Chlamydia Biology

From Genome to Disease

https://doi.org/10.21775/9781912530281

Edited by

Ming Tan

Departments of Microbiology and
Molecular Genetics, and Medicine
University of California
Irvine, CA
USA

Christine Sütterlin

Department of Developmental and Cell
Biology
University of California
Irvine, CA
USA

Johannes H. Hegemann

Institute of Functional Microbial
Genomics
Heinrich Heine University Düsseldorf
Düsseldorf
Germany

Caister Academic Press

Caister Academic Press
Norfolk, UK

www.caister.com

British Library Cataloguing-in-Publication Data
A catalogue record for this book is available from the British Library

ISBN: 978-1-912530-28-1 (paperback)
ISBN: 978-1-912530-29-8 (ebook)

Cover design: A chlamydial inclusion imaged by expansion microscopy. HeLa 229 cells were infected at an MOI of 1 with the dual fluorescent reporter strain *C. trachomatis* mCh(Gro_{L2}) and GFP($OmcA_{L2}$) (Cortina *et al.*, 2019, https://doi.org/10.1371/journal.pone.0217753) and imaged by expansion microscopy at 30 hours post infection, as described by Kunz *et al.* (2019, https://doi.org/10.3389/fcimb.2019.00276). The image shows a single inclusion containing a population of chlamydiae that is undergoing asynchronous RB-to-EB conversion, as detected by differential expression of mCherry from the early *GroESL* promoter (shown in magenta) and/or GFP from the late *omcAB* promoter (shown in cyan). This image was kindly provided by Tobias Kunz, Ralph Götz and Thomas Rudel (University of Würzburg, Germany), who used the dual fluorescent reporter strain generated by Maria-Eugenia Cortina and Isabelle Derré (University of Virginia).

Contents

Foreword

When Ming Tan, one of the editors and a contributor to this new volume on *Chlamydia*, called me earlier this summer to ask if I would be willing to write a few words about the book, I was delighted and honoured to have the opportunity to get a sneak preview of the contents. We chatted for a few minutes about new work that is under way on all aspects of chlamydiae biology, genetics and genomics, and host immune responses to chlamydiae. These are exciting times for *Chlamydia* researchers, and this excitement is reflected in each of the chapters that you, the reader, have before you. It is hard to believe that it has been nearly a decade since the last compendium of progress on *Chlamydia* research was published, and this volume provides a timely update. The editors, Drs Tan, Hegemann and Sütterlin, have done a superb job in organizing the book to provide cohesiveness between the broad areas covered. They have assembled a first-rate group of experts that nicely blends contributions from well-established, veteran researchers and highly talented newer investigators. This symbolic passing the mantel is, I think, extremely important for continued progress. The editors also have blended expertise in areas that run the gamut from clinician researchers to cell biologists and molecular geneticists. In short, there is something for everyone here.

The volume begins, quite appropriately, with chapters that describe both *Chlamydia trachomatis* and *C. pneumoniae* and human diseases they cause or are implicated in. It is most interesting to me that despite excellent physician awareness, quick and reliable diagnostic tests and effective treatment methods for *C. trachomatis* infection, the worldwide infection rates are increasing. This likely attests to the remarkable traits this microbe exhibits as a pathogen. We clearly need more research on the basic biology of chlamydiae to gain a better appreciation of early interactions with the host cell, intracellular survival strategies, host responses, and paths of transmission. It is very clear that during the past decade we have learned a great deal about chlamydiae and their interaction with the host. Several contributions are dedicated to informing us about how these pathogens invade, establish their intracellular niche, and co-opt host cell processes for their own benefit. There are chapters that combine expertise in chlamydiae biology (e.g. Hans Hegemann, Rey Carabeo, Kevin Hybiske, Isabelle Derré) and host cell biology (e.g. Christine Sütterlin). The propagation of chlamydiae is now known to be anything but straightforward and George Liechti, Patrick Viollier, Gilbert Greub and Tony Maurelli provide us with a very nice update on chlamydiae division and differentiation. Perhaps the most new knowledge developed during the past 10 years is in chlamydiae genetics and genomics. These topics are beautifully covered in three chapters. Christopher Rosario, Katelyn Soules, Scott Hefty and Ming Tan cover gene regulation; Colette O'Neill, Ian Clarke and Derek Fisher cover genetics; and Vitor Borges, Patrick Hyden, João Paulo Gomes and Thomas Rattei cover genomics. These chapters serve as an integrating theme for the book and help us to understand how new genetic and genomic information may have valuable practical applications. These chapters also showcase some of the newer investigators charged with moving this field forward. The response of the host to chlamydiae is not neglected in this volume. Cell-autonomous defence, an important feature of all intracellular infections, is discussed by George Häcker and Thomas Rudel. A discussion of innate immunity is led by Uma Nagarajan and adaptive immunity is discussed by Taylor Poston, Toni Darville and Ray Johnson. An update on the ever-elusive chlamydial vaccine is tackled in a chapter by Luis de la Maza and his colleagues. Simon Graspeuntner, Peter Timms, Jan Rupp and Bonnie Quigley provide an interesting chapter on the interface between chlamydiae and the microbiome. This is an area of particular interest to me, and one that could provide useful pathogenesis-related insights. We all know that there is much more to the study of chlamydiae than work on human disease. Konrad Sachse and Nicole Borel provide an overview of chlamydiae that are important as veterinary pathogens, and Alice Taylor-Brown, Tamara Halter, Adam Polkinghome and Matthias Horn give us valued insights on the ubiquity of chlamydiae in the biosphere. It would appear that wherever eukaryotic cells are found, chlamydiae are likely to – I almost wrote invade – but perhaps integrate in symbiotic relationships – may be a better way to put it. Finally, all chlamydiae must be studied in the context of the host cell and often work requiring animal models are needed. This is especially true for understanding pathogenesis of disease and protection from disease. Guangming Zhong, Alison Quayle, Ashok Aiyar and Tianyuan Zhang conclude the volume with a discussion of the pros and cons of animal models in the study of chlamydial infections and how these systems may be put to best use to study host responses and disease.

When Ming and I were on the phone talking about some of the newest work being done, his enthusiasm was infectious. I found I could almost bring back that feeling of excitement I had many years ago when I sat in Jim Moulder's lab. I recalled how Jim always said that he wanted to know the evolutionary adaptations chlamydiae have made to succeed in the intracellular environmental niche. I am sure that Jim would be pleased to know that we are getting closer to the goal thanks to the efforts of many, including those who contributed to this book. Please allow me to add my thanks to all of you. I know that this volume will be a wonderful addition to the literature. Keep up the good work!

Gerry Byrne
August 2019
Jamestown, RI

A note on usage

In the book, we have used the following terms with their respective formatting (Capitals and *italics*):

- chlamydia: single bacterium
- chlamydiae: multiple bacteria
- chlamydial: adjective (equivalent to 'bacterial')
- *Chlamydia*: Genus name
- *Chlamydiaceae, Parachlamydiaceae*: Family name
- *Chlamydiales*: Order name (includes *Chlamydia* and *Chlamydia*-related bacteria)
- *Chlamydiae*: Phylum name

For practical purposes *Chlamydiales* and *Chlamydiae* are equivalent because there is only one order, *Chlamydiales,* in the phylum *Chlamydiae.* We prefer *Chlamydiales* because *Chlamydiae* is easily confused with the generic term, chlamydiae.

The above italicization rules conform with the ASM style, which italicizes names of all bacterial taxa. However, some taxonomy nomenclature systems do not italicize names above family, i.e. they would use Chlamydiales and Chlamydiae for the order and phylum, respectively. We have deferred to authors of some chapters in the book on this usage.

More details on the usage of *Chlamydia* terms and taxonomy can be found in Bavoil, P., Kaltenboeck, B., and Greub, G. 2013. In *Chlamydia* veritas. Pathog. Dis. 67, 89–90.

MT: To my mother, Helen

JHH: To the most influential women in my life, Ursula and Angelika

CS: To my sons, Toby and Lucas

Chlamydia trachomatis Infections

1

Stephen Jordan[1]*, David Nelson[2] and William Geisler[3]

[1]Division of Infectious Diseases, Department of Medicine, Indiana University School of Medicine, Indianapolis, IN, USA.
[2]Department of Microbiology and Immunology, Indiana University School of Medicine, Indianapolis, IN, USA.
[3]Division of Infectious Diseases, Department of Medicine, University of Alabama at Birmingham, AL, USA.

*Correspondence: jordansj@iu.edu

https://doi.org/10.21775/9781912530281.01

Abstract

Chlamydia trachomatis is an obligate intracellular human bacterial pathogen that infects epithelial cells of the eye, oropharynx, urogenital, and anorectal mucosa and is responsible for the clinical diseases of trachoma, chlamydia (urogenital, oropharyngeal, and anorectal), and lymphogranuloma venereum. *C. trachomatis* infections occur worldwide and infection rates are increasing. *C. trachomatis* can cause chronic asymptomatic infections and can lead to complications, such as irreversible reproductive tract morbidity (e.g. tubal factor infertility) from genital chlamydia, and blindness from trachoma. Clinical manifestations of *C. trachomatis* infections range broadly from asymptomatic infection (most common) or relatively mild signs or symptoms (e.g. urethral or vaginal discharge) to more severe disease (e.g. pelvic inflammatory disease, epididymitis). Diagnosis of *C. trachomatis* infection cannot be accurately made by clinical signs or symptoms and requires identification of the pathogen, usually by nucleic acid detection. All *C. trachomatis* infections should be treated with antibiotics. For genital infections, reinfection should be ruled out by repeat testing approximately three months after treatment. Trachoma management involves a combination of four interventions at the community and individual levels: surgery for turned-in eyelashes (trichiasis), antibiotics, facial cleanliness, and environmental improvements.

Introduction

Members of the bacterial phylum Chlamydiae are obligate intracellular bacteria that can infect remarkably diverse hosts ranging from protozoans to humans. Despite their diverse host species, all chlamydiae share two key aspects of chlamydial biology: (1) they are obligate intracellular organisms that are dependent on their eukaryotic host for survival; and (2) they exhibit a biphasic developmental cycle. Three *Chlamydia* species are human pathogens: *C. trachomatis, C. pneumoniae,* and *C. psittaci*. This chapter will focus on *C. trachomatis* infections. *C. pneumoniae* and *C. psittaci* will be covered in Chapter 2, *Chlamydia pneumoniae* Infections, and Chapter 17, Recent Advances in Epidemiology, Pathology and Immunology of Veterinary Chlamydiae, respectively.

Infection with *C. trachomatis* results in three distinct clinical diagnoses: (1) trachoma; (2) chlamydia; (3) and lymphogranuloma venereum (LGV). As the term 'chlamydia' refers to both the organism and the infection, standard scientific nomenclature (capitalizing and italicizing the genus) is used to distinguish the genus '*Chlamydia*' from the clinical infection, 'chlamydia'. In this chapter, topics related to the basic biology, epidemiology, risk factors, natural history, and immunity of *C. trachomatis* infections will be reviewed, followed by chlamydial clinical manifestations, screening and diagnosis, and management. Clinical management includes treatment and prevention of reinfections. This chapter will predominantly focus on chlamydia and LGV. Trachoma will be discussed separately at the end of the chapter, given that the clinical manifestations and management of trachoma significantly differ from chlamydia and LGV.

In addition to providing a broad introduction to *C. trachomatis* infections, this chapter will highlight relevant clinical aspects that are essential for the basic scientist to understand when conducting translational research. The goal of this chapter is to combine the two seemingly disparate clinical and basic science chlamydial research fields, which we hope will foster an increase in translational research by non-clinical chlamydia researchers. To this end, each section begins with a list of three key clinically relevant questions (i.e. 'FAQs' for the chlamydial basic scientist) that are answered in the subsequent section. Our hope is that a deeper understanding of the clinical knowledge gaps in chlamydial research will foster new avenues of translational research.

C. trachomatis genetics and developmental cycle

- How do genetically similar *Chlamydia* serovars cause such distinct clinical diseases?
- Are 'persisting chlamydia' and 'persistent *Chlamydia*' the same thing?
- Why are *Chlamydia* infections so difficult to naturally clear?

The *C. trachomatis* chromosome is 1.0 Mb in size, and most natural isolates also carry a highly conserved 7.5 kb plasmid (Stephens *et al.*, 1998), for a total of approximately 900 genes. The plasmid encodes a non-coding RNA of unknown function and eight open reading frames that encode the plasmid glycoproteins Pgp1 to Pgp8. The plasmid is important for *Chlamydia* infectivity (Carlson *et al.*, 2008). Results from both mouse and nonhuman primate models indicate that the plasmid is a key virulence factor *in vivo* (Zhong, 2017b). Pgp3 may be a key plasmid-encoded virulence factor based on a murine study showing that a pgp3-deficient mutant was as pathologically attenuated as a plasmid-free strain (Liu *et al.*, 2014). Plasmid-free *C. trachomatis* isolates are infrequently detected in humans and the clinical implications of these strains are poorly understood.

Remarkably, *Chlamydia* strains that cause distinct diseases can share more than 99% genome identity (Carlson *et al.*, 2005; Thomson *et al.*, 2008). Some genes that determine chlamydial host range, infectivity and pathogenicity are localized to a region of high heterogeneity, called the 'plasticity zone' (PZ) (Read *et al.*, 2000). The PZ varies from 18 to 81 kb between different *Chlamydia* species and contains a number of virulence factors, which include the membrane attack complex/perforin protein, the tryptophan operon, phospholipase D, and the chlamydial cytotoxin (Read *et al.*, 2003). PZ genes will be described in more detail in Chapter 11, *Chlamydia* Genetics. The virulence factors that are associated with specific symptoms in humans, or that cause more invasive disease with LGV strains compared to non-LGV chlamydia strains, remain poorly understood.

The C. trachomatis developmental cycle initiates when the infectious (but relatively metabolically inert) form termed the elementary body (EB) binds to and triggers its own entry into a mucosal epithelial cell by a process that has been referred to as pathogen-mediated endocytosis (Hybiske and Stephens, 2007); see more details in Chapter 3, *Chlamydia* Adhesion and Invasion. The EB is deposited into a vacuole derived from the host plasma membrane, which is referred to as the inclusion, in which the pathogen resides for the remainder of the developmental cycle. EBs differentiate into metabolically active reticulate bodies (RB) inside the inclusion and undergo approximately 8–12 cycles of binary cell division before they differentiate back into EBs. Forty-eight to 72 hours post-infection, depending on the *Chlamydia* species, hundreds to thousands of daughter EBs are released upon near-simultaneous lysis of the inclusion and host cell plasma membrane or by extrusion of the entire inclusion (a specialized form of exocytosis). In the extracellular space, daughter EBs can then infect adjacent cells to re-initiate the developmental cycle (Moulder, 1991) (Fig. 1.1).

In some hostile *in vitro* conditions, *C. trachomatis* RBs adapt to the environmental stress by transitioning into a reversible altered growth state, termed chlamydial 'persistence.' Persistence is marked by a significant reduction in cellular metabolism and DNA replication, halting of EB production, and formation of large, aberrant forms that can be visualized by electron microscopy (Koehler *et al.*, 1997). Once the environmental stress has resolved, persistent forms revert back to replicating RBs and

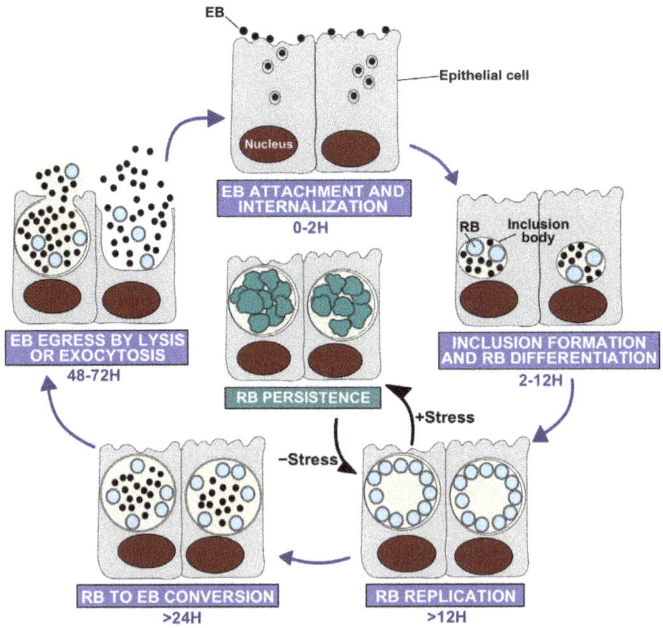

Figure 1.1 *Chlamydia* development cycle. Epithelial cell infection occurs with elementary body (EB) attachment and internalization within 2 hours, followed by formation of the inclusion body and differentiation into reticulate body (RB) forms by approximately 12 hours. RB replication begins at approximately 12 hours and continues throughout the development cycle. At 24 hours, RBs begin to differentiate into EBs in an asynchronous manner. Depending on the *Chlamydia* species, EBs are released by either cell lysis or exocytosis (extrusion) between 48 and 72 hours after infection. In environmental stress, RB replication halts and RBs differentiate into aberrant forms termed '*Chlamydia* persistence'. When the hostile condition is resolved, aberrant forms transform back into RBs and replication resumes.

the developmental cycle resumes. *C. trachomatis* persistence can be induced *in vitro* by depletion of amino acids (especially tryptophan), nutrients (e.g. iron), or exposure to beta-lactam antibiotics (e.g. penicillin, amoxicillin, etc.) or the cytokine interferon gamma (IFN-γ) (Mpiga and Ravaoarinoro, 2006). *In vitro*, persistent forms are more resistant to the first-line antibiotics used for chlamydia treatment, azithromycin (Wyrick and Knight, 2004) and doxycycline (Reveneau *et al.*, 2005), which raises concerns regarding *in vivo* treatment failures. If persistent forms do confer increased resistance to antibiotics *in vivo*, then they may increase risk for treatment failure (e.g. uncured infection), leading to *C. trachomatis* surviving within infected individuals and posing an ongoing transmission risk to sex partners. Indeed, inducing persistence in *C. muridarum*-infected mice with amoxicillin more than doubles the failure rate of azithromycin (Phillips Campbell *et al.*, 2014).

Primarily indirect evidence supports the hypothesis that chlamydial persistence occurs *in vivo*. Aberrant forms have been visualized in the gastrointestinal tract in pigs (Pospischil *et al.*, 2009) by electron microscopy and can be induced *in vivo* in mice by amoxicillin treatment (Phillips Campbell *et al.*, 2012). Several studies have reported serial antibiotic failures in a minority of women. In one study of 552 women, seven women were identified that had 3–10 repeat *C. trachomatis* infections with the same strain over a 2- to 5-year period. These infections were diagnosed by PCR testing, but the organisms were non-culturable (a characteristic of persistent forms) (Dean *et al.*, 2000). A second study using the same approach described 11 women who remained consistently infected with the same strain for up to three years, despite repeated treatments with azithromycin (Smith *et al.*, 2001). Another study observed aberrant *C. trachomatis* forms by electron microscopy in urethral and cervical specimens from sixteen men and women who were suspected azithromycin treatment failures (e.g. still infected two weeks post-treatment). This suggests that azithromycin treatment failure may be associated with *in vivo* persistence in humans (Bragina *et al.*, 2001). Azithromycin can induce persistence *in vitro* that is reversed when the drug is removed (Xue *et al.*, 2017). Further, it has been shown *in vitro* that after eight hours of azithromycin exposure the minimum inhibitory concentration of azithromycin for *C. trachomatis* is increased (Zheng *et al.*, 2015).

'Persistent *Chlamydia*' and 'persisting/chronic chlamydia' are not interchangeable terms; the former refers to the organism growth state and the latter to continuing infection. The term 'persistent chlamydia' (referring to the infection) should be used sparingly and never to indicate the state of persistence (persistent *Chlamydia* would be accurate), as the meaning of the term is ambiguous (i.e. does persistence refer to the infection, organism state, or both?) and its implication potentially inaccurate. For example, persistent chlamydia can be interpreted to refer to either chlamydial infection that is persisting over time (correct interpretation) or infection with chlamydial organisms that exist in a state of persistence *in vivo* (incorrect interpretation); the latter has yet to be determined. In contrast, 'persisting/chronic chlamydia' simply refers to ongoing or recurring chlamydial infection, with no reference to the persistent state. Restricting the term 'persistent' to only referencing the organism may help prevent confusion when describing persisting/chronic *C. trachomatis* infections.

In contrast to the traditional binary fission-mediated bacterial reproductive cycle, the biphasic developmental cycle may confer significant evolutionary advantages that allow chlamydial EBs and RBs to thrive in two very distinct environments: a hostile extracellular and the intracellular environment, respectively. Preservation of EB viability and structural integrity is critical to maintaining infectivity in the hostile extracellular environment. EBs do not have a classical cell wall and only form a narrow ring of peptidoglycan at the division plane in RBs, yet EBs are highly resilient to physical and chemical disruption (Liechti *et al.*, 2016). In contrast to most bacteria which rely on the cell wall to provide structural stability, extensive disulfide cross-linking of proteins in the outer membrane confers the structural integrity of EBs. Indeed, exposure of EBs to the strong detergent *N*-lauroyl sarcosine allows purification of the intact proteinaceous *Chlamydia* outer membrane complex (COMC) but does not alter EB shape (Caldwell *et al.*, 1981). The COMC contains extensively disulfide cross-linked outer membrane proteins, of which OmpA comprises up to 60% of the total COMC protein content. Other abundant EB proteins include the translocated domains of polymorphic proteins, which are located in the outer membrane where they may act as adhesins (Liu *et al.*, 2010; Wehrl *et al.*, 2004). Finally, EBs display a type III secretion system (TTSS) that allows them to inject pre-formed effectors into target host cells to mediate EB invasion and evasion of innate immunity (Fields *et al.*, 2003; Chen *et al.*, 2014; Clifton *et al.*, 2004). Further details on *Chlamydia* binding and entry and the TTSS are covered in Chapter 3, *Chlamydia* Adhesion and Invasion, and Chapter 7, Protein Secretion in *Chlamydia*, respectively.

In contrast to EBs, RBs are larger (approximately 1 μm in diameter), pleomorphic, and metabolically active. Beyond mediating RB metabolism, a broad range of TTSS effectors are deployed that divert host cell resources and circumvent intracellular innate immune sensing and killing. RBs manipulate multiple cellular processes, which include (1) intercepting sphingomyelin, cholesterol, and other nutrients; (2) inhibiting lysosome–inclusion fusion; (3) inhibiting apoptosis signalling to prolong cell survival; (4) altering the host transcriptome and proteome machinery; and (5) down-regulating MHC I and II surface expression to avoid detection by the immune system (Fields and Hackstadt, 2002). The precise mechanisms that chlamydiae use to interfere with many host processes are poorly understood, but it is believed that most RB–host interactions are mediated by inclusion membrane proteins and other proteins that RB secretes into the inclusion membrane or the host cell cytosol. Further details on the inclusion are provided in Chapter 4, The Chlamydial Inclusion, and Chapter 5, Interactions of the Chlamydial Inclusion with the Host Cell.

The biphasic development cycle has key evolutionary advantages in humans. EBs are highly resilient and infectious; a critical attribute that permits *C. trachomatis* to successfully infect multiple different mucosal sites. Also, the biphasic forms allow *C. trachomatis* to circumvent host immune defence mechanisms, which results in chronic infection with little-to-no symptoms elicited in the majority of infected individuals. For example, antibody responses that develop against EB antigens may be ineffective against RBs. In a similar fashion, EB survival will not be affected by cellular T-cell responses that target the intracellular RB. As a result, *C. trachomatis* is a highly formidable pathogen and difficult to eradicate, which has resulted in it becoming a worldwide major public health burden.

Biovars and serovars

- Does each serovar cause a distinct clinical manifestation?
- Which serovar is the most clinically relevant to use for *in vitro* and *in vivo* studies?
- Can an LGV serovar be used as an accurate model of chlamydia?

C. trachomatis infects mucosal epithelial cells of the eyes, oropharynx, genital tract, and anorectal tissues. Through molecular and serological typing techniques, *C. trachomatis* can be classified into two biovars: (1) trachoma strains, which cause either 'trachoma' (a distinct conjunctival infection) or the clinical syndrome of 'chlamydia' (non-ulcerative infection of the eye, oropharynx, genital tract, and anorectal tissues); and (2) LGV, which causes 'lymphogranuloma venereum' (LGV), an ulcerative infection of the genital tract and anorectal tissues. Classification of the biovars can be differentiated further, historically based on serological testing of the OmpA antigen encoded by the gene *ompA*, into 18 different major OmpA types, termed 'serovars', and minor variants. These 18 serovars can be further clustered into three classical clinical diseases (Wang and Grayston, 1991; Wang and Grayston, 1970): trachoma (serovars A, B, Ba, and C), chlamydia (serovars D, Da, E, F, G, H, I, Ia, J, and K), and LGV (serovars L1, L2, L2a, and L3). Modern OmpA identification techniques utilize PCR amplification of *ompA* and are usually only employed in research settings where the goals are to distinguish treatment failure from new infections or to track the transmission of specific *C. trachomatis* strains within sexual networks. Serovars D, E, and F are the most commonly identified trachoma biovars, and are responsible for up to 75% of urogenital infections (Morré *et al.*, 2000), and serovar L2 is the most commonly identified LGV biovar (Kuo *et al.*, 1983). Clinically, *C. trachomatis* OmpA types correlate poorly with clinical manifestations (i.e. no single serovar is consistently associated with a clinical sign/symptoms or disease severity). This was demonstrated in a large study of 1180 chlamydia-infected men and women, where signs and symptoms were not found to be strongly associated with the infecting OmpA type (Geisler *et al.*, 2003). The lack of correlation of clinical phenotypes with OmpA types may be explained by other virulence factors not linked to OmpA, including polymorphic outer membrane proteins, a putative cytotoxin, TTSS effectors, stress response proteins and proteins encoded by the cryptic plasmid (Byrne, 2010). Additionally, whole-genome phylogenetic studies of trachoma and LGV biovars have identified chimeric *ompA* alleles as well as instances in which *ompA* alleles have been exchanged within and between OmpA types and biovars (Harris *et al.*, 2012). As a result, *ompA* genotyping has little utility outside of the research arena and is rarely performed in clinical practice. One notable exception is the use of OmpA typing to determine if severe rectal chlamydial infections are caused by chlamydia or LGV strains, which are treated with different antibiotic durations (Workowski *et al.*, 2015).

When selecting serovars for research studies, the optimal approach would be to perform all translational research studies utilizing the homologous serovar (i.e. the same serovar the individual was infected with). When serovar-typing is not performed, an alternate approach is to utilize pools of the most common serovars (D-F for urogenital studies, L2 for LGV studies) to minimize the potential impact of serovar-specific responses. Importantly, although no specific serovar is associated with distinct clinical manifestations, the two biovar classes are clearly associated with distinct clinical manifestations (see the 'Clinical Manifestations and Complications' section). As a result, researchers should be cognizant that their model system could potentially utilize a clinically incompatible biovar, which may lead to a clinically inaccurate interpretation of study findings. For example, using L2 in an *in vitro* model of urogenital chlamydia would not be clinically accurate as L2 is associated with the disease LGV, not urogenital chlamydia.

Epidemiology

Incidence and prevalence rates

- How are chlamydia rates determined and what is a reportable infection?
- Why is the number of reported and estimated chlamydia cases so different?
- Are LGV rates as high as chlamydia?

C. trachomatis is the most prevalent bacterial sexually transmitted infection (STI) in the U.S. and worldwide, with estimates of > 1.7 million and > 130 million annual new infections, respectively (Newman *et al.*, 2015; CDC, 2018). In the U.S., chlamydia is a reportable infection to the Centers for Disease Control and Prevention (CDC), which means that all identified infections must be reported to the CDC so that national chlamydia rates can be monitored. In 2017, chlamydia infection rates were the highest ever reported to the CDC (CDC, 2018) (Fig. 1.2A). It is likely that many chlamydia cases remain undiagnosed, and the true annual chlamydia burden may be up to 2-fold higher than the number

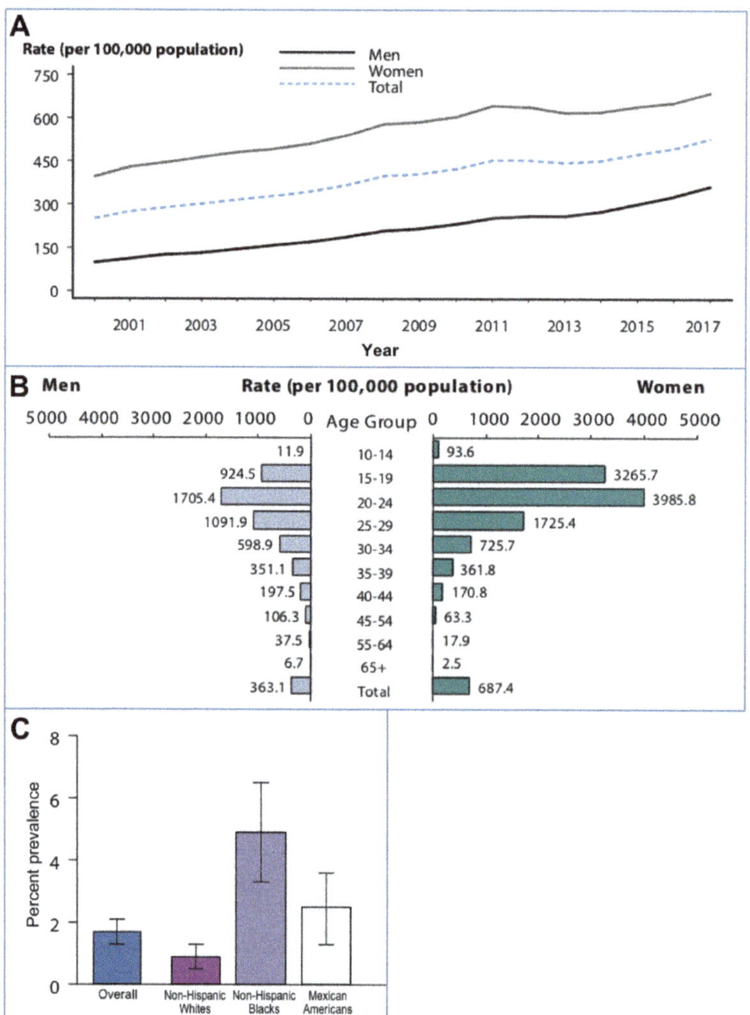

Figure 1.2 2017 CDC chlamydia surveillance data. (A) Chlamydia rates reported, by sex, from 2000 to 2017, showing rising rates over time. Rates are approx. twofold higher in women than in men. (b) 2017 chlamydia rates stratified by sex and age group. Highest rates occur in women aged < 25 years. (C) 2017 chlamydia rates stratified by race. Highest rates are seen in non-Hispanic black Americans. Error bars indicate 95% confidence intervals. Reproduced from the Centers for Disease Control and Prevention (2018).

of reported cases (CDC, 2013). The discrepancy between the number of reported and estimated chlamydia cases can likely be attributed to contributions from both the large burden of asymptomatic infections (CDC, 2018) and low compliance (< 50%) with screening recommendations in eligible women (Shih *et al.*, 2009). Chlamydia rates are highest in sexually active females 15–29 years of age (Fig. 1.2B); especially those aged 15–24. Rates in males follow a similar distribution, though they are about 50% lower. Based on U.S. national surveys from 1999–2008, the prevalence of chlamydia is < 2% in men and women aged 14–39, but 4.7% in women aged 14–24 (Torrone *et al.*, 2014). Prevalence rates may be > 10% in high-risk individuals.

LGV incidence and prevalence rates are less studied than chlamydia. In developed countries, LGV case numbers could be grossly underestimated due to the absence of serological or OmpA testing for suspected LGV infections, surveillance programmes, and established reporting mechanisms. In developing countries, only a few thousand annual cases are reported where LGV remains endemic: South America, the Caribbean, Southeast Asia, the Middle East and Africa (Stamm, 2008). In developed countries, LGV is rarely reported, except primarily in men who have sex with men (MSM), where the reported prevalence is < 1% (Ward *et al.*, 2009). Most of these cases appear to be clonally related to an outbreak of anorectal LGV from the L2b OmpA type, which occurred in 2003 in the Netherlands and has spread to other developed countries, likely following MSM sexual networks (Christerson *et al.*, 2010).

Risk factors for acquiring *Chlamydia trachomatis* infection

- Why is chlamydia twice as common in women, compared to men, and highest in African American/Hispanic women?
- How infectious is *Chlamydia trachomatis* and what is the transmission rate per sex act?
- Why are chlamydia rates increasing despite active national chlamydia screening programmes?

Prevalence rates vary considerably based on age, race/ethnicity, and other factors. In women, rates are highest in younger women and potential explanations are higher risk sexual behaviour (e.g. less frequent use of barrier protection, more frequent sexual partners, etc.), increased frequency of cervical ectopy (which increases the surface area of the cervical epithelial cells that EBs infect), hormonal effects from oral contraceptives, and less-developed mucosal immunity, compared to older women. Rates among women are approximately double that of men, which may reflect either greater susceptibility to infection than men and/or a more prolonged course of infection in asymptomatic women (explained further below). Hormone differences may also influence risk of *C. trachomatis* acquisition as a study that correlated the menstrual cycle stage with chlamydia prevalence found that *C. trachomatis* infection was more commonly detected during the high-progesterone luteal phase (Forcey *et al.*, 2014). Non-Hispanic African Americans have the highest chlamydia rates (Fig. 1.2C), which may be due to increased risk behaviours and/or immunogenetic factors. Increased susceptibility to reinfection in African Americans has been associated with HLA class II polymorphisms and IL-10 cytokine gene variants (Olson *et al.*, 2018; Wang *et al.*, 2005). Rates also vary by testing site and may exceed 10% in emergency rooms, sexually transmitted disease (STD) clinics, and youth correctional facilities. Risk factors for LGV in endemic areas likely parallel risk factors for chlamydia. In MSM, risk factors for LGV acquisition include reported unprotected receptive anal sex, multiple sex partners, HIV infection, drug use with sex (i.e. 'chemsex'), and engaging in group sex (Macdonald *et al.*, 2014). *C. trachomatis* is commonly detected as a co-pathogen with *Neisseria gonorrhoeae* in approximately 40% of cases (Datta *et al.*, 2007) and perhaps *Mycoplasma genitalium*. In the case of *N. gonorrhoeae*, *C. trachomatis* may promote co-infections by paralyzing anti-gonococcal neutrophils using the secreted protease CPAF (Rajeeve *et al.*, 2018). Bacterial vaginosis (BV) also appears to be a risk factor for acquiring chlamydia, and women with BV have an almost 2-fold increased risk for incident *C. trachomatis* infections (Bautista *et al.*, 2016; Wiesenfeld *et al.*, 2003). This association of BV with incident chlamydia was further shown in a study utilizing the antibiotic metronidazole (which has no anti-chlamydial activity) to prevent recurrent BV. In addition to reducing recurrent BV, metronidazole use unexpectedly reduced *C. trachomatis* incidence rates by 2-fold (Balkus *et al.*, 2016). The impact of the microbiota on chlamydia is discussed in more detail in Chapter 16, *Chlamydia* Infections and the Microbiota.

Although *C. trachomatis* EBs are highly infectious *in vitro*, the sexual transmission rate (i.e. rate per coital act) is thought to be relatively low and is highly variable, depending on risk factors that include age, the frequency of new partners, *C. trachomatis* prevalence, frequency of unprotected sex, etc. Multiple studies have attempted to estimate a transmission rate per sex act, which is reported as < 1% to 5.8% per sex act for both female-to-male and male-to-female urogenital chlamydia (Althaus *et al.*, 2012; Katz, 1992), but modelling studies suggest it may be as high as 13% per sex act in high–risk partnerships (Teng *et al.*, 2014; Tu *et al.*, 2011). Transmission rates per sex act for anorectal or oropharyngeal chlamydia are unknown.

Increased chlamydia testing has been proposed to explain increasing *C. trachomatis* infection rates, but this seems unlikely since gonorrhoea rates have not followed a similar trend (anogenital gonorrhoea testing is almost exclusively performed in parallel with chlamydia using a combo assay). Also, there is no recent evidence that increased chlamydia testing is actually being performed (CDC, 2018). One potential explanation for increasing chlamydia rates is that widespread treatment of *C. trachomatis* infections identified through screening programmes at an earlier stage of infection (before the development of protective immunity) could be arresting the development of effective immunity to chlamydia, a concept termed the 'arrested immunity' hypothesis. This hypothesis is supported by several observations. First, one murine study found that earlier clearance of chlamydia with antibiotics (i.e. shorter duration of infection) was associated with significantly less protection against chlamydia reinfection compared to mice who were allowed to naturally clear chlamydia (Su *et al.*, 1999). Second, a study of 200 women returning for treatment of a recent positive chlamydia test found that women who had active chlamydia infection cleared by antibiotics were 4-fold more likely to be reinfected within 6-months compared to women who had naturally cleared chlamydia before receiving antibiotics (Geisler *et al.*, 2013). Finally, a Canadian study reported a marked increase in chlamydia rates within five years of starting a national screening programme, which was driven by an approximately 40-fold increase in reinfection rates (Brunham *et al.*, 2005). Together, these studies suggest that screening programmes may be a risk factor for *C. trachomatis* reinfection by arresting development of natural immunity, which potentially explains why chlamydia rates continue to rise despite widespread surveillance programmes.

Natural course of infection and immunity

- Why do we know so little about the natural history of chlamydia in humans?
- Why study chlamydia in humans when we can use animal models?
- How is chlamydia naturally cleared and why do we not have a chlamydia vaccine?

The natural history of *C. trachomatis* infections in humans is not well understood, in part due to the ethical concerns that prohibit prospective studies of untreated chlamydia. Due to the risk of irreversible morbidity related to an untreated *C. trachomatis* infection, it is considered unethical to intentionally infect humans (which prohibits challenge studies) or to withhold treatment (which limits natural history studies). Natural clearance of chlamydia (i.e. clearance without antibiotic therapy) has been reported by multiple studies as far back as the late 1970s, when McCormack *et al.* reported that in 14 chlamydia-infected women, 50% were no longer infected when retested between 16 and 17 months later, despite denying receiving antibiotics (McCormack *et al.*, 1979). Similar findings were reported in a retrospective Colombian study of 82 chlamydia-infected women, in whom 50% cleared *C. trachomatis* at 1 year and > 90% had cleared at 3.5 years without treatment (Molano *et al.*, 2005). In one study, approximately 20% of women cleared *C. trachomatis* infection in the brief interval between screening and presenting for treatment of a positive chlamydia test (Geisler *et al.*, 2008; Parks *et al.*, 1997). Further, natural clearance may be a clinical correlate of protection (i.e. a clinical finding that correlates with a clinical outcome that may indicate protection against chlamydia) as one study found that women who naturally cleared chlamydia in the interval between screening and treatment visits were 4-fold less likely to be reinfected within 6 months than women who did not clear (Geisler *et al.*, 2013). These studies demonstrate that *C. trachomatis* is masterful at immune evasion because the majority of untreated women have chronic infection, which may take months to years to clear. In men, the duration of natural infection remains uncertain and the existing sparse studies are contradictory. One study of 14 men found that men were 2-fold more likely to naturally clear *C. trachomatis* infection in the 2-week interval between screening and treatment visits compared to women (36% vs. 16%) (Geisler *et al.*, 2008). In contrast, a mathematical modelling study suggests that men may take longer to clear infection than women (Lewis *et al.*, 2017).

The natural duration of oropharyngeal or anorectal chlamydia or LGV is unclear as studies are limited. In the interval between screening and treatment visits, the reported natural clearance rate is 43–50% for oropharyngeal chlamydia and 18–57% for anorectal chlamydia (Apewokin *et al.*, 2010; van Liere *et al.*, 2013). Natural clearance rates may be overestimated as studies that diagnose chlamydia by detecting *C. trachomatis* nucleic acids cannot distinguish contamination with nonviable organisms (i.e. *C. trachomatis* nucleic acids from infected semen, etc.) from natural clearance of an established infection. Indeed, one study found that 16% of women classified as having naturally cleared *C. trachomatis* infection were likely not infected, based on the fact that they were seronegative for the short-lived IgG isotype IgG3 (Bakshi *et al.*, 2017).

Given the ethical limitations inherent to human studies of natural clearance, animal Models of urogenital infection have been developed to characterize the mechanisms of natural clearance and protection against *Chlamydia* reinfection (further reviewed in Chapter 19, Animal Models of *Chlamydia trachomatis* Infection). The development of innate and adaptive immune responses is discussed in Chapter 13, *Chlamydia* and Innate Immunity, and Chapter 14, Adaptive Immunity to *Chlamydia trachomatis* infection, respectively. Animal models demonstrate that the cytokine IFN-γ, produced largely by CD4+ T-cells, is crucial to mediating natural clearance and protection against reinfection (Rank and Whittum Hudson, 2010). However, no animal model perfectly mimics the anatomy, histology and endocrinology of humans. Also, different *Chlamydia* species do not accurately reflect the immunopathogenesis of *C. trachomatis* infection in humans. Noted differences between models and humans include the *Chlamydia* species used, oestrous cycle duration, susceptibility of the upper genital tract to infection, and infection duration (i.e. time to immune-mediated *Chlamydia* clearance) (De Clercq *et al.*, 2013). Given that translation of animal models to humans is imperfect, it is important that observations from animal models be confirmed using human studies.

Chlamydia vaccine development is hindered by our incomplete understanding of how immune-mediated *C. trachomatis* clearance occurs in humans. Several human studies confirm the role of IFN-γ in protection against *C. trachomatis* reinfection. In high-risk commercial sex workers, IFN-γ production by *ex vivo* stimulated peripheral blood mononuclear cells (PBMCs), which contain T-cells, correlated with a decreased risk for subsequent chlamydia (Cohen *et al.*, 2005). In another study, IFN-γ production by CD4-enriched PBMCs from adolescents trended towards a reduced risk for *C. trachomatis* reinfection, though the study may have been underpowered (Barral et al., 2014). Also, a higher percentage of IFN-γ-producing CD4+ T-cells was observed in women with chlamydia who remained negative for one year after treatment (Russell *et al.*, 2016). Finally, an IFN-γ-positive CD4+ T-cell response on *ex vivo* stimulation with chlamydial antigens was more frequently found in women without *C. trachomatis* reinfection at a 3-month follow up visit, compared to reinfected women (Bakshi *et al.*, 2018).

The mechanism of IFN-γ-mediated clearance of *C. trachomatis* infection has been studied *in vitro*. IFN-γ inhibits *C. trachomatis* growth *in vitro* in a dose-dependent manner (Beatty *et al.*, 1993). This is mediated by IFN-γ upregulating expression of the enzyme indolamine 2,3-dioxygenase (IDO), which catabolizes tryptophan to kynurenine (Ibana *et al.*, 2011). As *C. trachomatis* is a tryptophan auxotroph (requires tryptophan to survive), its growth is arrested in the setting of cytosolic tryptophan depletion which, if prolonged, results in cell death (Beatty *et al.*, 1994). No *in vivo* evidence has yet confirmed this mechanism, but several observations support its role in natural clearance. First, cervical IFN-γ is produced in the setting of *C. trachomatis* infection in women (Aiyar *et al.*, 2014; Arno *et al.*, 1990). Second, a higher kynurenine–to-tryptophan ratio (a surrogate for tryptophan metabolism by IDO) was seen in vaginal secretions from chlamydia-infected women (Ziklo *et al.*, 2018). Third, women who naturally clear chlamydia have lower cervicovaginal tryptophan levels, compared to women with active chlamydia (Jordan *et al.*, 2017).

To circumvent IFN-γ-mediated clearance, urogenital strains (OmpA types D-K) encode a functional tryptophan synthase which can utilize indole to produce tryptophan, thereby counteracting IDO-mediated tryptophan catabolism. Indeed, *in vitro* indole supplementation is sufficient to protect urogenital *C. trachomatis* strains against inhibition by IFN-γ (Caldwell *et al.*, 2003). Several observations support the hypothesis that indole, a by-product of some vaginal microbes (some of which are associated with BV), may facilitate *C. trachomatis* persistence and explain the observed 2-fold increase in incident *C. trachomatis* infection rate in women with BV. First, one study found indole-producing organisms in the vagina of > 90% of chlamydia-infected women (Ziklo *et al.*, 2018). Also, eradicating BV organisms by prophylactic treatment resulted in an decrease in incident *C. trachomatis* infections (Balkus *et al.*, 2016). Finally, the growth of IFN-γ-treated *C. trachomatis* could be rescued by co-culturing with an indole-producing *Prevotella* species *in vitro*, but not with indole-negative *Prevotella* species (Ziklo *et al.*, 2016). Taken together, the ability of *C. trachomatis* to utilize available indole to restore tryptophan and escape clearance by IFN-γ may explain why *C. trachomatis* infections may be long-lasting in some women. This model in which the vaginal microbiota supports a *Chlamydia* infection by providing indole is discussed in more detail in Chapter 16, *Chlamydia* Infections and the Microbiota. Vaccine strategies aimed at eliciting strong IFN-γ mucosal responses will likely be needed to provide adequate protection against chlamydia and are discussed in Chapter 15, *Chlamydia* Vaccines.

Clinical manifestations and complications

Evaluation of clinical syndromes and the physical examination

- Can you diagnose chlamydia by physical examination?
- What does syndromic management mean?
- Can you rule out chlamydia infection with a normal physical examination?

No pathognomonic physical examination finding exists that distinguishes infection with *C. trachomatis* from other common sexually transmitted infections (STI), caused by *Neisseria gonorrhoeae, Mycoplasma genitalium* and *Trichomonas vaginalis.* As a result, a diagnosis of chlamydia cannot be made by physical exam alone. Instead, the evaluation of symptoms and physical exam findings is used to diagnose clinical syndromes (i.e. clinical manifestations that are characteristic of a disease) that may be compatible with *C. trachomatis* infections and inform treatment decisions; a process termed 'syndromic management.' In syndromic management, empiric treatment (*i.e.* treatment provided based on clinical suspicion or experience- not confirmatory testing) is directed by the diagnosed syndrome, not the pathogen, as STI test results are usually delayed given the lack of available highly sensitive point-of-care tests for common STI pathogens.

Syndrome evaluation involves elucidation of symptoms and objective findings of inflammation on physical examination. The physical exam involves a focused assessment of mucosal surfaces where *C. trachomatis* infection may be established (guided by reported sexual behaviours and symptoms) and organ systems where the infection may spread (e.g. lymph nodes, upper genital tract, etc.). The predominant role of the physical exam is to diagnose chlamydia-compatible syndromes so that empiric chlamydia treatment can be provided while awaiting confirmatory testing. The physical exam also allows differentiation of complicated chlamydia (upper genital tract disease or complications of lower genital tract infection, like a Bartholin's gland abscess or reactive arthritis) from uncomplicated chlamydia (asymptomatic or mildly symptomatic lower genital tract disease), which is important given that the treatment of each differs. Limitations of the physical exam are that it has low sensitivity and specificity for chlamydia and, therefore, cannot be used to rule out chlamydia. The former was demonstrated in a retrospective study of 577 asymptomatic women undergoing screening in which 94% had a normal physical exam (Geisler *et al.*, 2007). Of 68 women positive for chlamydia, only 13% had cervical discharge noted on exam.

Regarding the exam specificity, similar signs and symptoms may be elicited by infection with other pathogens, some of which can exist as co-infections with *C. trachomatis* (e.g. *N. gonorrhoeae*, etc.). In a study of 155 men with NGU presenting to an STD clinic, men were asked about specific urethritis symptoms, including urethral discharge, dysuria (pain with urination), burning/tingling, itching, meatal erythema (redness at the urethral opening), and/or lesions (abnormal skin changes). No single sign or symptom differentiated infection with *C. trachomatis* from other STI pathogens (Jordan and Nelson, unpublished data).

Syndromes commonly associated with chlamydia include cervicitis (inflammation of the cervix), urethritis (inflammation of the urethra), and pelvic inflammatory disease (PID) in women and urethritis in men. More specifically, chlamydia urethritis in men may be diagnosed initially as non-gonococcal urethritis (NGU) when Gram stain microscopy of urethral secretions reveal increased polymorphonuclear leucocytes (PMNs, a marker of inflammation) but an absence of gonococcal-like Gram-negative diplococci; however, not all clinics have Gram stain capabilities. Other signs that suggest inflammation include the presence of leucocyte esterase in urine specimens or the presence of ≥2 PMNs per 1000× oil immersion field (PMN/oif) on Gram stain of urethral secretions. Importantly, as other STI pathogens elicit similar syndromes, diagnostic testing is required to confirm chlamydia.

Urogenital chlamydia

- Can a specific symptom differentiate chlamydia from other genital infections?
- How common is asymptomatic chlamydia?
- Is treatment necessary for asymptomatic chlamydia?

In urogenital chlamydia, infection may involve specific areas of the lower genital tract and can then spread to upper genital tract structures. In women, the lower genital tract includes the vagina, urethra, Bartholin's glands, and cervix, and the upper genital tract includes the uterus, fallopian tubes, ovaries, and peritoneum. In men, the lower genital tract includes the urethra, and the upper urogenital tract includes the epididymis, testes, and prostate. Lower genital tract infection occurs first with *C. trachomatis* infection of the columnar epithelial cells of the cervix and/or urethra in women and the urethra in men, and then can spread to infect epithelial cells of the upper genital tract structures.

The incubation period of urogenital chlamydia is unknown but is estimated to be 1–3 weeks (Holmes *et al.*, 1967; Perroud and Miedzybrodzka, 1978). The majority of urogenital chlamydial infections are asymptomatic. A study of over 18,000 men and women across five countries in South America, Africa, and Asia found that, generally, > 80% of men and women with urogenital chlamydia reported no symptoms (Detels *et al.*, 2011). In the US, another study of > 1600 men and women with urogenital chlamydia reported that 84% of men and 75% of women were asymptomatic (Farley *et al.*, 2003).

In women, symptoms of lower urogenital tract chlamydia include vaginal discharge, vaginal itching, dysuria (pain on urination), dyspareunia (pain on intercourse), and inter-menstrual bleeding; Bartholin's gland infection may cause painful vulvar swelling/tenderness. Symptoms of upper genital tract chlamydia include non-specific pelvic and/or abdominal pain, nausea, vomiting and fever. Routine evaluation for urogenital chlamydia involves assessing for cervicitis and PID. To examine the cervix, a speculum is used to expand the vagina to visualize the cervix and to facilitate specimen collection for diagnostic testing. The cervix is inspected for signs of cervicitis, either the presence of cervical discharge (Fig. 1.3B) or cervical friability (i.e. 'easily induced' bleeding with gentle insertion of a swab into the cervical os). Evaluation for upper genital tract signs is performed by bimanual examination, in which one hand presses on the pelvis (to palpate the upper genital tract structures) while fingers of the other hand are inserted into the vagina to manoeuvre the cervix (to assess for cervical motion tenderness). PID is diagnosed if pain occurs with any of the following: (1) palpation of the uterus; (2) movement of the cervix; or (3) palpation of the adnexa (fallopian tube structures lateral to the uterus). Bartholin's gland infection typically presents as an exquisitely tender unilateral vulvar swelling. Cervicitis and/or PID may be present in women with chlamydia, but a normal exam is much more common. Indeed, one study found that approximately 90% of asymptomatic chlamydia-infected women had

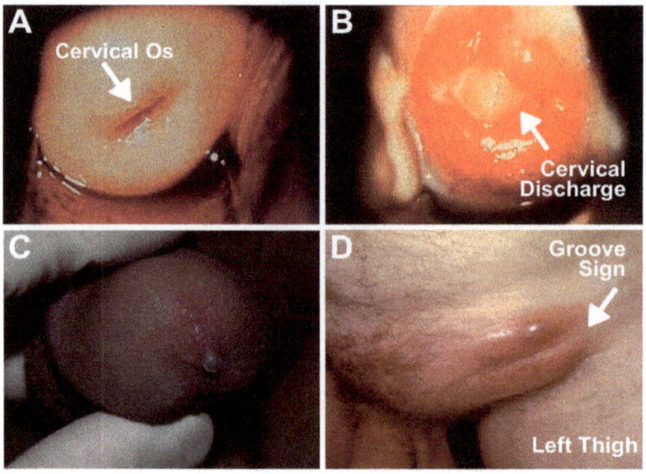

Figure 1.3 Clinical manifestations of chlamydia. (A) Normal-appearing cervix with no cervical discharge or erythema. (B) Mucopurulent cervical discharge in a chlamydia-infected woman. (C) Cloudy urethral discharge in a chlamydia-infected male. (D) Groove sign in a LGV-infected male. The 'groove' is formed by enlargement of the lymph nodes (buboes) above (inguinal nodes) and below (femoral nodes) the inguinal ligament. Reproduced courtesy of the University of Washington STD Prevention Training Center (A, B) and VisualDx (www.visualdx.com) (D). Figure 1.3C reproduced from Geisler, W.M. Diseases caused by Chlamydiae. In Goldman's Cecil Medicine, 24th edn, Goldman, L., and Schafer, A.I., eds, pp. 1916–1922. Copyright 2012, with permission from Elsevier.

a normal appearing cervix (Geisler *et al.*, 2007) (Fig. 1.3A).

In men, symptoms of urethritis include urethral discharge (clear, cloudy, or purulent) and urinary discomfort (pain [dysuria], tingling, and/or itching). A study of 194 men with urethritis symptoms found that only discharge and dysuria correlated with a clinical diagnosis of urethritis, based on urethral Gram stain evidence of inflammation or the presence of a urethral discharge on exam (Jordan *et al.*, 2018). Evaluation for urethritis involves assessing for the presence of urethral discharge (Fig. 1.3C) or erythema of the penile opening. However, *C. trachomatis* infection may elicit little-to-no inflammation in men, independent of urethritis symptoms. This was demonstrated by a study that attempted to correlate the PMN count from Gram stain smears of urethral swabs with a diagnosis of urogenital chlamydia in 13,520 symptomatic men (Rietmeijer and Mettenbrink, 2012). This study found that < 5% of symptomatic men with urogenital chlamydia had no PMNs/oif on urethral Gram stain and 16.3% had < 5; the PMN threshold previously CDC-recommended to diagnose urethritis. Another study found that 18% of men with chlamydia had < 5 PMNs/oif on Gram stain and 12% of men had no PMNs seen (Geisler *et al.*, 2005). As a result, the CDC lowered the PMN cut-off to diagnose urethritis from ≥ 5 to ≥ 2 PMNs/oif (Workowski *et al.*, 2015). Symptoms of epididymitis are typically unilateral scrotal or testicular pain and may also include scrotal erythema, testicular swelling and tenderness, and fever. Examination of the scrotal contents may reveal testicular swelling, erythema, warmth and tenderness, as well as epididymal swelling and tenderness. Other findings in epididymitis may include a urethral discharge with/without meatal erythema and/or a scrotal hydrocele.

Complications of chlamydial PID include chronic pelvic pain, tubo-ovarian abscess, tubal factor infertility (TFI), perihepatitis, and an increased risk for ectopic pregnancy, some of which may result from narrowing of the fallopian tube lumen and a loss of cilia on fallopian tube epithelial cells (Hvid *et al.*, 2007). PID is attributed to chlamydia in over 20% of cases (Jossens *et al.*, 1994), but up to 70% of women with PID test negative for gonorrhoea and/or chlamydia, which suggests that there is a large contribution from other pathogens (Ness *et al.*, 2002). Irreversible damage can occur from chronic inflammation and/or induced autoimmunity to *C. trachomatis* heat shock protein-60 (CtHSP60) (Neuer *et al.*, 2000), a highly conserved chaperone protein that shares considerable homology with known autoantigens (Jones *et al.*, 1993). Indeed, women with TFI were found to have antibodies reactive against CtHSP60 (Hjelholt *et al.*, 2011) and increased lymphocyte proliferation on exposure to CtHSP60 antigen, compared to healthy women (Witkin *et al.*, 1994). Tubo-ovarian abscess, a potentially life-threatening PID complication, may occur in up to 43% of women hospitalized with PID (Sorbye *et al.*, 2005). Tubo-ovarian abscesses elicit symptoms similar to PID and are diagnosed by pelvic imaging; if confirmed, surgical evaluation is indicated. Rarely, *C. trachomatis* may infect the peritoneum and it may also involve the liver capsule, causing perihepatitis with right-upper quadrant abdominal pain (termed the Fitz-Hugh-Curtis syndrome). In men, complications of epididymitis include chronic testicular pain, infertility, and rarely testicular abscess or prostatitis (infection of the prostate).

In pregnant women, urogenital chlamydia may be associated with premature rupture of membranes (breakage of the amniotic sac before labour onset), miscarriage, stillbirth, and preterm delivery (Chow *et al.*, 2009; Rours *et al.*, 2011). Neonatal infection can also occur from contact with infected maternal vaginal secretions during vaginal delivery and can result in neonatal conjunctivitis (eye infection, termed 'ophthalmia neonatorum') and/or neonatal pneumonia.

Urogenital chlamydia also increases the risk of acquisition of other infections (in addition to *N. gonorrhoeae*) and complications. Studies have demonstrated that chlamydia is associated with a 2-fold increase in the risk of HIV acquisition and transmission (Johnson and Lewis, 2008); the mechanism is likely multifactorial. Inflammatory cytokines from *C. trachomatis* infection may recruit the target cells of HIV (CXCR4+/CCR5+ CD4+ T-cells) to the genital mucosa (Ogendi *et al.*, 2017; Reighard *et al.*, 2010). Also, infection of U1 monocytes with LGV serovar L2 increased HIV replication by 9-fold (Ho *et al.*, 1995). An *in vitro* study found that cell-dependent HIV transmission to the basolateral side was increased in A2EN epithelial cells infected with *C. trachomatis* (Buckner *et al.*, 2016). Chlamydia is also associated with an increased risk for squamous cell cervical cancer (Anttila *et al.*, 2001; Zhu *et al.*, 2016). Pelvic inflammatory disease may increase the risk for ovarian disease by up to 70%, especially in Asian females (Zhou *et al.*, 2017).

Anorectal and oropharyngeal chlamydia

- Is the physical examination helpful for ruling out oral or anorectal chlamydia?
- Why are anorectal chlamydia rates the same in women with and without reported anal sex?
- Can anorectal chlamydia be acquired from oral inoculation of *Chlamydia*?

C. trachomatis can infect epithelial cells of the oropharynx and anorectal area in both men and women and usually results in asymptomatic infection. The physical exam is not sensitive enough to rule out chlamydia at those sites. Oropharyngeal chlamydia is uncommon, occurring in approximately 2% of high-risk men and women (Chan *et al.*, 2016), and symptoms are exceedingly rare and include pharyngeal redness and pain. Examination of the oropharynx is limited to inspection for pharyngeal erythema but is usually unremarkable.

In women, anorectal chlamydia is frequently reported in individuals who do not engage in anal sex, likely due to autoinoculation from infected vaginal secretions (Chandra *et al.*, 2018). Another possibility is that *C. trachomatis* disseminates through the bloodstream and colonizes the gastrointestinal tract, as demonstrated in mice infected with *C. muridarum* (Zhong, 2017). Although this has yet to be proven in humans, *C. trachomatis* organisms have been isolated from the blood (Cirino *et al.*, 2006; Zigangirova *et al.*, 2013). Another intriguing hypothesis is that *C. trachomatis* detected in rectal swabs may represent shedding from colonization of the gastrointestinal tract established through oral exposure to chlamydia (i.e. inoculation from oral sex with an infected partner), termed the 'oral hypothesis.' Indeed, gastrointestinal infection with various *Chlamydia* species has been widely reported to occur in various mammalian species (Rank and Yeruva, 2014). In mice, gastric inoculation with live *C. muridarum* will establish long-lasting gastrointestinal chlamydia infection, which can be detected by nucleic acid testing of rectal swabs (Wang *et al.*, 2018). No human data yet confirms the oral hypothesis. However, in a study of 196 men presenting to an STD clinic, *C. trachomatis* nucleic acids were detected in rectal swabs from two heterosexual men who denied ever engaging in anal sex, but both admitted to performing oral sex on a female partner (Batteiger *et al.*, 2019). These men were also negative for rectal *N. gonorrhoeae* and *M. genitalium*, commonly identified infections in men who have receptive anal sex.

In both men and women, *C. trachomatis* anorectal infections are usually asymptomatic but may result in anal pain, pruritus, bleeding, diarrhoea and/or an anorectal discharge. Examination of the anus and rectum for signs of proctitis, without the aid of an anoscope, is limited to inspection of the external tissues for discharge or lesions and assessment for pain on digital rectal examination (i.e. insertion of the index finger to palpate the lateral rectal walls and prostate). On exam, anorectal erythema and a discharge may suggest rectal chlamydia. Complications of oropharyngeal and anorectal chlamydia are not well described.

Reactive arthritis and associated manifestations

- Who is at risk for developing reactive arthritis?
- How common is reactive arthritis and is it linked to a specific *Chlamydia trachomatis* serovar?
- Is reactive arthritis due to disseminated *Chlamydia trachomatis* infection?

Reactive arthritis (formerly '*Reiter's syndrome*') is a rare complication of infection with *C. trachomatis* (and some other gastrointestinal pathogens [e.g. *Campylobacter*]) that may present with the classically described triad of urethritis, conjunctivitis, and arthritis. Sexually associated reactive arthritis (SARA) is most commonly attributed to *C. trachomatis* but is also caused by *N. gonorrhoeae* and *Ureaplasma urealyticum* (Carlin *et al.*, 2014). A review of three studies found that the estimated incidence of sexually associated reactive arthritis was 3.0–8.1% (Denison *et al.*, 2016). The arthritis typically involves one or more joints, and usually lasts <6 months, with the majority of symptoms resolving within one year, but bacterial cultures of synovial fluid are negative (Schmitt, 2017). Risk factors for reactive arthritis include male gender (2:1 predominance) and HLA-B27 genotype. The pathogenesis of reactive arthritis is poorly understood. Up to 80% of cases are linked to HLA-B27 (Sieper, 2001) and several hypotheses have been suggested to explain this association, including HLA-B27 misfolding hypothesis, arthritogenic peptide hypothesis, heavy chain homodimer hypothesis, persistent infection, and gastrointestinal microbiome (Schmitt, 2017). Reactive arthritis has not been definitely linked to a specific *C. trachomatis* serovar.

C. trachomatis (and *C. pneumoniae*) are increasingly being associated with reactive arthritis through recovery of *Chlamydia* DNA from synovial fluid/tissue in affected joints, raising concerns that disseminated *C. trachomatis* infection may be occurring. Further, *C. trachomatis* DNA can be isolated from peripheral blood mononuclear cells in subjects whose synovial fluid is PCR positive for *C. trachomatis* DNA (Carter *et al.*, 2009; Kuipers *et al.*, 1998). However, it is possible that *C. trachomatis* is merely associated with reactive arthritis (i.e. an innocent bystander), not the aetiological agent, as co-infections with other pathogens are increasingly being identified (Zeidler and Hudson, 2016). Also, *C. trachomatis* has been isolated from the synovial fluid in asymptomatic individuals (Schumacher *et al.*, 1999) and individuals with osteoarthritis (Carter *et al.*, 2009). Finally, studies have failed to convincingly demonstrate that antibiotic treatment cures reactive arthritis. Several limitations should be acknowledged in studies that demonstrate an association of *C. trachomatis* with reactive arthritis, including small sample size and lack of control groups, a lack of an agreed-upon clinical definition for reactive arthritis, and a lack of testing for other pathogens. Further translational research is needed into this area.

Reactive arthritis may be associated with extraarticular manifestations, especially urethritis. Genito-urinary manifestations include cystitis (bladder inflammation), salpingitis, and prostatitis. Skin manifestations include eruption of oral ulcers, erythema nodosum (painful target-shaped rash), keratoderma blennorrhagica (hyper-pigmented rash typically on palms and soles), and circinate balanitis (rash on the head of the penis). Rarely, neurological and cardiac disease manifestations may occur.

Lymphogranuloma venereum

- By what mechanism does LGV spread to the lymph nodes?
- Does asymptomatic LGV exist?
- How common is anorectal LGV in women?

The syndrome of LGV is distinct from chlamydia. LGV manifests initially as asymptomatic or painless ulcer(s) which can then progress to an inflammatory process, often severe in intensity, with characteristic lymph node enlargement and, ultimately, destruction of involved tissues (Stoner and Cohen, 2015). The mechanisms responsible for the distinct clinical manifestations of LGV, compared to those of chlamydia, remain poorly understood, but likely result in LGV dissemination through the lymphatic system (Mabey and Peeling, 2002). This is suggested by a number of observations. First, *C. trachomatis* serovars have been demonstrated to invade dendritic cells (Datta *et al.*, 2014; Gervassi *et al.*, 2004), which preferentially drain into the lymphatic circulation (Teijeira *et al.*, 2014). Second, dendritic cells infected by LGV serovar L2, but not serovar D, significantly upregulated the lymphatic homing receptor CCR7 (Armitage *et al.*, 2015). Also, serovars L2 and D both readily infect monocytes and dendritic cells, but only L2 maintains an active development cycle; serovar D becomes persistent (Datta *et al.*, 2014).

LGV can be divided into three stages termed primary, secondary, and tertiary stages (Stamm, 2008). In the primary stage (< 30 days after infection), infected individuals are largely asymptomatic, although a small painless genital ulcer may be noted (in men, this occurs in < 50%, in women, the frequency is not well described), which typically resolves within a few days. Accompanying symptoms of urethritis and/or cervicitis may also be present or there may be proctitis symptoms in anorectal infection. The secondary stage typically occurs 2–6 weeks after the primary stage and is characterized by enlargement and tenderness of the draining regional lymph nodes, typically the femoral and inguinal nodes. These symptoms correspond with migration of LGV through the lymphatic system to the lymph nodes (Schachter and Osoba, 1983), likely within macrophages (Miyagawa, 1956). Corresponding symptoms of systemic inflammation, including fevers, chills, myalgias (muscle tenderness), fatigue, and headache may also be present. On examination, marked enlargement of the lymph nodes (termed 'buboes') may be appreciated and the classic 'groove sign' may be seen (enlargement of both the inguinal and femoral lymph nodes on either side of the inguinal ligament) (Fig. 1.3D). The tertiary stage is characterized by regional tissue destruction and disruption of the lymphatic system integrity and drainage. Marked bubo purulence may result in abscess formation and ulceration of the overlying skin with spontaneous drainage and fistula formation. Bubo enlargement and tissue destruction may obliterate the lymphatic architecture and result in genital elephantiasis (massive genital enlargement from accumulated lymph fluid). Other late-stage manifestations of urogenital LGV include infertility and frozen pelvis (perineal fibrosis that fuses the organs in place).

For anorectal LGV, the distinct stages are less clear. Ulcers may be present and can bleed. The draining lymph nodes of the rectum are deep and may not be palpable on examination. As a result, lymph node enlargement and other signs of inflammation may not be clinically evident. Anorectal LGV may progress to rectal stricture, fistula, and perirectal abscess formation.

Limited data exists on the rate of asymptomatic LGV infection as widespread LGV screening is not performed and testing for LGV serovars is not routine. However, several studies have suggested that asymptomatic cases are common. A study of men who have sex with men (MSM) visiting an Amsterdam STI clinic found that 27% with anorectal LGV were asymptomatic. It is likely that the prevalence rate and LGV-associated symptom rate varies by mucosal site. This is supported by a U.K. study of 55 LGV-infected MSM, in which 27% ($n = 15$) were asymptomatic. Infection of the rectal site was the most common (80%), followed by urogenital (13%) and oropharyngeal (7%) sites (Saxon *et al.*, 2016). In developed countries, LGV infection rates in women are very low and associated with sex with bisexual men or men in endemic environments (Savage *et al.*, 2009; Ward *et al.*, 2007).

Diagnosis and screening for *C. trachomatis* infections

- Does a positive *C. trachomatis* nucleic acid amplification test (NAAT) always indicate an active chlamydial infection?
- Can a *C. trachomatis* NAAT distinguish between recent exposure and established infection?
- Why is empiric chlamydia treatment still used when a *C. trachomatis* NAAT is available?

Because the signs and symptoms of *C. trachomatis* infections are non-specific and frequently overlap with other STIs, diagnosis requires detection of the organism, either by culture isolation or identifying the presence of *C. trachomatis* nucleic acids in specimens. Until the mid-1980s, culture was considered the gold standard for *C. trachomatis* diagnosis. However, culture required specialized technical expertise and resources and the sensitivity was relatively low (approximately 60–80%) and highly laboratory- and technician-dependent. In the mid-1990s, NAATs became available and have significant advantages over prior culture and serological assays. NAATs can be performed on a wide range of specimens (vaginal, urethral, anorectal, oropharyngeal and urine), require little specialized training, and have both a high sensitivity (> 90%) and specificity (> 99%) for detecting *C. trachomatis*. NAAT is currently the CDC-recommended test of choice for detecting *C. trachomatis*. Given the high specificity of *C. trachomatis* NAATs and clinical importance of the infection, any positive NAAT should be presumed a true infection and treatment provided.

Multiple FDA-approved NAATs are available and many of these assays combine testing for *C. trachomatis* with *N. gonorrhoeae* and other pathogens. NAAT assays differ in their amplification methods and nucleic acid targets (Papp *et al.*, 2014). The types of nucleic acid amplification include polymerase chain reaction (PCR), strand displacement amplification (SDA), and transcription-mediated amplification (TMA). Nucleic acid targets include DNA within the cryptic plasmid and 23S ribosomal RNA. As they do not target *ompA*, commercial NAATs do not distinguish between *C. trachomatis* serovars. NAAT sensitivity varies by specimen type which may reflect differences in the load of *C. trachomatis* organisms in specimens from different body sites (Papp *et al.*, 2014). In men, first-catch urine (the first portion of the urine stream) and urethral swabs have near-equivalent sensitivity, but urine collection is less invasive (urethral swabs are frequently uncomfortable to obtain) and thus is the recommended specimen for NAAT screening. In women, vaginal swabs (swab inserted into the vagina) and cervical swabs (swab of the cervical os) have near-equivalent sensitivity. However, a vaginal swab can be self-collected (women cannot easily visualize or palpate their cervical os to collect a cervical swab) and thus is the recommended specimen for NAAT screening; both vaginal and cervical swabs have approximately a 10% increased sensitivity over urine to detect *C. trachomatis* organisms. No *C. trachomatis* NAATs have been approved by the FDA for oropharyngeal and rectal swabs but reported NAAT sensitivity of > 90% has been demonstrated with these specimens (Bachmann *et al.*, 2010; Schachter *et al.*, 2008).

C. trachomatis NAATs have some limitations as a diagnostic modality when compared with *Chlamydia* culture that should be considered. First, given that the diagnostic target is *C. trachomatis* nucleic acids, NAATs cannot distinguish between a viable *C. trachomatis* infection versus the presence of non-viable genetic material (e.g. presence of infected semen in the vagina), which can be differentiated by culture. Second, individuals treated with antibiotics often become culture negative prior to becoming NAAT negative, with culture negativity typically occurring within seven days of treatment and NAAT negativity taking up to a few weeks in some treated individuals (Williams *et al.*, 2014). Importantly, this increases the potential for false-positive tests if repeat NAAT testing is performed too early after treatment. Third, NAATs also amplify a specific target gene and may fail to detect any *C. trachomatis* organism that lacks that target (e.g. a deletion mutant); though it would still be culture-positive. This limitation of NAATs was highlighted in 2006 when a *C. trachomatis* isolate was detected in Sweden, termed the Swedish new variant, that had acquired a plasmid deletion that encompassed the target of two commonly used NAATs (Ripa and Nilsson, 2007). This variant was not detected by these two NAATs and had a selective advantage because of the failure to diagnose and treat infected individuals. As a result, areas where those commercial NAATs were commonly used saw the proportion of Swedish new variant *C. trachomatis* infections increase from 2–5% in 2003 to 48–65% in 2007 (Jurstrand *et al.*, 2013). In fact, it was estimated that up to 15,000 false-negative tests occurred in Sweden from 2003 to 2006 because of the use of those two commercial NAAT assays (Herrmann, 2007). In response, many modern NAAT assays amplify more than one target gene.

No highly sensitive commercially available rapid (< 1 hour) NAAT exists to diagnose *C. trachomatis* infections, though several assays are under development. Termed 'point of care' (POC) tests, rapid testing (< 1 hour) informs clinical treatment decisions within the same clinic visit (e.g. commonly used examples of POC tests include 'rapid flu' and 'rapid strep' tests). Several enzyme-based assays may be purchased commercially that are marketed as rapid (results within minutes), easy to use, home-based kits. However, their use is not recommended as they have an unacceptably low sensitivity and specificity to be considered a reliable diagnostic test (Schachter, 2016). Despite the widespread use of *Chlamydia* NAATs, syndromic management/empiric treatment continues to be the gold-standard approach to treatment of *C. trachomatis*-compatible syndromes. This is due to the lack of POC tests. As a result, immediate treatment decisions must be made before NAAT results are available.

Several serological tests also exist to detect *C. trachomatis* antibodies, but their use is not recommended due to key limitations. First, serological tests have low sensitivity to detect acute *C. trachomatis* infection as antibody responses may take weeks or longer to develop. Second, serological tests may be positive for months to years after *C. trachomatis* infection and, as a result, cannot distinguish a recent infection from a prior infection. Third, serological tests may have low specificity, depending on the antigen used, as they may cross-react with *C. pneumoniae* antigens. Also, commercially available antibody assays have lower sensitivity than research-based assays (Geisler *et al.*, 2012; Wills *et al.*, 2009). As a result of these limitations, the use of serological testing is mostly restricted to epidemiological surveys.

A presumptive LGV diagnosis can be made in any individual with an anorectal or urogenital clinical syndrome that is compatible with LGV and is accompanied by a positive *C. trachomatis* NAAT. Though several companies offer LGV serological testing (with LGV diagnosis confirmed by a complement fixation titre > 1 : 64 or microimmunofluorescence titres > 1 : 256), no FDA-approved LGV test is available, and confirmation of LGV can be difficult. In routine clinical practice, any individual with an LGV-compatible syndrome is generally treated empirically for LGV while awaiting confirmation of a *C. trachomatis* infection by NAAT. In an LGV-compatible clinical syndrome, a positive NAAT is presumed to be an LGV serovar, though the only way to confirm LGV is to perform *ompA* typing, which is restricted to a limited number of laboratories (Workowski *et al.*, 2015).

As the majority of chlamydial infections are asymptomatic, routine screening is recommended to detect incident infections. The CDC and European guidelines recommend annual urogenital chlamydia screening by NAAT in all sexually active women aged < 25 years and any older women at increased risk for infection (e.g. new/multiple sex partners, etc.) (Lanjouw *et al.*, 2016; Workowski *et al.*, 2015). Screening for urogenital chlamydia in men is not generally recommended, except in settings with a high chlamydia prevalence (e.g. STD clinics, correctional facilities, etc.) or populations with a high chlamydia burden (e.g. MSM). Though screening programmes have failed to curb rising chlamydia rates, they have likely contributed to the observed 50% reduction in chlamydia-associated sequalae, specifically pelvic inflammatory disease and ectopic pregnancy rates, over the last 10 years (CDC, 2018). Routine oropharyngeal chlamydia screening is not recommended. Anorectal chlamydia screening is recommended at least annually in MSM who engage in receptive anal sex. However, many clinicians screen all exposed sites (informed by reported sexual behaviours) for chlamydia in high-risk individuals as studies have demonstrated that extragenital chlamydia frequently goes undiagnosed if only urogenital screening is performed (van Liere *et al.*, 2014). No specific screening recommendations exist for LGV outside of routine chlamydia screening.

Clinical management

Management of chlamydia and LGV requires a multifaceted approach, including prompt treatment of detected infections and/or complications, patient education to reduce reinfection risk, partner treatment, and repeat screening to evaluate for reinfection.

Treatment of chlamydia and its complications

- Why choose azithromycin over doxycycline to treat chlamydia, and vice versa?
- Do we need new anti-chlamydial antibiotics?
- Why are chlamydia complications and LGV treated with a longer duration than uncomplicated chlamydia?

All detected chlamydial infections should be treated, even in asymptomatic individuals, to prevent transmission and to reduce the risk of chlamydia-associated sequelae (e.g. PID, infertility, etc.). The European guidelines and CDC-recommended first-line antibiotics for uncomplicated urogenital and oropharyngeal chlamydia are orally administered doxycycline (Lanjouw *et al.*, 2016; Workowski *et al.*, 2015) or azithromycin (Workowski *et al.*, 2015). Azithromycin is preferred by some clinicians and given as a single 1 g dose that can be directly observed, compared to seven days of doxycycline 100 mg twice-daily that requires individuals to be adherent to a more complicated treatment regimen. For urogenital chlamydia, a meta-analysis of 12 randomized controlled trials, most which used chlamydia culture testing, demonstrated that both regimens were equally efficacious, with cure rates of 97% for azithromycin and 98% for doxycycline (Lau and Qureshi, 2002). A subsequent randomized control trial of adolescent women in a correctional facility who had urogenital chlamydia diagnosed by NAAT reported similar high efficacy rates (Geisler *et al.*, 2015). The high efficacy of both azithromycin and doxycycline, combined with their broad spectrum of activity (e.g. activity against a number of other STI pathogens), is a strong argument against the need for new anti-chlamydial drugs. Also, co-infections with other STI are common in individuals with chlamydia which also argues for the continued use of broad-spectrum antibiotics as opposed to a narrow-spectrum anti-chlamydial antibiotic. Unlike some other STI that are developing antibiotic resistance (e.g. *Neisseria gonorrhoeae* and *Mycoplasma genitalium* infections), drug-resistant *C. trachomatis* isolates are rare. No resistance mutations to azithromycin or fluoroquinolones were identified in 113 *C. trachomatis* isolates from men with chlamydial urethritis, although azithromycin failed to eradicate *C. trachomatis* in 4% of cases (Deguchi *et al.*, 2018). When treatment failures occur, the cause is most likely due to host factors that contribute to subtherapeutic antibiotic levels (e.g. rapid metabolism by the liver, non-compliance, etc.). For oropharyngeal infection, insufficient data exists on the efficacy of these two regimens to preferentially recommend one over the over. For anorectal chlamydia, available studies suggest azithromycin has a lower efficacy than doxycycline, and that doxycycline may be preferred. A meta-analysis of retrospective studies showed that azithromycin efficacy was 83% compared to a doxycycline efficacy of approximately 99% (Kong *et al.*, 2015), although this has not yet been confirmed in a rectal chlamydia randomized control trial. A potential limitation of retrospective treatment studies is provider bias in the choice of antibiotic regimen and other factors that could influence treatment outcomes (e.g. infection severity, adherence with antibiotics, antibiotic duration of treatment, and socioeconomic factors of individuals), which would be better controlled for in a randomized control trial study design. However, one *in vitro* study suggests that the pharmacokinetic properties of azithromycin may be suboptimal compared to doxycycline in rectal tissue. Using immortalized rectal (Caco-2) and cervical (Hela) cell lines, one study found that four urogenital *C. trachomatis* serovars (D, E, F, G) had a 2–4-fold higher minimum-inhibitory concentration (MIC) to azithromycin (but not doxycycline) in rectal cells, compared to cervical cells (Foschi *et al.*, 2018). This suggests that azithromycin may not always achieve sufficient levels in anorectal tissues to clear *C. trachomatis* and, as a result, may be inferior to doxycycline for treating anorectal *C. trachomatis* infections. To confirm this, randomized control trials comparing azithromycin and doxycycline are under way. In chlamydia-infected pregnant women, azithromycin is the first-line treatment as doxycycline may permanently stain developing teeth in the foetus and is contraindicated. Chlamydia treatment recommendations are the same in individuals with or without HIV infection. Azithromycin and doxycycline are both well tolerated by most people. The most common side effects reported are gastrointestinal symptoms (nausea, diarrhoea, etc.), which can be minimized by taking the antibiotics with food. Treatment guidelines do not specify which antibiotic is preferred in specific populations, but several circumstances do inform treatment practices. Azithromycin is a single-dose therapy and is typically preferred in individuals where there is concern for medication non-adherence. In rectal chlamydia or when LGV is suspected, doxycycline is preferred. In circumstances when treatment failure is considered, the alternate regimen is typically provided and testing for other STI pathogens may be indicated. For chlamydia, the antibiotic choice or treatment duration is not influenced by the serovar.

Treatment regimens for chlamydia complications are different than for uncomplicated chlamydia, and hospitalization may be required for parenteral (e.g. non-oral route) antibiotics if there is concern that the standard oral treatment regimen has failed or will not be tolerated. Complicated chlamydia (e.g. upper genital tract involvement in men or women or reactive arthritis) is typically treated with a longer antibiotic course, depending on the diagnosis. Although no study has compared short- versus long-course treatment for complicated chlamydia, a longer course is provided because the infected tissue burden is higher with complicated chlamydia and the clinical course more severe. Treatment of reactive arthritis involves anti-inflammatory therapy, with choice of therapy determined by disease severity. In all cases of suspected reactive arthritis, active chlamydial infection should be ruled out by NAAT (and treated with antibiotics, if detected).

For treatment of LGV, the CDC-recommended regimen is doxycycline 100 mg twice-daily for 21 days. A longer duration is recommended due to the invasive nature of the disease, though a retrospective study suggests that a shorter course (7–14 days) may be adequate for some LGV infections (Simons *et al.*, 2018). Although azithromycin (1 g orally weekly for 21 days [3 doses]) may also be effective against LGV, its use is not currently recommended due to insufficient efficacy data. If buboes are present, incision and drainage should be performed for symptomatic relief and may help to prevent formation of inguinal/femoral ulcerations and fistulas.

Strategies to prevent reinfection

- When is a test-of-cure and a 3-month test for reinfection indicated?
- Are most reinfections with the same *C. trachomatis* strain?
- What is expedited partner therapy?

Chlamydia reinfections are common. A review of 38 studies found a reinfection rate of 14% with a peak of 20% within 12 months after treatment (Hosenfeld *et al.*, 2009). Using mathematical modelling, the peak rate of chlamydia reinfections in women 15–25 years old was estimated to occur between 2–5 months after treatment (Heijne *et al.*, 2013). As a result, the CDC recommends repeat NAAT testing at approximately three months after treatment to evaluate for reinfections. The majority of reinfections appear to be due to new infections (as opposed to treatment failures or reinfection with the same strain), as one study of 162 individuals found that 80% of reinfections were due to a heterologous *ompA* serovar (Kapil *et al.*, 2014). Routine test-of-cure (e.g. repeat *C. trachomatis* NAAT to ensure treatment success by being NAAT-negative) is not recommended, except in pregnant women or when non-adherence is suspected. This recommendation is due to the high efficacy of CDC recommended antibiotics to cure *C. trachomatis* infection, with cure rates of >97% (Geisler *et al.*, 2015). Strategies to prevent chlamydia reinfection also apply to preventing LGV reinfection. Test-of-cure testing should not be performed earlier than 3–4 weeks after treatment as false-positive results may occur due to the presence of persisting non-viable *C. trachomatis* nucleic in the urogenital tract. In a study of 123 women who self-collected weekly vaginal swabs after receiving azithromycin for NAAT-positive *C. trachomatis* urogenital, the median time to NAAT-negative was nine days and 28.5%, and 11.4% remained NAAT-positive at 14- and 21-days post-treatment, respectively (Williams *et al.*, 2014).

Counselling and education aimed at preventing reinfection should be provided. To minimize the risk of reinfection, sex partner(s) should be treated and the patient and partner(s) should abstain from sexual intercourse for seven days after treatment is initiated. Individuals should be instructed to refer their sex partners for evaluation, testing and treatment if they have had sexual contact within the past 60 days. The most recent sexual contact should receive testing and treatment, even if the time since the last sexual contact occurred > 60 days ago and the contact is asymptomatic. Partners of individuals with chlamydia should receive CDC-recommended treatment for uncomplicated chlamydia. Partners of individuals diagnosed with LGV should receive treatment for non-LGV chlamydia as long as they do not have clinical manifestations of LGV. The two primary approaches to partner treatment are partner referral and expedited partner therapy (EPT). Partner referral involves the clinician recommending to the patient that he/she inform his/her partner(s) that they may have been exposed to chlamydia and should go to a clinician for chlamydia testing and treatment. EPT is the practice of providing treatment to a patient's partner without the partner having to be seen by a clinician, thereby expediting treatment; this usually entails a clinician giving a packet to the patient to give to each partner that contains one dose of azithromycin (1 g) and possibly medication instructions, condoms, and/or an information sheet/pamphlet on chlamydia. EPT is widely accepted by most individuals and their partners and evidence suggests that it may decrease the rate of chlamydia reinfections (Shiely *et al.*, 2010). However, legal, financial, and regulatory hurdles have delayed widespread adoption. Currently in the U.S., EPT may be permitted in up to 42 states and appears to be either prohibited or only potentially allowable in seven states (see http://www.cdc.gov/STD/ept/default.htm for updated information).

Trachoma

- Are trachoma rates increasing similar to chlamydia?
- Is the trachoma mechanism of transmission similar to chlamydia and LGV?
- Does trachoma occur in developed countries?

Trachoma is a chronic keratoconjunctivitis (inflammation of the eye) caused by recurrent infection with *C. trachomatis* OmpA types A-C. Conjunctivitis is inflammation of the conjunctiva, a thin mucus membrane that lines the eye and inside of the eyelid. Trachoma is a public health problem in 42 countries and over 200 million people live in trachoma endemic areas and are at risk for trachoma blindness (WHO, 2016) (Fig. 1.4A). In highly endemic areas, over 60% of children < 5 years old may have active trachoma. Based on 2016 estimates, trachoma was responsible for visual impairment or blindness in approximately 1.9 million people in 2010 (Bourne *et al.*, 2013). Unlike chlamydia, the number of individuals with trachoma has decreased over the last few decades due to trachoma eradication campaigns. In 1993, the World Health Organization (WHO) launched the WHO Alliance for the Global Elimination of Trachoma by 2020.

Trachoma is highly infectious. Unlike chlamydia that is sexually transmitted, trachoma is transmitted through the spread of ocular and nasal secretions on fingers or contaminated fomites (objects that carry infection). Risk factors include poor access to clean water, overcrowding, poor hygiene (especially facial cleanliness), and increased contact with eye-seeking flies (Taylor *et al.*, 2014). As a result, trachoma is restricted to developing countries or regions in developed countries with poor community and personal hygiene. Repeated infections are a risk factor for blindness. Only a fraction of children will ultimately progress to blindness, which is likely a complex interaction between the total burden of infection and the immune response. Severe, persistent, or recurrent conjunctival inflammation is strongly associated with risk of scarring (Munoz *et al.*, 1999). Females are also at increased risk, which may result from increased exposure to young children, who are the major reservoir (Congdon *et al.*, 1993). Several studies have identified genetic polymorphisms related to immune function that are associated with trachomatous scarring. These include IFN-γ, tumour necrosis factor-α, and matrix metalloproteinase 9 (Hu *et al.*, 2013).

Clinical manifestations of trachoma

- Is asymptomatic infection with trachoma common?
- Are individuals with trachoma at risk for genitourinary or anorectal chlamydia and associated complications?
- Why are NAATs not used to diagnose trachoma?

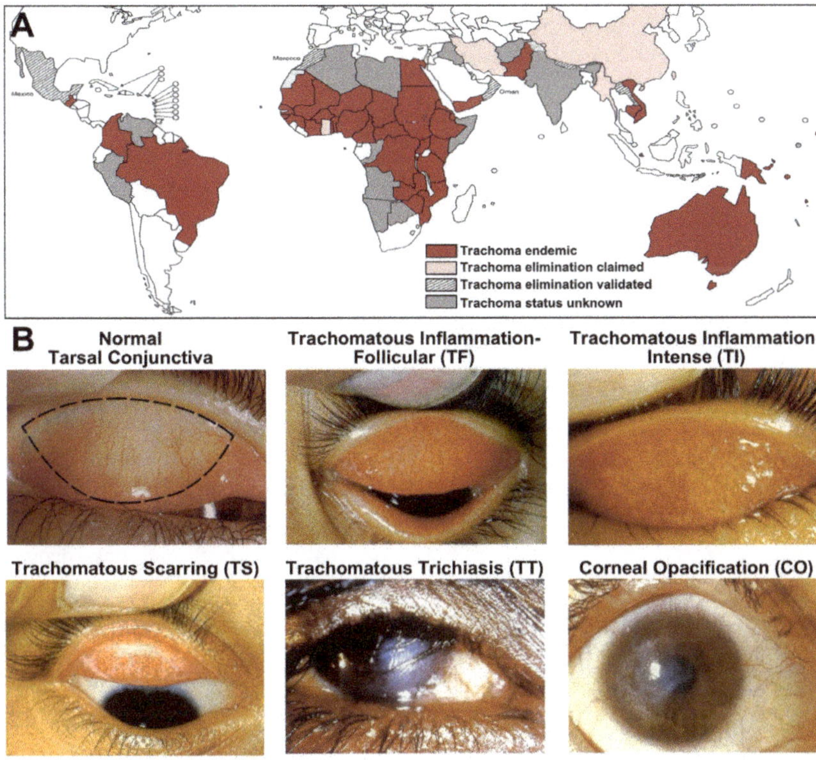

Figure 1.4 Trachoma distribution map and disease grading. (A) 2017 global distribution of trachoma. (B) Trachoma severity grading scale. Each eyelid is examined separately using binocular loupes (2.5× magnification) in adequate light by folding back the upper eyelid. The dashed line indicates the area to be examined. Adapted from www.trachomaatlas.org with permission from the International Trachoma Initiative (Decatur, GA, USA) (A) and the World Health Organization (B).

The clinical manifestations of trachoma can be divided into two phases: active trachoma (conjunctivitis) and cicatricial disease (conjunctival scarring). Active trachoma occurs most frequently in children and involves a mild, self-limiting follicular conjunctivitis, although the majority of infections are asymptomatic. Most symptoms typically represent chronic conjunctivitis and include discomfort, eye redness, sensitivity to light, and a mucopurulent eye discharge. With repeated trachoma episodes, cicatricial disease develops. Chronic conjunctival inflammation leads to scarring of the eyelid, which contracts and distorts the lid margin, and results in entropion (inward rolling of the eyelid) and trichiasis (eyelashes rubbing against the eye). Individuals with trichiasis have an increased risk of blindness as repeated eyelash abrasion (scraping) on the cornea leads to corneal oedema, opacification, and irreversible blindness. A simplified WHO trachoma grading system exists (available at http://www.who.int/trachoma/resources/SAFE_documents/en/) to stratify trachoma severity in the field, using 2.5× magnification in bright light (Fig. 1.4B). The normal conjunctiva is pink, smooth and transparent with large deep-lying blood vessels running vertically. In follicular trachomatous inflammation (TF), >5 follicles are visible (round pale swellings). In intense trachomatous inflammation (TI), the tarsal conjunctiva appears red, rough, and thick with numerous follicles which obscure more than half of the blood vessels. In trachoma scarring (TS), white lines (scarring) are easily visible. In trachoma trichiasis (TT), at least one eyelash rubs the globe (eyeball). In corneal opacification (CO), an opacified cornea obscures part of the pupillary margin.

Trachoma diagnosis is made by clinical examination, not by laboratory-based testing, and should be suspected in any individual with follicular conjunctivitis and/or conjunctival scarring living in a trachoma endemic area or at-risk for trachoma.

Trachoma is not associated with an increased risk for genitourinary or anorectal *C. trachomatis* infections due to cross-contamination, or associated chlamydia-related sequelae. This is because of existing distinct differences that make cross-contamination unlikely, including the infecting serovar (trachoma strains A–C in trachoma, chlamydia strains D–K in chlamydia) and transmission vector (non-sexual contact with infectious secretions in trachoma, sexual exposure in chlamydia). NAAT are not recommended as they are not cost-effective given the large number of people evaluated for trachoma in developing countries, and a current *C. trachomatis* infection may be absent, despite inflammation and progression towards trachomatous scarring (Burton *et al.*, 2015).

Management of trachoma

- Why is trachoma management different than chlamydia management?
- Why is azithromycin, not doxycycline, used for mass-drug administration?
- Is there a trachoma vaccine?

The approach to treatment and elimination of trachoma involves interventions at the community as well as individual level and can be stratified into four facets, termed the 'SAFE' strategy for trachoma control: **S**urgery for trichiasis, **A**ntibiotics for *C. trachomatis* infection, education on improving **F**acial cleanliness, and **E**nvironmental improvements. The reason that management approaches differ so significantly between trachoma and chlamydia is that the at-risk populations are so different due to the different transmission mechanisms. For example, with sexual transmission of *C. trachomatis*, only sexual contacts need screening and treatment (a term called 'contact tracing'); non-sexual contacts are not at-risk. However, in the case of trachoma, because infection is spread through poor hygiene, contact tracing would identify the at-risk populations as entire families and communities, so management requires interventions at the community level. Surveillance efforts help identify high density areas for deployment of the SAFE strategy. Given that trachoma rates are high in some regions with poor hygiene practices, antibiotic administration may be provided to entire regions (termed 'mass drug administration') with the goal of significantly reducing, and hopefully eliminating, the *C. trachomatis* reservoir. A review of 14 randomized trials estimated that antibiotic use was associated with a 20% decrease in the relative risk for trachoma (Evans and Solomon, 2011). Azithromycin is the preferred treatment due to the ease of a single dose regimen. Doxycycline is avoided as it is contraindicated in children (it can permanently discolour growing teeth) and can cause skin photohypersensitivity (e.g. rapidly developing sunburns in sun-exposed areas). Incidentally, mass drug azithromycin administration may also decrease the prevalence of chlamydia. This was shown in a retrospective study of almost 300 women presenting for antenatal care 10 months after receiving mass-drug administration with azithromycin in the Solomon Islands; a 40% age-adjusted reduction in chlamydia was observed (Marks *et al.*, 2016). To encourage facial cleanliness, intensive community-based health education programmes promote facial washing as children with grossly visible nasal/ocular discharge or flies on the face are over twice as likely to have active trachoma compared to children with clean faces (Ejere *et al.*, 2015). Environmental improvements include modification in the availability and use of water and latrines, improved fly control efforts, health education, and reduction in proximity to domesticated animals. When trichiasis occurs, surgery is indicated to prevent the progression of corneal abrasion and blindness.

No vaccine exists for trachoma. Early trachoma vaccine trials in the 1960s using inactivated whole EB demonstrated partial, short-term protection (Grayston and Wang, 1978). However, in subsequent non-human primate challenge studies, break-through infections appeared to exacerbate pathology in vaccinated animals (Wang *et al.*, 1967). As a result, trachoma vaccine development efforts largely abandoned whole organism vaccines in place of subunit vaccines (Mabey *et al.*, 2014). Recently, however, a plasmid-free trachoma strain has been developed for use as a live-attenuated vaccine and was demonstrated to confer protection without pathological responses in vaccinated non-human primates (Kari *et al.*, 2011), renewing hopes for a trachoma vaccine. Vaccines for chlamydia are discussed further in Chapter 15, *Chlamydia* Vaccines.

Conclusion

C. trachomatis has evolved elegant survival mechanisms that hijack host metabolic pathways and cellular resources, circumvent immune detection and associated inflammation, and maximize transmission efficiency. *C. trachomatis* infections can cause significant reproductive morbidity in women and remain a major public health concern. Capable of infecting multiple mucosal surfaces, *C. trachomatis* infections elicit protean clinical manifestations and controlling them requires a multifaceted approach that includes screening for asymptomatic infections, treating detected infections and their partners, and screening for reinfections. Although eradication campaigns are winning the public health fight against trachoma, the global burden of urogenital chlamydia is increasing. Importantly, this is happening despite the availability of highly effective antibiotics, highly sensitive screening tests (NAAT), and screening programmes designed to identify and treat asymptomatic infections in order to prevent complications and reduce chlamydia transmission; additional research is critically needed. Combating this threat will require further understanding into all aspects of the chlamydia developmental cycle and the host/pathogen interaction, which are detailed in the subsequent chapters.

References

Aiyar, A., Quayle, A.J., Buckner, L.R., Sherchand, S.P., Chang, T.L., Zea, A.H., Martin, D.H., and Belland, R.J. (2014). Influence of the tryptophan–indole-IFNγ axis on human genital *Chlamydia trachomatis* infection: role of vaginal co-infections. Front Cell Infect Micro *4*, 72. https://dx.doi.org/10.3389/fcimb.2014.00072.

Althaus, C.L., Turner, K.M., Schmid, B.V., Heijne, J.C., Kretzschmar, M., and Low, N. (2012). Transmission of *Chlamydia trachomatis* through sexual partnerships: a comparison between three individual-based models and empirical data. J Royal Society, Interface *9*, 136–146.

Anttila, T., Saikku, P., Koskela, P., Bloigu, A., Dillner, J., Ikaheimo, I., Jellum, E., Lehtinen, M., Lenner, P., Hakulinen, T., *et al.* (2001). Serotypes of *Chlamydia trachomatis* and risk for development of cervical squamous cell carcinoma. JAMA *285*, 47–51.

Apewokin, S.K., Geisler, W.M., and Bachmann, L.H. (2010). Spontaneous resolution of extragenital chlamydial and gonococcal infections prior to therapy. Sex Transm Dis *37*, 343–344. https://dx.doi.org/10.1097/OLQ.0b013e3181d73639.

Armitage, C.W., O'Meara, C.P., and Beagley, K.W. (2015). *Chlamydia pneumoniae* and *Chlamydia trachomatis* infection differentially modulates human dendritic cell line (MUTZ) differentiation and activation. Scand J Immunol *82*, 48–54. https://dx.doi.org/10.1111/sji.12295.

Arno, J.N., Ricker, V.A., Batteiger, B.E., Katz, B.P., Caine, V.A., and Jones, R.B. (1990). Interferon-gamma in endocervical secretions of women infected with *Chlamydia trachomatis*. J Infect Dis *162*, 1385–1389.

Bachmann, L.H., Johnson, R.E., Cheng, H., Markowitz, L., Papp, J.R., Palella, F.J., Jr., and Hook, E.W., 3rd (2010). Nucleic acid amplification tests for diagnosis of *Neisseria gonorrhoeae* and *Chlamydia trachomatis* rectal infections. J Clin Microbiol *48*, 1827–1832. https://dx.doi.org/10.1128/jcm.02398-09.

Bachmann, L.H., Richey, C.M., Waites, K., Schwebke, J.R., and Hook, E.W., 3rd (1999). Patterns of *Chlamydia trachomatis* testing and follow-up at a university hospital medical centre. Sex Transm Dis *26*, 496–499.

Bakshi, R., Gupta, K., Jordan, S.J., Brown, L.a.T., Press, C.G., Gorwitz, R.J., Papp, J.R., Morrison, S.G., Lee, J.Y., Morrison, R.P., *et al.* (2017). Immunoglobulin-based investigation of spontaneous resolution of *Chlamydia trachomatis* infection. J Infect Dis *215*, 1653–1656. https://dx.doi.org/10.1093/infdis/jix194.

Bakshi, R.K., Gupta, K., Jordan, S.J., Chi, X., Lensing, S.Y., Press, C.G., and Geisler, W.M. (2018). An adaptive *Chlamydia trachomatis*-specific IFN-gamma-producing CD4(+) T cell response is associated with protection against chlamydia reinfection in women. Front Immunol *9*, 1981. https://dx.doi.org/10.3389/fimmu.2018.01981.

Balkus, J.E., Manhart, L.E., Lee, J., Anzala, O., Kimani, J., Schwebke, J., Shafi, J., Rivers, C., Kabare, E., and Scott McClelland, R. (2016). Periodic presumptive treatment for vaginal infections may reduce the incidence of sexually transmitted bacterial infections. J Infect Dis *213*, 1932–1937. https://dx.doi.org/10.1093/infdis/jiw043.

Barral, R., Desai, R., Zheng, X., Frazer, L.C., Sucato, G.S., Haggerty, C.L., O'Connell, C.M., Zurenski, M.A., and Darville, T. (2014). Frequency of *Chlamydia trachomatis*-specific T cell interferon-γ and interleukin-17 responses in CD4-enriched peripheral blood mononuclear cells of sexually active adolescent females. J Repro Immunol *103*, 29–37. https://dx.doi.org/10.1016/j.jri.2014.01.002.

Batteiger, T.A., Jordan, S.J., Toh, E., Fortenberry, L., Williams, J.A., LaPradd, M., Katz, B., Fortenberry, J.D., Dodge, B., Arno, J.A., Batteiger, B.E., and Nelson, D.E. (2019). Detection of rectal Chlamydia trachomatis in heterosexual men who report cunnilingus. Sex Transm Dis *46*, 440–445. https://dx.doi.org/10.1097/OLQ.0000000000000998.

Bautista, C.T., Wurapa, E.K., Sateren, W.B., Morris, S.M., Hollingsworth, B.P., and Sanchez, J.L. (2016). Association of bacterial vaginosis With chlamydia and gonorrhea among women in the U.S. Army. Am J Prev Med *52*, 632–639. https://dx.doi.org/10.1016/j.amepre.2016.09.016.

Beatty, W.L., Belanger, T.A., Desai, A.A., Morrison, R.P., and Byrne, G.I. (1994). Tryptophan depletion as a mechanism of gamma interferon-mediated chlamydial persistence. Infect Immun *62*, 3705–3711.

Beatty, W.L., Byrne, G.I., and Morrison, R.P. (1993). Morphologic and antigenic characterization of interferon gamma-mediated persistent *Chlamydia trachomatis* infection *in vitro*. Proc Natl Acad Sci *90*, 3998–4002.

Bourne, R.R., Stevens, G.A., White, R.A., Smith, J.L., Flaxman, S.R., Price, H., Jonas, J.B., Keeffe, J., Leasher, J., Naidoo, K., *et al.* (2013). Causes of vision loss worldwide, 1990–2010: a systematic analysis. Lancet Global health *1*, e339–349. https://dx.doi.org/10.1016/s2214-109x(13)70113-x.

Bragina, E.Y., Gomberg, M.A., and Dmitriev, G.A. (2001). Electron microscopic evidence of persistent chlamydial infection following treatment. J Euro Acad Derm Vener *15*, 405–409.

Brunham, R.C., Pourbohloul, B., Mak, S., White, R., and Rekart, M.L. (2005). The unexpected impact of a *Chlamydia trachomatis* infection control program on susceptibility to reinfection. J Infect Dis *192*, 1836–1844. https://dx.doi.org/10.1086/497341.

Buckner, L.R., Amedee, A.M., Albritton, H.L., Kozlowski, P.A., Lacour, N., McGowin, C.L., Schust, D.J., and Quayle, A.J. (2016). *Chlamydia trachomatis* infection of endocervical epithelial cells enhances early HIV transmission events. PLoS ONE *11*, e0146663. https://dx.doi.org/10.1371/journal.pone.0146663.

Burton, M.J., Rajak, S.N., Hu, V.H., Ramadhani, A., Habtamu, E., Massae, P., Tadesse, Z., Callahan, K., Emerson, P.M., Khaw, P.T., *et al.* (2015). Pathogenesis of progressive scarring trachoma in Ethiopia and Tanzania and its implications for disease control: two cohort studies. PLoS Negl Trop Dis 9, e0003763. https://dx.doi.org/10.1371/journal.pntd.0003763.

Byrne, G.I. (2010). *Chlamydia trachomatis* strains and virulence: rethinking links to infection prevalence and disease severity. J Infect Dis *201 Suppl 2*, S126–133. https://dx.doi.org/10.1086/652398.

Caldwell, H.D., Kromhout, J., and Schachter, J. (1981). Purification and partial characterization of the major outer membrane protein of *Chlamydia trachomatis*. Infect Immun *31*, 1161–1176.

Caldwell, H.D., Wood, H., Crane, D., Bailey, R., Jones, R.B., Mabey, D., Maclean, I., Mohammed, Z., Peeling, R., Roshick, C., *et al.* (2003). Polymorphisms in *Chlamydia trachomatis* tryptophan synthase genes differentiate between genital and ocular isolates. J Clin Invest *111*, 1757–1769. https://dx.doi.org/10.1172/JCI17993.

Carlin, E.M., Ziza, J.M., Keat, A., and Janier, M. (2014). 2014 European Guideline on the management of sexually acquired reactive arthritis. Int J STD AIDS 25, 901–912. https://dx.doi.org/10.1177/0956462414540617.

Carlson, J.H., Porcella, S.F., McClarty, G., and Caldwell, H.D. (2005). Comparative genomic analysis of *Chlamydia trachomatis* oculotropic and genitotropic strains. Infect Immun 73, 6407–6418. https://dx.doi.org/10.1128/IAI.73.10.6407-6418.2005.

Carlson, J.H., Whitmire, W.M., Crane, D.D., Wicke, L., Virtaneva, K., Sturdevant, D.E., Kupko, J.J., 3rd, Porcella, S.F., Martinez-Orengo, N., Heinzen, R.A., *et al.* (2008). The *Chlamydia trachomatis* plasmid is a transcriptional regulator of chromosomal genes and a virulence factor. Infect Immun *76*, 2273–2283. https://dx.doi.org/10.1128/iai.00102-08.

Carter, J.D., Gerard, H.C., Espinoza, L.R., Ricca, L.R., Valeriano, J., Snelgrove, J., Oszust, C., Vasey, F.B., and Hudson, A.P. (2009). Chlamydiae as etiologic agents in chronic undifferentiated spondylarthritis. Arthritis Rheum 60, 1311–1316. https://dx.doi.org/10.1002/art.24431.

Centers for Disease Control and Prevention National Centre for HIV/AIDS, Viral Hepatitis, STD, and TB Prevention (U.S.) (2013). Incidence, Prevalence, and Cost of Sexually Transmitted Infections in the United States, 2013.

Centers for Disease Control and Prevention (2018). Sexually Transmitted Disease Surveillance 2017. Atlanta: US Department of Health and Human Services.

Chan, P.A., Robinette, A., Montgomery, M., Almonte, A., Cu-Uvin, S., Lonks, J.R., Chapin, K.C., Kojic, E.M., and Hardy, E.J. (2016). Extragenital infections caused by *Chlamydia trachomatis* and *Neisseria gonorrhoeae*: a review of the literature. Infect Dis Obstet Gynecol *2016*, 5758387. https://dx.doi.org/10.1155/2016/5758387.

Chandra, N.L., Broad, C., Folkard, K., Town, K., Harding-Esch, E.M., Woodhall, S.C., Saunders, J.M., Sadiq, S.T., and Dunbar, J.K. (2018). Detection of *Chlamydia trachomatis* in rectal specimens in women and its association with anal intercourse: a systematic review and meta-analysis. Sex Transm Infect *94*, 320–326. https://dx.doi.org/10.1136/sextrans-2017-053161.

Chen, Y.S., Bastidas, R.J., Saka, H.A., Carpenter, V.K., Richards, K.L., Plano, G.V., and Valdivia, R.H. (2014). The *Chlamydia trachomatis* type III secretion chaperone Slc1 engages multiple early effectors, including TepP, a tyrosine-phosphorylated protein required for the recruitment of CrkI-II to nascent inclusions and innate immune signalling. PLoS Pathog 10, e1003954. https://dx.doi.org/10.1371/journal.ppat.1003954.

Chow, J.M., Kang, M.S., Samuel, M.C., and Bolan, G. (2009). Assessment of the association of *Chlamydia trachomatis* infection and adverse perinatal outcomes with the use of population-based chlamydia case report registries and birth records. Pub Health Reports *124 Suppl 2*, 24–30. https://dx.doi.org/10.1177/00333549091240s205.

Christerson, L., de Vries, H.J., de Barbeyrac, B., Gaydos, C.A., Henrich, B., Hoffmann, S., Schachter, J., Thorvaldsen, J., Vall-Mayans, M., Klint, M., *et al.* (2010). Typing of lymphogranuloma venereum *Chlamydia trachomatis* strains. Emerg Infect Dis *16*, 1777–1779. https://dx.doi.org/10.3201/eid1611.100379.

Cirino, F., Webley, W.C., West, C., Croteau, N.L., Andrzejewski, C., Jr., and Stuart, E.S. (2006). Detection of *Chlamydia* in the peripheral blood cells of normal donors using *in vitro* culture, immunofluorescence microscopy and flow cytometry techniques. BMC Infect Dis 6, 23. https://dx.doi.org/10.1186/1471-2334-6-23.

Clifton, D.R., Fields, K.A., Grieshaber, S.S., Dooley, C.A., Fischer, E.R., Mead, D.J., Carabeo, R.A., and Hackstadt, T. (2004). A chlamydial type III translocated protein is tyrosine-phosphorylated at the site of entry and associated with recruitment of actin. Proc Natl Acad Sci *101*, 10166–10171. https://dx.doi.org/10.1073/pnas.0402829101.

Cohen, C.R., Koochesfahani, K.M., Meier, A.S., Shen, C., Karunakaran, K., Ondondo, B., Kinyari, T., Mugo, N.R., Nguti, R., and Brunham, R.C. (2005). Immunoepidemiologic profile of *Chlamydia trachomatis* infection: importance of heat-shock protein 60 and interferon-γ. J Infect Dis *192*, 591–599. https://dx.doi.org/10.1086/432070.

Congdon, N., West, S., Vitale, S., Katala, S., and Mmbaga, B.B. (1993). Exposure to children and risk of active trachoma in Tanzanian women. Am J Epidemiol *137*, 366–372.

Datta, B., Njau, F., Thalmann, J., Haller, H., and Wagner, A.D. (2014). Differential infection outcome of *Chlamydia trachomatis* in human blood monocytes and monocyte-derived dendritic cells. BMC Microbiol *14*, 209. https://dx.doi.org/10.1186/s12866-014-0209-3.

Datta, S.D., Sternberg, M., Johnson, R.E., Berman, S., Papp, J.R., McQuillan, G., and Weinstock, H. (2007). Gonorrhea and chlamydia in the United States among persons 14 to 39 years of age, 1999 to 2002. Annals Internal Med *147*, 89–96.

De Clercq, E., Kalmar, I., and Vanrompay, D. (2013). Animal models for studying female genital tract infection with *Chlamydia trachomatis*. Infect Immun *81*, 3060–3067. https://dx.doi.org/10.1128/IAI.00357-13.

Dean, D., Suchland, R.J., and Stamm, W.E. (2000). Evidence for long-term cervical persistence of *Chlamydia trachomatis* by omp1 genotyping. J Infect Dis *182*, 909–916. https://dx.doi.org/10.1086/315778.

Deguchi, T., Hatazaki, K., Ito, S., Kondo, H., Horie, K., Nakane, K., Mizutani, K., Tsuchiya, T., Yasuda, M., Yokoi, S., *et al*,. (2018). Macrolide and fluoroquinolone resistance is uncommon in clinical strains of Chlamydia trachomatis. J Infect Chemother *24*, 610–614. https://dx.doi.org/10.1016/j.jiac.2018.03.007.

Denison, H.J., Curtis, E.M., Clynes, M.A., Bromhead, C., Dennison, E.M., and Grainger, R. (2016). The incidence of sexually acquired reactive arthritis: a systematic literature review. Clin Rheumatol *35*, 2639–2648. https://dx.doi.org/10.1007/s10067-016-3364-0.

Detels, R., Green, A.M., Klausner, J.D., Katzenstein, D., Gaydos, C., Handsfield, H.H., Pequegnat, W., Mayer, K., Hartwell, T.D., and Quinn, T.C. (2011). The incidence and correlates of symptomatic and asymptomatic *Chlamydia trachomatis* and *Neisseria gonorrhoeae* infections in selected populations in five countries. Sex Transm Dis *38*, 503.

Ejere, H.O., Alhassan, M.B., and Rabiu, M. (2015). Face washing promotion for preventing active trachoma. Cochrane Database Syst Rev, Cd003659. https://dx.doi.org/10.1002/14651858.CD003659.pub4.

Evans, J.R., and Solomon, A.W. (2011). Antibiotics for trachoma. Cochrane Database Syst Rev, Cd001860. https://dx.doi.org/10.1002/14651858.CD001860.pub3.

Farley, T.A., Cohen, D.A., and Elkins, W. (2003). Asymptomatic sexually transmitted diseases: the case for screening. Prevent Med *36*, 502–509. https://dx.doi.org/10.1016/S0091-7435(02)00058-0.

Fields, K.A., and Hackstadt, T. (2002). The chlamydial inclusion: escape from the endocytic pathway. Ann Rev Cell Devel Biol *18*, 221–245. https://dx.doi.org/10.1146/annurev.cellbio.18.012502.105845.

Fields, K.A., Mead, D.J., Dooley, C.A., and Hackstadt, T. (2003). *Chlamydia trachomatis* type III secretion: evidence for a functional apparatus during early-cycle development. Mol Micro *48*, 671–683.

Forcey, D.S., Hocking, J.S., Tabrizi, S.N., Bradshaw, C.S., Chen, M.Y., Fehler, G., Nash, J.L., and Fairley, C.K. (2014). *Chlamydia* detection during the menstrual cycle: a cross-sectional study of women attending a sexual health service. PLoS One *9*, e85263. https://dx.doi.org/10.1371/journal.pone.0085263.

Foschi, C., Salvo, M., Cevenini, R., and Marangoni, A. (2018). *Chlamydia trachomatis* antimicrobial susceptibility in colorectal and endocervical cells. J Antimicrob Chemother *73*, 409–413. https://dx.doi.org/10.1093/jac/dkx392.

Geisler, W.M., Chow, J.M., Schachter, J., and McCormack, W.M. (2007). Pelvic examination findings and *Chlamydia trachomatis* infection in asymptomatic young women screened with a nucleic acid amplification test. Sex Transm Dis *34*, 335–338. https://dx.doi.org/10.1097/01.olq.0000240344.20665.63.

Geisler, W.M., Lensing, S.Y., Press, C.G., and Hook, E.W. (2013). Spontaneous resolution of genital *Chlamydia trachomatis* infection in women and protection from reinfection. J Infect Dis *207*, 1850–1856. https://dx.doi.org/10.1093/infdis/jit094.

Geisler, W.M., Morrison, S.G., Doemland, M.L., Iqbal, S.M., Su, J., Mancevski, A., Hook, E.W., and Morrison, R.P. (2012). Immunoglobulin-specific responses to *Chlamydia* elementary bodies in individuals with and at risk for genital chlamydial infection. J Infect Dis 206, 1836–1843. https://dx.doi.org/10.1093/infdis/jis621.

Geisler, W.M., Suchland, R.J., Whittington, W.L.H., and Stamm, W.E. (2003). The relationship of serovar to clinical manifestations of urogenital *Chlamydia trachomatis* infection. Sex Transm Dis *30*, 160–165.

Geisler, W.M., Uniyal, A., Lee, J.Y., Lensing, S.Y., Johnson, S., Perry, R.C.W., Kadrnka, C.M., and Kerndt, P.R. (2015). Azithromycin versus doxycycline for urogenital *Chlamydia trachomatis* infection. New Eng J Med *373*, 2512–2521. https://dx.doi.org/10.1056/NEJMoa1502599.

Geisler, W.M., Wang, C., Morrison, S.G., Black, C.M., Bandea, C.I., and Hook III, E.W. (2008). The Natural history of untreated *Chlamydia trachomatis* infection in the interval between screening and returning for treatment. Sex Transm Dis *35*, 119–123. https://dx.doi.org/10.1097/OLQ.0b013e318151497d.

Geisler, W.M., Yu, S., and Hook, E.W. (2005). Chlamydial and gonococcal infection in men without polymorphonuclear leucocytes on Gram stain: implications for diagnostic approach and management. Sex Transm Dis *32*, 630.

Gervassi, A., Alderson, M.R., Suchland, R., Maisonneuve, J.F., Grabstein, K.H., and Probst, P. (2004). Differential regulation of inflammatory cytokine secretion by human dendritic cells upon *Chlamydia trachomatis* infection. Infect Immun *72*, 7231–7239. https://dx.doi.org/10.1128/IAI.72.12.7231-7239.2004.

Grayston, J.T., and Wang, S.P. (1978). The potential for vaccine against infection of the genital tract with *Chlamydia trachomatis*. Sex Transm Dis *5*, 73–77.

Harris, S.R., Clarke, I.N., Seth-Smith, H.M., Solomon, A.W., Cutcliffe, L.T., Marsh, P., Skilton, R.J., Holland, M.J., Mabey, D., Peeling, R.W., *et al.* (2012). Whole-genome analysis of diverse *Chlamydia trachomatis* strains identifies phylogenetic relationships masked by current clinical typing. Nature Gen *44*, 413–419, s411. https://dx.doi.org/10.1038/ng.2214.

Heijne, J.C., Herzog, S.A., Althaus, C.L., Tao, G., Kent, C.K., and Low, N. (2013). Insights into the timing of repeated testing after treatment for *Chlamydia trachomatis*: data and modelling study. Sex Transm Infect *89*, 57–62. https://dx.doi.org/10.1136/sextrans-2011-050468.

Herrmann, B. (2007). A new genetic variant of *Chlamydia trachomatis*. Sex Transm Infect *83*, 253–254. https://dx.doi.org/10.1136/sti.2007.026260.

Ho, J.L., He, S., Hu, A., Geng, J., Basile, F.G., Almeida, M.G., Saito, A.Y., Laurence, J., and Johnson, W.D., Jr. (1995). Neutrophils from human immunodeficiency virus (HIV)-seronegative donors induce HIV replication from HIV-infected patients' mononuclear cells and cell lines: an *in vitro* model of HIV transmission facilitated by *Chlamydia trachomatis*. J Exp Med *181*, 1493–1505.

Holmes, K.K., Johnson, D.W., Floyd, T.M., and Kvale, P.A. (1967). Studies of venereal disease. II. Observations on the incidence, aetiology, and treatment of the postgonococcal urethritis syndrome. JAMA *202*, 467–473.

Hook, E.W., 3rd, Spitters, C., Reichart, C.A., Neumann, T.M., and Quinn, T.C. (1994). Use of cell culture and a rapid diagnostic assay for *Chlamydia trachomatis* screening. JAMA *272*, 867–870.

Hosenfeld, C.B., Workowski, K.A., Berman, S., Zaidi, A., Dyson, J., Mosure, D., Bolan, G., and Bauer, H.M. (2009). Repeat infection with chlamydia and gonorrhea among females: a systematic review of the literature. Sex Transm Dis *36*, 478–489. https://dx.doi.org/10.1097/OLQ.0b013e3181a2a933.

Hu, V.H., Holland, M.J., and Burton, M.J. (2013). Trachoma: protective and pathogenic ocular immune responses to *Chlamydia trachomatis*. PLoS Negl Trop Dis *7*, e2020. https://dx.doi.org/10.1371/journal.pntd.0002020.

Hvid, M., Baczynska, A., Deleuran, B., Fedder, J., Knudsen, H.J., Christiansen, G., and Birkelund, S. (2007). Interleukin-1 is the initiator of fallopian tube destruction during *Chlamydia trachomatis* infection. Cell Microbiol *9*, 2795–2803. https://dx.doi.org/10.1111/j.1462-5822.2007.00996.x.

Hybiske, K., and Stephens, R.S. (2007). Mechanisms of *Chlamydia trachomatis* entry into nonphagocytic cells. Infect Immun *75*, 3925–3934. https://dx.doi.org/10.1128/IAI.00106-07.

Ibana, J.A., Belland, R.J., Zea, A.H., Schust, D.J., Nagamatsu, T., AbdelRahman, Y.M., Tate, D.J., Beatty, W.L., Aiyar, A.A., and Quayle, A.J. (2011). Inhibition of indoleamine 2,3-dioxygenase activity by levo-1-methyl tryptophan blocks gamma interferon-induced *Chlamydia trachomatis* persistence in human epithelial cells. Infect Immun *79*, 4425–4437. https://dx.doi.org/10.1128/IAI.05659-11.

Johnson, L.F., and Lewis, D.A. (2008). The effect of genital tract infections on HIV-1 shedding in the genital tract: a systematic review and meta-analysis. Sex Transm Dis *35*, 946–959. 10.1097/OLQ.0b013e3181812d15.

Jones, D.B., Coulson, A.F., and Duff, G.W. (1993). Sequence homologies between hsp60 and autoantigens. Immunol Today *14*, 115–118. https://dx.doi.org/10.1016/0167-5699(93)90210-c.

Jordan, S.J., Aaron, K.J., Schwebke, J.R., Van Der Pol, B.J., and Hook, E.W., 3rd (2018). Defining the urethritis syndrome in men using patient reported symptoms. Sex Transm Dis *45*, e40-e42. https://dx.doi.org/10.1097/olq.0000000000000790.

Jordan, S.J., Olson, K.M., Barnes, S., Wilson, L.S., Berryhill, T.F., Bakshi, R., Brown, L.a.T., Press, C.G., and Geisler, W.M. (2017). Lower levels of cervicovaginal tryptophan are associated with natural clearance of chlamydia in women. J Infect Dis *215*, 1888–1892. https://dx.doi.org/10.1093/infdis/jix240.

Jossens, M.O., Schachter, J., and Sweet, R.L. (1994). Risk factors associated with pelvic inflammatory disease of differing microbial etiologies. Obstet Gyn *83*, 989–997.

Jurstrand, M., Fredlund, H., and Unemo, M. (2013). The new variant of *Chlamydia trachomatis* was present as early as 2003 in Orebro County, Sweden, but remained undetected until 2006. Sex Transm Infect *89*, 607–608. https://dx.doi.org/10.1136/sextrans-2013-051018.

Kapil, R., Press, C.G., Hwang, M.L., Brown, L., and Geisler, W.M. (2014). Investigating the epidemiology of repeat *Chlamydia trachomatis* detection after treatment using *Chlamydia trachomatis* Omp A genotyping. J Clin Micro *53*, 546–549. https://dx.doi.org/10.1128/JCM.02483-14.

Kari, L., Whitmire, W.M., Olivares-Zavaleta, N., Goheen, M.M., Taylor, L.D., Carlson, J.H., Sturdevant, G.L., Lu, C., Bakios, L.E., Randall, L.B., *et al.* (2011). A live-attenuated chlamydial vaccine protects against trachoma in nonhuman primates. J Exp Med *208*, 2217–2223. https://dx.doi.org/10.1084/jem.20111266.

Katz, B.P. (1992). Estimating transmission probabilities for chlamydial infection. Statistics Med *11*, 565–577.

Koehler, L., Nettelnbreker, E., Hudson, A.P., Ott, N., Gerard, H.C., Branigan, P.J., Schumacher, H.R., Drommer, W., and Zeidler, H. (1997). Ultrastructural and molecular analyses of the persistence of *Chlamydia trachomatis* (serovar K) in human monocytes. Microbial Path *22*, 133–142. https://dx.doi.org/10.1006/mpat.1996.0103.

Kong, F.Y.S., Tabrizi, S.N., Fairley, C.K., Vodstrcil, L.A., Huston, W.M., Chen, M., Bradshaw, C., and Hocking, J.S. (2015). The efficacy of azithromycin and doxycycline for the treatment of rectal chlamydia infection: a systematic review and meta-analysis. J Antimicro Chemo *70*, 1290–1297. https://dx.doi.org/10.1093/jac/dku574.

Kuipers, J.G., Jurgens-Saathoff, B., Bialowons, A., Wollenhaupt, J., Kohler, L., and Zeidler, H. (1998). Detection of *Chlamydia trachomatis* in peripheral blood leucocytes of reactive arthritis patients by polymerase chain reaction. Arthritis Rheum *41*, 1894–1895. https://dx.doi.org/10.1002/1529-0131(199810)41:10<1894::AID-ART24>3.0.CO;2-E.

Kuo, C.C., Wang, S.P., Holmes, K.K., and Grayston, J.T. (1983). Immunotypes of *Chlamydia trachomatis* isolates in Seattle, Washington. Infect Immun *41*, 865–868.

Lanjouw, E., Ouburg, S., de Vries, H.J., Stary, A., Radcliffe, K., and Unemo, M. (2016). 2015 European guideline on the management of *Chlamydia trachomatis* infections. Int J STD AIDS *27*, 333–348. https://dx.doi.org/10.1177/0956462415618837.

Lau, C.-Y., and Qureshi, A.K. (2002). Azithromycin versus doxycycline for genital chlamydial infections: a meta-analysis of randomized clinical trials. Sex Transm Dis *29*, 497–502.

Lewis, J., Price, M.J., Horner, P.J., and White, P.J. (2017). Genital *Chlamydia trachomatis* infections clear more slowly in men than women, but are less likely to become established. J Infect Dis *216*, 237–244. https://dx.doi.org/10.1093/infdis/jix283.

Liechti, G., Kuru, E., Packiam, M., Hsu, Y.P., Tekkam, S., Hall, E., Rittichier, J.T., VanNieuwenhze, M., Brun, Y.V., and Maurelli, A.T. (2016). Pathogenic *chlamydia* lack a classical sacculus but synthesize a narrow, mid-cell peptidoglycan ring, regulated by MreB, for cell division. PLoS Pathog *12*, e1005590. https://dx.doi.org/10.1371/journal.ppat.1005590.

Liu, X., Afrane, M., Clemmer, D.E., Zhong, G., and Nelson, D.E. (2010). Identification of *Chlamydia trachomatis* outer membrane complex proteins by differential proteomics. J Bac *192*, 2852–2860. https://dx.doi.org/10.1128/JB.01628-09.

Liu, Y., Huang, Y., Yang, Z., Sun, Y., Gong, S., Hou, S., Chen, C., Li, Z., Liu, Q., Wu, Y., *et al.* (2014). Plasmid-encoded Pgp3 is a major virulence factor for *Chlamydia muridarum* to induce hydrosalpinx in mice. Infect Immun *82*, 5327–5335. https://dx.doi.org/10.1128/IAI.02576-14.

Mabey, D., and Peeling, R.W. (2002). Lymphogranuloma venereum. Sexually transmitted infections *78*, 90–92.

Mabey, D.C.W., Hu, V., Bailey, R.L., Burton, M.J., and Holland, M.J. (2014). Towards a safe and effective chlamydial vaccine: lessons from the eye. Vaccine *32*, 1572–1578. https://dx.doi.org/10.1016/j.vaccine.2013.10.016.

Macdonald, N., Sullivan, A.K., French, P., White, J.A., Dean, G., Smith, A., Winter, A.J., Alexander, S., Ison, C., and Ward, H. (2014). Risk factors for rectal lymphogranuloma venereum in gay men: results of a multicentre case–control study in the UK. Sex Transm Infect *90*, 262–268. https://dx.doi.org/10.1136/sextrans-2013-051404.

Marks, M., Bottomley, C., Tome, H., Pitakaka, R., Butcher, R., Sokana, O., Kako, H., Solomon, A.W., and Mabey, D.C. (2016). Mass drug administration of azithromycin for trachoma reduces the prevalence of genital *Chlamydia trachomatis* infection in the Solomon Islands. Sex Transm Infect 92, 261–265. https://dx.doi.org/10.1136/sextrans-2015-052439.

McCormack, W.M., Alpert, S., McComb, D.E., Nichols, R.L., Semine, D.Z., and Zinner, S.H. (1979). Fifteen-month follow-up study of women infected with *Chlamydia trachomatis*. New Eng J Med *300*, 123–125. https://dx.doi.org/10.1056/NEJM197901183000305.

Miyagawa, Y. (1956). Studies on the characteristics of *Miyagawanella lymphogranulomatis*. Jap J Exp Med *26*, 157–159.

Molano, M., Meijer, C.J., Weiderpass, E., Arslan, A., Posso, H., Franceschi, S., Ronderos, M., Muñoz, N., and van den Brule, A.J. (2005). The natural course of *Chlamydia trachomatis* infection in asymptomatic Colombian women: a 5-year follow-up study. J Infect Dis *191*, 907–916. https://dx.doi.org/10.1086/428287.

Morré, S.A., Rozendaal, L., van Valkengoed, I.G., Boeke, A.J., van Voorst Vader, P.C., Schirm, J., de Blok, S., van Den Hoek, J.A., van Doornum, G.J., Meijer, C.J., *et al.* (2000). Urogenital *Chlamydia trachomatis* serovars in men and women with a symptomatic or asymptomatic infection: an association with clinical manifestations? J Clin Micro *38*, 2292–2296.

Moulder, J.W. (1991). Interaction of Chlamydiae and host cells *in vitro*. Microbiological reviews *55*, 143–190.

Mpiga, P., and Ravaoarinoro, M. (2006). *Chlamydia trachomatis* persistence: an update. Microbiol Research *161*, 9–19. https://dx.doi.org/10.1016/j.micres.2005.04.004.

Munoz, B., Bobo, L., Mkocha, H., Lynch, M., Hsieh, Y.H., and West, S. (1999). Incidence of trichiasis in a cohort of women with and without scarring. Int J Epidemiol *28*, 1167–1171.

Ness, R.B., Soper, D.E., Holley, R.L., Peipert, J., Randall, H., Sweet, R.L., Sondheimer, S.J., Hendrix, S.L., Amortegui, A., Trucco, G., *et al.* (2002). Effectiveness of inpatient and outpatient treatment strategies for women with pelvic inflammatory disease: results from the Pelvic Inflammatory Disease Evaluation and Clinical Health (PEACH) randomized trial. Am J Obstet Gynecol *186*, 929–937.

Neuer, A., Spandorfer, S.D., Giraldo, P., Dieterle, S., Rosenwaks, Z., and Witkin, S.S. (2000). The role of heat shock proteins in reproduction. Hum Repro Update *6*, 149–159.

Newman, L., Rowley, J., Vander Hoorn, S., Wijesooriya, N.S., Unemo, M., Low, N., Stevens, G., Gottlieb, S., Kiarie, J., and Temmerman, M. (2015). Global estimates of the prevalence and incidence of four curable sexually transmitted infections in 2012 based on systematic review and global reporting. PLoS ONE *10*, e0143304. https://dx.doi.org/10.1371/journal.pone.0143304.

Ogendi, B.M.O., Bakshi, R.K., Sabbaj, S., Brown, L., Lee, J.Y., Kapil, R., and Geisler, W.M. (2017). Distinct peripheral vs mucosal T-cell phenotypes in chlamydia-infected women. Am J Reprod Immunol *78*. https://dx.doi.org/10.1111/aji.12768.

Olson, K.M., Tang, J., Brown, L., Press, C.G., and Geisler, W.M. (2018). HLA-DQB1*06 is a risk marker for chlamydia reinfection in African American women. Genes Immun *201 Suppl 2*, S104. https://dx.doi.org/10.1038/s41435-018-0014-3.

Papp, J.R., Schachter, J., Gaydos, C., and Van Der Pol, B. (2014). Recommendations for the laboratory-based detection of *Chlamydia trachomatis* and *Neisseria gonorrhoeae*--2014. MMWR 63, 1–19.

Parks, K.S., Dixon, P.B., Richey, C.M., and Hook, E.W.I. (1997). Spontaneous clearance of *Chlamydia trachomatis* infection in untreated patients. Sex Transm Dis *24*, 229.

Perroud, H.M., and Miedzybrodzka, K. (1978). Chlamydial infection of the urethra in men. Br J Venereal Dis *54*, 45–49.

Phillips Campbell, R., Kintner, J., and Schoborg, R.V. (2014). Induction of the *Chlamydia muridarum* stress/persistence response increases azithromycin treatment failure in a murine model of infection. Antimicro Agents Chemo *58*, 1782–1784. https://dx.doi.org/10.1128/AAC.02097-13.

Phillips Campbell, R., Kintner, J., Whittimore, J., and Schoborg, R.V. (2012). *Chlamydia muridarum* enters a viable but non-infectious state in amoxicillin-treated

BALB/c mice. Microbes Infect *14*, 1177–1185. https://dx.doi.org/10.1016/j.micinf.2012.07.017.

Pospischil, A., Borel, N., Chowdhury, E.H., and Guscetti, F. (2009). Aberrant chlamydial developmental forms in the gastrointestinal tract of pigs spontaneously and experimentally infected with *Chlamydia suis*. Vet Micro *135*, 147–156. https://dx.doi.org/10.1016/j.vetmic.2008.09.035.

Rajeeve, K., Das, S., Prusty, B.K., and Rudel, T. (2018). *Chlamydia trachomatis* paralyses neutrophils to evade the host innate immune response. Nat Microbiol 3, 824–835. https://dx.doi.org/10.1038/s41564-018-0182-y.

Rank, R.G., and Whittum Hudson, J.A. (2010). Protective immunity to chlamydial genital infection: evidence from animal studies. J Infect Dis *201 Suppl 2*, S168–177. https://dx.doi.org/10.1086/652399.

Rank, R.G., and Yeruva, L. (2014). Hidden in plain sight: chlamydial gastrointestinal infection and its relevance to persistence in human genital infection. Infect Immun *82*, 1362–1371. https://dx.doi.org/10.1128/IAI.01244-13.

Read, T.D., Brunham, R.C., Shen, C., Gill, S.R., Heidelberg, J.F., White, O., Hickey, E.K., Peterson, J., Utterback, T., Berry, K., *et al.* (2000). Genome sequences of *Chlamydia trachomatis* MoPn and *Chlamydia pneumoniae* AR39. Nucleic Acids Res *28*, 1397–1406.

Read, T.D., Myers, G.S., Brunham, R.C., Nelson, W.C., Paulsen, I.T., Heidelberg, J., Holtzapple, E., Khouri, H., Federova, N.B., Carty, H.A., *et al.* (2003). Genome sequence of *Chlamydophila caviae* (*Chlamydia psittaci* GPIC): examining the role of niche-specific genes in the evolution of the *Chlamydiaceae*. Nucleic Acids Res *31*, 2134–2147.

Reighard, S.D., Sweet, R.L., Miguel, C.V., Miguel, R.D.V., Chivukula, M., Krishnamurti, U., and Cherpes, T.L. (2010). Endometrial leucocyte subpopulations associated with *Chlamydia trachomatis, Neisseria gonorrhoeae*, and *Trichomonas vaginalis* genital tract infection. YMOB *205*, 324.e321–324.e327. https://dx.doi.org/10.1016/j.ajog.2011.05.031.

Reveneau, N., Crane, D.D., Fischer, E., and Caldwell, H.D. (2005). Bactericidal activity of first-choice antibiotics against gamma interferon-induced persistent infection of human epithelial cells by *Chlamydia trachomatis*. Antimicrob Agents Chemother *49*, 1787–1793. https://dx.doi.org/10.1128/aac.49.5.1787-1793.2005.

Rietmeijer, C.A., and Mettenbrink, C.J. (2012). Recalibrating the Gram stain diagnosis of male urethritis in the era of nucleic acid amplification testing. Sex Transm Dis *39*, 18–20. https://dx.doi.org/10.1097/OLQ.0b013e3182354da3.

Ripa, T., and Nilsson, P.A. (2007). A *Chlamydia trachomatis* strain with a 377-bp deletion in the cryptic plasmid causing false-negative nucleic acid amplification tests. Sex Transm Dis *34*, 255–256. 10.1097/OLQ.0b013e31805ce2b9.

Rours, G.I., Duijts, L., Moll, H.A., Arends, L.R., de Groot, R., Jaddoe, V.W., Hofman, A., Steegers, E.A., Mackenbach, J.P., Ott, A., *et al.* (2011). *Chlamydia trachomatis* infection during pregnancy associated with preterm delivery: a population-based prospective cohort study. Eur J Epidem *26*, 493–502. https://dx.doi.org/10.1007/s10654-011-9586-1.

Russell, A.N., Zheng, X., O'Connell, C.M., Wiesenfeld, H.C., Hillier, S.L., Taylor, B.D., Picard, M.D., Flechtner, J.B., Zhong, W., Frazer, L.C., *et al.* (2016). Identification of *Chlamydia trachomatis* antigens recognized by T cells from highly exposed women who limit or resist genital tract Infection. J Infect Dis *214*, 1884–1892. https://dx.doi.org/10.1093/infdis/jiw485.

Savage, E.J., van de Laar, M.J., Gallay, A., van der Sande, M., Hamouda, O., Sasse, A., Hoffmann, S., Diez, M., Borrego, M.J., Lowndes, C.M., *et al.* (2009). Lymphogranuloma venereum in Europe, 2003–2008. Euro Surveill *14* (48), pii=19428. http://www.eurosurveillance.org/ViewArticle.aspx?ArticleId=19428.

Saxon, C., Hughes, G., and Ison, C. (2016). Asymptomatic lymphogranuloma venereum in men who have sex with men, United Kingdom. Emerg Infect Dis *22*, 112–116. https://dx.doi.org/10.3201/EID2201.141867.

Schachter, J. (2016). Point-of-care tests using enzyme detection to diagnose *Chlamydia trachomatis* infection do not work. But when they fail in clinical trials, they reappear under different names. Sex Transm Infect *92*, 406–407. https://dx.doi.org/10.1136/sextrans-2015-052508.

Schachter, J., Moncado, J., Liska, S., Shayevich, C., and Klausner, J.D. (2008). Nucleic acid amplification tests in the diagnosis of chlamydial and gonococcal infections of the oropharynx and rectum in men who have sex with men. Sex Transm Dis *35*, 637–642. https://dx.doi.org/10.1097/OLQ.0b013e31817bdd7e.

Schachter, J., and Osoba, A.O. (1983). Lymphogranuloma venereum. Br Med Bull *39*, 151–154.

Schmitt, S.K. (2017). Reactive Arthritis. Infect Dis Clin North Am *31*, 265–277. https://dx.doi.org/10.1016/j.idc.2017.01.002.

Schumacher, H.R., Jr., Arayssi, T., Crane, M., Lee, J., Gerard, H., Hudson, A.P., and Klippel, J. (1999). *Chlamydia trachomatis* nucleic acids can be found in the synovium of some asymptomatic subjects. Arthritis Rheumat *42*, 1281–1284. https://dx.doi.org/10.1002/1529-0131(199906)42:6<1281::AID-ANR27>3.0.CO;2-8.

Shiely, F., Hayes, K., Thomas, K.K., Kerani, R.P., Hughes, J.P., Whittington, W.L., Holmes, K.K., Handsfield, H.H., Hogben, M., and Golden, M.R. (2010). Expedited partner therapy: a robust intervention. Sex Transm Dis *37*, 602–607. https://dx.doi.org/10.1097/OLQ.0b013e3181e1a296.

Shih, S., Scholle, S., Irwin, K., Tao, G., Walsh, C., and Tun, W. (2009). Chlamydia screening among sexually active young female enrollees of health plans—United States, 2000–2007. MMWR Morb Mortal Wkly Rep *58*, 362–365.

Sieper, J. (2001). Pathogenesis of reactive arthritis. Cur Rheum Rep 3, 412–418.

Simons, R., Candfield, S., French, P., and White, J.A. (2018). Observed treatment responses to short-course doxycycline therapy for rectal lymphogranuloma venereum in men who have sex with men. Sex Transm Dis *45*, 406–408. https://dx.doi.org/10.1097/OLQ.0000000000000772.

Smith, A., Munoz, B., Hsieh, Y.H., Bobo, L., Mkocha, H., and West, S. (2001). OmpA genotypic evidence for persistent ocular *Chlamydia trachomatis* infection in Tanzanian village women. Ophthalmic Epi *8*, 127–135.

Sorbye, I.K., Jerve, F., and Staff, A.C. (2005). Reduction in hospitalized women with pelvic inflammatory disease in Oslo over the past decade. Acta Obstet Gyn Scand *84*, 290–296. https://dx.doi.org/10.1111/j.0001-6349.2005.00509.x.

Stamm, W.E. (2008). Lymphogranuloma Venereum. In Sexually Transmitted Diseases, S.P. Holmes KK, Stamm WE, ed. (McGraw Hill Medical), pp. 595–605.

Stephens, R.S., Kalman, S., Lammel, C., Fan, J., Marathe, R., Aravind, L., Mitchell, W., Olinger, L., Tatusov, R.L., Zhao, Q., *et al.* (1998). Genome sequence of an obligate intracellular pathogen of humans: *Chlamydia trachomatis*. Science *282*, 754–759.

Stoner, B.P., and Cohen, S.E. (2015). Lymphogranuloma venereum 2015: clinical presentation, diagnosis, and treatment. Clin Infect Dis *61 Suppl 8*, S865–873. https://dx.doi.org/10.1093/cid/civ756.

Su, H., Morrison, R., Messer, R., Whitmire, W., Hughes, S., and Caldwell, H.D. (1999). The effect of doxycycline treatment on the development of protective immunity in a murine model of chlamydial genital infection. J Infect Dis *180*, 1252–1258. https://dx.doi.org/10.1086/315046.

Taylor, H.R., Burton, M.J., Haddad, D., West, S., and Wright, H. (2014). Trachoma. Lancet *384*, 2142–2152. https://dx.doi.org/10.1016/s0140-6736(13)62182-0.

Teijeira, A., Russo, E., and Halin, C. (2014). Taking the lymphatic route: dendritic cell migration to draining lymph nodes. Sem Immunopath 36, 261–274. https://dx.doi.org/10.1007/s00281-013-0410-8.

Teng, Y., Kong, N., and Tu, W. (2014). Estimating age-dependent per-encounter *Chlamydia trachomatis* acquisition risk via a Markov-based state-transition model. J Clin Bioinform 4, 7. https://dx.doi.org/10.1186/2043-9113-4-7.

Thomson, N.R., Holden, M.T., Carder, C., Lennard, N., Lockey, S.J., Marsh, P., Skipp, P., O'Connor, C.D., Goodhead, I., Norbertzcak, H., *et al.* (2008). *Chlamydia trachomatis*: genome sequence analysis of lymphogranuloma venereum isolates. Genome Res *18*, 161–171. https://dx.doi.org/10.1101/gr.7020108.

Torrone, E., Papp, J., and Weinstock, H. (2014). Prevalence of *Chlamydia trachomatis* genital infection among persons aged 14–39 years - United States, 2007–2012. MMWR 63, 834–838.

Tu, W., Ghosh, P., and Katz, B.P. (2011). A stochastic model for assessing *Chlamydia trachomatis* transmission risk using longitudinal observational data. J Royal Stat Society *174*, 975–989. https://dx.doi.org/10.1111/j.1467-985X.2011.00691.x.

van Liere, G.A., Dukers-Muijrers, N.H., Wolffs, P.F., and Hoebe, C.J. (2013). P3.186 Substantial natural clearance of genital and extragenital *Chlamydia trachomatis* and *Neisseria gonorrhoeae* in STD clinic attendees. Sex Transm Dis. https://dx.doi.org/10.1136/sextrans-2013-051184.0643.

van Liere, G.A., Hoebe, C.J., and Dukers-Muijrers, N.H. (2014). Evaluation of the anatomical site distribution of chlamydia and gonorrhoea in men who have sex with men and in high-risk women by routine testing: cross-sectional study revealing missed opportunities for treatment strategies. Sex Transm Infect *90*, 58–60. https://dx.doi.org/10.1136/sextrans-2013-051248.

Wang, C., Tang, J., Geisler, W.M., Crowley-Nowick, P.A., Wilson, C.M., and Kaslow, R.A. (2005). Human leucocyte antigen and cytokine gene variants as predictors of recurrent *Chlamydia trachomatis* infection in high-risk adolescents. J Infect Dis *191*, 1084–1092. https://dx.doi.org/10.1086/428592.

Wang, L., Zhu, C., Zhang, T., Tian, Q., Zhang, N., Morrison, S., Morrison, R., Xue, M., and Zhong, G. (2018). Nonpathogenic colonization with chlamydia in the gastrointestinal tract as oral vaccination for inducing transmucosal protection. Infect Immun *86*, 919. https://dx.doi.org/10.1128/iai.00630-17.

Wang, S.-p., and Grayston, J.T. (1991). Three new serovars of *Chlamydia trachomatis*: Da, la, and L2a. J Infect Dis *163*, 403–405.

Wang, S.P., and Grayston, J.T. (1970). Immunologic relationship between genital TRIC, lymphogranuloma venereum, and related organisms in a new microtitre indirect immunofluorescence test. Am J Ophthalmol *70*, 367–374.

Wang, S.P., Grayston, J.T., and Alexander, E.R. (1967). Trachoma vaccine studies in monkeys. Am J Ophthalmol *63*, Suppl:1615–1630.

Ward, H., Alexander, S., Carder, C., Dean, G., French, P., Ivens, D., Ling, C., Paul, J., Tong, W., and White, J. (2009). The prevalence of lymphogranuloma venereum infection in men who have sex with men: results of a multicentre case finding study. Sex Transm Infect *85*, 173–175. https://dx.doi.org/10.1136/sti.2008.035311.

Ward, H., Martin, I., Macdonald, N., Alexander, S., Simms, I., Fenton, K., French, P., Dean, G., and Ison, C. (2007). Lymphogranuloma venereum in the United Kingdom. Clin Infect Dis *44*, 26–32. https://dx.doi.org/10.1086/509922.

Wehrl, W., Brinkmann, V., Jungblut, P.R., Meyer, T.F., and Szczepek, A.J. (2004). From the inside out—processing of the chlamydial autotransporter PmpD and its role in bacterial adhesion and activation of human host cells. Mol Microbiol *51*, 319–334. https://dx.doi.org/10.1046/j.1365-2958.2003.03838.x.

WHO (2016). Trachoma: fact sheet. 2016. Http://www.who.int/en/news-room/fact-sheets/detail/trachoma. Accessed September 2, 2018.

Wiesenfeld, H.C., Hillier, S.L., Krohn, M.A., Landers, D.V., and Sweet, R.L. (2003). Bacterial vaginosis is a strong predictor of *Neisseria gonorrhoeae* and *Chlamydia trachomatis* infection. Clin Infect Dis *36*, 663–668. https://dx.doi.org/10.1086/367658.

Williams, J.A., Ofner, S., Batteiger, B.E., Fortenberry, J.D., and Van Der Pol, B. (2014). Duration of polymerase chain reaction-detectable DNA after treatment of *Chlamydia trachomatis, Neisseria gonorrhoeae,* and *Trichomonas vaginalis* infections in women. Sex Transm Dis *41*, 215–219. https://dx.doi.org/10.1097/OLQ.0000000000000102.

Wills, G.S., Horner, P.J., Reynolds, R., Johnson, A.M., Muir, D.A., Brown, D.W., Winston, A., Broadbent, A.J., Parker, D., and McClure, M.O. (2009). Pgp3 antibody enzyme-linked immunosorbent assay, a sensitive and specific assay for seroepidemiological analysis of *Chlamydia trachomatis* infection. Clin Vac Immunol *16*, 835–843. https://dx.doi.org/10.1128/CVI.00021-09.

Witkin, S.S., Jeremias, J., Toth, M., and Ledger, W.J. (1994). Proliferative response to conserved epitopes of the *Chlamydia trachomatis* and human 60-kilodalton heat-shock proteins by lymphocytes from women with salpingitis. Am J Obstet Gynecol *171*, 455–460.

Workowski, K.A., Bolan, G.A., and Papp, J.R. (2015). Sexually transmitted diseases treatment guidelines, 2015., MMWR *64*, 1–137.

Wyrick, P.B., and Knight, S.T. (2004). Pre-exposure of infected human endometrial epithelial cells to penicillin *in vitro* renders *Chlamydia trachomatis* refractory to azithromycin. J Antimicro Chem *54*, 79–85. https://dx.doi.org/10.1093/jac/dkh283.

Xue, Y., Zheng, H., Mai, Z., Qin, X., Chen, W., Huang, T., Chen, D., and Zheng, L. (2017). An *in vitro* model of azithromycin-induced persistent *Chlamydia trachomatis* infection. FEMS Microbiol Let *364*. https://dx.doi.org/10.1093/femsle/fnx145.

Zeidler, H., and Hudson, A.P. (2016). Coinfection of Chlamydiae and other bacteria in reactive arthritis and spondyloarthritis: need for future research. Microorg *4*. https://dx.doi.org/10.3390/microorganisms4030030.

Zheng, H., Xue, Y., Bai, S., Qin, X., Lu, P., and Yang, B. (2015). Association of the *in vitro* susceptibility of clinical isolates of *Chlamydia trachomatis* with serovar and duration of antibiotic exposure. Sex Transm Dis *42*, 115–119. https://dx.doi.org/10.1097/OLQ.0000000000000241.

Zhong, G. (2017a). *Chlamydia* spreading from the genital tract to the gastrointestinal tract – a two-hit hypothesis. Trends Microbiol *26*, 611–623. https://dx.doi.org/10.1016/j.tim.2017.12.002.

Zhong, G. (2017). Chlamydial plasmid-dependent pathogenicity. Trends Microbiol *25*, 141–152. https://dx.doi.org/10.1016/j.tim.2016.09.006.

Zhou, Z., Zeng, F., Yuan, J., Tang, J., Colditz, G.A., Tworoger, S.S., Trabert, B., and Su, X. (2017). Pelvic inflammatory disease and the risk of ovarian cancer: a meta-analysis. Cancer Causes Control *28*, 415–428. https://dx.doi.org/10.1007/s10552-017-0873-3.

Zhu, H., Shen, Z., Luo, H., Zhang, W., and Zhu, X. (2016). *Chlamydia trachomatis* infection-associated risk of cervical cancer: a meta-analysis. Medicine (Baltimore) 95, e3077. https://dx.doi.org/10.1097/MD.0000000000003077.

Zigangirova, N.A., Rumyantseva, Y.P., Morgunova, E.Y., Kapotina, L.N., Didenko, L.V., Kost, E.A., Koroleva, E.A., Bashmakov, Y.K., and Petyaev, I.M. (2013). Detection of *C. trachomatis* in the serum of the patients with urogenital chlamydiosis. Biomed Res Int *489489*. https://dx.doi.org/10.1155/2013/489489.

Ziklo, N., Huston, W.M., Taing, K., Katouli, M., and Timms, P. (2016). *In vitro* rescue of genital strains of *Chlamydia trachomatis* from interferon-γ and tryptophan depletion with indole-positive, but not indole-negative Prevotella spp. BMC Microbiol *16*, 286. https://dx.doi.org/10.1186/s12866-016-0903-4.

Ziklo, N., Vidgen, M.E., Taing, K., Huston, W.M., and Timms, P. (2018). Dysbiosis of the vaginal microbiota and higher vaginal kynurenine/tryptophan ratio reveals an association with *Chlamydia trachomatis* genital infections. Front Cell Infect Microbiol *8*, 1. https://dx.doi.org/10.3389/fcimb.2018.00001.

Chlamydia pneumoniae Infections

2

Lee Ann Campbell[1] and David L. Hahn[2*]

[1]Department of Global Health, University of Washington, Seattle, WA, USA.
[2]Department of Family Medicine and Community Health, University of Wisconsin School of Medicine and Public Health, Madison, WI, USA.

*Correspondence: dlhahn@wisc.edu

https://doi.org/10.21775/9781912530281.02

Abstract

Chlamydia pneumoniae is an obligate intracellular *Chlamydia* organism that is an established cause of human acute upper and lower respiratory conditions, primarily bronchitis and pneumonia. Seroepidemiological studies indicate that the majority of individuals worldwide have been infected by the time adulthood is reached, and that reinfection and/or chronic infection is common. The majority of acute infections are asymptomatic or only mildly symptomatic below the threshold that medical care is sought. The known propensity for other *Chlamydia* species to cause chronic infection associated with chronic immunopathologic diseases early on raised the question of whether chronic *C. pneumoniae* infection, either ongoing or 'hit and run', could be causally associated with a number of chronic inflammatory diseases of unknown aetiology. Indeed, increasingly robust evidence implicates chronic *C. pneumoniae* infection as a potentially causal factor in chronic respiratory diseases (asthma, chronic bronchitis, chronic obstructive respiratory disease) and atherosclerotic cardiovascular diseases, cerebrovascular diseases, and lung cancers, among others. This chapter reviews historical aspects of the discovery of *C. pneumoniae*, its role in acute respiratory illnesses, and emerging evidence for the role of *C. pneumoniae* in the pathogenesis of selected chronic respiratory and vascular diseases.

History

The first *C. pneumoniae* isolate was obtained in 1965 from the conjunctiva of a primary school child who did not have active trachoma but was a participant in a trachoma vaccine trial in Taiwan carried out by Dr J. Thomas Grayston and colleagues (Alexander *et al.*, 1967). This isolate was designated TW-183 as it was the 183rd Taiwan *Chlamydia* isolate cultured from the eye in embryonated egg yolk sacs, the only method available at the time for culturing *Chlamydia* (Grayston and Wang, 1999). Initially, the isolate was thought to be a strain of *C. trachomatis*. However, several anomalies distinguishing TW-183 from *C. trachomatis* were noted. The appearance of the egg yolk following inoculation differed in comparison with other *C. trachomatis* isolates. At this time, the mouse toxicity prevention test was used to immunotype *C. trachomatis* isolates. The basis of this test was that intravenous inoculation of high titre *C. trachomatis* yolk sac suspensions was lethal to mice and death was prevented through immunization of mice with a non-lethal dose of the homologous serotype prior to challenge (Alexander *et al.*, 1967). TW-183 was not lethal in this test nor did it afford cross-protection to subsequent challenge with any of the *C. trachomatis* serotypes identified at that time. Thus, it was assumed that the organism represented a new *C. trachomatis* immunotype.

With the availability of cell culture methods for *C. trachomatis*, it became clear that TW-183 was not *C. trachomatis* based on the compact and dense inclusion morphology and the finding that the inclusions were glycogen negative. Because it shared inclusion characteristics with *C. psittaci*, TW-183 was then thought to be a human *C. psittaci* strain. Subsequently, another 'atypical' *Chlamydia* organism, designated IOL-207, was cultured by investigators in England from the conjunctiva of an Iranian child with trachoma (Dwyer *et al.*, 1972). Using the micro-immunofluorescence test (micro-IF test), which was developed by Dr San-Pin Wang in 1970 to facilitate immunotyping, Dwyer *et al.* reported that IOL-207 was serologically distinct from *C. trachomatis* and *C. psittaci* and similar to TW-183 (Dwyer *et al.*, 1972; Wang and Grayston, 1970). The micro-IF test enables determination of antibody titres based on reactivity of sera to formalinized elementary bodies of different *Chlamydia* species and *C. trachomatis* serovars on a glass slide (Wang and Grayston, 1970). It should be noted that the micro-IF test, which permitted differentiation of serotypes of *C. trachomatis* causing infection, proved equally invaluable in elucidating the disease associations of *C. pneumoniae* as well as in developing diagnostic criteria of acute and past or chronic infection. Although both TW183 and IOL-207 were obtained from conjunctival swabs, seroepidemiological studies of antibody against TW183 in Taiwanese families that were under surveillance for trachoma showed no evidence of an association of antibody against TW183 and eye infection. However, antibodies against TW183 were common in this group (Wang, 1986).

The first evidence indicating that the 'atypical' *C. psittaci* was a human respiratory pathogen was serological evidence that TW183 or a related organism was the etiologic agent of an epidemic of mild pneumonia in two communities in Finland (Saikku *et al.*, 1985). Unlike the typical transmission of *C. psittaci* to humans through an exposure to infected birds, there was no evidence of avian transmission due to the absence of bird vectors in the winter in the areas affected. The culture of eight isolates from throat swab specimens from a group of students with acute respiratory disease and serological evidence of infection at the University of Washington from January 1983 to June 1985 confirmed that this organism was a human respiratory pathogen (Grayston *et al.*, 1986). At this time, this pathogen was designated as *C. psittaci* strain TWAR, denoting the first two isolates by Dr Grayston's group, the ocular isolate TW-183 and the respiratory isolate AR-39.

In 1989, TWAR was established as a separate species of *Chlamydia*, *C. pneumoniae*, based on ultrastructure of the elementary body (EB), DNA analysis, and serology (Grayston, 1989). The yields of infectious organisms are highest at three days post-infection, when organisms are cultured at the optimal temperature of 35°C. The burst size (number of infectious units in the inclusion) of *C. pneumoniae* is significantly less than *C. trachomatis*. For example, when grown under similar conditions in HeLa cells, *C. pneumoniae* has an average burst size of 13–52 in comparison to *C. trachomatis* serovar G, which has a burst size of 500 (Kuo and Grayston, 1988; Sabet *et al.*, 1984).

Sequence similarity of *C. pneumoniae* isolates

Early analysis suggested that the human respiratory isolates were homogenous (Campbell *et al.*, 1987, 1990; Carter *et al.*, 1991; Gaydos *et al.*, 1992; Perez Melgosa *et al.*, 1991). Whole genome sequence comparisons of TW183, *C. pneumoniae* respiratory isolates (AR-39, CWL029, and J138), a coronary artery from a patient undergoing a heart transplant (A03) and a post-mortem brain tissue isolate from a patient with Alzheimer's disease (TOR-1) revealed >99% sequence similarity among respiratory and non-respiratory isolates (Myers *et al.*, 2009; Roulis *et al.*, 2015; Shirai *et al.*, 2000). This high clonality suggests that human infection is a relatively recent occurrence (Rattei *et al.*, 2007). However, two subgroups were identified, one consisting of TW-183 and the coronary artery isolate A03, and the other consisting of the Alzheimer's isolate (TOR-1) and the other respiratory isolates. The identification of two distinct clades suggests that recombination and mutation events have driven the evolution of *C. pneumoniae* strains (Roulis *et al.*, 2015).

To address genetic differences related to tissue tropism and evolutionary history, genome sequences were analysed from 18 clinical isolates from diverse geographical locales obtained from different anatomical sites including lung, blood, and coronary arteries of patients with disease (Weinmaier *et al.*, 2015). These studies showed evidence of recombination and low sequence diversity. Phylogenetic reconstruction, which also included TW-183 and respiratory isolates, demonstrated a clear separation of respiratory/ocular isolates from vascular isolates in the phylogenetic tree. Thus, SNP positions were determined that distinguished these groups from each other. Characteristic non-synonymous SNPs of the vascular isolates were identified in genes involved in the RB to EB transition, inclusion membrane formation, metabolism, and bacterial stress response. This study was the first to demonstrate genetic differences associated with the anatomical site of cultivation and paved the way to elucidate their functional role in tissue tropism and pathogenesis.

Host range

C. pneumoniae has been cultured or detected in specimens from a wide range of animals in the wild or captivity and by molecular methods in tissue specimens. These include horses, marsupials (e.g. koala and bandicoot), amphibians (e.g. frog and turtle), reptiles (e.g. iguana, chameleon and snake), and birds (Bodetti *et al.*, 2002; Hotzel *et al.*, 2001; Kabeya *et al.*, 2015; Kutlin *et al.*, 2007; Mitchell *et al.*, 2010; Rüegg *et al.*, 2015; Storey *et al.*, 1993). Similar to human infection, *C. pneumoniae* has been identified in various anatomical sites including the respiratory tract, liver, spleen, and conjunctiva (Berger *et al.*, 1999; Bodetti *et al.*, 2002; Hotzel *et al.*, 2001; Kutlin *et al.*, 2007; Mitchell *et al.*, 2010). Although plasmids have not been identified in human isolates, plasmids sharing 96.3 to 99.9% homology have been reported in animal isolates examined (Roulis *et al.*, 2014). Genetic and phylogenetic comparison of human isolates with animal isolates has revealed evidence that the animal lineages were basal to human lineages, a recent animal to human transmission event occurred, and at least two separate transmission events have occurred. (Mitchell *et al.*, 2010; Myers *et al.*, 2009; Rattei *et al.*, 2007). Whole genome sequence comparison of the koala isolate to human isolates indicated that the animal strain successfully adapted to humans through plasmid loss and gene decay, such that an animal reservoir was no longer required for transmission (Myers *et al.*, 2009).

Since the recognition of *C. pneumoniae* as an etiologic agent in acute respiratory disease, a broad spectrum of clinical manifestations and chronic inflammatory diseases have been associated with *C. pneumoniae*. This chapter will focus on the prevalence of infection, the use of animal models that mimic human infection to study *C. pneumoniae* pathogenesis, clinical diagnosis, and clinical spectrum of *C. pneumoniae* infections.

Clinical epidemiology

Acute respiratory illnesses

C. pneumoniae was identified as a cause for acute upper and lower respiratory tract illnesses in 1986 (Aldous *et al.*, 1992; Grayston *et al.*, 1993; Grayston *et al.*, 1986) although many *C. pneumoniae* infections are only mildly symptomatic or completely asymptomatic (Saikku *et al.*, 1985). The role of *C. pneumoniae* as a cause for acute upper and lower respiratory illnesses, including pharyngitis, sinusitis, bronchitis and pneumonia, has been confirmed in community practice-based (Hahn *et al.*, 1991), military (Coon *et al.*, 2011; Csángó *et al.*, 1997; Ekman *et al.*, 1993; Kleemola *et al.*, 1988), nursing home (Troy *et al.*, 1997), and hospitalized patient populations (McConnell *et al.*, 1994). Primary (first) infections predominate in children (Aldous *et al.*, 1992) and secondary (reinfection) infections are twice as common as primary infections in middle-aged and older adults (Thom *et al.*, 1994).

Reports of within family transmission of *C. pneumoniae* infections demonstrate that person to person spread tends to be slow – in the range of 2–3 weeks (Blasi *et al.*, 1994b; Ghosh *et al.*, 1992; Mordhorst *et al.*, 1992; Yamazaki *et al.*, 1990). Epidemics within closed communities have also been reported (Blasi *et al.*, 1994a; Conklin *et al.*, 2013; Ekman *et al.*, 1993; Pether *et al.*, 1989; Troy *et al.*, 1997). In Denmark, countrywide epidemics of *C. pneumoniae* infections had a periodicity of around six years, i.e. high incidence of infection lasted about 2.5 years, while periods of low incidence lasted 3+ years (Mordhorst *et al.*, 1990). Infectivity is thought to be primarily by contact with respiratory secretions – either directly or via aerosols – however, transfer from fomites on environmental surfaces with subsequent autoinoculation is also plausible (Falsey and Walsh, 1993). Infections can also be endemic and sporadic. Laboratory personnel working with *C. pneumoniae* should carefully follow infection protocols, as laboratory accidents can result in respiratory illnesses (Campbell and Kuo, 2009).

Diagnostic testing for *C. pneumoniae* infections is not commonly done in clinical practice, largely because of lack of readily available diagnostic tests. This may change with greater availability of multiplex PCR tests that check for multiple respiratory pathogens. As for many other pathogens, diagnosis of *C. pneumoniae* depends on direct organism identification, host immunologic response, or a combination (Centers-for-Disease-Control-and-Prevention, accessed June 23, 2018). The typical 5–10 day courses of antibiotics that are effective against *M. pneumoniae* and against typical pyogenic pathogens such as *Streptococcus pneumoniae* ('pneumococcus') and *Streptococcus pyogenes* ('Group A strep') were recognized early on to be of insufficient duration to achieve a satisfactory clinical response in many acute respiratory infections caused by *C. pneumoniae*. A minimum of two weeks of a tetracycline, a macrolide or – more recently – a quinolone are recommended, with the caveat that even longer treatment times may be required (Grayston *et al.*, 1990). Importantly, asymptomatic *C. pneumoniae* infections can persist after antibiotic treatment and resolution of acute respiratory illnesses (Conklin *et al.*, 2013; Hammerschlag *et al.*, 1992).

Chronic respiratory infection

A plethora of studies confirm that specific seroreactivity against *C. pneumoniae* begins to appear in human populations worldwide around age 5, increases to around 50% by young adulthood, and remains stable or even increases into old age (Aldous *et al.*, 1992; Asche *et al.*, 1993; Biendo *et al.*, 1994; Burney *et al.*, 1984; Darougar *et al.*, 1980; Karvonen *et al.*, 1992; Mahmoud *et al.*, 1994; Mordhorst *et al.*, 1986; Puolakkainen *et al.*, 1989; Stolk-Engelaar and Peeters, 1992; Wang and Grayston, 1990; Wang and Grayston, 1986). Antibodies against *Chlamydia* species wane after an acute primary (first) infection (Aldous *et al.*, 1992; Hahn *et al.*, 1991). Persistence of *C. pneumoniae* antibody is thus strong evidence for widespread repeated infections, or chronic infection, throughout the lifespan. *In vitro* data provide biologic plausibility for these chronic, persistent infections: in response to stressors such as tryptophan depletion, interferon-γ, and antibiotics, *C. pneumoniae* – like other *Chlamydia* species – can survive in a 'persistent' form, characterized by abnormal inclusion morphology, non-cultivability, and up-regulation of heat shock protein 60 (Beatty *et al.*, 1994). After removal of the immune stressor – or upon addition of an immunosuppressant such as a corticosteroid – *C. pneumoniae* is again cultivable. Reinfections and persistent infection as documented by seroepidemiology and human clinical studies may have significant implications for a variety of chronic inflammatory diseases of unknown aetiology (discussed below).

There are no accepted serological criteria for chronic infection. In asymptomatic young adults and older patients with lung cancer and chronic obstructive pulmonary disease (COPD), *C. pneumoniae* may only reside in mononuclear cells deep in the lung beyond the reach of the bronchoscope (Wu *et al.*, 2000), making direct organism identification impractical in most instances. These challenges to diagnosis of chronic lung infection have impeded research into potential chronic sequelae of acute *C. pneumoniae* infection.

Animal models of *C. pneumoniae* respiratory tract infection

Several animal models have been investigated for their susceptibility to *C. pneumoniae* respiratory infection including non-human primates, mice, and rabbits (Bell *et al.*, 1989; Holland *et al.*, 1990; Moazed *et al.*, 1996; Yang *et al.*, 1993). In these models, single or repeated inoculations are administered to simulate acute or chronic infection, respectively. Respiratory disease symptoms ranged from none (non-human primates) to pneumonia with bronchitis (rabbits and mice). Lung histopathology was less severe and of shorter duration in rabbits. In mice, human respiratory isolates were identified that differed in the severity of respiratory disease resulting from infection (Kaukoranta-Tolvanen *et al.*, 1993; Sommer *et al.*, 2009). Collectively, these models mimic respiratory disease in humans, which ranges from asymptomatic to mild to severe bronchitis and pneumonia (Gieffers *et al.*, 2004; Grayston *et al.*, 1990; Kleemola *et al.*, 1988). In all three animal models, infection disseminated to non-pulmonary sites (Bell *et al.*, 1989; Holland *et al.*, 1990; Moazed *et al.*, 1996; Yang *et al.*, 1995). This early observation raised the question whether *C. pneumoniae* infection of the human respiratory tract also disseminated to other sites, which subsequently was confirmed through the identification of the organism in cardiovascular tissue, non-cardiovascular tissue and peripheral blood mononuclear cells (PBMCs) (Boman *et al.*, 1998; Jackson *et al.*, 1997b; Kuo and Campbell, 2003).

Following acute infection, organisms were re-isolated from the nasopharynx or lung in all animal models; however, the recovery time-period varied. In non-human primates, the organism could be cultured for 5–8 weeks after infection from the nasopharynx. In mice, the organism can typically be cultured from the lung for approximately three weeks post infection depending on the inoculum titre and mouse strain (Kaukoranta-Tolvanen *et al.*, 1993; Murdin *et al.*, 2000a; Yang *et al.*, 1993). In contrast, the organism could only be cultured from the nasopharynx in New Zealand rabbits inoculated with high titres of the organism for two or three days after acute infection (Moazed *et al.*, 1996). Since the initial studies of *C. pneumoniae* infection in non-human primates, these animal models have not been pursued further.

Several common histopathological characteristics resulting from acute lung infection are observed in rabbits and mice. Infection results in a multifocal interstitial pneumonia. Lung histopathology is most severe early on with an inflammatory infiltrate consisting of polymorphonuclear leucocytes or heterophilic granulocytes, in mice and rabbits, respectively (Moazed *et al.*, 1996; Yang *et al.*, 1993). Later on in the infection, mononuclear leucocytes predominate and persist. A striking feature of acute infection is the accumulation of perivascular and peribronchial lymphoid cells (Moazed *et al.*, 1996; Yang *et al.*, 1993). The latter is distinct from *C. trachomatis* infection in which these foci occur only in chronic or repeated infection. In a subsequent study, the development of inflammatory pathology was monitored during the acute (the first two weeks post infection) and chronic stage (25–35 days post infection) (Jupelli *et al.*, 2013). Immunostaining analyses of the lymphoid aggregates identified these aggregates as inducible bronchus associated lymphoid tissue (iBALT) consisting of B, T and follicular dendritic cells (Jupelli *et al.*, 2013). iBALTs are thought to contribute to adaptive immunity, but may also contribute to lung pathology (Foo and Phipps, 2010; Hogg *et al.*, 2004; Rangel-Moreno *et al.*, 2006; Ulrichs *et al.*, 2004). The macrophage population present in the acute stage was predominantly M1 while M2 macrophages persisted in the chronic stage (Jupelli *et al.*, 2013). Significantly, collagen deposition indicating fibrosis was observed in the lung at 35 days post infection. Adoptive transfer experiments suggested that an excess of M1 macrophages in *C. pneumoniae*-infected lungs enhances inflammation, lung fibrosis and tissue damage (Jupelli *et al.*, 2013). These cumulative findings may be relevant to the association of *C. pneumoniae* with chronic lung diseases such as asthma and COPD. Significantly, a pathological feature of progression of COPD in humans is the infiltration of innate and adaptive inflammatory cells that form lymphoid follicles (Hogg *et al.*, 2004).

Evidence for partial immunity against subsequent *C. pneumoniae* infection has been observed in non-human primates and mice. In cynomolgus monkeys, decreased shedding was observed after the third inoculation of *C. pneumoniae* (Holland *et al.*, 1990). In mice, intranasal inoculation of *C. pneumoniae* or immunization with heat-killed organism followed by challenge with *C. pneumoniae* resulted in partial protection, as demonstrated by decreased bacterial burden or accelerated clearance of the organism from the lung (Murdin *et al.*, 2000a; Penttila *et al.*, 1998). However, repeated infection also increased the severity and duration of lung pathology (Penttila *et al.*, 1998).

Animal models have been useful for studying persistent *C. pneumoniae* infection. Evidence of *C. pneumoniae* persistence in human infection has been suggested by the presence of antigen and/or DNA in culture negative atherosclerotic, brain, and synovial tissue from patients with chronic disease (Carter and Hudson, 2017; Gerard *et al.*, 2006; Jackson *et al.*, 1997a; Ramirez, 1996). Aberrant RBs, similar to those seen in chlamydial persistence *in vitro* in cell culture, have been reported in human atheromatous tissue, and antigens that are suggested markers of persistence have been detected in patients with coronary atheroma undergoing heart transplantation (Borel *et al.*, 2006; Borel *et al.*, 2012; Borel *et al.*, 2008).

In rabbit and mouse models of *C. pneumoniae* pneumonitis, persistent lung infection is observed, in which the organism cannot be cultured, but antigen or DNA can be detected (Moazed *et al.*,

1996). Moreover, treatment of mice with doxycycline or azithromycin, using regimens comparable to those used for human infection, were effective based on the inability to culture the organism from lung tissue. However, *C. pneumoniae* DNA was detected in lung tissue in 25–77% of culture negative mice depending on the inoculating strain, and no differences were observed in histopathological findings in comparison to untreated mice (Malinverni *et al.*, 1995b). The presence of *C. pneumoniae* DNA could indicate either non-viable organisms or viable organism in a non-culturable state. Significantly, when mice were immunosuppressed by administration of cortisone at a time point when the lungs were culture negative but PCR positive, the organism could then be isolated from the lung (Laitinen *et al.*, 1996; Malinverni *et al.*, 1995a).

Mouse models have proven useful in understanding how *C. pneumoniae* elicits host inflammatory responses. *C. pneumoniae* is recognized by Toll-like receptors 2 and 4 (TLR2 and TLR4) and the nucleotide oligomerization domain (NOD) receptors (Naiki *et al.*, 2005; Rodriguez *et al.*, 2006; Shimada *et al.*, 2009). TLR/MyD88 signalling and Nod/RIP signalling play key roles in the host response to *C. pneumoniae* (Naiki *et al.*, 2005; Shimada *et al.*, 2009; Shimada *et al.*, 2012; Wantia *et al.*, 2011). Although TLR2–/– or TLR4–/– mice resolved *C. pneumoniae* infection, mice deficient in MyD88, the common TLR adaptor, were unable to clear infection due to delayed recruitment of PMNS as well as CD4+ and CD8+ cells to the lung. MyD88-deficient mice developed severe and persistent inflammation and had increased mortality (Naiki *et al.*, 2005). Nod1–/– or Nod2–/– mice exhibited delayed pulmonary clearance of *C. pneumoniae* (Shimada *et al.*, 2009). Mice that were deficient in Rip2, the downstream adaptor of Nod1 and Nod2, had decreased production of chemokines and cytokines, defective iNOS expression and NOS production, and delayed recruitment of neutrophils, collectively resulting in delayed clearance of the organism and enhanced lung pathology (Shimada *et al.*, 2009). In the adaptive response to *C. pneumoniae* infection, both CD4+ and CD8+ T cells played a role in bacterial clearance from the lung and in protection from infection (Rottenberg *et al.*, 1999).

Chronic respiratory conditions linked to *C. pneumoniae* infection

Asthma

In 1991 the Lung Division of the National Heart Lung and Blood Institute of the National Institutes of Health published the first expert consensus guideline on the management of asthma (National-Asthma-Education-Program, 1991). The guideline stated that 'the root cause of asthma is inflammation' and that bacterial infection played a minor role, if any role, in asthma (the non-infectious inflammatory hypothesis). The underlying cause(s) for asthma remain unknown, and 'there is no cure.' This non-infectious inflammatory hypothesis remains the dominant model, and asthma research currently focuses on a search for immunologic targets against which to fashion new anti-inflammatory agents (e.g. monoclonal antibodies). The model has three correlates: (1) asthma should be sub-categorized into 'phenotypes' based on differences in inflammatory cells recovered from the respiratory tract, usually by bronchoscopy, (2) asthma and COPD are entirely different entities that should be studied in strict isolation (Vermeire and Pride, 1991), and (3) COPD studies should exclude non-smokers and anyone with asthma. This 'splitting' approach (Vermeire and Pride, 1991) constitutes a *sine qua non* for research funding.

1991 was also the year of the first publication associating *C. pneumoniae* with wheezing, asthmatic bronchitis, and adult-onset asthma (Hahn *et al.*, 1991). The authors of this report suggested that *C. pneumoniae* might be a treatable – and even preventable – cause of asthma. Subsequent reports broadened the scope of *C. pneumoniae* infection as a potential cause of transient wheezing (Hahn, 1994), asthma in children (Cunningham *et al.*, 1998; Emre *et al.*, 1994), general adult asthma (Johnston and Martin, 2005), chronic bronchitis (Blasi *et al.*, 1997; Blasi *et al.*, 2002; Lorenz, 1999) and COPD (Hahn, 1999). It has even been suggested that *C. pneumoniae* might explain the worldwide asthma pandemic noted in the latter 20th century (Bone, 1991). Antibiotic treatment trials for asthma have yielded mixed results, possibly because of differences in patient enrolment criteria (Kew *et al.*, 2015; Reiter *et al.*, 2013; Stokholm *et al.*, 2015). Current best evidence supports long-term azithromycin treatment for severe, refractory asthma (Gibson *et al.*, 2017) and for COPD (Albert *et al.*, 2011).

Through the lens of *Chlamydia*, it seems inappropriate to split respiratory diseases into arbitrary subgroups, and subgroups into 'phenotypes' because simple logic and epidemiological evidence suggest otherwise. Logic says there is no assurance that inflammatory 'phenotypes' will remain stable over time (especially unlikely if they are manifestations of evolving infections). *Chlamydia* infections cause protean manifestations of histopathology – from neutrophil inflammation to mononuclear cell inflammation, even eosinophilia – depending on the phase of the infection (Kuo, 1988). Regarding the epidemiology of respiratory diseases, the prospective, population-based data show that asthma can evolve into COPD – i.e. asthma and COPD represent parts of a continuum, not distinct entities (Burrows, 1987; de Marco *et al.*, 2015; Silva *et al.*, 2004; Tai *et al.*, 2014). Lastly, population-based studies find that 6–18% of men with COPD and 30%–76% of women with COPD have never smoked (Lundbäck *et al.*, 2003; Rodriguez-Gonzalez Moro *et al.*, 2009; Zhou *et al.*, 2009). It has therefore been advocated that 'pragmatic' approaches are required to study *C. pneumoniae* and asthma (Webley and Hahn, 2017). A cardinal characteristic of pragmatic research is broad inclusion of all disease candidates to reflect 'real world' populations (Loudon *et al.*, 2015). The pragmatic approach allows for subgroup analyses to explore for underlying heterogeneity (Sluiter *et al.*, 1991), thus avoiding the unproven *a priori* assumptions inherent in the 'splitting' approach (Vermeire and Pride, 1991). The *C. pneumoniae*-asthma hypothesis challenges the basic assumptions of contemporary asthma research described above.

Chlamydia researchers interested in lung diseases should be aware of current approaches to asthma research, both to better interpret the existing asthma research literature and to inform future collaborations. Table 2.1 presents a series of studies

Table 2.1 Primary care-based studies relevant to *C. pneumoniae* and asthma

Reference	Study type	Main findings
Observational		
Hahn *et al.* (1991)	Prospective clinical and microbiologic study, *n* = 365	Nine (47%) of 19 patients with acute *C. pneumoniae* infection had bronchospasm during respiratory illness; four developed new-onset asthma. Positive dose–response association of *C. pneumoniae* antibody and wheezing: OR[a] = 3.5 (1.2–9.7) for MIF[b] titer ≥ 1:128.
Hahn *et al.* (1991)	Nested case–control, *n* = 142	Diagnosis of asthmatic bronchitis after illness was strongly associated with *C. pneumoniae* biomarkers: OR 12.0 (2.2–65) for antibody diagnostic for acute infection; OR 12.5 (2.5–62.5) for high titer non-diagnostic (MIF ≥ 1:128) antibody.
Hahn and Golubjatnikov (1994)	Prospective clinical and microbiologic study, *n* = 131	Two acute *C. pneumoniae* and one acute *M. pneumoniae* infections were associated with new-onset asthma.
Hahn and Golubjatnikov (1994)	Nested case–control, *n* = 131	*C. pneumoniae* seroreactivity (MIF≥1:16, and GMT) were strongly associated with adult-onset asthma (100%, 76.1) and asthmatic bronchitis (80.0%, 29.2) compared to non-wheezing respiratory illnesses (52.8%, 18.9), P < 0.001 and P = 0.0001, respectively.
Hahn and Beasley (1994)	Primary care asthma prevalence study, *n* = 14,127 (59 practices)	Active physician-diagnosed asthma = 6.1%, of which 46.3% was adult-onset. Possible undiagnosed asthma=3.3%, of which 56.7% was adult-onset.
Hahn (1995a)	Clinical description, *n* = 92	Asthma beginning after an acute respiratory illness (infectious asthma – IA) comprised 45% of patients, and was associated with worse pre- and post- bronchodilator pulmonary function despite shorter duration of asthma compared to non-IA (P < 0.05 for all comparisons).
Hahn *et al.* (1996)	Case–control, *n* = 70	Asthma onset within two years of testing was associated with *C. pneumoniae* IgA titers ≥ 1:10 (a proposed marker for chronic infection): 72% of cases v 44% of controls, adjusted OR 3.7 (1.2–11.5).

Table 2.1 Continued

Reference	Study type	Main findings
Hahn and McDonald (1998)	Prospective case series, $n=10$	Of 10 adults with *de novo* wheezing (first reported wheezing episode) five developed new-onset asthma, one developed chronic bronchitis (CB) and four resolved spontaneously. All had documented acute *C. pneumoniae* infections. The CB patient had *C. pneumoniae* cultured from his sputum six months after the index acute illness.
Hahn *et al.* (1998)	Case series, $n=3$	An adolescent and two adults – all with severe refractory asthma requiring chronic oral steroids – were *C. pneumoniae* seroreactive and were able to discontinue oral steroids after prolonged anti-chlamydial antibiotic treatment.
Hahn *et al.* (2000)	Case–control, $n=163$	*C. pneumoniae*-specific-IgG, -IgA, -IgE and CHSP60[c] were associated with asthma beginning after an acute respiratory infection.
Hahn and Peeling (2008)	Case–control, $n=138$	*C. pneumoniae*-specific HSP60 antibodies were associated with asthma (27% of cases v 8% of controls, $P<0.01$). For asthma, lower post-bronchodilator pulmonary function (indicating irreversible airflow limitation, i.e. development of COPD in asthma) was associated with specific HSP60 ($P<0.05$).
Hahn *et al.* (2012b)	Case-series/nested case–control, $n=66$	*C. pneumoniae*-specific IgE (Cp IgE) was associated with asthma compared with normal blood donor controls ($P=0.001$). 33 (50%) of 66 asthma patients were positive for Cp IgE and 16 (25%) were *C. pneumoniae* PCR positive in peripheral blood. Cp IgE was detected in 21% of mild intermittent asthma v 79% of severe persistent asthma (test for trend, $P=0.002$).
Interventional		
Hahn (1995b)	Open-label, before-after treatment trial, $n=46$	*C. pneumoniae* seroreactive patients with stable persistent asthma received anti-chlamydial antibiotics (median treatment four weeks, range 3–9). Asthma disappeared in four patients with proven *C. pneumoniae* infection treated soon after asthma onset. The remaining 42 were treated a mean of six years after asthma onset: three had complete remission and 18 had major persisting asthma improvement. Improvement was more likely in patients with early disease ($P=0.01$) and before the development of fixed obstruction ($P<0.01$).
Hahn *et al.* (2006)	RCT, $n=45$ (underpowered)	Six weekly doses of azithromycin improved asthma symptoms by 0.68 (0.1–1.3) units on a 5-point scale. Juniper Asthma Quality of Life Score improved by 0.25 (–0.3–0.8) units and rescue bronchodilator use decreased by 0.59 (–0.5–1.6) administrations. Baseline *C. pneumoniae* IgA antibodies were positively associated with worsening overall asthma symptoms at follow-up ($P=0.04$). Overall asthma symptom improvement attributable to azithromycin was 28% in high IgA participants versus 12% in low IgA participants (P for interaction $=0.27$).
Hahn *et al.* (2012a)	RCT with supplemental open-label (OL) [d] arm, $n=97$ (underpowered)	Treatment was 12 weekly doses of azithromycin or placebo followed by nine months of post-treatment observation. At baseline, open-label (OL arm participants[d] had significantly more severe persistent asthma than randomized participants (32% v 8%, $P=0.012$). The randomized arms showed no significant treatment effects. At one year compared with placebo, OL participants had improved symptoms ($P<0.02$), improved asthma control ($P<0.02$), and improved asthma quality of life (AQL) ($P=0.0006$). three subjects needed to be treated (number needed to treat – NNT) to achieve an AQL score change of ≥2 – a very large improvement. At one year, participants randomized to azithromycin v placebo were more likely to have an AQL score ≥ 1 unit increase (36% v 21%, NNT = 7, $P=0.35$).

[a]OR: odds ratio (95% confidence interval).
[b]MIF: microimmunofluorescence test using pooled IgM/IgA/IgG conjugate.
[c]CHSP60: chlamydial genus-specific heat shock protein 60 antibodies.
[d]For eligible participants unwilling to be randomized. Due to easy access to information via the Internet (www.clinicaltrials.gov and www.asthmastory.com) the word was already out on the street regarding the possible benefits of azithromycin in severe asthma when the trial was recruiting. The most severe and refractory study-eligible patients seeking enrollment did not want to risk receiving a placebo and instead opted for an open-label (OL) arm in which they could receive a prescription of azithromycin from their doctor and be followed for study outcomes in an identical way as those randomized.

of *C. pneumoniae* and asthma using the pragmatic ('lumping') approach, which were conducted in primary care settings in collaboration with *Chlamydia* researchers. Table 2.1 is meant as an illustrative example of primary care-basic science collaboration on 'real world' populations.

Animal models of *Chlamydia* infection in asthma

Epidemiological and clinical evidence motivated *in vitro* and animal model research into potential chlamydial pathogenesis in asthma. Collectively, the laboratory findings strongly support the biologic plausibility for a multifaceted causal role for chlamydial infection in asthma. *In vitro*, *C. pneumoniae* is capable of infecting virtually all relevant pulmonary cells (alveolar macrophages, ciliated bronchial epithelial cells, Type II alveolar pneumocytes, and bronchial smooth muscle cells) with subsequent production of a wide array of pro-inflammatory cytokines, most of which have been associated with asthma (Hahn, 2003). *C. pneumoniae* also infects lung immune cells (lung monocytes/macrophages and lung dendritic cells) resulting in altered function and implications for disease (Beagley *et al.*, 2009). The *in vitro* data have been reviewed elsewhere and will not be discussed in detail here (Hahn, 2004; Webley and Hahn, 2017). This section selectively highlights some of the main animal model findings; more information is available in recent reviews (Hansbro *et al.*, 2014; Webley and Hahn, 2017).

A neonatal mouse model inoculated with a mouse pneumonitis strain of *Chlamydia* (*Chlamydia muridarum*) has been extensively used as an experimental platform for the study of chlamydial 'infectious asthma.' Reasons for this choice were two-fold: neonatal mice are more susceptible to infection than adult mice due to a Th2 bias early in life, and *C. muridarum* naturally infects mice and was used as a surrogate for *C. pneumoniae*. It may be important to note, however, that although it is more difficult to create the 'asthma phenotype' by infecting adult mice, nevertheless the pulmonary damage created by infecting neonatal or juvenile mice persists into adulthood (Horvat *et al.*, 2010b; Jupelli *et al.*, 2011; Starkey *et al.*, 2013). Neonatal and adult mouse models using *C. pneumoniae* have largely confirmed and extended the results of the *C. muridarum* models (Crother *et al.*, 2011; Schröder *et al.*, 2008).

Pathologic hallmarks of asthma are pulmonary inflammation (variable cell composition) and bronchial hyper-responsiveness (manifested as reversible airway obstruction; 'twitchy lungs'). A secondary pathologic finding is lung fibrosis (manifested as irreversible airflow limitation). The secondary pathology is not a hallmark for asthma, which is characterized by complete reversibility of airway obstruction. Rather, irreversible airway obstruction is a hallmark for chronic obstructive pulmonary disease (COPD). This is an important distinction to understand: it was once thought that asthma and COPD were distinct, non-overlapping, clinical and pathologic entities. However, it is increasingly recognized that a significant amount of human obstructive airways disease manifests clinical and pathological characteristics of both asthma and COPD (termed the Asthma-COPD Overlap Syndrome, or ACOS) (Gibson and Simpson, 2009). Rather than representing three separate disease entities (asthma, ACOS and COPD), prospective clinical and epidemiological studies indicate that these three entities likely represent progressive stages of the same pathologic process, the underlying aetiologies of which are only now being unravelled (de Marco *et al.*, 2015; Hahn, 2002; Shirtcliffe *et al.*, 2012; Silva *et al.*, 2004).

Infection with either *C. muridarum* or *C. pneumoniae* in neonatal mouse models result in inflammatory lung pathology reminiscent of what is found in human asthma (Crother *et al.*, 2011; Horvat *et al.*, 2007; Horvat *et al.*, 2010a; Horvat *et al.*, 2010b; Patel and Webley, 2013; Schröder *et al.*, 2008). Horvat *et al.* demonstrated that *C. muridarum* infection in neonatal BALB/c mice, but not in adults, induced mixed T-cell responses that drove allergic airway disease pathology after ovalbumin (exogenous allergen) sensitization. Early life infection generated both Type 1 and Type 2 ovalbumin-specific helper T-cell cytokine and antibody responses and exacerbated hallmark features of asthma including mucous hypersecretion and airway hyperresponsiveness (Horvat *et al.*, 2007). In a follow-up study the same group demonstrated that *C. muridarum* infection during allergen sensitization drove neutrophilic allergic airways disease (Horvat *et al.*, 2010a). In this study, six-week old female BALB/c mice were infected with *C. muridarum*, challenged ('sensitized') with exogenous allergen (ovalbumin) or saline at time of infection, challenged again with ovalbumin on days 12–15 post-infection, and sacrificed on day 16. Current infection, but not resolved infection, attenuated hallmark features of allergy-induced asthma (pulmonary eosinophil influx, T-cell production of IL-5, mucous-secreting cell hyperplasia, and airway hyper-responsiveness). On the other hand, current infection induced robust ovalbumin-driven neutrophilic inflammation and IFN-γ release from T-cells reminiscent of human 'neutrophilic asthma' (one 'phenotype' of asthma described mainly in adults) (Horvat *et al.*, 2010a). Horvat *et al.* then went on to a more in depth comparison of neonatal (within 6 hours of birth), infant (3 weeks) and adult (6 week old) mice in a *C. muridarum* infection/ovalbumin sensitization model (Horvat *et al.*, 2010b). Neonatal and infant chlamydial infection, but not adult infection, enhanced the development of hallmark features of asthma including increased mucous-secreting cell numbers, IL-13 expression, and airway hyper-responsiveness. Notably neonatal, but not infant or adult infection, induced increases in alveolar diameter (a marker for possible fixed lung damage – 'emphysema') (Horvat *et al.*, 2010b).

Patel and Webley demonstrated that allergic sensitization to the infection itself could be demonstrated after *C. muridarum* infection in a mouse model (Patel and Webley, 2013). They infected neonatal (day 1) and adult (6 weeks) BALB/c mice and sensitized and challenged with ovalbumin seven weeks later. Adult mice cleared the infection ~30 days post-infection but neonates were still infected after 66 days. Infected neonatal mice produced increased amounts of Th2 cytokines (IL-4, IL-5, IL-10 and IL-13) whereas infected adult mice produced increased Th1 cytokines (IL-2, IFN-γ). Most significantly, infected neonates, but not infected adults, produced pathogen-specific IgE (in this model, specific IgE against *C. muridarum*) (Patel and Webley, 2013). This latter observation may be relevant to human asthma. *C. pneumoniae*-specific IgE can be demonstrated in children with serious chronic respiratory illnesses including asthma (Patel *et al.*, 2012). Up to 80% of adults encountered in primary care practice with severe asthma also demonstrate circulating *C. pneumoniae*-specific IgE in their peripheral blood (Hahn *et al.*, 2012b). Taken together, the findings from these animal models support a multifaceted role for *C. pneumoniae* infection in both childhood-onset asthma (COA: classically characterized by atopy and lung damage leading later in life to irreversible airflow obstruction) and in adult-onset asthma (AOA: associated with neutrophilic infiltrates, and less atopy).

It must be noted, however, that crystal clear distinctions between the pathogenesis of COA and AOA cannot be made; there are many areas of overlap, one of which is the presence of atopy demonstrated at all ages (Burrows *et al.*, 1989). For example, in the classic Tucson Study of Airway Obstructive Diseases (TESOAD), Burrows *et al.* reported that asthma at all ages was associated with total serum IgE reactivity; for children the sero-reactivity could be attributed to a battery of known aeroallergens but for adults the sero-reactivity could not be attributed to the known aeroallergens and was therefore assumed to a manifestation of one or more 'unknown' allergens (Burrows *et al.*, 1989). Lastly, macrolide therapy in a *C. muridarum* mouse model suppressed key features of experimental steroid-sensitive and steroid-insensitive asthma (Essilfie *et al.*, 2015). Whether antibiotic therapy is antimicrobial, purely anti-inflammatory (immunomodulatory) or a combination of both is currently controversial.

Comparable mouse models of *C. pneumoniae* infection largely confirm that *C. pneumoniae* infection creates the 'asthma phenotype' in dose- and age-dependent manners (Crother *et al.*, 2011; Schröder *et al.*, 2008). A novel finding from *C. pneumoniae* murine models is that adoptive transfer of lung dendritic cells from infected mice (who themselves have developed the 'asthma phenotype') will reproduce the same 'asthma phenotype' in recipients (Schröder *et al.*, 2008). Dendritic cells play a key role in immunoregulation within the lung, and chlamydial infection subverts dendritic cell function to promote Th2 immunity and airways hyper-responsiveness (Kaiko *et al.*, 2008). These findings suggest that a cell type that is relatively inaccessible and very difficult to sample (dendritic cells) could be responsible for initiating the 'inflammatory cascade' to produce human asthma (Hahn, 2013). The difficulty in proving direct causation should not deter clinical investigators from launching large, adequately powered pragmatic trials of long-term macrolides (such as azithromycin) in patients of all ages with severe asthma, particularly asthma that is refractory to usual treatments.

Other chronic respiratory conditions linked to *C. pneumoniae* infection

Chronic bronchitis and COPD

As mentioned above, chronic *C. pneumoniae* infection has been associated with chronic bronchitis and COPD (Blasi *et al.*, 1997; Blasi *et al.*, 2002; Hahn, 1999). In chronic bronchitis patients, (1) colonization of the sputum by *C. pneumoniae* was associated with lower lung function and greater airway bacterial colonization (Blasi *et al.*, 2002) and (2) positive blood PCR tests for *C. pneumoniae* (in 61 of 141 subjects) were associated with a significantly higher exacerbation rate (Blasi *et al.*, 2002). Interestingly, six weeks of azithromycin decreased blood PCR positivity from 100% to 29% at one year follow up in those treated for exacerbations (Blasi *et al.*, 2002). It is important to note that *C. pneumoniae* PCR positivity is common in healthy adult blood donors (~10–45%) (Boman *et al.*, 1998; Haranaga *et al.*, 2001; Karimi *et al.*, 2010; Kaul *et al.*, 2000) and that positive detections in the peripheral blood do not prove that the lung is infected. In patients with severe COPD, azithromycin has been found efficacious in improving symptoms and decreasing exacerbations (Albert *et al.*, 2011), but testing for *C. pneumoniae* was not done.

Bronchiolitis obliterans syndrome

Bronchiolitis obliterans syndrome is a complication associated with post-lung transplant pulmonary dysfunction and mortality. In one series, *C. pneumoniae* infection was reported in 25% of bronchoalveolar lung specimens obtained within 30 days of transplant from lung transplant recipients, and also in 25% of recipients studied later than 30 days post-transplant. In this study, *C. pneumoniae* infection, whether donor derived, *de novo*, or reactivated, was associated with early mortality, rejection, and bronchiolitis obliterans syndrome (Glanville *et al.*, 2005). A separate retrospective, observational cohort study of 107 lung transplant recipients with bronchiolitis obliterans syndrome reported that long-term azithromycin was associated with improved pulmonary function and survival, particularly in bronchiolitis obliterans syndrome with increased neutrophilia (testing for *C. pneumoniae* was not included in this study) (Vos *et al.*, 2010).

An uncontrolled treatment trial suggested that azithromycin may also be efficacious in post-bone marrow transplant bronchiolitis obliterans syndrome (Khalid *et al.*, 2005). A possible contribution of immunosuppression to pathogenesis of *C. pneumoniae*-associated chronic diseases, such as bronchiolitis obliterans syndrome, was suggested by a report showing that *C. pneumoniae* was detected by PCR in 10 of 11 pretreatment samples in patients with cyclosporine-associated gingival overgrowth (GO). Although azithromycin was effective in reducing GO, *C. pneumoniae* DNA remained detectable in all GO tissue samples, despite azithromycin treatment (Worm *et al.*, 2004). These intriguing reports, while not conclusive, do suggest that *C. pneumoniae* should be considered as a possible culprit in bronchiolitis obliterans syndrome –and related chronic inflammatory diseases of unknown aetiology – particularly in those with associated immunosuppression.

Cardiovascular infections

Atherosclerotic cardiovascular disease

Background

Atherosclerotic cardiovascular disease is the leading cause of mortality in the world (World Health Organization, 2011). Cardiovascular diseases resulting from atherosclerosis include ischaemic heart disease (also known as coronary heart disease or coronary artery disease), cerebrovascular disease, diseases of the aorta, and peripheral vascular disease. Ischaemic heart disease presents as stable and unstable angina, non-fatal myocardial infarction and coronary death (Wong, 2014). The most common presentation of cerebrovascular disease is ischaemic stroke (Hu *et al.*, 2017). Atherosclerosis is a disease of chronic inflammation that leads to blood vessel interior wall (intima) tissue damage and fibrosis (Geovanini and Libby, 2018; Libby, 2012; Ross, 1999). Early atherosclerotic lesion development results from blood vessel interior wall lining (endothelial) activation/injury leading to the expression of adhesion molecules and recruitment of monocytes and leucocytes that enter the intima

(Jang *et al.*, 1993; Libby, 2012; Pant *et al.*, 2013). Monocytes transform into tissue macrophages and internalize oxidized low-density lipoprotein (LDL) resulting in foam cell formation. Foam cells release cytokines, which further upregulate adhesion molecules on the endothelium and increase adherence and migration of inflammatory cells. The early atherosclerotic lesion, the fatty streak, consists of lipid-laden cells (monocyte-derived macrophages, macrophage-derived foam cells and T lymphocytes). Platelets accumulate at the site of endothelial injury and release platelet-derived growth factor that stimulates smooth muscle cell proliferation. Resulting secretion of collagen, elastin, and proteoglycans leads to the formation of fibrous tissue. The mature fibrolipid plaque is comprised of an acellular cholesterol/lipid core surrounded by a fibrous cap. Tissue proteolysis by matrix metalloproteinases promotes lesion vulnerability (Finn *et al.*, 2010). Conversion of a stable plaque to a vulnerable plaque can result in plaque rupture or erosion leading to thrombus formation, which can clinically manifest as an acute ischaemic event such as myocardial infarction, stroke, and/or cardiovascular death (Badimon and Vilahur, 2014; Geovanini and Libby, 2018; Libby, 2012).

Risk factors

Well-defined risk factors for the development of atherosclerosis include hypercholesterolemia, hypertension, hyperglycaemia and genetic factors, but collectively, these do not account for the total incidence of atherosclerosis (O'Connor *et al.*, 2001). The idea that infections play a role in the pathogenesis of atherosclerosis was described over a century ago. The first experimental evidence supporting this idea was the demonstration that Marek's disease virus, a chickenpox (herpes) virus, induced atherosclerosis in pathogen-free chickens, and immunization with turkey herpesvirus prior to infection prevented disease (Fabricant and Fabricant, 1999; Fabricant *et al.*, 1983). These elegant studies provided the impetus for investigations into the possible association of human pathogens with atherosclerosis. Cumulative evidence has suggested that infectious agents can contribute to atherosclerotic processes through direct or indirect mechanisms that ultimately promote chronic inflammation (Campbell and Rosenfeld, 2015). *C. pneumoniae* has been one of the most rigorously studied organisms as a risk factor for atherosclerosis. Compelling evidence of an association of *C. pneumoniae* and atherosclerosis is based on seroepidemiological studies, demonstration of the organism in atheromas, and *in vitro* and animal model studies demonstrating biological plausibility (Campbell and Rosenfeld, 2015; Rosenfeld and Campbell, 2011). Yet, causality of *C. pneumoniae* or any other infectious agent in human atherosclerosis has not yet been proved or refuted (Campbell and Rosenfeld, 2015; Grayston *et al.*, 2015).

Chlamydia pneumoniae infection and cardiovascular disease

Shortly after *C. pneumoniae* was recognized as a human respiratory pathogen, a landmark study by Saikku *et al.* reported an association of *C. pneumoniae* with chronic coronary heart disease and acute myocardial infarction based on the higher prevalence of *C. pneumoniae*-specific Immunoglobulin G antibody in patients versus control subjects (Saikku *et al.*, 1988). Subsequently, Dr Alan Shor identified pear-shaped structures by electron microscopy in coronary atheromas obtained at autopsy that were analogous to the *C. pneumoniae* EB ultrastructure, an impression that was confirmed by positive immunostaining results using *Chlamydia* genus and *C. pneumoniae*-species specific monoclonal antibodies (Shor *et al.*, 1992). Staining was not observed using a *C. trachomatis* species-specific MAb or a *C. psittaci* MAb reactive with a psittacine strain. Analysis of a larger number of these specimens that also included PCR found evidence of *C. pneumoniae* in over 50% of the coronary atheromas, supporting the seroepidemiological association of the organism with atherosclerosis (Kuo *et al.*, 1993).

As of 2004, over 50 seroepidemiological studies have demonstrated an association of *C. pneumoniae* with cardiovascular disease (coronary heart disease, carotid stenosis, aortic abdominal aneurysm, and stroke) (Kuo and Campbell, 2003). The average odds ratio in these studies were generally 2.0 and the risk was independent of other risk factors of atherosclerosis (Kuo and Campbell, 2003). However, there have also been seroepidemiological studies that have not demonstrated an association (Danesh *et al.*, 2000; Ridker *et al.*, 1999). Meta-analysis of 15 prospective studies focused only on the association of *Chlamydia pneumoniae* IgG titres and coronary heart disease demonstrated an odds ratio of 1.66, which was reduced to 1.22 when adjusted for smoking and socioeconomic status (Danesh *et al.*, 2000). Meta-analysis of 38 seroepidemiological studies from 1997–2000 demonstrated an overall odds ratio of 1.6 (Bloemenkamp *et al.*, 2003). Interpretation of seroepidemiological studies are confounded by the different assays used, the antibody measured (IgA vs. IgG), the varying level of expertise in interpreting the micro-IF test when it was the chosen assay, and the differences in study design.

Over sixty-five studies have observed *C. pneumoniae* in atheromas including coronary, carotid, pulmonary, iliac and femoral arteries and the aorta. The organism has been identified in fatty streaks as well as fibrolipid plaques and in atheromatous tissue from a wide age group ranging from young persons to the elderly and different races (Kuo and Campbell, 2003). Although a variety of pathogens have been detected in atherosclerotic lesions through identification of pathogen specific antigens or DNA, *C. pneumoniae* has been cultured from coronary and carotid atheromas (Campbell and Rosenfeld, 2015; Jackson *et al.*, 1997a; Maass *et al.*, 1998; Ramirez, 1996). The only other organism that has been isolated from atherosclerotic tissue (one patient) is *Enterobacter hormaechei* (Rafferty *et al.*, 2011).

Chlamydia pneumoniae infection and cerebrovascular disease

Case control studies investigating the association of *C. pneumoniae* with ischaemic stroke have yielded varying results. The recent IN2 study, the largest study done to date, was a prospective case–control study involving 1002 participants to address whether previous infections in general, and *C. pneumoniae* infection in particular, increased the risk of ischaemic stroke (Consoli *et al.*, 2015). Eleven stroke units in Italy participated, and controls were matched by age and sex with patients admitted to the hospital for acute ischaemic strokes. The IN2 study demonstrated that an infection seven days prior to the event, or seroprevalence of anti-*C. pneumoniae* IgA antibody significantly increased the risk of ischaemic stroke (Consoli *et al.*, 2015). Two meta-analyses have addressed the use of different methods in associating *C. pneumoniae* with cerebrovascular disease and cerebral infarction. The first analysed 52 published studies on cerebrovascular disease; an association with *C. pneumoniae* infection was indicated by serum-specific IgG, serum-specific IgA and by the presence of *C. pneumoniae* in peripheral blood cells (Chen *et al.*, 2013). Subgroup analyses demonstrated that the strongest *C. pneumoniae* association was with stroke from large artery atherosclerosis (Chen *et al.*, 2013). In the second meta-analysis analyzing 23 published papers, *C. pneumoniae* infection was found to be significantly associated with an increased risk of cerebral infraction (Su and Chen, 2014). *C. pneumoniae*-specific IgA was a stronger predictor than *C. pneumoniae*-specific IgG (Su and Chen, 2014).

C. pneumoniae infection and chronic inflammation

An unresolved question in the development of atherosclerotic lesions is what factors provide the constant stimuli promoting the chronic inflammatory processes in atherosclerosis. Although lipid modification and deposition are one source, there is considerable evidence suggesting that infection can also contribute to these processes (Rosenfeld and Campbell, 2011). A recent meta-analysis of 25 studies encompassing over 10,000 patients addressed whether *C. pneumoniae* contributed to chronic inflammation of atherosclerosis by determining whether *C. pneumoniae* seropositivity correlated with inflammatory markers (Filardo *et al.*, 2015). In patients with cardiovascular disease, several inflammatory markers were significantly increased and *C. pneumoniae*-specific antibodies were more prevalent. Importantly, high sensitivity C-reactive protein, IL-6, and fibrinogen levels were significantly higher in atherosclerotic patients who also had *C. pneumoniae*-specific IgA, suggesting that *C. pneumoniae* infection may enhance inflammation in patients with atherosclerotic cardiovascular disease (Filardo *et al.*, 2015).

Clinical trials with anti-chlamydial antibiotics in atherosclerotic diseases

The association of *C. pneumoniae* with atherosclerosis garnered considerable enthusiasm for the prospect that treatment with anti-chlamydial antibiotics would have benefits in secondary prevention of atherosclerotic vascular diseases. Secondary prevention refers to, e.g. treatment of patients with established coronary heart disease to prevent a second heart attack (myocardial infarction).

The concept also applies to patients with other atherosclerotic diseases such as stroke, peripheral vascular disease and aortic aneurysm. Pilot studies in patients with post-myocardial infarction or acute coronary syndrome were limited by small patient numbers, short treatment duration and follow-up times (Grayston, 2003). In these small trials, the treatment ranged from 3–7 days with azithromycin to 85 days with clarithromycin. Of these studies, three demonstrated beneficial effects, one had mixed results, and four were negative (Grayston, 2003). Beneficial effects were also observed in preliminary intervention studies of abdominal aortic aneurysm growth, carotid artery disease thickness, peripheral vascular disease symptoms and restenosis following stent placement in coronary arteries (Grayston, 2003). Patients were treated with roxithromycin for 28–30 days in all but one of these studies in which patients were treated with doxycycline for 28–30 days.

Overall, the small trial results underscored the need for larger clinical trials that addressed the limitations of the pilot studies. Because of the difficulty in treating chronic *Chlamydia* infections, treatment for longer duration was advocated (Grayston, 2003; Grayston *et al.*, 1999). Four well-designed clinical trials collectively enrolled over 20,000 patients. While three clinical trials (WIZARD, ACES, and CLARICOR) enrolled patients with stable coronary artery disease, one (PROVE-IT-TIMI) enrolled patients with acute coronary syndrome (Cannon *et al.*, 2005; Grayston *et al.*, 2005; Jespersen *et al.*, 2006; O'Connor *et al.*, 2003). Because short-term benefits were observed in patients treated with azithromycin for three months in the WIZARD trial, both ACES (600 mg azithromycin weekly for one year) and PROVE-IT-TIMI (gatifloxacin 400 mg for the first two weeks followed by a 10 day course, monthly for two years) had longer durations of antibiotic treatment (Cannon *et al.*, 2005; Grayston *et al.*, 2005). The CLARICOR trial assessed short-term treatment with clarithromycin (500 mg daily for two weeks) (Jespersen *et al.*, 2006). However, all four trials failed to demonstrate clinical benefits of antibiotic treatment.

Negative outcomes were also observed in two large-scale clinical trials of patients with peripheral artery disease. In the first study, 505 patients were randomized and treated with roxithromycin (300 mg daily for 28 days, with a mean follow up time of 2.1 years) (Joensen *et al.*, 2008). In the second study (Providence 1), 297 patients with intermittent claudication secondary to peripheral artery disease were treated with an eight-week regimen of rifalazil (25 mg once a week) or placebo, with a follow up time of 360 days (Jaff *et al.*, 2009). Collectively, these studies definitively showed that short or long term treatments with anti-chlamydial antibiotics were not efficacious in preventing secondary events in cardiovascular disease, nor did treatment provide any symptom improvement in peripheral artery disease or prevention of primary or secondary events.

Prior to completion of the large-scale, antibiotic treatment trials for coronary artery disease, the lead investigator of the ACES trial (Dr J.T. Grayston) predicted that, regardless of the outcome, the results would tend to be over interpreted with regard to causality. This would be problematic because the studies were not designed to address aetiology (Grayston, 2003). He envisaged that if the results were negative, interest in the association of *C. pneumoniae* with atherosclerotic processes would decrease although the trials would provide no insight into whether or not *C. pneumoniae* played a role. Sadly, he was correct in his assessment. The negative outcome of the clinical trials led many to conclude that *C. pneumoniae* had no role in atherosclerotic disease (Danesh, 2005). This misinterpretation has significantly impacted continued research support, not only for *C. pneumoniae* but also for the role of infectious agents generally in the pathology of atherosclerosis (Epstein *et al.*, 2009; Grayston *et al.*, 2015).

If the *C. pneumoniae* hypothesis is correct, there are potential caveats of the antibiotic treatment trials that warrant consideration (Anderson, 2005; Deniset and Pierce, 2010; Epstein *et al.*, 2009; Rosenfeld and Campbell, 2011; Taylor-Robinson and Boman, 2005). First, the design of the study was to treat patients who had advanced coronary artery disease, and intervention at that stage may be too late. Moreover, the events measured were likely due to plaque destabilization and rupture, which may be independent of atherosclerotic plaque progression due to infection. Second, the assumption was that long-term treatment with a single antibiotic effective against acute chlamydial infection would show efficacy in treatment of chronic disease. In patients with chronic *Chlamydia* induced reactive arthritis, treatment with combinations of antibiotics (doxycycline and rifampin or azithromycin and rifampin) given over six months resulted in improvement or complete resolution of symptoms in significant numbers of patients in comparison to subjects receiving placebos (Carter *et al.*, 2010). In this study, all subjects had *Chlamydia* PCR positive synovial tissue and/or PBMCs prior to treatment. Significant increases in the number of patients whose specimens were PCR negative were also observed following treatment. Interestingly, in patients undergoing treatment for coronary artery disease with a single agent (azithromycin), the organism was cultured from monocytes indicating that infection was not eradicated (Gieffers *et al.*, 2001). Lastly, it is possible that antibiotic treatment was ineffective due to pathogen burden with other pathogens not susceptible to anti-chlamydial antibiotics (Epstein *et al.*, 2009; Rosenfeld and Campbell, 2011). Thus, multiantibiotic trials might show benefit in secondary prevention of atherosclerotic diseases. Another possibility is that treating at the secondary prevention stage is 'too late' and intervention earlier in the natural history of atherosclerosis might be more successful.

Animal models of *C. pneumoniae* infection and atherosclerosis

Mice have been the most frequently used animal model in atherosclerosis research because of their small size, ease of breeding and ease of genetic manipulation (Getz and Reardon, 2012; Lee *et al.*, 2017). Thus, they have been frequently used for studying *C. pneumoniae* infection and atherosclerosis. Limitations of mouse models are those aspects that differ between atherosclerosis in humans and in mice. Mice do not develop unstable plaques or readily develop coronary artery atherosclerosis, but develop significant lesions in the aortic sinus, which is not the case in human atherosclerosis (Getz and Reardon, 2012; Lee *et al.*, 2017).

The effect of *C. pneumoniae* infection has been studied in various diet-induced animal models of atherosclerosis, including C57BL/6J mice and LDL receptor-deficient mice (LDLR–/–) that were fed a high fat/cholesterol diet, and New Zealand white rabbits that were fed a cholesterol diet (Blessing *et al.*, 2001; Hu *et al.*, 1999; Liu *et al.*, 2000; Muhlestein *et al.*, 1998). ApoE knockout mice, which develop atherosclerosis spontaneously in the absence of an atherogenic diet, have also been assessed (Burnett *et al.*, 2001; Moazed *et al.*, 1999; Moazed *et al.*, 1997; Rothstein *et al.*, 2001). These hyperlipidaemic animal models showed that repeated intranasal or intratracheal infection with *C. pneumoniae* resulted in dissemination of the organism to the vasculature, establishment of persistent infection in the aorta, detection of the organism in foam cells and acceleration of atherosclerotic lesion development. A single infection with *C. pneumoniae* also exacerbated atherosclerosis in ApoE knockout mice and enhanced foam cell induction and atherosclerosis production in a transgenic hyperlipidaemia-genetic hypertension rat model (Herrera *et al.*, 2003; Tumurkhuu *et al.*, 2018). Collectively, these studies provide the strongest evidence regarding the biological plausibility of a role for *C. pneumoniae* in atherosclerosis causation. In contrast, normolipidemic mice did not develop persistent infection or atherosclerosis of the aorta (Blessing *et al.*, 2000; Moazed *et al.*, 1997). However, aortic changes were observed following single or repeated inoculations in New Zealand White Rabbits without cholesterol feeding (Fong *et al.*, 1999b; Laitinen *et al.*, 1997).

The effect of antibiotic treatment on *Chlamydia pneumoniae* accelerated atherosclerosis

Significantly, treatment with two different regimens of azithromycin at doses equivalent to those administered for *C. pneumoniae* infection in humans had no effect on *C. pneumoniae* exacerbation of atherosclerosis in ApoE$^{-/-}$ mice, and organisms were detected by PCR in the aorta following treatment (Blessing *et al.*, 2005; Rothstein *et al.*, 2001). In both studies, antibiotic treatment was initiated after the last infection, suggesting that immediate treatment of acute *C. pneumoniae* infection might be required to prevent proatherogenic effects (Rothstein *et al.*, 2001). Indeed, anti-chlamydial therapy was efficacious in preventing *C. pneumoniae* accelerated atherosclerosis in New Zealand White rabbits when antibiotic treatment was initiated shortly after infection, although the organism was still detected in the vascular tissue in some animals (Fong, 2000; Fong *et al.*, 2002; Muhlestein *et al.*, 1998). Specifically, initiation of a 7-week course of azithromycin immediately following the third *C.*

pneumoniae inoculation in cholesterol fed rabbits prevented *C. pneumoniae*-accelerated maximal intimal thickening (Muhlestein *et al.*, 1998). Fong *et al.* compared early (5 days after the first of three *C. pneumoniae* inoculations) versus late (8 weeks after the first inoculation) azithromycin or clarithromycin treatment in rabbits fed standard chow. For both antibiotics, early treatment was highly effective in preventing lesions. Late azithromycin treatment was ineffective and late clarithromycin treatment was only partially effective (Fong *et al.*, 1999a; Fong *et al.*, 2002). Mouse models may prove useful in determining whether organisms detected following antibiotic treatment are viable through culture following immunosuppression or by transcriptional evidence in tissues. Future studies should also be directed at determining whether a combination of antimicrobial therapy administered later in infection decreases *C. pneumoniae* accelerated atherosclerosis and correlates with evidence of the organism in tissue.

Mechanisms contributing to *Chlamydia pneumoniae* accelerated atherosclerosis in mouse models

Mouse models have provided insight into pathways induced by *C. pneumoniae* infection that promote an increase in atherosclerotic lesion size. Through the use of knockout mice, it has been shown that the TNF-α p55 receptor, adhesion molecules (ICAM-1, E-selectin, and P-selectin), IL15, TLR2, TLR4, MyD88, IL-15, Nlrp3 and Caspase 1 contribute to *C. pneumoniae*-accelerated atherosclerosis (Campbell, 2010b; Campbell *et al.*, 2005; Chen *et al.*, 2010; Chen *et al.*, 2008; Naiki *et al.*, 2005; Tumurkhuu *et al.*, 2018). The innate immune system plays a role in inflammation resulting in atherosclerotic lesion progression through Toll receptors and nucleotide-binding domain and leucine-rich repeat pyrin domain containing 3 (NLRP3) inflammasome (Broderick *et al.*, 2015). In a series of studies, MyD88-dependent signalling from TLR2 and TLR4 pathways have been shown to play a role in both *C. pneumoniae*-accelerated atherosclerosis and lipid-mediated atherosclerosis (Bjorkbacka *et al.*, 2004; Michelsen *et al.*, 2004; Mullick *et al.*, 2005; Schröder *et al.*, 2008). Both *C. pneumoniae* and intracellular cholesterol activate the NLRP3 inflammasome, resulting in secretion of IL-1B (Duewell *et al.*, 2010; Tumurkhuu *et al.*, 2018). In turn, IL-1B inhibits ATP binding cassette transporter A1-mediated cholesterol efflux by downregulating ABCA1 expression (Tumurkhuu *et al.*, 2018). Ultimately, this results in the accumulation of intracellular lipids, foam cell formation, and lesion progression.

Mice deficient in CD8α were used to evaluate the role of CD8+ T cells in *C. pneumoniae*-induced atherosclerosis in C57Bl/6J mice fed a high fat diet (Zafiratos *et al.*, 2015). These studies were predicated on previous reports demonstrating that (i) upper genital tract pathology in the *C. muridarum* mouse model of genital tract infection was mediated by TNF-α-producing CD8+ T cells and (ii) CD8+ T cells contributed to the development of vulnerable atherosclerotic plaque in apoE-deficient mice (Kyaw *et al.*, 2013; Murthy *et al.*, 2011). In comparison to *C. pneumoniae*-infected wild type (WT) mice, infected CD8 knockout mice had significantly decreased atherosclerotic lesion area. Repletion of CD8 knockout mice with WT CD8+ T cells at the time of infection restored the lesion area to that observed with infected WT controls. These studies demonstrate that CD8+ T cells play a significant role in *C. pneumoniae* acceleration of atherosclerosis (Zafiratos *et al.*, 2015). Lastly, in mice, *C. pneumoniae* infection upregulates the expression of the scavenger receptor LOX-1, the major scavenger receptor on endothelial cells promoting uptake of oxidized low-density lipoprotein (Campbell, 2010a; Sawamura *et al.*, 1997). Activation of LOX-1 results in up-regulation of a variety of proatherogenic factors, including monocyte chemoattractant protein-1, matrix metalloproteinases, and TNF-α (Mehta *et al.*, 2006). Collectively, these studies demonstrate multiple pathways by which *C. pneumoniae* infection stimulates chronic inflammatory processes of atherosclerosis.

Vaccination to prevent atherosclerosis

An alternative approach has been to determine whether vaccination with appropriate antigens affords protection from *C. pneumoniae*-accelerated atherosclerosis. $Apob^{tm2Sgy}Ldlr^{tm1Her}/J$ mice (a strain that develops high levels of atherosclerosis on an atherogenic diet) were immunized with KDH-conjugated peptides derived from the MOMP and the putative outer membrane protein 5 of *C. pneumoniae*. Following a repetitive immunization strategy at different sites, mice were fed a high-fat diet and infected twice (Xia *et al.*, 2013). Immunization with a combined peptide containing both epitopes resulted in the greatest reduction in atherosclerotic lesion size in the aortic sinus and descending aorta. Intranasal vaccination with recombinant chlamydial protease-like activity factor (rCPAF) in combination with interleukin-12 prevented *C. pneumoniae* acceleration of atherosclerosis mice fed an atherogenic diet (Li *et al.*, 2018). Because CPAF is conserved in *Chlamydia* and intranasal vaccinations with CPAF and IL-12 protected against oviduct pathology in the *C. muridarum* model, the authors noted the potential of CPAF for use in a multivalent vaccine to protect against pathological consequences of chlamydial infection (Li *et al.*, 2018).

Other diseases associated with *C. pneumoniae*

C. pneumoniae infections have been described in association with a variety of acute and chronic inflammatory and neoplastic conditions of the skin, nervous system, cardiovascular system (other than atherosclerotic), and lung (other than asthma, chronic bronchitis and COPD). These descriptions are mostly derived from case reports and epidemiological associations; causality remains speculative at this time pending further research. Nevertheless, clinicians should include *C. pneumoniae* in the differential diagnosis of these and related inflammatory conditions that are acute, or subacute, as this may influence treatment choices. *C. pneumoniae* has been reported in association with skin wounds (King *et al.*, 2001), acne rosacea (Fernandez-Obregon and Patton, 2007), and primary cutaneous anaplastic large-cell lymphoma (in association with human Herpesvirus 8) (Caselli *et al.*, 2016). An association with cutaneous T-cell lymphoma could not be confirmed in one study (Rossler *et al.*, 2003). Case reports also implicate acute *C. pneumoniae* infection as one of many causes for myocarditis (Wesslén *et al.*, 1992), meningoencephalitis (Anton *et al.*, 2000; Elargoubi *et al.*, 2013), and lumbosacral meningoradiculitis (Michel *et al.*, 1992).

Regarding chronic inflammatory diseases of unknown cause, Alzheimer's disease and diabetes are associated epidemiologically through similar inflammatory patterns, including amyloid deposition, and are also associated with exogenous polymicrobial exposures, including *C. pneumoniae* (Miklossy and McGeer, 2016; Roubaud Baudron *et al.*, 2015). A serological association between lung cancer and *C. pneumoniae* infection was first reported by Laurila *et al.* in 1997 (Laurila *et al.*, 1997). More recently a meta-analysis of 12 studies involving 2595 lung cancer patients found that *C. pneumoniae* serological biomarkers were significantly associated with lung cancer in both cross-sectional and prospective studies (Zhan *et al.*, 2010). In a subsequent rat model, repeated intratracheal injections of *C. pneumoniae* were associated with lung cancer development in 15% of infected animals and in none of the uninfected group (Chu *et al.*, 2012). Taken together, these findings are provocative, and call for further investigation.

Future directions

An unresolved question is how to demonstrate unequivocally that *C. pneumoniae* infection is causally associated with coronary heart disease. Atherosclerosis may occur in young children, its prevalence increases with age, and lesion development progresses over decades. Similarly, *C. pneumoniae* infection occurs in early childhood, increases with day care or school attendance, and prevalence of infection increases with age (Kuo *et al.*, 1995). These facts preclude short-term studies to address any etiologic role of *C. pneumoniae* in human coronary heart disease. Whether antibiotics would be efficacious in treatment of patients with early atherosclerosis or in 'high-risk' individuals has not been determined as such studies would be difficult to design and execute. Although demonstration that vaccination against *C. pneumoniae* prevented or decreased atherosclerotic disease progress would prove causality, justification for vaccination in children to prevent putative atherosclerosis development would not be warranted, when causality has not been proven (Murdin *et al.*, 2000b). Second, a vaccine will be very difficult to test in humans due to the long-term course of atherosclerotic disease development. Grayston *et al.* have proposed longitudinal evaluation of families, an approach that was effective in studying the epidemiology of acute respiratory infections, utilizing recent advances in non-invasive imaging techniques and improved methods for diagnosing *C. pneumoniae* infection

(Grayston *et al.*, 2015). The latter permits the ability to monitor family members over time for *C. pneumoniae* infection through analysis of nasopharyngeal samples or PBMCs. The advancement and rapidity of DNA sequencing technology will enable identification of co-infecting organisms. The ultimate design of the proposed study is to correlate clinical disease with *C. pneumoniae* infection, airway microbiota, co-infections, and immune responses. This approach can also be applied to the genesis of obstructive airways diseases (asthma, chronic bronchitis and COPD). The major obstacle to the proposed studies is the expense of long-term observational studies.

References

Albert, R.K., Connett, J., Bailey, W.C., Casaburi, R., Cooper, J.A., Jr., Criner, G.J., Curtis, J.L., Dransfield, M.T., Han, M.K., Lazarus, S.C., *et al.* (2011). Azithromycin for prevention of exacerbations of COPD. The New England journal of medicine *365*, 689–698. doi:10.1056/NEJMoa1104623.

Aldous, M.B., Grayston, J.T., Wang, S.-P., and Foy, H.M. (1992). Seroepidemiology of *Chlamydia pneumoniae* TWAR infection in Seattle families, 1966–1979. The Journal of infectious diseases *166*, 646–649.

Alexander, E.R., Wang, S.P., and Grayston, J.T. (1967). Further classification of TRIC agents from ocular trachoma and other sources by the mouse toxicity prevention test. American journal of ophthalmology *63*, Suppl:1469–1478.

Anderson, J.L. (2005). Infection, antibiotics, and atherothrombosis--end of the road or new beginnings? New England Journal of Medicine *352*, 1706–1709 doi:10.1056/NEJMe058019.

Anton, E., Otegui, A., and Alonso, A. (2000). Meningoencephalitis and *Chlamydia pneumoniae* infection. Eur J Neurol *7*, 586.

Asche, L.V., Hutton, S.I., and Douglas, F.P. (1993). Serological evidence of the three chlamydial species in an Aboriginal community in the Northern Territory. Med J Australia *158*, 603–604.

Badimon, L., and Vilahur, G. (2014). Thrombosis formation on atherosclerotic lesions and plaque rupture. Journal of internal medicine *276*, 618–632 doi:10.1111/joim.12296.

Beagley, K.W., Huston, W.M., Hansbro, P.M., and Timms, P. (2009). Chlamydial infection of immune cells: Altered function and implications for disease. Critical Reviews in Immunology *29*, 275–305.

Beatty, W.L., Morrison, R.P., and Byrne, G. (1994). Persistent chlamydiae: from cell culture to a paradigm for chlamydial pathogenesis. Microbiological Reviews *58*, 686–699.

Bell, T.A., Kuo, C.C., Wang, S.P., and Grayston, J.T. (1989). Experimental infection of baboons (Papio cynocephalus anubis) with *Chlamydia pneumoniae* strain 'TWAR'. The Journal of infection *19*, 47–49.

Berger, L., Volp, K., Mathews, S., Speare, R., and Timms, P. (1999). *Chlamydia pneumoniae* in a free-ranging giant barred frog (Mixophyes iteratus) from Australia. Journal of clinical microbiology *37*, 2378–2380.

Biendo, M., Lefebvre, J.F., Fuentes, V., and Orfila, J. (1994). The prevalence of anti-*Chlamydia trachomatis* and anti-*Chlamydia pneumoniae* antibodies in Brazzaville. Bulletin de la Societe de Pathologie Exotique *87*, 85–88.

Bjorkbacka, H., Kunjathoor, V.V., Moore, K.J., Koehn, S., Ordija, C.M., Lee, M.A., Means, T., Halmen, K., Luster, A.D., Golenbock, D.T., *et al.* (2004). Reduced atherosclerosis in MyD88-null mice links elevated serum cholesterol levels to activation of innate immunity signaling pathways. Nature medicine *10*, 416–421 doi:10.1038/nm1008.

Blasi, F., Boschini, A., Cosentini, R., Legnani, D., Smacchia, C., Ghira, C., and Allegra, L. (1994a). Outbreak of *Chlamydia pneumoniae* infection in former injection-drug users. Chest *105*, 812–815.

Blasi, F., Cosentini, R., Damato, S., Denti, F., Fagetti, L., Molteni, L., Tarsia, P., Tien, T.V., and Allegra, L. (1997). *Chlamydia pneumoniae* chronic infection increases the risk of bacterial colonization in chronic bronchitis. Am J Resp Crit Care Med *155 (part 2 of 2 parts)*, A592.

Blasi, F., Cosentini, R., Denti, F., and Allegra, L. (1994b). Two family outbreaks of *Chlamydia pneumoniae* infection. Eur Resp J *7*, 102–104.

Blasi, F., Damato, S., Cosentini, R., Tarsia, P., Raccanelli, R., Centanni, S., and Allegra, L. (2002). *Chlamydia pneumoniae* and chronic bronchitis: association with severity and bacterial clearance following treatment. Thorax *57*, 672–676.

Blessing, E., Campbell, L.A., Rosenfeld, M.E., Chesebro, B., and Kuo, C.C. (2005). A 6 week course of azithromycin treatment has no beneficial effect on atherosclerotic lesion development in apolipoprotein E-deficient mice chronically infected with *Chlamydia pneumoniae*. The Journal of antimicrobial chemotherapy *55*, 1037–1040 doi:10.1093/jac/dki128.

Blessing, E., Campbell, L.A., Rosenfeld, M.E., Chough, N., and Kuo, C.C. (2001). *Chlamydia pneumoniae* infection accelerates hyperlipidemia induced atherosclerotic lesion development in C57BL/6J mice. Atherosclerosis *158*, 13–17.

Blessing, E., Lin, T.M., Campbell, L.A., Rosenfeld, M.E., Lloyd, D., and Kuo, C. (2000). *Chlamydia pneumoniae* induces inflammatory changes in the heart and aorta of normocholesterolemic C57BL/6J mice. Infection and immunity *68*, 4765–4768.

Bloemenkamp, D.G., Mali, W.P., Visseren, F.L., and van der Graaf, Y. (2003). Meta-analysis of sero-epidemiologic studies of the relation between *Chlamydia pneumoniae* and atherosclerosis: does study design influence results? American heart journal *145*, 409–417 doi:10.1067/mhj.2003.20.

Bodetti, T.J., Jacobson, E., Wan, C., Hafner, L., Pospischil, A., Rose, K., and Timms, P. (2002). Molecular evidence to support the expansion of the hostrange of Chlamydophila pneumoniae to include reptiles as well as humans, horses, koalas and amphibians. Systematic and applied microbiology *25*, 146–152 doi:10.1078/0723-2020-00086.

Boman, J., Soderberg, S., Forsberg, J., Birgander, L.S., Allard, A., Persson, K., Jidell, E., Kumlin, U., Juto, P., Waldenstrom, A., *et al.* (1998). High prevalence of *Chlamydia pneumoniae* DNA in peripheral blood mononuclear cells in patients with cardiovascular disease and in middle-aged blood donors. The Journal of infectious diseases *178*, 274–277.

Bone, R.C. (1991). Chlamydial pneumonia and asthma: a potentially important relationship. JAMA *266*, 265.

Borel, N., Mukhopadhyay, S., Kaiser, C., Sullivan, E.D., Miller, R.D., Timms, P., Summersgill, J.T., Ramirez, J.A., and Pospischil, A. (2006). Tissue MicroArray (TMA) analysis of normal and persistent Chlamydophila pneumoniae infection. BMC infectious diseases *6*, 152 doi:10.1186/1471-2334-6-152.

Borel, N., Pospischil, A., Dowling, R.D., Dumrese, C., Gaydos, C.A., Bunk, S., Hermann, C., Ramirez, J.A., and Summersgill, J.T. (2012). Antigens of persistent *Chlamydia pneumoniae* within coronary atheroma from patients undergoing heart transplantation. Journal of clinical pathology *65*, 171–177 doi:10.1136/jclinpath-2011-200270.

Borel, N., Summersgill, J.T., Mukhopadhyay, S., Miller, R.D., Ramirez, J.A., and Pospischil, A. (2008). Evidence for persistent *Chlamydia pneumoniae* infection of human coronary atheromas. Atherosclerosis *199*, 154–161 doi:10.1016/j.atherosclerosis.2007.09.026.

Broderick, L., De Nardo, D., Franklin, B.S., Hoffman, H.M., and Latz, E. (2015). The inflammasomes and autoinflammatory syndromes. Annual review of pathology *10*, 395–424 doi:10.1146/annurev-pathol-012414-040431.

Burnett, M.S., Gaydos, C.A., Madico, G.E., Glad, S.M., Paigen, B., Quinn, T.C., and Epstein, S.E. (2001). Atherosclerosis in apoE knockout mice infected with multiple pathogens. The Journal of infectious diseases *183*, 226–231 doi:10.1086/317938.

Burney, P., Forsey, T., Darougar, S., Sittampalam, Y., Booth, P., and Chamberlain, R. (1984). The epidemiology of chlamydial infections in childhood: a serological investigation. The International Journal of Epidemiology *13*, 491–495.

Burrows, B. (1987). The natural history of asthma. J Allergy Clin Immunol *80*, 375S–377S.

Burrows, B., Martinez, F., Halonen, M., Barbee, R.A., and Cline, M.G. (1989). Association of asthma with serum IgE levels and skin-test reactivity to allergens. New Engl J Med *320*, 271–277.

Campbell, L.A., Beryy, M.W., Heard, J., Rosenfeld, M.E., Kuo, C.-C. (2010a). *Chlamydia pneumoniae* induces expression of the scavenger receptor LOX-1 and Proatherogenic factors in C57BL/6J mice. Chlamydial Infections Proceedings of the Twelth International Symposium on Human Chlamydial Infections Schachter, J *et al.* (eds), p 161–164 San Francisco, CA.

Campbell, L.A., Jiang, S.-J., Ricks, J., Rosenfeld, M.E., Berry, M.W., Lee, A.W., Kuo, C.-C. (2010b). The role of adhesion moleculed in *C. pneumoniae* accelerated fatty streak development in hyperlipidemic C57BL/6J mice. Chlamydial Infections Proceedings of the Twelth International Symposium on Human Chlamydial Infections Schachter, J *et al.* (eds), p 479–482 San Francisco, CA.

Campbell, L.A., and Kuo, C.C. (2009). Cultivation and laboratory maintenance of *Chlamydia pneumoniae.* Current protocols in microbiology *Chapter 11*, Unit11B.11 doi:10.1002/9780471729259.mc11b01s12.

Campbell, L.A., Kuo, C.C., and Grayston, J.T. (1987). Characterization of the new Chlamydia agent, TWAR, as a unique organism by restriction endonuclease analysis and DNA–DNA hybridization. Journal of clinical microbiology *25*, 1911–1916.

Campbell, L.A., Kuo, C.C., and Grayston, J.T. (1990). Structural and antigenic analysis of *Chlamydia pneumoniae.* Infection and immunity *58*, 93–97.

Campbell, L.A., Nosaka, T., Rosenfeld, M.E., Yaraei, K., and Kuo, C.C. (2005). Tumor necrosis factor alpha plays a role in the acceleration of atherosclerosis by *Chlamydia pneumoniae* in mice. Infection and immunity 73, 3164–3165 doi:10.1128/IAI.73.5.3164-3165.2005.

Campbell, L.A., and Rosenfeld, M.E. (2015). Infection and Atherosclerosis Development. Archives of Medical Research *46*, 339–350 doi:https://doi.org/10.1016/j.arcmed.2015.05.006.

Cannon, C.P., Braunwald, E., McCabe, C.H., Grayston, J.T., Muhlestein, B., Giugliano, R.P., Cairns, R., and Skene, A.M. (2005). Antibiotic treatment of *Chlamydia pneumoniae* after acute coronary syndrome. The New England journal of medicine *352*, 1646–1654 doi:10.1056/NEJMoa043528.

Carter, J.D., Espinoza, L.R., Inman, R.D., Sneed, K.B., Ricca, L.R., Vasey, F.B., Valeriano, J., Stanich, J.A., Oszust, C., Gerard, H.C., *et al.* (2010). Combination antibiotics as a treatment for chronic Chlamydia-induced reactive arthritis: a double-blind, placebo-controlled, prospective trial. Arthritis and rheumatism *62*, 1298–1307 doi:10.1002/art.27394.

Carter, J.D., and Hudson, A.P. (2017). Recent advances and future directions in understanding and treating Chlamydia-induced reactive arthritis. Expert review of clinical immunology *13*, 197–206 doi:10.1080/1744666x.2017.1233816.

Carter, M.W., al-Mahdawi, S.A., Giles, I.G., Treharne, J.D., Ward, M.E., and Clark, I.N. (1991). Nucleotide sequence and taxonomic value of the major outer membrane protein gene of *Chlamydia pneumoniae* IOL-207. Journal of general microbiology *137*, 465–475 doi:10.1099/00221287-137-3-465.

Caselli, E., Borghi, A., Maritati, M., Gafa, R., Lanza, G., Di Luca, D., Virgili, A., and Contini, C. (2016). Relapses of primary cutaneous anaplastic large-cell lymphoma in a female immunocompetent patient with persistent chlamydophila pneumoniae and human herpesvirus 8 infection. Infect Agent Cancer *11*, 31 doi:10.1186/s13027-016-0079-4.

Centers-for-Disease-Control-and-Prevention (accessed June 23, 2018). *Chlamydia pneumoniae* Infection.

Chen, J., Zhu, M., Ma, G., Zhao, Z., and Sun, Z. (2013). *Chlamydia pneumoniae* infection and cerebrovascular disease: a systematic review and meta-analysis. BMC neurology *13*, 183 doi:10.1186/1471-2377-13-183.

Chen, S., Shimada, K., Zhang, W., Huang, G., Crother, T.R., and Arditi, M. (2010). IL-17A is proatherogenic in high-fat diet-induced and *Chlamydia pneumoniae* infection-accelerated atherosclerosis in mice. Journal of immunology (Baltimore, Md: 1950) *185*, 5619–5627 doi:10.4049/jimmunol.1001879.

Chen, S., Sorrentino, R., Shimada, K., Bulut, Y., Doherty, T.M., Crother, T.R., and Arditi, M. (2008). *Chlamydia pneumoniae*-induced foam cell formation requires MyD88-dependent and -independent signaling and is reciprocally modulated by liver X receptor activation. Journal of immunology (Baltimore, Md: 1950) *181*, 7186–7193.

Chu, D.J., Guo, S.G., Pan, C.F., Wang, J., Du, Y., Lu, X.F., and Yu, Z.Y. (2012). An experimental model for induction of lung cancer in rats by *Chlamydia pneumoniae*. Asian Pac J Cancer Prev *13*, 2819–2822.

Conklin, L., Adjemian, J., Loo, J., Mandal, S., Davis, C., Parks, S., Parsons, T., McDonough, B., Partida, J., Thurman, K., *et al.* (2013). Investigation of a *Chlamydia pneumoniae* Outbreak in a Federal Correctional Facility in Texas. Clin Infect Dis 57. https://dx.doi.org/10.1093/cid/cit357.

Consoli, D., Vidale, S., Aguglia, U., Bassi, P., Cavallini, A., Consoli, A., Galati, F., Guidetti, D., Micieli, G., Neri, G., *et al.* (2015). The influence of previous infections and anti*Chlamydia pneumoniae* seropositivity on functional outcome in ischaemic stroke patients: results from the IN2 study. Journal of neurology *262*, 1310–1316 doi:10.1007/s00415-015-7712-9.

Coon, R.G., Balansay, M.S., Faix, D.J., Hawksworth, A.W., Patterson, M.B., and Blair, P.J. (2011). Chlamydophila pneumoniae infection among Basic Underwater Demolition/SEAL (BUD/S) candidates, Coronado, California, July 2008. Mil Med *176*, 320–323.

Crother, T.R., Schroder, N.W., Karlin, J., Chen, S., Shimada, K., Slepenkin, A., Alsabeh, R., Peterson, E., and Arditi, M. (2011). *Chlamydia pneumoniae* Infection Induced Allergic Airway Sensitization Is Controlled by Regulatory T-Cells and Plasmacytoid Dendritic Cells. PloS one *6*, e20784. doi:10.1371/journal.pone.0020784.

Csángó, P.A., Haralstad, S., Pedersen, J.E., Jagars, G., and Føreland, I. (1997). Respiratory tract infection due to *Chlamydia pneumoniae* in military personnel. Scand J Infect Dis Suppl *104*, 26–29.

Cunningham, A.F., Johnston, S.L., Julious, S.A., Lampe, F.C., and Ward, M.E. (1998). Chronic *Chlamydia pneumoniae* infection and asthma exacerbations in children. Eur Resp J *11*, 345–349.

Danesh, J. (2005). Antibiotics in the prevention of heart attacks. Lancet *365*, 365–367 doi:10.1016/s0140-6736(05)17842-8.

Danesh, J., Whincup, P., Walker, M., Lennon, L., Thomson, A., Appleby, P., Wong, Y., Bernardes-Silva, M., and Ward, M. (2000). *Chlamydia pneumoniae* IgG titres and coronary heart disease: prospective study and meta-analysis. Bmj *321*, 208–213.

Darougar, S., Forsey, T., Brewerton, D.A., and Rogers, K.L. (1980). Prevalence of antichlamydial antibody in London blood donors. The British journal of venereal diseases *56*, 404–407.

de Marco, R., Marcon, A., Rossi, A., Anto, J.M., Cerveri, I., Gislason, T., Heinrich, J., Janson, C., Jarvis, D., Kuenzli, N., *et al.* (2015). Asthma, COPD and overlap syndrome: a longitudinal study in young European adults. Eur Respir J *46*, 671–679 doi:10.1183/09031936.00008615.

Deniset, J.F., and Pierce, G.N. (2010). Possibilities for therapeutic interventions in disrupting Chlamydophila pneumoniae involvement in atherosclerosis. Fundamental & clinical pharmacology *24*, 607–617 doi:10.1111/j.1472-8206.2010.00863.x.

Duewell, P., Kono, H., Rayner, K.J., Sirois, C.M., Vladimer, G., Bauernfeind, F.G., Abela, G.S., Franchi, L., Nunez, G., Schnurr, M., *et al.* (2010). NLRP3 inflammasomes are required for atherogenesis and activated by cholesterol crystals. Nature *464*, 1357–1361 doi:10.1038/nature08938.

Dwyer, R.S., Treharne, J.D., Jones, B.R., and Herring, J. (1972). Chlamydial infection. Results of micro-immunofluorescence tests for the detection of type-specific antibody in certain chlamydial infections. The British journal of venereal diseases *48*, 452–459.

Ekman, M.-R., Grayston, J.T., Visakorpi, R., Kleemola, M., Kuo, C.-c., and Saikku, P. (1993). An epidemic of infections due to *Chlamydia pneumoniae* in military conscripts. Clin Infect Dis *17*, 420–425.

Elargoubi, A., Verhoeven, P.O., Grattard, F., Stephan, J.L., Richard, O., and Pozzetto, B. (2013). Acute encephalitis associated to a respiratory infection due to Chlamydophila pneumoniae. Med Mal Infect 43, 345–349. doi: 10.1016/j.medmal.2013.06.015.

Emre, U., Roblin, P.M., Gelling, M., Dumornay, W., Rao, M., Hammerschlag, M.R., and Schachter, J. (1994). The association of *Chlamydia pneumoniae* infection and reactive airway disease in children. Arch Pediatr Adolesc Med *148*, 727–732.

Epstein, S.E., Zhu, J., Najafi, A.H., and Burnett, M.S. (2009). Insights into the role of infection in atherogenesis and in plaque rupture. Circulation *119*, 3133–3141 doi:119/24/3133 [pii] 10.1161/CIRCULATIONAHA.109.849455.

Essilfie, A.T., Horvat, J.C., Kim, R.Y., Mayall, J.R., Pinkerton, J.W., Beckett, E.L., Starkey, M.R., Simpson, J.L., Foster, P.S., Gibson, P.G., *et al.* (2015). Macrolide therapy suppresses key features of experimental steroid-sensitive and steroid-insensitive asthma. Thorax doi:10.1136/thoraxjnl-2014-206067.

Fabricant, C.G., and Fabricant, J. (1999). Atherosclerosis induced by infection with Marek's disease herpesvirus in chickens. American heart journal *138*, S465–468.

Fabricant, C.G., Fabricant, J., Minick, C.R., and Litrenta, M.M. (1983). Herpesvirus-induced atherosclerosis in chickens. Federation proceedings *42*, 2476–2479.

Falsey, A.R., and Walsh, E.E. (1993). Transmission of *Chlamydia pneumoniae*. The Journal of infectious diseases *168*, 493–496.

Fernandez-Obregon, A., and Patton, D.L. (2007). The role of *Chlamydia pneumoniae* in the aetiology of acne rosacea: response to the use of oral azithromycin. Cutis *79*, 163–167.

Filardo, S., Di Pietro, M., Farcomeni, A., Schiavoni, G., and Sessa, R. (2015). *Chlamydia pneumoniae*-Mediated Inflammation in Atherosclerosis: A Meta-Analysis. Mediators of inflammation *2015*, 378658 doi:10.1155/2015/378658.

Finn, A.V., Nakano, M., Narula, J., Kolodgie, F.D., and Virmani, R. (2010). Concept of vulnerable/unstable plaque. Arteriosclerosis, thrombosis, and vascular biology *30*, 1282–1292 doi:10.1161/atvbaha.108.179739.

Fong, I.W. (2000). Antibiotics effects in a rabbit model of *Chlamydia pneumoniae*-induced atherosclerosis. The Journal of infectious diseases *181 Suppl 3*, S514–518 doi:10.1086/315607.

Fong, I.W., Chiu, B., Viira, E., Jang, D., Fong, M.W., Peeling, R., and Mahony, J.B. (1999a). Can an antibiotic (macrolide) prevent *Chlamydia pneumoniae*-induced atherosclerosis in a rabbit model? Clinical and diagnostic laboratory immunology *6*, 891–894.

Fong, I.W., Chiu, B., Viira, E., Jang, D., and Mahony, J.B. (1999b). De Novo induction of atherosclerosis by *Chlamydia pneumoniae* in a rabbit model. Infection and immunity *67*, 6048–6055.

Fong, I.W., Chiu, B., Viira, E., Jang, D., and Mahony, J.B. (2002). Influence of clarithromycin on early atherosclerotic lesions after *Chlamydia pneumoniae* infection in a rabbit model. Antimicrobial agents and chemotherapy *46*, 2321–2326.

Foo, S.Y., and Phipps, S. (2010). Regulation of inducible BALT formation and contribution to immunity and pathology. Mucosal Immunology *3*, 537 doi:10.1038/mi.2010.52.

Gaydos, C.A., Quinn, T.C., Bobo, L.D., and Eiden, J.J. (1992). Similarity of *Chlamydia pneumoniae* strains in the variable domain IV region of the major outer membrane protein gene. Infection and immunity *60*, 5319–5323.

Geovanini, G.R., and Libby, P. (2018). Atherosclerosis and inflammation: overview and updates. Clinical science (London, England: 1979) *132*, 1243–1252 doi:10.1042/cs20180306.

Gerard, H.C., Dreses-Werringloer, U., Wildt, K.S., Deka, S., Oszust, C., Balin, B.J., Frey, W.H., 2nd, Bordayo, E.Z., Whittum-Hudson, J.A., and Hudson, A.P. (2006). Chlamydophila (Chlamydia) pneumoniae in the Alzheimer's brain. FEMS immunology and medical microbiology *48*, 355–366 doi:10.1111/j.1574-695X.2006.00154.x.

Getz, G.S., and Reardon, C.A. (2012). Animal models of atherosclerosis. Arteriosclerosis, thrombosis, and vascular biology *32*, 1104–1115 doi:10.1161/atvbaha.111.237693.

Ghosh, K., Frew, C.E., and Carrington, D. (1992). A family outbreak of *Chlamydia pneumoniae* infection. J Infection *25, Supplement 1*, 99–103.

Gibson, P.G., and Simpson, J.L. (2009). The overlap syndrome of asthma and COPD: what are its features and how important is it? BMJ *64*, 728–735.

Gibson, P.G., Yang, I.A., Upham, J.W., Reynolds, P.N., Hodge, S., James, A.L., Jenkins, C., Peters, M.J., Marks, G.B., Baraket, M., *et al.* (2017). Effect of azithromycin on asthma exacerbations and quality of life in adults with persistent uncontrolled asthma (AMAZES): a randomized, double-blind, placebo-controlled trial. Lancet *390*, 659–668 doi:10.1016/S0140-6736(17)31281-3.

Gieffers, J., Fullgraf, H., Jahn, J., Klinger, M., Dalhoff, K., Katus, H.A., Solbach, W., and Maass, M. (2001). *Chlamydia pneumoniae* infection in circulating human monocytes is refractory to antibiotic treatment. Circulation *103*, 351–356.

Gieffers, J., Rupp, J., Gebert, A., Solbach, W., and Klinger, M. (2004). First-choice antibiotics at subinhibitory concentrations induce persistence of *Chlamydia pneumoniae*. Antimicrobial agents and chemotherapy *48*, 1402–1405.

Glanville, A.R., Gencay, M., Tamm, M., Chhajed, P., Plit, M., Hopkins, P., and Aboyoun, C. (2005). *Chlamydia pneumoniae* infection after lung transplantation. Journal of Heart & Lung Transplantation *24*, 131–136.

Grayston, J.T. (2003). Antibiotic treatment of atherosclerotic cardiovascular disease. Circulation *107*, 1228–1230.

Grayston, J.T., Aldous, M., Easton, A., Wang, S.-p., Kuo, C.-c., Campbell, L.A., and Altman, J. (1993). Evidence that *Chlamydia pneumoniae* causes pneumonia and bronchitis. The Journal of infectious diseases *168*, 1231–1235.

Grayston, J.T., Belland, R.J., Byrne, G.I., Kuo, C.C., Schachter, J., Stamm, W.E., and Zhong, G. (2015). Infection with *Chlamydia pneumoniae* as a cause of coronary heart disease: the hypothesis is still untested. Pathogens and disease *73*, 1–9 doi:10.1093/femspd/ftu015.

Grayston, J.T., Campbell, L.A., Kuo, C.C., Mordhorst, C.H., Saikku, P., Thom, D.H., and Wang, S.P. (1990). A new respiratory tract pathogen: *Chlamydia pneumoniae* strain TWAR. The Journal of infectious diseases *161*, 618–625.

Grayston, J.T., Jackson, L.A., Kennedy, W.J., and Kronmal, R.A. (1999). Secondary prevention trials for coronary artery disease with antibiotic treatment for *Chlamydia pneumoniae*: design issues. American heart journal *138*, S545–549.

Grayston, J.T., Kronmal, R.A., Jackson, L.A., Parisi, A.F., Muhlestein, J.B., Cohen, J.D., Rogers, W.J., Crouse, J.R., Borrowdale, S.L., Schron, E., *et al.* (2005). Azithromycin for the secondary prevention of coronary events. The New England journal of medicine *352*, 1637–1645 doi:10.1056/NEJMoa043526.

Grayston, J.T., Kuo, C.C., Wang, S.P., and Altman, J. (1986). A new Chlamydia psittaci strain, TWAR, isolated in acute respiratory tract infections. The New England journal of medicine *315*, 161–168 doi:10.1056/nejm198607173150305.

Grayston, J.T., Kuo, C.-C., Campbell, L.A., Want, S.-P. (1989). *Chlamydia pneumoniae* sp. nov. for *Chlamydia* sp. Strain TWAR. Intern J Syst Bacteriol *39*, 88–90.

Grayston, J.T., and Wang, S.P. (1999). History of *Chlamydia pneumoniae* (TWAR). *Chlamydia pneumoniae* – The Lung and the Heart, Allegra, L, Blasi F (Eds), p 1–8 Springer, Milano, Italy.

Hahn, D., Bukstein, D., Luskin, A., and Zeitz, H. (1998). Evidence for *Chlamydia pneumoniae* infection in steroid-dependent asthma. Ann Allergy Asthma Immunol *80*, 45–49.

Hahn, D.L. (1994). Acute asthmatic bronchitis: A new twist to an old problem. J Fam Pract *39*, 431–435.

Hahn, D.L. (1995a). Infectious asthma: A reemerging clinical entity? J Fam Pract *41*, 153–157.

Hahn, D.L. (1995b). Treatment of *Chlamydia pneumoniae* infection in adult asthma: a before-after trial. J Fam Pract *41*, 345–351.

Hahn, D.L. (1999). *Chlamydia pneumoniae*, asthma and COPD: what is the evidence? Ann Allergy Asthma Immunol 83, 271–292.

Hahn, D.L. (2002). Evaluation and management of acute bronchitis. In 20 Common Problems in Respiratory Disorders, W.J. Hueston, ed. (New York: McGraw-Hill), pp. 141–153.

Hahn, D.L. (2003). *Chlamydia/Mycoplasma*: Do they cause new-onset asthma in adults? In Respiratory Infections in Allergy and Asthma, S.L. Johnston, and N.G. Papadopoulos, eds. (New York • Basel: Marcel Dekker, Inc.), pp. 645–662.

Hahn, D.L. (2004). Role of *Chlamydia pneumoniae* as an inducer of asthma. In *Chlamydia pneumoniae* Infection and Disease, H. Friedman, Y. Yamamoto, and M. Bendinelli, eds. (New York: Kluwer Academic/Plenum Publishers), pp. 239–262.

Hahn, D.L. (2013). A Cure for Asthma? What Your Doctor Is not Telling You, and Why (Durham, North Carolina: Peoples Pharmacy Press).

Hahn, D.L., Anttila, T., and Saikku, P. (1996). Association of *Chlamydia pneumoniae* IgA antibodies with recently symptomatic asthma. Epidemiol Infect *117*, 513–517.

Hahn, D.L., and Beasley, J.W. (1994). Diagnosed and possible undiagnosed asthma: a Wisconsin Research Network (WReN) study. J Fam Pract *38*, 373–379.

Hahn, D.L., Dodge, R., and Golubjatnikov, R. (1991). Association of *Chlamydia pneumoniae* (strain TWAR) infection with wheezing, asthmatic bronchitis and adult-onset asthma. JAMA *266*, 225–230.

Hahn, D.L., and Golubjatnikov, R. (1994). Asthma and chlamydial infection: a case series. J Fam Pract *38*, 589–595.

Hahn, D.L., Grasmick, M., Hetzel, S., and Yale, S. (2012a). Azithromycin for bronchial asthma in adults: an effectiveness trial. J Am Board Fam Med *25*, 442–459. doi:10.3122/jabfm.2012.04.110309.

Hahn, D.L., and McDonald, R. (1998). Can acute *Chlamydia pneumoniae* infection initiate chronic asthma? Ann Allergy Asthma Immunol *81*, 339–344.

Hahn, D.L., and Peeling, R.W. (2008). Airflow limitation, asthma, and *Chlamydia pneumoniae*-specific heat shock protein 60. Ann Allergy Asthma Immunol *101*, 614–618.

Hahn, D.L., Peeling, R.W., Dillon, E., McDonald, R., and Saikku, P. (2000). Serological markers for *Chlamydia pneumoniae* in asthma. Ann Allergy Asthma Immunol *84*, 227–233.

Hahn, D.L., Plane, M.B., Mahdi, O.S., and Byrne, G.I. (2006). Secondary outcomes of a pilot randomized trial of azithromycin treatment for asthma. PLoS Clin Trials *1*, e11 DOI: 10.1371/journal.pctr0010011.

Hahn, D.L., Schure, A., Patel, K., Childs, T., Drizik, E., and Webley, W. (2012b). *Chlamydia pneumoniae*-specific IgE is prevalent in asthma and is associated with disease severity. PloS one *7*, e35945. doi35910.31371/journal.pone.0035945.

Hammerschlag, M.R., Chirgwin, K., Roblin, P.M., Gelling, M., Dumornay, W., Mandel, L., and Schachter, J. (1992). Persistent infection with *Chlamydia pneumoniae* following acute respiratory illness. Clinical Infectious Diseases *14*, 178–182.

Hansbro, P.M., Starkey, M.R., Mattes, J., and Horvat, J.C. (2014). Pulmonary Immunity during Respiratory Infections in Early Life and the Development of Severe Asthma. Ann Am Thorac Soc *11 Suppl 5*, S297–302 doi:10.1513/AnnalsATS.201402-086AW.

Haranaga, S., Yamaguchi, H., Leparc, G.F., Friedman, H., and Yamamoto, Y. (2001). Detection of *Chlamydia pneumoniae* antigen in PBMNCs of healthy blood donors. Transfusion *41*, 1114–1119.

Herrera, V.L., Shen, L., Lopez, L.V., Didishvili, T., Zhang, Y.X., and Ruiz-Opazo, N. (2003). *Chlamydia pneumoniae* accelerates coronary artery disease progression in transgenic hyperlipidemia-genetic hypertension rat model. Molecular medicine (Cambridge, Mass) *9*, 135–142.

Hogg, J.C., Chu, F., Utokaparch, S., Woods, R., Elliott, W.M., Buzatu, L., Cherniack, R.M., Rogers, R.M., Sciurba, F.C., Coxson, H.O., *et al.* (2004). The nature of small-airway obstruction in chronic obstructive pulmonary disease. The New England journal of medicine *350*, 2645–2653 doi:10.1056/NEJMoa032158.

Holland, S.M., Taylor, H.R., Gaydos, C.A., Kappus, E.W., and Quinn, T.C. (1990). Experimental infection with *Chlamydia pneumoniae* in nonhuman primates. Infection and immunity *58*, 593–597.

Horvat, J.C., Beagley, K.W., Wade, M.A., Preston, J.A., Hansbro, N.C., Hickey, D.K., Kaiko, G.E., Gibson, P.G., Foster, P.S., and Hansbro, P.M. (2007). Neonatal chlamydial infection induces mixed T-cell responses that drive allergic airway disease. Am J Respir Crit Care Med *176*, 556–564.

Horvat, J.C., Starkey, M.R., Kim, R.Y., Beagley, K.W., Preston, J.A., Gibson, P.G., Foster, P.S., and Hansbro, P.H. (2010a). Chlamydial respiratory infection during allergen sensitization drives neutrophilic allergic airways disease. Journal of immunology (Baltimore, Md: 1950) doi:10.4049/jimmunol.0902287.

Horvat, J.C., Starkey, M.R., Kim, R.Y., Phipps, S., Gibson, P.G., Beagley, K.W., Foster, P.S., and Hansbro, P.H. (2010b). Early-life chlamydial lung infection enhances allergic airways disease through age-dependent differences in immunopathology. J Allergy Clin Immunol *125*, 617–625 doi: 10.1016/j.jaci.2009.10.018.

Hotzel, H., Grossmann, E., Mutschmann, F., and Sachse, K. (2001). Genetic characterization of a Chlamydophila pneumoniae isolate from an African frog and comparison to currently accepted biovars. Systematic and applied microbiology *24*, 63–66 doi:10.1078/0723-2020-00016.

Hu, H., Pierce, G.N., and Zhong, G. (1999). The atherogenic effects of chlamydia are dependent on serum cholesterol and specific to *Chlamydia pneumoniae*. The Journal of clinical investigation *103*, 747–753 doi:10.1172/jci4582.

Hu, X., De Silva, T.M., Chen, J., and Faraci, F.M. (2017). Cerebral Vascular Disease and Neurovascular Injury in Ischemic Stroke. Circulation research *120*, 449–471 doi:10.1161/circresaha.116.308427.

Jackson, L.A., Campbell, L.A., Kuo, C.C., Rodriguez, D.I., Lee, A., and Grayston, J.T. (1997a). Isolation of *Chlamydia pneumoniae* from a carotid endarterectomy specimen. The Journal of infectious diseases *176*, 292–295.

Jackson, L.A., Campbell, L.A., Schmidt, R.A., Kuo, C.C., Cappuccio, A.L., Lee, M.J., and Grayston, J.T. (1997b). Specificity of detection of *Chlamydia pneumoniae* in cardiovascular atheroma: evaluation of the innocent bystander hypothesis. The American journal of pathology *150*, 1785–1790.

Jaff, M.R., Dale, R.A., Creager, M.A., Lipicky, R.J., Constant, J., Campbell, L.A., and Hiatt, W.R. (2009). Anti-chlamydial antibiotic therapy for symptom improvement in peripheral artery disease: prospective evaluation of rifalazil effect on vascular symptoms of intermittent claudication and other endpoints in *Chlamydia pneumoniae* seropositive patients (PROVIDENCE-1). Circulation *119*, 452–458 doi:10.1161/circulationaha.108.815308.

Jang, I.K., Lassila, R., and Fuster, V. (1993). Atherogenesis and inflammation. European heart journal *14 Suppl K*, 2–6.

Jespersen, C.M., Als-Nielsen, B., Damgaard, M., Hansen, J.F., Hansen, S., Helo, O.H., Hildebrandt, P., Hilden, J., Jensen, G.B., Kastrup, J., *et al.* (2006). Randomised placebo controlled multicentre trial to assess short term clarithromycin for patients with stable coronary heart disease: CLARICOR trial. BMJ *332*, 22–27 doi:10.1136/bmj.38666.653600.55.

Joensen, J.B., Juul, S., Henneberg, E., Thomsen, G., Ostergaard, L., and Lindholt, J.S. (2008). Can long-term antibiotic treatment prevent progression of peripheral arterial occlusive disease? A large, randomized, double-blinded, placebo-controlled trial. Atherosclerosis *196*, 937–942 doi:10.1016/j.atherosclerosis.2007.02.025.

Johnston, S.L., and Martin, R.J. (2005). *Chlamydophila pneumoniae* and *Mycoplasma pneumoniae*. A role in asthma pathogenesis? Am J Respir Crit Care Med *172*, 1078–1089.

Jupelli, M., Murthy, A.K., Chaganty, B.K., Guentzel, M.N., Selby, D.M., Vasquez, M.M., Mustafa, S.B., Henson, B.M., Seidner, S.R., Zhong, G., *et al.* (2011). Neonatal chlamydial pneumonia induces altered respiratory structure and function lasting into adult life. Lab Invest *91*, 1530–1539 doi:10.1038/labinvest.2011.103.

Jupelli, M., Shimada, K., Chiba, N., Slepenkin, A., Alsabeh, R., Jones, H.D., Peterson, E., Chen, S., Arditi, M., and Crother, T.R. (2013). *Chlamydia pneumoniae* infection in mice induces chronic lung inflammation, iBALT formation, and fibrosis. PloS one *8*, e77447 doi:10.1371/journal.pone.0077447.

Kabeya, H., Sato, S., and Maruyama, S. (2015). Prevalence and characterization of Chlamydia DNA in zoo animals in Japan. Microbiology and immunology *59*, 507–515 doi:10.1111/1348-0421.12287.

Kaiko, G.E., Phipps, S., Hickey, D.K., Lam, C.E., Hansbro, P.M., Foster, P.S., and Beagley, K.W. (2008). Chlamydia muridarum infection subverts dendritic cell function to promote Th2 immunity and airways hyperreactivity. Journal of immunology (Baltimore, Md: 1950) *180*, 2225–2232.

Karimi, G., Samiei, S., Hatami, H., Gharehbaghian, A., Vafaiyan, V., and Namini, M.T. (2010). Detection of *Chlamydia pneumoniae* in peripheral blood mononuclear cells of healthy blood donors in Tehran Regional Educational Blood Transfusion Centre. Transfus Med *20*, 237–243 doi:10.1111/j.1365-3148.2010.01003.x.

Karvonen, M., Tuomilehto, J., Naukkarinen, A., and Saikku, P. (1992). The prevalence and regional distribution of antibodies against Chlamydia pneumonia (strain TWAR) in Finland in 1958. The International Journal of Epidemiology *21*, 391–397.

Kaukoranta-Tolvanen, S.-S.E., Laurila, A.L., Saikku, P., Leinonen, M., Liesirova, L., and Laitinen, K. (1993). Experimental infection of *Chlamydia pneumoniae* in mice. Microbial Pathogenesis *15*, 293–302 doi:https://doi.org/10.1006/mpat.1993.1079.

Kaul, R., Uphoff, J., Wiedeman, J., Yadlapalli, S., and Wenman, W.M. (2000). Detection of *Chlamydia pneumoniae* DNA in CD3+ lymphocytes from healthy blood donors and patients with coronary artery disease. Circulation *102*, 2341–2346.

Kew, K.M., Undela, K., Kotortsi, I., and Ferrara, G. (2015). Macrolides for chronic asthma. Cochrane Database Syst Rev, CD002997 doi:10.1002/14651858.CD002997.pub4.

Khalid, M., Al Saghir, A., Saleemi, S., Al Dammas, S., Zeitouni, M., Al Mobeireek, A., Chaudhry, N., and Sahovic, E. (2005). Azithromycin in bronchiolitis obliterans complicating bone marrow transplantation: a preliminary study. Eur Respir J *25*, 490–493.

King, L.E., Jr., Stratton, C.W., and Mitchell, W.M. (2001). *Chlamydia pneumoniae* and chronic skin wounds: a focused review. J Investig Dermatol Symp Proc *6*, 233–237 doi:10.1046/j.0022-202x.2001.00050.x.

Kleemola, M., Saikku, P., Visakorpi, R., Wang, S.P., and Grayston, J.T. (1988). Epidemics of pneumonia caused by TWAR, a new Chlamydia organism, in military trainees in Finland. The Journal of infectious diseases *157*, 230–236.

Kuo, C.-c. (1988). Chlamydia as pathogens. Host Response. In Microbiology of Chlamydia, A.L. Barron, ed. (Boca Raton, Florida: CRC Press, Inc.), pp. 193–208.

Kuo, C.C., and Campbell, L.A. (2003). Chlamydial infections of the cardiovascular system. Frontiers in bioscience: a journal and virtual library *8*, e36–43.

Kuo, C.C., and Grayston, J.T. (1988). Factors affecting viability and growth in HeLa 229 cells of Chlamydia sp. strain TWAR. Journal of clinical microbiology *26*, 812–815.

Kuo, C.C., Jackson, L.A., Campbell, L.A., and Grayston, J.T. (1995). *Chlamydia pneumoniae* (TWAR). Clinical microbiology reviews *8*, 451–461.

Kuo, C.C., Shor, A., Campbell, L.A., Fukushi, H., Patton, D.L., and Grayston, J.T. (1993). Demonstration of *Chlamydia pneumoniae* in atherosclerotic lesions of coronary arteries. The Journal of infectious diseases *167*, 841–849.

Kutlin, A., Roblin, P.M., Kumar, S., Kohlhoff, S., Bodetti, T., Timms, P., and Hammerschlag, M.R. (2007). Molecular characterization of Chlamydophila pneumoniae isolates from Western barred bandicoots. Journal of medical microbiology *56*, 407–417 doi:10.1099/jmm.0.46850-0.

Kyaw, T., Winship, A., Tay, C., Kanellakis, P., Hosseini, H., Cao, A., Li, P., Tipping, P., Bobik, A., and Toh, B.H. (2013). Cytotoxic and proinflammatory CD8+ T lymphocytes promote development of vulnerable atherosclerotic plaques in apoE-deficient mice. Circulation *127*, 1028–1039 doi:10.1161/circulationaha.112.001347.

Laitinen, K., Laurila, A., Pyhala, L., Leinonen, M., and Saikku, P. (1997). *Chlamydia pneumoniae* infection induces inflammatory changes in the aortas of rabbits. Infection and immunity *65*, 4832–4835.

Laitinen, K., Laurila, A.L., Leinonen, M., and Saikku, P. (1996). Reactivation of *Chlamydia pneumoniae* infection in mice by cortisone treatment. Infection and immunity *64*, 1488–1490.

Laurila, A.L., Antilla, T., Läärä, E., Bloigu, A., Virtamo, J., Albanes, D., Leinonen, M., and Saikku, P. (1997). Serological evidence of an association between *Chlamydia pneumoniae* infection and lung cancer. Int J Cancer (Pred Oncol) *74*, 31–34.

Lee, Y.T., Lin, H.Y., Chan, Y.W.F., Li, K.H.C., To, O.T.L., Yan, B.P., Liu, T., Li, G., Wong, W.T., Keung, W., *et al.* (2017). Mouse models of atherosclerosis: a historical perspective and recent advances. Lipids in Health and Disease *16*, 12 doi:10.1186/s12944-016-0402-5.

Li, W., Gudipaty, P., Li, C., Henderson, K.K., Ramsey, K.H., and Murthy, A.K. (2018). Intranasal immunization with recombinant chlamydial protease-like activity factor attenuates atherosclerotic pathology following *Chlamydia pneumoniae* infection in mice. Immunology and cell biology doi:10.1111/imcb.12192.

Libby, P. (2012). Inflammation in atherosclerosis. Arteriosclerosis, thrombosis, and vascular biology *32*, 2045–2051 doi:10.1161/atvbaha.108.179705.

Liu, L., Hu, H., Ji, H., Murdin, A.D., Pierce, G.N., and Zhong, G. (2000). *Chlamydia pneumoniae* infection significantly exacerbates aortic atherosclerosis in an LDLR-/- mouse model within six months. Molecular and cellular biochemistry *215*, 123–128.

Lorenz, J. (1999). *Chlamydia pneumoniae*: an important pathogen in chronic bronchitis. In *Chlamydia pneumoniae*: the lung and the heart, L. Allegra, and F. Blasi, eds. (Milano: Springer-Verlag, Italia), pp. 124–133.

Loudon, K., Treweek, S., Sullivan, F., Donnan, P., Thorpe, K.E., and Zwarenstein, M. (2015). The PRECIS-2 tool: designing trials that are fit for purpose. BMJ *350*, h2147 doi:10.1136/bmj.h2147.

Lundbäck, B., Lindberg, A., Lindström, M., Rönmark, E., Jonsson, A.C., Jönsson, E., Larsson, L.-G., Andersson, S., Sandström, T., and Larsson, K. (2003). Not 15 but 50% of smokers develop COPD? – Report from the Obstructive Lung Disease in Northern Sweden Studies. Respir Med *97*, 115–122.

Maass, M., Bartels, C., Engel, P.M., Mamat, U., and Sievers, H.H. (1998). Endovascular presence of viable *Chlamydia pneumoniae* is a common phenomenon in coronary artery disease. Journal of the American College of Cardiology *31*, 827–832.

Mahmoud, E.A., Elshibly, S., and Mårdh, P.-A. (1994). Seroepidemiologic study of *Chlamydia pneumoniae* and other chlamydial species in a hyperendemic area for trachoma in the Sudan. Am J Trop Med *51*, 489–494.

Malinverni, R., Kuo, C.C., Campbell, L.A., and Grayston, J.T. (1995a). Reactivation of *Chlamydia pneumoniae* lung infection in mice by cortisone. The Journal of infectious diseases *172*, 593–594.

Malinverni, R., Kuo, C.C., Campbell, L.A., Lee, A., and Grayston, J.T. (1995b). Effects of two antibiotic regimens on course and persistence of experimental *Chlamydia pneumoniae* TWAR pneumonitis. Antimicrobial agents and chemotherapy *39*, 45–49.

McConnell, C.T., Plouffe, J.F., File, T.M., Mueller, C.F., Wong, K.-H., Skelton, S.S., Marston, B.J., and Breiman, R.F. (1994). Radiographic appearance of *Chlamydia pneumoniae* (TWAR strain) respiratory infections. Radiology *192*, 819–824.

Mehta, J.L., Chen, J., Hermonat, P.L., Romeo, F., and Novelli, G. (2006). Lectin-like, oxidized low-density lipoprotein receptor-1 (LOX-1): a critical player in the development of atherosclerosis and related disorders. Cardiovascular research *69*, 36–45 doi:10.1016/j.cardiores.2005.09.006.

Michel, D., Antoine, J.C., Pozzetto, B., Gaudin, O.G., and Lucht, F. (1992). Lumbosacral meningoradiculitis associated with *Chlamydia pneumoniae* infection. J Neurol Neurosurg Psychiat *55*, 511.

Michelsen, K.S., Wong, M.H., Shah, P.K., Zhang, W., Yano, J., Doherty, T.M., Akira, S., Rajavashisth, T.B., and Arditi, M. (2004). Lack of Toll-like receptor 4 or myeloid differentiation factor 88 reduces atherosclerosis and alters plaque phenotype in mice deficient in apolipoprotein E. Proceedings of the National Academy of Sciences of the United States of America *101*, 10679–10684 doi:10.1073/pnas.0403249101.

Miklossy, J., and McGeer, P.L. (2016). Common mechanisms involved in Alzheimer's disease and type 2 diabetes: a key role of chronic bacterial infection and inflammation. Aging (Albany NY) *8*, 575–588 doi:10.18632/aging.100921.

Mitchell, C.M., Hutton, S., Myers, G.S.A., Brunham, R., and Timms, P. (2010). *Chlamydia pneumoniae* Is Genetically Diverse in Animals and Appears to Have Crossed the Host Barrier to Humans on (At Least) Two Occasions. PLOS Pathogens *6*, e1000903 doi:10.1371/journal.ppat.1000903.

Moazed, T.C., Campbell, L.A., Rosenfeld, M.E., Grayston, J.T., and Kuo, C.C. (1999). *Chlamydia pneumoniae* infection accelerates the progression of atherosclerosis in apolipoprotein E-deficient mice. The Journal of infectious diseases *180*, 238–241 doi:10.1086/314855.

Moazed, T.C., Kuo, C., Grayston, J.T., and Campbell, L.A. (1997). Murine models of *Chlamydia pneumoniae* infection and atherosclerosis. The Journal of infectious diseases *175*, 883–890.

Moazed, T.C., Kuo, C., Patton, D.L., Grayston, J.T., and Campbell, L.A. (1996). Experimental rabbit models of *Chlamydia pneumoniae* infection. The American journal of pathology *148*, 667–676.

Mordhorst, C.H., Wang, S.-P., Myhra, W., and Grayston, J.T. (1990). *Chlamydia pneumoniae*, strain TWAR, infections in Denmark 1975–1987. In Chlamydial Infections Proceedings of the Seventh International Symposium on Human Chlamydial Infections, W.R. Bowie, H.D. Caldwell, R.P. Jones, P.-A. Mardh, G.L. Ridgeway, J. Schacter, W.E. Stamm, and M.E. Ward, eds. (Cambridge: Cambridge University Press), pp. 418–421.

Mordhorst, C.H., Wang, S.P., and Grayston, J.T. (1986). Epidemic 'ornithosis' and TWAR infection, Denmark, 1976–85. In Chlamydial Infections: Proceedings of the Sixth International Symposium on Human Chlamydial Infections, D. Oriel, and G. Ridgeway, eds. (Cambridge: Cambridge University Press), pp. 325–328.

Mordhorst, C.H., Wang, S.P., and Grayston, J.T. (1992). Outbreak of *Chlamydia pneumoniae* infection in four farm families. Eur J Clin Microbiol Inf Dis *11*, 617–620.

Muhlestein, J.B., Anderson, J.L., Hammond, E.H., Zhao, L., Trehan, S., Schwobe, E.P., and Carlquist, J.F. (1998). Infection with *Chlamydia pneumoniae* accelerates the development of atherosclerosis and treatment with azithromycin prevents it in a rabbit model. Circulation *97*, 633–636.

Mullick, A.E., Tobias, P.S., and Curtiss, L.K. (2005). Modulation of atherosclerosis in mice by Toll-like receptor 2. The Journal of clinical investigation *115*, 3149–3156 doi:10.1172/jci25482.

Murdin, A.D., Dunn, P., Sodoyer, R., Wang, J., Caterini, J., Brunham, R.C., Aujame, L., and Oomen, R. (2000a). Use of a mouse lung challenge model to identify antigens protective against *Chlamydia pneumoniae* lung infection. The Journal of infectious diseases *181 Suppl 3*, S544–551 doi:10.1086/315605.

Murdin, A.D., Gellin, B., Brunham, R.C., Campbell, L.A., Christiansen, G., Deal, C.D., Jenson, H.B., Metcalf, B., Sankaran, B., Stephens, R.S., *et al.* (2000b). Collaborative multidisciplinary workshop report: progress toward a *Chlamydia pneumoniae* vaccine. The Journal of infectious diseases *181 Suppl 3*, S552–557 doi:10.1086/315601.

Murthy, A.K., Li, W., Chaganty, B.K., Kamalakaran, S., Guentzel, M.N., Seshu, J., Forsthuber, T.G., Zhong, G., and Arulanandam, B.P. (2011). Tumor necrosis factor alpha production from CD8+ T cells mediates oviduct pathological sequelae following primary genital Chlamydia muridarum infection. Infection and immunity *79*, 2928–2935 doi:10.1128/iai.05022-11.

Myers, G.S., Mathews, S.A., Eppinger, M., Mitchell, C., O'Brien, K.K., White, O.R., Benahmed, F., Brunham, R.C., Read, T.D., Ravel, J., *et al.* (2009). Evidence that human *Chlamydia pneumoniae* was zoonotically acquired. Journal of bacteriology *191*, 7225–7233 doi:10.1128/jb.00746-09.

Naiki, Y., Michelsen, K.S., Schroder, N.W., Alsabeh, R., Slepenkin, A., Zhang, W., Chen, S., Wei, B., Bulut, Y., Wong, M.H., *et al.* (2005). MyD88 is pivotal for the early inflammatory response and subsequent bacterial clearance and survival in a mouse model of *Chlamydia pneumoniae* pneumonia. The Journal of biological chemistry *280*, 29242–29249 doi:10.1074/jbc.M503225200.

National-Asthma-Education-Program (1991). Guidelines for the diagnosis and management of asthma. J Allergy Clin Immunol *88 (suppl)*, 425–534.

O'Connor, C.M., Dunne, M.W., Pfeffer, M.A., Muhlestein, J.B., Yao, L., Gupta, S., Benner, R.J., Fisher, M.R., and Cook, T.D. (2003). Azithromycin for the secondary prevention of coronary heart disease events: the WIZARD study: a randomized controlled trial. Jama *290*, 1459–1466 doi:10.1001/jama.290.11.1459.

O'Connor, S., Taylor, C., Campbell, L.A., Epstein, S., and Libby, P. (2001). Potential infectious etiologies of atherosclerosis: a multifactorial perspective. Emerging infectious diseases *7*, 780–788 doi:10.3201/eid0705.010503.

Pant, S., Deshmukh, A., GuruMurthy, G.S., Pothineni, N.V., Watts, T.E., Romeo, F., and Mehta, J.L. (2013). Inflammation and Atherosclerosis—Revisited. Journal of Cardiovascular Pharmacology and Therapeutics *19*, 170–178 doi:10.1177/1074248413504994.

Patel, K.K., Anderson, E.A., Salva, P.S., and Webley, W.C. (2012). The prevalence and identity of Chlamydia-specific IgE in children with asthma and other chronic respiratory symptoms. Respir Res *13*, 32 doi:1465-9921-13-32 [pii]
10.1186/1465–9921–13–32.

Patel, K.K., and Webley, W.C. (2013). Evidence of infectious asthma phenotype: Chlamydia-induced allergy and pathogen-specific IgE in a neonatal mouse model. PloS one *8*, e83453 doi:10.1371/journal.pone.0083453.

Penttila, J.M., Anttila, M., Puolakkainen, M., Laurila, A., Varkila, K., Sarvas, M., Makela, P.H., and Rautonen, N. (1998). Local immune responses to *Chlamydia pneumoniae* in the lungs of BALB/c mice during primary infection and reinfection. Infection and immunity *66*, 5113–5118.

Perez Melgosa, M., Kuo, C.C., and Campbell, L.A. (1991). Sequence analysis of the major outer membrane protein gene of *Chlamydia pneumoniae*. Infection and immunity *59*, 2195–2199.

Pether, J.V.S., Wang, S.-P., and Grayston, J.T. (1989). *Chlamydia pneumoniae*, strain TWAR, as the cause of an outbreak in a boys' school previously called psittacosis. Epidemiology and Infection *103*, 395–400.

Puolakkainen, M., Ukkonen, P., and Saikku, P. (1989). The seroepidemiology of Chlamydiae in Finland over the period 1971 to 1987. Epidemiol Infect *102*, 287–295.

Rafferty, B., Dolgilevich, S., Kalachikov, S., Morozova, I., Ju, J., Whittier, S., Nowygrod, R., and Kozarov, E. (2011). Cultivation of Enterobacter hormaechei from human atherosclerotic tissue. Journal of atherosclerosis and thrombosis *18*, 72–81.

Ramirez, J.A. (1996). Isolation of *Chlamydia pneumoniae* from the coronary artery of a patient with coronary atherosclerosis. The *Chlamydia pneumoniae*/Atherosclerosis Study Group. Annals of internal medicine *125*, 979–982.

Rangel-Moreno, J., Hartson, L., Navarro, C., Gaxiola, M., Selman, M., and Randall, T.D. (2006). Inducible bronchus-associated lymphoid tissue (iBALT) in patients with pulmonary complications of rheumatoid arthritis. The Journal of clinical investigation *116*, 3183–3194 doi:10.1172/jci28756.

Rattei, T., Ott, S., Gutacker, M., Rupp, J., Maass, M., Schreiber, S., Solbach, W., Wirth, T., and Gieffers, J. (2007). Genetic diversity of the obligate intracellular bacterium Chlamydophila pneumoniae by genome-wide analysis of single nucleotide polymorphisms: evidence for highly clonal population structure. BMC genomics *8*, 355 doi:10.1186/1471-2164-8-355.

Reiter, J., Demirel, N., Mendy, A., Gasana, J., Vieira, E.R., Colin, A.A., Quizon, A., and Forno, E. (2013). Macrolides for the long-term management of asthma – a meta-analysis of randomized clinical trials. Allergy *68*, 1040–1049 doi:10.1111/all.12199.

Ridker, P.M., Kundsin, R.B., Stampfer, M.J., Poulin, S., and Hennekens, C.H. (1999). Prospective study of *Chlamydia pneumoniae* IgG seropositivity and risks of future myocardial infarction. Circulation *99*, 1161–1164.

Rodriguez-Gonzalez Moro, J.M., Izquierdo, J.L., Anton, E., de Lucas, P., Martin, A., and Group, M.S. (2009).

Health-related quality of life in outpatient women with COPD in daily practice: the MUVICE Spanish study. Respir Med *103*, 1303–1312 doi:10.1016/j.rmed.2009.04.002.

Rodriguez, N., Wantia, N., Fend, F., Durr, S., Wagner, H., and Miethke, T. (2006). Differential involvement of TLR2 and TLR4 in host survival during pulmonary infection with *Chlamydia pneumoniae*. European journal of immunology *36*, 1145–1155 doi:10.1002/eji.200535152.

Rosenfeld, M.E., and Campbell, L.A. (2011). Pathogens and atherosclerosis: update on the potential contribution of multiple infectious organisms to the pathogenesis of atherosclerosis. Thrombosis and haemostasis *106*, 858–867 doi:10.1160/th11-06-0392.

Ross, R. (1999). Atherosclerosis--an inflammatory disease. The New England journal of medicine *340*, 115–126 doi:10.1056/nejm199901143400207.

Rossler, M.J., Rappl, G., Muche, M., Hasselmann, D.O., Sterry, W., Tilgen, W., and Reinhold, U. (2003). No evidence of skin infection with *Chlamydia pneumoniae* in patients with cutaneous T cell lymphoma. Clin Microbiol Infect *9*, 721–723.

Rothstein, N.M., Quinn, T.C., Madico, G., Gaydos, C.A., and Lowenstein, C.J. (2001). Effect of azithromycin on murine arteriosclerosis exacerbated by *Chlamydia pneumoniae*. The Journal of infectious diseases *183*, 232–238 doi:10.1086/317941.

Rottenberg, M.E., Gigliotti Rothfuchs, A.C., Gigliotti, D., Svanholm, C., Bandholtz, L., and Wigzell, H. (1999). Role of innate and adaptive immunity in the outcome of primary infection with *Chlamydia pneumoniae*, as analyzed in genetically modified mice. Journal of immunology (Baltimore, Md: 1950) *162*, 2829–2836.

Roubaud Baudron, C., Varon, C., Megraud, F., and Salles, N. (2015). Alzheimer's disease: the infectious hypothesis. Geriatr Psychol Neuropsychiatr Vieil *13*, 418–424 doi:10.1684/pnv.2015.0574.

Roulis, E., Bachmann, N., Polkinghorne, A., Hammerschlag, M., Kohlhoff, S., and Timms, P. (2014). Draft Genome and Plasmid Sequences of *Chlamydia pneumoniae* Strain B21 from an Australian Endangered Marsupial, the Western Barred Bandicoot. Genome announcements 2 doi:10.1128/genomeA.01223-13.

Roulis, E., Bachmann, N.L., Myers, G.S., Huston, W., Summersgill, J., Hudson, A., Dreses-Werringloer, U., Polkinghorne, A., and Timms, P. (2015). Comparative genomic analysis of human *Chlamydia pneumoniae* isolates from respiratory, brain and cardiac tissues. Genomics *106*, 373–383 doi:10.1016/j.ygeno.2015.09.008.

Rüegg, S.R., Regenscheit, N., Origgi, F.C., Kaiser, C., and Borel, N. (2015). Detection of *Chlamydia pneumoniae* in a collection of captive snakes and response to treatment with marbofloxacin. The Veterinary Journal *205*, 424–426 doi:https://doi.org/10.1016/j.tvjl.2015.05.007.

Sabet, S.F., Simmons, J., and Caldwell, H.D. (1984). Enhancement of *Chlamydia trachomatis* infectious progeny by cultivation of HeLa 229 cells treated with DEAE-dextran and cycloheximide. Journal of clinical microbiology *20*, 217–222.

Saikku, P., Leinonen, M., Mattila, K., Ekman, M.R., Nieminen, M.S., Makela, P.H., Huttunen, J.K., and Valtonen, V. (1988). Serological evidence of an association of a novel Chlamydia, TWAR, with chronic coronary heart disease and acute myocardial infarction. Lancet *2*, 983–986.

Saikku, P., Wang, S.P., Kleemola, M., Brander, E., Rusanen, E., and Grayston, J.T. (1985). An epidemic of mild pneumonia due to an unusual strain of Chlamydia psittaci. The Journal of infectious diseases *151*, 832–839.

Sawamura, T., Kume, N., Aoyama, T., Moriwaki, H., Hoshikawa, H., Aiba, Y., Tanaka, T., Miwa, S., Katsura, Y., Kita, T., *et al.* (1997). An endothelial receptor for oxidized low-density lipoprotein. Nature *386*, 73–77 doi:10.1038/386073a0.

Schröder, N.W.J., Crother, T.M., Naiki, Y., Chen, S., Wong, M.H., Yilmaz, A., Slepenkin, A., Schulte, D., Alsabeh, R., Doherty, T.M., *et al.* (2008). Innate immune responses during respiratory tract infection with a bacterial pathogen induce allergic airway sensitization. J Allergy Clin Immunol *122*, 595–602.

Shimada, K., Chen, S., Dempsey, P.W., Sorrentino, R., Alsabeh, R., Slepenkin, A.V., Peterson, E., Doherty, T.M., Underhill, D., Crother, T.R., *et al.* (2009). The NOD/RIP2 pathway is essential for host defenses against Chlamydophila pneumoniae lung infection. PLoS Pathog *5*, e1000379 doi:10.1371/journal.ppat.1000379.

Shimada, K., Crother, T.R., and Arditi, M. (2012). Innate immune responses to *Chlamydia pneumoniae* infection: role of TLRs, NLRs, and the inflammasome. Microbes and infection *14*, 1301–1307 doi:10.1016/j.micinf.2012.08.004.

Shirai, M., Hirakawa, H., Kimoto, M., Tabuchi, M., Kishi, F., Ouchi, K., Shiba, T., Ishii, K., Hattori, M., Kuhara, S., *et al.* (2000). Comparison of whole genome sequences of *Chlamydia pneumoniae* J138 from Japan and CWL029 from USA. Nucleic acids research *28*, 2311–2314.

Shirtcliffe, P., Marsh, S., Travers, J., Weatherall, M., and Beasley, R. (2012). Childhood asthma and GOLD-defined chronic obstructive pulmonary disease. Intern Med J *42*, 83–88 doi:10.1111/j.1445-5994.2010.02238.x.

Shor, A., Kuo, C.C., and Patton, D.L. (1992). Detection of *Chlamydia pneumoniae* in coronary arterial fatty streaks and atheromatous plaques. South African medical journal = Suid-Afrikaanse tydskrif vir geneeskunde *82*, 158–161.

Silva, G.E., Sherrill, D.L., Guerra, S., and Barbee, R.A. (2004). Asthma as a risk factor for COPD in a longitudinal study. Chest *126*, 59–65.

Sluiter, H.J., Koëter, G.H., de Monchy, J.G.R., Postma, D.S., de Vries, K., and Orie, N.G.M. (1991). The Dutch hypothesis (chronic non-specific lung disease) revisited. Eur Respir J *4*, 479–489.

Sommer, K., Njau, F., Wittkop, U., Thalmann, J., Bartling, G., Wagner, A., and Klos, A. (2009). Identification of high- and low-virulent strains of *Chlamydia pneumoniae* by their characterization in a mouse pneumonia model. FEMS immunology and medical microbiology *55*, 206–214 doi:10.1111/j.1574-695X.2008.00503.x.

Starkey, M.R., Nguyen, D.H., Kim, R.Y., Nair, P.M., Brown, A.C., Essifie, A.T., Horvat, J.C., and Hansbro,

P.M. (2013). Programming of the Lung in Early Life by Bacterial Infections Predisposes to Chronic Respiratory Disease. Clin Obstet Gynecol doi:10.1097/GRF.0b013e3182993a0c.

Stokholm, J., Chawes, B.L., Vissing, N.H., Bjarnadottir, E., Pedersen, T.M., Vinding, R.K., Schoos, A.M., Wolsk, H.M., Thorsteinsdottir, S., Hallas, H.W., *et al.* (2015). Azithromycin for episodes with asthma-like symptoms in young children aged 1–3 years: a randomised, double-blind, placebo-controlled trial. Lancet Respir Med doi:10.1016/S2213-2600(15)00500-7.

Stolk-Engelaar, M.V.M., and Peeters, M.F. (1992). Seroprevalence of *Chlamydia pneumoniae* in the normal population. Eur J Clin Microbiol Inf Dis *11*, 478.

Storey, C., Lusher, M., Yates, P., and Richmond, S. (1993). Evidence for *Chlamydia pneumoniae* of non-human origin. Journal of general microbiology *139*, 2621–2626 doi:10.1099/00221287-139-11-2621.

Su, X., and Chen, H.L. (2014). *Chlamydia pneumoniae* infection and cerebral infarction risk: a meta-analysis. International journal of stroke: official journal of the International Stroke Society *9*, 356–364 doi:10.1111/ijs.12248.

Tai, A., Tran, H., Roberts, M., Clarke, N., Wilson, J., and Robertson, C.F. (2014). The association between childhood asthma and adult chronic obstructive pulmonary disease. Thorax doi:10.1136/thoraxjnl-2013-204815.

Taylor-Robinson, D., and Boman, J. (2005). The failure of antibiotics to prevent heart attacks. BMJ *331*, 361–362 doi:10.1136/bmj.331.7513.361.

Thom, D.H., Grayston, J.T., Campbell, L.A., Kuo, C.-c., Diwan, V.K., and Wang, S.-P. (1994). Respiratory infection with *Chlamydia pneumoniae* in middle-aged and older adult outpatients. Eur J Clin Microbiol Infect Dis *13*, 785–792.

Troy, C.J., Peeling, R.W., Ellis, A.G., Hockin, J.C., Bennett, D.A., Murphy, M.R., and Spika, J.S. (1997). *Chlamydia pneumoniae* as a new source of infectious outbreaks in nursing homes. JAMA *277*, 1214–1218.

Tumurkhuu, G., Dagvadorj, J., Porritt, R.A., Crother, T.R., Shimada, K., Tarling, E.J., Erbay, E., Arditi, M., and Chen, S. (2018). *Chlamydia pneumoniae* Hijacks a Host Autoregulatory IL-1beta Loop to Drive Foam Cell Formation and Accelerate Atherosclerosis. Cell metabolism doi:10.1016/j.cmet.2018.05.027.

Ulrichs, T., Kosmiadi, G.A., Trusov, V., Jorg, S., Pradl, L., Titukhina, M., Mishenko, V., Gushina, N., and Kaufmann, S.H. (2004). Human tuberculous granulomas induce peripheral lymphoid follicle-like structures to orchestrate local host defence in the lung. The Journal of pathology *204*, 217–228 doi:10.1002/path.1628.

Vermeire, P.A., and Pride, N.B. (1991). A 'splitting' look at chronic nonspecific lung disease (CNSLD): common features but diverse pathogenesis. Eur Respir J *4*, 490–496.

Vos, R., Vanaudenaerde, B.M., Ottevaere, A., Verleden, S.E., De Vleeschauwer, S.I., Willems-Widyastuti, A., Wauters, S., Van Raemdonck, D.E., Nawrot, T.S., Dupont, L.J., *et al.* (2010). Long-term azithromycin therapy for bronchiolitis obliterans syndrome: divide and conquer? J Heart Lung Transplant *29*, 1358–1368 doi:10.1016/j.healun.2010.05.023.

Wang, S.-P., and Grayston, J.T. (1990). Population prevalence antibody to *Chlamydia pneumoniae*, strain TWAR. In Proceedings of the Seventh International Symposium on Human Chlamydial Infections, W.R. Bowie, H.D. Caldwell, R.P. Jones, P.-A. Mardh, G.L. Ridgway, J. Schachter, W.E. Stamm, and M.E. Ward, eds. (Cambridge: Cambridge University Press), pp. 402–405.

Wang, S.-P., and Grayston, J.T. (1986). Microimmunofluorescence serological studies with the TWAR organism. Chlamydial Infections Proceedings of the Sixth International Symposium on Human Chlamydial Infections, Oriel, D, Ridgwya, G, Schater, J, Taylor-Robinson, D, Ward, M (eds, p 329–33, Cambridge University Presss, Cambridge.

Wang, S.P., and Grayston, J.T. (1970). Immunologic relationship between genital TRIC, lymphogranuloma venereum, and related organisms in a new microtitre indirect immunofluorescence test. American journal of ophthalmology *70*, 367–374.

Wang, S.P., and Grayston, J.T. (1986). Microimmunofluorescence serological studies with the TWAR organism. In Chlamydial Infections: Proceedings of the Sixth International Symposium on Human Chlamydial Infections, D. Oriel, and G. Ridgeway, eds. (Cambridge: Cambridge University Press), pp. 329–332.

Wantia, N., Rodriguez, N., Cirl, C., Ertl, T., Durr, S., Layland, L.E., Wagner, H., and Miethke, T. (2011). Toll-like receptors 2 and 4 regulate the frequency of IFNgamma-producing CD4+ T-cells during pulmonary infection with *Chlamydia pneumoniae*. PloS one *6*, e26101 doi:10.1371/journal.pone.0026101.

Webley, W.C., and Hahn, D.L. (2017). Infection-mediated asthma: etiology, mechanisms and treatment options, with focus on *Chlamydia pneumoniae* and macrolides. Respir Res *18*, 98 doi:10.1186/s12931-017-0584-z.

Weinmaier, T., Hoser, J., Eck, S., Kaufhold, I., Shima, K., Strom, T.M., Rattei, T., and Rupp, J. (2015). Genomic factors related to tissue tropism in *Chlamydia pneumoniae* infection. BMC genomics *16*, 268 doi:10.1186/s12864-015-1377-8.

Wesslén, L., Påhlson, C., Friman, G., Fohlman, J., Lindquist, O., and Johansson, C. (1992). Myocarditis caused by *Chlamydia pneumoniae* (TWAR) and sudden unexpected death in a Swedish elite orienteer. Lancet *340*, 427–428.

Wong, N.D. (2014). Epidemiological studies of CHD and the evolution of preventive cardiology. Nature Reviews Cardiology *11*, 276 doi:10.1038/nrcardio.2014.26.

World Health Organization, Mendis, S., Puska, P., and Norrving, B. (2011). Global Atlas on Cardiovascular Disease Prevention and Control (Geneva, SWITZERLAND: World Health Organization).

Worm, H.C., Wirnsberger, G.H., Mauric, A., and Holzer, H. (2004). High prevalence of *Chlamydia pneumoniae* infection in cyclosporin A-induced post-transplant gingival overgrowth tissue and evidence for the possibility of persistent infection despite short-term treatment with azithromycin. Nephrol Dial Transplant *19*, 1890–1894.

Wu, L., Skinner, S.J.M., Lambie, N., Vuletic, J.C., Blasi, F., and Black, P.N. (2000). Immunohistochemical staining for *Chlamydia pneumoniae* is increased in lung tissue from subjects with chronic obstructive pulmonary disease. Am J Respir Crit Care Med *162*, 1148–1151.

Xia, M., Chen, D., Endresz, V., Faludi, I., Szabo, A., Gonczol, E., Kakkar, V., and Lu, X. (2013). Immunization of *Chlamydia pneumoniae* (Cpn)-Infected Apob(tm2Sgy) Ldlr(tm1Her)/J Mice with a Combined Peptide of Cpn Significantly Reduces Atherosclerotic Lesion Development. PloS one *8*, e81056 doi:10.1371/journal.pone.0081056.

Yamazaki, T., Nakada, H., Sakurai, N., Kuo, C.C., Wang, S.P., and Grayston, J.T. (1990). Transmission of *Chlamydia pneumoniae* in young children in a Japanese family. The Journal of infectious diseases *162*, 1390–1392.

Yang, Z.P., Kuo, C.C., and Grayston, J.T. (1993). A mouse model of *Chlamydia pneumoniae* strain TWAR pneumonitis. Infection and immunity *61*, 2037–2040.

Yang, Z.P., Kuo, C.C., and Grayston, J.T. (1995). Systemic dissemination of *Chlamydia pneumoniae* following intranasal inoculation in mice. The Journal of infectious diseases *171*, 736–738.

Zafiratos, M.T., Manam, S., Henderson, K.K., Ramsey, K.H., and Murthy, A.K. (2015). CD8+ T cells mediate *Chlamydia pneumoniae*-induced atherosclerosis in mice. Pathogens and disease 73 doi:10.1093/femspd/ftv052.

Zhan, P., Suo, L.J., Qian, Q., Shen, X.K., Qiu, L.X., Yu, L.K., and Song, Y. (2010). *Chlamydia pneumoniae* infection and lung cancer risk: A meta-analysis. Eur J Cancer doi: 10.1016/j.ejca.2010.11.003.

Zhou, Y., Wang, C., Yao, W., Chen, P., Kang, J., Huang, S., Chen, B., Wang, C., Ni, D., Wang, X., *et al.* (2009). COPD in Chinese nonsmokers. Eur Respir J *33*, 509–581.

Chlamydia Adhesion and Invasion

3

Matthew D. Romero[1†], Katja Mölleken[2†], Johannes H. Hegemann[2] and Rey A. Carabeo[1*]

[1]Department of Pathology and Microbiology, University of Nebraska Medical Center, Omaha, NE, USA.
[2]Institute of Functional Microbial Genomics, Heinrich Heine University Düsseldorf, Düsseldorf, Germany.

*Correspondence: rey.carabeo@unmc.edu

†Authors contributed equally

https://doi.org/10.21775/9781912530281.03

Abstract

Chlamydiae are obligate intracellular pathogens that require a specialized intracellular niche for their survival and replication. Hence, they possess an efficient means of gaining access to this niche. Adhesion and invasion of *Chlamydia* to host cells are the initiating processes of infection and pathogenesis, and thus it is important to characterize this process in detail. *In vivo*, the primary targets are the epithelial cells that line the mucosa of the ano-genital tract, conjunctiva, the respiratory tract, and the gut. These cells are non-phagocytic. In order for *Chlamydia* to invade these cells, they must induce their uptake by actively remodelling the cortical actin cytoskeleton and manipulating the endocytic machinery to facilitate the phagocytosis of infectious elementary bodies (EBs). Uptake is preceded by the stable adherence of EBs to the surface of epithelial cells. Stable adhesion is followed by the induction of signalling pathways that trigger a number of host cell processes, including remodelling of the cortical actin to facilitate uptake of the pathogen. Adhesion and the induction of signalling to the cytoskeletal remodelling machinery are shared by a number of chlamydial species, with *C. trachomatis* and *C. pneumoniae* being the best studied. Chlamydial adhesins stabilize the bacterium at the surface of the host cell via interaction with host cell surface receptors. This adhesion step does not need to be highly specific. It involves transient electrostatic interactions followed by more specific associations between chlamydial adhesins and specific receptors to trigger signalling within the host cell and initiate actin remodelling at the cell surface. Localized actin remodelling results in the formation of a variety of cell surface structures designed to internalize the invading bacteria. In this chapter, we will review the mechanisms of chlamydial adhesion and invasion, with a focus on bacterial and host factors that control these steps and that have been identified within the last decade.

Introduction

How is signalling at the plasma membrane during chlamydial adhesion transduced to the cytoskeleton remodelling machinery? Epithelial cells are naturally non-phagocytic. In order to ingest particles, they must be induced to do so by modulating their cytoskeleton remodelling machinery (Carabeo, 2011; Nogueira *et al.*, 2018). Indeed, a number of 'epitheliotropic' intracellular pathogens, e.g. *Salmonella, Listeria*, etc. have dedicated proteins to subvert cytoskeletal dynamics as part of their respective strategies to facilitate their uptake (Bonazzi and Cossart, 2006; Fredlund and Enninga, 2014).

Efficient invasion is crucial to pathogenesis because it gives bacteria access to protective intracellular niches, where they deploy additional subversive strategies and divert host cell processes to benefit the pathogen. Intracellular bacteria throughout evolution have acquired a collection of effector proteins that manipulate signalling to the cytoskeletal machinery to alter cytoskeleton dynamics. These include bacterial adhesin-host cell receptor interactions and translocation of effectors via specialized means (e.g. type III secretion system) (Deng *et al.*, 2017; Green and Mecsas, 2016).

The localized secretion of virulence factors and/or activation of receptors at the sites of adhesion are major determinants of localized activation of nucleation promoting factors (NPFs) of actin assembly, leading to the formation of actin-rich cell surface projections in the immediate vicinity of the invading bacteria. NPFs are found at lamellipodial protrusions, focal adhesions, podosomes, endosomal scissions and in the context of infection, at sites of invasion by microbial pathogens (Yu *et al.*, 2012; Grassart *et al.*, 2014; Truong *et al.*, 2013; Pizarro-Cerdá *et al.*, 2016). Binding of growth factors or pathogens to receptors triggers localized signalling at the plasma membrane to activate the Rho GTPase proteins Rac and Cdc42 and their downstream effectors, including nucleation promoting factors (NPFs), such as the WASP family of proteins (Derivery and Gautreau, 2010). Binding of Rac and Cdc42 to specific sites in the WASP family of proteins are necessary for the induction of actin remodelling (Rohatgi *et al.*, 1999). Membrane-localized phosphatidylinositol 4,5-bisphosphate molecules are also bound by WASPs via their pleckstrin homology (PH) domain (Miki *et al.*, 1996). WASPs also interact with the Src-homology 3 (SH3) domain via its proline-rich domain (Fukuoka *et al.*, 2001). These interactions, which occur at the plasma membrane relieve WASP from its autoinhibited conformation and reveal binding sites for actin and the Arp2/3 complex. A similar principle operates in the regulation of SCAR/WAVE, a relative of WASP. The major difference is that inhibition of SCAR/WAVE is via interaction with a number of proteins that make up the WAVE complex. They include Sra1/PIR12, Nap1, Abi1, and HSPC300 (Ibarra *et al.*, 2005). This multi–protein complex is disassembled in the presence of active Rac, and the activated state of the liberated SCAR/WAVE protein is further enhanced by its interaction with the host signalling protein IRSp53 (Miki *et al.*, 2000). As with WASP, the net result of SCAR/WAVE activation is the binding and activation of the Arp2/3 complex.

A common theme in NPF activation by intracellular bacteria during invasion is the modulation of Rac and/or Cdc42 GTPase activity, either directly through molecular mimicry or through the induction of signalling to activate these GTPases (Carabeo, 2011). The location of adhesion-receptor engagement and the highly localized translocation of bacterial effectors into the cytosol both determine where actin remodelling occurs. These two processes induce the invasion of *Listeria monocytogenes* and *Salmonella* Typhimurium, respectively. *Listeria* utilizes multiple adhesion molecules, e.g. InlA (internalin) and InlB, as well as accessory adhesion molecules (InlC, E, G, and H) on its surface to expand its cell tropism and induce efficient uptake of the pathogen (Bergmann *et al.*, 2002). Engagement of host cell receptors by either InlA or InlB stimulates signalling to the Rho GTPases Rac and/or Cdc42 to activate SCAR/WAVE, N-WASP and the Arp2/3 complex (Bierne *et al.*, 2001; Bierne *et al.*, 2005). *Salmonella* also targets and modulates host-signalling proteins to promote extensive actin remodelling to produce the trademark membrane ruffles. The virulence proteins SopE and SopE2 are secreted by *Salmonella* into the host cell cytosol to interact with and activate members of the small Rho GTPase signalling proteins (Rudolph *et al.*, 1999; Stender *et al.*, 2000).

As illustrated in the above examples, invasion is a pathogen-directed process, which is also exemplified *by Chlamydia* during its interactions with non-phagocytic cells. *Chlamydia* invasion is efficient as evidenced by the rapid kinetics of actin recruitment and internalization of elementary bodies (EBs) (Carabeo *et al.*, 2002). *In vivo*, the targets of *C. trachomatis* are the epithelial cells that line mucosal surfaces of the genital tract, the conjunctiva, and the rectum (Zhong, 2018; Hu *et al.*, 2013). As detailed below, *C. trachomatis* and *C. pneumoniae* deploy similar basic mechanisms to invade non-phagocytic cells. These include the engagement of cell surface receptors by specific bacterial adhesins and the translocation of effector proteins by specialized secretion systems to target host cell structural and signalling components, and to induce actin remodelling (Nogueira *et al.*, 2018).

Experimental approaches to study *Chlamydia* binding and entry

Chlamydia interactions with the host cell have much in common with other viral and bacterial pathogens that adhere to the surface of mammalian cells and invade to access their intracellular niche. These shared properties include non-specific electrostatic interaction of the pathogen with the host cell and the transition to a more stable temperature-dependent binding via intermolecular interactions between the pathogen's adhesin and the corresponding receptor on the host cell surface. Non-specific electrostatic interaction was demonstrated by exposing chlamydiae at the pre-adhesion stage to cations or polyanionic compounds, which disrupts pathogen binding (Hatch *et al.*, 1981; Su *et al.*, 1990). These initial studies revealed the multistage nature of *Chlamydia* adhesion, involving a temperature-independent electrostatic interaction, which was relatively non-specific and transient that transitions to a more stable temperature-dependent adhesion (Kuo and Grayston, 1976; Byrne, 1978; Carabeo and Hackstadt, 2001). Indeed, in this latter stage, *Chlamydia* could only be released by treatment with the trypsin protease (Byrne, 1978; Hatch *et al.*, 1981). The temperature-dependent shift in *Chlamydia* interaction with cells was confirmed when Carabeo *et al.* isolated a mutant line from a screen of chemically mutagenized Chinese hamster ovary K1 (CHO-K1) cells for resistance to *Chlamydia* infection (Carabeo and Hackstadt, 2001). Binding of *Chlamydia* to wild type parental CHO-K1 cells at 37°C was resistant to detachment by soluble heparin. However, *Chlamydia* adhesion to one CHO mutant line in particular could be disrupted by heparin, indicating the failure of adhesion to progress from reversible electrostatic interaction to the irreversible binding in mutant cells.

It has been more complicated to quantify the efficiency and dynamics of *Chlamydia* invasion compared to adhesion. Early studies relied on enumerating inclusions as a measure of successful invasion. However, inclusion formation is a process that is far removed from internalization, and thus can be affected by changes to post-adhesion stages of infection. In addition to invasion, inclusion formation is determined by the ability of the pathogen to evade lysosomal fusion, transport to the perinuclear region, express early genes to initiate development, and modify the inclusion membrane to maintain evasion of lysosomal fusion (Fields and Hackstadt, 2002). These determinants all depend on *Chlamydia* to transition from the relatively metabolically inert EB to the vegetative reticulate body (RB) form – itself a very complex process. Thus, older studies that quantified the inhibition of *Chlamydia* invasion by measuring inclusion numbers could not formally exclude post-invasion effects.

Subtil *et al.* and Carabeo *et al.* reported an invasion assay that quantified internalized and total cell-associated EBs using a combination of differential permeabilization and two antibodies directed against different *Chlamydia* spp. (Subtil *et al.*, 2004; Carabeo *et al.*, 2002). These experiments were performed soon after adhesion, with infection synchronized either by centrifugation or incubation at temperatures that inhibited uptake (< 15°C), followed by the addition of pre-warmed (37°C) cell culture media (Subtil *et al.*, 2004; Carabeo *et al.*, 2002). This approach enabled the quantification of the effects of inhibitors of signalling, cytoskeletal dynamics, chlamydial adhesion, etc. on invasion, and thus more conclusively implicate these processes in invasion. Supported by quantitative data, imaging-based approaches that monitored recruitment of targets of these inhibitors to the sites of entry (Subtil *et al.*, 2004; Carabeo *et al.*, 2004) soon followed, ushering in a change from descriptive to mechanistic research on adhesion and invasion.

The role of the host cytoskeleton and endocytic machinery in *Chlamydia* invasion

Actin cytoskeleton and remodelling during invasion

In vitro, *Chlamydia* displays a high degree of promiscuity with regards to infection of cultured cells, possibly due to a ubiquitous receptor expressed in many cell types or the utilization of multiple adhesin–receptor interactions to facilitate entry. Regardless of the mechanism used, the assembly of actin-rich structures at the base of and surrounding adhered *C. trachomatis* EB is observed (Carabeo, 2011; Carabeo *et al.*, 2002). Noteworthy are the rapid and transient dynamics of actin recruitment (Carabeo *et al.*, 2002), which, when inhibited by pharmacological agents that disrupts actin polymerization, concomitantly reduced invasion efficiency, highlighting the importance of actin-rich structures in chlamydial uptake (Carabeo, 2011; Carabeo *et al.*, 2002; Ford *et al.*, 2018). The use of the actin-destabilizing drug, cytochalasin D and globular actin-sequestering compound latrunculin B led to the loss of the microvillar structures, and markedly reduced invasion efficiencies (Carabeo, 2011; Carabeo *et al.*, 2002; Ford *et al.*, 2018). Interestingly, washout of cytochalasin D revealed the preferential re-establishment of the microvillar projections at the sites of EB adhesion, indicating a very localized re-induction of actin polymerization and remodelling in live-cell imaging experiments (Carabeo *et al.*, 2002). Enhanced green fluorescent protein (eGFP)-actin is specifically recruited in a rapid and transient manner to the site of chlamydial adhesion (Carabeo *et al.*, 2002), which may explain the formation of cell surface projections at the sites of invasion.

What mediates the localized actin recruitment that is often observed during chlamydial invasion? The signal that stimulates highly localized host actin remodelling may originate from the EB itself and involve the translocation of effectors to modulate host cell signalling at the site of invasion or the engagement of host cell receptors by adhesins on the surface of chlamydiae, or both. These events are likely to be associated with protein post-translational modification, such as phosphorylation, and an extensive network of protein–protein interactions that collectively mediate localized actin recruitment and remodelling, and the eventual uptake of the pathogen (Birkelund *et al.*, 1994; Fawaz *et al.*, 1997; Swanson *et al.*, 2007; Clifton *et al.*, 2005; Lane *et al.*, 2008). Elucidation of the signalling events that follow EB binding came from a combination of studies involving focused hypothesis-based approaches and large-scale screens. An RNA interference screen conducted by Elwell *et al.* detected a number of previously identified and novel cellular cytoskeleton-related genes that are necessary for *Chlamydia* invasion (Elwell *et al.*, 2008). The list of candidates included genes for Rac1 and/or Cdc42, phosphatidylinositol 3-kinase (PI3K) and the WAVE complex (Coombes and Mahony, 2002; Balañá *et al.*, 2005; Carabeo *et al.*, 2004; Subtil *et al.*, 2004). As will be detailed below, some of these signalling molecules were found to function in a signal transduction pathway centred on the chlamydial protein TarP (Lane *et al.*, 2008).

The dynamics of cortical actin are determined by the relative concentrations of monomeric globular (G-actin) and filamentous actin (F-actin), with points of regulation influencing this balance (Chhabra and Higgs, 2007). Nucleation promoting factors (NPFs), such as members of the WASP family of proteins, bind G-actin to effectively increase the local concentration of G-actin and to induce a conformational change in monomeric actin, which together results in the decrease in the thermodynamic barrier to actin nucleation. Actin remodelling mediated by N-WASP and WAVE2, and their regulators Rac and Cdc42 have been implicated in *Chlamydia* invasion (Balañá *et al.*, 2005; Carabeo *et al.*, 2004; Subtil *et al.*, 2004; Elwell *et al.*, 2008), likely through the formation of various actin-rich cell surface projections at the sites of *C. trachomatis* and *C. pneumoniae* adhesion (Coombes and Mahony, 2002; Carabeo *et al.*, 2002). Projection structures include hypertrophic microvilli, pockets, and pedestal-like structures. Ford *et al.* implicated Rac, N-WASP, and WAVE in the formation of macropinocytic-like structures during *C. trachomatis* uptake that require sorting nexin 9 (SNX9), which is a key regulator of clathrin-mediated and -independent endocytosis (Ford *et al.*, 2018), while Nans *et al.* showed that filopodia-mediated invasion (Nans *et al.*, 2015).

The formation of multiple distinct actin-rich structures during chlamydial invasion raises a number of questions. It remains to be seen what the biological relevance are of these different invasion mechanisms, i.e. mediated by microvilli/filopodia, pocket, and pedestal-like structures, and factors that determine their formation, and potential preferential utilization. Further studies are needed to determine functional relationships and their relevance to pathogenesis, such as host cell tropism.

Microtubule and intermediate filaments during *Chlamydia* invasion

The role of microtubules and intermediate filaments in *Chlamydia* invasion is unclear, with scant reports published. Ward and Murray reported inhibition of *Chlamydia* invasion upon treatment with vinblastine or vincristine, which are inhibitors of microtubule polymerization (Ward and

Murray, 1984). The RNAi screen conducted by Elwell *et al.* identified factors known to be involved in microtubule dynamics (Elwell *et al.*, 2008), but the same study did not distinguish invasion from potential post-invasion roles. There are reports that support a potential cooperation between the actin and microtubule networks in invasion by bacterial pathogens, such as *Campylobacter jejuni* (Bouwman *et al.*, 2013; Radhakrishnan and Splitter, 2012) suggesting a bacterial invasion mechanism that exploits functional cooperation between actin and microtubule networks.

Cooperation between the actin, microtubule, and intermediate filament networks has been reported (Fletcher and Mullins, 2010; Schoumacher *et al.*, 2010). This is best illustrated at focal adhesions, whose formation and turnover dynamics are determined by physical interactions between actin-rich stress fibres, microtubules and associated proteins that bridge the two (Gardel *et al.*, 2010). Furthermore, focal adhesions have been directly shown to serve as a site for keratin filament assembly (Windoffer *et al.*, 2006). The role of focal adhesions as a nexus of functional interactions between the three different cytoskeletal networks takes on increased importance for *Chlamydia* invasion. A recent report found that *C. trachomatis* invasion required β1-integrin (Stallmann and Hegemann, 2016), the focal adhesion kinase (Thwaites *et al.*, 2014), and vinculin (Thwaites *et al.*, 2015), which are all focal adhesion components. This is suggestive of the formation of a pseudo-focal adhesion signalling complex at the sites of invasion, where actin, microtubules and intermediate filament networks might cooperate to mediate efficient internalization of the pathogen.

The role of clathrin in *Chlamydia* uptake

The mechanism of chlamydial internalization remains unresolved because of several conflicting reports. For example, the role of clathrin is not clear. Clathrin is an essential factor controlling receptor-mediated endocytosis, and thus could have a role in *Chlamydia* invasion of non-phagocytic cells. Invasion by *C. trachomatis* serovar L2 [and *C. caviae* strain guinea pig inclusion conjunctivitis (GPIC)] was unaltered in cells expressing dominant negative forms of Eps15, a protein that facilitates the formation of clathrin structures by associating with the clathrin adaptor protein AP-2 (Boleti *et al.*, 1999; Benmerah *et al.*, 1999). More recent findings using RNA interference to reduce expression of proteins of the endocytic machinery led to a marginal, but statistically significant decrease in *C. trachomatis* serovar L2 invasion (Hybiske and Stephens, 2007), while entry of *C. pneumoniae* was clathrin-independent (Korhonen *et al.*, 2012), indicating species specificity for clathrin in invasion.

Interestingly, the role of clathrin in invasion may be influenced by the apico-basolateral polarization of epithelial cells. Earlier electron microscopy studies conducted by Wyrick *et al.* in human endometrial gland epithelial cells (HEGEC) and HeLa cells grown in a non-polarized configuration, demonstrated that adhered and internalized *Chlamydiae* were rarely associated with clathrin structures (Wyrick *et al.*, 1989). However, associations with clathrin-coated pit structures were observed when HEGECs were grown in a manner that promoted apico-basolateral polarization, indicating that clathrin-dependence may depend on cellular organization (Wyrick *et al.*, 1989). In addition to apical-basolateral polarization, clathrin requirement may also be dependent on the cell type and chlamydial species. Furthermore, the importance of the various routes of entry, e.g. filopodial capture, macropinocytosis, or microvilli-mediated engulfment might be different in polarized epithelial cells. With the advent of advanced quantitative imaging, e.g. super-resolution microscopy and high-resolution live-cell imaging, recruitment of clathrin components and subsequent assembly into coated pits should enable resolution of the role of clathrin in one or more routes of entry and may provide clues on possible clathrin-independent compensatory mechanisms.

Dynamin

Dynamin is 100 kilodalton protein that is localized via its pleckstrin-homology (PH) domain to the plasma membrane where it mediates membrane fission through its inherent GTPase activity (Sever *et al.*, 2000; Ferguson and De Camilli, 2012). Dynamin self-assembles into an oligomeric complex, forming a collar around the necks of nascent vesicles. GTP hydrolysis provides the energy that induces constriction and eventual vesicle scission and separation from the plasma membrane (Pawlowski, 2010). *C. pneumoniae* and *C. trachomatis* invasion exhibited susceptibility to dynamin inhibition. Depletion of dynamin 2, but not dynamin 1, reduced *C. trachomatis* uptake (Hybiske and Stephens, 2007). Consistent with this observation, *C. caviae* invasion was not affected in cells expressing the K44A dominant-negative mutant of dynamin 1 (Boleti *et al.*, 1999). *C. pneumoniae* entry was inhibited by the small molecule compound Dynasore (Fechtner *et al.*, 2016), which inhibits both dynamin 1 and 2 (Macia *et al.*, 2006). Furthermore, Dynasore was reported recently to reduce the levels of labile cholesterol at the plasma membrane and disrupted lipid rafts (Preta *et al.*, 2015). Thus, the Dynasore-related inhibition of *C. pneumoniae* invasion reported by Fechtner *et al.* may also be due to the loss of lipid raft microdomains. This becomes relevant in light of the reported role of the *C. pneumoniae* protein CPn0473, detailed below, in lipid raft-dependent invasion (Fechtner *et al.*, 2016). Dynasore's alternative mechanism of inhibiting *Chlamydia* invasion through its effects on microdomains also highlights the need for caution when interpreting the effects of inhibitors. Taken together, these reports point to a requirement for dynamin 2 in *C. trachomatis* and *C. pneumoniae* invasion. The role of dynamin 2 in *C. caviae* invasion remains to be investigated.

Bacterial adhesion proteins and their corresponding host cell receptors

In this section, we describe the potential roles of various chlamydial and host proteins in mediating adhesion and/or invasion of *C. trachomatis*, *C. pneumoniae*, and other chlamydial species (Table 3.1). Many studies used *C. trachomatis* serovar MoPn, which we will refer to by its new name, *C. muridarum*. In the following sections, it is

Table 3.1 Chlamydial adhesins and receptors

Adhesin	Species	Host receptor	References
OmcB	*C. trachomatis; C. pneumoniae*	Heparan sulfate proteoglycans (HSPG)	Su *et al.* (1996); Stephens *et al.* (2001); Fechtner *et al.* (2013)
Lipopolysaccharide	*C. trachomatis; C. pneumoniae*	CFTR (*C. trachomatis*)	Peterson *et al.* (1998); Fadel and Eley (2008); Ajonuma *et al.* (2010)
Major outer membrane protein	*C. trachomatis; C. pneumoniae*	HSPGs, mannose receptor/ M6PR	Zhang *et al.* (1992); Puolokkainen *et al.* (2005)
Ctad1/CT107	*C. trachomatis*	β1-integrin	Sonja *et al.* (2016)
–	*C. muridarum*	Fibroblast growth factor receptor (FGFR)	Kim *et al.* (2009)
–	*C. pneumoniae; C. muridarum*	Platelet-derived growth factor receptor (PDGFR)	Elwell *et al.* (2008)
–	*C. trachomatis*	Ephrin A2 receptor (EPHA2R)	Subbarayal *et al.* (2015)
–	*C. trachomatis; C. pneumoniae*	Protein disulfide isomerase (PDI)	Davis *et al.* (2002); Conant and Stephens 2007); Abromaitis *et al.* (2009)
Pmp21	*C. pneumoniae*	Epidermal growth factor receptor (EGFR)	Mölleken *et al.* (2013)
–	*C. trachomatis*	EGFR	Patel *et al.* (2014)
Pmp6	*C. pneumoniae*	–	Mölleken *et al.* (2010)
Pmp20	*C. pneumoniae*	–	Mölleken *et al.* (2010)
PmpA-PmpH	*C. trachomatis*	–	Becker *et al.* (2014)
CPn0473	*C. pneumoniae*	– (Raft-associated)	Fechtner *et al.* (2016)
Hsp60/GroEL-1	*C. pneumoniae*	–	Wuppermann *et al.* (2008)
ArtJ	*C. trachomatis; C. pneumoniae*	–	Soriani *et al.* (2010)
N-linked oligosaccharides	*C. trachomatis; C. pneumoniae*	–	Kuo *et al.* (2004); Lujan *et al.* (2018)
–	*C. pneumoniae*	Apolipoprotein E4	Gérard *et al.* (2008)

important to recognize that adhesins and receptors may be involved in adhesion only, and/or in triggering signalling to the cytoskeletal machinery to promote bacterial uptake. There are also reports that describe signalling pathways that may be more relevant to post-invasion processes, including reinforcing signals that promote host cell survival. Here, we will focus on entry-related aspects. There are a number of other chlamydial proteins that have been proposed to function as adhesins (Table 3.1). Below, we focus on those that have known receptors and/or characterized signalling pathways. The adhesin–receptor pairs relevant to invasion are not conserved among *Chlamydia* spp., which is interesting, but not surprising given the differences in the cell types infected *in vivo*.

Major outer membrane protein, OmcB, and glycosaminoglycans

The initial transient and reversible adhesion of a number of *Chlamydia* strains and species to host cells involves the negatively charged glycosaminoglycan (GAG) heparan sulfate (Moulder, 1991; Tiwari *et al.*, 2012). GAGs are composed of disaccharide units of hexuronic acid and hexosamine, forming unbranched chains. They are sulfated, with varying levels and position of sulfation (Sarrazin *et al.*, 2011). Heparin has a greater degree of sulfation modification per sugar residue (i.e. hexosamine) compared to heparan sulfate (Prydz, 2015). Furthermore, unlike heparin, which is restricted to mast cells, heparan sulfate is produced by many cell types, including epithelial cells (Sugahara and Kitagawa, 2002).When kept at low temperature to prevent uptake of the bacteria into host cells, *Chlamydia* association with the cell surface via GAGs was disrupted by soluble heparin molecules, illustrating the reversible nature of this interaction (Carabeo *et al.*, 2001; Kuo and Grayston, 1976).

Interestingly, heparin-dependence is not a conserved feature of *C. trachomatis* serovars, implying that there may be additional compensatory mechanisms that mediate the initial transient non–specific interactions. Heparan sulfate is required for adhesion of *C. trachomatis* lymphogranuloma venereum (LGV) serovars, but not of serovar E (Kuo and Grayston, 1976; Taraktchoglou *et al.*, 2001). This difference may be related to variations in the OmcB protein at the level of amino acid sequence. OmcB has been reported to be the primary heparin-binding protein present on the chlamydial surface (Fadel and Eley, 2007; Moelleken and Hegemann, 2008; Fechtner *et al.*, 2013). The XBBXBX motif (B and X representing basic and any amino acid residues, respectively) at the N-terminal portion of the protein, along with three amino acid residues immediately downstream, were crucial determinants of the interaction of OmcB with heparin, with proline in position 66 of OmcB being essential (Fechtner *et al.*, 2013) (Fig. 3.1). The Pro66 residue in heparin-requiring serovars is likely to be responsible for the differences in the secondary structure in this region of OmcB.

There is also added complexity to adhesion of some *C. trachomatis* serovars caused by varying degrees of sulfation of GAG on the cell surface. Using a panel of GAG sulfation-specific antibodies to inhibit adhesion of different serovars, Fechtner *et al.* revealed the relationship between OmcB polymorphisms at residues 66, 68, and 71 with the type of GAG recognized (Fechtner *et al.*, 2013). This specificity suggested that the role of host cell GAGs extended beyond simple charge interaction, and that it involves specific GAG configurations. This raises the possibility that the GAG type(s) may be an important determinant of cell tropism.

The chlamydial major outer membrane protein (MOMP) of *C. muridarum* was also shown to interact with GAGs and, like OmcB, also demonstrated some level of specificity for a particular type of GAG [e.g. heparan sulfate (HS) or chondroitin sulfate (CS)], as well as the degree of sulfation (Su *et al.*, 1996). Using recombinant MOMP fused to the *E. coli* maltose binding protein (MBP-MOMP), Su *et al.* demonstrated differences in binding of MBP-MOMP to various Chinese Hamster Ovary-K1 (CHO-K1) mutant cells with deficiencies in heparan sulfate biosynthesis. These mutant cell lines displayed a changed degree in GAG sulfation (mutant line psgD-606), complete lack of GAG biosynthesis, including HS and CS (mutant line psgA-745), and defects in HS but not CS biosynthesis (mutant line psgE-677). Furthermore, both proteins in recombinant forms were able to outcompete *Chlamydia* binding to cells. Consistent with these data, anti-*C. muridarum* MOMP monoclonal antibodies blocked *Chlamydia* binding and infection. With both OmcB and MOMP capable of interacting with GAGs, the possibility of functional compensation or redundancy becomes relevant. On

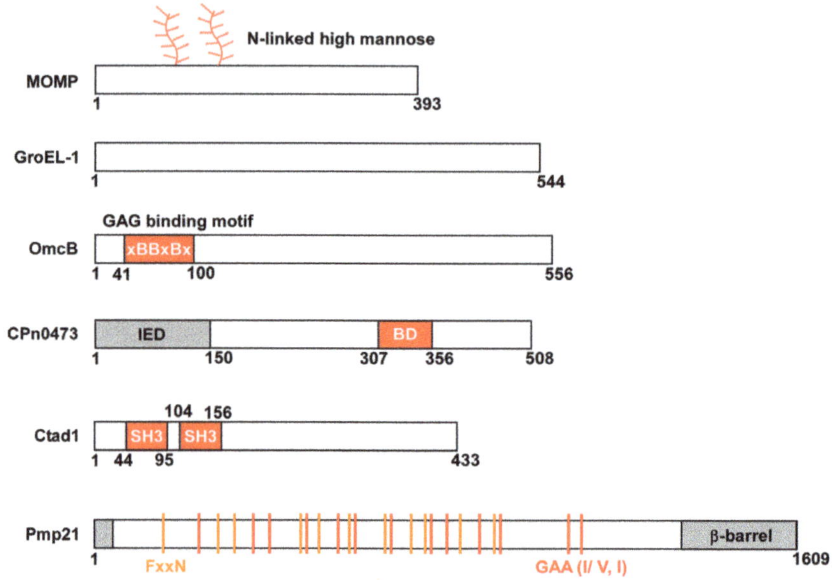

Figure 3.1 Domain organization and/or post-translational modification of candidate chlamydial adhesins. *C. trachomatis* MOMP is post-translationally modified by the addition of N-linked high mannose residues to mediate electrostatic interactions and binding to host mannose receptors, respectively. The origin of this post-translational modification has not been established. *C. pneumoniae* GroEL-1 binds to host cells via an unknown mechanism but can induce NF-κB signalling. The OmcB protein of all chlamydial species harbours an N-terminal GAG binding domain that mediates a reversible interaction with negatively charged GAGs on the host surface. *C. pneumoniae* CPn0473 possesses two functionally distinct domains involved in binding to host cell lipid rafts (BD) and enhancing infectivity (IED). *C. trachomatis* Ctad1 contains two bacterial SH3 domains which are important for the interaction with host β1-integrin receptors. Pmp21 from *C. pneumoniae*, an example of the autotransporter type V secreted Pmp proteins, binds to human EGFR. The short repetitive motifs FxxN and GG(A, I,V) are essential for structure and function of the protein.

the other hand, the different specificity of OmcB and MOMP for a variety of GAG isoforms might have consequences with regards to chlamydial cell/tissue tropism.

OmcB and MOMP of purified elementary bodies were reported to be post-translationally modified by N-linked high mannose oligosaccharides (Campbell and Kuo, 2006; Lujan *et al.*, 2018; Swanson and Kuo, 1991) (Fig. 3.1). Lujan *et al.* provided evidence for OmcB harbouring glycosylation modifications (Lujan *et al.*, 2018). In addition, both MOMP and OmcB exhibited reactivity to *Erythrina crista-galli* lectin in a modified Western blot experiment. This result indicates the presence of terminal N-acetyllactosamine glycosylation, which is recognized by galectin-1, a protein of 135 amino acid that is secreted into the extracellular milieu or remains associated with the cell surface. The authors therefore hypothesized that galectin-1 might be important for chlamydial adhesion. Indeed, galectin-1 bound to both chlamydial proteins and to host cells, bridging *Chlamydia* and glycosylation motifs on host proteins at the cell surface (Lujan *et al.*, 2018). This model is reminiscent of the GAG-mediated trimolecular bridge model proposed by Stephens *et al.*(Zhang and Stephens, 1992), in which heparan sulfate, rather than galectin-1, simultaneously binds an EB surface protein and a host cell receptor, bridging the two. There were other chlamydial proteins that are glycosylated but that did not show interaction with galectin-1 (Lujan *et al.*, 2018). Their roles are uncharacterized. An interesting question is how these chlamydial proteins are glycosylated. Based on the genome sequence, *Chlamydia* does not have the full glycosylation machinery (Stephens *et al.*, 1998; Thomson *et al.*, 2008). It is possible that this pathogen relies on host machinery for MOMP and/or OmcB modification with N-linked high-mannose glycosylation. Tunicamycin, an inhibitor of eukaryotic N-linked glycosylation of nascent polypeptide decreased MOMP glycosylation, which correlated with a decrease in the number of infectious progeny (Kuo *et al.*, 2004). The initial steps of N-linked glycosylation, including the addition of high-mannose residues take place in the endoplasmic reticulum (ER), with additional modification, trimming and terminal glycosylation, (i.e. the addition of N-acetylglucosamine, N-acetylneuraminic acid, etc.) occurring at the trans–Golgi network (Reynders *et al.*, 2011). N-linked high mannose modifications on MOMP and OmcB indicated no further modification by terminal glycosylation. This would be consistent with interaction with the ER-resident glycosylation machinery. Indeed, chlamydial inclusions interact with the ER involving the inclusion membrane protein, IncV (Stanhope *et al.*, 2017).

Galectin-1 may also trigger signalling to the cytoskeletal remodelling machinery, or perhaps the deployment of the type III secretion mechanism. A report from Su *et al.* directly investigated the role of MOMP, which in EBs is glycosylated (Su *et al.*, 1996). Recombinant unglycosylated MOMP, expressed as a fusion with the maltose-binding protein (MBP-MOMP) in and purified from *E. coli*, was able to bind to HeLa cells in a GAG-dependent manner at 37°C, but no internalization was observed, suggesting that the MOMP–GAG interaction was not sufficient to promote invasion.

Heat-shock protein 60/GroEL-1

Heat-shock protein 60 (Hsp60), also known as GroEL-1, is a candidate chlamydial adhesin (Fig. 3.1), as supported by evidence demonstrating association of HEp-2 cells with soluble and latex bead-conjugated recombinant GroEL-1 protein from *C. pneumoniae* (Wuppermann *et al.*, 2008). Paradoxically, when GroEL-1 was pre-bound to the cell surface, it did not block attachment of *C. pneumoniae* EBs, although their uptake was significantly reduced (Wuppermann *et al.*, 2008), indicating a more complex role of this putative adhesin in *Chlamydia* adhesion and uptake. The ability to bind to the cell surface was unique to GroEL-1 because the two *C. trachomatis* paralogues, GroEL-2 and GroEL-3, lacked this binding activity. How GroEL-1 is localized to the surface of elementary bodies is unclear, but may involve reabsorption of GroEL-1 from lysed chlamydial organisms.

There is precedent for GroEL as an adhesin because a number of bacterial and fungal pathogens utilize their GroEL homologue to interact with target cells. For example, *Helicobacter pylori* and *Haemophilus ducreyi* rely on GroEL for adhesion to target cells (Frisk *et al.*, 1998; Yamaguchi *et al.*, 1997). Localization of the Helicobacter GroEL homologue to the bacterial surface could be via an unidentified specialized secretion or autolysis followed by reabsorption by neighbouring intact bacteria (Vanet and Labigne, 1998; Phadnis *et al.*, 1996). The fungal pathogen *Histoplasma capsulatum* utilizes surface localized GroEL-like proteins to engage β2-integrin on the surface of macrophages (Long *et al.*, 2003). Interestingly, binding of soluble *C. pneumoniae* GroEL-1 to the cell surface caused the activation of nuclear factor-kappa B (NF-kB), which indicated modulation of host signalling (Wuppermann *et al.*, 2008). Further investigations are required to fully elucidate the role of GroEL-1 in *Chlamydia* attachment and invasion, as well as the nature of its signalling function.

Protein disulfide isomerase

Protein disulfide isomerase catalyses the remodelling of disulfide linkages by breaking and forming disulfide bonds between cysteine residues within and between proteins. Original disulfide linkages are not necessarily preserved during bond reformation. Rather, they are re-isomerized (Wilkinson and Gilbert, 2004). The protein disulfide isomerase (PDI) of the host cell is used by several *Chlamydia* species for adhesion (Abromaitis and Stephens, 2009; Conant and Stephens, 2007; Davis *et al.*, 2002). It was found associated with the host cell surface via electrostatic interaction. It did not appear to act as a conventional receptor for *Chlamydia* in that no direct interaction with *Chlamydia* EBs was observed. Thus, EB association with PDI might be indirect via a PDI binding partner located at the cell surface (Abromaitis and Stephens, 2009). Inhibition studies showed that the disulfide isomerase function of PDI was required for EB uptake, but not for adhesion (Davis *et al.*, 2002). Considering that EBs possess a heavily disulfide-linked outer membrane complex, the possibility that PDI targets this complex to remodel disulfide linkages is intriguing and suggests that PDI may promote initial interactions of EBs with the cell surface. Consistent with this was the observation by Carabeo *et al.* that re-isomerization of disulfide linkages following their reduction was essential for irreversible binding of chlamydiae to the host cell surface. Inhibition of the re-isomerization step by treatment with iodoacetamide (IAA) after reduction with dithiothreitol (DTT) did not significantly inhibit electrostatic interaction of *Chlamydia* with HeLa cells. However, this treatment reduced by approximately 50% the ability of *Chlamydia* to adhere in a manner resistant to heparin-mediated dissociation from the cell, which is a feature of the secondary irreversible binding step (Carabeo *et al.*, 2002). Thus, PDI may mediate proper reisomerization to enable progression of *Chlamydia* attachment from initial electrostatic interaction to stable binding, and trigger signal transduction and eventual uptake of the bacterium.

Disulfide linkage remodelling by PDI may also modulate the deployment of the type III secretion system (T3SS). Betts-Hampikian and Fields provided evidence that reduction of disulfide linkages changed the organization of the T3S machinery, specifically favouring the formation of the higher-order CdsF needle complex of increasing molecular weight (Betts-Hampikian and Fields, 2011). While the presence of other proteins besides CdsF was not explored, it remains possible that the newly formed needle complex as a result of disulfide linkage remodelling may contain additional T3S components. An additional intriguing possibility is the possible involvement of PDI in EB-to-RB conversion. While PDI has not been reported to localize to nascent inclusions, its association with EBs would lead to its co-internalization. Therefore, a disulfide linkage-isomerizing enzyme would be present *in situ* to mediate the conversion of the compact EB to the pleomorphic RB.

Mannose-6-phosphate/insulin-like growth factor 2 receptor

Engagement of insulin-like growth factor 2 receptor [IGF2R, also known as cation-independent mannose-6-phosphate receptor (CI-MPR)] leads to its internalization via ligand-dependent endocytosis, and thus could conceivably mediate bacterial uptake upon binding by the appropriate chlamydial adhesin. The involvement of IGF2R originated from the observation that *C. pneumoniae* infection was enhanced in the presence of exogenously added insulin-like growth factor 2 (IGF2) (Puolakkainen *et al.*, 2003). Interestingly, the addition of another ligand of IGF2R, namely mannose 6-phosphate (M6P) inhibited *C. pneumoniae* EB binding to host cells (Puolakkainen *et al.*, 2005). These opposing effects of two distinct IGF2R ligand on EB binding might appear contradictory. However, M6P and IGF2 have distinct binding sites in IGF2R, which could resolve this apparent contradiction. A possible model that unifies the role of IGF2R in *C. pneumoniae* adhesion and the enhancing effects of IGF2 involves a 'piggy-back' mechanism where an EB bound to IGF2R is internalized by IGF2-triggered endocytosis. This would require a non-overlapping and distinct binding sites for the chlamydial adhesin and IGF2. The chlamydial adhesin would be M6P-modified. Interestingly, the rate of internalization of IGF2R is enhanced by the binding of a multivalent ligand, such as β-glucuronidase (York *et al.*, 1999), with the addition of IGF2 further enhancing receptor uptake. Therefore, a spatially organized mannose-6-phosphate-containing chlamydial adhesin on the surface of the bacteria might mimic the multivalent β-glucuronidase ligand. This ligand-mediated uptake could be enhanced by the addition of IGF2 as observed by Puolakkainen *et al.* (Puolakkainen *et al.*, 2005).

Interestingly, the engagement of IGF2R by ligands influences the immediate intracellular fates of endocytosed receptors, including ubiquitination-dependent transport to the lysosome (Dahms *et al.*, 1989), raising the intriguing possibility that *Chlamydia* association with IGF2R may be a mechanism for evading the endolysosomal trafficking pathway. Therefore, a point of further investigation is the intracellular trafficking of *C. pneumoniae* immediately following IGF2R-dependent uptake.

Polymorphic membrane proteins

The polymorphic membrane proteins (Pmp) form the biggest protein family in *Chlamydiae* representing 13.6% and 17.5% of the coding capacity in *C. trachomatis* and *C. pneumoniae*, respectively (Rockey *et al.*, 2000). Initially identified in immunoblotting experiments of *C. abortus* (*Cab*) and *C. pneumoniae* as immunodominant antigens in infected hosts (Knudsen *et al.*, 1999; Longbottom *et al.*, 1998; Longbottom *et al.*, 1996), Pmps were later bioinformatically characterized and subdivided into six subtypes: A (PmpA), B (PmpB and PmpC), D (PmpD), E (PmpE and PmpF), G (PmpG and PmpI) and H (PmpH) (Grimwood and S. Stephens, 1999). While *C. trachomatis* expresses nine members (PmpA through PmpI), the Pmp family is expanded in the *C. pneumoniae* (17 to 28 genes, *pmp1* to *pmp21*), and in the two most virulent animal species *C. abortus* (12 to 17 genes) and *C. psittaci* (12 to 21) (including both full-length and truncated genes) (Van Lent *et al.*, 2016). Interestingly, the expansion of the Pmp family is solely due to expansion of the *pmpE/F* and *pmpG* gene subtypes, while the other subtypes are each represented by a single gene (Van Lent *et al.*, 2016).

Structurally all Pmps are predicted to be autotransporters with a N-terminal Sec-dependent leader sequence, followed by a passenger domain and a C-terminal β-barrel allowing the proteins to be transported onto the *Chlamydia* surface (Grimwood and Stephens, 1999); (Henderson and Lam, 2001) (Fig. 3.1). For several Pmps of different chlamydial species, surface localization was observed, accompanied by complex proteolytic processing (Crane *et al.*, 2006; Kiselev *et al.*, 2009; Mölleken *et al.*, 2010; Montigiani *et al.*, 2002; Swanson *et al.*, 2009; Tan *et al.*, 2010; Vandahl *et al.*, 2002; Wehrl *et al.*, 2004; Wheelhouse *et al.*, 2012; Qi *et al.*, 2011).

Pmps are characterized by the presence of multiple repeats of GGA (I, L, V) and FxxN tetrapeptide motifs in the N-terminal half (Fig. 3.1), which are both important for the structure as well as the function of these proteins (Becker and Hegemann, 2014; Luczak *et al.*, 2016; Mölleken *et al.*, 2010). Detailed studies of *C. pneumoniae* Pmp6, Pmp20 and Pmp21, which represent three different Pmp subtypes, ascertained that they act as adhesins and mediate attachment of EBs to human epithelial cells (Mölleken *et al.*, 2010). Similar results were obtained for all nine *C. trachomatis* Pmps, as soluble recombinant proteins bound to human cells and suppressed subsequent infections (Becker and Hegemann, 2014). More recently, the EGF growth factor receptor (EGFR) was identified as the receptor for the *C. pneumoniae* adhesin Pmp21 (Mölleken *et al.*, 2013). Mutagenesis experiments revealed that the two repetitive peptide motifs GGA (I, L, V) and FxxN were essential for adhesion function. This is likely because these motifs are critical for the oligomeric state of the protein, and only Pmp21 oligomers bind and interact with EGFR (Luczak *et al.*, 2016; Mölleken *et al.*, 2013). Upon binding to EGFR, recombinant Pmp21 triggers receptor-mediated endocytosis and intracellular signalling through MAP kinase, leading to ERK1/2 activation (Mölleken *et al.*, 2013). MAP kinases have been implicated in actin remodelling through their actions on the p16-Arc subunit of the actin nucleator Arp2/3 complex. Whether signalling to Arp2/3 is responsible for the actin remodelling associated with the Pmp21–EGFR interaction has not been investigated.

A *C. trachomatis* PmpD-null strain exhibited decreased adhesion to cells, indicating a role for Pmps in adhesion to host cells. Decreased adhesion correlated with attenuated virulence in a macaque ocular infection model (Kari *et al.*, 2014). It is noteworthy that the PmpD null mutation is compromised in adhesion despite the presence of other Pmps, and partial attenuation of infection by this mutant is limited to early time points post-infection (Kari *et al.*, 2014). Together, these findings indicate a role of PmpD at a specific stage in the infection, or compensation by other Pmps at later stages.

CPn0473

Although *Chlamydia* genomes are highly reduced, there are still species-specific proteins that account for tissue tropism and adaptation to specific hosts. Indeed *C. pneumoniae* encodes about 200 genes, which have no homologue in *C. trachomatis* (Roulis *et al.*, 2013). One of these is *CPn0473*. Fechtner *et al.* observed that the recombinant Cpn0473 protein bound to cholesterol- and sphingolipid-rich lipid raft domains in the host plasma membrane. Antibodies against Cpn0473 were able to reduce *C. pneumoniae* infection, but pretreatment of human cells with recombinant CPn0473 did not block infection, as would have been expected for a classical adhesin. Instead, Cpn0473 boosted infection in a dose-dependent manner at the level of bacterial internalization (Fechtner *et al.*, 2016). Binding and enhancement of invasion were mediated by the BD and IED domains, respectively (Fig. 3.1). The BD domain is a 50-amino acid region that is essential for Cpn0473 binding to cells, while the IED region resides in the first 150 amino acids of the protein. IED mediated the enhancement of infection by the purified recombinant Cpn0473 (Fechtner *et al.*, 2016). There is similarity to the enhancing effects of IGF2 on *C. pneumoniae* invasion (Puolakkainen *et al.*, 2005), thus invoking a possible interaction of Cpn0473 with a host cell receptor. The enhancing effect of CPn0473 on *C. pneumoniae* invasion required cholesterol- and sphingolipid-rich lipid raft domains, consistent with the previously reported role of this membrane domain in *C. pneumoniae* invasion (Stuart *et al.*, 2003; Korhonen *et al.*, 2012).

Subsequent work on CPn0473 revealed direct binding of the recombinant protein to the human plasma membrane. In mammalian cells, the internal and external leaflets of the plasma membrane possess different phospholipids. Phosphatidylserine (PS) is normally confined to the inner (cytoplasmic) membrane leaflet. In 2001, Goth and Stephens described that PS is transiently externalized upon infection with *C. pneumoniae* (Goth and Stephens, 2001). Interestingly, adhesion of recombinant CPn0473 to human cells induces PS externalization but unexpectedly not apoptosis (Galle *et al.*, 2019). Subsequently, it was shown that PS externalization is increased in human cells exposed to infectious *C. pneumoniae* cells expressing increased CPn0473 and reduced in exposure to *C. pneumoniae* expressing decreased CPn0473. CPn0473 also binds specifically to synthetic membranes carrying PS and stimulates pore formation. Asymmetric giant unilamellar vesicles (GUVs) in which PS is restricted to the inner leaflet reveal that CPn0473 induces PS externalization in the absence of other proteins. Thus, CPn0473 is a bacterial PS translocator capable of specific and apoptosis-independent PS externalization during infection, which facilitates uptake of the infectious chlamydial EB by the host cell. CPn0473 has therefore been renamed LIPP (for 'Lipid-dependent Internalization Promoting Protein) (Galle *et al.*, 2019). How exofacial PS localization supports EB internalization remains to be uncovered. *C. pneumoniae* may have invented PS translocase activity as a mechanism to expose a specific host-cell ligand, i.e., PS, to which the LIPP adhesin can then bind. Translocation of PS will also modulate the plasma membrane structure underneath the adhering EB. Specific PS translocation around the adhering EB increases the anionic charge density of the exofacial leaflet which will develop outward curvature at the edges of the invaginating membrane owing to electrostatic repulsion, thus supporting EB endocytosis (Galle *et al.*, 2019).

Ctad1 and β-integrin

For *C. trachomatis*, which targets polarized epithelial cells *in vivo*, invasion is mediated by the adhesin Ctad1 and its putative host cell receptor, β1-integrin (Stallmann and Hegemann, 2016) (Fig. 3.2). Ctad1 binding required two SH3-like domains at its N-terminus (Fig. 3.1). Incubation of cells with purified Ctad1 molecules immobilized on beads induced clustering of the receptor, phosphorylation of ERK1/2, and importantly uptake. Interestingly, β1-integrin is a glycoprotein with a terminal N-acetyllactosamine (Bellis, 2004), which may be the host cell component recognized by galectin-1 for binding β1-integrin and the bacterium, enabling chlamydial Ctad1 to bind to β1-integrin. In addition, the involvement of β1-integrin raises the possibility that cholesterol-rich microdomains (e.g. lipid rafts) at the plasma membrane may contribute to bacterial uptake. Endocytosis of antibody-clustered β1-integrin is highly dependent on plasma membrane cholesterol content (Vassilieva *et al.*, 2008). However, the role of cholesterol-rich microdomains in *C. trachomatis* invasion remains unresolved, with methyl-β-cyclodextrin treatment to disrupt cholesterol-rich microdomains yielding conflicting effects (Gabel *et al.*, 2004; Stuart *et al.*, 2003).

Receptor tyrosine kinases

EphrinA2 (EphA2) has been implicated in *Chlamydia* binding to host cells (Fig. 3.2). Receptor phosphorylation and activation of downstream signalling molecules, such as the phosphatidylinositol 3,4,5-triphosphate kinase (PI3K), occurred after adhesion of EBs (Subbarayal *et al.*, 2015). Depletion of EphA2 led to a 30% decrease in *C. trachomatis* invasion, indicating that other chlamydial adhesin-host cell receptor interactions also contributed to internalization. The chlamydial protein ligand for EphA2 has not been identified. Interestingly, Subbarayal *et al.* demonstrated a post-invasion function for EphA2, which they speculated was related to sustained activation of PI3K (Subbarayal *et al.*, 2015). Increased expression of EphA2 is dependent on PI3K induction. However, newly made EphA2 is retained at the chlamydial inclusion membrane, instead of being transported to the plasma membrane. The significance of this remains unclear, but it is possible that a post-invasion EphA2-mediated signalling occurs at the inclusion.

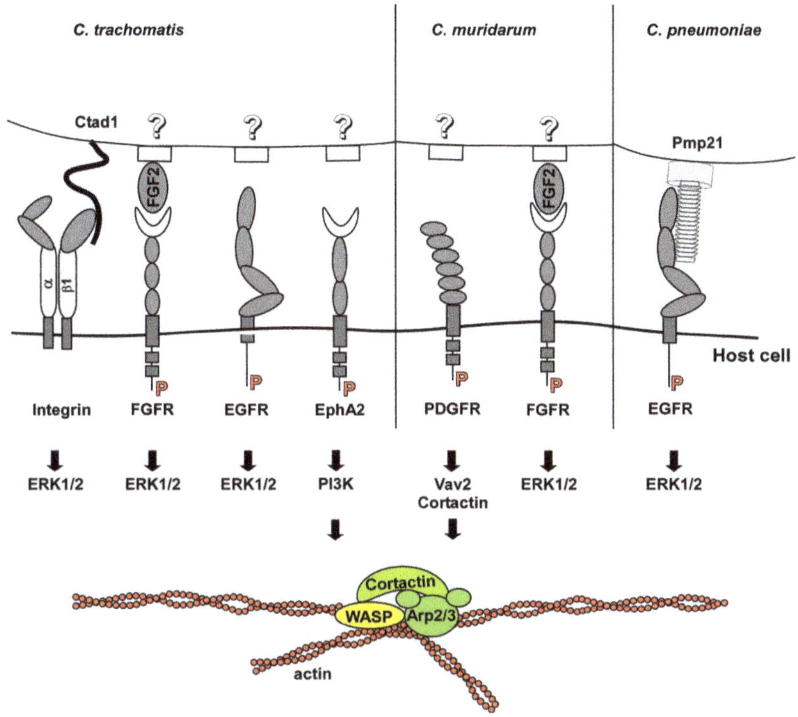

Figure 3.2 Host cell receptors that mediate *Chlamydia* attachment and entry. Examples of host cell receptors that induce signalling when engaged by *Chlamydia*. Various studies on *C. trachomatis* showed that the pathogen binds to or utilizes FGF2/FGFR, EGFR or the EphA2 for infection. All receptors are activated upon EB contact and induce intracellular signalling via ERK1/2 or PI3K pathway. Thus far no chlamydial protein interacting with these receptors could be identified although the FGF2/FGFR interaction of host and EB is dependent on GAGs. *C. trachomatis* Ctad1 interacts via two bacterial SH3 domain with host β1-integrin receptor. This binding leads to the activation of ERK1/2 signalling. In addition to FGF2/FGFR *C. muridarum* also uses PDGFR, which is activated upon EB binding and can activate cytoskeletal related proteins like Vav2 and Cortactin. No chlamydial interaction partner is known. Interaction of the *C. pneumoniae* Pmp21 protein with human EGFR leads to receptor activation and intracellular ERK1/2 signalling, which results in receptor-mediated endocytosis.

Platelet-derived growth factor receptor β (PDGFRβ) has been implicated in binding of *C. muridarum* (Fig. 3.2), with its depletion leading to decreased association of *C. muridarum* EBs with host cells (Elwell *et al.*, 2008). Inhibition of PDGFRβ kinase activity by the small molecule inhibitor AG1295 diminished phosphorylation-dependent activation of WAVE2, Vav, and cortactin, all molecules previously found to be necessary for *Chlamydia* invasion (Elwell *et al.*, 2008). The PDGFRβ pathway may be redundant to the FGF2-FGFR pathway, where FGF2 binds to EBs and acts as an adhesion bridge to FGFR (Kim *et al.*, 2011) (Fig. 3.2).

Only PDGFRβ, but not EGFR, FGFR, and EphrinA2 receptors described above has been functionally evaluated with regards to actin recruitment and remodelling. It remains to be seen if these signalling events, while triggered by chlamydial adhesion, have post-invasion roles, including the pre-conditioning of the host cell to become a more hospitable niche (i.e. more resistant to the induction of apoptosis).

Other adhesins and receptors

There are additional adhesins and receptors that have been functionally characterized at the levels of association with host cells and/or competitive inhibition of *Chlamydia* binding (Table 3.1). These include chlamydial lipopolysaccharide (LPS), which is thought to bind to the cystic fibrosis transmembrane conductance regulator (CFTR) (Peterson *et al.*, 1998; Ajonuma *et al.*, 2010), the bacterial protein ArtJ (Soriani *et al.*, 2010), and host apolipoprotein E4 (Gérard *et al.*, 2008). The corresponding binding partners for the latter two are not known. Signalling pathways associated with these receptor–adhesin interactions, their roles in chlamydial internalization, or how they affect the immediate intracellular fates of nascent *Chlamydia*-containing vacuoles have not been elucidated.

Secreted *Chlamydia* effectors relevant to invasion

Chlamydia invasion of non-phagocytic epithelial cells is a pathogen-directed process, that requires bacterial factors in addition to the adhesins associated with the invading EB (Elwell *et al.*, 2016). Indeed, invasion is mediated by chlamydial proteins that are pre-synthesized, localized within the EB, and translocated across the host cell plasma membrane, some by the type III secretion system (T3SS) (Dai and Li, 2014) (Table 3.2). These chlamydial effectors integrate into host cell signal transduction pathways to co-opt them and promote efficient EB uptake. A crucial target of these effectors is the actin remodelling machinery. The cortical actin network of non-phagocytic cell types, such as epithelial cells, is relatively static when compared to naturally phagocytic cells, such as macrophages. To facilitate uptake, *Chlamydia* likely induces dynamic remodelling of cortical actin to form cell surface projections designed to endocytose EBs adhered to the cell surface. Here, we describe key chlamydial effectors that induces actin remodelling.

TarP

The first chlamydial type III effector implicated in invasion was the translocated actin-recruiting protein (TarP) (Clifton *et al.*, 2004). TarP is conserved in all chlamydial species for which genome sequences are available (Clifton *et al.*, 2005). The open-reading frame (ORF) is large, at > 3 kilobases (kb) in length, encoding a protein with a predicted molecular weight of approximately 110 kilodaltons (kDa), but with an apparent gel mobility of ~150 kDa. The protein can be divided into two halves, with the C-terminal half being highly conserved amongst all *Chlamydia* orthologues (Clifton *et al.*, 2005). This domain consists of a proline-rich domain (PRD) and binding domains for host proteins, including actin (ABD), focal adhesion kinase (FAK; LD – leucine-aspartate motif), and vinculin (VBD). The N-terminal half of the protein is highly variable between the human pathogen *C. trachomatis* and veterinary non-trachomatis species. There are one to twelve copies of a tyrosine phosphorylation domain (phosphodomain) in *C. trachomatis* TarP, but none in non-trachomatis species, possibly indicative of *C. trachomatis* adaptation to human hosts (Thomson *et al.*, 2008; Carlson *et al.*, 2005; Clifton *et al.*, 2005; Clifton *et al.*, 2004).

C. trachomatis TarP was identified by virtue of its extensive phosphorylation after translocation from the EB into the host cell cytosol via T3SS (Clifton *et al.*, 2004). Phosphorylation preceded actin recruitment, and ectopic expression of TarP resulted in protein aggregates that colocalized with filamentous actin. Together, the data indicated a role in invasion. Lane *et al.* identified a number of host signalling molecules that interacted with TarP in a phosphorylation-dependent manner. These included two distinct guanine nucleotide exchange factor (GEF) specific to the Rac1 GTPase (Lane *et al.*, 2008). Carabeo *et al.* previously reported that

Table 3.2 Secreted factors implicated in invasion

Factors	Species	Host target/binding partner	Function	References
TarP	*C. trachomatis;*	FAK, Vinculin, Actin, Arp2/3, Sos1, Abi1, Eps8, Vav, p85 subunit of PI3K	Signalling; actin	Lane *et al.* (2008); Thwaites *et al.* (2014); Thwaites *et al.* (2015); Zrieq *et al.* (2017)
	C. pneumoniae	ADF/cofilin	Actin nucleation; cofilin displacement from F-actin	
TmeA	*C. trachomatis*	AHNAK (*C. trachomatis*)	–	Hower *et al.*, (2009); McKuen *et al.*, (2017)
CT166	*C. trachomatis; C. caviae; C. muridarum*	Ras family of small GTPases	GTPase inactivation by glucosylation	Belland *et al.* (2001); Thalmann *et al.* (2010); Bothe *et al.* (2015)

Rac1 signals to the WAVE2 complex to activate the Arp2/3 actin–nucleating complex (Carabeo *et al.*, 2007) (Fig. 3.3), which together with the TarP–GEF interactions functionally linked TarP to the Arp2/3 complex and actin remodelling. Interestingly, this phosphodomain signalling pathway is clearly dispensable in the invasion process of other species, such as *C. muridarum*, *C. caviae*, and *C. pneumoniae* (Clifton *et al.*, 2005).

Jewett *et al.* determined that purified TarP exhibited actin-nucleating activity *in vitro* (Jewett *et al.*, 2010) (Fig. 3.3). This activity was dependent on TarP oligomerization via its proline rich domain (PRD). Clustering of multiple TarP molecules brings multiple actin binding domains (ABD) in proximity, thereby increasing the local concentration of actin to overcome the thermodynamic barrier inherent to the formation of the actin trimer nucleus, which is absolutely required for actin polymerization. The requirement for PRD could be bypassed by the presence of additional ABDs in some TarP orthologues. This activity of TarP is reminiscent of the actin nucleating mechanism by NPFs, such as formins, mDia, Spire, Cordon-bleu, etc. (Dominguez, 2016). Nuclear magnetic resonance studies confirmed the ability of purified TarP to bind actin (Tolchard *et al.*, 2018).

A signalling function has been assigned to the C-terminal half of TarP. Sequence analysis identified an LD motif and vinculin binding domain, with the latter consisting of two or three vinculin binding sites (Thwaites *et al.*, 2014; Thwaites *et al.*, 2015). A role for TarP interaction with vinculin was confirmed by Whitewood *et al.* in a series of biochemical studies (Whitewood *et al.*, 2018). Functional characterization in cultured cells revealed either FAK- or vinculin-dependent actin recruitment when clustered at the plasma membrane. The requirement for these signalling pathways was further supported by observations of decreased invasion efficiencies of *C. caviae* in either FAK- or vinculin-knockout mouse embryo fibroblasts (MEFs) (Thwaites *et al.*, 2014; Thwaites *et al.*, 2015).

The vinculin-dependent pathway of TarP-mediated actin recruitment, which occurs in *C. trachomatis* and the non-trachomatis species *C. caviae* was found to be dependent on the Arp2/3 complex (Thwaites *et al.*, 2015). *Chlamydia* invasion is susceptible to inhibition by agents that disrupt the function of the Arp2/3 complex (Carabeo *et al.*, 2007; Thwaites *et al.*, 2014). Either overexpression of or titration by the verprolin/WASP homology 2/Acidic (WA) domain of N-WASP or treatment

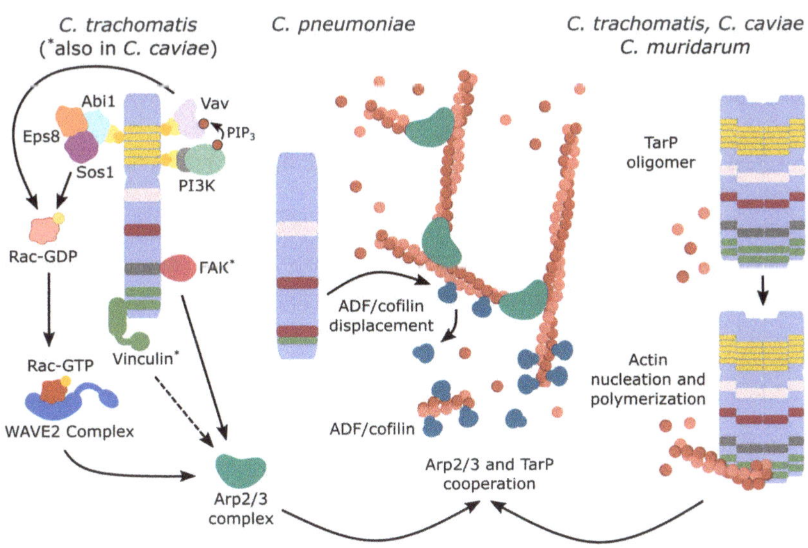

Figure 3.3 Various mechanisms of actin remodelling by TarP in invasion. TarP orthologues of *C. trachomatis* and *C. caviae* serve as a signalling scaffold to activate actin nucleating promoting factors and the Arp2/3 complex. TarP oligomerization mediates direct nucleation of linear actin polymers. Activated Arp2/3 complex binds linear actin to stimulate the polymerization of branched F-actin and a mesh-like actin network. The *C. pneumoniae* TarP orthologue displaces ADF/cofilin from polymerized F- actin to stabilize this network.

with the specific inhibitor CK-666 inhibited *C. trachomatis* invasion, supporting the signalling model, and on the surface contradicting the TarP actin nucleation model (Carabeo *et al.*, 2007; Thwaites *et al.*, 2014). However, high-resolution microscopy of the *Chlamydia* invasion site reveals a variety of cell surface projections that include elongated microvilli, pedestals, filopodia, and pockets that engulf invading EBs, hinting at different arrangements of the actin network, e.g. filamentous or branching that cooperate to produce these structures (Nans *et al.*, 2014). Indeed, Jewett *et al.* observed enhanced polymerization activity in the *in vitro* pyrene-actin assay when TarP and the Arp2/3 complex were both present in the reaction. This may indicate a level of cooperation between the two actin nucleation processes with filamentous actin produced from nucleation by TarP serving as the 'mother filament' from which the Arp2/3 complex initiates actin branching (Jewett *et al.*, 2006; Jewett *et al.*, 2010; Jiwani *et al.*, 2012) (Fig. 3.3).

As additional evidence supporting the mechanistic cooperation between the two TarP functions, Jewett *et al.* created a *C. trachomatis* L2 transformant that expressed a mutant version of TarP lacking the actin binding domain (ABD), which was predicted to disrupt actin polymerization when it oligomerizes with endogenous wild type TarP (Parrett *et al.*, 2016). Deletion of the ABD reduced invasion efficiency indicating that oligomers formed by the combination of wild type and ABD-deleted mutant TarP were functionally defective. The same study also demonstrated that a phosphodomain-deleted TarP had a similar effect on invasion efficiency, indicating that proper TarP function required both the N-terminal phosphodomain and C-terminal actin-binding domain.

C. pneumoniae possesses a TarP homologue (CPn0572). CPn0572, like other TarP homologues, can be divided into two halves, with the C-terminal half showing a higher degree of conservation with the other homologues, especially with their actin-binding domains, which are critical for the actin nucleating function of TarP. Indeed, Jewett *et al.* reported that CPn0572 nucleated actin in *in vitro* assays, albeit at a lower efficiency than other homologues (Jewett *et al.*, 2010) (Fig. 3.3). Two recent reports by Zrieq *et al.* and Braun *et al.* described a different actin-related function to CPn0572 (Zrieq *et al.*, 2017; Braun *et al.*, 2019). Both observed that expression of CPn0572 in yeast and mammalian cells was associated with F-actin patches and thickened actin cables, respectively. This change in F-actin morphology is related to CPn0572-mediated displacement of the actin depolymerizing factor cofilin. With cofilin displaced, the actin network was stabilized (Fig. 3.3). This activity was mediated by a domain of 59 amino acids that overlaps with the previously annotated actin-binding domain. Because this domain is conserved between chlamydial species, it would not be unexpected if other TarP homologues have an identical cofilin-displacing function to stabilize F-actin. Is cofilin displacement by CPn0572 relevant for invasion? The answer is not known, but an RNA interference screen conducted by Elwell *et al.* in Drosophila S2 cells identified a critical role for the ADF/cofilin homologue CFL1 in a *C. muridarum* infection (Elwell *et al.*, 2008). However, this screen did not distinguish between invasion and post-invasion requirements. CPn0572/TarP-mediated displacement of cofilin seems at odds with its requirement (Zrieq *et al.*, 2017), but it could signify the need for balanced actin polymerization and disassembly during invasion. Clearly, a more detailed investigation of TarP function during invasion is needed. With the advent of novel molecular genetic approaches in *Chlamydia* research, the full characterization of TarP is within the realm of possibility (Mueller *et al.*, 2016).

TmeA/CT694

TmeA is an EB-associated effector that is translocated into the host cell cytosol with the help of the T3S chaperone Slc1 (Chen *et al.*, 2014). It is encoded in an operon along with TmeB. Initial characterization showed its interaction with the host actin-bundling protein AHNAK, and when overexpressed in cells resulted in noticeable changes in stress fibres (Chen *et al.*, 2014) (Table 3.2). This activity of TmeA required its interaction with AHNAK. Furthermore, AHNAK localized to sites of EB adhesion. Together with its effect on actin structures, TmeA may have a role in invasion. In a genetic study using fluorescence-reported allelic exchange mutagenesis, McKuen *et al.* disrupted the *tmeA* open-reading frame and found it to be necessary for efficient invasion of HeLa cells, but in an AHNAK-independent manner (McKuen *et al.*, 2017). Indeed, AHNAK had no obvious role in invasion of cultured cells, with wildtype *C. trachomatis* still being capable of efficiently invading AHNAK-deficient mouse embryo fibroblasts (McKuen *et al.*, 2017). Importantly, the *tmeA* mutant exhibit reduced infectivity, with significant reduction in the yield of shed infectious particles in a mouse model of intra-vaginal infection (McKuen *et al.*, 2017). It remains to be seen if the invasion defect observed *in vitro* accounts for the reduced infectivity *in vivo*. The recruitment of AHNAK remains an intriguing observation with a potential post-invasion role. Perhaps AHNAK function in *Chlamydia* infection could be more prominent in an *in vivo* or three-dimensional cell culture context. *C. muridarum* and *C. suis* possess a homologue of TmeA, which could mean a similar mode of entry (Mojica *et al.*, 2015; Hower *et al.*, 2009).

CT166

The cytoskeleton-related role of CT166 was first reported by Belland *et al.* (Belland *et al.*, 2001). CT166 has homology to the TcdB and LifA/EfaA-related large cytotoxin (Bothe *et al.*, 2015). A putative glucosylation function has been assigned to CT166, which targets members of the Ras family of GTPases, based on the presence of a glucosylating motif (Asp/X/Asp – DXD). CT166 homologues of different *chlamydial species* are highly polymorphic. *C. trachomatis* serovar L2 lacks the entire gene, while the servar D homologue has a large deletion that is predicted to result in a truncated protein.). This truncated protein retains the amino acid residues (Asp/X/Asp – DXD) required for its glucosylating activity. The *C. muridarum* orthologue is predicted to encode a full-length toxin (Belland *et al.*, 2001. Inoculation of HeLa cells with *C. muridarum* or *C. trachomatis* serovar D at a high multiplicity of infection (MOI = 100) resulted in host cell cytotoxicity, which was in contrast to the lack of any visible effects in cells inoculated with *C. trachomatis* serovar L2. The species-dependent effects on the host cell cytoskeleton correlated with the presence of the DXD motif. Importantly, cytotoxicity was not inhibited by chloramphenicol, a bacterial protein synthesis inhibitor, indicative of an EB-associated preformed toxin.

CT166 is able to modify and inactivate the Ras GTPase, which is closely related to Rac and Cdc42, and mutating the DXD motif abrogated Ras inhibition (Thalmann *et al.*, 2010). The role of CT166 in *Chlamydia* invasion is unclear, with both DXD-containing and -deficient strains being able to efficiently invade non-phagocytic cells. It was speculated that CT166 inhibited Rac1 to shut down signalling to the actin remodelling machinery and to induce net disassembly of the F-actin network at the site of invasion to complete chlamydial uptake (Table 3.2). The dramatic morphological change observed by Belland *et al.* could be explained by the Rac-inhibiting function of CT166 (Belland *et al.*, 2001; Bothe *et al.*, 2015; Thalmann *et al.*, 2010), and may be related to the transient actin recruitment observed during invasion of host cells by *C. caviae*, another species harbouring a potentially active CT166 homologue (Thwaites *et al.*, 2015). It is tempting to speculate that CT166 acts to shut down signalling during invasion by toxin-harbouring strains. With Rac and/or Cdc42 being central to chlamydial invasion, it would not be surprising that CT166-targeting of these host proteins might be consequential to invasion. Because of the ability of CT166 to disrupt small GTPase function, the timing of deployment of this toxin is critical. Because Rac and/or Cdc42 are involved in invasion (Carabeo *et al.*, 2007; Lane *et al.*, 2008), premature activation of CT166 would effectively inhibit the small GTPases and thus, invasion (Thalmann *et al.*, 2010). Is there a toxin-independent mechanism of actin disassembly? Cofilin, also known as actin depolymerizing factor (ADF) could mediate actin disassembly. As discussed above, ADF/cofilin was found to be required for infection (Elwell *et al.*, 2008). While the invasion-specific function of cofilin was not addressed in this study, the basis for this apparent requirement could be related to the disassembly of the F-actin network that forms underneath the invading EBs.

Many questions about CT166 remain unanswered, including the basis for its association with EBs, how it modulates intracellular signalling molecules, and the regulation of the timing of toxin deployment during invasion. Is the toxin part of the heavily disulfide-linked outer membrane complex? Does its cytotoxic activity require toxin internalization; and if so, what is the mechanism of uptake? Finally, the role of the CT166 toxin in pathogenesis and host–pathogen interaction is ripe for future studies.

Concluding remarks and future research directions

The numerous chlamydial adhesins and receptors identified so far are indicative of promiscuous binding to host cells. However, a more detailed functional characterization must be conducted to strengthen the relevance of these adhesion–receptor pairs to adhesion and invasion. For example, invasion studies should be extended to investigations of the immediate trafficking of nascent *Chlamydia*-containing vacuoles away from the endolysosomal pathway. This is based on the known seamless transition at the molecular level of clathrin-dependent endocytosis and subsequent early endosome interaction of the newly formed vesicle. It has been reported that markers of early endosomes, such as Rab5, are recruited to clathrin-coated pits during their assembly (Vitale *et al.*, 1995; Galperin and Sorkin, 2003; Semerdjieva *et al.*, 2008). The relatively efficient uptake of *Chlamydia* by non-phagocytic cells might also indicate a similar seamless transition between uptake into the host cells and its immediate intracellular fate. Therefore, this aspect must be considered when evaluating candidate receptors and adhesion–receptor interactions. Does elimination of the receptor of interest or inhibiting its signalling activity affect the immediate intracellular fate of *Chlamydia*? This can be answered via live-cell imaging in conjunction with markers that indicate appropriate signalling and/or intracellular transport. In the past, successful invasion was indicated by the presence of inclusion. However, inclusion development is far removed from the process of bacterial uptake. When inhibited, a number of biological processes unrelated to invasion can prevent inclusion development, and thus be linked erroneously to the process of invasion.

Another aspect that needs to be considered when studying invasion and adhesin–receptor interactions is the possible post-invasion role of a specific signalling pathway. To determine the functional relevance of signalling triggered by chlamydial adhesion to the host cell surface is not straightforward. Instead of initiating pathogen uptake, some signalling might relate to the preconditioning of the host cell to support chlamydial growth. Induction of host cell resistance to apoptosis is an example (Kim *et al.*, 2011; Mehlitz *et al.*, 2010; Patel *et al.*, 2014). One possible way to distinguish invasion-related and -unrelated signalling is to focus on the induction of actin cytoskeleton remodelling, which is the convergence point of most, if not all invasion-related signalling. However, there will likely be exceptions, i.e. protein with functions related to both actin recruitment and preconditioning of the host cell. An example is TarP, which has been implicated in both invasion-related actin recruitment (Lane *et al.*, 2008) and apoptosis resistance, with the latter involving its interaction with the signalling protein Shc. Shc regulates cell survival genes in *Chlamydia*-infected cells. (Mehlitz *et al.*, 2010). And as mentioned above, one must also ask if a signalling pathway of interest leads to the proper intracellular trafficking of newly endocytosed *Chlamydia*.

The use of appropriate cell types should aid in identifying *bona fide* adhesin–receptor interaction. Whenever possible, adhesion and invasion should be evaluated in polarized epithelial cells, which are the primary target of *Chlamydia in vivo*. There are host cell receptors that are restricted in their localization to either the apical or basolateral surface, and as such have implications on the mode of infection. It has been reported that *C. trachomatis* serovar L2 preferentially infected from the basolateral surface, which may involve a different adhesin–receptor interaction and signalling pathways, while serovar E preferred adhesion to the apical surface (Davis and Wyrick, 1997; Schramm and Wyrick, 1995). The utilization of three-dimensional tissue culture models, such as epithelial raft cultures, organoids, and *ex vivo* organ culture is gaining momentum, which should further elucidate and offer clarity to the complex and varied means of EB binding and invasion. These alternative culture systems add a level of complexity that might distinguish *bona fide* adhesion and invasion events from artefacts associated with less sophisticated experimental models, e.g. non-epithelial cells, lack of apico-basolateral polarization, *etc.* It would not be surprising if particular receptor–adhesin combinations may be more relevant in the context of a three-dimensional tissue culture model than others.

There are also indications of signalling crosstalk. There are host-derived molecular participants that are common to signal transduction pathways utilized by host cell receptors and targeted by chlamydial effectors. To highlight one example, PDGFRβ signalling during invasion activates a number of signalling proteins that also interact with TarP, warranting further investigation (Elwell *et al.*, 2008). Does TarP recruit host signalling molecules to the plasma membrane to facilitate efficient activation by PDGFRβ? Or does PDGFRβ signalling activate these proteins to enhance their interaction either alone or in a multimeric complex with TarP? Another possibility is that PDGFRβ and TarP independently activate these proteins, implying molecular redundancy of the invasion process. Thus, multiple entry pathways may cooperate to induce uptake. Therefore, studying a single signalling pathway likely provides a very limited picture of the invasion process.

What is known so far is that there is a dynamic interplay between *Chlamydia* and the host cell via adhesion–receptor interactions to induce specific signalling pathways and targeting of host cell signalling via translocated type III effectors (Hybiske and Stephens, 2007; Mehlitz and Rudel, 2013; Subbarayal *et al.*, 2015). There are some species-specific differences in the mechanism, but the common outcome is the induction of remodelling of the cell surface to form various types of structures, such as hypertrophic microvilli, macropinocytic structures, and filopodia that are designed for engulfment of invading EBs by non-phagocytic epithelial cells. Decades of studies have identified some of the proteins involved, including the elucidation of their respective functions. We now have some insight as to the nature of the signalling pathways they induce. Together, years of investigation have laid the foundation for future studies to refine further models of early interactions between *Chlamydia* and its host epithelial cell. The advent of molecular genetic approaches should also facilitate structure–function analyses of adhesin–receptor interaction in the context of infection.

To summarize, significant progress has been made in elucidating in detail the mechanism of *Chlamydia* adhesion and invasion. Future studies on the invasion process will need to be tackled more comprehensively. Questions on signalling crosstalk and roles in determining intracellular fates, and the use of experimental models that better mimic *in vivo* conditions need to be employed. Functional evaluations of candidate adhesins and receptors should be investigated beyond demonstrations of interaction and binding competition. Their effects on the endolysosomal interactions (or lack thereof) of nascent inclusions should be evaluated. Kinetic studies may reveal more efficient invasion when multiple pathways cooperate; and possibilities of adhesin-receptor signalling having roles outside of invasion should not be discounted. A comprehensive approach to study invasion should generate novel insights into the biology, pathogenesis, and host interactions of *Chlamydia*.

References

Abromaitis, S., and Stephens, R.S. (2009). Attachment and Entry of *Chlamydia* Have Distinct Requirements for Host Protein Disulfide Isomerase. PLOS Pathogens 5, doi: e1000357.

Ajonuma, C.L., Fok, L.K., Ho, S.L., Chan, S.K.P., Chow, P.H., Tsang, L.L., Wong, Y.H.C., Chen, J., Li, S., Rowlands, K.D., *et al.* (2010). CFTR is required for cellular entry and internalization of *Chlamydia trachomatis.* Cell Biology International 34, 593–600. doi: 10.1042/CBI20090227.

Al-Zeer, M.A., Al-Younes, H.M., Kerr, M., Abu-Lubad, M., Gonzalez, E., Brinkmann, V., and Meyer, T.F. (2014). *Chlamydia trachomatis* remodels stable microtubules to coordinate Golgi stack recruitment to the chlamydial inclusion surface. Molecular Microbiology 94, 1285–1297. doi: 10.1111/mmi.12829.

Balañá, M.E., Niedergang, F., Subtil, A., Alcover, A., Chavrier, P., and Dautry-Varsat, A. (2005). ARF6 GTPase controls bacterial invasion by actin remodelling. Journal of Cell Science 118, 2201–2210. doi: 10.1242/jcs.02351.

Becker, E., and Hegemann, J.H. (2014). All subtypes of the Pmp adhesin family are implicated in chlamydial virulence and show species-specific function. MicrobiologyOpen 3, 544–556. doi: 10.1002/mbo3.

Belland, R.J., Scidmore, M.A., Crane, D.D., Hogan, D.M., Whitmire, W., McClarty, G., and Caldwell, H.D. (2001). *Chlamydia trachomatis* cytotoxicity associated with complete and partial cytotoxin genes. Proceedings of the National Academy of Sciences of the United States of America 98, 13984–13989.

Bellis, S.L. (2004). Variant glycosylation: an underappreciated regulatory mechanism for β1 integrins. Biochimica et Biophysica Acta (BBA) – Biomembranes 1663, 52–60. doi: 10.1016/S0005-2736(04)00095-1.

Bergmann, B., Raffelsbauer, D., Kuhn, M., Goetz, M., Hom, S., and Goebel, W. (2002). InlA- but not InlB-mediated internalization of *Listeria monocytogenes* by non-phagocytic mammalian cells needs the support of other internalins. Molecular Microbiology 43, 557–570.

Betts-Hampikian, H.J., and Fields, K.A. (2011). Disulfide Bonding within Components of the *Chlamydia* Type III Secretion Apparatus Correlates with Development. Journal of Bacteriology 193, 6950–6959. doi: 10.1128/JB.05163-11.

Bierne, H., Gouin, E., Roux, P., Caroni, P., Yin, H.L., and Cossart, P. (2001). A role for cofilin and LIM kinase in *Listeria*-induced phagocytosis. Journal of Cell Biology 155, 101–112.

Bierne, H., Miki, H., Innocenti, M., Scita, G., Gertler, F.B., Takenawa, T., and Cossart, P. (2005). WASP-related proteins, Abi1 and Ena/VASP are required for *Listeria* invasion induced by the Met receptor. Journal of Cell Science 118, 1537–1547. doi: 10.1242/jcs.02285.

Birkelund, S., Johnsen, H., and Christiansen, G. (1994). *Chlamydia trachomatis* serovar L2 induces protein tyrosine phosphorylation during uptake by HeLa cells. Infection and Immunity 62, 4900–4908.

Boleti, H., Benmerah, A., Ojcius, D.M., Cerf-Bensussan, N., and Dautry-Varsat, A. (1999). *Chlamydia* infection of epithelial cells expressing dynamin and Eps15 mutants: clathrin-independent entry into cells and dynamin-dependent productive growth. Journal of Cell Science 112, 1487.

Bonazzi, M., and Cossart, P. (2006). Bacterial entry into cells: A role for the endocytic machinery. FEBS Letters 580, 2962–2967. doi: 10.1016/j.febslet.2006.04.010.

Bothe, M., Dutow, P., Pich, A., Genth, H., and Klos, A. (2015). DXD Motif-Dependent and -Independent Effects of the *Chlamydia trachomatis* Cytotoxin CT166. Toxins 7, 621–637. doi: 10.3390/toxins7020621.

Bouwman, L.I., Niewold, P., and van Putten, J.P.M. (2013). Basolateral Invasion and Trafficking of Campylobacter jejuni in Polarized Epithelial Cells. PLoS ONE 8, e54759. doi: 10.1371/journal.pone.0054759.

Braun, C., Alcázar-Román, A.R., Laska, A., Mölleken, K., Fleig, U., and Hegemann, J.H. (2019). CPn0572, the *C. pneumoniae* orthologue of TarP, reorganizes the actin cytoskeleton via a newly identified F-actin binding domain and recruitment of vinculin. PLoS One 14(1):e0210403. doi: 10.1371/journal.pone.0210403.

Brunelle, B.W., Nicholson, T.L., and Stephens, R.S. (2004). Microarray-based genomic surveying of gene polymorphisms in *Chlamydia trachomatis*. Genome Biology 5, R42. doi: 10.1186/gb-2004-5-6-r42.

Byrne, G.I. (1978). Kinetics of phagocytosis of *Chlamydia* psittaci by mouse fibroblasts (L cells): separation of the attachment and ingestion stages. Infection and Immunity 19, 607–612.

Campbell, L.A., and Kuo, C.C. (2006). Interactions of *Chlamydia* with the host cells that mediate attachment and uptake. In: Bavoil, P.M., and Wyrick, P.B., editors. *Chlamydia*: Genomics and Pathogenesis. Wymondham, UK: Horizon Bioscience. pp. 505–522.

Campellone, K.G., Rankin, S., Pawson, T., Kirschner, M.W., Tipper, D.J., and Leong, J.M. (2004). Clustering of Nck by a 12-residue Tir phosphopeptide is sufficient to trigger localized actin assembly. Journal of Cell Biology 164, 407–416.

Carabeo, R. (2011). Bacterial subversion of host actin dynamics at the plasma membrane. Cellular Microbiology 13, 1460–1469. doi: 10.1111/j.1462-5822.2011.01651.x.

Carabeo, R.A., Dooley, C.A., Grieshaber, S.S., and Hackstadt, T. (2007). Rac interacts with Abi-1 and WAVE2 to promote an Arp2/3-dependent actin recruitment during chlamydial invasion. Cellular Microbiology 9, 2278–2288. doi: 10.1111/j.1462-5822.2007.00958.x.

Carabeo, R.A., Grieshaber, S.S., Fischer, E., and Hackstadt, T. (2002). *Chlamydia trachomatis* Induces Remodeling of the Actin Cytoskeleton during Attachment and Entry into HeLa Cells. Infection and Immunity 70, 3793–3803. doi: 10.1128/IAI.70.7.3793-3803.2002.

Carabeo, R.A., Grieshaber, S.S., Hasenkrug, A., Dooley, C., and Hackstadt, T. (2004). Requirement for the Rac GTPase in *Chlamydia trachomatis* Invasion of Non-phagocytic Cells. Traffic 5, 418–425. doi: 10.1111/j.1398-9219.2004.00184.x.

Carabeo, R.A., and Hackstadt, T. (2001). Isolation and characterization of a mutant Chinese hamster ovary cell line that is resistant to *Chlamydia* trachomatis infection at a novel step in the attachment process. Infection and Immunity 69, 5899–5904.

Carlson, J.H., Hughes, S., Hogan, D., Cieplak, G., Sturdevant, D.E., McClarty, G., Caldwell, H.D., and Belland, R.J. (2004). Polymorphisms in the *Chlamydia trachomatis* Cytotoxin Locus Associated with Ocular and Genital Isolates. Infection and Immunity 72, 7063–7072. doi: 10.1128/IAI.72.12.7063-7072.2004.

Carlson, J.H., Porcella, S.F., McClarty, G., and Caldwell, H.D. (2005). Comparative Genomic Analysis of *Chlamydia trachomatis* Oculotropic and Genitotropic Strains. Infection and Immunity 73, 6407–6418. doi: 10.1128/IAI.73.10.6407-6418.2005.

Carrasco, J.A., Tan, C., Rank, R.G., Hsia, R.C., and Bavoil, P.M. (2011). Altered developmental expression of polymorphic membrane proteins in penicillin-stressed *Chlamydia trachomatis*. Cell Microbiol 13, 1014–1025. doi: 10.1111/j.1462-5822.2011.01598.x.

Chen, Y.-S., Bastidas, R.J., Saka, H.A., Carpenter, V.K., Richards, K.L., Plano, G.V., and Valdivia, R.H. (2014). The *Chlamydia trachomatis* Type III Secretion Chaperone Slc1 Engages Multiple Early Effectors, Including TepP, a Tyrosine-phosphorylated Protein Required for the Recruitment of CrkI-II to Nascent Inclusions and Innate Immune Signaling. PLoS Pathogens 10, e1003954. https://doi.org/10.1371/journal.ppat.1003954.

Chhabra, E.S., and Higgs, H.N. (2007). The many faces of actin: matching assembly factors with cellular structures. Nature Cell Biology 9, 1110. doi: 10.1038/ncb1007-1110.

Clifton, D.R., Dooley, C.A., Grieshaber, S.S., Carabeo, R.A., Fields, K.A., and Hackstadt, T. (2005). Tyrosine Phosphorylation of the Chlamydial Effector Protein Tarp Is Species Specific and Not Required for Recruitment of Actin. Infection and Immunity 73, 3860–3868. doi: 10.1128/IAI.73.7.3860-3868.2005.

Clifton, D.R., Fields, K.A., Grieshaber, S.S., Dooley, C.A., Fischer, E.R., Mead, D.J., Carabeo, R.A., and Hackstadt, T. (2004). A chlamydial type III translocated protein is tyrosine-phosphorylated at the site of entry and associated with recruitment of actin. Proceedings of the National Academy of Sciences of the United States of America 101, 10166–10171. doi: 10.1073/pnas.0402829101.

Conant, C.G., and Stephens, R.S. (2007). *Chlamydia* attachment to mammalian cells requires protein disulfide isomerase. Cellular Microbiology 9, 222–232. doi: 10.1111/j.1462-5822.2006.00783.x.

Coombes, B.K., and Mahony, J.B. (2002). Identification of MEK- and phosphoinositide 3-kinase-dependent signalling as essential events during *Chlamydia pneumoniae* invasion of HEp2 cells. Cellular Microbiology 4, 447–460. https://doi.org/10.1046/j.1462-5822.2002.00203.x.

Crane, D.D., Carlson, J.H., Fischer, E.R., Bavoil, P., Hsia, R.C., Tan, C., Kuo, C.C., and Caldwell, H.D. (2006). *Chlamydia trachomatis* polymorphic membrane protein

D is a species-common pan-neutralizing antigen. Proc Natl Acad Sci U S A 103, 1894–1899. https://doi.org/10.1073/pnas.0508983103.

Dahms, N.M., Lobel, P., and Kornfeld, S. (1989). Mannose 6-phosphate receptors and lysosomal enzyme targeting. Journal of Biological Chemistry 264, 12115–12118.

Dai, W., and Li, Z. (2014). Conserved type III secretion system exerts important roles in *Chlamydia trachomatis*. International Journal of Clinical and Experimental Pathology 7, 5404–5414.

Davis, C.H., Raulston, J.E., and Wyrick, P.B. (2002). Protein Disulfide Isomerase, a Component of the Estrogen Receptor Complex, Is Associated with *Chlamydia trachomatis* Serovar E Attached to Human Endometrial Epithelial Cells. Infection and Immunity 70, 3413–3418. doi: 10.1128/IAI.70.7.3413-3418.2002.

Davis, C.H., and Wyrick, P.B. (1997). Differences in the association of *Chlamydia trachomatis* serovar E and serovar L2 with epithelial cells *in vitro* may reflect biological differences *in vivo*. Infection and Immunity 65, 2914–2924.

Deng, W., Marshall, N.C., Rowland, J.L., McCoy, J.M., Worrall, L.J., Santos, A.S., Strynadka, N.C.J., and Finlay, B.B. (2017). Assembly, structure, function and regulation of type III secretion systems. Nature Reviews Microbiology 15, 323. doi: 10.1038/nrmicro.2017.20.

Derivery, E., and Gautreau, A. (2010). Generation of branched actin networks: assembly and regulation of the N-WASP and WAVE molecular machines. BioEssays 32, 119–131.

Dominguez, R. (2016). The WH2 Domain and Actin Nucleation: Necessary but Insufficient. Trends in Biochemical Sciences 41, 478–490. doi: 10.1016/j.tibs.2016.03.004.

Dufrêne, Y.F. (2015). Sticky microbes: forces in microbial cell adhesion. Trends in Microbiology 23, 376–382. doi: 10.1002/bies.200900123.

Dumoux, M., Menny, A., Delacour, D., and Hayward, R.D. (2015). A *Chlamydia* effector recruits CEP170 to reprogram host microtubule organization. Journal of Cell Science 128, 3420–3434. doi: 10.1242/jcs.169318.

Elwell, C., Mirrashidi, K., and Engel, J. (2016). *Chlamydia* cell biology and pathogenesis. Nature reviews Microbiology 14, 385–400. doi: 10.1038/nrmicro.2016.30.

Elwell, C.A., Ceesay, A., Kim, J.H., Kalman, D., and Engel, J.N. (2008). RNA Interference Screen Identifies Abl Kinase and PDGFR Signaling in *Chlamydia trachomatis* Entry. PLOS Pathogens 4, e1000021.

Fadel, S., and Eley, A. (2007). *Chlamydia trachomatis* OmcB protein is a surface-exposed glycosaminoglycan-dependent adhesin. Journal of Medical Microbiology 56, 15–22. doi: 10.1099/jmm.0.46801-0.

Fadel, S., and Eley, A. (2008). Is lipopolysaccharide a factor in infectivity of *Chlamydia trachomatis*? Journal of Medical Microbiology 57, 261–266. doi:10.1099/jmm.0.47237-0.

Fawaz, F.S., van Ooij, C., Homola, E., Mutka, S.C., and Engel, J.N. (1997). Infection with *Chlamydia trachomatis* alters the tyrosine phosphorylation and/or localization of several host cell proteins including cortactin. Infection and Immunity 65, 5301–5308.

Fechtner, T., Galle, J.N., and Hegemann, J.H. (2016). The novel chlamydial adhesin *CPn0473* mediates the lipid raft-dependent uptake of *Chlamydia pneumoniae*. Cellular Microbiology 18, 1094–1105. doi: 10.1111/cmi.12569.

Fechtner, T., Stallmann, S., Moelleken, K., Meyer, K.L., and Hegemann, J.H. (2013). Characterization of the Interaction between the Chlamydial Adhesin OmcB and the Human Host Cell. Journal of Bacteriology 195, 5323–5333. doi: 10.1128/JB.00780-13.

Ferguson, S.M., De Camilli, P. (2012). Dynamin, a membrane-remodelling GTPase. Nature Reviews Molecular Cell Biology 13, 75–88. doi: 10.1038/nrm3266.

Fields, K.A., and Hackstadt, T. (2002). The chlamydial inclusion: escape from the endocytic pathway. Annual Reviews in Cell and Developmental Biology 18, 221–245.

Firat-Karalar, E.N., and Welch, M.D. (2011). New mechanisms and functions of actin nucleation. Current opinion in cell biology 23, 4–13. doi: 10.1016/j.ceb.2010.10.007.

Fletcher, D.A., and Mullins, R.D. (2010). Cell mechanics and the cytoskeleton. Nature 463, 485–492. doi: 10.1038/nature08908.

Ford, C., Nans, A., Boucrot, E., and Hayward, R.D. (2018). *Chlamydia* exploits filopodial capture and a macropinocytosis-like pathway for host cell entry. PLOS Pathogens 14. doi: 10.1371/journal.ppat.1007051.

Fredlund, J., and Enninga, J. (2014). Cytoplasmic access by intracellular bacterial pathogens. Trends in Microbiology 22, 128–137. doi: 10.1016/j.tim.2014.01.003.

Frisk, A., Ison, C.A., and Lagergård, T. (1998). GroEL Heat Shock Protein of Haemophilus ducreyi: Association with Cell Surface and Capacity To Bind to Eukaryotic Cells. Infection and Immunity 66, 1252–1257.

Fukuoka, M., Suetsugu, S., Miki, H., Fukami, K., Endo, T., and Takenawa, T. (2001). A novel neural Wiskott-Aldrich syndrome protein (N-WASP) binding protein, WISH, induces Arp2/3 complex activation independent of Cdc42. Journal of Cell Biology 152, 471–482.

Gabel, B.R., Elwell, C., van Ijzendoorn, S.C.D., and Engel, J.N. (2004). Lipid Raft-Mediated Entry Is Not Required for *Chlamydia trachomatis* Infection of Cultured Epithelial Cells. Infection and Immunity 72, 7367–7373. doi: 10.1128/IAI.72.12.7367-7373.2004.

Galle, J.N., Fechtner, T., Eierhoff, T., Römer, W., and Hegemann, J.H. (2019). A *Chlamydia pneumoniae* adhesin induces phosphatidylserine exposure on host cells. Nature Communications 10, 4644. doi.org/10.1038/s41467-019-12419-8.

Galperin, E., and Sorkin, A. (2003). Visualization of Rab5 activity in living cells by FRET microscopy and influence of plasma-membrane-targeted Rab5 on clathrin-dependent endocytosis. Journal of Cell Science 116, 4799–4810.

Gardel, M.L., Schneider, I.C., Aratyn-Schaus, Y., and Waterman, C.M. (2010). Mechanical Integration of Actin and Adhesion Dynamics in Cell Migration. Annual review of cell and developmental biology 26, 315–333. doi: 10.1146/annurev.cellbio.011209.

Gaydos, C.A., Summersgill, J.T., Sahney, N.N., Ramirez, J.A., and Quinn, T.C. (1996). Replication of *Chlamydia pneumoniae in vitro* in human macrophages, endothelial cells, and aortic artery smooth muscle cells. Infection and Immunity 64, 1614–1620.

Gérard, H.C., Fomicheva, E., Whittum-Hudson, J.A., and Hudson, A.P. (2008). Apolipoprotein E4 enhances attachment of Chlamydophila (*Chlamydia*) pneumoniae EBs to host cells. Microbial Pathogenesis 44, 279–285. doi: 10.1016/j.micpath.2007.10.002.

Goth, S.R., and Stephens, R.S. (2001). Rapid, transient phosphatidylserine externalization induced in host cells by infection with *Chlamydia* spp. Infection and Immunity 69, 1109–1119. doi: 10.1128/IAI.69.2.1109-1119.2001.

Grassart, A., Cheng, A.T., Hong, S.H., Zhang, F., Zenzer, N., Feng, Y., Briner, D.M., Davis, G.D., Malkov, D., and Drubin, D.G. (2014). Actin and dynamin2 dynamics and interplay during clathrin-mediated endocytosis. Journal of Cell Biology 205, 721–735. doi: 10.1083/jcb.201403041.

Green, E.R., and Mecsas, J. (2016). Bacterial Secretion Systems – An overview. Microbiology spectrum 4, doi: 10.1128/microbiolspec.VMBF-0012-2015.

Grieshaber, S.S., Grieshaber, N.A., and Hackstadt, T. (2003). *Chlamydia trachomatis* uses host cell dynein to traffic to the microtubule-organizing center in a p50 dynamitin-independent process. Journal of Cell Science 116, 3793–3802. doi: 10.1242/jcs.00695.

Grimwood, J., and Stephens, S.R. (1999). Computational Analysis of the Polymorphic Membrane Protein Superfamily of *Chlamydia trachomatis* and *Chlamydia pneumoniae*, Vol 4. doi: 10.1089/omi.1.1999.4.187.

Hatch, T.P., Vance, D.W. Jr, Al-Hossainy, E. (1981). Attachment of *Chlamydia* psittaci to formaldehyde-fixed and unfixed L cells. Journal of General Microbiology 125, 273–283.

Henderson, I.R., and Lam, A.C. (2001). Polymorphic proteins of *Chlamydia* spp.--autotransporters beyond the Proteobacteria. Trends Microbiol 9, 573–578. doi: 10.1016/S0966-842X(01)02234-X.

Hu, V.H., Holland, M.J., and Burton, M.J. (2013). Trachoma: protective and pathogenic ocular immune responses to *Chlamydia* trachomatis. PLoS Neglected Tropical Diseases 7(2):e2020. doi: 10.1371/journal.pntd.0002020.

Humphreys, D., Davidson, A.C., Hume, P.J., Makin, L.E., and Koronakis, V. (2013). Arf6 coordinates actin assembly through the WAVE complex, a mechanism usurped by *Salmonella* to invade host cells. Proceedings of the National Academy of Sciences 110, 16880–16885. doi: 10.1073/pnas.1311680110.

Hybiske, K., and Stephens, R.S. (2007). Mechanisms of *Chlamydia trachomatis* Entry into Nonphagocytic Cells. Infection and Immunity 75, 3925–3934. doi: 10.1128/IAI.00106-07.

Ibarra, N., Pollitt, A., and Insall, R.H. (2005). Regulation of actin assembly by SCAR/WAVE proteins. Biochemical Society Transactions 33, 1243–1246.

Jewett, T.J., Fischer, E.R., Mead, D.J., and Hackstadt, T. (2006). Chlamydial TARP is a bacterial nucleator of actin. Proceedings of the National Academy of Sciences of the United States of America 103, 15599–15604. doi: 10.1073/pnas.0603044103.

Jewett, T.J., Miller, N.J., Dooley, C.A., and Hackstadt, T. (2010). The Conserved Tarp Actin Binding Domain Is Important for Chlamydial Invasion. PLoS Pathogens 6, e1000997.

Jiwani, S., Ohr, R.J., Fischer, E.R., Hackstadt, T., Alvarado, S., Romero, A., and Jewett, T.J. (2012). *Chlamydia trachomatis* Tarp cooperates with the Arp2/3 complex to increase the rate of actin polymerization., *Chlamydia trachomatis* Tarp cooperates with the Arp2/3 complex to increase the rate of actin polymerization. Biochem Biophys Res Commun 420, 420, 816,–816–821. doi: 10.1016/j.bbrc.2012.03.080.

Johnson, C.M., and Fisher, D.J. (2014). Site-Specific, Insertional Inactivation of incA in *Chlamydia trachomatis* Using a Group II Intron. PLOS ONE 8. doi: 10.1371/journal. pone.0083989.

Kari, L., Southern, T.R., Downey, C.J., Watkins, H.S., Randall, L.B., Taylor, L.D., Sturdevant, G.L., Whitmire, W.M., and Caldwell, H.D. (2014). *Chlamydia trachomatis* polymorphic membrane protein D is a virulence factor involved in early host–cell interactions. Infection and Immunity 82, 2756–2762. doi: 10.1128/IAI.01686-14.

Kim, J.H., Jiang, S., Elwell, C.A., and Engel, J.N. (2011). *Chlamydia trachomatis* Co-opts the FGF2 Signaling Pathway to Enhance Infection. PLoS Pathogens 7. doi: 10.1371/journal.ppat.1002285.

Kiselev, A.O., Skinner, M.C., and Lampe, M.F. (2009). Analysis of pmpD expression and PmpD post-translational processing during the life cycle of *Chlamydia trachomatis* serovars A, D, and L2. PLoS One 4. doi: 10.1371/journal.pone.0005191.

Korhonen, J.T., Puolakkainen, M., Haveri, A., Tammiruusu, A., Sarvas, M., and Lahesmaa, R. (2012). *Chlamydia* pneumoniae entry into epithelial cells by clathrin-independent endocytosis. Microbial Pathogenesis 52, 57–64. doi:10.1016/j.micpath.2011.12.002.

Knudsen, K., Madsen, A.S., Mygind, P., Christiansen, G., and Birkelund, S. (1999). Identification of Two Novel Genes Encoding 97- to 99-Kilodalton Outer Membrane Proteins of *Chlamydia pneumoniae*. Infection and Immunity 67, 375–383.

Kuo, C.C., and Grayston, T. (1976). Interaction of *Chlamydia trachomatis* organisms and HeLa 229 cells. Infection and Immunity 13, 1103–1109.

Kuo, C.C., Lee, A., and Campbell, L.A. (2004). Cleavage of the N-linked oligosaccharide from the surfaces of *Chlamydia* species affects attachment and infectivity of the organisms in human epithelial and endothelial cells. Infection and Immunity 72, 6699–6701.

Lane, B.J., Mutchler, C., Al Khodor, S., Grieshaber, S.S., and Carabeo, R.A. (2008). Chlamydial Entry Involves TARP Binding of Guanine Nucleotide Exchange Factors. PLOS Pathogens 4. https://doi.org/10.1371/journal.ppat.1000014.

Long, K.H., Gomez, F.J., Morris, R.E., and Newman, S.L. (2003). Identification of Heat Shock Protein 60 as the Ligand on *Histoplasma capsulatum* That Mediates Binding to CD18 Receptors on Human Macrophages. The Journal of Immunology 170, 487. doi: https://doi.org/10.4049/jimmunol.170.1.487.

Longbottom, D., Russell, M., Dunbar, S.M., Jones, G.E., and Herring, A.J. (1998). Molecular Cloning and Characterization of the Genes Coding for the Highly Immunogenic Cluster of 90-Kilodalton Envelope Proteins from the *Chlamydia* psittaci Subtype That

Causes Abortion in Sheep. Infection and Immunity 66, 1317–1324.

Longbottom, D., Russell, M., Jones, G.E., Lainson, F.A., and Herring, A.J. (1996). Identification of a multigene family coding for the 90 kDa proteins of the ovine abortion subtype of *Chlamydia* psittaci. FEMS Microbiology Letters 142, 277–281.

Luczak, S.E.T., Smits, S.H.J., Decker, C., Nagel-Steger, L., Schmitt, L., and Hegemann, J.H. (2016). The *Chlamydia pneumoniae* Adhesin Pmp21 Forms Oligomers with Adhesive Properties. Journal of Biological Chemistry 291, 22806–22818. doi: 10.1074/jbc.M116.728915.

Lujan, A.L., Croci, D.O., Gambarte Tudela, J.A., Losinno, A.D., Cagnoni, A.J., Mariño, K.V., Damiani, M.T., and Rabinovich, G.A. (2018). Glycosylation-dependent galectin–receptor interactions promote *Chlamydia trachomatis* infection. Proceedings of the National Academy of Sciences. doi: 10.1073/pnas.1802188115.

Macia, E., Ehrlich, M., Massol, R., Boucrot, E., Brunner, C., and Kirchhausen, T. (2006). Dynasore, a cell-permeable inhibitor of dynamin. Developmental Cell 10, 839–850. doi: 10.1016/j.devcel.2006.04.002.

McKuen, M.J., Mueller, K.E., Bae, Y.S., and Fields, K.A. (2017). Fluorescence-Reported Allelic Exchange Mutagenesis Reveals a Role for *Chlamydia trachomatis* TmeA in Invasion That Is Independent of Host AHNAK. Infection and Immunity 85. doi: 10.1128/IAI.00640-17.

Mehlitz, A., Banhart, S., Mäurer, A.P., Kaushansky, A., Gordus, A.G., Zielecki, J., MacBeath, G., and Meyer, T.F. (2010). Tarp regulates early *Chlamydia*-induced host cell survival through interactions with the human adaptor protein SHC1. The Journal of Cell Biology 190, 143–157.: 10.1083/jcb.200909095.

Mehlitz, A., and Rudel, T. (2013). Modulation of host signalling and cellular responses by *Chlamydia*. Cell Communication and Signaling: CCS 11, 90–90. doi: 10.1186/1478-811X-11-90.

Miki, H., Miura, K., and Takenawa, T. (1996). N-WASP, a novel actin-depolymerizing protein, regulates the cortical cytoskeletal rearrangement in a PIP2-dependent manner downstream of tyrosine kinases. The EMBO Journal 15, 5326–5335.

Miki, H., Yamaguchi, H., Suetsugu, S., and Takenawa, T. (2000). IRSp53 is an essential intermediate between Rac and WAVE in the regulation of membrane ruffling. Nature 408, 732–735.

Mital, J., Lutter, E.I., Barger, A.C., Dooley, C.A., and Hackstadt, T. (2015). *Chlamydia trachomatis* inclusion membrane protein CT850 interacts with the dynein light chain DYNLT1 (Tctex1). Biochemical and biophysical research communications 462, 165–170. doi: 10.1016/j.bbrc.2015.04.116.

Moelleken, K., and Hegemann, J.H. (2008). The *Chlamydia* outer membrane protein OmcB is required for adhesion and exhibits biovar-specific differences in glycosaminoglycan binding. Molecular Microbiology 67, 403–419. doi: 10.1111/j.1365-2958.2007.06050.x.

Moelleken, K., Schmidt, E., and Hegemann, J.H. (2010). Members of the Pmp protein family of *Chlamydia pneumoniae* mediate adhesion to human cells via short repetitive peptide motifs. Mol Microbiol 78, 1004–1017. doi: 10.1111/j.1365-2958.2010.07386.x.

Molleken, K., Becker, E., and Hegemann, J.H. (2013). The *Chlamydia pneumoniae* invasin protein Pmp21 recruits the EGF receptor for host cell entry. PLoS Pathog 9, e1003325. doi: 10.1371/journal.ppat.1003325.

Montigiani, S., Falugi, F., Scarselli, M., Finco, O., PetraCca, R., Galli, G., Mariani, M., Manetti, R., Agnusdei, M., Cevenini, R., *et al.* (2002). Genomic Approach for Analysis of Surface Proteins in *Chlamydia pneumoniae*. Infection and Immunity 70, 368–379.

Moulder, J.W. (1991). Interaction of *Chlamydiae* and host cells *in vitro*. Microbiological Reviews 55, 143–190.

Mueller, K.E., Wolf, K., and Fields, K.A. (2016). Gene Deletion by Fluorescence-Reported Allelic Exchange Mutagenesis in *Chlamydia trachomatis*. mBio 7. doi: 10.1128/mBio.01817-15.

Müller, T., Becker, E., Stallmann, S., Waldhuber, A., Römmler-Dreher, F., Albrecht, S., Mohr, F., Hegemann, J.H., and Miethke, T. (2017). Vaccination with the polymorphic membrane protein A reduces *Chlamydia muridarum* induced genital tract pathology. Vaccine 35, 2801–2810. doi: 10.1016/j.vaccine.2017.04.017.

Nans, A., Ford, C., and Hayward, R.D. (2015). Host-pathogen reorganisation during host cell entry by *Chlamydia trachomatis*. Microbes and Infection/Institut Pasteur 17, 727–731. doi: 10.1016/j.micinf.2015.08.004.

Nans, A., Saibil, H.R., and Hayward, R.D. (2014). Pathogen–host reorganization during *Chlamydia* invasion revealed by cryo-electron tomography. Cellular Microbiology 16, 1457–1472. doi: 10.1111/cmi.12310.

Nogueira, A.T., Pedrosa, A.T., and Carabeo, R.A. (2018). Manipulation of the Host Cell Cytoskeleton by *Chlamydia*. In Biology of *Chlamydia*, G. Häcker, ed. (Cham: Springer International Publishing), pp. 59–80. doi: 10.1007/82_2016_10.

Pal, S., Tifrea, D.F., Zhong, G., and de la Maza, L.M. (2017). Transcervical inoculation with *Chlamydia trachomatis* induces infertility in HLA-DR4-transgenic and wild type mice. Infection and Immunity. doi: 10.1128/IAI.00722-17.

Parrett, C.J., Lenoci, R.V., Nguyen, B., Russell, L., and Jewett, T.J. (2016). Targeted Disruption of *Chlamydia trachomatis* Invasion by in Trans Expression of Dominant Negative Tarp Effectors. Frontiers in Cellular and Infection Microbiology 6, 84. doi: 10.3389/fcimb.2016.00084.

Patel, A.L., Chen, X., Wood, S.T., Stuart, E.S., Arcaro, K.F., Molina, D.P., Petrovic, S., Furdui, C.M., and Tsang, A.W. (2014). Activation of epidermal growth factor receptor is required for *Chlamydia trachomatis* development. BMC Microbiology 14, 277. doi: 10.1186/s12866-014–0277–4.

Pereira, S.F.F., Henriques, A.O., Pinho, M.G., de Lencastre, H., and Tomasz, A. (2009). Evidence for a dual role of PBP1 in the cell division and cell separation of Staphylococcus aureus. Molecular microbiology 72, 1423–1437. doi: 10.1111/j.1365-2958.2009.06687.x.

Peterson, E.M., de la Maza, L.M., Brade, L., and Brade, H. (1998). Characterization of a Neutralizing Monoclonal Antibody Directed at the Lipopolysaccharide of *Chlamydia pneumoniae*. Infection and Immunity 66, 3848–3855.

Phadnis, S.H., Parlow, M.H., Levy, M., Ilver, D., Caulkins, C.M., Connors, J.B., Dunn, B.E. (1996). Surface localization of *Helicobacter pylori* urease and a heat shock protein homolog requires bacterial autolysis. Infection and Immunity 64, 905–912.

Pizarro-Cerdá, J., Charbit, A., Enninga, J., Lafont, F., and Cossart, P. (2016). Manipulation of host membranes by the bacterial pathogens *Listeria*, Francisella, Shigella and Yersinia. Seminars in Cell and Developmental Biology 60, 155–167. doi: 10.1016/j.semcdb.2016.07.019.

Preta, G., Cronin, J.G., and Sheldon, I.M. (2015). Dynasore – not just a dynamin inhibitor. Cell Communication and Signalling 13, 24. doi: 10.1186/s12964-015-0102-1.

Prydz, K. (2015). Determinants of Glycosaminoglycan (GAG) Structure. Biomolecules 5, 2003–2022. doi: 10.3390/biom5032003.

Puolakkainen, M., Campbell, L.A., Lin, T.-M., Richards, T., Patton, D.L., and Kuo, C.-C. (2003). Cell-to-Cell Contact of Human Monocytes with Infected Arterial Smooth-Muscle Cells Enhances Growth of *Chlamydia pneumoniae*. The Journal of Infectious Diseases 187, 435–440. doi: 10.1086/368267.

Puolakkainen, M., Kuo, C.-C., and Campbell, L.A. (2005). *Chlamydia pneumoniae* Uses the Mannose 6-Phosphate/Insulin-Like Growth Factor 2 Receptor for Infection of Endothelial Cells. Infection and Immunity 73, 4620–4625. doi: 10.1128/IAI.73.8.4620-4625.2005.

Radhakrishnan, G.K., and Splitter, G.A. (2012). Modulation of host microtubule dynamics by pathogenic bacteria. Biomolecular concepts 3, 571–580. doi: 10.1515/bmc-2012-0030.

Reed, S.C.O., Serio, A.W., and Welch, M.D. (2012). Rickettsia parkeri invasion of diverse host cells involves an Arp2/3 complex, WAVE complex and Rho-family GTPase-dependent pathway. Cellular Microbiology 14, 529–545. doi: 10.1111/j.1462-5822.2011.01739.x.

Reynders, E., Foulquier, F., Annaert, W., and Matthijs, G. (2011). How Golgi glycosylation meets and needs trafficking: the case of the COG complex. Glycobiology. 21, 853–863. doi: 10.1093/glycob/cwq179.

Rockey, D.D., Lenart, J., and Stephens, R.S. (2000). Genome sequencing and our understanding of *Chlamydiae*. Infect Immun 68, 5473–5479.

Rohatgi, R., Ma, L., Miki, H., Lopez, M., Kirchhausen, T., Takenawa,T., and Kirschner, M.W. (1999). The interaction between N-WASP and the Arp2/3 complex links Cdc42-dependent signals to actin assembly. Cell 97, 221–231.

Roulis, E., Polkinghorne, A., and Timms, P. (2013). *Chlamydia* pneumoniae: modern insights into an ancient pathogen. Trends in Microbiology 21, 120–128. doi: 10.1016/j.tim.2012.10.009.

Rudolph, M.G., Weise, C., Mirold, S., Hillenbrand, B., Bader, B., Wittinghofer, A., and Hardt, W.D. (1999). Biochemical analysis of SopE from Salmonella typhimurium, a highly efficient guanosine nucleotide exchange factor for RhoGTPases. The Journal of Biological Chemistry 274, 30501–30509.

Rupp, J., Droemann, D., Goldmann, T., Zabel, P., Solbach, W., Vollmer, E., Branscheid, D., Dalhoff, K., and Maass, M. (2004). Alveolar epithelial cells type II are major target cells for *C. pneumoniae* in chronic but not in acute respiratory infection. FEMS Immunology & Medical Microbiology 41, 197–203. doi: 10.1016/j.femsim.2004.03.004.

Sarrazin, S., Lamanna, W.C., and Esko, J.D. (2011). Heparan Sulfate Proteoglycans. Cold Spring Harbor Perspectives in Biology 3. doi: 10.1101/cshperspect.a004952.

Schoumacher, M., Goldman, R.D., Louvard, D., and Vignjevic, D.M. (2010). Actin, microtubules, and vimentin intermediate filaments cooperate for elongation of invadopodia. The Journal of Cell Biology 189, 541–556. doi: 10.1083/jcb.200909113.

Schramm, N., and Wyrick, P.B. (1995). Cytoskeletal requirements in *Chlamydia trachomatis* infection of host cells. Infection and Immunity 63, 324–332.

Semerdjieva, S., Shortt, B., Maxwell, E., Singh, S., Fonarev, P., Hansen, J., Schiavo, G., Grant, B.D., and Smythe, E. (2008). Coordinated regulation of AP2 uncoating from clathrin-coated vesicles by rab5 and hRME-6. Journal of Cell Biology 183, 499–511. doi: 10.1083/jcb.200806016.

Sever, S., Damke, H., and Schmid, S.L. (2000). Dynamin:GTP controls the formation of constricted coated pits, the rate limiting step in clathrin-mediated endocytosis. Journal of Cell Biology 150, 1137–1148. doi: 10.1083/jcb.150.5.1137.

Shi, J., Scita, G., and Casanova, J.E. (2005). WAVE2 Signaling Mediates Invasion of Polarized Epithelial Cells by Salmonella typhimurium. Journal of Biological Chemistry 280, 29849–29855. doi: 10.1074/jbc.M500617200.

Soriani, M., Petit, P., Grifantini, R., Petra*Cca*, R., Gancitano, G., Frigimelica, E., Nardelli, F., Garcia, C., Spinelli, S., Scarabelli, G., *et al.* (2010). Exploiting Antigenic Diversity for Vaccine Design: the *Chlamydia* ArtJ paradigm. The Journal of Biological Chemistry 285, 30126–30138. doi: 10.1074/jbc.M110.11851.

Stallmann, S., and Hegemann, J.H. (2016). The *Chlamydia trachomatis* Ctad1 invasin exploits the human integrin β1 receptor for host cell entry. Cellular Microbiology 18, 761–775. doi: 10.1111/cmi.12549.

Stanhope, R., Flora, E., Bayne, C., and Derré, I. (2017). IncV, a FFAT motif-containing Chlamydia protein, tethers the endoplasmic reticulum to the pathogen-containing vacuole. Proceedings of the National Academy of Sciences U S A. 114, 12039–12044. doi: 10.1073/pnas.1709060114.

Stender, S., Friebel, A., Linder, S., Rohde, M., Mirold, S., and Hardt, W.D. (2000). Identification of SopE2 from Salmonella typhimurium, a conserved guanine nucleotide exchange factor for Cdc42 of the host cell. Molecular Microbiology 36, 1206–1221.

Stephens, R.S., Kalman, S., Lammel, C., Fan, J., Marathe, R., Aravind, L., Mitchell, W., Olinger, L., Tatusov, R.L., Zhao, Q., Koonin, E.V., and Davis, R.W. (1998). Genome sequence of an obligate intracellular pathogen of humans: *Chlamydia trachomatis*. Science. 282, 754–759.

Stephens, R.S., Koshiyama, K., Lewis, E., and Kubo, A. (2001). Heparin-binding outer membrane protein of *Chlamydiae*. Molecular Microbiology 40, 691–699. https://doi.org/10.1046/j.1365-2958.2001.02418.x.

Stuart, E.S., Webley, W.C., and Norkin, L.C. (2003). Lipid rafts, caveolae, caveolin-1, and entry by *Chlamydiae* into host cells. Experimental Cell Research 287, 67–78. https://doi.org/10.1016/S0014-4827(03)00059-4.

Su, H., Watkins, N.G., Zhang, Y.X., and Caldwell, H.D. (1990). *Chlamydia* trachomatis-host cell interactions: role of the chlamydial major outer membrane protein as an adhesin. Infection and Immunity 58, 1017–1025.

Su, H., Raymond, L., Rockey, D.D., Fischer, E., Hackstadt, T., and Caldwell, H.D. (1996). A recombinant *Chlamydia trachomatis* major outer membrane protein binds to heparan sulfate receptors on epithelial cells. Proceedings of the National Academy of Sciences of the United States of America 93, 11143–11148.

Subbarayal, P., Karunakaran, K., Winkler, A.-C., Rother, M., Gonzalez, E., Meyer, T.F., and Rudel, T. (2015). EphrinA2 Receptor (EphA2) Is an Invasion and Intracellular Signalling Receptor for *Chlamydia trachomatis*. PLOS Pathogens 11. doi: 10.1371/journal.ppat.1004846.

Subtil, A., Wyplosz, B., Balañá, M.E., and Dautry-Varsat, A. (2004). Analysis of *Chlamydia caviae* entry sites and involvement of Cdc42 and Rac activity. Journal of Cell Science 117, 3923–3933. doi: 10.1242/jcs.01247.

Sugahara, K., and Kitagawa, H. (2002). Heparin and heparan sulfate biosynthesis. IUBMB Life 54, 163–175.

Swanson, A.F., and Kuo, C.C. (1991). Evidence that the major outer membrane protein of *Chlamydia trachomatis* is glycosylated. Infection and Immunity 59, 2120–2125.

Swanson, K.A., Crane, D.D., and Caldwell, H.D. (2007). *Chlamydia trachomatis* Species-Specific Induction of Ezrin Tyrosine Phosphorylation Functions in Pathogen Entry. Infection and Immunity 75, 5669–5677. doi: 10.1128/IAI.01096-07.

Swanson, K.A., Taylor, L.D., Frank, S.D., Sturdevant, G.L., Fischer, E.R., Carlson, J.H., Whitmire, W.M., and Caldwell, H.D. (2009). *Chlamydia trachomatis* polymorphic membrane protein D is an oligomeric autotransporter with a higher-order structure. Infect Immun 77, 508–516. doi: 10.1128/IAI.01173-08.

Tan, C., Hsia, R.C., Shou, H., Carrasco, J.A., Rank, R.G., and Bavoil, P.M. (2010). Variable expression of surface-exposed polymorphic membrane proteins in *in vitro*-grown *Chlamydia trachomatis*. Cell Microbiol 12, 174–187. doi: 10.1111/j.1462-5822.2009.01389.x.

Tan, C., Hsia, R.C., Shou, H., Haggerty, C.L., Ness, R.B., Gaydos, C.A., Dean, D., Scurlock, A.M., Wilson, D.P., and Bavoil, P.M. (2009). *Chlamydia trachomatis*-infected patients display variable antibody profiles against the nine-member polymorphic membrane protein family. Infect Immun 77, 3218–3226. doi: 10.1128/IAI.01566-08.

Taraktchoglou, M., Pacey, A.A., Turnbull, J.E., and Eley, A. (2001). Infectivity of *Chlamydia trachomatis*Serovar LGV but Not E Is Dependent on Host Cell Heparan Sulfate. Infection and Immunity 69, 968–976. doi: 10.1128/IAI.69.2.968-976.2001.

Thalmann, J., Janik, K., May, M., Sommer, K., Ebeling, J., Hofmann, F., Genth, H., and Klos, A. (2010). Actin Re-Organization Induced by *Chlamydia trachomatis* Serovar D – Evidence for a Critical Role of the Effector Protein CT166 Targeting Rac. PLOS ONE 5, e9887. doi: 10.1371/journal.pone.0009887.

Thomson, N.R., Holden, M.T., Carder, C., Lennard, N., Lockey, S.J., Marsh, P., Skipp, P., O'Connor, C.D., Goodhead, I., Norbertzcak, H., Harris, B., Ormond, D., Rance, R., Quail, M.A., Parkhill, J., Stephens, R.S., and Clarke, I.N. (2008). *Chlamydia trachomatis*: genome sequence analysis of lymphogranuloma venereum isolates. Genome Research 18, 161–171.

Thwaites, T., Nogueira, A.T., Campeotto, I., Silva, A.P., Grieshaber, S.S., and Carabeo, R.A. (2014). The *Chlamydia* Effector TarP Mimics the Mammalian Leucine-Aspartic Acid Motif of Paxillin to Subvert the Focal Adhesion Kinase during Invasion. The Journal of Biological Chemistry 289, 30426–30442. doi: 10.1074/jbc.M114.604876.

Thwaites, T.R., Pedrosa, A.T., Peacock, T.P., and Carabeo, R.A. (2015). Vinculin Interacts with the *Chlamydia* Effector TarP Via a Tripartite Vinculin Binding Domain to Mediate Actin Recruitment and Assembly at the Plasma Membrane. Frontiers in Cellular and Infection Microbiology 5, 88. doi: 10.3389/fcimb.2015.00088.

Tiwari, V., Maus, E., Sigar, I.M., Ramsey, K.H., and Shukla, D. (2012). Role of heparan sulfate in sexually transmitted infections. Glycobiology 22, 1402–1412. doi: 10.1093/glycob/cws106.

Tolchard, J., Walpole, S.J., Miles, A.J., Maytum, R., Eaglen, L.A., Hackstadt, T., Wallace, B.A., and Blumenschein, T.M.A. (2018). The intrinsically disordered Tarp protein from *Chlamydia* binds actin with a partially preformed helix. Scientific Reports 8, 1960. doi: 10.1038/s41598-018-20290-8.

Truong, D., Brabant, D., Bashkurov, M., Wan, L.C., Braun, V., Heo, W.D., Meyer, T., Pelletier, L., Copeland, J., and Brumell, J.H. (2013). Formin-mediated actin polymerization promotes Salmonella invasion. Cellular Microbiology 15, 2051–2063. doi: 10.1111/cmi.12173.

Van Lent, S., De Vos, W.H., Huot Creasy, H., Marques, P.X., Ravel, J., Vanrompay, D., Bavoil, P., and Hsia, R.-c. (2016). Analysis of Polymorphic Membrane Protein Expression in Cultured Cells Identifies PmpA and PmpH of *Chlamydia* psittaci as Candidate Factors in Pathogenesis and Immunity to Infection. PLOS ONE 11. doi: 10.1371/journal.pone.0162392.

Vandahl, B.B., Pedersen, A.S., Gevaert, K., Holm, A., Vandekerckhove, J., Christiansen, G., and Birkelund, S. (2002). The expression, processing and localization of polymorphic membrane proteins in *Chlamydia pneumoniae* strain CWL029. BMC Microbiol 2, 36. doi: https://doi.org/10.1186/1471-2180-2-36.

Vasilevsky, S., Stojanov, M., Greub, G., and Baud, D. (2016). Chlamydialpolymorphicmembraneproteins: regulation, function and potential vaccine candidates. Virulence 7, 11–22. doi: 10.1080/21505594.2015.1111509.

Vassilieva, E.V., Gerner-Smidt, K., Ivanov, A.I., and Nusrat, A. (2008). Lipid rafts mediate internalization of β(1)-integrin in migrating intestinal epithelial cells. American Journal of Physiology – Gastrointestinal and Liver Physiology 295, G965-G976. doi: 10.1152/ajpgi.00082.2008.

Vitale, G., Alexandrov, K., Ullrich, O., Horiuchi, H., Giner, A., Dobson, C.,. Baykova, O., Gournier, H., Stenmark, H., and Zerial, M. (1995). The GDP/GTP cycle of Rab5 in the regulation of endocytotic membrane traffic. Cold Spring Harbor Symposium on Quantitative Biologly 60, 211–220.

Ward, M.E., and Murray, A. (1984). Control mechanisms governing the infectivity of *Chlamydia trachomatis* for HeLa cells: mechanisms of endocytosis. J Gen Microbiol 130, 1765–1780. doi: 10.1099/00221287-130-7-1765.

Wehrl, W., Brinkmann, V., Jungblut, P.R., Meyer, T.F., and Szczepek, A.J. (2004). From the inside out--processing of the Chlamydial autotransporter PmpD and its role in bacterial adhesion and activation of human host cells. Mol Microbiol 51, 319–334. doi: 10.1046/j.1365-2958.2003.03838.x.

Wesolowski, J., Weber, M.M., Nawrotek, A., Dooley, C.A., Calderon, M., St. Croix, C.M., Hackstadt, T., Cherfils, J., and Paumet, F. (2017). *Chlamydia* Hijacks ARF GTPases To Coordinate Microtubule Posttranslational Modifications and Golgi Complex Positioning. mBio 8. doi: 10.1128/mBio.02280-16.

Wheelhouse, N., Sait, M., Wilson, K., Aitchison, K., McLean, K., Smith, D.G., and Longbottom, D. (2012). Expression patterns of five polymorphic membrane proteins during the *Chlamydia* abortus developmental cycle. Vet Microbiol 160, 525–529. doi: 10.1016/j.vetmic.2012.06.017.

Whitewood, A.J., Singh, A.K., Brown, D.G., and Goult, B.T. (2018). Chlamydial virulence factor TarP mimics talin to disrupt the talin-vinculin complex. Febs Letters 592, 1751–1760. doi: 10.1002/1873-3468.13074.

Wilkinson, B., and Gilbert, H.F. (2004). Protein disulfide isomerase. Biochimica et Biophysica Acta. 1699:35–44.

Windoffer, R., Kölsch, A., Wöll, S., and Leube, R.E. (2006). Focal adhesions are hotspots for keratin filament precursor formation. The Journal of Cell Biology 173, 341–348. doi: 10.1083/jcb.200511124.

Wuppermann, F.N., Mölleken, K., Julien, M., Jantos, C.A., and Hegemann, J.H. (2008). *Chlamydia pneumoniae* GroEL1 Protein Is Cell Surface Associated and Required for Infection of HEp-2 Cells. Journal of Bacteriology 190, 3757–3767. doi: 10.1128/JB.01638-07.

Wyrick, P.B., Choong, J., Davis, C.H., Knight, S.T., Royal, M.O., Maslow, A.S., and Bagnell, C.R. (1989). Entry of genital *Chlamydia trachomatis* into polarized human epithelial cells. Infection and Immunity 57, 2378–2389.

Yamaguchi, H., Osaki, T., Kurihara, N., Taguchi, H., Hanawa, T., Yamamoto, T., and Kamiya, S. (1997). Heat-shock protein 60 homologue of Helicobacter pylori is associated with adhesion of H. pylori to human gastric epithelial cells. Journal of Medical Microbiology 46, 825–831. doi: 10.1099/00222615-46-10-825.

York, S.J., Arneson, L.S., Gregory, W.T., Dahms, N.M., and Kornfeld, S. (1999). The Rate of Internalization of the Mannose 6-Phosphate/Insulin-like Growth Factor II Receptor Is Enhanced by Multivalent Ligand Binding. Journal of Biological Chemistry 274, 1164–1171.

Yu, X., Zech, T., McDonald, L., Gonzalez, E.G., Li, A., Macpherson, I., Schwarz, J.P., Spence, H., Futó, K., Timpson, P., Nixon, C., Ma, Y., Anton, I.M., Visegrády, B., Insall, R.H., Oien, K., Blyth, K., Norman, J.C., and Machesky, L.M. (2012). N-WASP coordinates the delivery and F-actin-mediated capture of MT1-MMP at invasive pseudopods. Journal of Cell Biology 199, 527–544. doi: 10.1083/jcb.201203025.

Zhang, J.P., and Stephens, R.S. (1992). Mechanism of *C. trachomatis* attachment to eukaryotic host cells. Cell 69, 861–869.

Zhong G. (2018). *Chlamydia* Spreading from the Genital Tract to the Gastrointestinal Tract – A Two-Hit Hypothesis. Trends in Microbiology 26, 611–623. doi: 10.1016/j.tim.2017.12.002.

Zrieq, R., Braun, C., and Hegemann, J.H. (2017). The *Chlamydia pneumoniae* Tarp Ortholog *Cpn*0572 Stabilizes Host F-Actin by Displacement of Cofilin. Frontiers in Cellular and Infection Microbiology 7, 511. doi: 10.3389/fcimb.2017.00511.

The Chlamydial Inclusion

4

Kevin Hybiske[1]* and Dagmar Heuer[2]

[1]Division of Allergy and Infectious Diseases, University of Washington, Seattle, WA, USA.
[2]Unit of Sexually Transmitted Bacterial Pathogens, Robert Koch Institute, Berlin, Germany.

*Correspondence: khybiske@uw.edu

https://doi.org/10.21775/9781912530281.04

Abstract

A distinctive feature of *Chlamydia species* is that all of these bacterial species reside and replicate within an intracellular, host membrane-derived vacuole called the inclusion. Extensive modification of the inclusion membrane and lumen by *Chlamydia* is necessary to make it a hospitable home for the bacteria, which require nutrients and energy from the host cell. Given its central role in *Chlamydia* infection, the inclusion has been intensively studied for many decades. This chapter will summarize many of the important concepts that have been learned about the establishment and maintenance of the inclusion by *Chlamydia*. These include the mechanistic functions played by inclusion membrane proteins at the inclusion–host interface, and how *Chlamydia* scavenge and recruit critical nutrients and proteins from the host cell into the inclusion.

Introduction

All *Chlamydia* species reside within a modified vacuole, called the chlamydial inclusion, as their host intracellular niche for replication and developmental growth. The establishment and maintenance of the inclusion is an absolute requirement for *Chlamydia* infection – these bacteria are unable to replicate in either the host cytosol or the extracellular milieu. The inclusion provides a favourable intracellular environment for replication and development that is sequestered from lysosomal fusion, facilitates the acquisition of nutrients from the host cell, and protects against intracellular innate immune surveillance through the compartmentalization of bacterial molecules inside the infected host cell.

The *Chlamydia* inclusion is extensively modified by the bacteria in order to meet their changing needs over the 24–72 hour developmental cycle. *C. trachomatis* encodes ~51 predicted inclusion membrane proteins (Incs) – 7% of the *C. trachomatis* genome – that insert into the inclusion membrane and directly modulate functions of host cell factors (Table 4.1). The Inc protein family functions as the primary platform for orchestrating the recruitment of host factors and signalling events essential for the establishment and maintenance of the chlamydial intracellular niche (Elwell *et al.*, 2016; Mueller *et al.*, 2014).

Recent research advances have illuminated some of the underlying mechanisms governing critical functions mediated by the inclusion. This chapter will discuss the current state of information regarding *Chlamydia* inclusion biology and will provide a perspective on what major discoveries lie ahead. First, the general architecture of the inclusion will be discussed, including the important properties of its membrane and interior lumen. Second, we will discuss the molecular nature of Incs and the important roles played by these proteins, which are the primary type III effectors secreted by *Chlamydia*. Next, major early events that occur after the formation of the inclusion will be summarized, notably evasion from lysosomal trafficking and migration of the inclusion to the cell centre. We will look at how important nutrients necessary for chlamydial replication such as amino acids, ATP, lipids and cholesterol are delivered to the inclusion. Late

Table 4.1 *C. trachomatis* Incs and their known functions

Locus	Gene	Size (kDa)	Proposed function(s)	References
Conserved Incs				
CT005	incV	39.6	Endoplasmic reticulum contact sites	Stanhope *et al*. (2017)
CT006		20.2		
CT058		40.1		
CT134		15.1		
CT179		18.8		
CT195		40.9		
CT232	incB	12.2		
CT233	incC	18.4	Inclusion fusion; inclusion stability	Weber *et al*. (2017)
CT288		63.0		
CT324		33.6		
CT383		26.7	Inclusion fusion; inclusion stability	Weber *et al*. (2017)
CT440		12.6		
CT449		12.0		
CT483		12.9		
CT484		38.6		
CT565		16.0		
CT616		50.0		
CT618		27.9	Associates with lipid droplets	Saka *et al*. (2015)
CT642		32.2		
CT850		45.8	Interacts with dynein	Mital *et al*. (2015)
Non-conserved Incs				
CT036		45.3		
CT079		17.1		
CT101	mrcA	17.7	Inclusion stability, extrusion	Nguyen *et al*. (2018)
CT115	incD	14.9	Endoplasmic reticulum contact sites	Agaisse and Derré (2014); Derré *et al*. (2011)
CT116	incE	13.6	SNX5/6 recruitment; interference with retromer function	Aeberhard *et al*. (2015); Mirrashidi *et al*. (2015)
CT117	incF	10.4		
CT118	incG	17.4	Interacts with 14–3–3β	Scidmore and Hackstadt (2001)
CT119	incA	30.3	SNARE interactions; homotypic fusion of inclusions	Delevoye *et al*. (2004, 2008)
CT135		38.5		
CT147		162.1		
CT164		9.1		
CT192		28.0		
CT196		11.2		
CT214		59.8		
CT222		13.9		
CT223	ipam	29.4	Centrosome, microtubule assembly	Dumoux *et al*. (2015)
CT224		16.0		
CT225		13.3		
CT226		18.9		

Table 4.1 Continued

Locus	Gene	Size (kDa)	Proposed function(s)	References
CT227		14.1		
CT228		20.6	Actomyosin; MYPT1; extrusion	Lutter *et al*. (2013)
CT229	cpoS	23.4	Early endosomes; Rab4; inclusion and host cell stability	Rzomp *et al*. (2006); Weber *et al*. (2017)
CT249		12.2		
CT300		12.4		
CT345		13.1		
CT357		11.3		
CT358		18.9		
CT442	crpA	16.0		
CT529	cap1	31.0	Associates with lipid droplets	Saka *et al*. (2015)
CT789		10.0		
CT813	inaC	29.6	Interacts with ARF GTPases, 14–3–3	Kokes *et al*. (2015); Wesolowski *et al*. (2017)

events that occur within *Chlamydia* inclusions will be discussed, such as the asynchronous conversion of replicative RBs to infectious EBs, and the unique strategies employed by *Chlamydia* to exit host cells to begin their infectious cycles anew. Finally, we will present the emergence of modern experimental approaches for investigating *Chlamydia*–host interactions at the inclusion membrane, and what themes have developed from these large-scale studies.

Inclusion architecture and composition

Properties of the inclusion

The entirety of *Chlamydia*'s intracellular growth and development takes place within the membrane-bound inclusion (Fig. 4.1). Within the initial hours after a *Chlamydia* elementary body (EB) is internalized into an epithelial cell, the nascent inclusion is sequestered from late endosomal and lysosomal trafficking, and begins an approximately two day-long journey, during which it provides a suitable intracellular niche for *Chlamydia* reticulate body (RB) replication and developmental growth. Throughout the intracellular chlamydial developmental cycle, the inclusion harbours a compositional identity that is distinct from host organelles and vacuoles formed by other intracellular pathogens. The *Chlamydia* inclusion is therefore unique from those formed by other intracellular pathogens, many of which establish intracellular niches that are derived from or within pre-existing host compartments, such as the endoplasmic reticulum-based compartment of *Legionella pneumophila*, the late endosomal vacuole of *Coxiella burnetii*, and host *Listeria monocytogenes* and *Shigella flexneri*. Even though the inclusion does not acquire known markers from host organelles, recent findings have demonstrated that the inclusion is not completely 'walled off' from vesicular traffic within the host cell. Rather, numerous vesicular and non-vesicular communication routes with organelles are established and maintained throughout the chlamydial developmental cycle.

Inclusions across the *Chlamydia* genus show surprising diversity in shape and structure. *C. trachomatis* strains belonging to the LGV biovar grow within spheroid, distended inclusions (Moulder, 1991). *C. trachomatis* urogenital strains (D-K) inhabit ovoid inclusions that appear to exert less outward pressure on the host cell (Moulder, 1991). In contrast to these, *C. psittaci* inclusions are lobed and contoured, appearing to exert less outward physical pressure on the host cell and instead adopt a flexible or pliable ultrastructure (Moulder, 1991). The molecular basis for this spectrum in inclusion morphological diversity is not at all known; however, it is hypothesized that strain or species-specific Inc proteins may be responsible, or alternatively biophysical properties such as osmolarity.

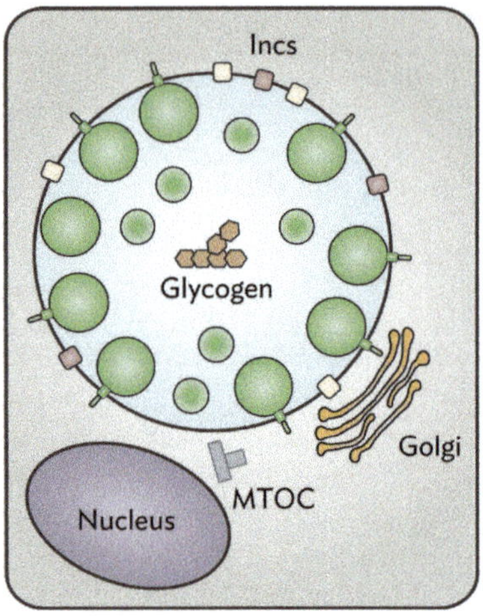

Figure 4.1 The *Chlamydia* inclusion. The inclusion sits in a perinuclear location adjacent to the microtubule organizing centre (MTOC) and in close association with Golgi mini-stacks. Most *Chlamydia* reticulate bodies (RBs) maintain close contact with the inclusion membrane and insert type III secretion needles across the inclusion membrane and into the host cytosol (Nans *et al*., 2015). *Chlamydia* elementary bodies (EBs) and glycogen deposits are additional features of the inclusion for some *Chlamydia* species. Inclusion membrane proteins (Incs) are important effectors secreted by type III secretion and inserted into the inclusion membrane.

The inclusion membrane is primarily made up of host-derived phospholipids, cholesterol, sphingolipids, host proteins, and chlamydial transmembrane proteins (Incs). Although the biogenesis of the inclusion membrane is not well understood, it is likely that the initial phospholipids are supplied from internalization of the host plasma membrane. Pathways for the acquisition of additional lipids required for inclusion expansion and chlamydial growth will be discussed later in this chapter. Physiologically, the inside of the *C. trachomatis* inclusion is in equilibrium with the ionic concentrations of the host cytosol. Experiments using membrane permeable, ion sensitive fluorescent dyes revealed that the inclusion membrane is freely permeable for H^+, Na^+, K^+, and Ca^{2+}, suggesting that there are no ionic gradients across this membrane (Grieshaber *et al.*, 2002). Additional experiments using fluorescent dyes and tracer molecules demonstrated that the *C. trachomatis* inclusion is permeable to molecules < 520 Da (Heinzen and Hackstadt, 1997; Kleba and Stephens, 2008). The protein composition of the inclusion lumen remains poorly defined compared to proteins on the inclusion membrane and soluble small molecules inside the inclusion. Light and transmission electron microscopy analysis of *Chlamydia* inclusions have established that the inclusion contains *Chlamydia* developmental forms and glycogen deposits (Moulder, 1991). Recent evidence has shown that lipid droplets and peroxisomes are translocated into inclusions (Boncompain *et al.*, 2014; Cocchiaro *et al.*, 2008; Kumar *et al.*, 2006), as also discussed in Chapter 5, Interactions of the Chlamydial Inclusion with the Host Cell.

Glycogen

A defining feature of the *C. trachomatis* inclusion is the luminal accumulation of glycogen deposits, which are readily visualized by iodine staining and brightfield microscopy (Moulder, 1991). In support of the hypothesis that *C. trachomatis* may use glycogen as an energy source, the *C. trachomatis* genome contains homologues of genes with functional roles in synthesizing glycogen from glucose-6-phosphate (*glgA*, *glgB*, *glgC*) (Stephens *et al.*, 1998). Furthermore, the genome contains glycogen phosphorylase (*glgP*) and the glycogen degrading enzyme glycogen hydrolase (*glgX*). It has been proposed that *C. trachomatis* traps glycogen inside inclusions as a strategy for glucose storage (Gehre *et al.*, 2016). *C. trachomatis* strains lacking the plasmid, or that contain a deletion in the transcriptional regulatory plasmid gene CDS6/*pgp4*, fail to accumulate glycogen in their inclusions (Carlson *et al.*, 2008; Miyashita *et al.*, 2001; Song *et al.*, 2013; Wang *et al.*, 2013). Plasmid-free *C. trachomatis* strains, as well as $CDS6^-$ mutants, orient in the periphery of inclusions with a vacuous 'hole' in the inclusion centre (Wang *et al.*, 2013). How this morphological phenotype relates to impaired glycogen accumulation is unclear. A recent study demonstrated that glycogen accumulation in the inclusion occurs by two distinct mechanisms (Gehre *et al.*, 2016). The first pathway is Gys1-mediated delivery of bulk glycogen from the host cytosol into the inclusion lumen. Gys1 is a host glycogen synthase enzyme that is capable of binding glycogen (Gehre *et al.*, 2016; Roach *et al.*, 2012), and recent work suggested that the translocation of glycogen-bound Gys1 into the *C. trachomatis* inclusion lumen serves as one route of glycogen accumulation (Gehre *et al.*, 2016). In the second mechanism, the host membrane transporter SLC35D2 is recruited to the inclusion membrane to facilitate the transport of UDP-glucose into the inclusion for its subsequent processing into glycogen by biosynthetic enzymes in the inclusion lumen (Gehre *et al.*, 2016). Although ample evidence supports the notion that *C. trachomatis* inclusions accumulate glycogen, it remains unclear why these strains do so. Glycogen-deficient *C. trachomatis* strains are viable and inclusions derived from *C. trachomatis* and *C. muridarum* appear to be the only ones that accumulate glycogen deposits (Moulder, 1991). Consequently, a popular hypothesis is that glycogen metabolism is an important trait for chlamydial growth in human reproductive tissues, possibly to cope with situations where other nutrient sources are limiting (Gehre *et al.*, 2016).

Homotypic fusion

A characteristic feature of *C. trachomatis* inclusions is that they are capable of homotypic fusion with each other, a phenomenon wherein two or more inclusions in the same host cell fuse their membrane bilayers with each other. This scenario is reproducibly observed in cell culture systems, for example, by infecting host cells with high *Chlamydia* titres to maximize the percentage of cells with multiple internalized *C. trachomatis* EBs (Ridderhof and Barnes, 1989). Each internalized EB forms its own distinct inclusion. Within a few hours after cell entry, these independent inclusions coalescence into a single one (Ridderhof and Barnes, 1989). Inclusion fusion was shown to require bacterial protein synthesis (Fields *et al.*, 2002) (Van Ooij *et al.*, 1998), host actin polymerization (Richards *et al.*, 2013), and is reliant on spatial migration to the cell centre by trafficking along microtubules (Ridderhof and Barnes, 1989; Schramm and Wyrick, 1995).

Unusual *C. trachomatis* clinical isolates with a non-fusogenic phenotype were determined to harbour a variety of mutations in the *incA* gene (Geisler *et al.*, 2001). It was therefore concluded that the inclusion membrane protein IncA plays a critical role in mediating homotypic fusion of inclusions. Curiously, the prevalence of *incA* mutants in circulating clinical strains from King County, Washington, US, suggests a fitness advantage associated with the loss of *incA*, despite their reduced growth rates compared to wildtype *C. trachomatis* strains (Geisler *et al.*, 2001). The basis of this potential advantage, and whether it is related to inclusion fusion, remains to be determined. In years since, the functional role of IncA in this process has been verified genetically (Johnson and Fisher, 2013; Suchland *et al.*, 2008), and two coiled-coil SNARE domains in IncA have been shown to mediate membrane fusion of *Chlamydia* inclusions (Delevoye *et al.*, 2008; Ronzone and Paumet, 2013; Ronzone *et al.*, 2014; Weber *et al.*, 2016).

The intrinsic propensity of *C. trachomatis* inclusions to fuse within host cells has been leveraged in recent years for genetic engineering purposes. Daniel Rockey, Walt Stamm, Robert Suchland, and co-workers have demonstrated the utility of inclusion fusion for generating intraspecies and interspecies recombinant *Chlamydia* strains, including crosses between *C. trachomatis*, *C. suis*, and *C. muridarum* (Jeffrey *et al.*, 2013; Suchland *et al.*, 2009). Intracellular DNA exchange between *Chlamydia* species and strains is possible with non-fusogenic *incA* mutant strains, indicating that occupation within the same inclusion compartment is not a requirement for interchlamydial DNA exchange (Jeffrey *et al.*, 2013).

Inclusion membrane proteins (Incs)

Incs are the molecular gateway for chlamydial manipulation of the host cell. Incs were originally identified in *C. psittaci* as proteins possessing a large predicted hydrophobic domain and localized to the inclusion membrane by immunofluorescence microscopy (Bannantine *et al.*, 1998; Rockey *et al.*, 1995). The Inc protein family is a source of significant genomic diversity across *Chlamydia* species and strains (Almeida *et al.*, 2012). *C. trachomatis* encodes ~51 predicted Incs (Table 4.1), in contrast to *C. pneumoniae* which encodes 107 predicted Incs (Dehoux *et al.*, 2011; Lutter *et al.*, 2012). All Incs are thought to be secreted out of chlamydiae and across the inclusion membrane by type III secretion (T3S) into the host cytosol, wherein they are subsequently inserted into the inclusion membrane by a mechanism that remains poorly defined. Incs are predicted to insert in the inclusion membrane via hydrophobic transmembrane domains in a hairpin orientation, and present short N-terminal domains and alpha helical C-terminal domains into the host cytosol (Fig. 4.2). Experimental evidence

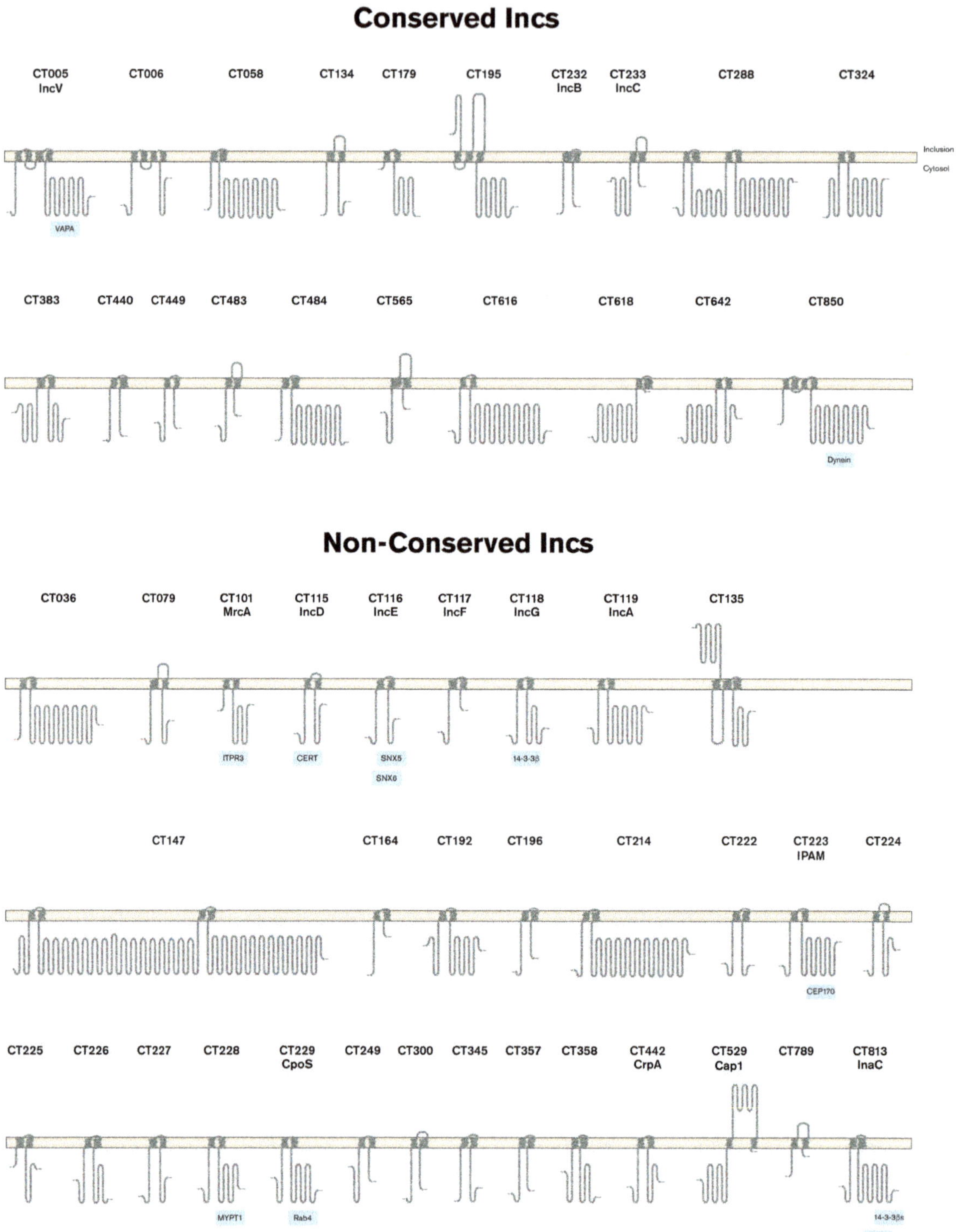

Figure 4.2 *C. trachomatis* Incs. Illustration depicts all validated *C. trachomatis* Incs, categorized as conserved (across *C. trachomatis*, *C. muridarum*, *C. caviae*, *C. felis*, and *C. pneumoniae*) or non-conserved based on computational analysis and experimental validation studies (Lutter *et al.*, 2012; Weber *et al.*, 2015). Predicted Inc protein topologies were computed using UniProt and Protter (Omasits *et al.*, 2014). Variable C-terminal lengths are indicated based on amino acid lengths. Known host interaction proteins are labelled in blue.

has shown that these sequences, in particular the C-terminal sequences of Incs, are important for promoting interactions with host proteins or with other Inc proteins (Elwell *et al.*, 2017; Gauliard *et al.*, 2015).

Incs lack sequence homology to characterized proteins, therefore homology to proteins or protein functional domains will likely need to be resolved at the structural level. Incs are expressed at different times in the chlamydial developmental cycle, suggesting unique functional roles elicited by specific members of this protein family. Many Incs are expressed early, and these are predicted to play important early functions during infection of host cells, such as inclusion biogenesis (Belland *et al.*, 2003; Nicholson *et al.*, 2003). Others are expressed at middle stages of infection, and may be important for inclusion maintenance or interference with apoptosis and innate immune signalling pathways (Belland *et al.*, 2003; Nicholson *et al.*, 2003). Incs with determined functions will be discussed in more detail elsewhere in this chapter.

In recent years, bioinformatic analysis of Incs across *Chlamydia* species has identified a set of highly conserved Incs, as well as many divergent Incs which may confer species specific tropisms (Table 4.1) (Agaisse and Derré, 2014; Dehoux *et al.*, 2011; Derré *et al.*, 2011; Lutter *et al.*, 2012). 'Conserved Incs' are defined as those which are conserved across five *Chlamydia* species – *C. trachomatis* (all biovars), *C. muridarum*, *C. pneumoniae*, *C. caviae*, and *C. felis* – and experimentally shown to be secreted into the inclusion membrane in *C. trachomatis* (Table 4.1) (Lutter *et al.*, 2012; Weber *et al.*, 2015). 'Non-conserved Incs' (Table 4.1) represent experimentally validated *C. trachomatis* inclusion membrane proteins that lack orthologues in one or more of the *Chlamydia* species listed above.

The early inclusion – towards the cell centre

Migration to the MTOC

Several ultrastructural studies have delineated the sequence of early events associated with *Chlamydia* infection of epithelial cells (Moulder, 1991; Shaw *et al.*, 2000). Within an hour after *C. trachomatis* EB internalization into epithelial cells, membrane-bound EBs are present near the cell periphery (Shaw *et al.*, 2000). After approximately two hours, many EBs have converted into metabolically active RBs, and by 12 hours bacteria have begun to replicate by binary fission (Shaw *et al.*, 2000). During the initial 2–6 hours of infection, inclusions are trafficked along microtubules from the cell periphery to the perinuclear microtubule organizing centre (MTOC). This phenomenon is well characterized for *C. trachomatis* inclusions, and somewhat for *C. pneumoniae* inclusions (Mital and Hackstadt, 2011). Transport of inclusions requires microtubule polymerization, Src family kinases, and the microtubule motor protein dynein (Clausen *et al.*, 1997; Grieshaber *et al.*, 2003; Mital *et al.*, 2010). Fibroblast cells with null mutations for the three Src family kinases Src, Fyn, and Yes, failed to traffic *C. trachomatis* and *C. pneumoniae* inclusions to the MTOC (Mital and Hackstadt, 2011). The conserved *C. trachomatis* Inc protein CT850 can directly bind the retrograde microtubule motor protein dynein light chain 1 (Mital *et al.*, 2015), and it was suggested that this interaction is important for facilitating inclusion trafficking to the MTOC. Based on microscopy data, additional Incs that cluster together in active Src family kinase 'microdomains' may also play a role in this process, including IncB, CT101, and CT222 (Mital *et al.*, 2010).

Interactions with the early endocytic pathway

Beginning at around four hours post-infection (hpi), transferrin-containing recycling endosomes concentrate around *C. trachomatis* inclusions (van Ooij *et al.*, 1997; Scidmore *et al.*, 1996). Using a yeast two-hybrid approach, it was discovered that the recycling endosome-localized Rab GTPase Rab4A was capable of directly interacting with the Inc CpoS/CT229 only in its active, GTP-bound state (Rzomp *et al.*, 2006). Recent work on *C. pneumoniae* inclusions has revealed that these inclusions are decorated with a progressive series of Rab proteins early after internalization into host cells, that facilitates the evasion of lysosomal fusion and the development of the inclusion into a recycling endosome-like compartment (Mölleken and Hegemann, 2017). *C. pneumoniae* inclusions recruit the early endosomal Rab GTPases Rab4, Rab5, and Rab7, and recycling endosome-associated proteins Rab11 and Rab14. While the early endosomal

Rab markers are lost within 30 minutes after EB internalization, Rab11 and Rab14 are retained on the inclusion membrane, thus marking the *C. pneumoniae* inclusion as 'recycling endosome-like' for intracellular membrane traffic (Mölleken and Hegemann, 2017). The disappearance of Rab7 from the inclusion is especially significant in that it prevents the acquisition of Golgi-derived transport vesicles trafficked to late endosomes, and also fusion with lysosomes. Furthermore, the conversion of the *C. pneumoniae* inclusion into a recycling endosome-like compartment may facilitate the migration of the inclusion to the perinuclear MTOC, where recycling endosomes are typically situated (Grant and Donaldson, 2009).

Evasion from lysosomes

A key feature of *Chlamydia* inclusions is the complete absence of canonical lysosomal and late endosomal markers, including transferrin receptor, M6PR, cathepsin D, v-H^+-ATPase, and LAMP-1 (Heinzen *et al.*, 1996). By contrast, phagocytosed objects are typically trafficked to late endosomes and lysosomes within hours (Conner and Schmid, 2003). As such, the early *Chlamydia* inclusion is thought to be sequestered from the default phagolysosomal trafficking pathway even though inclusions undergo homotypic fusion with each other. In the presence of chloramphenicol, to inhibit chlamydial protein synthesis, the fusion of nascent inclusions with lysosomes was found to be slow, taking at least 20 hours (Friis, 1972; Scidmore *et al.*, 1996). Lysosomal evasion may be an uncharacterized intrinsic feature of chlamydial cell wall protein(s), since purified *C. psittaci* cell walls were also trafficked to lysosomes with similarly slow kinetics as live *C. psittaci* (Eissenberg *et al.*, 1983; Levy and Moulder, 1982). Recent studies have revealed a potential molecular strategy by which *Chlamydia* inhibits fusion with lysosomes (Ronzone and Paumet, 2013; Ronzone *et al.*, 2014). IncA contains two coiled-coil, Soluble NSF Attachment Protein Receptor (SNARE)-like domains that are capable of interfering with SNARE-mediated membrane fusion, e.g. with late endosome or lysosomes (Ronzone and Paumet, 2013; Ronzone *et al.*, 2014). Complementation of an *C. trachomatis* IncA mutant with a series of full-length or truncated IncA constructs demonstrated that a 'functional core' consisting of SNARE-like domain 1 and part of SNARE-like domain 2 from IncA were necessary for homotypic fusion rescue (Ronzone *et al.*, 2014; Weber *et al.*, 2016).

The growing inclusion

For *Chlamydia trachomatis* L2, the volume of the inclusion and the number of bacteria inside it increase exponentially between 12 and 32 hpi (Lee *et al.*, 2018). At this mid-infection stage, about 70% of the predicted bacterial genome-encoded proteins are expressed (Østergaard *et al.*, 2016). The development inside a membrane-bound vacuole that protects *Chlamydia* species from host defence mechanisms comes at the price that essential nutrients have to be imported from the host cell to the bacteria. These metabolites include lipids, amino acids, nucleotides and cofactors (Haferkamp, 2018). This section summarizes our current view on mechanisms used by *C. trachomatis* to scavenge nutrients from the host cell.

'ATP parasitism' hypothesis revisited

In 1962, Moulder and co-workers postulated that *Chlamydia* spp. are not able to produce ATP by oxidation and hence depend on the import of ATP from the host cell. This ATP parasitism hypothesis was based on biochemical studies that failed to detect glycolytic enzyme activities capable of generating net ATP (Moulder 1991). It was further supported by others that showed that *Chlamydia* growth in L-cells was dependent on mitochondrial function and the characterization of an ADP/ATP translocase for the exchange of host cell ATP for bacterial ADP (Hatch *et al.*, 1982).

After publication of the first chlamydial genome in 1998, the ATP parasite hypothesis was revised as the genomic data provided revealing insights into the metabolic capacity of *C. trachomatis* (Stephens *et al.*, 1998). These genomic data indicated that *C. trachomatis* is able to generate ATP by glycolysis and oxidative phosphorylation using a minimal respiratory chain but lacks the hexokinase gene (Stephens *et al.*, 1998). Thus, *C. trachomatis* probably imports phosphorylated sugars from the host cell via the UhpC antiporter (Schwöppe *et al.*, 2002). Interestingly, the TCA cycle is incomplete in *C. trachomatis*, which confirms pre-genomic observations that acetyl-CoA generated by glycolysis and fatty acid degradation cannot enter the TCA cycle

(Weiss, 1967). In addition to endogenous ATP synthesis, *Chlamydia* can import ATP from the host cell cytosol by nucleotide transporters (Moulder, 1962; Saka and Valdivia, 2010; Tjaden *et al.*, 1999; Trentmann *et al.*, 2007). Interestingly, *in vitro* studies using axenic culture conditions showed that RBs and EBs differ in their ATP dependency suggesting that EBs are able to generate ATP whereas RBs scavenge ATP from the environment (Omsland *et al.*, 2012). Future studies are needed to analyse this metabolic difference under infection conditions.

Import of amino acids

Only small amounts of alanine, aspartate and glutamate are synthesized by the bacterium (Mehlitz *et al.*, 2017; Omsland *et al.*, 2012). Interestingly, the inclusion membrane is a barrier to molecules larger than 520 Da (Heinzen and Hackstadt, 1997). Thus, many amino acids are imported from the host cell, and these are used for bacterial protein biosynthesis and as carbon and/or energy sources. These amino acids could come from different sources such as lysosomal degradation of proteins, or proteolytic cleavage of host proteins by one or more bacterial proteases (Chen *et al.*, 2012; Ouellette *et al.*, 2011; Wu *et al.*, 2011; Zhong *et al.*, 2001). Currently it is not known how solutes including amino acids are transported across the inclusion membrane, but the identification of components of the human amino acid uptake system (SLC7A5 and SLC3A2) in the inclusion proteome implicates these as likely candidates (Aeberhard *et al.*, 2015).

In summary, it is generally accepted that *C. trachomatis* scavenges metabolites from the host cell and that the inclusion membrane represents a barrier for import of solutes. The characterization of these transport mechanisms will foster our understanding of chlamydial biology and are attractive targets for new anti-chlamydial drugs.

Transport and function of lipids

Lipids are important factors in the interplay between *Chlamydia* and its host. They control the formation and maturation of intracellular vesicles and serve as an energy source and as structural components of the bacterial cell and the inclusion membrane. *C. trachomatis*, like many bacteria, can synthesize most phospholipids with the exception of sphingomyelin, cholesterol and phosphatidylcholine. The identification of sphingomyelin and cholesterol inside chlamydial cells was surprising as these lipids are mainly produced by eukaryotic cells. The ensuing paragraphs describe the localization and trafficking of these host lipids in *C. trachomatis*-infected epithelial cells and their functions for *C. trachomatis.*

Sphingolipids

Sphingolipids are typical components of eukaryotes, where they are enriched in the outer leaflet of the plasma membrane. Most bacteria including *Chlamydia* lack enzymes for sphingolipid synthesis and therefore do not produce sphingolipids (Stephens *et al.*, 1998). Instead of sphingolipid synthesis, *Chlamydia* species have developed sophisticated mechanisms to scavenge sphingomyelin from their host cells.

In eukaryotic cells, *de novo* sphingolipid synthesis starts at the endoplasmic reticulum (ER). Ceramide, the precursor of sphingolipids, is produced at the ER and transported to the Golgi apparatus by two different mechanisms. Ceramide can be transferred to the Golgi apparatus by transport vesicles that bud of the ER and fuse with the *cis* side of the Golgi apparatus. This transport pathway is connected to the production of complex glycosphingolipids (D´Angelo *et al.*, 2013). Ceramide can also be extracted from ER membranes by the lipid transport protein CERT. This non-vesicular, CERT-dependent transport to the trans-Golgi apparatus contributes to the synthesis of sphingomyelin by cellular sphingomyelin synthase 2 (SMS2) (Hannun and Obeid, 2018).

In a landmark paper, Hackstadt and colleagues showed that *C. trachomatis* acquires fluorescently labelled sphingomyelin from the host cell (Hackstadt *et al.*, 1996). They demonstrated that purified EBs contain fluorescent sphingomyelin and concluded that bacteria gather Golgi-derived sphingomyelin instead of its precursor ceramide. Quantitative analysis indicated that chlamydial membranes contain up to 4% sphingolipids (Wylie *et al.*, 1997). Interestingly, proteins that are synthesized at the ER-Golgi compartment are not observed in the bacterial membrane leading to the hypothesis that a subpopulation of Golgi-derived exocytic vesicles is sequestered by *C. trachomatis.*

In recent years, our knowledge regarding the mechanisms *C. trachomatis* exploits to acquire sphingomyelin from their host cells has substantially increased. There are at least three different pathways to deliver sphingomyelin to *C. trachomatis* that are not redundant and contribute to distinct aspects of the infection. First, sphingomyelin can be transported in vesicles from Golgi mini-stacks to *C. trachomatis*. This requires the presence of the cellular GTPase Arf1, its guanine nucleotide exchange factor GBF1, and the Rab GTPases Rab14, Rab6A and Rab11A (Capmany and Damiani, 2010; Elwell *et al.*, 2011; Heuer *et al.*, 2009). Interestingly, presence of Rab6A, Rab11A and Rab14 seem to be important for *Chlamydia* progeny formation as shown by silencing of these proteins during *C. trachomatis* infections (Rejman Lipinski *et al.*, 2009; Capmany and Damiani, 2010). In contrast, Arf1 and GBF1 seem to contribute to stability of the inclusion by sphingomyelin delivery to the inclusion and actin stabilization around the inclusion but they seem to be dispensable for progeny formation (Elwell *et al.*, 2011; Rejman Lipinski *et al.*, 2009). Second, Rab39 mediated transport of sphingomyelin from multivesicular bodies (MVB) has been described, and the MVB marker CD63 has been detected by immunofluorescence inside the chlamydial inclusion; however, the functional relevance of this data are under debate (Beatty, 2006, 2008). Third, *C. trachomatis* intercepts non-vesicular ceramide transport for sphingolipid acquisition (Derré *et al.*, 2011; Elwell *et al.*, 2011). The early expressed Inc protein IncD directly binds the cellular ceramide transport protein CERT, which is relocated to the bacterial inclusion membrane (Agaisse and Derré, 2014; Derré *et al.*, 2011). Silencing or knockout of CERT dramatically reduced the uptake of sphingolipids inside *C. trachomatis* inclusions (Elwell *et al.*, 2011; Koch-Edelmann *et al.*, 2017). Molecular details of this interaction are discussed in Chapter 5, Interaction of the Chlamydial Inclusion with the Host Cell. Interestingly, SMS2 from the host cell was detected in the inclusion membrane by immunofluorescence microscopy suggesting that ceramide is converted into sphingomyelin at the inclusion membrane (Derré *et al.*, 2011; Elwell *et al.*, 2011). Additional experiments using a chemically modified ceramide derivative showed that sphingomyelin synthesis is a prerequisite for lipid uptake into the inclusion and into the bacteria (Banhart *et al.*, 2014). This modified ceramide lacks the hydroxyl group essential for the synthesis of sphingomyelin, and when added to *C. trachomatis*-infected cells, it was not transported to the bacteria inside the inclusion (Banhart *et al.*, 2014). Furthermore, this ceramide compound possessed a strong anti-chlamydial activity (Banhart *et al.*, 2014; Saied *et al.*, 2015). Thus, *C. trachomatis* targets vesicular and non-vesicular sphingolipid transport pathways, and by subversion of cellular proteins creates a new sphingomyelin-producing compartment inside the host cell, which is important for chlamydial growth.

Cholesterol

In the human body, cholesterol is an essential component of cell membranes and is enriched in the plasma membrane and in membranes of the Golgi apparatus, where it contributes to membrane rigidity, protein function and gene transcription. Like sphingolipids, cholesterol can be synthesized *de novo* in the ER or exogenous esterified cholesterol can be taken up by the cell, which involves vesicular and non-vesicular pathways. Vesicular transport is initiated by uptake of low density lipoprotein (LDL), an important cellular carrier of cholesterol, through receptor-mediated endocytosis. While the LDL receptor is recycled to the plasma membrane, LDL particles are transported via the endosomal system to lysosomes. Inside lysosomes, cholesterol is freed by lysosomal acid lipases. In addition to vesicular transport, cholesterol is transported by non-vesicular mechanisms from lysosomes to the smooth ER using oxysterol-binding protein-related proteins (Orps) and steroidogenic acute regulatory protein-related lipid transfer (StART) proteins (Olkkonen and Levine, 2004). Excess cholesterol is typically esterified and stored in specialized organelles called lipid droplets (LDs). These are comprised of esterified cholesterol, triacylglycerols, and phospholipids coated with perilipin proteins (Listenberger and Brown, 2008).

C. trachomatis takes up cholesterol from its host cell. In infected cells, *de novo* synthesized cholesterol can traffic from the Golgi apparatus to the chlamydial inclusion, which requires microtubules and bacterial protein synthesis (Carabeo *et al.*, 2003). In addition, cholesterol can be transported from multivesicular bodies and LDs. LDs are detected inside the inclusion, co-localizing with the bacterial protein Lda3, which is additionally found on cytosolic LDs (Cocchiaro *et al.*, 2008). A proteomic study on LDs isolated from infected cells identified the chlamydial proteins Cap1, CTL0882, and IncG, supporting the model that *C. trachomatis* manipulates the protein composition of LDs (Saka *et al.*, 2015). While the role of LD transport into the inclusion has yet to be elucidated, translocated LDs could deliver energy or membranes to the bacteria (Cocchiaro *et al.*, 2008). The recent identification of a putative bacterial carboxylic esterase (CT149) with cholesterol recognition motifs (Peters *et al.*, 2012) supports the idea that cholesterol could be freed from LDs. As a counterpoint to the above evidence, the role of LDs in *C. trachomatis* biology has recently been challenged by data showing that LDs are not important for chlamydial development (Sharma *et al.*, 2018).

While inhibition of MVB trafficking and LD formation affected bacterial growth, Golgi-derived cholesterol seemed dispensable for *C. trachomatis* replication. Furthermore, *C. trachomatis* growth in cholesterol-free mouse fibroblasts is moderately affected, suggesting that *C. trachomatis* can also use precursors of cholesterol synthesis (Gilk *et al.*, 2013). Future research on cholesterol transport and its role in *Chlamydia* biology will help to understand transport pathways and the significance of cholesterol in chlamydial infections.

Phosphatidylcholine

Chlamydia trachomatis does not synthesize phosphatidylcholine, but there is controversy about the specific mechanism used to import this lipid from the host cell. Phosphatidylcholine is found in the membranes of isolated EBs (Wylie *et al.*, 1997), but *C. trachomatis* lacks genes essential for phosphatidylcholine synthesis (Aktas *et al.*, 2010). Initial studies using radioactively labelled isoleucine, which is a precursor for *Chlamydia* specific branched chain fatty acids, showed that *C. trachomatis* modifies host-derived phospholipids with straight fatty acids by replacing this fatty acid with the chlamydia specific branched-chain fatty acid (Wylie *et al.*, 1997). It was further shown that the cellular enzyme calcium-dependent phospholipase A2 (cPLA2) deacylates cellular phosphatidylcholine. The generated lysophospholipid is taken up by *C. trachomatis* and reacylated with a bacteria-specific branched-chain fatty acid (Su *et al.*, 2004). Recent experiments that used mass spectrometry to reveal differences in molecular species of phosphatidylcholine in *C. trachomatis* infected cells or purified EBs have questioned the role of cPLA2 in scavenging phosphatidylcholine (Cox *et al.*, 2016). These observations are consistent with findings which showed by genetic and pharmacological evidence that cPLA2 is not required for *C. trachomatis* growth (Cox *et al.*, 2016; Vignola *et al.*, 2010; Yao *et al.*, 2015). Furthermore, chemical inhibition or silencing of the cellular phosphatidylcholine/cholesterol transporter CLA1 reduces phosphatidylcholine uptake into the bacteria, inclusion growth and bacterial progeny formation (Cox *et al.*, 2012, 2016). Additional experiments are needed to determine the contribution of phosphatidylcholine to inclusion expansion and to utilization by the replicating bacteria inside the inclusion.

Vesicular membrane trafficking and the inclusion

Membrane-enclosed organelles exchange proteins, lipids and other material via vesicles that traffic through the cytosol. This trafficking is tightly controlled to ensure specific fission and fusion of vesicles. Rab GTPases and SNARE proteins are key regulators of these transport processes. Rab proteins are peripheral membrane proteins that associate with membranes by prenyl groups. They control binding of distinct effector proteins, thereby orchestrating membrane trafficking events. Furthermore, they are important contributors to organellar identity (Behnia and Munro, 2005). The next paragraph summarizes findings regarding Rab, SNARE, and sorting nexin proteins during *C. trachomatis* infections.

Rab proteins on the *C. trachomatis* inclusion

A proteomic study of isolated inclusions identified 13 different Rab GTPases associated with inclusions (Aeberhard *et al.*, 2015). Five of these proteins, Rab1, Rab4, Rab6, Rab11 and Rab14, have previously been reported to be associated with *C. trachomatis* inclusions (Capmany and Damiani, 2010; Rejman Lipinski *et al.*, 2009; Rzomp *et al.*, 2003). These five proteins are involved in vesicular transport between different organelles. Rab1 is found on ER-derived vesicles, Rab4 and Rab11 have been shown to regulate recycling endosomes, which could carry iron to *C. trachomatis*, and Rab6, Rab11 and Rab14 have been implicated in the transport of sphingomyelin from the Golgi apparatus to the *C. trachomatis* inclusion (Capmany and Damiani, 2010; Ouellette and Carabeo, 2010; Rejman Lipinski *et al.*, 2009). Interestingly, while Rab11 and transferrin receptor (TfR)-positive vesicles are found in close proximity to the inclusion, they do not fuse with the inclusion (Ouellette and Carabeo, 2010). Accumulation of TfR-positive-recycling endosomes around the inclusion impairs chlamydial development suggesting that these interactions need to be tightly regulated (Ouellette and Carabeo, 2010).

For *Chlamydia*, it has been shown by ectopic expression of tagged Rab proteins that cellular Rab proteins are recruited to the chlamydial inclusions in a species-dependent manner (Rzomp *et al.*, 2003). Biochemical analysis showed that distinct Inc proteins (1) bind one specific Rab protein or (2) that a specific Inc protein binds multiple Rabs. For example, CT229 has been shown to interact with Rab4, and a recent screen for Inc–host interactions revealed that CT229 could also bind Rab1A/B and Rab35, while CT556 and CT642 co-immunoprecipitated with Rab11A and Rab8A, respectively (Alix *et al.*, 2011; Mirrashidi *et al.*, 2015). Interestingly, the absence of the multi-Rab binding protein CT229 induced premature host cell death in infected cells (Sixt *et al.*, 2017; Weber *et al.*, 2017).

In summary, *C. trachomatis* recruits a subset of cellular Rab proteins to the inclusion by secretion of specific chlamydial proteins. The function of many of these recruited Rab proteins are still under investigation. However, it seems that some of the recruited Rab proteins are important for establishment, maintenance and growth of the inclusion and the bacteria as they control the (1) transport of distinct cellular vesicles to the chlamydial inclusion, (2) cell signalling and (3) the recruitment of Rab-effectors to the chlamydial inclusion.

SNARE proteins and the *C. trachomatis* inclusion

In addition to the Rab proteins, SNARE proteins, which are a family of transmembrane proteins, control fusion of vesicles with their target membranes. At least three *C. trachomatis* Inc proteins, IncA (CT119), InaC (CT813), and IPAM (CT223), contain a SNARE-like domain (Delevoye *et al.*, 2008). Direct interactions of IncA with the known host SNAREs VAMP3, VAMP7 and VAMP8, as well as of CT813 with VAMP7 and VAMP8, were shown by co-immunoprecipitation of ectopically expressed proteins suggesting that these interactions could limit the fusion of vesicles that contain the respective SNARE proteins. Interestingly, ectopically expressed GFP-VAMP8 was found in the inclusion membrane of *C. trachomatis* D by cryo immune electron microscopy (Delevoye *et al.*, 2008). Furthermore, localization of the Golgi-associated SNARE proteins syntaxin (STX) 6 and 10, VAMP4 and GS15 to the chlamydial inclusion membrane has been observed by fluorescence microscopy (Kabeiseman *et al.*, 2013; Lucas *et al.*, 2015; Moore *et al.*, 2011; Pokrovskaya *et al.*, 2012). Although it is not completely understood how these proteins are recruited to the chlamydial inclusion, active bacterial synthesis is essential for the localization of these proteins to the chlamydial inclusion. In addition, a Golgi-targeting signal seemed to be involved in recruiting STX 6 to the inclusion (Kabeiseman *et al.*, 2013, 2014). Future studies on the nature of these associations and their species specificity will help to describe the functions of these proteins in distinct chlamydial infections.

Sorting nexin proteins on the *C. trachomatis* inclusion

The SNX proteins 1, 2, 5 and 6 are components of the retromer, a protein complex that is involved in recycling of proteins from endosomes to the Golgi apparatus (Gallon and Cullen, 2015). Five different sorting nexin (SNX) proteins have been identified in the proteome of isolated and purified inclusions (Aeberhard *et al.*, 2015). Immunofluorescence microscopy confirmed that SNX1, 2, 5 and 6 are recruited to the inclusion and to tubules extending from the inclusion (Aeberhard *et al.*, 2015; Mirrashidi *et al.*, 2015). SNX5 and SNX6 were shown to directly bind to IncE, and this binding competes with host cell proteins affecting the function of the retromer (Elwell *et al.*, 2017; Mirrashidi *et al.*, 2015; Paul *et al.*, 2017). Silencing of SNX5 using RNA interference promoted bacterial growth suggesting that SNX proteins normally restrict chlamydial propagation (Aeberhard *et al.*, 2015; Mirrashidi *et al.*, 2015). Thus, *C. trachomatis* recruits regulators of vesicular trafficking to its inclusion, which allows the bacteria to use vesicular trafficking routes such as the recycling pathway, while other trafficking pathways, such as retromer-dependent transport, are disturbed.

Golgi fragmentation

During chlamydial infection the structures of different organelles including mitochondria and the Golgi apparatus are affected (Chowdhury *et al.*, 2017; Heuer *et al.*, 2009). Infection with *C. trachomatis* induces Golgi fragmentation into Golgi mini-stacks that surround the bacterial inclusion (Heuer *et al.*, 2009). This fragmentation of the Golgi apparatus was first detected at 16 hours post-infection (hpi) by immunofluorescence microscopy using specific antibodies against distinct golgins including GM130, giantin and GPP130. Golgi repositioning required the cellular proteins Rab6A, Rab11A and Arf GTPases, as well as the formation of stable detyrosinated microtubules (Al-Zeer *et al.*, 2014; Rejman Lipinski *et al.*, 2009; Wesolowski *et al.*, 2017). The chlamydial protein InaC (CT813) has been shown to be important for Golgi mini-stack formation and repositioning by binding the cellular GTPases Arf1 and Arf4 (Kokes *et al.*, 2015; Wesolowski *et al.*, 2017). The involvement of other chlamydial proteins in this process is likely but has so for not been addressed.

Inhibition of Golgi fragmentation by either silencing of Rab6A and Rab11A, or by blocking detyrosination of microtubules, reduced bacterial progeny formation whereas stimulation of Golgi fragmentation enhanced bacterial infectivity (Al-Zeer *et al.*, 2014; Heuer *et al.*, 2009). Furthermore, *C. trachomatis* deficient for CT813 elicited smaller inclusions and produced fewer infectious progeny than wild type *C. trachomatis* (Wesolowski *et al.*, 2017) while the chemically mutagenized strain did not show any growth defect (Kokes *et al.*, 2015). This discrepancy is likely linked to the difference between targeted mutagenesis leading to a complete knockout of a gene and chemical mutagenesis, in which some functions of the protein may remain.

Maintenance of inclusion integrity

A long-standing dogma in the field, based on attempts to microinject *Chlamydia*-infected cells or to isolate cell-free inclusions, has been that *C. trachomatis* inclusion membranes are fragile (Gieffers *et al.*, 2003; Heinzen and Hackstadt, 1997). However, as described in a later section, the notion of inclusion incompatibility with experimental micromanipulation was overcome by recently by establishing procedures for subcellular fractionation and isolation of *C. trachomatis* inclusions from host cells (Aeberhard *et al.*, 2015).

Genetic studies using engineered knockout strains have shown that some Incs (IncC, CpoS, and CT383) are required for maintaining inclusion membrane integrity throughout the chlamydial developmental cycle (Weber *et al.*, 2017). In cells infected with any of these Inc knockout strains, inclusion membranes became prematurely permeable at around 18 hpi, after which *C. trachomatis* RBs diffused into the host cytosol and host cell apoptosis was induced (Weber *et al.*, 2017). The mechanisms by which these Incs regulate inclusion membrane stability, for example through interacting with other Incs or host protein targets, are not well established. A parallel study that investigated phenotypes associated with a chemically derived deletion mutant in CpoS/CT229 demonstrated that the loss of this Inc resulted in stimulator of interferon genes (STING) activation and rapid host cell death (Sixt *et al.*, 2017).

Maintenance of inclusion stability is also dependent on the recruitment of host cytoskeletal proteins to the inclusion membrane. It is now evident that *Chlamydia* inclusions dynamically recruit all members of the cytoskeleton family – actin filaments, microtubules, intermediate filaments, and septins – to their cytosolic surface throughout infection. Mesh-like scaffolds of actin and intermediate filaments are recruited to *C. trachomatis* inclusions in a developmentally regulated manner (Chin *et al.*, 2012; Kumar and Valdivia, 2008). Disruption of these rearrangements were shown to adversely affect inclusion integrity (Kumar and Valdivia, 2008). *C. trachomatis* inclusions have also been shown to preferentially recruit detyrosinated and acetylated 'stable' microtubules to their membrane surface and to regulate microtubule polymerization (Al-Zeer *et al.*, 2014; Dumoux *et al.*, 2015). Finally, heteromeric septin structures were demonstrated to be recruited to *C. trachomatis* inclusions, with detailed functions in the co-recruitment of actin networks and the promotion of *Chlamydia* extrusion (Chin *et al.*, 2012; Volceanov *et al.*, 2014).

The late inclusion

RB to EB conversion

Beginning at 18–24 hpi for *C. trachomatis*, *Chlamydia* RBs undergo developmental conversion to EBs. This process of conversion occurs asynchronously, with every RB within an inclusion tracking along its own independent program (Lee *et al.*, 2018). RB replication continues to occur after the onset of RB-EB conversion, and not every RB ends up converting to an EB prior to *Chlamydia* escape from the host cell (Abdelrahman and Belland, 2005; Moulder, 1991). The signals that govern RB-EB conversion are unknown, and have been hypothesized to be either an extrinsic (response to as-yet undefined environmental cues from the host cell, e.g. nutrient depletion) or an intrinsic cue (akin to an internal clock). It has been proposed that for *C. trachomatis*, type III secretion-mediated contact between RB and the inclusion membrane may function as a trigger for RB to EB differentiation (Wilson *et al.*, 2006). For most of the developmental cycle, *C. trachomatis* RBs are in close contact with the inner leaflet of the inclusion membrane. During this time, they project T3S apparatuses across the inclusion membrane bilayer and into the host cytosol (Nans *et al.*, 2015). Based on mathematical modelling, Bavoil and colleagues proposed a model wherein the loss of RB contacts with the inclusion membrane would lead to signalling events that promote differentiation to EB (Wilson *et al.*, 2006). More recently, Ming Tan, Christine Sütterlin and colleagues used a three dimensional transmission electron microscopy approach to quantitatively measure the volumetric sizes of RB, EB, and intermediate bodies (IB) at distinct stages of *C. trachomatis* development inside HeLa cells (Lee *et al.*, 2018). This study found that the size of RB varies early during infection and decreases significantly with successive rounds of replication (up to 6-fold reduction in mean volume). Once an RB reaches a size threshold, EBs begin to emerge within the inclusion. The model posits that RB size reduction is a trigger for EB conversion. The heterogeneity in starting RB sizes within the inclusion, and the variable amount of time it takes each RB to divide and reach this size threshold, are predicted to be responsible for the delay and asynchrony associated with RB-EB conversion (Lee *et al.*, 2018).

Exiting the cell by extrusion and lysis

At the end of their intracellular growth and conversion to EBs, chlamydiae are released from host cells by one of two mutually exclusive exit mechanisms: host cell lysis or membrane-encased extrusion of the chlamydial inclusion (Fig. 4.3) (Hybiske and Stephens, 2007). Data from several cell culture models indicates that lysis and extrusion are active at similar frequencies, yet mechanistically distinct (Hybiske and Stephens, 2007). Lysis is

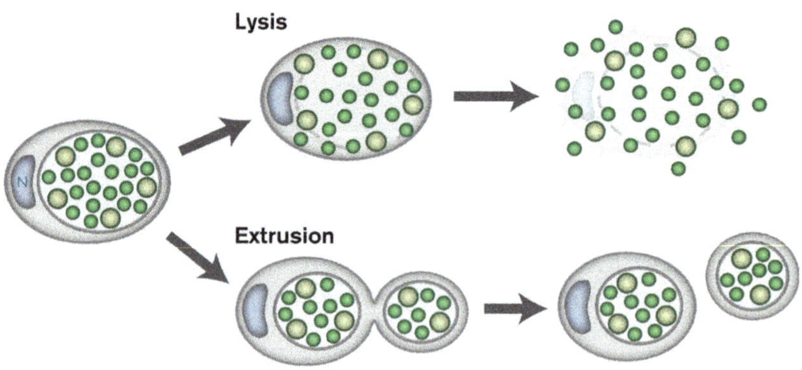

Figure 4.3 Exit strategies of *Chlamydia*. At the end of their developmental growth, *Chlamydia* exit epithelial cells by inducing either host cell lysis or extrusion of their inclusion. Host cell lysis consists of sequential disruptions of the inclusion, nucleus, and host plasma membranes. Inclusion breakdown is mediated by cysteine proteases; plasma membrane rupture is dependent on Ca^{2+} signalling. Extrusion is a slow, non-destructive exit strategy that morphologically and mechanistically resembles host cell cytokinesis. Extrusion requires cytoskeletal rearrangements on the cytosolic surface of the inclusion membrane, Rho GTPase signalling, myosin II, and chlamydial type III secreted effectors such as the Inc protein CT228. Importantly, the host cell remains intact and viable after releasing *Chlamydia* by the extrusion pathway.

approximately a 15-minute destructive process that consists of the sequential permeabilization of the inclusion, nuclear, and host plasma membranes (Hybiske and Stephens, 2007). Early experiments demonstrated that the permeabilization of these discrete membrane compartments were spaced approximately five minutes apart (Hybiske and Stephens, 2007). At the conclusion of lysis, there is release of chlamydial and intracellular debris into the extracellular milieu, including EBs that are free to infect new host cells. While inclusion lysis can be inhibited by treating cells with cysteine protease inhibitors, it remains unclear whether the implicated cysteine proteases are of chlamydial or host origin. In contrast, plasma membrane permeabilization was shown to be a calcium-dependent event (Hybiske and Stephens, 2007), but it is not known which specific calcium-mediated processes facilitate cell membrane disruption.

In stark contrast to lysis, extrusion-mediated exit of *Chlamydia* is a slow event that takes approximately two hours. It requires participation of numerous host factors and culminates in the release of chlamydiae within an encased compartment that is detached from the intact host cell (Fig. 4.3). Frequently, the host cell remains infected with a residual portion of the original inclusion (Hybiske and Stephens, 2007). Extrusions contain hundreds of EBs within a double membrane (inclusion and host plasma membrane), but organelles and other notable host structures are generally not present (Hybiske and Stephens, 2007; Zuck *et al.*, 2017). Extrusion is proposed to be triggered by RBs, since the process requires chlamydial protein synthesis and functional T3S (Chin *et al.*, 2012). Host actin polymerization, septins, RhoA GTPase activation, myosin II, and myosin phosphatase-targeting subunit 1 (MYPT1) have all been shown to play important roles for the extrusion mechanism (Hybiske and Stephens, 2007; Lutter *et al.*, 2013; Volceanov *et al.*, 2014; Zuck *et al.*, 2017). In addition to key host factors, two *C. trachomatis* Incs have been shown to be important for the extrusion pathway. CT228 modulates myosin II phosphorylation by recruiting MYPT1 to the inclusion membrane (Lutter *et al.*, 2013), and the Inc MrcA/CT101 promotes extrusion formation through functional interactions with host inositol-1,4,5-trisphosphate receptor 3 (Nguyen *et al.*, 2018). Although both exit mechanisms are driven by *Chlamydia*, the signal(s) for inducing them remain undefined. It has been speculated that cellular exit may be an example of bacterial altruism, wherein metabolically active, non-infectious RBs are the initiators of processes that only benefit EBs, which are the only chlamydial developmental form that can go on to infect new host cells (Hybiske and Stephens, 2008).

The route *Chlamydia* uses to exit host cells has important consequences for the fates of released EBs. EBs that escape by host cell lysis are released into a proinflammatory environment, surrounded by cellular debris and innate immune cells (Brunham and Rey-Ladino, 2005). These EBs must find a new host cell to infect quickly. In contrast, *Chlamydia* EBs that exist inside a membrane-encased extrusion are shielded from extracellular hazards, and are capable of sustaining their infectivity for prolonged periods compared to free EBs (Zuck *et al.*, 2017). Work by Kevin Hybiske and colleagues examined the intrinsic stability and distribution of EBs and RBs in isolated, cell-free extrusions, and found that extrusions provide long-term protective benefits for *Chlamydia* over prolonged times at 37°C (Zuck *et al.*, 2017). Bacteria within extrusions experienced significantly lower rates of decline in viability and infectivity compared to bacteria from mechanically disrupted extrusions, or purified EBs, and extrusions were capable of preserving EB infectivity for over 24 hours at 37°C (Zuck *et al.*, 2017). The same study showed that extrusions were heavily enriched for EBs compared to RBs, and the total number of bacteria and relative ratios remained unchanged in cell-free extrusions incubated over time. As such, the mechanisms by which extrusions sustain prolonged EB infectivity compared to free EBs remains an open question.

Extrusions display externalized phosphatidylserine, which seems to occur concurrent with or immediately after their release from host cells (Zuck *et al.*, 2017). A direct consequence of phosphatidylserine exposure is the recognition and engulfment of extrusions by macrophages and dendritic cells (Sherrid and Hybiske, 2017; Zuck *et al.*, 2017). After their phagocytosis into these antigen presenting cells, extrusions preserve *Chlamydia* EB viability and infectivity, and ultimately cause macrophage death and dendritic cell apoptosis (Sherrid and Hybiske, 2017; Zuck *et al.*, 2017).

Although extrusions have been observed *in vivo* (Doughri *et al.*, 1972; Shaw *et al.*, 2017), their exact roles in promoting chlamydial cell-to-cell spread and pathogenesis remain to be established. These remain major questions to be addressed by the field. Experimental evidence from cell culture models suggests that all three *C. trachomatis* biovars, *C. pneumoniae, C. psittaci, C. muridarum,* and *C. caviae* are capable of exiting host cells by both lysis and extrusion (Hybiske and Stephens, 2007; Zuck *et al.*, 2016). It is therefore attractive to speculate that extrusions convey unique advantages for chlamydial infection.

Interestingly, the process of chlamydial extrusion mechanism shares many features with malaria egress from liver hepatocytes (Burda *et al.*, 2017; Cowman and Kappe, 2006; Graewe *et al.*, 2011; Sturm *et al.*, 2006). The size and slow kinetics are very similar, but the mechanism for the formation of malarial extrusions is unknown. During the liver stage of malarial infection, *Plasmodium* sporozoites undergo rapid expansion within their parasitophorous vacuole followed by developmental conversion to merozoites. Thereafter, the vacuole membrane disperses and large, plasma membrane-encased packages (called 'merosomes') of merozoites are released into the hepatic vasculature (Baer *et al.*, 2007). *In vivo* imaging experiments have shown that the merosomes are delivered to the pulmonary microvasculature wherein they are sheared apart by microfluidic shear forces, and the onset of symptomatic malarial infection of red blood cells begins (Baer *et al.*, 2007). Thus, the egress of malaria from hepatocytes into extrusion-like merosomes facilitates the transition of the parasite from the liver to blood stages of infection, a critical bottleneck for malarial infection (Cowman and Kappe, 2006).

New methods to characterize the inclusion–host interface

In recent years, novel techniques have been developed that change our perception of inclusion development and maintenance, as well as our ideas about the role of bacterial effectors and their cellular interactions partners. The development of molecular tools to genetically modify *C. trachomatis* and *C. pneumoniae* has created opportunities to study the function of bacterial proteins and interacting factors during the infection process. Furthermore, the development of highly sensitive and quantitative mass spectrometric methods to study the protein composition of a complex biological sample have advanced our view of inclusion composition and the chlamydial effector interactome (Fig. 4.4).

Mass spectrometry-based proteomics of isolated *C. trachomatis* inclusions

The use of quantitative mass spectrometry to determine the proteome of pathogen-containing vacuoles has greatly expanded our understanding of these unique intracellular compartments (Banhart *et al.*, 2018; Herweg *et al.*, 2015). The technique requires the isolation of the pathogen-containing compartment in both high quality and quantity (Banhart *et al.*, 2018). Pathogen-containing vacuoles are very diverse and thus their biochemical isolation proved to be difficult. One of the earliest reports on an isolated pathogen-containing vacuole was the partial purification of inclusions from *C. psittaci* by a filter-based method (Matsumoto, 1981). Unfortunately, this fraction was contaminated with cellular debris and was therefore insufficiently pure for quantitative mass spectrometry-based proteomics. Using the same strain, early inclusions and phagosomes containing inactivated bacteria were isolated shortly thereafter (Zeichner, 1982, 1983). Recently, the first proteome of isolated and purified *C. trachomatis* inclusions was published (Aeberhard *et al.*, 2015). For inclusion isolation, *C. trachomatis*-infected and control cells were lysed at 24 hpi using a ball homogenizer. Freed inclusions were then further isolated on a Percol gradient followed by immunoseparation of the inclusion using an IncA-specific antibody (Aeberhard *et al.*, 2015). To control for co-purifying cellular debris, a SILAC based method for the proteome measurement was applied. This combination of label-free absolute and SILAC-based relative quantification to distinguish specific components of the bacterial compartment from co-purifying nonspecific proteins yielded a high confidence set of 351 HeLa cell proteins enriched in the inclusion fraction (Aeberhard *et al.*, 2015). Quantification of the associated cellular proteins showed that sorting nexins (SNX) especially SNX5 and SNX6 were highly enriched in this inclusion proteome suggesting specific and strong binding of these proteins to the inclusions. Furthermore, computational analysis of the host proteins

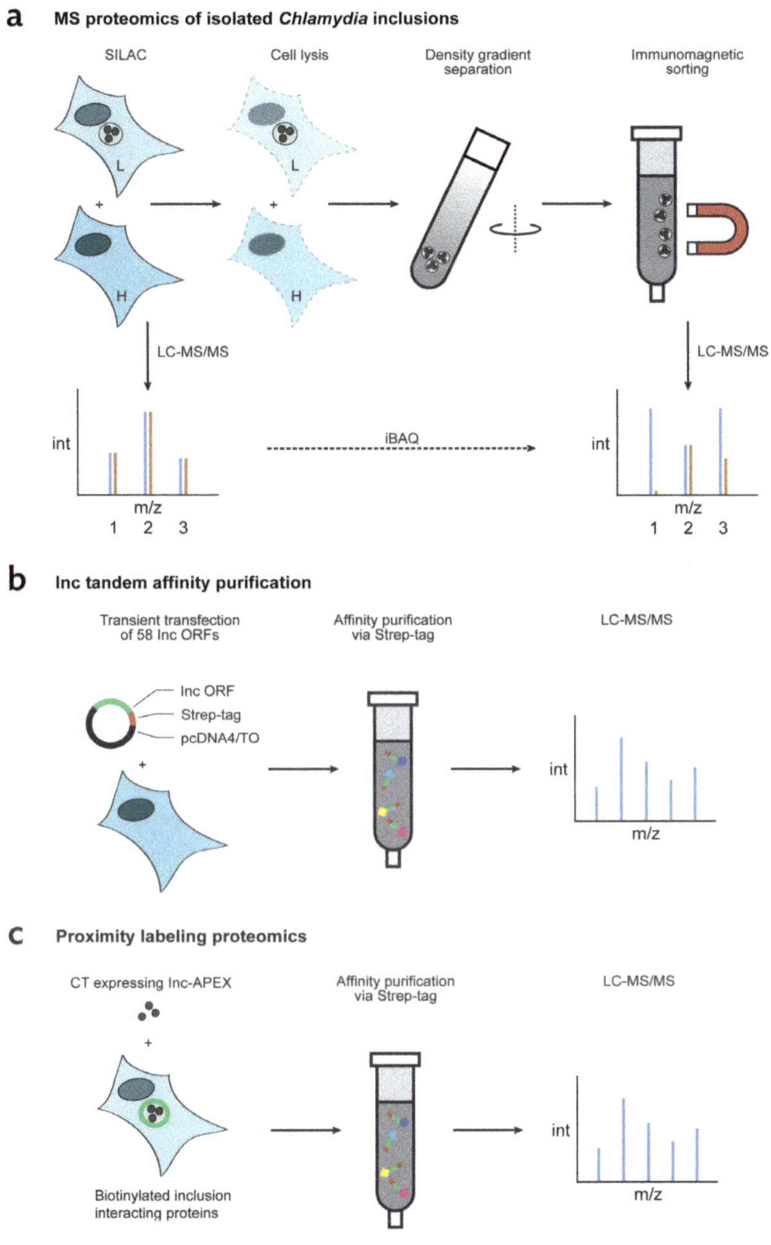

Figure 4.4 New methods for proteomic investigation of the *Chlamydia*–host interface. (a) Cells grown in medium supplemented with either light (L) amino acids or with amino acids labelled with stable heavy isotopes (H) are infected with *C. trachomatis* or left uninfected. Cell populations are mixed and lysed. After lysis, freed inclusions are purified by gradient centrifugation followed by immunoseparation with specific antibodies against bacterial IncA. After tryptic digestion of proteins, purified peptides are analysed by LC-MS/MS and SILAC ratios are calculated. For label-free quantification, full lysates from infected and mock-treated cells are prepared and protein samples are analysed by LC-MS/MS. Detailed study described in (Aeberhard *et al.*, 2015). (b) Inc proteins fused to the Strep-tag are ectopically expressed in 293T cells. Interacting cellular proteins are isolated by affinity purification and proteins samples are analysed by LC-MS/MS. Detailed study described in (Mirrashidi *et al.*, 2015) (c) Genetically modified *C. trachomatis* (CT) strains that express a specific Inc–APEX fusion protein are used to infect host cell in cell culture. After infection, cells are incubated with medium containing biotin. Expression of the engineered ascorbate peroxidase (APEX) fusion protein generates short-lived biotin-phenoxyl radicals that labels proteins in close proximity. After cell lysis, biotinylated proteins are purified by affinity purification and tryptic peptides are analysed by LC-MS/MS. Detailed studies described in (Dickinson *et al.*, 2019; Rucks *et al.*, 2017). Reprinted from Banhart *et al.*, 2018, with permission from Elsevier.

that were enriched in the proteome of isolated and purified inclusions supported the perception of the inclusion being an intracellular trafficking hub connected to distinct cellular vesicular pathways including the retromer.

Inc tandem affinity purification

Tandem affinity purification is a technique to identify protein–protein interactions. It uses a labelled protein as bait to capture and isolate interacting proteins. In 2002, this technique was combined with mass spectrometry to systematically analyse protein complexes in yeast (Gavin *et al.*, 2002). For *Chlamydia* species, a subset of *C. trachomatis* Inc proteins were subjected to affinity purification-mass spectroscopy. Chlamydial proteins were first fused to a streptavidin tag and expressed in uninfected embryonic kidney cells (HEK293T). Each fusion protein and its binding partners were then affinity purified after cell lysis (Mirrashidi *et al.*, 2015). Eluates were trypsin digested and subjected to LC-MS/MS. Specificity of interactions was assessed and scored by computational methods that used spectral counts and a two scoring metrics or protein abundance and specificity across all purifications. This approach identified 335 host proteins that were captured by 38 *C. trachomatis* Inc proteins. Among the identified cellular interaction partners were SNX5 and SNX6 that bind to bacterial IncE.

Proximity labelling proteomics

The advent of proximity-dependent proteomic technologies allows for the discovery of bacteria–host interactions in their native intracellular microenvironment, without the need for experimental manipulations such as subcellular fractionations or protein pull-downs. The BioID and APEX2 ascorbate peroxidase systems have been used extensively to probe protein interaction networks in subcellular regions of living cells (Hung *et al.*, 2016; Kim and Roux, 2016; Martell *et al.*, 2012; Varnaitė and MacNeill, 2016). A key advantage of APEX2 is that proteins of interest are genetically tagged to enable the covalent addition of biotin to fusion proteins within 20 nm of the tag, during rapid labelling reaction times of less than one minute. This allows *in situ* high spatial and temporal resolution of proteins in close proximity, when used in combination with mass spectrometry (Lam *et al.*, 2015; Lee *et al.*, 2016). Importantly, APEX2 has been adapted for use in combination with Inc proteins of *C. trachomatis*, through the expression and type III mediated secretion of APEX2-tagged Incs in infected cells (Dickinson *et al.*, 2019; Rucks *et al.*, 2017), and was recently used to generate a temporal map of host and chlamydial proteins enriched in close proximity to *C. trachomatis* inclusions at three stages of chlamydial developmental growth (Dickinson *et al.*, 2019). This study revealed that early 8 hpi inclusions had enriched associations with glycolytic enzymes, and that inclusion maturation corresponded with a shift to protein networks related to innate immune signalling, vesicular traffic, and the cytoskeleton. This study also highlighted interactions between *C. trachomatis* inclusions and endoplasmic reticulum exit sites, including a functional dependence of the early secretory pathway for *Chlamydia* developmental growth (Dickinson *et al.*, 2019).

Bioinformatic comparisons of the data sets generated by these landmark inclusion proteomics studies has led to important insight into the spectrum of host proteins associated with *Chlamydia* inclusions (Fig. 4.5). In total, 77 overlapping proteins were identified by the isolated inclusion and IncB-APEX approaches (Aeberhard *et al.*, 2015; Dickinson *et al.*, 2019), thus strengthening the likelihood that these protein associations are genuine. Cross-comparisons of these inclusion-based datasets with the Inc tandem affinity purification dataset (Mirrashidi *et al.*, 2015) results in 58 host proteins shared between those identified from isolated inclusions, and 29 proteins identified by proximity labelling (Dickinson *et al.*, 2019). Sixteen proteins were positively identified by all three approaches. Differences in technical approaches likely account for the low degree of overlap between the studies. For example, isolation of intact inclusions is closer to the *in vivo* situation than tandem affinity purification, but could miss weak inclusion–protein interactions and interactions that change over time. Tandem affinity purification should theoretically identify all possible interactions of the ectopically expressed proteins with the limitation that the proteins of interest are expressed out of context being e.g. being toxic to the cell or ectopic expression of a bacterial protein might result in false–positive interactions (Banhart *et al.*, 2018). Finally, while

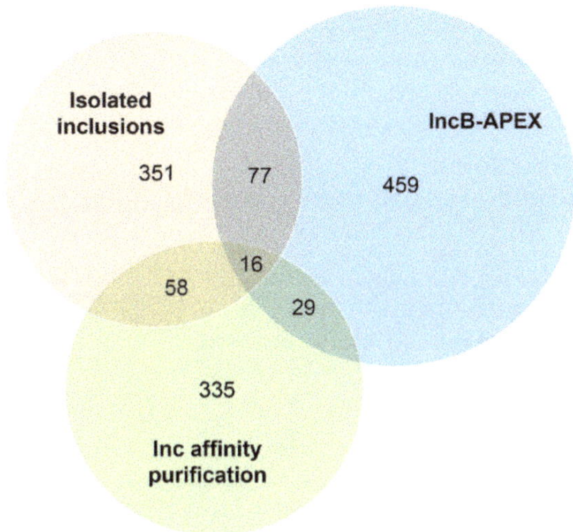

Figure 4.5 Bioinformatic comparisons of inclusion proteomics studies. Unique and overlapping host proteins identified by three major inclusion proteomics studies are represented by an Euler diagram. 351 proteins were identified in association with isolated inclusions (Aeberhard *et al.*, 2015), 459 proteins were identified by IncB-APEX (Dickinson *et al.*, 2019), and 335 proteins were affinity-purified with ectopically expressed Inc proteins (Mirrashidi *et al.*, 2015). The numbers of overlapping proteins identified by these discrete approaches, including those present in all three studies, are indicated.

proximity proteomics offers the important advantage of *in situ* labelling, it may mis-label proteins that are not actually present at the inclusion membrane. While tremendous new insight has been gained from these proteomics efforts, future expansion and development of the technical approaches should provide clearer resolution for the protein composition of *Chlamydia* inclusions.

Chlamydia development without an inclusion?

All *Chlamydia* species identified to date develop inside the inclusion, which is a membrane-bound vacuole inside a eukaryotic host cell. In biology, membrane bilayers serve as barriers to separate different compartments or organelles. They prevent diffusion of nutrients including ions, glucose, ATP, amino acids, proteins and other molecules. An intriguing thought question is thus: if provided the essential nutrients and factors, could *Chlamydia* replication and development proceed without an inclusion? In this final section, we will speculate on this hypothesis based on current knowledge, and discuss our current view on the importance of the inclusion and inclusion membrane.

In the light of this idea, different functions of the inclusion and the inclusion membrane must be taken into consideration: (1) defence against anti-chlamydial mechanisms of the host cell, (2) metabolism and (3) signalling. It has been shown that the integrity of the inclusion is essential to prevent the induction of cellular anti-chlamydial pathways (Feeley *et al.*, 2017; Sixt *et al.*, 2017). In an inclusion-free, and by extension host cell-free system, this first functional property may not be necessary.

It is clear that several distinct metabolic pathways take place inside the inclusion lumen and at the inclusion membrane, including metabolism of glycogen and lipids. From these data, we can speculate that the inclusion and the inclusion membrane provide some enzymatic functions important for *Chlamydia* metabolism. In an inclusion-free system, the proper concentrations of these nutrients – still incompletely defined – would have to be supplemented to the bacteria. Some advances in this area have been made by Omsland and co-workers, wherein a cell-free culture system was derived for enabling metabolic activity in *C. trachomatis* RBs and EBs (Omsland *et al.*, 2012). Ongoing efforts are necessary to generate a culture system fully capable of supporting chlamydial replication and developmental switching.

The inclusion membrane plays major functions in manipulating host cell signalling (Elwell *et al.*, 2016). In *Chlamydia* infections, the T3SS of membrane-localized RBs perforate the inclusion membrane resulting in RB-inclusion membrane contact sites, and detachment of RBs from the inclusion membrane has been predicted to trigger RB-to-EB conversion (Wilson *et al.*, 2006). Recently, observations made using quantitative three-dimensional electron microscopy suggested a new model, that re-differentiation was triggered by size of RBs (Lee *et al.*, 2018). This last potential property of the inclusion, and the availability of essential nutrients, will be critical for the establishment of an *in vitro*, inclusion-free *Chlamydia* developmental system but we propose that *Chlamydia* researches will sooner or later overcome these obstacles.

Summary and outlook

Chlamydia species have developed sophisticated mechanisms to interact with their host cells allowing the formation of their intracellular compartment, the inclusion. The inclusion is a defining feature of *Chlamydia*, and its establishment and maintenance are requirements for successful chlamydial infection of the host cell. Despite their highly streamlined genomes, Chlamydiae have preserved a high percentage of genetic resources – 7–10% for *C. trachomatis* – for type III mediated secretion of Incs and soluble effector proteins into the inclusion membrane and host cell. These proteins are predicted to fulfil critical processes for *Chlamydia*, for example recruiting host proteins and organelles to the inclusion membrane and manipulating host cell signalling pathways. The past decade has witnessed a veritable explosion of discoveries related to inclusion biology and the molecular mechanisms of *Chlamydia*–host interactions. These advances have been greatly facilitated by breakthroughs in genetic manipulation of *Chlamydia* (Hooppaw and Fisher, 2015), that now make possible: the deletion of Incs (Johnson and Fisher, 2013), tagging of Incs with split fluorescent proteins (Wang *et al.*, 2018), ectopic inducible expression of Incs (Wickstrum *et al.*, 2013), and knockdown of endogenous chlamydial proteins (Ouellette, 2018). Despite these great achievements, the genetic manipulation of this important pathogen still poses many challenges. Targeted knockouts are not easily accomplished, and many genes are predicted to be refractory to genetic disruption (Kokes *et al.*, 2015; Stephens *et al.*, 1998). The identification of proteins enriched in the host cell-derived inclusion proteome (Aeberhard *et al.*, 2015) and by tandem affinity purification of Inc-interacting proteins (Mirrashidi *et al.*, 2015) represent useful starting points for characterizing novel proteins and their functions for inclusion biology and beyond.

Despite the field's progress, many important details surrounding the chlamydial inclusion remain unanswered. We lack a detailed temporal, spatial, and compositional map of the inclusion, as well as a comprehensive understanding of the host cell pathways manipulated by Incs on the inclusion membrane. How do the protein and phospholipid compositions differ across *Chlamydia* species, and what are the defining, critical features shared by all chlamydial inclusions? These and other important questions await discovery, and their answers will reveal major new insights into *Chlamydia* biology and pathogenesis.

References

Abdelrahman, Y.M., and Belland, R.J. (2005). The chlamydial developmental cycle. FEMS Microbiol. Rev. *29*, 949–959.

Aeberhard, L., Banhart, S., Fischer, M., Jehmlich, N., Rose, L., Koch, S., Laue, M., Renard, B.Y., Schmidt, F., and Heuer, D. (2015). The Proteome of the Isolated *Chlamydia trachomatis* Containing Vacuole Reveals a Complex Trafficking Platform Enriched for Retromer Components. PLoS Pathog. *11*, e1004883.

Agaisse, H., and Derré, I. (2014). Expression of the effector protein IncD in *Chlamydia trachomatis* mediates recruitment of the lipid transfer protein CERT and the endoplasmic reticulum-resident protein VAPB to the inclusion membrane. Infect. Immun. *82*, 2037–2047.

Aktas, M., Wessel, M., Hacker, S., Klüsener, S., Gleichenhagen, J., and Narberhaus, F. (2010). Phosphatidylcholine biosynthesis and its significance in bacteria interacting with eukaryotic cells. Eur. J. Cell Biol. *89*, 888–894.

Alix, E., Mukherjee, S., and Roy, C.R. (2011). Subversion of membrane transport pathways by vacuolar pathogens. J. Cell Biol. *195*, 943–952.

Almeida, F., Borges, V., Ferreira, R., Borrego, M.J., Gomes, J.P., and Mota, L.J. (2012). Polymorphisms in inc proteins and differential expression of inc genes among *Chlamydia trachomatis* strains correlate with invasiveness and tropism of lymphogranuloma venereum isolates. J. Bacteriol. *194*, 6574–6585.

Al-Zeer, M.A., Al-Younes, H.M., Kerr, M., Abu-Lubad, M., Gonzalez, E., Brinkmann, V., and Meyer, T.F. (2014). *Chlamydia trachomatis* remodels stable microtubules to coordinate Golgi stack recruitment to the chlamydial inclusion surface. Mol. Microbiol. *94*, 1285–1297.

Baer, K., Klotz, C., Kappe, S.H.I., Schnieder, T., and Frevert, U. (2007). Release of hepatic Plasmodium yoelii merozoites into the pulmonary microvasculature. PLoS Pathog. 3, e171.

Banhart, S., Saied, E.M., Martini, A., Koch, S., Aeberhard, L., Madela, K., Arenz, C., and Heuer, D. (2014). Improved plaque assay identifies a novel anti-Chlamydia ceramide derivative with altered intracellular localization. Antimicrob. Agents Chemother. *58*, 5537–5546.

Banhart, S., Rose, L., Aeberhard, L., Koch-Edelmann, S., and Heuer, D. (2018). *Chlamydia trachomatis* and its interaction with the cellular retromer. Int. J. Med. Microbiol. *308*, 197–205.

Bannantine, J.P., Rockey, D.D., and Hackstadt, T. (1998). Tandem genes of Chlamydia psittaci that encode proteins localized to the inclusion membrane. Mol. Microbiol. *28*, 1017–1026.

Beatty, W.L. (2006). Trafficking from CD63-positive late endocytic multivesicular bodies is essential for intracellular development of *Chlamydia trachomatis*. J. Cell Sci. *119*, 350–359.

Beatty, W.L. (2008). Late endocytic multivesicular bodies intersect the chlamydial inclusion in the absence of CD63. Infect. Immun. *76*, 2872–2881.

Behnia, R., and Munro, S. (2005). Organelle identity and the signposts for membrane traffic. Nature *438*, 597–604.

Belland, R.J., Zhong, G., Crane, D.D., Hogan, D., Sturdevant, D., Sharma, J., Beatty, W.L., and Caldwell, H.D. (2003). Genomic transcriptional profiling of the developmental cycle of *Chlamydia trachomatis*. Proc. Natl. Acad. Sci. U. S. A. *100*, 8478–8483.

Boncompain, G., Müller, C., Meas-Yedid, V., Schmitt-Kopplin, P., Lazarow, P.B., and Subtil, A. (2014). The intracellular bacteria Chlamydia hijack peroxisomes and utilize their enzymatic capacity to produce bacteria-specific phospholipids. PLoS One *9*, e86196.

Brunham, R.C., and Rey-Ladino, J. (2005). Immunology of Chlamydia infection: implications for a *Chlamydia trachomatis* vaccine. Nat. Rev. Immunol. *5*, 149–161.

Burda, P.-C., Caldelari, R., and Heussler, V.T. (2017). Manipulation of the Host Cell Membrane during Plasmodium Liver Stage Egress. MBio *8*, e00139-17.

Capmany, A., and Damiani, M.T. (2010). *Chlamydia trachomatis* intercepts Golgi-derived sphingolipids through a Rab14-mediated transport required for bacterial development and replication. PLoS One *5*, e14084.

Carabeo, R.A., Mead, D.J., and Hackstadt, T. (2003). Golgi-dependent transport of cholesterol to the *Chlamydia trachomatis* inclusion. Proc. Natl. Acad. Sci. U. S. A. *100*, 6771–6776.

Carlson, J.H., Whitmire, W.M., Crane, D.D., Wicke, L., Virtaneva, K., Sturdevant, D.E., Kupko, J.J., 3rd, Porcella, S.F., Martinez-Orengo, N., Heinzen, R.A., *et al.* (2008). The *Chlamydia trachomatis* plasmid is a transcriptional regulator of chromosomal genes and a virulence factor. Infect. Immun. *76*, 2273–2283.

Chen, A.L., Johnson, K.A., Lee, J.K., Sütterlin, C., and Tan, M. (2012). CPAF: a Chlamydial protease in search of an authentic substrate. PLoS Pathog. *8*, e1002842.

Chin, E., Kirker, K., Zuck, M., James, G., and Hybiske, K. (2012). Actin recruitment to the Chlamydia inclusion is spatiotemporally regulated by a mechanism that requires host and bacterial factors. PLoS One *7*, e46949.

Chowdhury, S.R., Reimer, A., Sharan, M., Kozjak-Pavlovic, V., Eulalio, A., Prusty, B.K., Fraunholz, M., Karunakaran, K., and Rudel, T. (2017). Chlamydia preserves the mitochondrial network necessary for replication via microRNA-dependent inhibition of fission. J. Cell Biol. *216*, 1071–1089.

Clausen, J.D., Christiansen, G., Holst, H.U., and Birkelund, S. (1997). *Chlamydia trachomatis* utilizes the host cell microtubule network during early events of infection. Mol. Microbiol. *25*, 441–449.

Cocchiaro, J.L., Kumar, Y., Fischer, E.R., Hackstadt, T., and Valdivia, R.H. (2008). Cytoplasmic lipid droplets are translocated into the lumen of the *Chlamydia trachomatis* parasitophorous vacuole. Proc. Natl. Acad. Sci. U. S. A. *105*, 9379–9384.

Conner, S.D., and Schmid, S.L. (2003). Regulated portals of entry into the cell. Nature *422*, 37–44.

Cowman, A.F., and Kappe, S.H.I. (2006). Microbiology. Malaria's stealth shuttle. Science *313*, 1245–1246.

Cox, J.V., Naher, N., Abdelrahman, Y.M., and Belland, R.J. (2012). Host HDL biogenesis machinery is recruited to the inclusion of *Chlamydia trachomatis*-infected cells and regulates chlamydial growth. Cell. Microbiol. *14*, 1497–1512.

Cox, J.V., Abdelrahman, Y.M., Peters, J., Naher, N., and Belland, R.J. (2016). *Chlamydia trachomatis* utilizes the mammalian CLA1 lipid transporter to acquire host phosphatidylcholine essential for growth. Cell. Microbiol. *18*, 305–318.

D'Angelo, G., Capasso, S., Sticco, L., and Russo, D. (2013). Glycosphingolipids: synthesis and functions. FEBS J. *280*, 6338–6353.

Dehoux, P., Flores, R., Dauga, C., Zhong, G., and Subtil, A. (2011). Multi-genome identification and characterization of chlamydiae-specific type III secretion substrates: the Inc proteins. BMC Genomics *12*, 109.

Delevoye, C., Nilges, M., Dautry-Varsat, A., and Subtil, A. (2004). Conservation of the biochemical properties of IncA from *Chlamydia trachomatis* and Chlamydia caviae: oligomerization of IncA mediates interaction between facing membranes. J. Biol. Chem. *279*, 46896–46906.

Delevoye, C., Nilges, M., Dehoux, P., Paumet, F., Perrinet, S., Dautry-Varsat, A., and Subtil, A. (2008). SNARE protein mimicry by an intracellular bacterium. PLoS Pathog. *4*, e1000022.

Derré, I., Swiss, R., and Agaisse, H. (2011). The lipid transfer protein CERT interacts with the Chlamydia inclusion protein IncD and participates to ER-Chlamydia inclusion membrane contact sites. PLoS Pathog. *7*, e1002092.

Dickinson, M.S., Anderson, L.N., Webb-Robertson, B.-J.M., Hansen, J.R., Smith, R.D., Wright, A.T., and Hybiske, K. (2019). Proximity-dependent proteomics of the *Chlamydia trachomatis* inclusion membrane reveals functional interactions with endoplasmic reticulum exit sites. PLoS Pathog. *15*, e1007698.

Doughri, A.M., Storz, J., and Altera, K.P. (1972). Mode of entry and release of chlamydiae in infections of intestinal epithelial cells. J. Infect. Dis. *126*, 652–657.

Dumoux, M., Menny, A., Delacour, D., and Hayward, R.D. (2015). A Chlamydia effector recruits CEP170 to reprogram host microtubule organization. J. Cell Sci. *128*, 3420–3434.

Eissenberg, L.G., Wyrick, P.B., Davis, C.H., and Rumpp, J.W. (1983). Chlamydia psittaci elementary body envelopes: ingestion and inhibition of phagolysosome fusion. Infect. Immun. *40*, 741–751.

Elwell, C., Mirrashidi, K., and Engel, J. (2016). Chlamydia cell biology and pathogenesis. Nat. Rev. Microbiol. *14*, 385–400.

Elwell, C.A., Jiang, S., Kim, J.H., Lee, A., Wittmann, T., Hanada, K., Melancon, P., and Engel, J.N. (2011). *Chlamydia trachomatis* co-opts GBF1 and CERT to acquire host sphingomyelin for distinct roles during intracellular development. PLoS Pathog. *7*, e1002198.

Elwell, C.A., Czudnochowski, N., von Dollen, J., Johnson, J.R., Nakagawa, R., Mirrashidi, K., Krogan, N.J., Engel, J.N., and Rosenberg, O.S. (2017). Chlamydia interfere with an interaction between the mannose-6-phosphate receptor and sorting nexins to counteract host restriction. Elife *6*, e22709.

Feeley, E.M., Pilla-Moffett, D.M., Zwack, E.E., Piro, A.S., Finethy, R., Kolb, J.P., Martinez, J., Brodsky, I.E., and Coers, J. (2017). Galectin-3 directs antimicrobial guanylate binding proteins to vacuoles furnished with bacterial secretion systems. Proc. Natl. Acad. Sci. U. S. A. *114*, E1698–E1706.

Fields, K.A., Fischer, E., and Hackstadt, T. (2002). Inhibition of fusion of *Chlamydia trachomatis* inclusions at 32 degrees C correlates with restricted export of IncA. Infect. Immun. *70*, 3816–3823.

Friis, R.R. (1972). Interaction of L cells and Chlamydia psittaci: entry of the parasite and host responses to its development. J. Bacteriol. *110*, 706–721.

Gallon, M., and Cullen, P.J. (2015). Retromer and sorting nexins in endosomal sorting. Biochem. Soc. Trans. *43*, 33–47.

Gauliard, E., Ouellette, S.P., Rueden, K.J., and Ladant, D. (2015). Characterization of interactions between inclusion membrane proteins from *Chlamydia trachomatis*. Front. Cell. Infect. Microbiol. *5*, 13.

Gavin, A.-C., Bösche, M., Krause, R., Grandi, P., Marzioch, M., Bauer, A., Schultz, J., Rick, J.M., Michon, A.-M., Cruciat, C.-M., *et al.* (2002). Functional organization of the yeast proteome by systematic analysis of protein complexes. Nature *415*, 141–147.

Gehre, L., Gorgette, O., Perrinet, S., Prevost, M.-C., Ducatez, M., Giebel, A.M., Nelson, D.E., Ball, S.G., and Subtil, A. (2016). Sequestration of host metabolism by an intracellular pathogen. Elife *5*, e12552.

Geisler, W.M., Suchland, R.J., Rockey, D.D., and Stamm, W.E. (2001). Epidemiology and clinical manifestations of unique *Chlamydia trachomatis* isolates that occupy nonfusogenic inclusions. J. Infect. Dis. *184*, 879–884.

Gieffers, J., Tamplin, V., Maass, M., Belland, R.J., and Caldwell, H.D. (2003). Micromanipulation of the *Chlamydia pneumoniae* inclusion: implications for cloning and host–pathogen interactions. FEMS Microbiol. Lett. *226*, 45–49.

Gilk, S.D., Cockrell, D.C., Luterbach, C., Hansen, B., Knodler, L.A., Ibarra, J.A., Steele-Mortimer, O., and Heinzen, R.A. (2013). Bacterial colonization of host cells in the absence of cholesterol. PLoS Pathog. *9*, e1003107.

Graewe, S., Rankin, K.E., Lehmann, C., Deschermeier, C., Hecht, L., Froehlke, U., Stanway, R.R., and Heussler, V. (2011). Hostile takeover by Plasmodium: reorganization of parasite and host cell membranes during liver stage egress. PLoS Pathog. *7*, e1002224.

Grant, B.D., and Donaldson, J.G. (2009). Pathways and mechanisms of endocytic recycling. Nat. Rev. Mol. Cell Biol. *10*, 597–608.

Grieshaber, S., Swanson, J.A., and Hackstadt, T. (2002). Determination of the physical environment within the *Chlamydia trachomatis* inclusion using ion–selective ratiometric probes. Cell. Microbiol. *4*, 273–283.

Grieshaber, S.S., Grieshaber, N.A., and Hackstadt, T. (2003). *Chlamydia trachomatis* uses host cell dynein to traffic to the microtubule-organizing centre in a p50 dynamitin-independent process. J. Cell Sci. *116*, 3793–3802.

Hackstadt, T., Rockey, D.D., Heinzen, R.A., and Scidmore, M.A. (1996). *Chlamydia trachomatis* interrupts an exocytic pathway to acquire endogenously synthesized sphingomyelin in transit from the Golgi apparatus to the plasma membrane. EMBO J. *15*, 964–977.

Haferkamp, I. (2018). Crossing the border – Solute entry into the chlamydial inclusion. Int. J. Med. Microbiol. *308*, 41–48.

Hannun, Y.A., and Obeid, L.M. (2018). Sphingolipids and their metabolism in physiology and disease. Nat. Rev. Mol. Cell Biol. *19*, 175–191.

Hatch, T.P., Al-Hossainy, E., and Silverman, J.A. (1982). Adenine nucleotide and lysine transport in *Chlamydia psittaci*. J. Bacteriol. *150*, 662–670.

Heinzen, R.A., and Hackstadt, T. (1997). The *Chlamydia trachomatis* parasitophorous vacuolar membrane is not passively permeable to low-molecular-weight compounds. Infect. Immun. *65*, 1088–1094.

Heinzen, R.A., Scidmore, M.A., Rockey, D.D., and Hackstadt, T. (1996). Differential interaction with endocytic and exocytic pathways distinguish parasitophorous vacuoles of Coxiella burnetii and *Chlamydia trachomatis*. Infect. Immun. *64*, 796–809.

Herweg, J.-A., Hansmeier, N., Otto, A., Geffken, A.C., Subbarayal, P., Prusty, B.K., Becher, D., Hensel, M., Schaible, U.E., Rudel, T., *et al.* (2015). Purification and proteomics of pathogen-modified vacuoles and membranes. Front. Cell. Infect. Microbiol. *5*, 48.

Heuer, D., Rejman Lipinski, A., Machuy, N., Karlas, A., Wehrens, A., Siedler, F., Brinkmann, V., and Meyer, T.F. (2009). Chlamydia causes fragmentation of the Golgi compartment to ensure reproduction. Nature *457*, 731–735.

Hooppaw, A.J., and Fisher, D.J. (2015). A Coming of Age Story: Chlamydia in the Post-Genetic Era. Infect. Immun. *84*, 612–621.

Hung, V., Udeshi, N.D., Lam, S.S., Loh, K.H., Cox, K.J., Pedram, K., Carr, S.A., and Ting, A.Y. (2016). Spatially resolved proteomic mapping in living cells with the engineered peroxidase APEX2. Nat. Protoc. *11*, 456–475.

Hybiske, K., and Stephens, R.S. (2007). Mechanisms of host cell exit by the intracellular bacterium Chlamydia. Proc. Natl. Acad. Sci. U. S. A. *104*, 11430–11435.

Hybiske, K., and Stephens, R.S. (2008). Exit strategies of intracellular pathogens. Nat. Rev. Microbiol. *6*, 99–110.

Jeffrey, B.M., Suchland, R.J., Eriksen, S.G., Sandoz, K.M., and Rockey, D.D. (2013). Genomic and phenotypic characterization of *in vitro*-generated *Chlamydia trachomatis* recombinants. BMC Microbiol. *13*, 142.

Johnson, C.M., and Fisher, D.J. (2013). Site-specific, insertional inactivation of incA in *Chlamydia trachomatis* using a group II intron. PLoS One *8*, e83989.

Kabeiseman, E.J., Cichos, K., Hackstadt, T., Lucas, A., and Moore, E.R. (2013). Vesicle-associated membrane protein 4 and syntaxin 6 interactions at the chlamydial inclusion. Infect. Immun. *81*, 3326–3337.

Kabeiseman, E.J., Cichos, K.H., and Moore, E.R. (2014). The eukaryotic signal sequence, YGRL, targets the chlamydial inclusion. Front. Cell. Infect. Microbiol. *4*, 129.

Kim, D.I., and Roux, K.J. (2016). Filling the Void: Proximity-Based Labelling of Proteins in Living Cells. Trends Cell Biol. *26*, 804–817.

Kleba, B., and Stephens, R.S. (2008). Chlamydial effector proteins localized to the host cell cytoplasmic compartment. Infect. Immun. *76*, 4842–4850.

Koch-Edelmann, S., Banhart, S., Saied, E.M., Rose, L., Aeberhard, L., Laue, M., Doellinger, J., Arenz, C., and Heuer, D. (2017). The cellular ceramide transport protein CERT promotes Chlamydia psittaci infection and controls bacterial sphingolipid uptake. Cell. Microbiol. *19*, e12752.

Kokes, M., Dunn, J.D., Granek, J.A., Nguyen, B.D., Barker, J.R., Valdivia, R.H., and Bastidas, R.J. (2015). Integrating chemical mutagenesis and whole-genome sequencing as a platform for forward and reverse genetic analysis of Chlamydia. Cell Host Microbe *17*, 716–725.

Kumar, Y., and Valdivia, R.H. (2008). Actin and intermediate filaments stabilize the *Chlamydia trachomatis* vacuole by forming dynamic structural scaffolds. Cell Host Microbe *4*, 159–169.

Kumar, Y., Cocchiaro, J., and Valdivia, R.H. (2006). The obligate intracellular pathogen *Chlamydia trachomatis* targets host lipid droplets. Curr. Biol. *16*, 1646–1651.

Lam, S.S., Martell, J.D., Kamer, K.J., Deerinck, T.J., Ellisman, M.H., Mootha, V.K., and Ting, A.Y. (2015). Directed evolution of APEX2 for electron microscopy and proximity labelling. Nat. Methods *12*, 51–54.

Lee, J.K., Enciso, G.A., Boassa, D., Chander, C.N., Lou, T.H., Pairawan, S.S., Guo, M.C., Wan, F.Y.M., Ellisman, M.H., Sütterlin, C., *et al.* (2018). Replication-dependent size reduction precedes differentiation in *Chlamydia trachomatis*. Nat. Commun. *9*, 45.

Lee, S.-Y., Kang, M.-G., Park, J.-S., Lee, G., Ting, A.Y., and Rhee, H.-W. (2016). APEX Fingerprinting Reveals the Subcellular Localization of Proteins of Interest. Cell Rep. *15*, 1837–1847.

Levy, N.J., and Moulder, J.W. (1982). Attachment of cell walls of Chlamydia psittaci to mouse fibroblasts (L cells). Infect. Immun. 37, 1059–1065.

Listenberger, L.L., and Brown, D.A. (2008). Lipid droplets. Curr. Biol. *18*, R237–R238.

Lucas, A.L., Ouellette, S.P., Kabeiseman, E.J., Cichos, K.H., and Rucks, E.A. (2015). The trans-Golgi SNARE syntaxin 10 is required for optimal development of *Chlamydia trachomatis*. Front. Cell. Infect. Microbiol. *5*, 68.

Lutter, E.I., Martens, C., and Hackstadt, T. (2012). Evolution and conservation of predicted inclusion membrane proteins in chlamydiae. Comp. Funct. Genomics *2012*, 362104.

Lutter, E.I., Barger, A.C., Nair, V., and Hackstadt, T. (2013). *Chlamydia trachomatis* inclusion membrane protein CT228 recruits elements of the myosin phosphatase pathway to regulate release mechanisms. Cell Rep. 3, 1921–1931.

Martell, J.D., Deerinck, T.J., Sancak, Y., Poulos, T.L., Mootha, V.K., Sosinsky, G.E., Ellisman, M.H., and Ting, A.Y. (2012). Engineered ascorbate peroxidase as a genetically encoded reporter for electron microscopy. Nat. Biotechnol. *30*, 1143–1148.

Matsumoto, A. (1981). Isolation and electron microscopic observations of intracytoplasmic inclusions containing Chlamydia psittaci. J. Bacteriol. *145*, 605–612.

Mehlitz, A., Eylert, E., Huber, C., Lindner, B., Vollmuth, N., Karunakaran, K., Goebel, W., Eisenreich, W., and Rudel, T. (2017). Metabolic adaptation of *Chlamydia trachomatis* to mammalian host cells. Mol. Microbiol. *103*, 1004–1019.

Mirrashidi, K.M., Elwell, C.A., Verschueren, E., Johnson, J.R., Frando, A., Von Dollen, J., Rosenberg, O., Gulbahce, N., Jang, G., Johnson, T., *et al.* (2015). Global Mapping of the Inc-Human Interactome Reveals that Retromer Restricts Chlamydia Infection. Cell Host Microbe *18*, 109–121.

Mital, J., and Hackstadt, T. (2011). Diverse requirements for SRC-family tyrosine kinases distinguish chlamydial species. MBio *2*, e00031-11.

Mital, J., Miller, N.J., Fischer, E.R., and Hackstadt, T. (2010). Specific chlamydial inclusion membrane proteins associate with active Src family kinases in microdomains that interact with the host microtubule network. Cell. Microbiol. *12*, 1235–1249.

Mital, J., Lutter, E.I., Barger, A.C., Dooley, C.A., and Hackstadt, T. (2015). *Chlamydia trachomatis* inclusion membrane protein CT850 interacts with the dynein light chain DYNLT1 (Tctex1). Biochem. Biophys. Res. Commun. *462*, 165–170.

Miyashita, N., Matsumoto, A., Fukano, H., Niki, Y., and Matsushima, T. (2001). The 7.5-kb common plasmid is unrelated to the drug susceptibility of *Chlamydia trachomatis*. J. Infect. Chemother. *7*, 113–116.

Mölleken, K., and Hegemann, J.H. (2017). Acquisition of Rab11 and Rab11-Fip2-A novel strategy for *Chlamydia pneumoniae* early survival. PLoS Pathog. *13*, e1006556.

Moore, E.R., Mead, D.J., Dooley, C.A., Sager, J., and Hackstadt, T. (2011). The trans-Golgi SNARE syntaxin 6 is recruited to the chlamydial inclusion membrane. Microbiology *157*, 830–838.

Moulder, J.W. (1962). Some basic properties of the psittacosis-lymphogranuloma venereum group of agents. Structure and chemical composition of isolated particles. Ann. N. Y. Acad. Sci. *98*, 92–99.

Moulder, J.W. (1991). Interaction of chlamydiae and host cells *in vitro*. Microbiol. Rev. *55*, 143–190.

Mueller, K.E., Plano, G.V., and Fields, K.A. (2014). New frontiers in type III secretion biology: the Chlamydia perspective. Infect. Immun. *82*, 2–9.

Nans, A., Kudryashev, M., Saibil, H.R., and Hayward, R.D. (2015). Structure of a bacterial type III secretion system in contact with a host membrane *in situ*. Nat. Commun. *6*, 10114.

Nguyen, P.H., Lutter, E.I., and Hackstadt, T. (2018). *Chlamydia trachomatis* inclusion membrane protein MrcA interacts with the inositol 1,4,5-trisphosphate receptor type 3 (ITPR3) to regulate extrusion formation. PLoS Pathog. *14*, e1006911.

Nicholson, T.L., Olinger, L., Chong, K., Schoolnik, G., and Stephens, R.S. (2003). Global stage-specific gene regulation during the developmental cycle of *Chlamydia trachomatis*. J. Bacteriol. *185*, 3179–3189.

Olkkonen, V.M., and Levine, T.P. (2004). Oxysterol binding proteins: in more than one place at one time? Biochem. Cell Biol. *82*, 87–98.

Omasits, U., Ahrens, C.H., Müller, S., and Wollscheid, B. (2014). Protter: interactive protein feature visualization and integration with experimental proteomic data. Bioinformatics *30*, 884–886.

Omsland, A., Sager, J., Nair, V., Sturdevant, D.E., and Hackstadt, T. (2012). Developmental stage-specific metabolic and transcriptional activity of *Chlamydia trachomatis* in an axenic medium. Proc. Natl. Acad. Sci. U. S. A. *109*, 19781–19785.

van Ooij, C., Apodaca, G., and Engel, J. (1997). Characterization of the *Chlamydia trachomatis* vacuole and its interaction with the host endocytic pathway in HeLa cells. Infect. Immun. *65*, 758–766.

Østergaard, O., Follmann, F., Olsen, A.W., Heegaard, N.H., Andersen, P., and Rosenkrands, I. (2016). Quantitative Protein Profiling of *Chlamydia trachomatis* Growth Forms Reveals Defence Strategies Against Tryptophan Starvation. Mol. Cell. Proteomics *15*, 3540–3550.

Ouellette, S.P. (2018). Feasibility of a Conditional Knockout System for Chlamydia Based on CRISPR Interference. Front. Cell. Infect. Microbiol. *8*, 59.

Ouellette, S.P., and Carabeo, R.A. (2010). A Functional Slow Recycling Pathway of Transferrin is Required for Growth of Chlamydia. Front. Microbiol. *1*, 112.

Ouellette, S.P., Dorsey, F.C., Moshiach, S., Cleveland, J.L., and Carabeo, R.A. (2011). Chlamydia species-dependent differences in the growth requirement for lysosomes. PLoS One *6*, e16783.

Paul, B., Kim, H.S., Kerr, M.C., Huston, W.M., Teasdale, R.D., and Collins, B.M. (2017). Structural basis for the hijacking of endosomal sorting nexin proteins by *Chlamydia trachomatis*. Elife *6*, e22311.

Peters, J., Onguri, V., Nishimoto, S.K., Marion, T.N., and Byrne, G.I. (2012). The *Chlamydia trachomatis* CT149 protein exhibits esterase activity *in vitro* and catalyses cholesteryl ester hydrolysis when expressed in HeLa cells. Microbes Infect. *14*, 1196–1204.

Pokrovskaya, I.D., Szwedo, J.W., Goodwin, A., Lupashina, T.V., Nagarajan, U.M., and Lupashin, V.V. (2012). *Chlamydia trachomatis* hijacks intra-Golgi COG complex-dependent vesicle trafficking pathway. Cell. Microbiol. *14*, 656–668.

Rejman Lipinski, A., Heymann, J., Meissner, C., Karlas, A., Brinkmann, V., Meyer, T.F., and Heuer, D. (2009). Rab6 and Rab11 regulate *Chlamydia trachomatis* development and golgin-84-dependent Golgi fragmentation. PLoS Pathog. *5*, e1000615.

Richards, T.S., Knowlton, A.E., and Grieshaber, S.S. (2013). *Chlamydia trachomatis* homotypic inclusion fusion is promoted by host microtubule trafficking. BMC Microbiol. *13*, 185.

Ridderhof, J.C., and Barnes, R.C. (1989). Fusion of inclusions following superinfection of HeLa cells by two serovars of *Chlamydia trachomatis*. Infect. Immun. *57*, 3189–3193.

Roach, P.J., Depaoli-Roach, A.A., Hurley, T.D., and Tagliabracci, V.S. (2012). Glycogen and its metabolism: some new developments and old themes. Biochem. J *441*, 763–787.

Rockey, D.D., Heinzen, R.A., and Hackstadt, T. (1995). Cloning and characterization of a Chlamydia psittaci gene coding for a protein localized in the inclusion membrane of infected cells. Mol. Microbiol. *15*, 617–626.

Ronzone, E., and Paumet, F. (2013). Two coiled-coil domains of *Chlamydia trachomatis* IncA affect membrane fusion events during infection. PLoS One *8*, e69769.

Ronzone, E., Wesolowski, J., Bauler, L.D., Bhardwaj, A., Hackstadt, T., and Paumet, F. (2014). An α-helical core encodes the dual functions of the chlamydial protein IncA. J. Biol. Chem. *289*, 33469–33480.

Rucks, E.A., Olson, M.G., Jorgenson, L.M., Srinivasan, R.R., and Ouellette, S.P. (2017). Development of a Proximity Labelling System to Map the *Chlamydia trachomatis* Inclusion Membrane. Front. Cell. Infect. Microbiol. *7*, 40.

Rzomp, K.A., Scholtes, L.D., Briggs, B.J., Whittaker, G.R., and Scidmore, M.A. (2003). Rab GTPases are recruited to chlamydial inclusions in both a species-dependent and species-independent manner. Infect. Immun. *71*, 5855–5870.

Rzomp, K.A., Moorhead, A.R., and Scidmore, M.A. (2006). The GTPase Rab4 interacts with *Chlamydia trachomatis* inclusion membrane protein CT229. Infect. Immun. *74*, 5362–5373.

Saied, E.M., Banhart, S., Bürkle, S.E., Heuer, D., and Arenz, C. (2015). A series of ceramide analogues modified at the 1-position with potent activity against the intracellular growth of *Chlamydia trachomatis*. Future Med. Chem. *7*, 1971–1980.

Saka, H.A., and Valdivia, R.H. (2010). Acquisition of nutrients by Chlamydiae: unique challenges of living in an intracellular compartment. Curr. Opin. Microbiol. *13*, 4–10.

Saka, H.A., Thompson, J.W., Chen, Y.-S., Dubois, L.G., Haas, J.T., Moseley, A., and Valdivia, R.H. (2015). *Chlamydia trachomatis* Infection Leads to Defined Alterations to the Lipid Droplet Proteome in Epithelial Cells. PLoS One *10*, e0124630.

Schramm, N., and Wyrick, P.B. (1995). Cytoskeletal requirements in *Chlamydia trachomatis* infection of host cells. Infect. Immun. *63*, 324–332.

Schwöppe, C., Winkler, H.H., and Neuhaus, H.E. (2002). Properties of the glucose-6-phosphate transporter from *Chlamydia pneumoniae* (HPTcp) and the glucose-6-phosphate sensor from Escherichia coli (UhpC). J. Bacteriol. *184*, 2108–2115.

Scidmore, M.A., and Hackstadt, T. (2001). Mammalian 14–3–3beta associates with the *Chlamydia trachomatis* inclusion membrane via its interaction with IncG. Mol. Microbiol. *39*, 1638–1650.

Scidmore, M.A., Rockey, D.D., Fischer, E.R., Heinzen, R.A., and Hackstadt, T. (1996). Vesicular interactions of the *Chlamydia trachomatis* inclusion are determined by chlamydial early protein synthesis rather than route of entry. Infect. Immun. *64*, 5366–5372.

Sharma, M., Recuero-Checa, M.A., Fan, F.Y., and Dean, D. (2018). *Chlamydia trachomatis* regulates growth and development in response to host cell fatty acid availability in the absence of lipid droplets. Cell. Microbiol. *20*, e12801.

Shaw, E.I., Dooley, C.A., Fischer, E.R., Scidmore, M.A., Fields, K.A., and Hackstadt, T. (2000). Three temporal classes of gene expression during the *Chlamydia trachomatis* developmental cycle. Mol. Microbiol. *37*, 913–925.

Shaw, J.H., Behar, A.R., Snider, T.A., Allen, N.A., and Lutter, E.I. (2017). Comparison of Murine Cervicovaginal Infection by Chlamydial Strains: Identification of Extrusions Shed In vivo. Front. Cell. Infect. Microbiol. *7*, 18.

Sherrid, A.M., and Hybiske, K. (2017). *Chlamydia trachomatis* Cellular Exit Alters Interactions with Host Dendritic Cells. Infect. Immun. *85*, e00046-17.

Sixt, B.S., Bastidas, R.J., Finethy, R., Baxter, R.M., Carpenter, V.K., Kroemer, G., Coers, J., and Valdivia, R.H. (2017). The *Chlamydia trachomatis* Inclusion Membrane

Protein CpoS Counteracts STING-Mediated Cellular Surveillance and Suicide Programs. Cell Host Microbe *21*, 113–121.

Song, L., Carlson, J.H., Whitmire, W.M., Kari, L., Virtaneva, K., Sturdevant, D.E., Watkins, H., Zhou, B., Sturdevant, G.L., Porcella, S.F., *et al.* (2013). *Chlamydia trachomatis* plasmid-encoded Pgp4 is a transcriptional regulator of virulence-associated genes. Infect. Immun. *81*, 636–644.

Stanhope, R., Flora, E., Bayne, C., and Derré, I. (2017). IncV, a FFAT motif-containing Chlamydia protein, tethers the endoplasmic reticulum to the pathogen-containing vacuole. Proc. Natl. Acad. Sci. U. S. A. *114*, 12039–12044.

Stephens, R.S., Kalman, S., Lammel, C., Fan, J., Marathe, R., Aravind, L., Mitchell, W., Olinger, L., Tatusov, R.L., Zhao, Q., *et al.* (1998). Genome sequence of an obligate intracellular pathogen of humans: *Chlamydia trachomatis*. Science *282*, 754–759.

Sturm, A., Amino, R., van de Sand, C., Regen, T., Retzlaff, S., Rennenberg, A., Krueger, A., Pollok, J.-M., Menard, R., and Heussler, V.T. (2006). Manipulation of host hepatocytes by the malaria parasite for delivery into liver sinusoids. Science *313*, 1287–1290.

Su, H., McClarty, G., Dong, F., Hatch, G.M., Pan, Z.K., and Zhong, G. (2004). Activation of Raf/MEK/ERK/cPLA2 signalling pathway is essential for chlamydial acquisition of host glycerophospholipids. J. Biol. Chem. *279*, 9409–9416.

Suchland, R.J., Jeffrey, B.M., Xia, M., Bhatia, A., Chu, H.G., Rockey, D.D., and Stamm, W.E. (2008). Identification of concomitant infection with *Chlamydia trachomatis* IncA-negative mutant and wild-type strains by genomic, transcriptional, and biological characterizations. Infect. Immun. *76*, 5438–5446.

Suchland, R.J., Sandoz, K.M., Jeffrey, B.M., Stamm, W.E., and Rockey, D.D. (2009). Horizontal transfer of tetracycline resistance among Chlamydia spp. *in vitro*. Antimicrob. Agents Chemother. *53*, 4604–4611.

Tjaden, J., Winkler, H.H., Schwöppe, C., Van Der Laan, M., Möhlmann, T., and Neuhaus, H.E. (1999). Two nucleotide transport proteins in *Chlamydia trachomatis*, one for net nucleoside triphosphate uptake and the other for transport of energy. J. Bacteriol. *181*, 1196–1202.

Trentmann, O., Horn, M., van Scheltinga, A.C.T., Neuhaus, H.E., and Haferkamp, I. (2007). Enlightening energy parasitism by analysis of an ATP/ADP transporter from chlamydiae. PLoS Biol. *5*, e231.

Van Ooij, C., Homola, E., Kincaid, E., and Engel, J. (1998). Fusion of *Chlamydia trachomatis*-containing inclusions is inhibited at low temperatures and requires bacterial protein synthesis. Infect. Immun. *66*, 5364–5371.

Varnaitė, R., and MacNeill, S.A. (2016). Meet the neighbours: Mapping local protein interactomes by proximity-dependent labelling with BioID. Proteomics *16*, 2503–2518.

Vignola, M.J., Kashatus, D.F., Taylor, G.A., Counter, C.M., and Valdivia, R.H. (2010). cPLA2 regulates the expression of type I interferons and intracellular immunity to *Chlamydia trachomatis*. J. Biol. Chem. *285*, 21625–21635.

Volceanov, L., Herbst, K., Biniossek, M., Schilling, O., Haller, D., Nölke, T., Subbarayal, P., Rudel, T., Zieger, B., and Häcker, G. (2014). Septins arrange F-actin-containing fibres on the *Chlamydia trachomatis* inclusion and are required for normal release of the inclusion by extrusion. MBio *5*, e01802–e01814.

Wang, X., Hybiske, K., and Stephens, R.S. (2018). Direct visualization of the expression and localization of chlamydial effector proteins within infected host cells. Pathog. Dis. *76*, fty011.

Wang, Y., Cutcliffe, L.T., Skilton, R.J., Persson, K., Bjartling, C., and Clarke, I.N. (2013). Transformation of a plasmid-free, genital tract isolate of *Chlamydia trachomatis* with a plasmid vector carrying a deletion in CDS6 revealed that this gene regulates inclusion phenotype. Pathog. Dis. *67*, 100–103.

Weber, M.M., Bauler, L.D., Lam, J., and Hackstadt, T. (2015). Expression and localization of predicted inclusion membrane proteins in *Chlamydia trachomatis*. Infect. Immun. *83*, 4710–4718.

Weber, M.M., Noriea, N.F., Bauler, L.D., Lam, J.L., Sager, J., Wesolowski, J., Paumet, F., and Hackstadt, T. (2016). A Functional Core of IncA Is Required for *Chlamydia trachomatis* Inclusion Fusion. J. Bacteriol. *198*, 1347–1355.

Weber, M.M., Lam, J.L., Dooley, C.A., Noriea, N.F., Hansen, B.T., Hoyt, F.H., Carmody, A.B., Sturdevant, G.L., and Hackstadt, T. (2017). Absence of Specific *Chlamydia trachomatis* Inclusion Membrane Proteins Triggers Premature Inclusion Membrane Lysis and Host Cell Death. Cell Rep. *19*, 1406–1417.

Weiss, E. (1967). Transaminase activity and other enzymatic reactions involving pyruvate and glutamate in Chlamydia (psittacosis-trachoma group). J. Bacteriol. *93*, 177–184.

Wesolowski, J., Weber, M.M., Nawrotek, A., Dooley, C.A., Calderon, M., St Croix, C.M., Hackstadt, T., Cherfils, J., and Paumet, F. (2017). Chlamydia Hijacks ARF GTPases To Coordinate Microtubule Posttranslational Modifications and Golgi Complex Positioning. MBio *8*, e02280-16.

Wickstrum, J., Sammons, L.R., Restivo, K.N., and Hefty, P.S. (2013). Conditional gene expression in *Chlamydia trachomatis* using the tet system. PLoS One *8*, e76743.

Wilson, D.P., Timms, P., McElwain, D.L.S., and Bavoil, P.M. (2006). Type III secretion, contact-dependent model for the intracellular development of chlamydia. Bull. Math. Biol. *68*, 161–178.

Wu, X., Lei, L., Gong, S., Chen, D., Flores, R., and Zhong, G. (2011). The chlamydial periplasmic stress response serine protease cHtrA is secreted into host cell cytosol. BMC Microbiol. *11*, 87.

Wylie, J.L., Hatch, G.M., and McClarty, G. (1997). Host cell phospholipids are trafficked to and then modified by *Chlamydia trachomatis*. J. Bacteriol. *179*, 7233–7242.

Yao, J., Dodson, V.J., Frank, M.W., and Rock, C.O. (2015). *Chlamydia trachomatis* Scavenges Host Fatty Acids for Phospholipid Synthesis via an Acyl-Acyl Carrier Protein Synthetase. J. Biol. Chem. *290*, 22163–22173.

Zeichner, S.L. (1982). Isolation and characterization of phagosomes containing Chlamydia psittaci from L cells. Infect. Immun. *38*, 325–342.

Zeichner, S.L. (1983). Isolation and characterization of macrophage phagosomes containing infectious and heat-inactivated Chlamydia psittaci: two phagosomes with different intracellular behaviours. Infect. Immun. *40*, 956–966.

Zhong, G., Fan, P., Ji, H., Dong, F., and Huang, Y. (2001). Identification of a chlamydial protease-like activity factor responsible for the degradation of host transcription factors. J. Exp. Med. *193*, 935–942.

Zuck, M., Sherrid, A., Suchland, R., Ellis, T., and Hybiske, K. (2016). Conservation of extrusion as an exit mechanism for Chlamydia. Pathog. Dis. *74*, ftw093.

Zuck, M., Ellis, T., Venida, A., and Hybiske, K. (2017). Extrusions are phagocytosed and promote Chlamydia survival within macrophages. Cell. Microbiol. *19*, e12683.

Interactions of the Chlamydial Inclusion with the Host Cell

5

Christine Sütterlin[1]* and Isabelle Derré[2]

[1]Department of Developmental and Cell Biology, University of California, Irvine, CA, USA.
[2]Department of Microbiology, Immunology and Cancer Biology, University of Virginia, Charlottesville, VA, USA.

*Correspondence: suetterc@uci.edu

https://doi.org/10.21775/9781912530281.05

Abstract

Bacteria of the *Chlamydia* genus are obligate intracellular pathogens that reside at all times within a membrane-bound cytoplasmic compartment called the chlamydial inclusion. This replicative niche is necessary for the chlamydial developmental cycle and depends on many interactions with the host cell. In this chapter, we will discuss how these bacteria orchestrate the intricate relationship of the inclusion with most of the cellular elements of their host cell. In the first section, we will focus on the cytoskeleton and summarize our knowledge on how and why the infection alters F-actin, intermediate filaments and microtubules. We will also discuss the interaction of the inclusion with the centrosome, including its potentially negative consequences for the host cell. In the second section, we will describe how interactions between the inclusion and the endoplasmic reticulum, the Golgi and mitochondria are established, and how they promote the infection. In the last section, we will focus on associations of the inclusion with lipid droplets and peroxisomes, which have both been found to translocate into the inclusion. Overall, this description shows that chlamydiae are master manipulators of their host cell.

Introduction

Chlamydiae are obligate intracellular bacteria that infect epithelium mucosa of the host. All *Chlamydia* species share a biphasic developmental cycle that occurs strictly within the lumen of an intracellular compartment known as the inclusion (see Chapter 4, The Chlamydial Inclusion). The inclusion membrane is originally derived from the plasma membrane, but is quickly remodelled to avoid trafficking down the endocytic pathway and clearance of the bacteria. This is achieved through the use of a Type III secretion system (T3SS, described in great detail in Chapter 7, Protein Secretion in *Chlamydia*), which translocates effector molecules either into the host cytosol or the inclusion membrane. The subset of proteins that are embedded in the inclusion membrane are referred to as Inclusion membrane proteins (Inc proteins). Incs display an overall conserved structure, with a bilobed hydrophobic domain (often found in the N-terminal part) that is embedded in the inclusion membrane, flanked by tails facing the host cytosol. Inc proteins facilitate the interaction of the inclusion with cellular factors and organelles, thereby allowing chlamydiae to manipulate the host cell and to establish a productive infection from within the safe haven of the inclusion. Incs constitute 5–10% of the *Chlamydia* proteome, reflecting their importance to the intracellular infection (Dehoux *et al.*, 2011; Lutter *et al.*, 2012).

Complementary approaches have shed light on the molecular mechanisms that mediate the interactions of the inclusion with cellular structures and organelles. For example, electron and fluorescence microscopy-based studies showed that the inclusion is located in the cell centre. Genome-wide

RNAi screens have identified host factors that are essential for *Chlamydia* intracellular replication. Moreover, interaction screens revealed host cell proteins that bind *Chlamydia* effector proteins. Finally, genetic approaches, taking advantage of newly developed tools to genetically manipulate chlamydiae, have established essential roles of *Chlamydia* factors in these interactions.

In this chapter, unless specifically stated, we will review how the *Chlamydia trachomatis* inclusion interacts with the three key cytoskeletal factors (F-actin, intermediate filaments and microtubules) as well as the centrosome. We will also describe the association of the inclusion with cellular organelles that are recruited to its periphery (Golgi, endoplasmic reticulum and mitochondria) and with those that translocate into its lumen (lipid droplets and peroxisomes). A graphical representation of these interactions is shown in Fig. 5.1, and specific microscopic illustrations are provided throughout the chapter. We will summarize in the text and in Table 5.1 how these interactions are established

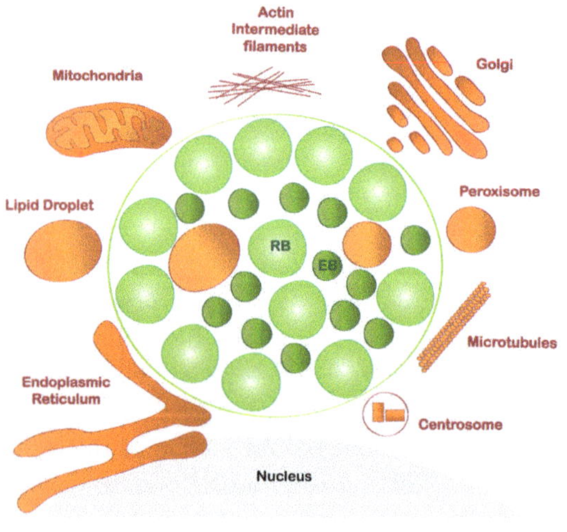

Figure 5.1 The *Chlamydia* inclusion associates with numerous cellular elements. Cartoon depicting interactions of host cytoskeletal elements (filamentous actin, intermediate filaments and microtubules), the centrosome and cellular organelles (endoplasmic reticulum, Golgi, mitochondria, lipid droplets and peroxisome) with the *Chlamydia* inclusion. Most cellular elements are recruited to the vicinity of the inclusion, while some translocate into the inclusion lumen. Cellular components and organelles are depicted in orange. The inclusion membrane and the *Chlamydia* are shown in green (RB, reticulate bodies, light green and EB, elementary bodies, dark green).

and maintained during the infection, and how they promote the infection.

Interaction of the inclusion with the host cell cytoskeleton and the centrosome

In this first section, we will discuss the interplay of the inclusion with F-actin, intermediate filaments and microtubules, which are the main components of the host cell cytoskeleton. These cytoskeletal elements each form filamentous structures that are assembled from monomeric proteins as building blocks and fulfil critical functions, ranging from structural support for cell architecture to roles in vesicular trafficking. We will also describe how the infection affects the centrosome, the major microtubule organizing centre (MTOC) of the cell.

F-Actin

The actin cage around the mid-stage *C. trachomatis* inclusion

Chlamydia manipulates the actin cytoskeleton during the entire developmental cycle. Here, we will focus on the actin cage that surrounds a chlamydial inclusion during the mid-stage of the intracellular infection. The role of regulated actin reorganization in bacterial entry and exit from the host cell will be discussed in Chapter 3, *Chlamydia* Adhesion and Invasion, and Chapter 4, The Chlamydial Inclusion, respectively.

Although actin stress fibres and cortical actin are unaffected in a *Chlamydia*-infected cell (Campbell *et al.*, 1989; Kumar and Valdivia, 2008), the inclusion is encased by a compact uniform ring of actin filaments, as first described by the Valdivia lab (Kumar and Valdivia, 2008) (Fig. 5.2A). This unique F-actin arrangement, which is commonly referred to as an actin cage or coat, was also visualized in cells expressing the GFP-tagged F-actin binding domain of moesin (Kumar and Valdivia, 2008). Additional fluorescence and electron microscopy analyses revealed that thick actin filaments around the inclusion were beginning to form at 20 hours post-infection (hpi) in *C. trachomatis* LGV-L2-infected cells (Chin *et al.*, 2012; Kumar and Valdivia, 2008). The strength and prevalence of this actin arrangement increased with the progressing *Chlamydia* LGV-L2 infection,

Table 5.1 Characteristics of host and bacterial factors involved in a physical or functional link between the inclusion and the cytoskeleton, the centrosome and cellular organelles

Bacterial factor	Host factor	Characteristics and proposed role	References
Actin			
InaC/CT813		Actin cage formation	Kokes *et al.* (2015)
	Formin	Nucleation of the actin cage	Chin *et al.* (2012)
	Septins	Actin cage formation	Chin *et al.* (2012); Volceanov *et al.* (2014)
Intermediate filaments vimentin, **keratin-8 keratin-18**			
	F-actin	Vimentin cage formation	Kumar and Valdivia (2008)
	RhoA	Vimentin cage formation	Kumar and Valdivia (2008)
	O-GlcNAc transferase	Vimentin dimerization and cage formation	Tarbet *et al.* (2018)
Microtubules			
	Dynein	Inclusion movement along microtubules	Clausen *et al.* (1997); Grieshaber *et al.* (2003); Mital *et al.* (2015)
CT850	Tctex1	Inclusion movement along microtubules	Mital *et al.* (2015)
	RhoA; ROCK	Post-translational modification of microtubules	Al-Zeer *et al.* (2014)
	Arf1/4	Post-translational modification of microtubules	Wesolowski *et al.* (2017)
CopN		Microtubule depolymerization in *C. pneumoniae*-infected cells	Archuleta *et al.* (2011); Huang *et al.* (2008); Nawrotek *et al.* (2014); Campanacci *et al.* (2019)
Centrosome			
	λ-tubulin	Microtubule nucleation	Mital *et al.* (2010)
CT850	Tctex1	Centrosome recruitment to the inclusion	Mital *et al.* (2015); Mital *et al.* (2010)
CT223/IPAM	CEP170	Centrosome recruitment to inclusion, microtubule reorganization	Alzhanov *et al.* (2009); Dumoux *et al.* (2015a)
CT288	CCDC146	Localize to inclusion membrane	Almeida *et al.* (2018)
	Cdk1; Plk4	Kinase inactive form blocked *Chlamydia*-induced centrosome amplification	Johnson *et al.* (2009)
CPAF		Reduced rate of centrosome amplification when mutated	Brown *et al.* (2014)
Endoplasmic reticulum			
	Calreticulin; DP1; Rtn4B	Inclusion membrane localized	Dumoux *et al.* (2012)
IncD	CERT	Localize to ER–inclusion MCS; sphingomyelin acquisition; (CERT only); bacterial replication (CERT only)	Agaisse and Derré (2014); Derré *et al.* (2011); Elwell *et al.* (2011)
IncV	VAPA, VAPB	ER–inclusion MCS localized; ER tethering to the inclusion; bacterial replication (VAPA/B only)	Derré *et al.* (2011); Elwell *et al.* (2011); Stanhope *et al.* (2017)
	STIM1	ER–inclusion MCS and inclusion microdomains localized; extrusion	Agaisse and Derré (2015); Nguyen *et al.* (2018)
CT101/MrcA	ITPR3	Localize to inclusion microdomains; extrusion	Nguyen *et al.* (2018)
Golgi			
	Rab6, Rab11	Golgi fragmentation	Rejman Lipinski *et al.* (2009)
	Actin	Golgi recruitment to the inclusion	Kokes and Valdivia (2015)
	Micro-tubules	Golgi recruitment to the inclusion	Al-Zeer *et al.* (2014)
InaC/CT813		Golgi recruitment to the inclusion	Kokes and Valdivia (2015)

Table 5.1 Continued

Bacterial factor	Host factor	Characteristics and proposed role	References
Mitochondria			
	Kinesin	Mitochondria recruitment to *C. psittaci* inclusion	Escalante-Ochoa *et al.* (1999)
	miR-30c	Overexpressed in *C. trachomatis*-infected cells, bacterial replication via mitochondria morphology and integrity	Chowdhury *et al.* (2017)
	Drp1	Downregulated in *C. trachomatis*-infected cells, bacterial replication via mitochondria morphology and integrity	Chowdhury *et al.* (2017)
	Tom40, Tom22	*C. caviae* replication	Derré *et al*. (2007)
Lipid droplets			
CT156 (Lda1), CT163 (Lda2), CT527		Localized to lipid droplets	Kumar *et al.* (2006)
CT473 (Lda3)		Docking of lipids droplet to inclusion, removal of ADRP coat protein	Cocchiaro *et al.* (2008); Kumar *et al.* (2006)
Cap1/CT529, CT618, IncG		Localized to lipid droplets	Saka *et al.* (2015)
	DGAT1/2	Bacterial replication	Saka *et al.* (2015)
	Long chain acyl-CoA synthetase	Bacterial replication	Kumar *et al.* (2006); Recuero-Checa *et al.* (2016)
CT775	hACBD6	Localize to lipid droplets, CT775 acyl transferase activity regulated by hACBD6	Soupene *et al.* (2015)
Peroxisomes			
	Pex19	Inclusion size	Boncompain *et al.* (2014)

Proteins listed on the same line have been shown to interact and, unless noted, share the same characteristics and proposed role. Please note that there are strain/species-specific differences for most of the interactions listed below and refer to the text to determine if they are conserved beyond *C. trachomatis*.

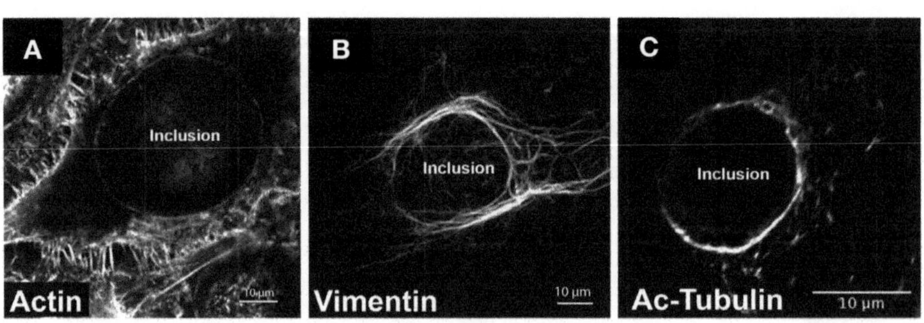

Figure 5.2 *Chlamydia* inclusion associates with the host cytoskeleton. A. Confocal images of HeLa cells infected with *C. trachomatis* LGV-L2 for 30h and stained for F-actin. Image courtesy of Lee Dolan and Raphael Valdivia. B. Confocal images of HeLa cells expressing Vimentin-mEmerald and infected with *C. trachomatis* LGV-L2 for 30h. Image courtesy of Lee Dolan and Raphael Valdivia. C. Confocal image of HeLa cells infected with *C. trachomatis* LGV-L2 for 24h and stained for Acetylated-Tubulin. Image courtesy of Jordan Weslowski and Fabienne Paumet. In each image the cytoskeletal elements prominently wrap around the inclusion.

but the precise timing differed between investigators, which may be due to different actin detection methods. While Kumar and Valdivia observed an actin cage in 95% of infected HeLa cells at 30 hpi using fluorescent phalloidin to label actin (Kumar and Valdivia, 2008), Chin *et al.* used LifeAct-GFP to detect a prevalent actin rearrangement in only 55% of infected HeLa cells at 68 hpi (Chin *et al.*, 2012). Live imaging in these cells demonstrated that this actin cage is highly dynamic and that it is actively recruited to the inclusion (Chin *et al.*, 2012).

Regulation of actin cage formation by bacterial and host factors

Treatment with the antibiotics, chloramphenicol or rifampin, or the putative T3SS inhibitor C1 prevented actin cage assembly (Chin *et al.*, 2012). This suggests that F-actin reorganization in an infected cell is likely controlled by bacterial proteins that are either in the inclusion membrane or secreted into the host cell cytosol. In order to identify such chlamydial factors, Kokes and colleagues screened an arrayed library of chemically induced mutants for the absence of the actin cage around the inclusion (Kokes *et al.*, 2015). This microscopy-based screen resulted in the isolation of a *C. trachomatis* strain with a nonsense mutation in the Inc protein InaC/CT813 (Table 5.1). The *inaC* mutant was no longer able to induce F-actin rearrangement around the inclusion, a phenotype that was reverted by expression of the wild-type allele of *inaC*, definitively linking phenotype to genotype (Kokes *et al.*, 2015).

On the host side, pharmacological approaches, together with RNAi knockdown experiments, were used to examine well known actin regulatory factors for a role in F-actin cage formation. For example, actin nucleation at the inclusion was characterized by using small molecule inhibitors to block the two actin nucleating factors, Arp2/3 and formin, which promote nucleation of branched and linear actin filaments, respectively. While inhibition of Arp2/3 activation with CK-548 and CK-666 had no effect, blocking the activity of formin with SMIFH2 prevented cage formation (Chin *et al.*, 2012) (Table 5.1). Additional regulatory factors include the small GTPase RhoA, myosin II and Src family kinase signalling (Chin *et al.*, 2012; Kumar and Valdivia, 2008). Furthermore, selected septins, which are filament-forming small GTPases, were found to contribute to actin cage formation (Chin *et al.*, 2012; Volceanov *et al.*, 2014) (Table 5.1).

Role of the actin cage during infection

At least three distinct functions have been attributed to the actin cage around the mid-stage *C. trachomatis* inclusion. An obvious role is to provide mechanical support to the chlamydial inclusion. This is illustrated by the fact that treatment of cells with the actin depolymerizing drugs cytochalasin D and latrunculin A resulted in increased sensitivity of the inclusion to detergent and led to the presence of bacteria in the host cell cytosol (Kumar and Valdivia, 2008). Another function is the recruitment and positioning of Golgi membranes around the inclusion, exemplified by the inability of the *inaC* mutant strain, which does not induce F-actin cage formation, to recruit Golgi membranes to the inclusion (Kokes *et al.*, 2015). Lastly, the actin cage is necessary for the formation of networks made up of septins (Volceanov *et al.*, 2014) or intermediate filaments (Kumar and Valdivia, 2008), as will be discussed in the next section of this chapter.

In spite of these seemingly critical functions of actin reorganization around the inclusion, the overall significance of the actin cage for a *Chlamydia* infection is unclear. This mainly stems from studies with the *inaC* mutant strain, which showed that the absence of the actin cage did not alter chlamydial growth or inclusion integrity in the standard cell culture infection model (Kokes *et al.*, 2015). Furthermore, not all chlamydial species induce actin reorganization (Chin *et al.*, 2012). While an actin cage was prominently seen around *C. trachomatis* LGV-L2 inclusions, it was less pronounced in infections with *C. trachomatis* serovar D and *C. caviae* (Chin *et al.*, 2012). In addition, *C. muridarum* inclusions displayed minimal actin recruitment with a punctate rather than a ring-like pattern. Together these results suggest that alternative mechanisms are likely to support the inclusion in these species.

Intermediate filaments

The intermediate filament network around the chlamydial inclusion

In 1989, Campbell and colleagues first reported that the cytokeratin meshwork, normally present around the nucleus, was reorganized to encase the nucleus and the inclusion of a *C. trachomatis* (LGV-L2)-infected cell (Campbell *et al.*, 1989). Immunofluorescence analysis by the Valdivia lab confirmed these initial reports and showed that vimentin-, keratin-8- and keratin-18-positive intermediate filaments assembled around the inclusion (Kumar and Valdivia, 2008) (Fig. 5.2B). When infected, a reporter cell line expressing vimentin-GFP showed a similar phenotype. The association of these intermediate filaments with the inclusion increased with progression of the developmental cycle and was seen in about 70% of the infected cells at 30 hpi (hours post infection).

Factors that influence the vimentin cage

Vimentin recruitment to the inclusion was investigated with pharmacological approaches (Kumar and Valdivia, 2008). In conditions that interfered with actin polymerization or RhoA activity, vimentin reorganization was not observed, suggesting that assembly of this intermediate filament around the inclusion depended on an intact actin cage (Table 5.1). Interestingly, these effects were reversible because the vimentin cage reformed shortly after the removal of the actin depolymerizing drug. In contrast, microtubule depolymerization did not alter vimentin reorganization in an infected cell, but it is not known if the recruitment of keratin-8 and -18 follows the same requirements. Furthermore, it is unclear which chlamydial factor(s) induce the recruitment of these intermediate filaments to the inclusion.

A glycosylation-dependent mechanism may be important for vimentin reorganization in *Chlamydia*-infected cells, as recently revealed by Tarbet and colleagues (Tarbet *et al.*, 2018). In uninfected cells, vimentin filament assembly was found to be dependent on the site-specific post-translational modification of vimentin with O-linked-β-N-acetylglucosamine (O-GlcNAc). More specifically, a vimentin glycosylation site mutant or inhibition of the N-acetylglucosamine transferase with the specific inhibitor 5SGlcNAc affected vimentin dimerization and induced severe alterations in vimentin filament morphology. Interestingly, prevention of vimentin O-GlcNAc modification with these same methods led to perturbation of the vimentin cage around the inclusion, suggesting that *Chlamydia*-induced vimentin reorganization requires vimentin O-GlcNAc modification (Table 5.1).

It was originally proposed that the intermediate filament cage is remodelled to accommodate the growing inclusion (Kumar and Valdivia, 2008). The proposed mechanism involved cleavage of vimentin, keratin-8 and keratin-18 by the chlamydial protease CPAF. However, there are now doubts about this model because Chen *et al.* demonstrated that keratins-8 and -18 are not cleaved in intact infected cells (Chen *et al.*, 2012). There is controversy about whether vimentin is cleaved by CPAF in infected cells. When precautions were taken to prevent proteolysis by CPAF during cell lysate preparation, researchers have either detected no vimentin cleavage or only a small amount of cleavage that occurred very late in the intracellular infection (≥ 48 hours post infection) (Chen *et al.*, 2012; Snavely *et al.*, 2014).

Role of the vimentin cage during infection

As observed for F-actin, the intermediate filament cage is critical for the maintenance and stabilization of the inclusion. Inclusions in infected vimentin (–/–) MEFs showed increased sensitivity to detergent treatment as an indication of an unstable inclusion (Kumar and Valdivia, 2008). Further evidence for vimentin providing structural support to the inclusion came from cells that were defective in vimentin O-GlcNAc modifications (Tarbet *et al.*, 2018). Infection of these cells resulted in smaller inclusions with compromised integrity, as demonstrated by an increase in the number of bacteria released into the cytosol. Finally, the vimentin cage was found to be required for proper actin reorganization around the inclusion because F-actin rings appeared disordered in vimentin (–/–) MEFs (Kumar and Valdivia, 2008). These results suggest that F-actin and intermediate filament cages cooperate to provide structural support for the inclusion.

Microtubules

Transport of the inclusion to the cell centre

The first indication of an important link between microtubules and the inclusion came from experiments by Schramm and Wyrick, who tested the effects of microtubule depolymerizing compounds on the chlamydial infection (Schramm and Wyrick, 1995). Treatment of infected cells with nocodazole or colchicine early in the infection prevented inclusion fusion and led to cells with multiple inclusions. This phenotype suggested that the clustering and fusion of inclusions might depend on transport of inclusions along microtubules towards the centrosome in the centre of the cell. However, at that time the authors failed to detect convincing co-localization between nascent inclusions and microtubules.

Several studies have provided additional support for microtubule- and dynein-dependent positioning of the inclusion in the cell centre (Clausen *et al.*, 1997; Grieshaber *et al.*, 2003; Mital *et al.*, 2015). The Birkelund and Hackstadt labs both demonstrated that between 2 and 5 hpi, *C. trachomatis* LGV-L2 and serovar D inclusions move towards the microtubule minus end where they eventually cluster (Clausen *et al.*, 1997; Grieshaber *et al.*, 2003). This process, which occurred at a rate that was comparable to dynein-dependent transport of vesicles along microtubules, was blocked by inhibitors of bacterial protein synthesis, microtubule polymerization or dynein motor activity (Table 5.1), (Clausen *et al.*, 1997; Grieshaber *et al.*, 2003). Furthermore, dynein was detected on *C. trachomatis* LGV-L2 inclusions using immunofluorescence microscopy (Grieshaber *et al.*, 2003). Interfering with dynactin, which mediates cargo binding to dynein, did not affect inclusion transport, suggesting that a chlamydial factor, potentially an Inc protein, may link the inclusion to the dynein motor. Overall, these findings led to the accepted model that the inclusion reaches the cell centre by dynein-mediated transport along microtubules, although a definitive demonstration of this idea is still missing. It is not yet known how the inclusion is kept close to the microtubule organizing centre (MTOC) and if/how it is unloaded from the microtubule motor upon arrival. Surprisingly, preventing the perinuclear inclusion positioning through RNAi-mediated depletion of the dynein light chain Tctex1 did not reduce infectious progeny (Mital *et al.*, 2015) (Table 5.1). These results raise the question of the significance of inclusion positioning in the cell centre, at least in an *in vitro* infection of non-polarized cells.

A cage made up of post-translationally modified microtubules around the midstage inclusion

At the mid-stage of the infection, a network of microtubules forms around the *C. trachomatis* inclusion, with microtubule plus ends extending towards the plasma membrane (Al-Younes *et al.*, 2011; Al-Zeer *et al.*, 2014; Dumoux *et al.*, 2015a). As the infection progresses, tubulin subunits within these microtubules are increasingly modified by detyrosination and acetylation, which are modifications known to stabilize microtubules (Al-Zeer *et al.*, 2014). Immunofluorescence microscopy showed that these post-translationally modified microtubules were restricted to the close proximity of the inclusion (Fig. 5.2C). The post-translational modification of these cage microtubules may be conserved among different *Chlamydia* species because modified microtubules were also detected around *C. pneumoniae* and *C. muridarum* inclusions (Al-Zeer *et al.*, 2014).

Several groups have investigated the mechanism of microtubule modification in *Chlamydia*-infected cells. Al-Zeer and colleagues reported that microtubule modification required the activity of RhoA and ROCK, two regulators of F-actin dynamics that also control microtubule modification and stabilization (Al-Zeer *et al.*, 2014; Heng *et al.*, 2012) (Table 5.1). Furthermore, Paumet and colleagues found that enhanced modification of microtubules required the recruitment and activation of the small GTPases Arf1 and Arf4 at the inclusion (Wesolowski *et al.*, 2017) (Table 5.1). The *inaC/ct813* mutant, which is defective in recruiting these two small GTPases to the inclusion (Kokes *et al.*, 2015), lacked enhanced microtubule modification (Wesolowski *et al.*, 2017). A similar phenotype was also seen in infected cells depleted of Arf1 and Arf4 by RNAi. At this point, it is not known how these small GTPases, which normally function in protein transport, enhance microtubule modification.

Role of the microtubule cage around the inclusion

The microtubule cage encasing the inclusion is important for infection. When Al-Zeer and colleagues induced microtubule depolymerization and disruption of the cage in time windows between 10 and 25 hpi or 25 to 48 hpi, bacterial progeny was reduced (Al-Zeer *et al.*, 2014). Interestingly, this defect was not due to inclusion destabilization because inclusion morphology was normal in the absence of the microtubule cage (Al-Zeer *et al.*, 2014). It may be caused by defects in microtubule-dependent recruitment of proteins, such as the microtubule-binding proteins LC3 and MAP1, but it is not known if/how these proteins promote a productive infection (Al-Younes *et al.*, 2011). The reduced progeny could also result from defects in the recruitment of Golgi fragments to the inclusion because modified stable microtubules were required for Golgi reorganization and positioning (Al-Zeer *et al.*, 2014), as will be discussed in the Golgi section of this chapter.

Role of *C. pneumoniae* CopN in microtubule depolymerization

In a *C. pneumoniae* infection, an additional relationship between microtubule dynamics and the infection has been reported (Huang *et al.*, 2008). Using heterologous systems in yeast and mammalian cells, Huang and colleagues reported that expression of CopN, a proposed component of the T3S translocation machinery and T3S effector (Fields and Hackstadt, 2000), promoted microtubule depolymerization and caused cell cycle arrest (Huang *et al.*, 2008) (Table 5.1). The role of CopN during infection was investigated using a chemical biology approach. Two small molecules were found to counteract CopN-induced cell cycle arrest and inhibited the *C. pneumoniae* infection (Huang *et al.*, 2008). It is unclear how CopN activity has beneficial effects on the infection. This protein was found to sequester α/β–tubulin heterodimers *in vitro*, which led to a significant decrease in microtubule nucleation. A recent study confirmed a possible role of CopN in microtubule nucleation, although a mechanism was proposed that involved preventing the binding of microtubule protofilaments (Campanacci *et al.*, 2019). An alternative mechanism focused on the interaction of CopN with the enzyme fructose biphosphate aldolase A, which was found to limit the infection (Ishida *et al.*, 2014). Interestingly, only CopN from *C. pneumoniae* and *C. abortus*, but not from *C. trachomatis*, affected microtubule dynamics and cell cycle progression in yeast (Huang *et al.*, 2008). It is not known if CopN is able to induce disassembly of the modified microtubules that surround the inclusion.

The centrosome

The centrosome functions as the major microtubule organizing centre (MTOC) of the cell by nucleating the radial array of microtubules in interphase and spindle microtubules in mitosis. This non-membrane bound organelle contains about 500 proteins. In a newly formed daughter cell, it is organized into two centrioles, which are microtubule-based barrels, and the pericentriolar material, which is arranged in concentric layers around the centrioles. During centrosome duplication in S-phase, each centriole serves as a template for the formation of a daughter centriole. To maintain proper centriole number, it is critical that each centriole duplicates only once per cell cycle, with one daughter centriole forming per mother centriole template.

During most of the chlamydial developmental cycle, the centrosome has been detected in close association with the inclusion. As discussed above, this proximity is established early during the intracellular infection when the nascent inclusion moves from the cell periphery to the cell centre (Clausen *et al.*, 1997; Grieshaber *et al.*, 2003) and is maintained as the infection progresses. Early on, *Chlamydia* may use its link to the centrosome to establish its intracellular niche in the metabolite-rich cell centre. However, it is not clear why the centrosome is subsequently mislocalized to the inclusion. As discussed below three specific protein pairs, each made up of a centrosomal protein from the host cell and an Inc protein, have been described, but surprisingly only one of these complexes seems to be required for establishing the physical proximity between these two organelles.

Molecular mechanisms that link the centrosome to the inclusion

Inclusion microdomains

Inclusion membrane microdomains were defined by Hackstadt and colleagues as small subsections of the inclusion membrane where active Src-family kinases colocalize with a set of specific Inc proteins (IncB, IncC, CT101, CT222, CT223/IPAM, CT224, CT228, CT288 and CT850) (Lutter *et al.*, 2013; Mital *et al.*, 2010; Weber *et al.*, 2015). Inclusion microdomains also stained positive for γ-tubulin, a tubulin isoform that is necessary for microtubule nucleation at the centrosome (Mital *et al.*, 2010) (Table 5.1). Here we will focus on how specific microdomain Inc proteins associate with centrosomal proteins to recruit the centrosome to the inclusion.

CT850 and Tctex1

In interphase and mitosis, the centrosome, detected with antibodies to γ-tubulin, is in close apposition to CT850- and CT101-containing microdomains (Mital *et al.*, 2010). When expressed ectopically, CT850 localized to the centrosome (Mital *et al.*, 2010). This specific localization motivated Mital *et al.* to perform a Yeast Two-Hybrid screen, which identified the dynein light chain Tctex1, a subunit of the dynein/dynactin complex, as a CT850 interactor (Mital *et al.*, 2015). Tctex1 binding to CT850 was necessary for recruitment of the centrosome to the inclusion because the centrosome was separated from the inclusion in *C. trachomatis*-infected cells depleted of Tctex1 (Table 5.1). Interestingly, Tctex1-depletion had no effect on chlamydial infectivity (Mital *et al.*, 2015), raising the question about the significance of centrosome recruitment to the inclusion.

CT223/IPAM and CEP170

The Hayward lab identified an interaction between the microdomain Inc protein CT223/IPAM and the centrosomal protein CEP170 (Dumoux *et al.*, 2015a). A possible connection between CT223/IPAM and the microtubule cytoskeleton was proposed based on sequence homology of CT223/IPAM with microtubule- and centrosome-interacting proteins. Pull-down assays showed that CT223/IPAM interacted with the mother centriole-specific protein CEP170 (Dumoux *et al.*, 2015a) (Table 5.1). In uninfected cells, ectopically expressed CT223/IPAM localized to the centrosome (Alzhanov *et al.*, 2009; Dumoux *et al.*, 2012). However, RNAi-mediated CEP170 depletion did not interfere with this localization, suggesting that another host protein is necessary for CT223/IPAM recruitment to the centrosome. It is currently not clear why depletion of CEP170 from infected cells resulted in changes in the microtubule network around the inclusion and decreased infectivity. Ectopic expression of CT223/IPAM also caused defects in cytokinesis, but the mechanism is not known (Alzhanov *et al.*, 2009).

CT288 and CCDC146

The Mota Lab showed that the microdomain Inc protein CT288 interacts with the human centrosomal protein CCDC146 (coiled-coil domain-containing protein 146) (Almeida *et al.*, 2018) (Table 5.1). This interaction was observed by Yeast-2 Hybrid analysis, co-immunoprecipitation after co-expression in uninfected mammalian cells and when CT288 was exogenously expressed in *C. trachomatis* (Almeida *et al.*, 2018). In uninfected cells, CCDC146 co-localized with γ-tubulin at the centrosome, but in infected cells it was recruited to the entire circumference of the inclusion. A *ct288* mutant was still able to recruit CCDC146 to the inclusion suggesting that redundant mechanisms for CCDC146 recruitment must exist. At this point it is unclear how the CT228/CCDC146 pair contributes to inclusion–centrosome association.

Centrosome abnormalities in *Chlamydia*-infected cells

C. trachomatis infection causes centrosome dysregulation. These centrosome abnormalities were first observed by Grieshaber and colleagues and included the recruitment of the centrosome to the inclusion membrane, leading to the centrosome being pulled away from its normal perinuclear position (Grieshaber *et al.*, 2006). Furthermore, about 60% of infected cells contained supernumerary, or extra, centrosomes, which were seen as multiple γ-tubulin-positive foci (Fig. 5.3). Centrosome abnormalities were still observed five days after clearance of the infection with rifampin, suggesting that infection is not required to maintain them and leading the authors to propose that the effects of *Chlamydia* on the centrosome may have lasting negative effects on the host cell (Grieshaber *et al.*, 2006).

Chlamydia may produce centrosome abnormalities by at least two distinct mechanisms. Johnson *et al.* provided evidence supporting *Chlamydia*-induced dysregulation of centrosome duplication. Interfering with regulators of centrosome duplication, such as the Cdk1 and Plk4 protein kinases, or by blocking S-phase progression prevented *Chlamydia*-induced centrosome amplification (Johnson *et al.*, 2009) (Table 5.1). This finding suggested that *Chlamydia* induces centrosome overduplication. In

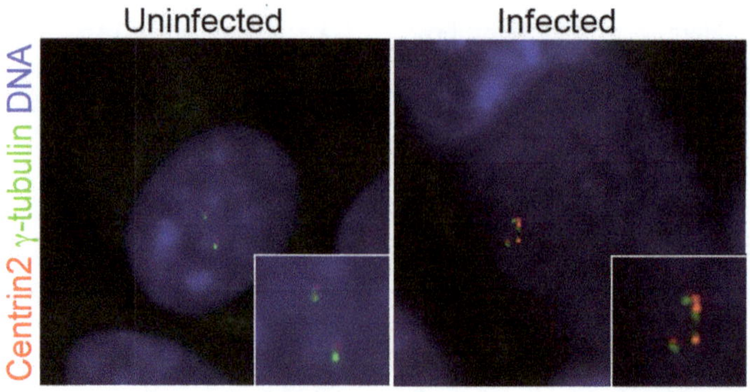

Figure 5.3 *Chlamydia*-infected cells show centrosome defects. Immunofluorescence images of uninfected control or *C. trachomatis* L2-infected HeLa cells at 36 hours post infection. Cells were stained with antibodies to the centrosomal marker protein centrin2 (red) and γ-tubulin (green) and with the DNA dye Hoechst 33342 (blue). The inset in the bottom right corner of each image is a magnified view of the centrioles of each cell. Image courtesy of Kevin Wang and Christine Sütterlin.

confirmation of this idea, Johnson *et al.* observed that infected cells contained mostly immature centrioles, which typically result from centriole overduplication (Johnson *et al.*, 2009). Another possible mechanism to generate supernumerary centrosomes is through defects in cytokinesis, which lead to an increase in mature centrioles (Brown *et al.*, 2012). Several studies described a defect in cytokinesis in *Chlamydia*-infected cells, as measured by an increase in multinucleated cells (Brown *et al.*, 2012; Sun *et al.*, 2011). However, the relative contribution of each these mechanisms to the amplified centrosome phenotype is unclear and has to be studied in more detail.

In a separate study, Brown and colleagues did not detect extra centrosomes in cells infected with a *C. trachomatis* mutant strain lacking the chlamydial protease CPAF (rst17 and M532 strains), suggesting that CPAF activity may be required for centrosome amplification (Brown *et al.*, 2014) (Table 5.1). The specific role of CPAF in centrosome dysregulation has not been determined. More details about this protease are discussed in Chapter 8, The Chlamydial Protease CPAF.

In addition to the increase in centrosome number, infected cells contained abnormal spindles, which lead to micronuclei and chromosome segregation errors (Grieshaber *et al.*, 2006). Spindle defects may be caused by two different mechanism (Knowlton *et al.*, 2011). Supernumerary centrosomes can lead to spindle defects. While uninfected cells are generally able to maintain spindle bipolarity by clustering centrosomes (Quintyne *et al.*, 2005), Knowlton *et al.* observed that *Chlamydia*-infected cells had lost the ability to cluster their extra centrosomes. Defects in the spindle checkpoint, which ensures normal mitotic progression and error-free chromosome segregation, can also lead to chromosome segregation errors. Knowlton *et al.* used the fluorescent cell cycle reporter system FUCCI to show that *Chlamydia*-infected cells progressed through mitosis at a faster rate than control cells, suggesting that the spindle checkpoint may not be activated. The mechanism of mitotic dysregulation is not known. It had been previously proposed to be due to cleavage of the mitotic regulators cyclin B1 and securin (Balsara *et al.*, 2006; Knowlton *et al.*, 2011), but on re-evaluation, there is no evidence that these proteins undergo proteolysis in intact infected cells (Chen *et al.*, 2012; Grieshaber and Grieshaber, 2014; Snavely *et al.*, 2014) (discussed in Chapter 8, The Chlamydial Protease CPAF).

The extent of centrosome dysregulation and the resulting defects in spindle formation and ultimately multinucleation vary between chlamydial species (Brown *et al.*, 2014). Centrosome abnormalities were most pronounced in *C. trachomatis* serovars LGV-L2, D and G (Grieshaber *et al.*, 2006), but were also seen, to a lesser extent, in *C. caviae* and *C. muridarum*-infected cells. Interestingly, abnormal spindle formation and multinucleation were less severe with these latter chlamydial species.

It is unclear why *Chlamydia* induces centrosome abnormalities because cells in which centrosome amplification was blocked produced similar numbers of progeny as control cells (Johnson *et al.*,

2009). It is possible that centrosome amplification may promote the infection in polarized cells, in which microtubules are organized differently from typical tissue culture cells, or *in vivo*. Alternatively, it may be the result of a bystander effect.

Potential consequences of centrosome dysregulation for the host cell

The effect of *C. trachomatis* infection on the centrosome may explain the link between chlamydial infection and cervical cancer because centrosome abnormalities lead to chromosome instability, which are typically associated with cancer (Godinho and Pellman, 2014). *C. trachomatis* infection has been linked to cervical cancer, either as an independent risk factor or as co-factor with human papillomavirus (Karim *et al.*, 2018; Zhu *et al.*, 2016). An association with ovarian cancer has also been reported. The idea that *C. trachomatis* may contribute to carcinogenesis is based on epidemiological analyses (Fortner *et al.*, 2019; Zhu *et al.*, 2016) and has been further supported by tissue culture and mouse studies. Knowlton *et al.* showed that infection with *C. trachomatis* was able to transform 3T3 cells in soft agar and to disrupt normal growth behaviour and regulation (Knowlton *et al.*, 2013). This study also found that *C. trachomatis* infects replicating cells *in vivo*, and that it promotes cell proliferation within the murine cervix. As confocal analysis of these tissues indicated elements of *C. trachomatis*-induced chromosomal instability, it is possible that *Chlamydia*-induced centrosome dysregulation contributes to cellular transformation and carcinogenesis.

Membrane-bound organelles that are actively recruited to the inclusion

The endoplasmic reticulum (ER)

The ER is the largest membrane-bound organelle of the cell. It is organized as interconnected sheets and tubules that extend throughout the cytoplasm and plays a critical role in protein and lipid synthesis, two functions that an obligate intracellular bacterium with incomplete biosynthetic pathways may want to target.

In 1988, Peterson and de la Maza showed by electron microscopy that smooth and rough endoplasmic reticulum vesicles were in close contact with the inclusion (Peterson and de la Maza, 1988). A similar observation was made twenty years later by Giles and Wyrick (Giles and Wyrick, 2008). These original findings most likely described structures that are now referred to as pathogen synapses, ER-inclusion membrane contact sites and inclusion microdomains. Here, we will describe the architecture and the molecular composition of these structures, how they compare to each other and how they may contribute to *Chlamydia* replication.

Pathogen synapses

Pathogen synapses were first described in a study by the Hayward lab that provided compelling support for an interaction between the inclusion and the ER (Dumoux *et al.*, 2012; Dumoux *et al.*, 2015a). Using electron tomography and immunolabelling microscopy, the authors identified small sections of the inclusion membrane that had an RB and rough ER structures on either side and that were connected by arrays of T3SS (Fig. 5.4A). Furthermore, confocal microscopy showed that ER proteins, such as Calreticulin, DP1 and Rtn4B, were in the vicinity of the inclusion (Dumoux *et al.*, 2012) (Table 5.1). Despite the presence of these general ER markers, the characterization of the pathogen synapse has been mostly limited to ultrastructural analysis, and their specific molecular composition has not yet been determined.

ER-inclusion membrane contact sites

ER-inclusion membrane contact sites (MCS) are specific subsections of the inclusion, where the ER is maintained in close proximity to the inclusion membrane without fusion of these membranes (Fig. 5.4B). Chlamydial MCS share morphological and molecular similarities with MCS that have been described between the ER and other organelles in uninfected cells (Prinz, 2014).

Independent studies, respectively led by J. Engel and I. Derré, identified CERT, a ceramide transfer protein at ER-Golgi MCS, as a key host factor required for *Chlamydia* replication. Both groups showed by confocal microscopy that CERT associated with the inclusion in small patches that were also positive for the ER resident proteins VAPA and VAPB (Derré *et al.*, 2011; Elwell *et al.*, 2011) (Fig. 5.4C and Table 5.1). Derré and

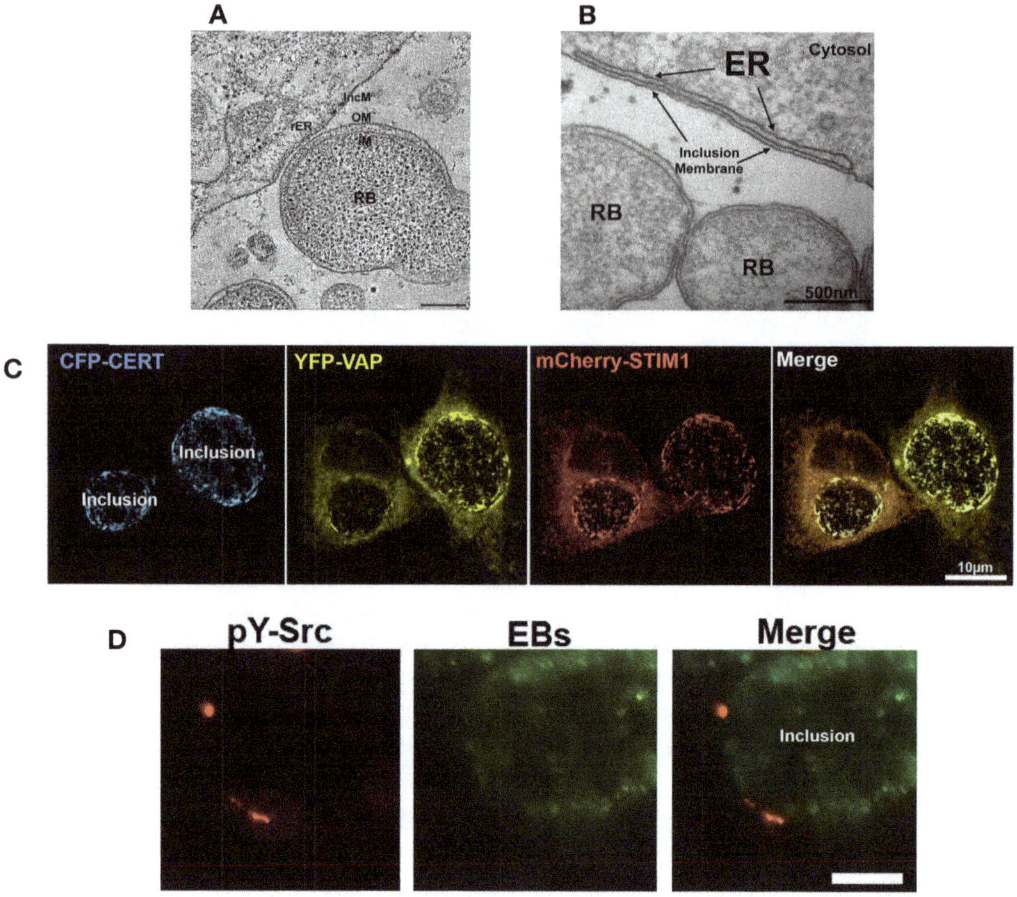

Figure 5.4 Inclusion interaction with the endoplasmic reticulum. A. Electron micrograph of HeLa cells infected with *C. trachomatis* LGV-L2 for 24h. The samples were high pressure frozen, and tomograms were recorded from 200 nm thick sections of freeze-substituted, embedded samples. The tomogram shows a pathogen synapse connecting the reticulate body (RB) inner (IM) and outer (OM) membranes, the inclusion membrane (IncM) and the rough endoplasmic reticulum (rER) recruited to the cytoplasmic face of the inclusion. Widening of the periplasmic space is evident at the hemisphere of the RB in contact with the inclusion membrane, accommodating an array of type III secretion systems. Scale bar, 200nm. Image courtesy of Maud Dumoux, Dan Clare, Helen Saibil and Richard Hayward. B. Electron micrograph of HeLa cells infected with *C. trachomatis* LGV-L2 for 24h. A close up of the close apposition of an endoplasmic reticulum (ER) tubule to the inclusion membrane characteristic of endoplasmic reticulum (ER)-inclusion membrane contact site is shown. RB: reticulate body. Scale Bar 500 nm. Image courtesy of Isabelle Derré. C. Confocal micrographs of HeLa cells co-expressing CFP-CERT (blue), YFP-VAP (Yellow) and mCherry-STIM1 (Red) and infected with *C. trachomatis* LGV-L2 for 24h. The merged image to the right shows the co-localization of the three markers at endoplasmic reticulum (ER)-inclusion membrane contact sites (MCS). Scale Bar: 10 µm. Image courtesy of Isabelle Derré. D. Confocal micrographs of HeLa cells infected with *C. trachomatis* LGV-L2 for 24h and stained with anti-active Src-family kinase (pY-Src, red) and anti-MOMP antibodies (EBs, green). An active Src kinase-positive microdomains in close apposition of the inclusion is shown on the merged image to the right. Scale Bar: 5 µm. Image courtesy of Phu Hai Nguyen and Ted Hackstadt.

colleagues subsequently used electron microscopy and immunogold labelling to demonstrate that these small patches corresponded to ER structures adjacent to the inclusion membrane (Derré *et al.*, 2011). ER-inclusion MCS also contained STIM1, an ER calcium sensor and known component of ER-plasma membrane (PM) MCS in uninfected cells (Agaisse and Derré, 2015) (Fig. 5.4C and Table 5.1). Moreover, STIM1, CERT and VAP associated with the inclusion membrane and colocalized throughout the developmental cycle (Agaisse and Derré, 2015).

Additional biochemical, microscopy and *Chlamydia* genetics approaches identified specific host and bacterial factors present in the ER-inclusion MCS (Agaisse and Derré, 2014; Derré *et al.*, 2011; Stanhope *et al.*, 2017). Specifically, IncD was found to recruit CERT to the inclusion membrane by interacting with its PH domain, while the FFAT motif of CERT mediated CERT binding to VAP on the ER membrane (Agaisse and Derré, 2014; Derré *et al.*, 2011) (Table 5.1). CERT partially co-localized with the sphingomyelin synthase SMS2, and ceramide was also detected in the inclusion membrane (Elwell *et al.*, 2011). These findings led to the proposal that the IncD/CERT/VAP complex may facilitate the non-vesicular transfer of ceramide to the inclusion membrane for on-site conversion into sphingomyelin, a host lipid essential for *Chlamydia* replication. IncV was also found to contribute to the formation of ER-inclusion MCS (Table 5.1). This protein tethered the ER to the inclusion membrane by mimicking two eukaryotic VAP binding motifs (i.e. FFAT motifs) that promoted binding to VAP (Stanhope *et al.*, 2017). However, while an *incV* mutant displayed reduced VAP recruitment to the inclusion, ER-inclusion MCS still formed and progeny production was normal, suggesting the existence of redundant mechanisms. Altogether, these data indicate that *Chlamydia* actively induces the formation of ER-inclusion MCS by tethering the ER to the inclusion membrane and recruiting host factors that could facilitate lipid acquisition.

ER-inclusion MCS have only been detected by electron microscopy in *C. trachomatis* infected cells (Derré *et al.*, 2011). However, the Elwell study, which produced similar results regarding CERT and VAP, was performed with *C. muridarum* (Elwell *et al.*, 2011), suggesting that these contact sites are likely to be conserved between these two species. In addition, the Heuer lab detected CERT on the surface of *C. psittaci* inclusion (Koch-Edelmann *et al.*, 2017), but it is unknown if there is an association between the *C. psittaci* inclusion and the ER.

Inclusion microdomains

As discussed earlier (section on microtubules), the inclusion membrane contains microdomains that are enriched in specific Inc proteins and activated Src-family kinases Src and Fyn (Lutter *et al.*, 2013; Mital *et al.*, 2010; Weber *et al.*, 2015) (Fig. 5.4D). Here, we focus on a report by the Hackstadt lab that suggests a role for ER-microdomain contacts in chlamydial extrusion (Nguyen *et al.*, 2018). Nguyen and colleagues used Yeast-2 Hybrid analysis, fluorescence microscopy, host gene silencing and *Chlamydia* mutagenesis to reveal an interaction between CT101, a microdomain Inc (renamed to MrcA), and the inositol trisphosphate receptor 3 (ITPR3) receptor, a calcium channel in the ER that controls Ca^{2+} release from the ER in response to inositol 1,4,5-trisphophate IP3 (Table 5.1). ITPR3 co-localized with inclusion microdomains that were positive for the active Src-family kinase (Nguyen *et al.*, 2018). ITPR3 depletion or *ct101/mrcA* inactivation caused no growth defect. However, components of the myosin phosphoregulatory pathway (namely pT853-MYPT1 and pS19-MLC2; Lutter *et al.*, 2012) were no longer recruited to the inclusion, leading to a significant extrusion defect. Extrusion was also disrupted by Ca^{2+} depletion or by removal of STIM1, the ER Ca^{2+} sensor previously identified as an ER-inclusion MCS component. STIM1 partially colocalized with CT101/MrcA and pY464-MCLK, another component of the myosin phosphoregulatory pathway (Nguyen *et al.*, 2018). These results are consistent with a model in which CT101/MrcA recruits ITPR3 to inclusion microdomains, which may then induce a local change in Ca^{2+} concentrations to modulate the myosin phosphoregulatory pathway and promote extrusion. It is unclear how STIM1 fits into this model.

Comparison of pathogen synapses, ER-inclusion MCS and inclusion microdomains

Based on electron microscopy, pathogen synapses and ER-inclusion MCS display a similar morphology and may be the same structures (Agaisse and Derré, 2015; Derré *et al.*, 2011; Dumoux *et al.*, 2012; Dumoux *et al.*, 2015b). While ER membranes have not yet been detected in proximity to Src family kinase-positive inclusion microdomains, the association of two integral ER resident proteins (ITPR3 and STIM1) with inclusion microdomains indicates that the ER is likely to be in close proximity. By confocal microscopy, ER-inclusion MCS and inclusion microdomains both manifest as small patches on the surface of the inclusion membrane, although their pattern and numbers were slightly different (Agaisse and Derré, 2015; Derré *et al.*,

2011; Mital *et al.*, 2010; Nguyen *et al.*, 2018). Nevertheless, Nguyen and colleagues showed partial co-localization between proteins found in ER-inclusion MCS and components of inclusion microdomains. These pairs include STIM1 and CT101/MrcA, CERT and active Src-family kinases, and STIM1 and a component of the myosin phosphoregulatory pathway (pY464-MCLK) (Nguyen *et al.*, 2018). Therefore, the literature suggests that there is at least some structural and molecular overlap between the ER domains that interact with the inclusion membrane. Future studies will determine the extent of this overlap and if it extends to functionality.

Role of ER–inclusion association for the infection

It is not clear how ER-inclusion proximity promotes infection. A role in cholesterol trafficking to the inclusion has been proposed because IncB-positive microdomains were found to be enriched in the cholesterol dye filipin (Mital *et al.*, 2010). Moreover, depletion of CERT, VAPA and VAPB or SMS2 by RNAi led to a significant growth defect in *C. trachomatis* and *C. muridarum* infection (Derré *et al.*, 2011; Elwell *et al.*, 2011). Pharmacological inactivation of CERT by HPA-12 also prevented *C. muridarum* growth and sphingomyelin acquisition (Elwell *et al.*, 2011). These results suggested that ER–inclusion interaction contribute to sphingomyelin acquisition by these two species. However, it is not clear yet if this phenomenon extends to *C. psittaci*. CERT inactivation by either HPA-12 or CRISPR/Cas9-mediated knockout led to smaller inclusions and decreased *C. psittaci* infectious progeny (Koch-Edelmann *et al.*, 2017). However, sphingomyelin accumulation in *C. psittaci* inclusions was higher in CERT-KO cells. This phenotype was reversed with the addition of HPA-12, suggesting that an additional HPA-12 sensitive factor promotes sphingomyelin acquisition in *C. psittaci* infection. Although the *C. psittaci* inclusion recruits CERT, it is not known if it has interactions with the ER and how such interactions could contribute to the phenotypes described by Koch-Edelmann and colleagues.

ER–inclusion interactions have been implicated in additional functions during a chlamydial infection. As discussed above, they may play a role in exit of chlamydiae from the host cell by extrusion (Nguyen *et al.*, 2018). Furthermore, the Hayward lab has used a pharmacological approach to investigate the role of the ER in *Chlamydia* infection. The authors showed that Aerolysin, a drug that globally affects ER function, disrupted inclusion biogenesis and infectious particle production (Dumoux *et al.*, 2012). They discussed that this effect could be due to a defect in lipid acquisition or modulation of the ER stress response and proposed that interaction with the ER might be involved in antigen presentation. In addition, this group hypothesized that the ER could also serve as a 'receiver membrane' to assist in the folding and insertion of Inc proteins into the inclusion membrane.

The Golgi apparatus

Close interactions between Golgi membranes and the inclusion

The Golgi apparatus is a membrane-bound organelle localized in the cell centre, next to the nucleus and the centrosome. It is composed of flattened cisternae that are organized into polarized stacks. 40–80 such stacks form the single interconnected Golgi ribbon that is typical for a mammalian cell in interphase. In mitosis, Golgi membranes lose their perinuclear position and are converted into tubular-vesicular elements that are dispersed throughout the cytosol. Key functions of the Golgi are the transport, modification and sorting of proteins and lipids, which, as discussed for the ER, are processes that could be beneficial for *Chlamydia*.

The original observation that the inclusion may interact with the Golgi dates back to 1989 when Campbell and colleagues noted 'scattered stain associated with the inclusion' in *Chlamydia*-infected cells that were labelled with the Golgi stain C6-NBD-Ceramide (Campbell *et al.*, 1989). While the authors concluded at that time that '*dispersal of the complex... may have limited functional significance*', the 'scattered stain' has since been extensively characterized and shown to be important for the infection.

In 1996, the Hackstadt lab used C6-NBD-Ceramide as a probe to visualize the Golgi of infected cells. The authors reported that after internalization, the *Chlamydia* inclusion migrated to a perinuclear region of the cell next to the Golgi (Hackstadt *et al.*, 1996) (Fig. 5.5A). This result confirmed the earlier findings by Campbell *et al.* and provided the

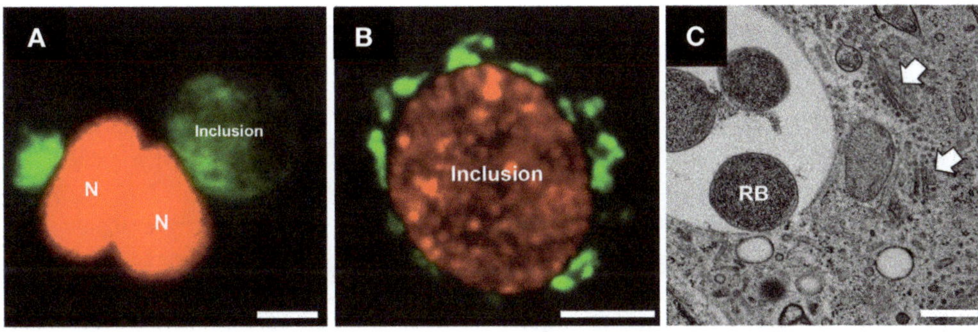

Figure 5.5 Inclusion association with the Golgi. A. Micrographs of HeLa cells infected with *C. trachomatis* LGV-L2 for 24h labelled with C6-NBD-Ceramide for 1h (green) and the DNA dye Hoechst (red). Note the accumulation of the fluorescent probe at the Golgi of the uninfected cell to the left and in the inclusion of the infected cell to the right. N= Nucleus. Scale Bar: 10 µm. Image courtesy of Isabelle Derré. B. Confocal micrograph of HeLa cells expressing a GFP–Golgi construct and infected with mCherry-expressing *C. trachomatis* LGV-L2 for 24h. Note the presence of Golgi ministacks surrounding the inclusion. Scale Bar: 10 µm. Image courtesy of Rachel Ende and Isabelle Derré. C. Electron micrograph of HeLa cells infected with *C. trachomatis* LGV-L2 for 24h. Note the presence of two Golgi ministacks (white arrows) in close vicinity of the inclusion. RB: Reticulate body. Scale Bar: 500 nm. Image courtesy of Michael Laue and Dagmar Heuer.

first solid evidence for physical proximity between the inclusion and the Golgi. Years later, Heuer and colleagues used confocal and live cell microscopy to unambiguously demonstrate that Golgi membranes of infected cells were extensively reorganized around the inclusion (Fig. 5.5B). Electron microscopy further showed that the Golgi membranes that were detected by immunofluorescence microscopy corresponded to intact and seemingly functional Golgi ministacks (Heuer *et al.*, 2009) (Fig. 5.5C). These results indicated that during an infection, the interconnected Golgi ribbon is fragmented into mini-stacks that are then recruited to the inclusion.

Mechanism of Golgi reorganization around the inclusion

The mechanism by which *Chlamydia* promotes Golgi fragmentation has been extensively studied. Rab6 and Rab11, two small GTPases that were previously detected at the inclusion, are necessary for *Chlamydia*-induced Golgi fragmentation because their depletion prevented Golgi fragmentation upon infection (Rejman Lipinski *et al.*, 2009) (Table 5.1). The F-actin cage around the inclusion is an additional factor that promotes Golgi reorganization in infected cells because the *inaC* mutant, which is unable to induce actin cage formation, was defective in Golgi rearrangement (Kokes *et al.*, 2015) (Table 5.1). Furthermore, stable microtubules were necessary for Golgi repositioning (Table 5.1).

The *Chlamydia* protease CPAF had also been invoked in *Chlamydia*-induced Golgi fragmentation through effects on the Golgi matrix component Golgin-84 (Christian *et al.*, 2011). However, these results were later disproven because there was no evidence of CPAF-mediated Golgin–84 cleavage in intact cells (Chen *et al.*, 2012). In addition, Golgi fragmentation and the recruitment of the resulting fragments were still observed in cell infected with the CPAF null mutant strain (Snavely *et al.*, 2014) (discussed in more detail in Chapter 8, The Chlamydial Protease CPAF, for more details). Although Golgi repositioning around the inclusion is recognized as a hallmark of *C. trachomatis* infected epithelial cells, the molecular mechanisms involved in this process remain only partially understood.

Role of Golgi reorganization in lipid acquisition

Using confocal microscopy and thin layer chromatography, the Hackstadt group showed that in infected cells, sphingomyelin, which was synthesized from fluorescent C6-NBD-Ceramide in the Golgi, accumulated in the inclusion membrane and the cell wall of the bacteria (Hackstadt *et al.*, 1996; Hackstadt *et al.*, 1995). This process was inhibited by Brefeldin A (BFA), a small molecule inhibitor that blocks protein transport by inducing relocalization of the Golgi into the ER. This result provided the first evidence that Golgi-derived sphingomyelin reaches the inclusion by vesicular transport. Incubation of infected cells with fluorescently labelled C6-NBD-ceramide has since become a standard assay in the field to study vesicle-mediated sphingomyelin transport to the inclusion. This assay subsequently facilitated the identification of several regulators of sphingomyelin acquisition in HeLa cells as well as in polarized epithelial cells (Moore *et al.*, 2008). For example, depletion of Syntaxin 10 increases lipid acquisition (Lucas *et al.*, 2015), whereas genetic inactivation of Rab14 (Capmany and Damiani, 2010), the Src family kinase Fyn (Mital and Hackstadt, 2011), CERT (Elwell *et al.*, 2011) and VAMP4 (Kabeiseman *et al.*, 2013) or pharmacological inhibition of dynamin (Gurumurthy *et al.*, 2014) prevented sphingomyelin trafficking. Furthermore, Rab6 and Rab11 depletion, which disrupted *Chlamydia*-induced Golgi fragmentation, also decreased sphingomyelin acquisition (Rzomp *et al.*, 2003). As experimentally induced Golgi fragmentation corrected this phenotype, these experiments demonstrated that Golgi reorganization is required for sphingomyelin acquisition (Rejman Lipinski *et al.*, 2009).

Host cell-derived sphingolipids are critical for *Chlamydia* growth. The Engel lab was first to show that *Chlamydia* was unable to replicate in a cell line carrying a mutation in the first enzyme of the sphingolipid biosynthetic pathway (van Ooij *et al.*, 2000). Similarly, Rab6 and Rab11 depletion, as discussed above, led to a decrease in the production of infectious progeny (Rejman Lipinski *et al.*, 2009). To the contrary, BFA, which reduced sphingomyelin trafficking to the inclusion, led to smaller inclusions without affecting the number of infectious particles recovered (Hackstadt *et al.*, 1996). These seemingly contradictory observations were explained 15 years later by the Engel lab, which showed that two distinct pathways contribute to sphingomyelin acquisition (Elwell *et al.*, 2011). A vesicular-, Golgi-, GBF1/ARF1-dependent and BFA-sensitive pathway was essential for inclusion membrane growth and stability but not for bacterial replication, while a non-vesicular pathway involving CERT, VAPA, VAPB and SMS2 contributed to bacterial replication (Derré *et al.*, 2011; Elwell *et al.*, 2011) (See section above on the ER). In the future, it would be interesting to determine if the specific contribution of each pathway is due to a difference in the nature of the sphingomyelin molecule that is being delivered and/or the physical location of its delivery.

Altogether, the currently available data point towards a role for the interaction between the inclusion and the Golgi as an essential step for sphingomyelin acquisition through Golgi-derived vesicles. While most studies on vesicular transport of sphingomyelin to the inclusion have been performed in *C. trachomatis* infection, sphingomyelin was also trafficked to the inclusion of *C. pneumoniae* and *C. psittaci*-infected cells (Koch-Edelmann *et al.*, 2017; Wolf and Hackstadt, 2001). However, it is not known if Golgi repositioning around the inclusion occurs during infection with *Chlamydia* species besides *C. trachomatis*, and if so, if it contributes to sphingomyelin acquisition.

Finally, association of the *C. trachomatis* inclusion with the Golgi has been implicated in cholesterol acquisition (Carabeo *et al.*, 2003). Carabeo and colleagues showed that the inclusion membrane was positive for the cholesterol probe filipin. Both radiolabelled cholesterol precursor ($[^{14}C]$ squalene) and low-density lipoproteins (LDL-$[^{3}H]$ cholesteryl linoleate) accumulated in EBs, indicating that both *de novo*-synthesized and LDL-derived cholesterol traffic to the inclusion. This process required *Chlamydia* protein synthesis and was inhibited by nocodazole or BFA suggesting that it was driven by the bacteria and was dependent on microtubules and functional Golgi membranes. Surprisingly, mutant cell lines defective in synthesis or transport of cholesterol still supported *Chlamydia* growth. While it is likely that there are compensatory mechanisms for cholesterol acquisition through these two pathways, it has not been possible to unambiguously determine if Golgi-derived cholesterol is necessary for *Chlamydia* replication. As described for sphingomyelin, it is not known if cholesterol acquisition is shared by other *Chlamydia* species.

Mitochondria

Mitochondria are organelles that are surrounded by two membranes, extend throughout the cytosol of a eukaryotic cell and constantly undergo fusion and fission. These organelles are critical for supplying energy to the cell via oxidative phosphorylation and also play key roles in lipid synthesis and apoptosis.

An association between *Chlamydia* infection and mitochondria has been predominantly studied in the context of apoptosis. This particular host–pathogen interaction is mostly conserved among different chlamydial species and will be discussed in detail in Chapter 6, *Chlamydia* and Cell-Autonomous Defence: Apoptosis and Autophagy. Here, we will describe four exciting interactions between the inclusion and mitochondria that are not conserved between chlamydial species.

Recruitment of mitochondria to the *C. psittaci* inclusion

Morphological studies with *C. psittaci* infection revealed a close physical relationship between the inclusion and mitochondria. Using electron microscopy, Friis and colleagues were the first to report the localization of mitochondria in close proximity to the inclusion (Friis, 1972). This observation was later confirmed by Matsumoto *et al.* and Peterson and de la Maza through additional EM studies and by Escalante-Ochoa *et al.* through immunofluorescence microscopy experiments (Escalante-Ochoa *et al.*, 1999; Matsumoto *et al.*, 1991; Peterson and de la Maza, 1988). The localization of mitochondria around the inclusion required the activity of kinesin, a (+) end-directed microtubule motor protein because inhibiting kinesin function with blocking antibodies prevented mitochondria recruitment (Escalante-Ochoa *et al.*, 1999) (Table 5.1). To date, no kinesin-binding chlamydial protein has been identified, and therefore, it is not known if and how the activity of this molecular motor is controlled in an infected cell.

The *C. psittaci* infection was found to depend on the positioning of mitochondria next to the inclusion. If blocked, there was delayed RB to EB conversion and an overall defect in the developmental cycle (Escalante-Ochoa *et al.*, 1999). Peterson and de la Maza proposed in 1988 that close contact between mitochondria and the inclusion could facilitate the transfer of metabolites to the RB (Peterson and de la Maza, 1988), but potential mechanisms have not been determined. Surprisingly, the recruitment of mitochondria to the inclusion was not observed in cells infected with *C. trachomatis* or *C. pneumoniae* (Friis, 1972), suggesting that other pathways must provide RBs with host cell derived metabolites, such as ATP. Alternatively, *C. psittaci* may acquire a metabolite other than ATP that is not important for *C. trachomatis* and *C. pneumoniae.*

Protection of mitochondrial organization by *C. trachomatis*

The *C. trachomatis* infection requires the maintenance of mitochondrial integrity (Chowdhury *et al.*, 2017). In a recent study, the Rudel lab described the up-regulation of the microRNA miR-30c in *C. trachomatis*-infected cells. This miRNA has been previously linked to the down-regulation of the mitochondria fission protein Drp1 (Table 5.1). Infected cells with increased miR-30c expression displayed decreased Drp1 levels and slightly elongated mitochondria. The significance of miR-30c up-regulation was examined by inducibly depleting miR-30c from infected cells. Upon miR-30c removal, mitochondria of infected cell appeared fragmented, ATP production was low and chlamydial growth was severely reduced. Consistent with these results, Drp1-overexpressing cells were unable to support *Chlamydia* infection. Interestingly, *Chlamydia*-induced up-regulation of miR-30c was able to protect mitochondrial integrity against other stress stimuli, such as H_2O_2. These results suggest that *C. trachomatis* actively protects the morphology and function of mitochondria of its host cell. It is not known if this miR-30c-dependent pathway is also active in other chlamydial species.

C. pneumoniae and mitochondrial function

C. pneumoniae infection was found to be enhanced by reduced mitochondria function (Kading *et al.*, 2017). The Rupp lab reported significant defects in mitochondria function in *C. pneumoniae*-infected cells, which was measured as an increase in ROS levels, decrease in respiration and hyperpolarization of mitochondrial membranes. The effect of mitochondria dysregulation on the infection was subsequently examined by disrupting mitochondrial ATP synthase subunits. These conditions had no effect on the infection rate and inclusion size, but led to a significant increase in chlamydial growth as measured in genome copy numbers assays. Enhanced chlamydial growth was also observed under hypoxic conditions, which is a low oxygen condition that reduces mitochondria function. Hypoxia led to enhanced metabolic activity of the bacteria, which may ultimately enhance their growth. These findings led the authors to conclude that impairment of mitochondrial function is beneficial for a *C. pneumoniae* infection. This is in contrast to *C. trachomatis* infection, which actively prevents changes to mitochondrial integrity and function by miR-30c up-regulation.

C. caviae infection and mitochondrial protein import

An RNA interference screen in Drosophila S2 cells found that loss of several mitochondrial proteins negatively affected *C. caviae* infection (Derré *et al.*, 2007). Among these proteins were two components of the outer mitochondrial membrane import machinery (Tom40 and Tom22), which are necessary for the transport of nuclear-encoded proteins across the outer mitochondrial membrane. Loss of Tom40 or Tom22 resulted in smaller inclusions and reduced progeny in insect and mammalian cells, however the mechanism remains unknown (Table 5.1). It is surprising that this effect was specific for *C. caviae*, and not observed in *C. trachomatis*-infected cells.

Conclusion

Many aspects about the link between *Chlamydia* and mitochondria appear to be species-specific. These differences include mitochondrial recruitment to the inclusion and effects on mitochondrial organization and function. A comprehensive comparison of phenotypes in different chlamydial strains and species would help establish conserved versus non-conserved effects of *Chlamydia* infection on mitochondrial organization and function, and improve our understanding of the molecular mechanism involved.

Organelles that have been detected inside the inclusion

Lipid droplets

Localization of lipid droplets in infected cells

Lipid droplets are neutral lipid-rich organelles that are surrounded by a phospholipid monolayer and are closely linked to energy and lipid metabolism (Saka and Valdivia, 2012).

In a collaboration, the Valdivia and Hackstadt labs used live and electron microscopy to show that lipid droplets not only docked to the inclusion membrane, but also translocated into the inclusion, where they ultimately associated with RBs (Cocchiaro *et al.*, 2008) (Fig. 5.6A). This was the first report of a whole organelle inside the lumen of the inclusion and suggested that the inclusion is

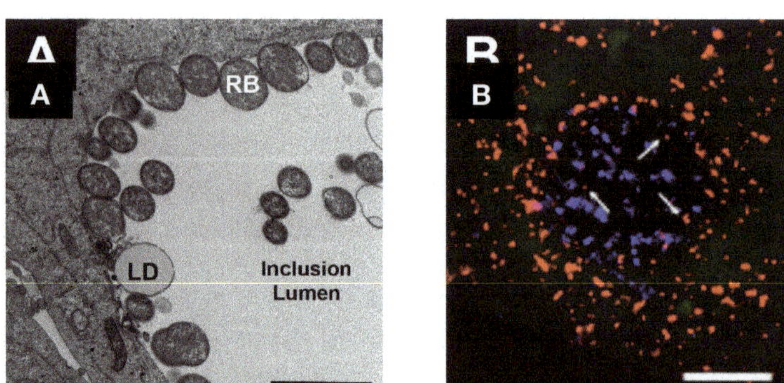

Figure 5.6 Lipid droplets and peroxisomes translocate into the lumen of the inclusion. A. Electron micrograph of HeLa cells infected with *C. trachomatis* LGV-L2 for 18h in the presence of malachite green to preserve lipid structures. Note the presence of a lipid droplet (LD) in the inclusion lumen. RB: Reticulate Body. Scale Bar: 2 μm. Image courtesy of Elizabeth Fisher and Ted Hackstadt. B. Confocal micrograph of HeLa cells expressing cytosolic GFP (green) were infected with *C. trachomatis* LGV-L2 for 20h and stained with anti-HSP60 (*Chlamydia*, blue) and anti-ALDP (Peroxisome, red). The white arrows point to peroxisomes that have translocated into the lumen of the inclusion. Scale Bar: 5 μm. Image courtesy of Agathe Subtil.

able to take up entire organelles. During the time course of the infection, lipid droplets expanded and increased in number as determined by measurements of cholesteryl ester levels and by fluorescence microscopy (Kumar *et al.*, 2006; Saka *et al.*, 2015). Furthermore, a proteomic analysis revealed that their protein composition changed during the time course of the infection, and that there was an enrichment in proteins related to lipid metabolism, biosynthesis and lipid droplet function (Saka *et al.*, 2015).

Chlamydial proteins linked to lipid droplets

Several chlamydial proteins have been found to associate with lipid droplets. In studies using yeast as a heterologous system, the Valdivia lab identified four chlamydial proteins that associated with lipid droplets (Lda: lipid droplet associated): CT156 (Lda1), CT163 (Lda2), CT257and CT473 (Lda3) (Kumar *et al.*, 2006). Of these, Lda3 (CT473) was proposed to control the docking of lipid droplets to the inclusion membrane and the removal of their protective coat protein ADRP (Cocchiaro *et al.*, 2008) (Table 5.1). Another proteomic study showed that the inclusion membrane proteins Cap1 (CTL0791/CT529), CTL0882/CT618 and IncG (CTL0373/CT118) also localized to lipid droplets (Saka *et al.*, 2015) (Table 5.1). The affinity of these Inc proteins for lipid droplets may be an indicator of their potential roles in lipid droplet recruitment to the inclusion membrane. CT775, a lysophosphatidylcholine acyltransferase (LPCAT) has also been detected on lipid droplets and its proposed role is discussed below (Soupene *et al.*, 2015). Overall, additional experiments are required to confirm and characterize the potential function of the *Chlamydia* proteins that display tropism for lipid droplets.

Role of inclusion-lipid droplet interaction

Several studies indicate that lipid droplets are critical for the infection. *Chlamydia* replication was reduced in diacylglycerol-acyltransferases 1 and 2 (DGAT1/2) knockout MEFs in which lipid droplet formation is prevented (Saka *et al.*, 2015) (Table 5.1). Furthermore, ST-669, a broad-spectrum antiviral compound that associates with lipid droplets, disrupted *C. trachomatis* growth in Vero cells but not in McCoy cells (Sandoz *et al.*, 2014), although the mode of action of ST-669 is unknown. ST-699 also negatively affected *C. muridarum* and *C. caviae* infection in Vero cells. Finally, two studies used the pharmacological inhibitor Triacsin C, which targets the long chain acyl-CoA synthetases (ACSLs) that mediate triacylglyceride and cholesterol ester biosynthesis and thereby blocks lipid droplet formation. Triacsin C treatment negatively impacted the infection (Kumar *et al.*, 2006; Recuero-Checa *et al.*, 2016). However, Triacsin C still induced a growth phenotype in DGAT1/2 MEFs, suggesting that the effect of the drug may be caused by inhibition of ACSLs, but not by the absence of lipid droplets (Recuero-Checa *et al.*, 2016).

Lipid droplets may promote the chlamydial infection by enhancing lipid acquisition. Although lipid droplets translocate into the lumen of the inclusion, they do not accumulate over time suggesting that they are consumed after translocation (Cocchiaro *et al.*, 2008). This result led Cocchiaro and colleagues to propose that lipid droplets could serve as a source of lipids for *Chlamydia.* For example, lipid droplets are enriched in phosphatidylcholine (PC) and lyso-PC, two lipids that cannot be synthesized by chlamydiae, but which make up 40% of their phospholipid content. As a major lipid storage organelle, lipid droplets may also supply neutral lipids to the bacteria. However, it is unclear how *Chlamydia* would degrade neutral lipids as it does not encode any triacylglycerol lipases. Furthermore, lipid droplets may contribute to the recruitment of acyl-coA synthases into the inclusion, although the presence of these enzymes in the inclusion in the absence of lipid droplets suggests the existence of redundant mechanisms (Recuero-Checa *et al.*, 2016). Finally, lipid droplets inside the inclusion may promote fatty acid remodelling (Soupene *et al.*, 2015). Soupene and colleagues showed that CT775, a chlamydial lysophosphatidylcholine acyltransferase (LPCAT) which reacylates host lysophophatidylcholine to generate *Chlamydia*-specific phosphatidylcholine, associated with lipid droplets (Table 5.1). Furthermore, the human acyl-CoA binding protein hACBD6, regulated CT775 acyl transferase activity *in vitro* and was recruited to lipid droplets upon infection. hACBD6-positive lipid droplets were detected in the cytosol and inside the inclusion.

Based on these results the authors proposed a model in which *Chlamydia* infection induces the association of hACBD6 with lipid droplets prior to their translocation into the lumen of the inclusion. Once inside the inclusion, hACBD6 would modulate the LPCAT activity of CT775, supposedly by buffering the concentration of acyl-CoA, thereby allowing for CT775-mediated host lipid remodelling.

Peroxisomes

Peroxisomes are membrane-bound cytosolic organelles with a key role in lipid metabolism, such as breakdown of very long chain fatty acids via beta-oxidation.

Localization of peroxisomes in infected cells

The Subtil lab showed that *C. trachomatis* infection induces changes in the localization of peroxisomes. Immunofluorescence staining revealed that these small cytosolic organelles are recruited to the vicinity of the inclusion. Furthermore, peroxisomes have been detected inside about 50% of inclusions, suggesting that similar to lipid droplets, these organelles translocate into the inclusion (Boncompain *et al.*, 2014) (Fig. 5.6B).

Mechanism and role of peroxisome–inclusion interaction

It is unclear how *Chlamydiae* 'hijack' peroxisomes. This host–pathogen interaction has been linked to changes in lipid composition of the host cell and chlamydiae (Boncompain *et al.*, 2014). Boncompain and colleagues found that the overall abundance of cellular phospholipids was not altered by *C. trachomatis* infection, but that the levels of plasmalogens, which are rare host cell lipids synthesized in peroxisomes, were decreased. Furthermore, purified chlamydiae contained host-derived plasmalogens as well as *Chlamydia*-modified lipid species. These results are consistent with a transfer of peroxisome-specific lipids to the bacteria, which may be essential for *Chlamydia* replication. In agreement, the authors also showed that in tissue culture, inclusions were smaller in cells lacking peroxisomes due to a mutation in the *pex19* gene (Table 5.1). However, there was no defect in bacterial progeny, but it was proposed that peroxisomes might be critical for an *in vivo* infection.

Challenges to the detection of organelle markers inside the inclusion

As discussed, markers of several organelles have been detected inside the chlamydial inclusion, raising the intriguing possibility that either specific host proteins or entire organelles can translocate into this bacterial compartment. However, a recent study compared the detection of various organelle markers inside the inclusion by live or fixed cell microscopy (Kokes and Valdivia, 2015). For example, a GFP-tagged form of the ER marker Sec61 was observed inside the inclusion when cells were fixed with different fixatives, however, this signal was not observed in live imaging experiments. Thus, cell fixation may significantly enhance the signal for these markers inside the inclusion, raising the question of possible fixation-induced artefacts for some markers. This study also demonstrates the importance of validating inclusion–organelle association by live cell imaging. While lipid droplets have been detected inside the inclusion by live cell imaging, this approach has not been taken for peroxisomes.

Conclusions

In this chapter, we have summarized our current knowledge of interaction of the chlamydial inclusion with host cell structures and organelles. We elaborated on how *Chlamydia*, mostly through specific Inc proteins, recruits cytoskeletal elements and organelles to the inclusion for mechanical support and for the acquisition of lipids and metabolites. Many of these host–pathogen interactions appear to enhance the chlamydial infection.

Pharmacological approaches and RNAi have been instrumental in identifying host cell proteins that directly contribute to or regulate these interactions. In contrast, only few chlamydial counterparts of these host factors are known. In general, bacterial genetics provide an invaluable toolbox to understand how bacteria manipulate their host cell, but the lack of such genetic approaches hampered *Chlamydia* research for a long time and is the main reason for an incomplete understanding of the function of many chlamydial genes. We predict that with the recent development of genetic methods, including targeted gene disruption and genetic screens, we will be able to move forward at a faster rate and obtain insights into the chlamydial factors that manipulate the host cell.

Another limitation for studying the host–pathogen interactions described in this chapter is their interdependency. For example, if F-actin cannot be recruited to the inclusion, the vimentin cage, Golgi positioning and the post-translational modification of microtubules are all defective. However, a better understanding of these interactions on a molecular level is likely to help us resolve this issue.

To date, we have a limited understanding of the temporal and spatial regulation of these interactions during the developmental cycle. Is the expression of both a chlamydial protein and its interacting host factor sufficient to produce a phenotype, or are other regulatory mechanisms involved? Also, once a phenotype has been produced, does it require continued interactions for its maintenance? Are these interactions dynamic and how is the spatial positioning of the cytoskeletal elements and organelles around the inclusion regulated? These are interesting questions that are likely to be addressed in the near future.

The functional implications of many of these interactions are incompletely understood. Many studies have been performed in tissue culture cells, such as HeLa cells, which may not faithfully represent the *in vivo* environment. Thus, there is a clear need to analyse all the interactions that we have discussed in this chapter in more physiologically relevant systems such as polarized cells, 3D organoids or tissue explants. Furthermore, it is not known if a small difference (< 5-fold) in progeny, as observed upon disruption of many of these interactions, will alter the course of an *in vivo* infection.

Nevertheless, tremendous progress has been made over the past 30 years, and it will be exciting to observe the advances that are likely to occur in the near future. A better understanding of the molecular mechanisms governing the interactions between the *Chlamydia* inclusion and cellular structures and organelles will not only expand our knowledge of how obligate intracellular bacteria establish their replication niche but may also highlight fundamental cell biology processes.

References

Agaisse, H., and Derré, I. (2014). Expression of the effector protein IncD in *Chlamydia trachomatis* mediates recruitment of the lipid transfer protein CERT and the endoplasmic reticulum-resident protein VAPB to the inclusion membrane. Infect Immun *82*, 2037–2047.

Agaisse, H., and Derré, I. (2015). STIM1 is a novel component of ER-*Chlamydia trachomatis* inclusion membrane contact sites. PLoS One *10*, e0125671.

Al-Younes, H.M., Al-Zeer, M.A., Khalil, H., Gussmann, J., Karlas, A., Machuy, N., Brinkmann, V., Braun, P.R., and Meyer, T.F. (2011). Autophagy-independent function of MAP-LC3 during intracellular propagation of *Chlamydia trachomatis*. Autophagy *7*, 814–828.

Al-Zeer, M.A., Al-Younes, H.M., Kerr, M., Abu-Lubad, M., Gonzalez, E., Brinkmann, V., and Meyer, T.F. (2014). *Chlamydia trachomatis* remodels stable microtubules to coordinate Golgi stack recruitment to the chlamydial inclusion surface. Mol Microbiol *94*, 1285–1297.

Almeida, F., Luis, M.P., Pereira, I.S., Pais, S.V., and Mota, L.J. (2018). The Human Centrosomal Protein CCDC146 Binds *Chlamydia trachomatis* Inclusion Membrane Protein CT288 and Is Recruited to the Periphery of the *Chlamydia*-Containing Vacuole. Front Cell Infect Microbiol *8*, 254.

Alzhanov, D.T., Weeks, S.K., Burnett, J.R., and Rockey, D.D. (2009). Cytokinesis is blocked in mammalian cells transfected with *Chlamydia trachomatis* gene CT223. BMC Microbiol *9*, 2.

Archuleta, T.L., Du, Y., English, C.A., Lory, S., Lesser, C., Ohi, M.D., Ohi, R., and Spiller, B.W. (2011). The *Chlamydia* effector chlamydial outer protein N (CopN) sequesters tubulin and prevents microtubule assembly. J Biol Chem *286*, 33992–33998.

Balsara, Z.R., Misaghi, S., Lafave, J.N., and Starnbach, M.N. (2006). *Chlamydia trachomatis* infection induces cleavage of the mitotic cyclin B1. Infect Immun *74*, 5602–5608.

Boncompain, G., Muller, C., Meas-Yedid, V., Schmitt-Kopplin, P., Lazarow, P.B., and Subtil, A. (2014). The intracellular bacteria *Chlamydia* hijack peroxisomes and utilize their enzymatic capacity to produce bacteria-specific phospholipids. PLoS One *9*, e86196.

Brown, H.M., Knowlton, A.E., and Grieshaber, S.S. (2012). Chlamydial infection induces host cytokinesis failure at abscission. Cell Microbiol *14*, 1554–1567.

Brown, H.M., Knowlton, A.E., Snavely, E., Nguyen, B.D., Richards, T.S., and Grieshaber, S.S. (2014). Multinucleation during *C. trachomatis* infections is caused by the contribution of two effector pathways. PLoS One *9*, e100763.

Campanacci, V., Urvoas, A., Cantos-Fernandes, S., Aumont-Nicaise, M., Arteni, A.A., Velours, C., Valerio-Lepiniec, M., Dreier, B., Pluckthun, A., Pilon, A., *et al.* (2019). Insight into microtubule nucleation from tubulin-capping proteins. Proc Natl Acad Sci U S A *116*, 9859–9864.

Campbell, S., Richmond, S.J., and Yates, P.S. (1989). The effect of *Chlamydia trachomatis* infection on the host cell cytoskeleton and membrane compartments. J Gen Microbiol *135*, 2379–2386.

Capmany, A., and Damiani, M.T. (2010). *Chlamydia trachomatis* intercepts Golgi-derived sphingolipids through a Rab14-mediated transport required for bacterial development and replication. PLoS One *5*, e14084.

Carabeo, R.A., Mead, D.J., and Hackstadt, T. (2003). Golgi-dependent transport of cholesterol to the *Chlamydia trachomatis* inclusion. Proc Natl Acad Sci U S A *100*, 6771–6776.

Chen, A.L., Johnson, K.A., Lee, J.K., Sutterlin, C., and Tan, M. (2012). CPAF: a Chlamydial protease in search of an authentic substrate. PLoS Pathog *8*, e1002842.

Chin, E., Kirker, K., Zuck, M., James, G., and Hybiske, K. (2012). Actin recruitment to the *Chlamydia inclusion* is spatiotemporally regulated by a mechanism that requires host and bacterial factors. PLoS One *7*, e46949.

Chowdhury, S.R., Reimer, A., Sharan, M., Kozjak-Pavlovic, V., Eulalio, A., Prusty, B.K., Fraunholz, M., Karunakaran, K., and Rudel, T. (2017). *Chlamydia* preserves the mitochondrial network necessary for replication via microRNA-dependent inhibition of fission. J Cell Biol *216*, 1071–1089.

Christian, J.G., Heymann, J., Paschen, S.A., Vier, J., Schauenburg, L., Rupp, J., Meyer, T.F., Hacker, G., and Heuer, D. (2011). Targeting of a chlamydial protease impedes intracellular bacterial growth. PLoS Pathog *7*, e1002283.

Clausen, J.D., Christiansen, G., Holst, H.U., and Birkelund, S. (1997). *Chlamydia trachomatis* utilizes the host cell microtubule network during early events of infection. Mol Microbiol *25*, 441–449.

Cocchiaro, J.L., Kumar, Y., Fischer, E.R., Hackstadt, T., and Valdivia, R.H. (2008). Cytoplasmic lipid droplets are translocated into the lumen of the *Chlamydia trachomatis* parasitophorous vacuole. Proc Natl Acad Sci U S A *105*, 9379–9384.

Dehoux, P., Flores, R., Dauga, C., Zhong, G., and Subtil, A. (2011). Multi-genome identification and characterization of *chlamydiae*-specific type III secretion substrates: the Inc proteins. BMC Genomics *12*, 109.

Derré, I., Pypaert, M., Dautry-Varsat, A., and Agaisse, H. (2007). RNAi screen in *Drosophila* cells reveals the involvement of the Tom complex in *Chlamydia* infection. PLoS Pathog *3*, 1446–1458.

Derré, I., Swiss, R., and Agaisse, H. (2011). The lipid transfer protein CERT interacts with the *Chlamydia* inclusion protein IncD and participates to ER-*Chlamydia* inclusion membrane contact sites. PLoS Pathog *7*, e1002092.

Dumoux, M., Clare, D.K., Saibil, H.R., and Hayward, R.D. (2012). *Chlamydiae* assemble a pathogen synapse to hijack the host endoplasmic reticulum. Traffic *13*, 1612–1627.

Dumoux, M., Menny, A., Delacour, D., and Hayward, R.D. (2015a). A *Chlamydia* effector recruits CEP170 to reprogram host microtubule organization. J Cell Sci *128*, 3420–3434.

Dumoux, M., Nans, A., Saibil, H.R., and Hayward, R.D. (2015b). Making connections: snapshots of chlamydial type III secretion systems in contact with host membranes. Current Opin Microbiol *23*, 1–7.

Elwell, C.A., Jiang, S., Kim, J.H., Lee, A., Wittmann, T., Hanada, K., Melancon, P., and Engel, J.N. (2011). *Chlamydia trachomatis* co-opts GBF1 and CERT to acquire host sphingomyelin for distinct roles during intracellular development. PLoS Pathog *7*, e1002198.

Escalante-Ochoa, C., Ducatelle, R., Charlier, G., De Vos, K., and Haesebrouck, F. (1999). Significance of host cell kinesin in the development of *Chlamydia psittaci*. Infect Immun *67*, 5441–5446.

Fields, K.A., and Hackstadt, T. (2000). Evidence for the secretion of *Chlamydia trachomatis* CopN by a type III secretion mechanism. Mol Microbiol *38*, 1048–1060.

Fortner, R.T., Terry, K.L., Bender, N., Brenner, N., Hufnagel, K., Butt, J., Waterboer, T., and Tworoger, S.S. (2019). Sexually transmitted infections and risk of epithelial ovarian cancer: results from the Nurses' Health Studies. Br J Cancer *120*, 855–860.

Friis, R.R. (1972). Interaction of L cells and *Chlamydia psittaci*: entry of the parasite and host responses to its development. J Bacteriol *110*, 706–721.

Giles, D.K., and Wyrick, P.B. (2008). Trafficking of chlamydial antigens to the endoplasmic reticulum of infected epithelial cells. Microbes Infect *10*, 1494–1503.

Godinho, S.A., and Pellman, D. (2014). Causes and consequences of centrosome abnormalities in cancer. Philos Trans R Soc Lond B Biol Sci *369*.

Grieshaber, S.S., and Grieshaber, N.A. (2014). The role of the chlamydial effector CPAF in the induction of genomic instability. Pathog Dis *72*, 5–6.

Grieshaber, S.S., Grieshaber, N.A., and Hackstadt, T. (2003). *Chlamydia trachomatis* uses host cell dynein to traffic to the microtubule-organizing centre in a p50 dynamitin-independent process. J Cell Sci *116*, 3793–3802.

Grieshaber, S.S., Grieshaber, N.A., Miller, N., and Hackstadt, T. (2006). *Chlamydia trachomatis* causes centrosomal defects resulting in chromosomal segregation abnormalities. Traffic *7*, 940–949.

Gurumurthy, R.K., Chumduri, C., Karlas, A., Kimmig, S., Gonzalez, E., Machuy, N., Rudel, T., and Meyer, T.F. (2014). Dynamin-mediated lipid acquisition is essential for *Chlamydia trachomatis* development. Mol Microbiol *94*, 186–201.

Hackstadt, T., Rockey, D.D., Heinzen, R.A., and Scidmore, M.A. (1996). *Chlamydia trachomatis* interrupts an exocytic pathway to acquire endogenously synthesized sphingomyelin in transit from the Golgi apparatus to the plasma membrane. EMBO *15*, 964–977.

Hackstadt, T., Scidmore, M.A., and Rockey, D.D. (1995). Lipid metabolism in *Chlamydia trachomatis*-infected cells: directed trafficking of Golgi-derived sphingolipids to the chlamydial inclusion. Proc Natl Acad Sci U S A *92*, 4877–4881.

Heng, Y.W., Lim, H.H., Mina, T., Utomo, P., Zhong, S., Lim, C.T., and Koh, C.G. (2012). TPPP acts downstream of RhoA-ROCK-LIMK2 to regulate astral microtubule organization and spindle orientation. J Cell Sci *125*, 1579–1590.

Heuer, D., Rejman Lipinski, A., Machuy, N., Karlas, A., Wehrens, A., Siedler, F., Brinkmann, V., and Meyer, T.F. (2009). *Chlamydia* causes fragmentation of the Golgi compartment to ensure reproduction. Nature *457*, 731–735.

Huang, J., Lesser, C.F., and Lory, S. (2008). The essential role of the CopN protein in *Chlamydia pneumoniae* intracellular growth. Nature *456*, 112–115.

Ishida, K., Matsuo, J., Yamamoto, Y., and Yamaguchi, H. (2014). *Chlamydia pneumoniae* effector chlamydial outer protein N sequesters fructose bisphosphate aldolase A, providing a benefit to bacterial growth. BMC Microbiol *14*, 330.

Johnson, K.A., Tan, M., and Sutterlin, C. (2009). Centrosome abnormalities during a *Chlamydia trachomatis* infection are caused by dysregulation of the normal duplication pathway. Cell Microbiol *11*, 1064–1073.

Kabeiseman, E.J., Cichos, K., Hackstadt, T., Lucas, A., and Moore, E.R. (2013). Vesicle-associated membrane protein 4 and syntaxin 6 interactions at the chlamydial inclusion. Infect Immun *81*, 3326–3337.

Kading, N., Kaufhold, I., Muller, C., Szaszak, M., Shima, K., Weinmaier, T., Lomas, R., Conesa, A., Schmitt-Kopplin, P., Rattei, T., *et al.* (2017). Growth of *Chlamydia pneumoniae* is enhanced in cells with impaired mitochondrial function. Front Cell Infect Microbiol *7*, 499.

Karim, S., Souho, T., Benlemlih, M., and Bennani, B. (2018). Cervical Cancer Induction Enhancement Potential of *Chlamydia trachomatis*: A Systematic Review. Curr Microbiol *75*, 1667–1674.

Knowlton, A.E., Brown, H.M., Richards, T.S., Andreolas, L.A., Patel, R.K., and Grieshaber, S.S. (2011). *Chlamydia trachomatis* infection causes mitotic spindle pole defects independently from its effects on centrosome amplification. Traffic *12*, 854–866.

Knowlton, A.E., Fowler, L.J., Patel, R.K., Wallet, S.M., and Grieshaber, S.S. (2013). *Chlamydia* induces anchorage independence in 3T3 cells and detrimental cytologic defects in an infection model. PLoS One *8*, e54022.

Koch-Edelmann, S., Banhart, S., Saied, E.M., Rose, L., Aeberhard, L., Laue, M., Doellinger, J., Arenz, C., and Heuer, D. (2017). The cellular ceramide transport protein CERT promotes *Chlamydia psittaci* infection and controls bacterial sphingolipid uptake. Cell Microbiol *19*.

Kokes, M., Dunn, J.D., Granek, J.A., Nguyen, B.D., Barker, J.R., Valdivia, R.H., and Bastidas, R.J. (2015). Integrating chemical mutagenesis and whole-genome sequencing as a platform for forward and reverse genetic analysis of *Chlamydia*. Cell Host Microbe *17*, 716–725.

Kokes, M., and Valdivia, R.H. (2015). Differential Translocation of Host Cellular Materials into the *Chlamydia trachomatis* Inclusion Lumen during Chemical Fixation. PLoS One *10*, e0139153.

Kumar, Y., Cocchiaro, J., and Valdivia, R.H. (2006). The obligate intracellular pathogen *Chlamydia trachomatis* targets host lipid droplets. Curr Biol *16*, 1646–1651.

Kumar, Y., and Valdivia, R.H. (2008). Actin and intermediate filaments stabilize the *Chlamydia trachomatis* vacuole by forming dynamic structural scaffolds. Cell Host Microbe *4*, 159–169.

Lucas, A.L., Ouellette, S.P., Kabeiseman, E.J., Cichos, K.H., and Rucks, E.A. (2015). The trans-Golgi SNARE syntaxin 10 is required for optimal development of *Chlamydia trachomatis*. Front Cell Infect Microbiol *5*, 68.

Lutter, E.I., Barger, A.C., Nair, V., and Hackstadt, T. (2013). *Chlamydia trachomatis* inclusion membrane protein CT228 recruits elements of the myosin phosphatase pathway to regulate release mechanisms. Cell Rep *3*, 1921–1931.

Lutter, E.I., Martens, C., and Hackstadt, T. (2012). Evolution and conservation of predicted inclusion membrane proteins in *chlamydiae*. Comp Funct Genome *2012*, 362104.

Matsumoto, A., Bessho, H., Uehira, K., and Suda, T. (1991). Morphological studies of the association of mitochondria with chlamydial inclusions and the fusion of chlamydial inclusions. J Electron Microsc (Tokyo) *40*, 356–363.

Mital, J., and Hackstadt, T. (2011). Role for the SRC family kinase Fyn in sphingolipid acquisition by chlamydiae. Infect Immun *79*, 4559–4568.

Mital, J., Lutter, E.I., Barger, A.C., Dooley, C.A., and Hackstadt, T. (2015). *Chlamydia trachomatis* inclusion membrane protein CT850 interacts with the dynein light chain DYNLT1 (Tctex1). Biochem Biophys Res Commun *462*, 165–170.

Mital, J., Miller, N.J., Fischer, E.R., and Hackstadt, T. (2010). Specific chlamydial inclusion membrane proteins associate with active Src family kinases in microdomains that interact with the host microtubule network. Cell Microbiol *12*, 1235–1249.

Moore, E.R., Fischer, E.R., Mead, D.J., and Hackstadt, T. (2008). The chlamydial inclusion preferentially intercepts basolaterally directed sphingomyelin-containing exocytic vacuoles. Traffic *9*, 2130–2140.

Nawrotek, A., Guimaraes, B.G., Velours, C., Subtil, A., Knossow, M., and Gigant, B. (2014). Biochemical and structural insights into microtubule perturbation by CopN from *Chlamydia pneumoniae*. J Biol Chem *289*, 25199–25210.

Nguyen, P.H., Lutter, E.I., and Hackstadt, T. (2018). *Chlamydia trachomatis* inclusion membrane protein MrcA interacts with the inositol 1,4,5-trisphosphate receptor type 3 (ITPR3) to regulate extrusion formation. PLoS Pathog *14*, e1006911.

Peterson, E.M., and de la Maza, L.M. (1988). *Chlamydia* parasitism: ultrastructural characterization of the interaction between the chlamydial cell envelope and the host cell. J Bacteriol *170*, 1389–1392.

Prinz, W.A. (2014). Bridging the gap: membrane contact sites in signalling, metabolism, and organelle dynamics. J Cell Biol *205*, 759–769.

Quintyne, N.J., Reing, J.E., Hoffelder, D.R., Gollin, S.M., and Saunders, W.S. (2005). Spindle multipolarity is prevented by centrosomal clustering. Science *307*, 127–129.

Recuero-Checa, M.A., Sharma, M., Lau, C., Watkins, P.A., Gaydos, C.A., and Dean, D. (2016). *Chlamydia trachomatis* growth and development requires the activity of host Long-chain Acyl-CoA Synthetases (ACSLs). Sci Rep *6*, 23148.

Rejman Lipinski, A., Heymann, J., Meissner, C., Karlas, A., Brinkmann, V., Meyer, T.F., and Heuer, D. (2009). Rab6 and Rab11 regulate *Chlamydia trachomatis* development and golgin-84-dependent Golgi fragmentation. PLoS Pathog *5*, e1000615.

Rzomp, K.A., Scholtes, L.D., Briggs, B.J., Whittaker, G.R., and Scidmore, M.A. (2003). Rab GTPases are recruited to chlamydial inclusions in both a species-dependent and species-independent manner. Infect Immun *71*, 5855–5870.

Saka, H.A., Thompson, J.W., Chen, Y.S., Dubois, L.G., Haas, J.T., Moseley, A., and Valdivia, R.H. (2015). *Chlamydia trachomatis* Infection Leads to Defined Alterations to the Lipid Droplet Proteome in Epithelial Cells. PLoS One *10*, e0124630.

Saka, H.A., and Valdivia, R. (2012). Emerging roles for lipid droplets in immunity and host–pathogen interactions. Annu Rev Cell Dev Biol *28*, 411–437.

Sandoz, K.M., Valiant, W.G., Eriksen, S.G., Hruby, D.E., Allen, R.D., 3rd, and Rockey, D.D. (2014). The broad-spectrum antiviral compound ST-669 restricts chlamydial inclusion development and bacterial growth and localizes to host cell lipid droplets within treated cells. Antimicrob Agents Chemother *58*, 3860–3866.

Schramm, N., and Wyrick, P.B. (1995). Cytoskeletal requirements in *Chlamydia trachomatis* infection of host cells. Infect Immun *63*, 324–332.

Snavely, E.A., Kokes, M., Dunn, J.D., Saka, H.A., Nguyen, B.D., Bastidas, R.J., McCafferty, D.G., and Valdivia, R.H. (2014). Reassessing the role of the secreted protease CPAF in *Chlamydia trachomatis* infection through genetic approaches. Pathog Dis *71*, 336–351.

Soupene, E., Wang, D., and Kuypers, F.A. (2015). Remodelling of host phosphatidylcholine by *Chlamydia* acyltransferase is regulated by acyl-CoA binding protein ACBD6 associated with lipid droplets. Microbiologyopen *4*, 235–251.

Stanhope, R., Flora, E., Bayne, C., and Derré, I. (2017). IncV, a FFAT motif-containing *Chlamydia* protein, tethers the endoplasmic reticulum to the pathogen-containing vacuole. Proc Natl Acad Sci U S A *114*, 12039–12044.

Sun, H.S., Wilde, A., and Harrison, R.E. (2011). *Chlamydia trachomatis* inclusions induce asymmetric cleavage furrow formation and ingression failure in host cells. Mol Cell Biol *31*, 5011–5022.

Tarbet, H.J., Dolat, L., Smith, T.J., Condon, B.M., O'Brien, E.T., 3rd, Valdivia, R.H., and Boyce, M. (2018). Site-specific glycosylation regulates the form and function of the intermediate filament cytoskeleton. Elife *7*.

van Ooij, C., Kalman, L., van, I., Nishijima, M., Hanada, K., Mostov, K., and Engel, J.N. (2000). Host cell-derived sphingolipids are required for the intracellular growth of *Chlamydia trachomatis*. Cell Microbiol *2*, 627–637.

Volceanov, L., Herbst, K., Biniossek, M., Schilling, O., Haller, D., Nolke, T., Subbarayal, P., Rudel, T., Zieger, B., and Hacker, G. (2014). Septins arrange F-actin-containing fibres on the *Chlamydia trachomatis* inclusion and are required for normal release of the inclusion by extrusion. MBio *5*, e01802–01814.

Weber, M.M., Bauler, L.D., Lam, J., and Hackstadt, T. (2015). Expression and localization of predicted inclusion membrane proteins in *Chlamydia trachomatis*. Infect Immun *83*, 4710–4718.

Wesolowski, J., Weber, M.M., Nawrotek, A., Dooley, C.A., Calderon, M., St Croix, C.M., Hackstadt, T., Cherfils, J., and Paumet, F. (2017). *Chlamydia* Hijacks ARF GTPases To Coordinate Microtubule Posttranslational Modifications and Golgi Complex Positioning. MBio *8*.

Wolf, K., and Hackstadt, T. (2001). Sphingomyelin trafficking in *Chlamydia pneumoniae*-infected cells. Cell Microbiol *3*, 145–152.

Zhu, H., Shen, Z., Luo, H., Zhang, W., and Zhu, X. (2016). *Chlamydia Trachomatis* Infection-Associated Risk of Cervical Cancer: A Meta-Analysis. Medicine (Baltimore) *95*, e3077.

Chlamydia and Cell-autonomous Defence: Apoptosis and Autophagy

6

Georg Häcker[1]* and Thomas Rudel[2]

[1]Faculty of Medicine, Institute of Medical Microbiology and Hygiene, University of Freiburg, Freiburg, Germany.
[2]Department of Microbiology, University of Würzburg Biocenter, Würzburg, Germany.

*Correspondence: georg.haecker@uniklinik-freiburg.de

https://doi.org/10.21775/9781912530281.06

Abstract

In this chapter, we will discuss two major mechanisms of cell-autonomous defence against chlamydial infection in mammalian cells: apoptosis and autophagy. Both have been implicated in chlamydial infection numerous times, and there is substantial evidence that *Chlamydia* has established ways to escape these two defence reactions. We will describe the molecular mechanisms of apoptosis and autophagy and review the literature concerning their roles during a chlamydial infection. We will also discuss how these mechanisms may be triggered during an infection to defend the host cell, and how they are subverted by the infecting bacteria. *Chlamydia* can inhibit apoptosis in infected cells: experimental stimuli that would normally induce apoptosis in human cells fail to do that in *Chlamydia*-infected cells. It is generally assumed that *Chlamydia* needs to inhibit apoptosis to keep its host cell alive, although this has not been experimentally substantiated. Autophagy can also be used as a defence mechanism against intracellular bacteria. *Chlamydia* uses at least two mechanisms to avoid recognition by the autophagy machinery. Both apoptosis and autophagy could defend a human cell against chlamydial infection, but the bacteria appear to have evolved mechanisms to escape these defence mechanisms.

Introduction

One of the challenges that *Chlamydia* faces is the anti-bacterial defence reaction that the infected cell can muster. Not only professional immune cells, but probably all nucleated mammalian cells have the capacity to detect invading microorganisms and to respond with a panel of defence reactions against the invader. One of these mechanisms is apoptosis, a form of 'programmed cell death' or cellular suicide. A cell can activate its apoptotic programme, which will kill the infected cell: within a dead cell *Chlamydia* cannot complete its developmental cycle, and apoptotic cells are taken up by macrophages, which are equipped to digest bacteria. Death of an infected cell would thus block chlamydial replication. A second defence reaction is autophagy, whose main function is to regulate the turnover of cellular components including organelles. This leads to degradation of cellular material, which is then recycled. It is well documented that the autophagy system can be used to get rid of intracellular pathogens that have invaded a human cell (Gomes and Dikic, 2014). Through autophagy, intracellular bacteria can be targeted to lysosomes and destroyed. Both apoptosis and autophagy therefore have the potential to defend a human cell against infection with *Chlamydia*.

Apoptosis and autophagy are complex cell biological mechanisms, and it has not been easy to work out their molecular intersection with the developmental cycle of *Chlamydia*. The bacterial species of the genus *Chlamydia*, which infect animal hosts with a sophisticated apoptosis apparatus have, where tested, been found to be able to inhibit apoptosis. How *Chlamydia* inhibits apoptosis has still not been settled. There are two principal possibilities. First, *Chlamydia* may express a bacterial apoptosis inhibitor that can be translocated into the host cell and that can interfere with apoptotic signalling. Second, the infection may tamper with the host cell apoptotic signalling by altering the expression or activity of host pro- and anti-apoptotic proteins. There is no published example of the first possibility, while numerous effects of the second sort have been reported, and sometimes refuted again (see below). Apoptosis during chlamydial infection is thus a still changing and moving field, and our understanding of this interaction is likely to benefit from recent and future improvements in our understanding of the cell biology of apoptosis.

In mammals, specialized cells from the innate and adaptive immune system combat infections. In addition, most cells of the body have some capacity to defend themselves against pathogens. This ability is collectively referred to as 'cell-autonomous defence'. Mechanisms of cell-autonomous defence include the initiation of apoptosis, targeting of the pathogen to phagolysosomes or autophagosomes, and NF-κB- and IFN-dependent expression of cytokines and antimicrobial effectors (Randow *et al.*, 2013). Each *Chlamydia* species is well-adapted to the immune system of their respective host but is susceptible to defence mechanisms of divergent host species (for instance human versus mouse). These differences in the response to infection can be explained by the differences in IFN-signalling between human and murine cells and the specific subversion strategies of chlamydial species. The mechanism underlying clearance of heterologous *Chlamydia* infections depends on autophagy targeting the chlamydial inclusion for autolysosomal degradation.

In this chapter we will focus on apoptosis and autophagy during a *Chlamydia* infection. We will discuss our current understanding of mechanisms of autophagy and apoptosis, and how they may be used by an infected human cell to combat the infection. We will also present the current knowledge of how *Chlamydia* can overcome these defence mechanisms.

What is apoptosis?

Apoptosis is a form of cell death that results from activation of a dedicated signalling pathway. Proteolysis by aspartate proteases termed caspases is a core signalling mechanism and is required for the downstream effects of apoptosis (Hengartner, 2000). In all cases of apoptosis the activity of 'effector' caspases can be detected. The main effector caspase is caspase-3; a role for caspase-7 and caspase-6 is assumed but not actually very clear. Caspase-3 has to be activated by another caspase, an 'initiator' caspase. There are two initiator caspases that are regulated by different upstream signalling events, caspases 8 and 9: caspase-8 is activated following the engagement of a death receptor such as TNFR1 or CD95/Fas. Caspase-9 becomes activated in the apoptosome, a cytosolic complex that forms upon release of cytochrome *c* from its normal localization in the mitochondrial intermembrane space (mitochondrial apoptosis). Initiator caspases can directly cleave and activate caspase-3. Caspase-8 can cleave caspase-3 directly but can also cleave an activator of mitochondrial apoptosis, the Bcl-2-family protein Bid (see below). In this case caspase-3 is activated in the sequence of caspase-8-Bid-caspase-9-caspase-3 (Scaffidi *et al.*, 1998).

Of the two forms of apoptosis, the much more common form naturally occurring in an organism is mitochondrial apoptosis. A critical step in mitochondrial apoptosis is the release of cytochrome *c*. This release is regulated by the activity of Bcl-2-family proteins (Strasser and Vaux, 2018). The Bcl-2-family consists of about 15 members that fall into three separate groups that are either pro- or anti-apoptotic (two are pro-, one is anti-apoptotic); the three groups can also be defined structurally since they share certain characteristic domains (Chipuk *et al.*, 2010). One pro-apoptotic Bcl-2-family group encompasses the initial triggers of mitochondrial apoptosis. Because of a particular structural feature, the proteins in this group are called BH3-only proteins. BH3-only proteins start mitochondrial apoptosis by activating members of the second pro-apoptotic group of Bcl-2-proteins. This second group consists of two functionally equivalent proteins, Bax and Bak. During their activation, Bax and Bak form homo-oligomers and insert into the outer mitochondrial membrane (OMM). This event is believed to form a pore in the membrane that releases cytochrome *c* and other intermembrane space proteins; the precise nature of such a pore is unknown. The greatest difference between Bax and Bak is that Bax is kept cytosolic and probably cycles between OMM and cytosol while Bak is always inserted in the OMM, where it is kept inactive by a mitochondrial porin, VDAC2 (a membrane-inserted protein with a β-barrel structure) (Naghdi and Hajnoczky, 2016). Bax and Bak can be neutralized by direct binding to proteins from the third group of Bcl-2-proteins, the anti-apoptotic group. These proteins comprise five members (Bcl-2, Bcl-X_L, Bcl-w, Mcl-1 and A1/Bfl-1), which are, some exclusively, some mostly tail-anchored proteins that are inserted into the OMM. Table 6.1 shows the allocation of Bcl-2-proteins to these groups for further reference.

One way of inducing mitochondrial apoptosis is the down-regulation of Bax/Bak-inactivators of the Bcl-2 family. This has been most extensively documented for Mcl-1. The rapid turnover of Mcl-1 is regulated by a number of ubiquitinating and deubiquitinating enzymes and by proteasomal degradation, often depending on its phosphorylation status. Major cellular kinase pathways target Mcl-1, most notably the PI3 kinase/AKT-pathway, which stabilizes Mcl-1 as one aspect of its anti-apoptotic function (Mojsa *et al.*, 2014).

In addition to the pure loss of inhibitors, and partly overlapping with this function, Bax/Bak are activated by members of the BH3-only protein group (Bim, Bid, Puma, Bmf, Noxa, Bad, Bik, Hrk) (Chipuk *et al.*, 2010). Their pro-apoptotic function has two molecular aspects. All BH3-only proteins can bind to and thereby inactivate anti-apoptotic Bcl-2-proteins, relieving their inhibition on Bax/Bak. However, great differences exist in terms of their binding affinity (Chen *et al.*, 2005). The two most broadly binding BH3-only proteins are Bim and Bid: they can bind all anti-apoptotic Bcl-2-proteins. Others such as Bmf are more restricted; Bmf is for instance inferior to Bim in binding to Mcl-1 at least in some assays. Noxa is very restricted and can bind only to Mcl-1 and A1. By neutralizing anti-apoptotic Bcl-2-proteins in this way, BH3-only proteins can activate Bax/Bak. In addition, at least three BH3-only proteins (Bim, Bid and Puma) can directly bind and activate Bax and Bak, probably through transient binding, which induces a conformational change that allows the insertion of Bax/Bak into the OMM. There are two BH3-only proteins, that do not seem to localize to mitochondria, which may arguably challenge their purported role as BH3-only proteins: Bik is localized in the ER-membrane while Bad, although not very closely studied, does not localize to mitochondria in cell-free-systems (Wilfling *et al.*, 2012). All other BH3-only proteins are tail-anchored in the OMM via a C-terminal alpha-helix.

There are largely two ways to activate mitochondrial apoptosis. One is the down-regulation of anti-apoptotic proteins, the other is the up-regulation or activation of (pro-apoptotic) BH3-only proteins. Although theoretically possible, changes in the abundance of Bax/Bak are unusual and do not seem to play a role of substance. BH3-only proteins can be regulated transcriptionally. An example of this is the up-regulation of Puma and Noxa by the transcription factor p53, typically following DNA damage (Villunger *et al.*, 2003). Bid is different in that its inactive form is cytosolic, but once it is activated by caspase–8-mediated cleavage, this active form inserts into the OMM (Youle and Strasser, 2008). Activation of Bim, and very likely also of Bmf, involves modifications to its binding partner dynein light chain (DLC) 1/2 (Singh *et al.*, 2018). The abundance and possibly activation of Bim is further regulated by MAPK kinase-signalling. ERK signals regulate the ubiquitination and degradation of Bim and are anti-apoptotic. JNK signalling, in contrast, is pro-apoptotic through Bim and may act by disrupting the Bim-DLC-interaction.

Table 6.1 Bcl-2-proteins

Category	Specific proteins
Initiators of apoptosis (BH3-only proteins)	Bim, Noxa, Puma, Bmf, Bid, Bik, Hrk
Effectors of mitochondrial cytochrome c-release activated by BH3-only proteins	Bax and Bak
Inhibitors of apoptosis	Bcl-2, Bcl-XL, Mcl-1,Bcl-w, A1/Bfl-1

There are clearly many undiscovered details of situations where BH3-only proteins are activated post-translationally. More generally, quite a number of Bcl-2-family proteins can replace each other, and there are few assays to measure their activity individually. It is therefore extremely difficult to understand the role each Bcl-2-family protein plays in a given situation of apoptosis. Nevertheless, we do know much about the molecular processes that drive or inhibit apoptosis and can in many situations now probe mechanisms with some precision.

Apoptosis is difficult to study *in vivo*. Although many studies show apoptotic cells in tissue sections, the general understanding is that apoptotic cells are rapidly engulfed and degraded by either neighbouring cells or, probably more commonly, by macrophages (Kawane *et al.*, 2003). *In vitro*, a number of approaches can be used to detect either the activation of the apoptotic pathway or its downstream effects. A good way of measuring apoptosis is the activation of caspase-3. Caspase-3-activation involves its proteolytic processing, which generates a neo-epitope that can be detected using antibodies directed against end-fragments of the proteolytic product and detection by flow cytometry or microscopy. Another assay uses the exposure of the membrane phospholipid phosphatidylserine (PS), which, in the absence of an apoptotic stimulus, is almost exclusively found in the inner membrane leaflet, but which is 'flipped' to the extracellular membrane leaflet during apoptosis. PS binds the protein annexin V, and binding of fluorescently labelled annexin V to cells undergoing apoptosis often provides a good way of detecting these cells. Later on, the loss of membrane integrity (which however only indicates cell death and is not specific for apoptosis) is also commonly used. Less frequently used methods include the detection of nuclear morphological changes and DNA damage.

Why may apoptosis be relevant to chlamydial infection?

Chlamydia can only grow inside host cells. This immediately suggests that apoptosis could be used as an efficient defence reaction by the host cell. Indeed, when *C. trachomatis*-infected HeLa cells were experimentally forced to undergo apoptosis by massive over-expression of the BH3-only protein Bim, chlamydial replication was – fairly unsurprisingly – prevented (Ying *et al.*, 2008b). Does this mean that *Chlamydia* requires the ability to block apoptosis? Not necessarily: the regular lifespan of an epithelial cell is probably substantially longer than the two days or so required for completion of the chlamydial developmental cycle. We do not know whether some species of *Chlamydia* may reside in their host cells much longer and not simply replicate as fast as possible. It is possible that some long-term infections require keeping host cells alive beyond their normal lifespan. In this case, an anti-apoptotic activity may be needed to extend the normal life span of the host cell. It is, however, not obvious that anti-apoptosis is relevant to acute infection of humans.

The alternative is that *Chlamydia* triggers an apoptotic response that would kill the infected cell but *Chlamydia* has developed a means to inhibit apoptosis as a countermeasure. There is in fact a substantial number of data about viruses that do exactly that (Galluzzi *et al.*, 2008), so could *Chlamydia* do the same? In this scenario, apoptosis acts as the self-defence mechanism alluded to above. Rather than permitting the pathogen to replicate, which would generate progeny of infectious particles and spread to other cells when released, the cell kills itself. The infectious agent cannot replicate and is probably taken up by a phagocyte, which is normally equipped to digest the agent together with the remnants of the apoptotic host cell. As already mentioned, the initiation of apoptosis is a common response of a cell to viral infection. There may not be much that viruses and *Chlamydia* have in common, but both depend on the integrity of the host cell for their replication.

In the case of a number of viruses, pro-apoptotic pathways have been elucidated. Viruses can be recognized by pattern recognition receptors (PRR) that typically detect viral nucleic acids. PRR are better known for activating major signalling pathways such as NF-κB, type I interferon and MAP kinase pathways. However, PRR are also potent activators of the apoptosis machinery. This potential seems to have been found for every PRR that has been tested. Examples are toll-like receptors (Aliprantis *et al.*, 1999; Weber *et al.*, 2010), RIG-like helicases (Besch *et al.*, 2009) and the DNA recognition system cGAS/STING (Gulen *et al.*, 2017). How the decision is made for a PRR to activate signalling or to promote apoptosis is not clear, but may depend on signal strength, expression of anti-apoptotic proteins, cell type and other factors.

Now looking at *Chlamydia*, it is possible that chlamydial recognition by PRR could generate a pro-apoptotic signal. PRR certainly recognize chlamydiae, and several different receptors have been found to be involved. In epithelial cells, the recognition of *Chlamydia* has been shown to occur through NOD1 (Buchholz and Stephens, 2008) and by cGAS (Zhang *et al.*, 2014), leading to activation of STING. NOD1 is a cytosolic receptor for peptidoglycan fragments that can detect the presence of cytosolic bacteria (Philpott *et al.*, 2014). This suggests that chlamydial peptidoglycan reaches the cytosol of the infected cells. NOD1 has at least weak pro-apoptotic activity (Inohara *et al.*, 1999) and may therefore be involved in the activation of apoptosis during infection. STING is most often activated by cyclic dinucleotides that are produced by cGAS upon cytosolic recognition of DNA (Li and Chen, 2018). cGAS/STING may therefore recognize chlamydial DNA that somehow reaches the cytosol. It has also been suggested that chlamydial cyclic dinucleotides, generated as second messengers in bacterial physiology, leak out of the inclusion and directly stimulate STING (Barker *et al.*, 2013). STING has clear pro-apoptotic potential (Gulen *et al.*, 2017); STING activation may therefore be a pro-apoptotic signal during chlamydial infection that has to be countered by *Chlamydia*. Recognition of chlamydial material that required the TLR3/4-sigalling adapter TRIF in epithelial cells has also been described (Derbigny *et al.*, 2007); TLR3/4 also have pro-apoptotic potential (Ruckdeschel *et al.*, 2004).

The recognition of *Chlamydiae* thus has the potential, through the engagement of PRR, to induce apoptosis. There are other potential pathways that might trigger mitochondrial apoptosis in *Chlamydia*-infected cells. Apoptosis is notoriously easily initiated by changes in a cell's environment or interior milieu: imbalances in cellular nutrients, depletion of host cell materials, accumulation of metabolites and many more changes can induce apoptosis (Vaux *et al.*, 1994). Quite a number of such changes have been shown or postulated to occur during chlamydial infection. This suggests that a pro-apoptotic signal is generated during the infection but no apoptosis is seen. It may well be the case that the strong chlamydial anti-apoptotic activity (see below) masks any pro-apoptotic effect, as indeed would be expected. There are some older studies that suggest that there is a pro-apoptotic effect (Ojcius *et al.*, 1998; Perfettini *et al.*, 2003). However, a later study concluded that *Chlamydia*-infected cells do not undergo apoptosis, even though they display some morphological changes of apoptosis, because the changes were independent of caspases (Ying *et al.*, 2006). Similarly, infection with *C. pneumoniae* was able to kill human cells, with some features of apoptosis, but in a caspase-independent manner. It should be noted that very high bacterial numbers were required for this form of cell death (Dumrese *et al.*, 2005). Generally, with bacterial numbers in a range of about one per host cell, there is no cell death, apoptotic or otherwise.

Apoptosis inhibition by chlamydial infection

When human epithelial cells are infected with *Chlamydia*, they become resistant to experimental stimuli that would otherwise induce apoptosis in the infected cell. All stimuli tested that induce mitochondrial apoptosis in human cells, such as cytotoxic agents used for tumour therapy, are inhibited in their apoptosis-inducing activity. The most plausible interpretation of this broad inhibition of apoptosis seems to be that *Chlamydia* has to inhibit apoptosis triggered by its recognition but that this activity happens to extend to the inhibition of other stimuli.

This anti-apoptotic activity was for the first time described for HeLa cells infected with *C. trachomatis* serovar L2 (Fan *et al.*, 1998). Similar findings were soon reported for *C. pneumoniae* (Fischer *et al.*, 2001; Rajalingam *et al.*, 2001), *C. muridarum*, *C. psittaci*, *C. caviae* (Zhong *et al.*, 2006) as well as infections of other cell types (fibroblasts) (Ying *et al.*, 2008a). All members of the *Chlamydiaceae* family tested have been found capable of inhibiting apoptosis. Apoptosis inhibition has been thoroughly investigated on a molecular level only for *C. trachomatis* L2. All the results discussed below have been obtained with that strain. It is therefore possible that different chlamydial strains and species use different mechanisms to inhibit apoptosis. Personally, we believe this is unlikely. It is not straightforward for a bacterium to evolve the potent apoptosis-inhibition that has been shown for all tested chlamydial species. There is no reason to assume that the various species have developed their individual mechanisms after they have split from whatever the first ancestor was that had anti-apoptotic activity. However, we cannot exclude the possibility.

We think it is of significance that all tested *Chlamydiaceae* but not *Chlamydia*-related bacteria (which are reviewed in Chapter 18, Thinking Outside the *Chlamydia* Box), except for *Simkania negevensis* (Karunakaran *et al.*, 2011) appear to be able to inhibit apoptosis. For example, *Waddlia chondrophila* grows well in human host cells (its natural host is unknown) but has no anti-apoptotic activity (Dille *et al.*, 2015), at least when tested with the strong apoptosis-inducer staurosporine. Although *W. chondrophila* has been detected using sensitive assays in human infections (Baud *et al.*, 2014), it seems very unlikely that this epidemiology of very few infected human hosts could sustain the species. Rather, *W. chondrophila* likely has a different, possibly non-mammalian host and infects humans accidentally. This is an important consideration if we pursue the line of thought put forward above. It is pure speculation but the lack of apoptosis inhibition may make it more difficult or impossible to establish itself permanently in a mammalian population. *Chlamydia*-related bacteria that grow in protozoa appear to have no anti-apoptotic activity (Sixt *et al.*, 2012). This makes sense since their natural hosts have no known apoptosis machinery.

Molecular mechanism of apoptosis inhibition by *Chlamydia*

While a number of reports have investigated the question how *Chlamydia* inhibits apoptosis, there is curiously little agreement between the proposed mechanisms. The data seem to agree on the point that only mitochondrial apoptosis is inhibited while direct caspase-8-mediated apoptosis (see above) is not (Fischer *et al.*, 2004a). However, all stimuli of mitochondrial apoptosis tested to date were blocked by chlamydial infection. It therefore seems safe to suggest that a general block in mitochondrial apoptosis is imposed by chlamydial infection.

Regarding the mechanism of apoptosis inhibition, there are two principal possibilities. One is that *Chlamydia* carries an inhibitor of apoptosis that is expressed during infection and perhaps localizes to mitochondria where it could block signal transduction in the Bcl-2-family. Alternatively, *Chlamydia* may manipulate the expression and/or activity of host cell proteins and thereby inhibit apoptosis. For instance, the infection could upregulate an anti-apoptotic factor such as Bcl-2, thereby inducing broad inhibition of apoptosis.

Most published studies have focused on the latter possibility. Within this area, several studies reported an analysis of upstream signalling pathways. Thus, the activation of PI3K during infection has been reported – PI3K can be anti-apoptotic in some situations, most likely through a stabilizing effect on Mcl-1 (Mojsa *et al.*, 2014). The transcription factor p53 can have a pro-apoptotic activity (Lowe *et al.*, 1993), and the levels of p53 decrease during chlamydial infection (Gonzalez *et al.*, 2014; Siegl *et al.*, 2014). Other kinase pathways are also activated and can have anti-apoptotic function (e.g. Mehlitz *et al.*, 2010; Subbarayal *et al.*, 2015). It is very difficult to measure the anti-apoptotic activity of major signalling pathways in mammalian cells because there are only limited ways of activating such signalling pathways directly; most experiments rely on triggering some cellular receptor that triggers, for instance, the ERK pathway and then test the effect of chemical inhibition of ERK. In most studies, a surrogate marker of apoptosis such as the levels of a pro- or anti-apoptotic protein is then tested. We therefore cannot be certain about the contribution of these upstream signalling pathways to apoptosis inhibition by *Chlamydia*. However, decades in the field of apoptosis research make us confident that none of these signals can explain the strong anti-apoptotic effect of chlamydial infection. That does not exclude the possibility that such pathways contribute to apoptosis-inhibition. They, however, do not explain the breadth of the activity.

Two further hypotheses have been put forward that directly concern changes within the Bcl-2-family system at mitochondria.

1 The first hypothesis is based on the strong up-regulation of Mcl-1 by *Chlamydia* infection

(Rajalingam *et al.*, 2008). This up-regulation is impressive by Western blotting starting at about 24 h post-infection and may involve transcriptional regulation as well as stabilization by deubiquitylation (Fischer *et al.*, 2017). High levels of Mcl-1 have the potential to protect broadly against apoptosis, as shown in Mcl-1-transgenic mice (Campbell *et al.*, 2010). Targeting Mcl-1 by RNAi showed that its loss enhanced apoptosis in small but not in large inclusions in infected HeLa cells, and not at later stages of the infection, when apoptosis inhibition is Mcl-1-independent (Rajalingam *et al.*, 2008).

2. The second hypothesis centres around the loss of pro-apoptotic BH3-only proteins during infection, most notably Bim. Bim levels decrease during infection (in this report lysates were prepared in Laemmli buffer containing 2% SDS) (Fischer *et al.*, 2004b). Later, it was proposed that the loss of Bim is an extraction artefact of infected cells and that only 8M urea during extraction can prevent its loss (Chen *et al.*, 2012). However, a re-analysis detected loss of Bim in extracts from infected cells prepared using 8M urea (Dille *et al.*, 2015). Thus, the issue of Bim loss therefore remains unresolved.

The loss of Bim alone is again insufficient to explain the complete block of apoptosis observed, and although some data on additional BH3-only proteins have been presented (Ying *et al.*, 2005) this may not be sufficiently substantiated. While these alterations to Bcl-2-family members may contribute, they are insufficient to explain how *Chlamydia* inhibits apoptosis.

Changes in host cell apoptosis-regulatory proteins that could explain the anti-apoptotic effect of *Chlamydia* would have to be profound, and would be expected to be fairly obvious. From molecular studies we know that the levels of neither the Bax/Bak-group nor of the anti-apoptotic Bcl-2-protein group (with the exception of Mcl-1) are substantially altered during infection. We further know that the anti-apoptotic activity prevents the activation of Bax and of Bak (the published analysis is limited to the finding that a certain N-terminal conformational change in these proteins, which is seen during their activation, is prevented in *Chlamydia*-infected cells (Fischer *et al.*, 2004b). There is no obvious candidate for a host cell protein that could have this effect. We now in fact have the tools to probe this question more deeply than before. In particular new small molecule Bcl-2 antagonists that specifically inhibit an individual anti-apoptotic protein, such as Bcl-2 (Souers *et al.*, 2013), Mcl-1 (Kotschy *et al.*, 2016) or combinations of Bcl-2-family proteins (Oltersdorf *et al.*, 2005), will be valuable for studying the molecular basis of chlamydial anti-apoptosis in greater detail.

The second possibility, namely that a chlamydial protein (or other component) may be relevant to the inhibition of apoptosis, has to our knowledge not been stringently explored. Many viruses, especially large DNA viruses, carry dedicated anti-apoptotic proteins, including Bcl-2-homologues, to inhibit host cell apoptosis (Galluzzi *et al.*, 2008). However, the acquisition of such inhibitory factors is much easier for a virus than it is for a bacterium. Viruses often 'steal' host cell genes and modify them for their purpose, and the host cell has several of these potent apoptosis-inhibitors. Although gene transfer between animal cells and bacteria is not unheard of (Dunning Hotopp, 2011), it is certainly not common, and for instance Bcl-2-like proteins have not been detected in the *Chlamydia* genome. A protein with homology to mammalian anti-apoptotic proteins therefore seems unlikely. Future research should consider the possibility that a secreted *Chlamydia*-derived factor makes an infected human cell resistant to apoptosis. The existence and nature of such an inhibitor is completely speculative, but there seems to be no other explanation for the anti-apoptotic effect of *Chlamydia* through a known host apoptosis-modulator.

What is autophagy?

Autophagy is a highly conserved and regulated process of cell homeostasis by which cytoplasmic material is delivered to and degraded in lysosomes. It is induced under diverse cellular stress conditions as a cytoprotective response to prevent cell damage and promote cell survival. Three different types of autophagy have been described. Chaperone-mediated autophagy facilitates the degradation of intracellular proteins by chaperone-guided direct translocation of substrate proteins into the lysosome (Cuervo and Wong, 2014). During microautophagy, cytoplasmic cargo is engulfed by the lysosome in a process similar to endocytosis (Li *et al.*, 2012). Macroautophagy is characterized by the sequestration of cytosolic components into typical double-membrane vesicles, the so-called autophagosomes which subsequently fuse with lysosomes to mediate cargo degradation (Dikic and Elazar, 2018). We will focus on macroautophagy because it is most relevant in the context of bacterial infections (Gomes and Dikic, 2014). Together with the ubiquitin proteasome system (UPS), autophagy is an essential cellular mechanism to maintain protein homeostasis (Labbadia and Morimoto, 2015). The UPS is used to proteolytically degrade short-lived, misfolded and damaged proteins, whereas the autophagy–lysosomal pathway removes larger particles like protein aggregates, dysfunctional organelles and invading bacteria (Dikic, 2017).

The main function of autophagy in an uninfected cell is to provide essential metabolites during starvation and excessive cellular stress. Autophagy was therefore considered to be a non-specific process. This view changed recently when more and more examples demonstrated that autophagy specifically delivers selected substrates for autophagic degradation (Farre and Subramani, 2016). Similar to proteins targeted for proteasomal degradation, selective autophagy requires labelling of the substrate with ubiquitin. However, ubiquitination of substrates does not always lead to their degradation but sometimes changes the biological function of the substrate. In this case, ubiquitin has a role of a signalling molecule. Ubiquitin is covalently linked to lysine residues of target proteins in a multistep process by ubiquitin-activating E1 and ubiquitin-conjugating E2 enzymes together with E3 ubiquitin ligases. As ubiquitin itself harbours lysine residues, chains of repeatedly attached ubiquitin molecules can be formed. The fate of the target proteins can be very diverse, depending on the length of the ubiquitin chain, and by which of its seven lysine residues ubiquitin molecules are linked to each other (Grumati and Dikic, 2018). Attachment of ubiquitin molecules can be reversed by deubiquitinating enzymes (DUBs). Ubiquitination and deubiquitination serve as regulatory mechanisms to control the so-called ubiquitin code – the specific ubiquitin chain length and linkage – and thereby the fate of targets (Mevissen and Komander, 2017).

The autophagic machinery of eukaryotic cells consists of eighteen autophagy-related (ATG) proteins and five additional factors that are grouped into five functional complexes (Klionsky *et al.*, 2011). Autophagosomes are generated at phagophore assembly sites (PAS), which are specific subcellular locations that are derived from membranes of various source organelles, including the ER and mitochondria (Hamasaki *et al.*, 2013), plasma membrane (Ravikumar *et al.*, 2010), Golgi (van der Vaart *et al.*, 2010) and recycling endosomes (Puri *et al.*, 2013). The serine/threonine kinase mechanistic target of rapamycin (mTOR) is a main regulator of cell homeostasis and is part of the mTOR complex (mTORC) 1 (Saxton and Sabatini, 2017). mTORC1 negatively regulates the UNC-51-like kinase (ULK) complex, which initiates autophagosome biogenesis. Upon cellular stress conditions, the ULK complex is dephosphorylated and thereby activated, so that it translocates to the PAS to which other components of the autophagy machinery are also recruited, including the phosphatidylinositol-3-kinase (PtdIns3K) complex (Yang and Klionsky, 2010). Subsequent generation of the lipid phosphatiylinositol-3-phosphate (PtdIns3P) at the PAS leads to the localization of additional factors required for phagophore formation. Elongation and closure of the phagophore completes the formation process resulting in a double-membrane coated vacuole called an autophagosome. This process of autophagosome formation is mediated by two ubiquitin-like conjugation systems involved in the conjugation of phosphatidylethanolamine (PE) to members of the ubiquitin-like ATG8 family (Yang and Klionsky, 2010). ATG8 proteins comprise the light-chain 3 (LC3) and γ-aminobutyric acid receptor-associated protein (GABARAP) subfamilies, with their members LC3A, LC3B, LC3C, GABARAP, GABARAP-like 1 (L1), GABARAP L2. After their synthesis, ATG8 proteins are post-translationally modified by the ATG8/LC3 conjugation system. First, they are processed at their C-termini by the cysteine protease ATG4, subsequently activated by the E1-like enzyme ATG7 and then conjugated to PE, associated with the phagophore membrane, by the E2-like enzyme ATG3. In complex with ATG16L1, the ATG12–ATG5 conjugate functions as an E3-like ligase. Thereby ATG8 proteins in general, and LC3 in particular, are transferred from a cytosolic (LC3-I) to a lipidated, membrane bound (LC3-II) state.

ATG8 proteins mediate selective recognition of ubiquitinated substrates by interacting with autophagy receptors or adaptors. These adaptor proteins, including neighbour of BRCA1 gene 1 (NBR1) (Kirkin *et al.*, 2009), p62 (Pankiv *et al.*, 2007) and optineurin (OPTN) (Wild *et al.*, 2011), are ubiquitin binding proteins, which harbour a LC3 interacting region (LIR)-motif for the interaction with ATG8 proteins (Rogov *et al.*, 2014). The autophagosome then matures by stripping off ATG proteins. The further transport of autophagosomes towards lysosomes is thought to occur in a microtubule-dependent manner facilitated by LC3. Fusion then creates autolysosomes, the compartments where acidic hydrolyses degrade the autophagic cargo to recycle them back to the cytoplasm.

There is no standard assay to monitor autophagy in infected cells. Autophagy is a highly dynamic process and the number of autophagosomes at any given time point represents the balance of autophagosome formation and their conversion into autolysosomes. This means that an increased number of autophagosomes in an infected cell could result from the induction of autophagosome formation or the inhibition of downstream processes. Measurements of autophagic activity and determination of the impact of intracellular pathogens on this process requires careful investigation and quantification of intermediate structures. For example if autophagosome formation is blocked, the numbers of all autophagic structures decrease. In contrast, interference with any step downstream of autophagosome formation increases the numbers of autophagosomes while decreasing the numbers of autolysosomes. The most traditional method is the detection of autophagosomes and autolysosomes by electron microscopy (Mizushima *et al.*, 2010). Autophagosomes are enclosed by a double-membrane and contain undigested material (bacteria, mitochondria, ER membranes). Autolysosomes are more difficult to define since they have a single limiting membrane and partially digested material. The detection of autophagy by electron microscopy has been more and more replaced by fluorescence microscopy and immunoblot assays that are easier to apply for most researchers.

LC3 has been widely used as a maker of autophagosomes. LC3 is processed at the C-terminus by ATG4 to LC3-I which is in the cytosol. LC3-I is subsequently conjugated with PE to become LC3-II, which associates with the inner and outer membrane of the autophagosome. Endogenous LC3 or GFP-LC3 is detected by fluorescence microscopy either as diffuse cytoplasmic staining or as punctate structures representing autophagosomes. The conversion of LC3-I to LC3-II can also be detected by immunoblotting since LC3-II, in spite of its higher molecular weight, migrates faster than LC3-I by SDS-PAGE. Both, fluorescence microscopy and immunoblotting are well suited detection methods to quantify autophagosomes. Detection and quantification of autolysosomes are more difficult and assays depending on the degradative and acidic nature of the autolysosome are applied. For example, LC3-II and other autophagy markers like p62 are degraded in the autolysosome, and this is investigated by comparing two samples with and without lysosomal inhibitor treatment, e.g. Bafilomycin A1, an inhibitor of vacuolar ATPase required for lysosomal function. Another assay depends on the quenching of the GFP fluorescence in the acidic environment of the autolysosome by monitoring the green and red (autophagosome) or red only (autolysosome) fluorescence of an RFP–GFP-LC3 fusion (Mizushima *et al.*, 2010).

How does autophagy function in cell-autonomous defence?

An important function of autophagy is the cell-autonomous defence against intracellular pathogens. This selective, ubiquitin-dependent induction of autophagy efficient in delivering pathogens to the autophagosome has been termed xenophagy. The term 'xenophagy' has been introduced to discriminate the process of selective autophagy of foreign material ('xeno') like infecting microbes from the autophagy of host material like mitochondria ('mitophagy') or ribosomes ('ribophagy'). After the initial description in *Rickettsia*-infected neutrophils (Rikihisa, 1984), it took 20 years until xenophagy was recognized as a crucial cell-autonomous defence mechanism against pathogenic bacteria like *Mycobacterium tuberculosis* or group A streptococci (Gutierrez *et al.*, 2004; Nakagawa *et al.*, 2004). Like other types of selective autophagy, xenophagy depends on the core autophagy machinery (Gomes and Dikic, 2014).

One of the best studied examples is the clearance of *Salmonella* by an infection-induced host autophagy response. Cytosolic *Salmonella* or damaged *Salmonella*-containing vacuoles are decorated with ubiquitin (Birmingham *et al.*, 2006). Subsequently the autophagy receptors p62 (Zheng *et al.*, 2009), nuclear dot protein 52 kDa (NDP52) (Thurston *et al.*, 2009), and OPTN (Wild *et al.*, 2011) associate with ubiquitinated *Salmonella* and mediate binding to LC3-positive autophagosomes. NDP52 recruits the protein kinase TANK-binding kinase 1 (TBK1), which in turn phosphorylates OPTN enhancing its affinity to LC3. The autophagy system is even able to sense damaged bacteria-containing vacuoles. Host glycans lining the inner leaflet of the bacteria-containing vacuolar membrane are exposed on damaged *Salmonella*-containing vacuoles to the host cell cytosol and recognized by galectin-8 which recruits NDP52 in a ubiquitin-independent manner (Thurston *et al.*, 2012). The galectin-8/NDP52-marked membranes from *Salmonella*-containing vacuoles together with the ubiquitinated and OPTN/NDP52/p62-decorated *Salmonella* are then recognized by LC3 and targeted to the autophagosome for degradation.

Why may autophagy be relevant for *Chlamydia*?

Chlamydia as obligate intracellular bacteria acquire many metabolites from the host cell, which may create acute metabolite shortages that induce nutrient stress and autophagy. For example, *Chlamydia* take up amino acids from the host cell (Mehlitz *et al.*, 2017) and thereby may induce starvation signalling via mTORC1, a master regulator that drives anabolic cell growth. mTORC1 senses amino acid levels and couples amino acid availability to growth or autophagy (Inoki *et al.*, 2012; Jewell *et al.*, 2013). In addition, uptake and depletion of host ATP (Omsland *et al.*, 2012) and the induction of reactive oxygen species (ROS) (Abdul-Sater *et al.*, 2010) by replicating *Chlamydia* are efficient inducers of autophagy (Filomeni *et al.*, 2015; Kim *et al.*, 2011).

Although *Chlamydia* does not reside in the cytosol of host cells for extended times, a leaky or perforated inclusion membrane during late stage infections or as a consequence of cell-autonomous immune attack may offer targets for autophagic removal (Fischer and Rudel, 2018a). In this regard, the maintenance of inclusion integrity may be a major defence mechanism of *Chlamydia*. *Chlamydia* uses a type III secretion system (T3SS) to secrete inclusion membrane (Inc) proteins into the membrane of the inclusion. These Inc proteins, together with other proteins secreted via the T3SS, function in orchestrating the intracellular adaptation of *Chlamydia* (Moore and Ouellette, 2014). However, Inc proteins may also have a structural role in maintaining the inclusion membrane integrity. This has recently been proposed in a study in which mutations in three different Inc proteins led to premature inclusion lysis, recognition by the autophagosomal system and the induction of host cell death (Weber *et al.*, 2017). Incs are discussed in detail in Chapter 4, The Chlamydial Inclusion. However, it is currently not known if the autophagy machinery recognizes damaged chlamydial inclusions or only the chlamydial particles released after inclusion damage.

Interaction of *Chlamydia* with the autophagic pathway

Chlamydia has established ways to evade autophagosomal degradation. Early studies revealed that *C. trachomatis* inclusions do not fuse with autophagosomes in human epithelial cells, as the autophagosome-specific marker monodansylcadaverine (MDC) does not accumulate within inclusions (Al-Younes *et al.*, 2004). Nevertheless, an interaction with the autophagy pathway was shown by the detection of the autophagosome-linked protein LC3 in close proximity to the inclusion (Al-Younes *et al.*, 2004). Increased LC3-II levels revealed that autophagy is strongly induced in *C. trachomatis*-infected human epithelial cells (Pachikara *et al.*, 2009). LC3 also plays an autophagy-independent role in *Chlamydia* infection as it mediates interaction with the microtubule network and might promote the transport of nutrient-rich vesicles to the inclusion (Al-Younes *et al.*, 2011). Earlier work by the same group proposed autophagy as mechanism to provide nutrients to *Chlamydia* because inhibition of autophagy impaired chlamydial intracellular survival (Al-Younes *et al.*, 2004).

The adaptation of *Chlamydia* species to the host specific autophagy system is a fascinating example of co-evolution. For example, comparative studies revealed that EB production for *C. trachomatis* but not *C. muridarum* was reduced in ATG5$^{+/+}$ mouse fibroblasts compared to ATG5$^{-/-}$ cells. This result demonstrates that the mouse pathogen *C. muridarum* was able to overcome the autophagic defence, whereas the human pathogen *C. trachomatis* was susceptible (Yasir *et al.*, 2011). In human epithelial cells, in contrast, *C. muridarum* inclusions attracted several components of the autophagy pathway. However, *C. trachomatis* inclusions were resistant to ubiquitination, which prevented the association of the autophagy adaptors p62 and NDP52 and the autophagosomal marker LC3 with the inclusion (Haldar *et al.*, 2016). The family of guanylate binding proteins (GBPs), involved in the induction of macroautophagy, appear to be central regulators that target non-adapted *Chlamydia* for destruction (Al-Zeer *et al.*, 2013; Haldar *et al.*, 2015; Haldar *et al.*, 2016). These studies, in which host cells are infected with non-adapted chlamydial species are very informative since they reveal the host factors important for the response against intracellular *Chlamydia*.

Although the chlamydial factors mediating the evasion of host autophagy remain unknown, *Chlamydia* appears to subvert host autophagy by at least two mechanisms: (i) keeping the inclusion intact (Weber *et al.*, 2017); and (ii) interfering with initial ubiquitination of the inclusion. Two deubiquitinases (DUBs) have been identified in *C. trachomatis* (Misaghi *et al.*, 2006) and one of these effectors called Cdu1 is secreted into the inclusion membrane with its catalytic site facing the host cytosol (Fischer *et al.*, 2017). Expression of both DUBs is first detectable around 12 to 16 hours post infection and remains high during the replicative phase of the bacteria until the end of the cycle. *C. trachomatis* inclusions are not ubiquitinated, but inclusions of chlamydial mutants lacking a functional Cdu1 protein are highly ubiquitinated. This ubiquitination of inclusions did not require IFNγ stimulation, the potent inducer of autophagic clearance of intracellular pathogens (Fischer *et al.*, 2017). These observations suggest that *Chlamydia* utilizes its DUBs to prevent the constitutive ubiquitination of its inclusion (Fischer and Rudel, 2018b). With the Cdu1 mutant, there was insufficient ubiquitination of the inclusion to induce autophagic clearance during infection of human epithelial cells, but its growth was restricted by low doses of IFNγ that had no effect on wild type *Chlamydia*. A possible explanation for this observation is that IFNγ induces the clearance of Cdu1 mutants by activating the autophagy machinery to assemble on the ubiquitinated inclusion surface whereas the deubiquitinated inclusions of wild type *Chlamydia* are not recognized in this manner. *C. muridarum* expresses a Cdu1 homologue that exhibits only 33% identity to the *C. trachomatis* Cdu1 protein. Therefore, it is possible that the two DUBs possess distinct substrate specificities. If so, the cell-autonomous effector mechanisms restricting *C. muridarum* replication in human cells may also limit growth of *C. trachomatis* Cdu1 mutants, which are not protected from inclusion ubiquitination. In this case, the *C. muridarum* DUB should not be able to functionally complement the *C. trachomatis* Cdu1 mutant. The identification of the respective Cdu1 substrates and the E3 ubiquitin ligases mediating inclusion ubiquitination are therefore critical for understanding how *Chlamydia* hide inside their host cell. Since the growth restriction of the *C. trachomatis* Cdu1 mutant required IFNγ stimulation but the ubiquitination of the inclusion formed by the Cdu1 mutant was independent of IFNγ stimulation, an alternative role of Cdu1 in protecting the inclusion from autophagolysosomal degradation has to be considered. For *C. caviae* an interesting alternative mechanism has been proposed. When *C. caviae* infects human epithelial cells, the chlamydial entry site is locally and transiently ubiquitinated. In this case, the deubiquitinase ChlaOTU clears ubiquitination during chlamydial entry and thereby may protect the nascent inclusion from early autophagolysosomal degradation (Furtado *et al.*, 2013).

Concluding remarks

Although both, apoptosis and autophagy play major roles in many different cellular processes not connected to infection, such as development and cell stress management, their function as cell-autonomous defence mechanisms is of utmost importance for the prevention of infection with intracellular pathogens. This is particularly evident by the fact that multiple strategies of intracellular pathogens aim to prevent apoptosis and autophagy (particularly xenophagy) induction. In the case of *Chlamydia* infection, recent research supports a central role for apoptosis and autophagy in cellular defence, but there are still many unanswered questions. It is still not clear how inhibition of apoptosis is initiated and maintained by *Chlamydia* throughout the replicative phase of the bacteria in cultured cells. Further, it is unknown if and where inhibition of apoptosis occurs in infected mammals, and whether it is required for an effective acute and/or persistent infection *in vivo*. Similarly, xenophagy appears to be prevented by *Chlamydia*, but the picture is far from clear. An emerging common strategy of *Chlamydia* appears to be the interference with autophagy initiation, for instance by secreted chlamydial deubiquitinases. Keeping the inclusion intact has been suggested as another mechanism to prevent autophagy and even death of the infected cell, which may point to autophagy-dependent cell death (ADCD), a specialized form of cell death induced by autophagy under conditions when cellular apoptosis is prevented (Bialik *et al.*, 2018). Autophagy as a consequence of leaky inclusions may thereby induce cell death even in the infected, apoptosis-resistant cell.

There is clearly a need for further research to define the role of apoptosis and autophagy under relevant infection conditions. Since apoptosis and autophagy are deregulated in many tumour cell lines, the development of more sophisticated human infection models is particularly relevant for research with the human pathogenic *Chlamydia* species. This may be a challenging but also rewarding aim for future explorations of these cell-autonomous defence mechanisms in *Chlamydia*-infected cells.

References

Abdul-Sater, A.A., Said-Sadier, N., Lam, V.M., Singh, B., Pettengill, M.A., Soares, F., Tattoli, I., Lipinski, S., Girardin, S.E., Rosenstiel, P., *et al.* (2010). Enhancement of reactive oxygen species production and chlamydial infection by the mitochondrial Nod-like family member NLRX1. J Biol Chem *285*, 41637–41645.

Al-Younes, H.M., Al-Zeer, M.A., Khalil, H., Gussmann, J., Karlas, A., Machuy, N., Brinkmann, V., Braun, P.R., and Meyer, T.F. (2011). Autophagy-independent function of MAP-LC3 during intracellular propagation of Chlamydia trachomatis. Autophagy *7*, 814–828.

Al-Younes, H.M., Brinkmann, V., and Meyer, T.F. (2004). Interaction of Chlamydia trachomatis serovar L2 with the host autophagic pathway. Infect Immun *72*, 4751–4762.

Al-Zeer, M.A., Al-Younes, H.M., Lauster, D., Abu Lubad, M., and Meyer, T.F. (2013). Autophagy restricts Chlamydia trachomatis growth in human macrophages via IFNG-inducible guanylate binding proteins. Autophagy *9*, 50–62.

Aliprantis, A.O., Yang, R.B., Mark, M.R., Suggett, S., Devaux, B., Radolf, J.D., Klimpel, G.R., Godowski, P., and Zychlinsky, A. (1999). Cell activation and apoptosis by bacterial lipoproteins through toll-like receptor-2. Science *285*, 736–739.

Barker, J.R., Koestler, B.J., Carpenter, V.K., Burdette, D.L., Waters, C.M., Vance, R.E., and Valdivia, R.H. (2013). STING-dependent recognition of cyclic di-AMP mediates type I interferon responses during Chlamydia trachomatis infection. mBio *4*, e00018–00013.

Baud, D., Goy, G., Osterheld, M.C., Croxatto, A., Borel, N., Vial, Y., Pospischil, A., and Greub, G. (2014). Role of Waddlia chondrophila placental infection in miscarriage. Emerg Infect Dis *20*, 460–464.

Besch, R., Poeck, H., Hohenauer, T., Senft, D., Hacker, G., Berking, C., Hornung, V., Endres, S., Ruzicka, T., Rothenfusser, S., *et al.* (2009). Proapoptotic signalling induced by RIG-I and MDA-5 results in type I interferon-independent apoptosis in human melanoma cells. J Clin Invest *119*, 2399–2411.

Bialik, S., Dasari, S.K., and Kimchi, A. (2018). Autophagy-dependent cell death – where, how and why a cell eats itself to death. J Cell Sci *131*, pii.jcs215152.

Birmingham, C.L., Smith, A.C., Bakowski, M.A., Yoshimori, T., and Brumell, J.H. (2006). Autophagy controls Salmonella infection in response to damage to the Salmonella-containing vacuole. J Biol Chem *281*, 11374–11383.

Buchholz, K.R., and Stephens, R.S. (2008). The cytosolic pattern recognition receptor NOD1 induces inflammatory IL-8 during *C. trachomatis* infection. Infect Immun *76*, 3150–3155.

Campbell, K.J., Bath, M.L., Turner, M.L., Vandenberg, C.J., Bouillet, P., Metcalf, D., Scott, C.L., and Cory, S. (2010). Elevated Mcl-1 perturbs lymphopoiesis, promotes transformation of haematopoietic stem/progenitor cells, and enhances drug resistance. Blood *116*, 3197–3207.

Chen, A.L., Johnson, K.A., Lee, J.K., Sutterlin, C., and Tan, M. (2012). CPAF: a Chlamydial protease in search of an authentic substrate. PLoS pathogens *8*, e1002842.

Chen, L., Willis, S.N., Wei, A., Smith, B.J., Fletcher, J.I., Hinds, M.G., Colman, P.M., Day, C.L., Adams, J.M., and Huang, D.C. (2005). Differential targeting of prosurvival Bcl-2 proteins by their BH3-only ligands allows complementary apoptotic function. Mol Cell *17*, 393–403.

Chipuk, J.E., Moldoveanu, T., Llambi, F., Parsons, M.J., and Green, D.R. (2010). The BCL-2 family reunion. Molecular cell 37, 299–310.

Cuervo, A.M., and Wong, E. (2014). Chaperone-mediated autophagy: roles in disease and ageing. Cell Res *24*, 92–104.

Derbigny, W.A., Hong, S.C., Kerr, M.S., Temkit, M., and Johnson, R.M. (2007). Chlamydia muridarum infection elicits a beta interferon response in murine oviduct epithelial cells dependent on interferon regulatory factor 3 and TRIF. Infect Immun *75*, 1280–1290.

Dikic, I. (2017). Proteasomal and Autophagic Degradation Systems. Annu Rev Biochem *86*, 193–224.

Dikic, I., and Elazar, Z. (2018). Mechanism and medical implications of mammalian autophagy. Nat Rev Mol Cell Biol *19*, 349–364.

Dille, S., Kleinschnitz, E.M., Kontchou, C.W., Nolke, T., and Hacker, G. (2015). In contrast to Chlamydia trachomatis, Waddlia chondrophila grows in human cells without inhibiting apoptosis, fragmenting the Golgi apparatus, or diverting post-Golgi sphingomyelin transport. Infect Immun *83*, 3268–3280.

Dumrese, C., Maurus, C.F., Gygi, D., Schneider, M.K., Walch, M., Groscurth, P., and Ziegler, U. (2005). Chlamydia pneumoniae induces aponecrosis in human aortic smooth muscle cells. BMC Microbiol *5*, 2.

Dunning Hotopp, J.C. (2011). Horizontal gene transfer between bacteria and animals. Trends in Genetics: TIG *27*, 157–163.

Fan, T., Lu, H., Hu, H., Shi, L., McClarty, G.A., Nance, D.M., Greenberg, A.H., and Zhong, G. (1998). Inhibition of Apoptosis in Chlamydia-infected Cells: Blockade of Mitochondrial Cytochrome c Release and Caspase Activation. J Exp Med *187*, 487–496.

Farre, J.C., and Subramani, S. (2016). Mechanistic insights into selective autophagy pathways: lessons from yeast. Nat Rev Mol Cell Biol *17*, 537–552.

Filomeni, G., De Zio, D., and Cecconi, F. (2015). Oxidative stress and autophagy: the clash between damage and metabolic needs. Cell Death Differ *22*, 377–388.

Fischer, A., Harrison, K.S., Ramirez, Y., Auer, D., Chowdhury, S.R., Prusty, B.K., Sauer, F., Dimond, Z., Kisker, C., Scott Hefty, P., *et al.* (2017). Chlamydia trachomatis-containing vacuole serves as deubiquitination platform to stabilize Mcl-1 and to interfere with host defence. eLife *6*, pii.e21465.

Fischer, A., and Rudel, T. (2018a). Safe haven under constant attack-The Chlamydia-containing vacuole. Cell Microbiol *20*, e12940.

Fischer, A., and Rudel, T. (2018b). Subversion of Cell-Autonomous Host Defence by Chlamydia Infection. Curr Top Microbiol Immunol *412*, 81–106.

Fischer, S.F., Harlander, T., Vier, J., and Hacker, G. (2004a). Protection against CD95-induced apoptosis by chlamydial infection at a mitochondrial step. Infect Immun *72*, 1107–1115.

Fischer, S.F., Schwarz, C., Vier, J., and Hacker, G. (2001). Characterization of Antiapoptotic Activities of *Chlamydia pneumoniae* in Human Cells. Infect Immun *69*, 7121–7129.

Fischer, S.F., Vier, J., Kirschnek, S., Klos, A., Hess, S., Ying, S., and Hacker, G. (2004b). Chlamydia Inhibit Host Cell Apoptosis by Degradation of Proapoptotic BH3-only Proteins. J Exp Med *200*, 905–916.

Furtado, A.R., Essid, M., Perrinet, S., Balana, M.E., Yoder, N., Dehoux, P., and Subtil, A. (2013). The chlamydial OTU domain-containing protein ChlaOTU is an early type III secretion effector targeting ubiquitin and NDP52. Cell Microbiol *15*, 2064–2079.

Galluzzi, L., Brenner, C., Morselli, E., Touat, Z., and Kroemer, G. (2008). Viral control of mitochondrial apoptosis. PLoS pathogens *4*, e1000018.

Gomes, L.C., and Dikic, I. (2014). Autophagy in antimicrobial immunity. Mol Cell *54*, 224–233.

Gonzalez, E., Rother, M., Kerr, M.C., Al-Zeer, M.A., Abu-Lubad, M., Kessler, M., Brinkmann, V., Loewer, A., and Meyer, T.F. (2014). Chlamydia infection depends on a functional MDM2–p53 axis. Nature Commun *5*, 5201.

Grumati, P., and Dikic, I. (2018). Ubiquitin signalling and autophagy. J Biol Chem *293*, 5404–5413.

Gulen, M.F., Koch, U., Haag, S.M., Schuler, F., Apetoh, L., Villunger, A., Radtke, F., and Ablasser, A. (2017). Signalling strength determines proapoptotic functions of STING. Nature Commun *8*, 427.

Gutierrez, M.G., Master, S.S., Singh, S.B., Taylor, G.A., Colombo, M.I., and Deretic, V. (2004). Autophagy is a defence mechanism inhibiting BCG and Mycobacterium tuberculosis survival in infected macrophages. Cell *119*, 753–766.

Haldar, A.K., Foltz, C., Finethy, R., Piro, A.S., Feeley, E.M., Pilla-Moffett, D.M., Komatsu, M., Frickel, E.M., and Coers, J. (2015). Ubiquitin systems mark pathogen-containing vacuoles as targets for host defence by guanylate binding proteins. Proc Natl Acad Sci U S A *112*, E5628–5637.

Haldar, A.K., Piro, A.S., Finethy, R., Espenschied, S.T., Brown, H.E., Giebel, A.M., Frickel, E.M., Nelson, D.E., and Coers, J. (2016). Chlamydia trachomatis Is Resistant to Inclusion Ubiquitination and Associated Host Defence in Gamma Interferon-Primed Human Epithelial Cells. MBio *7*, pii.e01417-16.

Hamasaki, M., Furuta, N., Matsuda, A., Nezu, A., Yamamoto, A., Fujita, N., Oomori, H., Noda, T., Haraguchi, T., Hiraoka, Y., *et al.* (2013). Autophagosomes form at ER-mitochondria contact sites. Nature *495*, 389–393.

Hengartner, M.O. (2000). The biochemistry of apoptosis. Nature *407*, 770–776.

Inohara, N., Koseki, T., del Peso, L., Hu, Y., Yee, C., Chen, S., Carrio, R., Merino, J., Liu, D., Ni, J., *et al.* (1999). Nod1, an Apaf-1-like activator of caspase-9 and nuclear factor-kappaB. J Biol Chem *274*, 14560–14567.

Inoki, K., Kim, J., and Guan, K.L. (2012). AMPK and mTOR in cellular energy homeostasis and drug targets. Annu Rev Pharmacol Toxicol *52*, 381–400.

Jewell, J.L., Russell, R.C., and Guan, K.L. (2013). Amino acid signalling upstream of mTOR. Nat Rev Mol Cell Biol *14*, 133–139.

Karunakaran, K., Mehlitz, A., and Rudel, T. (2011). Evolutionary conservation of infection-induced cell death inhibition among Chlamydiales. PloS one *6*, e22528.

Kawane, K., Fukuyama, H., Yoshida, H., Nagase, H., Ohsawa, Y., Uchiyama, Y., Okada, K., Iida, T., and Nagata, S. (2003). Impaired thymic development in mouse embryos deficient in apoptotic DNA degradation. Nat Immunol *4*, 138–144.

Kim, J., Kundu, M., Viollet, B., and Guan, K.L. (2011). AMPK and mTOR regulate autophagy through direct phosphorylation of Ulk1. Nat Cell Biol *13*, 132–141.

Kirkin, V., Lamark, T., Sou, Y.S., Bjorkoy, G., Nunn, J.L., Bruun, J.A., Shvets, E., McEwan, D.G., Clausen, T.H., Wild, P., *et al.* (2009). A role for NBR1 in autophagosomal degradation of ubiquitinated substrates. Mol Cell *33*, 505–516.

Klionsky, D.J., Baehrecke, E.H., Brumell, J.H., Chu, C.T., Codogno, P., Cuervo, A.M., Debnath, J., Deretic, V., Elazar, Z., Eskelinen, E.L., *et al.* (2011). A comprehensive glossary of autophagy-related molecules and processes (2nd edition). Autophagy *7*, 1273–1294.

Kotschy, A., Szlavik, Z., Murray, J., Davidson, J., Maragno, A.L., Le Toumelin-Braizat, G., Chanrion, M., Kelly, G.L.,

Gong, J.N., Moujalled, D.M., *et al.* (2016). The MCL1 inhibitor S63845 is tolerable and effective in diverse cancer models. Nature *538*, 477–482.

Labbadia, J., and Morimoto, R.I. (2015). The biology of proteostasis in ageing and disease. Annu Rev Biochem *84*, 435–464.

Li, T., and Chen, Z.J. (2018). The cGAS-cGAMP-STING pathway connects DNA damage to inflammation, senescence, and cancer. J Exp Med.

Li, W.W., Li, J., and Bao, J.K. (2012). Microautophagy: lesser-known self-eating. Cell Mol Life Sci *69*, 1125–1136.

Lowe, S.W., Schmitt, E.M., Smith, S.W., Osborne, B.A., and Jacks, T. (1993). p53 is required for radiation-induced apoptosis in mouse thymocytes. Nature *362*, 847–849.

Mehlitz, A., Banhart, S., Maurer, A.P., Kaushansky, A., Gordus, A.G., Zielecki, J., Macbeath, G., and Meyer, T.F. (2010). Tarp regulates early Chlamydia-induced host cell survival through interactions with the human adaptor protein SHC1. J Cell Biol *190*, 143–157.

Mehlitz, A., Eylert, E., Huber, C., Lindner, B., Vollmuth, N., Karunakaran, K., Goebel, W., Eisenreich, W., and Rudel, T. (2017). Metabolic adaptation of Chlamydia trachomatis to mammalian host cells. Mol Microbiol *103*, 1004–1019.

Mevissen, T.E.T., and Komander, D. (2017). Mechanisms of Deubiquitinase Specificity and Regulation. Ann Rev Biochem *86*, 159–192.

Misaghi, S., Balsara, Z.R., Catic, A., Spooner, E., Ploegh, H.L., and Starnbach, M.N. (2006). Chlamydia trachomatis-derived deubiquitinating enzymes in mammalian cells during infection. Mol Microbiol *61*, 142–150.

Mizushima, N., Yoshimori, T., and Levine, B. (2010). Methods in mammalian autophagy research. Cell *140*, 313–326.

Mojsa, B., Lassot, I., and Desagher, S. (2014). Mcl-1 ubiquitination: unique regulation of an essential survival protein. Cells 3, 418–437.

Moore, E.R., and Ouellette, S.P. (2014). Reconceptualizing the chlamydial inclusion as a pathogen-specified parasitic organelle: an expanded role for Inc proteins. Front Cell Infect Microbiol 4, 157.

Naghdi, S., and Hajnoczky, G. (2016). VDAC2-specific cellular functions and the underlying structure. Biochem Biophys Acta *1863*, 2503–2514.

Nakagawa, I., Amano, A., Mizushima, N., Yamamoto, A., Yamaguchi, H., Kamimoto, T., Nara, A., Funao, J., Nakata, M., Tsuda, K., *et al.* (2004). Autophagy defends cells against invading group A Streptococcus. Science *306*, 1037–1040.

Ojcius, D.M., Souque, P., Perfettini, J.L., and Dautry-Varsat, A. (1998). Apoptosis of epithelial cells and macrophages due to infection with the obligate intracellular pathogen Chlamydia psittaci. J Immunol *161*, 4220–4226.

Oltersdorf, T., Elmore, S.W., Shoemaker, A.R., Armstrong, R.C., Augeri, D.J., Belli, B.A., Bruncko, M., Deckwerth, T.L., Dinges, J., Hajduk, P.J., *et al.* (2005). An inhibitor of Bcl-2 family proteins induces regression of solid tumours. Nature *435*, 677–681.

Omsland, A., Sager, J., Nair, V., Sturdevant, D.E., and Hackstadt, T. (2012). Developmental stage-specific metabolic and transcriptional activity of Chlamydia trachomatis in an axenic medium. Proc Natl Acad Sci U S A *109*, 19781–19785.

Pachikara, N., Zhang, H., Pan, Z., Jin, S., and Fan, H. (2009). Productive Chlamydia trachomatis lymphogranuloma venereum 434 infection in cells with augmented or inactivated autophagic activities. FEMS Microbiol Lett *292*, 240–249.

Pankiv, S., Clausen, T.H., Lamark, T., Brech, A., Bruun, J.A., Outzen, H., Overvatn, A., Bjorkoy, G., and Johansen, T. (2007). p62/SQSTM1 binds directly to Atg8/LC3 to facilitate degradation of ubiquitinated protein aggregates by autophagy. J Biol Chem *282*, 24131–24145.

Perfettini, J.L., Ojcius, D.M., Andrews, C.W., Jr., Korsmeyer, S.J., Rank, R.G., and Darville, T. (2003). Role of proapoptotic BAX in propagation of Chlamydia muridarum (the mouse pneumonitis strain of Chlamydia trachomatis) and the host inflammatory response. J Biol Chem *278*, 9496–9502.

Philpott, D.J., Sorbara, M.T., Robertson, S.J., Croitoru, K., and Girardin, S.E. (2014). NOD proteins: regulators of inflammation in health and disease. Nat Rev Immunol *14*, 9–23.

Puri, C., Renna, M., Bento, C.F., Moreau, K., and Rubinsztein, D.C. (2013). Diverse autophagosome membrane sources coalesce in recycling endosomes. Cell *154*, 1285–1299.

Rajalingam, K., Al Younes, H., Muller, A., Meyer, T.F., Szczepek, A.J., and Rudel, T. (2001). Epithelial cells infected with Chlamydophila pneumoniae (Chlamydia pneumoniae) are resistant to apoptosis. Infect Immun *69*, 7880–7888.

Rajalingam, K., Sharma, M., Lohmann, C., Oswald, M., Thieck, O., Froelich, C.J., and Rudel, T. (2008). Mcl-1 is a key regulator of apoptosis resistance in Chlamydia trachomatis-infected cells. PloS one *3*, e3102.

Randow, F., MacMicking, J.D., and James, L.C. (2013). Cellular self-defence: how cell-autonomous immunity protects against pathogens. Science *340*, 701–706.

Ravikumar, B., Moreau, K., Jahreiss, L., Puri, C., and Rubinsztein, D.C. (2010). Plasma membrane contributes to the formation of pre-autophagosomal structures. Nat Cell Biol *12*, 747–757.

Rikihisa, Y. (1984). Glycogen autophagosomes in polymorphonuclear leucocytes induced by rickettsiae. Anat Rec *208*, 319–327.

Rogov, V., Dotsch, V., Johansen, T., and Kirkin, V. (2014). Interactions between autophagy receptors and ubiquitin-like proteins form the molecular basis for selective autophagy. Molecular cell *53*, 167–178.

Ruckdeschel, K., Pfaffinger, G., Haase, R., Sing, A., Weighardt, H., Hacker, G., Holzmann, B., and Heesemann, J. (2004). Signalling of apoptosis through TLRs critically involves toll/IL-1 receptor domain-containing adapter inducing IFN-beta, but not MyD88, in bacteria-infected murine macrophages. J Immun *173*, 3320–3328.

Saxton, R.A., and Sabatini, D.M. (2017). mTOR Signalling in Growth, Metabolism, and Disease. Cell *168*, 960–976.

Scaffidi, C., Fulda, S., Srinivasan, A., Friesen, C., Li, F., Tomaselli, K.J., Debatin, K.M., Krammer, P.H., and Peter, M.E. (1998). Two CD95 (APO-1/Fas) signalling pathways. EMBO J *17*, 1675–1687.

Siegl, C., Prusty, B.K., Karunakaran, K., Wischhusen, J., and Rudel, T. (2014). Tumour suppressor p53 alters host cell metabolism to limit Chlamydia trachomatis infection. Cell reports 9, 918–929.

Singh, P.K., Weber, A., and Hacker, G. (2018). The established and the predicted roles of dynein light chain in the regulation of mitochondrial apoptosis. Cell cycle, 1–11.

Sixt, B.S., Hiess, B., Konig, L., and Horn, M. (2012). Lack of effective anti-apoptotic activities restricts growth of Parachlamydiaceae in insect cells. PloS one 7, e29565.

Souers, A.J., Leverson, J.D., Boghaert, E.R., Ackler, S.L., Catron, N.D., Chen, J., Dayton, B.D., Ding, H., Enschede, S.H., Fairbrother, W.J., *et al.* (2013). ABT-199, a potent and selective BCL-2 inhibitor, achieves antitumour activity while sparing platelets. Nat Med *19*, 202–208.

Strasser, A., and Vaux, D.L. (2018). Viewing BCL2 and cell death control from an evolutionary perspective. Cell Death Diff 25, 13–20.

Subbarayal, P., Karunakaran, K., Winkler, A.C., Rother, M., Gonzalez, E., Meyer, T.F., and Rudel, T. (2015). EphrinA2 receptor (EphA2) is an invasion and intracellular signalling receptor for Chlamydia trachomatis. PLoS pathogens *11*, e1004846.

Thurston, T.L., Ryzhakov, G., Bloor, S., von Muhlinen, N., and Randow, F. (2009). The TBK1 adaptor and autophagy receptor NDP52 restricts the proliferation of ubiquitin-coated bacteria. Nat Immunol *10*, 1215–1221.

Thurston, T.L., Wandel, M.P., von Muhlinen, N., Foeglein, A., and Randow, F. (2012). Galectin 8 targets damaged vesicles for autophagy to defend cells against bacterial invasion. Nature *482*, 414–418.

van der Vaart, A., Griffith, J., and Reggiori, F. (2010). Exit from the Golgi is required for the expansion of the autophagosomal phagophore in yeast Saccharomyces cerevisiae. Mol Biol Cell *21*, 2270–2284.

Vaux, D.L., Haecker, G., and Strasser, A. (1994). An evolutionary perspective on apoptosis. Cell *76*, 777–779.

Villunger, A., Michalak, E.M., Coultas, L., Mullauer, F., Bock, G., Ausserlechner, M.J., Adams, J.M., and Strasser, A. (2003). p53- and Drug-Induced Apoptotic Responses Mediated by BH3-Only Proteins Puma and Noxa. Science *302*, 1036–1038.

Weber, A., Kirejczyk, Z., Besch, R., Potthoff, S., Leverkus, M., and Hacker, G. (2010). Proapoptotic signalling through Toll-like receptor-3 involves TRIF-dependent activation of caspase-8 and is under the control of inhibitor of apoptosis proteins in melanoma cells. Cell death and differentiation *17*, 942–951.

Weber, M.M., Lam, J.L., Dooley, C.A., Noriea, N.F., Hansen, B.T., Hoyt, F.H., Carmody, A.B., Sturdevant, G.L., and Hackstadt, T. (2017). Absence of Specific Chlamydia trachomatis Inclusion Membrane Proteins Triggers Premature Inclusion Membrane Lysis and Host Cell Death. Cell Rep *19*, 1406–1417.

Wild, P., Farhan, H., McEwan, D.G., Wagner, S., Rogov, V.V., Brady, N.R., Richter, B., Korac, J., Waidmann, O., Choudhary, C., *et al.* (2011). Phosphorylation of the autophagy receptor optineurin restricts Salmonella growth. Science *333*, 228–233.

Wilfling, F., Weber, A., Potthoff, S., Vogtle, F.N., Meisinger, C., Paschen, S.A., and Hacker, G. (2012). BH3-only proteins are tail-anchored in the outer mitochondrial membrane and can initiate the activation of Bax. Cell Death Diff *19*, 1328–1336.

Yang, Z., and Klionsky, D.J. (2010). Mammalian autophagy: core molecular machinery and signalling regulation. Curr Opin Cell Biol *22*, 124–131.

Yasir, M., Pachikara, N.D., Bao, X., Pan, Z., and Fan, H. (2011). Regulation of chlamydial infection by host autophagy and vacuolar ATPase-bearing organelles. Infect Immun *79*, 4019–4028.

Ying, S., Christian, J.G., Paschen, S.A., and Hacker, G. (2008a). Chlamydia trachomatis can protect host cells against apoptosis in the absence of cellular Inhibitor of Apoptosis Proteins and Mcl-1. Microbes and infection/ Institut Pasteur *10*, 97–101.

Ying, S., Fischer, S.F., Pettengill, M., Conte, D., Paschen, S.A., Ojcius, D.M., and Hacker, G. (2006). Characterization of Host Cell Death Induced by Chlamydia trachomatis. Infect Immun *74*, 6057–6066.

Ying, S., Pettengill, M., Latham, E.R., Walch, A., Ojcius, D.M., and Hacker, G. (2008b). Premature apoptosis of Chlamydia-infected cells disrupts chlamydial development. J Infect Dis *198*, 1536–1544.

Ying, S., Seiffert, B.M., Hacker, G., and Fischer, S.F. (2005). Broad degradation of proapoptotic proteins with the conserved Bcl-2 homology domain 3 during infection with Chlamydia trachomatis. Infect Immun *73*, 1399–1403.

Youle, R.J., and Strasser, A. (2008). The BCL-2 protein family: opposing activities that mediate cell death. Nat Rev Mol Cell Biol *9*, 47–59.

Zhang, Y., Yeruva, L., Marinov, A., Prantner, D., Wyrick, P.B., Lupashin, V., and Nagarajan, U.M. (2014). The DNA sensor, cyclic GMP-AMP synthase, is essential for induction of IFN-beta during Chlamydia trachomatis infection. J Immunol *193*, 2394–2404.

Zheng, Y.T., Shahnazari, S., Brech, A., Lamark, T., Johansen, T., and Brumell, J.H. (2009). The adaptor protein p62/ SQSTM1 targets invading bacteria to the autophagy pathway. J Immunol *183*, 5909–5916.

Zhong, Y., Weininger, M., Pirbhai, M., Dong, F., and Zhong, G. (2006). Inhibition of staurosporine-induced activation of the proapoptotic multidomain Bcl-2 proteins Bax and Bak by three invasive chlamydial species. J Infect *53*, 408–414.

Protein Secretion in *Chlamydia*

7

Agathe Subtil[1]* and Richard D. Hayward[2]*

[1]Institut Pasteur, CNRS UMR3691, Paris, France.
[2]Department of Pathology, University of Cambridge, Cambridge, UK.

*Correspondence: asubtil@pasteur.fr; rdh24@cam.ac.uk

https://doi.org/10.21775/9781912530281.07

Abstract

Protein secretion is fundamental to microbial physiology and underlies virulence-associated processes including adhesion, invasion, intracellular replication and the delivery of exotoxins. Protein secretion is key to the interaction with the environment, and is thus an essential feature of the biology of the Chlamydiaceae, which interact with a eukaryotic host throughout their developmental cycle. Here we review the mechanisms of secretion relevant to the pathogenesis of the Chlamydiaceae. The major secretion pathways are described, including the current knowledge of the structure, mechanism and function of the associated macromolecular machineries. Specifically, we review the role of the type 2, 3 and 5 secretion pathways in the interactions between these bacteria and their host cell, and we discuss the emerging role for vesicles released from the outer membrane of the intracellular reticulate bodies. We review how new genetic techniques and chemical inhibitors have advanced our understanding of the contribution of different secretory pathways to the infection process and allowed the definitive identification of secretion substrates. We also discuss how state-of-the-art imaging has contributed to further understanding the structure and mechanism of the associated macromolecular machineries. We consider the future opportunities for translating this knowledge into the development of vaccines, novel antimicrobials and diagnostic tools, and discuss the implications beyond the field of chlamydial research.

Introduction

Prokaryotes must secrete proteins across the cell envelope to display structures on their surface, to assemble the macromolecular machinery necessary for motility and conjugation, to mediate essential interactions with their environment, and to communicate and compete with other organisms. Protein secretion is also critical for the virulence of pathogenic bacteria to position adhesins on the envelope surface, to deploy exotoxins and to directly translocate virulence effector proteins into eukaryotic target cells. Protein secretion is best characterized in Gram-negative bacteria in which the envelope consists of an archetypal energized inner membrane (IM) and an asymmetric outer membrane (OM), the latter comprising an inner lipid monolayer and an outer leaflet of lipopolysaccharide. The IM and OM both contain proteins and are separated by a periplasmic space. Secretion often occurs as a two-step process whereby proteins are exported initially into the periplasm prior to their transfer across the OM to the envelope surface or extracellular milieu, e.g. type 2 secretion (T2S) and type 5 secretion (T5S) (Cianciotto and White, 2017; Fan *et al.*, 2016). Alternatively, secretion can occur directly (i.e. without a periplasmic intermediate) via conduits that traverse both membranes and the intervening periplasmic space, e.g. type 1 secretion (T1S). Effector proteins can also be translocated directly across three membranes, the third being a membrane of the eukaryotic host cell, which is a key interface during host–pathogen interaction, e.g. type 3 secretion (T3S) (Galan and Waksman, 2018). Conventionally, 'secretion' refers to the transport of proteins across the bacterial envelope, whereas 'translocation' describes the delivery of toxins and effectors across an additional eukaryotic membrane. In some cases, e.g. T3S, secretion and translocation are coupled and we will refer to both steps in this review.

Consequently, studying protein secretion is key to understanding the mechanisms of bacterial virulence as well as physiology, since this process underpins the display of bacterial surface antigens, the assembly of adhesins on the envelope surface, and the delivery of protein toxins and effectors, each of which represent targets for the design and development of novel interventions. Factors present on the bacterial surface, which are delivered into the surrounding environment, or that are translocated into the host cell are often antigens presented to the host immune system, and consequently represent important vaccine targets. Although the secretion system substrates encoded by different bacteria are diverse, leading to their disparate antigenic properties and effects on the host, the structure and mechanisms of the secretion systems themselves are conserved. Molecules that act as secretion system inhibitors are therefore potentially viable novel antimicrobials with a broad host range (Charro and Mota, 2015). The development of such alternative antimicrobials is a key strategic priority given the increasing incidence of antibiotic resistance.

In this chapter, we will provide an overview of the current knowledge of protein secretion in *Chlamydia* and highlight some of the outstanding questions driving current research. We acknowledge that some aspects have recently been reviewed elsewhere and we will refer to these sources, which provide a greater level of specific detail, where appropriate.

Secretion mechanisms in *Chlamydia*

No less than seven different secretion mechanisms have now been identified in Gram-negative bacteria (Maffei *et al.*, 2017). While some secretion systems such as tripartite pumps required for the export of exotoxins and drug efflux utilize only a few components embedded in the envelope, others are composed of more than 15 components that assemble into complex macromolecular machines. In addition to these protein-assisted translocation devices, the release of OM vesicles (OMVs) is increasingly becoming recognized as a broadly relevant secretion process, and was recently proposed as 'type zero' secretion (Guerrero-Mandujano *et al.*, 2017). We will review the evidence for the contribution of the known secretion pathways (T2S, T3S and T5S) to secretion in *Chlamydiae*, and discuss the emerging evidence that shedding of OMVs might also allow for protein export (Fig. 7.1). Proteins identified to be secreted by *C. trachomatis* are summarized in Table 7.1. For a description of the methods used to study protein secretion in bacteria more generally, including a description of different reporter systems that have been or could be applied to *Chlamydia* in the future, we refer the reader to a recent comprehensive review (Maffei *et al.*, 2017).

The Sec translocon

The export of periplasmic proteins across the IM occurs via an ancestral mechanism employing the Sec translocon. The bacterial and archaeal Sec translocon is composed of SecYEG (the integral membrane channel), SecA (the peripheral membrane ATPase), and SecD and SecF (the auxiliary release proteins). Homologues of SecY (CT510 in *C. trachomatis* serovar D/UW-3/CX), SecE (CT321), SecA (CT701) and a SecD/F fusion (CT448) are present in *Chlamydia* genomes. Proteins targeted to the Sec translocon carry an N-terminal stretch of hydrophobic amino acids called the signal peptide, which can be predicted from the sequence. In *Escherichia coli*, a cytoplasmic chaperone, SecB, assists post-translational transport. Although no SecB homologue is present in the *C. trachomatis* genome, it is also lacking in other bacteria, and it remains possible that an as yet uncharacterized protein might substitute this function to engage and pilot substrates with a signal peptide to the SecYEG translocation platform.

In many other bacteria, an alternative to the Sec pathway, called the twin arginine translocation (Tat) pathway, can translocate folded proteins across the inner membrane; however, there is no evidence for the existence of this mechanism in *C. trachomatis*.

Given the essential nature of this conserved pathway in other organisms, small molecule inhibitors targeting the Sec translocon are expected to severely impact chlamydial fitness (Segers and

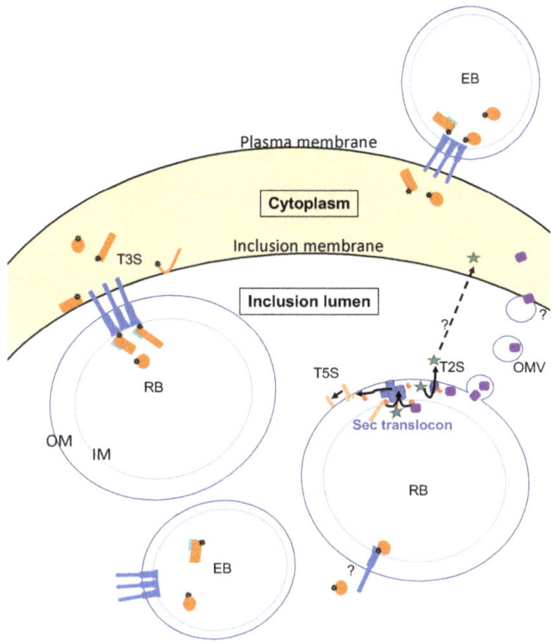

Figure 7.1 Secretion mechanisms in *Chlamydiaceae*. Proteins that possess an N-terminal signal peptide (red dot) are translocated via the Sec translocon across the inner membrane (IM) into the periplasm, where signal peptidases can cleave the peptide. From this location, there are several possibilities. T2S substrates (green) are translocated across the outer membrane (OM) via a dedicated multiprotein machinery, and the translocated proteins reach the inclusion lumen. It is not known whether some T2S substrates reach the host cytoplasm, and if so, the mechanism remains to be determined. In contrast to the multiprotein T2S machinery, T5S (pink) is a largely autonomous process that delivers a 'passenger domain' across the outer membrane, sometimes followed by proteolytic cleavage. The released fragments can remain associated with the bacterial outer membrane, or be free within the inclusion lumen. Outer membrane vesicles (OMVs) are a likely mechanism for the secretion of a subset of periplasmic proteins (purple) into the inclusion lumen. Whether this pathway, or an alternative mechanism, is used to deliver a subset of Sec-translocon substrates into the host cytoplasm remains to be determined. The secretion of T3S substrates is better understood, because this multiprotein machinery can translocate proteins directly to the host cytosol, from either an EB at the entry step (top right) or an RB during intracellular replication (left). T3S substrates (orange) have an N-terminal secretion signal (black star). Their secretion can be facilitated by chaperone proteins (light blue). Detection of T3S substrates within the inclusion suggests that the T3S machinery might also deliver substrates directly into the inclusion lumen.

Anne, 2011). A screen of over 40,000 chemically defined compounds led to the identification of one compound, MDQA, which significantly inhibited *C. trachomatis* multiplication (Sandoz *et al.*, 2012). Bacteria resistant to this compound were isolated and possessed suppressor mutations in *secY*, demonstrating that a functional Sec translocon is indeed essential for chlamydial growth.

While some proteins exported via the Sec pathway are destined for the periplasm, others will eventually also traverse the outer membrane. This additional step can be mediated by T2S or T5S, or alternatively by incorporation into outer membrane vesicles (OMVs), which is currently a less well-documented route.

Shedding of OMVs, a secretion pathway in chlamydiae?

OMV production has been described during the growth of a variety of bacteria *in vivo* and *in vitro* (Guerrero-Mandujano *et al.*, 2017; Jan, 2017). In general, OMVs are formed due to the intrinsic instability of the OM, which leads to stochastic deformation, pinching and extrusion of the membrane that can incorporate lipopolysaccharide, lipids, membrane proteins and periplasmic content into vesicles of 50–250 nm diameter. This extrusion process and its regulation remain poorly understood, however the composition of OMVs suggests that sorting occurs to select at least some of their constituents, implying that this process is

Table 7.1 Secreted proteins of *Chlamydiaceae* (other than Inc proteins and Pmps)

Protein	Name	Location	Evidence for secretion	Suspected secretion mechanism	Putative function
C. trachomatis					
CT042	GlgX	Inclusion lumen and inclusion membrane	Functional T3S signal and IF against endogenous protein (Gehre *et al.*, 2016)	T3S	Glycogen debranching enzyme
CT049	PlsI	Inclusion lumen	IF against endogenous protein (Jorgensen and Valdivia, 2008)	Unknown (not Sec dependent)	–
CT050	PlsII	Inclusion lumen	IF against endogenous protein (Jorgensen and Valdivia, 2008)	Unknown (not Sec dependent)	–
CT082		Bacterial surface	Functional T3S signal, enhanced secretion in heterologous system in co-expression with CT584 (Hovis *et al.*, 2013; Pais *et al.*, 2013)	T3S	Unknown, probably not an effector
CT089	CopN	Inclusion membrane	Functional T3S signal and IF against endogenous protein (Fields and Hackstadt, 2000)	T3S	Exported regulator of T3S. Binds tubulin (*C. pneumoniae* only)
CT142		Inclusion lumen	Functional T3S signal and IF against tagged protein (da Cunha *et al.*, 2014; da Cunha *et al.*, 2017)	T3S	–
CT143		Inclusion lumen	Functional T3S signal and IF against tagged and endogenous protein (da Cunha *et al.*, 2014; da Cunha *et al.*, 2017; Patton *et al.*, 2018)	T3S	–
CT144		Inclusion lumen	Functional T3S signal and IF against tagged and endogenous protein (da Cunha *et al.*, 2014; da Cunha *et al.*, 2017; Patton *et al.*, 2018)	T3S	–
CT156	Lda1	Inclusion membrane periphery	IF against endogenous protein (Kumar *et al.*, 2006)	Unknown	Unknown, tropism for lipid droplets
CT163	Lda2	Inclusion membrane periphery	IF against endogenous protein (Kumar *et al.*, 2006)	Unknown	Unknown, tropism for lipid droplets
CT166		Unknown	Indirect: biological activity on host substrate (Thalmann *et al.*, 2010)	Unknown (not Sec dependent)	Rac1 glucosylation
CT248	GlgP	Inclusion membrane?	Indirect: functional T3S signal (Gehre *et al.*, 2016) and enrichment in inclusion membrane fraction (Saka *et al.*, 2011)	T3S	Glycogen phosphorylase
CT311		Cytoplasm and nucleus	IF against endogenous protein (Lei *et al.*, 2011)	Unknown, Sec dependent	–
CT456	TarP	Cytoplasm	Functional T3S signal, phosphorylation in the host cytosol (Clifton *et al.*, 2004)	T3S	Actin remodelling

Table 7.1 Continued

Protein	Name	Location	Evidence for secretion	Suspected secretion mechanism	Putative function
CT473	Lda3	Inclusion membrane periphery	IF against endogenous protein (Kumar *et al*., 2006)	Unknown	Unknown, tropism for lipid droplets
CT529	Cap1	Inclusion membrane	Functional T3S signal (Subtil *et al*., 2005), IF against endogenous protein (Fling *et al*., 2001)	T3S	–
CT578	CopB	Inclusion membrane	Functional T3S signal and IF against endogenous protein (Fields *et al*., 2005)	T3S	T3S Translocator (EB only?) (Chellas-Gery *et al*., 2011)
CT619			Indirect: functional T3S and domain shared with T3S effectors CT620 and CT621(Muschiol *et al*., 2011)	T3S	–
CT620		Inclusion lumen and cytosol	Functional T3S signal, IF against endogenous protein (Muschiol *et al*., 2011)	T3S	–
CT621		Inclusion lumen, cytosol, nucleus	Functional T3S signal, IF against endogenous protein, detection in nuclear fractions (Hobolt-Pedersen *et al*., 2009; Muschiol *et al*., 2011)	T3S	–
CT622		Cytosol and nucleus	Functional T3S signal, IF against endogenous and tagged protein (Cossé *et al*., 2018; Gong *et al*., 2011)	T3S	–
CT666	CdsF	T3S needle and inclusion membrane	TEM and IF against endogenous protein (Betts *et al*., 2008)	T3S	–
CT694	TmeA		Functional T3S signal, IF against endogenous protein, secretion in fusion with TEM (Hower *et al*., 2009; Mueller and Fields, 2015)	T3S	Role in invasion (McKuen *et al*., 2017)
CT695		Inclusion membrane (early and late)	Functional T3S signal, IF against endogenous protein, secretion in fusion with TEM (Hower *et al*., 2009; Mueller and Fields, 2015)	T3S	–
CT711			Indirect: functional T3S and domain shared with T3S effectors CT620 and CT621 (Muschiol *et al*., 2011)	T3S	–
CT712			Indirect: functional T3S and domain shared with T3S effectors CT620 and CT621(Muschiol *et al*., 2011)	T3S	–
CT737	NUE	Cytoplasm and nucleus	Functional T3S signal, detection in nuclear fractions (Pennini *et al*., 2010)	T3S	Histone methyl transferase
CT795		Cytoplasm	IF against endogenous protein (Qi *et al*., 2011)	Unknown, Sec dependent	–
CT798	GlgA	Inclusion lumen and cytoplasm	Functional T3S signal (Gehre *et al*., 2016), IF against endogenous protein (Lu *et al*., 2013)	T3S	Glycogen Synthase

Table 7.1 Continued

Protein	Name	Location	Evidence for secretion	Suspected secretion mechanism	Putative function
CT823	HtrA	Cytoplasm	IF against endogenous protein (Wu *et al.*, 2011)	Unknown, Sec dependent	Serine protease
CT847			Indirect: secreted by *Yersinia* T3S machinery and co-immunoprecipitated with host protein GCIP (Chellas-Géry *et al.*, 2007)	T3S	–
CT858	CPAF	Inclusion lumen and cytoplasm	IF against endogenous and tagged protein (Bauler and Hackstadt, 2014; Zhong *et al.*, 2001; Prusty *et al.*, 2018)	Unknown, Sec dependent	Protease
CT861	CopB2		Functional T3S signal and similarity with CopB (Fields *et al.*, 2005)	T3S	Role in T3S effector translocation? (Chellas-Gery *et al.*, 2011)
CT866	GlgB	Inclusion lumen	Indirect: functional T3S (Gehre *et al.*, 2016) and defect in the glycogen formation in GlgB mutant strain (Nguyen and Valdivia, 2012)	T3S	–
CT867	ChlaDUB2/ Cdu2	Inclusion membrane and plasma membrane	Split-GFP technology (Wang *et al.*, 2018)	T3S (A.S. unpublished)	Deubiquitinase (Misaghi *et al.*, 2006)
CT868	ChlaDUB1/ Cdu1	Inclusion membrane	IF against tagged protein and split-GFP technology (Fischer *et al.*, 2017; Misaghi *et al.*, 2006; Wang *et al.*, 2018)	T3S (A.S. unpublished)	Deubiquitinase (Misaghi *et al.*, 2006)
CT875	TepP	Cytoplasm	Interaction with Slc1, phosphorylation in host cytosol (Chen *et al.*, 2014)	T3S	Scaffolding protein (Carpenter *et al.*, 2017)
Pgp3 (pORF5)		Cytoplasm and bacterial surface	IF against endogenous protein (Li *et al.*, 2008)	Unknown	Neutralizing action against anti-microbial peptide (Hou *et al.*, 2015)
C. caviae					
CCA00037 (CT671 orthologue)		Cytoplasm	IF against endogenous protein (Subtil *et al.*, 2005)	T3S	T3S ruler? CT671 interacts with CT670 (Lorenzini *et al.*, 2010)
C. pneumoniae					
CPn0572 (TarP orthologue)		Cytoplasm	IF against endogenous protein (Zrieq *et al.*, 2017)	T3S	Actin remodelling (Zrieq *et al.*, 2017)
CPn0712		Inclusion membrane and cytoplasm	IF against endogenous protein (Müller *et al.*, 2008)	Unknown	–

Table 7.1 Continued

Protein	Name	Location	Evidence for secretion	Suspected secretion mechanism	Putative function
CPn0796		Cytoplasm	IF against endogenous protein (Vandahl *et al*., 2005)	T5S	–
CPn0797			IF against endogenous protein (Dong *et al*., 2006)	Unknown	–
CPn0809 (CT578 orthologue)		Bacterial surface	IF against endogenous protein and functional T3S signal (Engel *et al*., 2016; Müller *et al*., 2008; Subtil *et al*., 2005)	T3S	T3S Translocator
CPn1020 (CT861 orthologue)		Endoplasmic reticulum?	IF against endogenous protein and functional T3S signal (Müller *et al*., 2008; Subtil *et al*., 2005)	T3S	–
C. psittaci					
G5Q_0070	SINC	Nuclear membrane	IF and TEM against endogenous protein (Mojica *et al*., 2015)	T3S	–

We did not include in this list the candidate T3S substrates (based on the detection of a functional T3S signal; da Cunha *et al*., 2014; Subtil *et al*., 2005) that have not yet been validated by complementary approaches. For a list of Inc proteins see (Dehoux *et al*., 2011; Weber *et al*., 2016). *C. trachomatis* serovar D/UW-3/CX, *C. caviae* GPIC, *C. pneumoniae* CWL029 and *C. psittaci* Cal10 strains were used as reference genomes.

not simply stochastic (Guerrero-Mandujano *et al.*, 2017; Jan, 2017).

Due to the predominantly intracellular nature of the chlamydial developmental cycle, and the fragility of replicative RBs, biochemical approaches to isolate these vesicles are extremely challenging. Membranous structures that could correspond to OMVs, are observed at the ultrastructural level within inclusions. Their abundance in *C. trachomatis* inclusions increased upon treatment with the antibiotic azithromycin (Giles *et al.*, 2006; Wyrick *et al.*, 1999).

While OMVs may have relevant physiological functions within the inclusion lumen, they could also constitute a mechanism for the translocation of chlamydial antigens into the host cytoplasm. This could proceed via the fusion of extruded OMVs with the inclusion membrane, or their engulfment by this membrane into an invaginated endocytic-like structure (Giles *et al.*, 2006). At present, the hypothesis that chlamydial OMV content can be delivered to the host cytoplasm is mainly based on ultrastructural imaging of chemically fixed cell sections and the localization of some bacterial surface antigens, such as major outer membrane protein (MOMP) or lipopolysaccharide, on the surface of infected cells following immunogold labelling of fixed samples (Wyrick *et al.*, 1999). Further studies should now be facilitated by genetic techniques that enable epitope tagging of specific proteins during infection and advances in higher resolution imaging techniques, particularly super-resolution light microscopy, cryo-electron microscopy (cryoEM) and tomography.

Type 2 secretion

Proteins exported into the periplasm by the Sec translocon can subsequently be secreted across the OM by a dedicated machinery called the T2S system. This is a sophisticated double-membrane-spanning apparatus comprised of 12–15 distinct components, which are conserved amongst a large number of pathogenic and non-pathogenic Gram-negative bacteria (Korotkov *et al.*, 2012). Genes encoding homologues of GspD (CT572), GspE (CT571), GspF (CT570) and GspG (CT569) are present in *C. trachomatis,* as well as in other *Chlamydia* spp. Note that we noticed that GspG was wrongly annotated in all *C. trachomatis* genomes: a GTG codon located 15 codons upstream of the annotated site encodes a protein that matches the full length of GspG homologues in other bacterial species. Despite the presence of these core constituents, other expected T2S components are less evident and remain to be experimentally identified. T2S substrates are challenging to predict from their sequence alone since the features recognized by T2S machineries have not been elucidated and appear to vary between different bacterial species (Korotkov *et al.*, 2012).

To date no T2S substrate has been unequivocally identified in *Chlamydia*. The Valdivia laboratory has isolated a mutant in the T2S ATPase GspE, which exhibits a significant attenuation in infectivity (Nguyen and Valdivia, 2012). It remains to be seen whether this phenotype is an indirect consequence of the impaired assembly of the OM or other virulence-associated machinery, or whether T2S substrates are themselves key to sustain bacterial development. Of course, these two possibilities are also not mutually exclusive. Several chlamydial proteins of unknown function possess a signal peptide and are thus possible T2S substrates. The analysis of such proteins will now be expedited by the isolation of the *C. trachomatis gspE* mutant, in which T2S substrates would be predicted to accumulate in the periplasmic space and could be analysed by quantitative proteomics.

Finally, it is important to consider that unlike T3S, T2S only transports substrates across the OM, and therefore does not provide direct access to the host cytoplasm. T2S substrates are thus expected to remain associated with the bacterial surface or to remain entrapped within the inclusion lumen, unless invagination of the inclusion membrane can deliver some of its content to the cytoplasm (Giles *et al.*, 2006).

Type 5 secretion

T5S refers generically to the secretion of a large family of bacterial surface proteins called autotransporters (Bernstein, 2015). A T5S substrate has an N-terminal signal peptide that directs its export into the periplasm via the Sec translocon, and two major domains, an N-terminal extracellular or 'passenger' domain and a C-terminal domain that forms a β-barrel residing in the OM. The passenger domain of a T5S substrate often functions as an adhesin, but can also be cytotoxic, or possess other virulence-associated activities. Passenger domains are highly diverse in size and sequence and often exceed 100 kDa in size. Following their transport across the OM, many passenger domains are subsequently released from the cell surface by proteolytic cleavage and in some instances this occurs by autoproteolysis. More recent evidence suggests that the heterooligomeric barrel assembly machine (Bam) complex, which catalyses the fundamental integration of β-barrel proteins into the OM, is also required for insertion of the autotransporter β-barrel domain, indicating that secretion driven by autotransport may in fact not be as 'autonomous' as previously believed (Bernstein, 2015).

Chlamydia express a family of polymorphic membrane proteins (Pmps) that are exposed on the bacterial surface and are often involved in adhesion to the host cell. These vary in number between bacterial species, but all possess a canonical signal peptide and a C-terminal domain predicted to form a β-barrel structure. In addition, several Pmps are known to undergo proteolytic processing, a characteristic feature of autotransporters. Pmps are thus likely T5S substrates, and their expression and putative function are described in Chapter 3, *Chlamydia* Adhesion and Invasion.

A cluster of six related *C. pneumoniae* genes, *cpn0794–cpn0799*, have also been investigated as potential T5S substrates. These genes lack orthologues in *C. trachomatis*. Each of the encoded proteins except Cpn0795 contain predicted signal peptides. In the first study of this group of proteins, CPn0796 distribution was restricted to the bacteria at early times of infection whereas it was detected in the cytoplasm of the cell harbouring larger inclusions (Vandahl *et al.*, 2005). Only full length CPn0796 was present in EB extracts, while a cleavage product appeared during infection. A possible β-barrel could be predicted in the C-terminal region of Cpn0796 preceded by a putative alpha helical domain, which are strong indicators that this protein is an autotransporter. None of the other five proteins of the cluster were found to be secreted into the host cell cytoplasm in this initial study. Of those, only CPn0795 contained the putative C-terminal β-barrel, but the protein did not have an N-terminal signal peptide. However, a later study detected CPn0797 in the cytosol, indicating it might also be secreted, although possibly via a T5S-independent mechanism (Dong *et al.*, 2006). As with T2S, T5S transports substrates either onto the surface of the bacteria or into the inclusion lumen as a result of proteolytic processing. T5S substrates must exploit additional as yet uncharacterized processes to translocate from the lumen of the inclusion across the inclusion membrane into the host cell cytoplasm.

How are proteins with signal peptides secreted in *Chlamydia*?

A subset of chlamydial proteins with signal peptides have been detected in the inclusion lumen and/or in the host cytoplasm (Chen *et al.*, 2010, Wu *et al.*, 2011, Lei *et al.*, 2011). As stated earlier, the secretion signal targets substrates to the Sec translocon to be delivered into the periplasm. Transport across the OM requires a second step. With the exception of T5S substrates, this second step has not been defined for any of the known Sec-dependent secreted proteins of *Chlamydia*.

The prominent substrate in this category of secreted proteins is the chlamydial virulence-associated protease CPAF (discussed in more detail in Chapter 8, The Chlamydial Protease CPAF). It contains a signal peptide, and the functionality of this sequence was demonstrated using a reporter gene fusion construct (*phoA*), which was sufficient to drive export across the IM into the periplasm of wild-type *E. coli* (Chen *et al.*, 2010). While CPAF has been detected in the host cytoplasm by immunofluorescence, including by epitope-tagging (Bauler and Hackstadt, 2014; Heuer *et al.*, 2003; Zhong *et al.*, 2001), and by electron microscopy (Dille *et al.*, 2014; Heuer *et al.*, 2003), its ability to reach the cytoplasm has recently been questioned (Prusty *et al.*, 2018). It is, however, largely accepted that CPAF is present in the lumen of the inclusion, but its mechanism of secretion remains unclear.

As explained above, OMVs constitute one possible mechanism of protein export beyond the periplasm. It could also account for the release of CPAF into the inclusion lumen assuming that the membrane of OMVs is unstable, or even to its delivery into the host cytoplasm if OMVs can fuse with the inclusion membrane or be translocated intact into the host cytoplasm. Noticeably, CPAF was no longer detected by immunofluorescence in the cytoplasm of cells infected with the T2S-deficient *C. trachomatis* strain encoding impaired GspE (Snavely *et al.*, 2014). This finding reinforced the view that CPAF translocation beyond the OM required the activity of the T2S. This requirement could, however, be indirect, for instance by impairing the secretion of a protein required for the specific incorporation of CPAF into OMVs or a more generalized reduction in the release of OMVs from the RB outer membrane.

A second substrate of the Sec machinery, another protease called HtrA, has been detected in the cytoplasm of *Chlamydia*-infected cells (Wu *et al.*, 2011). HtrA is not exclusively virulence-related since it is critical to the stress response in many bacteria, acting as a serine protease and as a chaperone. While a specific inhibitor of the protease activity was lethal to *C. trachomatis* development (Gloeckl *et al.*, 2013), and a point mutation in the *htrA* gene was linked to a developmental defect (Marsh *et al.*, 2015), the exact role of this protease in the developmental cycle remains undetermined. As with CPAF, the N-terminal secretion signal of HtrA is sufficient to direct the export of a PhoA reporter protein into the periplasm in *E. coli* (Wu *et al.*, 2011). How HtrA is secreted is unknown, but this is also not a question restricted to the *Chlamydia* field. Extracellular HtrA has been described in many bacterial species, and the mechanism of export beyond the periplasm remains a matter of debate (Backert *et al.*, 2018). OMVs are proposed as a mode of HtrA secretion in several bacterial species, and it remains feasible as for CPAF, that OMVs might mediate HtrA transport to the inclusion lumen, and possibly into the cytoplasm during chlamydial infection. This raises the question whether all periplasmic proteins can be potentially be secreted via OMVs. To address this question, Zhong and colleagues compared the localization of HtrA to that of three other periplasmic proteins during a *C. trachomatis* infection; thioredoxin (CT539), protein disulphide isomerase (CT783) and the periplasmic component of an ABC transporter YtgA (CT067). Intriguingly, only HtrA was detected in the infected cell cytoplasm, indicating that either there is a directed mechanism enabling the specific translocation of secreted HtrA from the inclusion lumen, or that the content of OMVs is controlled by sorting in the bacterial periplasm (Wu *et al.*, 2011).

Finally CT311, a protein of unknown function that contains a functional peptide signal for Sec-dependent export into the periplasm, was similarly detected in the host cytoplasm (Lei *et al.*, 2011) and can also be translocated into the nucleus later during infection (Lei *et al.*, 2013).

Bioinformatic analysis suggests that other proteins with a potential N-terminal signal sequence are present within the *C. trachomatis* genome. The studies of CPAF, HtrA and CT311 indicate that these might also represent secreted chlamydial proteins, which have the potential to reach the infected cell cytoplasm.

Type 3 secretion

T3S systems are macromolecular assemblies in the bacterial envelope responsible for the translocation of proteins from the bacterial cytoplasm directly into the eukaryotic host cell (Galan and Waksman, 2018). They are widespread amongst Gram-negative bacterial pathogens. The core components comprising the envelope-spanning machinery are evolutionarily related to the hook-basal body complex of the bacterial flagellum (Diepold and Armitage, 2015). Archetypally, the T3S comprises oligomeric ring complexes located in the IM and OM, connected by a trans-periplasmic rod. A polymeric 'needle' protrudes from the external face of the OM complex, which connects at its distal end to a transmembrane 'translocon' complex of unknown structure embedded in a eukaryotic cell membrane (Fig. 7.2). The T3S assembly therefore provides a continuous conduit for effector protein delivery linking the cytosol of the bacterium to that of the host cell and consequently there are no periplasmic export intermediates. T3S substrates dock at a cytosolic platform that includes an ATPase, to which they are piloted by binding to substrate-specific chaperones. Substrates have an N-terminal export signal defined by its secondary structure rather than the primary sequence for Sec-dependent export (Galan and Waksman, 2018).

T3S is the only recognized mechanism present in *Chlamydia* to translocate proteins directly into the host cell. In contrast to other bacteria, no mutants in T3S components have been described to date, possibly since T3S activity is essential for the progression of the developmental cycle and is implicated in both cell entry and intracellular replication (Mueller *et al.*, 2014). T3S effector proteins can be translocated into the host cell cytoplasm or subsequently transit to cellular compartments such as the nucleus (e.g. Clifton *et al.*, 2004; Pennini *et al.*, 2010). A further large family of ~40–50 inclusion proteins (Incs), which are all T3S substrates, are associated with the membrane of the inclusion (Dehoux *et al.*, 2011; Elwell *et al.*, 2016; Subtil *et al.*, 2001). Finally, an increasing number of effectors apparently reside within the lumen of the inclusion, suggesting that this subset of T3S substrates might

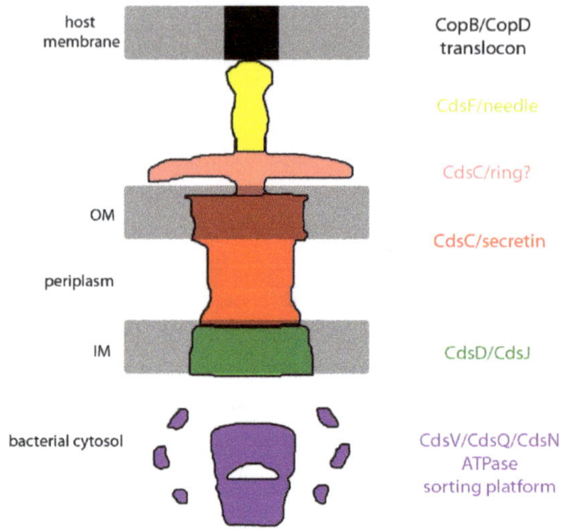

Figure 7.2 Schematic representation of the *C. trachomatis* T3S system based on subtomogram averaging (Nans *et al.*, 2015), showing the ATPase/sorting complex (CdsV/CdsQ/CdsN; purple) at the interface of the bacterial inner membrane (IM), the oligomeric inner membrane complex (CdsD/CdsJ; green), the secretin (CdsC; red) embedded in the bacterial outer membrane (OM) with the outer membrane ring (possibly an extension of CdsC; pink). The needle (CdsF; yellow) connects the secretin to the translocon complex (CopB/CopD; black) of unknown structure, which is embedded in the host plasma membrane or inclusion membrane.

not be delivered across the inclusion membrane (da Cunha *et al.*, 2017; Gehre *et al.*, 2016; Muschiol *et al.*, 2011).

Secreted proteins as vaccine candidates

Recent advances in vaccine development are comprehensively described in Chapter 15, *Chlamydia* Vaccines. Here we will briefly discuss secreted chlamydial proteins that have generated promising results when used as antigens in different immunization regimens.

Secreted proteins that become exposed but remain anchored to the surface of EBs or RBs are clear candidates for the generation of neutralizing antibodies, as well as being processed as targets of cell-mediated immunity. Indeed, Pmp-derived antigens trigger potent immune responses in a number of experimental systems (Desclozeaux *et al.*, 2017; Muller *et al.*, 2017; Pal *et al.*, 2017). Surface exposed components of the T3S machinery, including components of the translocon (Bulir *et al.*, 2016) and the needle subunit CdsF (Koroleva *et al.*, 2017), have also been documented to stimulate protective immunity.

Chlamydial antigens that may be translocated into the host cytoplasm have been recognized as immunodominant antigens, with a strong potential for the induction of protective immunity. These include CPAF (Murthy *et al.*, 2009), TARP (Wang *et al.*, 2009), Pgp3 (Donati *et al.*, 2003), and CopN (Tammiruusu *et al.*, 2007) (Table 7.1). In addition, many inclusion-associated proteins are highly immunogenic (Li *et al.*, 2008) and show potential for vaccine development (Fling *et al.*, 2001; Starnbach *et al.*, 2003). The T3S chaperone Slc1, which remains within the bacterial cytoplasm, induces a CD4+ Th1 response and is proposed to induce protective immunity (Meoni *et al.*, 2009).

Since an ideal vaccine should target both the extracellular and intracellular stages of the chlamydial developmental cycle, combinations of antigens expressed at different stages are now being tested (O'Meara *et al.*, 2017). Ultimately, this will require the generation of substantial quantities of high-purity antigens, emphasizing the importance of the continued biochemical and structural analyses of both chlamydial secretion systems and their associated effector substrates.

Mechanistic insights into T3S

Structure of the apparatus

Chlamydial genomes encode the conserved components of the T3S apparatus resident in other Gram-negative bacteria. These are predicted to form the core complex, comprising the oligomeric IM ring (CdsD/J), the cytoplasmic C–ring complex including the associated ATPase (CdsN), the OM secretin channel (CdsC) that appears to have evolved separately from other systems, the subunit (CdsF) that polymerizes to assemble the external needle, and the translocon proteins (CopB, CopD) (Fig. 7.2). Current knowledge of the assembly and activity of *Chlamydia* T3S has been reviewed in detail by Ferrell and Fields (2016). Here we will extend this by focusing on the most recent structural advances, and highlight current outstanding questions relating to T3S activity.

Even prior to the formal identification of T3S, seminal freeze-fracture and thin section electron microscopy (EM) studies by Matsumoto (e.g. Matsumoto, 1981) described regularly spaced projections on a limited surface of the EB, and subsequently immunogold labelling demonstrated that CdsF was exposed on the EB surface (Betts *et al.*, 2008). This correlation implies that the previously observed projections were in fact the first described T3S complexes in bacteria. More recently, cryoEM tomography of *C. trachomatis* and *Chlamydia*-related bacteria also revealed T3S-like densities spanning the EB envelope, which were restricted to one hemisphere of the bacterial surface (Nans *et al.*, 2014; Pilhofer *et al.*, 2014). In *C. trachomatis*, this pole is characterized by an expansion of the periplasmic space to ~28 nm compared to ~14 nm at the opposing pole. The expanded periplasmic space accommodates 14–20 T3S complexes per EB. These form a semi-ordered array with an average spacing of ~57 nm. The widening of the periplasmic space arises from the remodelling of the IM, consistent with the rigidity of the highly cross-linked OM, with each T3S complex originating at a specific concave deformation of the IM. This polar distribution of T3S complexes might be unique to *Chlamydiae*, as equivalently analysed envelope-embedded T3S machineries in other pathogenic bacteria are uniformly distributed around the envelope perimeter (Nans *et al.*, 2014).

CryoEM tomography and sub-tomogram averaging also permitted the detailed analysis of the *in situ* structure of the T3S in the envelope of the EB (Nans *et al.*, 2015). The overall architecture of the chlamydial T3S complex at 3.3 nm resolution differed from that in other bacteria since the membrane-embedded core is more elongated, with a pronounced convex curvature, a length of 34 nm and a diameter from 14 to 20 nm. A ring-like structure surrounds the needle on the face of the OM, which likely corresponds to an additional domain of the secretin CdsC, since it is significantly larger than T3S secretins present in other bacteria. Additionally, no pilotin, chaperone-like lipoproteins required for secretin insertion in the OM, or other predicted OM proteins reside within the operons encoding T3S components. Density is present at the IM interface with the membrane-embedded core characteristic of the cytoplasmic domain of the export apparatus (CdsV), along with the cytoplasmic sorting platform (CdsQ)-ATPase (CdsN) complex, although this is less distinct, reflecting an increased level of disorder. Consistent with labelling approaches, no peptidoglycan layer could be resolved (Nans *et al.*, 2015).

Since EBs could also be frequently visualized in contact with the plasma membrane of host cells by cryoEM tomography (Nans *et al.*, 2014), it was also possible to capture the T3S complex in contact with a host membrane for the first time in any biological system (Nans *et al.*, 2015). Analogous sub-tomogram averaging at 3.8 nm resolution allowed the cross comparison of the membrane-contact and cell-free states of the EB T3S *in situ*. There were several significant differences when compared to the apparatus visualized in cell-free EBs. The length of the envelope–spanning complex was ~4.5 nm shorter in the host contact state, and this contact-associated compression was accompanied by a stabilization of the cytoplasmic sorting platform and the ATPase proximal to the bacterial IM. Conformational changes were observed in the needle, and a pore-like density was also evident proximal to the needle tip in the host cell membrane, implying the assembly of a continuum between the EB cytosol, T3S complex and the host cell. However, this structure could not be definitively identified in the absence of chlamydial mutants lacking the predicted CopB/CopD translocon components. Nevertheless, these data revealed intriguing new insights into the 'pump-action' conformational changes that occur during T3S activation upon contact with the host cell membrane, although it remains to be determined whether these are restricted to *Chlamydia* or more conserved facets of the T3S mechanism (Nans *et al.*, 2015).

When RBs were visualized in contact with the inclusion membrane in infected cell sections by electron tomography following high pressure freezing and freeze-substitution, macromolecular complexes were also evident spanning both bacterial membranes (Dumoux *et al.*, 2012). These complexes were in contact with the luminal face of the inclusion membrane at sites co-incident with the recruitment of the rough endoplasmic reticulum on the cytoplasmic face of the inclusion. As in EBs, the dimensions of these complexes were consistent with T3S, with each RB containing a polar array of 20–100 T3S associated with a widening of the periplasmic space. Needle structures could also be defined extending from the RB surface, in some places connecting the bacterial OM to the inclusion membrane. These structures could also be immunogold-labelled using an anti-CdsF antibody. The arrays, in contact with four membranes (bacterial IM and OM, inclusion membrane and host endoplasmic reticulum) *in situ*, were termed 'pathogen synapses' and likely represent portals for the translocation of T3S effectors, particularly Incs. Significantly, T3S complexes could only be observed by tomography when RBs were in contact with the inclusion membrane connected to the rough endoplasmic reticulum, suggesting an essential functional link between these bacterial and host structures (Dumoux *et al.*, 2012). Common structural features therefore intriguingly exist between T3S assemblies in EBs and RBs (Dumoux *et al.*, 2015).

Phylogenetic analysis has revealed that the last common ancestor of pathogenic and symbiotic *Chlamydiae* possessed a T3S apparatus and several of the effector proteins present in modern isolates. Ancient *Chlamydiae* therefore appear to be the originators of mechanisms for the successful hijack of eukaryotic host cells (Horn *et al.*, 2004). In support of this notion, there is only limited evidence for horizontal gene transfer either between chlamydial strains or with other species of bacteria. This is in contrast to other species where mobile genetic elements are a driving force in the development of pathogenic traits (Bobay and Ochman, 2017). In particular, the genes encoding the T3S components, effectors and their cognate chaperones are dispersed in multiple small clusters throughout the genome (Peters *et al.*, 2007). In other pathogens the T3S and a core effector cohort have been acquired during a single horizontal gene transfer event and are located together within a single chromosomal pathogenicity island or on a plasmid (Hueck, 1998). Intriguingly, although considered non-motile, *C. trachomatis* has retained a number of orphan T3S-related flagellar genes, in addition to the homologues comprising the virulence-associated T3S system. These are predicted to encode the IM ring FliF (CT719), the traffic ATPase FliI (CT717) and its regulator FliH (CT718), a core component of the export sorting platform in the inner membrane FlhA (CT060), and a homologue of the flagella-associated sigma factor FliA (CT061) (Peters *et al.*, 2007; Stone *et al.*, 2010). CT061 has been shown to act as an alternative sigma factor σ^{28} that transcribes a subset of late genes in *Chlamydia* (Yu and Tan, 2003). σ^{28} is discussed in more detail in Chapter 10, *Chlamydia* Gene Regulation. However, the function of these additional flagellar-related proteins at present remains unknown.

Salicylidene acylhydrazides were initially identified as inhibitors of the T3S system in *Yersinia* species, and were additionally demonstrated to have an inhibitory effect on chlamydial infection. Since their initial use and classification as T3S inhibitors, it has become apparent that these compounds also have the capacity to chelate iron (Slepenkin *et al.*, 2007) and to bind chlamydial protoporphyrinogen oxidase (HemG) (Engstrom *et al.*, 2013). It is now accepted that the dramatic impact of these compounds on chlamydial development likely represents a pleiotropic or combinatorial effect rather than a specific inhibition of the T3S system.

At present, there are therefore no methods to specifically inhibit T3S in *Chlamydia*. The lack of chlamydial T3S inhibitors has made it difficult to validate effector substrates and to study the role of T3S during the intracellular infection. From work in other bacteria, several alternative T3S inhibitor candidates have been identified (Grishin *et al.*, 2018; Johnson *et al.*, 2009; Zigangirova *et al.*, 2016). If validated, these compounds would represent valuable experimental tools to decipher the role of T3S during the *Chlamydia* infection cycle, and could represent promising new anti-chlamydial agents.

Regulation of T3S activity

These structural and functional insights into chlamydial T3S complexes have informed subsequent hypotheses and have also resulted in the reassessment of previous data. The T3S must be activated in EBs and RBs and recognize substrates specific to different phases of the infection cycle. Moreover, since the T3S is present in both EBs and RBs, what happens to T3S assemblies during the profound membrane reorganization during bacterial differentiation remains unknown. Regulation could occur either by disassembling the entire machinery, or by inactivation or reprogramming of the system via the modification or substitution of key components.

One aspect of this regulation concerns the unique chlamydial OM, which is stabilized by extensive disulphide bonding between the unusually cysteine-rich surface proteins. The Fields laboratory established that this disulphide bonding also extended to structural components of the T3S complex including CdsD and CdsC, but most significantly occurred within the needle subunit CdsF (Betts-Hampikian and Fields, 2011). As with the OM proteins, the extent of disulphide bonding correlated with the progression of the developmental cycle as these bonds were in a reduced state during replication of the RBs. It was also apparent that CdsF became re-oxidized into higher order oligomers upon the conversion of RBs back to EBs. These features, which are apparently also unique to *Chlamydiae*, indicate that the redox state is potentially an important regulator of T3S complex assembly and activity. Interfering with this redox process could inhibit chlamydial T3S and consequently provide an alternative route to developing specific antimicrobial compounds.

Although T3S systems were originally defined as activated on contact with a eukaryotic membrane, the physiologically relevant signals that trigger their activation remain incompletely understood (Hueck, 1998). In many bacteria, eukaryotic lipids including cholesterol seem important factors for T3S activation (Lafont and van der Goot, 2005), while other *in vitro* triggers such as Congo Red or calcium chelation can stimulate T3S in the absence of host cells (Hueck, 1998). In *Chlamydia*, the Hackstadt laboratory showed that TARP secretion by the T3S could be induced, albeit at a moderate level, by treating *C. trachomatis* EBs with serum or by the addition of liposomes enriched in sphingolipids and cholesterol (Jamison and Hackstadt, 2008). These conditions could be exploited to examine conformational changes that accompany T3S complex activation and to prospect for novel effectors. By contrast, the signals that activate T3S in RBs, or potentially suppress T3S in newly differentiated EBs prior to their release from the infected cell, are not known.

It is not known whether the protein composition of the chlamydial T3S remains the same throughout the developmental cycle. This issue has been raised because the chlamydial T3S genes uniquely include duplicated copies of the translocon proteins and orphan flagella genes that could potentially substitute for structural and regulatory components of the T3S (Chellas-Gery *et al.*, 2011). Although the structures of the T3S complex in EBs and RBs appear to be similar (Dumoux *et al.*, 2015), it is not known if their protein composition is identical. A change could induce a switch in substrate specificity, for example from effectors required for entry to those that control inclusion. Substitution of key components might also determine specificity for plasma membrane versus inclusion membrane targeting. These translocon and flagella-like proteins can be studied by testing if they can complement predicted T3S homologues in other bacteria, or by looking for phenotypic effects associated with mutations in these loci.

Regulation might also be achieved by complete or partial disassembly and reassembly of the T3S complex during the developmental cycle (Ferrell and Fields, 2016; Dumoux *et al.*, 2015). Indeed, electron tomography of early stages of interaction between EBs and the host cell revealed that the T3S complex was apparently disassembled following bacterial internalization, with an associated loss of EB polarity (Nans *et al.*, 2014; Dumoux *et al.*, 2015). T3S complex dynamics now requires further investigation throughout the developmental cycle, since the T3S system must be reassembled and reactivated again to enable the translocation of newly synthesized effectors such as Incs during nascent inclusion biogenesis. More generally, the fate of trans–periplasmic complexes directing T2S and T3S is unclear during differentiation from EBs to RBs. Similarly, within the inclusion it is unclear how the RB containing up to 100 T3S complexes transitions to an EB containing 14–20 complexes, while polarity apparently remains evident (Dumoux *et al.*, 2015). Related to this is the fate of T3S complexes during the formation of aberrant bodies during chlamydial persistence. These enlarged bacteria have significantly deformed membranes and lack apparent polarity yet maintain inclusion integrity and presumably must therefore continue to translocate effector proteins, although T3S complexes are yet to be visualized in this state.

Detachment of RBs from the inclusion membrane has been proposed as a signal to trigger re-differentiation into EBs (Wilson *et al.*, 2009). However, new evidence suggests that dividing RBs are apparently not restricted to the periphery of the inclusion (Lee *et al.*, 2017). Whether these luminal RBs also express T3S machinery is not known. Active T3S from RBs that are not associated with the inclusion membrane could account for detection of several T3S substrates in the inclusion lumen (da Cunha *et al.*, 2017; Gehre *et al.*, 2016; Muschiol *et al.*, 2011). If, as in other T3S systems, contact with a host-derived membrane is required to trigger secretion, some of the membranous structures frequently observed in the inclusion lumen could potentially induce T3S. The nature of these membranous structures still remains unclear; some could be OMVs, but others could correspond to vesicles imported from the host cytoplasm, as it has become clear that exchange between the inclusion and the cytoplasm is not unidirectional, and host-derived material is translocated into the inclusion lumen (Boncompain *et al.*, 2014; Cocchiaro *et al.*, 2008; Gehre *et al.*, 2016). Interestingly, several T3S substrates have now been detected on the surface of different Gram-negative bacteria before being translocated into a mammalian cell (Akopyan *et al.*, 2011; Tejeda-Dominguez *et al.*, 2017). Thus, it is possible that chlamydiae could secrete T3S substrates into the inclusion lumen rather than via the T3S translocon pore into the host cell.

Substrates of T3S: what, when and how?

Methods for the identification of T3S substrates

Chlamydial effectors translocated into the inclusion membrane and the host cytoplasm are the principal weaponry by which the bacteria manipulate fundamental host cell processes and circumvent host defence mechanisms. Therefore, their identification, as an initial step towards functional studies, has been the subject of intense investigation over the last two decades. Despite the sequencing of chlamydial genomes, genetic context provides little information about T3S effectors since atypically compared to other bacteria the genes encoding T3S components and substrates are generally scattered in *Chlamydiales*, even if some hot spots of T3S effectors are present as gene clusters (Table 7.1) (Subtil *et al.*, 2000; Subtil *et al.*, 2005). Putative effectors also show extremely limited sequence similarity to other bacterial or eukaryotic proteins, meaning that bioinformatic approaches using sequence or structural databases have made limited impact.

Prospecting for chlamydial T3S effectors has been facilitated by one feature very specific to the major family of chlamydial effectors, the Inc proteins, and by two common features of T3S substrates first described in *Salmonella*, *Shigella* and *Yersinia* (Galán *et al.*, 2014):

1. Incs are T3S substrates characterized by one or several predicted bilobed hydrophobic domain(s) in their sequence, which are proposed to act as an anchor in the inclusion membrane exposing the terminal domains to the host cytoplasm (Bannantine *et al.*, 2000). Between 60 and 100 putative Incs can be identified bioinformatically in chlamydial genomes based on this feature (Dehoux *et al.*, 2011; Lutter *et al.*, 2012). About 40 have been confirmed to localize to the inclusion using antibodies or more recently by exploiting newly developed *Chlamydia* expression vectors to epitope-tag Incs (Weber *et al.*, 2015).
2. T3S substrates are directed to the secretion machinery by an N-terminal signal within the first 20–25 amino acids. Although this domain is highly variable in sequence and can tolerate significant changes, it can be useful for T3S effector identification (Galán *et al.*, 2014).
3. A second motif, present at least in some T3S substrates within the first 100 residues of the protein, after the secretion signal, serves as a binding site for substrate-specific chaperones (Galán *et al.*, 2014).

For a long time, these characteristics, coupled with the use of secretion assays employing heterologous T3S systems in other bacteria, were the only tools available to search for potential T3S substrates in *Chlamydia*. Validation of a candidate effector required the generation of specific antibodies to detect its localization in the inclusion membrane or host cytoplasm. Often, these approaches were limited by the low abundance of T3S substrates in infected cells, rendering them beneath the detection threshold of immunofluorescence. The opportunity to express tagged proteins in wild-type *C. trachomatis* now substantially facilitates this effector validation step (Maffei *et al.*, 2017).

Several machine-learning approaches have been employed to identify N-terminal T3S signals. These methods differ in terms of algorithms, the curated data sets and the precise features used, and reach different levels of prediction performance. Several of these tools were recently compared in other bacteria (An *et al.*, 2016). All predictors were flawed to various degrees, due to the generation of false positives, where secretion signals are identified in non-secreted proteins, and false negatives when documented secretion substrates were not predicted. Thus, these tools enable the initial prediction of candidate proteins, but subsequent experimental validation is essential. In comparison to other pathogens, these approaches in *Chlamydia* appear to perform somewhat poorly, showing a rather large number of false negatives when tested on Inc proteins (Dehoux *et al.*, 2011). The reasons for these inaccuracies are not known, but under these circumstances heterologous secretion assays in a surrogate host remain the most sensitive assay to determine whether the amino-terminal sequence of a given chlamydial protein is sufficient to function as a T3S signal. In addition to being used to specifically assess whether one candidate protein might be a T3S substrate, these assays have also been applied to perform genome-wide screens (da Cunha *et al.*, 2014; Subtil *et al.*, 2005). Heterologous secretion assays often use chimeric constructs, in which the N-terminal secretion signal is fused to a reporter protein that does not require chaperone binding for secretion. In some instances, the corresponding full-length chlamydial proteins were secreted by surrogate bacteria, either because their secretion can occur in a chaperone-independent manner, or because this function could be replaced by an endogenous chaperone.

Chaperone binding sites cannot as yet be predicted from the substrate primary sequence, but since the T3S chaperones share structural similarity as well as some biochemical properties, another strategy to find novel T3S substrates has been to first identify chlamydial chaperones, and then use these as bait to identify novel substrates. For instance, the best studied chlamydial chaperone, Slc1 (CT043), was used to pull-down bound chlamydial proteins, a strategy that led to the identification of a novel T3S substrate, TepP (Chen *et al.*, 2014). Mota and colleagues adopted a more systematic approach to identify novel T3S chaperone–substrate complexes (Pais *et al.*, 2013).

Novel strategies to identify chlamydial proteins translocated into the host cytoplasm have now been enabled by the expansion of the chlamydial genetic toolbox that allows the generation of fusions with reporter tags. Since T3S and Sec-dependent secretion signals reside at the N-terminus, it is important to maintain this region of the protein intact. Correspondingly, a C-terminal FLAG-tag and a reporter tag that becomes specifically phosphorylated in the eukaryotic cytosol have been successfully introduced into Incs (Bauler and Hackstadt, 2014). Larger proteins, such as the enzyme APEX2, have also been successfully used to tag Inc proteins (Rucks *et al.*, 2017). Furthermore, over the last 10 years, the development of fluorogenic beta lactamase (TEM1) substrates has allowed the tracing of effectors fused to beta lactamase. This assay is very sensitive and is amenable to live microscopy, circumventing some of the previous issues with antibody detection thresholds and fixation artefacts (Maffei *et al.*, 2017). This technique has been adapted for *C. trachomatis* by the Fields laboratory (Mueller and Fields, 2015), although a disadvantage of the TEM1-based assay is that it does not provide spatial resolution. Even more recently the Stephens laboratory reported the adaptation of a split-green fluorescent protein (GFP) system for use in *Chlamydia* (Wang *et al.*, 2018). While less sensitive than the TEM1-based approach, and with a lower temporal resolution, this strategy is interesting as it allows the localization of the secreted protein in infected host cells. Despite these substantial advances, the use of *C. trachomatis*-modified plasmids enhances the dose and likely alters the kinetics of effector expression, even if the endogenous promoter is employed. A combination of approaches therefore remains optimal. Also, one limitation is that not all tags are compatible with T3S. Some proteins like the GFP and GST (glutathione S-transferase), which fold rapidly, block T3SS (Jaumouillé *et al.*, 2008; Riordan *et al.*, 2008). We have also observed that Inc proteins fused to a carboxyterminal SNAP or Halo tag were not expressed at the inclusion membrane (Subtil unpublished).

T3S chaperones and their role in the secretion process

Molecular chaperones bind transiently to one or several proteins to prevent premature or inappropriate intra- and inter–molecular interactions. Unlike other housekeeping chaperones, such as the GroEL/Hsp60 and DnaK/Hsp70 protein family, T3S-associated chaperones lack nucleotide-binding or hydrolysis activity but specifically pilot cognate T3S substrates to the cytoplasmic export platform. They exhibit no discernible sequence similarity to one another or to other types of chaperone, but share common features such as a small size, an acidic pI, and a predicted amphiphilic alpha-helix within the C-terminal region (Galan *et al.*, 2014; Parsot *et al.*, 2003). Based on tertiary structure and their binding specificity, T3S chaperones have been divided into three classes:

1. Class I chaperones are sub-divided into class IA chaperones that bind a single effector and class IB chaperones that bind multiple effectors. They share limited sequence similarity to each other, but display a conserved three-dimensional structure and normally form homodimers. In the absence of its class IA chaperone the cognate effector protein is usually degraded, whereas this is often not the case for effectors bound by the multicargo class IB chaperones. It is now well accepted that the primary function of these chaperones is to target their cognate effectors to the T3S apparatus, as the absence of these chaperones results in the lack of secretion of their corresponding effectors. Furthermore, removal of the effector chaperone-binding domain similarly prevents secretion.
2. Class II chaperones usually interact with the components of the T3S translocon and in some cases with the subunits of the needle. These chaperones are characterized by the presence of tetratricopeptide repeats, and while it is clear that they stabilize and protect their substrates from detrimental interactions, their precise role in T3S is still unclear.
3. Other chaperones are sometimes referred to class III chaperones. Initially dedicated to the flagellar export system, which is proposed to be the ancestor of the T3S system, this group now tends to include all the chaperones that

cannot be defined as class I or II (Parsot *et al.*, 2003).

Amino acid sequence analysis has suggested that *C. trachomatis* encodes at least three putative class I T3S chaperones, namely Slc1 (CT043), Scc1 (CT088) and Scc4 (CT663, SycE/CesT-like). Unlike their counterparts in other pathogenic bacteria, homotypic interactions among chlamydial class I T3S chaperones were not detected by two hybrid screens in yeast, but heterotypic Scc1/Scc4 complexes were observed (Spaeth *et al.*, 2009). In addition, orthologues in *C. pneumoniae* were shown to interact with CopN and to facilitate its secretion in a heterologous system (Silva-Herzog *et al.*, 2011). Slc1 was defined by independent groups to be a multicargo chaperone (Table 7.2) (Brinkworth *et al.*, 2011; Chen *et al.*, 2014; Pais *et al.*, 2013). The secretion of Slc1-chaperoned T3S effectors has mostly been studied very early during infection, but might feasibly also occur later in the developmental cycle, as shown at least for CT695, which is enriched on the inclusion membrane (Mueller and Fields, 2015).

While it was not apparent based on its primary sequence, the predicted structure and biophysical properties of MscC (CT260) revealed features consistent with a class I chaperone, which were confirmed experimentally (Spaeth *et al.*, 2009). MscC is a multicargo chaperone, and its two known early effector substrates Cap1 and CT618 are present on the inclusion membrane a few hours post infection.

CT635 was also recently identified as a potential T3S chaperone (Cossé *et al.*, 2018). CT635 is small (17 kDa) with a slightly acidic pI (6.5). It binds within the first 100 amino acids of the T3S effector CT622. However, co-expression of CT635 and CT622 did not enhance secretion of CT622 in the heterologous secretion assay using *Shigella* as a surrogate host, so the definitive classification of CT635 as a novel T3S chaperone awaits further experimental validation.

Database similarity searches revealed three *C. trachomatis* class II chaperones: CT274 (LcrH/SycD-like), Scc2 (CT576) and Scc3 (CT862) (Fields *et al.*, 2005; Slepenkin *et al.*, 2005; Spaeth *et al.*, 2009). Heterotypic interactions between CT274 and a small acidic protein CT668 protein were detected, suggesting that CT668 might also be a chaperone (Spaeth *et al.*, 2009). CT274 also interacted with CT161, a homologue of the secreted protein Lda2 and a putative T3S substrate (da Cunha *et al.*, 2014; Spaeth *et al.*, 2009).

The protein CT584 had initially been proposed to be the chlamydial needle tip protein based on its

Table 7.2 T3S chaperones and associated T3S substrates

Protein	Class	Secreted proteins
Chaperone		
Scc1 (CT088)/Scc4 (CT663)	I	CopN (Silva-Herzog *et al.*, 2011)
McsC (CT260)	I	CT529, CT618, CT225 (Spaeth *et al.*, 2009)
Slc1 (CT043)	I	TarP (CT456), TepP (CT875), TmeA (CT694), CT695 (Brinkworth *et al.*, 2011; Chen *et al.*, 2014; Pais *et al.*, 2013)
Scc2 (CT576)	II	CopB and CopD (Fields *et al.*, 2005; Slepenkin *et al.*, 2005; Spaeth *et al.*, 2009)
Scc3 (CT862)	II	CopN (Fields *et al.*, 2005; Slepenkin *et al.*, 2005; Spaeth *et al.*, 2009)
CT274/(CT668?)	II	CT161? (da Cunha *et al.*, 2014; Spaeth *et al.*, 2009)
CdsE/CdsG (CT665/CT667)	III	CdsF (Betts *et al.*, 2008; Spaeth *et al.*, 2009)
Putative chaperone		
CT584	III	CT082 (Pais *et al.*, 2013) CdsF, CdsN (Stone *et al.*, 2012)
CT635	I	CT622 (Cossé *et al.*, 2018)

For consistency the *C. trachomatis* D/UW-3/CX name is indicated, even if some of the experimental data were obtained in *C. pneumoniae*. All these chaperones have homologues in several *Chlamydiales* families except CT584, that is only found in *Chlamydiaceae*.

biophysical properties (Markham *et al.*, 2009). The orthologous protein in *C. pneumoniae* (Cpn0803) was shown to interact with components of the T3S apparatus (Stone *et al.*, 2012) and the expression of CT584 facilitated the secretion of a candidate T3S substrate, CT082 (Pais *et al.*, 2013). CT584 is thus classified as a candidate class III chaperone. CT082 could be a T3S substrate, although it might not be an effector and could instead be involved in assembly of the apparatus or part of a related substructure (Pais *et al.*, 2013).

Finally, the proteins CdsE and CdsG, which interact with each other and whose genes are just adjacent to the one coding for the needle protein CdsF, very likely act as chaperones for CdsF (Betts *et al.*, 2008; Spaeth *et al.*, 2009)

The roles of T3S chaperones have been extensively experimentally addressed in other genetically tractable bacteria, and these are consequently assumed to also hold true in *Chlamydiae*. The current dogma is that T3S chaperones stabilize their cognate T3S substrates and maintain them in a secretion-competent state (Parsot *et al.*, 2003). In addition, in *Yersinia*, *Salmonella* and *Shigella*, some chaperones contribute to the regulation of the expression of T3S genes at the transcriptional level (Parsot *et al.*, 2003), a function potentially retained by *Chlamydia* Scc4 (see below).

Many open questions still remain about T3S chaperones in *Chlamydia*. Firstly, it is likely that additional chaperones are still to be defined in *C. trachomatis* among the remaining proteins of unknown function. It is also likely that not all T3S effectors require a chaperone for efficient secretion in *Chlamydia*, especially given that T3S chaperones are often dispensable in other bacteria and act to enhance the efficiency of effector secretion. In addition, *Chlamydia* is somewhat atypical in that many effectors are synthesized when the bacteria are already in contact with the host cell and so do not need to be stored in a secretion-primed pool. In this respect, T3S in *Chlamydia* shares similarities to that mediated by the SPI-2 T3S apparatus that controls the intracellular replication of *Salmonella*. In the RB, the translocation of T3S substrates such as Incs could be coupled to translation and consequently chaperones might not be essential. However, inclusion-associated CT529 and CT618 do bind the MscC chaperone, suggesting that both chaperone-dependent and -independent modes of T3S might occur simultaneously during intracellular replication.

Contribution of T3S chaperones to regulation of gene expression

The regulation of gene expression will be discussed in detail in Chapter 10, *Chlamydia* Gene Regulation. However, we understand comparatively little about the regulation of T3S at a genetic level in comparison to other pathogenic bacteria. As in other systems, there may be a coupling role for T3S chaperones. For example, Scc4, a small protein that interacts with two subunits of RNA polymerase and suppresses its activity could potentially act to link T3S and transcriptional regulation (Rao *et al.*, 2009), as it also serves as a chaperone for the T3S substrate CopN (Silva-Herzog *et al.*, 2011). Two studies have subsequently provided evidence that the dual functions of Scc4 may link T3S to the regulation of gene expression in *Chlamydia* (Hanson *et al.*, 2015; Shen *et al.*, 2015). However, the exact timing of this activity in relation to the bacterial developmental cycle is not yet clear. Shen and colleagues suggested that Scc4 inhibited early transcription once released from CopN (Shen *et al.*, 2015), which does not correlate with the increased levels of transcription observed at that stage of the cycle. Since Scc1 interacts with Scc4, one explanation could be that Scc1 might initially be present in sufficient quantity to sequester Scc4 and prevent the inhibition of RNA polymerase. Consequently, transcriptional regulation by Scc4 might take place later in the developmental cycle (Ferrell and Fields, 2016; Hanson *et al.*, 2015).

Translocation of Incs into the inclusion membrane

A key outstanding question concerns the mechanism of translocation of the Incs into the inclusion membrane. This large family of T3S effectors are characterized by one or more hydrophobic regions, with the majority predicted to adopt a unidirectional hairpin-like topology in the inclusion membrane, allowing the exposure of the termini in the host cell cytoplasm where some engage host targets. How these hydrophobic effectors, which are unique to *Chlamydiae*, insert into the inclusion membrane following their transit through the T3S translocon complex without aggregating in the cytosol remains unknown (Dumoux *et al.*, 2015).

One possibility is that the host rough endoplasmic reticulum, which is specifically recruited to the cytoplasmic face of the inclusion at sites proximal to RBs containing assembled T3S arrays forming 'pathogen synapses', acts as a receiver membrane for this hydrophobic cargo. Since the inclusion membrane forms extensive contacts with the endoplasmic reticulum, membrane contact sites or fusion of vesicles derived from the endoplasmic reticulum with the nascent inclusion might then mediate Inc integration. Notwithstanding this, it remains unknown whether partially unfolded Incs emerging from the T3S translocon are competent to spontaneously insert into membranes or require either specific lipids or additional host machinery like the Sec61 translocon or the GET complex involved in the integration of eukaryotic transmembrane proteins, also resident in the endoplasmic reticulum (Dumoux *et al.*, 2015). Bacterial two-hybrid analysis indicated that Inc interactions could be reconstituted in *E. coli* membranes and that some apparently specific homo- and heterotypic Inc–Inc interactions could be retained (Gauliard *et al.*, 2015). The potential for Inc-Inc oligomerization adds yet more intriguing questions regarding how sub-complexes are assembled and maintained within the inclusion membrane and how this oligomerization is mediated. The answers to these questions currently await further experiments, which will undoubtedly now be facilitated by the ability to tag and analyse Inc proteins translocated during infection, coupled to proximity labelling and the application of mass spectrometry analysis (Rucks *et al.*, 2017). Fundamental questions also remain about the mechanism of Inc integration that can only be addressed following the purification of full-length Incs at quantities sufficient to allow the study and reconstitution of Inc interactions with model membranes using biophysical and structural biology approaches.

Conclusion

The past few years have seen significant advances in our understanding of protein secretion in the *Chlamydiaceae*, which at least in part have been driven by a more comprehensive understanding of genomic architecture and genetic diversity and the development of computational tools to identify putative export components, their substrates and regulators. In addition, the expansion of the genetic toolkit has enabled the generation of bacterial mutants and allowed heterologous protein expression. This has revolutionized the analysis of effectors and will facilitate understanding the contribution of the different secretory pathways to bacterial virulence and the control of the developmental cycle. Finally, the application of high-resolution imaging, particularly with the recent advances in electron microscopy and tomography coupled to the increasing accessibility of live imaging has substantially enhanced our understanding of the bacterial envelope, the structure and dynamics of the T3S, enabled the localization of effectors and illuminated the surprisingly extensive interplay between the inclusion and intracellular compartments.

As with all the best science, there are now many new intriguing unanswered questions. These include but are not restricted to the following:

- how are some proteins translocated into the host cytosol independently of the T3S pathway?
- to what extent do bacterial OMVs contribute to the transport of proteins into the inclusion lumen? Can OMV-entrapped proteins reach the host cytoplasm, and if so how?
- what is the repertoire of T2S substrates and their roles?
- what is the composition of the T3S apparatus in EBs versus RBs and to what extent is the apparatus assembled and disassembled during the developmental cycle? Why are there two sets of translocon-related proteins and to what extent does the redox state regulate T3S assembly?
- how are Incs integrated into the inclusion membrane and how does the T3S deal with this uniquely hydrophobic cargo?
- is there differential expression of genes in RBs that are in contact with the inclusion membrane versus luminal bacteria and what determines whether a T3S substrate is translocated across the inclusion membrane into the host cytosol or remains in the lumen?
- how are the different secretion machineries regulated along the developmental cycle? Are they active in EBs?

The ability to manipulate the *C. trachomatis* genome will undoubtedly facilitate the functional study of secreted proteins. However, the phenotypes observed in deletion mutants will need to be interpreted cautiously because intricate layers of regulation appear to link the expression and secretion of many effector proteins. For instance, the deletion of CPAF increased the abundance of several T3S substrates in total cell lysates (Patton *et al.*, 2016), indicating that some of the phenotypes observed in CPAF deficient strains could actually result from an indirect effect on other effector proteins. Another illustration of such confounding factors arose recently from the genetic disruption of the T3S effector CT622 (Cossé *et al.*, 2018). Loss of expression of this single T3S effector severely affected the viability of elementary bodies, infectivity (with a defect in TARP secretion), and inclusion growth. The direct contribution of CT622 to each of these defects remains to be determined, some of which might indirectly result from the production of defective EBs in the absence of CT622.

Naturally, each technique and advance has associated limitations, and so the challenge will be to combine these approaches to address the outstanding unsolved questions relating to protein secretion in *Chlamydia*. This in turn is likely to deliver new fundamental understanding of the biology of *Chlamydia* that may be translated into the development of vaccines, the design of small molecule inhibitors and superior diagnostic tools.

Acknowledgements

We thank Jaime Mota for critical review of Table 7.1. Current work in the Hayward laboratory is funded by a Medical Research Council Project Grant (N000846/1). The Subtil laboratory is supported by the Institut Pasteur, the Centre National pour la Recherche Scientifique and by the European Research Council (N° 282046).

References

Akopyan, K., Edgren, T., Wang-Edgren, H., Rosqvist, R., Fahlgren, A., Wolf-Watz, H., and Fallman, M. (2011). Translocation of surface-localized effectors in type III secretion. Proc Natl Acad Sci U S A *108*, 1639–1644. https://doi.org/10.1073/pnas.1013888108.

An, Y., Wang, J., Li, C., Leier, A., Marquez-Lago, T., Wilksch, J., Zhang, Y., Webb, G.I., Song, J., and Lithgow, T. (2016). Comprehensive assessment and performance improvement of effector protein predictors for bacterial secretion systems III, IV and VI. Briefings in Bioinformatics, bbw100. https://doi.org/10.1093/bib/bbw100.

Backert, S., Bernegger, S., Skórko-Glonek, J., and Wessler, S. (2018). Extracellular HtrA serine proteases: An emerging new strategy in bacterial pathogenesis. Cellular Microbiology *20*, e12845. doi doi:10.1111/cmi.12845.

Bannantine, J.P., Griffiths, R.S., Viratyosin, W., Brown, W.J., and Rockey, D.D. (2000). A secondary structure motif predictive of protein localization to the chlamydial inclusion membrane. Cell Microbiol 2, 35–47. doi.

Bauler, L.D., and Hackstadt, T. (2014). Expression and Targeting of Secreted Proteins from *Chlamydia trachomatis*. J Bacteriol *196*, 1325–1334. https://doi.org/10.1128/JB.01290-13.

Bernstein, H.D. (2015). Looks can be deceiving: recent insights into the mechanism of protein secretion by the autotransporter pathway. Mol Microbiol 97, 205–215. doi doi:10.1111/mmi.13031.

Betts, H.J., Twiggs, L.E., Sal, M.S., Wyrick, P.B., and Fields, K.A. (2008). Bioinformatic and biochemical evidence for the identification of the type III secretion system needle protein of *Chlamydia trachomatis*. J Bacteriol *190*, 1680–1690. doi.

Betts-Hampikian H.J., and Fields, K.A. (2011). Disulfide bonding within components of the *Chlamydia* type III secretion apparatus correlates with development. J Bacteriol *193*, 6950–6959. https://doi.org/10.1128/JB.05163-11.

Bobay, L.M., and Ochman, H. (2017). The evolution of bacterial genome architecture. Front Genet *8*, 72. https://doi.org/10.3389/fgene.2017.00072.

Boncompain, G., Muller, C., Meas-Yedid, V., Schmitt-Kopplin, P., Lazarow, P.B., and Subtil, A. (2014). The intracellular bacteria *Chlamydia* hijack peroxisomes and utilize their enzymatic capacity to produce bacteria-specific phospholipids. PLoS One *9*, e86196. https://doi.org/10.1371/journal.pone.0086196.

Brinkworth, A.J., Malcolm, D.S., Pedrosa, A.T., Roguska, K., Shahbazian, S., Graham, J.E., Hayward, R.D., and Carabeo, R.A. (2011). *Chlamydia trachomatis* Slc1 is a type III secretion chaperone that enhances the translocation of its invasion effector substrate TARP. Mol Microbiol *82*, 131–144. https://doi.org/10.1111/j.1365-2958.2011.07802.x.

Bulir, D.C., Liang, S., Lee, A., Chong, S., Simms, E., Stone, C., Kaushic, C., Ashkar, A., and Mahony, J.B. (2016). Immunization with chlamydial type III secretion antigens reduces vaginal shedding and prevents fallopian tube pathology following live C. muridarum challenge. Vaccine *34*, 3979–3985. https://doi.org/10.1016/j.vaccine.2016.06.046.

Carpenter, V., Chen, Y.S., Dolat, L., and Valdivia, R.H. (2017). The Effector TepP Mediates Recruitment and Activation of Phosphoinositide 3-Kinase on Early *Chlamydia trachomatis* Vacuoles. mSphere *2*. https://doi.org/10.1128/mSphere.00207-17.

Charro, N., and Mota, L.J. (2015). Approaches targeting the type III secretion system to treat or prevent bacterial infections. Expert Opin Drug Discov *10*, 373–387. https://doi.org/10.1517/17460441.2015.1019860.

Chellas-Géry, B., Linton, C.N., and Fields, K.A. (2007). Human GCIP interacts with CT847, a novel *Chlamydia trachomatis* type III secretion substrate, and is degraded in a tissue-culture infection model. Cellular

Microbiology *9*, 2417–2430. https://doi.org/10.1111/j.1462-5822.2007.00970.x.

Chellas-Gery, B., Wolf, K., Tisoncik, J., Hackstadt, T., and Fields, K.A. (2011). Biochemical and localization analyses of putative type III secretion translocator proteins CopB and CopB2 of *Chlamydia trachomatis* reveal significant distinctions. Infect Immun *79*, 3036–3045. https://doi.org/10.1128.IAI.00159-11.

Chen, D., Lei, L., Lu, C., Flores, R., DeLisa, M.P., Roberts, T.C., Romesberg, F.E., and Zhong, G. (2010). Secretion of the chlamydial virulence factor CPAF requires the Sec-dependent pathway. Microbiology *156*, 3031–3040. https://doi.org/10.1099/mic.0.040527-0.

Chen, Y.-S., Bastidas, R.J., Saka, H.A., Carpenter, V.K., Richards, K.L., Plano, G.V., and Valdivia, R.H. (2014). The *Chlamydia trachomatis* Type III Secretion Chaperone Slc1 Engages Multiple Early Effectors, Including TepP, a Tyrosine-phosphorylated Protein Required for the Recruitment of CrkI-II to Nascent Inclusions and Innate Immune Signalling. PLoS pathogens *10*, e1003954. https://doi.org/10.1371/journal.ppat.1003954.

Cianciotto, N.P., and White, R.C. (2017). Expanding role of type II secretion in bacterial pathogenesis and beyond. Infect Immun *85*, e00014–17. https://doi.org/10.1128/IAI.00014-17.

Clifton, D.R., Fields, K.A., Grieshaber, S.S., Dooley, C.A., Fischer, E.R., Mead, D.J., *et al.* (2004). A chlamydial type III translocated protein is tyrosine-phosphorylated at the site of entry and associated with the recruitment of actin. Proc Natl Acad Sci USA *101*, 10166–10171. doi.

Cocchiaro, J.L., Kumar, Y., Fischer, E.R., Hackstadt, T., and Valdivia, R.H. (2008). Cytoplasmic lipid droplets are translocated into the lumen of the *Chlamydia trachomatis* parasitophorous vacuole. Proc Natl Acad Sci U S A *105*, 9379–9384. doi.

Cossé, M.M., Barta, M.L., Fisher, D.J., Oesterlin, L.K., Niragire, B., Perrinet, S., Millot, G.A., Hefty, P.S., and Subtil, A. (2018). The Loss of Expression of a Single Type 3 Effector (CT622) Strongly Reduces *Chlamydia trachomatis* Infectivity and Growth. Front Cell Inf Microbiol *8*, 145. https://doi.org/10.3389/fcimb.2018.00145.

da Cunha, M., Milho, C., Almeida, F., Pais, S.V., Borges, V., Mauricio, R., Borrego, M.J., Gomes, J.P., and Mota, L.J. (2014). Identification of type III secretion substrates of *Chlamydia trachomatis* using *Yersinia enterocolitica* as a heterologous system. BMC Microbiol *14*, 40. https://doi.org/10.1186/1471-2180-14-40.

da Cunha, M., Pais, S.V., Bugalhão, J.N., and Mota, L.J. (2017). The *Chlamydia trachomatis* type III secretion substrates CT142, CT143, and CT144 are secreted into the lumen of the inclusion. PLoS ONE *12*, e0178856. https://doi.org/10.1371/journal.pone.0178856.s007.

Dehoux, P., Flores, R., Dauga, C., Zhong, G., and Subtil, A. (2011). Multi-genome identification and characterization of chlamydiae-specific type III secretion substrates: the Inc proteins. BMC Genomics *12*, 109. https://doi.org/10.1186/1471-2164-12-109.

Desclozeaux, M., Robbins, A., Jelocnik, M., Khan, S.A., Hanger, J., Gerdts, V., Potter, A., Polkinghorne, A., and Timms, P. (2017). Immunization of a wild koala population with a recombinant Chlamydia pecorum Major Outer Membrane Protein (MOMP) or Polymorphic Membrane Protein (PMP) based vaccine: New insights into immune response, protection and clearance. PLoS One *12*, e0178786. https://doi.org/10.1371/journal.pone.0178786.

Diepold, A., and Armitage, J.P. (2015). Type III secretion systems: the bacterial flagellum and the injectisome. Philos Trans R Soc Lond B Biol Sci *370*. https://doi.org/10.1098/rstb.2015.0020.

Dille, S., Herbst, K., Volceanov, L., Nolke, T., Kretz, O., and Hacker, G. (2014). Golgi fragmentation and sphingomyelin transport to *Chlamydia trachomatis* during penicillin-induced persistence do not depend on the cytosolic presence of the chlamydial protease CPAF. PLoS One *9*, e103220. https://doi.org/10.1371/journal.pone.0103220.

Donati, M., Sambri, V., Comanducci, M., Di Leo, K., Storni, E., Giacani, L., Ratti, G., and Cevenini, R. (2003). DNA immunization with pgp3 gene of *Chlamydia trachomatis* inhibits the spread of chlamydial infection from the lower to the upper genital tract in C3H/HeN mice. Vaccine *21*, 1089–1093. doi.

Dong, F., Flores, R., Chen, D., Luo, J., Zhong, Y., Wu, Z., and Zhong, G. (2006). Localization of the hypothetical protein Cpn0797 in the cytoplasm of *Chlamydia pneumoniae*-infected host cells. Infect Immun *74*, 6479–6486. doi.

Dumoux, M., Clare, D.K., Saibil, H.R., and Hayward, R.D. (2012). *Chlamydiae* assemble a pathogen synapse to hijack the host endoplasmic reticulum. Traffic *13*, 1612–1627. https://doi.org/10.1111/tra.12002.

Dumoux, M., Nans, A., Saibil, H.R., and Hayward, R.D. (2015). Making connections: snapshots of chlamydial type III secretion systems in contact with host membranes. Curr Opin Microbiol *23*, 1–7. https://doi.org/10.1016/j.mib.2014.09.019.

Elwell, C., Mirrashidi, K., and Engel, J. (2016). Chlamydia cell biology and pathogenesis. Nat Rev Microbiol *14*, 385–400. https://doi.org/10.1038/nrmicro.2016.30.

Engel, A.C., Herbst, F., Kerres, A., Galle, J.N., and Hegemann, J.H. (2016). The Type III Secretion System-Related CPn0809 from *Chlamydia pneumoniae*. PLoS One *11*, e0148509. https://doi.org/10.1371/journal.pone.0148509.

Engstrom, P., Nguyen, B.D., Normark, J., Nilsson, I., Bastidas, R.J., Gylfe, A., Elofsson, M., Fields, K.A., Valdivia, R.H., Wolf-Watz, H., *et al.* (2013). Mutations in hemG Mediate Resistance to Salicylidene Acylhydrazides, Demonstrating a Novel Link between Protoporphyrinogen Oxidase (HemG) and *Chlamydia trachomatis* Infectivity. J Bacteriol *195*, 4221–4230. https://doi.org/10.1128/JB.00506-13.

Elwell, C., Mirrashidi, K., and Engel, J. (2016). *Chlamydia* cell biology and pathogenesis. Nat Rev Microbiol *14*, 385–400. https://doi.org/10.1038/nrmicro.2016.30.

Fan, E., Chauhan, N., Udatha, D.B., Leo, J.C., and Linke, D. (2016). Type V secretion systems in bacteria. Microbiol Spectr *4*. https://doi.org/10.1128/microbiolspec.VMBF-0009-2015.

Ferrell, J.C., and Fields, K.A. (2016). A working model for the type III secretion mechanism in *Chlamydia*. Microbes Infect *18*, 84–92. https://doi.org/10.1016/j.micinf.2015.10.006.

Fields, K., and Hackstadt, T. (2000). Evidence for the secretion of *Chlamydia trachomatis* CopN by a type III secretion mechanism. Mol Microbiol *38*, 1048–1060. doi.

Fields, K.A., Fischer, E.R., Mead, D.J., and Hackstadt, T. (2005). Analysis of putative *Chlamydia trachomatis* chaperones Scc2 and Scc3 and their use in the identification of type III secretion substrates. J Bacteriol *187*, 6466–6478. https://doi.org/10.1128/jb.187.18.6466-6478.2005.

Fling, S.P., Sutherland, R.A., Steele, L.N., Hess, B., D'Orazio, S.E., Maisonneuve, J., Lampe, M.F., Probst, P., and Starnbach, M.N. (2001). CD8+ T cells recognize an inclusion membrane-associated protein from the vacuolar pathogen *Chlamydia trachomatis*. Proc Natl Acad Sci U S A *98*, 1160–1165. doi.

Fields, K.A., Fischer, E.R., Mead, D.J., and Hackstadt, T. (2005). Analysis of putative *Chlamydia trachomatis* chaperones Scc2 and Scc3 and their use in the identification of type III secretion substrates. J Bacteriol *187*, 6466–6478. https://doi.org/10.1128/jb.187.18.6466-6478.2005.

Fischer, A., Harrison, K.S., Ramirez, Y., Auer, D., Chowdhury, S.R., Prusty, B.K., Sauer, F., Dimond, Z., Kisker, C., Scott Hefty, P., *et al.* (2017). *Chlamydia trachomatis*-containing vacuole serves as deubiquitination platform to stabilize Mcl-1 and to interfere with host defence. eLife *6*. https://doi.org/10.7554/eLife.21465.

Fling, S.P., Sutherland, R.A., Steele, L.N., Hess, B., D'Orazio, S.E., Maisonneuve, J., Lampe, M.F., Probst, P., and Starnbach, M.N. (2001). CD8+ T cells recognize an inclusion membrane-associated protein from the vacuolar pathogen *Chlamydia trachomatis*. Proc Natl Acad Sci U S A *98*, 1160–1165. doi.

Galan, J.E., Lara-Tejero, M., Marlovits, T.C., and Wagner, S. (2014). Bacterial type III secretion systems: specialized nanomachines for protein delivery into target cells. Annu Rev Microbiol *68*, 415–438. https://doi.org/10.1146/annurev-micro-092412-155725.

Galán, J.E., Lara-Tejero, M., Marlovits, T.C., and Wagner, S. (2014). Bacterial Type III Secretion Systems: Specialized Nanomachines for Protein Delivery into Target Cells. Annu Rev Microbiol *68*, 415–438. https://doi.org/10.1146/annurev-micro-092412-15572.

Galan, J.E., and Waksman, G. (2018). Protein-injection machines in bacteria. Cell *172*, 1306–1318. https://doi.org/10.1016/j.cell.2018.01.034.

Gauliard, E., Ouellette, S.P., Rueden, K.J., and Ladant, D. (2015). Characterization of interactions between inclusion membrane proteins from *Chalmydia trachomatis*. Front Cell Infect Microbiol 5, 13. https://doi.org/10.3389/fcimb.2015.00013.

Gehre, L., Gorgette, O., Perrinet, S., Prevost, M.C., Ducatez, M., Giebel, A.M., Nelson, D.E., Ball, S.G., and Subtil, A. (2016). Sequestration of host metabolism by an intracellular pathogen. Elife *5*, e12552. https://doi.org/10.7554/eLife.12552.

Giles, D.K., Whittimore, J.D., LaRue, R.W., Raulston, J.E., and Wyrick, P.B. (2006). Ultrastructural analysis of chlamydial antigen-containing vesicles everting from the *Chlamydia trachomatis* inclusion. Microbes Infect *8*, 1579–1591. https://doi.org/10.1016/j.micinf.2006.01.018.

Gloeckl, S., Ong, V.A., Patel, P., Tyndall, J.D., Timms, P., Beagley, K.W., Allan, J.A., Armitage, C.W., Turnbull, L., Whitchurch, C.B., *et al.* (2013). Identification of a serine protease inhibitor which causes inclusion vacuole reduction and is lethal to *Chlamydia trachomatis*. Mol Microbiol *89*, 676–689. https://doi.org/10.1111/mmi.12306.

Gong, S., Lei, L., Chang, X., Belland, R., and Zhong, G. (2011). *Chlamydia trachomatis* secretion of hypothetical protein CT622 into host cell cytoplasm via a secretion pathway that can be inhibited by the type III secretion system inhibitor compound 1. Microbiology *157*, 1134–1144. https://doi.org/10.1099/mic.0.047746-0.

Grishin, A.V., Luyksaar, S.I., Kapotina, L.N., Kirsanov, D.D., Zayakin, E.S., Karyagina, A.S., and Zigangirova, N.A. (2018). Identification of chlamydial T3SS inhibitors through virtual screening against T3SS ATPase. Chemical biology & drug design *91*, 717–727. https://doi.org/10.1111/cbdd.13130.

Guerrero-Mandujano, A., Hernandez-Cortez, C., Ibarra, J.A., and Castro-Escarpulli, G. (2017). The outer membrane vesicles: Secretion system type zero. Traffic *18*, 425–432. https://doi.org/10.1111/tra.12488.

Hanson, B.R., Slepenkin, A., Peterson, E.M., and Tan, M. (2015). *Chlamydia trachomatis* Type III Secretion Proteins Regulate Transcription. J Bacteriol *197*, 3238–3244. https://doi.org/10.1128/jb.00379-15.

Heuer, D., Brinkmann, V., Meyer, T.F., and Szczepek, A.J. (2003). Expression and translocation of chlamydial protease during acute and persistent infection of the epithelial HEp-2 cells with Chlamydophila (Chlamydia) pneumoniae. Cellular Microbiology 5, 315–322. doi.

Hobolt-Pedersen, A.-S., Christiansen, G., Timmerman, E., Gevaert, K., and Birkelund, S. (2009). Identification of *Chlamydia trachomatis* CT621, a protein delivered through the type III secretion system to the host cell cytoplasm and nucleus. FEMS Immunol Med Microbiol 57, 46–58. https://doi.org/10.1111/j.1574-695X.2009.00581.x.

Horn, M., Collingro, A., Schmitz-Esser, S., Beier, C.L., Purkhold, U., Fartmann, B. *et al.* (2004). Illuminating the evolutionary history of *Chlamydiae*. Science *304*, 728–730. doi.

Hou, S., Dong, X., Yang, Z., Li, Z., Liu, Q., and Zhong, G. (2015). Chlamydial plasmid-encoded virulence factor Pgp3 neutralizes the antichlamydial activity of human cathelicidin LL-37. Infect Immun, IAI.00746–00715. https://doi.org/10.1128/IAI.00746-15.

Hovis, K.M., Mojica, S., McDermott, J.E., Pedersen, L., Simhi, C., Rank, R.G., Myers, G.S., Ravel, J., Hsia, R.C., and Bavoil, P.M. (2013). Genus-optimized strategy for the identification of chlamydial type III secretion substrates. Pathog Dis *69*, 213–222. https://doi.org/10.1111/2049-632x.12070.

Hower, S., Wolf, K., and Fields, K.A. (2009). Evidence that CT694 is a novel *Chlamydia trachomatis* T3S substrate capable of functioning during invasion or early cycle development. Mol Microbiol 72, 1423–1437. doi.

Hueck, C.J., (1998). Type III protein secretion systems in bacterial pathogens of animals and plants. Microbiol Mol Biol Rev *62*, 379–433. doi.

Jamison, W.P., and Hackstadt, T., (2008). Induction of type III secretion by cell-free *Chlamydia trachomatis* elementary bodies. Microb Pathog *45*, 435–440. https://doi.org/10.1016/j.micpath.2008.10.002.

Jan, A.T. (2017). Outer Membrane Vesicles (OMVs) of Gram-negative Bacteria: A Perspective Update. Front Microbiol *8*, 1053. https://doi.org/10.3389/fmicb.2017.01053.

Jaumouillé, V., Francetic, O., Sansonetti, P.J., and Tran Van Nhieu, G. (2008). Cytoplasmic targeting of IpaC to the bacterial pole directs polar type III secretion in Shigella. The EMBO Journal *27*, 447–457. https://doi.org/10.1038/sj.emboj.7601976.

Jorgensen, I., and Valdivia, R.H. (2008). Pmp-Like Proteins Pls1 and Pls2 Are Secreted into the Lumen of the *Chlamydia trachomatis* Inclusion. Infect Immun *76*, 3940–3950. https://doi.org/10.1128/IAI.00632-08.

Johnson, D.L., Stone, C.B., Bulir, D.C., Coombes, B.K., and Mahony, J.B. (2009). A novel inhibitor of Chlamydophila pneumoniae protein kinase D (PknD) inhibits phosphorylation of CdsD and suppresses bacterial replication. BMC Microbiol *9*, 218. https://doi.org/10.1186/1471-2180-9-218.

Koroleva, E.A., Kobets, N.V., Shcherbinin, D.N., Zigangirova, N.A., Shmarov, M.M., Tukhvatulin, A.I., Logunov, D.Y., Naroditsky, B.S., and Gintsburg, A.L. (2017). Chlamydial Type III Secretion System Needle Protein Induces Protective Immunity against Chlamydia muridarum Intravaginal Infection. BioMed research international *2017*, 3865802. https://doi.org/10.1155/2017/3865802.

Korotkov, K.V., Sandkvist, M., and Hol, W.G.J. (2012). The type II secretion system: biogenesis, molecular architecture and mechanism. Nature Rev Microbiol *10*, 336. https://doi.org/10.1038/nrmicro2762.

Kumar, Y., Cocchiaro, J., and Valdivia, R.H. (2006). The obligate intracellular pathogen *Chlamydia trachomatis* targets host lipid droplets. Curr Biol *16*, 1646–1651. doi.

Lafont, F., and van der Goot, F.G. (2005). Oiling the key hole. Mol Microbiol *56*, 575–7. doi.

Lee, J.K., Enciso, G.A., Boassa, D., Chander, C.N., Lou, T.H., Pairawan, S.S., Guo, M.C., Wan, F.Y.M., Ellisman, M.H., Sütterlin, C., *et al.* (2017). Replication-dependent size reduction precedes differentiation in *Chlamydia trachomatis*. Nature Comm, 1–9. https://doi.org/10.1038/s41467-017-02432-0.

Lei, L., Dong, X., Li, Z., and Zhong, G. (2013). Identification of a Novel Nuclear Localization Signal Sequence in *Chlamydia trachomatis*-Secreted Hypothetical Protein CT311. PLoS ONE *8*, e64529. https://doi.org/10.1371/journal.pone.0064529.g006.

Lei, L., Qi, M., Budrys, N., Schenken, R., and Zhong, G. (2011). Localization of *Chlamydia trachomatis* hypothetical protein CT311 in host cell cytoplasm. Microb Pathog *51*, 101–109. https://doi.org/10.1016/j.micpath.2011.05.002.

Li, Z., Chen, C., Chen, D., Wu, Y., Zhong, Y., and Zhong, G. (2008). Characterization of fifty putative inclusion membrane proteins encoded in the *Chlamydia trachomatis* genome. Infect Immun *76*, 2746–2757. doi.

Li, Z., Chen, D., Zhong, Y., Wang, S., and Zhong, G. (2008). The Chlamydial Plasmid-Encoded Protein pgp3 Is Secreted into the Cytosol of Chlamydia-Infected Cells. Infect Immun *76*, 3415–3428. https://doi.org/10.1128/IAI.01377-07.

Lorenzini, E., Singer, A., Singh, B., Lam, R., Skarina, T., Chirgadze, N.Y., Savchenko, A., and Gupta, R.S. (2010). Structure and protein–protein interaction studies on *Chlamydia trachomatis* protein CT670 (YscO Homologue). J Bacteriol *192*, 2746–2756. https://doi.org/10.1128/jb.01479-09.

Lu, C., Lei, L., Peng, B., Tang, L., Ding, H., Gong, S., Li, Z., Wu, Y., and Zhong, G. (2013). *Chlamydia trachomatis* GlgA is secreted into host cell cytoplasm. PLoS One *8*, e68764. https://doi.org/10.1371/journal.pone.0068764.

Lutter, E.I., Martens, C., and Hackstadt, T. (2012). Evolution and Conservation of Predicted Inclusion Membrane Proteins in Chlamydiae. Comparative and Functional Genomics *2012*, 1–13. https://doi.org/10.1155/2012/362104.

Maffei, B., Francetic, O., and Subtil, A. (2017). Tracking Proteins Secreted by Bacteria: What's in the Toolbox? Front Cell Infect Microbiol *7*, 221. https://doi.org/10.3389/fcimb.2017.00221.

Markham, A.P., Jaafar, Z.A., Kemege, K.E., Middaugh, C.R., and Hefty, P.S. (2009). Biophysical characterization of *Chlamydia trachomatis* CT584 supports its potential role as a type III secretion needle tip protein. Biochemistry (Mosc) *48*, 10353–10361. https://doi.org/10.1021/bi901200y.

Marsh, J.W., Wee, B.A., Tyndall, J.D., Lott, W.B., Bastidas, R.J., Caldwell, H.D., Valdivia, R.H., Kari, L., and Huston, W.M. (2015). A *Chlamydia trachomatis* strain with a chemically generated amino acid substitution (P370L) in the cthtrA gene shows reduced elementary body production. BMC Microbiol *15*, 194. https://doi.org/10.1186/s12866-015-0533-2.

Matsumoto, A., (1981). Electron microscopic observations of surface projections and related intracellular structures of *Chlamydia* organisms. J Electron Microsc *30*, 315–320.

McKuen, M.J., Mueller, K.E., Bae, Y.S., and Fields, K.A. (2017). Fluorescence-Reported Allelic Exchange Mutagenesis Reveals a Role for *Chlamydia trachomatis* TmeA in Invasion That Is Independent of Host AHNAK. Infect Immun *85*. https://doi.org/10.1128/iai.00640-17.

Meoni, E., Faenzi, E., Frigimelica, E., Zedda, L., Skibinski, D., Giovinazzi, S., Bonci, A., Petracca, R., Bartolini, E., Galli, G., *et al.* (2009). CT043, a protective antigen that induces a CD4+ Th1 response during *Chlamydia trachomatis* infection in mice and humans. Infect Immun *77*, 4168–4176. https://doi.org/10.1128/IAI.00344-09.

Misaghi, S., Balsara, Z.R., Catic, A., Spooner, E., Ploegh, H.L., and Starnbach, M.N. (2006). *Chlamydia trachomatis*-derived deubiquitinating enzymes in mammalian cells during infection. Mol Microbiol *61*, 142–150. doi.

Mojica, S.A., Hovis, K.M., Frieman, M.B., Tran, B., Hsia, R.-C., Ravel, J., Jenkins-Houk, C., Wilson, K.L., and Bavoil, P.M. (2015). SINC, a type III secreted protein of Chlamydia psittaci, targets the inner nuclear membrane of infected cells and uninfected neighbors. Mol Biol Cell *26*, 1918–1934. https://doi.org/10.1091/mbc.E14-11-1530.

Mueller, K.E., and Fields, K.A. (2015). Application of β-Lactamase Reporter Fusions as an Indicator of Effector Protein Secretion during Infections with the Obligate Intracellular Pathogen *Chlamydia trachomatis*. PloS One *10*, e0135295. https://doi.org/10.1371/journal.pone.0135295.

Mueller, K.E., Plano, G.V., and Fields, K.A. (2014). New frontiers in type III secretion biology: the *Chlamydia* perspective. Infect Immun *82*, 2–9. https://doi.org/10.1128/IAI.00917-13.

Müller, N., Sattelmacher, F., Lugert, R., and Gross, U. (2008). Characterization and intracellular localization of putative *Chlamydia pneumoniae* effector proteins. Med Microbiol Immunol *197*, 387–396. https://doi.org/10.1007/s00430-008-0097-y.

Muller, T., Becker, E., Stallmann, S., Waldhuber, A., Rommler-Dreher, F., Albrecht, S., Mohr, F., Hegemann, J.H., and Miethke, T. (2017). Vaccination with the polymorphic membrane protein A reduces Chlamydia muridarum induced genital tract pathology. Vaccine 35, 2801–2810. https://doi.org/10.1016/j.vaccine.2017.04.017.

Murthy, A.K., Guentzel, M.N., Zhong, G., and Arulanandam, B.P. (2009). Chlamydial protease-like activity factor--insights into immunity and vaccine development. J Reprod Immunol *83*, 179–184. https://doi.org/10.1016/j.jri.2009.05.007.

Muschiol, S., Boncompain, G., Vromman, F., Dehoux, P., Normark, S., Henriques-Normark, B., and Subtil, A. (2011). Identification of a family of type III secreted effectors conserved in pathogenic *Chlamydiae*. Infect Immun *79*, 571–580. doi.

Nans, A., Saibil, H.R., and Hayward, R.D. (2014). Pathogen-host reorganisation during *Chlamydia* invasion revealed by cryo-electron tomography. Cell Microbiol *16*, 1457–1472. doi: 10.1111/cmi.12310.

Nans, A., Kudryashev, M., Saibil, H.R., and Hayward, R.D. (2015). Structure of a bacterial type III secretion system in contact with a host membrane *in situ*. Nat Comm *6*, 10114. https://doi.org/10.1038/ncomms10114.

Nguyen, B.D., and Valdivia, R.H. (2012). Virulence determinants in the obligate intracellular pathogen *Chlamydia trachomatis* revealed by forward genetic approaches. Proc Natl Acad Sci U S A *109*, 1263–1268. https://doi.org/10.1073/pnas.1117884109.

O'Meara, C.P., Armitage, C.W., Andrew, D.W., Kollipara, A., Lycke, N.Y., Potter, A.A., Gerdts, V., Petrovsky, N., and Beagley, K.W. (2017). Multistage vaccines containing outer membrane, type III secretion system and inclusion membrane proteins protects against a Chlamydia genital tract infection and pathology. Vaccine *35*, 3883–3888. https://doi.org/10.1016/j.vaccine.2017.05.063.

Pais, S.V., Milho, C., Almeida, F., and Mota, L.J. (2013). Identification of novel type III secretion chaperone-substrate complexes of *Chlamydia trachomatis*. PLoS One *8*, e56292. https://doi.org/10.1371/journal.pone.0056292.

Pal, S., Favaroni, A., Tifrea, D.F., Hanisch, P.T., Luczak, S.E.T., Hegemann, J.H., and de la Maza, L.M. (2017). Comparison of the nine polymorphic membrane proteins of *Chlamydia trachomatis* for their ability to induce protective immune responses in mice against a C. muridarum challenge. Vaccine *35*, 2543–2549. https://doi.org/10.1016/j.vaccine.2017.03.070.

Parsot, C., Hamiaux, C., and Page, A.L. (2003). The various and varying roles of specific chaperones in type III secretion systems. Curr Opin Microbiol *6*, 7–14. doi.

Patton, M.J., McCorrister, S., Grant, C., Westmacott, G., Fariss, R., Hu, P., Zhao, K., Blake, M., Whitmire, B., Yang, C., *et al.* (2016). Chlamydial Protease-Like Activity Factor and Type III Secreted Effectors Cooperate in Inhibition of p65 Nuclear Translocation. MBio 7. https://doi.org/10.1128/mBio.01427-16.

Patton, M.J., Chen, C.Y., Yang, C., McCorrister, S., Grant, C., Westmacott, G., Yuan, X.Y., Ochoa, E., Fariss, R., Whitmire, W.M., *et al.* (2018). Plasmid Negative Regulation of CPAF Expression Is Pgp4 Independent and Restricted to Invasive *Chlamydia trachomatis* Biovars. MBio *9*. https://doi.org/10.1128/mBio.02164-17.

Pennini, M.E., Perrinet, S., Dautry-Varsat, A., and Subtil, A. (2010). Histone methylation by NUE, a novel nuclear effector of the intracellular pathogen *Chlamydia trachomatis*. PLoS Pathog *6*, e1000995. https://doi.org/10.1371/journal.ppat.1000995.

Peters, J., Wilson, D.P., Myers, G., Timms, P., and Bavoil, P.M. (2007). Type III secretion a la *Chlamydia*. Trends Microbiol *15*, 241–251. doi.

Pilhofer, M., Aistleitner, K., Ladinsky, M.S., Koenig, L., Horn, M., and Jensen, G.J. (2014). Architecture and host interface of environmental *Chlamydiae* revealed by electron cryotomography. Environ Microbiol *16*, 417–429. https://doi.org/10.1111/1462-2920.12299.

Prusty, B.K., Chowdhury, S.R., Gulve, N., and Rudel, T. (2018). Peptidase Inhibitor 15 (PI15) Regulates Chlamydial CPAF Activity. Front Cell Infect Microbiol *8*, 183. https://doi.org/10.3389/fcimb.2018.00183.

Qi, M., Lei, L., Gong, S., Liu, Q., DeLisa, M.P., and Zhong, G. (2011). *Chlamydia trachomatis* Secretion of an Immunodominant Hypothetical Protein (CT795) into Host Cell Cytoplasm. J Bacteriol *193*, 2498–2509. https://doi.org/10.1128/JB.01301-10.

Rao, X., Deighan, P., Hua, Z., Hu, X., Wang, J., Luo, M., Wang, J., Liang, Y., Zhong, G., Hochschild, A., *et al.* (2009). A regulator from *Chlamydia trachomatis* modulates the activity of RNA polymerase through direct interaction with the beta subunit and the primary sigma subunit. Genes Dev *23*, 1818–1829. https://doi.org/10.1101/gad.1784009.

Riordan, K.E., Sorg, J.A., Berube, B.J., and Schneewind, O. (2008). Impassable YscP substrates and their impact on the Yersinia enterocolitica type III secretion pathway. J Bacteriol *190*, 6204–6216. https://doi.org/10.1128/jb.00467-08.

Rucks, E.A., Olson, M.G., Jorgenson, L.M., Srinivasan, R.R., and Ouellette, S.P. (2017). Development of a Proximity Labelling System to Map the *Chlamydia trachomatis*

Inclusion Membrane. Front Cell Inf Microbiol 7, 949. https://doi.org/10.1371/journal.pone.0076743.

Saka, H.A., Thompson, J.W., Chen, Y.S., Kumar, Y., Dubois, L.G., Moseley, M.A., and Valdivia, R.H. (2011). Quantitative proteomics reveals metabolic and pathogenic properties of *Chlamydia trachomatis* developmental forms. Mol Microbiol *82*, 1185–1203. https://doi.org/10.1111/j.1365-2958.2011.07877.x.

Sandoz, K.M., Eriksen, S.G., Jeffrey, B.M., Suchland, R.J., Putman, T.E., Hruby, D.E., Jordan, R., and Rockey, D.D. (2012). Resistance to a Novel Antichlamydial Compound Is Mediated through Mutations in *Chlamydia trachomatis* secY. Antimicrob Agents Chemother *56*, 4296–4302. https://doi.org/10.1128/AAC.00356-12.

Segers, K., and Anne, J. (2011). Traffic jam at the bacterial sec translocase: targeting the SecA nanomotor by small-molecule inhibitors. Chem Biol *18*, 685–698. https://doi.org/10.1016/j.chembiol.2011.04.007.

Shen, L., Macnaughtan, M.A., Frohlich, K.M., Cong, Y., Goodwin, O.Y., Chou, C.W., LeCour, L., Jr., Krup, K., Luo, M., and Worthylake, D.K. (2015). Multipart Chaperone-Effector Recognition in the Type III Secretion System of *Chlamydia trachomatis*. J Biol Chem *290*, 28141–28155. https://doi.org/10.1074/jbc.M115.670232.

Silva-Herzog, E., Joseph, S.S., Avery, A.K., Coba, J.A., Wolf, K., Fields, K.A., and Plano, G.V. (2011). Scc1 (CP0432) and Scc4 (CP0033) function as a type III secretion chaperone for CopN of *Chlamydia pneumoniae*. J Bacteriol *193*, 3490–3496. https://doi.org/10.1128/jb.00203-11.

Slepenkin, A., de la Maza, L.M., and Peterson, E.M. (2005). Interaction between components of the type III secretion system of Chlamydiaceae. J Bacteriol *187*, 473–479. https://doi.org/10.1128/jb.187.2.473-479.2005.

Slepenkin, A., Enquist, P.A., Hagglund, U., de la Maza, L.M., Elofsson, M., and Peterson, E.M. (2007). Reversal of the antichlamydial activity of putative type III secretion inhibitors by iron. Infect Immun 75, 3478–3489. doi.

Snavely, E.A., Kokes, M., Dunn, J.D., Saka, H.A., Nguyen, B.D., Bastidas, R.J., McCafferty, D.G., and Valdivia, R.H. (2014). Reassessing the role of the secreted protease CPAF in *Chlamydia trachomatis* infection through genetic approaches. Pathog Dis *71*, 336–351. https://doi.org/10.1111/2049-632x.12179.

Spaeth, K.E., Chen, Y.-S., and Valdivia, R.H. (2009). The Chlamydia type III secretion system C-ring engages a chaperone-effector protein complex. PLoS Pathog *5*, e1000579. https://doi.org/10.1371/journal.ppat.1000579.

Starnbach, M.N., Loomis, W.P., Ovendale, P., Regan, D., Hess, B., Alderson, M.R., and Fling, S.P. (2003). An inclusion membrane protein from *Chlamydia trachomatis* enters the MHC class I pathway and stimulates a CD8+ T cell response. J Immunol *171*, 4742–4749. doi.

Stone, C.B., Bulir, D.C., Gilchrist, J.D., Toor, R.K., and Mahoney, J.B. (2010). Interactions between flagellar and type III secretion proteins in *Chlamydia pneumoniae*. BMC Microbiol *10*, 18. https://doi.org/10.1186/1471-2180-10-18.

Stone, C.B., Sugiman-Marangos, S., Bulir, D.C., Clayden, R.C., Leighton, T.L., Slootstra, J.W., Junop, M.S., and Mahony, J.B. (2012). Structural characterization of a novel *Chlamydia pneumoniae* type III secretion-associated protein, Cpn0803. PLoS One *7*, e30220. https://doi.org/10.1371/journal.pone.0030220.

Subtil, A., Blocker, A., and Dautry-Varsat, A. (2000). Type III secretion system in< i> Chlamydia</i> species: identified members and candidates. Microbes and Infection 2, 367–369. doi.

Subtil, A., Delevoye, C., Balañá, M.E., Tastevin, L., Perrinet, S., and Dautry-Varsat, A. (2005). A directed screen for chlamydial proteins secreted by a type III mechanism identifies a translocated protein and numerous other new candidates. Mol Microbiol 56, 1636–1647. doi.

Tammiruusu, A., Penttila, T., Lahesmaa, R., Sarvas, M., Puolakkainen, M., and Vuola, J.M. (2007). Intranasal administration of chlamydial outer protein N (CopN) induces protection against pulmonary *Chlamydia pneumoniae* infection in a mouse model. Vaccine 25, 283–290. https://doi.org/10.1016/j.vaccine.2006.07.043.

Tejeda-Dominguez, F., Huerta-Cantillo, J., Chavez-Dueñas, L., and Navarro-Garcia, F. (2017). A Novel Mechanism for Protein Delivery by the Type 3 Secretion System for Extracellularly Secreted Proteins. mBio *8*. https://doi.org/10.1128/mBio.00184-17.

Thalmann, J., Janik, K., May, M., Sommer, K., Ebeling, J., Hofmann, F., Genth, H., and Klos, A. (2010). Actin Re-Organization Induced by *Chlamydia trachomatis* Serovar D – Evidence for a Critical Role of the Effector Protein CT166 Targeting Rac. PLoS ONE *5*, e9887. https://doi.org/10.1371/journal.pone.0009887.s006.

Vandahl, B.B., Stensballe, A., Roepstorff, P., Christiansen, G., and Birkelund, S. (2005). Secretion of Cpn0796 from *Chlamydia pneumoniae* into the host cell cytoplasm by an autotransporter mechanism. Cell Microbiol 7, 825–836. https://doi.org/10.1111/j.1462-5822.2005.00514.x.

Wang, J., Chen, L., Chen, F., Zhang, X., Zhang, Y., Baseman, J., Perdue, S., Yeh, I.T., Shain, R., Holland, M., *et al.* (2009). A chlamydial type III-secreted effector protein (Tarp) is predominantly recognized by antibodies from humans infected with *Chlamydia trachomatis* and induces protective immunity against upper genital tract pathologies in mice. Vaccine *27*, 2967–2980. https://doi.org/10.1016/j.vaccine.2009.02.095.

Wang, X., Hybiske, K., and Stephens, R.S. (2018). Direct visualization of the expression and localization of chlamydial effector proteins within infected host cells. Pathog Dis *76*. https://doi.org/10.1093/femspd/fty011.

Weber, M.M., Bauler, L.D., Lam, J., and Hackstadt, T. (2015). Expression and Localization of Predicted Inclusion Membrane Proteins in *Chlamydia trachomatis*. Infect Immun *83*, 4710–4718. https://doi.org/10.1128/iai.01075-15.

Wilson, D.P., Whittum-Hudson, J.A., Timms, P., and Bavoil, P.M. (2009). Kinematics of intracellular chlamydiae provide evidence for contact-dependent development. J Bacteriol *191*, 5734–5742. https://doi.org/10.1128/jb.00293-09.

Wu, X., Lei, L., Gong, S., Chen, D., Flores, R., and Zhong, G. (2011). The chlamydial periplasmic stress response serine protease cHtrA is secreted into host cell cytosol. BMC Microbiol *11*, 87. https://doi.org/10.1186/1471-2180-11-87.

Wyrick, P.B., Knight, S.T., Paul, T.R., Rank, R.G., and Barbier, C.S. (1999). Persistent chlamydial envelope antigens in antibiotic-exposed infected cells trigger neutrophil chemotaxis. J Infect Dis *179*, 954–966. https://doi.org/10.1086/314676.

Yu, H.H., and Tan, M. (2003). Sigma28 RNA polymerase regulates *hctB,* a late developmental gene in *Chlamydia*. Mol Microbiol *50*, 577–584. doi.

Zhong, G., Fan, P., Ji, H., Dong, F., and Huang, Y. (2001). Identification of a chlamydial protease-like activity factor responsible for the degradatin of host transcription factors. J Exp Med *193*, 935–942. doi.

Zigangirova, N.A., Kost, E.A., Didenko, L.V., Kapotina, L.N., Zayakin, E.S., Luyksaar, S.I., Morgunova, E.Y., Fedina, E.D., Artyukhova, O.A., Samorodov, A.V., *et al.* (2016). A small-molecule compound belonging to a class of 2,4-disubstituted 1,3,4-thiadiazine-5-ones inhibits intracellular growth and persistence of *Chlamydia trachomatis*. J Med Microbiol *65*, 91–98. https://doi.org/10.1099/jmm.0.000189.

Zrieq, R., Braun, C., and Hegemann, J.H. (2017). The *Chlamydia pneumoniae* Tarp Orthologue CPn0572 Stabilizes Host F-Actin by Displacement of Cofilin. Front Cell Infect Microbiol *7*, 511. https://doi.org/10.3389/fcimb.2017.00511

The Chlamydial Protease CPAF

8

Julie A. Brothwell[1], Christine Sütterlin[2*], Thomas Rudel[3] and Ming Tan[4]

[1]Department of Microbiology and Molecular Genetics, University of California, Irvine, CA, USA.
[2]Department of Developmental and Cell Biology, University of California, Irvine, CA, USA.
[3]Department of Microbiology, University of Würzburg Biocenter, Würzburg, Germany.
[4]Departments of Microbiology and Molecular Genetics, and Medicine, University of California, Irvine, CA, USA.

*Correspondence: suetterc@uci.edu

https://doi.org/10.21775/9781912530281.08

Abstract

CPAF is a conserved chlamydial protease, but our understanding of its role in a *Chlamydia* infection is incomplete and evolving. CPAF is expressed as an inactive enzyme that requires secretion, homodimerization, and proteolytic processing for conversion into the active protease. It cleaves or degrades many chlamydial and host proteins, but it has been challenging to distinguish *in vivo* substrates that are true biological targets from *in vitro* substrates that do not undergo proteolysis during a *Chlamydia* infection. The subcellular localization of active CPAF in a *Chlamydia*-infected cell is also controversial because it is uncertain whether it is primarily localized in the chlamydial inclusion during infection or also present in the host cytoplasm. CPAF activity must be regulated to account for the observed differences between *in vitro* substrate susceptibility and *in vivo* proteolysis, but the control mechanisms have not been completely defined. Multiple functions have been proposed for CPAF. The strongest data indicate that CPAF modulates the host immune response by down-regulating the type I interferon response and by inhibiting neutrophil activation. *In vivo*, CPAF is necessary to maintain the *Chlamydia* infection in the genital tract of mice. CPAF is also a protective antigen that is being studied as a vaccine candidate. Thus, despite the controversy surrounding the true substrates of CPAF, CPAF remains an important virulence factor of *Chlamydia*.

Introduction

Chlamydial proteasome/protease-like activity factor (CPAF) is a conserved *Chlamydia*-specific protease that is encoded by all sequenced *Chlamydia* spp. (Dong *et al.*, 2005; Shaw *et al.*, 2002). The gene for CPAF is also found in some *Chlamydia*-related bacteria, such as *Waddlia* and the *Parachlamydiaceae*, but not in *Simkania* (Collingro *et al.*, 2011). However, orthologues of CPAF are not present in other bacteria (Zhong *et al.*, 2001).

Since its discovery in 2001 (Zhong *et al.*, 2001), CPAF has gained notoriety as an increasing number of host and then chlamydial proteins were identified as substrates. This in turn fuelled efforts to understand its function, localization within an infected cell and regulation. However, the discovery of CPAF activity in lysates of infected cells revealed that the cleavage or degradation of these proteins had not always occurred within a *Chlamydia*-infected cell (Chen *et al.*, 2012). Thus, the challenge is to identify the substrates that undergo proteolysis during the intracellular infection and to elucidate the biological significance of this proteolysis.

The protease

Chlamydia encodes more than two dozen proteases (Zhong, 2009), but more work has been done to study CPAF than all the other proteases combined. CPAF was first identified by Guangming Zhong and colleagues as a chlamydial protease that degraded two MHC transcriptional regulators in the host cell (Zhong *et al.*, 2001; Zhong *et al.*, 1999; Zhong *et al.*, 2000). As will be discussed, CPAF has since been found to cleave or degrade multiple host and chlamydial proteins. However, the interpretation of these results is challenging because of the paucity of evidence for cleavage or degradation of susceptible proteins within a *Chlamydia*-infected cell (Chen *et al.*, 2012).

CPAF is an atypical serine protease. The catalytic site is formed by H105 and S499 and a water molecule that is coordinated by H105 and E558 (Chen *et al.*, 2009; Huang *et al.*, 2008). BLAST analysis showed that CPAF shares homology with the superfamily of Tail-specific proteases. The catalytic triad itself is most similar in structure to the active site of tricorn proteases, which are serine proteases that form hexamers and act as proteasomes (Chen *et al.*, 2009). However, CPAF is resistant to the serine protease inhibitors found in standard, commercial protease inhibitor cocktails, such as PMSF (Jorgensen *et al.*, 2011; Zhong *et al.*, 2000), and, for this reason, is considered to be an atypical serine protease.

CPAF processing

CPAF is produced and proteolytically processed within a *Chlamydia*-infected host cell. CPAF is first expressed during the midstage of the developmental cycle (Belland *et al.*, 2003; Dong *et al.*, 2004b). The CPAF gene (CT858 in *Chlamydia trachomatis* serovar D) encodes a 70kDa protein, but full-length CPAF is not usually detected because it is rapidly cleaved as early as 16 hours post infection into a 29kDa N-terminal fragment (CPAFn) and a 35kDa C-terminal fragment (CPAFc) (Dong *et al.*, 2004a; Dong *et al.*, 2004b; Shaw *et al.*, 2002). CPAF does not appear to be cleaved by any of the other 28 predicted proteases in *C. trachomatis* (Chen *et al.*, 2010a). Instead, CPAF mediates its own proteolytic processing from the inactive proenzyme, or zymogen, into the active enzyme.

The events that lead to autoprocessing have been delineated in a series of biochemical and structural studies by Guangming Zhong and colleagues (Chen *et al.*, 2010a; Dong *et al.*, 2004a; Huang *et al.*, 2008; Yang *et al.*, 2015). Their data strongly support a model in which two molecules of full-length CPAF homodimerize when in close proximity and proteolytically cleave each other to produce the active enzyme (Fig. 8.1) (Dong *et al.*, 2004a; Huang *et al.*, 2008). The processing step removes a 40 amino acid internal inhibitory peptide, which otherwise occludes the active catalytic site, and produces the two main CPAF fragments, which remain associated as a heterodimer (Huang *et al.*, 2008). Active CPAF is thus a pair of two heterodimers. This model of proximity-based homodimerization is supported by the efficient processing of ectopically expressed CPAF with an N-terminal oligomerization domain in uninfected cells (Paschen *et al.*, 2008).

CPAF autoprocessing is coupled to its secretion out of the bacterium. Secretion and processing of CPAF were both altered in a *C. trachomatis* strain defective for type 2 secretion (T2S) system, which is conserved machinery used by Gram-negative bacteria to secrete folded proteins from the periplasm to the external environment (see Chapter 7, Protein Secretion in *Chlamydia*, for details on *Chlamydia* T2S) (Costa *et al.*, 2015). This strain, which has a mutation in the T2S ATPase, produced no CPAF cleavage fragments or CPAF activity, although there was a small amount of a protein corresponding to the size of full-length CPAF (Snavely *et al.*, 2014). CPAF was only detected in the chlamydiae within the inclusion and was not seen in the host cytoplasm. These differences from the parental strain indicate that secretion of CPAF is necessary for its processing and activation.

CPAF expression and processing were both downregulated during *Chlamydia* cell culture persistence, which is an altered growth state. The CPAF gene was not transcribed during *Chlamydia* persistence induced by interferon-γ or ampicillin (Rodel *et al.*, 2012; Wang *et al.*, 2011). CPAF protein levels were also lower in persistence induced by penicillin (Dille *et al.*, 2014). In addition, CPAF processing was partially inhibited, and the zymogen form was readily apparent during penicillin-induced persistence (Dille *et al.*, 2014). Effects on CPAF in other models of persistence have not been investigated. These results indicate that CPAF expression and processing require active chlamydial growth.

CPAF inhibitors

Of the commercially available protease and proteasome inhibitors, CPAF proteolytic activity is only blocked by the antibiotic lactacystin and its active form, clasto-lactacystin (Chen *et al.*, 2012; Zhong *et al.*, 2001). Lactacystin irreversibly inhibits CPAF

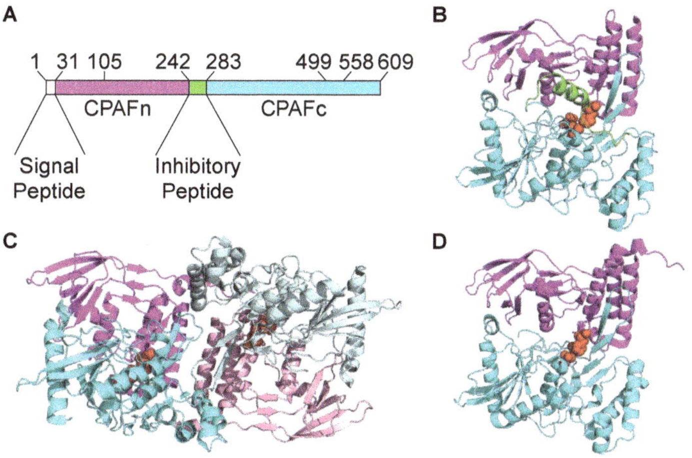

Figure 8.1 Important sequence features of CPAF. The crystal structure of *C. trachomatis* serovar D CPAF has been solved and mutational analysis was used to confirm interactions and functions of the structure. (A) Schematic of full-length CPAF showing N-terminal (CPAFn; magenta) and C-terminal (CPAFc; cyan) fragments, and the signal (aa 1–31; white) and inhibitory (aa 243–283; green) peptides that are removed upon secretion. The catalytic triad is formed by H105, S499, and E558, which are indicated. (B) The crystal structure of the zymogen form of CPAF was obtained using a E558A mutant (PDB ID 3DPN). The colour scheme is the same as in panel A. Note how the inhibitory peptide (green) obstructs the catalytic site residues (H105, S499, and E558; shown as red space fill). (C) Processing of the inhibitory peptide in wild type CPAF occurs by three cleavage events and requires homodimerization of two full-length CPAF proteins (one mature dimer shown at magenta and cyan and the other as lighter shades of magenta and cyan); shown for mutant PDB ID 3DOR. (D) Removal of the inhibitory peptide (red) exposes the catalytic site to substrates. For further structural details, see Huang, *et al.*, 2008.

enzymatic activity by covalently binding to S499 in the catalytic site (Huang *et al.*, 2008). Lactacystin and clasto-lactocystin are cell-permeable, so they can be used to inhibit CPAF in *Chlamydia*-infected cells. However, they are expensive, which has limited their routine use for chlamydial lysate preparation. Since lactacystin and clasto-lactacystin are eukaryotic proteasome inhibitors, they also have additional effects on the host cell because cells must be pre-treated prior to cell processing.

Two peptide-based CPAF inhibitors have been described. The tetrapeptide inhibitor WEHD-fmk, which is typically used to inhibit caspases, inhibits CPAF enzymatic activity *in vitro* (Christian *et al.*, 2011). A more specific non-covalent CPAF inhibitor has been developed that adds a polybasic tract to the sequence of the internal inhibitory peptide of full-length CPAF to make it cell-permeable (Bednar *et al.*, 2011). This peptide inhibitor blocked the *in vitro* activity of CPAF from *C. trachomatis* and *Chlamydia muridarum*, but not *Chlamydia caviae* (Jorgensen *et al.*, 2011), which may be due to differences in the amino acid sequence. It has also been reported to inhibit CPAF-mediated proteolysis within *C. trachomatis*-infected cells (Jorgensen *et al.*, 2011). However, the substrates that were used to monitor proteolysis are now known to be cleaved by residual CPAF activity in the lysates (Chen *et al.*, 2012), as will be discussed in the next section. Thus, it is not clear if this cell-permeable inhibitory peptide blocked proteolysis inside an infected cell, or during lysate preparation. Both WEHD-fmk and the internal inhibitory peptide appear to have off-target effects on the infected cell because they inhibited chlamydial growth to a greater extent than a CPAF null mutant strain (Christian *et al.*, 2011; Hou *et al.*, 2013; Jorgensen *et al.*, 2011; Snavely *et al.*, 2014). In this regard, WEHD-fmk is a known caspase-1 inhibitor (Christian *et al.*, 2011), and the internal inhibitory peptide induces caspase-1 dependent cell death of the infected host cell (Jorgensen *et al.*, 2011).

CPAF substrates

CPAF has proteolytic activity against many host and chlamydial proteins. Contrary to some descriptions, this activity is selective and not indiscriminate or non-specific. For instance, incubation of recombinant CPAF with EB lysates caused the proteolysis of susceptible proteins but not broad degradation of other proteins (Jorgensen *et al.*, 2011). The end product of proteolysis is specific for each substrate and ranges from complete degradation to cleavage fragments resistant to further proteolysis. It is currently not possible to predict if a protein is a CPAF substrate from its amino acid sequence because the cleavage site recognized by CPAF is poorly understood. However, an *in vitro* proteomics study suggests that cleavage is most favourable after an aliphatic-X-small amino acid sequence motif (Biniossek *et al.*, 2016). Alternatively, it is hypothesized that the cleavage site may be determined by secondary structural elements, although a structural recognition motif has not been identified (Kumar and Valdivia, 2008).

We now know that the susceptibility of a protein to CPAF-mediated proteolysis does not necessarily mean that the proteolysis is biologically significant. Chen *et al.* demonstrated that proteolysis of 11 published CPAF substrates was prevented by blocking CPAF activity during preparation of lysates of the *Chlamydia*-infected cells (Chen *et al.*, 2012). Thus the reported proteolysis of these host proteins was an *in vitro* phenomenon that did not occur within the *Chlamydia*-infected cell. This general finding has been confirmed by many research groups (Grieshaber and Grieshaber, 2014; Hou *et al.*, 2013; Patton *et al.*, 2016; Rajalingam *et al.*, 2008; Snavely *et al.*, 2014).

There are a number of reasons for the presence of CPAF activity in *Chlamydia*-infected cell lysates. CPAF activity can be readily induced by detaching a monolayer of infected cells from a plastic support with either trypsinization or mechanical scraping (Johnson *et al.*, 2015), which are routine methods to prepare cell lysates for subsequent analysis. CPAF induction by trypsinization occurs even in the absence of host cell lysis (Johnson *et al.*, 2015). One possibility is that manipulation of infected cells, particularly at late times when the inclusion is large, could cause loss of inclusion stability. CPAF is active in the presence of detergents, such as NP-40, deoxycholate and SDS, which are found in RIPA buffer (Chen *et al.*, 2012). In addition, CPAF remains active at 4°C and is not blocked by inhibitors of typical serine proteases, which explains why established precautions such as keeping chlamydial lysates on ice and adding standard protease inhibitor cocktails did not prevent CPAF-mediated proteolysis (Chen *et al.*, 2012; Johnson *et al.*, 2015; Jorgensen *et al.*, 2011; Zhong *et al.*, 2000). Thus, a 'perfect' storm of a robust bacterial protease, multiple potential substrates, an undefined mechanism for inducing CPAF activity and ineffective precautions for inhibiting CPAF activity, caused the *Chlamydia* research community to be misled about the true targets of this protease.

Table 8.1 Methods used to prevent CPAF-mediated proteolysis during lysate preparation

Method	Notes	References
Lyse infected cells in Laemmli sample buffer	Most reliable method. Suitable for SDS-PAGE but not for mass spectrometry or enzymatic assays	Rajeeve *et al.* (2018)
Lyse infected cells with 8M urea	8M urea must be prepared fresh. Not 100% reliable, so lysates should be checked to confirm that CPAF activity has been inhibited	Chen *et al.* (2012); Hou *et al.* (2013); Johnson *et al.* (2015)
Lyse infected cells with boiling-hot 1–2% SDS	Not 100% reliable, so lysates should be checked to confirm that CPAF activity has been inhibited	Johnson *et al.* (2015); Patton *et al.* (2016); Snavely *et al.* (2014)
Lactocystin or clasto-lactocystin (CPAF inhibitors)	Pretreat infected cells with the inhibitor for one hour prior to harvest, and also include the inhibitor in the lysis buffer. Not 100% reliable, so lysates should be checked to confirm that CPAF activity has been inhibited	Chen *et al.* (2012); Johnson *et al.* (2015); Zhong *et al.* (2001)
CPAF null mutant	Currently only available for *C. trachomatis*. Lack of CPAF activity reveals whether proteolysis is CPAF-dependent, but does not differentiate between proteolysis that occurred during infection versus during lysate preparation	Snavely *et al.* (2014)

The problem of residual CPAF activity has been managed by taking measures to inhibit CPAF in chlamydial lysates (Table 8.1). If the lysates are to be analysed by SDS-PAGE, the most reliable method is to directly lyse the infected cells in Laemmli sample buffer (Rajeeve *et al.*, 2018). Lysates for mass spectrometry or enzymatic assays can be harvested in freshly prepared 8M urea or boiling-hot 1–2% SDS, although these methods do not inhibit CPAF with 100% reliability (Johnson *et al.*, 2015). Thus it is necessary to confirm that the treated lysates do not cause proteolysis when added to a recombinant substrate or an uninfected cell lysate (Johnson *et al.*, 2015). Alternatively, lysates can be treated with CPAF inhibitors, but this method is not commonly used as current inhibitors are expensive.

As shown in Table 8.2, CPAF substrates can be broadly divided into three classes:

1. Chlamydial substrates
2. Intracellular host substrates: host proteins within an infected host cell that are cleaved or degraded by CPAF produced by this cell
3. Extracellular host substrates: host proteins that are cleaved or degraded by CPAF released from another infected cell

In evaluating whether a published CPAF substrate is a true *in vivo* target, it is important to assess whether the proteolysis was solely based on Western blot analysis of cell lysates, whether effective precautions were taken during lysate preparation to inhibit CPAF activity, and whether the proteolysis has been confirmed in intact host cells by methods such as immunofluorescence microscopy that do not involve lysate preparation.

Chlamydial substrates

A number of chlamydial proteins have been proposed as CPAF substrates. Active CPAF is most likely to be present outside of the bacterium because CPAF processing and activation requires its secretion (Snavely *et al.*, 2014). Thus, chlamydial proteins that are *in vivo* CPAF targets are not expected to be in the bacterial cytosol, but are more likely secreted into the inclusion or host cytoplasm, or found in external domains of chlamydial outer membrane proteins. It has not been ascertained if there is active CPAF in the periplasm to cleave or degrade periplasmic proteins.

OmcB is a conserved outer membrane protein of EBs that has been reported to be a chlamydial CPAF substrate (Hou *et al.*, 2013). Partial OmcB cleavage was detected in Western blots of lysates prepared with 8M urea to inhibit CPAF. This proteolysis was attributed to CPAF because recombinant OmcB was cleaved by CPAF, but not by two other chlamydial proteases, HtrA and the tail-specific protease CT441. In addition, there is evidence that OmcB is cleaved in *Chlamydia*-infected cells because its C-terminal fragment predominantly localized to the host cytoplasm, while its N-terminal fragment was restricted to the chlamydial inclusion (Qi *et al.*, 2011). However, OmcB is predicted to be a single pass transmembrane protein with its N-terminus exposed to the inclusion lumen and its C-terminus in the chlamydial periplasm (Mygind *et al.*, 1998; Stephens *et al.*, 2001). It is therefore not clear how CPAF could cleave the C-terminus of OmcB and release it into the host cytoplasm. Stronger evidence that CPAF is responsible for OmcB processing could be obtained with immunofluorescence studies showing that there is no OmcB cleavage in cells infected with a CPAF null mutant or a T2S null mutant.

To find CPAF targets on a global scale, McClarty, Caldwell and colleagues used a proteomics approach to identify proteins that are differentially expressed during infection with the *C. trachomatis* CPAF mutant compared to the wild type strain. Nine chlamydial proteins (seven type 3 secretion (T3S) effectors and two T3S structural proteins) had > 2-fold higher expression in the CPAF mutant during infection (Patton *et al.*, 2016). Precautions were taken to inhibit CPAF by preparing the lysates in hot 2% SDS (Patton *et al.*, 2016). This relatively small number of differentially expressed proteins indicates that CPAF did not cause widespread proteolysis of chlamydial proteins. It is not known if these nine chlamydial proteins are direct or indirect targets of CPAF. Confidence in these chlamydial proteins as *in vivo* CPAF substrates would be enhanced by data showing that they undergo proteolysis in intact infected cells.

Valdivia and colleagues tested more than 300 potential chlamydial effectors, of which nine were either cleaved or degraded by *Chlamydia*-infected cell lysates (Table 8.2) (Jorgensen *et al.*, 2011). Precautions were not taken to inhibit CPAF activity in the lysates because it was not yet known that

Table 8.2 Published CPAF substrates

Cellular process	Protein	Initial studies reporting protein as CPAF substrate	Type of CPAF substrate	Immunoblot data when CPAF was inhibited during lysate preparation	Data supporting the protein as an *in vivo* CPAF target (in intact cells or in an animal)	Follow-up studies
Immune response	RFX5	Zhong *et al*. (2000)	Intracellular host	No proteolysis	–	Chen *et al*. (2012)
	USF-1	Zhong *et al*. (1999)	Intracellular host	No proteolysis	–	Patton *et al*. (2016)
	CD1d	Kawana *et al*. (2007)	Intracellular host	No data	–	–
	p65/RelA	Christian *et al*. (2010)	Intracellular host	No proteolysis	–	Chen *et al*. (2012); Patton *et al*. (2016)
	LL-37	Tang *et al*. (2015)	Extracellular host	Not relevant	Cell-free proteolysis assay; CPAF inhibited anti-chlamydial activity of LL-37	Tang *et al*. (2015)
	C3b	Yang *et al*. (2016b)	Extracellular host	Not relevant	Cell-free proteolysis assay; CPAF inhibited anti-chlamydial activity of complement	Yang *et al*. (2016b)
	FPR2	Rajeeve *et al*. (2018)	Extracellular host	Not relevant	Cell-free proteolysis assay; FPR2 cleaved from surface of neutrophils by CPAF; mice studies with FPR2-KO mice and CPAF null mutant	Rajeeve *et al*. (2018)
Cell death	Bik	Dong *et al*. (2005)	Intracellular host	No proteolysis	–	Chen *et al*. (2012)
	Bim	Dong *et al*. (2005)	Intracellular host	No proteolysis	–	Chen *et al*. (2012); Rajalingam *et al*. (2008)
	Puma	Dong *et al*. (2005)	Intracellular host	No proteolysis	–	Chen *et al*. (2012); Rajalingam *et al*. (2008)
	BH3-only	Pirbhai *et al*. (2006)	Intracellular host	No proteolysis	–	Rajalingam *et al*. (2008)
	PARP	Matsuo *et al*. (2013); Paschen *et al*. (2008); Yu *et al*. (2010)	Intracellular host	No data	–	–
	HMGB1	Yu *et al*. (2010)	Intracellular host	No data	–	–
Stress response	HIF-1α	Rupp *et al*. (2007)	Intracellular host	No data	–	–
	PI15	Prusty *et al*. (2018)	Intracellular host	Proteolysis	Proteolysis in chlamydial lysates by immunoblot; co-localization with CPAF in inclusion lumen	Prusty *et al*. (2018)

Golgi organization	Golgin-84	Christian *et al*. (2011); Heuer *et al*. (2009)	Intracellular host	No proteolysis	–	Chen *et al*. (2012)
Cytoskeleton	CK8	Dong *et al*. (2004b); Savijoki *et al*. (2008)	Intracellular host	No proteolysis	–	Chen *et al*. (2012)
	CK18	Savijoki *et al*. (2008)	Intracellular host	No proteolysis	–	Chen *et al*. (2012)
	Vimentin	Kumar and Valdivia (2008); Savijoki *et al*. (2008)	Intracellular host	Proteolysis in one study; no proteolysis in other studies	Proteolysis at late times in chlamydial lysates by immunoblot; relocalization on inclusion rupture	Chen *et al*. (2012); Snavely *et al*. (2014); Hou *et al*. (2013); Patton *et al*. (2016)
	β-tubulin	Savijoki *et al*. (2008)	Intracellular host	No data	–	–
	LAP-1	Snavely *et al*. (2014)	Intracellular host	Proteolysis	Proteolysis at late times in chlamydial lysates by immunoblot; relocalization on inclusion rupture	Snavely *et al*. (2014)
	Nectin-1	Sun and Schoborg (2009)	Intracellular host	No proteolysis	–	Chen *et al*. (2012)
Cell cycle	Cyclin B1	Balsara *et al*. (2006); Brown *et al*. (2012); Paschen *et al*. (2008)	Intracellular host	No proteolysis	–	Chen *et al*. (2012)
	Securin	Brown *et al*. (2012)	Intracellular host	No proteolysis	–	Grieshaber and Grieshaber (2014)
Chlamydial proteins	CPAF	Paschen *et al*. (2008)	Chlamydial	Proteolysis	Proteolysis in chlamydial lysates by immunoblot	Chen *et al*. (2012)
	OmcB	Hou *et al*. (2013)	Chlamydial	Proteolysis	Proteolysis in chlamydial lysates by immunoblot	Hou *et al*. (2013)
	CT005	Jorgensen *et al*. (2011)	Chlamydial	No data	–	–
	IncD	Jorgensen *et al*. (2011)	Chlamydial	No data	–	–
	IncE	Jorgensen *et al*. (2011)	Chlamydial	No data	–	–
	IncC	Jorgensen *et al*. (2011)	Chlamydial	No data	–	–
	CT288	Jorgensen *et al*. (2011)	Chlamydial	No data	–	–
	CT694	Jorgensen *et al*. (2011)	Chlamydial	No data	–	–
	CT695	Jorgensen *et al*. (2011)	Chlamydial	No data	–	–
	CT813	Jorgensen *et al*. (2011)	Chlamydial	No data	–	–
	TARP	Jorgensen *et al*. (2011)	Chlamydial	No data	–	–

residual CPAF activity could cause *in vitro* proteolysis. Thus, more data are needed to determine if these chlamydial proteins are *in vivo* substrates. These nine proteins were inclusion membrane proteins or effectors, but there was no overlap with the Patton *et al.* study described above (Patton *et al.*, 2016).

Intracellular host substrates

It is no longer clear if host proteins inside an infected cell are cleaved or degraded by CPAF. The proteomics study discussed above provided strong evidence that few, if any, host proteins are intracellular CPAF substrates (Patton *et al.*, 2016). Only 0.1% (6 of 6,228) of host proteins had a > 2-fold increase in expression levels when infected with a CPAF mutant compared to the wild type strain (Patton *et al.*, 2016). These six host proteins all have roles in the innate immune response, but there were no data to indicate that their reduced expression was due to proteolysis by CPAF.

There is uncertainty about whether the intermediate filament protein vimentin is an intracellular CPAF substrate. It is now known that most if not all of the vimentin cleavage that was initially reported (Kumar and Valdivia, 2008) was due to *in vitro* proteolysis from CPAF in the chlamydial lysates (Chen *et al.*, 2012; Hou *et al.*, 2013; Patton *et al.*, 2016; Snavely *et al.*, 2014). Chen *et al.* detected no vimentin cleavage products at 24, 36 or 48 hours post infection with lysates that were verified to be free of CPAF activity (Chen *et al.*, 2012). Hou *et al.* also failed to observe vimentin cleavage when infected cells were lysed with 8M urea to inhibit CPAF (Hou *et al.*, 2013). In addition, the McClarty and Caldwell proteomics study found that vimentin levels did not differ between the CPAF mutant and wildtype strain (Patton *et al.*, 2016). In contrast, Snavely *et al.* detected a small amount of cleavage products for vimentin and the nuclear envelope protein LAP1 in infected cell lysates prepared at very late times (≥ 48 hours post infection) with hot 1% SDS to inhibit CPAF (Snavely *et al.*, 2014).

How can these different results be reconciled? One explanation is that vimentin and LAP1 cleavage was due to residual CPAF activity in the chlamydial lysates despite the use of hot SDS, which is not always reliable for inhibiting CPAF activity (Johnson *et al.*, 2015). Another possibility is that this cleavage only occurred at very late times in the intracellular infection. To support this model, the Snavely *et al.* study used an innovative fluorescent microscopy method to obtain time lapse images showing diffuse relocalization of vimentin and LAP1 within minutes of inclusion rupture (Snavely *et al.*, 2014). Unfortunately, there was no negative control to show that protein relocalization was specific for CPAF substrates. Nevertheless, these findings suggest that CPAF is primarily in the inclusion until it is released into the host cytoplasm by inclusion rupture just before the host cell is lysed. It is thus possible that CPAF-mediated proteolysis is a terminal event in the intracellular infection that could affect many susceptible proteins.

A host protein called PI15, which is a serine protease inhibitor, was recently reported to be cleaved in a CPAF-dependent manner (Prusty *et al.*, 2018). PI15 cleavage was detected in chlamydial lysates from infected cells lysed in Laemmli sample buffer to prevent *in vitro* proteolysis. Proteolysis of PI15 was proposed to occur in the chlamydial inclusion because of data showing that PI15 was transported into the inclusion and partially co-localized with CPAF. Whether CPAF causes proteolysis of other host proteins that are translocated into the inclusion membrane is unknown.

Extracellular host substrates

There are now several reports indicating that host proteins are cleaved or degraded by CPAF outside of the *Chlamydia*-infected cell. These models still propose that an infected cell is the source of active CPAF. However, the protease is hypothesized to be released, perhaps upon cell lysis, in an enzymatically active form that causes proteolysis of host proteins in the extracellular environment. These published targets include extracellular host effectors and a surface receptor on uninfected host cells.

CPAF has been shown to cleave soluble innate immunity effectors. In a cell-free proteolysis assay, recombinant CPAF cleaved cathelicidin LL-37, which is an antimicrobial peptide (Tang *et al.*, 2015). Recombinant CPAF also cleaved two complement components, C3 and factor B, which are present in human serum (Yang *et al.*, 2016b). Furthermore, the anti-chlamydial activity of LL-37 and human serum were both neutralized by recombinant CPAF (Tang *et al.*, 2015; Yang *et al.*, 2016b).

Another recently reported extracellular CPAF substrate is a surface receptor on host neutrophils called FPR2 (formyl protein receptor 2) (Rajeeve *et al.*, 2018). FPR2, and its paralogue FPR1, were degraded by recombinant CPAF *in vitro*, although FPR2 was a more susceptible substrate. However, only FPR2, and not FPR1, was detected in the supernatant of neutrophils that were incubated with EB preparations or treated with recombinant CPAF. These results suggest that extracellular CPAF released from *Chlamydia*-infected cells cleaves FPR2 on the neutrophil surface.

There are challenges in verifying that an extracellular protein is an *in vivo* CPAF substrate. The *in vivo* proteolysis of these host proteins cannot be replicated in the standard *Chlamydia* cell culture infection model. Assays showing that the protein is cleaved or degraded by recombinant CPAF only demonstrate that it is an *in vitro* substrate, but do not indicate if the cleavage is biologically relevant. Animal studies with a CPAF null mutant and mice defective for specific pathways could help provide evidence that CPAF is necessary for a proposed function. However, the caveats about *in vitro* proteolysis during cell lysate preparation also apply to tissue preparation for protein analysis. Thus, the best assays, if available, would be imaging studies that do not require tissue manipulation.

Localization

CPAF is not usually detected within chlamydiae because it is secreted out of the bacterium soon after its production (Shaw *et al.*, 2002; Snavely *et al.*, 2014). The protease reaches the chlamydial periplasm through the Sec general secretory pathway (Chen *et al.*, 2010b), which transports unfolded proteins from the bacterial cytoplasm across the inner membrane into the periplasm. The first 31 amino acids are sufficient for secretion and contain a signal peptidase I cleavage site to release the signal sequence once CPAF is secreted into the periplasm (Shaw *et al.*, 2002). CPAF secretion is blocked by arylomycin C16, an inhibitor of the Sec secretory pathway (Chen *et al.*, 2010b).

CPAF likely folds in the periplasm, but it is not known how it is then secreted across the outer membrane and out of the bacterium. Valdivia and colleagues have suggested that CPAF is a T2S substrate because there was no CPAF secretion in a T2S defective strain (Snavely *et al.*, 2014). However, Subtil and Hayward postulate that this T2S requirement does not prove that CPAF secretion occurs via T2S and could instead be due to another mechanism that is dependent on T2S (discussed in more detail in Chapter 7, Protein Secretion in *Chlamydia*).

There is controversy over the localization of CPAF after its secretion out of chlamydiae. Initial microscopy analyses suggested that when CPAF is secreted out of the bacterium, it is translocated across the inclusion membrane into the host cell cytoplasm. This idea was supported by the detection of CPAF in the host cytoplasm of *C. trachomatis*-infected cells by immunofluorescence microscopy at late times in the infection (Shaw *et al.*, 2002; Zhong *et al.*, 2001). In addition, *C. pneumoniae* CPAF appeared to show progressive translocation from the inclusion to the host cell cytoplasm (Heuer *et al.*, 2003). The localization of CPAF to the host cytoplasm was further corroborated by electron microscopy studies with *C. pneumoniae* (Heuer *et al.*, 2003) and *C. trachomatis* (Dille *et al.*, 2014). The mechanism by which CPAF reaches the host cytoplasm is not known, but has been speculated to involve translocation across the inclusion membrane by outer membrane vesicles (OMVs) (Chen *et al.*, 2010b).

More recently, the cytoplasmic localization of CPAF has been challenged (Bavoil and Byrne, 2014). As discussed, Snavely *et al.* used time lapse fluorescent imaging to propose that CPAF activity is largely restricted to the inclusion until it is released into the host cytoplasm by inclusion rupture late in the infection (Snavely *et al.*, 2014). Furthermore, Rudel and colleagues reported that fixation conditions affected CPAF localization (Prusty *et al.*, 2018). In infected cells fixed with 4% paraformaldehyde at 37°C, CPAF was only detected in the inclusion, but when this fixation was performed at room temperature, the protease was also seen in the host cytoplasm (Prusty *et al.*, 2018). CPAF was also restricted to the inclusion when cells were fixed with ice cold methanol (Prusty *et al.*, 2018). These results demonstrate that alternative CPAF localization patterns can be obtained solely by altering the fixation conditions. While Zhong and colleagues always detected CPAF in the cytoplasm of *C. trachomatis*-infected cells fixed with 4% paraformaldehyde, they only performed the fixations at room temperature and not at 37°C (Yang *et al.*, 2015). Fixation may have a general effect on inclusion membrane permeability because it greatly increased the detection of endoplasmic reticulum markers within the chlamydial inclusion compared to live cells (Kokes and Valdivia, 2015). Since immunogold electron microscopy studies also require chemical fixation of cells, they could also have fixation artefacts. Whether CPAF localizes to the inclusion and/or the host cytoplasm could be resolved by live cell imaging of fluorescently tagged CPAF in *Chlamydia*-infected cells.

Regulation of CPAF-dependent proteolysis

The presence of CPAF inside an infected cell without obvious proteolysis of susceptible substrates has been called the CPAF paradox (Tan and Sutterlin, 2014). This discrepancy strongly suggests that CPAF is regulated in some manner. Several different control mechanisms have been proposed, including regulation of CPAF enzymatic activity and compartmentalization of the protease within the infected cell.

Regulation of enzymatic activity

CPAF activity could be controlled at the intramolecular autoprocessing step, which is necessary for conversion of the zymogen into the active protease. A pulse-chase study showed that the zymogen was processed into its cleaved fragments within four hours (Dong *et al.*, 2004b). Processing requires secretion, and, therefore, could be controlled at this step, although such a regulatory mechanism has not been described. There is evidence, however, of a host factor that modulates CPAF processing in a concentration-dependent manner (Prusty *et al.*, 2018). This protein, human serine peptidase inhibitor PI15, bound to the CPAF zymogen in co-immunoprecipitation studies with *Chlamydia*-infected cells and also with recombinant proteins. In addition, PI15 partially co-localized with CPAF in the lumen of the chlamydial inclusion. These findings suggest that CPAF processing, and its regulation by PI15, occur in the inclusion lumen.

Another potential regulatory mechanism would be to regulate the enzymatic activity of mature CPAF with a host factor or a secreted chlamydial effector. One approach for identifying a CPAF regulator would be to search for proteins that physically interact with CPAF. CPAF structure–function data (Chen *et al.*, 2009; Huang *et al.*, 2008) could be useful for identifying candidate inhibitors that bind to the catalytic site of CPAF.

Regulation of physical location

Proteolysis by CPAF could also be regulated by controlling the physical location of CPAF relative to its substrates. As discussed, there should be no active CPAF in the chlamydial cytoplasm because CPAF processing requires secretion out of the bacterium (Chen *et al.*, 2010b; Snavely *et al.*, 2014). If CPAF is restricted to the inclusion (Prusty *et al.*, 2018), it could explain why susceptible host proteins in the cytosol are not cleaved or degraded in an infected cell. It has been suggested that CPAF may have access to these proteins when the inclusion ruptures, prior to disruption of the host cell plasma membrane at the end of the intracellular infection (Snavely *et al.*, 2014).

Even if CPAF is translocated across the inclusion membrane into the host cytoplasm, its ability to come in contact with host proteins could still be regulated (Chen *et al.*, 2010a). For instance, CPAF might be restricted to a subcellular compartment within the host cytoplasm. Alternatively, individual host proteins could be sequestered from CPAF by their location, or by physical interactions with proteins that shield them from CPAF.

Perhaps the best evidence for CPAF regulation is that the proteolysis of host proteins observed upon ectopic expression of CPAF in uninfected cells (Christian *et al.*, 2010; Paschen *et al.*, 2008) resembles the *in vitro* proteolysis by residual CPAF activity in infected cell lysates (Chen *et al.*, 2012). These similarities suggest that a control mechanism to keep CPAF activity in check may be absent when this protease is exogenously expressed in uninfected cells.

Function

The function of CPAF in a *Chlamydia* infection has been a subject of much interest and investigation (Conrad *et al.*, 2013; Hacker, 2014; Tan and Sutterlin, 2014). As a protease with many potential substrates, CPAF could be critical for modulating the levels of proteins involved in specific bacterial or host processes. Two general approaches have been used to study CPAF function. Traditionally, the role of CPAF in an infected cell has been extrapolated from the normal, cellular function of each host substrate. This approach is heavily dependent on the correct identification of *in vivo* substrates and will lead to faulty conclusions if a protein is only an *in vitro* substrate. More recently, a CPAF null mutant has been used to reveal phenotypic differences in an infected cell or animal that require CPAF (Snavely *et al.*, 2014). This genetic approach has many strengths and has been used to support or refute functions attributed to CPAF. However, because CPAF likely has multiple substrates, it does not identify the specific CPAF target responsible for a given phenotype. Furthermore, a null mutant does not distinguish between a direct or an indirect effect of CPAF, and does not reveal if a direct target is an *in vitro* or *in vivo* substrate.

The CPAF null mutant

Valdivia and colleagues identified two CPAF null mutants from a *C. trachomatis* serovar L2 library chemically mutagenized with EMS (ethyl-methanesulfonate) (Snavely *et al.*, 2014). They used lateral gene transfer to reduce, but not eliminate, the number of additional mutations introduced into the genome by the mutagenesis procedure (Snavely *et al.*, 2014). The most widely used recombinant strain is RST17, which codes for a nonsense mutation in CPAF at position 294 (of 601 amino acids) and does not produce detectable CPAF or CPAF activity (Snavely *et al.*, 2014). Because of the recombination steps, RST17 has resistance markers to rifampin, spectinomycin and trimethoprim. RST17 is often used together with RST5, which contains the wildtype CPAF allele and serves as a complemented strain. RST5 is nearly isogenic with RST17 outside of the CPAF allele and has the same resistance markers.

There is currently no *C. muridarum* CPAF mutant. A limitation of RST17 is that it is not optimal for mouse studies because it is a human strain and therefore cannot be used to model a chronic infection in mice. The *C. trachomatis* and *C. muridarum* CPAF homologues share 82% amino acid identity, but it has not been confirmed that they have the same substrate repertoire. The inability to cleave the orthologous substrate proteins in another host has been observed for a related chlamydial protease, Tsp (CT441), and one of its substrates, host p65 (Lad *et al.*, 2007). Recent advances in genetic manipulation of *Chlamydia* should make it possible to generate a *C. muridarum* CPAF mutant without background mutations or antibiotic resistance markers.

CPAF is not necessary for a *Chlamydia* infection in cell culture. The CPAF null mutant was able to infect host cells, establish an inclusion and progress through the developmental cycle to produce infectious EBs (Snavely *et al.*, 2014). Compared to the wildtype strain, the CPAF mutant produced 3-fold fewer EBs (Snavely *et al.*, 2014). However, to put this in perspective, a *C. trachomatis* infection typically produces about 1,000 EBs per cell (Lee *et al.*, 2018). A 3-fold reduction would still generate about 300 EBs per infected cell, indicating that the basic machinery for RB replication and EB production is intact in the absence of CPAF.

Putative CPAF functions with good experimental support

In this section, we review data about proposed functions of CPAF.

Inhibition of type I interferon response and IL-6 and IL-8 expression

The McClarty group showed that CPAF inhibited nuclear translocation of the NF-κB subunit p65, which downregulated the type I interferon response and expression of IL-6 and IL-8 (Patton *et al.*, 2016). This study is noteworthy for a number of experimental strengths:

1. A complementation approach, using the CPAF null mutant and a strain complemented with the CPAF gene, provided strong genetic evidence that CPAF is necessary to inhibit p65 nuclear translocation.
2. Multiple assays, including Western blots and immunofluorescence microscopy, were used to measure CPAF-dependent effects.
3. Functional assays, measuring decreased secretion of IFN-β, IL-6 and IL-8 in cell culture supernatants, were used to determine the effect of CPAF on NF-κB-dependent gene expression.

The mechanism by which CPAF inhibits p65 translocation is not known. CPAF did not appear to cleave p65, and the authors speculate that CPAF may have an indirect role by regulating chlamydial T3S effectors, which in turn blocks p65 nuclear translocation (Patton *et al.*, 2016). It is not clear why the findings of this study differ from Snavely *et al.*, who reported that CPAF was not necessary for inhibition of NF-κB-dependent gene expression, as measured with a NF-κB luciferase reporter assay (Snavely *et al.*, 2014).

Inhibition of neutrophil activation

Using the CPAF null mutant and a complemented strain expressing wild type CPAF, Rudel and colleagues provided compelling evidence that CPAF inhibits neutrophil activation (Rajeeve *et al.*, 2018). CPAF was necessary to prevent the formation of neutrophil extracellular traps (NETs), which are networks of DNA fibres that bind bacteria and other pathogens. FPR2, a surface receptor important for neutrophil signalling and activation, underwent CPAF–dependent cleavage and down-regulation. Mice infected with the CPAF mutant had a reduction in chlamydial load in a transcervical genital tract model of infection. The reduced chlamydial load for the CPAF mutant was rescued in mice that were either depleted of neutrophils or that lacked FPR2, providing evidence that CPAF targets FPR2 to inhibit neutrophil activation. This study stands out because it describes a complete mechanism of CPAF function from release of CPAF by infected cells to cause proteolysis of a surface protein on uninfected neutrophils and downstream effects in inhibiting NET formation and neutrophil activation. FPR2 is present on the surface of other host cells, including epithelial cells of the respiratory and intestinal tracts (Duvall and Levy, 2016), but it is not known if CPAF also has effects on these cells.

Genomic instability and multinucleation

Scott Grieshaber and colleagues used the *C. trachomatis* CPAF null mutant to show that CPAF is necessary for formation of extra centrosomes and multipolar spindles, which can lead to chromosome segregation errors and early exit from mitosis (Brown *et al.*, 2014). CPAF was also necessary for the induction of multinucleation, which is a consequence of cytokinesis defects. These CPAF functions may be conserved in other *Chlamydia* spp. because *C. muridarum* and *C. caviae* also induced extra centrosomes and early mitotic exit. These other chlamydial species also induced multipolar spindles and multinucleation, but at a lower prevalence than *C. trachomatis.* The authors suggest that this difference is due to the absence of a CPAF-independent defect, such as centrosome mislocalization, from *C. muridarum* and *C. caviae.* The mechanism by which CPAF causes these features of genomic instability is not known.

Genital tract infection

CPAF contributes to tissue tropism in the mouse genital tract (Rajeeve *et al.*, 2018; Yang *et al.*, 2016a). Yang *et al.* found that the *C. trachomatis* CPAF null mutant was cleared more rapidly than the wild type strain from the lower genital tract, but not the upper genital tract, of mice lacking innate immunity. However, Rajeeve *et al.* measured a 3-fold reduction in chlamydial load in the uterine horns of mice infected with the CPAF mutant, which demonstrates a role for CPAF in upper genital tract infection (Rajeeve *et al.*, 2018). Vaccination of mice with CPAF protected against upper genital infection (Cong *et al.*, 2007; Murphey *et al.*, 2006). These studies indicate that CPAF has an important role during genital tract infection.

CPAF as a vaccine candidate

CPAF is an antigen that elicits a protective immune response (Murthy *et al.*, 2009). High anti-CPAF titres are observed in patients with previous *C. trachomatis* urogenital or ocular infection and prior *C. pneumoniae* pneumonia (Sharma *et al.*, 2004; Skwor *et al.*, 2010; Susnea *et al.*, 2011), and higher titres of anti-CPAF antibodies correlate with more severe disease sequelae (Skwor *et al.*, 2010; van Ess *et al.*, 2019). CPAF from animal *Chlamydia* spp. are also conserved immunodominant antigens (Forsbach-Birk *et al.*, 2013; Kari *et al.*, 2013; Li *et al.*, 2007). Neutralizing antibody to CPAF is directed against the central portion of the protein and blocks processing into the active enzyme (Sharma *et al.*, 2005), which suggests that full-length CPAF is accessible as an antibody target. Antibody against CPAF is also protective against upper genital infection in mice (Cong *et al.*, 2007; Murphey *et al.*, 2006). The role of CPAF in adaptive immunity is discussed in more depth in Chapter 14, Adaptive Immunity to *Chlamydia trachomatis* Infection.

These antigenic properties make CPAF a potential vaccine candidate (Murthy *et al.*, 2009). Immunization with recombinant CPAF, or adoptive transfer of CPAF-responsive T cells, protected against both genital and respiratory infection (Li *et al.*, 2013). In mice infected with *C. muridarum,* intranasal CPAF vaccination increased chlamydial clearance and protected against complications of genital disease, such as fallopian tube pathology and infertility (Li *et al.*, 2007; Murthy *et al.*, 2007; Murthy *et al.*, 2011). Addition of CPAF to PBMCs from *C. trachomatis* infected women resulted in higher secretion of TNF-α and IL-1β (Cheong *et al.*, 2019), which may at least partially explain the reduced pathology in vaccinated mice. Intranasal vaccination also increased anti-CPAF vaginal IgA levels, demonstrating that vaccination with CPAF can drive class switching and induce protection at distant mucosal sites (Cong *et al.*, 2007). In addition, restimulation of $CD4^+$ T-cells from vaccinated mice released high levels of IFN-γ, which has been shown to promote early infection clearance in adoptive transfer experiments (Li *et al.*, 2008; Murthy *et al.*, 2006). Intriguingly, vaccination with CPAF from *C. pneumoniae* reduced pathologic changes associated with atherosclerosis in a mouse model (Li *et al.*, 2018). More details about CPAF and *Chlamydia* vaccine development are in Chapter 15, *Chlamydia* Vaccines.

Proposed CPAF functions that require further validation

In this section we review additional functions of CPAF that have been proposed.

OmcB processing

CPAF has been reported to mediate the proteolytic processing of OmcB (Hou *et al.*, 2013), which could potentially affect binding between EBs and the host cell because OmcB is a chlamydial adhesin (Fadel and Eley, 2007; Moelleken and Hegemann, 2008). OmcB processing could also have effects on the host immune response because the C-terminal OmcB fragment is an immunodominant CD8+ T cell antigen (Gervassi *et al.*, 2004). Infection with the CPAF null mutant could be used to verify that OmcB processing requires CPAF. The CPAF null mutant also provides a means to test if CPAF-dependent OmcB processing is necessary for binding of EBs to the host cell in cell culture and for production of antibodies against OmcB in animal studies.

Proteolysis of an antimicrobial peptide and complement factors

Active CPAF is proposed to be released from lysed *Chlamydia*-infected cells to cleave extracellular host defence molecules to prevent their anti-chlamydial effects (Conrad *et al.*, 2013). This model is based on the ability of recombinant CPAF to cleave and neutralize the antichlamydial activities of the antimicrobial peptide LL-37 and the complement factors, C3 and factor B (Tang *et al.*, 2015; Yang *et al.*, 2016b). *Chlamydia* infection in a mouse model is cleared by activation of the alternative complement cascade (Bode *et al.*, 2012), so inactivation of complement by CPAF would be advantageous for chlamydiae. Genetic proof could be obtained by showing that the CPAF null mutant has lower infectivity in mice and that this deficit is rescued in mice deficient for LL-37 or C3 and factor B.

Increased expression of T3S proteins

Infection with the CPAF mutant selectively upregulated several structural and effector T3S proteins (Patton *et al.*, 2016). The significance of this finding is not clear, but the authors speculated that CPAF regulates T3S by directly targeting T3S effectors for proteolysis in the host cytoplasm. This model awaits further experimental testing to verify that these chlamydial T3S proteins are *in vivo* CPAF substrates and that their proteolysis modulates T3S.

Proposed CPAF functions that have been disproven

CPAF had been proposed to cause reorganization of the Golgi apparatus, resistance to apoptosis, and protection from reinfection (Christian *et al.*, 2011; Dong *et al.*, 2005; Jorgensen *et al.*, 2011; Pirbhai *et al.*, 2006), but newer data show that CPAF is not required to produce these phenotypes. Doubts about CPAF regulating Golgi function and apoptosis were raised when the proteolysis CPAF substrates that are known to regulate Golgi function and apoptosis was found to be *in vitro* phenomena (Chen *et al.*, 2012). Golgi reorganization, apoptosis resistance and protection from superinfection were still observed during infection with the CPAF null mutant, which provided compelling genetic evidence that these are not CPAF-dependent phenomena (Snavely *et al.*, 2014). CPAF had also been postulated to regulate adaptive immunity by cleaving transcription factors required for expression of antigen presentation molecules (Zhong *et al.*, 1999; Zhong *et al.*, 2000), but this proteolysis was shown to only occur during lysate preparation (Chen *et al.*, 2012). CPAF had also been proposed to be required for inclusion integrity and remodelling (Kumar and Valdivia, 2008), but the CPAF null mutant is still able to form and maintain an inclusion (Snavely *et al.*, 2014). Thus, CPAF is not necessary for these phenotypes, which must therefore be mediated by other mechanisms within a *Chlamydia*-infected cell.

CPAF: looking ahead

Our understanding of the role of CPAF during the *Chlamydia* infection continues to evolve. A major challenge is to elucidate the biological relevance of this potent protease, including its full repertoire of *in vivo* targets and the mechanisms that regulate their proteolysis. CPAF is not necessary for cell culture infections, and there is no evidence of a redundant protein. Nevertheless, CPAF is likely to have one or more important functions because it is conserved in all *Chlamydia* and *Chlamydia*-related bacteria.

CPAF has broad relevance to *Chlamydia* researchers because of its potential to compromise any analysis of chlamydial or host proteins from *Chlamydia*-infected cells. It is now abundantly clear that assays measuring proteolysis of CPAF substrates must include precautions to ensure that there is no *in vitro* proteolysis. This caveat even applies to studies that do not focus on CPAF, such as proteomics studies or *in vitro* reporter assays, which can be affected by experimental manipulations that induce CPAF-mediated proteolysis.

Confidence in the correct identification of CPAF substrates would be greatly enhanced by the wider use of alternative methods to document proteolysis within a *Chlamydia*-infected cell. Microscopy is likely to be the main method to convincingly show that putative CPAF substrates undergo proteolysis in an intact cell. For example, immunofluorescence microscopy may detect protein cleavage if the fragments have different subcellular localization patterns, as was observed with *C. trachomatis* OmcB (Qi *et al.*, 2011). Protein degradation could be detected by the loss of immunofluorescence signal. Proteolysis could be monitored during the course of infection in live cells using a fluorescently tagged version of the proposed protein substrate that is expressed ectopically in *Chlamydia*-infected cells or by a transformed chlamydial strain. Furthermore, loss of a FRET signal may provide a suitable assay to detect cleavage of a tagged substrate in a live infected cell. These methods could then be repeated with the CPAF null mutant to show that CPAF is necessary for the observed proteolysis.

There is a critical need to determine where CPAF is located in an infected cell. Since fixation conditions affect CPAF localization (Prusty *et al.*, 2018), live cell microscopy could be used to follow the location and secretion of fluorescently tagged CPAF over the course of the developmental cycle. Live cell imaging could also be used to investigate if experimental manipulation alters CPAF localization.

Determining how the protease activity of CPAF is regulated is critical for understanding how susceptible substrates undergo proteolysis at the appropriate time and place. If CPAF is confined to the inclusion, the protease could be viewed as a ticking time bomb that explodes and triggers host cell death when the inclusion ruptures (Kerr *et al.*, 2017; Snavely *et al.*, 2014). Interventions that cause premature detonation could then serve as a novel therapeutic approach to prevent *Chlamydia* replication, at the cost of killing the infected host cell.

There is undoubtedly more to learn about the function of CPAF during a *Chlamydia* infection. The CPAF null mutant is an important genetic tool, but by itself, it does not reveal which CPAF substrates are responsible for particular phenotypes. The contribution of an individual CPAF substrate could be studied by designing a proteolysis-resistant version of the protein. Alternatively, the contribution of a specific cellular pathway could be investigated by testing if the phenotype of a CPAF null mutant is rescued in cells or mice defective in that pathway (Rajeeve *et al.*, 2018). CPAF may be challenging to study, but exciting discoveries about the full spectrum of its functions lie ahead.

References

Balsara, Z.R., Misaghi, S., Lafave, J.N., and Starnbach, M.N. (2006). *Chlamydia trachomatis* infection induces cleavage of the mitotic cyclin B1. Infect. Immun. 74, 5602–5608.

Bavoil, P.M., and Byrne, G.I. (2014). Analysis of CPAF mutants: new functions, new questions (The ins and outs of a chlamydial protease). Pathog. Dis. *71*, 287–291.

Bednar, M.M., Jorgensen, I., Valdivia, R.H., and McCafferty, D.G. (2011). Chlamydia protease-like activity factor

(CPAF): characterization of proteolysis activity in vitro and development of a nanomolar affinity CPAF zymogen-derived inhibitor. Biochemistry *50*, 7441–7443.

Belland, R.J., Zhong, G., Crane, D.D., Hogan, D., Sturdevant, D., Sharma, J., Beatty, W.L., and Caldwell, H.D. (2003). Genomic transcriptional profiling of the developmental cycle of *Chlamydia trachomatis*. Proc. Natl. Acad. Sci. U S A *100*, 8478–8483.

Biniossek, M.L., Niemer, M., Maksimchuk, K., Mayer, B., Fuchs, J., Huesgen, P.F., McCafferty, D.G., Turk, B., Fritz, G., Mayer, J., Haecker, G., Mach, L., and Schilling, O. (2016). Identification of protease specificity by combining proteome-derived peptide libraries and quantitative proteomics. Mol. Cell Proteomics *15*, 2515–2524.

Bode, J., Dutow, P., Sommer, K., Janik, K., Glage, S., Tummler, B., Munder, A., Laudeley, R., Sachse, K.W., and Klos, A. (2012). A new role of the complement system: C3 provides protection in a mouse model of lung infection with intracellular *Chlamydia psittaci*. PLoS ONE *7*, e50327.

Brown, H.M., Knowlton, A.E., and Grieshaber, S.S. (2012). Chlamydial infection induces host cytokinesis failure at abscission. Cell Microbiol. *14*, 1554–1567.

Brown, H.M., Knowlton, A.E., Snavely, E., Nguyen, B.D., Richards, T.S., and Grieshaber, S.S. (2014). Multinucleation during *C. trachomatis* infections is caused by the contribution of two effector pathways. PLoS ONE *9*, e100763.

Chen, A.L., Johnson, K.A., Lee, J.K., Sutterlin, C., and Tan, M. (2012). CPAF: a chlamydial protease in search of an authentic substrate. PLoS Pathogens *8*, e1002842.

Chen, D., Chai, J., Hart, P.J., and Zhong, G. (2009). Identifying catalytic residues in CPAF, a *Chlamydia*-secreted protease. Arch. Biochem. Biophys. *485*, 16–23.

Chen, D., Lei, L., Flores, R., Huang, Z., Wu, Z., Chai, J., and Zhong, G. (2010a). Autoprocessing and self-activation of the secreted protease CPAF in *Chlamydia*-infected cells. Microb. Pathog. *49*, 164–173.

Chen, D., Lei, L., Lu, C., Flores, R., DeLisa, M.P., Roberts, T.C., Romesberg, F.E., and Zhong, G. (2010b). Secretion of the chlamydial virulence factor CPAF requires the Sec-dependent pathway. Microbiology *156*, 3031–3040.

Cheong, H.C., C.Lee, Y.Q., Cheok, Y.Y., Shankar, E.M., Sabet, N.S., G.Tan, M.Y., Movahed, E., Yeow, T.C., Sulaiman, S., Wong, W.F., Looi, C.Y., Gupta, R., Hassan, J., Arulanandam, B., and AbuBakar, S. (2019). CPAF, HSP60 and MOMP antigens elicit pro-inflammatory cytokines production in the peripheral blood mononuclear cells from genital *Chlamydia trachomatis*-infected patients. Immunobiology *224*, 34–41.

Christian, J., Vier, J., Paschen, S.A., and Hacker, G. (2010). Cleavage of the NF-kappaB family protein p65/RelA by the chlamydial protease-like activity factor (CPAF) impairs proinflammatory signaling in cells infected with Chlamydiae. J. Biol. Chem. *285*, 41320–41327.

Christian, J.G., Heymann, J., Paschen, S.A., Vier, J., Schauenburg, L., Rupp, J., Meyer, T.F., Hacker, G., and Heuer, D. (2011). Targeting of a chlamydial protease impedes intracellular bacterial growth. PLoS Pathog. *7*, e1002283.

Collingro, A., Tischler, P., Weinmaier, T., Penz, T., Heinz, E., Brunham, R.C., Read, T.D., Bavoil, P.M., Sachse, K., Kahane, S., Friedman, M.G., Rattei, T., Myers, G.S., and Horn, M. (2011). Unity in variety - the pan-genome of the *Chlamydiae*. Mol. Biol. Evol. 3253–3270.

Cong, Y., Jupelli, M., Guentzel, M.N., Zhong, G., Murthy, A.K., and Arulanandam, B.P. (2007). Intranasal immunization with chlamydial protease-like activity factor and CpG deoxynucleotides enhances protective immunity against genital *Chlamydia muridarum* infection. Vaccine *25*, 3773–3780.

Conrad, T., Yang, Z., Ojcius, D., and Zhong, G. (2013). A path forward for the chlamydial virulence factor CPAF. Microbes Infect. *15*, 1026–1032.

Costa, T.R., Felisberto-Rodrigues, C., Meir, A., Prevost, M.S., Redzej, A., Trokter, M., and Waksman, G. (2015). Secretion systems in Gram-negative bacteria: structural and mechanistic insights. Nat. Rev. Microbiol. *13*, 343–359.

Dille, S., Herbst, K., Volceanov, L., Nolke, T., Kretz, O., and Hacker, G. (2014). Golgi fragmentation and sphingomyelin transport to *Chlamydia trachomatis* during penicillin-induced persistence do not depend on the cytosolic presence of the chlamydial protease CPAF. PLoS ONE *9*, e103220.

Dong, F., Pirbhai, M., Xiao, Y., Zhong, Y., Wu, Y., and Zhong, G. (2005). Degradation of the proapoptotic proteins Bik, Puma, and Bim with Bcl-2 domain 3 homology in *Chlamydia trachomatis*-infected cells. Infect. Immun. *73*, 1861–1864.

Dong, F., Pirbhai, M., Zhong, Y., and Zhong, G. (2004a). Cleavage-dependent activation of a chlamydia-secreted protease. Mol. Microbiol. *52*, 1487–1494.

Dong, F., Su, H., Huang, Y., Zhong, Y., and Zhong, G. (2004b). Cleavage of host keratin 8 by a chlamydia-secreted protease. Infect. Immun. *72*, 3863–3868.

Duvall, M.G., and Levy, B.D. (2016). DHA- and EPA-derived resolvins, protectins, and maresins in airway inflammation. Eur. J. Pharmacol. *785*, 144–155.

Fadel, S., and Eley, A. (2007). *Chlamydia trachomatis* OmcB protein is a surface-exposed glycosaminoglycan-dependent adhesin. J. Med. Microbiol. *56*, 15–22.

Forsbach-Birk, V., Foddis, C., Simnacher, U., Wilkat, M., Longbottom, D., Walder, G., Benesch, C., Ganter, M., Sachse, K., and Essig, A. (2013). Profiling antibody responses to infections by *Chlamydia abortus* enables identification of potential virulence factors and candidates for serodiagnosis. PLoS ONE *8*, e80310.

Gervassi, A.L., Grabstein, K.H., Probst, P., Hess, B., Alderson, M.R., and Fling, S.P. (2004). Human CD8+ T cells recognize the 60-kDa cysteine-rich outer membrane protein from *Chlamydia trachomatis*. J. Immunol. *173*, 6905–6913.

Grieshaber, S.S., and Grieshaber, N.A. (2014). The role of the chlamydial effector CPAF in the induction of genomic instability. Pathog. Dis. *72*, 5–6.

Hacker, G. (2014). The chlamydial protease CPAF: important or not, important for what? Microbes Infect. *16*, 367–370.

Heuer, D., Brinkmann, V., Meyer, T.F., and Szczepek, A.J. (2003). Expression and translocation of chlamydial protease during acute and persistent infection of the epithelial HEp-2 cells with *Chlamydophila (Chlamydia) pneumoniae*. Cell Microbiol. *5*, 315–322.

Heuer, D., Rejman Lipinski, A., Machuy, N., Karlas, A., Wehrens, A., Siedler, F., Brinkmann, V., and Meyer, T.F. (2009). *Chlamydia* causes fragmentation of the Golgi compartment to ensure reproduction. Nature *457*, 731–735.

Hou, S., Lei, L., Yang, Z., Qi, M., Liu, Q., and Zhong, G. (2013). *Chlamydia trachomatis* outer membrane complex protein B (OmcB) is processed by the protease CPAF. J. Bacteriol. *195*, 951–957.

Huang, Z., Feng, Y., Chen, D., Wu, X., Huang, S., Wang, X., Xiao, X., Li, W., Huang, N., Gu, L., Zhong, G., and Chai, J. (2008). Structural basis for activation and inhibition of the secreted *Chlamydia* protease CPAF. Cell Host Microbe *4*, 529–542.

Johnson, K.A., Lee, J.K., Chen, A.L., Tan, M., and Sutterlin, C. (2015). Induction and inhibition of CPAF activity during analysis of *Chlamydia*-infected cells. Pathog. Dis. *73*, 1–8.

Jorgensen, I., Bednar, M.M., Amin, V., Davis, B.K., Ting, J.P., McCafferty, D.G., and Valdivia, R.H. (2011). The *Chlamydia* protease CPAF regulates host and bacterial proteins to maintain pathogen vacuole integrity and promote virulence. Cell Host Microbe *10*, 21–32.

Kari, L., Bakios, L.E., Goheen, M.M., Bess, L.N., Watkins, H.S., Southern, T.R., Song, L., Whitmire, W.M., Olivares-Zavaleta, N., and Caldwell, H.D. (2013). Antibody signature of spontaneous clearance of *Chlamydia trachomatis* ocular infection and partial resistance against re-challenge in a nonhuman primate trachoma model. PLoS Negl. Trop. Dis. *7*, e2248.

Kawana, K., Quayle, A.J., Ficarra, M., Ibana, J.A., Shen, L., Kawana, Y., Yang, H., Marrero, L., Yavagal, S., Greene, S.J., Zhang, Y.X., Pyles, R.B., Blumberg, R.S., and Schust, D.J. (2007). CD1d degradation in *Chlamydia trachomatis*-infected epithelial cells is the result of both cellular and chlamydial proteasomal activity. J. Biol. Chem. *282*, 7368–7375.

Kerr, M.C., Gomez, G.A., Ferguson, C., Tanzer, M.C., Murphy, J.M., Yap, A.S., Parton, R.G., Huston, W.M., and Teasdale, R.D. (2017). Laser-mediated rupture of chlamydial inclusions triggers pathogen egress and host cell necrosis. Nat. Commun. *8*, 14729.

Kokes, M., and Valdivia, R.H. (2015). Differential translocation of host cellular materials into the *Chlamydia trachomatis* inclusion lumen during chemical fixation. PLoS ONE *10*, e0139153.

Kumar, Y., and Valdivia, R.H. (2008). Actin and intermediate filaments stabilize the *Chlamydia trachomatis* vacuole by forming dynamic structural scaffolds. Cell Host Microbe *4*, 159–169.

Lad, S.P., Yang, G., Scott, D.A., Wang, G., Nair, P., Mathison, J., Reddy, V.S., and Li, E. (2007). Chlamydial CT441 is a PDZ domain-containing tail-specific protease that interferes with the NF-kappaB pathway of immune response. J. Bacteriol. *189*, 6619–6625.

Lee, J.K., Enciso, G.A., Boassa, D., Chander, C.N., Lou, T.H., Pairawan, S.S., Guo, M.C., F.Wan, Y.M., Ellisman, M.H., Sutterlin, C., and Tan, M. (2018). Replication-dependent size reduction precedes differentiation in *Chlamydia trachomatis*. Nat. Commun. *9*, 45.

Li, W., Gudipaty, P., Li, C., Henderson, K.K., Ramsey, K.H., and Murthy, A.K. (2018). Intranasal immunization with recombinant chlamydial protease-like activity factor attenuates atherosclerotic pathology following *Chlamydia pneumoniae* infection in mice. Immunol. Cell Biol.

Li, W., Guentzel, M.N., Seshu, J., Zhong, G., Murthy, A.K., and Arulanandam, B.P. (2007). Induction of cross-serovar protection against genital chlamydial infection by a targeted multisubunit vaccination approach. Clin. Vaccine Immunol. *14*, 1537–1544.

Li, W., Murthy, A.K., Guentzel, M.N., Seshu, J., Forsthuber, T.G., Zhong, G., and Arulanandam, B.P. (2008). Antigen-specific CD4+ T cells produce sufficient IFN-gamma to mediate robust protective immunity against genital *Chlamydia muridarum* infection. J. Immunol. *180*, 3375–3382.

Li, W., Murthy, A.K., Lanka, G.K., Chetty, S.L., Yu, J.J., Chambers, J.P., Zhong, G., Forsthuber, T.G., Guentzel, M.N., and Arulanandam, B.P. (2013). A T cell epitope-based vaccine protects against chlamydial infection in HLA-DR4 transgenic mice. Vaccine *31*, 5722–5728.

Matsuo, J., Nakamura, S., Ito, A., Yamazaki, T., Ishida, K., Hayashi, Y., Yoshida, M., Takahashi, K., Sekizuka, T., Takeuchi, F., Kuroda, M., Nagai, H., Hayashida, K., Sugimoto, C., and Yamaguchi, H. (2013). Protochlamydia induces apoptosis of human HEp-2 cells through mitochondrial dysfunction mediated by chlamydial protease-like activity factor. PLoS ONE *8*, e56005.

Moelleken, K., and Hegemann, J.H. (2008). The *Chlamydia* outer membrane protein OmcB is required for adhesion and exhibits biovar-specific differences in glycosaminoglycan binding. Mol. Microbiol. *67*, 403–419.

Murphey, C., Murthy, A.K., Meier, P.A., Guentzel, N.M., Zhong, G., and Arulanandam, B.P. (2006). The protective efficacy of chlamydial protease-like activity factor vaccination is dependent upon CD4+ T cells. Cell Immunol. *242*, 110–117.

Murthy, A.K., Chambers, J.P., Meier, P.A., Zhong, G., and Arulanandam, B.P. (2007). Intranasal vaccination with a secreted chlamydial protein enhances resolution of genital *Chlamydia muridarum* infection, protects against oviduct pathology, and is highly dependent upon endogenous gamma interferon production. Infect. Immun. *75*, 666–676.

Murthy, A.K., Cong, Y., Murphey, C., Guentzel, M.N., Forsthuber, T.G., Zhong, G., and Arulanandam, B.P. (2006). Chlamydial protease-like activity factor induces protective immunity against genital chlamydial infection in transgenic mice that express the human HLA-DR4 allele. Infect. Immun. *74*, 6722–6729.

Murthy, A.K., Guentzel, M.N., Zhong, G., and Arulanandam, B.P. (2009). Chlamydial protease-like activity factor--insights into immunity and vaccine development. J. Reprod. Immunol. *83*, 179–184.

Murthy, A.K., Li, W., Guentzel, M.N., Zhong, G., and Arulanandam, B.P. (2011). Vaccination with the defined chlamydial secreted protein CPAF induces robust protection against female infertility following repeated genital chlamydial challenge. Vaccine *29*, 2519–2522.

Mygind, P., Christiansen, G., and Birkelund, S. (1998). Topological analysis of *Chlamydia trachomatis* L2 outer membrane protein 2. J. Bacteriol. *180*, 5784–5787.

Paschen, S.A., Christian, J.G., Vier, J., Schmidt, F., Walch, A., Ojcius, D.M., and Hacker, G. (2008). Cytopathicity of *Chlamydia* is largely reproduced by expression of a single chlamydial protease. J. Cell Biol. *182*, 117–127.

Patton, M.J., McCorrister, S., Grant, C., Westmacott, G., Fariss, R., Hu, P., Zhao, K., Blake, M., Whitmire, B., Yang, C., Caldwell, H.D., and McClarty, G. (2016). Chlamydial protease-like activity factor and type III secreted effectors cooperate in inhibition of p65 nuclear translocation. MBio 7.

Pirbhai, M., Dong, F., Zhong, Y., Pan, K.Z., and Zhong, G. (2006). The secreted protease factor CPAF is responsible for degrading pro-apoptotic BH3-only proteins in *Chlamydia trachomatis*-infected cells. J. Biol. Chem. *281*, 31495–31501.

Prusty, B.K., Chowdhury, S.R., Gulve, N., and Rudel, T. (2018). Peptidase inhibitor 15 (PI15) regulates chlamydial CPAF activity. Front. Cell Infect. Microbiol. *8*, 183.

Qi, M., Gong, S., Lei, L., Liu, Q., and Zhong, G. (2011). A *Chlamydia trachomatis* OmcB C-terminal fragment is released into the host cell cytoplasm and is immunogenic in humans. Infect. Immun. *79*, 2193–2203.

Rajalingam, K., Sharma, M., Lohmann, C., Oswald, M., Thieck, O., Froelich, C.J., and Rudel, T. (2008). Mcl-1 is a key regulator of apoptosis resistance in *Chlamydia trachomatis*-infected cells. PLoS ONE 3, e3102.

Rajeeve, K., Das, S., Prusty, B.K., and Rudel, T. (2018). *Chlamydia trachomatis* paralyses neutrophils to evade the host innate immune response. Nat. Microbiol. *3*, 824–835.

Rodel, J., Grosse, C., Yu, H., Wolf, K., Otto, G.P., Liebler-Tenorio, E., Forsbach-Birk, V., and Straube, E. (2012). Persistent *Chlamydia trachomatis* infection of HeLa cells mediates apoptosis resistance through a *Chlamydia* protease-like activity factor-independent mechanism and induces high mobility group box 1 release. Infect. Immun. *80*, 195–205.

Rupp, J., Gieffers, J., Klinger, M., van Zandbergen, G., Wrase, R., Maass, M., Solbach, W., Deiwick, J., and Hellwig-Burgel, T. (2007). *Chlamydia pneumoniae* directly interferes with HIF-1alpha stabilization in human host cells. Cell Microbiol. *9*, 2181–2191.

Savijoki, K., Alvesalo, J., Vuorela, P., Leinonen, M., and Kalkkinen, N. (2008). Proteomic analysis of *Chlamydia pneumoniae*-infected HL cells reveals extensive degradation of cytoskeletal proteins. FEMS Immunol. Med. Microbiol. *54*, 375–384.

Sharma, J., Bosnic, A.M., Piper, J.M., and Zhong, G. (2004). Human antibody responses to a *Chlamydia*-secreted protease factor. Infect. Immun. 72, 7164–7171.

Sharma, J., Dong, F., Pirbhai, M., and Zhong, G. (2005). Inhibition of proteolytic activity of a chlamydial proteasome/protease-like activity factor by antibodies from humans infected with *Chlamydia trachomatis*. Infect. Immun. 73, 4414–4419.

Shaw, A.C., Vandahl, B.B., Larsen, M.R., Roepstorff, P., Gevaert, K., Vandekerckhove, J., Christiansen, G., and Birkelund, S. (2002). Characterization of a secreted *Chlamydia* protease. Cell Microbiol. *4*, 411–424.

Skwor, T., Kandel, R.P., Basravi, S., Khan, A., Sharma, B., and Dean, D. (2010). Characterization of humoral immune responses to chlamydial HSP60, CPAF, and CT795 in inflammatory and severe trachoma. Invest. Ophthalmol. Vis. Sci. *51*, 5128–5136.

Snavely, E.A., Kokes, M., Dunn, J.D., Saka, H.A., Nguyen, B.D., Bastidas, R.J., McCafferty, D.G., and Valdivia, R.H. (2014). Reassessing the role of the secreted protease CPAF in *Chlamydia trachomatis* infection through genetic approaches. Pathog. Dis. *71*, 336–351.

Stephens, R.S., Koshiyama, K., Lewis, E., and Kubo, A. (2001). Heparin-binding outer membrane protein of chlamydiae. Mol. Microbiol. *40*, 691–699.

Sun, J., and Schoborg, R.V. (2009). The host adherens junction molecule nectin-1 is degraded by chlamydial protease-like activity factor (CPAF) in *Chlamydia trachomatis*-infected genital epithelial cells. Microbes Infect. *11*, 12–19.

Susnea, I., Bunk, S., Wendel, A., Hermann, C., and Przybylski, M. (2011). Biomarker candidates of *Chlamydophila pneumoniae* proteins and protein fragments identified by affinity-proteomics using FTICR-MS and LC-MS/MS. J. Am. Soc. Mass Spectrom. 22, 784–788.

Tan, M., and Sutterlin, C. (2014). The *Chlamydia* protease CPAF: Caution, Precautions And Function. Pathog. Dis.

Tang, L., Chen, J., Zhou, Z., Yu, P., Yang, Z., and Zhong, G. (2015). *Chlamydia*-secreted protease CPAF degrades host antimicrobial peptides. Microbes Infect. *17*, 402–408.

van Ess, E.F., Eck-Hauer, A., Land, J.A., Morre, S.A., and Ouburg, S. (2019). Combining individual *Chlamydia trachomatis* IgG antibodies MOMP, TARP, CPAF, OMP2, and HSP60 for tubal factor infertility prediction. Am. J. Reprod. Immunol. *81*, e13091.

Wang, J., Frohlich, K.M., Buckner, L., Quayle, A.J., Luo, M., Feng, X., Beatty, W., Hua, Z., Rao, X., Lewis, M.E., Sorrells, K., Santiago, K., Zhong, G., and Shen, L. (2011). Altered protein secretion of *Chlamydia trachomatis* in persistently infected human endocervical epithelial cells. Microbiology *157*, 2759–2771.

Yang, Z., Tang, L., Sun, X., Chai, J., and Zhong, G. (2015). Characterization of CPAF critical residues and secretion during *Chlamydia trachomatis* infection. Infect. Immun. *83*, 2234–2241.

Yang, Z., Tang, L., Shao, L., Zhang, Y., Zhang, T., Schenken, R., Valdivia, R., and Zhong, G. (2016a). The *Chlamydia*-secreted protease CPAF promotes chlamydial survival in the mouse lower genital tract. Infect. Immun. *84*, 2697–2702.

Yang, Z.S., Tang, L.L., Zhou, Z.G., and Zhong, G.M. (2016b). Neutralizing antichlamydial activity of complement by chlamydia-secreted protease CPAF. Microbes Infect. *18*, 669–674.

Yu, H., Schwarzer, K., Forster, M., Kniemeyer, O., Forsbach-Birk, V., Straube, E., and Rodel, J. (2010). Role of high-mobility group box 1 protein and poly(ADP-ribose) polymerase 1 degradation in *Chlamydia trachomatis*-induced cytopathicity. Infect. Immun. *78*, 3288–3297.

Zhong, G. (2009). Killing me softly: chlamydial use of proteolysis for evading host defenses. Trends Microbiol. *17*, 467–474.

Zhong, G., Fan, P., Ji, H., Dong, F., and Huang, Y. (2001). Identification of a chlamydial protease-like activity factor responsible for the degradation of host transcription factors. J. Exp. Med. *193*, 935–942.

Zhong, G., Fan, T., and Liu, L. (1999). *Chlamydia* inhibits interferon gamma-inducible major histocompatibility complex class II expression by degradation of upstream stimulatory factor 1. J. Exp. Med. *189*, 1931–1938.

Zhong, G., Liu, L., Fan, T., Fan, P., and Ji, H. (2000). Degradation of transcription factor RFX5 during the inhibition of both constitutive and interferon gamma-inducible major histocompatibility complex class I expression in *Chlamydia*-infected cells. J. Exp. Med. *191*, 1525–1534.

Chlamydia Cell Division and Differentiation

9

George W. Liechti[1]*, Patrick H. Viollier[2], Gilbert Greub[3] and Anthony T. Maurelli[4]

[1]Department of Microbiology and Immunology, Uniformed Services University of the Health Sciences, Bethesda, MD, USA.
[2]Department of Microbiology and Molecular Medicine, Institute of Genetics and Genomics of Geneva, University of Geneva, Geneva, Switzerland.
[3]Institute of Microbiology, University of Lausanne and University Hospital, Lausanne, Switzerland.
[4]Department of Environmental and Global Health and Emerging Pathogens Institute, University of Florida, Gainesville, FL, USA.

*Correspondence: george.liechti@usuhs.edu

https://doi.org/10.21775/9781912530281.09

Abstract

In this chapter we review the current state of our knowledge regarding the molecular processes by which *Chlamydia* species divide and differentiate between their replicative and infectious forms. We will (1) give a general introduction to cell division in Gram-negative bacteria, (2) compare this process with the process of cell division in *Chlamydia*, (3) address the 'chlamydial anomaly' by discussing our collective knowledge of peptidoglycan synthesis in *Chlamydia*, (4) describe our current understanding of the localization and kinetics of peptidoglycan assembly, (5) compare binary fission and polarized division as models of *Chlamydia* replication, (6) review what is known about the elementary body (EB) to reticulate body (RB) transition, and (7) evaluate the current models for RB to EB differentiation. We will also describe similarities and differences observed regarding division, peptidoglycan and persistence among the different members of the Chlamydiae phylum.

Introduction

Members of the Chlamydiae utilize a biphasic developmental cycle consisting of an extracellular, infectious form (elementary body, EB) and an intracellular, replicative form (reticulate body, RB). This unique adaptation to an intracellular environment has prompted curiosity and confusion since the first representatives of the Chlamydiae phylum were identified over a century ago. Progress in understanding how members of the Chlamydiae carry out basic biological processes, such as cell division and differentiation/conversion between developmental forms, has been modest compared to advances that have been made in other microbes, largely due to the inability to grow and propagate these obligate intracellular pathogens in an extracellular environment. Scientists have spent significant time and resources over the last 10 years adapting common molecular approaches for use in genetically intractable *Chlamydia* species. These persistent efforts have recently begun to bear fruit, shedding light on a number of the physiological and metabolic processes utilized by these microbes that have confounded the field for decades. Thanks largely to advances in molecular, genetic, and imaging techniques, notable progress has been made in the elucidation and evaluation of the complex molecular machinery utilized by members of the Chlamydiae phylum to undergo cell division.

Meanwhile, recent discoveries in *Escherichia coli* and *Bacillus subtilis* have provided insight into how members of the *Chlamydia* genus carry out these intricate, cellular processes utilizing only a fraction of the components required by other bacterial species. Many of the most successful antibiotics in use today target the ability of bacterial cells to divide, highlighting the importance of understanding chlamydial replication. Thus, the identification of key structural and regulatory components essential for the division and propagation of pathogenic microbes remains essential in the pursuit and development of new therapeutics.

Cell division complex/ components

Cell division in Gram-negative bacteria

The main purpose of the cell division complex is to provide the biosynthetic capacity to polymerize and remodel peptidoglycan (PG), a large macromolecule that in most bacteria comprises the cell wall or 'sacculus', and acts as an exoskeleton enveloping the bacterial cell and protecting it from osmotic lysis (Egan and Vollmer, 2013). PG synthesis dictates the direction in which a bacterial cell will grow, as well as its eventual shape. PG also promotes bacterial division. Nascent PG biosynthesis is spatially and temporally restricted within most bacterial cells via two molecular complexes, each relying on its own principal, filament-forming cytoskeletal protein: the FtsZ complex ('divisome'), which is associated with septal PG synthesis required during cell division, and the MreB complex ('elongasome'), which is generally associated with bacterial cell wall synthesis at other locations (Fig. 9.1). In both systems, filament-forming proteins (FtsZ or MreB) are organizers of the PG assembly machinery and act on the cytoplasmic side of the inner membrane. As these protein monomers polymerize or 'treadmill' along the membrane, the assembly complexes travel with them, effectively directing localized PG synthesis.

At the end of the cell division cycle, in a process referred to as septation, highly localized, mid-cell (septal) PG synthesis results in an invagination in

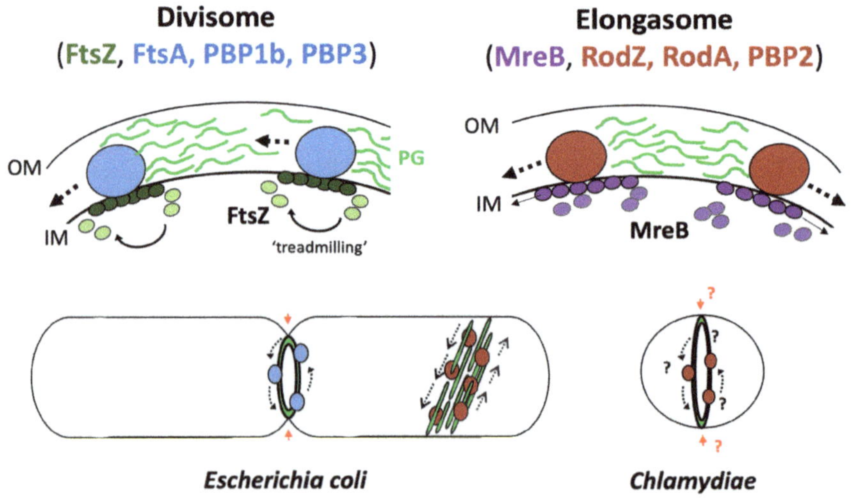

Figure 9.1 Rod-shaped bacteria utilize two molecular complexes for peptidoglycan biosynthesis. The 'divisome' (blue) and the 'elongasome' (red) are associated with cell division and elongation, respectively. These complexes are governed by the cytoplasmic, filament-forming proteins FtsZ and MreB. Both complexes associate with the polymerized form of either FtsZ or MreB and the progression of these protein polymers across the inner leaflet of the inner membrane (solid arrows) results in the movement of the synthesis complex (dashed arrows) and subsequent, localized peptidoglycan synthesis along the same tract (green). In most bacteria, FtsZ is recruited to mid-cell where localized peptidoglycan synthesis is required for a dividing cell to undergo septation. As the Chlamydiae lack FtsZ but still require peptidoglycan synthesis in order to complete septation, it has been proposed that the elongasome (also referred to as the Rod complex) functions as the division complex in members of the Chlamydiae, and that MreB acts as the septal organizer.

the cell envelope along the plane of constriction. This circumferential constriction at the septal plane is under tight spatio-temporal control by the septal organizing protein FtsZ, which structurally resembles the eukaryotic cytoskeletal protein tubulin. Perturbation of the bacterial cytoskeleton with depolymerizing drugs or inhibition of PG biosynthetic proteins with antibiotics such as penicillin prevents constriction at the cell division plane and thus blocks cell division (Jacquier *et al.*, 2015b). Upon synthesis of septal PG, the bacterial inner membrane and cytoplasm constrict at the leading edge of the septal invagination. In addition to the cytoplasmic membrane, Gram-negative bacteria also maintain an essential protective outer membrane decorated with lipopolysaccharides that cover the PG layer and prevent the penetration of noxious, hydrophobic compounds (Silhavy *et al.*, 2010). As such, cell division in Gram-negative bacteria also requires the invagination and constriction of this outer membrane, and this can only be accomplished if PG is simultaneously degraded at the trailing edge of the septum. Protein complexes bridge the outer and inner membranes at this trailing edge, and promote coordinated invagination of outer and inner membranes during septation.

Removal of septal PG during septation requires the action of hydrolytic enzymes. Many of these enzymes are also required for the assembly of trans-envelope protein structures such as the type III secretion system for protein secretion, transporters for the uptake of peptides and other metabolites, and efflux pumps for the expulsion of small toxic molecules. Intricate regulatory mechanisms that ensure and restrict PG hydrolysis at the division site have been described in Gram-negative bacteria (Frirdich and Gaynor, 2013; Singh *et al.*, 2015; Srivastava *et al.*, 2018; Uehara and Bernhardt, 2011; Uehara and Park, 2008).

Chlamydiae cell division

Compared to most other bacterial lineages, the genomes of all members of the Chlamydiae reveal a remarkably reduced and seemingly minimal complement of cell division genes important for binary fission. Chlamydiae are exceptional in that their genomes do not encode an orthologue of the tubulin-like FtsZ protein, which almost all other bacteria utilize to organize PG biosynthetic and hydrolytic enzymes in space and time (Fig. 9.2). While FtsZ and its interaction partners appear to be absent in all of the Chlamydiae, most of the enzymes known to be involved in PG biosynthesis and processing appear to be encoded in chlamydial genomes (Jacquier *et al.*, 2015b). Owing to the limited genetic tractability of most chlamydial species, past research on predicted chlamydial orthologues of cell division proteins have been examined by testing if they can complement *E. coli* mutants. These systems tend to work well for enzymes whose activities

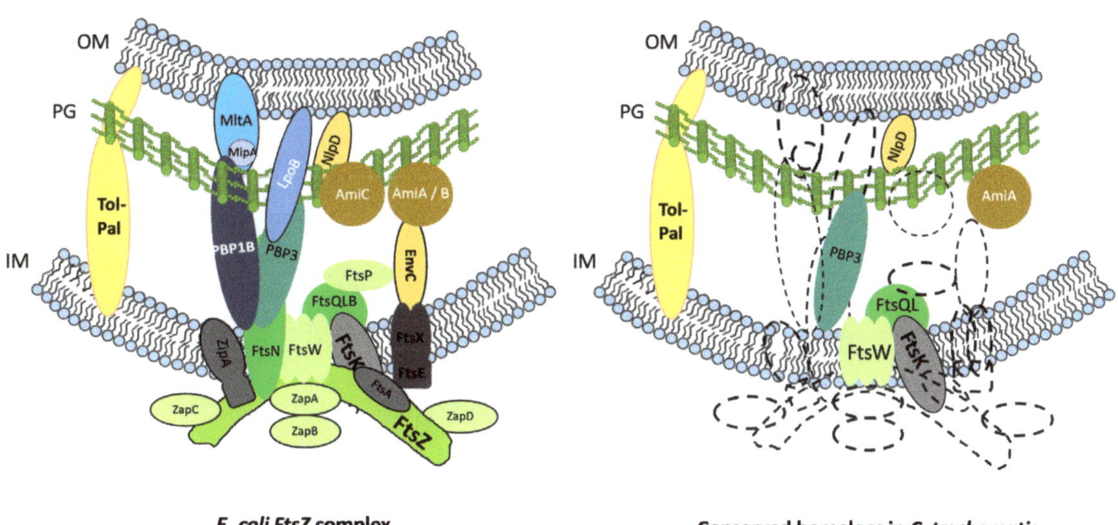

Figure 9.2 Members of the Chlamydiae lack the majority of the 'early divisome' components present in most Gram-negative bacteria.

do not require association with native partner proteins, and/or can use the substrate provided in *in vitro* assays or heterologous *in vivo* systems. This is the case for most enzymes of the PG precursor (Lipid II, see below) biosynthesis pathway, as well as for hydrolytic enzymes that cleave either amide (amidases), peptide (endopeptidases or carboxypeptidases) or glycosidic (lytic transglycosylases) bonds in PG (Jacquier *et al.*, 2015b). It has not yet been possible to demonstrate the biosynthetic activity of the predicted chlamydial orthologues that carry out the polymerization of PG from the Lipid II precursor via trans-peptidation (TP) and trans-glycosylation (TG) reactions. These activities usually reside within polypeptides that are part of the PG biosynthetic machinery, and it appears that TP and TG activities either require the integrity of the entire complex or depend on other factors that are not met in *in vitro* assays or in heterologous host systems. Similarly, the regulatory action(s) of the cytoskeletal elements of these complexes (such as FtsZ and MreB) are often highly specialized and heterologous complementation experiments with these proteins are often unsuccessful.

The cell division complex in the Chlamydiae

Genomic studies investigating the distinctive specializations of Chlamydiae are providing new insights into the evolution of bacterial cell division in osmotically stable environments. One notable genomic feature of the Chlamydiae is the absence of a homologue of the essential cell division gene, *ftsZ*, as well as of homologues of many of the regulatory proteins (ZipA, ZapA-D, MinC and FtsEX) that bind FtsZ in *E. coli* (Fig. 9.2). Since *Chlamydia* are capable of dividing without FtsZ, other organization systems must exist to spatially restrict PG synthesis and remodel it at the division site during septation. Strong cytologic evidence in *Chlamydia* implicates the cytoskeletal protein MreB, a bacterial actin-like protein and principal regulatory component of the Rod Complex, and its membrane anchor protein, RodZ (Fig. 9.3). The Rod system is generally associated with PG synthesis on bacterial side walls and is responsible for cell elongation in rod-shaped bacteria (Divakaruni *et al.*, 2007). Components of this complex (MreB, RodZ, NlpD, FtsQ) localize to the division plane in *Chlamydia* and in *Chlamydia*-related bacteria (Frandi *et al.*, 2014; Jacquier *et al.*, 2014; Kemege *et al.*, 2015; Ouellette *et al.*, 2015). Interactions between MreB/RodZ and the homologues of the chlamydial division machinery have been detected using bacterial two-hybrid assays performed in *E. coli* (Gaballah *et al.*, 2011; Ouellette *et al.*, 2015; Ouellette *et al.*, 2014). In its filamentous form MreB promotes localized PG synthesis. Inhibitors of MreB polymerization effectively halt *Chlamydia* replication, lead to the disintegration of the MreB/RodZ complex at the septum, and inhibit cell division in *Chlamydia* and '*Chlamydia*-like' bacteria

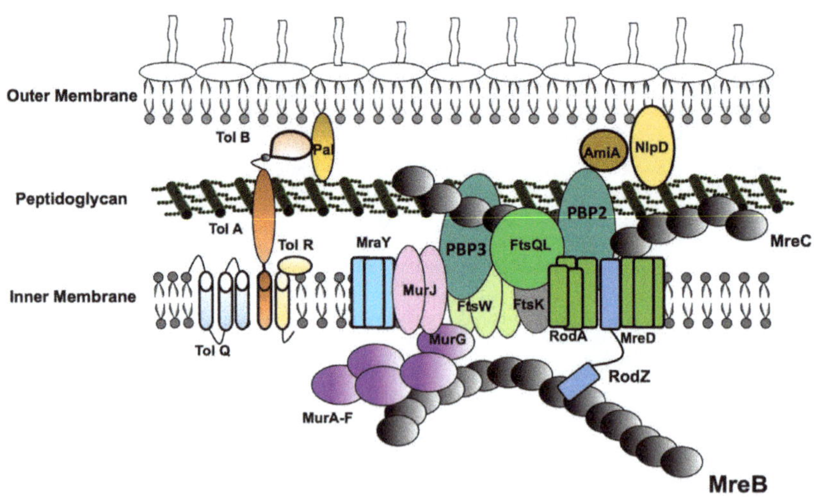

Figure 9.3 The Chlamydiae Rod complex. *Chlamydia* species encode the majority of the molecular components of the Rod system in addition to the Tol–Pal complex.

(Jacquier *et al.*, 2014; Liechti *et al.*, 2016; Ouellette *et al.*, 2012). Remarkably, this also occurs when Lipid II production in '*Chlamydia*-like'-bacteria is inhibited by the addition of fosfomycin, a compound that inhibits the first enzyme in the Lipid II biosynthesis pathway (Fig. 9.4), but not when TP activity is inhibited with penicillin (Jacquier *et al.*, 2014). The notion that Lipid II plays an important role in spatiotemporal regulation of the division machinery in '*Chlamydia*-like' bacteria is supported by recent findings in other bacterial systems, but has not yet been explored in the members of the *Chlamydia* genus due to their innate resistance to fosfomycin (McCoy *et al.*, 2003).

FtsZ not only regulates PG synthesis in *E. coli*, but also the PG-cleaving amidases via the ABC transporter complex FtsEX (Yang *et al.*, 2011) (Fig. 9.2). Engagement of FtsEX by FtsZ at the division plane triggers amidase activity through dislocation of an auto-inhibitory loop domain that occludes the amidase active site. This inhibitory loop ensures that amidases are by default in an inactive state and that their activation occurs in a topologically constrained manner. Interestingly, the components of this amidase activation pathway do not appear to be present in *Chlamydia* and '*Chlamydia*-like' bacteria: their genomes encode amidases but they lack the inhibitory loop domain. Heterologous complementation revealed that the chlamydial amidases are active in their native state (Klockner *et al.*, 2014), and questions remain as to the mechanism(s) that control amidase abundance/action and restrict its activity in space and/or time in *Chlamydia* and '*Chlamydia*-like' bacteria.

Chlamydia **peptidoglycan and its role in cell division**

Initial observations

The question of whether *Chlamydia* species contain PG in their cell walls has been the subject of a highly contentious debate spanning the last 80 years. Early successes with penicillin in the treatment of *Chlamydia* psittacosis and ornithoses

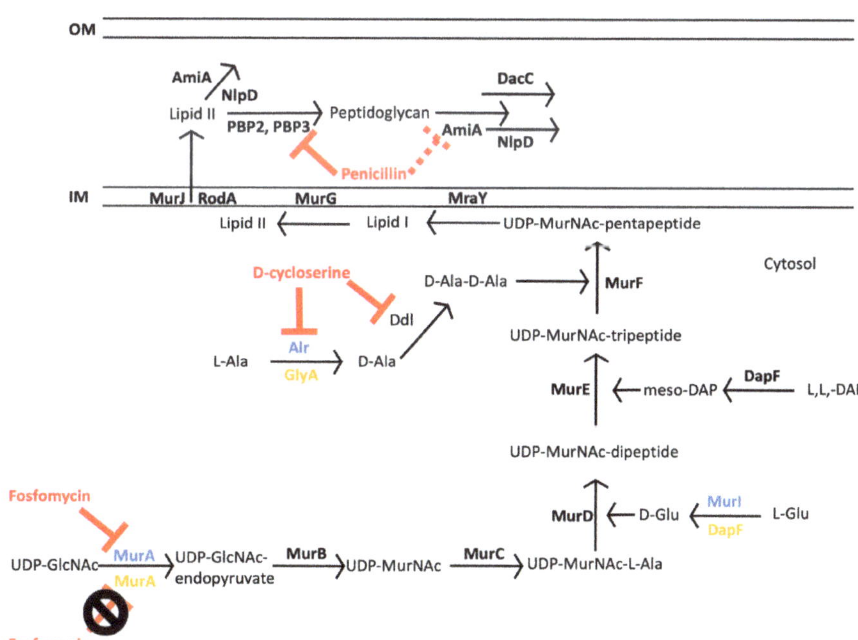

Figure 9.4 Peptidoglycan biosynthesis pathway comparison between *W. chondrophila* and *Chlamydiaceae*. Enzymes present only in *W. chondrophila* (Alr and MurI) are depicted in blue; the enzymes replacing Alr and MurI in *Chlamydiaceae* (GlyA and DapF) are depicted in orange. Steps inhibited by d-cycloserine, fosfomycin, and penicillin are shown in red. Dotted red line indicates that penicillin inhibits the carboxypeptidase activity of AmiA (which is bifunctional) but not the amidase activity. Fosfomycin-sensitive MurA is depicted in blue for *W. chondrophila*, and fosfomycin-resistant MurA in orange for *Chlamydiaceae*. Figure adapted from Jacquier *et al.* (2015b) by permission of Oxford University Press.

in the early 1940s (Heilman, 1944) were largely influential in the ultimate reclassification of 'Psittacosis and Lymphogranuloma Group viruses' as 'bacteria'. Of even greater significance is that these studies pre-dated the initial discovery of bacterial PG (Work, 1961) and the finding that penicillin targets the synthesis of the major muropeptide component of bacterial cell walls (Park and Strominger, 1957). While early results were promising, penicillin was less effective for treatment of lymphogranuloma venereum, trachoma, and inclusion conjunctivitis infections as treatment did not always result in bacterial clearance. In the early 1940s and 1950s, researchers noted that in some cases treatment with penicillin was capable of halting and even clearing infections in mice and in chick embryos (Heilman, 1944; Meiklejohn *et al.*, 1946; Wiseman *et al.*, 1946). However, it was later concluded that treatment of infections due to certain Psittacosis and Lymphogranuloma Group 'viruses' resulted in a carrier state, rather than actual clearance (Bedson, 1945; Early and Morgan, 1946; Meyer and Eddie, 1947). *In vitro* treatment of 'non-multiplying virus' appeared to have no effect on infectivity (Eaton *et al.*, 1948), hinting that the target of penicillin was only present in the microbe's intracellular, replicative form. Later studies utilizing electron microscopy (EM) noted morphological changes in the 'virus' upon penicillin treatment (Gogolak and Weiss, 1950; Hurst *et al.*, 1953; Weiss, 1950). Furthermore, PG-targeting antibiotics were subsequently shown to inhibit division of chlamydiae propagated in tissue culture cells (Hurst *et al.*, 1953; Lin and Moulder, 1966; Moulder *et al.*, 1963; Weiss, 1950). Inhibition of chlamydial cell division by the PG biosynthesis inhibitors penicillin (Matsumoto and Manire, 1970; Tajima *et al.*, 1959; Tamura and Manire, 1968; Weiss, 1950) and d-cycloserine (Moulder *et al.*, 1963) resulted in what has been termed an 'aberrant' phenotype characterized by an arrest of cell division and a significant expansion in the size of the intracellular RBs. These aberrant forms were later referred to as aberrant bodies or aberrant RBs. Subsequent studies revealed that DNA replication proceeds unabated under these aberrance-inducing conditions (Lambden *et al.*, 2006), indicating that, like *E. coli*, cell division and DNA replication are de-coupled in members of the *Chlamydia* genus. This also holds true for *Waddlia chondrophila*, a *Chlamydia*-related bacterium, following penicillin or fosfomycin treatment (A. Scherler and G. Greub, unpublished observation).

It is noteworthy that all known inhibitors of *Chlamydia* cell division also prevent the generation of new infectious EBs, indicating that the processes of *Chlamydia* division and differentiation are either linked or that they utilize overlapping components (Jacquier *et al.*, 2015b). A recent study has reported that RBs exhibit cell size reduction over the course of the developmental cycle, and the authors proposed a model whereby a minimal size threshold must be reached by an RB prior to its conversion to an EB (Lee *et al.*, 2018). Cell size in other bacterial species is often regulated, directly or indirectly, by enzymes that alter the PG sacculus and/or the major division-associated protein FtsZ (Chien *et al.*, 2012; Taheri-Araghi *et al.*, 2015). As the members of the Chlamydiae phylum lack FtsZ and frequently also the PG sacculus, the mechanism by which they regulate cell size is likely to be novel. Without a sacculus, a number of variables could theoretically affect the timing of cell division, and subsequently, the size of the daughter cells. Rates of metabolism, nutrient uptake/processing, PG synthesis/degradation could all potentially influence the rate of cell growth prior to division and the rate at which the division processes ultimately terminates.

Peptidoglycan precursor pathway in *Chlamydia* species

Early studies showed that PG-targeting antibiotics such as penicillin are effective at inhibiting *Chlamydia* cell division, however, confirming the existence of PG in pathogenic *Chlamydia* was problematic. Trace amounts of muramic acid (MurNAc, Fig. 9.4), a molecule unique to bacterial peptidoglycan, were detected by some researchers (Jenkin, 1960; Perkins and Allison, 1963; Su, 1985) but not others (Barbour *et al.*, 1982; Fox *et al.*, 1990; Garrett *et al.*, 1974; Manire and Tamura, 1967; Tamura and Manire, 1967). This paradox, a bacterium susceptible to PG-targeting antibiotics despite lacking canonical PG, was dubbed the '*Chlamydia* anomaly' by Moulder (Moulder, 1993) and. investigators were consequently obligated to explain how bacterial cellular integrity and cell division could be preserved in the absence of a canonical PG sacculus. It was proposed that disulfide cross-linked envelope proteins carried out the osmoprotective role of PG for *Chlamydia* in its extracellular (EB) state (Everett and Hatch, 1995).

The successful sequencing of the *C. trachomatis* genome revealed that the pathogen encodes almost all of genes associated with the PG biosynthesis pathway (Fig. 9.4), which constitutes ~1.8% of the genome (Stephens *et al.*, 1998). Many of these homologues appear within the division/cell wall (*dcw*) operon, a chromosomal locus highly conserved throughout Gram-negative bacteria and comprised of cell wall biosynthesis and cell division-associated genes. Subsequent genetic and biochemical studies confirmed the functionality of the enzymes encoded in the genome, either by enzymatic assays or genetic complementation in surrogate (*E. coli*) model systems (De Benedetti *et al.*, 2014; Hesse *et al.*, 2003; Klockner *et al.*, 2014; Liechti *et al.*, 2018; McCoy *et al.*, 2006; McCoy and Maurelli, 2005; McCoy *et al.*, 2003; Patin *et al.*, 2009; Patin *et al.*, 2012). The first of these enzymes to be functionally characterized was UDP-N-acetylglucosamine 1-carboxyvinyltransferase, MurA (McCoy *et al.*, 2003), which suggested that *C. trachomatis* possesses the enzymatic machinery to synthesize muramic acid. Subsequent characterization of the enzymes responsible for the addition of amino acids to muramic acid; UDP-N-acetylmuramate–alanine ligase, MurC (Hesse *et al.*, 2003), D-alanine-D-alanine ligase, Ddl (McCoy and Maurelli, 2005), UDP-N-acetylmuramoyl-L-alanyl-D-glutamate-diaminopimelate ligase, MurE (Patin *et al.*, 2009), UDP-N-acetylmuramoyl-tripeptide-D-alanyl-D-alanine ligase, MurF (Patin *et al.*, 2012), and N-acetylmuramoyl-L-alanine amidase, AmiA (Klockner *et al.*, 2014) confirmed the presence of a minimal PG biosynthetic pathway (Fig. 9.4). Transcriptomic data indicated that the majority of these genes were expressed after the EB-to-RB transition (Albrecht *et al.*, 2011; Albrecht *et al.*, 2010; Belland *et al.*, 2003b), which is consistent with a role in RB division. However, several key enzymatic activities did not appear to be encoded in the *Chlamydia* genome, raising the question of whether *Chlamydia* possess a complete PG biosynthesis pathway.

PG is a unique polymer in that it utilizes a number of non-canonical amino acids: meso-diaminopimelic acid (mDAP) a metabolic intermediate of the lysine biosynthesis pathway, and the D-stereoisomers of glutamic acid and alanine (Fig. 9.4). Curiously, the canonical enzymes involved in the synthesis of these key amino acids (N-succinyl-aminoketopimelate aminotransferase, tetrahydrodipicolinate succinylase, succinyl-diaminopimelate desuccinylase, glutamate racemase, and alanine racemase) are not encoded in the genomes of members of the *Chlamydia* genus. It was later discovered that *Chlamydia* make use of the aminotransferase pathway, also present in cyanobacteria (Hudson *et al.*, 2006), for mDAP biosynthesis (McCoy *et al.*, 2006). For d-amino acid biosynthesis, *Chlamydia* encode enzymes with dual function ('moonlighting') activities: Chlamydial serine hydroxymethyltransferase (GlyA) and diaminopimelate epimerase (DapF) are capable of synthesizing d-alanine and d-glutamate, respectively (De Benedetti *et al.*, 2014; Liechti *et al.*, 2018). *Chlamydia* also appears to lack homologues for canonical PG synthases (transglycosylases), which gave rise to the assumption that *Chlamydia* maintains a 'glycanless' cell wall (Ghuysen and Goffin, 1999). However, the structure of PG was resolved for both pathogenic *Chlamydia* (Packiam *et al.*, 2015) and *Chlamydia*-related bacteria (Pilhofer *et al.*, 2013), and analysis of muropeptide fragments from *Chlamydia*-infected cell lysates indicate that muramic acid-containing glycan chains are clearly present (Packiam *et al.*, 2015). Subsequent studies indicated that while β-lactam antibiotics can prevent *Chlamydia* cell division and differentiation, *Chlamydia* is still capable of forming uncrosslinked PG in the presence of these antibiotics. As this antibiotic class targets the activity of penicillin-binding proteins (PBPs), this result indicates that these proteins are not essential for transglycosylase activity in *Chlamydia* and hint that *Chlamydia* utilizes a different enzyme for carrying out transglycosylation during PG synthesis (Liechti *et al.*, 2016; Liechti *et al.*, 2014). Recent discoveries in other model systems suggest that components of the Rod complex (Meeske *et al.*, 2016; Taguchi, 2019) may provide a mechanism by which *Chlamydia* species carry out this transglycosylase activity in the absence of canonical PG synthases. The reliance on these non-canonical enzymatic activities for PG biosynthesis is now one of the defining characteristics of chlamydial physiology.

Peptidoglycan abundance and localization in *Chlamydia*

In the presence of β-lactam antibiotics, *Chlamydia* species fail to divide but do not lyse, and remain viable within an intracellular niche. The resulting enlarged aberrant bodies are incapable of differentiating into infectious EBs until the antibiotic is removed and PG transpeptidase activity is restored. Thus, it appears that the primary function of PG in these microbes is RB division, and not maintenance of osmotic stability. Imaging and localization studies indicate that pathogenic *Chlamydia* species lack a canonical sacculus and maintain PG about their septa (Liechti *et al.*, 2016). The visualization of *Chlamydia* PG in fixed cells was first accomplished by Brown *et al*, utilizing antiserum raised against cell walls purified from mycobacteria (Brown and Rockey, 2000). The authors reported that multiple *Chlamydia* species appeared to possess septal-localizing 'SEP antigen' in their replicative forms. More recently, Liechti *et al.* confirmed that four pathogenic *Chlamydia* species (*C. trachomatis, C. caviae, C. muridarum,* and *C. psittaci*) all lack classical sacculi, and instead maintain PG either at their division septa or at 'polar disks', possibly reflecting septal PG remnants in cells that have recently divided (Liechti *et al.*, 2014, 2016). Liechti *et al.* utilized a novel labelling technique in which modified D-alanine dipeptide analogues are added to growth media. These modified dipeptides are taken up, first by host cells and then intracellular *Chlamydia,* and subsequently inserted into the Lipid II stem peptide via the MurF enzyme during new PG synthesis (Fig. 9.4). Similar to the observations of Brown and Rockey (2000), PG was only observed in RBs, and labelled PG was first observed ~8 hours post infection (hpi) in *C. trachomatis* (Liechti *et al.*, 2016; Liechti *et al.*, 2014), roughly coinciding with the up-regulation of PG biosynthesis and cell division genes associated with EB to RB transition (Albrecht *et al.*, 2010, 2011; Belland *et al.*, 2003b). Images of late-stage, intracellular growth suggest that the muropeptide is degraded upon transition of *Chlamydia* RBs to the infectious EB form, implicating PG endopeptidases in this event. This observation was independently confirmed utilizing a reporter cell assay and mass spectrometry. Lysates from cells containing *Chlamydia* RBs activate NFκB signalling in an HEK293 NOD2 reporter cell line (NOD2 is an intracellular receptor for bacterial PG fragments) while lysates of cells containing only EBs do not (Packiam *et al.*, 2015). Interestingly, a D-amino-acid-based labelling methodology similar to that used by Liechti *et al.* revealed that '*Chlamydia*-like' species also possess PG, but that it appears to form a canonical saccular structure that completely covers the surface of the bacterium, demonstrating a significant physiological variance in the evolution of the *Chlamydiae* (Pilhofer *et al.*, 2013).

It is tempting to speculate about this evolutionary divergence with regards to PG abundance and localization in these organisms. Recent work on the growth and replication of PG-inhibited 'L-form' bacteria suggests that under osmotically stable/intracellular conditions, the loss of a sacculus is well tolerated by both Gram-positive and Gram-negative bacteria (Mercier *et al.*, 2014, 2016; Rojas *et al.*, 2018). Gram-negative bacteria likely rely on the inherent strength of the outer membrane, which is a load-bearing structure capable of approximately the same stiffness and tensile strength as a PG sacculus (Rojas *et al.*, 2018). Both membrane proteins and lipopolysaccharide (LPS) contribute to the strength of the Gram-negative outer membrane, which is consistent with previous observations indicating that LPS is an essential membrane component of PG-less *Chlamydia* EBs (Nguyen *et al.*, 2011). *Chlamydia* does not encode a homologue of Braun's lipoprotein (Braun, 1975; Torti and Park, 1976), a key lipoprotein anchor that tethers the outer membrane of Gram-negative bacteria to the PG sacculus (Stephens *et al.*, 1998). As *Chlamydia* species lack sacculi, this is not entirely surprising, and other PG-binding lipoproteins that reside in the inner leaflet of the outer membrane such as Pal (see below) may provide an anchoring function at the division septum.

The role of peptidoglycan in *Chlamydia* cell division

Members of the *Chlamydiaceae* family likely restrict PG to what will eventually become the division septum in order to complete the septation process across both membranous envelope layers. The fact that PG is retained by the *Chlamydiaceae* would appear to indicate an inherent benefit of the efficient division and septation conferred via a muropeptide-based mechanism. This mechanism is in contrast to the often erratic PG-independent mechanism employed by mycoplasma and by L-form bacteria that make use of enhanced phospholipid trafficking (Mercier *et al.*, 2013). One potential benefit for retention of a division complex in an osmotically stable environment is the ability to ensure the even distribution of genetic material between emergent daughter cells. As inhibition of PG biosynthesis also prevents EB formation, it is likely that PG also aids in the constriction of the bacterial cell membranes, resulting in the reduction in overall cell volume that is known to occur prior to the transition from an RB to an EB (Lee *et al.*, 2018). The Tol–Pal complex (Figs. 9.1 and 9.2) is an essential element in Gram-negative cell division, responsible for linking the cell division apparatus with the bacterial outer membrane (Gerding *et al.*, 2007). Pal (peptidoglycan-associated lipoprotein) binds mDAP in the Gram-negative PG stem peptide, providing a point of interaction between the outer membrane and the division complex, and making it essential for septation. *Chlamydia* encodes homologues to Tol–Pal complex subunits (Stephens *et al.*, 1998) supporting the hypothesis that the point of interaction between PG and the outer membrane in *Chlamydia* is localized to the division septum. The continual synthesis and degradation of septal PG is thus likely required for constriction of the division plane, which is the final step of *Chlamydia* cell division (Abdelrahman *et al.*, 2016). The rate of PG synthesis is directly regulated by the Lipid II flippase, MurJ (Fig. 9.4) (Monteiro *et al.*, 2018). In *E. coli*, MurJ 'flips' PG monomers (Lipid II) across the inner membrane into the periplasmic space. MurJ is recruited to the septum simultaneously with MraY and MurG, which are both also required for Lipid II production (Liu *et al.*, 2018). *Chlamydia* encode a MurJ orthologue but its role in the PG biosynthesis pathway remains to be characterized.

Most of the Lipid II biosynthesis genes are expressed beginning 6–8 hours post infection in *C. trachomatis*, roughly coinciding with EB to RB transition (Belland *et al.*, 2003b). Interruption of Lipid II biosynthesis early in the developmental cycle of *C. trachomatis* via the chemical inhibitor D-cycloserine results in the formation of enlarged aberrant bodies (Moulder *et al.*, 1963). This finding indicates that PG is essential for cell division in RBs, but it may not be essential for the increase in cell size associated with the transition between EBs and RBs. Visualization studies investigating the kinetics of *Chlamydia* PG assembly and degradation suggest that the PG biosynthesis and degradation pathways are uncoupled (Liechti *et al.*, 2016). That is, inhibition of synthesis does not prevent the subsequent degradation of the septal PG.

Early studies investigating mechanisms of cell division arrest in *Chlamydia* aberrant bodies noted that treatment with interferon gamma results in an inhibition in *Chlamydia* cell division and development (Byrne *et al.*, 1986; Carlin *et al.*, 1989; Rapoza *et al.*, 1991). Transcriptional analysis indicated that interferon treatment does not appear to significantly alter expression of any genes essential for *Chlamydia* PG biosynthesis or assembly (Belland *et al.*, 2003a). Subsequent studies have described a tryptophan-dependent mechanism in which gamma interferon inhibits translation of PG biosynthetic enzymes or elements of the PG assembly complex (Bonner *et al.*, 2014; Ouellette *et al.*, 2016). Gamma interferon induces production of indoleamine 2,3-dioxygenase, which degrades intracellular and extracellular tryptophan pools (Shimizu *et al.*, 1978). Recently it has been noted that PG biosynthetic enzymes such as GlmV and MraY in the Chlamydiales and DapA and LysC in the Chlamydiaceae (Bonner *et al.*, 2014) contain an abundance of tryptophan residues compared to their orthologues in other bacteria. It was hypothesized that under tryptophan-depletion conditions, the successful translation of these proteins may be delayed, if not halted, limiting PG biosynthetic capacity. It has also been reported that gamma interferon treatment results in translational stalling of mRNA for RodZ, a component of the Rod complex (Fig. 9.3) (Ouellette *et al.*, 2016). RodZ is thought to direct localization of MreB, which in turn, directs PG biosynthesis. These findings illustrate steps in the PG synthesis and assembly pathways that may be susceptible to interferon-induced cell division arrest.

What we can learn from *Chlamydia*-like organisms

Scientific investigation of the Chlamydiae phylum, many of whose members diverged from the *Chlamydiaceae* family over 700 million years ago, has been largely limited to a few model organisms belonging to the *Waddliaceae*, *Parachlamydiaceae* and *Simkaniaceae* families (reviewed in Jacquier *et al.*, 2015b) and discussed in more detail in Chapter 18, Thinking Outside the *Chlamydia* Box. Nevertheless, these studies have demonstrated that the majority of the conclusions derived from research on *C. trachomatis* and *C. pneumoniae* hold true for all members of the Chlamydiae. Despite some structural differences, such as a five-layered cell membrane in *Rhabdochlamydiaceae* (Corsaro *et al.*, 2007) and a star-like shape intrinsic to the EB form in Criblamydiaceae (Lienard *et al.*, 2011; Thomas *et al.*, 2006), the overall conservation of the divisome genes among all the different family level lineages examined to date suggests that the underlying molecular mechanisms of cell division has also been preserved across the Chlamydiae.

Differences between *Chlamydia*-related bacteria and members of the *Chlamydiaceae* have directly aided in the exploration and deciphering of the complete Chlamydiae divisome. For example, the rapid replication time and larger cell size of *Waddlia chondrophila* (a member of the *Waddliaceae*) make it an ideal model organism for investigating the cell biology of cell division, although the genetic intractability limits functional analyses. Fortunately, pharmacological approaches can be used instead to reveal the requirement of Lipid II for cell division, which is only possible because *W. chondrophila* is intrinsically susceptible to fosfomycin (Jacquier *et al.*, 2014). In contrast, most other members of the *Chlamydiaceae* are resistant to this MurA inhibitor (Fig. 9.4). RodZ precedes MreB in localizing to the septal site in *W. chondrophila*. This recruitment is dependent on the biosynthesis of Lipid II by MurA and is inhibited by fosfomycin (Jacquier *et al.*, 2014). Additionally, in *W. chondrophila* the crosslinking of Lipid II peptides by the Pbp3 enzyme effectively disperses RodZ away from the division plane (Jacquier *et al.*, 2014). Furthermore, the *Waddlia* genome encodes a putative PG-binding endopeptidase (NlpD) that localizes to the bacterial septum in a Lipid II-dependent manner, suggesting that localized PG synthesis directs NlpD to the septum (Frandi *et al.*, 2014). The Pal component of the Tol–Pal complex also binds PG (Jacquier *et al.*, 2015a) and is localized to the *W. chondrophila* septum, allowing for symmetrical binary fission to proceed. The investigation of these biological processes in members of the *Parachlamydiaceae* has also demonstrated the importance of PG in the division of members of the Chlamydiae phylum, as fosfomycin treatment results in termination in cell division and the formation of aberrant bodies (Pilhofer *et al.*, 2013). Thus, *Parachlamydiaceae* and *Waddliaceae* represent good alternative models for the continued investigation of chlamydial cell division, although differences in the division and developmental cycle likely exist across chlamydial genera.

Mechanism of *Chlamydia* replication

Rate(s) of replication

The length of time *Chlamydia* requires to complete its developmental cycle is strain and species specific, varying from ~40 to 72 hours (Miyairi *et al.*, 2006). Numerous factors influence rates of propagation in tissue culture, including cell lines used, per cent serum, and the addition of non-essential amino acids and/or cycloheximide. For *C. trachomatis*, the time from the initial attachment of an EB to the cell surface to the lysis of the infected cell is approximately 48 hours. (Hybiske and Stephens, 2007). Several groups have measured the average rate of *C. trachomatis* replication. Timms *et al.* calculated chlamydial doubling time at ~3 hours based on genome copy number increase and examination of inclusions 8–40 hours post infection (Mathews *et al.*, 1999). Hackstadt *et al.* calculated an average doubling time of ~2 hours based on genome copy number (Shaw *et al.*, 2000) and Lambden *et al.* estimated that chromosomal DNA was amplified 100–200-fold throughout the developmental cycle (Lambden *et al.*, 2006). These findings are largely consistent with approximately 8–10 rounds of *Chlamydia* RB replication (Tan and Bavoil, 2012). These initial studies have recently been confirmed by Lee *et al.*, who utilized 3D electron microscopy to accurately count all of the bacteria residing within inclusions at various time points and categorize them by developmental stage (RB, dividing RB, intermediate body (IB), and EB). Based on this technique and supported by qPCR and infectious EB counts, they estimated a doubling time of ~1.8 hours for *C. trachomatis* RBs between 12 and 24 hours post infection (hpi) (Lee *et al.*, 2018). As *Chlamydia* division occurs much more slowly than in the majority of other Gram-negative bacteria, this long generation time may account for the unique features of the division process inherent to *Chlamydia* species.

Binary fission

Binary fission is the most common form of reproduction in bacteria. It is defined by a parent cell initially doubling in size and subsequently splitting into two equal-sized daughter cells in coordination with DNA replication and chromosome segregation. This symmetric division process is the predominant mechanism used to describe division in pathogenic *Chlamydia* (Brown and Rockey, 2000; Friis, 1972; Gaylord, 1954; Higashi, 1965; Lee *et al.*, 2018; Matsumoto, 1981a, b; Matsumoto and Manire, 1970) and '*Chlamydia*-like' organisms (Chua *et al.*, 2005; Draghi *et al.*, 2004; Greub and Raoult, 2002; Thomas *et al.*, 2006). The earliest electron micrographs of the *Chlamydia* division process support the initial conclusion that *Chlamydia* species divide by binary fission (Gaylord, 1954; Higashi, 1965; Matsumoto, 1981a, b) and were used by Moulder to argue that these organisms should be classified as bacteria and not viruses (Moulder, 1964). Binary fission is an attractive division model as it accounts for not only the rates of nucleic acid and protein synthesis calculated for *C. psittaci* growth within mammalian cells (Schechter, 1966) but also that the division process appears to give rise to daughter cells of approximately the same size (Lee *et al.*, 2018). Lee *et al.* examined inclusion contents via 3D electron microscopy to determine how well a binary fission model of cell division accounts for size variance exhibited by daughter cells during the division of *C. trachomatis*. The boundaries for all RBs within two inclusions 24 hpi were defined by 3D EM and a daughter/parent volume ratio was calculated for each to establish whether parental cells would give rise to equivalently sized daughter cells, effectively testing the fit of a binary fission model of cell propagation. It was determined that the size divergence between daughter cells in *Chlamydia* (11%) was only slightly larger than that observed in *E. coli* (4%), which is known to undergo symmetric, binary cell division. These data largely fit a binary fission division process, as opposed to a budding or asymmetric division process, which would have resulted in much larger ratios (Lee *et al.*, 2018).

Observations inconsistent with 'symmetric' binary fission

While binary fission has been the favoured model for describing the *Chlamydia* division process for ~80 years, additional mechanisms of replication have been proposed. Researchers commenting on some of the first micrographs displaying the *Chlamydia* developmental cycle concluded that while two daughter cells are generally observed emerging from a single parent cell, they could be comprised of either equal or unequal portions of the cell parent (Friis, 1972; Gaylord, 1954; Higashi, 1965). Instances of 'RB budding' or the formation of 'mini-cells'/'miniature RBs' have been reported for *C. psittaci* (Falcieri *et al.*, 1979; Gaylord, 1954; Tanami and Yamada, 1973), *C. trachomatis* (Falcieri *et al.*, 1979; Hackstadt *et al.*, 1996), *C. pneumoniae* (Wolf *et al.*, 2000), and *C. muridarum* (Phillips *et al.*, 1984). Newly formed daughter cells from *C. trachomatis* RBs comprised of equal and unequal portions of the parent cell have also been observed in endocervical samples obtained from infected human patients (Lewis *et al.*, 2014; Swanson *et al.*, 1975).

The frequency of these findings, which span multiple strains/species of *Chlamydia* and *Chlamydia*-related bacteria, as well as the fact that this morphological feature has been observed in clinical samples, appear to indicate a relatively common and universal phenomenon. The preponderance of data accumulated to date would seem to rule out the possibility that these observations are simply an anomaly or representative of an isolated, 'abnormal', laboratory-passaged strain. Yet these observations of *Chlamydia* cells apparently undergoing 'asymmetric' or unequal division have been met with a healthy degree of scepticism. A common critique is that these studies were all carried out utilizing fixed cells and electron microscopy. It is therefore unclear whether these 'unequal portions' represent true developmental processes or if they are simply artefacts inherent to either (i) visualizing bacteria undergoing division in single cross-sections or (ii) fixation. A comparison of various fixation methods in different members of the Chlamydiae phylum clearly showed that fixation artefacts occur more commonly in some species, such as *Estrella lausanensis*, *Criblamydia sequanensis* and *Waddlia chondrophila* (Rusconi *et al.*, 2013), likely as a result of slight differences in cell wall composition. As members of the Chlamydiales lack a classical cell wall altogether, fixation may have an even greater effect on cellular morphology. Alternatively, these images may instead represent snapshots of a novel, highly dynamic division process. Assuming the apparent asymmetry presented in micrographs is reflective of a genuine morphological form, it remains uncertain whether parental cells eventually distribute their cytoplasmic contents equally or unequally between newly formed daughter cells.

'Polarized cell division'

More recently, investigators have begun to examine the processes by which *Chlamydia* initiates daughter cell formation and completes septation, the final step in the division process. Evidence of an initial, unequal distribution of parental cell material, defined as 'polarized cell division' by Abdelrahman *et al.*(Abdelrahman *et al.*, 2016) (Fig. 9.5), has been observed in *C. trachomatis* by two groups, each utilizing molecular approaches

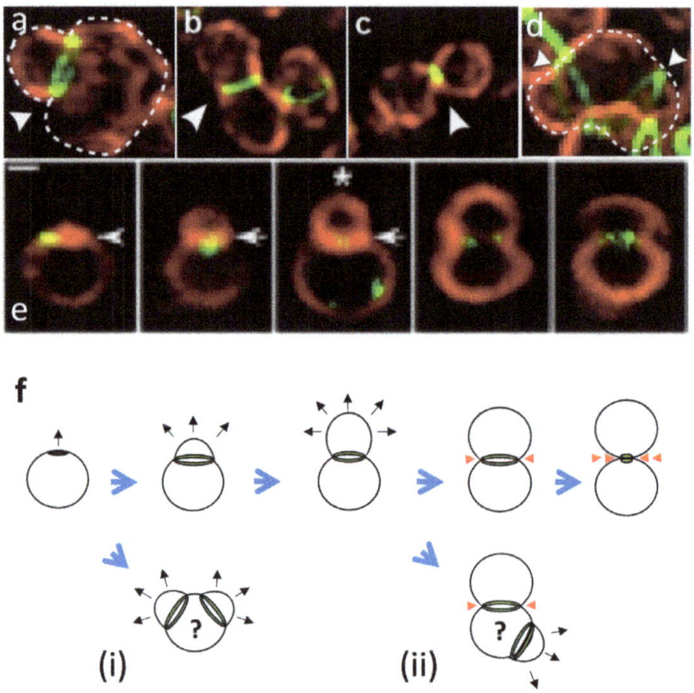

Figure 9.5 Polarized growth/division in *C. trachomatis*. (a, d) Maximum intensity projections of 0.5 μm thick planes of *C. trachomatis*-infected cells (18 hpi) that were incubated with the peptidoglycan-labelling reagent EDA-DA, an analogue of d-alanine-d-alanine. Arrowheads indicate areas of peptidoglycan staining. Dashed lines indicate the boundaries of individual bacteria. Examples of RBs containing multiple peptidoglycan rings are outlined and marked with arrowheads for clarity. Peptidoglycan is shown in green, and the major *C. trachomatis* outer membrane protein (MOMP) is shown in red. (e) *C. trachomatis* expressing GFP-FtsQ (green) and labelled with Bodipy-sphingomyelin (red) in living cells. Images are selected intermediates (as opposed to time lapse of an individual cell) and are representative of all observations. Arrowhead indicates region of intense BODIPY-sphingomyelin fluorescence at one pole of a round cell prior to division. Arrows indicate a sphingomyelin-rich membrane separating a nascent daughter cell from a mother cell. (f) Graphical representation illustrating the polarized division model. Spatial (i) and temporal (ii) regulation of polar growth may not be tightly controlled during the division cycle, potentially explaining the presence of multiple division planes within single bacterial cells. Black arrows indicate areas of cell expansion and red arrowheads illustrate septation, the terminal step in cell division. Figure adapted from Liechti *et al.* (2016) and Abdelrahman *et al.* (2016). The article by Liechti *et al.* (2016) is an open access article, free of all copyright, and may be freely reproduced, distributed, transmitted, modified, built upon, or otherwise used by anyone for any lawful purpose. See: https://creativecommons.org/publicdomain/zero/1.0/. The article by Abdelrahmen is an open access article distributed under the terms of the Creative Commons Attribution License, which permits unrestricted use, distribution, and reproduction in any medium, provided the original author and source are credited. See: http://creativecommons.org/licenses/by/4.0/

combined with fluorescence microscopy (Abdelrahman *et al.*, 2016; Liechti *et al.*, 2016). Unlike previous studies that relied on electron micrographs and single imaging planes, both Liechti and Abdelrahman provided composite images generated from numerous cross sections, thereby significantly reducing the likelihood that these results are artefacts attributable to the orientation of the bacteria within a single, imaging cross section.

In their investigation of PG localization in *Chlamydia* RBs, Liechti *et al.* observed that while the vast majority of *C. trachomatis* RBs in mid-stage (16–18 hpi) inclusions appear to maintain their PG at mid-cell (consistent with canonical binary fission), newly divided daughter cells appeared to orient their PG in a structure similar to a polar patch or disk near where the division plane had existed in the parent cell (Liechti *et al.*, 2016). They also noted that in a subset of RBs, PG rings did not localize directly to the mid-cell but appear to orient closer to one cell 'pole' than to the other. It was proposed that this uneven localization of PG rings could represent an additional, asymmetric division process, akin to budding, resulting in daughter cells of different sizes, with the smaller of the daughter cells potentially differentiating into IBs and eventually EBs (Liechti *et al.*, 2016). In some instances, RBs with multiple PG-containing division planes were also observed (Liechti *et al.*, 2016), consistent with previous observations of more than one set of membrane invaginations being present within a single RB (Lewis *et al.*, 2014; Swanson *et al.*, 1975).

When examining EB to RB transition and initial RB cell division events (6–12 hpi), Abdelrahman *et al.* observed via immunofluorescence microscopy that several membrane associated components of the *Chlamydia* membrane (major outer membrane protein, lipo-oligosaccharide, sphingomyelin), heat-shock protein 60 (HSP60), and the cell division protein FtsQ exhibited polar localization in newly forming RBs (Abdelrahman *et al.*, 2016). This initial orientation of the division complex adjacent to a cell pole is in line with previous observations of cell division proteins remaining at the nascent cell pole in other bacterial species (Thanbichler, 2010). Others have also noted that both RBs and EBs exhibit cell polarity with an asymmetric distribution of type III secretion system (T3SS) structures (Chen *et al.*, 2014; Matsumoto, 1981b, 1988; Nans *et al.*, 2014) as well as various inclusion membrane proteins and the major outer membrane protein (MOMP) on/in RBs (Scidmore-Carlson *et al.*, 1999). In light of their observations, Abdelrahman *et al.* proposed a 'polarized' division model that is similar to the budding mode of division described previously for a subset of *Planctomycetes* that, like *Chlamydia* species, lack an FtsZ homologue (Fuerst, 1995; Lee *et al.*, 2009; Santarella-Mellwig *et al.*, 2013). Such unipolar growth or budding is also prevalent in alpha-proteobacteria (Brown *et al.*, 2012). According to this polarized division model, *Chlamydia* initially organizes the components of the cell division machinery at a single cell pole, rather than at mid-cell. In this case, an asymmetric expansion of the cell envelope from this polar-localized patch or, alternatively, the migration of polar-localized components to mid-cell, ensues. The initiation of this process is denoted by the extrusion of the cell membrane until the parent cell's contents are equally distributed on either side of the division septum. Unlike the model proposed by Liechti *et al.*, the polarized division model results in two daughter cells that are approximately the same size. While Abdelrahman's observations of the initial RB division event immediately following the EB to RB transition appear to fit this model, it is important to note that similar observations at later time points may represent division complexes that have yet to disassemble and re-assemble in the wake of medial (symmetric) division.

In the presence of inhibitors of the cell division protein MreB (A22), fatty acid biosynthesis (AFN1252), and acquisition of phosphatidylcholine (Glyburide), these polarized cell division intermediates collapse, and polarized MOMP disperses (Abdelrahman *et al.*, 2016). β-lactam antibiotics (penicillin and carbenicillin) do not prevent asymmetric membrane expansion and only inhibit the completion of septation, suggesting that the role of crosslinked PG in the division process is not the establishment of cell polarity but rather constriction of the *Chlamydia* division septum. The biosynthesis of Lipid II acts as the driving factor for septation in '*Chlamydia*-like' microbes and in other bacteria (Monteiro *et al.*, 2018), and it may play a role in directing septation in *Chlamydia* as well. Liechti *et al.* reported that the PG ring dissociates rapidly upon treatment with MreB polymerization inhibitors (A22 and MP265) whereas PG labelling is retained in cells treated with ampicillin, albeit in an altered, disjointed orientation (Liechti *et al.*, 2016). Whether PG intermediates or uncrosslinked PG (i.e. monomeric PG precursors with or without the carrier lipid moiety, bactoprenol) play an active role in the initial polar expansion of the cell membrane remains an open question.

Abdelrahman *et al.* noted that division of *Chlamydia* undergoing a second round of cell division occurs perpendicular to the first division plane, similar to division in *Staphylococcus aureus* (Turner *et al.*, 2010; Tzagoloff and Novick, 1977), suggesting that division plane formation may be spatially regulated. Additionally, they observed that ~ 5% of RBs undergoing their first division event exhibited polar membrane expansion from more than one location simultaneously, indicating that spatial regulation of polar growth may not be tightly controlled in *C. trachomatis* (Abdelrahman *et al.*, 2016).

Reconciling observations

Unlike Liechti *et al.* and Abdelrahman *et al.*, Lee *et al.* did not report any instances of unequal daughter cells or multiple division planes within single RBs consistent with the binary fission model (Lee *et al.*, 2018). The notable absence of asymmetric RBs in the Lee *et al.* study may indicate that these asymmetric RBs are more prevalent at early time points, as observed by Abdelrahman *et al.*(Abdelrahman *et al.*, 2016). Alternatively, they may represent short-lived division intermediates, particularly later in the *Chlamydia* developmental cycle, if the initial polar growth/extrusion event occurs rapidly compared to the division event. Liechti *et al.* reported that RBs apparently undergoing asymmetric division are relatively rare in mid/late stage inclusions and that the vast majority of RBs appear to have PG septal division planes at mid-cell (Liechti *et al.*, 2016). It is important to recognize that all of these studies provide only snapshots of what are likely highly dynamic processes. Studies investigating how the kinetics of these processes align with actual changes in cell morphology will very likely affect how these data are interpreted in the future.

Inconsistencies with regard to whether RBs are capable of forming multiple division planes simultaneously are more difficult to reconcile. Multiple division plane formation has been previously reported in *Chlamydia*-infected cells treated with β-lactam antibiotics (Matsumoto and Manire, 1970). Lewis (Lewis *et al.*, 2014), Liechti (Liechti *et al.*, 2016), and Abdelrahman (Abdelrahman *et al.*, 2016) have all reported evidence of either multiple, septal PG rings by fluorescent approaches or early stages of division plane formation simultaneously within single RBs, resembling the two ledges of the 'piecrust' seen at the division plane of staphylococci (Turner *et al.*, 2010). Lee *et al.* did not report instances of RBs containing multiple invaginations, which would suggest multiple division plane formation (Lee *et al.*, 2018). Given the findings of Abdelrahman *et al.* that PG appears to only be essential for septal constriction, it could be that these instances actually represent rare aberrant forms with alterations or deficiencies in the PG biosynthesis/remodelling pathways in which these stained rings later merge into a single septal disc of polymerized PG, as seen in staphylococci (Lund *et al.*, 2018; Turner *et al.*, 2010). Alternatively, if 1) the *Chlamydia* cell division machinery and PG play an active role in both chromosomal partitioning and septation and, 2) these processes are temporally decoupled, then these observations may simply illustrate instances where RBs begin forming new bulges upon chromosomal partitioning prior to the completion of septation (Fig. 9.5f). As all studies to date have been conducted in fixed cells, it remains unknown whether RBs containing multiple division planes give rise to more than two daughter cells, each containing a full chromosome and the structural components necessary for normal RB growth and development.

In summary, *Chlamydia* species appear to divide by symmetric, binary fission, but the process is unlike what is observed in other bacterial species. Most bacteria orient their cell division complexes at mid cell just prior to the terminal event in the division process (septation), and this process is dependent on FtsZ. *Chlamydia* species lack FtsZ, and, as a result, the mechanism by which they establish where 'mid-cell' is, is different. In contrast to the FtsZ-dependent mechanism, *Chlamydia* initially orients its cell division complex at a single cell pole early in the division process. As the cell enlarges, the cell membrane expands asymmetrically from that pole, in a process akin to budding. However, unlike budding, septation does not occur until the cell contents are equally distributed on either side of the division complex.

Models for *Chlamydia* differentiation

Reticulate body to elementary body conversion

Approximately 24 hours into the chlamydial developmental cycle, RBs begin converting into EBs in an asynchronous manner. While the mechanism underlying this conversion is currently unknown, the basic stages of the process have been methodically defined. DNA condensation is one of the more obvious features of conversion, with IBs and EBs exhibiting condensed nucleoids (Costerton *et al.*, 1975) and has been linked to chlamydial histone-like proteins (Barry *et al.*, 1992; Hackstadt *et al.*, 1991). An additional feature of conversion is the alteration in configuration and composition of the chlamydial outer membrane proteins. The membranes of EBs contain a number of highly cross-linked (disulfide bonded), cysteine-rich outer membrane proteins that provide the bacterium with the rigidity needed for maintaining cell integrity in an extracellular environment (Bavoil *et al.*, 1984; Hackstadt *et al.*, 1985; Hatch *et al.*, 1984; Hatch *et al.*, 1986; Newhall and Jones, 1983). Cysteine-rich membrane proteins specific to EBs include outer membrane proteins OmcA and OmcB (Clarke *et al.*, 1988; Hatch *et al.*, 1984), which interact with the major outer membrane protein (OmpA or MOMP) to form the cross-linked bacterial outer membrane complex. Temporal gene expression (discussed in more detail in Chapter 10, *Chlamydia* Gene Regulation) is a hallmark of differentiation in *Chlamydia* species with 'early' genes being associated with establishing the inclusion, 'midcycle' genes being involved in RB growth and replication, and 'late' genes, which are expressed at advanced stages of the developmental cycle (Belland *et al.*, 2003b), encoding elements specific to the RB to EB transition. Many late genes, such as the genes encoding for histone-like proteins Hc1 and Hc2 and outer membrane proteins OmcA and OmcB share a common regulatory element, the transcriptional repressor EUO (Rosario *et al.*, 2014; Rosario and Tan, 2012), which binds to the promoters of these late genes, preventing their full expression until late in the developmental cycle.

Contact-dependent model of differentiation

Given the large difference in size between RBs and EBs and notable differences in their transcription profiles (Belland *et al.*, 2003b), it has been postulated that some external factor or signal acts as a trigger for the initiation of differentiation from RB to EB. The distribution of RBs, IBs and EBs within *Chlamydia* inclusions is not random and size appears to be a determining factor with regard to bacterial cell positioning within inclusions. Studies based largely on transmission electron microscopy noted that RB size does not appear to be uniform and that larger RBs often appear to reside closer to the inclusion membrane, with smaller RBs, IBs and EBs clustering more towards the inclusion interior (Matsumoto, 1981a, b). Wilson *et al.* (Wilson *et al.*, 2006) and Rank *et al.* (Rank *et al.*, 2011) both used this observation as the foundation for a contact-dependent model of *Chlamydia* differentiation. This contact-dependent model was put forward as a means of explaining the observation that as more bacteria filled the inclusion, RBs accumulated along the inclusion membrane whereas EBs were more likely to be found in the inclusion interior. According to this model, loss of RB contact with the inclusion membrane, whether by spatial constraints of the inclusion or some other factor, was proposed to eventually lead to the inactivation of the *Chlamydia* T3SS, which then triggers RB to EB differentiation (Bavoil *et al.*, 2000). The 'contact-dependent model' provides a rational explanation for the observations that *C. trachomatis* RBs that do not maintain contact with the inclusion membrane are displaced to the interior of the inclusion and eventually undergo differentiation into EBs (Hoare *et al.*, 2008; Wilson *et al.*, 2006; Wilson *et al.*, 2009). It is important to note that this is an untested mathematical model based on correlative observations and that these observations do not appear to hold true for *C. pneumoniae*. As such, there may be more than one mechanism of differentiation among *Chlamydia* species (see below).

Reticulate body size control model of differentiation

By examining the composition of numerous inclusions at various stages of development utilizing 3D EM, Lee *et al.* showed that there is heterogeneity in RB size within inclusions, and subsequent computational modelling showed that this RB size heterogeneity could produce asynchronous conversion (Lee *et al.*, 2018). While significant heterogeneity in RB size was reported, the average RB size reduced six-fold during the intracellular developmental cycle, providing a rational explanation for the asynchronous conversion of RBs to EBs inherent to pathogenic *Chlamydia* species. Lee *et al.*'s observations support a 'size control model' for RB to EB conversion in which replication-dependent RB size reduction has the potential to produce delayed and asynchronous conversion (Lee *et al.*, 2018). The assumptions for this model are 1) that RBs gradually get smaller through replication and 2) that there is a size threshold for the conversion of RBs into EBs. Lee *et al.* observed that RBs within early stage inclusions on average were larger than those within later stage inclusions, and that the appearance of small RBs coincided with the detection of EBs. They reasoned that if RB size decreases uniformly in a population, then the RB-to-EB conversion should occur simultaneously for all bacteria within an inclusion, but if there is RB size heterogeneity, a variable number of replication cycles will be required to reach the proper conversion size, causing conversion to be asynchronous (Fig. 9.6). This computational study did not seek to test the contact-dependent model and relied on RB size averages; therefore, no correlation can be made with the size of RBs and their distribution within the inclusion. The authors did note, however, that some areas within the inclusion appeared to contain a higher percentage of dividing RBs than others (Lee *et al.*, 2018). Additionally, Lee *et al.* found that while the number of chlamydiae increased proportionately with inclusion volume at all time points, the total volume of chlamydiae within inclusions did not keep pace with inclusion growth. This finding appears to challenge the notion that spatial constraints within the inclusion result in the physical crowding of RBs at the membrane, causing loss of contact and resulting in differentiation, as was predicted previously (Hoare *et al.*, 2008).

The contact-dependent model and the replication-dependent RB size reduction model are not mutually exclusive. RBs in contact with the inclusion membrane could be larger because of enhanced growth between division events as a result of their proximity to the membrane and available nutrients. In contrast, loss of contact with the inclusion membrane could limit access to nutrients and ATP, potentially leading to smaller interior RBs that can convert into EBs. Thus, although RB size has been proposed as an intrinsic signal for RB-to-EB conversion, it may be controlled by external signals.

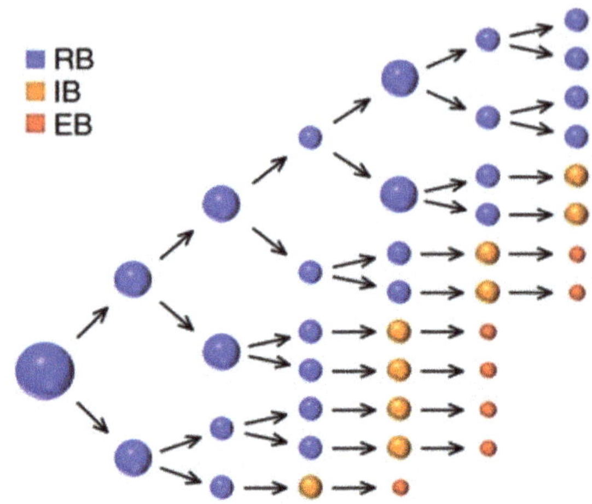

Figure 9.6 Replication-dependent size reduction precedes differentiation in *C. trachomatis*. Proposed model in which the size of an RB determines whether it can differentiate into an EB or continue to replicate. RBs become progressively smaller because they divide, on average, at less than twice their starting size, but they can only convert into an EB below a given size threshold. These two elements of RB size control ensure that the RB population expands before conversion occurs. The figure demonstrates how the absence of a strong cell size control mechanism prior to replication would result in RB size heterogeneity. This would provide an effective model for asynchronous conversion in *Chlamydia* RBs by varying the number of replication cycles required to reach the conversion size threshold. Figure adapted from Lee *et al.* (2018). This article is licensed under a Creative Commons Attribution 4.0 International License, which permits use, sharing, adaptation, distribution and reproduction in any medium or format, as long as you give appropriate credit to the original author(s) and the source. See: http://creativecommons.org/licenses/by/4.0/

Conclusions and future perspectives

Questions remain

Despite the advances in the field summarized in this chapter, we are only beginning to appreciate the intricacies of these fundamental and dynamic processes. Many important questions about peptidoglycan biology remain in an organism that 1) lacks FtsZ and utilizes only a Rod system for PG biosynthesis and cell division, and 2) only assembles a polar/septal peptidoglycan ring.

1 The septal ring and the mechanism behind its positioning: How is expansion of the septum into the lateral dimension restricted? How is new PG synthesis initiated/nucleated after each round of cell division, at mid-cell and after the EB to RB transition? What is the role of the PG synthesis module (i.e. the Rod Complex, Fig. 9.3) that normally drives lateral PG synthesis in other bacteria? What determines the site of divisome assembly? While MreB is thought to spatially regulate PG synthesis, and RodZ/RodA regulate MreB localization, what determines RodZ localization within RBs, and is MurJ involved? Divisome assembly and disassembly are dynamic processes in other microbes and it remains unclear how dynamic they are in pathogenic *Chlamydia* species.
2 Division: Does a bacterium undergoing binary fission have to initiate septal formation at mid-cell? What signals control PG synthesis, septum formation and subsequent division? If the initiation of *Chlamydia* cell division requires first establishing polarity within the cell, then how is this polarity attained? If the initiation of polar division in *Chlamydia* is followed very swiftly by chromosomal segregation in the newly forming daughter cell, how does this process occur and how is it regulated?
3 Link between cell division and differentiation: As PG appears to be necessary for constriction of the cell envelope at the terminus of division (Abdelrahman *et al.*, 2016), it may also be indispensable for constriction (size and membrane reduction) of the entire cell that occurs upon the RB to EB transition, similar to LPS (Nguyen *et al.*, 2011). As PG-based constriction would likely need to take place uniformly and not exclusively at the septum, a potential mechanism for such envelope reduction solely based on PG is not readily apparent. *Chlamydia* exhibit RB size heterogeneity but what factors determine when an RB is ready to divide and how large it is when division occurs? In some bacterial systems, nutrient availability plays a direct role in the size of the microbe upon division (Weart *et al.*, 2007) whereas in others, surface area and/or phospholipid synthesis and transport directly regulate cell size/propagation, particularly in osmotically stable environments (Harris and Theriot, 2016; Mercier *et al.*, 2013). Potential candidate triggers, such as DNA replication/segregation, the rate of Lipid II biosynthesis, and ATP concentration gradients within individual inclusions, await further exploration.

Current and future challenges

While remarkable progress has been made within the last few years concerning the investigation of *Chlamydia* division and differentiation, several technical challenges remain. As the vast majority of cellular components of the *Chlamydia* division apparatus are likely essential, gene inactivation/deletion approaches are unlikely to provide a reliable means of exploring the intricacies of the division process. The recent adaptation of CRISPRi technology for use in *C. trachomatis* is highly encouraging (Ouellette, 2018), as knockdown/depletion-based approaches (described in more detail in Chapter 11, *Chlamydia* Genetics) offer a more reliable and controlled method for the examination of the effects of individual components of these complex molecular machines. Additionally, one limitation of using *Waddlia* and other *Chlamydia*-related bacteria as model organisms to examine these processes is the absence of genetic tools, and a key challenge for the future is the development of methodologies for the genetic manipulation of *W. chondrophila*. In conjunction with its innate susceptibility to fosfomycin, future genetic approaches would allow for the use of this bacterium as a model organism for the investigation of the effects of Lipid II biosynthesis on cell division/size regulation. Additionally, the recent development of a shuttle plasmid system for *C. pneumoniae* (Shima *et al.*, 2018) will allow for the subsequent evaluation of this microbe's division process, potentially highlighting similarities and differences between those observed in *C. trachomatis* and *W. chondrophila*.

To date the vast majority of imaging-based studies have utilized fixed, permeabilized cells, providing sufficient resolution to note morphological features but not to precisely assess the temporal and dynamic reality of the underlying processes. It is remarkable to consider that almost our entire working knowledge of the functionality of these systems is based on protein–protein interaction assays conducted in surrogate systems, and snapshots of structural processes generated using crude localization data obtained from either EM micrographs, polyclonal immuno-labelling or tagged fusion proteins expressed well above physiological levels. What is sorely needed to make further progress in the characterization of these systems is what has become standard practice in other, more genetically tractable microbial systems: (i) chromosomal allelic exchange utilizing fluorescent fusion proteins coupled with (ii) live-cell imaging. Gene deletion by allelic exchange has recently been achieved in *C. trachomatis* (Johnson and Fisher, 2013; Mueller *et al.*, 2016) and these techniques could be utilized for the replacement of chromosomal alleles with fluorescent gene fusions (see Chapter 11, *Chlamydia* Genetics). The fact that *Chlamydia* resides within an intracellular environment largely limits the use of some approaches used in other systems for tracking dynamic cellular processes (TIRF, TIRF-SIM, etc.). However, recent advances in Light Sheet Fluorescence Microscopy (LSFM) and instantaneous structured illumination microscopy (iSIM) may allow for the successful tracking of these dynamic processes in chlamydiae growing and dividing within living cells.

Acknowledgements

The opinions or assertions contained herein are those of the authors and are not to be construed as official or as reflecting the views of the Department of Defense or the Uniformed Services University. Research on *Chlamydia* in the laboratory of ATM is supported by award R01AI12330 from the National Institute of Allergy and Infectious Diseases and GL is supported by USU Faculty Startup Award.

References

Abdelrahman, Y., Ouellette, S.P., Belland, R.J., and Cox, J.V. (2016). Polarized cell division of *Chlamydia trachomatis*. PLoS Pathog *12*, e1005822.

Albrecht, M., Sharma, C.M., Dittrich, M.T., Muller, T., Reinhardt, R., Vogel, J., and Rudel, T. (2011). The transcriptional landscape of *Chlamydia pneumoniae*. Genome Biol *12*, R98.

Albrecht, M., Sharma, C.M., Reinhardt, R., Vogel, J., and Rudel, T. (2010). Deep sequencing-based discovery of the *Chlamydia trachomatis* transcriptome. Nucleic Acids Res *38*, 868–877.

Barbour, A.G., Amano, K., Hackstadt, T., Perry, L., and Caldwell, H.D. (1982). *Chlamydia trachomatis* has penicillin-binding proteins but not detectable muramic acid. J Bacteriol *151*, 420–428.

Barry, C.E., 3rd, Hayes, S.F., and Hackstadt, T. (1992). Nucleoid condensation in *Escherichia coli* that express a chlamydial histone homolog. Science *256*, 377–379.

Bavoil, P., Ohlin, A., and Schachter, J. (1984). Role of disulfide bonding in outer membrane structure and permeability in *Chlamydia trachomatis*. Infect Immun *44*, 479–485.

Bavoil, P.M., Hsia, R., and Ojcius, D.M. (2000). Closing in on *Chlamydia* and its intracellular bag of tricks. Microbiology *146* (Pt 11), 2723–2731.

Bedson, S.P.a.M., H.B. (1945). Penicillin in Experimental Psittacosis of Mice. Lancet 2, 394–397.

Belland, R.J., Nelson, D.E., Virok, D., Crane, D.D., Hogan, D., Sturdevant, D., Beatty, W.L., and Caldwell, H.D. (2003a). Transcriptome analysis of chlamydial growth during IFN-gamma-mediated persistence and reactivation. Proc Natl Acad Sci U S A *100*, 15971–15976.

Belland, R.J., Zhong, G., Crane, D.D., Hogan, D., Sturdevant, D., Sharma, J., Beatty, W.L., and Caldwell, H.D. (2003b). Genomic transcriptional profiling of the developmental cycle of *Chlamydia trachomatis*. Proc Natl Acad Sci U S A *100*, 8478–8483.

Bonner, C.A., Byrne, G.I., and Jensen, R.A. (2014). *Chlamydia* exploit the mammalian tryptophan-depletion defense strategy as a counter-defensive cue to trigger a survival state of persistence. Front Cell Infect Microbiol *4*, 17.

Braun, V. (1975). Covalent lipoprotein from the outer membrane of *Escherichia coli*. Biochim Biophys Acta *415*, 335–377.

Brown, P.J., de Pedro, M.A., Kysela, D.T., Van der Henst, C., Kim, J., De Bolle, X., Fuqua, C., and Brun, Y.V. (2012). Polar growth in the Alphaproteobacterial order Rhizobiales. Proc Natl Acad Sci U S A *109*, 1697–1701.

Brown, W.J., and Rockey, D.D. (2000). Identification of an antigen localized to an apparent septum within dividing chlamydiae. Infect Immun *68*, 708–715.

Byrne, G.I., Lehmann, L.K., and Landry, G.J. (1986). Induction of tryptophan catabolism is the mechanism for gamma-interferon-mediated inhibition of intracellular *Chlamydia psittaci* replication in T24 cells. Infect Immun *53*, 347–351.

Carlin, J.M., Borden, E.C., Sondel, P.M., and Byrne, G.I. (1989). Interferon-induced indoleamine 2,3-dioxygenase activity in human mononuclear phagocytes. J Leukoc Biol *45*, 29–34.

Chen, Y.S., Bastidas, R.J., Saka, H.A., Carpenter, V.K., Richards, K.L., Plano, G.V., and Valdivia, R.H. (2014). The *Chlamydia trachomatis* type III secretion chaperone Slc1 engages multiple early effectors, including TepP, a tyrosine-phosphorylated protein required for the recruitment of CrkI-II to nascent inclusions and innate immune signaling. PLoS Pathog *10*, e1003954.

Chien, A.C., Hill, N.S., and Levin, P.A. (2012). Cell size control in bacteria. Curr Biol *22*, R340–349.

Chua, P.K., Corkill, J.E., Hooi, P.S., Cheng, S.C., Winstanley, C., and Hart, C.A. (2005). Isolation of *Waddlia malaysiensis*, a novel intracellular bacterium, from fruit bat (*Eonycteris spelaea*). Emerg Infect Dis *11*, 271–277.

Clarke, I.N., Ward, M.E., and Lambden, P.R. (1988). Molecular cloning and sequence analysis of a developmentally regulated cysteine-rich outer membrane protein from *Chlamydia trachomatis*. Gene *71*, 307–314.

Corsaro, D., Thomas, V., Goy, G., Venditti, D., Radek, R., and Greub, G. (2007). 'Candidatus Rhabdochlamydia crassificans', an intracellular bacterial pathogen of the cockroach *Blatta orientalis* (Insecta: Blattodea). Syst Appl Microbiol *30*, 221–228.

Costerton, J.W., Poffenroth, L., Wilt, J.C., and Kordova, N. (1975). Ultrastructural studies of *Chlamydia psittaci* 6BC in situ in yolk sac explants and L cells: a comparison with gram-negative bacteria. Can J Microbiol *21*, 1433–1447.

De Benedetti, S., Buhl, H., Gaballah, A., Klockner, A., Otten, C., Schneider, T., Sahl, H.G., and Henrichfreise, B. (2014). Characterization of serine hydroxymethyltransferase GlyA as a potential source of D-alanine in *Chlamydia pneumoniae*. Front Cell Infect Microbiol *4*, 19.

Divakaruni, A.V., Baida, C., White, C.L., and Gober, J.W. (2007). The cell shape proteins MreB and MreC control cell morphogenesis by positioning cell wall synthetic complexes. Mol Microbiol *66*, 174–188.

Draghi, A., 2nd, Popov, V.L., Kahl, M.M., Stanton, J.B., Brown, C.C., Tsongalis, G.J., West, A.B., and Frasca, S., Jr. (2004). Characterization of "Candidatus piscichlamydia salmonis" (order Chlamydiales), a chlamydia-like bacterium associated with epitheliocystis in farmed Atlantic salmon (Salmo salar). J Clin Microbiol *42*, 5286–5297.

Early, R.L., and Morgan, H.R. (1946). Studies on the chemotherapy of viruses in the psittacosis-lymphogranuloma venereum group; effect of certain chemotherapeutic agents on psittacosis virus (6BC strain) infections in mice. J Immunol *53*, 251–257.

Eaton, M.D., Dozois, T.F., and et al. (1948). Studies on inhibition of the agents of murine and feline pneumonitis by penicillin. J Immunol *58*, 251–267.

Egan, A.J., and Vollmer, W. (2013). The physiology of bacterial cell division. Ann N Y Acad Sci *1277*, 8–28.

Everett, K.D., and Hatch, T.P. (1995). Architecture of the cell envelope of *Chlamydia psittaci* 6BC. J Bacteriol *177*, 877–882.

Falcieri, E., Cevenini, R., Landini, M.P., and Donati, M. (1979). The replication cycle of *Chlamydia trachomatis* and Chlamydia psittaci: ultrastructural analysis. Boll Ist Sieroter Milan *58*, 395–405.

Fox, A., Rogers, J.C., Gilbert, J., Morgan, S., Davis, C.H., Knight, S., and Wyrick, P.B. (1990). Muramic acid is not detectable in *Chlamydia psittaci* or *Chlamydia trachomatis* by gas chromatography-mass spectrometry. Infect Immun *58*, 835–837.

Frandi, A., Jacquier, N., Theraulaz, L., Greub, G., and Viollier, P.H. (2014). FtsZ-independent septal recruitment and function of cell wall remodelling enzymes in chlamydial pathogens. Nat Commun *5*, 4200.

Friis, R.R. (1972). Interaction of L cells and *Chlamydia psittaci*: entry of the parasite and host responses to its development. J Bacteriol *110*, 706–721.

Frirdich, E., and Gaynor, E.C. (2013). Peptidoglycan hydrolases, bacterial shape, and pathogenesis. Curr Opin Microbiol *16*, 767–778.

Fuerst, J.A. (1995). The planctomycetes: emerging models for microbial ecology, evolution and cell biology. Microbiology *141 (Pt 7)*, 1493–1506.

Gaballah, A., Kloeckner, A., Otten, C., Sahl, H.G., and Henrichfreise, B. (2011). Functional analysis of the cytoskeleton protein MreB from *Chlamydophila pneumoniae*. PLoS One *6*, e25129.

Garrett, A.J., Harrison, M.J., and Manire, G.P. (1974). A search for the bacterial mucopeptide component, muramic acid, in *Chlamydia*. J Gen Microbiol *80*, 315–318.

Gaylord, W.H., Jr. (1954). Intracellular forms of meningopneumonitis virus. J Exp Med *100*, 575–580.

Gerding, M.A., Ogata, Y., Pecora, N.D., Niki, H., and de Boer, P.A. (2007). The trans-envelope Tol-Pal complex is part of the cell division machinery and required for proper outer-membrane invagination during cell constriction in *E. coli*. Mol Microbiol *63*, 1008–1025.

Ghuysen, J.M., and Goffin, C. (1999). Lack of cell wall peptidoglycan versus penicillin sensitivity: new insights into the chlamydial anomaly. Antimicrob Agents Chemother *43*, 2339–2344.

Gogolak, F.M., and Weiss, E. (1950). The effect of antibiotics on agents of the psittacosis-lymphogranuloma group. J Infect Dis *87*, 264–274.

Greub, G., and Raoult, D. (2002). Crescent bodies of *Parachlamydia acanthamoeba* and its life cycle within *Acanthamoeba polyphaga* : an electron micrograph study. Appl Environ Microbiol *68*, 3076–3084.

Hackstadt, T., Baehr, W., and Ying, Y. (1991). *Chlamydia trachomatis* developmentally regulated protein is homologous to eukaryotic histone H1. Proc Natl Acad Sci U S A *88*, 3937–3941.

Hackstadt, T., Rockey, D.D., Heinzen, R.A., and Scidmore, M.A. (1996). *Chlamydia trachomatis* interrupts an exocytic pathway to acquire endogenously synthesized sphingomyelin in transit from the Golgi apparatus to the plasma membrane. EMBO J *15*, 964–977.

Hackstadt, T., Todd, W.J., and Caldwell, H.D. (1985). Disulfide-mediated interactions of the chlamydial major outer membrane protein: role in the differentiation of Chlamydiae? J Bacteriol *161*, 25–31.

Harris, L.K., and Theriot, J.A. (2016). Relative rates of surface and volume synthesis set bacterial cell size. Cell *165*, 1479–1492.

Hatch, T.P., Allan, I., and Pearce, J.H. (1984). Structural and polypeptide differences between envelopes of infective and reproductive life cycle forms of *Chlamydia* spp. J Bacteriol *157*, 13–20.

Hatch, T.P., Miceli, M., and Sublett, J.E. (1986). Synthesis of disulfide-bonded outer membrane proteins during the developmental cycle of *Chlamydia psittaci* and *Chlamydia trachomatis*. J Bacteriol *165*, 379–385.

Heilman, F.R.a.H., W. E. (1944). Penicillin in the treatment of experimental psittacosis. Proe Staff Meetings Mayo Clinic *19*, 204–207.

Hesse, L., Bostock, J., Dementin, S., Blanot, D., Mengin-Lecreulx, D., and Chopra, I. (2003). Functional and biochemical analysis of *Chlamydia trachomatis* MurC, an enzyme displaying UDP-N-acetylmuramate:amino acid ligase activity. J Bacteriol *185*, 6507–6512.

Higashi, N. (1965). Electron Microscopic Studies on the Mode of Reproduction of Trachoma Virus and

Psittacosis Virus in Cell Cultures. Exp Mol Pathol *4*, 24–39.

Hoare, A., Timms, P., Bavoil, P.M., and Wilson, D.P. (2008). Spatial constraints within the chlamydial host cell inclusion predict interrupted development and persistence. BMC Microbiol *8*, 5.

Hudson, A.O., Singh, B.K., Leustek, T., and Gilvarg, C. (2006). An LL-diaminopimelate aminotransferase defines a novel variant of the lysine biosynthesis pathway in plants. Plant Physiol *140*, 292–301.

Hurst, E.W., Landquist, J.K., Melvin, P., Peters, J.M., Senior, N., Silk, J.A., and Stacy, G.J. (1953). The therapy of experimental psittacosis and lymphogranuloma venereum (inguinale) II. The activity of quinoxaline-1:4-dioxide and substituted and related compounds, with a note on the morphological changes induced in lymphogranuloma virus by these compounds and by antibiotics. Br J Pharmacol Chemother *8*, 297–305.

Hybiske, K., and Stephens, R.S. (2007). Mechanisms of host cell exit by the intracellular bacterium *Chlamydia*. Proc Natl Acad Sci U S A *104*, 11430–11435.

Jacquier, N., Frandi, A., Pillonel, T., Viollier, P., and Greub, G. (2014). Cell wall precursors are required to organize the chlamydial division septum. Nat Commun *5*, 3578.

Jacquier, N., Frandi, A., Viollier, P.H., and Greub, G. (2015a). Disassembly of a medial transenvelope structure by antibiotics during intracellular division. Chem Biol *22*, 1217–1227.

Jacquier, N., Viollier, P.H., and Greub, G. (2015b). The role of peptidoglycan in chlamydial cell division: towards resolving the chlamydial anomaly. FEMS Microbiol Rev *39*, 262–275.

Jenkin, H.M. (1960). Preparation and properties of cell walls of the agent of meningopneumonitis. J Bacteriol *80*, 639–647.

Johnson, C.M., and Fisher, D.J. (2013). Site-specific, insertional inactivation of *incA* in *Chlamydia trachomatis* using a group II intron. PLoS One *8*, e83989.

Kemege, K.E., Hickey, J.M., Barta, M.L., Wickstrum, J., Balwalli, N., Lovell, S., Battaile, K.P., and Hefty, P.S. (2015). *Chlamydia trachomatis* protein CT009 is a structural and functional homolog to the key morphogenesis component RodZ and interacts with division septal plane localized MreB. Mol Microbiol *95*, 365–382.

Klockner, A., Otten, C., Derouaux, A., Vollmer, W., Buhl, H., De Benedetti, S., Munch, D., Josten, M., Molleken, K., Sahl, H.G., *et al.* (2014). AmiA is a penicillin target enzyme with dual activity in the intracellular pathogen *Chlamydia pneumoniae*. Nat Commun *5*, 4201.

Lambden, P.R., Pickett, M.A., and Clarke, I.N. (2006). The effect of penicillin on *Chlamydia trachomatis* DNA replication. Microbiology *152*, 2573–2578.

Lee, J.K., Enciso, G.A., Boassa, D., Chander, C.N., Lou, T.H., Pairawan, S.S., Guo, M.C., Wan, F.Y.M., Ellisman, M.H., Sutterlin, C., *et al.* (2018). Replication-dependent size reduction precedes differentiation in *Chlamydia trachomatis*. Nat Commun *9*, 45.

Lee, K.C., Webb, R.I., and Fuerst, J.A. (2009). The cell cycle of the planctomycete *Gemmata obscuriglobus* with respect to cell compartmentalization. BMC Cell Biol *10*, 4.

Lewis, M.E., Belland, R.J., AbdelRahman, Y.M., Beatty, W.L., Aiyar, A.A., Zea, A.H., Greene, S.J., Marrero, L., Buckner, L.R., Tate, D.J., *et al.* (2014). Morphologic and molecular evaluation of *Chlamydia trachomatis* growth in human endocervix reveals distinct growth patterns. Front Cell Infect Microbiol *4*, 71.

Liechti, G., Kuru, E., Packiam, M., Hsu, Y.P., Tekkam, S., Hall, E., Rittichier, J.T., VanNieuwenhze, M., Brun, Y.V., and Maurelli, A.T. (2016). Pathogenic *Chlamydia* lack a classical sacculus but synthesize a narrow, mid-cell peptidoglycan ring, regulated by mreb, for cell division. PLoS Pathog *12*, e1005590.

Liechti, G., Singh, R., Rossi, P.L., Gray, M.D., Adams, N.E., and Maurelli, A.T. (2018). *Chlamydia trachomatis dapF* encodes a bifunctional enzyme capable of both d-glutamate racemase and diaminopimelate epimerase activities. MBio *9*, e00204–00218.

Liechti, G.W., Kuru, E., Hall, E., Kalinda, A., Brun, Y.V., VanNieuwenhze, M., and Maurelli, A.T. (2014). A new metabolic cell-wall labelling method reveals peptidoglycan in *Chlamydia trachomatis*. Nature *506*, 507–510.

Lienard, J., Croxatto, A., Prod'hom, G., and Greub, G. (2011). *Estrella lausannensis*, a new star in the *Chlamydiales* order. Microbes Infect *13*, 1232–1241.

Lin, H.S., and Moulder, J.W. (1966). Patterns of response to sulfadiazine, D-cycloserine and D-alanine in members of the psittacosis group. J Infect Dis *116*, 372–376.

Liu, X., Meiresonne, N.Y., Bouhss, A., and den Blaauwen, T. (2018). FtsW activity and lipid II synthesis are required for recruitment of MurJ to midcell during cell division in *Escherichia coli*. Mol Microbiol *109*, 855-884.

Lund, V.A., Wacnik, K., Turner, R.D., Cotterell, B.E., Walther, C.G., Fenn, S.J., Grein, F., Wollman, A.J., Leake, M.C., Olivier, N., *et al.* (2018). Molecular coordination of *Staphylococcus aureus* cell division. Elife *7*, e32057.

Manire, G.P., and Tamura, A. (1967). Preparation and chemical composition of the cell walls of mature infectious dense forms of meningopneumonitis organisms. J Bacteriol *94*, 1178–1183.

Mathews, S.A., Volp, K.M., and Timms, P. (1999). Development of a quantitative gene expression assay for *Chlamydia trachomatis* identified temporal expression of sigma factors. FEBS Lett *458*, 354–358.

Matsumoto, A. (1981a). Electron microscopic observations of surface projections and related intracellular structures of *Chlamydia* organisms. J Electron Microsc (Tokyo) *30*, 315–320.

Matsumoto, A. (1981b). Isolation and electron microscopic observations of intracytoplasmic inclusions containing *Chlamydia psittaci*. J Bacteriol *145*, 605–612.

Matsumoto, A. (1988). Structural characteristics of chlamydial bodies. In Microbiology of Chlamydia, A.L. Barron, ed. (Boca Raton, FL: CRC Press), pp. pp. 21–45.

Matsumoto, A., and Manire, G.P. (1970). Electron microscopic observations on the effects of penicillin on the morphology of *Chlamydia psittaci*. J Bacteriol *101*, 278–285.

McCoy, A.J., Adams, N.E., Hudson, A.O., Gilvarg, C., Leustek, T., and Maurelli, A.T. (2006). L,L-diaminopimelate aminotransferase, a trans-kingdom enzyme shared by *Chlamydia* and plants for synthesis of diaminopimelate/lysine. Proc Natl Acad Sci U S A *103*, 17909–17914.

McCoy, A.J., and Maurelli, A.T. (2005). Characterization of *Chlamydia* MurC-Ddl, a fusion protein exhibiting D-alanyl-D-alanine ligase activity involved in peptidoglycan synthesis and D-cycloserine sensitivity. Mol Microbiol *57*, 41–52.

McCoy, A.J., Sandlin, R.C., and Maurelli, A.T. (2003). In vitro and in vivo functional activity of *Chlamydia* MurA, a UDP-N-acetylglucosamine enolpyruvyl transferase involved in peptidoglycan synthesis and fosfomycin resistance. J Bacteriol *185*, 1218–1228.

Meeske, A.J., Riley, E.P., Robins, W.P., Uehara, T., Mekalanos, J.J., Kahne, D., Walker, S., Kruse, A.C., Bernhardt, T.G., and Rudner, D.Z. (2016). SEDS proteins are a widespread family of bacterial cell wall polymerases. Nature *537*, 634–638.

Meiklejohn, G., Wagner, J.C., and Beveridge, G.W. (1946). Studies on the chemotherapy of viruses in the psittacosis-lymphogranuloma group; effect of penicillin and sulfadiazine on ten strains in chick embryos. J Immunol *54*, 1–8.

Mercier, R., Kawai, Y., and Errington, J. (2013). Excess membrane synthesis drives a primitive mode of cell proliferation. Cell *152*, 997–1007.

Mercier, R., Kawai, Y., and Errington, J. (2014). General principles for the formation and proliferation of a wall-free (L-form) state in bacteria. Elife 3.

Mercier, R., Kawai, Y., and Errington, J. (2016). Wall proficient *E. coli* capable of sustained growth in the absence of the Z-ring division machine. Nat Microbiol *1*, 16091.

Meyer, K.F., and Eddie, B. (1947). The knowledge of human virus infections of animal origin. J Am Med Assoc *133*, 822–828.

Miyairi, I., Mahdi, O.S., Ouellette, S.P., Belland, R.J., and Byrne, G.I. (2006). Different growth rates of *Chlamydia trachomatis* biovars reflect pathotype. J Infect Dis *194*, 350–357.

Monteiro, J.M., Pereira, A.R., Reichmann, N.T., Saraiva, B.M., Fernandes, P.B., Veiga, H., Tavares, A.C., Santos, M., Ferreira, M.T., Macario, V., *et al.* (2018). Peptidoglycan synthesis drives an FtsZ-treadmilling-independent step of cytokinesis. Nature *554*, 528–532.

Moulder, J.W. (1964). The psittacosis group as bacteria (New York,: Wiley).

Moulder, J.W. (1993). Why is *Chlamydia* sensitive to penicillin in the absence of peptidoglycan? Infect Agents Dis *2*, 87–99.

Moulder, J.W., Novosel, D.L., and Officer, J.E. (1963). inhibition of the growth of agents of the psittacosis group by d-cycloserine and its specific reversal by d-alanine. J Bacteriol *85*, 707–711.

Mueller, K.E., Wolf, K., and Fields, K.A. (2016). Gene deletion by fluorescence-reported allelic exchange mutagenesis in *Chlamydia trachomatis*. MBio 7.

Nans, A., Saibil, H.R., and Hayward, R.D. (2014). Pathogen-host reorganization during *Chlamydia* invasion revealed by cryo-electron tomography. Cell Microbiol *16*, 1457–1472.

Newhall, W.J., and Jones, R.B. (1983). Disulfide-linked oligomers of the major outer membrane protein of Chlamydiae. J Bacteriol *154*, 998–1001.

Nguyen, B.D., Cunningham, D., Liang, X., Chen, X., Toone, E.J., Raetz, C.R., Zhou, P., and Valdivia, R.H. (2011). Lipooligosaccharide is required for the generation of infectious elementary bodies in *Chlamydia trachomatis*. Proc Natl Acad Sci U S A *108*, 10284–10289.

Ouellette, S.P. (2018). Feasibility of a conditional knockout system for *Chlamydia* Based on CRISPR Interference. Front Cell Infect Microbiol *8*, 59.

Ouellette, S.P., Karimova, G., Subtil, A., and Ladant, D. (2012). *Chlamydia* co-opts the rod shape-determining proteins MreB and Pbp2 for cell division. Mol Microbiol *85*, 164–178.

Ouellette, S.P., Rueden, K.J., AbdelRahman, Y.M., Cox, J.V., and Belland, R.J. (2015). Identification and Partial Characterization of Potential FtsL and FtsQ Homologs of *Chlamydia*. Front Microbiol *6*, 1264.

Ouellette, S.P., Rueden, K.J., Gauliard, E., Persons, L., de Boer, P.A., and Ladant, D. (2014). Analysis of MreB interactors in *Chlamydia* reveals a RodZ homolog but fails to detect an interaction with MraY. Front Microbiol *5*, 279.

Ouellette, S.P., Rueden, K.J., and Rucks, E.A. (2016). Tryptophan codon-dependent transcription in *Chlamydia pneumoniae* during gamma interferon-mediated tryptophan limitation. Infect Immun *84*, 2703–2713.

Packiam, M., Weinrick, B., Jacobs, W.R., Jr., and Maurelli, A.T. (2015). Structural characterization of muropeptides from *Chlamydia trachomatis* peptidoglycan by mass spectrometry resolves "chlamydial anomaly". Proc Natl Acad Sci U S A *112*, 11660–11665.

Park, J.T., and Strominger, J.L. (1957). Mode of action of penicillin. Science *125*, 99–101.

Patin, D., Bostock, J., Blanot, D., Mengin-Lecreulx, D., and Chopra, I. (2009). Functional and biochemical analysis of the *Chlamydia trachomatis* ligase MurE. J Bacteriol *191*, 7430–7435.

Patin, D., Bostock, J., Chopra, I., Mengin-Lecreulx, D., and Blanot, D. (2012). Biochemical characterisation of the chlamydial MurF ligase, and possible sequence of the chlamydial peptidoglycan pentapeptide stem. Arch Microbiol *194*, 505–512.

Perkins, H.R., and Allison, A.C. (1963). Cell-wall constituents of rickettsiae and psittacosis-lymphogranuloma organisms. J Gen Microbiol *30*, 469–480.

Phillips, D.M., Swenson, C.E., and Schachter, J. (1984). Ultrastructure of *Chlamydia trachomatis* infection of the mouse oviduct. J Ultrastruct Res *88*, 244–256.

Pilhofer, M., Aistleitner, K., Biboy, J., Gray, J., Kuru, E., Hall, E., Brun, Y.V., VanNieuwenhze, M.S., Vollmer, W., Horn, M., *et al.* (2013). Discovery of chlamydial peptidoglycan reveals bacteria with murein sacculi but without FtsZ. Nat Commun *4*, 2856.

Rank, R.G., Whittimore, J., Bowlin, A.K., and Wyrick, P.B. (2011). In vivo ultrastructural analysis of the intimate relationship between polymorphonuclear leukocytes and the chlamydial developmental cycle. Infect Immun *79*, 3291–3301.

Rapoza, P.A., Tahija, S.G., Carlin, J.P., Miller, S.L., Padilla, M.L., and Byrne, G.I. (1991). Effect of interferon on a primary conjunctival epithelial cell model of trachoma. Invest Ophthalmol Vis Sci *32*, 2919–2923.

Rojas, E.R., Billings, G., Odermatt, P.D., Auer, G.K., Zhu, L., Miguel, A., Chang, F., Weibel, D.B., Theriot, J.A., and Huang, K.C. (2018). The outer membrane is an essential load-bearing element in Gram-negative bacteria. Nature, 559, 617–621.

Rosario, C.J., Hanson, B.R., and Tan, M. (2014). The transcriptional repressor EUO regulates both subsets of *Chlamydia* late genes. Mol Microbiol *94*, 888–897.

Rosario, C.J., and Tan, M. (2012). The early gene product EUO is a transcriptional repressor that selectively regulates promoters of *Chlamydia* late genes. Mol Microbiol *84*, 1097–1107.

Rusconi, B., Lienard, J., Aeby, S., Croxatto, A., Bertelli, C., and Greub, G. (2013). Crescent and star shapes of members of the *Chlamydiales* order: impact of fixative methods. Antonie Van Leeuwenhoek *104*, 521–532.

Santarella-Mellwig, R., Pruggnaller, S., Roos, N., Mattaj, I.W., and Devos, D.P. (2013). Three-dimensional reconstruction of bacteria with a complex endomembrane system. PLoS Biol *11*, e1001565.

Schechter, E.M. (1966). Synthesis of nucleic acid and protein in L cells infected with the agent of meningopneumonitis. J Bacteriol *91*, 2069–2080.

Scidmore-Carlson, M.A., Shaw, E.I., Dooley, C.A., Fischer, E.R., and Hackstadt, T. (1999). Identification and characterization of a *Chlamydia trachomatis* early operon encoding four novel inclusion membrane proteins. Mol Microbiol *33*, 753–765.

Shaw, E.I., Dooley, C.A., Fischer, E.R., Scidmore, M.A., Fields, K.A., and Hackstadt, T. (2000). Three temporal classes of gene expression during the *Chlamydia trachomatis* developmental cycle. Mol Microbiol *37*, 913–925.

Shima, K., Wanker, M., Skilton, R.J., Cutcliffe, L.T., Schnee, C., Kohl, T.A., Niemann, S., Geijo, J., Klinger, M., Timms, P., *et al.* (2018). The Genetic Transformation of *Chlamydia pneumoniae*. mSphere *3(5)*, e00412–18.

Shimizu, T., Nomiyama, S., Hirata, F., and Hayaishi, O. (1978). Indoleamine 2,3-dioxygenase. Purification and some properties. J Biol Chem *253*, 4700–4706.

Silhavy, T.J., Kahne, D., and Walker, S. (2010). The bacterial cell envelope. Cold Spring Harb Perspect Biol 2, a000414.

Singh, S.K., Parveen, S., SaiSree, L., and Reddy, M. (2015). Regulated proteolysis of a cross-link-specific peptidoglycan hydrolase contributes to bacterial morphogenesis. Proc Natl Acad Sci U S A *112*, 10956–10961.

Srivastava, D., Seo, J., Rimal, B., Kim, S.J., Zhen, S., and Darwin, A.J. (2018). A proteolytic complex targets multiple cell wall hydrolases in *Pseudomonas aeruginosa*. MBio *9(4)*, e00972–18.

Stephens, R.S., Kalman, S., Lammel, C., Fan, J., Marathe, R., Aravind, L., Mitchell, W., Olinger, L., Tatusov, R.L., Zhao, Q., *et al.* (1998). Genome sequence of an obligate intracellular pathogen of humans: *Chlamydia trachomatis*. Science *282*, 754–759.

Su, H., Zhang, Y. X., and Li, R. (1985). Presence of muramic acid in *Chlamydia trachomatis* proved by liquid chromatography-mass spectrometry. Kexue Tongbao *30*, 695–699.

Swanson, J., Eschenbach, D.A., Alexander, E.R., and Holmes, K.K. (1975). Light and electron microscopic study of *Chlamydia trachomatis* infection of the uterine cervix. J Infect Dis *131*, 678–687.

Taguchi, A., Welsh, M.A., Marmont, L.S., Lee, W., Sjodt, M., Kruse, A.C., Kahne, D., Bernhardt, T.G.* and Walker, S. (2019). FtsW is a peptidoglycan polymerase that is functional only in complex with its cognate penicillin-binding protein. Nat. Microbiol. 4, 587–594.

Taheri-Araghi, S., Bradde, S., Sauls, J.T., Hill, N.S., Levin, P.A., Paulsson, J., Vergassola, M., and Jun, S. (2015). Cell-size control and homeostasis in bacteria. Curr Biol *25*, 385–391.

Tajima, M., Samejima, T., and Nomura, Y. (1959). Morphology of meningopneumonitis virus exposed to penicillin as observed with the electron microscope. J Bacteriol *77*, 23–34.

Tamura, A., and Manire, G.P. (1967). Preparation and chemical composition of the cell membranes of developmental reticulate forms of meningopneumonitis organisms. J Bacteriol *94*, 1184–1188.

Tamura, A., and Manire, G.P. (1968). Effect of penicillin on the multiplication of meningopneumonitis organisms (*Chlamydia psittaci*). J Bacteriol *96*, 875–880.

Tan, M., and Bavoil, P.M. (2012). Intracellular pathogens I : *Chlamydiales* (Washington, DC: ASM Press).

Tanami, Y., and Yamada, Y. (1973). Miniature cell formation in *Chlamydia psittaci*. J Bacteriol *114*, 408–412.

Thanbichler, M. (2010). Synchronization of chromosome dynamics and cell division in bacteria. Cold Spring Harb Perspect Biol *2*, a000331.

Thomas, V., Casson, N., and Greub, G. (2006). *Criblamydia sequanensi*, a new intracellular *Chlamydiales* isolated from Seine river water using amoebal co-culture. Environ Microbiol *8*, 2125–2135.

Torti, S.V., and Park, J.T. (1976). Lipoprotein of gram-negative bacteria is essential for growth and division. Nature *263*, 323–326.

Turner, R.D., Ratcliffe, E.C., Wheeler, R., Golestanian, R., Hobbs, J.K., and Foster, S.J. (2010). Peptidoglycan architecture can specify division planes in *Staphylococcus aureus*. Nat Commun *1*, 26.

Tzagoloff, H., and Novick, R. (1977). Geometry of cell division in *Staphylococcus aureus*. J Bacteriol *129*, 343–350.

Uehara, T., and Bernhardt, T.G. (2011). More than just lysins: peptidoglycan hydrolases tailor the cell wall. Curr Opin Microbiol *14*, 698–703.

Uehara, T., and Park, J.T. (2008). Growth of *Escherichia coli*: significance of peptidoglycan degradation during elongation and septation. J Bacteriol *190*, 3914–3922.

Weart, R.B., Lee, A.H., Chien, A.C., Haeusser, D.P., Hill, N.S., and Levin, P.A. (2007). A metabolic sensor governing cell size in bacteria. Cell *130*, 335–347.

Weiss, E. (1950). The effect of antibiotics on agents of the psittacosis-lymphogranuloma group. I. The effect of penicillin. J Infect Dis *87*, 249–263.

Wilson, D.P., Timms, P., McElwain, D.L., and Bavoil, P.M. (2006). Type III secretion, contact-dependent model for the intracellular development of *Chlamydia*. Bull Math Biol *68*, 161–178.

Wilson, D.P., Whittum-Hudson, J.A., Timms, P., and Bavoil, P.M. (2009). Kinematics of intracellular chlamydiae provide evidence for contact-dependent development. J Bacteriol *191*, 5734–5742.

Wiseman, R.W., Meiklejohn, G., *et al.* (1946). Studies on the chemotherapy of viruses in the psittacosis-lymphogranuloma group; effect of penicillin and sulfadiazine on seven strains in mice. J Immunol *54*, 9–16.

Wolf, K., Fischer, E., and Hackstadt, T. (2000). Ultrastructural analysis of developmental events in *Chlamydia pneumoniae*-infected cells. Infect Immun *68*, 2379–2385.

Work, E. (1961). The mucopeptides of bacterial cell walls. A review. J Gen Microbiol *25*, 169–189.

Yang, D.C., Peters, N.T., Parzych, K.R., Uehara, T., Markovski, M., and Bernhardt, T.G. (2011). An ATP-binding cassette transporter-like complex governs cell-wall hydrolysis at the bacterial cytokinetic ring. Proc Natl Acad Sci U S A *108*, E1052–1060.

Chlamydia Gene Regulation

10

Christopher J. Rosario[1], Katelyn Soules[2], P. Scott Hefty[2] and Ming Tan[3]*

[1]Department of Microbiology and Molecular Genetics, University of California, Irvine, CA, USA.
[2]Department of Molecular Biosciences, University of Kansas, Lawrence, KS, USA.
[3]Departments of Microbiology and Molecular Genetics, and Medicine, University of California, Irvine, CA, USA.

*Correspondence: mingt@uci.edu

https://doi.org/10.21775/9781912530281.10

Abstract

A defining feature of the *Chlamydia* developmental cycle is the expression of chlamydial genes in three main temporal classes. Early genes are the only genes transcribed at the beginning of the intracellular infection, but the mechanism for their selective expression is unknown. The large majority of chlamydial genes are midcycle genes, which are transcribed during the midstage of the developmental cycle at the time of reticulate body (RB) growth and replication. Gene expression in midcycle is proposed to be activated by higher DNA supercoiling levels and to be regulated by a DNA gyrase and two DNA topoisomerases, which are enzymes that control DNA supercoiling. Late genes are up-regulated late in the developmental cycle, when RB to elementary body (EB) conversion takes place. There are two subsets of late genes that are respectively transcribed by σ^{66} RNA polymerase, which is the major chlamydial RNA polymerase, or an alternative RNA polymerase called σ^{28} RNA polymerase. However, both subsets of late genes are repressed by the transcription factor EUO, which is thus known as the master regulator of late gene expression. Transcription is then proposed to be silenced in EBs by two histone-like proteins, Hc1 and Hc2, which condense DNA. *Chlamydia* gene expression is also controlled by transcription factors that homeostatically regulate the levels of amino acids, nucleotides, metal ions and the major heat shock proteins. Regulators of chlamydial gene expression with less understood roles include a second alternative sigma factor, σ^{54}, putative regulators of each of the three forms of chlamydial RNA polymerase, and the transcription factors, ChxR and GrgA. In addition, there are potentially many sRNAs that control gene expression by binding to target transcripts. Collectively, these mechanisms allow *Chlamydia* to regulate the expression of its genes in both a pre-programmed manner and in response to external signals.

Introduction

Gene expression in bacteria can be regulated at a number of different steps, but transcription is the most important control point (Feklistov *et al.*, 2014). Transcripts can also be regulated post-transcriptionally by mechanisms that alter mRNA stability (Gottesman and Storz, 2011). Another major regulatory step is translation, where each mRNA is translated into its encoded protein by the ribosome. Protein function can then be altered by post-translational modifications, which include phosphorylation, glycosylation and disulfide crosslinking.

Most of the work on chlamydial gene regulation has focused on understanding transcriptional regulation because of its pivotal role in controlling gene expression in *Chlamydia* (Tan, 2006; Tan, 2012). In this chapter, we will first introduce the main mechanisms that bacteria use to regulate transcription. This repertoire includes RNA polymerase to transcribe RNA from DNA, and transcription factors and DNA structure to modulate transcription from individual genes or groups of genes. We will then discuss how *Chlamydia* utilizes this same basic toolbox to co-ordinately express its genes in temporal waves during the chlamydial developmental cycle and in response to external cues.

The transcriptional machinery: RNA polymerase

RNA polymerase in other bacteria

The bacterial enzyme that transcribes RNA from DNA is called RNA polymerase. Its catalytically active portion, known as core enzyme, polymerizes nucleotide triphosphates (NTPs) into RNA, but lacks DNA sequence specificity. Core enzyme is made up of two copies of the alpha (α) subunit, a beta (β) subunit and a beta-prime (β') subunit. An additional specificity subunit, called a sigma (σ) factor is necessary for RNA polymerase to recognize promoters, which are specific DNA sequences upstream of the gene where RNA polymerase binds and initiates transcription (Feklistov *et al.*, 2014).

Bacteria produce different forms of RNA polymerase that contain the same core enzyme bound to a different sigma factor (Feklistov *et al.*, 2014). For example, the major form of RNA polymerase in *E. coli* is σ^{70} RNA polymerase, which transcribes most vegetative genes. In addition, *E. coli* encodes a number of alternative sigma factors, each with its own promoter specificity and target genes. This mechanism provides a means to reprogram the transcriptional machinery to transcribe different sets of target genes in response to environmental and physiologic cues.

RNA polymerase in *Chlamydia*

Chlamydia utilizes the same basic transcriptional machinery as other bacteria. Each of the subunits of chlamydial core enzyme subunits is well conserved with its homologue in other bacteria (Kalman *et al.*, 1999; Lonetto *et al.*, 1992; Read *et al.*, 2003; Stephens *et al.*, 1998). *Chlamydia* and *E. coil* core enzyme are functionally equivalent and can substitute for each other in cell-free transcription assays (Schaumburg and Tan, 2003).

Chlamydia has three sigma factors, which allows it to produce three different forms of RNA polymerase. The major chlamydial RNA polymerase consists of core enzyme bound to σ^{66}, which is a 66-kDa homologue of the major *E. coli* sigma factor, σ^{70} (Engel and Ganem, 1990; Koehler *et al.*, 1990). σ^{66} RNA polymerase transcribes the large majority of chlamydial genes and appears to be present at all times, including in RBs and EBs (Saka *et al.*, 2011; Skipp *et al.*, 2016). *Chlamydia* also encodes two alternative sigma factors, σ^{28} and σ^{54}. From work done in *C. trachomatis*, σ^{28} RNA polymerase transcribes a subset of late genes (Rosario and Tan, 2015; Yu *et al.*, 2006b). So far, seven σ^{28}-dependent genes have been identified in *C. trachomatis* (Table 10.1) (Rosario and Tan, 2015; Yu *et al.*, 2006b). Neither the biological role of *Chlamydia* σ^{54} nor its complement of target genes have been elucidated.

The promoter specificity of each of the three chlamydial sigma factors is conserved with its homologues in other bacteria. Early work on *Chlamydia* transcription raised the question of whether chlamydial σ^{66} RNA polymerase recognized different promoters from *E. coli* and other bacteria

Table 10.1 *C. trachomatis* genes regulated by the alternative sigma factors σ^{28} and σ^{54}

Sigma	Gene	References for functional data
σ^{28}	*hctB*	Yu *et al.* (2006a); Yu *et al.* (2006b)
	tsp	Yu *et al.* (2006b)
	dnaK	Rosario and Tan (2015); Yu *et al.* (2006b)
	bioY	Rosario and Tan (2015); Yu *et al.* (2006b)
	pgk	Rosario and Tan (2015); Yu *et al.* (2006b)
	tlyC_1	Yu *et al.* (2006b)
	ct415	Rosario and Tan (2015)
σ^{54}	*ct652.1*	Mathews and Timms (2000)
	ct683	Mathews and Timms (2000)

(Mathews and Sriprakash, 1994; Stephens, 1990). However, functional studies have demonstrated that the optimal promoter sequence for *C. trachomatis* σ^{66} RNA polymerase is very similar to the consensus *E. coli* σ^{70} promoter sequence (Schaumburg and Tan, 2003; Tan, 2006). Similarly, the promoter recognized by *C. trachomatis* σ^{28} RNA polymerase is well conserved with the bacterial σ^{28} promoter sequence (Yu *et al.*, 2006a; Yu *et al.*, 2006b). The third chlamydial sigma factor, σ^{54}, has not been studied as well as σ^{66} and σ^{28}, but promoter sequences resembling the consensus bacterial σ^{54} promoter have been identified in the *C. trachomatis* genome (Mathews and Timms, 2000).

There are, however, two main differences between RNA polymerase in *Chlamydia* and other bacteria. Chlamydial RNA polymerase appears to lack an omega subunit, which is an accessory, and non-essential, subunit that plays a role in RNA polymerase assembly and transcriptional response to nutrient deprivation (Bhardwaj *et al.*, 2018; Mathew and Chatterji, 2006). It has been suggested that core enzyme lacking the omega subunit may bind better to an alternative sigma factor (Gunnelius *et al.*, 2014). The other major difference is that *Chlamydia* RNA polymerase only utilizes two alternative sigma factors. By comparison, other bacteria can have up to 100 alternative sigma factors (Paget, 2015), including heat shock, stationary phase and extracytoplasmic function (ECF) sigma factors, that are not present in *Chlamydia*. This difference may be a reflection of the relatively stable intracellular environment that *Chlamydia* encounters, compared to other bacteria that have to adapt to changing external conditions.

In summary, *Chlamydia* contains the same basal transcriptional machinery as other bacteria. The major and two minor forms of chlamydial RNA polymerases are similar to their homologues in other bacteria. Each form of chlamydial RNA polymerase recognizes its own promoter sequence, which allows it to transcribe its own subset of target genes. Almost all of the functional studies on chlamydial RNA polymerase and its regulation have been performed with *C. trachomatis*, and mainly with serovar L2. However, RNA polymerase is conserved among *Chlamydia* spp., and it is likely that fundamental properties such as promoter specificity are well conserved too.

Transcriptional regulation by transcription factors

Like other bacteria, *Chlamydia* uses transcription factors to modulate the ability of RNA polymerase to transcribe target genes (Table 10.2). All *Chlamydia* spp. are predicted to have 12–15 transcription factors, most of which have orthologues in other bacteria (Domman and Horn, 2015). An exception is EUO, which is a regulator of the developmental cycle that is conserved among all *Chlamydia* and *Chlamydia*-related bacteria, but not found in other bacteria (Rosario *et al.*, 2014; Rosario and Tan, 2012; Zhang *et al.*, 1998; Zhang *et al.*, 2000). GrgA is only found in *Chlamydia* but its *Chlamydia*-specific function is not known (Bao *et al.*, 2012; Desai *et al.*, 2018). Except for TrpR and ArgR, chlamydial transcription factors are present in all *Chlamydia* spp., which suggests that they have important roles in chlamydial biology (Domman and Horn, 2015). Eight chlamydial transcription factors have been validated with functional studies (Table 10.2), but it is now doubtful that another chlamydial protein DcrA is a bone fide transcription factor. Many well-conserved families of bacterial transcription factors are absent from *Chlamydia* (Domman and Horn, 2015). Specific chlamydial transcription factors will be discussed in more detail below.

This small number of chlamydial transcription factors is not surprising because it is characteristic of obligate intracellular bacteria, which have a reduced genome and face limited environmental changes within an infected host cell (Madan Babu *et al.*, 2006). In contrast, *E. coli* has more than 300 transcription factors because it encounters a dynamic range of environments and growth conditions (Yamamoto *et al.*, 2014). *Pseudomonas* utilizes 475 transcription factors to live in diverse environments from water to soil, and within plants and animals (Galan-Vasquez *et al.*, 2011).

The majority of chlamydial transcription factors are transcriptional repressors. A repressor down-regulates transcription by binding to its operator, which is a specific DNA sequence (Payankaulam *et al.*, 2010). The operator is located near the promoter of a target gene such that repressor-operator binding interferes with RNA polymerase–promoter interactions (Seshasayee *et al.*, 2011). The same operator sequence can be present at the promoters of multiple genes, allowing a repressor to co-ordinately regulate a group, or regulon, of target genes.

Table 10.2 *Chlamydia* transcription factors

Transcription factor	Function	Notes	Co-factor	Conservation among *Chlamydia*	References
EUO	Master regulator of late gene expression	Mechanism of derepression not determined		Conserved; but only found in *Chlamydia* and *Chlamydia*-related bacteria	Rosario *et al.* (2014); Rosario and Tan (2012); Zhang *et al.* (1998); Zhang *et al.* (2000)
TrpR	Tryptophan biosynthesis		Tryptophan is co-repressor	Present in some *Chlamydia* spp.	Akers and Tan (2006); Carlson *et al.* (2006; Wood *et al.* (2003)
ArgR	Arginine transport		Arginine is co-repressor	Present in some *Chlamydia* spp.	Schaumburg and Tan (2006)
YtgR	Iron transport	Fusion protein with YtgC	Iron is co-repressor	Conserved	Akers *et al.* (2011; Thompson *et al.* (2012)
NrdR	Nucleotide triphosphate (NTP) transport		dNTP is co-repressor	Conserved	Case *et al.* (2011)
HrcA	Stress response expression of heat shock proteins	Co-repressor augments repression but not required for HrcA activity	GroEL is co-repressor	Conserved	Hanson and Tan (2015); Wilson and Tan (2002); Wilson *et al.* (2005)
GrgA	Activator of σ^{66}- and σ^{28}-dependent transcription	Binds DNA non-specifically		Conserved, but only found in *Chlamydia* and not *Chlamydia*-related bacteria	Bao *et al.* (2012); Desai *et al.* (2018)
ChxR	Activator	Regulates virulence genes		Conserved	Koo *et al.* (2006); Yang *et al.* (2017)
CtcC	σ^{54} activator	Activator function not yet confirmed: phosphorylated by CtcB		Conserved	Koo and Stephens (2003)
DksA	No functional studies			Conserved	Domman and Horn (2015)
PhoU	No functional studies			Conserved	Domman and Horn (2015)
YebC	No functional studies			Conserved	Domman and Horn (2015)
YbjN	No functional studies			Conserved	Domman and Horn (2015)

Another class of transcription factors are the transcriptional activators, which upregulate transcription of specific target genes (Lee *et al.*, 2012). An activator binds its cognate DNA sequence, located upstream of, or adjacent to, its target promoters. Contact between the activator and RNA polymerase helps recruit RNA polymerase to the promoter or enhances its ability to initiate transcription. As with a repressor, an activator can co-ordinately control a regulon of target genes, each containing the activator binding site. Specific examples of chlamydial repressors and activators will be discussed below.

Transcriptional regulation by changes in DNA structure

Structural changes to the DNA template can have a dramatic effect on transcription by altering its interactions with the transcriptional machinery. In

Chlamydia, two distinct DNA structural alterations have global effects on gene expression. The first is DNA supercoiling, which is the topological state of the DNA helix. As will be discussed, changes in DNA supercoiling appear to be used as a mechanism to globally activate gene expression during RB replication. The second is DNA condensation, which is a characteristic feature of EBs that may mediate transcriptional silencing in this extracellular form of chlamydiae.

DNA supercoiling

DNA supercoiling refers to the twisting of the DNA double helix around its own axis (Dorman and Dorman, 2016). DNA can be in its intrinsically relaxed state, or can be overwound or underwound, which are defined as positive or negative supercoiling, respectively. The level of bacterial DNA supercoiling is generally regulated by enzymes called topoisomerases that include DNA gyrase, which increases negative supercoiling, and DNA topoisomerases I and IV, which relax DNA by removing negative supercoils. Most cellular DNA, including chlamydial DNA, is negatively supercoiled, and so for simplicity, we will refer to negative supercoiling as supercoiling in this chapter.

Chlamydia utilizes three topoisomerases to control the supercoiling level of its DNA (Orillard and Tan, 2015). From studies with *C. trachomatis* and *C. pneumoniae*, chlamydial DNA gyrase is a two-subunit enzyme that increases DNA negative supercoiling by hydrolyzing ATP (Ameyama *et al.*, 2003; Orillard and Tan, 2015). In contrast, *C. trachomatis* DNA topoisomerase I relaxes DNA in an ATP-independent manner (Orillard and Tan, 2015). *C. trachomatis* DNA topoisomerase IV is a two-subunit enzyme for decatenating DNA during DNA replication, but it has also been shown to relax DNA (Orillard and Tan, 2015). These three enzymes serve as opposing mediators of DNA supercoiling and are conserved between *Chlamydia* and other bacteria, and among *Chlamydia* spp.

Topoisomerases alter the supercoiling state of the entire genome, but the effects on gene expression are gene-specific. The main reason for this specificity is that only a subset of promoters are responsive to changes in supercoiling levels. (Dorman and Dorman, 2016). From work in other bacteria, it is known that DNA supercoiling can activate an individual promoter through several mechanisms. For example, supercoiling levels can optimize promoter alignment, thereby increasing RNA polymerase recruitment (Ahmed *et al.*, 2016), or increase transcription by promoting DNA strand separation during transcription initiation and by enhancing promoter clearance (Revyakin *et al.*, 2004).

Supercoiling provides a mechanism to co-ordinately regulate a large number of genes through their supercoiling-responsive promoters. For example, changes in supercoiling co-ordinately regulate the expression of hundreds of genes in *E. coli* and *Haemophilus influenzae* (Gmuender *et al.*, 2001; Peter *et al.*, 2004). As will be discussed, DNA supercoiling may have an even larger role in *Chlamydia* as a global regulator of gene expression during the developmental cycle.

DNA condensation

In a bacterium, DNA is condensed into a nucleoid (Dillon and Dorman, 2010; Joyeux, 2015). This compaction of DNA reduces its volume so that it only occupies a portion of the bacterial cell. DNA condensation is mediated by a number of non-specific mechanisms, including macromolecular crowding and DNA charge neutralization. Bacterial DNA can also be condensed by nucleoid-associated proteins, which bind DNA and form DNA-protein-DNA bridges. Negative DNA supercoiling also contributes to DNA compaction, but only to a modest extent.

DNA binding and condensation by nucleoid-associated proteins can have major effects on transcription. A nucleoid-associated protein may cause transcriptional repression by preventing RNA polymerase from binding to a promoter or by keeping bound RNA polymerase in an inactive transcriptional complex (Joyeux, 2015). These effects on transcription can be global but can also act on individual promoters. In addition, nucleoid-associated proteins can affect transcription indirectly by constraining, or trapping, negative supercoils, which decreases the effective DNA supercoiling level (Dillon and Dorman, 2010).

Chlamydia appear to lack most of the canonical nucleoid-associated proteins found in other bacteria. The only nucleoid-associated protein that can be readily identified from the *Chlamydia* genome is HU, which binds DNA and constrains DNA supercoiling, thereby preventing binding of

RNA polymerase and transcription factors (Berger *et al.*, 2016; Grove, 2011). HU is composed of two subunits (Drlica and Rouviere-Yaniv, 1987), but *Chlamydia* only encodes the beta subunit and appears to lack the alpha subunit. However, *Chlamydia* HU-beta could be functional because *E. coli* HU-beta forms homodimers that bind DNA, albeit with different DNA-binding properties than an HU–alpha/HU-beta heterodimer (Pinson *et al.*, 1999). The role of *Chlamydia* HU-beta as a DNA binding protein and a regulator of gene expression has not been studied. *Chlamydia* does not have identifiable homologues of other common bacterial nucleoid-associated proteins, such as IHF, Fis, and H-NS.

Interestingly, *Chlamydia* has two nucleoid-associated proteins, Hc1 and Hc2, which are not found in other bacteria (Hackstadt, 1991; Perara *et al.*, 1992; Tao *et al.*, 1991). They have been called histone-like proteins because they are similar to eukaryotic histones in binding DNA without sequence specificity. Hc1 and Hc2 are encoded by *hctA* and *hctB*, respectively, and are conserved in all *Chlamydia* spp., but are not homologous to other bacterial nucleoid-associated proteins. The role of Hc1 and Hc2 in mediating DNA condensation in EBs is supported by experimental evidence showing that they each bind and compact DNA (Barry *et al.*, 1992; Barry *et al.*, 1993; Brickman *et al.*, 1993). However, only Hc1, and not Hc2, is encoded by the *Chlamydia*-related bacteria (Collingro *et al.*, 2011). Thus Hc1 may be the primary histone-like protein for condensing chlamydial DNA in EBs, while Hc2 may have a more specialized role. Consistent with this hypothesis, Hc1 expression in *E. coli* produced a more compact nucleoid than Hc2 (Barry *et al.*, 1992; Brickman *et al.*, 1993).

Regulation of transcription during the *Chlamydia* developmental cycle

The defining features of the intracellular *Chlamydia* infection are its developmental cycle and the expression of chlamydial genes in temporal waves (Abdelrahman and Belland, 2005). *Chlamydia* are unusual bacteria because they convert between an elementary body (EB), which is an infectious but non-replicating form, and a reticulate body (RB), which is the metabolically active, dividing form that is only found within a eukaryotic host cell. Transcription in three main temporal groups was first described by the Hackstadt group in an RT-PCR analysis of 8% of the genes in *C. trachomatis* (Shaw *et al.*, 2000). This general temporal pattern was confirmed at a genome-wide level with DNA microarrays in *C. trachomatis* (Belland *et al.*, 2003; Nicholson *et al.*, 2003) and *C. pneumoniae* (Maurer *et al.*, 2007). More recently, transcriptional profiling has been performed with RNA-seq, although this analysis was limited to two time points in the developmental cycle (Humphrys *et al.*, 2013).

The three temporal groups of chlamydial genes correspond to different phases of the developmental cycle. This correlation is best shown by a microarray study by Belland *et al.*, which reported genome-wide transcript levels and transmission electron micrographs at the same time points (Belland *et al.*, 2003). Early genes were transcribed within a few hours of EB entry into the host cell, but before EB-to-RB conversion. Midcycle genes were transcribed during the RB stage when RBs were undergoing multiple rounds of replication. Late genes were upregulated late in the developmental cycle, when RBs were converting into EBs, starting at about 24 hours post infection (hpi). Chlamydial genes differed in when they were first transcribed, but once expressed, their transcripts were detected for the rest of the developmental cycle. However, an important caveat for this study, and all other chlamydial transcriptome analyses, is that only steady-state transcript levels were measured. Thus, it is not clear if all chlamydial genes continue to be transcribed, or whether there are also temporal patterns of transcriptional shut-off.

Initiation of transcription at the start of the intracellular infection

An EB becomes transcriptionally active within an hour of entering a host cell (Belland *et al.*, 2003; Humphrys *et al.*, 2013), but the mechanism is not known. Transcription in EBs is presumed to be silenced by DNA condensation mediated by the histone-like proteins, as will be discussed in more detail below. It would thus be tempting to postulate that DNA is decondensed within one hour of EB entry. However, electron micrographs clearly indicate that chlamydiae are still in the form of EBs at 1 hpi, and their DNA remains condensed as shown by their electron-dense appearance (Belland *et al.*, 2003; Shaw *et al.*, 2000).

Further evidence that the onset of transcription precedes overall DNA decondensation comes from work showing that binding of Hc1 and Hc2 to DNA is regulated by a small metabolite, MEC (2-C-methyl-o-erythritol 2,4-cyclodiphosphate) (Grieshaber *et al.*, 2004). MEC disrupts interactions between Hc1 and Hc2 and DNA and induces DNA decondensation, and thus it is unlikely to be present in EBs. Nor is it likely to be produced in the first hours of the intracellular infection because it is synthesized by isoprenoid biosynthesis (Isp) genes that are first expressed between 3 and 8 hpi (Belland *et al.*, 2003; Grieshaber *et al.*, 2004). These findings indicate that transcription in EBs precedes, and thus does not require, general DNA decondensation. This paradox could be explained if there is a transitional period in which there is selective disassociation of the histone-like proteins from early genes to allow them to be transcribed.

Early genes

At the start of the intracellular infection, there is selective transcription of early genes, while midcycle and late genes are not yet transcribed. Of the almost 900 genes in *C. trachomatis*, 29 genes were transcribed by 1 hpi and an additional 200 were transcribed by 3 hpi in the Belland microarray study (Belland *et al.*, 2003). These two groups of genes have been labelled as immediate early and early genes, respectively, although it is not known if they are regulated by the same or by different mechanisms. An RNA-seq study detected transcripts for up to 400 genes at 1 hpi, but the methodology did not distinguish between newly synthesized transcripts and transcripts that were already present in EBs (Humphrys *et al.*, 2013). As a consequence, the number of early genes identified by this study is not known. EBs contain carry-over transcripts from the previous developmental cycle, but these transcripts do not appear to be translated into proteins and their significance is unknown (Belland *et al.*, 2003).

Early genes are important for establishing the chlamydial inclusion and the intracellular infection. They encode subunits of RNA polymerase, 16S ribosomal RNA, ribosomal proteins, and DNA polymerase, which are the basic machineries for transcription, translation, and DNA replication (Belland *et al.*, 2003). Early genes also encode regulators of later temporal classes of genes, such as DNA gyrase, which is a midcycle regulator (Niehus *et al.*, 2008; Orillard and Tan, 2015), and the late regulator EUO (Rosario and Tan, 2012), which will be discussed in more detail. The major heat shock proteins, DnaK, GroEL, and GroES, and a large number of inclusion membrane proteins are also expressed from early genes (Belland *et al.*, 2003).

The mechanism for the selective transcription of early genes is not known. They are scattered around the chlamydial genome and not clustered into domains that might be differentially transcribed. Early genes are transcribed by σ^{66} RNA polymerase (Cheng and Tan, 2012), which does not provide selectivity because it is the major form of chlamydial RNA polymerase that also transcribes midcycle and late genes (Niehus *et al.*, 2008; Rosario and Tan, 2012). In addition, early genes are not known to be regulated by a transcription factor. One possible explanation is that early genes might be constitutively expressed, unlike midcycle and late genes, which are not initially transcribed because they are actively regulated, as will be discussed.

Midcycle genes

Midcycle genes make up the largest group of chlamydial genes. In the Belland microarray study, about 650 *C. trachomatis* genes were first transcribed in midcycle (Belland *et al.*, 2003). They include many housekeeping genes involved in metabolism and growth, such as enzymes of glycolysis and gluconeogenesis. They also encode the major outer membrane protein MOMP and the inclusion membrane protein IncA, which is involved in homotypic fusion of chlamydial inclusions.

The developmental expression of midcycle genes does not appear to be due to selective recognition by a particular form of RNA polymerase, nor regulation by a transcription factor. Midcycle promoters are only known to be transcribed by σ^{66} RNA polymerase (Case *et al.*, 2010; Niehus *et al.*, 2008), which also transcribes other temporal classes of genes (Cheng and Tan, 2012; Rosario and Tan, 2012). Transcription factors that control specific midcycle promoters have been identified, and will be discussed below, but none have been found to be a general regulator of midcycle transcription. Furthermore, it is unlikely that a single transcription factor can regulate all the midcycle genes because very few bacterial transcription factors, as listed in the transcription factor database RegulonDB, regulate more than a hundred target genes (Santos-Zavaleta *et al.*, 2018).

Instead, it has been proposed that midcycle genes are activated by increased DNA supercoiling (Tan, 2012). This model is based on the related findings that chlamydial supercoiling levels are temporally regulated and that midcycle promoters are supercoiling-responsive. From studies with the chlamydial plasmid, *C. trachomatis* DNA was found to be relatively relaxed at early times, became progressively more negatively supercoiled in midcycle, and returned to its initial, relaxed state at late times and in EBs (Niehus *et al.*, 2008). Furthermore, promoters for midcycle genes, but not late genes, were more active at higher negative supercoiling levels (Niehus *et al.*, 2008). In addition, a subset of early genes that are upregulated in midcycle also had promoters that were activated by increased supercoiling (Cheng and Tan, 2012). Collectively, these data support a model in which increased DNA supercoiling in midcycle upregulates transcription of chlamydial genes that have supercoiling-responsive promoters.

An attractive feature of this supercoiling model is that it has the potential to activate a very large number of genes. Supercoiling upregulates promoters by inducing favourable conformational changes in their structure. Thus, in principle, supercoiling could co-ordinately regulate many genes, beyond the number of genes controlled by a single transcription factor. In studies showing supercoiling-dependent transcription in *E. coli* and *H. influenzae*, hundreds of genes, representing 10% of the genome, were transcribed in a supercoiling-dependent manner (Gmuender *et al.*, 2001; Peter *et al.*, 2004). Thus supercoiling is used in other bacteria to regulate a large network of genes, but there is no precedent for it to regulate the majority of genes in a bacterium. Supercoiling-responsive promoters have so far been shown for 12 *C. trachomatis* genes (Case *et al.*, 2010; Cheng and Tan, 2012; Niehus *et al.*, 2008; Orillard and Tan, 2015). Further testing will need to be done to confirm if supercoiling is the mechanism that upregulates hundreds of chlamydial genes in midcycle.

The temporal regulation of chlamydial DNA supercoiling may be due to the changing growth state of chlamydiae during the developmental cycle. Rapidly growing bacterial cells have higher supercoiling levels because they have a higher ATP/ADP ratio, which increases DNA gyrase activity (Dorman and Dorman, 2016). Thus the superhelical density of a reporter plasmid in *E. coli* was highest at −0.05 during logarithmic growth but decreased to −0.03 in stationary phase, indicative of a change from a supercoiled to a more relaxed topological state (Hatfield and Benham, 2002). Similarly, the superhelical density of the *C. trachomatis* plasmid decreased from a range of −0.063 to −0.077 at 24 hpi, to a range of −0.028 to −0.035 at 46 hpi (Niehus *et al.*, 2008), corresponding to a transition from negatively supercoiled DNA in RBs to more relaxed DNA in EBs.

DNA supercoiling levels during the chlamydial developmental cycle are also controlled by *Chlamydia*-specific changes in the expression of the topoisomerase genes (Orillard and Tan, 2015). In other bacteria, DNA supercoiling levels are homeostatically controlled by the counteracting activities of DNA gyrase, which increases supercoiling, and DNA topoisomerase I to relax DNA (Dorman and Dorman, 2016). In *C. trachomatis*, however, DNA gyrase is expressed from an early gene but DNA topoisomerase I and IV are transcribed later, as midcycle genes (Belland *et al.*, 2003). This temporal regulation of topoisomerase expression initially provides unopposed gyrase activity to increase DNA supercoiling to its peak in midcycle, at which time topoisomerase I and IV are expressed to counterbalance gyrase. There is also an important difference in the feedback control of the chlamydial gyrase promoter in response to DNA supercoiling levels. Gyrase promoters in other bacteria are generally downregulated by increased supercoiling, as a mechanism of negative feedback control (Menzel and Gellert, 1983), but the *C. trachomatis* gyrase promoter is unique in being activated by higher negative supercoiling (Orillard and Tan, 2015). This unusual property provides a mechanism of positive feedback, in which increasing supercoiling promotes more gyrase expression, which further increases supercoiling. Thus, the normal control of DNA supercoiling levels, with balanced gyrase and topoisomerase activity and feedback regulation of gyrase expression, has been genetically altered in *Chlamydia*.

We propose the following model to explain how a combination of growth-dependent and *Chlamydia*-specific effects on topoisomerase expression and activity combine to temporally regulate DNA supercoiling levels during the developmental cycle (Orillard and Tan, 2015). At the start of the developmental cycle, when supercoiling levels are low and RB growth has not yet begun, unopposed DNA gyrase expression may be the dominant mechanism to make chlamydial DNA more negatively supercoiled. Increasing supercoiling provides positive feedback to further upregulate the gyrase promoter and enhance gyrase expression. Higher supercoiling levels then activate midcycle genes, on a genome-wide basis through their supercoiling-responsive promoters, to promote RB growth and additional growth-dependent up-regulation of gyrase activity. Collectively, these mechanisms provide a means for chlamydial DNA to transition from its initial relaxed state to a highly negatively supercoiled state in midcycle.

We propose that the main switch that causes a decrease in supercoiling levels from their peak in midcycle is the expression of DNA topoisomerase I and IV from midcycle genes (Belland *et al.*, 2003). The DNA relaxing activities of these topoisomerases are likely to be additive, but may also be sequential because transcripts for topoisomerase I are detected before those of topoisomerase IV (Belland *et al.*, 2003; Orillard and Tan, 2015). Once supercoiling levels start to decrease, they inhibit the supercoiling-sensitive gyrase promoter, and this positive feedback decreases gyrase expression and further reduces the level of DNA supercoiling. At the same time, decreased RB growth and potentially lower ATP/ADP ratios may also inhibit DNA gyrase activity, leading to the resetting of chlamydial DNA supercoiling to its initial low level by the end of the developmental cycle. This supercoiling model of midcycle gene regulation is supported by data from *in vitro* studies and measurements of chlamydial DNA supercoiling levels (Case *et al.*, 2010; Cheng and Tan, 2012; Niehus *et al.*, 2008; Orillard and Tan, 2015), but awaits validation in intracellular chlamydiae.

The role of DNA supercoiling in controlling chlamydial gene expression indicates that the three chlamydial topoisomerases are likely to be key regulators of the developmental cycle and attractive therapeutic targets. DNA gyrase from *C. trachomatis* and *C. pneumoniae*, and *C. trachomatis* topoisomerase IV, have been shown to be inhibited by fluoroquinolones (Ameyama *et al.*, 2003; Orillard and Tan, 2015), which are a class of antibiotics used to treat *Chlamydia* infections (Workowski *et al.*, 2015). The main chlamydial target of the fluoroquinolones may be DNA gyrase because it was 2.5 to 5 times more sensitive than topoisomerase IV (Orillard and Tan, 2015). Furthermore, fluoroquinolone resistance in *C. trachomatis* has been linked with mutations in the gene for the gyrase subunit *gyrA* (Dessus-Babus *et al.*, 1998). The conventional model of fluoroquinolone action is that they inhibit DNA gyrase and DNA topoisomerase IV, thereby blocking DNA replication and leading to DNA fragmentation (Aldred *et al.*, 2014). In *Chlamydia*, however, the important role of supercoiling in activating midcycle gene expression suggests that the fluoroquinolones may have a second and *Chlamydia*-specific mechanism of action by inhibiting progression of the developmental cycle. Although topoisomerase I is another potential anti-chlamydial target, there are no inhibitors of this bacterial enzyme in clinical use.

Late genes

Late genes are upregulated at or after 24 hpi in *C. trachomatis* serovar L2, at the time when RB-to-EB conversion is first detected (Belland *et al.*, 2003; Nicholson *et al.*, 2003). Belland *et al.* identified 26 late genes and Nicholson *et al.* reported 75 late or very late genes in their respective DNA microarray studies (Belland *et al.*, 2003; Nicholson *et al.*, 2003). This temporal group includes genes with known roles in RB-to-EB conversion and EB function. For example, *hctA* and *hctB* are late genes that encode the two histone-like proteins, Hc1 and Hc2, which condense DNA in EBs (Barry *et al.*, 1992; Brickman *et al.*, 1993). Another late gene *omcB* encodes the 60 kDa cysteine-rich outer membrane protein, which is an adhesin that is important for binding of an EB to a new host cell (Moelleken and Hegemann, 2008). Additional late gene products include putative disulfide oxidoreductases and membrane thiol proteases, which may have roles in formation of the highly crosslinked outer membrane complex that protects an EB from the extracellular environment (Liu *et al.*, 2010). The likely reason for the temporal regulation of late genes is to prevent RBs from prematurely converting into an EB before the RB population has had the time to expand.

Late genes are transcribed as two distinct subsets by different forms of chlamydial RNA polymerase.

One subset of late genes, which includes *hctA* and *omcB*, is transcribed by the major chlamydial RNA polymerase, σ^{66} RNA polymerase (Niehus *et al.*, 2008; Rosario and Tan, 2012). A second subset, which includes *hctB*, is transcribed by the alternative polymerase, σ^{28} RNA polymerase (Yu *et al.*, 2006b; Yu and Tan, 2003). This polymerase specificity is determined by the promoter of each late gene because the two forms of chlamydial polymerase recognize mutually exclusive promoter sequences (Schaumburg and Tan, 2003; Tan, 2006; Yu *et al.*, 2006a). These findings about late genes demonstrate that a temporal class of chlamydial genes can be regulated by more than one transcriptional mechanism (Yu and Tan, 2003). However, the two subsets of late genes cannot be distinguished by their transcriptional patterns (Belland *et al.*, 2003; Nicholson *et al.*, 2003).

Transcriptional studies with σ^{28} RNA polymerase have identified σ^{28} as a late regulator in *C. trachomatis*. So far, seven genes have been identified as σ^{28}-regulated genes (Table 10.1) (Rosario and Tan, 2015; Yu *et al.*, 2006b; Yu and Tan, 2003). All seven contain a σ^{28}-dependent promoter that is upregulated at late times (Rosario *et al.*, 2014; Rosario and Tan, 2015). However, this temporal pattern was not immediately obvious because some of these genes are also transcribed by σ^{66} RNA polymerase at other times in the developmental cycle, as will be discussed further (Rosario and Tan, 2015). σ^{28} is conserved among all *Chlamydia*, but is absent and appears to have been lost from *Chlamydia*-related bacteria (Domman and Horn, 2015).

The transcription factor EUO has been recognized as the master regulator of late gene expression in *Chlamydia* because it represses both subsets of late genes (Rosario *et al.*, 2014; Rosario and Tan, 2012). EUO binds an AT-rich operator and blocks transcription (Zhang *et al.*, 1998; Zhang *et al.*, 2000). It is able to repress σ^{66}-dependent and σ^{28}-dependent promoters as long as they contain its binding sequence, which is called the EUO operator (Rosario *et al.*, 2014; Rosario and Tan, 2012). This mechanism allows EUO to selectively repress late genes even though they are transcribed by two different forms of RNA polymerase.

Late genes are not regulated by DNA supercoiling. *C. trachomatis* late promoters were not responsive to supercoiling changes *in vitro* (Case *et al.*, 2010; Niehus *et al.*, 2008). This property makes sense because late promoters are only transcribed at late times, when supercoiling levels are low (Niehus *et al.*, 2008).

We propose the following model to account for the temporal expression of late genes. EUO is transcribed from an immediate early promoter by 1 hpi (Belland *et al.*, 2003), and is thus available from early on in the developmental cycle to block expression of late genes (Rosario *et al.*, 2014; Rosario and Tan, 2012). This temporal regulation may prevent RBs from prematurely converting into EBs before the RB population has expanded. EUO does not affect transcription of σ^{66}-dependent early and midcycle genes because they lack EUO operators (Rosario and Tan, 2012). At the end of midcycle, repression by EUO is relieved to allow transcription of a subset of late genes by σ^{66} RNA polymerase and a second subset by σ^{28} RNA polymerase. Neither the reason for these two subsets of late chlamydial genes, nor the mechanism of EUO derepression, is known. Nevertheless, these regulatory control mechanisms are likely to be important because late gene expression is a critical step in the developmental cycle that is temporally and functionally linked to the production of EBs. EUO, as a regulator of late genes, has the potential to control RB-to-EB conversion, although this role awaits experimental validation.

Combining temporal expression patterns

Chlamydia has expanded its repertoire of temporal expression patterns by transcribing some genes from more than one promoter. For instance, *bioY*, which encodes a biotin transporter, is transcribed by σ^{66} RNA polymerase from a supercoiling-regulated midcycle promoter and by σ^{28} RNA polymerase from an EUO-regulated late promoter (Rosario and Tan, 2015). This example illustrates how tandem promoters can allow a gene to be regulated by different forms of chlamydial RNA polymerase and multiple mechanisms of temporal regulation. As a consequence, however, the use of the temporal expression pattern of a gene to predict its underlying regulatory mechanism has limitations if the gene is transcribed from tandem promoters.

Transcriptional silencing in EBs

After these waves of temporal gene expression, the developmental cycle transitions to a transcriptionally inactive phase. EBs have the potential for transcription because they contain all the subunits of σ^{66} RNA polymerase (Saka *et al.*, 2011; Skipp *et al.*, 2016). However, they are considered to be transcriptionally silent because they incorporate very low levels of NTPs into RNA (Barry *et al.*, 1993; Moulder, 1991).

Silencing of transcription in EBs is hypothesized to be due to chromatin condensation caused by the histone-like proteins, Hc1 and Hc2. Expression of either Hc1 or Hc2 in *E. coli* inhibited overall transcription and translation (Barry *et al.*, 1993; Pedersen *et al.*, 1994; Pedersen *et al.*, 1996). Each histone-like protein also inhibited transcription by T7 RNA polymerase *in vitro* (Pedersen *et al.*, 1994; Pedersen *et al.*, 1996). These effects are consistent with a model in which binding of Hc1 and Hc2 proteins prevents access of RNA polymerase to promoters. This mechanism is likely to be global in nature, affecting all genes in the genome and all three forms of chlamydial RNA polymerase.

Summing up transcriptional regulation during the developmental cycle

Chlamydia controls the temporal expression of its genes via multiple mechanisms of transcriptional regulation, including alternative forms of RNA polymerase, transcription factors, DNA supercoiling, nucleoid-associated proteins and tandem promoters (Table 10.3). These mechanisms constitute the standard toolbox of transcriptional regulators in bacteria, but *Chlamydia* applies them in its own way to control the timing of gene expression during the developmental cycle. This toolbox also has *Chlamydia*-specific features, such as DNA supercoiling as a global mechanism of gene regulation, positive feedback control of gyrase expression to temporally regulate supercoiling levels, and DNA condensation and transcriptional silencing to limit gene expression in EBs.

These control mechanisms are linked together in a chain in which the regulator of a temporal class of genes is expressed in an earlier temporal class (Table 10.3). For example, DNA gyrase, which is a midcycle regulator, is expressed from an early gene. EUO and σ^{28}, which are late regulators, are expressed from an early and a midcycle gene, respectively. The histone-like proteins, Hc1 and Hc2, which prevent transcription in EBs, are expressed as late genes. And σ^{66} RNA polymerase is loaded into EBs ready to transcribe genes in the next developmental cycle. Thus the developmental cycle can be viewed as a circle or an endless relay, with each temporal stage handing off to the next stage.

Temporal gene regulation in *Chlamydia* is likely to be due to a combination of intrinsic and extrinsic controls. Some of the temporal expression patterns are genetically programmed, such as midcycle

Table 10.3 Regulators of temporal gene expression during the chlamydial developmental cycle

Stage	Regulator	Function	Stage when regulator is expressed	References
Early	None identified			
Midcycle	DNA gyrase	Increases negative supercoiling	Early	Ameyama *et al.* (2003); Orillard and Tan (2015)
	DNA topoisomerase I	Relaxes negatively supercoiled DNA	Midcycle	Orillard and Tan (2015)
	DNA topoisomerase IV	Relaxes negatively supercoiled DNA	Midcycle	Orillard and Tan (2015)
Late	EUO	Represses late gene expression	Early	Rosario *et al.* (2014); Rosario and Tan (2012)
	σ^{28}	Transcribes a subset of late genes	Midcycle	Rosario and Tan (2015); Yu *et al.* (2006b)
EB	Hc1, Hc2	Condenses DNA; Inhibits transcription	Late	Barry *et al.* (1992); Barry *et al.* (1993); Brickman *et al.* (1993)

genes with supercoiling-dependent promoters, or late genes with an operator for EUO to bind and repress promoter activity. It is anticipated, however that temporal gene expression will also be regulated by external signals that allow chlamydiae to sense their environment.

Sigma factors and their potential regulators

σ^{54}

All *Chlamydia* and *Chlamydia*-related bacteria encode another sigma factor, σ^{54}, but its biological role is unknown. *Chlamydia* σ^{54} activity has not yet been confirmed, but is likely because it contains a conserved RpoN box (ARRTVAKYR) that is responsible for σ^{54} promoter recognition (Mathews and Timms, 2000; Taylor *et al.*, 1996). Mathews and Timms used the distinctive sequence and spacing of the consensus bacterial σ^{54} promoter (Barrios *et al.*, 1999) to identify potential σ^{54} targets in the *C. trachomatis* genome (Mathews and Timms, 2000). The strongest experimental support is for CT652.1 and CT683 because their *in vivo* transcription start sites have been mapped immediately downstream of predicted σ^{54} promoters (Table 10.1). Moreover, σ^{54} promoter sequences have been identified upstream of the orthologous genes for CT652.1 and CT683 in *C. pneumoniae* and *C. muridarum* (Mathews and Timms, 2000). However, chlamydial σ^{54} promoters have not yet been experimentally verified with transcription assays.

A distinctive feature of σ^{54} is its absolute requirement for an activator (Bush and Dixon, 2012). From work in other bacteria, it is known that σ^{54} can bind a target promoter, but it does not initiate transcription until there is ATP hydrolysis by the activator bound at an upstream DNA sequence called an enhancer. The chlamydial protein CtcC is predicted to be the activator of σ^{54} (Koo and Stephens, 2003) because of its sequence similarity to *E. coli* NtrC, which is an activator of *E. coli* σ^{54} (Bush and Dixon, 2012). Furthermore, CtcC is likely to be regulated by CtcB, which is a kinase that phosphorylates CtcC (Koo and Stephens, 2003) and a homologue of *E. coli* NtrB that activates NtrC. CtcC is unusual in lacking an enhancer-binding domain, but there is precedent as *Helicobacter* FlgR activates σ^{54} even though it does not have an enhancer-binding domain (Brahmachary *et al.*, 2004; Koo and Stephens, 2003).

There are many unanswered questions about the role of σ^{54} in *Chlamydia*. It is not known if σ^{54} is a temporal regulator, although the genes for σ^{54} (*ct609*), CtcC (*ct468*) and CtcB (*ct467*) are all transcribed as midcycle genes (Belland *et al.*, 2003). Canonical roles of σ^{54} in regulating nitrogen fixation (Garcia *et al.*, 1977) and flagellar motility (Riordan and Mitra, 2017), are unlikely in *Chlamydia* because this bacterium lacks nitrogen fixation genes and a flagellum. In addition, the signals that induce CtcB to phosphorylate CtcC and activate σ^{54} are not known. Progress in understanding how σ^{54} regulates chlamydial gene expression would benefit from the development of a σ^{54} transcription assay, which is an approach that has been helpful for studying σ^{66} and σ^{28}.

CdsZ, a potential regulator of σ^{54}

CdsZ has been proposed to be a regulator of *Chlamydia* σ^{54} (Barta *et al.*, 2015). Barta and colleagues showed that *C. trachomatis* CdsZ (CT398) interacted with σ^{54} in a bacterial two-hybrid assay and *in vitro* affinity pull-down studies, and with σ^{54} and core enzyme in affinity purification studies that used *Chlamydia*-infected cells. The crystal structure of CdsZ revealed high structural similarity to FlgZ, a regulator of flagellar synthesis genes in *Helicobacter* that prevents the rapid turnover of σ^{54} (Pereira *et al.*, 2011). These findings provide strong evidence that CdsZ physically interacts with, and likely regulates, σ^{54}, although the exact regulatory role remains to be determined. Interestingly, CdsZ also interacted with two type 3 secretion (T3S) proteins, FliH and CdsL, which suggests that it may link T3S to σ^{54}-dependent transcription (Barta *et al.*, 2015).

RsbW, a potential regulator of σ^{28} or σ^{66}

RsbW is an anti-sigma factor, which is a protein that binds and inhibits a specific sigma factor. In *Bacillus*, RsbW binds to an alternative sigma factor called σ^B thereby preventing σ^B from binding core enzyme (Hecker *et al.*, 2007). However, RsbW can change partners and instead bind RsbV, freeing up σ^B to bind core enzyme to form σ^B RNA polymerase. This partner switching mechanism is controlled by the phosphorylation state of RsbV – only unphosphorylated RsbV binds RsbW, and thus when RsbV is phosphorylated, RsbW is free to bind σ^B. This pathway is controlled by a phosphatase RsbU and by RsbW, which has a second function as a kinase for RsbV (Voelker *et al.*, 1995).

The Rsb pathway is present in *Chlamydia* although some components, such as RsbV, appear to be duplicated (Hua *et al.*, 2006; Thompson *et al.*, 2015). *C. trachomatis* RsbU has been confirmed to be a phosphatase and shown to dephosphorylate one of the RsbV paralogues, RsbV1, but not RsbV2 (Thompson *et al.*, 2015). Unlike *Bacillus* RsbU, which is a cytosolic protein, chlamydial RsbU is predicted to be a transmembrane protein with a periplasmic sensor domain and a cytosolic phosphatase domain (Hua *et al.*, 2006). CT589 appears to be a paralogue of RsbU, but it is not certain that it is a phosphatase because it lacks two important residues necessary for phosphatase function (Hua *et al.*, 2006). RsbW has been shown to bind and phosphorylate RsbV1 and RsbV2 (Hua *et al.*, 2006), although it bound RsbV1 with greater affinity (Thompson *et al.*, 2015). This association of RsbW to RsbV1 and RsbV2 was unusual because it required ATP (Thompson *et al.*, 2015), which is not a feature of RsbW-RsbV binding in *Bacillus*. It is not known why there are two RsbV proteins, but the *in vitro* differences in their susceptibility to dephosphorylation by RsbU and binding affinity for RsbW suggest that they are not functionally equivalent. Interestingly, an RsbV1 deletion strain of *C. trachomatis* has been generated (Thompson *et al.*, 2015), which indicates that RsbV1 is not essential and suggests that there may be redundancy between RsbV1 and RsbV2. One possibility is that the Rsb pathway in *Chlamydia* has parallel arms that respond to different upstream signals but converge on a single RsbW anti-sigma factor.

There is controversy about the sigma factor that is the target for RsbW in *Chlamydia*. Douglas and Hatch showed that RsbW bound σ^{28} in a pull-down assay (Douglas and Hatch, 2006). However, Hua and colleagues did not detect interactions between RsbW and any of the three chlamydial sigma factors in a yeast two-hybrid study (Hua *et al.*, 2006). More recently, Thompson and colleagues showed that RsbW interacted with σ^{66}, but not σ^{28} or σ^{54}, in a bacterial two-hybrid assay and confirmed a greater interaction with σ^{66} by surface plasmon resonance analysis (Thompson *et al.*, 2015). They also reported changes in transcription of selected σ^{66} target genes in mutant strains overexpressing RsbW or RsbV1, or lacking RsbV1. However, there is no precedent for the Rsb pathway to regulate the major sigma factor in other bacteria (Hecker *et al.*, 2007). Thompson *et al.* have proposed that the Rsb pathway serves as a brake to modulate σ^{66} RNA polymerase activity, perhaps in an ATP-dependent manner (Thompson *et al.*, 2015).

Scc4, a regulator of σ^{66} RNA polymerase

Scc4 is an inhibitor of σ^{66} RNA polymerase. *C. trachomatis* Scc4 (CT663) interacted with both the σ^{66} and β subunits of RNA polymerase in bacterial two-hybrid and *in vitro* pull-down assays (Rao *et al.*, 2009). The specific interaction sites were narrowed down to region 4 of σ^{66}, and the β-flap tip helix of the β subunit, which are sites involved in recognizing the −35 element of the σ^{66} promoter (Rao *et al.*, 2009). Consistent with this structure–function analysis, Scc4 inhibited *in vitro* transcription by σ^{66} RNA polymerase but not σ^{28} RNA polymerase.

Scc4 provides a second potential mechanism, in addition to CdsZ, to link T3S and gene expression (Betts *et al.*, 2008; Hefty and Stephens, 2007). Scc4 is a T3S chaperone and forms a complex with another T3S chaperone Scc1 and with CopN, which is a secreted T3S effector as well as the 'plug' for the T3S channel (Hanson *et al.*, 2015; Shen *et al.*, 2015; Silva-Herzog *et al.*, 2011). The ability of Scc4 to inhibit transcription was antagonized by Scc1 and CopN (Hanson *et al.*, 2015). Shen and colleagues have proposed that CopN is secreted when EBs make contact with the host cell plasma membrane, which frees up Scc4 to inhibit transcription of σ^{66}-dependent genes (Shen *et al.*, 2015), perhaps as a mechanism for the selective transcription of early genes.

Chlamydial transcription factors not linked with the developmental cycle

Repressors that control nutrient biosynthesis and transport

Chlamydia utilizes a number of transcription factors to regulate the levels of specific nutrients during homeostasis (Table 10.2). These transcription factors are aporepressors, which means that they are inactive unless bound by a co-factor (Payankaulam *et al.*, 2010), and they repress genes involved in either biosynthesis or import of the co-factor. This mechanism allows feedback control in which the repressor senses and regulates the levels of the co-factor, which can be an amino acid, nucleotide or metal ion (Akers *et al.*, 2011; Case *et al.*, 2011; Schaumburg and Tan, 2006; Thompson *et al.*, 2012). For example, the TrpR repressor requires tryptophan as a co-factor, which is thus called its co-repressor (Akers and Tan, 2006; Carlson *et al.*, 2006). In elegant studies with *C. trachomatis*-infected cells, McClarty and colleagues showed that tryptophan availability regulated the ability of TrpR to bind and repress tryptophan synthase genes (Carlson *et al.*, 2006; Wood *et al.*, 2003). These findings provide experimental support for a model in which high tryptophan levels allow TrpR to bind tryptophan and repress tryptophan synthase genes, which limits further production of tryptophan. In the absence of tryptophan, TrpR is inactive and unable to repress, which allows tryptophan synthase to be expressed so that it can convert indole to tryptophan.

The stress response regulator, HrcA

HrcA is a transcription factor that controls the expression of the main chlamydial heat shock proteins, DnaK, GroEL and GroES (Wilson and Tan, 2002; Wilson and Tan, 2004), which have been strongly implicated in chlamydial pathogenesis (LaVerda *et al.*, 1999; Stephens, 2003). In *C. trachomatis*, HrcA represses the genes for these heat shock proteins, but its repression is enhanced by GroEL. Thus, GroEL acts as a co-repressor and autoregulates its own expression (Chen *et al.*, 2011; Wilson *et al.*, 2005). HrcA has been shown by chromatin immunoprecipitation (ChIP) to bind and regulate its target promoters within intracellular chlamydiae in response to temperature changes (Hanson and Tan, 2015). However, chlamydiae are unlikely to encounter temperature shifts in a mammalian host cell, and the physiologic stressors that regulate HrcA in a *Chlamydia*-infected cell are not known.

ChxR

ChxR (Chlamydial expression regulator) is a chlamydial orthologue of OmpR response regulators that are widely used in other bacteria to control gene expression in response to changes in environmental osmolality (Koo *et al.*, 2006). ChxR is considered to be an atypical OmpR regulator because it does not require phosphorylation to activate transcription. Furthermore, it does not have an identifiable sensor kinase, and the OmpR phosphorylation site is substantially modified in ChxR, with the canonical phosphorylated aspartate (D54) replaced by a glutamate.

C. trachomatis ChxR has been confirmed to be a transcription factor that activated specific promoters in *lacZ* reporter assays performed in *E. coli* (Koo *et al.*, 2006). The specific DNA sequence that ChxR binds to has been defined with gel shift assays (Hickey *et al.*, 2011). ChxR autoregulates its own expression via multiple ChxR binding sites that are upstream of the *chxR* gene (Koo *et al.*, 2006). Putative ChxR target genes that are differentially expressed in a *C. trachomatis* ChxR null mutant strains have also been identified. Five proteins, CT005, CT214, CT565, CT694, and CT695, were consistently downregulated in three different ChxR null mutant strains, providing strong evidence that they are regulated by ChxR. However, it is not known if they are direct or indirect targets of ChxR (Yang *et al.*, 2017). Unfortunately, there was no overlap between the proposed target genes identified with the *lacZ* reporter assays and the ChxR null mutant strains, leaving uncertainty about the genes that make up the ChxR regulon.

ChxR has been proposed to regulate the expression of virulence genes, although the precise role remains to be determined. It is not an essential gene because *C. trachomatis* null mutant strains had no detectable growth defect in cell culture, but there was decreased chlamydial shedding in a mouse infection (Yang *et al.*, 2017).

GrgA

GrgA (General regulator of genes A) is an activator of promoters transcribed by σ^{66} and σ^{28} RNA polymerases (Bao *et al.*, 2012; Desai *et al.*, 2018). While it makes different physical interactions with both σ^{66} and σ^{28}, it bound σ^{66} with higher affinity than σ^{28}, which suggests that it may preferentially bind and activate σ^{66} (Desai *et al.*, 2018). GrgA binds DNA in a sequence-nonspecific manner, which is reminiscent of histones, and thus could potentially activate many chlamydial genes (Bao *et al.*, 2012). So far, GrgA has been shown to activate five chlamydial promoters, including σ^{66}-dependent promoters of early, midcycle and late genes (Bao *et al.*, 2012; Desai *et al.*, 2018). Thus GrgA does not appear to be a temporal regulator. GrgA is conserved among all *Chlamydia* but not present in *Chlamydia*-related bacteria (Domman and Horn, 2015).

Pgp4 and Pgp5

Pgp4 and Pgp5 are plasmid-encoded proteins that are proposed to be transcriptional regulators. A genome-wide microarray analysis of a *C. trachomatis* Pgp4 deletion mutant revealed differential transcription of 39 genes, including the plasmid gene *pgp3*, which encodes a virulence factor (Zhong, 2017), and chromosomal genes such as *glgA*, which encodes glycogen synthase (Song *et al.*, 2013). Transcript levels for most, but not all, of these genes were decreased in the Pgp4 deletion strain, consistent with Pgp4 functioning as a positive regulator. However, Pgp4 did not bind its target promoters, so it is not clear if it is a transcription factor or if it acts indirectly through another factor. Selected target genes were also upregulated in a *C. muridarum* Pgp4 deletion mutant (Liu *et al.*, 2014).

C. muridarum Pgp5 is postulated to be a suppressor of transcription because six genes that were downregulated in the Pgp4 deletion strain were transcribed at higher levels in a Pgp5 deletion strain (Liu *et al.*, 2014). These results suggest that Pgp4 and Pgp5 have antagonistic roles in regulating the same genes, with Pgp5 acting as a negative regulator of transcription. In this study, *C. trachomatis* Pgp5 also decreased transcription of target genes, but caused a smaller reduction in promoter activity than *C. muridarum* Pgp5.

DcrA

DcrA (CT296) has been proposed as a transcription factor because it complemented an *E. coli* mutant strain that lacked Fur, which is an iron-dependent repressor, and bound the *E. coli* consensus Fur operator and chlamydial promoter sequences (Rau *et al.*, 2005; Wyllie and Raulston, 2001). However, a structural analysis of CT296 revealed similarity to a non-heme Fe(II) 2-oxoglutarate-dependent enzyme, but not to Fur proteins or known transcription factors (Kemege *et al.*, 2011). This study also failed to reproduce CT296 binding to reported target promoters and complementation of *E. coli* Fur deficient strains. Thus, there is now considerable doubt that CT296 is a Fur homologue.

Regulation of chlamydial gene expression by small RNAs

Small RNAs (sRNAs) are short, highly structured, non-coding RNAs that post-transcriptionally regulate expression of specific target genes (Storz, 2002). They are typically between 50 and 500 nucleotides in length and can be divided into 2 classes. A *cis*-encoded sRNA is transcribed from the complementary strand of its target gene and functions as an antisense RNA to regulate the stability of its target mRNA (Gottesman and Storz, 2011). *Trans*-encoded sRNAs also bind their target transcript, but either block translation by binding to the ribosomal binding site, or promote translation by altering the structure of the ribosomal binding site (Gottesman and Storz, 2011).

sRNAs can be identified with a number of methods (Gottesman, 2005; Storz *et al.*, 2004). These techniques include labelling of sRNAs and detection by gel electrophoresis, mapping of transcripts to non-coding regions of the genome, bioinformatic identification of conserved intergenic regions, microarray analysis of intergenic regions, and deep-sequencing of transcriptomes (Abdelrahman *et al.*, 2011; Huttenhofer and Vogel, 2006; Sharma and Vogel, 2009). Albrecht and colleagues used deep-sequencing studies to identify 43 potential sRNAs in *C. trachomatis* and 75 potential sRNAs in *C. pneumoniae* (Albrecht *et al.*, 2011; Albrecht *et al.*, 2010). Nine *C. trachomatis* and 13 *C. pneumoniae* sRNA candidates were confirmed with Northern blots.

In general, targets of *cis*-encoded sRNAs are easy to identify because they are transcribed from the same gene, but in the complementary direction. It is more difficult to identify targets of *trans*-encoded sRNA because they have limited sequence complementarity and are not encoded in the same genomic location.

The first confirmed sRNA–target pair in *Chlamydia* was the *trans*-encoded sRNA IhtA and its target, the histone-like protein Hc1 (Grieshaber *et al.*, 2004). IhtA was identified in a genetic screen in *E. coli* where this 120 nt sRNA rescued the lethal effects of Hc1 expression (Grieshaber *et al.*, 2004). IhtA inhibited translation of Hc1, but not Hc2, in an *E. coli* co-expression assay, but did not affect Hc1 transcript levels (Grieshaber *et al.*, 2006a; Grieshaber *et al.*, 2006b). IhtA is transcribed from early times and is proposed to prevent translation of HctA until late in the developmental cycle (Grieshaber *et al.*, 2006a).

A small 80 nt transcript sRNA called pL2-sRNA1 is likely to be a *cis*-encoded sRNA because it is highly abundant (> 10,000 reads) and is encoded on the complementary strand of its putative target, pL2–07a, on the chlamydial plasmid (Albrecht *et al.*, 2010).

AbdelRahman, Rose, and Belland used an intergenic microarray and northern blotting to identify sRNAs in *C. trachomatis* serovar D that were differentially expressed during IFN-γ- and carbenicillin-induced persistence and reactivation (Abdelrahman *et al.*, 2011). One of these sRNA, CTIG270 appears to be a *cis*-encoded sRNA for *ftsI*, which is involved in peptidoglycan synthesis. In co-expression studies in *E. coli*, CTIG270 prevented transcription and translation of *ftsI*. In addition, there was a correlation between the presence of CTIG270 and decreased *ftsI* transcript levels at late times in the *Chlamydia* developmental cycle.

sRNAs are likely to provide an additional layer of transcriptional regulation in *Chlamydia* that is tailored to individual target genes. Much remains to be discovered because the repertoire of sRNAs in the chlamydial genome has not been completely defined. Transcriptome studies for identifying sRNAs have so far been limited to a few time points in the developmental cycle. It is also not known whether these sRNAs are conserved between *Chlamydia* spp. and strains.

New genetic tools and approaches for investigating mechanisms of gene regulation

Progress in understanding gene expression in *Chlamydia* will depend more and more on methods for transforming chlamydiae and manipulating their genetic content. Most of our knowledge so far about chlamydial gene regulation has come from transcriptional studies performed in cell-free or heterologous bacterial systems. The advantage of genetic approaches is that they can provide important information about the role of a regulator within chlamydiae, during the developmental cycle and at a genome-wide level. The range of chlamydial genetic approaches now available are discussed in more detail in Chapter 11, *Chlamydia* Genetics.

An obvious approach will be to alter the expression of a regulatory factor and then measure changes in global gene expression or effects on the chlamydial infection and developmental cycle. The regulator could be expressed at higher or lower levels, or at times that are different from its normal expression pattern during the developmental cycle. For example, a Tet-On expression system is available to inducibly express a protein in *C. trachomatis* (Wickstrum *et al.*, 2013). As an alternative approach, a CRISPR interference method has been developed to selectively deplete a chlamydial protein by blocking its transcription (Ouellette, 2018). This method uses an inactivated Cas9 protein (dCas9) with a guide RNA complementary to a specific promoter to generate an inducible, custom repressor that blocks transcription of the gene of interest. CRISPR interference has been used in *C. trachomatis* to repress IncA expression, but could in principle be used to selectively and reversibly decrease protein levels of any chlamydial protein (Ouellette, 2018). Another approach is to create a knockout strain that no longer encodes a regulatory factor. This approach has been used in *C. trachomatis* to show that Pgp4, Pgp5 and ChxR are transcriptional regulators and to identify target genes (Liu *et al.*, 2014; Song *et al.*, 2013; Yang *et al.*, 2017). These methods may not work if the regulator is essential or if its expression levels are intrinsically regulated.

These approaches to alter the expression of a chlamydial regulatory factor can be paired with genome-wide transcriptional profiling methods, such as RNA-seq (Humphrys *et al.*, 2013), to identify target genes. However, additional studies will be needed to distinguish between direct targets and indirect targets that are regulated via an additional factor. Chromatin immunoprecipitation (ChIP) can be used to detect target genes bound by the transcription factor within chlamydiae (Hanson and Tan, 2016; Hanson and Tan, 2018), as has been done with HrcA, a stress response transcriptional repressor (Hanson and Tan, 2015). Alternatively, tried-and-true cell-free assays, such as electrophoretic mobility shift assays and *in vitro* transcription assays can be used to test if putative regulators are transcription factors that directly regulate their target genes (Rosario and Tan, 2012; Tan, 2006).

Genetic approaches can also be used to investigate if the activity of transcriptional regulators can be controlled by post-translational modifications. For example, chlamydial transcriptional regulators such as RsbV1 and CtcC are regulated by phosphorylation of specific residues (Hua *et al.*, 2006; Koo and Stephens, 2003; Thompson *et al.*, 2015). Thus these regulators are themselves controlled by phosphatases and kinases, as part of a phosphorelay signalling pathway. If phosphorylation sites can be identified, mutant forms with modifications at the critical residues, such as an alanine to mimic an unphosphorylated residue or a glutamic acid to mimic a phosphorylated residue, can be expressed in *Chlamydia*. The modified regulator may disrupt a signalling pathway by out-competing the native protein or by acting as a dominant negative form that interacts with normal binding partners.

This discussion has focused on the regulatory proteins, but in addition, *cis*-acting regulatory DNA sequences can be investigated with genetic approaches. For example, a promoter region together with its *cis*-acting sequences could be cloned upstream of a reporter gene on a plasmid and then subject to a mutational analysis in which the effects of sequence and spacing alterations on reporter activity are measured. Multiple fluorescent proteins as well as β-galactosidase have been shown to function as reporter genes in *Chlamydia* (Bauler and Hackstadt, 2014; Wang *et al.*, 2011; Wickstrum *et al.*, 2013). A limitation of this approach is that a separate *Chlamydia* strain will have to be generated for each promoter mutation, which is labour-intensive with current chlamydial transformation methods that are relatively inefficient.

Final words

The challenge is to use these genetic approaches and functional studies to gain a fuller understanding of gene regulation during the intracellular *Chlamydia* infection. The models for the activation of midcycle gene expression by DNA supercoiling, and the regulation of late gene expression by EUO and σ^{28}, provide insights into the temporal control of chlamydial genes. However they are based on *in vitro* data and await confirmation of their *in vivo* function from studies of chlamydiae within infected cells. There are additional questions about the roles of gene regulation in critical events of the developmental cycle. For instance, what is the mechanism and the signal that initiates transcription within one hour of EB entry into a host cell? How is repression by EUO relieved, and does the resulting expression of late genes control or mediate RB-to-EB conversion? What are the biological roles of transcriptional regulators such as σ^{54}, Scc4, RsbW, ChxR, GrgA, Pgp4, Pgp5 and sRNAs? How conserved are the mechanisms of gene regulation among *Chlamydia* spp., and what is the balance between intrinsic programming of gene expression and extrinsic regulation by environmental signals? In the coming years, we will undoubtedly learn more about the key roles that the regulators of *Chlamydia* gene expression play in the developmental cycle and the intracellular infection.

References

Abdelrahman, Y.M., and Belland, R.J. (2005). The chlamydial developmental cycle. FEMS Microbiol. Rev. *29*, 949–959.

Abdelrahman, Y.M., Rose, L.A., and Belland, R.J. (2011). Developmental expression of non-coding RNAs in *Chlamydia trachomatis* during normal and persistent growth. Nucleic Acids Res. *39*, 1843–1854.

Ahmed, W., Menon, S., Karthik, P.V., and Nagaraja, V. (2016). Autoregulation of topoisomerase I expression by supercoiling sensitive transcription. Nucleic Acids Res. *44*, 1541–1552.

Akers, J.C., HoDac, H., Lathrop, R.H., and Tan, M. (2011). Identification and functional analysis of CT069 as a novel transcriptional regulator in *Chlamydia*. J. Bacteriol. *193*, 6123–6131.

Akers, J.C., and Tan, M. (2006). Molecular mechanism of tryptophan-dependent transcriptional regulation in *Chlamydia trachomatis*. J. Bacteriol. *188*, 4236–4243.

Albrecht, M., Sharma, C.M., Dittrich, M.T., Muller, T., Reinhardt, R., Vogel, J., and Rudel, T. (2011). The transcriptional landscape of *Chlamydia pneumoniae*. Genome Biol. *12*, R98.

Albrecht, M., Sharma, C.M., Reinhardt, R., Vogel, J., and Rudel, T. (2010). Deep sequencing-based discovery of the *Chlamydia trachomatis* transcriptome. Nucleic Acids Res. *38*, 868–877.

Aldred, K.J., Kerns, R.J., and Osheroff, N. (2014). Mechanism of quinolone action and resistance. Biochemistry *53*, 1565–1574.

Ameyama, S., Shinmura, Y., and Takahata, M. (2003). Inhibitory activities of quinolones against DNA gyrase of *Chlamydia pneumoniae*. Antimicrob. Agents Chemother. *47*, 2327–2329.

Bao, X., Nickels, B.E., and Fan, H. (2012). *Chlamydia trachomatis* protein GrgA activates transcription by contacting the nonconserved region of sigma66. Proc. Natl. Acad. Sci. U S A *109*, 16870–16875.

Barrios, H., Valderrama, B., and Morett, E. (1999). Compilation and analysis of sigma(54)-dependent promoter sequences. Nucleic Acids Res. *27*, 4305–4313.

Barry, C., Hayes, S., and Hackstadt, T. (1992). Nucleoid condensation in *Escherichia coli* that express a chlamydial histone homolog. Science *256*, 377–379.

Barry, C.E., 3rd, Brickman, T.J., and Hackstadt, T. (1993). Hc1-mediated effects on DNA structure: a potential regulator of chlamydial development. Mol. Microbiol. *9*, 273–283.

Barta, M.L., Battaile, K.P., Lovell, S., and Hefty, P.S. (2015). Hypothetical protein CT398 (CdsZ) interacts with sigma(54) (RpoN)-holoenzyme and the type III secretion export apparatus in *Chlamydia trachomatis*. Protein Sci. *24*, 1617–1632.

Bauler, L.D., and Hackstadt, T. (2014). Expression and targeting of secreted proteins from *Chlamydia trachomatis*. J. Bacteriol. *196*, 1325–1334.

Belland, R.J., Zhong, G., Crane, D.D., Hogan, D., Sturdevant, D., Sharma, J., Beatty, W.L., and Caldwell, H.D. (2003). Genomic transcriptional profiling of the developmental cycle of *Chlamydia trachomatis*. Proc. Natl. Acad. Sci. U S A *100*, 8478–8483.

Berger, M., Gerganova, V., Berger, P., Rapiteanu, R., Lisicovas, V., and Dobrindt, U. (2016). Genes on a wire: the nucleoid-associated protein HU insulates transcription units in *Escherichia coli*. Sci. Rep. *6*, 31512.

Betts, H.J., Twiggs, L.E., Sal, M.S., Wyrick, P.B., and Fields, K.A. (2008). Bioinformatic and biochemical evidence for the identification of the type III secretion system needle protein of *Chlamydia trachomatis*. J. Bacteriol. *190*, 1680–1690.

Bhardwaj, N., Syal, K., and Chatterji, D. (2018). The role of omega-subunit of *Escherichia coli* RNA polymerase in stress response. Genes Cells *23*, 357–369.

Brahmachary, P., Dashti, M.G., Olson, J.W., and Hoover, T.R. (2004). *Helicobacter pylori* FlgR is an enhancer-independent activator of sigma54-RNA polymerase holoenzyme. J. Bacteriol. *186*, 4535–4542.

Brickman, T.J., Barry, C.E. III, and Hackstadt, T. (1993). Molecular cloning and expression of *hctB* encoding a strain-variant chlamydial histone-like protein with DNA-binding activity. J. Bacteriol. *175*, 4274–4281.

Bush, M., and Dixon, R. (2012). The role of bacterial enhancer binding proteins as specialized activators of sigma54-dependent transcription. Microbiol. Mol. Biol. Rev. *76*, 497–529.

Carlson, J.H., Wood, H., Roshick, C., Caldwell, H.D., and McClarty, G. (2006). *In vivo* and *in vitro* studies of *Chlamydia trachomatis* TrpR:DNA interactions. Mol. Microbiol. *59*, 1678–1691.

Case, E.D., Akers, J.C., and Tan, M. (2011). CT406 encodes a chlamydial ortholog of NrdR, a repressor of ribonucleotide reductase. J. Bacteriol. *193*, 4396–4404.

Case, E.D., Peterson, E.M., and Tan, M. (2010). Promoters for *Chlamydia* type III secretion genes show a differential response to DNA supercoiling that correlates with temporal expression pattern. J. Bacteriol. *192*, 2569–2574.

Chen, A.L., Wilson, A.C., and Tan, M. (2011). A *Chlamydia*-specific C-terminal region of the stress response regulator HrcA modulates its repressor activity. J. Bacteriol. *193*, 6733–6741.

Cheng, E., and Tan, M. (2012). Differential effects of DNA supercoiling on *Chlamydia* early promoters correlate with expression patterns in midcycle. J. Bacteriol. *194*, 3109–3115.

Collingro, A., Tischler, P., Weinmaier, T., Penz, T., Heinz, E., Brunham, R.C., Read, T.D., Bavoil, P.M., Sachse, K., Kahane, S., Friedman, M.G., Rattei, T., Myers, G.S., and Horn, M. (2011). Unity in variety - the pan-genome of the *Chlamydiae*. Mol. Biol. Evol. 3253–3270.

Desai, M., Wurihan, W., Di, R., Fondell, J.D., Nickels, B.E., Bao, X., and Fan, H. (2018). Role for GrgA in regulation of sigma(28)-dependent transcription in the obligate intracellular bacterial pathogen *Chlamydia trachomatis*. J. Bacteriol. *200*, 1–11.

Dessus-Babus, S., Bebear, C.M., Charron, A., Bebear, C., and de Barbeyrac, B. (1998). Sequencing of gyrase and topoisomerase IV quinolone-resistance-determining regions of *Chlamydia trachomatis* and characterization of quinolone-resistant mutants obtained *in vitro*. Antimicrob. Agents Chemother. *42*, 2474–2481.

Dillon, S.C., and Dorman, C.J. (2010). Bacterial nucleoid-associated proteins, nucleoid structure and gene expression. Nat. Rev. Microbiol. *8*, 185–195.

Domman, D., and Horn, M. (2015). Following the footsteps of chlamydial gene regulation. *Mol. Biol. Evol.*

Dorman, C.J., and Dorman, M.J. (2016). DNA supercoiling is a fundamental regulatory principle in the control of bacterial gene expression. Biophys. Rev. *8*, 209–220.

Douglas, A., and Hatch, T.P. (2006). The phosphorelay system in *Chlamydia trachomatis*. In Chlamydial Infections: Proceedings of the Eleventh International Symposium on Human Chlamydial Infections. M. Chernesky, editor, Niagara-on-the-Lake, Canada. 177–180.

Drlica, K., and Rouviere-Yaniv, J. (1987). Histonelike proteins of bacteria. Microbiol. Rev. *51*, 301–319.

Engel, J., and Ganem, D. (1990). A PCR-based approach to cloning sigma factors from eubacteria and its application to the isolation of a sigma70 homolog from *Chlamydia trachomatis*. J. Bacteriol. *172*, 2447–2455.

Feklistov, A., Sharon, B.D., Darst, S.A., and Gross, C.A. (2014). Bacterial sigma factors: a historical, structural, and genomic perspective. Annu. Rev. Microbiol. *68*, 357–376.

Galan-Vasquez, E., Luna, B., and Martinez-Antonio, A. (2011). The Regulatory Network of *Pseudomonas aeruginosa*. Microb. Inform. Exp. *1*, 3.

Garcia, E., Bancroft, S., Rhee, S.G., and Kustu, S. (1977). The product of a newly identified gene, gInF, is required for synthesis of glutamine synthetase in *Salmonella*. Proc. Natl. Acad. Sci. U S A 74, 1662–1666.

Gmuender, H., Kuratli, K., Di Padova, K., Gray, C.P., Keck, W., and Evers, S. (2001). Gene expression changes triggered by exposure of *Haemophilus influenzae* to novobiocin or ciprofloxacin: combined transcription and translation analysis. Genome Res. *11*, 28–42.

Gottesman, S. (2005). Micros for microbes: non-coding regulatory RNAs in bacteria. Trends Genet. *21*, 399–404.

Gottesman, S., and Storz, G. (2011). Bacterial small RNA regulators: versatile roles and rapidly evolving variations. Cold Spring Harb. Perspect. Biol. 3.

Grieshaber, N.A., Fischer, E.R., Mead, D.J., Dooley, C.A., and Hackstadt, T. (2004). Chlamydial histone-DNA interactions are disrupted by a metabolite in the methylerythritol phosphate pathway of isoprenoid biosynthesis. Proc. Natl. Acad. Sci. U S A *101*, 7451–7456.

Grieshaber, N.A., Grieshaber, S.S., Fischer, E.R., and Hackstadt, T. (2006a). A small RNA inhibits translation of the histone-like protein Hc1 in *Chlamydia trachomatis*. Mol. Microbiol. *59*, 541–550.

Grieshaber, N.A., Sager, J.B., Dooley, C.A., Hayes, S.F., and Hackstadt, T. (2006b). Regulation of the *Chlamydia trachomatis* histone H1-like protein Hc2 is IspE dependent and IhtA independent. J. Bacteriol. *188*, 5289–5292.

Grove, A. (2011). Functional evolution of bacterial histone-like HU proteins. Curr. Issues Mol. Biol. *13*, 1–12.

Gunnelius, L., Hakkila, K., Kurkela, J., Wada, H., Tyystjarvi, E., and Tyystjarvi, T. (2014). The omega subunit of the RNA polymerase core directs transcription efficiency in cyanobacteria. Nucleic Acids Res. *42*, 4606–4614.

Hackstadt, T (1991). Purification and N-terminal amino acid sequences of *Chlamydia trachomatis* histone analogs. J. Bacteriol. *173*, 7046–7049.

Hanson, B.R., Slepenkin, A., Peterson, E.M., and Tan, M. (2015). *Chlamydia trachomatis* type III secretion proteins regulate transcription. J. Bacteriol. *197*, 3238–3244.

Hanson, B.R., and Tan, M. (2015). Transcriptional regulation of the *Chlamydia* heat shock stress response in an intracellular infection. Mol. Microbiol. *97*, 1158–1167.

Hanson, B.R., and Tan, M. (2016). Intra-ChIP: studying gene regulation in an intracellular pathogen. Curr. Genet. *62*, 547–551.

Hanson, B.R., and Tan, M. (2018). Using Intra-ChIP to measure protein–DNA interactions in intracellular pathogens. Methods Mol. Biol. *1689*, 147–155.

Hatfield, G.W., and Benham, C.J. (2002). DNA topology-mediated control of global gene expression in *Escherichia coli*. Annu. Rev. Genet. 36, 175–203.

Hecker, M., Pane-Farre, J., and Volker, U. (2007). SigB-dependent general stress response in *Bacillus subtilis* and related gram-positive bacteria. Annu. Rev. Microbiol. *61*, 215–236.

Hefty, P.S., and Stephens, R.S. (2007). Chlamydial type III secretion system is encoded on ten operons preceded by sigma 70-like promoter elements. J. Bacteriol. *189*, 198–206.

Hickey, J.M., Weldon, L., and Hefty, P.S. (2011). The atypical OmpR/PhoB response regulator ChxR from *Chlamydia trachomatis* forms homodimers in vivo and binds a direct repeat of nucleotide sequences. J. Bacteriol. *193*, 389–398.

Hua, L., Hefty, P.S., Lee, Y.J., Lee, Y.M., Stephens, R.S., and Price, C.W. (2006). Core of the partner switching signalling mechanism is conserved in the obligate intracellular pathogen *Chlamydia trachomatis*. Mol. Microbiol. 59, 623–636.

Humphrys, M.S., Creasy, T., Sun, Y., Shetty, A.C., Chibucos, M.C., Drabek, E.F., Fraser, C.M., Farooq, U., Sengamalay, N., Ott, S., Shou, H., Bavoil, P.M., Mahurkar, A., and Myers, G.S. (2013). Simultaneous transcriptional profiling of bacteria and their host cells. PLoS ONE *8*, e80597.

Huttenhofer, A., and Vogel, J. (2006). Experimental approaches to identify non-coding RNAs. Nucleic Acids Res. *34*, 635–646.

Joyeux, M. (2015). Compaction of bacterial genomic DNA: clarifying the concepts. J. Phys. Condens. Matter. *27*, 383001.

Kalman, S., Mitchell, W., Marathe, R., Lammel, C., Fan, J., Hyman, R.W., Olinger, L., Grimwood, J., Davis, R.W., and Stephens, R.S. (1999). Comparative genomes of *Chlamydia pneumoniae* and *C. trachomatis*. Nat. Genet. *21*, 385–389.

Kemege, K.E., Hickey, J.M., Lovell, S., Battaile, K.P., Zhang, Y., and Hefty, P.S. (2011). Ab initio structural modeling of and experimental validation for *Chlamydia trachomatis* protein CT296 reveal structural similarity to Fe(II) 2-oxoglutarate-dependent enzymes. J. Bacteriol. *193*, 6517–6528.

Koehler, J.E., Burgess, R.R., Thompson, N.E., and Stephens, R.S. (1990). *Chlamydia trachomatis* RNA polymerase major sigma subunit. Sequence and structural comparison of conserved and unique regions with *Escherichia coli* sigma 70 and *Bacillus subtilis* sigma 43. J. Biol. Chem. *265*, 13206–13214.

Koo, I.C., and Stephens, R.S. (2003). A developmentally regulated two-component signal transduction system in *Chlamydia*. J. Biol. Chem. *278*, 17314–17319.

Koo, I.C., Walthers, D., Hefty, P.S., Kenney, L.J., and Stephens, R.S. (2006). ChxR is a transcriptional activator in Chlamydia. Proc. Natl. Acad. Sci. U S A *103*, 750–755.

LaVerda, D., Kalayoglu, M.V., and Byrne, G.I. (1999). Chlamydial heat shock proteins and disease pathology: new paradigms for old problems? Infect. Dis. Obstet. Gynecol. 7, 64–71.

Lee, D.J., Minchin, S.D., and Busby, S.J. (2012). Activating transcription in bacteria. Annu. Rev. Microbiol. *66*, 125–152.

Liu, X., Afrane, M., Clemmer, D.E., Zhong, G., and Nelson, D.E. (2010). Identification of *Chlamydia trachomatis* outer membrane complex proteins by differential proteomics. J. Bacteriol. *192*, 2852–2860.

Liu, Y., Chen, C., Gong, S., Hou, S., Qi, M., Liu, Q., Baseman, J., and Zhong, G. (2014). Transformation of *Chlamydia muridarum* reveals a role for Pgp5 in suppression of plasmid-dependent gene expression. J. Bacteriol. *196*, 989–998.

Lonetto, M., Gribskov, M., and Gross, C.A. (1992). The sigma 70 family: sequence conservation and evolutionary relationships. J. Bacteriol. *174*, 3843–3849.

Madan Babu, M., Teichmann, S.A., and Aravind, L. (2006). Evolutionary dynamics of prokaryotic transcriptional regulatory networks. J. Mol Biol. *358*, 614–633.

Mathew, R., and Chatterji, D. (2006). The evolving story of the omega subunit of bacterial RNA polymerase. Trends Microbiol. *14*, 450–455.

Mathews, S.A., and Sriprakash, K.S. (1994). The RNA polymerase of *Chlamydia trachomatis* has a flexible sequence requirement at the -10 and -35 boxes of its promoters. J. Bacteriol. *176*, 3785–3789.

Mathews, S.A., and Timms, P. (2000). Identification and mapping of sigma-54 promoters in *Chlamydia trachomatis*. J. Bacteriol. *182*, 6239–6242.

Maurer, A.P., Mehlitz, A., Mollenkopf, H.J., and Meyer, T.F. (2007). Gene expression profiles of *Chlamydophila pneumoniae* during the developmental cycle and iron depletion-mediated persistence. PLoS Pathog. *3*, e83.

Menzel, R., and Gellert, M. (1983). Regulation of the genes for *E. coli* DNA gyrase: homeostatic control of DNA supercoiling. Cell *34*, 105–113.

Moelleken, K., and Hegemann, J.H. (2008). The *Chlamydia* outer membrane protein OmcB is required for adhesion and exhibits biovar-specific differences in glycosaminoglycan binding. Mol. Microbiol. *67*, 403–419.

Moulder, J.W. (1991). Interaction of chlamydiae and host cells *in vitro*. Microbiol. Rev. *55*, 143–190.

Nicholson, T.L., Olinger, L., Chong, K., Schoolnik, G., and Stephens, R.S. (2003). Global stage-specific gene regulation during the developmental cycle of *Chlamydia trachomatis*. J. Bacteriol. *185*, 3179–3189.

Niehus, E., Cheng, E., and Tan, M. (2008). DNA supercoiling-dependent gene regulation in *Chlamydia*. J. Bacteriol. *190*, 6419–6427.

Orillard, E., and Tan, M. (2015). Functional analysis of three topoisomerases that regulate DNA supercoiling levels in *Chlamydia*. Mol. Microbiol. *99*, 484–496.

Ouellette, S.P. (2018). Feasibility of a conditional knockout system for *Chlamydia* based on CRISPR Interference. Front. Cell Infect. Microbiol. *8*, 59.

Paget, M.S. (2015). Bacterial sigma factors and anti-sigma factors: structure, function and distribution. Biomolecules *5*, 1245–1265.

Payankaulam, S., Li, L.M., and Arnosti, D.N. (2010). Transcriptional repression: conserved and evolved features. Curr. Biol. *20*, R764–771.

Pedersen, L.B., Birkelund, S., and Christiansen, G. (1994). Interaction of the *Chlamydia trachomatis* histone H1-like protein (Hc1) with DNA and RNA causes repression of transcription and translation *in vitro*. Mol. Microbiol. *11*, 1085–1098.

Pedersen, L.B., Birkelund, S., and Christiansen, G. (1996). Purification of recombinant *Chlamydia trachomatis* histone H1-like protein Hc2, and comparative functional analysis of Hc2 and Hc1. Mol. Microbiol. *20*, 295–311.

Perara, E., Ganem, D., and Engel, J. (1992). A developmentally regulated chlamydial gene with apparent homology to eukaryotic histone H1. Proc. Natl. Acad. Sci. U S A *89*, 2125–2129.

Pereira, L.E., Tsang, J., Mrazek, J., and Hoover, T.R. (2011). The zinc-ribbon domain of *Helicobacter pylori* HP0958: requirement for RpoN accumulation and possible roles of homologs in other bacteria. Microb. Inform. Exp. *1*, 1–10.

Peter, B.J., Arsuaga, J., Breier, A.M., Khodursky, A.B., Brown, P.O., and Cozzarelli, N.R. (2004). Genomic transcriptional response to loss of chromosomal supercoiling in *Escherichia coli*. Genome Biol. *5*, R87.

Pinson, V., Takahashi, M., and Rouviere, J.-Yaniv (1999). Differential binding of the *Escherichia coli* HU, homodimeric forms and heterodimeric form to linear, gapped and cruciform DNA. J. Mol. Biol. *287*, 485–497.

Rao, X., Deighan, P., Hua, Z., Hu, X., Wang, J., Luo, M., Wang, J., Liang, Y., Zhong, G., Hochschild, A., and Shen, L. (2009). A regulator from *Chlamydia trachomatis* modulates the activity of RNA polymerase through direct interaction with the beta subunit and the primary sigma subunit. Genes Dev. *23*, 1818–1829.

Rau, A., Wyllie, S., Whittimore, J., and Raulston, J.E. (2005). Identification of *Chlamydia trachomatis* genomic sequences recognized by chlamydial divalent cation-dependent regulator A (DcrA). J. Bacteriol. *187*, 443–448.

Read, T.D., Myers, G.S., Brunham, R.C., Nelson, W.C., Paulsen, I.T., Heidelberg, J., Holtzapple, E., Khouri, H., Federova, N.B., Carty, H.A., Umayam, L.A., Haft, D.H., Peterson, J., Beanan, M.J., White, O., Salzberg, S.L., Hsia, R.C., McClarty, G., Rank, R.G., Bavoil, P.M., and Fraser, C.M. (2003). Genome sequence of *Chlamydophila caviae* (*Chlamydia psittaci* GPIC): examining the role of niche-specific genes in the evolution of the Chlamydiaceae. Nucleic Acids Res. *31*, 2134–2147.

Revyakin, A., Ebright, R.H., and Strick, T.R. (2004). Promoter unwinding and promoter clearance by RNA polymerase: detection by single-molecule DNA nanomanipulation. Proc. Natl. Acad. Sci. U S A *101*, 4776–4780.

Riordan, J.T., and Mitra, A. (2017). Regulation of *Escherichia coli* Pathogenesis by Alternative Sigma Factor N. EcoSal Plus. 7.

Rosario, C.J., Hanson, B.R., and Tan, M. (2014). The transcriptional repressor EUO regulates both subsets of *Chlamydia* late genes. Mol. Microbiol. *94*, 888–897.

Rosario, C.J., and Tan, M. (2012). The early gene product EUO is a transcriptional repressor that selectively regulates promoters of *Chlamydia* late genes. Mol. Microbiol. *84*, 1097–1107.

Rosario, C.J., and Tan, M. (2015). Regulation of *Chlamydia* gene expression by tandem promoters with different temporal patterns. J. Bacteriol. *198*, 363–369.

Saka, H.A., Thompson, J.W., Chen, Y.S., Kumar, Y., Dubois, L.G., Moseley, M.A., and Valdivia, R.H. (2011). Quantitative proteomics reveals metabolic and pathogenic properties of *Chlamydia trachomatis* developmental forms. Mol. Microbiol. *82*, 1185–1203.

Santos-Zavaleta, A., Salgado, H., Gama-Castro, S., Sanchez-Perez, M., Gomez-Romero, L., Ledezma-Tejeida, D., Garcia-Sotelo, J.S., Alquicira-Hernandez, K., Muniz-Rascado, L.J., Pena-Loredo, P., Ishida-Gutierrez, C., Velazquez-Ramirez, D.A., Del Moral-Chavez, V., Bonavides-Martinez, C., Mendez-Cruz, C.F., Galagan, J., and Collado-Vides, J. (2018). RegulonDB v 10.5: tackling challenges to unify classic and high throughput knowledge of gene regulation in *E. coli* K-12. Nucleic Acids Res.

Schaumburg, C.S., and Tan, M. (2003). Mutational analysis of the *Chlamydia trachomatis dnaK* promoter defines the optimal -35 promoter element. Nucleic Acids Res. *31*, 551–555.

Schaumburg, C.S., and Tan, M. (2006). Arginine-dependent gene regulation via the ArgR repressor is species specific in *Chlamydia*. J. Bacteriol. *188*, 919–927.

Seshasayee, A.S., Sivaraman, K., and Luscombe, N.M. (2011). An overview of prokaryotic transcription factors: a summary of function and occurrence in bacterial genomes. Subcell. Biochem. *52*, 7–23.

Sharma, C.M., and Vogel, J. (2009). Experimental approaches for the discovery and characterization of regulatory small RNA. Curr. Opin. Microbiol. *12*, 536–546.

Shaw, E.I., Dooley, C.A., Fischer, E.R., Scidmore, M.A., Fields, K.A., and Hackstadt, T. (2000). Three temporal classes of gene expression during the *Chlamydia trachomatis* developmental cycle. Mol. Microbiol. *37*, 913–925.

Shen, L., Macnaughtan, M.A., Frohlich, K.M., Cong, Y., Goodwin, O.Y., Chou, C.W., LeCour, L., Jr., Krup, K.L., Luo, M., and Worthylake, D.K. (2015). Multipart chaperone-effector recognition in the type III secretion system of *Chlamydia trachomatis*. J. Biol. Chem. *290*, 28141–28155.

Silva-Herzog, E., Joseph, S.S., Avery, A.K., Coba, J.A., Wolf, K., Fields, K.A., and Plano, G.V. (2011). Scc1 (CP0432) and Scc4 (CP0033) function as a type III secretion chaperone for CopN of *Chlamydia pneumoniae*. J. Bacteriol. *193*, 3490–3496.

Skipp, P.J., Hughes, C., McKenna, T., Edwards, R., Langridge, J., Thomson, N.R., and Clarke, I.N. (2016). Quantitative proteomics of the infectious and replicative forms of *Chlamydia trachomatis*. PLoS ONE *11*, e0149011.

Song, L., Carlson, J.H., Whitmire, W.M., Kari, L., Virtaneva, K., Sturdevant, D.E., Watkins, H., Zhou, B., Sturdevant, G.L., Porcella, S.F., McClarty, G., and Caldwell, H.D. (2013). *Chlamydia trachomatis* plasmid-encoded Pgp4 is a transcriptional regulator of virulence-associated genes. Infect. Immun. *81*, 636–644.

Stephens, R.S (1990). Molecular Genetics of *Chlamydia*. *In* Chlamydial Infections 1990. W.R.e.a. Bowie, editor. Cambridge University Press, Cambridge, UK. 63–72.

Stephens, R.S. (2003). The cellular paradigm of chlamydial pathogenesis. Trends Microbiol. *11*, 44–51.

Stephens, R.S., Kalman, S., Lammel, C., Fan, J., Marathe, R., Aravind, L., Mitchell, W., Olinger, L., Tatusov, R.L., Zhao, Q., Koonin, E.V., and Davis, R.W. (1998). Genome sequence of an obligate intracellular pathogen of humans: *Chlamydia trachomatis*. Science *282*, 754–759.

Storz, G. (2002). An expanding universe of noncoding RNAs. Science *296*, 1260–1263.

Storz, G., Opdyke, J.A., and Zhang, A. (2004). Controlling mRNA stability and translation with small, noncoding RNAs. Curr. Opin. Microbiol. 7, 140–144.

Tan, M. (2006). Regulation of gene expression. *In Chlamydia*: Genomics and Pathogenesis. P.M. Bavoil and P.B. Wyrick, editors. Horizon Bioscience, Wymondham, UK. 103–131.

Tan, M. (2012). Temporal gene regulation during the chlamydial development cycle. *In* Intracellular Pathogens I: *Chlamydiales*. M. Tan and P.M. Bavoil, editors. ASM Press, Washington, D.C. 149–169.

Tao, S., Kaul, R., and Wenman, W.M. (1991). Identification and nucleotide sequence of a developmentally regulated gene encoding a eukaryotic histone H1-like protein from *Chlamydia trachomatis*. J. Bacteriol. *173*, 2818–2822.

Taylor, M., Butler, R., Chambers, S., Casimiro, M., Badii, F., and Merrick, M. (1996). The RpoN-box motif of the RNA polymerase sigma factor sigma N plays a role in promoter recognition. Mol. Microbiol. *22*, 1045–1054.

Thompson, C.C., Griffiths, C., Nicod, S.S., Lowden, N.M., Wigneshweraraj, S., Fisher, D.J., and McClure, M.O. (2015). The Rsb phosphoregulatory network controls availability of the primary sigma factor in *Chlamydia trachomatis* and influences the kinetics of growth and development. PLoS Pathog. *11*, e1005125.

Thompson, C.C., Nicod, S.S., Malcolm, D.S., Grieshaber, S.S., and Carabeo, R.A. (2012). Cleavage of a putative metal permease in *Chlamydia trachomatis* yields an iron-dependent transcriptional repressor. Proc. Natl. Acad. Sci. U S A *109*, 10546–10551.

Voelker, U., Dufour, A., and Haldenwang, W.G. (1995). The *Bacillus subtilis* rsbU gene product is necessary for RsbX-dependent regulation of sigma B. J. Bacteriol. *177*, 114–122.

Wang, Y., Kahane, S., Cutcliffe, L.T., Skilton, R.J., Lambden, P.R., and Clarke, I.N. (2011). Development of a transformation system for *Chlamydia trachomatis*: restoration of glycogen biosynthesis by acquisition of a plasmid shuttle vector. PLoS Pathog. *7*, e1002258.

Wickstrum, J., Sammons, L.R., Restivo, K.N., and Hefty, P.S. (2013). Conditional gene expression in *Chlamydia trachomatis* using the tet system. *PLoS ONE 8*, e76743.

Wilson, A.C., and Tan, M. (2002). Functional analysis of the heat shock regulator HrcA of *Chlamydia trachomatis*. *J. Bacteriol. 184*, 6566–6571.

Wilson, A.C., and Tan, M. (2004). Stress response gene regulation in *Chlamydia* is dependent on HrcA-CIRCE interactions. J. Bacteriol. *186*, 3384–3391.

Wilson, A.C., Wu, C.C., Yates, J.R., 3rd, and Tan, M. (2005). Chlamydial GroEL autoregulates its own expression through direct interactions with the HrcA repressor protein. J. Bacteriol. *187*, 7535–7542.

Wood, H., Fehlner-Gardner, C., Berry, J., Fischer, E., Graham, B., Hackstadt, T., Roshick, C., and McClarty, G. (2003). Regulation of tryptophan synthase gene expression in *Chlamydia trachomatis*. Mol. Microbiol. *49*, 1347–1359.

Workowski, K.A., Bolan, G.A., Centers for Disease Control and Prevention (2015). Sexually transmitted diseases treatment guidelines, 2015. MMWR Recomm. Rep. *64*, 1–137.

Wyllie, S., and Raulston, J.E. (2001). Identifying regulators of transcription in an obligate intracellular pathogen: a metal-dependent repressor in *Chlamydia trachomatis*. Mol. Microbiol. *40*, 1027–1036.

Yamamoto, K., Watanabe, H., and Ishihama, A. (2014). Expression levels of transcription factors in *Escherichia coli*: growth phase- and growth condition-dependent variation of 90 regulators from six families. Microbiology *160*, 1903–1913.

Yang, C., Kari, L., Sturdevant, G.L., Song, L., Patton, M.J., Couch, C.E., Ilgenfritz, J.M., Southern, T.R., Whitmire, W.M., Briones, M., Bonner, C., Grant, C., Hu, P., McClarty, G., and Caldwell, H.D. (2017). *Chlamydia trachomatis* ChxR is a transcriptional regulator of virulence factors that function in in vivo host-pathogen interactions. Pathog. Dis. 75.

Yu, H.H.Y., Di Russo, E.G., Rounds, M.A., and Tan, M. (2006a). Mutational analysis of the promoter recognized by *Chlamydia* and *Escherichia coli* sigma 28 RNA polymerase. J. Bacteriol. *188*, 5524–5531.

Yu, H.H.Y., Kibler, D., and Tan, M. (2006b). *In silico* prediction and functional validation of sigma 28-regulated genes in *Chlamydia* and *Escherichia coli*. J. Bacteriol. *188*, 8206–8212.

Yu, H.H.Y., and Tan, M. (2003). Sigma 28 RNA polymerase regulates *hctB*, a late developmental gene in *Chlamydia*. Mol. Microbiol. *50*, 577–584.

Zhang, L., Douglas, A.L., and Hatch, T.P. (1998). Characterization of a *Chlamydia psittaci* DNA binding protein (EUO) synthesized during the early and middle phases of the developmental cycle. Infect. Immun. *66*, 1167–1173.

Zhang, L., Howe, M.M., and Hatch, T.P. (2000). Characterization of *in vitro* DNA binding sites of the EUO protein of *Chlamydia psittaci*. Infect. Immun. *68*, 1337–1349.

Zhong, G. (2017). Chlamydial plasmid-dependent pathogenicity. Trends Microbiol. *25*, 141–152.

Chlamydia Genetics

Colette E. O'Neill[1], Ian N. Clarke[1] and Derek J. Fisher[2*]

11

[1]Molecular Microbiology Group, Department of Clinical and Experimental Science, University of Southampton, Southampton, UK.
[2]Department of Microbiology, Southern Illinois University, Carbondale, IL, USA.

*Correspondence: fisher@micro.siu.edu

https://doi.org/10.21775/9781912530281.11

Abstract

Since the seminal publication in 2011 from Wang and colleagues on plasmid-based gene transfer and the introduction of competence in *Chlamydia trachomatis*, numerous significant advances have been made in the field of chlamydial genetics using their transformation technique. Collectively, we can now generate targeted insertions and gene deletions in the chromosome, transform four chlamydial species with a variety of different shuttle vectors, construct random chromosomal mutants using chemical and transposon mutagenesis methods, and even generate conditional gene knockdowns using CRISPRi technology. In this chapter, we seek to provide background on how the field of chlamydial genetics grew, where it currently rests, and the needs that remain. While the transformation problem has been overcome, the low transformation frequencies and obligatory intracellular lifestyle are a significant bottleneck. Consequently, mutant construction remains tedious and operon structure and gene essentiality complicate both mutant construction and analysis. Pitfalls that arise when new technologies are brought to bear in a field starved of experimental approaches long practised for other bacteria are also discussed along with standardization of protocols. It is our hope that this approach will yield significant and reproducible insights into the molecular details underlying the pathogenesis of *Chlamydia*.

Introduction

Research on *Chlamydia* is challenging due to its obligate intracellular nature and bi-phasic developmental cycle, and this is particularly so in the field of genetics. For many years the absence of a simple, cell-free culture system and the lack of a reproducible means by which to stably introduce foreign DNA into the microorganism have hampered progress. Alternative 'surrogate' systems have been developed to study gene function where chlamydial proteins are expressed and analysed in other bacterial hosts, and while these approaches have provided some advances there are clear limitations.

The two main challenges in developing chlamydial genetics were getting *Chlamydia* to readily take up foreign DNA (competence) and designing vectors for delivering the DNA. These two challenges will be discussed first, followed by more in-depth discussion on the pathways that illuminated the possibility of chlamydial genetics, the currently available tools, and our perception of the remaining needs and limitations.

Bacterial competence in *Chlamydia*

Many species of bacteria can take up DNA through a natural process known as 'competence'. Early molecular evidence for DNA transfer in *Chlamydia* comes from phylogenetic analysis (Fitch *et al.*, 1993) and the discovery of naturally occurring recombinants in the *ompA* gene (Hayes *et al.*, 1994b; Lampe *et al.*, 1993). Recombination has subsequently been developed to allow genetic crosses between different strains and species of *Chlamydia* following the direct, but random, exchange/recombination of genomes (DeMars and Weinfurter, 2008; DeMars *et al.*, 2007). Alternative ways of introducing DNA are by bacterial conjugation or by transfer via bacteriophages in a process known as transduction. To perform precise genetic manipulation of bacteria, vectors based on plasmids and/or bacteriophages are required. Biological studies and genomic analyses of Chlamydiae have revealed that there are few such extrachromosomal genetic elements (Lusher *et al.*, 1991), consistent with the niche intracellular lifestyle of the Chlamydiae. Unfortunately, Chlamydiae lack conjugative plasmids and transducing bacteriophages, which are critical tools for genetic methods developed in the late 1970s and 1980s to manipulate free-living bacteria such as *Escherichia coli*.

Chlamydia do have the ability to take up DNA and this process occurs when two different parental strains infect the same host cell, as has been demonstrated experimentally (DeMars *et al.*, 2007). It is well known that inclusions formed by two different strains in a multiply infected host cell will fuse to make one large mature inclusion (Bernkopf *et al.*, 1962; Blyth and Taverne, 1972; Campbell *et al.*, 1989; Kramer and Gordon, 1971) and presumably it is at this stage that genetic exchange takes place. Because of their obligate intracellular developmental cycle, a culture of *Chlamydia*-infected host cells should be considered not as a single homogenous entity, but as a set of microcolonies growing in parallel. When a cell monolayer is infected with fewer EBs than cells, each mature inclusion can be viewed as a colony and most likely represents the outcome of infection with a single EB (Lambden *et al.*, 2006b). An EB, once it has differentiated into an RB, will undergo multiple divisions before differentiating back to the EB form (Lambden *et al.*, 2006b). If chlamydial cell division was similar to that of free-living bacteria, then the most competent chlamydial form would be the RB growing in the exponential phase (Chung *et al.*, 1989). Unfortunately for researchers, the chlamydial genome at this critical stage is at its most inaccessible point for exposure to foreign DNA (and hence transforming vectors) since it is protected by the membrane barrier of the host cell, the inclusion membrane and the RB cell envelope (inner and outer membranes). These multiple barriers hinder delivery of DNA via approaches commonly used for eukaryotic cells such as transfection or electroporation. Thus, the prospects of transforming RBs with foreign DNA inside a cell are challenging. Now that an *in vitro* system has been developed for keeping purified RBs metabolically active (Haider *et al.*, 2010; Omsland *et al.*, 2012) it might be possible to target cell-free RBs for transformation. To make this approach practical, methods would have to be developed to replicate RBs under axenic conditions and to reintroduce them into a host cell, which is currently not possible. Therefore, from a pragmatic viewpoint, the most accessible target for transformation is the EB.

This infectious form of *Chlamydia* can be purified in high quantity (Skipp *et al.*, 2005, 2016) and seemed to be a good starting point to make competent bacteria since even if they are not naturally competent, competence can be induced in bacteria using a variety of protocols (Yoshida and Sato, 2009). Regardless of which form of chlamydia was selected as the most suitable transformation target, there were still significant barriers to overcome. For example, modification of plasmid DNA can affect transformation frequencies (Macaluso and Mettus, 1991). It has been demonstrated that chlamydial DNA has low level Dcm methylation (Wagar *et al.*, 1992), although it is not known whether *Chlamydia* carries a unique restriction modification system. Knowledge and understanding of the methylation systems in *Chlamydia* would help in working out optimal ways of preparing DNA for transformation. It is not known whether there are uptake sequences or specific DNA transporter proteins. Furthermore, in the case of EBs, an additional and much overlooked complication is the issue of infectivity. A seminal study by Rosanna Peeling (Peeling and Brunham, 1991) showed that even in highly purified preparations the ratio of infectious to non-infectious EBs can be as low as 1:100 to 1:500 and differs between species and serovars. This presents a significant challenge because low competency levels will be further compounded if greater than 99% of the microorganisms in a preparation are not infectious. The problem of dead EBs is exemplified by a study on phage infectivity (Skilton *et al.*, 2007). Whether the infectivity ratio varies with different methods of preparing EBs and could perhaps be improved is currently unknown. Thus, there are many problems associated with making this bacterium competent and to date, there is no easily applied means of measuring *Chlamydia* competency and transformation frequencies. One is left to perform blind attempts at transformation as not all strains are able to form plaques in cell culture assays and flow-cytometry analysis of first passage, fluorescent-transformants has not been validated. Consequently, rigorous optimization of transformation efficiency remains challenging.

Vector design

The chlamydial plasmid as a suitable vector or plasmid backbone

Early phylogenetic studies have indicated that *Chlamydia* are distant from other eubacteria (Weisburg *et al.*, 1986) and the unique developmental cycle hinted at complex and divergent systems of gene regulation. At the time of vector design, many differentially expressed promoters could have been selected for vector construction and there were legitimate concerns about what would work in Chlamydiae. The first specific studies on gene regulation in *Chlamydia* showed intricate and sophisticated control systems, which bore little resemblance to the standard regulons of free-living bacteria (Lambden *et al.*, 1990; Stephens *et al.*, 1988). Consensus promoters in *Chlamydia* appeared to be different from those of other bacterial species and there seemed to be no standard rules to define promoters and transcriptional regulators. Developmental gene regulation is controlled through the use of a series or cascade of sigma factors (Mathews *et al.*, 1999) and there are many other regulatory control loops that affect development of *Chlamydia*. Thus, if a vector was to be designed for *Chlamydia* transformation then there would be compelling scientific evidence that it should employ bespoke chlamydial controls and should be designed to coordinate expression of selectable markers with gene regulation throughout the developmental cycle. In view of the apparent complexity of chlamydial gene regulation an arbitrary decision was made to use two simple constitutive promoters from free living bacteria that had no external regulatory controls (Wang *et al.*, 2011). The assumption was that such 'core' promoters, shown to function in different bacterial species, could work within most prokaryotic environments and would therefore be useful to drive the expression of two different selectable markers in *Chlamydia*. We now appreciate that the chlamydial transcription machinery and regulation mechanisms largely adhere to the rules learned from the study of other bacteria (discussed in more detail in Chapter 10, *Chlamydia* Gene Regulation).

Prior attempts to transform *Chlamydia* with *Escherichia coli* shuttle vectors were not successful, presumably due to failure of the vector to replicate (unpublished). So, selecting a plasmid as a vector for *C. trachomatis* was left to Hobson's choice, the endogenous chlamydial plasmid, which plays a role in regulating various aspects of chlamydial growth and virulence (reviewed in Zhong, 2017). *C. trachomatis* and several other chlamydial species have a 7.5 kb plasmid, which encodes eight genes (CDS1 through CDS8) and for which phylogenetic studies indicate a common evolutionary origin (Seth-Smith *et al.*, 2009). Comparative studies both within *C. trachomatis* and with plasmids from other chlamydial species (Seth-Smith *et al.*, 2009; Thomas *et al.*, 1997) suggest that CDS1 is functionally redundant with CDS2 since CDS1 carries truncating mutations and is a partial duplication of CDS2, which is the most highly conserved plasmid gene. However, efforts to make the chlamydial plasmid replicate in *E. coli* failed. To reproducibly generate large amounts for transformation studies, the chlamydial plasmid was turned into a 'shuttle vector', pBR325::L2, by cloning it into a standard *E. coli* vector, pBR325 (within the chlamydial CDS1), which carried an *E. coli* origin of replication and conferred both ampicillin and chloramphenicol resistance in *E. coli* (Wang *et al.*, 2011). To avoid the potential reduction in transformation rate owing to non-native DNA methylation patterns, the plasmid vector DNA was prepared in Dcm-/Dam- *E. coli* hosts. This was because in early optimization experiments using this same shuttle vector prepared in standard *E. coli* without the methylase mutations, no transformants were ever recovered (unpublished observation, Ian Clarke; suggested by Binet and Maurelli, 2009).

Selectable markers and promoters

The choice of selectable markers to use in chlamydial transformation is restricted by clinical priorities. Azithromycin and tetracycline are the frontline antibiotics currently used for treatment (CDC, 2015). While there is evidence that spontaneous azithromycin resistance occurs *in vitro* (Binet and Maurelli, 2007) and that tetracycline resistance mutants exist amongst *Chlamydia suis* isolates (Sandoz and Rockey, 2010), these antibiotics could not be used as selectable markers even though they would appear the best biological choices. Erythromycin is also effective and a reserved treatment for specific conditions and so using this as a selectable marker was also not possible. In contrast, penicillin and chloramphenicol are not recommended to treat chlamydial infections (CDC, 2015). These antibiotics have single gene resistance markers and are simple and easy to use in bacterial genetics. A list of selectable markers that have been used for chlamydial transformation with recombinant plasmids is shown in Table 11.1. The data in Table 11.1 are not intended to cover all markers used to date but illustrates the first reported transformation selection marker of a particular strain, and more refined selection protocols using the same antibiotic may have subsequently been developed.

Penicillin treatment of *Chlamydia*-infected cells has a dramatic effect. When added immediately after infection of host cells, penicillin blocks the developmental cycle (Lambden *et al.*, 2006a). EBs can develop into RBs, but RB division is inhibited while chromosomal DNA replication continues. Under these conditions RBs will undergo one or two divisions and then start to increase rapidly in size. Typically, a mature inclusion will hold two or four such aberrant or enlarged RBs, which remain non-infectious and are eventually lost when host cells lyse. We and others (Chacko *et al.*, 2015; Goellner *et al.*, 2006; Skilton *et al.*, 2009) have previously shown that it was possible for the bacteria to continue to divide and produce EBs if penicillin is removed from these cultures. The formation of viable, but non-infectious bacteria in the presence of penicillin was exploited to select for transformants (Wang *et al.*, 2011). The rationale was that transformed *C. trachomatis* would, under penicillin selection, have a normal developmental cycle due to the production of β-lactamase to hydrolyse penicillin. Untransformed EBs would become non-infectious aberrant RBs that would be lost on subsequent passage in culture. The gamble was whether β lactamase would be expressed in *C. trachomatis* as a periplasmic protein and retain its enzymatic activity. *C. trachomatis* L2, a well-characterized fast growing isolate, was used as the recipient host and EBs were transformed using a calcium-chloride based transformation protocol, which remains the method of choice for introducing foreign, recombinant DNA vectors into *Chlamydia*. Following antibiotic selection, normal inclusions were recovered, albeit at very low frequency, and therefore needed amplification by several passages. Purified, transformed EBs had β lactamase activity, and western blotting indicated that the β-lactamase signal peptide was cleaved (Wang *et al.*, 2011). Observing penicillin-resistant inclusions answered many questions including whether *C. trachomatis* could be stably transformed and express foreign genes. Extensive proof that these represented genuine stable transformants was provided by Southern blotting of the recovered supercoiled recombinant shuttle vector. A key factor for the field in adopting the technique reported by Wang *et al.* was the initial repetition in two independent laboratories in Israel and the UK prior to publication (Wang *et al.*, 2011).

Recombination amongst clinical isolates as support for the feasibility of genetics

Typing of *C. trachomatis* strains obtained from clinical cases into serotypes A-K and LGV L1-L3 was classically performed using antisera specific to the major outer membrane protein, MOMP, encoded by the *ompA* gene (listed in some early literature as *omp1*). With the advent of molecular biology and DNA sequencing using the Sanger method, investigators began to assess the full nucleotide sequence of *ompA* within and across serovars A-K and LGV L1-L3. As early as 1993, Lampe *et al.* (Lampe *et al.*, 1993) reported variation in the *ompA* sequence across serovar I strains, with an apparent insertion of a 12 nucleotide stretch that encoded a VDIV amino acid sequence identical to that found in serovar H isolates and not previously reported for serovar I isolates. The authors speculated that recombination between serovar I and H isolates might have been responsible for the variant MOMP sequence within the serovar I isolate. Soon afterwards, Hayes *et al.* reported an *ompA* sequence that appeared to be a hybrid of the LGV L1 and L2 *ompA* genes indicating that DNA exchange and recombination was occurring between serovars (Hayes *et al.*, 1994a). These findings led to a prescient statement that 'efforts to manipulate the *Chlamydia* genome *in vitro* by recombination should be intensified' (Hayes *et al.*, 1994b). It would take over a decade before such *in vitro* experiments would appear in the literature.

Numerous follow-up studies directly or indirectly supported these early findings of recombination amongst clinical isolates (Brunham *et al.*, 1994; Frost *et al.*, 1995; Hayes *et al.*, 1994b). Prior to the publication of the first chlamydial genome by the Stephens group (Stephens *et al.*, 1998b), two reports described the initial identification and characterization of the chlamydial *recA* gene, which plays a pivotal role in DNA recombination in other bacteria (Hintz *et al.*, 1995; Zhang *et al.*, 1995). The first chlamydial genome sequence from a *C. trachomatis* serovar D isolate confirmed the presence of *recA* along with other genes encoding for DNA repair and recombination systems (Stephens *et al.*, 1998a). Subsequent annotations of the genomes of all sequenced *Chlamydia* species have come to the same overall conclusion – *Chlamydia* should be capable of recombination if foreign DNA could be acquired by the bacterium.

Multiple studies using large collections of clinical isolates during the early 2000s continued to show strong evidence of recombination, including recombination outside of the *ompA* locus (Gomes *et al.*, 2004; Gomes *et al.*, 2007; Lysen *et al.*, 2004; Millman *et al.*, 2004; Millman *et al.*, 2001; Read *et al.*, 2000a; Read *et al.*, 2003). The advent of next generation sequencing technology, allowing for analysis of complete genomes from numerous strains, has further documented the fairly robust recombinatorial capacity of *C. trachomatis* (Hadfield *et al.*, 2017; Harris *et al.*, 2012). And finally, in 2007, Demars *et al.* described recombination between *Chlamydia* strains *in vitro* (DeMars *et al.*, 2007). Following co-infection experiments with *C. trachomatis* serovar L1 strains marked with chromosomal point mutations conferring resistance to different antibiotics, the DeMars group reported obtaining dual-resistant strains at rates of ~10^4 higher than could be obtained from infection with a single strain. Their case was proven by subsequent co-infection experiments using L1 and D serovars followed by whole genome sequencing of recombinant, dual-resistant progeny (DeMars and Weinfurter, 2008). Sequencing revealed transfer of regions of DNA in excess of 123 bp, which amounts to ~10% of the genome. As these strains lack bacteriophages, the authors proposed conjugation or transformation as the most likely mechanisms with the latter being the most probable based upon genome analyses, which show an absence of conjugation-related genes. Cross-species DNA transfer was demonstrated in 2009 when Suchland *et al.* showed that co-infections with a tetracycline-resistant *Chlamydia suis* (*tet*(C) marker) strain and *C. muridarum* or *C. trachomatis* could give rise to tetracycline-resistant, *tet*(C)-positive hybrid *C. muridarum* and *C. trachomatis* strains (Suchland *et al.*, 2009). Interestingly, tetracycline-resistant recombinant *C. caviae* strains could not be obtained following co-infection, possibly hindered by the sequence divergence of *C. suis* and *C. caviae* compared to the highly interrelated *C. suis/muridarum/trachomatis* (Donati *et al.*, 2014). The *in vitro* recombination experiments devised in these studies have been subsequently leveraged by numerous groups to obtain chlamydial mutants and have proven particularly important for studies developing mutants through high-content chemical mutagenesis (Chen *et al.*, 2014; Nguyen and Valdivia, 2012).

Early successes in genetics

Chemical mutagenesis

Despite their well-known genetic intractability, there were some early successes in genetic modification of the Chlamydiae. In 1983, the first successful mutagenesis of a chlamydial species was reported. Temperature sensitive mutants of *Chlamydia psittaci* (now reclassified as *Chlamydia abortus*) were identified after N-Methyl-N'-Nitro-N-Nitrosoguanidine (NTG) mutagenesis (Rodolakis and Souriau, 1983), and inoculation of pregnant ewes with one of these mutants was shown to give long-lasting protection against spontaneous abortion, resulting in the development of an effective live vaccine that is still used today (genome sequence reported by Burall *et al.*, 2009). However, no further reports of mutagenesis of *Chlamydia* were made until 2011, by which time sufficient genome analysis techniques were available to allow the potential of chemical mutagenesis in the Chlamydiae to be fully realized (Kari *et al.*, 2011). Using an approach known as TILLING (targeting-induced local lesions in genomes) adopted from plant genetics, Kari *et al.* used low-level ethyl methanesulfonate (EMS) mutagenesis to generate Chlamydiae that contained less than one mutation per genome. Target regions were amplified by PCR and the amplicons were heat-denatured, slowly re-annealed, and then digested by the endonuclease CEL-1, which cuts DNA at the 3′ end of a mismatched DNA strand, enabling identification of mutated loci. This study produced the first library of *Chlamydia* mutants that could be screened for phenotypic changes using a reverse-genetics approach. As proof of principle, the authors identified a *trpB* null mutant, and showed that it was unable to synthesize tryptophan from the precursor indole (Kari *et al.*, 2011). *de novo* sequencing of the mutant showed that no other mutations had occurred, indicating the phenotypic change was due to the single point mutation in *trpB*. The ability to compare known mutants with their isogenic partner was a significant step forwards in chlamydial genetics, particularly given the difficulties experienced by those trying to transform *Chlamydia*. The TILLING report was shortly followed by a second chemical-based mutagenesis study that leveraged EMS to generate a higher content mutant library (3–20 transitions per genome) and which was screened for mutants with altered glycogen-production morphology (Nguyen and Valdivia, 2012). In this study, natural lateral gene transfer via back-crossing with the wild type strain *in vitro*, as pioneered by Demars *et al.* (DeMars *et al.*, 2007), was used to isolate the gene mutations responsible for the phenotype. Collectively, researchers now had two approaches for making genotype/phenotype connections, albeit in a non-targeted and laborious fashion requiring an identifiable phenotype and genome sequencing of strains.

Vector-based transformation approaches

The first apparently successful transformation was reported in 1994 when Tam and co-workers designed a prototype chlamydial shuttle vector, pPBW100, and used electroporation to introduce it into *C. trachomatis* EBs (Tam *et al.*, 1994). The plasmid pPBW100 consists of the 7.5 kb chlamydial plasmid, from a urogenital serovar E strain of *C. trachomatis*, fused to the *E. coli* plasmid pBGS9, which has the pBR322 origin of replication and both β-lactam and kanamycin resistance genes. Additionally, a chloramphenicol resistance gene (chloramphenicol acetyl transferase [*cat*]) expressed from a chlamydial promoter was incorporated for selection of transformed *C. trachomatis*. Although chloramphenicol-resistant inclusions were identified, these were transient and typically lost after four passages. The authors suggested that this may have been due to incompatibility of the *E. coli* and chlamydial plasmids used to construct the shuttle vector, activation of a restriction/modification system resulting in degradation of pPBW100 or to regulation of the chlamydial promoter for *cat*, resulting in insufficient expression to achieve stability. Current information on plasmid maintenance indicates that the deletion in pBW100 between the *PstI* and *BglII* sites used to construct the shuttle vector resulted in the loss of ~500 bp from *pgp1* (CDS3), which is predicted to be a helicase and we now know is essential for plasmid maintenance (Liu *et al.*, 2014; Song *et al.*, 2013). Nevertheless, this work demonstrated that *Chlamydia* could accept and, at least transiently, express proteins from exogenous plasmid DNA.

Fifteen years later, transformation via electroporation was again reported, but this time for *C. psittaci* strain 6BC (Binet and Maurelli, 2009). The single copy 16S rRNA gene of *C. psittaci* was cloned into a pUC-derivative vector containing a *cat* chloramphenicol-resistance gene and propagated in *E. coli*. The 16S rRNA gene was modified to contain two point mutations conferring spectinomycin (C1192T) and kasugamycin (A794G) resistance. Two additional non-selective mutations (C1071T and A72G [23S rRNA]) were included in the recombination template that had never been described in *C. psittaci* before to exclude the possibility of spontaneous emergence of resistance. The vector DNA was then transformed into *C. psittaci* at various concentrations, lengths of homology, and in both circular and linearized formats. They found that the most successful combination was 10 μg of circular plasmid DNA prepared in a DNA methylase-deficient *E. coli*. The high degree of homology between the 16S rRNA gene carried by the *E. coli* plasmids and that of the *C. psittaci* chromosome likely enabled allelic exchange (although gene conversion remains a possibility), which was easily detected by the selection of stable spectinomycin and kasugamycin resistant strains. The plasmid bearing the recombining sequence did not contain a chlamydial plasmid backbone and was lost within a single passage, approximating an early suicide vector – thus resistance could only be attributed to chromosomal modification. This was the first application of a transformation protocol to genetically manipulate *Chlamydia* in a stable manner, but the method has not been adopted, presumably due to the need for high level containment for *C. psittaci* and the limited application of such a system where mutant selection relies on generation of a drug-resistant strain by replacement of a wild-type allele. The ability to introduce a stable, replicating plasmid to a chlamydial species remained elusive.

The transformation barrier was finally breached in 2011 when a new shuttle vector pBR325::L2 was devised (Wang *et al.*, 2011). Similar to pPBW100, this new vector utilized the *C. trachomatis* native plasmid to carry *E. coli*-derived genes into EBs; however, in this vector the *pgp1* gene was uninterrupted. The transformation protocol was greatly simplified by using a modification of the standard calcium-chloride based chemical transformation protocol frequently used in other bacteria, negating the need for any specialized electroporation equipment. An initial, selection-free passage was carried out to allow EBs to recover from the transformation treatment, and then penicillin was used to select transformants. As discussed above, any untransformed *Chlamydia* grew aberrantly, consistent with penicillin-induced persistence (Matsumoto and Manire, 1970), resulting in large, non-infectious RBs which did not differentiate into EBs and thus were lost upon passage. Conversely, transformed *C. trachomatis* progressed along the normal developmental pathway, resulting in the formation of EBs which could be passaged many times. Identification of transformants required at least two passages under selection, due to the low transformation efficiency of the system – a limitation that remains today. Nevertheless, transformants expressed both β lactamase and chloramphenicol acetyl transferase from *E. coli* promoters and were stable over at least eight passages.

A second shuttle vector, pGFP::SW2, was constructed which consisted of the SW2 (new variant from a *C. trachomatis* serovar E strain with a 377 bp deletion in CDS1) plasmid backbone, and additionally contained the engineered red-shifted green fluorescent protein (RSGFP) gene, fused to CAT and expressed from a constitutive neisserial promoter (Wang *et al.*, 2011). This was a significant advance as it allowed rapid screening for the presence of transformants by fluorescence microscopy (Fig. 11.1). It also proved that plasmid construction and stable transformation could be performed with different plasmid backbones and in different strains. This study provided invaluable insights into chlamydial genetics. Firstly, it showed that the native plasmid is eliminated by the presence of the competing shuttle vector which provides an advantage under selection, suggesting that only a single extrachromosomal replicon can be maintained at any one time. Secondly, proteins of non-chlamydial origin can be expressed from foreign promoters in *C. trachomatis*. This is important as it suggested that genes could be placed under control of previously

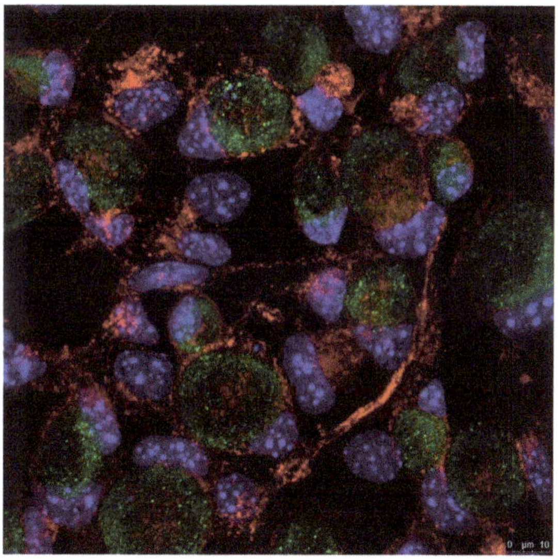

Figure 11.1 Confocal microscope image of mature **C. trachomatis* SWFP- transformed with plasmid pSW2::GFP in infected McCoy cells at 48 hours post infection. Inclusions fluoresce green and the cells were counter-stained using wheat germ agglutinin Alexa Fluor® 594 staining the plasma membrane (red) and DAPI to identify the cell nucleus (blue). Use of fluorescent-protein expressing strains has opened up avenues for real-time analysis of chlamydial growth and infection, monitoring of transformation success, and even facilitates titering in the absence of costly antibodies. **C. trachomatis* SWFP- (Wang *et al.*, 2013) is a naturally occurring plasmid-free urogenital tract serovar F isolate from Sweden that is widely used as a recipient host for transformation with chlamydial plasmids.

characterized inducible promoters indicating that, while phylogenetically divergent, *Chlamydia* follow the basic 'rules' of regulation. Thirdly, prior studies had found that plasmid-minus strains of *C. trachomatis* and *C. muridarum*, unlike typical plasmid-positive isolates, do not produce glycogen (Carlson *et al.*, 2008; O'Connell *et al.*, 2007). The Wang *et al.* study proved that restoration of the plasmid in a naturally plasmid-free strain restored the glycogen-accumulation phenotype and thus confirmed the role of the plasmid in regulation of carbon storage (Wang *et al.*, 2011).

Gerard *et al.* (Gerard *et al.*, 2013) also reported successful transformation of Chlamydiae, although they used an approach in which a dendrimer-enabled system helped to introduce the vector. Dendrimers are synthetic molecules that have shown promise in gene delivery methodologies, and of particular relevance Poly(amidoamine) (PAMAM) dendrimers have been used to deliver plasmid DNA during mouse transfection experiments (Maruyama-Tabata *et al.*, 2000; Navarro and Tros de Ilarduya, 2009). The authors constructed a vector based on an *E. coli*/yeast shuttle vector to enable plasmid construction and amplification in *Saccharomyces cerevisiae* and combined the shuttle vector with the native *C. trachomatis* L2 plasmid with and without the *gfp* gene to yield plasmids pGFP-EG(KG) and pCL2-EG(KG), respectively. In a departure from standard cloning protocols, the authors removed all antibiotic resistance genes from these vectors. Dendrimer–plasmid complexes were then incubated with HEp-2 cells infected with *Chlamydia pneumoniae* strain AR-39, which does not have an endogenous plasmid. High transformation efficiencies (at around 80%) were reported, but it was not possible to eliminate the background of untransformed *Chlamydia* using this vector due to the lack of selective markers, which is a major limitation. Infection of cells via limiting dilution with mixed cultures and/or fluorescence-activated cell sorting (FACS) could potentially have circumvented this background problem. Additionally, it is now known that tropism barriers exist between chlamydial species making it unclear if the *C. trachomatis* plasmid could be stably maintained in other species at all. When *C. muridarum* was transformed with a plasmid of *C. trachomatis* origin, acquisition of a stable cross-species plasmid mutant was dependent upon recombination between the gene CDS2 from the native plasmid with the incoming plasmid's CDS2 creating a hybrid vector (Wang *et al.*, 2014). There have been no independent reports of the dendrimer methodology being repeated in *Chlamydia* elsewhere at the time of writing, perhaps indicating the difficulty in formulating the dendrimer–DNA complexes. While a similar approach has been reported for another obligate intracellular bacterium, *Anaplasma phagocytophilum* (Oki *et al.*, 2015), chemical transformation with CaCl2 remains the gold standard for introducing DNA into *Chlamydia*.

Current genetic tools

Due to the efforts described in the preceding sections, chlamydial researchers are now poised to tackle fundamental biological questions often beginning with the 'simple' decision of how they wish to manipulate the bacterium to address their hypothesis. While a set of genetic methods now exists including an array of shuttle vectors, transposon and intron-based insertion mutagenesis methods, CRISPRi, allelic exchange, and Cre-lox gene deletions, many limitations remain (discussed below).

Although reports for various shuttle vectors and targeted gene inactivation methods have largely been focused on *C. trachomatis* L2, inroads into non-L2 serovars and other species have commenced and are likely to expand as we understand more about plasmid replication and maintenance mechanisms. In addition, chemical mutagenesis approaches, currently reported for *C. abortus* (Rodolakis and Souriau, 1983), *C. muridarum* (Rajaram *et al.*, 2015), and *C. trachomatis* L2 (Kari *et al.*, 2011; Nguyen and Valdivia, 2012), are in theory applicable to all chlamydial species and strains. In combination with complementation approaches enabled via transformation, chemical mutants can be rigorously evaluated for genotype-phenotype linkages. The following sections will highlight the generic shuttle vector systems and the approaches for mutating the chromosome available at the time of publishing. Validated selection markers for recombinant strain construction are listed in Table 11.1 with their description of primary use, and a list of chromosomal mutants constructed via targeted mutagenesis is provided in Table 11.2.

Table 11.1 Antibiotics used to select *Chlamydia* transformed with a plasmid shuttle vector (that replicates in two host species)

Class	Antibiotic	Species/strain	Concentration/ conditions used during selection	Primary reference
β-lactam	Penicillin G*	*C. trachomatis*, L2/434/Bu	10 units/ml[a]	Wang *et al.* (2011)
		C. trachomatis L2P (naturally occurring plasmid free strain) also known as L2R	10 units/ml	Wang *et al.* (2014); Wang *et al.* (2013)
		C. trachomatis SWFP (naturally occurring plasmid free urogenital strain)	10 units/ml	Wang *et al.* (2013)
		C. trachomatis A2497 (plasmid cured ocular strain)	10 units/ml	Song *et al.* (2014)
		C. muridarum Nigg II	10 units/ml	Skilton *et al.* (2018)
	Ampicillin	*C. muridarum*, *Nigg II* (plasmid cured)	5 μg/ml in initial rounds of selection increased to 20 μg/ml	Liu *et al.* (2014)
		C. muridarum, Nigg II *CMUT3* (plasmid cured)	5 μg/ml in initial rounds of selection increased to 20 μg/ml	Gong *et al.* (2013)
		C. trachomatis L2R *C. trachomatis* L2/434/Bu	1 μg/ml in initial round of selection increased to 10 μg/ml	Wickstrum *et al.* (2013)
Nitrobenzenes	Chloramphenicol	*C. trachomatis*, L2/434/Bu	5 μg/ml	Tam *et al.* (1994)
		C. trachomatis, E/UW-5/Cx		Tam *et al.* (1994)
		C. trachomatis L2P- (L2R)	0.1–0.2 μg/ml	Xu *et al.* (2013)
		C. pneumoniae	0.2–0.5 μg/ml	Shima *et al.* (2018)
		LPCoLN (Koala)		
		C. pneumoniae CV-6 (human cardiac)		
		C. pneumoniae IOL207 (human respiratory)		
		C. felis 02DC26		
		C. felis Cello		
		C. felis NI		
Aminoacyl-nucleoside	Blasticidin	*C. trachomatis* D/UW3/Cx	5 μg/ml for 24 hours and then 20 μg/ml	Ding *et al.* (2013)
Aminoglycoside	Spectinomycin	*C. trachomatis* L2 (LGV 434)	500 μg/ml	Mueller *et al.* (2016)

We have listed the primary publication where a specific host and vector combination were used for the first time, with the antibiotic selection specified. In several cases vectors have subsequently been improved and the transformation/selection protocols subject to minor modification (such as incremental increases in antibiotic concentrations when selecting for and propagating transformants). In the interests of clarity these secondary changes have not been included.

[a]Penicillin is measured in internationally recognized units, hence 1 mg pure sodium penicillin G = 1667 units; therefore, 10 units/ml = ~6 μg/ml.

Table 11.2 Chromosomal gene insertion and deletion mutants (*C. trachomatis* L2 LGV 434)

Method	Gene	Selection marker	Primary reference
Group II intron (TargeTron)	*incA*	*bla*	Johnson and Fisher (2013); Weber *et al.* (2016)
	rsbV1 and *incA*	*bla (rsbV1), aadA (incA)*	Lowden *et al.* (2015)
	rsbV1	*aadA*	Thompson *et al.* (2015)
	hsp60_2 and *hsp60_3*	*aadA*	Illingworth *et al.* (2017)
	tepP	*bla*	Carpenter *et al.* (2017)
	cpoS	*bla*	Sixt *et al.* (2017)
	incC, ct229, ct288, ct383, ct449, ct005, ct179, ct224, ct813, ct850	*bla*	Weber *et al.* (2017)
	ct288	*aadA*	Almeida *et al.* (2018)
	ct622	*aadA*	Cosse *et al.* (2018)
	ct228	*aadA*	Shaw *et al.* (2018)
	mrcA	*bla*	Nguyen *et al.* (2018)
Transposon	*cdu1*	*bla*	Fischer *et al.* (2017)
Allelic exchange (FRAEM)	*trpA, ctl0063 (tmeA), ctl0064, ctl0065*	*bla*	Mueller *et al.* (2016)
FRAEM with Cre lox	*tmeA*	*bla*	Keb *et al.* (2018)

Shuttle vectors to introduce genes into *Chlamydia*

The currently available shuttle vectors have been extensively, and recently, reviewed elsewhere (Bastidas and Valdivia, 2016) and so we will focus on broader themes pertaining to vector construction and vector/host considerations when designing experiments. While most *Chlamydia* species encode a highly conserved native plasmid, the difficulty in genetically manipulating and growing the organism has led researchers to perform plasmid construction primarily using *E. coli*. Numerous studies have shown that the chlamydial plasmid impacts virulence through both direct contributions from plasmid-encoded genes and through regulation of chromosomal genes (Carlson *et al.*, 2008). Thus, one of two approaches are necessary when designing a vector: either the chlamydial plasmid must be included as part of the final vector backbone or a vector in a different incompatibility group must be used. As non-chlamydial plasmids that will replicate in *Chlamydia* are not known, all reported replication vectors include a combination of the chlamydial plasmid with an *E. coli* compatible vector. This combination leads to vector constructs that are large, often exceeding 10 kb, which makes cloning work challenging. While the upwards size limit of a shuttle vector in terms of what *Chlamydia* can stably maintain is unknown, increased plasmid size will likely decrease what is already a poor transformation efficiency (Hanahan, 1983). Construction of a mini-replicon chlamydial plasmid backbone containing the essential plasmid maintenance genes *pgp*1, 2, 6, and 8 (coding sequence only; Gong *et al.*, 2013; Song *et al.*, 2013) would reduce plasmid size, but at the expense of the plasmid-encoded virulence genes.

Use of chlamydial plasmid-based shuttle vectors eventually leads to loss of the native plasmid either through competition by selection of the recombinant plasmid or by recombination between the incoming shuttle vector and the native plasmid (Wang *et al.*, 2011). Use of natural plasmid-less clinical isolates (Wang *et al.*, 2013) or lab-derived plasmid-cured strains (Skilton *et al.*, 2018) for transformation alleviates the concern about plasmid-recombination although one is then assuming that a plasmid-free strain can stably maintain the incoming shuttle vector. Another important consideration when choosing the strain for transformation comes from recent studies that have shown that for the greatest likelihood of success the chlamydial plasmid backbone should match the species/serovar destined for transformation to ensure plasmid maintenance (Song *et al.*, 2014; Wang *et al.*, 2014). However, at least within *C. trachomatis*, the SW2 plasmid-derived vector appears to have a universal *C. trachomatis* host-strain spectrum in that it can transform and replicate in all biovars (Wang *et al*, 2011; O'Neill *et al.*, 2018).

While most studies currently employ an L2 plasmid backbone, limited successes with plasmid-based genetics in other species and serovars indicate that the foreign (non-chlamydial) regulatory elements and genes used in L2 genetics will function outside of L2 strains (but might be restricted to *C. trachomatis*; Wang *et al.*, 2014). For those wishing to embark upon chlamydial genetics research, our recommendation would be to start with the most tractable system, i.e. transformation of a plasmid-free *C. trachomatis* L2 isolate with one of the GFP-expressing shuttle vectors. From a host-cell consideration, cell line usage appears to vary amongst labs and reflects the cell line typically used by the individual research group. An important point to keep in mind when working with the shuttle vectors is that the cloned genes of interest will be present in multiple copies reflecting the plasmid copy number and non-native expression will occur if the gene is cloned behind a foreign promoter. Both situations will lead to gene dosage effects.

A wide range of shuttle vectors are now employed in the field, and calcium-chloride mediated transformation is the method of choice across labs for introducing vectors. The plasmids offer the ability to express foreign proteins including a variety of fluorescent proteins (Agaisse and Derre, 2013), which are useful for isolating transformants and following bacterial development. Ectopic expression of tagged proteins has also been used to study chlamydial protein secretion and to identify type III secretion system substrates (Bauler and Hackstadt, 2014; Mueller and Fields, 2015; Wang *et al.*, 2018). A variety of multiple cloning sites (MCS) are provided across the vectors, although restriction enzyme-independent modern cloning techniques are largely making these classic vector elements optional. In addition, the expression of a gene can be controlled by using its native promoter, constitutive chlamydial or foreign promoters (exemplified by selection cassettes), or via the inducible Tet-On expression system. For the Tet-On system, gene expression is induced by addition of anhydrotetracycline (ATc) to remove the TetR repressor from the gene operator site (Wickstrum *et al.*, 2013). While effective, Tet-regulation is not a binary system and promoter leakiness as well as inclusion-to-inclusion induction variability has been observed (Ouellette, 2018, and unpublished observations), possibly due to vector design, including alternative numbers of TetR operator sites and their respective location to the promoter and/or the metabolic state of the bacterium (Wickstrum *et al.*, 2013). In addition, some fetal bovine serum (FBS) batches contain low levels of tetracycline due to commercial use in the cattle industry (tetracycline-free FBS can be purchased).

Approaches to constructing chromosomal mutants

Unlike the relative wealth of shuttle vector options for *Chlamydia*, approaches for generating mutations in the chromosome are more limited and labour intensive. In addition, there are likely to be a limited number of nonessential genes in *Chlamydia* because it is an obligate intracellular bacterium with a reduced genome. Thus, the failure to obtain a chromosomal mutant may be difficult to interpret – did the approach fail or is the gene essential? A recently described inducible shuttle-vector based CRISPR interference (CRISPRi) system for *C. trachomatis* may allow the study of essentiality and essential gene function to be addressed through gene knockdown (Ouellette, 2018). In this approach, a catalytically inactive version of the nuclease Cas9, known as dCas9, is targeted to the 5′ end of the gene of interest by a guide RNA (gRNA) molecule. Regulated co-expression of the dCas9 and gRNA from the shuttle vector leads to binding of dCas9 to the 5′ end of the target gene resulting in steric hindrance of the transcriptional machinery and loss of gene expression.

Random mutagenesis approaches using the chemical mutagens EMS, NTG, and/or N-ethyl-N-nitrosourea (ENU) have been used to make libraries of *Chlamydia* carrying 'random' mutations. At least three large collections of chemical mutants have been reported, and the phenotypes explored have included glycogen production, plaque morphology, and temperature sensitive growth (Brothwell *et al.*, 2016; Kokes *et al.*, 2015; Nguyen and Valdivia, 2012). While highly useful, the random nature of the approach and the introduction of multiple mutations per genome necessitate a fair amount of 'back-end' work. Full genome sequencing and strain crosses are a prerequisite for experimental rigour that are not feasible for all laboratories. A recent report of a Himar1 transposon-mutant (*cdu1*::Tn *bla*) (Fischer *et al.*, 2017) suggest that a new tool for random mutagenesis is on the horizon for both *C. trachomatis* and *C. muridarum*, but experimental details are lacking for this approach.

While the function of chlamydial plasmid genes have been individually queried by construction of shuttle vectors lacking each CDS ('drop-out' cloning) (Gong *et al.*, 2013; Song *et al.*, 2013), limited studies to date have reported generation of targeted chromosomal mutations. Currently, two approaches are available: intron-based insertional mutagenesis (Johnson and Fisher, 2013) or gene deletion via allelic-exchange (Fluorescence-reported allelic exchange mutagenesis or FRAEM) (Mueller *et al.*, 2016).

Intron mutagenesis: The intron-based approach results in an insertional-mutant. This method utilizes a group II intron that inserts into genes via a retrotransposition mechanism that uses site-specific recognition (which can be modified) between the intron and the target sequence leading to intron insertion at the targeted site. This approach was initially described by the Lambowitz group (Zhong *et al.*, 2003) and marketed by Sigma for other bacterial species as TargeTron™. It was adapted for use in *Chlamydia* by placing a chlamydial promoter from CT398 (CdsZ), (a constitutive gene; Barta *et al.*, 2015) upstream of the intron, which contained the penicillin resistance gene *blaZ* (Johnson and Fisher, 2013). This approach, with different selection markers, has been used successfully to inactivate multiple genes in *C. trachomatis* (Table 11.2). While Sigma has stopped providing the algorithm needed to re-target the intron, the company Targetronics, LLC provides a similar insertion prediction service and in theory one could design an in-house algorithm (Perutka *et al.*, 2004). Similar to the vector devised by Binet and Maurelli (Binet and Maurelli, 2009), Johnson and Fisher (Johnson and Fisher, 2013) did not incorporate the chlamydial plasmid in their vector design, and it was therefore lost upon passage (i.e. it is not a shuttle vector), enabling ampicillin selection for only those bacteria in which the intron had inserted. While off-target intron insertions for *C. trachomatis* have not been reported, whole genome sequencing or Southern blotting could be performed to ensure a single intron insertion.

FRAEM: Classical allelic exchange, known to be feasible based on documented natural recombination in *Chlamydia*, was first reported in 2016 (Mueller *et al.*, 2016). The Fields group developed a conditional lethal shuttle vector where the essential *pgp6* gene was under the control of the Tet-On system. The vector encoded both *mCherry* and *gfp* along with the *bla* gene flanked by the 5′ and 3′ sequences of the gene targeted for deletion. Upon initial transformation in the presence of ATc, *C. trachomatis* L2 transformants would appear as penicillin-resistant bacteria that show red and green fluorescence. The removal of ATc leads to loss of *pgp6* expression and subsequent loss of the plasmid and resistance to penicillin unless the plasmid integrates into the chromosome via homologous recombination at the targeted gene. A double crossover event would then generate a gene deletion mutant resistant to penicillin and maintaining green fluorescence. Hence deletion mutants can be differentiated from single-crossover mutants by their fluorescence profile. The advantage of such an approach over a classical suicide vector approach for allelic exchange is that it allows for separation of the transformation event from the recombination event, which is significant when trouble shooting genetics in *Chlamydia*. The *pgp6*-inducible shuttle vector system was also shown to have utility for generating plasmid-free strains. In the presence of penicillin selection, the native plasmid would be lost with the shuttle vector being maintained. Following removal of ATc, the shuttle vector was no longer maintained leading to formation of a plasmid-free strain (Table 11.2).

Future goals and remaining challenges

While the age of chlamydial genetics is clearly upon us, critical needs remain and future goals from a genetics standpoint should include the construction of improved tools (such as marker recycling approaches) and broad application of those tools across *Chlamydia* species as the majority of genetic efforts have been applied to *C. trachomatis* L2. We have found that there are species- and strain-based differences in selection robustness (unpublished) requiring analysis of drug-selection conditions for each strain prior to transformation experiments. In addition, there are public-health based restrictions on selection marker usage (*bla* usage is restricted in *C. trachomatis* serovars D-K in the United States). As already discussed for transformation of *C. muridarum* with a *C. trachomatis*-based shuttle vector (Wang *et al.*, 2014), species/plasmid host-range problems exist and the majority of shuttle vectors are built from the L2 plasmid. And perhaps the greatest barrier remains the poor efficiency of transformation, which seems to differ between species and strains (unpublished, personal experience), and remains difficult to quantify in the absence of axenic culturing conditions.

As the genetic landscape has opened up, the field has been constrained by the number of selection markers available, limiting our ability to construct chromosomal mutants with multiple gene disruptions in the presence of complementing vectors. Based on publication numbers and our experience, the *bla* and the aminoglycoside 3′-adenyltransferase *aadA* markers (and to a lesser extent, *cat*) are fairly robust making them the current field 'favourites' for selection of vectors and gene deletion/insertion mutants. The ability to recycle selection markers by deleting them after mutant selection, allowing for repeated use of a marker within the same strain would be a boon to researchers. The Cre-*lox* (Dale and Ow, 1991; Marx and Lidstrom, 2002) system has been reported by the Fields group for use with their FRAEM system (Keb *et al.*, 2018) and the flippase-flippase recognition target (Flp-FRT) system (Posfai *et al.*, 1994) could also be adapted for use in *Chlamydia*. In addition to marker recycling, these systems can also be used for genome editing approaches where genome sections are removed following the insertion of the appropriate Cre/Flp-recognized DNA sequences at sites flanking the genes of interest (Enyeart *et al.*, 2013). Ideally, these recombination systems would be employed from a conditional-suicide vector such as that described by the Fields group for use in FRAEM (Mueller *et al.*, 2016).

The stream-lined genome and the obligate intracellular nature of *Chlamydia* also presents a challenge as many genes are likely essential for chlamydial growth. Inactivation of an essential gene would be lethal. Systems for generating conditional lethal mutants can overcome this limitation by engineering a situation where the gene can be turned off in a controlled manner allowing for analysis of the condition-dependent null mutant. Conditional-lethal mutants can be obtained using shuttle vectors with the target gene 'A' behind a regulatable promoter in a chromosomal 'A' mutant background, by gene knockdown via siRNA (Mishra *et al.*, 2012) or CRISPRi (Ouellette, 2018), and/or using regulatable protein inactivation via expression of RNA aptamers (Chen *et al.*, 2015). While a targeted, conditional lethal gene mutant has yet to be reported, a substantial collection of temperature-sensitive mutants has been derived using chemical mutagenesis (Brothwell *et al.*, 2016). Currently, the method of generating chemically derived temperature-sensitive mutants is not easily adaptable for the field at-large and will not allow for the targeted interrogation of a selected gene.

An additional draw-back for strain construction is the size of available *E. coli-Chlamydia* shuttle vectors. The current collection generally exceeds 10 kb prior to insertion of study-related genetic elements making cloning cumbersome and effort-intensive and potentially reducing transformation efficiency. The field would benefit from the development of smaller shuttle-vectors (~5 kb) commonly employed in other bacterial genetic systems. Complicating the design of smaller vectors is our limited understanding of plasmid replication and plasmid incompatibility grouping for *Chlamydia*, and the need to maintain the native plasmid for full virulence in *C. trachomatis* and *C. muridarum*. Screening of broad host range vectors for maintenance in chlamydial species could facilitate discovery of vectors that would be easier to manipulate at the bench leading to more efficient construction of recombinant strains.

Collectively, the current state of genetics has taught us that approaches successful in other bacteria are likely applicable to *Chlamydia* following some species-specific modifications. Consequently, mini-Tn7 gene cargo insertions (Bao *et al.*, 1991; Choi *et al.*, 2005), recombineering (Datta *et al.*, 2006), CRISPR-Cas9 (Jiang *et al.*, 2013), Tn-seq (van Opijnen *et al.*, 2009), etc. should be possible. Unfortunately, transformation efficiency is extremely poor and robust methods for selecting or screening for mutants following the initial round of transformation are lacking. These limitations are especially troublesome for transposon approaches where a high number of mutants are typically desired for downstream experiments. While high transformation efficiency (~80%) was found for DNA complexed with a PAMAM-dendrimer, the use of this method in *Chlamydia* has only been reported by a single lab (Gerard *et al.*, 2013). To improve DNA introduction efficiency, further insights into the physiological processes mediating chlamydial transformation and/or alternative methods of DNA delivery such as transduction, discussed below, are needed. As for improved methods to quickly identify and obtain clones, fluorescent reporter systems coupled with single-cell isolation approaches such as FACS or microscopy-guided picking of single inclusions would appear to hold promise. Development of axenic culturing conditions would be ideal, as demonstrated for the selection of *Coxiella burnetii* mutants (Beare *et al.*, 2012), but for a variety of reasons axenic culturing of *Chlamydia* presents a tremendously difficult challenge (Omsland *et al.*, 2012) and this protocol remains to be repeated independently.

Genetics outside *C. trachomatis*

Plasmid based transformation studies have been extended to *C. muridarum* (Liu *et al.*, 2014; Song *et al.*, 2014; Wang *et al.*, 2014). Interestingly, plasmid-free *C. muridarum* cannot be transformed by the *C. trachomatis* shuttle vectors (Wang *et al.*, 2014). This observation was initially presented as an example of 'transformation tropism' i.e. that there is a physical barrier to DNA transforming *C. muridarum* with the *C. trachomatis* plasmid (Song *et al.*, 2014). However, it has been shown that by using the *C. trachomatis* shuttle vector pSW2GFP to transform wild-type, plasmid-bearing *C. muridarum* it was possible to select transformants resistant to penicillin (Wang *et al.*, 2014). In these transformants, recombination had occurred between the native *C. muridarum* plasmid, pNigg, and the *C. trachomatis* shuttle vector. Mapping and sequencing of the plasmids present in the recombinant strains that survived selection allowed identification of the minimal sequence needed to allow replication of the *C. trachomatis* shuttle plasmid pSW2GFP in *C. muridarum.* The transforming *C. trachomatis* plasmid had exchanged an approximately 1 kb region with pNigg covering CDS2 and short adjacent sequences. The resultant plasmid pSW2NiggCDS2 could be maintained under penicillin selection when transformed into plasmid-cured *C. muridarum.* These experiments prove that (1) the concept of a 'transformation tropism' is incorrect and that wild type, plasmid-bearing *C. muridarum* can acquire penicillin resistance through the *C. trachomatis* shuttle plasmid pSW2GFP, and (2) that the tropism barrier is not at transformation, but at either plasmid replication and/or maintenance. We do not know the molecular mechanisms that prevent replication of the chlamydial plasmid from one chlamydial species in another. It is possible that there are strict species-level plasmid requirements such as the need for additional replication proteins supplied by the host, but not enough information exists to make such conclusions.

Recently the transformation of *C. pneumoniae* was described (Shima *et al.*, 2018); this approach was based on the plasmid-mediated calcium chloride-based transformation system described for *Chlamydia trachomatis* (Wang *et al.*, 2011). A plasmid shuttle vector that was derived from the equine *C. pneumoniae* isolate N16 (Storey *et al.*, 1993), from the UK, was engineered in a similar fashion to the *C. trachomatis* shuttle vectors except that selection of transformants was performed with chloramphenicol (Table 11.1). Using this equine *C. pneumoniae* plasmid construct, it was first possible to transform an Australian koala *C. pneumoniae* isolate. All tested human isolates of *C. pneumoniae* (which are naturally plasmid free) could also be stably transformed by using the same plasmid shuttle vector (Shima *et al.*, 2018). Moreover, it was possible to overcome the cross–species barrier to genetic transmission (observed in *C. trachomatis*) between several isolates of two chlamydial species (*C. pneumoniae* and *C. felis*) without the need for plasmid recombination (Shima *et al.*, 2018).

Importantly, the results were replicated in two independent laboratories in Lübeck/Germany and Southampton/UK, working with two genetically different human *C. pneumoniae* isolates CV-6 (cardiac) and IOL-207 (respiratory) (Shima *et al.*, 2018). In addition, the work incorporated and built upon remarkable veterinary fieldwork performed over the past three decades making use of rare *C. pneumoniae* isolates from the UK and Australia, which have been used to develop the shuttle vector and refine transformation protocols (Shima *et al.*, 2018). This work underscores the importance of archiving and preserving chlamydial isolates, a practice which is being lost in the age of high-throughput genomics and nucleic acid amplification testing of clinical specimens *in lieu* of classical culturing methods.

Bacteriophage

As previously discussed, the remaining obstacle to the application of genetic technology in *Chlamydia* is the low transformation frequency of EBs. Bacteriophage may offer an alternative and potentially attractive option to the plasmid for developing low copy vectors and for obtaining higher transformation frequencies by using transduction approaches i.e. packaging and infection. The developmental cycle allows the growing RBs to occupy an econiche that is isolated from the world of free-living bacteria and which shields them from exposure to most bacteriophages in the environment. Nevertheless, bacteriophages that infect *Chlamydia* species (the so-called 'Chp-like' phages) have been described. These phages all belong to the virus family *Microviridae* for which the prototype is the very well characterized T = 1 bacteriophage ΦX174 (Fane *et al.*, 2006). The *Microviridae* has a deep evolutionary divide between the ΦX174-like phages and the Chp-like phages, although genome structure and replication strategies are conserved (Brentlinger *et al.*, 2002). A large number of related viruses have now been described mainly from metagenomic studies and this latter group, including the Chp-like phages, is collectively known as the *Gokushovirinae* (Doore and Fane, 2016).

The first 'chlamydiaphage' to be characterized, named Chp1, was discovered by chance during transmission EM analysis of cells infected with *C. psittaci* from an outbreak of ornithosis in ducks (Bacon *et al.*, 1986). Chp1 replicates in the RBs of *C. psittaci* and is a 22 nm icosahedral virus with a buoyant density of 1.37 g/ml (Storey *et al.*, 1989a). Unusually, it forms crystalline arrays within RBs causing them to distend, preventing EB formation. Phage-infected *C. psittaci* contain double stranded (ds) phage DNA that is thought to be the replicative form (RF) of Chp1 (Storey *et al.*, 1989a). Hybridization studies with the Chp1 genome showed that the phage dsDNA does not integrate into the chlamydial chromosome. Polyacrylamide gel electrophoresis of purified Chp1 revealed three major structural proteins VP1, VP2 and VP3, with molecular weights of 75, 30 and 16.5 kDa, respectively (Storey *et al.*, 1989b). The genome of Chp1 is 4,877 nucleotides encoding five major open reading frames (ORFs). These ORFs all start with an ATG codon that in each case is preceded by a ribosome binding site. N-terminal sequencing of purified VP1, VP2 and VP3 allowed their respective open reading frames in the genome sequence to be assigned. The mechanisms and techniques to package DNA in the prototype *E. coli* bacteriophage ΦFX174 are well defined (Fane *et al.*, 2006), so it should be possible to develop similar technology for the Chp-like phages enabling high frequency delivery of *in vitro* packaged DNA by transduction. Unfortunately, Chp1 is no longer available because it was lost in a freezer breakdown (C. Storey – personal communication).

A set of phages related to Chp1 (ΦFCPG1, ΦFCPAR39, Chp2, Chp3, Chp4) has been discovered (Garner *et al.*, 2004; Hsia *et al.*, 2000; Liu *et al.*, 2000; Read *et al.*, 2000b; Sait *et al.*, 2011), of which the best characterized is Chp2 that infects *C. abortus*. EM of thin sections of *Chlamydia*-infected cells showed that Chp2 particles were located to membranous structures surrounding RBs which appeared to be lysed suggesting that Chp2 is cytopathic for RBs (Everson *et al.*, 2002). Chp2 ds circular RF DNA was purified and sequenced. The Chp2 genome consists of 4,567 base pairs and encodes five major ORFs with a similar overall organization to the Chp1 genome. Sequence comparisons indicate that VP1, 2, and 3 from Chp1 and Chp2 share amino acid sequence identities of 48.3%, 27.3% and 26%, respectively. Amino acid residues in the ΦFX174 procapsid known to mediate interactions between the viral coat protein and internal scaffolding proteins are conserved in the Chp2 VP1 and VP3 proteins suggesting that VP3 performs a scaffolding-like function, but has evolved into a structural protein (Clarke *et al.*, 2004). Interestingly, ORF4 of ΦFCPAR39 (*C. pneumoniae* phage) is also found integrated into the host chromosome (Read *et al.*, 2000b). This is unusual since *Microviruses* do not have an integrative stage in their replication cycle, so this is likely to be a fortuitous recombination.

While the host range of *Chlamydia* phages has not been extensively studied (Everson *et al.*, 2003), none have been reported to infect *C. trachomatis* or

C. muridarum. The development of simple robust transformation vectors and protocols for two of the main susceptible host species *C. pneumoniae* and *C. felis* will allow detailed studies on the genetics of the chlamydiaphages. Such studies will open up the possibility of generating a second *Chlamydia*-specific replicon in addition to the existing shuttle-vectors, which if successful, will advance our capabilities for genetic transformation. Transduction has been demonstrated with ΦX174; all that is necessary for that system is to identify the stage III origin of replication and to provide procapsids and bacterial extracts (Hayashi M, 1988). Chlamydiaphage procapsids are easy to purify in abundance (Clarke *et al.*, 2004) and construction of a vector incorporating the stage III origin and a selectable marker e.g. *bla* is theoretically possible and would amount to a ~2 kb vector leaving approximately 2.4 kb unused phage genome capacity for packaging foreign DNA that could be delivered with high efficiency by transduction.

What we have learned and moving forward

The genome is 'flexible'

Somewhat surprising given the obligatory nature of *Chlamydia* is the number of genes that have been proven dispensable for growth. At least 19 genes have been inactivated via targeted approaches (Table 11.2) and loss of activity verified using a variety of approaches. While nonsense mutations obtained through chemical mutagenesis have been found in > 84 ORFs, most have not been verified for loss of activity (Kari *et al.*, 2011; Kokes *et al.*, 2015; Rajaram *et al.*, 2015). These findings are unexpected given the reduced and streamlined genomes of these bacteria, which contain few apparent pseudogenes and rather short intergenic regions. Superficially, this would seem to indicate that the genes remaining are essential for bacterial growth and survival. It is certainly possible, and seems likely, that the cell culture conditions used and cell culture in general, impacts the type of mutations that can be obtained. As a consequence, a modest loss of fitness in cell culture is not likely to prevent isolation of said mutant, but could be devastating in the natural host where the immune system, competition from resident microflora, etc. would exist. As of yet, studies comparing the fitness of mutants grown in cell culture versus animal infection models have largely been limited to plasmid positive and plasmid negative strains and a nonsense CPAF mutant (Carlson *et al.*, 2008; O'Connell *et al.*, 2007; Yang *et al.*, 2016). Future work should be performed to determine the set of genes required for survival *in vivo* compared to growth *in vitro*. This information could be used for design of a live attenuated vaccine strain.

Many chlamydial genes appear to be regulated as part of operons, thus it is also important to consider that polar effects (where a mutation in gene A leads to alterations in expression of gene B often due to their presence in an operon) will complicate mutant selection and phenotype analysis. Limited, direct operon mapping in *Chlamydia* has been performed for select genes such as the type III secretion system (Hefty and Stephens, 2007). In the absence of such data, operon structure can be inferred from analysis of gene organization using the annotated genomes or from analysis of data generated by deep sequencing of chlamydial transcriptomes as performed by Albrecht *et. al* (Albrecht *et al.*, 2010) to identify genome-wide transcriptional start sites. Information on gene operon status will prove important for selecting genes to target for inactivation and approaches to generate mutations within operons without inducing polarity will need to be developed for use in *Chlamydia*. Additionally, genes essential for *in vitro* growth will require the use of gene knockdown or conditional lethal approaches. And finally, virulence systems, such as the type III secretion system, that are likely essential for *Chlamydia*'s intracellular growth will be difficult to genetically dissect in the absence of axenic growth conditions.

Concerns as the field moves forward

The need to grow *Chlamydia* in cells in culture brings a high level of variability to studies. Laboratories use different cell culture conditions, media, and cell lines to propagate *Chlamydia*. Even cell cultures that are purportedly the same may have clonal differences. Does this matter when studying *Chlamydia*? In a broader context the field has been aware of and has accommodated this variability since routine isolation in cell culture began some 50 years ago, and it is inevitable that there will be differences between laboratories and studies. However, within the field of genetics the tolerance limits for describing new mutants by comprehensive phenotyping of growth kinetics as well as inclusion morphology are becoming exacting. This is particularly so for subtle and complex phenotypes. Thus, good science needs reproducible and robust methods. The application of good science ensures that research is built on the best possible evidence. The use of reliable and repeatable methods also improves efficiency and confidence in data. Sharing data, resources, and results speeds up the pace of research, advancing discovery and making the field attractive to young scientists. The publication of full data sets in open access journals, allowing other researchers to access all the data, is becoming more common and there are now several journals available where such activities are standard practice.

A new open access database Chlambase.org (https://chlambase.org/; Putman *et al.*, 2019) is a promising resource that fulfils a significant gap in chlamydial bioinformatics and comparative genetics. It has already solved at least one enduring problem, the ability to cross-reference the differently annotated genomes of L2 and D with respect to the numbering of CDSs. While data and annotation incompatibilities are relatively straightforward to resolve, there are different growth and biological characteristics between natural isolates of *Chlamydia* and thus the description and definition of chlamydial growth characteristics are a case in point for accurate determination of the properties of mutants. It is now appropriate to move away from describing phenotypes based on the observation of inclusion morphology at a single time point in the developmental cycle. Growth kinetics based on IFUs in combination with qPCR determination of DNA replication gives a more accurate definition of the mutant phenotype (Skilton *et al.*, 2018). Performing these analyses constitutes a great deal more work, but ultimately quantitative data are more meaningful than the selective use of a handful of infectivity time points.

In addition to random mutagenesis of the genome, it is now possible to make specifically targeted chromosomal gene knock outs and to delete plasmid genes. We are entering a phase in chlamydial genetics when a wealth of possibilities are available, yet there are frequently unrealized pitfalls in assigning specific functions to single gene deletions. Given the complex and compact nature of the chlamydial genome it should be expected that random mutagenesis and single gene deletions may cause genomic changes outside the target areas. These will include compensatory mutations as well as unintended consequences such as polar effects and unknowing deletion of uncharacterized control sequences. Mutations and gene deletions may also weaken the bacterium in different physiological conditions and give rise to unstable progeny. Thus, mutants should be subject to rigorous clonal purification together with complete genome sequence determination and/or complementation, if possible, and any new vector constructs should be provided with their complete sequences rather than vague descriptions. And while the development of plasmid-based transformation systems provides a novel way of investigating gene function by complementation, the plasmid has variable copy number both between strains, during the development cycle, and under different conditions (Pickett *et al.*, 2000; Tam *et al.*, 1992). Thus, altered copy number can lead to over-expression of regulated genes which may cripple a host or could result in gain of function for specific changes leading to confounding phenotypes.

Despite the stated concerns arising from our newfound technologies, the future of chlamydial research in the age of genetics looks exceedingly bright (Fig. 11.1).

References

Agaisse, H., and Derre, I. (2013). A *C. trachomatis* cloning vector and the generation of *C. trachomatis* strains expressing fluorescent proteins under the control of a *C. trachomatis* promoter. PLoS One *8*. e57090.

Albrecht, M., Sharma, C.M., Reinhardt, R., Vogel, J., and Rudel, T. (2010). Deep sequencing-based discovery of the *Chlamydia trachomatis* transcriptome. Nucleic Acids Res *38*, 868–877.

Almeida, F., Luis, M.P., Pereira, I.S., Pais, S.V., and Mota, L.J. (2018). The Human Centrosomal Protein CCDC146 Binds *Chlamydia trachomatis* Inclusion Membrane Protein CT288 and Is Recruited to the Periphery of the *Chlamydia*-Containing Vacuole. Front Cell Infect Microbiol *8*, 254.

Bacon, E.J., Richmond, S.J., Wood, D.J., Stirling, P., Bevan, B.J., and Chalmers, W.S.K. (1986). Serological detection of phage infection in *Chlamydia psittaci* recovered from ducks. The Veterinary Record *119*, 618–620.

Bao, Y., Lies, D.P., Fu, H., and Roberts, G.P. (1991). An improved Tn7-based system for the single-copy insertion of cloned genes into chromosomes of Gram-negative bacteria. Gene *109*, 167–168.

Barta, M.L., Battaile, K.P., Lovell, S., and Hefty, P.S. (2015). Hypothetical protein CT398 (CdsZ) interacts with sigma(54) (RpoN)-holoenzyme and the type III secretion export apparatus in *Chlamydia trachomatis*. Protein Sci *24*, 1617–1632.

Bastidas, R.J., and Valdivia, R.H. (2016). Emancipating *Chlamydia*: Advances in the Genetic Manipulation of a Recalcitrant Intracellular Pathogen. Microbiol Mol Biol Rev *80*, 411–427.

Bauler, L.D., and Hackstadt, T. (2014). Expression and Targeting of Secreted Proteins from *Chlamydia trachomatis*. J Bacteriol *196*, 1325–1334.

Beare, P.A., Larson, C.L., Gilk, S.D., and Heinzen, R.A. (2012). Two systems for targeted gene deletion in *Coxiella burnetii*. Appl Environ Microbiol *78*, 4580–4589.

Bernkopf, H., Mashiah, P., and Becker, Y. (1962). Correlation between morphological and biochemical changes and the appearance of infectivity in FL cell cultures infected with trachoma agent. Annals of the New York Academy of Sciences *98*, 62–81.

Binet, R., and Maurelli, A.T. (2007). Frequency of development and associated physiological cost of azithromycin resistance in *Chlamydia psittaci* 6BC and *C. trachomatis* L2. Antimicrob Agents Chemother *51*, 4267–4275.

Binet, R., and Maurelli, A.T. (2009). Transformation and isolation of allelic exchange mutants of *Chlamydia psittaci* using recombinant DNA introduced by electroporation. Proceedings Of The National Academy Of Sciences Of The United States Of America *106*, 292–297.

Blyth, W.A., and Taverne, J. (1972). Some consequences of the multiple infection of cell cultures by TRIC organisms. The Journal of hygiene *70*, 33–37.

Brentlinger, K.L., Hafenstein, S., Novak, C.R., Fane, B.A., Borgon, R., McKenna, R., and Agbandje-McKenna, M. (2002). *Microviridae*, a family divided: Isolation, characterization and genome sequence of a phi MH2K, a bacteriophage of the obligate intracellular parasitic bacterium *Bdellovibrio bacteriovorus*. J. Bacteriol.*184*, 1089–1094.

Brothwell, J.A., Muramatsu, M.K., Toh, E., Rockey, D.D., Putman, T.E., Barta, M.L., Hefty, P.S., Suchland, R.J., and Nelson, D.E. (2016). Interrogating Genes That Mediate *Chlamydia trachomatis* Survival in Cell Culture Using Conditional Mutants and Recombination. J Bacteriol *198*, 2131–2139.

Brunham, R., Yang, C.L., Maclean, I., Kimani, J., Maitha, G., and Plummer, F. (1994). *Chlamydia-Trachomatis* from Individuals in A Sexually-Transmitted Disease Core Group Exhibit Frequent Sequence Variation in the Major Outer-Membrane Protein (Omp1) Gene. Journal of Clinical Investigation *94*, 458–463.

Burall, L.S., Rodolakis, A., Rekiki, A., Myers, G.S.A., and Bavoil, P.M. (2009). Genomic Analysis of an Attenuated *Chlamydia abortus* Live Vaccine Strain Reveals Defects in Central Metabolism and Surface Proteins. Infection and Immunity *77*, 4161–4167.

Campbell, S., Richmond, S.J., and Yates, P. (1989). The development of *Chlamydia trachomatis* inclusions within the host eukaryotic cell during interphase and mitosis. Journal of general microbiology *135*, 1153–1165.

Carlson, J.H., Whitmire, W.M., Crane, D.D., Wicke, L., Virtaneva, K., Sturdevant, D.E., Kupko, J.J., 3rd, Porcella, S.F., Martinez-Orengo, N., Heinzen, R.A., *et al.* (2008). The *Chlamydia trachomatis* plasmid is a transcriptional regulator of chromosomal genes and a virulence factor. Infect Immun *76*, 2273–2283.

Carpenter, V., Chen, Y.S., Dolat, L., and Valdivia, R.H. (2017). The Effector TepP Mediates Recruitment and Activation of Phosphoinositide 3-Kinase on Early *Chlamydia trachomatis* Vacuoles. mSphere *2*, e00207-17.

CDC (2015). 2015 Sexually Transmitted Diseases Treatment Guidelines: Chlamydial infections. Centers for Disease Control and Prevention, https://www.cdc.gov/std/tg2015/chlamydia.htm.

Chacko, A., Beagley, K.W., Timms, P., and Huston, W.M. (2015). Human *Chlamydia pneumoniae* isolates demonstrate ability to recover infectivity following penicillin treatment whereas animal isolates do not. FEMS microbiology letters *362*.

Chen, L., Rashid, F., Shah, A., Awan, H.M., Wu, M., Liu, A., Wang, J., Zhu, T., Luo, Z., and Shan, G. (2015). The isolation of an RNA aptamer targeting to p53 protein with single amino acid mutation. Proc Natl Acad Sci U S A *112*, 10002–10007.

Chen, Y.-S., Bastidas, R.J., Saka, H.A., Carpenter, V.K., Richards, K.L., Plano, G.V., and Valdivia, R.H. (2014). The *Chlamydia trachomatis* Type III Secretion Chaperone Slc1 Engages Multiple Early Effectors, Including TepP, a Tyrosine-phosphorylated Protein Required for the Recruitment of CrkI-II to Nascent Inclusions and Innate Immune Signalling. PLoS Pathogens *10*, e1003954.

Choi, K.H., Gaynor, J.B., White, K.G., Lopez, C., Bosio, C.M., Karkhoff-Schweizer, R.R., and Schweizer, H.P. (2005). A Tn7-based broad-range bacterial cloning and expression system. Nat Methods *2*, 443–448.

Chung, C.T., Niemela, S.L., and Miller, R.H. (1989). One-step preparation of competent *Escherichia coli*: transformation and storage of bacterial cells in the same solution. Proc Natl Acad Sci U S A *86*, 2172–2175.

Clarke, I.N., Cutcliffe, L.T., Everson, J.S., Garner, S.A., Lambden, P.R., Pead, P.J., Pickett, M.A., Brentlinger, K.L., and Fane, B.A. (2004). Chlamydiaphage Chp2, a skeleton in the phiX174 closet: scaffolding protein and procapsid identification. J Bacteriol *186*, 7571–7574.

Cosse, M.M., Barta, M.L., Fisher, D.J., Oesterlin, L.K., Niragire, B., Perrinet, S., Millot, G.A., Hefty, P.S., and Subtil, A. (2018). The Loss of Expression of a Single Type 3 Effector (CT622) Strongly Reduces *Chlamydia trachomatis* Infectivity and Growth. Front Cell Infect Microbiol *8*, 145.

Dale, E.C., and Ow, D.W. (1991). Gene transfer with subsequent removal of the selection gene from the host genome. Proc Natl Acad Sci U S A *88*, 10558–10562.

Datta, S., Costantino, N., and Court, D.L. (2006). A set of recombineering plasmids for Gram-negative bacteria. Gene *379*, 109–115.

DeMars, R., and Weinfurter, J. (2008). Interstrain gene transfer in *Chlamydia trachomatis in vitro*: Mechanism and significance. J. Bacteriol.*190*, 1605–1614.

DeMars, R., Weinfurter, J., Guex, E., Lin, J., and Potucek, Y. (2007). Lateral gene transfer *in vitro* in the intracellular pathogen *Chlamydia trachomatis*. J. Bacteriol.*189*, 991–1003.

Ding, H., Gong, S., Tian, Y., Yang, Z., Brunham, R., and Zhong, G. (2013). Transformation of sexually transmitted infection-causing serovars of *Chlamydia trachomatis* using Blasticidin for selection. PLoS One *8*, e80534.

Donati, M., Huot-Creasy, H., Humphrys, M., Di Paolo, M., Di Francesco, A., and Myers, G.S. (2014). Genome Sequence of *Chlamydia suis* MD56, Isolated from the Conjunctiva of a Weaned Piglet. Genome Announc *2(3):* e00425–14.

Doore, S.M., and Fane, B.A. (2016). The microviridae: Diversity, assembly, and experimental evolution. Virology *491*, 45–55.

Enyeart, P.J., Chirieleison, S.M., Dao, M.N., Perutka, J., Quandt, E.M., Yao, J., Whitt, J.T., Keatinge-Clay, A.T., Lambowitz, A.M., and Ellington, A.D. (2013). Generalized bacterial genome editing using mobile group II introns and Cre-lox. Mol Syst Biol 9, 685.

Everson, J.S., Garner, S.A., Fane, B., Liu, B.-L., Lambden, P.R., and Clarke, I.N. (2002). Biological Properties and Cell Tropism of Chp2, a Bacteriophage of the Obligate Intracellular Bacterium *Chlamydophila abortus*. J. Bacteriol.*184*, 2748–2754.

Everson, J.S., Garner, S.A., Lambden, P.R., Fane, B.A., and Clarke, I.N. (2003). Host range of chlamydiaphages fCPAR39 and Chp3. J. Bacteriol.*185*, 6490–6492.

Fane, B., Brentlinger, K.L., Burch, A.D., Chen, M., Hafenstein, S., Moore, E., Novak, C.R., and Uchiyama, A. (2006). øX174 *et al.*, the *Microviridae*. In The Bacteriophages, R. Calendar, ed. (Oxford: Oxford University Press), pp. 129–145.

Fischer, A., Harrison, K.S., Ramirez, Y., Auer, D., Chowdhury, S.R., Prusty, B.K., Sauer, F., Dimond, Z., Kisker, C., Scott Hefty, P., *et al.* (2017). *Chlamydia trachomatis*-containing vacuole serves as deubiquitination platform to stabilize Mcl-1 and to interfere with host defence. Elife *6*, e21465.

Fitch, W.M., Peterson, E.M., and de la Maza, L.M. (1993). Phylogenetic analysis of the outer-membrane-protein genes of Chlamydiae, and its implication for vaccine development. MolBiolEvol *10*, 892–913.

Frost, E.H., Deslandes, S., Gendron, D., Bourgaux-Ramoisy, D., and Bourgaux, P. (1995). Variation outside variable segments of the major outer membrane protein distinguishes trachoma from urogenital isolates of the same serovar of *Chlamydia trachomatis*. Genitourinary Medicine *71*, 18–23.

Garner, S., Everson, J.S., Lambden, P.R., Fane, B., and Clarke, I.N. (2004). Isolation, molecular characterization and genome sequence of a bacteriophage (Chp3) from *Chlamydophila pecorum*. Virus Genes *28*, 207–214.

Gerard, H.C., Mishra, M.K., Mao, G., Wang, S., Hali, M., Whittum-Hudson, J.A., Kannan, R.M., and Hudson, A.P. (2013). Dendrimer-enabled DNA delivery and transformation of *Chlamydia pneumoniae*. Nanomedicine: nanotechnology, biology, and medicine *9*, 996–1008.

Goellner, S., Schubert, E., Liebler-Tenorio, E., Hotzel, H., Saluz, H.P., and Sachse, K. (2006). Transcriptional response patterns of *Chlamydophila psittaci* in different *in vitro* models of persistent infection. Infect Immun *74*, 4801–4808.

Gomes, J.P., Bruno, W.J., Borrego, M.J., and Dean, D. (2004). Recombination in the genome of *Chlamydia trachomatis* involving the polymorphic membrane protein C gene relative to *ompA* and evidence for horizontal gene transfer. J Bacteriol *186*, 4295–4306.

Gomes, J.P., Bruno, W.J., Nunes, A., Santos, N., Florindo, C., Borrego, M.J., and Dean, D. (2007). Evolution of *Chlamydia trachomatis* diversity occurs by widespread interstrain recombination involving hotspots. Genome Research *17*, 50–60.

Gong, S., Yang, Z., Lei, L., Shen, L., and Zhong, G. (2013). Characterization of *Chlamydia trachomatis* plasmid-encoded open reading frames. J Bacteriol *195*, 3819–3826.

Hadfield, J., Harris, S.R., Seth-Smith, H.M.B., Parmar, S., Andersson, P., Giffard, P.M., Schachter, J., Moncada, J., Ellison, L., Vaulet, M.L.G., *et al.* (2017). Comprehensive global genome dynamics of *Chlamydia trachomatis* show ancient diversification followed by contemporary mixing and recent lineage expansion. Genome research *27*, 1220–1229.

Haider, S., Wagner, M., Schmid, M.C., Sixt, B.S., Christian, J.G., Hacker, G., Pichler, P., Mechtler, K., Muller, A., Baranyi, C., *et al.* (2010). Raman microspectroscopy reveals long-term extracellular activity of Chlamydiae. Mol Microbiol *77*, 687–700.

Hanahan, D. (1983). Studies on transformation of *Escherichia coli* with plasmids. J Mol Biol *166*, 557–580.

Harris, S.R., Clarke, I.N., Seth-Smith, H.M., Solomon, A.W., Cutcliffe, L.T., Marsh, P., Skilton, R.J., Holland, M.J., Mabey, D., Peeling, R.W., *et al.* (2012). Whole-genome analysis of diverse *Chlamydia trachomatis* strains identifies phylogenetic relationships masked by current clinical typing. Nature genetics *44*, 413–419, S411.

Hayashi M., A.A., Richardson, D.L., Jr., Hayashi, M.N. (1988). The Bacteriophages: Biology of the Bacteriophage fX174, Vol 1 (New York: Plenum).

Hayes, L.J., Yearsley, P., Treharne, J.D., Ballard, R.A., Fehler, G.H., and Ward, M.E. (1994a). Evidence for naturally occurring recombination in the gene encoding the major outer membrane protein of lymphogranuloma venereum isolates of *Chlamydia trachomatis* Infection and Immunity *62*, 5659–5663.

Hayes, L.J., Yearsley, P., Treharne, J.D., Ballard, R.A., Fehler, G.H., and Ward, M.E. (1994b). Evidence for naturally occurring recombination in the gene encoding the major outer membrane protein of lymphogranuloma venereum isolates of *Chlamydia trachomatis*. Infection and Immunity *62*, 5659–5663.

Hefty, P.S., and Stephens, R.S. (2007). Chlamydial type III secretion system is encoded on ten operons preceded by sigma 70-like promoter elements. J. Bacteriol.*189*, 198–206.

Hintz, N.J., Ennis, D.G., Liu, W.F., and Larsen, S.H. (1995). The *recA* gene of *Chlamydia trachomatis*: cloning, sequence, and characterization in *Escherichia coli*. FEMS microbiology letters *127*, 175–180.

Hsia, R.C., Ting, L.M., and Bavoil, P.M. (2000). Microvirus of *Chlamydia psittaci* strain Guinea pig Inclusion Conjunctivitis: isolation and molecular characterization. Microbiology-Uk *146*, 1651–1660.

Illingworth, M., Hooppaw, A.J., Ruan, L., Fisher, D.J., and Chen, L. (2017). Biochemical and Genetic Analysis of the *Chlamydia* GroEL Chaperonins. J Bacteriol *199*, e00844-16.

Jiang, W., Bikard, D., Cox, D., Zhang, F., and Marraffini, L.A. (2013). RNA-guided editing of bacterial genomes using CRISPR-Cas systems. Nat Biotechnol *31*, 233–239.

Johnson, C.M., and Fisher, D.J. (2013). Site-specific, insertional inactivation of *incA* in *Chlamydia trachomatis* using a group II intron. PLoS One *8*, e83989.

Kari, L., Goheen, M.M., Randall, L.B., Taylor, L.D., Carlson, J.H., Whitmire, W.M., Virok, D., Rajaram, K., Endresz, V., McClarty, G., *et al.* (2011). Generation of targeted *Chlamydia trachomatis* null mutants. ProcNatlAcadSciUSA *108*, 7189–7193.

Keb, G., Hayman, R., and Fields, K.A. (2018). Floxed-Cassette Allelic Exchange Mutagenesis Enables Markerless Gene Deletion in *Chlamydia trachomatis* and Can Reverse Cassette-Induced Polar Effects. J Bacteriol *200*, e00479–18.

Kokes, M., Dunn, J.D., Granek, J.A., Nguyen, B.D., Barker, J.R., Valdivia, R.H., and Bastidas, R.J. (2015). Integrating chemical mutagenesis and whole-genome sequencing as a platform for forward and reverse genetic analysis of *Chlamydia*. Cell Host Microbe *17*, 716–725.

Kramer, M.J., and Gordon, F.B. (1971). Ultra structural analysis of the effects of penicillin and chlortetracycline on the development of genital tract *Chlamydia* Infection and Immunity 3, 333–341.

Lambden, P.R., Everson, J.S., Ward, M.E., and Clarke, I.N. (1990). Sulfur-rich proteins of *Chlamydia trachomatis*: developmentally regulated transcription of polycistronic mRNA from tandem promoters. Gene *87*, 105–112.

Lambden, P.R., Pickett, M.A., and Clarke, I.N. (2006a). The effect of penicillin on *Chlamydia trachomatis* DNA replication. Microbiology *152*, 2573–2578.

Lambden, P.R., Pickett, M.A., and Clarke, I.N. (2006b). The effect of penicillin on *Chlamydia trachomatis* DNA replication. Microbiology *152*, 2573–2578.

Lampe, M.F., Suchland, R.J., and Stamm, W.E. (1993). Nucleotide sequence of the variable domains within the major outer membrane protein gene from serovariants of *Chlamydia trachomatis*. Infection and Immunity *61*, 213–219.

Liu, B.L., Everson, J.S., Fane, B., Giannikopoulou, P., Vretou, E., Lambden, P.R., and Clarke, I.N. (2000). Molecular characterization of a bacteriophage (Chp2) from *Chlamydia psittaci*. Journal of Virology *74*, 3464–3469.

Liu, Y., Chen, C., Gong, S., Hou, S., Qi, M., Liu, Q., Baseman, J., and Zhong, G. (2014). Transformation of *Chlamydia muridarum* reveals a role for Pgp5 in suppression of plasmid-dependent gene expression. J Bacteriol *196*, 989–998.

Lowden, N.M., Yeruva, L., Johnson, C.M., Bowlin, A.K., and Fisher, D.J. (2015). Use of aminoglycoside 3′ adenyltransferase as a selection marker for *Chlamydia trachomatis* intron-mutagenesis and *in vivo* intron stability. BMC Res Notes *8*, 570.

Lusher, M., Storey, C.C., and Richmond, S.J. (1991). Extrachromosomal elements of the genus *Chlamydia*. In Advances in gene technology, vol 2, P.J.Greenaway, ed. (Elsevier Science, New York), pp. 261–285.

Lysen, M., Osterlund, A., Rubin, C.J., Persson, T., Persson, I., and Herrmann, B. (2004). Characterization of *ompA* genotypes by sequence analysis of DNA from all detected cases of *Chlamydia trachomatis* infections during 1 year of contact tracing in a Swedish county. Journal of Clinical Microbiology *42*, 1641–1647.

Macaluso, A., and Mettus, A.M. (1991). Efficient transformation of *Bacillus thuringiensis* requires nonmethylated plasmid DNA. J Bacteriol *173*, 1353–1356.

Maruyama-Tabata, H., Harada, Y., Matsumura, T., Satoh, E., Cui, F., Iwai, M., Kita, M., Hibi, S., Imanishi, J., Sawada, T., *et al.* (2000). Effective suicide gene therapy *in vivo* by EBV-based plasmid vector coupled with polyamidoamine dendrimer. Gene therapy *7*, 53–60.

Marx, C.J., and Lidstrom, M.E. (2002). Broad-host-range cre-lox system for antibiotic marker recycling in Gram-negative bacteria. Biotechniques *33*, 1062–1067.

Mathews, S.A., Volp, K.M., and Timms, P. (1999). Development of a quantitative gene expression assay for *Chlamydia trachomatis* identified temporal expression of sigma factors. Federation of European Biochemical Societies Letters *458*, 354–358.

Matsumoto, A., and Manire, G.P. (1970). Electron microscopic observations on the effects of penicillin on the morphology of *Chlamydia psittaci* J. Bacteriol.*101*, 278–285.

Millman, K., Black, C.M., Johnson, R.E., Stamm, W.E., Jones, R.B., Hook, E.W., Martin, D.H., Bolan, G., Tavaré, S., and Dean, D. (2004). Population-Based Genetic and Evolutionary Analysis of *Chlamydia trachomatis* Urogenital Strain Variation in the United States. J. Bacteriol.*186*, 2457–2465.

Millman, K.L., Tavare, S., and Dean, D. (2001). Recombination in the *ompA* gene but not the *omcB* gene of *Chlamydia* contributes to serovar-specific differences in tissue tropism, immune surveillance, and persistence of the organism. J. Bacteriol.*183*, 5997–6008.

Mishra, M.K., Gerard, H.C., Whittum-Hudson, J.A., Hudson, A.P., and Kannan, R.M. (2012). Dendrimer-enabled modulation of gene expression in *Chlamydia trachomatis*. Mol Pharm *9*, 413–421.

Mueller, K.E., and Fields, K.A. (2015). Application of beta-Lactamase Reporter Fusions as an Indicator of Effector Protein Secretion during Infections with the Obligate Intracellular Pathogen *Chlamydia trachomatis*. PLoS One *10*, e0135295.

Mueller, K.E., Wolf, K., and Fields, K.A. (2016). Gene Deletion by Fluorescence-Reported Allelic Exchange Mutagenesis in *Chlamydia trachomatis*. MBio *7*, e01817–01815.

Navarro, G., and Tros de Ilarduya, C. (2009). Activated and non-activated PAMAM dendrimers for gene delivery *in vitro* and *in vivo*. Nanomedicine: nanotechnology, biology, and medicine *5*, 287–297.

Nguyen, B.D., and Valdivia, R.H. (2012). Virulence determinants in the obligate intracellular pathogen *Chlamydia trachomatis* revealed by forward genetic approaches. Proceedings of the National Academy of Sciences of the United States of America *109*, 1263–1268.

Nguyen, P.H., Lutter, E.I., and Hackstadt, T. (2018). *Chlamydia trachomatis* inclusion membrane protein MrcA interacts with the inositol 1,4,5-trisphosphate receptor type 3 (ITPR3) to regulate extrusion formation. PLoS Pathog *14*, e1006911.

O'Connell, C.M., Ingalls, R.R., Andrews, C.W., Jr., Scurlock, A.M., and Darville, T. (2007). Plasmid-deficient *Chlamydia muridarum* fail to induce immune pathology and protect against oviduct disease. J Immunol *179*, 4027–4034.

Oki, A.T., Seidman, D., Lancina, M.G., 3rd, Mishra, M.K., Kannan, R.M., Yang, H., and Carlyon, J.A. (2015). Dendrimer-enabled transformation of *Anaplasma phagocytophilum*. Microbes Infect *17*, 817–822.

Omsland, A., Sager, J., Nair, V., Sturdevant, D.E., and Hackstadt, T. (2012). Developmental stage-specific metabolic and transcriptional activity of *Chlamydia trachomatis* in an axenic medium. Proc Natl Acad Sci U S A *109*, 19781–19785.

Ouellette, S.P. (2018). Feasibility of a Conditional Knockout System for *Chlamydia* Based on CRISPR Interference. Front Cell Infect Microbiol *8*, 59.

Peeling, R.W., and Brunham, R.C. (1991). Neutralization of *Chlamydia trachomatis*: Kinetics and Stoichiometry. Infect Immun *59*, 2624–2630.

Perutka, J., Wang, W., Goerlitz, D., and Lambowitz, A.M. (2004). Use of computer-designed group II introns to disrupt *Escherichia coli* DExH/D-box protein and DNA helicase genes. J Mol Biol *336*, 421–439.

Pickett, M.A., Everson, J.S., and Clarke, I.N. (2000). Determination of Chlamydial plasmid copy number using a fluorescent, 5′-exonuclease (Taqman) assay. European Society for *Chlamydia* Research *Proceedings of Fourth Meeting, Helsinki, Finland. 20–23 August 2000*, 45–45.

Posfai, G., Koob, M., Hradecna, Z., Hasan, N., Filutowicz, M., and Szybalski, W. (1994). *In vivo* excision and amplification of large segments of the *Escherichia coli* genome. Nucleic Acids Res *22*, 2392–2398.

Putman, T., Hybiske, K., Jow, D., Afrasiabi, C., Lelong, S., Cano, M.A., Wu, C., and Su, A.I. (2019). ChlamBase: a curated model organism database for the *Chlamydia* research community. Database (Oxford) *2019*.

Rajaram, K., Giebel, A.M., Toh, E., Hu, S., Newman, J.H., Morrison, S.G., Kari, L., Morrison, R.P., and Nelson, D.E. (2015). Mutational Analysis of the *Chlamydia muridarum* Plasticity Zone. Infect Immun *83*, 2870–2881.

Read, T.D., Brunham, R.C., Shen, C., Gill, S.R., Heidelberg, J.F., White, O., Hickey, E.K., Peterson, J., Utterback, T., Berry, K., *et al.* (2000a). Genome sequences of *Chlamydia trachomatis* MoPn and *Chlamydia pneumoniae* AR39. Nucleic Acids Research *28*, 1397–1406.

Read, T.D., Brunham, R.C., Shen, C., Gill, S.R., Heidelberg, J.F., White, O., Hickey, E.K., Peterson, J., Utterback, T., Berry, K., *et al.* (2000b). Genome sequences of *Chlamydia trachomatis* MoPn and *Chlamydia pneumoniae* AR39. Nucleic Acids Research *28*, 1397–1406.

Read, T.D., Myers, G.S., Brunham, R.C., Nelson, W.C., Paulsen, I.T., Heidelberg, J., Holtzapple, E., Khouri, H., Federova, N.B., Carty, H.A., *et al.* (2003). Genome sequence of *Chlamydophila caviae* (*Chlamydia psittaci* GPIC): examining the role of niche-specific genes in the evolution of the Chlamydiaceae. Nucleic Acids Research *31*, 2134–2147.

Rodolakis, A., and Souriau, A. (1983). Response of ewes to temperature-sensitive mutants of *Chlamydia psittaci* (var ovis) obtained by NTG mutagenesis. Annales de recherches veterinaires Annals of veterinary research *14*, 155–161.

Sait, M., Livingstone, M., Graham, R., Inglis, N.F., Wheelhouse, N., and Longbottom, D. (2011). Identification, sequencing and molecular analysis of Chp4, a novel chlamydiaphage of *Chlamydophila abortus* belonging to the family Microviridae. J Gen Virol *92*, 1733–1737.

Sandoz, K.M., and Rockey, D.D. (2010). Antibiotic resistance in Chlamydiae FutureMicrobiol *5*, 1427–1442.

Seth-Smith, H.M.B., Harris, S.R., Persson, K., Marsh, P., Barron, A., Bignell, A., Bjartling, C., Clark, L., Cutcliffe, L.T., Lambden, P.R., *et al.* (2009). Co-evolution of genomes and plasmids within *Chlamydia trachomatis* and the emergence in Sweden of a new variant strain. Bmc Genomics *10*, 239.

Shaw, J.H., Key, C.E., Snider, T.A., Sah, P., Shaw, E.I., Fisher, D.J., and Lutter, E.I. (2018). Genetic Inactivation of *Chlamydia trachomatis* Inclusion Membrane Protein CT228 Alters MYPT1 Recruitment, Extrusion Production, and Longevity of Infection. Front Cell Infect Microbiol *8*, 415.

Shima, K., Wanker, M., Skilton, R.J., Cutcliffe, L.T., Schnee, C., Kohl, T.A., Niemann, S., Geijo, J., Klinger, M., Timms, P., *et al.* (2018). The Genetic Transformation of *Chlamydia pneumoniae*. mSphere 3, e00412–18.

Sixt, B.S., Bastidas, R.J., Finethy, R., Baxter, R.M., Carpenter, V.K., Kroemer, G., Coers, J., and Valdivia, R.H. (2017). The *Chlamydia trachomatis* Inclusion Membrane Protein CpoS Counteracts STING-Mediated Cellular Surveillance and Suicide Programs. Cell Host Microbe *21*, 113–121.

Skilton, R.J., Cutcliffe, L.T., Barlow, D., Wang, Y., Salim, O., Lambden, P.R., and Clarke, I.N. (2009). Penicillin induced persistence in *Chlamydia trachomatis*: high quality time lapse video analysis of the developmental cycle. PLoSOne *4*, e7723.

Skilton, R.J., Cutcliffe, L.T., Pickett, M.A., Lambden, P.R., Fane, B.A., and Clarke, I.N. (2007). Intracellular parasitism of chlamydiae: specific infectivity of chlamydiaphage Chp2 in *Chlamydophila abortus* J. Bacteriol.*189*, 4957–4959.

Skilton, R.J., Wang, Y., O'Neill, C., Filardo, S., Marsh, P., Benard, A., Thomson, N.R., Ramsey, K.H., and Clarke, I.N. (2018). The *Chlamydia muridarum* plasmid revisited: new insights into growth kinetics. Wellcome Open Res 3, 25.

Skipp, P., Robinson, J., O'Connor, C.D., and Clarke, I.N. (2005). Shotgun proteomic analysis of *Chlamydia trachomatis*. Proteomics *5*, 1558–1573.

Skipp, P.J., Hughes, C., McKenna, T., Edwards, R., Langridge, J., Thomson, N.R., and Clarke, I.N. (2016). Quantitative Proteomics of the Infectious and Replicative Forms of *Chlamydia trachomatis*. PLoS One *11*, e0149011.

Song, L., Carlson, J.H., Whitmire, W.M., Kari, L., Virtaneva, K., Sturdevant, D.E., Watkins, H., Zhou, B., Sturdevant, G.L., Porcella, S.F., *et al.* (2013). *Chlamydia trachomatis* plasmid-encoded Pgp4 is a transcriptional regulator of virulence-associated genes. Infect Immun *81*, 636–644.

Song, L., Carlson, J.H., Zhou, B., Virtaneva, K., Whitmire, W.M., Sturdevant, G.L., Porcella, S.F., McClarty, G., and Caldwell, H.D. (2014). Plasmid-mediated transformation tropism of chlamydial biovars. Pathog Dis *70*, 189–193.

Stephens, R.S., Kalman, S., Lammel, C., Fan, J., Marathe, R., Aravind, L., Mitchell, W., Olinger, L., Tatusov, R.L., Zhao, Q., *et al.* (1998a). Genome sequence of an obligate intracellular pathogen of humans: *Chlamydia trachomatis* Science *282*, 754–759.

Stephens, R.S., Kalman, S., Lammel, C., Fan, J., Marathe, R., Aravind, L., Mitchell, W., Olinger, L., Tatusov, R.L., Zhao, Q., *et al.* (1998b). Genome sequence of an obligate intracellular pathogen of humans: *Chlamydia trachomatis*. Science *282*, 754–759.

Stephens, R.S., Wagar, E.A., and Edman, U. (1988). Developmental Regulation of Tandem Promoters for the Major Outer-Membrane Protein Gene of *Chlamydia trachomatis*. J. Bacteriol.*170*, 744–750.

Storey, C., Lusher, M., Yates, P., and Richmond, S. (1993). Evidence for *Chlamydia pneumoniae* of non-human origin. Journal of general microbiology *139*, 2621–2626.

Storey, C.C., Lusher, M., and Richmond, S.J. (1989a). Analysis of the complete nucleotide sequence of Chp1, a phage which infects avian *Chlamydia psittaci*. Journal of General Virology *70*, 3381–3390.

Storey, C.C., Lusher, M., Richmond, S.J., and Bacon, J. (1989b). Further characterization of a bacteriophage recovered from an avian strain of *Chlamydia psittaci*. Journal of General Virology *70*, 1321–1327.

Suchland, R.J., Sandoz, K.M., Jeffrey, B.M., Stamm, W.E., and Rockey, D.D. (2009). Horizontal transfer of tetracycline resistance among *Chlamydia* spp. *in vitro*. Antimicrobial Agents and Chemotherapy *53*, 4604–4611.

Tam, J.E., Davis, C.H., Thresher, R.J., and Wyrick, P.B. (1992). Location of the origin of replication for the 7.5-kb *Chlamydia trachomatis* plasmid. Plasmid *27*, 231–236.

Tam, J.E., Davis, C.H., and Wyrick, P.B. (1994). Expression of recombinant DNA introduced into *Chlamydia trachomatis* by electroporation. Canadian Journal of Microbiology *40*, 583–591.

Thomas, N.S., Lusher, M., Storey, C.C., and Clarke, I.N. (1997). Plasmid diversity in *Chlamydia*. Microbiology *143*, 1847–1854.

Thompson, C.C., Griffiths, C., Nicod, S.S., Lowden, N.M., Wigneshweraraj, S., Fisher, D.J., and McClure, M.O. (2015). The Rsb Phosphoregulatory Network Controls Availability of the Primary Sigma Factor in *Chlamydia trachomatis* and Influences the Kinetics of Growth and Development. PLoS Pathog *11*, e1005125.

van Opijnen, T., Bodi, K.L., and Camilli, A. (2009). Tn-seq: high-throughput parallel sequencing for fitness and genetic interaction studies in microorganisms. Nat Methods *6*, 767–772.

Wagar, E.A., Safarians, S., and Pang, M. (1992). Analysis of genomic DNA from *Chlamydia trachomatis* for Dam and Dcm methylation. FEMS Microbiology Letters *98*, 161–168.

Wang, X., Hybiske, K., and Stephens, R.S. (2018). Direct visualization of the expression and localization of chlamydial effector proteins within infected host cells. Pathog Dis *76*.

Wang, Y., Cutcliffe, L.T., Skilton, R.J., Ramsey, K.H., Thomson, N.R., and Clarke, I.N. (2014). The genetic basis of plasmid tropism between *Chlamydia trachomatis* and *Chlamydia muridarum*. Pathog Dis *72*, 19–23.

Wang, Y., Kahane, S., Cutcliffe, L.T., Skilton, R.J., Lambden, P.R., and Clarke, I.N. (2011). Development of a transformation system for *Chlamydia trachomatis*: restoration of glycogen biosynthesis by acquisition of a plasmid shuttle vector. PLoS Pathog *7*, e1002258.

Wang, Y., Kahane, S., Cutcliffe, L.T., Skilton, R.J., Lambden, P.R., Persson, K., Bjartling, C., and Clarke, I.N. (2013). Genetic transformation of a clinical (genital tract), plasmid-free isolate of *Chlamydia trachomatis*: engineering the plasmid as a cloning vector. PLoS One *8*, e59195.

Weber, M.M., Lam, J.L., Dooley, C.A., Noriea, N.F., Hansen, B.T., Hoyt, F.H., Carmody, A.B., Sturdevant, G.L., and Hackstadt, T. (2017). Absence of Specific *Chlamydia trachomatis* Inclusion Membrane Proteins Triggers Premature Inclusion Membrane Lysis and Host Cell Death. Cell Rep *19*, 1406–1417.

Weber, M.M., Noriea, N.F., Bauler, L.D., Lam, J.L., Sager, J., Wesolowski, J., Paumet, F., and Hackstadt, T. (2016). A Functional Core of IncA Is Required for *Chlamydia trachomatis* Inclusion Fusion. J Bacteriol *198*, 1347–1355.

Weisburg, W.G., Hatch, T.P., and Woese, C.R. (1986). Eubacterial origin of *Chlamydia*. J. Bacteriol.*167*, 570–574.

Wickstrum, J., Sammons, L.R., Restivo, K.N., and Hefty, P.S. (2013). Conditional Gene Expression in *Chlamydia trachomatis* Using the Tet System. PLoS One *8*, e76743.

Xu, S., Battaglia, L., Bao, X., and Fan, H. (2013). Chloramphenicol acetyltransferase as a selection marker for chlamydial transformation. BMC Res Notes *6*, 377.

Yang, Z., Tang, L., Shao, L., Zhang, Y., Zhang, T., Schenken, R., Valdivia, R., and Zhong, G. (2016). The *Chlamydia*-Secreted Protease CPAF Promotes Chlamydial Survival in the Mouse Lower Genital Tract. Infect Immun *84*, 2697–2702.

Yoshida, N., and Sato, M. (2009). Plasmid uptake by bacteria: a comparison of methods and efficiencies. Appl Microbiol Biotechnol *83*, 791–798.

Zhang, D.J., Fan, H., McClarty, G., and Brunham, R.C. (1995). Identification of the *Chlamydia trachomatis* RecA-encoding gene. Infection and Immunity *63*, 676–680.

Zhong, G. (2017). Chlamydial Plasmid-Dependent Pathogenicity. Trends Microbiol *25*, 141–152.

Zhong, J., Karberg, M., and Lambowitz, A.M. (2003). Targeted and random bacterial gene disruption using a group II intron (targetron) vector containing a retrotransposition-activated selectable marker. Nucleic Acids Res *31*, 1656–1664.

Chlamydia Genomics

12

Vítor Borges[1], Patrick Hyden[2], João Paulo Gomes[1] and Thomas Rattei[2*]

[1]Bioinformatics Unit, Department of Infectious Diseases, National Institutes of Health, Lisbon, Portugal.
[2]Department of Microbiology and Ecosystem Science, University of Vienna, Vienna, Austria.

*Correspondence: thomas.rattei@univie.ac.at

https://doi.org/10.21775/9781912530281.12

Abstract

The field of *Chlamydia* genomics has seen a rampant development, from the very first release of the whole genome sequence of a *C. trachomatis* strain to the current availability of hundreds of genomes from multiple *Chlamydia* species. While this scientific advance has revealed fascinating patterns of adaptation to the intracellular environment, it has also disclosed the main genetic mechanisms driving *Chlamydia* evolution as well as many genes of unknown function that lack orthologues in the microbial world. It has allowed advances not only in the understanding of fundamental aspects of the epidemiology and biology of these unique intracellular bacteria but also in deciphering the molecular basis of phenotypic differences, such as host and tissue tropism, invasiveness and ecological success. Improvements and decreased cost of next-generation sequencing are coupled with advances in mutagenesis and proteome and transcriptome profiling. There is a critical need for multidisciplinary efforts to gain insights into the biological role of specific genes, to identify the within-patient *Chlamydia* adaptive traits during infection, and to decipher the *Chlamydia*–host interactive players that may ultimately lead to the development of new therapies.

Introduction

The term 'genomics' covers the sequencing and analysis of complete, or nearly complete, genomic sequences. It comprises all relevant laboratory and computational techniques, including DNA isolation, DNA amplification and sequencing, genome assembly, annotation and sequence analysis. The ability to sequence complete genomes has also enabled the study of other molecular targets, such as RNA transcripts ('transcriptomics'), proteins ('proteomics') and metabolites ('metabolomics'). This chapter starts by introducing the techniques and specific challenges for the sequencing of *Chlamydiales* genomes, while summarizing the key computational techniques for genome analysis and the present status of sequences in public repositories. It also provides an overview of the evolutionary dynamics of *Chlamydiales* genomes, from ancestry to host speciation, and discusses the current knowledge about the population structure, highlighting the main genetic events and genome targets underlying chlamydial evolution and diversification. In this regard, it also emphasizes the impact of genome changes on both phenotype diversification among same-niche infecting strains and on tissue tropism and ecological success. This chapter further explains why and how genomics acts as the basis for downstream high-throughput 'omics' studies in the *Chlamydiales*. Finally, it anticipates the application of genome sequencing as a tool to scrutinize *Chlamydia* evolution during infection, by providing relevant literature examples for other pathogens showing how novel genomics approaches may help deciphering the *Chlamydia*-pathogen 'arms race'.

Sequencing and characteristics of Chlamydiales genomes

Chlamydiales genomes

Most *Chlamydiales* genomes consist of a single, circular chromosome, although many species, particularly from the genus *Chlamydia*, also carry a small amount of genetic material on a plasmid. Due to the limited read length of DNA sequencing methods, genomic DNA is randomly fragmented by physical or enzymatic techniques. The necessary amount of DNA is traditionally obtained from *Chlamydiales* culture within laboratory host cell lines. To speed up the sequencing process, e.g. from clinical samples, and to avoid mutations introduced during laboratory cultivation, culture-free sequencing techniques for *Chlamydiales* have been developed (e.g. Seth-Smith *et al.*, 2013). The highest number (164) of sequenced genomes are available for *Chlamydia trachomatis*, reflecting its status as a highly prevalent cause of human disease (Table 12.1). 58 of all listed genomes contain a plasmid sequence of a few kB in length, but these short additional elements may not have been identified in assemblies derived from whole genome shotgun sequencing.

The sequencing of genomes requires a workflow of three main steps: (i) isolation and preparation of genomic DNA, (ii) sequencing of genomic DNA, and (iii) computational assembly, annotation and analysis. Genomic DNA can be derived from cultures or obtained directly from patient isolates. Patient isolates provide too little *Chlamydia*-specific DNA, but this hurdle is being successfully surpassed by applying random DNA amplification techniques, such as Multiple Displacement Amplification (MDA) (e.g. Putman *et al.*, 2013) or targeted genome enrichment procedures (e.g. Seth-Smith *et al.*, 2013).

The assembly of whole-genome shotgun sequencing reads into genomic contigs (contiguous sequences from overlapping reads), scaffolds (contigs arranged with each other by using additional information, such as mate pairs or optical maps) and finished chromosomes is typically performed by using *de novo* assembly techniques. For a few genomes, comparative assembly, which is a combination of *de novo* assembly and alignment-based mapping to a reference genome, has been used (e.g. Weinmaier *et al.*, 2015). Mixtures of strains, as they may occur in patient isolates with multiple infections, can be separated and quantified by read mapping and variant analysis workflows, such as VarCap (Zojer *et al.*, 2017).

Genome sequences are deposited into databases of the International Nucleotide Sequence Database

Table 12.1 Genome records of *Chlamydia*, as of November 2018, based on NCBI GenBank (https://www.ncbi.nlm.nih.gov)

Species	Number of chromosome sequences	Number of plasmid sequences	Median length (Mb)	Median protein-coding sequence count	Median % G+C	Selected references
Chlamydia abortus	66	0	1.16	962	39.9	Seth-Smith *et al.* (2017b)
Chlamydia avium	2	1	1.05	880	36.8	Sachse *et al.* (2014)
Chlamydia caviae	1	1	1.18	976	39.2	Read *et al.* (2003)
Chlamydia felis	1	1	1.17	974	39.4	Azuma *et al.* (2006)
Chlamydia gallinacea	2	2	1.07	902	37.9	Sachse *et al.* (2014)
Chlamydia ibidis	2	0	1.18	956	38.4	Vorimore *et al.* (2013)
Chlamydia muridarum	28	1	1.07	899	40.3	Yeruva *et al.* (2014)
Chlamydia pecorum	12	1	1.11	933	41.1	Sait *et al.* (2014)
Chlamydia pneumoniae	12	2	1.23	1064	40.6	Weinmaier *et al.* (2015)
Chlamydia psittaci	67	30	1.18	989	39.1	Van Lent *et al.* (2012)
Chlamydia suis	33	4	1.09	910	42.1	Seth-Smith *et al.* (2017a)
Chlamydia trachomatis	164	58	1.05	902	41.3	Harris *et al.* (2012)
Unclassified *Chlamydia* sp.	36	4	1.20	996	39.3	Staub *et al.* (2018)

Collaboration (INSDC) as part of the publication process (Karsch-Mizrachi *et al.*, 2018). The INSDC contains a taxonomy database, which categorizes and numbers all taxonomy terms that are used for sequence submissions. The taxonomy terms are maintained and adapted by voluntary maintainers, representing the scientific community, according to their typical use in the scientific literature. Remarkably, the recent suggestion to combine the genera *Chlamydia* and *Chlamydophila* into one single genus *Chlamydia* is already reflected in most genome records of the former genus *Chlamydophila* (Pannekoek *et al.*, 2016). However, the INSDC member databases (NCBI GenBank, European Nucleotide Archive and DNA Databank of Japan) do not yet scientifically maintain or update genome records, but only serve as long-term archives. Therefore, the NCBI Reference Sequences project (RefSeq) has started to re-annotate selected genomes once per year (Haft *et al.*, 2018). The gene and function prediction for RefSeq genomes is based on the NCBI prokaryotic genome annotation pipeline PGAP, which combines *ab initio* gene prediction algorithms with homology-based methods (Tatusova *et al.*, 2016). However, a major problem for the use of *Chlamydiales* genomes in comparative research has not been solved by PGAP: the nomenclature of gene names and function descriptions. e.g. *Chlamydia trachomatis* and *Chlamydia pneumoniae* have similar but not identical genes/proteins. For example, Tarp was initially identified from *Chlamydia trachomatis* L2, and the L2 protein is still the best characterized one (Carabeo *et al.*, 2002). Tarp and its homologue from *Chlamydia pneumoniae* are similar but have functional differences. It is likely that many genes will have species-specific functions – and these should be reflected in the nomenclature and functional description. A solution would be a species-specific nomenclature, e.g. Ctr-Tarp versus Cpn-Tarp, but this would contradict the idea of species-nonspecific gene symbols that easily indicate orthologues between genomes. Therefore, our vocabulary of gene names needs to be updated and extended, in order to reasonably reflect function. Additionally, systematic locus tags from RefSeq records (for example CT0123) should be used in publications to unambiguously identify genes and gene products. During re-annotation, locus tags and gene names need to be conserved.

Chlamydia-related bacteria have more variability in their genome size than the members of the genus *Chlamydia* (Table 12.2). Although the number of completely sequenced genomes of *Chlamydia*-related bacteria is still comparatively low, they have tremendously contributed to our understanding of *Chlamydiales* evolution and pathogenicity. Remarkably, the recent development of metagenome assembly and binning techniques, which are described below, has allowed us to sequence *Chlamydiales* genomes directly from environmental samples without cultivation (Morowitz *et al.*, 2011; Parks *et al.*, 2018; Tully *et al.*, 2018). Metagenome assembly links sequencing reads into coherent genomic segments, contigs and scaffolds,

Table 12.2 Genomes of *Chlamydia*-related bacteria, as of November 2018, based on NCBI GenBank (https://www.ncbi.nlm.nih.gov)

Genus/order	Number of chromosome sequences	Number of plasmid sequences	Median length (Mb)	Median protein-coding sequence count	Median % G+C	Selected references
Candidatus Protochlamydia	3	1	2.6	1917	34.6	Horn *et al.* (2004)
Candidatus Rubidus	1	0	2.8	2446	32.6	Bou Khalil *et al.* (2016)
Candidatus Similichlamydia	2	0	0.9	758	41.6	Taylor-Brown *et al.* (2017)
Criblamydia	1	0	3.0	2422	38.2	Bertelli *et al.* (2014)
Neochlamydia	3	0	2.5	1882	38.0	Domman *et al.* (2014)
Parachlamydia	6	0	3.1	2543	40.3	Parks *et al.* (2018)
Simkania	2	1	2.6	2355	41.6	Collingro *et al.* (2011)
Waddlia	3	1	2.1	1823	43.8	Tully *et al.* (2018)

and is required to reconstruct genomes from metagenomes ('binning'). Simulations show that greater length and coverage of contigs/scaffolds reduce the risk of chimeric sequences. However, strain diversity (presence of multiple strains of a species in a sample) can either prevent successful assembly or lead to co-assembly of related strains, and there is therefore the need for forthcoming long-read sequencing methods (Turaev and Rattei, 2016). Genome-centric metagenomics is made possible by efficient binning approaches. Three independent information sources are used in modern binning programs: (i) sequence similarity (e.g. to universal or lineage-specific marker genes), (ii) sequence composition (e.g. tetranucleotide frequencies), and (iii) the variation of sequencing coverage in multiple samples (Turaev and Rattei, 2016). These novel computational techniques will substantially expand our knowledge of *Chlamydia*-related bacteria in the coming years, and will generate data that will allow us to isolate and characterize yet unknown species, genera or even families and orders. These bacteria are discussed in more detail in Chapter 1, *Chlamydia*-related Bacteria.

For the phylum *Chlamydiae*, the database eggnog 5.0 (Huerta-Cepas *et al.*, 2019) lists 2514 clusters of orthologous groups (chlaNOG). These clusters, representing proteins that have been inherited from the same gene in the last common ancestor, were derived from 21 representative complete genome sequences of different *Chlamydiae*. Only 324 of these chlaNOG are represented in all 21 genomes and thereby constitute the set of *Chlamydia* core protein-coding genes. The majority of chlaNOG contain genes from two to ten genomes, indicating a substantial number of genes of the *Chlamydia* pangenome that are not ubiquitous for all strains.

Virulence genes

One of the most important questions in the analysis of *Chlamydiales* genomes is the prediction of virulence genes. These genes are discussed in detail in different chapters of this book. However, the annotation of newly sequenced genomes has often inspired and enabled their further investigation. To understand what the genome tells us about virulence, we need to emphasize the complexity of this problem and the limits of computational methods and resources. Most protein sequences in public databases are computationally predicted and have not been experimentally validated. The prediction of protein-coding genes is the automated detection of start- and stop-codon, followed by a computational decision to designate this open-reading-frame (ORF) as a gene or not. In some cases, additional evidence on protein or transcript level is available to verify the existence of the gene, and has also been used to assess the accuracy of protein gene prediction (Tripp *et al.*, 2015). As computational predictions of operon structures are not sufficiently confident so far, the genome and protein databases do not group predicted genes into predicted operons.

To assign functions to a gene, the protein would have to be extensively studied. In many cases, sequence similarity to genes that have been studied gives a good first indication about their function. The UniProtKB database links experimental data to protein sequences, which allows genes to be annotated with their functions. Roughly 1% of the UniProtKB, the 'SwissProt' section, are manually reviewed by experts in this way (UniProt, 2019). If one compares the number of genes of a well-studied *E. coli* genome found in this database with a well-studied *C. trachomatis* strain (Table 12.3), it becomes clear that the existence of most *C. trachomatis* genes is only supported by homologues that have been confirmed to be real genes in other strains or species. A large proportion of chlamydial genes remain as 'predicted only', while most *E. coli* proteins will at least have been confirmed to exist at the protein level. In addition, the function of proteins is mainly provided from homology inference. A large proportion of genes in *C. trachomatis* are hypothetical proteins of unknown function.

Genomes from the genus *Chlamydia* contain a 'plasticity zone' with increased genetic diversity between strains and species (Nunes *et al.*, 2013; Read *et al.*, 2003; Voigt *et al.*, 2012). Other variable regions include *ompA* and the genes for the polymorphic membrane proteins (Pmps). It has been hypothesized that genes within plasticity zones are important for host and tissue specificity (Read *et al.*, 2000). Recently, a core genome multilocus sequence typing approach (cgMLST) of *C. trachomatis* found that these regions vary among the serovars according to their tissue tropism. A mutagenesis study in *C. muridarum* revealed that defective variants of genes in the *C. muridarum*

Table 12.3 Comparison of the numbers of genes for an *E. coli* strain and a *C. trachomatis* strain present in the UniProtKB database

	Escherichia coli strain K-12	*Chlamydia trachomatis* strain D/UW-3/Cx
Total number of predicted genes	4350	897
Evidence at protein level	2,934	25
Evidence at transcript level	165	13
Inferred from homology	702	467
Predicted only	549	392
Manually reviewed	4,475	483
Function from literature	1,864	9
Function from homology	1,137	332
No function annotated	1,630	556

Almost all chlamydial genes are predicted and get the functional assignment by homology. The genes of the model organism *E. coli* in turn are extensively studied in experiments.

plasticity zone result in less infective phenotypes (Versteeg *et al.*, 2018).

Various virulence factors were identified by *in vivo* or *in vitro* screenings using random mutagenesis and knockout strategies. The applied molecular methods are discussed in Chapter 11, *Chlamydia* Genetics and Chapter 19, Animal Models of *Chlamydia trachomatis* Infection. Type III secretion systems (T3SS) and their effectors are introduced in Chapter 7, Protein Secretion in *Chlamydia*, inclusion proteins in Chapter 4, The Chlamydial Inclusion, and the protease CPAF in Chapter 8, The Chlamydial Protease CPAF. The database EffectiveDB (Eichinger *et al.*, 2016) provides specific computational approaches for the prediction of functional Type III secretion systems, which in *Chlamydiales* genomes are often arranged in three to four clusters spread over the genome (Hefty and Stephens, 2007). This is unusual because T3SS structural genes in other bacteria are often found on plasmids or pathogenicity islands together with their effectors (Tampakaki *et al.*, 2004). *In-silico* methods such as machine learning were used in EffectiveDB to identify novel effectors with a common N-terminal secretion signal (Arnold *et al.*, 2009; Eichinger *et al.*, 2016; Hovis *et al.*, 2013). In addition, inclusion membrane proteins (Inc proteins) play a substantial role in virulence. These proteins share a bilobed hydrophobic domain, which has been used for *in-silico* predictions (Dehoux *et al.*, 2011).

During the last 20 years the field of *Chlamydia* genomics has not only been established, but became one of the main pillars of chlamydial molecular and clinical research. It greatly facilitates other research approaches, such as genetics, molecular biology, cell biology and epidemiology. The recent development of long read sequencing techniques, which partially can be used at point of care (e.g. Nanopore MinION sequencing; Jain *et al.*, 2016), has the potential to massively expand the number of sequenced genomes and to provide strain resolution even from patient material with multiple infections.

Evolutionary dynamics of *Chlamydia* genomes

From ancestral to host speciation and population structure

The evolution of *Chlamydia* is a function of its interaction with multiple eukaryotic hosts over the course of millions of years. While the time scale of the emergence and diversification of *Chlamydiales* cannot be precisely defined, it is believed that the evolutionary separation of the human pathogen *Chlamydia trachomatis* as a species occurred in parallel with the emergence of modern human lineages (about 6 million years ago) (Clarke, 2004; Horn *et al.*, 2004; Ochman *et al.*, 2010; Steiper and Young, 2006). The evolutionary transition from a free-living to an obligate intracellular state (intrinsic to all *Chlamydia*) involved a massive loss of genes that became non-essential in the intracellular habitat. These are mainly those genes that are involved in biosynthetic functions required to produce metabolites that are available in the host cell cytoplasm (Andersson and Andersson, 1999; Stephens *et al.*, 1998; Zomorodipour and Andersson, 1999). This genome reduction, which is characteristic of other obligate intracellular bacteria, has led to a stabilized genome size among *Chlamydiales* of about 1–1.2 Mb with about 900–1100 genes (Merhej *et al.*, 2009; Nunes and Gomes, 2014; Sachse *et al.*, 2015). Typical for *Chlamydia* genomes is a high protein-coding density, few pseudogenes and few non-essential genes, which is in contrast to other obligate intracellular pathogens, such as *Rickettsia prowazekii*, that have a high content of noncoding sequences likely representing DNA in the process of being eliminated (Andersson and Kurland, 1998) (Andersson and Andersson, 1999). *Chlamydia* also present a nearly overlapping core-genome [i.e. genes shared by (virtually) all strains of a given species] and pan-genome [i.e. all genes described in a given species], with just a few mobile genetic elements described so far (Nunes and Gomes, 2014). The most relevant non-chromosomal element is a ~7.5kb plasmid, which co-evolved with the chromosome and is present in several chlamydial species, including *C. trachomatis*, *C. pneumoniae* and *C. psittaci*. Noteworthy, the genome contraction in intracellular bacteria may have also relied on recurrent evolutionary bottlenecks that accelerated the DNA degenerative process, in a phenomenon commonly known as Muller's ratchet (Felsenstein, 1974; Moran, 1996; Muller, 1964) (Andersson and Hughes, 1996). In fact, bottleneck-derived bacterial populations present small effective sizes, being less able to purge slightly deleterious mutations compared to free-living bacteria (Andersson and Andersson, 1999).

Whole-genome sequencing to study hundreds of *Chlamydia* genomes has made it possible to reach important phylogenetic inferences, revealing an evolutionary separation of species based on differences in their biological properties, such as host range, tissue tropism and disease outcomes (Fig. 12.1A) (Hadfield *et al.*, 2017; Harris *et al.*, 2012; Jelocnik *et al.*, 2015; Nunes and Gomes, 2014; Pannekoek *et al.*, 2010; Pannekoek *et al.*, 2008; Stephens *et al.*, 2009; Voigt *et al.*, 2012; Weinmaier *et al.*, 2015). This is remarkably exemplified by the phylodynamics within the *C. trachomatis* species, for which strain clustering generally reflects adaptation to the organ/cell-type that they preferentially infect (ocular conjunctiva, genitalia and lymph nodes) (Harris *et al.*, 2012; Joseph *et al.*, 2012; Nunes *et al.*, 2008). Also, the genome-based *C. trachomatis* species tree unveils traces of contemporary mixing and recent lineage expansion. After the ancient diversification, four main clades have appeared over the last few thousand years (Hadfield *et al.*, 2017; Harris *et al.*, 2012) (Fig. 12.1B): (i) a LGV lineage (serovars L1 to L3) that likely evolved prior to the separation of the ocular (serovars A to C) and the non-LGV genital strains (serovars D to K) and that is marked by a recent expansion of a L2b clone; (ii) two well-separated non-LGV genital lineages, recently designated as T1 (essentially composed of clinically prevalent serovars E and F), and T2 (composed of non-E/F genital serovars) and; (iii) an ocular lineage that constitutes a sub-cluster that shares a more recent ancestor with the T2 clade (Hadfield *et al.*, 2017; Harris *et al.*, 2012). It is possible that more genome data may challenge the historical assumption that the core genomic backbones of the main lineages reflect an adaptation to their respective niche. For example, a recent study of trachoma-associated strains in Australia revealed that they did not descend from the classical ocular clade, but instead from T1 and T2 clade ancestors that acquired genetic markers from the ocular genetic backbone (Andersson *et al.*, 2016). Curiously, despite the lack of congruence between the *ompA*- and genome-based phylogenies, the *ompA* genotype classification *per se* shows a strong correlation with tissue tropism and disease outcome, thus constituting a strong predictor of a strain's genome backbone (i.e. the position in the species tree) (Harris *et al.*, 2012).

Besides genomic features, there are other drivers that have influenced the population structure and worldwide epidemiological picture of *C. trachomatis*. For instance, the genetic conservation within trachoma-causing lineages is believed to be a consequence of geographical constraints (Andersson *et al.*, 2016; Hadfield *et al.*, 2017). The recent spread of a clonal LGV lineage (L2b) has been linked to behaviour-related factors [e.g. networks of men who have sex with men (MSM)] (de Vrieze and de Vries, 2014; Savage *et al.*, 2009; Thomson *et al.*, 2008). On the other hand, the parallel segregation of the most successful non-LGV genital *C. trachomatis* serovars likely relies on their 'most fitted'

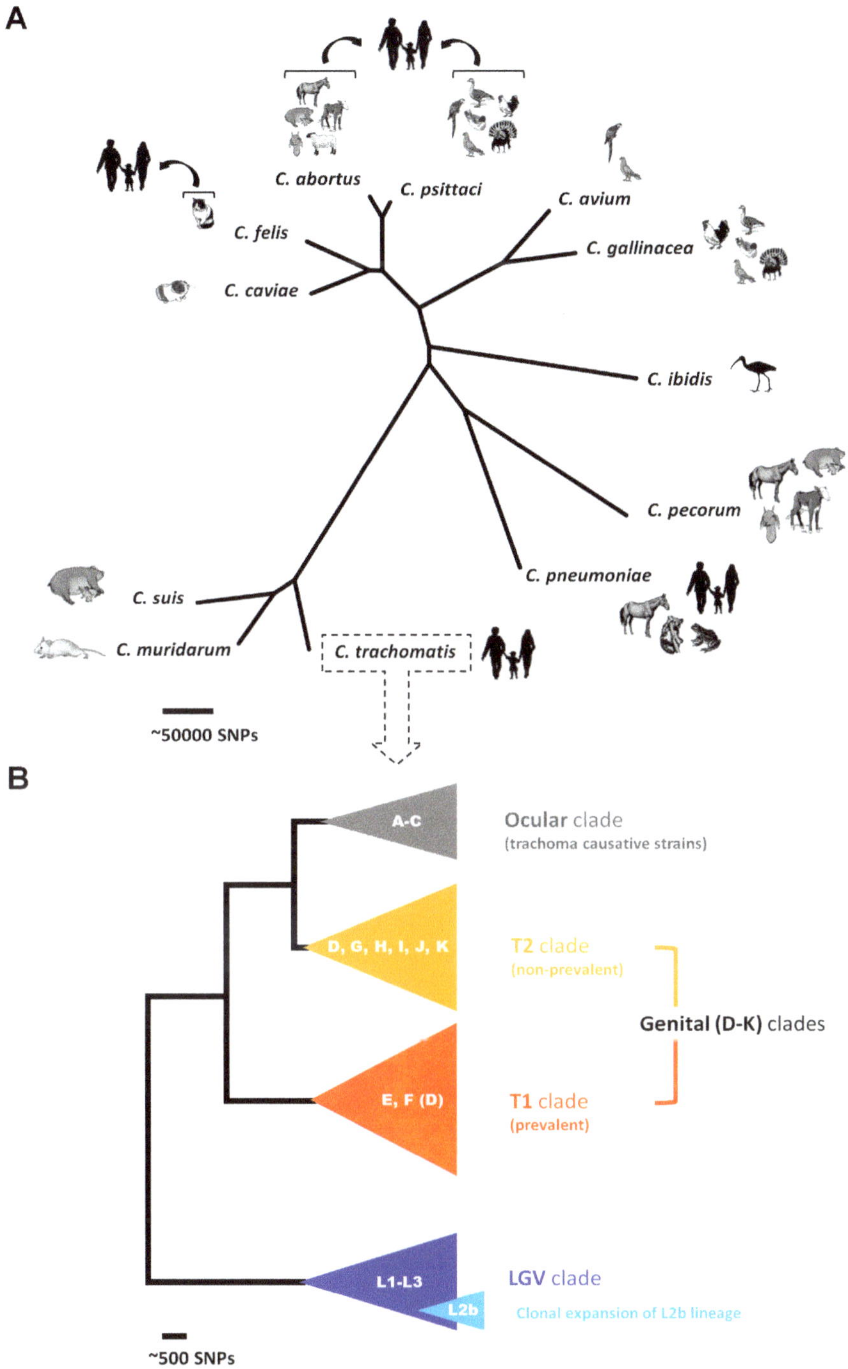

Figure 12.1 Phylogenetic reconstruction of the *Chlamydia* genus. A. Maximum Likelihood tree constructed based on 406734 core-genome variant sites detected between representative genomes of 12 *Chlamydia* species (NCBI GenBank records as of November 2018; see Table 12.1). Natural hosts are highlighted for each chlamydial species and zoonotic transmission (already proven for *C. psittaci*, *C. abortus* and to a lesser extension *C. felis*) is indicated with brackets (modified and updated from Nunes and Gomes, 2014). A dashed arrow pointing to the phylogenetic species tree shown below marks *C. trachomatis*. B. *C. trachomatis* species tree. Schematic representation of the *C. trachomatis* genome-based phylogenetic tree showing strains' segregation into four main clades, according to tissue tropism and/or ecological success. The main tree topology (and scale) was retrieved from Hadfield *et al*, 2017, where minor branches were collapsed in non-scaled triangles. Serovars mostly associated with each clade are indicated inside each respective collapsed triangle.

genomic backbone (Nunes *et al*, 2013; Harris *et al*, 2012; Hadfield *et al*, 2017). Still, considering that serovar E has been the most prevalent serovar causing sexually transmitted infections since *C. trachomatis* serotyping was implemented in the 1980s, this plausible hypothesis is hard to demonstrate as there is no data available supporting an ancestral 'selective sweep' (i.e. the rapid increase in prevalence of a lineage with advantageous genetic features).

Main genetic events and genome targets driving evolution and diversification

The genome evolution of *Chlamydia* is consistent with their obligate intracellular lifestyle. Indeed, this ecological restriction, where several barriers (chlamydial membranes, inclusion membrane, and host cell plasma membrane) separate bacterial DNA from the extracellular environment, imposes strong constraints on DNA exchange, which is virtually restricted to recombination between different *Chlamydia* species and strains during a co-infection. As such, and different from other bacteria, horizontal genetic exchange with other species has not been a relevant factor in the evolution of *Chlamydia*. Exceptions are: (i) the transposable tetracycline element identified in *C. suis* strains (Dugan *et al.*, 2004), which may have contributed to high rates of tetracycline resistance in this species (Joseph *et al.*, 2016); and (ii) a handful of distinct but related chlamydiaphages in different *Chlamydia* species (Everson *et al.*, 2002; Garner *et al.*, 2004; Hsia *et al.*, 2000; Liu *et al.*, 2000; Read *et al.*, 2000; Richmond *et al.*, 1982; Sait *et al.*, 2011), which may be involved in chlamydial pathogenesis by enhancing host inflammatory response or by promoting bacterial persistence (Rank *et al.*, 2009; Salim *et al.*, 2008).

Still, the *Chlamydiales* genomes have undergone other genetic events that have had a major impact on the chromosomal architecture and have led to deviations from the background synteny and/or sequence conservation. Examples include the symmetric inversion around the axes of the replication origin, the replication terminus that affects the so-called 'plasticity zone' and the expansion of paralogous genes (Nunes *et al.*, 2013; Read *et al.*, 2003; Voigt *et al.*, 2012). The plasticity zone is hypervariable in size and gene content [e.g. ~12kb, 11 genes for *C. abortus* (Thomson *et al.*, 2005), ~40kb, ~17 genes for *C. pecorum* (Sait *et al.*, 2014), and ~50kb, 45–49 genes for *C. trachomatis* (Read *et al.*, 2000; Stephens *et al.*, 1998)], and harbours genes important for chlamydial biology and adaptation, such as the phospholipase D enzyme family (Nelson *et al.*, 2006), the cytotoxin gene (McClarty *et al.*, 2007), and the tropism-associated tryptophan biosynthesis operon (Fehlner-Gardiner *et al.*, 2002; McClarty *et al.*, 2007).

The biological importance of these genetic events for *Chlamydiales* evolution is also illustrated by the heterogeneous paralogous family of polymorphic membrane proteins (Pmps), which are large autotransporter proteins with well-known roles as adhesins and antigens (Coler *et al.*, 2009; Crane *et al.*, 2006; Gomes *et al.*, 2005; Luczak *et al.*, 2016; Molleken *et al.*, 2010; Pal *et al.*, 2017; Tan *et al.*, 2010; Wehrl *et al.*, 2004; Becker and Hegemann, 2014). A comparison of Pmps within and between species showed that these proteins are highly variable in size, sequence and number. For example, there are nine *pmp* genes in both *C. trachomatis* and *C. muridarum* and 21 in *C. pneumoniae* and *C. psittaci* (Read *et al.*, 2000; Read *et al.*, 2003; Stephens *et al.*, 1998; Voigt *et al.*, 2012). In addition, Pmps have variable numbers of repetitive sequence motifs GGAI and FXXN (often flanked by cysteine residues) that may be key for adhesion and infection (Gomes *et al.*, 2006; Grimwood and Stephens, 1999; Molleken *et al.*, 2010).

Despite the impact of these discrete mutational events during *Chlamydia* evolution, it is well established that homologous recombination and point mutations are the primary evolutionary driving forces for diversification. As stated above, the heterogeneous *Chlamydia* genomic signatures resulted from co-evolution with different hosts (at inter-species level) and distinct anatomical sites (at intra-species level, with emphasis for *C. trachomatis*). The accumulation of single nucleotide polymorphisms (SNPs) is the basis for *Chlamydia* tree topology (700 orthologous genes shared among the species) (Voigt *et al.*, 2012), as well as for the evolutionary radiation of *C. trachomatis* species into main clades consistent with disease properties and/or ecological success (Fig. 12.1B) (Hadfield *et al.*, 2017; Harris *et al.*, 2012; Nunes *et al.*, 2008). On the other hand, homologous recombination is a major driver of diversity within species by acting on the SNP-containing regions towards the generation of mosaic genomic structures. The contribution of recombination and mutation to diversity not only varies between species (Ferreira *et al.*, 2014; Hadfield *et al.*, 2017; Joseph *et al.*, 2011; Joseph *et al.*, 2012), but is also determined by ecological and population-related constraints of strains within a species. This is exemplified by *C. trachomatis*, where a higher recombination flow is observed between strains that infect the same tissue, as markedly observed for diverse and broadly tissue-infecting non-LGV genital strains (Hadfield *et al.*, 2017; Harris *et al.*, 2012; Joseph *et al.*, 2011; Joseph *et al.*, 2012). In fact, LGV and ocular strains face not only a harder barrier for cell-to-cell contact due to the specificity of their respective infection niche, but also low prevalence and effective population size yielding a low likelihood of co-infection with either same or distinct niche-infecting strains. Due to this 'allopatric speciation' (i.e. an evolutionary scenario where biological populations of the same species become isolated from each other so that their ability to exchange DNA is impaired), LGV and ocular populations may be more prone to accumulation of slightly deleterious mutations ('Muller's ratchet') that have had fewer opportunities to be purged by recombination (Muller, 1964). The contribution of *Chlamydia* niche-restriction to the Muller's ratchet effect can be illustrated by the observed overrepresentation of non-synonymous mutations underlying *C. trachomatis* serovar radiation (polymorphism) than *C. trachomatis* species separation (divergence) (Borges *et al.*, 2012).

The SNP-based genomic background does not show a random distribution throughout the *Chlamydia* chromosome. In fact, SNP density has been found to be higher in genes coding for host-interacting proteins, with an emphasis for: (i) MOMP, the strongest antigen within chlamydial species, which displays widespread diversity and recombinant character; (ii) Pmps, the largest outer membrane protein family, which displays important antigenic and adhesion properties; and, (iii) effectors and substrates of secretion systems, such as inclusion membrane proteins, which are pivotal for host-cell manipulation and subversion of host-cellular functions (Almeida *et al.*, 2012; Borges *et al.*, 2012; Ferreira *et al.*, 2014; Harris *et al.*, 2012; Jeffrey *et al.*, 2010; Joseph *et al.*, 2012; Nunes *et al.*, 2008; Roulis *et al.*, 2015; Sait *et al.*, 2014; Weinmaier *et al.*, 2015; Wolff *et al.*, 2015). High SNP density is also observed in genes coding for proteins with unknown function ('hypothetical proteins'), a group of genes that is massively represented in *Chlamydiales* genomes, with most of them lacking homology outside *Chlamydia*. With the advance of functional studies, some hypothetical proteins have been assigned to the category of host-interacting proteins, namely effectors and Incs (Subtil *et al.*, 2005) (McClure *et al.*, 2017; Pruneda *et al.*, 2018; Sixt *et al.*, 2017; Sixt and Valdivia, 2016; Valdivia, 2008; Valdivia and Bastidas, 2018), and some of them are strong antigens (Fling *et al.*, 2001; Follmann *et al.*, 2008; Patton *et al.*, 2014; Picard *et al.*, 2012; Sharma *et al.*, 2006; Starnbach *et al.*, 2003).

Impact of genome changes on tissue tropism and ecological success

In the field of infectious diseases, an important area of research has been the development of therapeutic or prophylactic measures that target microbial mechanisms associated with specific disease outcomes. Considering that the genetic diversity within *Chlamydia* is shaped by tissue-specific selective pressures (such as concomitant microflora, physiological features and targeted local immune responses), the detection of genes presenting molecular 'fingerprints' of tissue tropism, ecological success or specific disease outcomes has been an important research goal. Most of the efforts to elucidate the genetic basis of phenotypic differences has been focused on *C. trachomatis* (Abdelsamed *et al.*, 2013; Bachmann *et al.*, 2014; Byrne, 2010; Nunes *et al.*, 2013), although work has also been done in other *Chlamydia* species (Jelocnik *et al.*, 2015; Morrison *et al.*, 2018; Roulis *et al.*, 2015; Weinmaier *et al.*, 2015). Of these phenotypes, there has been particular interest in studying the functional and genomic basis of tissue tropism.

The best-studied genetic models of *Chlamydia* tissue tropism involve the tryptophan synthase operon (*trpRBA*) and the cytotoxin locus, which are both located in the plasticity zone (Caldwell *et al.*, 2003; Carlson *et al.*, 2004; Fehlner-Gardiner *et al.*, 2002). Tryptophan is an essential amino acid that is depleted by the host IFN-γ immune response (a primary anti-chlamydial human immune response). Genital strains have a functional tryptophan synthase, which enables them to synthesize tryptophan from exogenous indole, which can be obtained from the abundant competing flora of the genital tract (Caldwell *et al.*, 2003; Fehlner-Gardiner *et al.*, 2002; Ziklo *et al.*, 2016). This scavenging mechanism allows genital strains to survive under conditions of tryptophan starvation. In contrast, ocular strains have a disruption of one of the subunits of the tryptophan synthase (alpha and beta, encoded by *trpA* and *trpB*, respectively) but may not need this enzyme because most ocular microflora do not produce indole (Caldwell *et al.*, 2003). The phenotypic impact of the *trpRBA* operon was ultimately assessed by transformation and complementation assays with an ocular strain that enabled it to produce tryptophan from indole (O'Neill *et al.*, 2018). This genetic signature also remarkably exemplifies how concomitant microflora may impact the evolutionary pathway of microorganisms, and is discussed in more detail in Chapter 16, *Chlamydia* Infections and the Microbiota.

C. trachomatis strains have major differences in a cytotoxin gene that are associated with differences in tissue tropism. This gene is present, but at different copy numbers, in the plasticity zone of *Chlamydia* species (Bachmann *et al.*, 2014; Lo *et al.*, 2012; Mojica *et al.*, 2011; Nunes and Gomes, 2014; Read *et al.*, 2000; Read *et al.*, 2003; Voigt *et al.*, 2012). In *C. trachomatis*, the cytotoxin (encoded by the CT166 gene) is thought to be linked to the ability of chlamydial strains to promote the disassembly of cytoskeleton actin filaments, which may influence bacterial internalization ('cytophatic effect') and niche specificity (Bothe *et al.*, 2015; Caldwell *et al.*, 2003; Carabeo, 2011; Thalmann *et al.*, 2010). Both ocular and non-LGV genital strains retain remnants of a larger ancestral cytotoxin gene, but CT166 is absent in LGV strains (Carlson *et al.*, 2004). However, only the non-LGV genital isolates encode a cytotoxin that has both the UDP-glucose binding domain and the functional glycosyltransferase domain that are characteristic of a functional cytotoxin (Belland *et al.*, 2001; Carlson *et al.*, 2004; McClarty *et al.*, 2007).

There are many phylogenomic signatures that have the potential to be linked to *C. trachomatis* tropism, although functional support is lacking. In a general perspective, hundreds of genes show mutational patterns that correlate with specific clades, with about 16%, 50%, 2%, and 80% of genes phylogenetically segregating into the ocular, prevalent genital (clade T1), genital (clade T2), and LGV clades, respectively. Notably, about one third of the *C. trachomatis* genes exclusively segregate with the LGV clade. Due to the evolutionary separation of clade T1 (leaving the remaining non-LGV genital strains clustered with ocular strains), only about 1% of the *C. trachomatis* genes present a tree topology perfectly reflecting the segregation of the three disease groups (ocular, genital, and LGV) (Ferreira *et al.*, 2014). This is somehow surprising considering that the genital strains (clades T1 and T2) are subjected to similar niche-specific selective pressures, but have the opportunity to recombine. Another curious genetic feature associated with clade T1 (and likely with its expansion) is the observation that about 20% of *C. trachomatis* genes show co-segregation between this clade and LGV strains. These data, together with multiple other antigenic and genetic features (Nunes *et al.*, 2010), should be considered when interpreting the genetic basis behind the high ecological success of clade T1 strains. In addition, it may suggest that LGV and T1 clade strains co-evolved before LGV segregation (Ferreira *et al.*, 2014), contradicting the assumption of a more recent expansion of clade T1 (Hadfield *et al.*, 2017).

Multiple studies have focused on the identification of genetic targets of *Chlamydiales* evolutionary adaptation (reviewed in Abdelsamed *et al.*, 2013; Bachmann *et al.*, 2014; Nunes *et al.*, 2013). Among the best candidates are effectors secreted by the type 3 secretion system (T3SS). Translocated actin-recruiting protein (Tarp) is a T3S effector with genetic traits most likely associated with chlamydial tropism. Tarp contains distinct functional regions, such as the C- terminal actin binding domain and the N-terminal tyrosine-rich repeat domain, which vary between strains from different disease groups (both in number and sequence). These domains are likely to mediate differences in the host cell actin-driven uptake of *Chlamydia* and in the strength of interactions with proteins from immune signalling pathways (Lutter *et al.*, 2010; Mehlitz *et al.*, 2010). It is curious that the T3SS effectors CT694 (TmeA, translocated membrane-associated effector A) and CT875 (TepP, Translocated early phosphoprotein), which act on the establishment of the nascent inclusions upon host cell invasion (as Tarp) (Bastidas *et al.*, 2013; Bullock *et al.*, 2012; Chen *et al.*, 2014; Scidmore, 2011; Valdivia, 2008; McKuen *et al*, 2017), also reveal clade-specific genetic traits. In fact, while CT875 (similarly to Tarp) phylogeny defines the three main disease groups (ocular, genital and LGV), CT694 perfectly segregates the four main *C. trachomatis* phylogenetic lineages, and presents signs of positive selection during the evolutionary separation of the clinically prevalent T1 clade strains (Borges *et al*., 2012; Ferreira *et al*., 2014). These strong genetic/phylogenetic signals shared by CT694, TepP and Tarp may constitute a good example on how similar selective pressures acting on distinct proteins (likely engaged in related biological processes) may leave concordant traces of adaptation. One may speculate that these proteins present discreet changes in the *modus operandi* (data for Tarp support this assumption) depending on the infected tissue.

Another group of proteins that has raised particular interest are the Inc proteins (also generally transported by T3S), which are known to subvert host cell functions while decorating the cytosolic face of the chlamydial inclusion (Delevoye *et al*., 2008; Derre *et al*., 2011; Lutter *et al*., 2013; Rzomp *et al*., 2006; Scidmore and Hackstadt, 2001; Sixt *et al*., 2017). Incs have been linked to the unique tropism and invasiveness of LGV strains, since a huge number of LGV-specific mutations (most affecting their cytoplasmic side) are found within genes encoding for Incs (Almeida *et al*., 2012; Borges *et al*., 2012; Ferreira *et al*., 2014). Genetic features of Pmps (especially the more phylogenetically informative PmpE, PmpF and PmpH) have also been associated with chlamydial phenotypic properties, namely clusters of mutations correlating with the segregation of strains by tissue tropism or ecological success (Borges *et al*., 2012; Gomes *et al*., 2006). It might be hypothesized that these mutations may fine-tune the Pmp role in antigenic variation and host-cell adhesion, thus impacting phenotypic diversity (Carlson *et al*., 2005; Coler *et al*., 2009; Crane *et al*., 2006; Gomes *et al*., 2006; Grimwood and Stephens, 1999; Nunes *et al*., 2007; Tan *et al*., 2009). The existence of clade-specific pseudogenes (i.e. a gene inactivated due to an internal frameshift mutation) should also be highlighted because protein truncations in strains from one disease group, but not from another group, may suggest that some molecular pathways are essential (or discarded) for the infection of specific niches. This scenario is similar to what has been observed for the *trpRBA* operon and the cytotoxin. Among the most noteworthy examples are CT037 (coding for a conserved hypothetical protein), which is a pseudogene only in prevalent genital (clade T1) and LGV strains (Ferreira *et al*., 2014) and the potential 'ocular-specific' pseudogenes CT058 (coding for an Inc protein) (Almeida *et al*., 2012), CT105 (coding for T3SS effector) and CT374/aacX (coding for a cytoplasmic-membrane arginine:agmatine antiporter) (Borges *et al*., 2012; da Cunha *et al*., 2014; Giles *et al*., 2009).

The 7.5 kb *C. trachomatis* plasmid plays an important role in the pathogenesis of infection and disease (Carlson *et al*., 2008; Gong *et al*., 2013; Kari *et al*., 2011; Matsumoto *et al*., 1998; O'Connell *et al*., 2011; Ramsey *et al*., 2014; Song *et al*., 2014; Sturdevant *et al*., 2010; Wang *et al*., 2013). It is expected that the plasmid may influence phenotypic diversity of *Chlamydia* strains by means of polymorphisms that can be observed in genomic and transcriptomic datasets. In fact, despite its high genetic conservation, the phylogenetic profile of the plasmid and the chromosome are similar, which suggests that they have co-evolved during niche-specific adaptation (Harris *et al*., 2012; Seth-Smith *et al*., 2009). Also, early plasmid-mediated transformation assays seemed to indicate that the ability of the plasmid to replicate may be biovar-specific, or even strain-specific, corroborating its potential role in strain-specific phenotypes (Song *et al*., 2014; Wang *et al*., 2014). Of note, a more recent study developed a plasmid vector potentially showing no strain–specific barrier (O'Neill *et al*., 2018). Most importantly, the plasmid gene *pgp4* has been shown to function as a transcriptional regulator of multiple chromosomal virulence genes (Carlson *et al*., 2008; O'Connell *et al*., 2011; Song *et al*., 2014). These include some of the most polymorphic *C. trachomatis* genes (e.g. CT049, CT050 and CT051) as well as genes displaying mutational profiles consistent with tissue tropism. The latter is exemplified by CT144, which exhibits variants that could be linked to rectal tropism of some genital strains (Christerson *et al*., 2012; Jeffrey *et al*., 2010). The recent observation that the plasmid-dependent regulation of the virulence-associated CPAF was restricted to a *C. trachomatis* macrophage-tropic strain provides further evidence that the plasmid can mediate tropism-specific phenotypes (Patton *et al*., 2018).

Impact of genome changes on phenotype diversification among same-niche infecting strains

Although the main diversity within *C. trachomatis* species is represented by the different serovars (Fig. 12.1B), genomic data from same-serovar strains has revealed subtle polymorphisms as well as novel recombination-driven genomic mosaics (Borges and Gomes, 2015; Harris *et al.*, 2012). In addition to host genetic and/or immunological factors, disease severity and differential phenotypic signatures may be due to genetic differences between *C. trachomatis* strains, even if they belong to the same phylogenetic clade (Abdelsamed *et al.*, 2013; Asner *et al.*, 2014; Roberts *et al.*, 2018). For instance, a pool of genes carrying subtle genomic variations was implicated in virulence dissimilarities between two *C. trachomatis* ocular strains in nonhuman primates (Kari *et al.*, 2008). A recent genome-wide association study identified strains within the T1 and T2 genital clades that cause trachoma, which was likely due to the acquisition of *ompA* alone or together with *pmpEFGH* from typical ocular strains (Andersson *et al.*, 2016). Genomic studies of the LGV clade have revealed multiple SNPs and indels that differentiate those strains causing the classic bubonic LGV from those that are associated with the anorectal syndrome (L2b genotype) (Borges and Gomes, 2015; Thomson *et al.*, 2008). Also, the expansion of the L2b lineage is marked by protein-changing mutations, with Tarp as the main target, which may justify dissimilar disease outcomes even among L2b strains (Borges and Gomes, 2015; de Vries *et al.*, 2015; Korhonen *et al.*, 2012; Marcotte *et al.*, 2014; Pendle and Gowers, 2012; Savage *et al.*, 2009).

Genomics as the basis for downstream high-throughput 'omics' studies in *Chlamydia*

Genomics, transcriptomics and proteomics (so called 'omics' approaches) have for long constituted relevant and complementary approaches for the understanding of *Chlamydia* biology. While earlier studies focused on individual (or classes of) genes/proteins by low-throughput methodologies (PCR/Sanger sequencing for genomics, RT-PCR for quantitative analysis of gene expression and Western Blot for proteomics), large-scale 'omics' approaches have been used only since the late 90s. In fact, the first chlamydial genome (*C. trachomatis* serovar D strain) was released in 1998, followed by a large-scale quantitative analysis of the *C. pneumoniae* proteome in 2001 and the microarray-based transcriptome analysis of *C. trachomatis* serovar D strain in 2003 (Belland *et al.*, 2003a; Stephens *et al.*, 1998; Vandahl *et al.*, 2001). Besides the remarkable gains in the understanding of *Chlamydia* biology and diversity at the genomic level, the genome era unequivocally triggered the generation of more comprehensive transcriptomic and proteomic data. This approach essentially yielded, in initial stages, unprecedented knowledge about the transcript and protein content/levels throughout the different stages of the unique biphasic developmental cycle of *Chlamydia* during normal or persistent growth (e.g. Belland *et al.*, 2003a,b; Maurer *et al.*, 2007; Miura *et al.*, 2008; Nicholson *et al.*, 2003; Ouellette *et al.*, 2006; Shaw *et al.*, 1999,2000; Skipp *et al.*, 2005; Vandahl *et al.*, 2004).

These large-scale approaches have also helped to identify early expressed genes involved in adhesion, protein secretion and inclusion formation, and late gene products involved in host cell lysis and pre-loading newly formed infectious EBs with machinery needed for invasion of new host cells (Belland *et al.*, 2003a; Nicholson *et al.*, 2003). It has also led to a better understanding that the *Chlamydia* persistent state is likely a defensive mechanism against host immune system, rather than a strict response to stress conditions (Belland *et al.*, 2003b; Ouellette *et al.*, 2006).

By the 2010s, novel and more complex outputs could be obtained due to significant advances in RNA-sequencing and mass spectrometry, which represented a tremendous turning point for the specificity and sensitivity of transcriptomics and proteomics. For RNA sequencing, the major outputs involved transcript and transcriptional start sites mapping, identification of massively expressed small RNAs, whose role is far from being deciphered, and insights into the differential transcriptional landscape between and within *Chlamydia* species (Albrecht *et al.*, 2011; Albrecht *et al.*, 2010; Beder and Saluz, 2018; Ferreira *et al.*, 2014; Humphrys *et al.*, 2013).

In the field of global proteome profiling, the main gains relied not only on reinforcing and/or consolidating the functional and/or temporal protein roles previously hypothesized from low-throughput techniques (Brinkworth *et al.*, 2018; Ostergaard *et al.*, 2016; Saka *et al.*, 2011; Skipp *et al.*, 2016), but also on the identification of potential vaccine candidates through large-scale immunoprofiling approaches through protein chip arrays (Cruz-Fisher *et al.*, 2011; Hufnagel *et al.*, 2018; Susnea *et al.*, 2011). More recently, a comparative study combining proteome expression profiling with mutagenesis strengthened previous findings (obtained through transcriptomics) on the virulence role of the *C. trachomatis* plasmid (Patton *et al.*, 2018). It also provided additional clues about the existence of distinctive regulation dynamics underlying the macrophage-tropism of LGV strains (Patton *et al.*, 2018).

Finally, the advances from 'omics' approaches are now allowing a step forward in understanding *Chlamydia*–host interactions by enabling the simultaneous profiling of chlamydial and host factors and molecular pathways (Hop *et al.*, 2017; Humphrys *et al.*, 2013; Rother *et al.*, 2018; Westermann *et al.*, 2017). This dual focus may ultimately contribute to the development of host-based therapies. For example, targeting inosine-5-monophosphate dehydrogenase, an enzyme in guanine nucleotide biosynthesis, efficiently inhibits *C. trachomatis* growth *in vitro* and *in vivo* (Rother *et al.*, 2018). This finding illustrates that the dependence of obligate intracellular pathogens on essential metabolites from the host cell can be exploited to develop new pharmaceutical approaches to treat *Chlamydia* infections.

Chlamydia evolution during infection: genome sequencing as a tool to scrutinize the host–pathogen 'arms race'

One of the hottest research fields in infectious diseases in the era of whole-genome sequencing is the understanding of how microbial pathogens adapt to the host during the infectious process in response to host-associated factors, such as specific immune responses, drug pressures, competing flora and physiological conditions. In fact, this adaptation underlies a co-evolution process, or 'host–pathogen arms race', that continuously shapes the pathogen genome, which may alter its virulence (enhancing it or attenuating it) and transmissibility due to the selection for survival of the fittest phenotypes. As such, the identification of the genetic mechanisms and factors that mediate pathogen adaptation may facilitate the development of new prophylactic and therapeutic strategies. On the other hand, studies focusing on the dynamics of intra-host pathogen populations may disclose transmission chains, which may support the understanding of local and global epidemiology of infectious diseases (Didelot *et al.*, 2016). These studies typically required the collection of multiple samples from the same patient either by simultaneously sampling (e.g. from two distinct anatomical sites) or by tracking evolution longitudinally (e.g. for a long-term disease). Multiple mutational mechanisms, such as acquisition of SNP, homologous recombination, phase variation and horizontal gene transfer, have been described as drivers of within-host pathogen diversity. These may constitute the genetic basis, for instance, for the acquisition of anti-microbial resistance, breach of anatomical barriers (e.g. from colonization to invasiveness) and the emergence of persistent phenotypes, which may impact the disease outcomes. Advantageous phenotypes thus arise upon the emergence of subpopulations (carrying beneficial mutations) whose relative proportion is highly dependent on the specific selective pressures imposed though the host–pathogen 'arms race' taking place at local and/or circumstantial levels. For example, whole-genome sequencing of multiple clones of *S. aureus* from infected patients with nasal colonization revealed an enrichment of mutations in pathogenesis-associated genes (e.g. surface antigens and toxin production) in infecting bacteria when comparing with nasal-colonizing bacteria) (Young *et al.*, 2017). Regarding persistence, important evidence could be found when tracking the long-term evolution of opportunistic pathogens (e.g. *Burkholderia dolosa*, *Pseudomonas aeruginosa* or *Mycobacterium absessus*) in patients with cystic fibrosis, revealing patterns of convergent evolution affecting, for instance, genes linked to regulation, lipopolysaccharide biosynthesis, antigenicity or drug resistance (Bryant *et al.*, 2013; Lieberman *et al.*, 2014; Marvig *et al.*, 2015; Viberg *et al.*, 2017). Given the increasing interest arising from this type of studies, cutting-edge methodologies are being developed towards the targeted sequencing of microbial genomes directly from clinical samples. This approach minimizes population bias associated with the *in vitro* culture and experimental constraints imposed by non-cultivable, highly fastidious or obligate intracellular bacteria, as *Chlamydia* (Arora *et al.*, 2016; Brown *et al.*, 2015; Brown and Christiansen, 2017; Christiansen *et al.*, 2014; Hadfield *et al.*, 2017; Pinto *et al.*, 2016; Putman *et al.*, 2013; Seth-Smith *et al.*, 2013; Votintseva *et al.*, 2017).

When compared with other important pathogens, very modest achievements have been reported so far for *Chlamydia*, where the major goal has been to determine the sequence of the entire genomes directly from clinical samples, rather than the study of population-based and genetic dynamics during the infectious process. Indeed, besides the few studies reporting whole genome sequencing for some species without bacterial enrichment in cell lines (Bachmann *et al.*, 2015; Christiansen *et al.*, 2014; Hadfield *et al.*, 2017; Jelocnik *et al.*, 2015; Putman *et al.*, 2013; Seth-Smith *et al.*, 2013; Taylor-Brown *et al.*, 2017), only one study has focused on understanding the adaptive pathway of *C. trachomatis* during long-term infection (Suchland *et al.*, 2017). These findings provided the feasibility for *in vivo* microevolutionary studies in such tricky organisms, which may open up research avenues to better understand *Chlamydia* biology and disease outcomes. One may anticipate that those studies will be potentially informative about population and genetic traits underlying several adaptive scenarios, such as: (i) the emergence of persistent forms of *Chlamydia* during infection (e.g. triggered by antibiotic therapy or host immunological response) (Panzetta *et al.*, 2018); (ii) the expansion of the infection from the lower to the upper genital tract, with impact on disease severity (e.g. tubal infertility, salpingitis and ectopic pregnancy) (Bebear and de Barbeyrac, 2009); and, (iii) the ability to infect extra-genital anatomical sites (e.g. through the invasion of synovial tissue leading to reactive arthritis) (Carter and Hudson, 2017). In parallel, *in vitro* studies should not be discarded considering that *Chlamydia* is able to gradually adapt to the *in vitro* environment through genetic and transcriptomic changes (Borges *et al.*, 2015). This occurs in a time frame that would allow relevant adaptive scenarios, such as the induction of persistence, to be mimicked. In fact, experimental evolution has been validated as proof of concept for the interpretation of *in vivo*-derived data for multiple pathogens (Barrick and Lenski, 2013; Barrick *et al.*, 2009).

Outlook

Despite the huge advances in the genomics field, some important milestones still need to be reached in order to understand most of the genotypic-phenotypic links addressed throughout this chapter. In particular, unveiling the functions of important targets of diversity and adaptation (proteins with unknown function, Incs and effectors, etc) comes up as a mandatory mission in the *Chlamydia* field. In order to make maximal use of the exponentially growing collection of sequenced genomes, our vocabulary of gene names needs to be updated and extended, in order to reasonably reflect function within and across species. Recent developments in genetic manipulation and mutagenesis in *Chlamydia* (discussed in Chapter 11, *Chlamydia* Genetics) provide new tools to answer some key scientific questions: (i) which are the genetic determinants of the differential host preference within the *Chlamydia* genus?; (ii) what is the molecular basis underlying the ability of same-species *Chlamydia* strains to infect different human tissues, e.g. why do some *C. trachomatis* strains preferentially cause mucosal infections while others are lymphotropic?; (iii) what are the genetic features behind the ecological success of *C. trachomatis* clade T1 strains?; and, (iv) are there genetic features that may enable the epidemic *C. trachomatis* L2b strains to exhibit wider tropism and transmission skills than other LGV strains? An illustrative example of the application of genetic manipulation and mutagenesis to understand these hot questions would be to introduce clade-specific genetic fingerprints into *Chlamydia* from other clades and then to check for phenotypic changes.

In another perspective, it is expected that advances in the 'omics' field will also allow to understand how *Chlamydia* shape its genome and transcriptome during the infection process and to identify which host mechanisms are critical for chlamydial survival. The identification of the *Chlamydia*-host interacting factors will constitute a huge step forward towards the understanding of the biological context behind pathogenicity-associated phenotypes (e.g. persistence, tropism expansion or antibiotic resistance) and, ultimately, for the development of innovative pathogen- and/or host-based therapeutic strategies.

References

Abdelsamed, H., Peters, J., and Byrne, G.I. (2013). Genetic variation in *Chlamydia trachomatis* and their hosts: impact on disease severity and tissue tropism. Future microbiology *8*, 1129–1146. 10.2217/fmb.13.80.

Albrecht, M., Sharma, C.M., Dittrich, M.T., Muller, T., Reinhardt, R., Vogel, J., and Rudel, T. (2011). The transcriptional landscape of *Chlamydia pneumoniae*. Genome biology *12*, R98. 10.1186/gb-2011-12-10-r98.

Albrecht, M., Sharma, C.M., Reinhardt, R., Vogel, J., and Rudel, T. (2010). Deep sequencing-based discovery of the *Chlamydia trachomatis* transcriptome. Nucleic Acids Res *38*, 868–877. 10.1093/nar/gkp1032.

Almeida, F., Borges, V., Ferreira, R., Borrego, M.J., Gomes, J.P., and Mota, L.J. (2012). Polymorphisms in inc proteins and differential expression of inc genes among *Chlamydia trachomatis* strains correlate with invasiveness and tropism of lymphogranuloma venereum isolates. J Bacteriol *194*, 6574–6585. 10.1128/jb.01428-12.

Andersson, D.I., and Hughes, D. (1996). Muller's ratchet decreases fitness of a DNA-based microbe. Proc Natl Acad Sci U S A *93*, 906–907.

Andersson, J.O., and Andersson, S.G. (1999). Insights into the evolutionary process of genome degradation. Current opinion in genetics & development *9*, 664–671.

Andersson, P., Harris, S.R., Seth Smith, H.M., Hadfield, J., O'Neill, C., Cutcliffe, L.T., Douglas, F.P., Asche, L.V., Mathews, J.D., Hutton, S.I., *et al.* (2016). *Chlamydia trachomatis* from Australian Aboriginal people with trachoma are polyphyletic composed of multiple distinctive lineages. Nature communications *7*, 10688. 10.1038/ncomms10688.

Andersson, S.G., and Kurland, C.G. (1998). Reductive evolution of resident genomes. Trends Microbiol *6*, 263–268.

Arias, C.A., Panesso, D., McGrath, D.M., Qin, X., Mojica, M.F., Miller, C., Diaz, L., Tran, T.T., Rincon, S., Barbu, E.M., *et al.* (2011). Genetic basis for *in vivo* daptomycin resistance in enterococci. The New England journal of medicine *365*, 892–900. 10.1056/NEJMoa1011138.

Arnold, R., Brandmaier, S., Kleine, F., Tischler, P., Heinz, E., Behrens, S., Niinikoski, A., Mewes, H.W., Horn, M., and Rattei, T. (2009). Sequence-based prediction of type III secreted proteins. PLoS Pathog *5*, e1000376. 10.1371/journal.ppat.1000376.

Arora, N., Schuenemann, V.J., Jager, G., Peltzer, A., Seitz, A., Herbig, A., Strouhal, M., Grillova, L., Sanchez-Buso, L., Kuhnert, D., *et al.* (2016). Origin of modern syphilis and emergence of a pandemic Treponema pallidum cluster. Nature microbiology *2*, 16245. 10.1038/nmicrobiol.2016.245.

Asner, S.A., Morre, S.A., Bochud, P.Y., and Greub, G. (2014). Host factors and genetic susceptibility to infections due to intracellular bacteria and fastidious organisms. Clinical microbiology and infection: the official publication of the European Society of Clinical Microbiology and Infectious Diseases *20*, 1246–1253. 10.1111/1469-0691.12806.

Azarian, T., Daum, R.S., Petty, L.A., Steinbeck, J.L., Yin, Z., Nolan, D., Boyle-Vavra, S., Hanage, W.P., Salemi, M., and David, M.Z. (2016). Intrahost Evolution of Methicillin-Resistant Staphylococcus aureus USA300 Among Individuals With Recurring Skin and Soft-Tissue Infections. The Journal of infectious diseases *214*, 895–905. 10.1093/infdis/jiw242.

Azuma, Y., Hirakawa, H., Yamashita, A., Cai, Y., Rahman, M.A., Suzuki, H., Mitaku, S., Toh, H., Goto, S., Murakami, T., *et al.* (2006). Genome sequence of the cat pathogen, Chlamydophila felis. DNA Res *13*, 15–23. 10.1093/dnares/dsi027.

Bachmann, N.L., Polkinghorne, A., and Timms, P. (2014). Chlamydia genomics: providing novel insights into chlamydial biology. Trends Microbiol *22*, 464–472. 10.1016/j.tim.2014.04.013.

Bachmann, N.L., Sullivan, M.J., Jelocnik, M., Myers, G.S., Timms, P., and Polkinghorne, A. (2015). Culture-independent genome sequencing of clinical samples reveals an unexpected heterogeneity of infections by Chlamydia pecorum. J Clin Microbiol *53*, 1573–1581. 10.1128/jcm.03534-14.

Barrick, J.E., and Lenski, R.E. (2013). Genome dynamics during experimental evolution. Nature reviews Genetics *14*, 827–839. 10.1038/nrg3564.

Barrick, J.E., Yu, D.S., Yoon, S.H., Jeong, H., Oh, T.K., Schneider, D., Lenski, R.E., and Kim, J.F. (2009). Genome evolution and adaptation in a long-term experiment with Escherichia coli. Nature *461*, 1243–1247. 10.1038/nature08480.

Bastidas, R.J., Elwell, C.A., Engel, J.N., and Valdivia, R.H. (2013). Chlamydial intracellular survival strategies. Cold Spring Harbour perspectives in medicine *3*, a010256. 10.1101/cshperspect.a010256.

Bebear, C., and de Barbeyrac, B. (2009). Genital *Chlamydia trachomatis* infections. Clinical microbiology and infection: the official publication of the European Society of Clinical Microbiology and Infectious Diseases *15*, 4–10. 10.1111/j.1469-0691.2008.02647.x.

Beder, T., and Saluz, H.P. (2018). Virulence-related comparative transcriptomics of infectious and non-infectious chlamydial particles. BMC Genomics *19*, 575. 10.1186/s12864-018-4961-x.

Belland, R.J., Nelson, D.E., Virok, D., Crane, D.D., Hogan, D., Sturdevant, D., Beatty, W.L., and Caldwell, H.D. (2003a). Transcriptome analysis of chlamydial growth during IFN-gamma-mediated persistence and reactivation. Proc Natl Acad Sci U S A *100*, 15971–15976. 10.1073/pnas.2535394100.

Belland, R.J., Scidmore, M.A., Crane, D.D., Hogan, D.M., Whitmire, W., McClarty, G., and Caldwell, H.D. (2001). *Chlamydia trachomatis* cytotoxicity associated with complete and partial cytotoxin genes. Proc Natl Acad Sci U S A *98*, 13984–13989. 10.1073/pnas.241377698.

Belland, R.J., Zhong, G., Crane, D.D., Hogan, D., Sturdevant, D., Sharma, J., Beatty, W.L., and Caldwell, H.D. (2003b). Genomic transcriptional profiling of the developmental cycle of *Chlamydia trachomatis*. Proc Natl Acad Sci U S A *100*, 8478–8483. 10.1073/pnas.1331135100.

Bertelli, C., Goesmann, A., and Greub, G. (2014). Criblamydia sequanensis Harbours a Megaplasmid Encoding Arsenite Resistance. Genome Announc 2. 10.1128/genomeA.00949-14.

Borges, V., and Gomes, J.P. (2015). Deep comparative genomics among *Chlamydia trachomatis* lymphogranuloma venereum isolates highlights genes potentially involved in pathoadaptation. Infection, genetics and evolution: journal of molecular epidemiology and evolutionary genetics in infectious diseases *32*, 74–88. 10.1016/j.meegid.2015.02.026.

Borges, V., Nunes, A., Ferreira, R., Borrego, M.J., and Gomes, J.P. (2012). Directional evolution of *Chlamydia trachomatis* towards niche-specific adaptation. J Bacteriol *194*, 6143–6153. 10.1128/jb.01291-12.

Borges, V., Pinheiro, M., Antelo, M., Sampaio, D.A., Vieira, L., Ferreira, R., Nunes, A., Almeida, F., Mota, L.J., Borrego, M.J., *et al.* (2015). *Chlamydia trachomatis* In Vivo to In Vitro Transition Reveals Mechanisms of Phase Variation and Down-Regulation of Virulence Factors. PLoS One *10*, e0133420. 10.1371/journal.pone.0133420.

Bothe, M., Dutow, P., Pich, A., Genth, H., and Klos, A. (2015). DXD motif-dependent and -independent effects of the *Chlamydia trachomatis* cytotoxin CT166. Toxins 7, 621–637. 10.3390/toxins7020621.

Bou Khalil, J.Y., Benamar, S., Baudoin, J.P., Croce, O., Blanc-Tailleur, C., Pagnier, I., Raoult, D., and La Scola, B. (2016). Developmental Cycle and Genome Analysis of 'Rubidus massiliensis,' a New Vermamoeba vermiformis Pathogen. Front Cell Infect Microbiol *6*, 31. 10.3389/fcimb.2016.00031.

Brinkworth, A.J., Wildung, M.R., and Carabeo, R.A. (2018). Genomewide Transcriptional Responses of Iron-Starved *Chlamydia trachomatis* Reveal Prioritization of Metabolic Precursor Synthesis over Protein Translation. mSystems 3. 10.1128/mSystems.00184-17.

Brown, A.C., Bryant, J.M., Einer-Jensen, K., Holdstock, J., Houniet, D.T., Chan, J.Z., Depledge, D.P., Nikolayevskyy, V., Broda, A., Stone, M.J., *et al.* (2015). Rapid Whole-Genome Sequencing of Mycobacterium tuberculosis Isolates Directly from Clinical Samples. J Clin Microbiol *53*, 2230–2237. 10.1128/jcm.00486-15.

Brown, A.C., and Christiansen, M.T. (2017). Whole-Genome Enrichment Using RNA Probes and Sequencing of *Chlamydia trachomatis* Directly from Clinical Samples. Methods in molecular biology (Clifton, NJ) *1616*, 1–22. 10.1007/978-1-4939-7037-7_1.

Bryant, J.M., Grogono, D.M., Greaves, D., Foweraker, J., Roddick, I., Inns, T., Reacher, M., Haworth, C.S., Curran, M.D., Harris, S.R., *et al.* (2013). Whole-genome sequencing to identify transmission of Mycobacterium abscessus between patients with cystic fibrosis: a retrospective cohort study. Lancet (London, England) *381*, 1551–1560. 10.1016/s0140-6736(13)60632-7.

Bullock, H.D., Hower, S., and Fields, K.A. (2012). Domain analyses reveal that *Chlamydia trachomatis* CT694 protein belongs to the membrane-localized family of type III effector proteins. The Journal of biological chemistry *287*, 28078–28086. 10.1074/jbc.M112.386904.

Byrne, G.I. (2010). *Chlamydia trachomatis* strains and virulence: rethinking links to infection prevalence and disease severity. The Journal of infectious diseases *201 Suppl 2*, S126–133. 10.1086/652398.

Caldwell, H.D., Wood, H., Crane, D., Bailey, R., Jones, R.B., Mabey, D., Maclean, I., Mohammed, Z., Peeling, R., Roshick, C., *et al.* (2003). Polymorphisms in *Chlamydia trachomatis* tryptophan synthase genes differentiate between genital and ocular isolates. The Journal of clinical investigation *111*, 1757–1769. 10.1172/jci17993.

Carabeo, R. (2011). Bacterial subversion of host actin dynamics at the plasma membrane. Cell Microbiol *13*, 1460–1469. 10.1111/j.1462-5822.2011.01651.x.

Carabeo, R.A., Grieshaber, S.S., Fischer, E., and Hackstadt, T. (2002). *Chlamydia trachomatis* induces remodelling of the actin cytoskeleton during attachment and entry into HeLa cells. Infect Immun *70*, 3793–3803.

Carlson, J.H., Hughes, S., Hogan, D., Cieplak, G., Sturdevant, D.E., McClarty, G., Caldwell, H.D., and Belland, R.J. (2004). Polymorphisms in the *Chlamydia trachomatis* cytotoxin locus associated with ocular and genital isolates. Infect Immun 72, 7063–7072. 10.1128/iai.72.12.7063-7072.2004.

Carlson, J.H., Porcella, S.F., McClarty, G., and Caldwell, H.D. (2005). Comparative genomic analysis of *Chlamydia trachomatis* oculotropic and genitotropic strains. Infect Immun *73*, 6407–6418. 10.1128/iai.73.10.6407-6418.2005.

Carlson, J.H., Whitmire, W.M., Crane, D.D., Wicke, L., Virtaneva, K., Sturdevant, D.E., Kupko, J.J., 3rd, Porcella, S.F., Martinez-Orengo, N., Heinzen, R.A., *et al.* (2008). The *Chlamydia trachomatis* plasmid is a transcriptional regulator of chromosomal genes and a virulence factor. Infect Immun *76*, 2273–2283. 10.1128/iai.00102-08.

Carter, J.D., and Hudson, A.P. (2017). Recent advances and future directions in understanding and treating Chlamydia-induced reactive arthritis. Expert review of clinical immunology *13*, 197–206. 10.1080/1744666x.2017.1233816.

Chen, Y.S., Bastidas, R.J., Saka, H.A., Carpenter, V.K., Richards, K.L., Plano, G.V., and Valdivia, R.H. (2014). The *Chlamydia trachomatis* type III secretion chaperone Slc1 engages multiple early effectors, including TepP, a tyrosine-phosphorylated protein required for the recruitment of CrkI-II to nascent inclusions and innate immune signalling. PLoS Pathog *10*, e1003954. 10.1371/journal.ppat.1003954.

Christerson, L., Bom, R.J., Bruisten, S.M., Yass, R., Hardick, J., Bratt, G., Gaydos, C.A., Morre, S.A., and Herrmann, B. (2012). *Chlamydia trachomatis* strains show specific clustering for men who have sex with men compared to heterosexual populations in Sweden, the Netherlands, and the United States. J Clin Microbiol *50*, 3548–3555. 10.1128/jcm.01713-12.

Christiansen, M.T., Brown, A.C., Kundu, S., Tutill, H.J., Williams, R., Brown, J.R., Holdstock, J., Holland, M.J., Stevenson, S., Dave, J., *et al.* (2014). Whole-genome enrichment and sequencing of *Chlamydia trachomatis* directly from clinical samples. BMC infectious diseases *14*, 591. 10.1186/s12879-014-0591-3.

Clarke, J. (2004). Predicting bacterial cause in infectious conjunctivitis: Chlamydia needs to be taken into account. BMJ *329*, 625; author reply 625. 10.1136/bmj.329.7466.625-a.

Coler, R.N., Bhatia, A., Maisonneuve, J.F., Probst, P., Barth, B., Ovendale, P., Fang, H., Alderson, M., Lobet, Y., Cohen, J., *et al.* (2009). Identification and characterization of novel recombinant vaccine antigens for immunization against genital *Chlamydia trachomatis*. FEMS immunology and medical microbiology *55*, 258–270. 10.1111/j.1574-695X.2008.00527.x.

Collingro, A., Tischler, P., Weinmaier, T., Penz, T., Heinz, E., Brunham, R.C., Read, T.D., Bavoil, P.M., Sachse,

K., Kahane, S., *et al.* (2011). Unity in variety--the pan-genome of the Chlamydiae. Mol Biol Evol *28*, 3253–3270. 10.1093/molbev/msr161.

Crane, D.D., Carlson, J.H., Fischer, E.R., Bavoil, P., Hsia, R.C., Tan, C., Kuo, C.C., and Caldwell, H.D. (2006). *Chlamydia trachomatis* polymorphic membrane protein D is a species-common pan-neutralizing antigen. Proc Natl Acad Sci U S A *103*, 1894–1899. 10.1073/pnas.0508983103.

Cruz-Fisher, M.I., Cheng, C., Sun, G., Pal, S., Teng, A., Molina, D.M., Kayala, M.A., Vigil, A., Baldi, P., Felgner, P.L., *et al.* (2011). Identification of immunodominant antigens by probing a whole *Chlamydia trachomatis* open reading frame proteome microarray using sera from immunized mice. Infect Immun *79*, 246–257. 10.1128/iai.00626-10.

da Cunha, M., Milho, C., Almeida, F., Pais, S.V., Borges, V., Mauricio, R., Borrego, M.J., Gomes, J.P., and Mota, L.J. (2014). Identification of type III secretion substrates of *Chlamydia trachomatis* using Yersinia enterocolitica as a heterologous system. BMC microbiology *14*, 40. 10.1186/1471-2180-14-40.

de Vries, H.J., Zingoni, A., Kreuter, A., Moi, H., and White, J.A. (2015). 2013 European guideline on the management of lymphogranuloma venereum. Journal of the European Academy of Dermatology and Venereology: JEADV *29*, 1–6. 10.1111/jdv.12461.

de Vrieze, N.H., and de Vries, H.J. (2014). Lymphogranuloma venereum among men who have sex with men. An epidemiological and clinical review. Expert review of anti-infective therapy *12*, 697–704. 10.1586/14787210.2014.901169.

Dehoux, P., Flores, R., Dauga, C., Zhong, G., and Subtil, A. (2011). Multi-genome identification and characterization of chlamydiae-specific type III secretion substrates: the Inc proteins. BMC Genomics *12*, 109. 10.1186/1471-2164-12-109.

Delevoye, C., Nilges, M., Dehoux, P., Paumet, F., Perrinet, S., Dautry-Varsat, A., and Subtil, A. (2008). SNARE protein mimicry by an intracellular bacterium. PLoS Pathog *4*, e1000022. 10.1371/journal.ppat.1000022.

Derre, I., Swiss, R., and Agaisse, H. (2011). The lipid transfer protein CERT interacts with the Chlamydia inclusion protein IncD and participates to ER-Chlamydia inclusion membrane contact sites. PLoS Pathog *7*, e1002092. 10.1371/journal.ppat.1002092.

Didelot, X., Walker, A.S., Peto, T.E., Crook, D.W., and Wilson, D.J. (2016). Within-host evolution of bacterial pathogens. Nature reviews Microbiology *14*, 150–162. 10.1038/nrmicro.2015.13.

Domman, D., Collingro, A., Lagkouvardos, I., Gehre, L., Weinmaier, T., Rattei, T., Subtil, A., and Horn, M. (2014). Massive expansion of Ubiquitination-related gene families within the Chlamydiae. Mol Biol Evol *31*, 2890–2904. 10.1093/molbev/msu227.

Dugan, J., Rockey, D.D., Jones, L., and Andersen, A.A. (2004). Tetracycline resistance in Chlamydia suis mediated by genomic islands inserted into the chlamydial inv-like gene. Antimicrobial agents and chemotherapy *48*, 3989–3995. 10.1128/aac.48.10.3989-3995.2004.

Eichinger, V., Nussbaumer, T., Platzer, A., Jehl, M.A., Arnold, R., and Rattei, T. (2016). EffectiveDB--updates and novel features for a better annotation of bacterial secreted proteins and Type III, IV, VI secretion systems. Nucleic Acids Res *44*, D669–674. 10.1093/nar/gkv1269.

Everson, J.S., Garner, S.A., Fane, B., Liu, B.L., Lambden, P.R., and Clarke, I.N. (2002). Biological properties and cell tropism of Chp2, a bacteriophage of the obligate intracellular bacterium Chlamydophila abortus. J Bacteriol *184*, 2748–2754.

Fehlner-Gardiner, C., Roshick, C., Carlson, J.H., Hughes, S., Belland, R.J., Caldwell, H.D., and McClarty, G. (2002). Molecular basis defining human *Chlamydia trachomatis* tissue tropism. A possible role for tryptophan synthase. The Journal of biological chemistry *277*, 26893–26903. 10.1074/jbc.M203937200.

Felsenstein, J. (1974). The evolutionary advantage of recombination. Genetics *78*, 737–756.

Ferreira, R., Antelo, M., Nunes, A., Borges, V., Damiao, V., Borrego, M.J., and Gomes, J.P. (2014). In silico scrutiny of genes revealing phylogenetic congruence with clinical prevalence or tropism properties of *Chlamydia trachomatis* strains. G3 (Bethesda, Md) *5*, 9–19. 10.1534/g3.114.015354.

Fling, S.P., Sutherland, R.A., Steele, L.N., Hess, B., D'Orazio, S.E., Maisonneuve, J., Lampe, M.F., Probst, P., and Starnbach, M.N. (2001). CD8+ T cells recognize an inclusion membrane-associated protein from the vacuolar pathogen *Chlamydia trachomatis*. Proc Natl Acad Sci U S A *98*, 1160–1165. 10.1073/pnas.98.3.1160.

Follmann, F., Olsen, A.W., Jensen, K.T., Hansen, P.R., Andersen, P., and Theisen, M. (2008). Antigenic profiling of a *Chlamydia trachomatis* gene-expression library. The Journal of infectious diseases *197*, 897–905. 10.1086/528378.

Garner, S.A., Everson, J.S., Lambden, P.R., Fane, B.A., and Clarke, I.N. (2004). Isolation, molecular characterization and genome sequence of a bacteriophage (Chp3) from Chlamydophila pecorum. Virus genes *28*, 207–214. 10.1023/B:VIRU.0000016860.53035.f3.

Giles, T.N., Fisher, D.J., and Graham, D.E. (2009). Independent inactivation of arginine decarboxylase genes by nonsense and missense mutations led to pseudogene formation in *Chlamydia trachomatis* serovar L2 and D strains. BMC evolutionary biology *9*, 166. 10.1186/1471-2148-9-166.

Giulieri, S.G., Baines, S.L., Guerillot, R., Seemann, T., Goncalves da Silva, A., Schultz, M., Massey, R.C., Holmes, N.E., Stinear, T.P., and Howden, B.P. (2018). Genomic exploration of sequential clinical isolates reveals a distinctive molecular signature of persistent Staphylococcus aureus bacteraemia. Genome medicine *10*, 65. 10.1186/s13073-018-0574-x.

Gomes, J.P., Hsia, R.C., Mead, S., Borrego, M.J., and Dean, D. (2005). Immunoreactivity and differential developmental expression of known and putative *Chlamydia trachomatis* membrane proteins for biologically variant serovars representing distinct disease groups. Microbes Infect *7*, 410–420. 10.1016/j.micinf.2004.11.014.

Gomes, J.P., Nunes, A., Bruno, W.J., Borrego, M.J., Florindo, C., and Dean, D. (2006). Polymorphisms in the nine polymorphic membrane proteins of *Chlamydia trachomatis* across all serovars: evidence for serovar Da recombination and correlation with tissue tropism.

J Bacteriol *188*, 275–286. 10.1128/jb.188.1.275-286.2006.

Gong, S., Yang, Z., Lei, L., Shen, L., and Zhong, G. (2013). Characterization of *Chlamydia trachomatis* plasmid-encoded open reading frames. J Bacteriol *195*, 3819–3826. 10.1128/jb.00511-13.

Grimwood, J., and Stephens, R.S. (1999). Computational analysis of the polymorphic membrane protein superfamily of *Chlamydia trachomatis* and *Chlamydia pneumoniae*. Microbial & comparative genomics *4*, 187–201. 10.1089/omi.1.1999.4.187.

Hadfield, J., Harris, S.R., Seth-Smith, H.M.B., Parmar, S., Andersson, P., Giffard, P.M., Schachter, J., Moncada, J., Ellison, L., Vaulet, M.L.G., *et al.* (2017). Comprehensive global genome dynamics of *Chlamydia trachomatis* show ancient diversification followed by contemporary mixing and recent lineage expansion. Genome Res *27*, 1220–1229. 10.1101/gr.212647.116.

Haft, D.H., DiCuccio, M., Badretdin, A., Brover, V., Chetvernin, V., O'Neill, K., Li, W., Chitsaz, F., Derbyshire, M.K., Gonzales, N.R., *et al.* (2018). RefSeq: an update on prokaryotic genome annotation and curation. Nucleic Acids Res *46*, D851-D860. 10.1093/nar/gkx1068.

Harris, S.R., Clarke, I.N., Seth-Smith, H.M., Solomon, A.W., Cutcliffe, L.T., Marsh, P., Skilton, R.J., Holland, M.J., Mabey, D., Peeling, R.W., *et al.* (2012). Whole-genome analysis of diverse *Chlamydia trachomatis* strains identifies phylogenetic relationships masked by current clinical typing. Nat Genet *44*, 413–419, S411. 10.1038/ng.2214.

Hefty, P.S., and Stephens, R.S. (2007). Chlamydial type III secretion system is encoded on ten operons preceded by sigma 70-like promoter elements. J Bacteriol *189*, 198–206. 10.1128/JB.01034-06.

Honsa, E.S., Cooper, V.S., Mhaissen, M.N., Frank, M., Shaker, J., Iverson, A., Rubnitz, J., Hayden, R.T., Lee, R.E., Rock, C.O., *et al.* (2017). RelA Mutant Enterococcus faecium with Multiantibiotic Tolerance Arising in an Immunocompromised Host. MBio *8*. 10.1128/mBio.02124-16.

Hop, H.T., Arayan, L.T., Reyes, A.W.B., Huy, T.X.N., Min, W., Lee, H.J., Son, J.S., and Kim, S. (2017). Simultaneous RNA-seq based transcriptional profiling of intracellular Brucella abortus and B. abortus-infected murine macrophages. Microbial pathogenesis *113*, 57–67. 10.1016/j.micpath.2017.10.029.

Horn, M., Collingro, A., Schmitz-Esser, S., Beier, C.L., Purkhold, U., Fartmann, B., Brandt, P., Nyakatura, G.J., Droege, M., Frishman, D., *et al.* (2004). Illuminating the evolutionary history of chlamydiae. Science *304*, 728–730. 10.1126/science.1096330.

Hovis, K.M., Mojica, S., McDermott, J.E., Pedersen, L., Simhi, C., Rank, R.G., Myers, G.S., Ravel, J., Hsia, R.C., and Bavoil, P.M. (2013). Genus-optimized strategy for the identification of chlamydial type III secretion substrates. Pathog Dis *69*, 213–222. 10.1111/2049-632X.12070.

Hsia, R.C., Ting, L.M., and Bavoil, P.M. (2000). Microvirus of chlamydia psittaci strain guinea pig inclusion conjunctivitis: isolation and molecular characterization. Microbiology *146 (Pt 7)*, 1651–1660. 10.1099/00221287-146-7-1651.

Huerta-Cepas, J., Szklarczyk, D., Heller, D., Hernández-Plaza, A., Forslund, S.K., Cook, H., Mende, D.R., Letunic, I., Rattei, T., Jensen, L.J., von Mering, C., and Bork, P. (2019). eggNOG 5.0: a hierarchical, functionally and phylogenetically annotated orthology resource based on 5090 organisms and 2502 viruses. Nucleic Acids Res *47*(D1):D309-D314. 10.1093/nar/gky1085.

Hufnagel, K., Lueong, S., Willhauck-Fleckenstein, M., Hotz-Wagenblatt, A., Miao, B., Bauer, A., Michel, A., Butt, J., Pawlita, M., Hoheisel, J.D., *et al.* (2018). Immunoprofiling of *Chlamydia trachomatis* using whole-proteome microarrays generated by on-chip *in situ* expression. Sci Rep *8*, 7503. 10.1038/s41598-018-25918-3.

Humphrys, M.S., Creasy, T., Sun, Y., Shetty, A.C., Chibucos, M.C., Drabek, E.F., Fraser, C.M., Farooq, U., Sengamalay, N., Ott, S., *et al.* (2013). Simultaneous transcriptional profiling of bacteria and their host cells. PLoS One *8*, e80597. 10.1371/journal.pone.0080597.

Jain, M., Olsen, H.E., Paten, B., and Akeson, M. (2016). The Oxford Nanopore MinION: delivery of nanopore sequencing to the genomics community. Genome biology *17*, 239. 10.1186/s13059-016-1103-0.

Jeffrey, B.M., Suchland, R.J., Quinn, K.L., Davidson, J.R., Stamm, W.E., and Rockey, D.D. (2010). Genome sequencing of recent clinical *Chlamydia trachomatis* strains identifies loci associated with tissue tropism and regions of apparent recombination. Infect Immun *78*, 2544–2553. 10.1128/iai.01324-09.

Jelocnik, M., Bachmann, N.L., Kaltenboeck, B., Waugh, C., Woolford, L., Speight, K.N., Gillett, A., Higgins, D.P., Flanagan, C., Myers, G.S., *et al.* (2015). Genetic diversity in the plasticity zone and the presence of the chlamydial plasmid differentiates Chlamydia pecorum strains from pigs, sheep, cattle, and koalas. BMC Genomics *16*, 893. 10.1186/s12864-015-2053-8.

Joseph, S.J., Didelot, X., Gandhi, K., Dean, D., and Read, T.D. (2011). Interplay of recombination and selection in the genomes of *Chlamydia trachomatis*. Biology direct *6*, 28. 10.1186/1745-6150-6-28.

Joseph, S.J., Didelot, X., Rothschild, J., de Vries, H.J., Morre, S.A., Read, T.D., and Dean, D. (2012). Population genomics of *Chlamydia trachomatis*: insights on drift, selection, recombination, and population structure. Mol Biol Evol *29*, 3933–3946. 10.1093/molbev/mss198.

Joseph, S.J., Marti, H., Didelot, X., Read, T.D., and Dean, D. (2016). Tetracycline Selective Pressure and Homologous Recombination Shape the Evolution of Chlamydia suis: A Recently Identified Zoonotic Pathogen. Genome Biol Evol *8*, 2613–2623. 10.1093/gbe/evw182.

Kari, L., Goheen, M.M., Randall, L.B., Taylor, L.D., Carlson, J.H., Whitmire, W.M., Virok, D., Rajaram, K., Endresz, V., McClarty, G., *et al.* (2011). Generation of targeted *Chlamydia trachomatis* null mutants. Proc Natl Acad Sci U S A *108*, 7189–7193. 10.1073/pnas.1102229108.

Kari, L., Whitmire, W.M., Carlson, J.H., Crane, D.D., Reveneau, N., Nelson, D.E., Mabey, D.C., Bailey, R.L., Holland, M.J., McClarty, G., *et al.* (2008). Pathogenic diversity among *Chlamydia trachomatis* ocular strains in nonhuman primates is affected by subtle genomic variations. The Journal of infectious diseases *197*, 449–456. 10.1086/525285.

Karsch-Mizrachi, I., Takagi, T., Cochrane, G., and International Nucleotide Sequence Database, C. (2018). The international nucleotide sequence database collaboration. Nucleic Acids Res *46*, D48-D51. 10.1093/nar/gkx1097.

Korhonen, S., Hiltunen-Back, E., and Puolakkainen, M. (2012). Genotyping of *Chlamydia trachomatis* in rectal and pharyngeal specimens: identification of LGV genotypes in Finland. Sexually transmitted infections *88*, 465–469. 10.1136/sextrans-2011–050458.

Lees, J.A., Kremer, P.H., Manso, A.S., Croucher, N.J., Ferwerda, B., Seron, M.V., Oggioni, M.R., Parkhill, J., Brouwer, M.C., van der Ende, A., *et al.* (2017). Large scale genomic analysis shows no evidence for pathogen adaptation between the blood and cerebrospinal fluid niches during bacterial meningitis. Microbial genomics *3*, e000103. 10.1099/mgen.0.000103.

Lieberman, T.D., Flett, K.B., Yelin, I., Martin, T.R., McAdam, A.J., Priebe, G.P., and Kishony, R. (2014). Genetic variation of a bacterial pathogen within individuals with cystic fibrosis provides a record of selective pressures. Nat Genet *46*, 82–87. 10.1038/ng.2848.

Liu, B.L., Everson, J.S., Fane, B., Giannikopoulou, P., Vretou, E., Lambden, P.R., and Clarke, I.N. (2000). Molecular characterization of a bacteriophage (Chp2) from Chlamydia psittaci. Journal of virology *74*, 3464–3469.

Lo, C.C., Xie, G., Bonner, C.A., and Jensen, R.A. (2012). The alternative translational profile that underlies the immune-evasive state of persistence in Chlamydiaceae exploits differential tryptophan contents of the protein repertoire. Microbiology and molecular biology reviews: MMBR *76*, 405–443. 10.1128/mmbr.05013–11.

Luczak, S.E., Smits, S.H., Decker, C., Nagel-Steger, L., Schmitt, L., and Hegemann, J.H. (2016). The *Chlamydia pneumoniae* Adhesin Pmp21 Forms Oligomers with Adhesive Properties. The Journal of biological chemistry *291*, 22806–22818. 10.1074/jbc.M116.728915.

Lutter, E.I., Barger, A.C., Nair, V., and Hackstadt, T. (2013). *Chlamydia trachomatis* inclusion membrane protein CT228 recruits elements of the myosin phosphatase pathway to regulate release mechanisms. Cell Rep *3*, 1921–1931. 10.1016/j.celrep.2013.04.027.

Lutter, E.I., Bonner, C., Holland, M.J., Suchland, R.J., Stamm, W.E., Jewett, T.J., McClarty, G., and Hackstadt, T. (2010). Phylogenetic analysis of *Chlamydia trachomatis* Tarp and correlation with clinical phenotype. Infect Immun *78*, 3678–3688. 10.1128/IAI.00515–10.

Marcotte, T., Lee, Y., Pandori, M., Jain, V., and Cohen, S.E. (2014). Lymphogranuloma venereum causing a persistent genital ulcer. Sexually transmitted diseases *41*, 280–282. 10.1097/olq.0000000000000104.

Marvig, R.L., Sommer, L.M., Molin, S., and Johansen, H.K. (2015). Convergent evolution and adaptation of Pseudomonas aeruginosa within patients with cystic fibrosis. Nat Genet *47*, 57–64. 10.1038/ng.3148.

Matsumoto, A., Izutsu, H., Miyashita, N., and Ohuchi, M. (1998). Plaque formation by and plaque cloning of *Chlamydia trachomatis* biovar trachoma. J Clin Microbiol *36*, 3013–3019.

Maurer, A.P., Mehlitz, A., Mollenkopf, H.J., and Meyer, T.F. (2007). Gene expression profiles of Chlamydophila pneumoniae during the developmental cycle and iron depletion-mediated persistence. PLoS Pathog *3*, e83. 10.1371/journal.ppat.0030083.

McClarty, G., Caldwell, H.D., and Nelson, D.E. (2007). Chlamydial interferon gamma immune evasion influences infection tropism. Current opinion in microbiology *10*, 47–51. 10.1016/j.mib.2006.12.003.

McClure, E.E., Chavez, A.S.O., Shaw, D.K., Carlyon, J.A., Ganta, R.R., Noh, S.M., Wood, D.O., Bavoil, P.M., Brayton, K.A., Martinez, J.J., *et al.* (2017). Engineering of obligate intracellular bacteria: progress, challenges and paradigms. Nature reviews Microbiology *15*, 544–558. 10.1038/nrmicro.2017.59.

Mehlitz, A., Banhart, S., Maurer, A.P., Kaushansky, A., Gordus, A.G., Zielecki, J., Macbeath, G., and Meyer, T.F. (2010). Tarp regulates early Chlamydia-induced host cell survival through interactions with the human adaptor protein SHC1. J Cell Biol *190*, 143–157. 10.1083/jcb.200909095.

Merhej, V., Royer-Carenzi, M., Pontarotti, P., and Raoult, D. (2009). Massive comparative genomic analysis reveals convergent evolution of specialized bacteria. Biology direct *4*, 13. 10.1186/1745–6150–4–13.

Miura, K., Toh, H., Hirakawa, H., Sugii, M., Murata, M., Nakai, K., Tashiro, K., Kuhara, S., Azuma, Y., and Shirai, M. (2008). Genome-wide analysis of Chlamydophila pneumoniae gene expression at the late stage of infection. DNA Res *15*, 83–91. 10.1093/dnares/dsm032.

Mojica, S., Huot Creasy, H., Daugherty, S., Read, T.D., Kim, T., Kaltenboeck, B., Bavoil, P., and Myers, G.S. (2011). Genome sequence of the obligate intracellular animal pathogen Chlamydia pecorum E58. J Bacteriol *193*, 3690. 10.1128/jb.00454–11.

Molleken, K., Schmidt, E., and Hegemann, J.H. (2010). Members of the Pmp protein family of *Chlamydia pneumoniae* mediate adhesion to human cells via short repetitive peptide motifs. Molecular microbiology *78*, 1004–1017. 10.1111/j.1365–2958.2010.07386.x.

Moran, N.A. (1996). Accelerated evolution and Muller's rachet in endosymbiotic bacteria. Proc Natl Acad Sci U S A *93*, 2873–2878.

Morowitz, M.J., Denef, V.J., Costello, E.K., Thomas, B.C., Poroyko, V., Relman, D.A., and Banfield, J.F. (2011). Strain-resolved community genomic analysis of gut microbial colonization in a premature infant. Proc Natl Acad Sci U S A *108*, 1128–1133. 10.1073/pnas.1010992108.

Morrison, S.G., Giebel, A.M., Toh, E.C., Spencer, H.J., 3rd, Nelson, D.E., and Morrison, R.P. (2018). Chlamydia muridarum Genital and Gastrointestinal Infection Tropism Is Mediated by Distinct Chromosomal Factors. Infect Immun *86*. 10.1128/iai.00141–18.

Muller, H.J. (1964). THE RELATION OF RECOMBINATION TO MUTATIONAL ADVANCE. Mutation research *106*, 2–9.

Nelson, D.E., Crane, D.D., Taylor, L.D., Dorward, D.W., Goheen, M.M., and Caldwell, H.D. (2006). Inhibition of chlamydiae by primary alcohols correlates with the strain-specific complement of plasticity zone phospholipase D genes. Infect Immun *74*, 73–80. 10.1128/iai.74.1.73–80.2006.

Nicholson, T.L., Olinger, L., Chong, K., Schoolnik, G., and Stephens, R.S. (2003). Global stage-specific gene regulation during the developmental cycle of *Chlamydia trachomatis*. J Bacteriol *185*, 3179–3189.

Nunes, A., Borrego, M.J., and Gomes, J.P. (2013). Genomic features beyond *Chlamydia trachomatis* phenotypes: what do we think we know? Infection, genetics and evolution: journal of molecular epidemiology and evolutionary genetics in infectious diseases *16*, 392–400. 10.1016/j.meegid.2013.03.018.

Nunes, A., and Gomes, J.P. (2014). Evolution, phylogeny, and molecular epidemiology of Chlamydia. Infection, genetics and evolution: journal of molecular epidemiology and evolutionary genetics in infectious diseases *23*, 49–64. 10.1016/j.meegid.2014.01.029.

Nunes, A., Gomes, J.P., Mead, S., Florindo, C., Correia, H., Borrego, M.J., and Dean, D. (2007). Comparative expression profiling of the *Chlamydia trachomatis* pmp gene family for clinical and reference strains. PLoS One *2*, e878. 10.1371/journal.pone.0000878.

Nunes, A., Nogueira, P.J., Borrego, M.J., and Gomes, J.P. (2008). *Chlamydia trachomatis* diversity viewed as a tissue-specific coevolutionary arms race. Genome biology *9*, R153. 10.1186/gb-2008-9-10-r153.

Nunes, A., Nogueira, P.J., Borrego, M.J., and Gomes, J.P. (2010). Adaptive evolution of the *Chlamydia trachomatis* dominant antigen reveals distinct evolutionary scenarios for B- and T-cell epitopes: worldwide survey. PLoS One *5*. 10.1371/journal.pone.0013171.

O'Connell, C.M., AbdelRahman, Y.M., Green, E., Darville, H.K., Saira, K., Smith, B., Darville, T., Scurlock, A.M., Meyer, C.R., and Belland, R.J. (2011). Toll-like receptor 2 activation by *Chlamydia trachomatis* is plasmid dependent, and plasmid-responsive chromosomal loci are coordinately regulated in response to glucose limitation by *C. trachomatis* but not by C. muridarum. Infect Immun *79*, 1044–1056. 10.1128/iai.01118-10.

O'Neill, C.E., Skilton, R.J., Pearson, S.A., Filardo, S., Andersson, P., and Clarke, I.N. (2018). Genetic Transformation of a *C. trachomatis* Ocular Isolate With the Functional Tryptophan Synthase Operon Confers an Indole-Rescuable Phenotype. Front Cell Infect Microbiol *8*, 434. 10.3389/fcimb.2018.00434.

Ochman, H., Worobey, M., Kuo, C.H., Ndjango, J.B., Peeters, M., Hahn, B.H., and Hugenholtz, P. (2010). Evolutionary relationships of wild hominids recapitulated by gut microbial communities. PLoS biology *8*, e1000546. 10.1371/journal.pbio.1000546.

Ostergaard, O., Follmann, F., Olsen, A.W., Heegaard, N.H., Andersen, P., and Rosenkrands, I. (2016). Quantitative Protein Profiling of *Chlamydia trachomatis* Growth Forms Reveals Defence Strategies Against Tryptophan Starvation. Molecular & cellular proteomics: MCP *15*, 3540–3550. 10.1074/mcp.M116.061986.

Ouellette, S.P., Hatch, T.P., AbdelRahman, Y.M., Rose, L.A., Belland, R.J., and Byrne, G.I. (2006). Global transcriptional upregulation in the absence of increased translation in Chlamydia during IFNgamma-mediated host cell tryptophan starvation. Molecular microbiology *62*, 1387–1401. 10.1111/j.1365-2958.2006.05465.x.

Pal, S., Favaroni, A., Tifrea, D.F., Hanisch, P.T., Luczak, S.E.T., Hegemann, J.H., and de la Maza, L.M. (2017). Comparison of the nine polymorphic membrane proteins of *Chlamydia trachomatis* for their ability to induce protective immune responses in mice against a C. muridarum challenge. Vaccine *35*, 2543–2549. 10.1016/j.vaccine.2017.03.070.

Pannekoek, Y., Dickx, V., Beeckman, D.S., Jolley, K.A., Keijzers, W.C., Vretou, E., Maiden, M.C., Vanrompay, D., and van der Ende, A. (2010). Multi locus sequence typing of Chlamydia reveals an association between Chlamydia psittaci genotypes and host species. PLoS One *5*, e14179. 10.1371/journal.pone.0014179.

Pannekoek, Y., Morelli, G., Kusecek, B., Morre, S.A., Ossewaarde, J.M., Langerak, A.A., and van der Ende, A. (2008). Multi locus sequence typing of Chlamydiales: clonal groupings within the obligate intracellular bacteria *Chlamydia trachomatis*. BMC microbiology *8*, 42. 10.1186/1471-2180-8-42.

Pannekoek, Y., Qi-Long, Q., Zhang, Y.Z., and van der Ende, A. (2016). Genus delineation of Chlamydiales by analysis of the percentage of conserved proteins justifies the reunifying of the genera Chlamydia and Chlamydophila into one single genus Chlamydia. Pathog Dis *74*. 10.1093/femspd/ftw071.

Panzetta, M.E., Valdivia, R.H., and Saka, H.A. (2018). Chlamydia Persistence: A Survival Strategy to Evade Antimicrobial Effects in-vitro and in-vivo. Frontiers in microbiology *9*, 3101. 10.3389/fmicb.2018.03101.

Parks, D.H., Chuvochina, M., Waite, D.W., Rinke, C., Skarshewski, A., Chaumeil, P.A., and Hugenholtz, P. (2018). A standardized bacterial taxonomy based on genome phylogeny substantially revises the tree of life. Nat Biotechnol *36*, 996–1004. 10.1038/nbt.4229.

Patton, D.L., Teng, A., Randall, A., Liang, X., Felgner, P.L., and de la Maza, L.M. (2014). Whole genome identification of *C. trachomatis* immunodominant antigens after genital tract infections and effect of antibiotic treatment of pigtailed macaques. Journal of proteomics *108*, 99–109. 10.1016/j.jprot.2014.05.009.

Patton, M.J., Chen, C.Y., Yang, C., McCorrister, S., Grant, C., Westmacott, G., Yuan, X.Y., Ochoa, E., Fariss, R., Whitmire, W.M., *et al.* (2018). Plasmid Negative Regulation of CPAF Expression Is Pgp4 Independent and Restricted to Invasive *Chlamydia trachomatis* Biovars. MBio *9*. 10.1128/mBio.02164-17.

Pendle, S., and Gowers, A. (2012). Reactive arthritis associated with proctitis due to *Chlamydia trachomatis* serovar L2b. Sexually transmitted diseases *39*, 79–80. 10.1097/OLQ.0b013e318235b256.

Picard, M.D., Cohane, K.P., Gierahn, T.M., Higgins, D.E., and Flechtner, J.B. (2012). High-throughput proteomic screening identifies *Chlamydia trachomatis* antigens that are capable of eliciting T cell and antibody responses that provide protection against vaginal challenge. Vaccine *30*, 4387–4393. 10.1016/j.vaccine.2012.01.017.

Pinto, M., Borges, V., Antelo, M., Pinheiro, M., Nunes, A., Azevedo, J., Borrego, M.J., Mendonca, J., Carpinteiro, D., Vieira, L., *et al.* (2016). Genome-scale analysis of the non-cultivable Treponema pallidum reveals extensive within-patient genetic variation. Nature microbiology *2*, 16190. 10.1038/nmicrobiol.2016.190.

Pruneda, J.N., Bastidas, R.J., Bertsoulaki, E., Swatek, K.N., Santhanam, B., Clague, M.J., Valdivia, R.H., Urbe, S., and Komander, D. (2018). A Chlamydia effector combining deubiquitination and acetylation activities induces Golgi fragmentation. Nature microbiology *3*, 1377–1384. 10.1038/s41564-018-0271-y.

Putman, T.E., Suchland, R.J., Ivanovitch, J.D., and Rockey, D.D. (2013). Culture-independent sequence analysis of *Chlamydia trachomatis* in urogenital specimens identifies regions of recombination and in-patient sequence mutations. Microbiology *159*, 2109–2117. 10.1099/mic.0.070029–0.

Ramsey, K.H., Schripsema, J.H., Smith, B.J., Wang, Y., Jham, B.C., O'Hagan, K.P., Thomson, N.R., Murthy, A.K., Skilton, R.J., Chu, P., *et al.* (2014). Plasmid CDS5 influences infectivity and virulence in a mouse model of *Chlamydia trachomatis* urogenital infection. Infect Immun *82*, 3341–3349. 10.1128/iai.01795–14.

Rank, R.G., Bowlin, A.K., Cane, S., Shou, H., Liu, Z., Nagarajan, U.M., and Bavoil, P.M. (2009). Effect of Chlamydiaphage phiCPG1 on the course of conjunctival infection with 'Chlamydia caviae' in guinea pigs. Infect Immun 77, 1216–1221. 10.1128/IAI.01109–08.

Read, T.D., Brunham, R.C., Shen, C., Gill, S.R., Heidelberg, J.F., White, O., Hickey, E.K., Peterson, J., Utterback, T., Berry, K., *et al.* (2000). Genome sequences of *Chlamydia trachomatis* MoPn and *Chlamydia pneumoniae* AR39. Nucleic Acids Res *28*, 1397–1406.

Read, T.D., Myers, G.S., Brunham, R.C., Nelson, W.C., Paulsen, I.T., Heidelberg, J., Holtzapple, E., Khouri, H., Federova, N.B., Carty, H.A., *et al.* (2003). Genome sequence of Chlamydophila caviae (Chlamydia psittaci GPIC): examining the role of niche-specific genes in the evolution of the Chlamydiaceae. Nucleic Acids Res *31*, 2134–2147.

Richmond, S.J., Stirling, P., and Ashley, C.R. (1982). Virus infecting the reticulate bodies of an avian strain of Chlamydia psittaci. FEMS Microbiol Lett *14*, 31–36.

Roberts, C.H., Ouburg, S., Preston, M.D., de Vries, H.J.C., Holland, M.J., and Morre, S.A. (2018). Pathway-Wide Genetic Risks in Chlamydial Infections Overlap between Tissue Tropisms: A Genome-Wide Association Scan. Mediators of inflammation *2018*, 3434101. 10.1155/2018/3434101.

Rother, M., Gonzalez, E., Teixeira da Costa, A.R., Wask, L., Gravenstein, I., Pardo, M., Pietzke, M., Gurumurthy, R.K., Angermann, J., Laudeley, R., *et al.* (2018). Combined Human Genome-wide RNAi and Metabolite Analyses Identify IMPDH as a Host-Directed Target against Chlamydia Infection. Cell host & microbe *23*, 661–671.e668. 10.1016/j.chom.2018.04.002.

Roulis, E., Bachmann, N.L., Myers, G.S., Huston, W., Summersgill, J., Hudson, A., Dreses-Werringloer, U., Polkinghorne, A., and Timms, P. (2015). Comparative genomic analysis of human *Chlamydia pneumoniae* isolates from respiratory, brain and cardiac tissues. Genomics *106*, 373–383. 10.1016/j.ygeno.2015.09.008.

Rzomp, K.A., Moorhead, A.R., and Scidmore, M.A. (2006). The GTPase Rab4 interacts with *Chlamydia trachomatis* inclusion membrane protein CT229. Infect Immun *74*, 5362–5373. 10.1128/IAI.00539-06.

Sachse, K., Bavoil, P.M., Kaltenboeck, B., Stephens, R.S., Kuo, C.C., Rossello-Mora, R., and Horn, M. (2015). Emendation of the family Chlamydiaceae: proposal of a single genus, Chlamydia, to include all currently recognized species. Syst Appl Microbiol *38*, 99–103. 10.1016/j.syapm.2014.12.004.

Sachse, K., Laroucau, K., Riege, K., Wehner, S., Dilcher, M., Creasy, H.H., Weidmann, M., Myers, G., Vorimore, F., Vicari, N., *et al.* (2014). Evidence for the existence of two new members of the family Chlamydiaceae and proposal of Chlamydia avium sp. nov. and Chlamydia gallinacea sp. nov. Syst Appl Microbiol *37*, 79–88. 10.1016/j.syapm.2013.12.004.

Sait, M., Livingstone, M., Clark, E.M., Wheelhouse, N., Spalding, L., Markey, B., Magnino, S., Lainson, F.A., Myers, G.S., and Longbottom, D. (2014). Genome sequencing and comparative analysis of three Chlamydia pecorum strains associated with different pathogenic outcomes. BMC Genomics *15*, 23. 10.1186/1471-2164-15-23.

Sait, M., Livingstone, M., Graham, R., Inglis, N.F., Wheelhouse, N., and Longbottom, D. (2011). Identification, sequencing and molecular analysis of Chp4, a novel chlamydiaphage of Chlamydophila abortus belonging to the family Microviridae. The Journal of general virology *92*, 1733–1737. 10.1099/vir.0.031583-0.

Saka, H.A., Thompson, J.W., Chen, Y.S., Kumar, Y., Dubois, L.G., Moseley, M.A., and Valdivia, R.H. (2011). Quantitative proteomics reveals metabolic and pathogenic properties of *Chlamydia trachomatis* developmental forms. Molecular microbiology *82*, 1185–1203. 10.1111/j.1365-2958.2011.07877.x.

Salim, O., Skilton, R.J., Lambden, P.R., Fane, B.A., and Clarke, I.N. (2008). Behind the chlamydial cloak: the replication cycle of chlamydiaphage Chp2, revealed. Virology 377, 440–445. 10.1016/j.virol.2008.05.001.

Savage, E.J., van de Laar, M.J., Gallay, A., van der Sande, M., Hamouda, O., Sasse, A., Hoffmann, S., Diez, M., Borrego, M.J., Lowndes, C.M., *et al.* (2009). Lymphogranuloma venereum in Europe, 2003–2008. Euro surveillance: bulletin Europeen sur les maladies transmissibles = European communicable disease bulletin *14*.

Scidmore, M.A. (2011). Recent advances in Chlamydia subversion of host cytoskeletal and membrane trafficking pathways. Microbes Infect *13*, 527–535. 10.1016/j.micinf.2011.02.001.

Scidmore, M.A., and Hackstadt, T. (2001). Mammalian 14–3–3beta associates with the *Chlamydia trachomatis* inclusion membrane via its interaction with IncG. Molecular microbiology *39*, 1638–1650.

Seth-Smith, H.M., Harris, S.R., Persson, K., Marsh, P., Barron, A., Bignell, A., Bjartling, C., Clark, L., Cutcliffe, L.T., Lambden, P.R., *et al.* (2009). Co-evolution of genomes and plasmids within *Chlamydia trachomatis* and the emergence in Sweden of a new variant strain. BMC Genomics *10*, 239. 10.1186/1471-2164-10-239.

Seth-Smith, H.M., Harris, S.R., Skilton, R.J., Radebe, F.M., Golparian, D., Shipitsyna, E., Duy, P.T., Scott, P., Cutcliffe, L.T., O'Neill, C., *et al.* (2013). Whole-genome sequences of *Chlamydia trachomatis* directly from clinical samples without culture. Genome Res *23*, 855–866. 10.1101/gr.150037.112.

Seth-Smith, H.M., Wanninger, S., Bachmann, N., Marti, H., Qi, W., Donati, M., di Francesco, A., Polkinghorne, A., and Borel, N. (2017a). The Chlamydia suis Genome Exhibits High Levels of Diversity, Plasticity, and Mobile Antibiotic Resistance: Comparative Genomics of a Recent Livestock Cohort Shows Influence of Treatment

Regimes. Genome Biol Evol *9*, 750–760. 10.1093/gbe/evx043.

Seth-Smith, H.M.B., Buso, L.S., Livingstone, M., Sait, M., Harris, S.R., Aitchison, K.D., Vretou, E., Siarkou, V.I., Laroucau, K., Sachse, K., *et al.* (2017b). European Chlamydia abortus livestock isolate genomes reveal unusual stability and limited diversity, reflected in geographical signatures. BMC Genomics *18*, 344. 10.1186/s12864-017-3657-y.

Sharma, J., Zhong, Y., Dong, F., Piper, J.M., Wang, G., and Zhong, G. (2006). Profiling of human antibody responses to *Chlamydia trachomatis* urogenital tract infection using microplates arrayed with 156 chlamydial fusion proteins. Infect Immun *74*, 1490–1499. 10.1128/iai.74.3.1490-1499.2006.

Shaw, A.C., Christiansen, G., and Birkelund, S. (1999). Effects of interferon gamma on *Chlamydia trachomatis* serovar A and L2 protein expression investigated by two-dimensional gel electrophoresis. Electrophoresis *20*, 775–780. 10.1002/(sici)1522-2683(19990101)20:4/5<775::Aid-elps775>3.0.Co;2-z.

Shaw, E.I., Dooley, C.A., Fischer, E.R., Scidmore, M.A., Fields, K.A., and Hackstadt, T. (2000). Three temporal classes of gene expression during the *Chlamydia trachomatis* developmental cycle. Molecular microbiology *37*, 913–925.

Sixt, B.S., Bastidas, R.J., Finethy, R., Baxter, R.M., Carpenter, V.K., Kroemer, G., Coers, J., and Valdivia, R.H. (2017). The *Chlamydia trachomatis* Inclusion Membrane Protein CpoS Counteracts STING-Mediated Cellular Surveillance and Suicide Programs. Cell host & microbe *21*, 113–121. 10.1016/j.chom.2016.12.002.

Sixt, B.S., and Valdivia, R.H. (2016). Molecular Genetic Analysis of Chlamydia Species. Annual review of microbiology *70*, 179–198. 10.1146/annurev-micro-102215-095539.

Skipp, P., Robinson, J., O'Connor, C.D., and Clarke, I.N. (2005). Shotgun proteomic analysis of *Chlamydia trachomatis*. Proteomics *5*, 1558–1573. 10.1002/pmic.200401044.

Skipp, P.J., Hughes, C., McKenna, T., Edwards, R., Langridge, J., Thomson, N.R., and Clarke, I.N. (2016). Quantitative Proteomics of the Infectious and Replicative Forms of *Chlamydia trachomatis*. PLoS One *11*, e0149011. 10.1371/journal.pone.0149011.

Song, L., Carlson, J.H., Zhou, B., Virtaneva, K., Whitmire, W.M., Sturdevant, G.L., Porcella, S.F., McClarty, G., and Caldwell, H.D. (2014). Plasmid-mediated transformation tropism of chlamydial biovars. Pathog Dis *70*, 189–193. 10.1111/2049-632x.12104.

Starnbach, M.N., Loomis, W.P., Ovendale, P., Regan, D., Hess, B., Alderson, M.R., and Fling, S.P. (2003). An inclusion membrane protein from *Chlamydia trachomatis* enters the MHC class I pathway and stimulates a CD8+ T cell response. Journal of immunology (Baltimore, Md: 1950) *171*, 4742–4749.

Staub, E., Marti, H., Biondi, R., Levi, A., Donati, M., Leonard, C.A., Ley, S.D., Pillonel, T., Greub, G., Seth-Smith, H.M.B., *et al.* (2018). Novel Chlamydia species isolated from snakes are temperature-sensitive and exhibit decreased susceptibility to azithromycin. Sci Rep *8*, 5660. 10.1038/s41598-018-23897-z.

Steiper, M.E., and Young, N.M. (2006). Primate molecular divergence dates. Molecular phylogenetics and evolution *41*, 384–394. 10.1016/j.ympev.2006.05.021.

Stephens, R.S., Kalman, S., Lammel, C., Fan, J., Marathe, R., Aravind, L., Mitchell, W., Olinger, L., Tatusov, R.L., Zhao, Q., *et al.* (1998). Genome sequence of an obligate intracellular pathogen of humans: *Chlamydia trachomatis*. Science *282*, 754–759.

Stephens, R.S., Myers, G., Eppinger, M., and Bavoil, P.M. (2009). Divergence without difference: phylogenetics and taxonomy of Chlamydia resolved. FEMS immunology and medical microbiology *55*, 115–119. 10.1111/j.1574-695X.2008.00516.x.

Sturdevant, G.L., Kari, L., Gardner, D.J., Olivares-Zavaleta, N., Randall, L.B., Whitmire, W.M., Carlson, J.H., Goheen, M.M., Selleck, E.M., Martens, C., *et al.* (2010). Frameshift mutations in a single novel virulence factor alter the *in vivo* pathogenicity of *Chlamydia trachomatis* for the female murine genital tract. Infect Immun *78*, 3660–3668. 10.1128/IAI.00386-10.

Subtil, A., Delevoye, C., Balana, M.E., Tastevin, L., Perrinet, S., and Dautry-Varsat, A. (2005). A directed screen for chlamydial proteins secreted by a type III mechanism identifies a translocated protein and numerous other new candidates. Molecular microbiology *56*, 1636–1647. 10.1111/j.1365-2958.2005.04647.x.

Suchland, R.J., Dimond, Z.E., Putman, T.E., and Rockey, D.D. (2017). Demonstration of Persistent Infections and Genome Stability by Whole-Genome Sequencing of Repeat-Positive, Same-Serovar *Chlamydia trachomatis* Collected From the Female Genital Tract. The Journal of infectious diseases *215*, 1657–1665. 10.1093/infdis/jix155.

Susnea, I., Bunk, S., Wendel, A., Hermann, C., and Przybylski, M. (2011). Biomarker candidates of Chlamydophila pneumoniae proteins and protein fragments identified by affinity-proteomics using FTICR-MS and LC-MS/MS. Journal of the American Society for Mass Spectrometry *22*, 784–788. 10.1007/s13361-011-0082-3.

Tampakaki, A.P., Fadouloglou, V.E., Gazi, A.D., Panopoulos, N.J., and Kokkinidis, M. (2004). Conserved features of type III secretion. Cell Microbiol *6*, 805–816. 10.1111/j.1462-5822.2004.00432.x.

Tan, C., Hsia, R.C., Shou, H., Carrasco, J.A., Rank, R.G., and Bavoil, P.M. (2010). Variable expression of surface-exposed polymorphic membrane proteins in *in vitro*-grown *Chlamydia trachomatis*. Cell Microbiol *12*, 174–187. 10.1111/j.1462-5822.2009.01389.x.

Tan, C., Hsia, R.C., Shou, H., Haggerty, C.L., Ness, R.B., Gaydos, C.A., Dean, D., Scurlock, A.M., Wilson, D.P., and Bavoil, P.M. (2009). *Chlamydia trachomatis*-infected patients display variable antibody profiles against the nine-member polymorphic membrane protein family. Infect Immun *77*, 3218–3226. 10.1128/iai.01566-08.

Tatusova, T., DiCuccio, M., Badretdin, A., Chetvernin, V., Nawrocki, E.P., Zaslavsky, L., Lomsadze, A., Pruitt, K.D., Borodovsky, M., and Ostell, J. (2016). NCBI prokaryotic genome annotation pipeline. Nucleic Acids Res *44*, 6614–6624. 10.1093/nar/gkw569.

Taylor-Brown, A., Pillonel, T., Bridle, A., Qi, W., Bachmann, N.L., Miller, T.L., Greub, G., Nowak, B., Seth-Smith, H.M.B., Vaughan, L., *et al.* (2017). Culture-independent genomics of a novel chlamydial pathogen of fish provides new insight into host-specific adaptations utilized by these intracellular bacteria. Environ Microbiol *19*, 1899–1913. 10.1111/1462-2920.13694.

Thalmann, J., Janik, K., May, M., Sommer, K., Ebeling, J., Hofmann, F., Genth, H., and Klos, A. (2010). Actin re-organization induced by *Chlamydia trachomatis* serovar D-evidence for a critical role of the effector protein CT166 targeting Rac. PLoS One *5*, e9887. 10.1371/journal.pone.0009887.

Thomson, N.R., Holden, M.T., Carder, C., Lennard, N., Lockey, S.J., Marsh, P., Skipp, P., O'Connor, C.D., Goodhead, I., Norbertzcak, H., *et al.* (2008). *Chlamydia trachomatis*: genome sequence analysis of lymphogranuloma venereum isolates. Genome Res *18*, 161–171. 10.1101/gr.7020108.

Thomson, N.R., Yeats, C., Bell, K., Holden, M.T., Bentley, S.D., Livingstone, M., Cerdeno-Tarraga, A.M., Harris, B., Doggett, J., Ormond, D., *et al.* (2005). The Chlamydophila abortus genome sequence reveals an array of variable proteins that contribute to interspecies variation. Genome Res *15*, 629–640. 10.1101/gr.3684805.

Tripp, H.J., Sutton, G., White, O., Wortman, J., Pati, A., Mikhailova, N., Ovchinnikova, G., Payne, S.H., Kyrpides, N.C., and Ivanova, N. (2015). Towards a standard in structural genome annotation for prokaryotes. Stand Genomic Sci *10*, 45. 10.1186/s40793-015-0034-9.

Tully, B.J., Graham, E.D., and Heidelberg, J.F. (2018). The reconstruction of 2,631 draft metagenome-assembled genomes from the global oceans. Sci Data *5*, 170203. 10.1038/sdata.2017.203.

Turaev, D., and Rattei, T. (2016). High definition for systems biology of microbial communities: metagenomics gets genome-centric and strain-resolved. Curr Opin Biotechnol 39, 174–181. 10.1016/j.copbio.2016.04.011.

UniProt, C. (2019). UniProt: a worldwide hub of protein knowledge. Nucleic Acids Res *47*, D506-D515. 10.1093/nar/gky1049.

Valdivia, R.H. (2008). Chlamydia effector proteins and new insights into chlamydial cellular microbiology. Current opinion in microbiology *11*, 53–59. 10.1016/j.mib.2008.01.003.

Valdivia, R.H., and Bastidas, R.J. (2018). The Expanding Molecular Genetics Tool Kit in Chlamydia. J Bacteriol *200*. 10.1128/jb.00590-18.

Van Lent, S., Piet, J.R., Beeckman, D., van der Ende, A., Van Nieuwerburgh, F., Bavoil, P., Myers, G., Vanrompay, D., and Pannekoek, Y. (2012). Full genome sequences of all nine Chlamydia psittaci genotype reference strains. J Bacteriol *194*, 6930–6931. 10.1128/JB.01828-12.

Vandahl, B.B., Birkelund, S., and Christiansen, G. (2004). Genome and proteome analysis of Chlamydia. Proteomics *4*, 2831–2842. 10.1002/pmic.200400940.

Vandahl, B.B., Birkelund, S., Demol, H., Hoorelbeke, B., Christiansen, G., Vandekerckhove, J., and Gevaert, K. (2001). Proteome analysis of the *Chlamydia pneumoniae* elementary body. Electrophoresis *22*, 1204–1223. 10.1002/1522-2683()22:6<1204::Aid-elps1204>3.0.Co;2-m.

Versteeg, B., Bruisten, S.M., Pannekoek, Y., Jolley, K.A., Maiden, M.C.J., van der Ende, A., and Harrison, O.B. (2018). Genomic analyses of the *Chlamydia trachomatis* core genome show an association between chromosomal genome, plasmid type and disease. BMC Genomics *19*, 130. 10.1186/s12864-018-4522-3.

Viberg, L.T., Sarovich, D.S., Kidd, T.J., Geake, J.B., Bell, S.C., Currie, B.J., and Price, E.P. (2017). Within-Host Evolution of Burkholderia pseudomallei during Chronic Infection of Seven Australasian Cystic Fibrosis Patients. MBio *8*. 10.1128/mBio.00356-17.

Voigt, A., Schofl, G., and Saluz, H.P. (2012). The Chlamydia psittaci genome: a comparative analysis of intracellular pathogens. PLoS One *7*, e35097. 10.1371/journal.pone.0035097.

Vorimore, F., Hsia, R.C., Huot-Creasy, H., Bastian, S., Deruyter, L., Passet, A., Sachse, K., Bavoil, P., Myers, G., and Laroucau, K. (2013). Isolation of a New Chlamydia species from the Feral Sacred Ibis (Threskiornis aethiopicus): Chlamydia ibidis. PLoS One *8*, e74823. 10.1371/journal.pone.0074823.

Votintseva, A.A., Bradley, P., Pankhurst, L., Del Ojo Elias, C., Loose, M., Nilgiriwala, K., Chatterjee, A., Smith, E.G., Sanderson, N., Walker, T.M., *et al.* (2017). Same-Day Diagnostic and Surveillance Data for Tuberculosis via Whole-Genome Sequencing of Direct Respiratory Samples. J Clin Microbiol *55*, 1285–1298. 10.1128/jcm.02483-16.

Wang, Y., Cutcliffe, L.T., Skilton, R.J., Persson, K., Bjartling, C., and Clarke, I.N. (2013). Transformation of a plasmid-free, genital tract isolate of *Chlamydia trachomatis* with a plasmid vector carrying a deletion in CDS6 revealed that this gene regulates inclusion phenotype. Pathog Dis *67*, 100–103. 10.1111/2049-632x.12024.

Wang, Y., Cutcliffe, L.T., Skilton, R.J., Ramsey, K.H., Thomson, N.R., and Clarke, I.N. (2014). The genetic basis of plasmid tropism between *Chlamydia trachomatis* and Chlamydia muridarum. Pathog Dis *72*, 19–23. 10.1111/2049-632x.12175.

Wehrl, W., Brinkmann, V., Jungblut, P.R., Meyer, T.F., and Szczepek, A.J. (2004). From the inside out--processing of the Chlamydial autotransporter PmpD and its role in bacterial adhesion and activation of human host cells. Molecular microbiology *51*, 319–334. 10.1046/j.1365-2958.2003.03838.x.

Weinmaier, T., Hoser, J., Eck, S., Kaufhold, I., Shima, K., Strom, T.M., Rattei, T., and Rupp, J. (2015). Genomic factors related to tissue tropism in *Chlamydia pneumoniae* infection. BMC Genomics *16*, 268. 10.1186/s12864-015-1377-8.

Westermann, A.J., Barquist, L., and Vogel, J. (2017). Resolving host–pathogen interactions by dual RNA-seq. PLoS Pathog *13*, e1006033. 10.1371/journal.ppat.1006033.

Wolff, B.J., Morrison, S.S., Pesti, D., Ganakammal, S.R., Srinivasamoorthy, G., Changayil, S., Weil, M.R., MacCannell, D., Rowe, L., Frace, M., *et al.* (2015). Chlamydia psittaci comparative genomics reveals intraspecies variations in the putative outer membrane and type III secretion system genes. Microbiology *161*, 1378–1391. 10.1099/mic.0.000097.

Yeruva, L., Myers, G.S., Spencer, N., Creasy, H.H., Adams, N.E., Maurelli, A.T., McChesney, G.R., Cleves, M.A., Ravel, J., Bowlin, A., *et al.* (2014). Early microRNA expression profile as a prognostic biomarker for the development of pelvic inflammatory disease in a mouse model of chlamydial genital infection. MBio 5, e01241–01214. 10.1128/mBio.01241-14.

Young, B.C., Wu, C.H., Gordon, N.C., Cole, K., Price, J.R., Liu, E., Sheppard, A.E., Perera, S., Charlesworth, J., Golubchik, T., *et al.* (2017). Severe infections emerge from commensal bacteria by adaptive evolution. eLife *6*. 10.7554/eLife.30637.

Ziklo, N., Huston, W.M., Hocking, J.S., and Timms, P. (2016). *Chlamydia trachomatis* Genital Tract Infections: When Host Immune Response and the Microbiome Collide. Trends Microbiol *24*, 750–765. 10.1016/j.tim.2016.05.007.

Zojer, M., Schuster, L.N., Schulz, F., Pfundner, A., Horn, M., and Rattei, T. (2017). Variant profiling of evolving prokaryotic populations. PeerJ *5*, e2997. 10.7717/peerj.2997.

Zomorodipour, A., and Andersson, S.G. (1999). Obligate intracellular parasites: Rickettsia prowazekii and *Chlamydia trachomatis*. FEBS letters *452*, 11–15.

Chlamydia and Innate Immunity

13

Uma M. Nagarajan[1*], Breanna J. Turman[1], Michael R. Knittler[2] and Andreas Klos[3]

[1]University of North Carolina, Chapel Hill, NC, USA.
[2]Institute of Immunology, Friedrich Loeffler Institute, Federal Research Institute for Animal Health, Greifswald–Isle of Riems, Germany.
[3]Hannover Medical School, Institute for Medical Microbiology and Hospital Epidemiology, Hannover, Germany.

*Correspondence: nagaraja@email.unc.edu

https://doi.org/10.21775/9781912530281.13

Abstract

As an obligate intracellular bacterium, *Chlamydia* is detected by intracellular and extracellular host proteins to mount an innate immune response. The extracellular innate detectors in body fluids include complement and antimicrobial peptides, while the intracellular detectors are the pathogen recognition receptors. Engagement of these receptors with *Chlamydia* lead to intracellular signalling resulting in cytokine/chemokine secretion. Innate immune cells, such as neutrophils, NK cells, and macrophages, come to the site of infection in response to the chemokine gradients and also produce cytokines such as IFN-γ, which can reduce bacterial burden by activation of interferon-response genes and GTPases. The innate immune response to *Chlamydia* is ineffective at clearing the infection, a finding which has long suggested that the pathogen has evolved means, at the humoral, extracellular and cytosolic levels, to subvert this host response. However, the innate response during *Chlamydia* infection is also associated with tissue pathology. Therefore, a better understanding of the key mediators of chlamydial recognition and early host responses can provide us with potential targets to mitigate host pathology.

Introduction

Due to its obligate intracellular lifestyle, chlamydiae have developed close interactions with specific host proteins for their uptake, survival, growth, and inhibition of host apoptosis. Not surprisingly, the eukaryotic host has evolved means to sense chlamydiae in body fluids and at the cellular level. Anti-microbial peptides and complement present in body fluid are bactericidal. Further, activation of the complement system causes the generation of soluble chemoattractant inflammatory mediators and immune-modulators, resulting in chlamydial killing. At the cellular level, chlamydiae are recognized by pattern recognition receptors (PRRs), which activate intracellular signalling events leading to chemokine/cytokine synthesis and secretion by the epithelial cells. Chemokines recruit 'first responder' immune cells, such as neutrophils, NK cells, and monocytes, which reduce bacterial load. Despite the presence of a significant arsenal of antimicrobial effectors, innate immune responses cannot clear the intracellular chlamydial infection. In fact, in some instances, this response can actually be exploited by chlamydiae to benefit their own uptake and transmission. The innate immune response can also result in collateral damage, suggesting a dichotomy between protection and damage to the host tissue. Therefore, insights into innate immune responses to *Chlamydia* are essential for understanding interactions of the pathogen with its host that contribute to both chlamydial control and host pathology.

In this chapter, we will discuss the innate immune molecules in the body fluid and cell-bound

PRRs, which interact with chlamydiae before and after their entry into the host cell. Next, we will discuss the cell-signalling outcome of *Chlamydia*–PRR interactions, which lead to cytokine/chemokine responses. We will outline the cell-intrinsic outcome of a major cytokine, IFN-γ, and its effect on chlamydial growth. Finally, we will conclude the chapter by discussing how innate immune signalling is integral to pathogenesis and by highlighting how chlamydial proteins or effectors could potentially subvert some of these innate mechanisms of host defence for their survival.

Innate responses in the body fluid

Physical barriers, such as mucus and secretions serve as the first line of defence against chlamydial infection. Secretions in the mucosal surface contain antimicrobial peptides that can be bactericidal. In body fluids, complement forms a cornerstone of innate immunity and is strongly intertwined with other essential components of host defence against pathogens.

Antimicrobial peptides

Defensins, protegrins, and cathelicidins are small peptides (< 45 aa) with antimicrobial properties, also known as host defence peptides (reviewed in Di Francesco *et al.*, 2013). They are preformed and secreted by leukocytes upon activation, but some are also constitutively secreted in the genitourinary and respiratory tracts. Antimicrobial peptides function mainly by disrupting microbial membranes, but they can also modulate innate immune responses via their pro- or anti-inflammatory activities or by recruiting immune cells (reviewed by Di Francesco *et al.*, 2013). Defensins are induced by cytokines such as TNFα (Zhao *et al.*, 2015b) and by stimulation of innate immune receptors (Romano Carratelli *et al.*, 2009). Human defensin HNP2 and protegrin PG-1 reduced chlamydial infection in McCoy cells in a serum-independent manner, with protegrins being particularly effective (Yasin *et al.*, 1996). Likewise, cathelicidin was found to be effective against multiple chlamydial strains (Donati *et al.*, 2005). In a cohort in Brazil, lower levels of human β-defensins were found in cervicovaginal fluid of women with chlamydial infection compared to uninfected women (Noda-Nicolau *et al.*, 2017). However, the relationship between antimicrobial peptides and *Chlamydia* is unclear because higher levels of cathelicidin LL-37 were found in vaginal washes of women with *C. trachomatis* genital infection (Hou *et al.*, 2018).

Complement system

The complement system is composed of almost 50 serum factors, their cleavage products, receptors, and soluble or cell-bound regulators (Freeley *et al.*, 2016; Hajishengallis *et al.*, 2017; Klos *et al.*, 2013; West *et al.*, 2018)]. There are three main activation pathways of the complement cascade. The 'classical' pathway is mainly triggered by immune complexes. The 'Mannan-Binding-Lectin' and the 'alternative' pathway are activated by surface components of pathogens (such as LPS or lectins) or parts of necrotic host cells (such as mitochondria). All three pathways cause cleavage of the central complement factor C3. Some complement cleavage products (e.g. C3b, C5b) bind covalently to the pathogen surface to become part of complement convertases and promote subsequent steps of the cascade, serve as a tag to improve recognition by leukocytes or promote direct pathogen cell lysis. Others (e.g. C3a or C5a) bind specific host cell receptors, such as C3aR or C5aR, thereby mediating intracellular signalling and responses. Besides serving as an immediate first line of defence, complement also triggers, enhances, and orchestrates the development of an effective adaptive B- and T-cell response. Though complement activation is a protective host response, local or systemic overactivation can be harmful (reviewed by Freeley *et al.*, 2016; Hajishengallis *et al.*, 2017; Klos *et al.*, 2013; West *et al.*, 2018).

During *C. trachomatis* infection, complement serves as a first line of defence and is activated by elementary bodies (EB) (Megran *et al.*, 1985). Complement was essential for *in vitro* neutralization of chlamydial infectivity by one monoclonal antibody in cell culture (Fedorko *et al.*, 1987). Additionally, neutralization of *C. trachomatis* infectivity by human sera was enhanced by early complement components (Lin *et al.*, 1992). Activation of the alternative pathway was also observed by the binding of the complement factor properdin to EBs of *Chlamydia pneumoniae* and C3b and C9 deposition with concomitant protection from infection in cell culture (Cortes *et al.*, 2011). Moreover, a recent study showed the opsonins iC3b and C3c, which are degradation products of C3, are deposited on the *C. trachomatis* EB surface and accelerate EB uptake and phagocytosis by monocytes. Therefore, complement activation and opsonization of EBs by iC3b deposition help monocytes remove EBs from the extracellular space, reducing infection of nearby epithelial cells.

The role of complement in chlamydial pathogenesis has been studied using gene knockout mouse models. Initial studies showed that a C5-deficient mouse strain was no more sensitive to intranasal *C. muridarum* infection than the control strain (Williams *et al.*, 1984). Moreover, mice given cobra venom factor (CVF), which is a stable C3 convertase and consumes C3, were not additionally susceptible to *Chlamydia* infection (Williams *et al.*, 1984). However, these initial results were misleading as a *single dose* of CVF likely caused only transient and ineffective consumption of C3. There are now several animal studies clearly demonstrating an important role for complement in chlamydial infection (Fig. 13.1). In a mouse lung infection model using intranasal delivery of a low dose of the *C. psittaci* DC15 strain (from a calf abort), high, early, and long-lasting complement activation was detected (Bode *et al.*, 2012). Noteworthy, in this model early complement effector functions are not important in the defence against *C. psittaci* during the first one to two weeks post infection. However, after that period, complement is highly protective, as C3-deficient mice either hardly recover or succumb to pneumonia, with a 50-fold decrease in survival compared to wild-type mice. In contrast, there was almost no effect of C5 or C5aR1 deficiency.

Intranasal infection of C3aR$^{-/-}$ mice with *C. psittaci* DC15 revealed that missing C3a-signalling (mainly on cells of myeloid origin) is less critical than complete lack of C3; nonetheless, C3a/C3aR are still indispensable for effective protection late in the course of *C. psittaci* pneumonia (Bode *et al.*, 2012). The presence of C3a/C3aR significantly increased chlamydial clearance, shortened severe acute disease, prevented weight loss, and resulted in an improved clinical outcome, with higher survival rate in mice. Intriguingly, draining lymph nodes in lungs are smaller in infected C3aR$^{-/-}$ mice, likely due to reduced numbers of CD19^{+} B cells and *Chlamydia*-specific IFN-γ producing CD4^{+} and

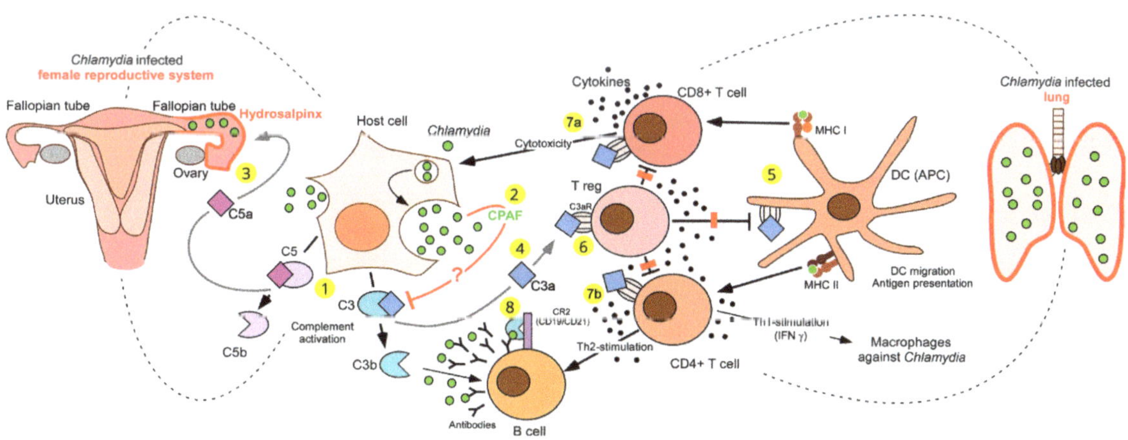

Figure 13.1. Complement-dependent regulation of the immune response against chlamydial infections in the female genital tract (left) and in the lung (right). Centre: During complement activation by *Chlamydia* (green circles), inactive complement factors C3 and C5 are cleaved into immunomodulatory C3a and C3b, and the inflammatory mediator C5a and C5b, respectively (1). CPAF may inhibit C3 activation and thereby also inhibit downstream C5 (2). Left: C5a generated from enzymatic cleavage of C5 is likely the inflammatory mediator aggravating tissue pathology and development of hydrosalpinx in *C. muridarum* infection (3). Right: In lung infection with *C. psittaci* (DC15) early, high, and long-lasting complement activation occurs with an increase of C3a (4). C3 and C3a/C3aR are indispensable for effective protection and chlamydial clearance (4), as well as DC maturation and migration (5). In addition, C3aR-expressing Treg (6), or direct effects of C3a on CD8^{+} (7a) and CD4^{+} T cells (7b) leading to IFN-γ release, or cytotoxicity, could play a role. Moreover, C3b binding to EBs could augment antibody production via CD19/CD21 on B cells (8).

CD8+ cells. Transfer of hyperimmune serum to C3aR$^{-/-}$ mice during ongoing acute infection had a very limited protective effect. Thus, in addition to the (C3b-dependent) humoral response, other C3a-dependent functions, most likely the improved T-cell response, seem to be critical for recovery and better survival of wild-type mice in primary infection (Dutow *et al.*, 2014). With extracellular bacteria, C3a acts as a weak pro-inflammatory mediator, but in the context of *C. psittaci* chlamydial infection C3a acts as a powerful immune modulator of the adaptive immune response – linking innate with adaptive immunity. So far, similar functions of C3/C3aR in mice have mainly been described for viral infections (Rattan *et al.*, 2017).

In contrast to lung infection, lack of C3 (upstream of C5) does not affect susceptibility of mice to *C. muridarum* intravaginal infection, direct infection of oviducts, or chlamydial induction of hydrosalpinx (Yang *et al.*, 2014). However, mice lacking C5 fail to develop hydrosalpinx, a pathological hallmark of tubal infertility, while 42% of wild-type mice develop hydrosalpinx. Most likely, the inflammatory mediator C5a (via C5aR), or, less likely, C5 to C9 are detrimental in this setting. Rarely, enzymatic activation of C5 bypassing C3 can occur (e.g. by thrombin) during over-activated coagulation (Huber-Lang *et al.*, 2006), which could explain the phenotype of C5$^{-/-}$ mice.

Complement is activated early and can influence pathology and disease outcome in chlamydial infection, potentially depending on the involved organ and the chlamydial species. On the one hand, C3 is highly protective in mouse lung infection by *C. psittaci* (Bode *et al.*, 2012; Dutow *et al.*, 2014) – and to a lower degree also by *C. pneumoniae* (Klos, unpublished). C3a/C3aR signalling induces an appropriate protective T-cell response to overcome lung infection. Additionally, C3b is needed to give rise to an effective antibody response. On the other hand, local over-activation of complement, in particular of C5, augments the development of hydrosalpinx induced by *C. muridarum* (Yang *et al.*, 2014). If this is also the case in humans, it might be possible to interfere therapeutically with C5a/C5aR-specific complement inhibitors (reviewed by Ricklin *et al.*, 2018) without hampering the overall protective function of complement.

Interactions with innate immune sensors on the surface and in the cytoplasm of epithelial cells

Once chlamydiae have successfully escaped the humoral innate immune response and infected an epithelial cell, they occupy an intracellular niche of an inclusion, which escapes phagolysosome fusion. However, the infection is 'visible' to membrane-bound and cytosolic innate immune system receptors called pattern recognition receptors (PRRs), which recognize conserved molecular patterns on pathogens. Fig. 13.2 summarizes multiple PRRs that sense chlamydiae or cellular alterations caused by their presence in an infected cell.

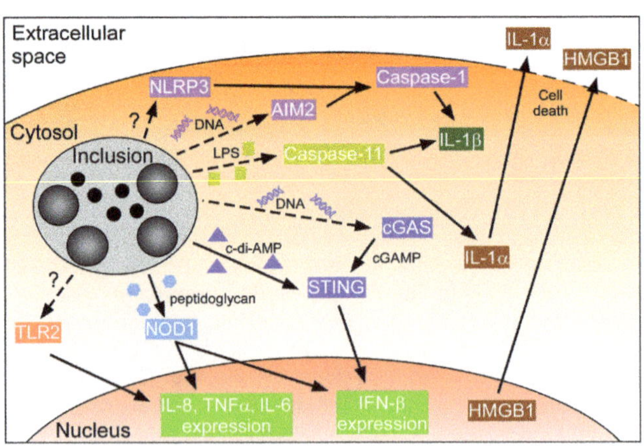

Figure 13.2. Schematic of multiple pattern recognition receptors that sense chlamydial infections. Solid line indicates confirmed ligands, while dotted lines indicate predicted chlamydial ligands. Details are provided in the text.

The Toll-like receptors

Toll-like receptors (TLRs), which are a subset of PRRs, are transmembrane proteins expressed on multiple cell types and serve as the sentinels of the innate immune system. TLR activation leads to interactions with accessory molecules in the host cytosol, resulting in intracellular signalling cascades. The outcome of these signalling events is expression of chemokine/cytokine genes, followed by immune cell influx to the infection site.

At least 13 TLRs have been identified which bind to unique molecular patterns in bacteria and virus (reviewed in Kawai and Akira, 2006). Of the TLRs, recognition of chlamydiae by TLR2 on macrophages and epithelial cells has been reported for *C. pneumoniae* (Netea *et al.*, 2002; Prebeck *et al.*, 2001), *C. muridarum* (Darville *et al.*, 2003) and *C. trachomatis* (Erridge *et al.*, 2004). TLR2 is expressed on the plasma membrane and endosomes of multiple cells types and its activation by *C. muridarum* induced gene expression of IL-6, TNFα and IL-8 in macrophages (Darville *et al.*, 2003). TLR2 recognition by *C. pneumoniae* in macrophages has also been shown to play a role in foam cell formation, which mediates the development of atherosclerotic lesions (Cao *et al.*, 2007). In HEK293T cells transfected with GFP-tagged TLR2, TLR2 and its signalling adaptor MyD88 co-localized with the *C. trachomatis* inclusion membrane (O'Connell *et al.*, 2006), suggesting that there may be sampling of chlamydial ligands by TLR2 on the inclusion.

TLR2 can recognize cell wall components in Gram-positive bacteria and bacterial lipoproteins (Quesniaux *et al.*, 2004; Takeuchi *et al.*, 1999). However, it remains unclear which chlamydial component is recognized by TLR2. Plasmid-cured strains of *C. muridarum* (O'Connell *et al.*, 2007) and *C. trachomatis* (O'Connell *et al.*, 2011) failed to induce cytokine stimulation in a TLR2-dependent manner in HEK reporter cells transfected with TLR2, which suggests a role for the plasmid in the generation or decoration of membrane proteins, which could serve as a TLR2 ligand. Of the proteins encoded by the plasmid, Pgp3 (or pORF5), is a major virulence factor. Although one study showed that pORF5 activates TLR2 (Zhou *et al.*, 2013), two other reports showed that purified recombinant Pgp3 failed to stimulate epithelial cells directly (Bas *et al.*, 2008; Hou *et al.*, 2018), suggesting that it is less likely to be a TLR2 ligand in this form. A recombinant form of a chlamydial lipoprotein called macrophage infectivity potentiator (MIP), which is expressed from the chlamydial chromosome, was shown to induce inflammatory cytokines through TLR2/TLR6 and CD14 (Bas *et al.*, 2008). However, *E.coli*-expressed MIP or Pgp3 used in the studies described above can have modifications that may not exist in the native protein, which may cause TLR2 activation. Therefore, it is essential to study TLR2 activation by native Pgp3 or MIP. Importantly, native *C. trachomatis* MOMP was shown to stimulate TLR2, but not TLR4, which conventionally recognizes LPS, in HEK reporter cells (Massari *et al.*, 2013). In addition, MOMP has been observed in the ER in infected cells, suggesting that MOMP could be available for recognition by TLR2 outside the inclusion (Giles and Wyrick, 2008). Based on the findings that the chlamydial plasmid and/or native MOMP contributes to TLR2 signalling, it would be of interest to determine if a plasmid-dependent regulator modifies MOMP in a manner that is recognized by host TLR2. These studies are certainly feasible with recent mass spectrometric methods for determining plasmid-dependent changes in post translational modification, such as glycosylation and lipidation of MOMP.

TLR4, is another TLR that has been well studied during *Chlamydia* infection. LPS from Gram-negative bacteria, such as *E. coli*, *Salmonella* spp. and *Yersinia* spp., contains hexacylated lipid A that is well recognized by TLR4 (reviewed by Park and Lee, 2013), leading to induction of a potent cytokine response in macrophages (Frendéus *et al.*, 2001). However, *Chlamydia* spp., like *Legionella* spp. and *Francisella* spp., produce weakly stimulatory LPS (Ingalls *et al.*, 1995) and failed to stimulate TLR4 in macrophages (Darville *et al.*, 2003). *C. pneumoniae* stimulate human vascular smooth muscle cell proliferation, through TLR4, which was confirmed using diphosphoryl lipid A, a competitive TLR4 antagonist (Sasu *et al.*, 2001). Recombinant Hsp60 from *C. pneumoniae* has also been shown to stimulate TLR4 in a MyD88-dependent manner to induce TNF-α (Bulut *et al.*, 2002). In *C. pneumoniae* about 50% of Hsp60 is associated with the surface of infectious EBs, which suggests that it could directly stimulate TLR4 on the plasma membrane of macrophages.

Besides the above-mentioned TLRs, TLR3 has been shown to contribute to an IFN-β response during *C. muridarum* infection in mouse oviduct epithelial cells (Johnson, 2004). The ligand for TLR3 is dsRNA, which is detected in the host cytosol during viral infection. The presence of dsRNA during *Chlamydia* infection or localization of TLR3 to the chlamydial inclusion have not been demonstrated. However, one report showed that TLR3 in keratinocytes sensed self-RNA from damaged cells (Lai *et al.*, 2009). Therefore, it is possible that TLR3 functions in an indirect manner to affect host responses during *Chlamydia* infection. It should be noted that the contribution of TLR3 to chlamydial sensing and IFN-β is partial because the DNA-sensing pathway (next section) contributes to the bulk of IFN-β expression in epithelial cells during infection. Other intracellular TLRs, such as TLR7 and TLR9, which recognize ssRNA and CpG DNA, respectively, have a minimal contribution in chlamydial recognition (Nagarajan *et al.*, 2011).

The contribution of TLRs in pathogenesis has been studied, but the results are contradictory. In a *C. muridarum* genital infection of TLR2 knockout (TLR2 KO) mice, there were significantly lower inflammatory cytokine responses, including TNF-α and MIP-2, and less oviduct pathology in comparison to wild-type mice (Darville *et al.*, 2003), confirming the role of TLR2 in cytokine production and chlamydial pathogenesis. However, Dong *et al.* reported no impact of TLR2 on oviduct pathology, despite observing a similar reduction in cytokine production in TLR2 KO mice. These differences in results could be caused by using mice not fully backcrossed to a C57BL/6 background or from using different *C. muridarum* isolates (Ramsey *et al.*, 2009). Contrary to the results for genital infection, lung infection of TLR2 KO mice with *C. muridarum* resulted in more severe disease compared to wild-type mice (He *et al.*, 2011), suggesting a tissue-specific role for TLR2 in disease outcome. These data emphasize the 'location-specific' nature of innate immune receptors, and cautions researchers to be careful about extrapolating data from one tissue site to another. TLR3 has also recently been shown to mildly contribute to resolution of infection in the mouse model of genital infection (Carrasco *et al.*, 2018; Kumar *et al.*, 2019). Additionally, it was shown that TLR3 contributes to epithelial integrity in infected epithelial cells in an IFN-β-independent manner, using cells derived from wildtype and gene knockout mice (Kumar *et al.*, 2019). These studies need to be confirmed using polarized cells and TLR3-deficient cells complemented with TLR3 to avoid confounding results from using two independent cell types. TLR7 and TLR9 have no significant contribution in chlamydial pathogenesis, consistent with their minimal role in chlamydial recognition (Nagarajan *et al.*, 2011).

Nucleotide-binding oligomerization domain-like Receptors or NOD-like receptors (NLRs)

Owing to its strict intracellular developmental nature, the likelihood of chlamydial ligands being sensed by cytosolic sensors is very high. Nod-like receptors (NLRs) are cytosolic sensors of which there are two classes whose roles have been tested during *Chlamydia* infection: NLRC (with the CARD domain) or NLRP (with a PYRIN domain). Within the NLRC protein class, the Nucleotide oligomerization domain (NOD) proteins are cytosolic proteins that sense bacterial peptidoglycan-derived components, γ-d-glutamyl-meso-diaminopimelic acid (Chamaillard *et al.*, 2003) and muramyl dipeptide (Girardin *et al.*, 2003). Long before the presence of intact peptidoglycan in *Chlamydia* was discovered (Liechti *et al.*, 2014; Packiam *et al.*, 2015), it was shown that NOD1 deficient fibroblasts and HeLa cells depleted for NOD1 expression had reduced cytokine production in response to infection (Buchholz and Stephens, 2008; Prantner *et al.*, 2010; Welter-Stahl *et al.*, 2006). These data suggested that precursors of peptidoglycan in chlamydiae are recognized by NOD1. Both NOD1 and NOD2 also play a predominant role in recognizing *C. pneumoniae* in endothelial cells (Opitz *et al.*, 2005). Contrary to the results from NOD1-deficient cells, mice deficient for NOD1 showed no difference in cytokine induction, infection course, or pathology during *C. muridarum* when compared to wildtype mice (Welter-Stahl *et al.*, 2006), suggesting that chlamydial recognition by NOD1 is not absolutely essential for the immune response during *in vivo* infection.

The NLR protein family studied most intensely in chlamydial pathogenesis are the NLRPs, as they are involved in the activation of inflammatory caspase-1 (reviewed in Ting *et al.*, 2008). The NLRPs are the major cytosolic pathogen-sensing systems, which function in parallel to classical

PRR recognition pathways. They culminate in the assembly of a large multiprotein complex, called the 'inflammasome,' which consists of multimers of the protein 'Apoptosis-associated speck-like protein containing a CARD' or ASC. Assembly of the inflammasome results in cleavage of inactive pro-caspase-1 to enzymatically active caspase-1. Caspase-1 in turn cleaves inactive pro-IL-1β and pro-IL-18 to their mature forms. Brunham and colleagues were the first to show infection of HeLa cells with several biovars of *Chlamydia* spp. stimulated caspase-1 activation and induced IL-18 secretion (Lu *et al.*, 2000). Subsequent studies in HeLa cells infected with *C. trachomatis* confirmed these findings, and siRNA knockdown studies demonstrated a role for the NLRP3 inflammasome in caspase-1 activation (Abdul-Sater *et al.*, 2009). Caspase-1 activation in resting mouse macrophages infected with *C. muridarum* is also NLRP3- and ASC-dependent. However, in macrophages pre-exposed to TLR4 or TLR2 ligand, caspase-1 activation does not require NLRP3, but requires ASC (Nagarajan *et al.*, 2012). More recently, Finnety *et al.* showed the AIM2 inflammasome, which samples DNA, also activates caspase-1 in *C. muridarum* -infected macrophages that were pre-exposed to TLR4 ligand (Finethy *et al.*, 2015). Further, caspase-1 activation in macrophages did not require bacterial growth but required viable *C. muridarum* (Prantner *et al.*, 2009) suggesting that an early chlamydial effector is required for host caspase-1 activation.

The caspase-1 pathway activated by the inflammasome described above is defined as the 'canonical' inflammasome pathway. LPS from Gram-negative bacteria can also activate another caspase, namely caspase-11, by directly binding to it. Caspase-11 activation leads to cell death by 'pyroptosis', an inflammatory cell death accompanied by the release of active IL-1β (Hagar *et al.*, 2013). This pathway is defined as the 'non-canonical' inflammasome pathway (Kayagaki *et al.*, 2011). There is evidence that *Chlamydia* activates this pathway (Finethy *et al.*, 2015) although, the stimulation is relatively weak, compared to observations with other Gram-negative bacteria (Broz *et al.*, 2012; Kayagaki *et al.*, 2011).

Caspase-1 activation was shown to contribute to oviduct pathology in the mouse model of genital *C. muridarum* infection (Cheng *et al.*, 2008). However, the inflammasome adaptors ASC and NLRP3, which are essential for caspase-1 activation *in vitro*, did not contribute to oviduct pathology (Nagarajan *et al.*, 2012). Because ASC is required for caspase-1 activation, this dichotomy could not be explained until recently. Caspase-1 knockout mice used in the previous study were shown to be inadvertently doubly deficient for caspase-1 and caspase-11 (Kayagaki *et al.*, 2011). Our results show that caspase-11 deficiency was sufficient to recapitulate the combined absence of both caspase-1 and caspase-11 on oviduct pathology (Allen *et al.*, 2019). Relative to WT mice, caspase-11 deficient mice were protected from oviduct and uterine pathology. Further, when caspase-11 expression was restored in mice deficient for both caspase-1 and caspase-11, by expressing a caspase-11 transgene, the mice were not significantly protected from oviduct pathology during infection. These data suggest that caspase-11 has a greater impact on oviduct pathology than caspase-1. As chlamydial LPS is a weak inducer of both canonical and non-canonical inflammasome (Yang *et al.*, 2019), caspase-11 activation could occur independently of LPS or require an additional stimulus present during *in vivo* infection.

Cytoplasmic DNA/RNA sensors

Because chlamydiae are restricted inside an inclusion, one would not expect the host cell to sense chlamydial DNA or RNA. However, DNA sensors contribute to the IFN-β response against *Chlamydia* infection. IFN-β is an antiviral cytokine made in response to the presence of viral DNA/RNA in the cytosol, but a similar response is also observed with intracellular bacterial infection. Unlike its antiviral role, IFN-β is not protective during *Chlamydia* infection and exacerbates pathology (Nagarajan *et al.*, 2008). The central adaptor for DNA sensing, stimulator of interferon genes (STING) was essential for IFN-β induction during *C. muridarum* infection in both mouse and human epithelial cells (Prantner *et al.*, 2010). STING was also found to localize in close proximity to the inclusion (Prantner *et al.*, 2010) and as specks in the cytosol; a result of direct sensing of chlamydial cyclic di-AMP produced by *C. trachomatis* EBs (Barker *et al.*, 2013). DNA is also sensed upstream of STING activation by the DNA sensor cyclic GAMP synthase (cGAS) during infection with multiple *C. trachomatis* serovars (Zhang *et al.*, 2014). cGAS was found localized on the cytosolic side of the chlamydial inclusion membrane, suggesting that DNA is likely sensed there.

The source of cytosolic DNA recognized by cGAS during infection remains unclear. We speculate that the cytosolic DNA sensed by cGAS is of chlamydial origin because mitochondrial damage has not been observed at the time of DNA sensing, and halting chlamydial growth with chloramphenicol treatment abrogates IFN-β expression (Prantner and Nagarajan, 2009). The question still remains if DNA-sensing is a direct or indirect effect of infection and whether chlamydial DNA enters the host cytosol. Multiple studies have linked IFN-β expression and bacterial secretion systems (Crimmins *et al.*, 2008; Roux *et al.*, 2007; Stanley *et al.*, 2007; Stetson and Medzhitov, 2006), suggesting that bacterial secretion structures may allow passage of DNA from bacteria to the host cells. A study by Manzanillo *et al.* showed the ESX-1 secretion system of *Mycobacterium tuberculosis* causes enough permeabilization of the phagosomal membrane to allow DNA sensing by host cytosolic recognition pathways. (Manzanillo *et al.*, 2012). Inhibition of the type III secretion system (T3SS) during chlamydial infection reduces the chlamydial-induced IFN-β response (Prantner and Nagarajan, 2009) suggesting that the chlamydial T3SS could function similarly to ESX-1 to permeabilize the inclusion membrane. Contact points between chlamydial RBs and the inclusion membrane (Wilson *et al.*, 2006), likely via T3SS, could act as permeabilization points allowing nucleic acids to leak into the cytosol and to be recognized by host sensors. This T3SS leakage mechanism could also deliver other chlamydial ligands such as RNA in minute quantities to the host cytosol, where they may be recognized by the innate immune sensors.

Signalling and gene expression in response to *Chlamydia* infection

Signalling

Recognition of *Chlamydia* by host cell sensors activates a plethora of signalling events, which lead to expression and secretion of chemokines and cytokines. These events mark the beginning of the cellular innate immune response to *Chlamydia* infection.

Activation of the mitogen-activated protein kinase (MAPK) family, NFκB, and IRF3 are the primary signalling events after chlamydial recognition. MAPK is comprised of three members: extracellular signal-regulated kinase (ERK), p38, and c-Jun NH_2-terminal kinase (JNK). Together, these three molecules form an extensive signalling network called the MAPK pathway that responds to growth factors, cytokines, and cell stress. ERK responds to growth factors to control cell proliferation, differentiation, and migration, while activation of JNK and/or p38 can lead to cell proliferation, differentiation, apoptosis, and expression of inflammatory genes (reviewed by Kim and Choi, 2010).

Given the robust inflammatory response to *Chlamydia* infection, it is no surprise that MAPK signalling occurs during infection. The outcome of MAPK signalling, and the mediators required for signalling, differ by chlamydial species. During *C. pneumoniae* infection, IL-8 expression in endothelial cells requires both ERK and p38 (Krull *et al.*, 2004). Furthermore, GM-CSF expression in *C. pneumoniae* infected cells is reliant on p38 (Krüll *et al.*, 2006). *C. trachomatis* induces a slightly different MAPK signature with expression of IL-8 being dependent upon ERK signalling (Buchholz and Stephens, 2007). p38 increased phosphorylation of JNK to aid in activation of the pro-inflammatory AP-1 transcription factor in *C. trachomatis*-infected epithelial cells and regulates secretion of type I interferons in *C. trachomatis*-infected fibroblasts and macrophages infected with *C. muridarum* (Olive *et al.*, 2014; Prantner, 2010; Vignola *et al.*, 2010).

Activation of the pro-inflammatory transcription factor NFκB is another key signalling event associated with MAPK activation. NFκB is expressed in immune cells as well as epithelial cells and positively controls the expression of multiple pro-inflammatory genes, including the cytokines IL-1, IL-6, IL-8, and TNF-α and the chemokines CXCL1, and CXCL10 (reviewed by Liu *et al.*, 2017). In an unactivated cell, the protein IκB is bound to NFκB to inhibit its activity. However, following activation of many pattern recognition receptors and TNF superfamily members, IκB is phosphorylated and directed for degradation. NFκB is released, dimerizes and translocates to the nucleus, where it activates expression of pro-inflammatory genes (reviewed by Lawrence, 2009).

NFκB activation in epithelial cells early in *Chlamydia* infection is conserved among *C. trachomatis*, *C. pneumoniae*, *C. muridarum*, and *C. pecorum* (Gencay *et al.*, 2003; Leonard *et al.*, 2017; Welter-Stahl *et al.*, 2006). Despite early activation, *in vitro* studies suggest that NFκB activation is inhibited at later time points during infection (Lad *et al.*, 2007; Le Negrate *et al.*, 2008). Analysis of cytokines secreted in the *C. muridarum* mouse model support NFκB activation at three hours post infection (Rank *et al.*, 2010). However, evidence of NFκB activation at later time points in this study is confounded by immune cell influx to the genital tract beginning as early as 12 hours post infection (Rank *et al.*, 2010). To study temporal activation of NFκB *in vivo* would require isolation of intact infected epithelial cells or other defined cell populations and direct measurement of phosphorylated or translocated p65. The biological benefit for early activation and potential suppression of NFκB at later timepoints remains elusive. Early suppression or late activation of NFκB in infected cells by chemical or genetic means may shed light on the significance of temporal NFκB regulation by *Chlamydia* during infection.

Besides NFκB, IRF3 is a pro-inflammatory transcription factor downstream of the pattern recognition receptors TLR3, TLR4, STING, and NOD2, and is essential for IFN-β expression (reviewed by Zhao *et al.*, 2015a). Activation of IRF3 by any of these pathways leads to IRF3 dimerization and translocation to the nucleus (reviewed by Zhao *et al.*, 2015a). During *Chlamydia* infection, IRF3 is responsible for early IFN-β expression in mouse macrophages (Nagarajan *et al.*, 2005) and primary mouse epithelial cells (Derbigny *et al.*, 2007; Hu *et al.*, 2015). This conclusion is supported by observations that IRF3 knockout mice had lower IFN-β levels in genital secretions compared to wild-type mice (Prantner *et al.*, 2011). Importantly, despite few differences in IFN-β expression over the course of *C. muridarum* infection and similar infection kinetics, IRF3 knockout mice exhibit more severe pathology in the uterine horns than their wildtype counterparts (Prantner *et al.*, 2011). This suggests that IRF3 may function in other cell signalling pathways during *Chlamydia* infection to affect genital tract pathology (Di Paolo *et al.*, 2013).

Gene expression

Activation of innate immune receptors and signalling pathways by *Chlamydia* leads to a myriad of changes in gene expression, including increased expression and secretion of chemokines and cytokines. Changes in the expression of chemokines and cytokines can occur at the transcriptional or translational levels, while secretion is usually regulated post-translationally.

Numerous studies have examined cytokine and chemokine expression in response to *Chlamydia* infection. Though there is slight variation amongst the chlamydial species, *in vitro* infected epithelial cells increase expression of the cytokines IL-6, IL-12, TNF-α, IFN-β, IL-1α, IL-8, and GM-CSF (Dessus-Babus *et al.*, 2000; Johnson, 2004; Rasmussen *et al.*, 1997) and chemokines such as CXCL1, MIP-2/CXCL2, and RANTES/CCL5 (Dessus-Babus *et al.*, 2000; Johnson, 2004; Rasmussen *et al.*, 1997). Monitoring of cytokine and chemokine levels in *C. muridarum* infected mice supports a role for these cytokines *in vivo* and expands the number of cytokines secreted to include those secreted primarily by immune cells such as IL-1β and IFN-γ (Darville *et al.*, 2001; Prantner, 2010; Rank *et al.*, 2010). Cytokines are the major mediators of pathogenesis and will be discussed in detail with innate immune cells in 'Innate immune cells and their role in pathogenesis', below. Of the cytokines IFN-γ is a critical player in chlamydial control and will be first discussed in detail.

IFN-γ-induced cell autonomous self-defence mechanisms

IFN-γ induces expression of proteins involved in cell-intrinsic immunity against several pathogens (MacMicking, 2012), including *Chlamydia*. Specifically, four families of IFN-γ inducible proteins including IFN-inducible GTPases, indoleamine-2,3-dioxygenase (IDO), IFN-γ inducible nitric oxide synthase (iNOS/NOS2), and NADPH oxidase (NOXs), are major participants in the anti-microbial attack against *Chlamydia* and play important roles in the clearance of chlamydial infections in cell culture and the mouse model (reviewed in Finethy and Coers, 2016; McClarty *et al.*, 2007; Rottenberg *et al.*, 2002).

Interferon-inducible GTPases

Two families of interferon-inducible GTPases have been described: the immunity-related GTPases (IRGs) and the guanylate-binding proteins (GBPs) (Kim *et al.*, 2012). The IRGs constitute a family of IFN-inducible GTPases that play an important role in cell-intrinsic immunity against a number of intracellular bacterial infections (Pilla-Moffett *et al.*, 2016). The IRGs include cytosolic 'effector' GKSs (glycine–lysine–serine class IRGs) and membrane-bound 'regulator' IRGMs (glycine–methionine–serine class IRGs) (Al-Zeer *et al.*, 2009; Coers *et al.*, 2008; Haldar *et al.*, 2013; MacMicking, 2012). Mouse cells exposed to IFN-γ induce high levels of GKS and IRGM. In uninfected cells, the GKS protein is predominantly in the cytosol in its GDP-bound form. In infected cells, GKSs translocate and bind to the inclusion membrane in their active GTP-bound form. Conversely, induced IRGMs were found in abundance on 'self-membranes' (e.g. on lipid droplets (LDs) or mitochondria) in host cells but at low levels on 'non-self' membranes, such as the chlamydial inclusion membrane (Haldar *et al.*, 2013). Since IRGMs are negative regulators of GKSs (and GBPs), IRGM-positive cellular 'self-membranes and -compartments' are protected against GKS-binding, whereas IRGM-negative inclusions serve as targets for GKS-/GBP-binding.

The IRG family significantly differs between mice and humans. Mice have >20 IRGs that are the main anti-chlamydial effectors (Nelson *et al.*, 2005), while in humans there is a single IRG, which is ubiquitously expressed and not inducible by IFN-γ (Bekpen *et al.*, 2005). In mice, genetic variation in the GKS family members Irgb10 and/or IIGP2 (also referred to as Irgm2) confer differential susceptibility to *C. psittaci* infection (Bernstein-Hanley *et al.*, 2006b; Miyairi *et al.*, 2007). DBA/2J mice, which express lower levels of Irgb10 than C57BL/6J mice, develop higher chlamydial burdens than their C57BL/6J counterparts and ultimately succumb to infection (Miyairi *et al.*, 2007). GKSs are also important in *C. trachomatis* infection of mouse cells. Irga6, Irgb10, and GBP1/2 all localize to *C. trachomatis* inclusion membranes following IFN-γ treatment (Al-Zeer *et al.*, 2009; Bernstein-Hanley *et al.*, 2006a; Bernstein-Hanley *et al.*, 2006b; Coers *et al.*, 2008; Haldar *et al.*, 2013). Targeting of the inclusion membrane by GKSs is important to keep *C. trachomatis* infection in check, as mouse embryonic fibroblasts (MEFs) deficient for Irgb10 are resistant to IFN-γ induced inhibition of *C. trachomatis*. The role of another GKS family member Irga6 in anti-chlamydial defence is still a little unclear. Though Irga6 associates with the inclusion membrane, Coers and colleagues found that Irga6 depletion did not interfere with *C. trachomatis* infection of mouse cells in the presence of IFN-γ (Coers *et al.*, 2008). In contrast, other studies show a role for Irga6 in controlling *C. trachomatis* infection and growth (Al-Zeer *et al.*, 2009; Nelson *et al.*, 2005). Further studies on specific membrane recognition and binding are needed to reconcile differences in observations of particular GKSs, such as Irga6. However, it is clear that GKSs play an important role in controlling chlamydial growth within infected cells and regulating the IFN-γ inducible immune response to chlamydial infection.

GBPs constitute the second family of interferon-inducible GTPases and are members of the dynamin-related large GTPase family, which assist in inflammasome activation by lysis of pathogen containing vacuoles (PV), bacteriolysis, or directly promoting formation of the inflammasome. Tietzel and co-workers (Tietzel *et al.*, 2009) showed that GBP1 and 2 interact with the membrane of *C. trachomatis* inclusions and prevent bacterial growth in epithelial cells, whereas GBP depletion decreased the inhibitory IFN-γ effect leading to improved bacterial development with larger inclusions. This kind of GBP1 recruitment by inclusions was also seen in infected dendritic cells (Fiegl *et al.*, 2013) and macrophages (Al-Zeer *et al.*, 2013). GBPs also physically associated with *C. trachomatis* inclusions in mouse fibroblasts (Haldar *et al.*, 2013; Lindenberg *et al.*, 2017). In *C. muridarum* infection of mouse macrophages, loss of GBPs led to reduction of pyroptosis, but GBPs were precluded from the chlamydial inclusion (Finethy *et al.*, 2015). Therefore, GBP-dependent inflammasome activation is independent of inclusion lysis and suggests that there is no link between GBPs and inclusion rupture (Finethy *et al.*, 2015; Haldar *et al.*, 2015). These data reveal distinct differences between the adaptability of *C. muridarum* and human chlamydial strains in infected mouse cells.

Mechanism of IRG and GBP cooperation to induce auto-/xenophagy

To mount an effective anti-chlamydial defence, IRGs often cooperate directly with GBPs. GBP localization depends on a functional IRG system (Haldar *et al.*, 2013). Contrary to the original hypothesis that GBPs and IRGs disrupt pathogen-containing vacuoles, it is now believed that these proteins function via membrane tethering. Al-Zeer and colleagues showed that GBP1 and 2 control alterations in the auto-/xenophagy pathway leading to a rerouting of inclusions to lysosomal degradation (Al-Zeer *et al.*, 2013). Auto-/xenophagy is a proteolytic degradation process in which cells remove intracellular (auto) as well as microbial material and compartments (xeno) that eliminates invading pathogens. The molecular mechanism is covered in more detail in Chapter 6, *Chlamydia* and Cell-autonomous Defence: Apoptosis and Autophagy. Interestingly, factors involved in xenophagy also activate GBPs, revealing considerable cross talk between the two pathways. Moreover, loading of IRGs on inclusions seems to depend on a functional auto-/xenophagic flux (Al-Zeer *et al.*, 2009; Khaminets *et al.*, 2010; Zhao *et al.*, 2008). In this context, recruitment of IRGs and GBPs to the inclusion membrane also requires expression of the autophagy-related (ATG) genes ATG3 and ATG5 (Haldar *et al.*, 2015). Surprisingly, induction of auto-/xenophagy without the involvement of IFN-γ signalling has no detectable effect on *C. trachomatis* growth in infected MEFs. This demonstrates a crucial role of IRGs in labelling chlamydial inclusions for auto/xenophagic elimination (Pachikara *et al.*, 2009). The link between IFN-γ signalling and xenophagy is the ubiquitination of chlamydial inclusions via IRG-dependent recruitment of E3 ubiquitin ligases (Haldar *et al.*, 2015). Ubiquitin decoration of chlamydial compartments creates physical binding sites for ubiquitin adaptor protein p62/Sequestosome 1 (SQSTM1), which in turn enhances further ubiquitination of inclusions and recruitment of GBPs. The chlamydial inclusion is now tagged for xenolysosomal elimination. Together, these data suggest the IFN-γ-mediated self-defence of infected cells against chlamydiae requires functional auto-/xenophagic degradation pathways, in addition to GBPs and IRGs.

Similar to GBP recruitment, activation of the ubiquitination pathway is also host- and *Chlamydia*-specific. *C. muridarum* blocks ubiquitination of its inclusion by blocking the IRGs in mouse cells. But in human epithelial cells, IFNγ-priming decorates mouse-specific *C. muridarum* inclusions with ubiquitin-binding proteins like LC3 and p62. In contrast, the inclusions of human-specific *C. trachomatis* do not get ubiquitinated in these cells. In human cells, unlike in mouse cells, ubiquitination of inclusions is independent of IRGs. These results suggest that *C. trachomatis* evolved a host cell-specific mechanism to avoid ubiquitination of its inclusion (Haldar *et al.*, 2016). The chlamydial mediator of this novel defence mechanism is not known. Interestingly, a similar protection is observed during *C. muridarum* infection of mouse epithelial cells. It has been proposed that *C. muridarum* prevents IFN-γ-induced xenophagic degradation of its inclusions in mouse epithelial cells by cytosolic secretion of the bacterial TC438 protein. Remarkably, this chlamydial effector is related to the UDP-glycosyltransferase domain of the large clostridial toxin (LCT) as well as the cysteine protease domain of YopT from *Yersinia pestis*, both of which are known to inactivate GTPase activity by removal of the lipid modification on the enzyme. *C. muridarum* might use cytosol-localized TC438 to block IRGs and downstream targeting of the inclusions to xenophagic degradation (Nelson *et al.*, 2005; Read *et al.*, 2000). Clearly, the findings that IRGs and GBPs mediated defence mechanisms are counteracted by *Chlamydia* in a host-specific manner, explains why multiple means of innate effectors are constantly evolving.

Role of indoleamine-2,3-dioxygenase in *C. trachomatis* infection

In human cells, the anti-chlamydial effect of IFN-γ is predominantly mediated by indoleamine-2,3-dioxygenase (IDO), a heme-containing enzyme (Byrne *et al.*, 1986), that catabolizes the essential amino acid tryptophan (Trp) (Shimizu *et al.*, 1978). Cellular IDO expression is induced by IFN-γ and leads to nutritional depletion of Trp in host cells. A partial Trp pathway consisting of gene products encoded by the operon assemblage *trpR*/*trpEb*/*trpEa* is maintained by *C. trachomatis* allowing tryptophan to be produced to some extent from indole. The source of indole is not the human host, but rather comes from indole-producing organisms that can co-exist in the tissue occupied by genital strains (please also see Chapter 16, *Chlamydia* Infections and the Microbiota, for details. Nevertheless, Trp depletion is detrimental for chlamydial growth and development (Voigt *et al.*, 2012). *C. trachomatis* can respond to Trp depletion in cell culture by adopting a viable, non-replicating phenotype (Beatty *et al.*, 1994). Importantly, IDO-mediated bacteriostatic effects on *Chlamydia* can be reversed by external addition of Trp. IDO is also important in the defence against infection by *C. pneumoniae*, another Trp-auxotroph (Njau *et al.*, 2009). Though IFN-γ/IDO-mediated Trp reduction was shown to generate signals sensed by cellular pathways linked to autophagy (Metz *et al.*, 2012), the extent that this process is involved in the xenophagic elimination of chlamydial structures after Trp depletion is not known. Furthermore, expression of IDO is inducible in many different types of human cells (Hwu *et al.*, 2000; Moon *et al.*, 2015), but not in mouse epithelial cells primed with IFN-γ (Roshick *et al.*, 2006). These data suggest unique species-specific adaptations of chlamydial strains, and one can speculate that the IDO response may play a role in establishing chronic infection in humans, which is not observed in mice. Thus, the question remains open if the IDO response benefits the host or the pathogen.

IFN-γ-inducible nitric oxide synthase (iNOS/NOS2) and NADPH oxidase (NOXs)

IFN-γ-inducible isoform of nitric oxide synthase 2 (termed iNOS and/or NOS2) is a critical player in the inhibition of intracellular bacterial growth in macrophages by producing the antibacterial and immunomodulatory compound nitric oxide (NO). Further, NO can combine with Reactive Oxygen Species (ROS) to form Reactive Nitrogen Species (RNS), which directly damage the bacterial membrane inside phagosomes. There are a number of *in vitro* and *in vivo* studies showing a correlation between the induction of NO and the inhibition of intracellular chlamydial growth (Chen *et al.*, 1996; Igietseme, 1996; Igietseme *et al.*, 1998; Igietseme *et al.*, 1997; Woods *et al.*, 1994). During *C. muridarum* infection in mice, iNOS/NOS2 activity reduced disseminated infection and protected mice from hydrosalpinx, despite similar bacterial burdens (Ramsey *et al.*, 2001a). Additional studies show that reduction of NO production in macrophages by incomplete inhibition of NOS2 in mice was sufficient to abolish susceptibility to *C. muridarum* lung infection (Huang *et al.*, 2002). iNOS/NOS2 was also required for control of *C. pneumoniae* infection, as iNOS$^{-/-}$ mice showed higher bacterial burden compared with wildtype mice (Rottenberg *et al.*, 1999), suggesting that iNOS/NOS2 is an important anti-chlamydial effector.

In addition to reactive nitrogen species (RNS), chlamydial infection also elicits the production of short-lived ROS in infected macrophages (Rajaram and Nelson, 2015). Induction of ROS during respiratory burst of macrophages is mediated via IFN-γ-dependent activation of nicotinamide adenine dinucleotide phosphate (NADPH) oxidases (NOXs) (Bedard and Krause, 2007; Casbon *et al.*, 2012). NOXs are membrane-associated multicomponent enzymes that catalyze electron transfer from NADPH to O2. Surprisingly, *C. trachomatis* has the ability to target NOX activity and shut it down to delay the host response to infection (Boncompain *et al.*, 2010). Although the regulatory NOX subunit Rac is relocated to chlamydial inclusions, the exact molecular subversion mechanism is not known (Boncompain *et al.*, 2010). Finally, the observation that ROS repress the formation of RNS by iNOS/NOS2 suggests that the balance between ROS and RNS might be critical in disease development during chlamydial infections (Ramsey *et al.*, 2001b; Rusconi and Greub, 2011).

In conclusion, the highly complex and multifaceted functions of IFN-γ-induced cellular self-defence (IRGs, GBPs, IDO, iNOS/NOS2, and NOX) in the innate immune response (Fig. 13.3) against chlamydial infections are only partly understood. Strain-, species- and cell type-specific differences in *Chlamydia*/host cell interactions suggest that further investigations are required to clarify early self-defence mechanisms in infected cells. Such studies may shed light on some of the differences in infection course and pathology observed between animal models and human patients.

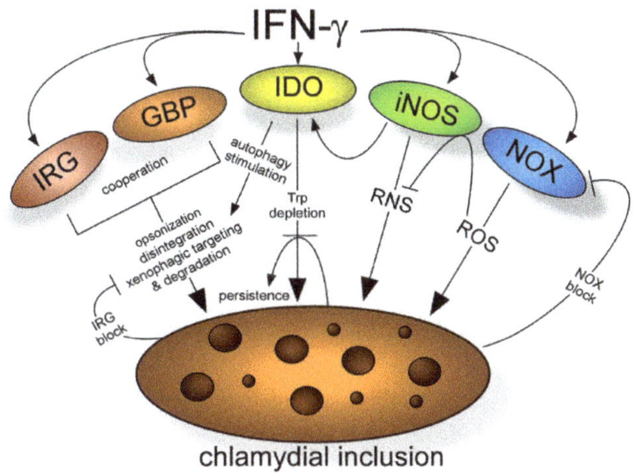

Figure 13.3. IFN-γ-induced cellular self-defence in the innate immune response against chlamydial infections cause recognition, opsonization and degradation of bacterial compartments.

Innate immune cells and their role in pathogenesis

NK cells, neutrophils and matrix metalloproteinases

Animal studies in mice and guinea pigs have shown that pathology following *Chlamydia* infections is largely attributed to a robust host innate immune response. The first set of innate immune cells that arrive at the site of infection are NK cells and neutrophils. NK cells produce high levels of IFNγ, which is essential for activation of innate cells and to promote interferon-mediated cell autonomous killing. NK cells appear in the genital tract as early as 12 h after infection and their depletion leads to an exacerbated course of infection with a diminished Th1 response (Tseng and Rank, 1998). NK cells also play an important role in host defence against chlamydial lung infection, as their depletion resulted in decreased Th1 and TH17 cells and correlated with reduced IFN-γ, IL-12, IL-17 and IL-22 (Li *et al.*, 2016).

Neutrophils, the other early responder to infection, can also be found in the submucosa as early as 12 hours post infection and in the epithelium by 24 hours post infection (Rank *et al.*, 2010). Though neutrophils alone are unable to clear the infection, neutrophil depletion by a monoclonal antibody leads to increased *Chlamydia* burden and prolonged infection *in vivo* (Barteneva *et al.*, 1996). Infection control by neutrophils is likely partially mediated by direct interactions of PMNs with infected cells (Rank *et al.*, 2008; Rank *et al.*, 2011). Neutrophils surround and dislodge infected epithelial cells from the epithelial layers. Dislodged cells would likely burst in the genital tract lumen releasing free EBs that may be more easily eliminated by phagocytes than those sequestered in inclusions. Surprisingly, neutrophils were also observed inside infected epithelial cells in the *C. muridarum* genital tract model, suggesting neutrophils invade epithelial cells to engulf and eliminate intracellular *Chlamydia* (Rank *et al.*, 2011). Further, IFN-γ activated neutrophils play a significant role in antibody-mediated killing of *Chlamydia*, emphasizing its protective role in secondary infection (Naglak *et al.*, 2017). Neutrophil-mediated control of infection is likely executed via production of reactive oxygen species (ROS), antimicrobial molecules, and proteases in response to infection. These molecules are effective at killing engulfed *Chlamydia* but are also implicated in neutrophil-mediated pathology.

Although neutrophils have some role in chlamydial killing, their association with pathology is overwhelming. In the mouse genital tract model, neutrophil influx to the oviduct is correlated with oviduct pathology and hydrosalpinx formation following *C. muridarum* infection (Shah *et al.*, 2005). In an ocular guinea pig model, depletion of neutrophils was associated with decreased pathology (Lacy *et al.*, 2011). However, neutrophil depletion also resulted in a reduced T cell response. Therefore, it is difficult to attribute decreases in pathology solely to neutrophil depletion in this study. Neutrophils and the innate immune response were further implicated in mice infected with plasmid-cured *C. muridarum* strains, which developed significantly less oviduct pathology than mice infected with wild-type *C. muridarum* (Lei *et al.*, 2014; O'Connell *et al.*, 2007). Reduction in pathology was associated with reduced neutrophil influx, and significant decreases in multiple cytokines and chemokines such as GM-CSF, CXCL1, CXCL2, and TNFα throughout the course of infection (Frazer *et al.*, 2011). A similar association of neutrophils with oviduct pathology was observed in mice infected with *C. muridarum* missing the plasmid-encoded virulence factor Pgp3 (Liu *et al.*, 2014).

At least one of the ways neutrophils affect pathology is through extracellular release of matrix metalloproteinases (MMPs). MMPs are enzymes that rearrange the extracellular matrix, process chemokines and cytokines, release growth and signalling factors, and aid in leukocyte chemotaxis through tissues (Page-McCaw *et al.*, 2007). MMP expression is increased during *Chlamydia* infection (Ramsey *et al.*, 2005). Specifically, MMP-9 is highly expressed in response to infection and is predominantly found in mouse neutrophils. Deletion of MMP-9 decreased oviduct pathology in *C. muridarum* infected mice, suggesting that MMP-9 has a role in causing immunopathology (Imtiaz *et al.*, 2007). This observation is further supported in humans, as MMP-9 is highly expressed in *C. trachomatis*-infected tissue, and polymorphisms in MMP-9 are associated with protection from scarring following ocular *C. trachomatis* infection (El-Asrar *et al.*, 2000; Natividad *et al.*, 2006).

Cytokine(s) and chemokines in tissue damage

Cytokines/chemokines, produced from innate (and adaptive) immune cells in response to pathogen detection, play the integral role in interaction and communication between cells. The decision to secrete pro-inflammatory molecules or induce cell death is driven by other factors in the microenvironment of an infected cell, but both can lead to tissue pathology. As discussed previously in this chapter, TNFα is one of the predominant cytokines secreted during *Chlamydia* infection. TNFα signalling induces NFκB activation and subsequent chemokine/cytokine secretion or cell death in epithelial and immune cells. In humans, polymorphisms in TNFα are associated with increased risk of tissue damage following *C. trachomatis* infection (Ohman *et al.*, 2009). Furthermore, *C. muridarum* infected mice lacking TNF-receptor 1, TNF-receptor 2, or TNFα are protected from oviduct pathology (Kamalakaran *et al.*, 2013; Murthy *et al.*, 2013). TNFR1 knockout mice exhibit lower incidence of hydrosalpinx than TNFR2 knockout mice, suggesting that TNFR1 signalling may contribute to tissue damage more than TNFR2 (Murthy *et al.*, 2013). No difference in chlamydial clearance was observed in the lower genital tract among any of the knockout mice (Kamalakaran *et al.*, 2013; Murthy *et al.*, 2013). This suggests that although TNFα is not absolutely essential for infection resolution, TNFα does aid tissue damage and pathology in *Chlamydia* infection. TNFα-induced secretion of chemokines and neutrophil activation could account for some of the TNF-mediated pathology. However, this is unlikely to be the sole mechanism as TNF-producing $CD8^+$ T cells have also been implicated in pathology (Murthy *et al.*, 2011), which is discussed in detail in Chapter 14, Adaptive Immunity to *Chlamydia trachomatis* Infection.

Other inflammatory cytokines associated with neutrophil recruitment, such as IL-1, have been implicated in the tissue-damaging response during *Chlamydia* infection. There are two distinct forms of IL-1: IL-1α and IL-1β. Both bind and transduce signal via the IL-1 receptor (IL-1R). IL-1R signalling can also synergize with TNFR signalling in epithelial cells (Janes *et al.*, 2006). IL-1α is expressed by some immune cells, constitutively expressed by epithelial cells, and is released upon necrotic cell death (reviewed by Dinarello, 2018). Once released from an epithelial cell, IL-1α acts as a damage-associated molecular pattern to alert the immune system to potential tissue damage or infection. IL-1β is primarily expressed by immune cells and its expression is induced following TLR and/or IL-1R activation. Signalling via the IL-1R can lead to the activation of MAPKs, including p38, to induce expression and secretion of pro-inflammatory molecules (Hvid *et al.*, 2007). IL-1R signalling is modulated by IL-1 receptor antagonist

(IL-1RA) (Dinarello, 2018). IL-1RA competes with IL-1α and IL-1β for binding to the IL-1R; thus shutting down pro-inflammatory signalling via IL-1R (Dinarello, 2018).

Mice deficient in IL-1R1 have delayed bacterial clearance of *C. muridarum* infection but develop less severe oviduct pathology than infected wildtype mice (Nagarajan *et al.*, 2012). In contrast, mice deficient in IL-1RA clear *C. muridarum* infection faster than wildtype mice, but develop more severe oviduct pathology. Studies of IL-1β deficient mice further support a role for IL-1 in mediating oviduct pathology (Prantner *et al.*, 2009). Similar to IL-1R-deficient mice, deletion of IL-1β results in a prolonged infection, but protection from oviduct pathology. However, recent studies from our lab suggest that IL-1α rather than IL-1β is a significant contributor to pathology. *In vitro* studies in a fallopian tube model suggest that IL-1 may directly induce tissue damage at the epithelial cell level (Hvid *et al.*, 2007). Together, these results suggest that IL-1 signalling plays dual functions in both clearance of chlamydial infection, as well as development of pathology following infection.

miRNAs and pathogenesis

Recently, a growing appreciation for the role of micro RNA (miRNA) in pathology following *Chlamydia* infection has arisen. miRNAs are small, single-stranded RNA molecules that regulate gene expression by binding to the 3′ untranslated region of messenger RNA to prevent translation. miRNAs are differentially expressed during *Chlamydia* infection *in vitro* (Chowdhury *et al.*, 2017) and *in vivo* (Derrick *et al.*, 2016; Derrick *et al.*, 2017; Derrick *et al.*, 2013; Gupta *et al.*, 2015; Yeruva *et al.*, 2014; Yeruva *et al.*, 2017). Multiple miRNAs involved in regulating the immune response, including cytokine and chemokine secretion and PRR expression, are differentially expressed during *C. muridarum* genital tract infection. However, the particular miRNA signature observed in *C. muridarum* infected mice is dependent on the strain and timepoint studied (Gupta *et al.*, 2015; Yeruva *et al.*, 2014; Yeruva *et al.*, 2017). miRNA regulation of the immune response has been implicated in human conjunctival *C. trachomatis* infection as well (Derrick *et al.*, 2013). Differential expression of miRNAs that regulate genes involved in fibrosis and epithelial cell differentiation was observed in adult trachoma patients from The Gambia. Specifically, miR-147b, a negative regulator of inflammation, and miR-1285, a regulator of cell proliferation and fibrosis, were up-regulated in patients of this cohort with trachomatous scarring, suggesting that these miRNAs may keep inflammation and fibrosis in check during and following *Chlamydia* infection to influence the development of pathology (Derrick *et al.*, 2013). However, a subsequent study of trachoma in children from Guinea-Bissou, West Africa did not observe differential expression of miR-147b and miR-1285, but identified a negative correlation between miR-155, which regulates immune cell function (reviewed in (Alivernini *et al.*, 2017) and inflammation, and a positive correlation between miR-184, a miRNA involved in cellular proliferation and differentiation (Nagosa *et al.*, 2017; Yu *et al.*, 2008) and inflammation (Derrick *et al.*, 2016). Age differences between the two cohorts may account for the divergent findings, and miRNA responses to *Chlamydia* may also be influenced by other host and environmental factors. Despite contrasting findings, these studies show a clear role for miRNAs in regulating the immune response to *Chlamydia* infection.

Chlamydia infection leads to a robust host innate immune response characterized by chemokine and cytokine secretion by the epithelial cells and large numbers of neutrophils infiltrating to the site of infection. Although the host response is meant to curb bacterial replication and aid in infection clearance, the innate response is ineffective against *Chlamydia* and can ultimately cause more damage to the host than the bacteria. Key players in causing immunopathology have been identified. However, more studies are required to determine the specific molecular mechanisms by which these interact to induce tissue damage and pathology following *Chlamydia* infection.

Chlamydial subversion of innate immune response

Several studies suggest that it is likely that an ineffective anti-microbial innate response during infection is the result of chlamydial subversion (reviewed by Fischer and Rudel, 2018a,b). From the time *Chlamydia* is deposited at the mucosal surface, it is equipped to compromise the ability of innate immune response to eradicate infection. The infectious EB is reasonably invisible to the immune system due to its inert membrane, which could give *Chlamydia* a better chance to establish epithelial cell infection and avoid immediate clearance by the immune system. However, an inert EB is not enough to protect the intracellular chlamydiae from the barrage of attacks from the innate immune response. Thus, *Chlamydia* has evolved more active mechanisms to subvert the innate immune response in the extracellular space.

At the extracellular level, cell lysis associated with chlamydial release can also release chlamydial effectors to neutralize soluble innate mediators and promote additional rounds of infection. Among the more well-studied active subversion tactics is by the chlamydial protease CPAF (chlamydial protease/proteasome-like activity factor). CPAF is discussed in detail in Chapter 8, The Chlamydial Protease CPAF. Released CPAF could inactivate antimicrobial peptides, complement, and innate immune cells to promote a subsequent round of infection. Indeed, purified recombinant CPAF can degrade the antimicrobial peptide LL-37 *in vitro* (Tang *et al.*, 2015). Additionally, *in vitro* preincubation of human serum with CPAF from *C. trachomatis* led to cleavage of C3 and factor B resulting in decreased C3-dependent anti-chlamydial activity (Yang *et al.*, 2016b). Further, chlamydial CPAF mutants have reduced survival and ascension in the mouse genital tract, suggesting that inactivation of LL-37 and complement by CPAF could promote chlamydial survival and pathogenesis (Yang *et al.*, 2016a). In addition to CPAF, other chlamydial proteins such as Pgp3 could also be released during cell lysis and subvert innate immunity. Pgp3 can bind to human LL-37 secreted by genital tract epithelial cells and infiltrating neutrophils and neutralize its anti-chlamydial activity (Hou *et al.*, 2018).

Chlamydia can also subvert killing by neutrophils. Recent studies by Rudel and colleagues show that *Chlamydia* paralyzes neutrophils by selective cleavage of a protein formyl peptide receptor (FPR2) by chlamydial protease CPAF (Rajeeve *et al.*, 2018). This results in suppression of neutrophil oxidative bursts and extended survival of chlamydiae in the neutrophil. These studies contrast with previous studies that showed significant killing of chlamydiae by neutrophils (Register *et al.*, 1986). One possibility for this difference is that Rajeeve *et al.* used supernatants containing CPAF released from epithelial cells infected with wildtype or CPAF mutant chlamydial strains, instead of purified EBs, to demonstrate that CPAF directly suppressed neutrophil activation. This suggestion is supported by the observation of Rajeeve *et al.* that purified EBs induced significantly more neutrophil extracellular traps (NETs) than EB preps containing secreted CPAF. Together, these results suggest that neutrophils are likely to be effective in engulfing free EBs, but with every round of infection, cell lysis, and CPAF release, the activity of newly recruited neutrophils is compromised.

Once inside the epithelial cells, *Chlamydia* remain susceptible to recognition by innate immune receptors and cell-autonomous clearance mechanisms. Subversion of intracellular recognition is exemplified by weak activation of PRRs by chlamydial ligands. Chlamydial LPS is a weak stimulant for TLR4 compared to *Salmonella* and *Neisseria* LPS (Ingalls *et al.*, 1995). *In vitro* studies showed that chlamydial LPS is also a weak stimulant of caspase-11(Yang *et al.*, 2019). Therefore, we posit that either chlamydial effectors are intrinsically weak stimulants of PRRs or *Chlamydia* actively inhibits certain host responses. Active inhibition of intracellular host responses by *Chlamydia* is supported by observations of downregulated expression of inducible NO synthase (iNOS/NOS2) in *C. trachomatis*-infected human cells. The infection simultaneously upregulates ornithine decarboxylase, diverting the iNOS/NOS2 substrate l-arginine toward the synthesis of polyamines. Consequently, the antimicrobial NO production of infected cells is blocked (Abu-Lubad *et al.*, 2014). The specific chlamydial effectors driving this subversion are not yet known.

Evidence of active inhibition at signalling levels is also demonstrated by the inhibition of NFκB activation during *C. trachomatis* infection. NFκB inhibition was proposed to be the result of cleavage of the NFκB p65 subunit by two chlamydial proteases, CT441 and CPAF (Christian *et al.*, 2010; Lad *et al.*, 2007). However, two later studies showed p65 cleavage is dependent on cell lysis conditions and is likely experimentally induced in the cell lysates (Chen *et al.*, 2012; Johnson *et al.*, 2015). A second mechanism of NFκB inhibition was proposed, whereby the chlamydial deubiquitinase ChlaDub1 prevents degradation of IκB and subsequent NFκB translocation into the nucleus (Le Negrate *et al.*, 2008).

Active regulation of cell death pathways is yet another mechanism for chlamydial manipulation of the innate immune response. The chlamydial inclusion membrane protein CpoS, a type III secreted effector, negatively impacts STING-mediated IFN-β and cell-death pathways (Sixt *et al.*, 2017; Weber *et al.*, 2017). Prevention of cell death by p53 degradation (Siegl *et al.*, 2014) or induction of cell death late in the chlamydial developmental cycle (Sixt *et al.*, 2018) by *Chlamydia* to promote its survival can also have direct consequences on the innate host response. Further, the presence of chlamydial Pgp3 in the host cytosol (Li *et al.*, 2008) and the association of chlamydial plasmid with oviduct pathology, suggest that Pgp3 may also manipulate host immune responses via interactions with host signalling intermediates in the cytosol. Together, these effectors, known and unknown, actively regulate the host response on multiple levels to create an environment conducive to chlamydial infection, growth, and replication.

Conclusion

The eukaryotic innate immune system has evolved in a complex manner. The multifaceted and broad nature of the immune system allows the host to clear or at least control most infections for host survival. In some cases of control, a type of 'homeostatic duelling' develops between the host and pathogen in which the host is continually fighting the pathogen and the pathogen continually subverts its attempts, but the end result is stalemate with minimal damage to the host or pathogen. *Chlamydia* is clearly one such pathogen that has co-evolved with its host to maintain its growth and replication, despite an arsenal of innate immune mediators, and to cause minimal host damage in most people. However, a proportion of women develop long term sequalae from infection, largely due to an increased innate immune response. Thus, understanding host–pathogen interactions and determining the contribution of the innate immune response to pathology during infection can help to develop therapeutics that, when given along with antibiotics, can prevent excessive tissue damage in infected women. An additional benefit in studying innate responses to infection lies in using this knowledge for immune modulation in vaccine development or to improve a protective response during a secondary infection. Developing a vaccine for *Chlamydia* infection is the 'Holy Grail' for the field. Although a number of vaccine candidates has been shown to be effective, vaccine efficacy in humans is dependent on the use of an effective adjuvant. A number of these adjuvants engage the innate immune system (reviewed in Del Giudice *et al.*, 2018). A classic example of how the right adjuvant can elicit a protective response to *Chlamydia* is evident from elegant vaccine studies by Stary *et al.* in which UV-killed *C. muridarum* only produced a protective immune response in the presence of a TLR7 stimulating adjuvant (Stary *et al.*, 2015). Therefore, modulating the innate immune system towards more effective clearance and the development of a protective adaptive response can shift immune homeostasis towards protection instead of pathology, and provide a new approach to prevent chlamydial disease.

References

Abdul-Sater, A.A., Koo, E., Hacker, G., and Ojcius, D.M. (2009). Inflammasome-dependent caspase-1 activation in cervical epithelial cells stimulates growth of the intracellular pathogen Chlamydia trachomatis. J Biol Chem *284*, 26789–26796.

Abu-Lubad, M., Meyer, T.F., and Al-Zeer, M.A. (2014). Chlamydia trachomatis inhibits inducible NO synthase in human mesenchymal stem cells by stimulating polyamine synthesis. J Immunol *193*, 2941–2951.

Allen, J., 4th, Gyorke, G., Tripathy, M., Zhang, Y., Montgomery, S., and Nagarajan, U.M. (2019). Caspase-11 contributes to oviduct pathology during C. murdairum genital infection in mice. Infect Immun *87*, 8.

Al-Zeer, M.A., Al-Younes, H.M., Braun, P.R., Zerrahn, J., and Meyer, T.F. (2009). IFN-gamma-inducible Irga6 mediates host resistance against Chlamydia trachomatis via autophagy. PLoS One *4*, e4588.

Al-Zeer, M.A., Al-Younes, H.M., Lauster, D., Abu Lubad, M., and Meyer, T.F. (2013). Autophagy restricts Chlamydia trachomatis growth in human macrophages via IFNG-inducible guanylate binding proteins. Autophagy *9*, 50–62.

Alivernini, S., Gremese, E., McSharry, C., Tolusso, B., Ferraccioli, G., McInnes, I.B., and Kurowska-Stolarska, M. (2017). MicroRNA-155-at the Critical Interface of Innate and Adaptive Immunity in Arthritis. Front Immunol *8*, 1932.

Barker, J.R., Koestler, B.J., Carpenter, V.K., Burdette, D.L., Waters, C.M., Vance, R.E., and Valdivia, R.H. (2013). STING-Dependent Recognition of Cyclic di-AMP Mediates Type I Interferon Responses during Chlamydia trachomatis Infection. MBio *4*.

Barteneva, N., Theodor, I., Peterson, E.M., and de la Maza, L.M. (1996). Role of neutrophils in controlling early stages of a Chlamydia trachomatis infection. Infection and Immunity *64*, 4830–4833.

Bas, S., Neff, L., Vuillet, M., Spenato, U., Seya, T., Matsumoto, M., and Gabay, C. (2008). The proinflammatory cytokine response to Chlamydia trachomatis elementary bodies in human macrophages is partly mediated by a lipoprotein, the macrophage infectivity potentiator, through TLR2/TLR1/TLR6 and CD14. J Immunol *180*, 1158–1168.

Beatty, W.L., Belanger, T.A., Desai, A.A., Morrison, R.P., and Byrne, G.I. (1994). Tryptophan depletion as a mechanism of gamma interferon-mediated chlamydial persistence. Infect Immun *62*, 3705–3711.

Bedard, K., and Krause, K.H. (2007). The NOX family of ROS-generating NADPH oxidases: physiology and pathophysiology. Physiol Rev *87*, 245–313.

Bekpen, C., Hunn, J.P., Rohde, C., Parvanova, I., Guethlein, L., Dunn, D.M., Glowalla, E., Leptin, M., and Howard, J.C. (2005). The interferon-inducible p47 (IRG) GTPases in vertebrates: loss of the cell autonomous resistance mechanism in the human lineage. Genome Biology *6*, R92.

Bernstein-Hanley, I., Balsara, Z.R., Ulmer, W., Coers, J., Starnbach, M.N., and Dietrich, W.F. (2006a). Genetic analysis of susceptibility to Chlamydia trachomatis in mouse. Genes and immunity *7*, 122–129.

Bernstein-Hanley, I., Coers, J., Balsara, Z.R., Taylor, G.A., Starnbach, M.N., and Dietrich, W.F. (2006b). The p47 GTPases Igtp and Irgb10 map to the Chlamydia trachomatis susceptibility locus Ctrq-3 and mediate cellular resistance in mice. Proceedings of the National Academy of Sciences of the United States of America *103*, 14092–14097.

Bode, J., Dutow, P., Sommer, K., Janik, K., Glage, S., Tummler, B., Munder, A., Laudeley, R., Sachse, K.W., and Klos, A. (2012). A new role of the complement system: C3 provides protection in a mouse model of lung infection with intracellular Chlamydia psittaci. PLoS One *7*, e50327.

Boncompain, G., Schneider, B., Delevoye, C., Kellermann, O., Dautry-Varsat, A., and Subtil, A. (2010). Production of reactive oxygen species is turned on and rapidly shut down in epithelial cells infected with Chlamydia trachomatis. Infection and immunity *78*, 80–87.

Broz, P., Ruby, T., Belhocine, K., Bouley, D.M., Kayagaki, N., Dixit, V.M., and Monack, D.M. (2012). Caspase-11 increases susceptibility to Salmonella infection in the absence of caspase-1. Nature *490*, 288–291.

Buchholz, K.R., and Stephens, R.S. (2007). The extracellular signal-regulated kinase/mitogen-activated protein kinase pathway induces the inflammatory factor interleukin-8 following Chlamydia trachomatis infection. Infection and immunity *75*, 5924–5929.

Buchholz, K.R., and Stephens, R.S. (2008). The cytosolic pattern recognition receptor NOD1 induces inflammatory interleukin-8 during Chlamydia trachomatis infection. Infection and immunity *76*, 3150–3155.

Bulut, Y., Faure, E., Thomas, L., Karahashi, H., Michelsen, K.S., Equils, O., Morrison, S.G., Morrison, R.P., and Arditi, M. (2002). Chlamydial heat shock protein 60 activates macrophages and endothelial cells through Toll-like receptor 4 and MD2 in a MyD88-dependent pathway. J Immunol *168*, 1435–1440.

Byrne, G.I., Lehmann, L.K., and Landry, G.J. (1986). Induction of tryptophan catabolism is the mechanism for gamma-interferon-mediated inhibition of intracellular Chlamydia psittaci replication in T24 cells. Infect Immun *53*, 347–351.

Cao, F., Castrillo, A., Tontonoz, P., Re, F., and Byrne, G.I. (2007). Chlamydia pneumoniae-Induced Macrophage Foam Cell Formation Is Mediated by Toll-Like Receptor 2. Infect Immun *75*, 753–759.

Carrasco, S.E., Hu, S., Imai, D.M., Kumar, R., Sandusky, G.E., Yang, X.F., and Derbigny, W.A. (2018). Toll-like receptor 3 (TLR3) promotes the resolution of Chlamydia muridarum genital tract infection in congenic C57BL/6N mice. PLoS One *13*, e0195165.

Casbon, A.J., Long, M.E., Dunn, K.W., Allen, L.A., and Dinauer, M.C. (2012). Effects of IFN-gamma on intracellular trafficking and activity of macrophage NADPH oxidase flavocytochrome b558. J Leukoc Biol *92*, 869–882.

Chamaillard, M., Hashimoto, M., Horie, Y., Masumoto, J., Qiu, S., Saab, L., Ogura, Y., Kawasaki, A., Fukase, K., Kusumoto, S., *et al.* (2003). An essential role for NOD1 in host recognition of bacterial peptidoglycan containing diaminopimelic acid. Nat Immunol *4*, 702–707.

Chen, A.L., Johnson, K.A., Lee, J.K., Sutterlin, C., and Tan, M. (2012). CPAF: a Chlamydial protease in search of an authentic substrate. PLoS Pathog *8*, e1002842.

Chen, B., Stout, R., and Campbell, W.F. (1996). Nitric oxide production: a mechanism of Chlamydia trachomatis inhibition in interferon-gamma-treated RAW264.7 cells. FEMS immunology and medical microbiology *14*, 109–120.

Cheng, W., Shivshankar, P., Li, Z., Chen, L., Yeh, I.T., and Zhong, G. (2008). Caspase-1 contributes to Chlamydia trachomatis-induced upper urogenital tract inflammatory pathologies without affecting the course of infection. Infection and immunity *76*, 515–522.

Chowdhury, S.R., Reimer, A., Sharan, M., Kozjak-Pavlovic, V., Eulalio, A., Prusty, B.K., Fraunholz, M., Karunakaran, K., and Rudel, T. (2017). Chlamydia preserves the mitochondrial network necessary for replication via microRNA-dependent inhibition of fission. J Cell Biol *216*, 1071–1089.

Christian, J., Vier, J., Paschen, S.A., and Hacker, G. (2010). Cleavage of the NF-kappaB family protein p65/RelA by the chlamydial protease-like activity factor (CPAF) impairs proinflammatory signaling in cells infected with Chlamydiae. J Biol Chem *285*, 41320–41327.

Coers, J., Bernstein-Hanley, I., Grotsky, D., Parvanova, I., Howard, J.C., Taylor, G.A., Dietrich, W.F., and Starnbach, M.N. (2008). Chlamydia muridarum evades growth restriction by the IFN-gamma-inducible host resistance factor Irgb10. J Immunol *180*, 6237–6245.

Cortes, C., Ferreira, V.P., and Pangburn, M.K. (2011). Native properdin binds to Chlamydia pneumoniae and promotes complement activation. Infection and immunity *79*, 724–731.

Crimmins, G.T., Herskovits, A.A., Rehder, K., Sivick, K.E., Lauer, P., Dubensky, T.W., Jr., and Portnoy, D.A. (2008). Listeria monocytogenes multidrug resistance transporters activate a cytosolic surveillance pathway of innate immunity. Proc Natl Acad Sci U S A *105*, 10191–10196.

Darville, T., Andrews, C.W., Jr., Sikes, J.D., Fraley, P.L., and Rank, R.G. (2001). Early local cytokine profiles in strains of mice with different outcomes from chlamydial genital tract infection. Infection and immunity *69*, 3556–3561.

Darville, T., O'Neill, J.M., Andrews, C.W., Jr., Nagarajan, U.M., Stahl, L., and Ojcius, D.M. (2003). Toll-like receptor-2, but not Toll-like receptor-4, is essential for development of oviduct pathology in chlamydial genital tract infection. J Immunol *171*, 6187–6197.

Del Giudice, G., Rappuoli, R., and Didierlaurent, A.M. (2018). Correlates of adjuvanticity: A review on adjuvants in licensed vaccines. Semin Immunol.

Derbigny, W.A., Hong, S.C., Kerr, M.S., Temkit, M., and Johnson, R.M. (2007). Chlamydia muridarum infection elicits a beta interferon response in murine oviduct epithelial cells dependent on interferon regulatory factor 3 and TRIF. Infection and immunity *75*, 1280–1290.

Derrick, T., Last, A.R., Burr, S.E., Roberts, C.H., Nabicassa, M., Cassama, E., Bailey, R.L., Mabey, D.C.W., Burton, M.J., and Holland, M.J. (2016). Inverse relationship between microRNA-155 and-184 expression with increasing conjunctival inflammation during ocular Chlamydia trachomatis infection. Bmc Infect Dis *16*.

Derrick, T., Ramadhani, A.M., Mtengai, K., Massae, P., Burton, M.J., and Holland, M.J. (2017). miRNAs that associate with conjunctival inflammation and ocular Chlamydia trachomatis infection do not predict progressive disease. Pathogens and disease *75*, ftx016.

Derrick, T., Roberts, C., Rajasekhar, M., Burr, S.E., Joof, H., Makalo, P., Bailey, R.L., Mabey, D.C., Burton, M.J., and Holland, M.J. (2013). Conjunctival MicroRNA expression in inflammatory trachomatous scarring. PLoS Negl Trop Dis *7*, e2117.

Dessus-Babus, S., Knight, S.T., and Wyrick, P.B. (2000). Chlamydial infection of polarized HeLa cells induces PMN chemotaxis but the cytokine profile varies between disseminating and non-disseminating strains. Cell Microbiol 2, 317–327.

Di Francesco, A., Favaroni, A., and Donati, M. (2013). Host defense peptides: general overview and an update on their activity against Chlamydia spp. Expert Rev Anti Infect Ther *11*, 1215–1224.

Di Paolo, N.C., Doronin, K., Baldwin, L.K., Papayannopoulou, T., and Shayakhmetov, D.M. (2013). The transcription factor IRF3 triggers 'defensive suicide' necrosis in response to viral and bacterial pathogens. Cell Rep 3, 1840–1846.

Dinarello, C.A. (2018). Overview of the IL-1 family in innate inflammation and acquired immunity. Immunol Rev *281*, 8–27.

Donati, M., Di Leo, K., Benincasa, M., Cavrini, F., Accardo, S., Moroni, A., Gennaro, R., and Cevenini, R. (2005). Activity of cathelicidin peptides against Chlamydia spp. Antimicrobial agents and chemotherapy *49*, 1201–1202.

Dutow, P., Fehlhaber, B., Bode, J., Laudeley, R., Rheinheimer, C., Glage, S., Wetsel, R.A., Pabst, O., and Klos, A. (2014). The complement C3a receptor is critical in defense against Chlamydia psittaci in mouse lung infection and required for antibody and optimal T cell response. J Infect Dis *209*, 1269–1278.

El-Asrar, A.M., Geboes, K., Al-Kharashi, S.A., Al-Mosallam, A.A., Missotten, L., Paemen, L., and Opdenakker, G. (2000). Expression of gelatinase B in trachomatous conjunctivitis. Br J Ophthalmol *84*, 85–91.

Erridge, C., Pridmore, A., Eley, A., Stewart, J., and Poxton, I.R. (2004). Lipopolysaccharides of Bacteroides fragilis, Chlamydia trachomatis and Pseudomonas aeruginosa signal via toll-like receptor 2. J Med Microbiol *53*, 735–740.

Fedorko, D.P., Clark, R.B., Nachamkin, I., and Dalton, H.P. (1987). Complement dependent *in vitro* neutralization of Chlamydia trachomatis by a subspecies-specific monoclonal antibody. Med Microbiol Immunol *176*, 225–228.

Fiegl, D., Kagebein, D., Liebler-Tenorio, E.M., Weisser, T., Sens, M., Gutjahr, M., and Knittler, M.R. (2013). Amphisomal route of MHC class I cross-presentation in bacteria-infected dendritic cells. J Immunol *190*, 2791–2806.

Finethy, R., and Coers, J. (2016). Sensing the enemy, containing the threat: cell-autonomous immunity to Chlamydia trachomatis. FEMS microbiology reviews *40*, 875–893.

Finethy, R., Jorgensen, I., Haldar, A.K., de Zoete, M.R., Strowig, T., Flavell, R.A., Yamamoto, M., Nagarajan, U.M., Miao, E.A., and Coers, J. (2015). Guanylate binding proteins enable rapid activation of canonical and noncanonical inflammasomes in Chlamydia-infected macrophages. Infection and immunity *83*, 4740–4749.

Fischer, A., and Rudel, T. (2018a). Safe haven under constant attack-The Chlamydia-containing vacuole. Cell Microbiol *20*, e12940.

Fischer, A., and Rudel, T. (2018b). Subversion of Cell-Autonomous Host Defense by Chlamydia Infection. Curr Top Microbiol Immunol *412*, 81–106.

Frazer, L.C., O'Connell, C.M., Andrews, C.W., Zurenski, M.A., and Darville, T. (2011). Enhanced Neutrophil Longevity and Recruitment Contribute to the Severity of Oviduct Pathology during Chlamydia muridarum Infection. Infection and Immunity *79*, 4029–4041.

Freeley, S., Kemper, C., and Le Friec, G. (2016). The 'ins and outs' of complement-driven immune responses. Immunol Rev *274*, 16–32.

Frendéus, B., Wachtler, C., Hedlund, M., Fischer, H., Samuelsson, P., Svensson, M., and Svanborg, C. (2001). Escherichia coli P fimbriae utilize the Toll-like receptor 4 pathway for cell activation. Molecular Microbiology *40*, 37–51.

Gencay, M.M., Tamm, M., Glanville, A., Perruchoud, A.P., and Roth, M. (2003). Chlamydia pneumoniae activates epithelial cell proliferation via NF-kappaB and the glucocorticoid receptor. Infect Immun *71*, 5814–5822.

Giles, D.K., and Wyrick, P.B. (2008). Trafficking of chlamydial antigens to the endoplasmic reticulum of infected epithelial cells. Microbes and infection/Institut Pasteur.

Girardin, S.E., Boneca, I.G., Viala, J., Chamaillard, M., Labigne, A., Thomas, G., Philpott, D.J., and Sansonetti, P.J. (2003). Nod2 is a general sensor of peptidoglycan through muramyl dipeptide (MDP) detection. J Biol Chem *278*, 8869–8872.

Gupta, R., Arkatkar, T., Yu, J.J., Wali, S., Haskins, W.E., Chambers, J.P., Murthy, A.K., Bakar, S.A., Guentzel, M.N., and Arulanandam, B.P. (2015). Chlamydia muridarum infection associated host MicroRNAs in the murine genital tract and contribution to generation of host immune response. Am J Reprod Immunol *73*, 126–140.

Hagar, J.A., Powell, D.A., Aachoui, Y., Ernst, R.K., and Miao, E.A. (2013). Cytoplasmic LPS activates caspase-11: implications in TLR4-independent endotoxic shock. Science *341*, 1250–1253.

Hajishengallis, G., Reis, E.S., Mastellos, D.C., Ricklin, D., and Lambris, J.D. (2017). Novel mechanisms and functions of complement. Nature immunology *18*, 1288–1298.

Haldar, A.K., Foltz, C., Finethy, R., Piro, A.S., Feeley, E.M., Pilla-Moffett, D.M., Komatsu, M., Frickel, E.M., and Coers, J. (2015). Ubiquitin systems mark pathogen-containing vacuoles as targets for host defense by guanylate binding proteins. Proc Natl Acad Sci U S A *112*, E5628–5637.

Haldar, A.K., Piro, A.S., Finethy, R., Espenschied, S.T., Brown, H.E., Giebel, A.M., Frickel, E.M., Nelson, D.E., and Coers, J. (2016). Chlamydia trachomatis Is Resistant to Inclusion Ubiquitination and Associated Host Defense in Gamma Interferon-Primed Human Epithelial Cells. MBio 7.

Haldar, A.K., Saka, H.A., Piro, A.S., Dunn, J.D., Henry, S.C., Taylor, G.A., Frickel, E.M., Valdivia, R.H., and Coers, J. (2013). IRG and GBP host resistance factors target aberrant, 'non-self' vacuoles characterized by the missing of 'self' IRGM proteins. PLoS Pathog *9*, e1003414.

He, X., Nair, A., Mekasha, S., Alroy, J., O'Connell, C.M., and Ingalls, R.R. (2011). Enhanced Virulence of *Chlamydia muridarum* Respiratory Infections in the Absence of TLR2 Activation. PLoS ONE *6*, e20846.

Hou, S., Sun, X., Dong, X., Lin, H., Tang, L., Xue, M., and Zhong, G. (2018). Chlamydial plasmid-encoded virulence factor Pgp3 interacts with human cathelicidin peptide LL-37 to modulate immune response. Microbes and infection/Institut Pasteur.

Hu, S., Hosey, K.L., and Derbigny, W.A. (2015). Analyses of the pathways involved in early- and late-phase induction of IFN-beta during *C. muridarum* infection of oviduct epithelial cells. Plos One *10*, e0119235.

Huang, J., DeGraves, F.J., Lenz, S.D., Gao, D., Feng, P., Li, D., Schlapp, T., and Kaltenboeck, B. (2002). The quantity of nitric oxide released by macrophages regulates Chlamydia-induced disease. Proc Natl Acad Sci U S A *99*, 3914–3919.

Huber-Lang, M., Sarma, J.V., Zetoune, F.S., Rittirsch, D., Neff, T.A., McGuire, S.R., Lambris, J.D., Warner, R.L., Flierl, M.A., Hoesel, L.M., *et al.* (2006). Generation of C5a in the absence of C3: a new complement activation pathway. Nature medicine *12*, 682–687.

Hvid, M., Baczynska, A., Deleuran, B., Fedder, J., Knudsen, H.J., Christiansen, G., and Birkelund, S. (2007). Interleukin-1 is the initiator of Fallopian tube destruction during Chlamydia trachomatis infection. Cell Microbiol *9*, 2795–2803.

Hwu, P., Du, M.X., Lapointe, R., Do, M., Taylor, M.W., and Young, H.A. (2000). Indoleamine 2,3-dioxygenase production by human dendritic cells results in the inhibition of T cell proliferation. J Immunol *164*, 3596–3599.

Igietseme, J.U. (1996). Molecular mechanism of T-cell control of Chlamydia in mice: role of nitric oxide *in vivo*. Immunology *88*, 1–5.

Igietseme, J.U., Perry, L.L., Ananaba, G.A., Uriri, I.M., Ojior, O.O., Kumar, S.N., and Caldwell, H.D. (1998). Chlamydial infection in inducible nitric oxide synthase knockout mice. Infection and immunity *66*, 1282–1286.

Igietseme, J.U., Uriri, I.M., Chow, M., Abe, E., and Rank, R.G. (1997). Inhibition of intracellular multiplication of human strains of Chlamydia trachomatis by nitric oxide. Biochemical and biophysical research communications *232*, 595–601.

Imtiaz, M.T., Distelhorst, J.T., Schripsema, J.H., Sigar, I.M., Kasimos, J.N., Lacy, S.R., and Ramsey, K.H. (2007). A role for matrix metalloproteinase-9 in pathogenesis of urogenital Chlamydia muridarum infection in mice. Microbes and infection/Institut Pasteur *9*, 1561–1566.

Ingalls, R., Rice, P., Qureshi, N., Takayama, K., Lin, J., and Golenbock, D. (1995). The inflammatory cytokine response to Chlamydia trachomatis infection is endotoxin mediated. Infection and immunity *63*, 3125–3130.

Janes, K.A., Gaudet, S., Albeck, J.G., Nielsen, U.B., Lauffenburger, D.A., and Sorger, P.K. (2006). The response of human epithelial cells to TNF involves an inducible autocrine cascade. Cell *124*, 1225–1239.

Johnson, K.A., Lee, J.K., Chen, A.L., Tan, M., and Sutterlin, C. (2015). Induction and inhibition of CPAF activity during analysis of Chlamydia-infected cells. Pathog Dis *73*, 1–8.

Johnson, R.M. (2004). Murine oviduct epithelial cell cytokine responses to Chlamydia muridarum infection include interleukin-12-p70 secretion. Infection and immunity *72*, 3951–3960.

Kamalakaran, S., Chaganty, B.K.R., Gupta, R., Guentzel, M.N., Chambers, J.P., Murthy, A.K., and Arulanandam, B.P. (2013). Vaginal chlamydial clearance following primary or secondary infection in mice occurs independently of TNF-alpha. Front Cell Infect Mi 3.

Kawai, T., and Akira, S. (2006). TLR signaling. Cell Death Differ *13*, 816–825.

Kayagaki, N., Warming, S., Lamkanfi, M., Vande Walle, L., Louie, S., Dong, J., Newton, K., Qu, Y., Liu, J., Heldens, S., *et al.* (2011). Non-canonical inflammasome activation targets caspase-11. Nature *479*, 117–121.

Khaminets, A., Hunn, J.P., Konen-Waisman, S., Zhao, Y.O., Preukschat, D., Coers, J., Boyle, J.P., Ong, Y.C., Boothroyd, J.C., Reichmann, G., *et al.* (2010). Coordinated loading of IRG resistance GTPases on to the Toxoplasma gondii parasitophorous vacuole. Cell Microbiol *12*, 939–961.

Kim, B.H., Shenoy, A.R., Kumar, P., Bradfield, C.J., and MacMicking, J.D. (2012). IFN-inducible GTPases in host cell defense. Cell Host Microbe *12*, 432–444.

Kim, E.K., and Choi, E.J. (2010). Pathological roles of MAPK signaling pathways in human diseases. Biochim Biophys Acta *1802*, 396–405.

Klos, A., Wende, E., Wareham, K.J., and Monk, P.N. (2013). International Union of Basic and Clinical Pharmacology. [corrected]. LXXXVII. Complement peptide C5a, C4a, and C3a receptors. Pharmacol Rev *65*, 500–543.

Krüll, M., Bockstaller, P., Wuppermann, F.N., Klucken, A.C., Mühling, J., Schmeck, B., Seybold, J., Walter, C., Maass, M., Rosseau, S., *et al.* (2006). Mechanisms of Chlamydophila pneumoniae–Mediated GM-CSF Release in Human Bronchial Epithelial Cells. American Journal of Respiratory Cell and Molecular Biology *34*, 375–382.

Krull, M., Kramp, J., Petrov, T., Klucken, A.C., Hocke, A.C., Walter, C., Schmeck, B., Seybold, J., Maass, M., Ludwig, S., *et al.* (2004). Differences in cell activation by Chlamydophila pneumoniae and Chlamydia trachomatis infection in human endothelial cells. Infection and Immunity *72*, 6615–6621.

Kumar, R., Gong, H., Liu, L., Ramos-Solis, N., Seye, C.I., and Derbigny, W.A. (2019). TLR3 deficiency exacerbates the loss of epithelial barrier function during genital tract Chlamydia muridarum infection. PLoS One *14*, e0207422.

Lacy, H.M., Bowlin, A.K., Hennings, L., Scurlock, A.M., Nagarajan, U.M., and Rank, R.G. (2011). Essential role for neutrophils in pathogenesis and adaptive immunity in Chlamydia caviae ocular infections. Infection and immunity *79*, 1889–1897.

Lad, S.P., Li, J., da Silva Correia, J., Pan, Q., Gadwal, S., Ulevitch, R.J., and Li, E. (2007). Cleavage of p65/RelA of the NF-kappaB pathway by Chlamydia. Proceedings of the National Academy of Sciences of the United States of America *104*, 2933–2938.

Lai, Y., Di Nardo, A., Nakatsuji, T., Leichtle, A., Yang, Y., Cogen, A.L., Wu, Z.R., Hooper, L.V., Schmidt, R.R., von Aulock, S., *et al.* (2009). Commensal bacteria regulate Toll-like receptor 3-dependent inflammation after skin injury. Nat Med *15*, 1377–1382.

Lawrence, T. (2009). The nuclear factor NF-kappaB pathway in inflammation. Cold Spring Harb Perspect Biol *1*, a001651.

Le Negrate, G., Krieg, A., Faustin, B., Loeffler, M., Godzik, A., Krajewski, S., and Reed, J.C. (2008). ChlaDub1 of Chlamydia trachomatis suppresses NF-kappa B activation and inhibits I kappa B alpha ubiquitination and degradation. Cell Microbiol *10*, 1879–1892.

Lei, L., Chen, J., Hou, S., Ding, Y., Yang, Z., Zeng, H., Baseman, J., and Zhong, G. (2014). Reduced live organism recovery and lack of hydrosalpinx in mice infected with plasmid-free Chlamydia muridarum. Infection and immunity *82*, 983–992.

Leonard, C.A., Schoborg, R.V., and Borel, N. (2017). Productive and Penicillin-Stressed Chlamydia pecorum Infection Induces Nuclear Factor Kappa B Activation and Interleukin-6 Secretion In Vitro. Front Cell Infect Microbiol *7*, 180.

Li, J., Dong, X., Zhao, L., Wang, X., Wang, Y., Yang, X., Wang, H., and Zhao, W. (2016). Natural killer cells regulate Th1/Treg and Th17/Treg balance in chlamydial lung infection. J Cell Mol Med *20*, 1339–1351.

Li, Z., Chen, D., Zhong, Y., Wang, S., and Zhong, G. (2008). The chlamydial plasmid-encoded protein pgp3 is secreted into the cytosol of Chlamydia-infected cells. Infection and immunity *76*, 3415–3428.

Liechti, G.W., Kuru, E., Hall, E., Kalinda, A., Brun, Y.V., VanNieuwenhze, M., and Maurelli, A.T. (2014). A new metabolic cell-wall labelling method reveals peptidoglycan in Chlamydia trachomatis. Nature *506*, 507–510.

Lin, J.S., Yan, L.L., Ho, Y., and Rice, P.A. (1992). Early complement components enhance neutralization of Chlamydia trachomatis infectivity by human sera. Infection and immunity *60*, 2547–2550.

Lindenberg, V., Molleken, K., Kravets, E., Stallmann, S., Hegemann, J.H., Degrandi, D., and Pfeffer, K. (2017). Broad recruitment of mGBP family members to Chlamydia trachomatis inclusions. PLoS One *12*, e0185273.

Liu, T., Zhang, L., Joo, D., and Sun, S.C. (2017). NF-kappaB signaling in inflammation. Signal Transduct Target Ther 2.

Liu, Y., Huang, Y., Yang, Z., Sun, Y., Gong, S., Hou, S., Chen, C., Li, Z., Liu, Q., Wu, Y., *et al.* (2014). Plasmid-encoded Pgp3 is a major virulence factor for Chlamydia muridarum to induce hydrosalpinx in mice. Infection and immunity *82*, 5327–5335.

Lu, H., Shen, C., and Brunham, R.C. (2000). Chlamydia trachomatis infection of epithelial cells induces the activation of caspase-1 and release of mature IL-18. J Immunol *165*, 1463–1469.

MacMicking, J.D. (2012). Interferon-inducible effector mechanisms in cell-autonomous immunity. Nat Rev Immunol *12*, 367–382.

Manzanillo, P.S., Shiloh, M.U., Portnoy, D.A., and Cox, J.S. (2012). Mycobacterium Tuberculosis Activates the DNA-Dependent Cytosolic Surveillance Pathway within Macrophages. Cell host & microbe *11*, 469–480.

Massari, P., Toussi, D.N., Tifrea, D.F., and de la Maza, L.M. (2013). Toll-like receptor 2-dependent activity of native major outer membrane protein proteosomes of Chlamydia trachomatis. Infection and immunity *81*, 303–310.

McClarty, G., Caldwell, H.D., and Nelson, D.E. (2007). Chlamydial interferon gamma immune evasion influences infection tropism. Current opinion in microbiology *10*, 47–51.

Megran, D.W., Stiver, H.G., and Bowie, W.R. (1985). Complement activation and stimulation of chemotaxis by Chlamydia trachomatis. Infection and immunity *49*, 670–673.

Metz, R., Rust, S., Duhadaway, J.B., Mautino, M.R., Munn, D.H., Vahanian, N.N., Link, C.J., and Prendergast, G.C. (2012). IDO inhibits a tryptophan sufficiency signal that stimulates mTOR: A novel IDO effector pathway targeted by D-1-methyl-tryptophan. Oncoimmunology *1*, 1460–1468.

Miyairi, I., Tatireddigari, V.R., Mahdi, O.S., Rose, L.A., Belland, R.J., Lu, L., Williams, R.W., and Byrne, G.I. (2007). The p47 GTPases Iigp2 and Irgb10 regulate innate immunity and inflammation to murine Chlamydia psittaci infection. J Immunol *179*, 1814–1824.

Moon, Y.W., Hajjar, J., Hwu, P., and Naing, A. (2015). Targeting the indoleamine 2,3-dioxygenase pathway in cancer. Journal for immunotherapy of cancer *3*, 51.

Murthy, A., Thomas, J., and Manam, S. (2013). TNF-alpha receptors 1 and 2 contribute to the development of upper genital tract pathology following primary genital chlamydia muridarum infection in mice. J Immunol *190*.

Murthy, A.K., Li, W., Chaganty, B.K., Kamalakaran, S., Guentzel, M.N., Seshu, J., Forsthuber, T.G., Zhong,

G., and Arulanandam, B.P. (2011). Tumor necrosis factor alpha production from CD8+ T cells mediates oviduct pathological sequelae following primary genital Chlamydia muridarum infection. Infection and immunity *79*, 2928–2935.

Nagarajan, U.M., Ojcius, D.M., Stahl, L., Rank, R.G., and Darville, T. (2005). Chlamydia trachomatis induces expression of IFN-gamma-inducible protein 10 and IFN-beta independent of TLR2 and TLR4, but largely dependent on MyD88. J Immunol *175*, 450–460.

Nagarajan, U.M., Prantner, D., Sikes, J.D., Andrews, C.W., Jr., Goodwin, A.M., Nagarajan, S., and Darville, T. (2008). Type I interferon signaling exacerbates Chlamydia muridarum genital infection in a murine model. Infection and immunity *76*, 4642–4648.

Nagarajan, U.M., Sikes, J., Prantner, D., Andrews, C.W., Jr., Frazer, L., Goodwin, A., Snowden, J.N., and Darville, T. (2011). MyD88 Deficiency Leads to Decreased NK Cell Gamma Interferon Production and T Cell Recruitment during Chlamydia muridarum Genital Tract Infection, but a Predominant Th1 Response and Enhanced Monocytic Inflammation Are Associated with Infection Resolution. Infection and immunity *79*, 486–498.

Nagarajan, U.M., Sikes, J.D., Yeruva, L., and Prantner, D. (2012). Significant Role of IL-1 Signaling, but Limited Role of Inflammasome Activation, in Oviduct Pathology during Chlamydia muridarum Genital Infection. J Immunol *188*, 2866–2875.

Naglak, E.K., Morrison, S.G., and Morrison, R.P. (2017). Neutrophils Are Central to Antibody-Mediated Protection against Genital Chlamydia. Infection and immunity *85*.

Nagosa, S., Leesch, F., Putin, D., Bhattacharya, S., Altshuler, A., Serror, L., Amitai-Lange, A., Nasser, W., Aberdam, E., Rouleau, M., *et al.* (2017). microRNA-184 Induces a Commitment Switch to Epidermal Differentiation. Stem Cell Reports *9*, 1991–2004.

Natividad, A., Cooke, G., Holland, M.J., Burton, M.J., Joof, H.M., Rockett, K., Kwiatkowski, D.P., Mabey, D.C., and Bailey, R.L. (2006). A coding polymorphism in matrix metalloproteinase 9 reduces risk of scarring sequelae of ocular Chlamydia trachomatisinfection. BMC Medical Genetics *7*, 40.

Nelson, D.E., Virok, D.P., Wood, H., Roshick, C., Johnson, R.M., Whitmire, W.M., Crane, D.D., Steele-Mortimer, O., Kari, L., McClarty, G., *et al.* (2005). Chlamydial IFN-gamma immune evasion is linked to host infection tropism. Proceedings of the National Academy of Sciences of the United States of America *102*, 10658–10663.

Netea, M.G., Kullberg, B.J., Galama, J.M., Stalenhoef, A.F., Dinarello, C.A., and Van der Meer, J.W. (2002). Non-LPS components of Chlamydia pneumoniae stimulate cytokine production through Toll-like receptor 2-dependent pathways. European Journal of Immunology 32, 1188–1195.

Njau, F., Geffers, R., Thalmann, J., Haller, H., and Wagner, A.D. (2009). Restriction of Chlamydia pneumoniae replication in human dendritic cell by activation of indoleamine 2,3-dioxygenase. Microbes Infect *11*, 1002–1010.

Noda-Nicolau, N.M., Bastos, L.B., Bolpetti, A.N., Pinto, G.V.S., Marcolino, L.D., Marconi, C., Ferreira, C.S.T., Polettini, J., Vieira, E.P., and da Silva, M.G. (2017). Cervicovaginal Levels of Human beta-Defensin 1, 2, 3, and 4 of Reproductive-Aged Women With Chlamydia trachomatis Infection. Journal of lower genital tract disease *21*, 189–192.

O'Connell, C.M., AbdelRahman, Y.M., Green, E., Darville, H.K., Saira, K., Smith, B., Darville, T., Scurlock, A.M., Meyer, C.R., and Belland, R.J. (2011). Toll-Like Receptor 2 Activation by Chlamydia trachomatis Is Plasmid Dependent, and Plasmid-Responsive Chromosomal Loci Are Coordinately Regulated in Response to Glucose Limitation by *C. trachomatis* but Not by *C. muridarum*. Infection and immunity *79*, 1044–1056.

O'Connell, C.M., Ingalls, R.R., Andrews, C.W., Jr., Scurlock, A.M., and Darville, T. (2007). Plasmid-deficient Chlamydia muridarum fail to induce immune pathology and protect against oviduct disease. J Immunol *179*, 4027–4034.

O'Connell, C.M., Ionova, I.A., Quayle, A.J., Visintin, A., and Ingalls, R.R. (2006). Localization of TLR2 and MyD88 to Chlamydia trachomatis inclusions. Evidence for signaling by intracellular TLR2 during infection with an obligate intracellular pathogen. Journal of Biological Chemistry *281*, 1652–1659.

Ohman, H., Tiitinen, A., Halttunen, M., Lehtinen, M., Paavonen, J., and Surcel, H.M. (2009). Cytokine polymorphisms and severity of tubal damage in women with Chlamydia-associated infertility. J Infect Dis *199*, 1353–1359.

Olive, A.J., Haff, M.G., Emanuele, M.J., Sack, L.M., Barker, J.R., Elledge, S.J., and Starnbach, M.N. (2014). Chlamydia trachomatis-induced alterations in the host cell proteome are required for intracellular growth. Cell host & microbe *15*, 113–124.

Opitz, B., Forster, S., Hocke, A.C., Maass, M., Schmeck, B., Hippenstiel, S., Suttorp, N., and Krull, M. (2005). Nod1-Mediated Endothelial Cell Activation by Chlamydophila pneumoniae. Circulation Research *96*, 319–326.

Pachikara, N., Zhang, H., Pan, Z., Jin, S., and Fan, H. (2009). Productive Chlamydia trachomatis lymphogranuloma venereum 434 infection in cells with augmented or inactivated autophagic activities. FEMS Microbiol Lett *292*, 240–249.

Packiam, M., Weinrick, B., Jacobs, W.R., Jr., and Maurelli, A.T. (2015). Structural characterization of muropeptides from Chlamydia trachomatis peptidoglycan by mass spectrometry resolves 'chlamydial anomaly'. Proceedings of the National Academy of Sciences of the United States of America *112*, 11660–11665.

Page-McCaw, A., Ewald, A.J., and Werb, Z. (2007). Matrix metalloproteinases and the regulation of tissue remodelling. Nat Rev Mol Cell Biol *8*, 221–233.

Park, B.S., and Lee, J.O. (2013). Recognition of lipopolysaccharide pattern by TLR4 complexes. Exp Mol Med *45*, e66.

Pilla-Moffett, D., Barber, M.F., Taylor, G.A., and Coers, J. (2016). Interferon-Inducible GTPases in Host Resistance, Inflammation and Disease. J Mol Biol *428*, 3495–3513.

Prantner, D., Darville, T., and Nagarajan, U.M. (2010). Stimulator of IFN gene is critical for induction of IFN-beta during Chlamydia muridarum infection. J Immunol *184*, 2551–2560.

Prantner, D., Darville, T., Sikes, J.D., Andrews, C.W., Jr., Brade, H., Rank, R.G., and Nagarajan, U.M. (2009). Critical role for interleukin-1beta (IL-1beta) during Chlamydia muridarum genital infection and bacterial replication-independent secretion of IL-1beta in mouse macrophages. Infection and immunity *77*, 5334–5346.

Prantner, D., Darville, T., and Nagarajan, U.M. (2010). Stimulator of IFN Gene is Critical for Induction of IFN-β during Chlamydia muridarum infection. J Immunol *184*, 2551–2560.

Prantner, D., and Nagarajan, U.M. (2009). Role for the chlamydial type III secretion apparatus in host cytokine expression. Infection and immunity *77*, 76–84.

Prantner, D., Sikes, J.D., Hennings, L., Savenka, A.V., Basnakian, A.G., and Nagarajan, U.M. (2011). Interferon regulatory transcription factor 3 protects mice from uterine horn pathology during Chlamydia muridarum genital infection. Infect Immun *79*, 3922–3933.

Prebeck, S., Kirschning, C., Durr, S., da Costa, C., Donath, B., Brand, K., Redecke, V., Wagner, H., and Miethke, T. (2001). Predominant role of toll-like receptor 2 versus 4 in Chlamydia pneumoniae-induced activation of dendritic cells. J Immunol *167*, 3316–3323.

Quesniaux, V.J., Nicolle, D.M., Torres, D., Kremer, L., Guerardel, Y., Nigou, J., Puzo, G., Erard, F., and Ryffel, B. (2004). Toll-like receptor 2 (TLR2)-dependent-positive and TLR2-independent-negative regulation of proinflammatory cytokines by mycobacterial lipomannans. Journal of Immunology *172*, 4425–4434.

Rajaram, K., and Nelson, D.E. (2015). Chlamydia muridarum infection of macrophages elicits bactericidal nitric oxide production via reactive oxygen species and cathepsin B. Infect Immun *83*, 3164–3175.

Rajeeve, K., Das, S., Prusty, B.K., and Rudel, T. (2018). Chlamydia trachomatis paralyses neutrophils to evade the host innate immune response. Nat Microbiol *3*, 824–835.

Ramsey, K.H., Miranpuri, G.S., Sigar, I.M., Ouellette, S., and Byrne, G.I. (2001a). Chlamydia trachomatis persistence in the female mouse genital tract: inducible nitric oxide synthase and infection outcome. Infect Immun *69*, 5131–5137.

Ramsey, K.H., Sigar, I.M., Rana, S.V., Gupta, J., Holland, S.M., and Byrne, G.I. (2001b). Role for inducible nitric oxide synthase in protection from chronic Chlamydia trachomatis urogenital disease in mice and its regulation by oxygen free radicals. Infection and immunity *69*, 7374–7379.

Ramsey, K.H., Sigar, I.M., Schripsema, J.H., Denman, C.J., Bowlin, A.K., Myers, G.A., and Rank, R.G. (2009). Strain and virulence diversity in the mouse pathogen Chlamydia muridarum. Infection and immunity *77*, 3284–3293.

Ramsey, K.H., Sigar, I.M., Schripsema, J.H., Shaba, N., and Cohoon, K.P. (2005). Expression of matrix metalloproteinases subsequent to urogenital Chlamydia muridarum infection of mice. Infect Immun *73*, 6962–6973.

Rank, R.G., Lacy, H.M., Goodwin, A., Sikes, J., Whittimore, J., Wyrick, P.B., and Nagarajan, U.M. (2010). Host chemokine and cytokine response in the endocervix within the first developmental cycle of Chlamydia muridarum. Infection and immunity *78*, 536–544.

Rank, R.G., Whittimore, J., Bowlin, A.K., Dessus-Babus, S., and Wyrick, P.B. (2008). Chlamydiae and polymorphonuclear leukocytes: unlikely allies in the spread of chlamydial infection. FEMS immunology and medical microbiology *54*, 104–113.

Rank, R.G., Whittimore, J., Bowlin, A.K., and Wyrick, P.B. (2011). In Vivo Ultrastructural Analysis of the Intimate Relationship between Polymorphonuclear Leukocytes and the Chlamydial Developmental Cycle. Infection and immunity *79*, 3291–3301.

Rasmussen, S.J., Eckmann, L., Quayle, A.J., Shen, L., Zhang, Y.X., Anderson, D.J., Fierer, J., Stephens, R.S., and Kagnoff, M.F. (1997). Secretion of proinflammatory cytokines by epithelial cells in response to Chlamydia infection suggests a central role for epithelial cells in chlamydial pathogenesis. J Clin Invest *99*, 77–87.

Rattan, A., Pawar, S.D., Nawadkar, R., Kulkarni, N., Lal, G., Mullick, J., and Sahu, A. (2017). Synergy between the classical and alternative pathways of complement is essential for conferring effective protection against the pandemic influenza A(H1N1) 2009 virus infection. PLoS Pathog *13*, e1006248.

Read, T.D., Brunham, R.C., Shen, C., Gill, S.R., Heidelberg, J.F., White, O., Hickey, E.K., Peterson, J., Utterback, T., Berry, K., *et al.* (2000). Genome sequences of Chlamydia trachomatis MoPn and Chlamydia pneumoniae AR39. Nucleic Acids Res *28*, 1397–1406.

Register, K.B., Morgan, P.A., and Wyrick, P.B. (1986). Interaction between Chlamydia spp. and human polymorphonuclear leukocytes *in vitro*. Infection and immunity *52*, 664–670.

Ricklin, D., Mastellos, D.C., Reis, E.S., and Lambris, J.D. (2018). The renaissance of complement therapeutics. Nature reviews Nephrology *14*, 26–47.

Romano Carratelli, C., Mazzola, N., Paolillo, R., Sorrentino, S., and Rizzo, A. (2009). Toll-like receptor-4 (TLR4) mediates human beta-defensin-2 (HBD-2) induction in response to Chlamydia pneumoniae in mononuclear cells. FEMS immunology and medical microbiology *57*, 116–124.

Roshick, C., Wood, H., Caldwell, H.D., and McClarty, G. (2006). Comparison of gamma interferon-mediated antichlamydial defense mechanisms in human and mouse cells. Infection and immunity *74*, 225–238.

Rottenberg, M.E., Gigliotti Rothfuchs, A.C., Gigliotti, D., Svanholm, C., Bandholtz, L., and Wigzell, H. (1999). Role of innate and adaptive immunity in the outcome of primary infection with Chlamydia pneumoniae, as analyzed in genetically modified mice. J Immunol *162*, 2829–2836.

Rottenberg, M.E., Gigliotti-Rothfuchs, A., and Wigzell, H. (2002). The role of IFN-gamma in the outcome of chlamydial infection. Current opinion in immunology *14*, 444–451.

Roux, C.M., Rolan, H.G., Santos, R.L., Beremand, P.D., Thomas, T.L., Adams, L.G., and Tsolis, R.M. (2007). Brucella requires a functional Type IV secretion system to elicit innate immune responses in mice. Cell Microbiol *9*, 1851–1869.

Rusconi, B., and Greub, G. (2011). Chlamydiales and the innate immune response: friend or foe? FEMS Immunol Med Microbiol *61*, 231–244.

Sasu, S., LaVerda, D., Qureshi, N., Golenbock, D.T., and Beasley, D. (2001). Chlamydia pneumoniae and chlamydial heat shock protein 60 stimulate proliferation of human vascular smooth muscle cells via toll-like receptor 4 and p44/p42 mitogen-activated protein kinase activation. Circulation Research *89*, 244–250.

Shah, A.A., Schripsema, J.H., Imtiaz, M.T., Sigar, I.M., Kasimos, J., Matos, P.G., Inouye, S., and Ramsey, K.H. (2005). Histopathologic changes related to fibrotic oviduct occlusion after genital tract infection of mice with Chlamydia muridarum. Sexually transmitted diseases *32*, 49–56.

Shimizu, T., Nomiyama, S., Hirata, F., and Hayaishi, O. (1978). Indoleamine 2,3-dioxygenase. Purification and some properties. J Biol Chem *253*, 4700–4706.

Siegl, C., Prusty, B.K., Karunakaran, K., Wischhusen, J., and Rudel, T. (2014). Tumor suppressor p53 alters host cell metabolism to limit Chlamydia trachomatis infection. Cell reports *9*, 918–929.

Sixt, B.S., Bastidas, R.J., Finethy, R., Baxter, R.M., Carpenter, V.K., Kroemer, G., Coers, J., and Valdivia, R.H. (2017). The Chlamydia trachomatis Inclusion Membrane Protein CpoS Counteracts STING-Mediated Cellular Surveillance and Suicide Programs. Cell host & microbe *21*, 113–121.

Sixt, B.S., Nunez-Otero, C., Kepp, O., Valdivia, R.H., and Kroemer, G. (2018). Chlamydia trachomatis fails to protect its growth niche against pro-apoptotic insults. Cell Death Differ.

Stanley, S.A., Johndrow, J.E., Manzanillo, P., and Cox, J.S. (2007). The Type I IFN response to infection with Mycobacterium tuberculosis requires ESX-1-mediated secretion and contributes to pathogenesis. J Immunol *178*, 3143–3152.

Stary, G., Olive, A., Radovic-Moreno, A.F., Gondek, D., Alvarez, D., Basto, P.A., Perro, M., Vrbanac, V.D., Tager, A.M., Shi, J., *et al.* (2015). VACCINES. A mucosal vaccine against Chlamydia trachomatis generates two waves of protective memory T cells. Science *348*, aaa8205.

Stetson, D.B., and Medzhitov, R. (2006). Recognition of cytosolic DNA activates an IRF3-dependent innate immune response. Immunity *24*, 93–103.

Takeuchi, O., Hoshino, K., Kawai, T., Sanjo, H., Takada, H., Ogawa, T., Takeda, K., and Akira, S. (1999). Differential roles of TLR2 and TLR4 in recognition of Gram-negative and Gram-positive bacterial cell wall components. Immunity *11*, 443–451.

Tang, L., Chen, J., Zhou, Z., Yu, P., Yang, Z., and Zhong, G. (2015). Chlamydia-secreted protease CPAF degrades host antimicrobial peptides. Microbes and infection/Institut Pasteur *17*, 402–408.

Tietzel, I., El-Haibi, C., and Carabeo, R.A. (2009). Human guanylate binding proteins potentiate the anti-chlamydia effects of interferon-gamma. PLoS One *4*, e6499.

Ting, J.P.Y., Willingham, S.B., and Bergstralh, D.T. (2008). NLRs at the intersection of cell death and immunity. Nat Rev Immunol *8*, 372–379.

Tseng, C.T., and Rank, R.G. (1998). Role of NK cells in early host response to chlamydial genital infection. Infection and immunity *66*, 5867–5875.

Vignola, M.J., Kashatus, D.F., Taylor, G.A., Counter, C.M., and Valdivia, R.H. (2010). cPLA2 regulates the expression of type I interferons and intracellular immunity to chlamydia trachomatis. J Biol Chem.

Voigt, A., Schofl, G., and Saluz, H.P. (2012). The Chlamydia psittaci genome: a comparative analysis of intracellular pathogens. PLoS One *7*, e35097.

Weber, M.M., Lam, J.L., Dooley, C.A., Noriea, N.F., Hansen, B.T., Hoyt, F.H., Carmody, A.B., Sturdevant, G.L., and Hackstadt, T. (2017). Absence of Specific Chlamydia trachomatis Inclusion Membrane Proteins Triggers Premature Inclusion Membrane Lysis and Host Cell Death. Cell reports *19*, 1406–1417.

Welter-Stahl, L., Ojcius, D.M., Viala, J., Girardin, S., Liu, W., Delarbre, C., Philpott, D., Kelly, K.A., and Darville, T. (2006). Stimulation of the cytosolic receptor for peptidoglycan, Nod1, by infection with Chlamydia trachomatis or Chlamydia muridarum. Cell Microbiol *8*, 1047–1057.

West, E.E., Kolev, M., and Kemper, C. (2018). Complement and the Regulation of T Cell Responses. Annual review of immunology *36*, 309–338.

Williams, D.M., Schachter, J., Weiner, M.H., and Grubbs, B. (1984). Antibody in host defense against mouse pneumonitis agent (murine Chlamydia trachomatis). Infection and immunity *45*, 674–678.

Wilson, D.P., Timms, P., McElwain, D.L., and Bavoil, P.M. (2006). Type III secretion, contact-dependent model for the intracellular development of chlamydia. Bull Math Biol *68*, 161–178.

Woods, M.L., 2nd, Mayer, J., Evans, T.G., and Hibbs, J.B., Jr. (1994). Antiparasitic effects of nitric oxide in an *in vitro* murine model of Chlamydia trachomatis infection and an *in vivo* murine model of Leishmania major infection. Immunology series *60*, 179–195.

Yang, C., Briones, M., Chiou, J., Lei, L., Patton, M.J., Ma, L., McClarty, G., and Caldwell, H.D. (2019). Chlamydia trachomatis Lipopolysaccharide Evades the Canonical and Noncanonical Inflammatory Pathways To Subvert Innate Immunity. MBio *10*.

Yang, Z., Conrad, T., Zhou, Z., Chen, J., Dutow, P., Klos, A., and Zhong, G. (2014). Complement factor C5 but not C3 contributes significantly to hydrosalpinx development in mice infected with Chlamydia muridarum. Infection and immunity *82*, 3154–3163.

Yang, Z., Tang, L., Shao, L., Zhang, Y., Zhang, T., Schenken, R., Valdivia, R., and Zhong, G. (2016a). The Chlamydia-Secreted Protease CPAF Promotes Chlamydial Survival in the Mouse Lower Genital Tract. Infection and immunity *84*, 2697–2702.

Yang, Z., Tang, L., Zhou, Z., and Zhong, G. (2016b). Neutralizing antichlamydial activity of complement by chlamydia-secreted protease CPAF. Microbes and infection/Institut Pasteur *18*, 669–674.

Yasin, B., Harwig, S.S., Lehrer, R.I., and Wagar, E.A. (1996). Susceptibility of Chlamydia trachomatis to protegrins and defensins. Infection and immunity *64*, 709–713.

Yeruva, L., Myers, G.S.A., Spencer, N., Creasy, H.H., Adams, N.E., Maurelli, A.T., McChesney, G.R., Cleves, M.A., Ravel, J., Bowlin, A., *et al.* (2014). Early MicroRNA Expression Profile as a Prognostic Biomarker for the Development of Pelvic Inflammatory Disease in a Mouse Model of Chlamydial Genital Infection (vol 5, e01241, 2014). Mbio *5*.

Yeruva, L., Pouncey, D.L., Eledge, M.R., Bhattacharya, S., Luo, C.Q., Weatherford, E.W., Ojcius, D.M., and Rank, R.G. (2017). MicroRNAs Modulate Pathogenesis Resulting from Chlamydial Infection in Mice. Infection and Immunity *85*.

Yu, J., Ryan, D.G., Getsios, S., Oliveira-Fernandes, M., Fatima, A., and Lavker, R.M. (2008). MicroRNA-184 antagonizes microRNA-205 to maintain SHIP2 levels in epithelia. Proc Natl Acad Sci U S A *105*, 19300–19305.

Zhang, Y., Yeruva, L., Marinov, A., Prantner, D., Wyrick, P.B., Lupashin, V., and Nagarajan, U.M. (2014). The DNA sensor, cyclic GMP-AMP synthase, is essential for induction of IFN-beta during Chlamydia trachomatis infection. J Immunol *193*, 2394–2404.

Zhao, G.N., Jiang, D.S., and Li, H.L. (2015a). Interferon regulatory factors: at the crossroads of immunity, metabolism, and disease. Bba-Mol Basis Dis *1852*, 365–378.

Zhao, X., Zhu, D., Ye, J., Li, X., Wang, Z., Zhang, L., and Xu, W. (2015b). The potential protective role of the combination of IL-22 and TNF-alpha against genital tract Chlamydia trachomatis infection. Cytokine *73*, 66–73.

Zhao, Z., Fux, B., Goodwin, M., Dunay, I.R., Strong, D., Miller, B.C., Cadwell, K., Delgado, M.A., Ponpuak, M., Green, K.G., *et al.* (2008). Autophagosome-independent essential function for the autophagy protein Atg5 in cellular immunity to intracellular pathogens. Cell Host Microbe *4*, 458–469.

Zhou, H., Huang, Q., Li, Z., Wu, Y., Xie, X., Ma, K., Cao, W., Zhou, Z., Lu, C., and Zhong, G. (2013). PORF5 plasmid protein of Chlamydia trachomatis induces MAPK-mediated pro-inflammatory cytokines via TLR2 activation in THP-1 cells. Sci China Life Sci *56*, 460–466.

Adaptive Immunity to *Chlamydia trachomatis* Infection

14

Taylor B. Poston[1], Toni Darville[1] and Raymond M. Johnson[2]*

[1]Department of Pediatrics, University of North Carolina at Chapel Hill, Chapel Hill, NC, USA.
[2]Department of Medicine, Yale University School of Medicine, New Haven, CT, USA.

*Correspondence: raymond.johnson@yale.edu

https://doi.org/10.21775/9781912530281.14

Abstract

Understanding mechanics of mucosal adaptive immunity may facilitate development of a *Chlamydia trachomatis* vaccine. This chapter details mechanisms involved in determining whether an individual's adaptive immune response (antigen-specific B and T cell response) to a chlamydial genital tract infection or urogenital vaccine results in asymptomatic clearance (protection) or disease (urethritis/salpingitis/infertility). Over 30 years of research have revealed that adaptive responses to *C. trachomatis* are more complex than the presence or absence of antigen-specific T cell production of interferon-gamma (IFN-γ) (Th1 immunity) and B cell production of neutralizing antibody against elementary bodies. Basic research is beginning to define adaptive immunity to *Chlamydia* infections as functioning in tissues rather than in lymph nodes and spleens. Central to local tissue immunity against the epithelial-tropic *C. trachomatis* serovars (A-K) are B cells, T cells, and highly specialized immune infrastructures localized immediately beneath the reproductive tract epithelium. These subepithelial immune structures, called memory lymphocyte clusters, harbour plasma B cells and tissue resident memory T cells that are likely to be critical to protective immunity. While this story is incomplete, we try to present a working narrative of how chlamydial-specific mucosal adaptive immunity may be relevant to vaccine development. We touch on trachoma adaptive immunity to highlight how the eye differs from the genital tract as that informs the possibility of a universal urogenital/trachoma vaccine. Adaptive immunity mechanisms in *Chlamydia*-associated reactive arthritis, *C. trachomatis* serovar L_{1-3} (causative agent of lymphogranuloma venereum), and other Chlamydiae (e.g. *C. psittaci*) are unique and cannot be extrapolated from *C. trachomatis* urogenital tract data.

Introduction

Reproductive tracts are exposed to the environment and the body fluids of other individuals during sexual encounters. Prolonged replication of *C. trachomatis* urogenital serovars in mucosal epithelia increases the probability of transmission to multiple sexual partners. *Chlamydiae* are obligate intracellular bacterial pathogens that cannot be cleared from the reproductive tract by innate defence mechanisms in their natural hosts. In addition, convincing research shows that antibodies are not sufficient to clear chlamydial genital tract infections. Though innate immunity dramatically reduces the level of bacterial shedding (see Chapter 13, *Chlamydia* and Innate Immunity), preservation of the reproductive tract epithelium during *C. trachomatis* infections falls principally on the adaptive T cell response. Although self-clearance of *C. trachomatis* takes a long time, humans are capable of self-clearing chlamydial genital tract infections and develop resistance to reinfection, similar to mice and guinea pigs (Geisler *et al.*, 2013; Morre *et al.*, 2002; van den Brule *et al.*, 2002). A working paradigm for protective immunity must account for the underlying mechanisms that finally resolve infection to facilitate development of an efficacious vaccine. In the coming sections we (1) cover protective immunity as defined in the mouse model, (2) individually address CD4 T cells, CD8 T cells and B cells, incorporating human data as they relate to protective immunity and immunopathology, and (3) contrast cytokine polarization with tissue resident immunity as working paradigms for understanding *Chlamydia* pathogenesis.

Animal model data tell us what to look for in human studies

Animal model data are necessary to understand T cell-mediated immunity since human investigations need a mechanistic context for interpreting results. For example, IFN-γ knockout mice cannot clear *Chlamydia muridarum* genital tract infections so it is logical to study whether human IFN-γ responses correlate with *C. trachomatis* acquisition and clearance. However, it is not possible to identify a cohort of humans with IFN-γ deficiencies and study whether their natural course of *C. trachomatis* genital tract infection is altered compared to the general population.

It is important to pick the right animal model. Human *C. trachomatis* strains infect mice but can be cleared solely by innate mechanisms, as demonstrated with CD4 T cell depletion (Morrison *et al.*, 2011) and recombination activating gene 1 (*Rag*) knockout mice lacking B and T cells (Sturdevant and Caldwell, 2014). If the goal is to make a T cell vaccine, infecting mice with human *C. trachomatis* serovars may not be an optimal model. Thus, the majority of mechanistic T cell data come from the *C. muridarum* mouse model that requires T cells for clearance (Rank *et al.*, 1985). While this model has some limitations, it reproduces human genital tract infections with respect to occurrence of an ascending infection from a vaginal inoculum. Mouse manipulation is limited to a medroxyprogesterone injection one week prior to experimental infection, preserving murine uterine epithelium that otherwise cyclically involutes and rejuvenates in a rapid < 1-week cycle. Medroxyprogesterone is also a common method of birth control used by women 15–24 years of age, who are at high risk for *C. trachomatis* infection. An ascending *C. muridarum* infection reaches the oviducts (mouse version of Fallopian tubes) and causes salpingitis, which in a subset of mice progresses to scarring and oviduct dilatation or hydrosalpinx (pathology also seen in women). Repeated infections in mice cause reduced fertility and infertility (Igietseme *et al.*, 2009). A similar narrative describes repeated urogenital *C. trachomatis* infections in women. *Chlamydia muridarum* is closely related to *C. trachomatis* with gene-for-gene synteny except for a region known as the plasticity zone that encodes genes necessary for evading species-specific IFN-γ induced innate immunity that differs between mice and humans (Nelson *et al.*, 2005). Mice, unlike humans, have interferon-inducible p47 GTPases, a cell intrinsic defence mechanism that relies on altering endosomal trafficking. Conversely, human urogenital *C. trachomatis* strains, unlike *C. muridarum*, have tryptophan synthesis operons for synthesizing their own tryptophan to circumvent IFN-γ-induced host cell tryptophan depletion mediated by indoleamine 2,3-dioxygenase (IDO) (Nelson *et al.*, 2005). Curiously, the lack of the *trp* operon in *C. muridarum* implies that tryptophan depletion does not apply selective pressure in mice. To our knowledge, there are no major species-specific differences in cellular machinery related to T cell biology, including antigen processing, antigen presentation molecules (MHC/HLA class Ia, Ib, and II), and T cell receptor (TCR) recognition and activation pathways. Therefore, it is reasonable to extrapolate T cell mechanisms elucidated in the *C. muridarum* mouse model to the human condition.

Data from the *C. muridarum* model identify CD4 T cells as the most important and reliable effector for controlling genital tract infection and preventing immunopathology (Morrison and Caldwell, 2002). However, there are caveats to that statement. To understand the caveats, it is necessary to examine basic adaptive immune mechanisms and ask which antigen-presentation molecules and co-receptor(s) are associated with T cell recognition of infected cells, protective immunity and immunopathology. Data based on experiments using gene-deficient mice revealed the following (Morrison *et al.*, 1995):

1 MHC class II is required to clear *C. muridarum* genital tract infections, as MHC class II deficiency leads to persistent infection. CD4 is

the T cell co-receptor for MHC class II. Co-receptors on T cells bind MHC molecules on the target cells presenting chlamydial peptides bound and exposed in an outer surface groove. Co-receptors, in this case CD4, are needed for antigen recognition events through the T cell receptor complex; CD4 must bind MHC class II for CD4 T cells to be activated.

2. MHC class Ia and class Ib molecules are entirely dispensable, and their absence has negligible effects on bacterial clearance kinetics. The T cell co-receptor for MHC class Ia is CD8, and CD8 does not bind class Ib molecules. T cells that recognize antigens bound to MHC class Ib molecules lack a known co-receptor.
3. The 'CD4 knockout mouse caveat'. The MHC class I and class II knockout mouse data predict that CD4 knockout mice would have persistent genital tract infections because their remaining T cells lack the co-receptor for binding MHC class II molecules. However, CD4 knockout mice clear infection with a slight one-week delay. This discordant result has been overlooked or minimized, but is potentially an important finding (Morrison *et al.*, 1995).

These data demonstrate that MHC class II molecules, not MHC class Ia and Ib, are *unequivocally* necessary to clear infection (Morrison *et al.*, 1995), and have been presumptively interpreted as CD4 T cells clearing infection with little or no contribution from CD8 T cells. However, the 'CD4 knockout mouse caveat' will be revisited during examination of the role of CD8 T cells in chlamydial immune protection and immunopathology. Consistent with the mouse model data, MHC class II is linked to *C. trachomatis* genital tract infection outcomes in humans. HLA-DQ alleles (human MHC class II) have been associated with fertility outcomes (Kinnunen *et al.*, 2002; Ness *et al.*, 2004) and susceptibility to reinfection (Olson *et al.*, 2018; Wang *et al.*, 2005).

CD4 T cells

Data from experiments in wild type mice strongly support CD4 T cell recognition of MHC class II-presented chlamydial peptides as the major effector cell type and immune event responsible for clearing *C. muridarum* from the genital tract. Mice deficient in MHC class II molecules or directly depleted of CD4 T cells with antibody are severely compromised in their ability to clear primary genital infections (Landers *et al.*, 1991; Morrison and Morrison, 2005a). Adoptively transferring chlamydial-specific CD4 T cells into nude mice that lack functional T cells resolves chronic infections (Igietseme *et al.*, 1993). While CD4 T cells mediate bacterial clearance, one cannot conclude from existing data that there is no role for CD8 T cells. CD8 T cell expansion during a primary response can be dependent on helper CD4 T cells; therefore, absence of CD4 T cells due to antibody-depletion or MHC class II deficiency likely also precludes any CD8 T cell response (Jennings *et al.*, 1991). An adaptive T cell response without a helper T cell component is an experimental tool rather than a physiological event in nature, and therefore the conclusion that CD8 T cells play little or no role in protective immunity is not definitively supported by existing data.

Efficient vaccine development requires a surrogate biomarker for protective immunity. Chlamydial-specific CD4 T cell responses are evident in individuals with active or prior infections (Barral *et al.*, 2014; Jordan *et al.*, 2017). Generic demonstration of a CD4 T cell IFN-γ response specific for *C. trachomatis* elementary bodies (EB) *is not predictive of protection* in humans (details below). However, more refined analyses in two studies provide proof-in-principle that surrogate biomarkers for vaccine efficacy are possible:

Study 1 in Kenyan female sex workers showed that a peripheral blood mononuclear cell (PBMC) IFN-γ response to *chlamydial* HSP60 or, independently, an IL-13 response to purified EB correlated with resistance to reinfection (Cohen *et al.*, 2005). Unless *C. trachomatis* HSP60 is the subunit vaccine antigen (unlikely) or the vaccine is live-attenuated, a PBMC IFN-γ response to HSP60 is not relevant to vaccine assessment. The IFN-γ response rate to EB was 40% (69 out of ~170 evaluable individuals with adequate assay data); the IL-13 response to EB was 20% (35 out of ~170 evaluable individuals). The incidence of reinfection in EB/IFN-γ+ individuals was the same as the cohort as a whole (risk ratio = 1); the incidence of reinfection in EB/IL-13+ individuals was 5-fold lower than the cohort as a whole (risk ratio = 0.2). It is reasonable to assume female sex workers in this study were repeatedly challenged with *C. trachomatis* exposures. *If* you were sure that *C. trachomatis* antigen 'X' was protective, *perhaps* demonstrating a vaccine-antigen-specific IL-13 PBMC response post-vaccination would be a surrogate biomarker for protective immunity. *By crude extrapolation*, 20% of women who had the EB/IL-13 vaccine response would get infected upon chlamydial exposure; 80% would not. This is a relatively uncomplicated assay that may have additional predictive value as IL-13 appears to be a biomarker for mucosal T cell subsets (more later).

Study 2 recently published by Bakshi *et al.* (Bakshi *et al.*, 2018) is similar to study 1 but more refined. It analysed T cell subsets (CD4 *vs.*CD8), and responses to EBs and specific *C. trachomatis* antigens including MOMP. Women with incident *C. trachomatis* infections were enrolled and immune parameters performed at follow -up visits were then correlated with the incidence of reinfection going forward. *Uninfected status* on return clinic visit is a surrogate for protective immunity; for clarity, we use the 3-month follow up visit for this discussion. As in study 1, chlamydial-specific IFN-γ responses to EBs did not correlate with reduced risk of reinfection. A CD4, but not CD8, IFN-γ response to MOMP correlated with a 2-fold reduction in likelihood of reinfection. While 60% of the *uninfected* subjects had a CD4 IFN-γ response to MOMP, 40% of *uninfected* subjects did not, arguing for presence of non-MOMP protective antigens, limited re-exposure to *C. trachomatis*, or some combination thereof. A CD4 MOMP IFN-γ response did not reflect solid protection as 33% of reinfected women had CD4 MOMP IFN-γ responses. Reinfections in women with MOMP-specific CD4 IFN-γ responses were in part explained by an IFN-γ *monokine* response shown in mice to be nonprotective (Yu *et al.*, 2010), i.e. T cells that produce IFN-γ without additional cytokines (IL-2 or TNFα). Advanced T cell subset analysis for multifunctional CD4 T cell responses against any *C. trachomatis* antigen (EB, MOMP, or PgP3) showed that 70% of *uninfected* subjects had a multifunctional CD4 response (IFN-γ plus IL-2/TNFα); 20% of reinfected individuals had a multifunctional CD4 IFN-γ/IL-2 response. *If* you were sure that *C. trachomatis* antigen 'X' was protective, *perhaps* demonstrating a vaccine-antigen-specific CD4 IFNγ/IL-2 response post-vaccination would be a surrogate biomarker for protective immunity. *By crude extrapolation*, 20% of women who had the CD4 IFNγ/IL-2 response would get infected upon *C. trachomatis* exposure; 80% would not. A similar correlation of multifunctional CD4 IFN-γ T cells with protective immunity exists in the *C. muridarum* model (Yu *et al.*, 2010; Yu *et al.*, 2012; Yu *et al.*, 2011). Thus, mouse experiments and human investigations show that multifunctional CD4 T cells are important for protection.

In summary, more data are needed regarding the CD4 T cell subset response profile including, but not limited to, cytokine polarization and antigen specificity to facilitate vaccine development. It is important to remember that human investigations based on infected individuals, by definition, evaluate a mucosal immune response that will differ from conventional subunit vaccine injections to skeletal muscle. While the *crude analysis* above suggests that we have plausible surrogate biomarkers for protection, an understanding of how local mucosal immunity works, raises significant concerns that we do not. Recent work demonstrating that mucosal CD4 T cells can be divided into subsets using parameters other than cytokine polarization (Johnson and Brunham, 2016; Johnson *et al.*, 2017) may be a step forward towards practicable surrogate biomarkers for protective mucosal immunity (upcoming section on tissue resident memory T cells).

CD8 T cells

Chlamydial-specific CD8 responses occur in humans and primates (Gervassi *et al.*, 2003; Van Voorhis *et al.*, 1996). In mice, the CD8 T cell response amounts to about 10% of the overall T cell response in the genital tract and draining lymph node (Jiang *et al.*, 2017b; Kelly and Rank, 1997). The reason(s) why *C. muridarum*, an intracellular pathogen, elicits a limited CD8 T cell response is unclear. One plausible explanation comes from the murine *C. muridarum* immunoproteome. Sequenced peptides eluted from MHC class I and II molecules purified from infected dendritic cells yielded one chlamydial peptide bound to MHC class Ia, while multiple chlamydial peptides have been eluted from MHC class II molecules (Karunakaran *et al.*, 2015; Yu *et al.*, 2009). Because dendritic cells are the critical antigen presenting cell type for naïve T cell expansion during primary responses, the anaemic CD8 T cell response may reflect limited MHC class I processing of chlamydial antigens by dendritic cells. Murine dendritic cells are semi-permissive for chlamydial replication; *C. muridarum* infection of bone marrow-derived dendritic cells showed increasing numbers of inclusions over time, but the number of recoverable infectious chlamydiae was less than the input inoculum and decreased over time. It was postulated that this discordance between increasing inclusion number and decreasing infectious progeny represented a non-replicative persistent state (Rey-Ladino *et al.*, 2007). Similar findings have been observed with the replication of urogenital *C. trachomatis* serovars in human dendritic cells (Datta *et al.*, 2014). If inclusions within dendritic cells represent a 'persistent state', then chlamydial gene expression and protein synthesis would be limited in quantity and diversity [see *C. trachomatis* gene expression patterns in productive and persistent states (Belland *et al.*, 2003a; Belland *et al.*, 2003b)], and the array of antigens presented by dendritic cells would be potentially discordant with antigens presented by productively infected epithelial cells where chlamydiae complete a full developmental-cycle; i.e. express all 900 genes.

The murine chlamydial-specific CD8 T cell response is important despite its limited size. In the *C. muridarum* model, CD8 T cells are principally responsible for infection-associated immunopathology, including infertility (Igietseme *et al.*, 2009). Important studies have shown that CD8-mediated immunopathology is dependent on TNFα production (Murthy *et al.*, 2011) and chlamydial antigen specificity (Manam *et al.*, 2013; Vlcek *et al.*, 2016). CD8 mediated immunopathology is not a bystander event, it is part and parcel of the antigen-specific CD8 T cell response to infection. CD8 T cells on the whole cause immunopathology; however, it is unclear if all CD8 T cell subsets are immunopathologic. Chlamydial-specific CD8 T cell clones isolated from infected humans and mice are atypical compared to those active against viruses because the majority of chlamydial-specific T cell clones are not conventionally restricted by MHC class I molecules, (i.e. TCRs recognize chlamydial peptides bound to molecules other than MHC class I) (Gervassi *et al.*, 2003; Johnson *et al.*, 2014; Matyszak and Gaston, 2004). There is not a similar phenomenon associated with MHC class II presentation for the CD4 T cell subset. Broad conclusions regarding chlamydial-specific CD8 T cell function require better understanding of these two major subsets: a minority CD8 T cell subset restricted by MHC class Ia (conventional CD8 T cells), and a majority 'unrestricted' CD8 T cell subset (unconventional CD8 T cells). The ontogeny, regional distribution, and function of 'unrestricted' CD8 T cells likely differs dramatically from that of conventional MHC class Ia restricted CD8 T cells. In one human study, the investigators noted that 'unrestricted' CD8 T cell clones recognized infected target cells at earlier time points post-infection than conventional CD8 T cells, a phenotype potentially advantageous for clearing chlamydial infection (Matyszak and Gaston, 2004). Among published murine data, adoptive transfer of a chlamydial-specific CD8 T cell clone into nude mice with chronic infections led to bacterial clearance in a slight majority (Igietseme *et al.*, 1994). Recalling the 'CD4 knockout mouse caveat', a live attenuated trachoma vaccine induced protective immunity in primates that mapped to MHC class II (Kari *et al.*, 2011), however, subsequent antibody depletion studies in the same previously vaccinated monkeys showed that protective response was mediated by CD8 T cells (Olivares-Zavaleta *et al.*, 2014), rather than the expected result of CD4 T cells. This finding implies an unconventional MHC class II-restricted CD8 T cell subset may be the relevant effector for this protective experimental vaccine. Similarly, the 'CD4 knockout mouse caveat' itself is most simply explained by a protective MHC class II-restricted CD8 T cell response. These results emphasize that chlamydial-specific CD8 T cell responses require additional investigations that may yield important insights into immunopathology and protective immunity.

T cell effector mechanisms

The ability of a CD4 or CD8 T cell to recognize an infected cell does not in-and-of-itself cause bacterial clearance. Some event (effector mechanism) triggered by T cell recognition must physically damage and kill chlamydiae. Based on mechanisms of viral clearance, the logical event would be for T cells, CD4 or CD8, to kill chlamydial-infected cells. Target cell lysis destroys the chlamydial incubator causing the death of non-infectious replicating RBs and exposing intracellular infectious EBs to extracellular effector mechanisms such as antibody, complement, and phagocytes capable of mediating EB demise. Most CD8 and CD4 T cells are capable of cytolysis. The question is whether cytolysis, rapidly via perforin or more slowly by induction of apoptotic pathways, is a relevant effector mechanism for clearing chlamydiae from reproductive tract epithelium. The simple answer from studies in the mouse model is no. Mice deficient in perforin, apoptosis receptor–ligand pairs, or both, do not show any deficiency in their ability to clear genital tract infections (Perry *et al.*, 1999). These results have practical implications for vaccine development. For example, measuring vaccine-induced chlamydial-specific PBMC cytolytic capacity is unlikely to be a useful surrogate biomarker for vaccine efficacy.

Nitric oxide

T cell-mediated clearance of chlamydiae from the genital tract requires a physical assault on intracellular bacteria. In the mid-1990s, the search for an effector mechanism led to investigation of inducible nitric oxide synthetase (iNOS) and nitric oxide as the relevant effector mechanism. iNOS is transcriptionally upregulated in epithelial cells by IFN-γ, and further upregulated by T cell-epithelial cell physical interaction (Igietseme *et al.*, 1996). Induced nitric oxide kills chlamydiae residing within mouse and human epithelial cells (Igietseme, 1996; Igietseme *et al.*, 1997). iNOS and nitric oxide production seemed to be the answer. Chlamydial-specific T cells, especially Th1 CD4 T cells (polarized to make IFN-γ), bathe the reproductive tract epithelium in IFN-γ to induce epithelial production of nitric oxide. Nitric oxide, the nitrogenous version of hydrogen peroxide, sterilizes intracellular chlamydiae throughout the reproductive tract. It was postulated that the goal of vaccination was a robust Th1 response with levels of IFN-γ sufficient to purge chlamydiae with a deadly fog of nitric oxide. This was an appealing model as one can readily imagine a cytokine like IFN-γ acting at a distance to induce epithelial nitric oxide production without T cells having to track down and eliminate every infected epithelial cell, and perhaps accomplish this without physically interacting with epithelial cells at all. In retrospect, this model is discordant with the kinetics of IFN-γ levels within the genital tract (Darville *et al.*, 2001b). IFN-γ in genital tract secretions becomes undetectable 1–2 weeks prior to final bacterial clearance. In the last stages of bacterial clearance, the absence of nitric oxide fog would presumably leave reproductive tract epithelium susceptible to infection by viable bacteria still detectable in the lower reproductive tract. The model as narrated above was put to rest when independent investigators showed that *iNOS* knockout mice were not dramatically compromised in their ability to clear *C. muridarum* genital tract infections (Igietseme *et al.*, 1998; Ramsey *et al.*, 1998). That result argued that nitric oxide was either irrelevant or there was a second redundant chlamydiacidal mechanism that compensates when nitric oxide is absent.

The chlamydial clearance model required two modifications. The first is that it is unlikely that T cells act at a distance as this is contrary to their behaviour as we otherwise know it. T cells interact with target cells. The likely reason it takes mice at least a month, and humans 6–12 months or more, to clear primary genital tract infections is that T cells must track down every last infected epithelial cell, a logical consequence of the inability of anti-chlamydial antibodies and innate mechanisms to clear infections. Reproductive tract epithelial cells at baseline have very low expression levels of MHC molecules, especially MHC class II. Infected epithelial cells have been wrongly assumed to also have low MHC class II expression and thus to be invisible to CD4 T cells. However, increased expression of mouse and human epithelial MHC class II molecules is induced by inflammatory cytokine milieus that contain IFN-γ, and has been demonstrated in infected mouse and human epithelial cells (Abu el-Asrar *et al.*, 2001; Jayarapu *et al.*, 2009; Mabey *et al.*, 1991). Furthermore, experiments *in vitro* showed that chlamydial-specific CD4 and CD8 T cells recognized infected epithelial cells and terminated chlamydiae within them (Jayarapu *et al.*, 2010; Jayarapu *et al.*, 2009; Johnson *et al.*, 2014). The timing of mouse CD4 T cell clone recognition of infected epithelial cells was influenced by IFN-γ (earlier recognition), and varied by T cell clone, with earliest detectable recognition 12 h post infection in IFN-γ pretreated epithelial cells. Full recognition/activation occurred 16–18 hours post-infection, too late to destroy the *C. muridarum* epithelial incubator before non-infectious RBs transition to infectious EBs (Jayarapu *et al.*, 2009).

Plac8

The second modification to the chlamydial clearance model is that there are redundant mechanisms for terminating chlamydiae. Experiments with CD4 T cell clones and infected epithelial cells *in vitro* identified two independent mechanisms. The first mechanism was dependent on iNOS and nitric oxide production, and seemed to be universally used by chlamydial-specific CD4 T cell clones. When iNOS was inhibited, a subset of T cell clones could still terminate chlamydiae within infected epithelial cells. That iNOS-independent termination event was dependent on T cell degranulation (Jayarapu *et al.*, 2010). Microarray studies comparing these iNOS-dependent versus -independent T cell subsets identified a protein called Plac8 that localizes to granules in neutrophils, and is broadly expressed in haematopoietic cells (Ledford *et al.*, 2007); *Plac8* mRNA transcription is also inducible in epithelial cells (Johnson and Kerr, 2015). *Plac8*-deficient neutrophils had a reduced ability to kill Gram-negative bacteria, and *Plac8* knockout mice had increased susceptibility to bacterial peritonitis (Ledford *et al.*, 2007). These observations led to chlamydial genital tract challenge experiments in *Plac8* knockout mice (Johnson *et al.*, 2012a). *Plac8* knockout mice cleared genital tract infections with a slight delay, but *Plac8* knockout mice treated with an iNOS inhibitor (i.e. genetically deficient in *Plac8* and pharmacologically deficient in nitric oxide) were almost entirely unable to clear *C. muridarum* from the genital tract over eight weeks and had exacerbated damage to their oviducts and uterine horns. Wild type mice treated with iNOS inhibitor cleared infections under the same conditions.

Plac8 and nitric oxide are both relevant to clearance of bacteria from the genital tract

These experiments revealed two potent and redundant T cell-associated pathways for clearing chlamydiae from genital tract epithelium and resolved a longstanding conundrum in chlamydial pathogenesis. The first T cell clearance mechanism is based on iNOS and nitric oxide production; the second mechanism is independent of nitric oxide and requires *Plac8* (Johnson *et al.*, 2012a). Mice deficient in iNOS clear chlamydial infections, because they still have the *Plac8*-dependent mechanism (and *vice versa*). It is unknown whether these two mechanisms are additive, synergistic, or antagonistic. One scenario is that they have a temporal relationship to clearance, with the iNOS/nitric oxide mechanism dominant when IFN-γ is detectable in the genital tract, and the *Plac8*-dependent mechanism becoming dominant during the last stages of bacterial clearance after IFN-γ wanes.

T cell-mediated iNOS-dependent killing of intracellular chlamydiae appears to be mechanistically straightforward. *Chlamydia*-specific T cells produce IFN-γ in a paracrine fashion, inducing iNOS transcription in epithelial cells, which is further boosted by the physical interaction between T cell and epithelial target, leading to bactericidal concentrations of intracellular nitric oxide, a nitrogen-based free radical mechanism that indiscriminately damages proteins, nucleic acids and lipids. The second mechanism is relatively IFN-γ independent and requires T cell degranulation, and likely Plac8. The nature of the physical damage that kills intracellular chlamydiae by this degranulation-dependent mechanism is unknown.

T cell-mediated termination of chlamydiae has an overarching and satisfying theme. T cells recognize infected epithelial cells and 'tell them' to kill themselves and anything in them. During this process, toxic molecules reach high concentrations inside infected epithelial cells to maximize chlamydial damage. When those epithelial cells die, the toxic molecules are released into an infinite-dilution void (reproductive tract lumen) thereby minimizing damage to the reproductive tract, a far less damaging approach than seen in pelvic inflammatory disease where activated neutrophils unload free radicals, proteases, and elastases at the epithelial interface causing irreversible damage. Based on the 'toxic bag' model of chlamydial termination (Fig. 14.1), T cell-mediated cytolysis via perforin channels would lead to decreased termination of chlamydiae due to leakage of toxic effector molecules through perforin-induced membrane pores, and that is the case (Johnson *et al.*, 2013). Chlamydial-specific CD4 T cells from immune perforin-knockout mice are more potent terminators of intracellular chlamydiae than similar immune CD4 T cells from wild type mice. Fig. 14.2 shows unpublished CD4 and CD8 T cell killing

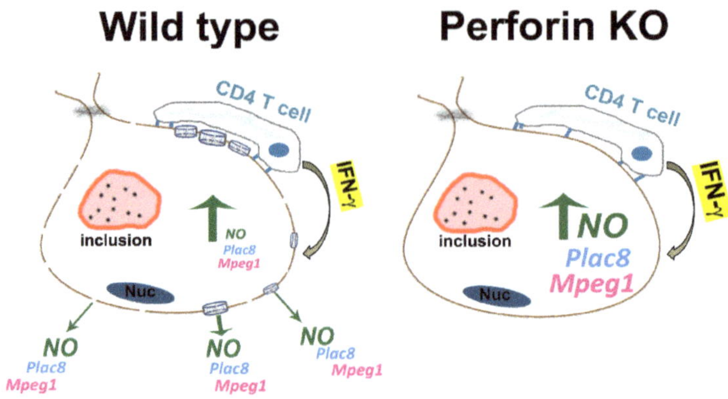

Figure 14.1 Toxic bag model. Wild type T cell insert perforin-channels into the target cell plasma membrane. Experimentally, immune T cells from perforin knockout mice are more potent terminators of intracellular chlamydiae than immune wild type T cells, presumably because KO T cells cannot punch perforin holes in the target cell plasma membrane, an event that diminishes the effective intracellular concentration of effector molecules (nitric oxide, Plac8, Mpeg1, etc.). Mpeg1 is also known as perforin-2 (Fields *et al.*, 2013).

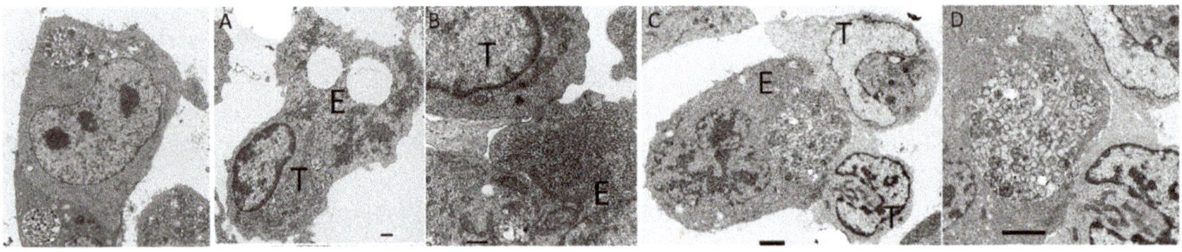

Figure 14.2 T cell-mediated termination of intracellular chlamydiae at the single cell level. The first panel shows an epithelial cell 16 hours post-infection without T cell contact (note healthy appearing nucleus with nucleoli and a relatively homogeneous grey cytoplasm). (A and B) *Chlamydia*-specific CD4T cell interaction with infected epithelial cells [note that granular osmium staining deposits throughout the epithelial cytoplasm that may originate at the epithelial point of contact with the T cell (B)]. (C and D) *Chlamydia*-specific CD8T cell interaction with infected epithelial cells [note that the nucleus is pyknotic and the inclusion body appears distressed (D), whereas the cytoplasm looks unaffected, i.e. similar to the first panel]. E=epithelial cell; T=T cell.

of infected epithelial cells *in vitro* using electron microscopy (*unpublished data*, RM Johnson and RC Brunham). During carefully choreographed experiments mixing T cells with infected epithelial cells, it was only possible to capture an epithelial inclusion and the attacking T cell in the same plane for the CD8 clone. Upon T cell recognition, the bactericidal epithelial suicide events are likely rapid on the time scale of minutes. It is highly unlikely that one can observe such events in cross sections of infected mouse genital tracts as targeted epithelial cells likely release from the basal lamina into the reproductive tract lumen. It is noteworthy that the epithelial suicide event induced by CD4 and CD8 T cells may be different. The CD4 event appears to cause coagulation of the target cell cytosol (Fig. 14.2A and B), while the CD8 event resembles necrosis (panel C), with some specificity for the inclusion body (panel D). Our inability to capture a CD4 T cell and an epithelial inclusion in the same plane and the gross coagulation of the cytosol suggests that the CD4-mediated event is rapid and cataclysmic. Very little is known about the intracellular events (cascades) triggered during T cell-epithelial cell termination of intracellular bacterial pathogens. An interesting facet of T cell-mediated clearance is that ultimately it is infected epithelial cells that produce/activate the effector molecules that kill chlamydiae.

Cytokine polarization

For two decades chlamydial pathogenesis has been described using the cytokine polarization Th1, Th2,

... Thx framework (Mosmann *et al.*, 1986). In this paradigm the combination of dendritic cell type presenting antigen plus the local environmental cytokine milieu, determined principally by the nature of the invading pathogen and the resulting innate response, lead to T cell differentiation into specific effector types that make specific cytokines. For intracellular pathogens, CD4 T cell differentiation is typically towards production of IFN-γ and is known as T helper type 1, Th1. For extracellular bacteria and parasites, CD4 T cell specialization is typically towards cytokines that facilitate potent B cell (IL-4) and eosinophil responses (IL-5) and is known as Th2. There are additional cytokine patterns associated with other T effector types, e.g. Th17 is prominent with some bacterial infections, such as *Neisseria gonorrheoae* and autoimmune arthritis. It is unclear whether the Th1/Th2/Thx paradigm is sufficient to explain new data or guide chlamydial vaccine development.

The utility of this cytokine-based paradigm has been starkly demonstrated by adoptive transfer of chlamydial-specific Th1 (IFN-γ) and Th2 clones (IL-4) into nude mice chronically infected with *C. muridarum* (Hawkins *et al.*, 2002). Th1 recipient mice cleared infections, while Th2 recipient mice did not. Th1 recipients had a Th1-type antibody response (IgG2a), while Th2 mice had a Th2-type antibody response (IgG1) and an IgA response without obvious benefit. While the antigen-specificity of the T cell clones was not controlled, it provided a demonstration of the basic premise of the Th1/Th2 paradigm with Th1 being protective, while Th2 being ineffective against *Chlamydia* infection. In the *C. trachomatis* L2 mouse model, naïve T cells from a *C. trachomatis* L2-specific TCR transgenic mouse were polarized *in vitro* to Th1, Th2, and Th17 phenotypes prior to being adoptively transferred into mice that were subsequently challenged intravenously with human *C. trachomatis* serovar L2. Th1 recipients had lower chlamydial burdens in the spleen than control mice, while Th2 recipients had higher splenic burdens than control mice (Gondek *et al.*, 2009). Inoculation route and infected-spleen-readout caveats aside, this study demonstrated detrimental effects of Th2 immunity.

IFN-γ is critical for T cell-mediated clearance of chlamydiae from the genital tract. IFN-γ knockout mice have persistent *C. muridarum* genital tract shedding and a severe disseminated infection (Cotter *et al.*, 1997; Wang *et al.*, 1999). IFN-γ is required for induction of epithelial nitric oxide production. Because IFN-γ regulates expression of MHC class II on epithelial cells, and MHC class II is critical to clearance, it is also plausible that IFN-γ plays a central role in all MHC class II restricted T cell clearance mechanisms. However, IFN-γ in-and-of-itself is not enough. Mice immunized with the protective chlamydial antigen PmpG in CpG adjuvant have a monokine IFN-γ Th1 response. In contrast, mice immunized with PmpG in DDA/TDB adjuvant have a multifunctional Th1 response (IFN-γ + TNFα + IL-17). CpG + PmpG-vaccinated mice showed no protection during genital tract *C. muridarum* challenges while DDA/TDB/PmpG mice were significantly protected in head-to-head studies using the same antigen, same mouse, same route of vaccination and challenge (Yu *et al.*, 2010). TNFα is likely the critical second cytokine based on its association with the protective phenotype in a follow-up study (Yu *et al.*, 2011), and TNFα has a plausible mechanism for synergy with IFN-γ. The iNOS promoter has IFN-γ and TNFα responsive regulatory elements (Xie *et al.*, 1993); TNFα synergizes with IFN-γ to enhance transcription of iNOS that, by extension, boosts nitric oxide production and killing of intracellular chlamydiae.

The Th1/Th2 paradigm has major limitations even with data going back more than a decade. C57BL/6 mice have a Th1 dominant response and relatively rapid clearance with less immunopathology, compared to BALB/c and C3H/HeN mice. However, none of these mouse strains have reliably detectable Th2 responses in their draining iliac lymph nodes during genital tract infections (Darville *et al.*, 2001a). While C57BL/6 mice have significantly less immunopathology than other strains, they still have a 60–80% rate of hydrosalpinx. Furthermore, studies have shown that TNFα is a dual-edged cytokine with a likely role in vaccine-induced protection (Yu *et al.*, 2011) and a definite role in CD8-mediated immunopathology (Murthy *et al.*, 2011). TNFα has no definitive home in the Th1/Th2/Th3/Th9/Th17/Th22/Th33 paradigm. It is not possible to describe protective immunity and immunopathology based on Th1/Th2 cytokine polarization if there is no Th2 response in mice strains demonstrating delayed clearance and enhanced immunopathology, and a Th_x assignment for TNFα.

IL-13 may be a cytokine-based surrogate marker for protective immunity, but it is not specifically associated with an effector mechanism for clearance. In a prospective study of Kenyan female sex workers, absence of incident infections correlated with a PBMC IL-13, with or without IFN-γ, response to purified EB, while ubiquitous PBMC IFN-γ responses to EB did not correlate with protection (Cohen *et al.*, 2005). These results are counterintuitive because IL-13 is a prototypical Th2 cytokine and IFN-γ is the quintessential Th1 cytokine. There are two plausible explanations for these data: 1) protected individuals' PBMCs contained separate populations of Th1 and Th2 polarized T cells or 2) protected individuals had T cells that produced both IFN-γ and IL-13. Neither explanation is compatible with the Th1/Th2 paradigm because it would violate the concept of polarization or mutual exclusivity, respectively.

Mucosal T cell phenotypes and tissue resident T cell immunity

The epithelial surface area of the gut is estimated to be between that of a badminton court and a tennis court (Helander and Fandriks, 2014). There are no published estimates of the epithelial surface area of the female urogenital tract, but perhaps it may cover a ping pong table. Utilizing T cells ~20 microns in diameter, the adaptive immune system is tasked with defending the reproductive epithelium from all comers. To do this, the immune system appears to have adopted the strategy used by reigning powers operating in hostile environments. Build outposts housing defenders on the contested frontier, and build enough of them close enough together to reach critical mass wherein external threats cannot gain a foothold to threaten the empire; think Roman Empire or *Dances with Wolves*. There is an inescapable economy to this strategy. A large army of T cells does not have to continuously defend the entire territory. Instead, when attacked, epithelial sentries call for help using chemokines to recruit a regional host defence response from the closest outposts. In this way, vast areas can be protected with fewer T cells expending energy only as needed. However, until you have assembled a critical mass of outposts, the empire remains at risk.

The *Chlamydia* field has for many years considered lymphoid follicles of the eyelid and genital tract to be the defensive outposts of these epithelial surfaces. These outposts are readily visible on the underside of the eyelid in trachoma and as a 'strawberry' cervix in urogenital *C. trachomatis* infections. Recently, viral pathogenesis researchers descriptively labelled this immune infrastructure 'memory lymphocyte clusters' (MLC), and demonstrated their importance in local immunity mediated by tissue resident memory T cells (Trm) (Iijima and Iwasaki, 2014, 2015). However, Morrison and Morrison previously noted the appearance of lymphoid follicles in the mouse reproductive tract during later stages of *C. muridarum* genital tract infections, and proposed that 'persistence of CD4$^+$ T-cell clusters long after infection had resolved (day 70) may provide for a readily mobilizable T-cell response by which previously infected animals can quickly respond to and control a secondary infectious challenge' (Morrison and Morrison, 2000). Historically, this is the first data-driven statement of a local immunity hypothesis mediated by Trm.

Chlamydial-specific T cells localized in these follicles are unique compared to T cells circulating in the blood or to those residing in peripheral lymph nodes. The microenvironment of the reproductive tract and conjunctiva of the eye differs from the systemic immune compartment, and those differences are reflected in unique antibody classes (IgA), novel infrastructure (MLC), and a cytokine milieu that includes abundant latent TGFβ (Granstein *et al.*, 1990; Jones *et al.*, 2003). In the setting of inflammation, such as *C. trachomatis* infections, the pool of latent-TGFβ is processed into active TGFβ that has profound influence on local adaptive immune responses. A well-known function of active TGFβ is B cell immunoglobulin class switching to IgA. Local IgA production is readily demonstrable in eyes (Bailey *et al.*, 1993) and vaginal fluid (Brunham *et al.*, 1983) of chlamydial infected humans.

Extrapolating from viral pathogenesis data, it is reasonable to postulate that *Chlamydia*-specific T cells localized to chlamydial-specific memory lymphocyte clusters (c-MLC) are Trm. In genital tracts of *C. muridarum* immune mice, c-MLC in the endometrium are predominantly populated by CD4 T cells, with CD8 T cells localized to the lamina propria or intraepithelial spaces in the vagina (Morrison and Morrison, 2000); mice infected with human *C. trachomatis* serovars do not form c-MLC or have a local IgA response (Morrison *et al.*, 2011). In human trachoma, c-MLC are populated by a mix of CD4 and CD8 T cells (Abu el-Asrar *et al.*, 1998). The first *Chlamydia*-specific Trm T cell clones came from mice that self-cleared *C. muridarum* genital tract infections (Johnson *et al.*, 2017). *Chlamydia*-specific CD4 Trm T cell clones were selectively expanded from spleens of immune mice by restricting antigen presentation to immune B cells. Selective B cell-APC (antigen presenting cell) driven expansion of Trm *ex vivo* suggests an intimate and critical relationship between immune B cells and Trm. The presence of *Chlamydia*-specific Trm in the spleen suggests that this organ banks memory *Chlamydia-specific* T cells from systemic and mucosal immune events. Non-trafficking T cells with Trm transcriptomes have previously been documented in the spleen and draining lymph nodes of mice infected with lymphocytic choriomeningitis virus (Beura *et al.*, 2018; Schenkel *et al.*, 2014).

Chlamydial-specific Trm clones provided an opportunity to gain insights into Trm immunobiology, and supported the hypothesis that mucosal CD4 T cells are different from systemic T cells (22). This study revealed the following:

1 CD4 Trm include a subset of CD4 T cells polarized to produce both IFN-γ and IL-13 (CD4γ_{13}).
2 CD4γ_{13}T cells represent 5–15% of the chlamydial-specific IFN-γ response in the genital tracts of immune mice challenged with *C. muridarum*, whether the pre-existing immunity was generated with a protective PmpG + DDA/TDB vaccine or prior genital tract infection. CD4γ_{13}T cells were present but rare in the spleens of immune mice.
3 CD4γ_{13} production of IL-13 in early expansion cultures was lost unless active TGF-β was included in the culture medium and, unlike systemic CD4 T cell clones, addition of TGF-β enhanced Trm expansion.
4 The CD4γ_{13}T cell transcriptome included an enhanced mRNA signal for Lrrc32 (Leucine-rich repeat containing 32), the receptor for latent TGF-β, implying that CD4γ_{13}T cells utilize TGF-β for homeostasis.
5 Chlamydial-specific CD4 T cell clones, systemic and Trm, did not follow Th1/2 rules for differentiation markers or cytokine polarization patterns. All chlamydial-specific CD4 clones produced IFN-γ (Th1) and TNFα upon activation. Four of the five clones (1 systemic, 4 Trm) expressed Tbet (Th1 marker) and all five expressed the Th2 differentiation marker *Gata3* at the mRNA and protein level. One CD4γ_{13}T cell clone was polarized to produce IFN-γ, IL-2, TNFα, IL-13, IL-4, and IL-5 – two quintessential Th1 cytokines and three quintessential Th2 cytokines. That unconventional CD4γ_{13}T cell clone also violated Th1/2 rules by producing IFN-γ without Tbet. The only Th1/2 rule relevant to chlamydial-specific CD4 T cells was that production of IL-4 and IL-5 was only seen in the CD4γ_{13}T cell clone lacking Tbet. Thus, during mucosal T cell responses, expression of Tbet may prevent differentiation into an IL-4/5 producing T cell. Additionally, the CD4γ_{13} clone with IFN-γ, IL-2, TNFα, IL-13, IL-4, IL-5 expression was not protective after adoptive transfer. It is plausible that an IL-4/5 phenotype represents an immunopathologic CD4 phenotype, although calling it Th2 is problematic, since it produced IFN-γ and IL-2 (Th1 cytokines) at much higher levels than IL-4 and IL-5 (Th2 cytokines). None of the CD4 T cell clones in this study could be described with a Th1/2 narrative.
6 The transcription factor that appears to uniquely identify CD4γ_{13}T cells is *Fhl2*. The core CD4γ_{13} transcription factor differentiation pattern appears to be *Gata3* and *Fhl2* positive.
7 Adoptive transfer of one CD4γ_{13}T cell clone (IFN-γ, IL-2, TNFα, IL-13, IL-17, IL-22) into C57BL/6 mice completely prevented oviduct immunopathology without accelerating clearance. Thus, the $CD4_{\gamma13}$ clone mitigated infection-associated pathology by a mechanism not based on accelerated clearance. A non-Trm Gata3 + /IFN-γ CD4 T cell clone showed protection from immunopathology to a lesser extent, also without accelerating clearance.
8 CD79a+ B220(−) memory B cells expanded in the genital tract during the course of *C. muridarum* genital tract infections.

Based on these data, IL-13 was postulated to be a biomarker for a unique subset of CD4 Trm rather than an indicator of Th1/Th2 polarization. The *C. trachomatis* -specific Trm data from this initial study has implications for understanding human *C. trachomatis* pathogenesis and facilitating rational vaccine development.

In 2013, immunohistochemistry of *C. trachomatis*-infected women clearly showed that endometrial c-MLC contained *Gata3*+ CD4 T cells (Vicetti Miguel *et al.*, 2013). In the context of the Th1/2 paradigm, the *Gata3* finding was interpreted as demonstrating a Th2 response, perhaps turning the entire IFN-γ protection paradigm that grew out of mouse studies on its head. The authors postulated that protective immunity and the goal of vaccination was a healing Th2 IL-4 T cell response rather than a Th1 IFN-γ response (Vicetti Miguel and Cherpes, 2012; Vicetti Miguel *et al.*, 2013). However, our recent and previous mouse studies showed *Gata3* expression by all IFN-γ-producing murine chlamydial-specific CD4 T cell clones tested to date (Johnson and Brunham, 2017; Johnson *et al.*, 2017). Regardless of Trm phenotype or method of derivation, the murine CD4 T cell response to *C. muridarum* genital tract infections appears to be consistently *Gata3*+ (Th2) and IFN-γ producing (Th1).

Discovery of the CD4γ_{13}T cell subset provides a plausible Trm-based explanation for the Kenyan sex worker study where EB stimulated PBMC production of IL-13 and IFN-γ predicted resistance to *C. trachomatis* incident infections (Cohen *et al.*, 2005). Chlamydial-specific IL-13 may have been a biomarker for circulating CD4γ_{13}T cells. In the *C. muridarum* Trm study, CD4γ_{13}T cells completely prevented oviduct pathology when adoptively transferred into naïve mice prior to infection, a scenario analogous to a hypothetical *C. trachomatis* vaccine that generates CD4γ_{13}T cells in naïve humans.

From mouse and human T cell studies, it is reasonable to conclude that the Th1/2 paradigm does not adequately describe protective immunity and immunopathology parameters relevant to vaccine development. Arguments that support moving beyond the Th1/Th2 paradigm include:

1 As a single parameter, T cell production of IFN-γ to EB has not predicted protection from incident infections in the *C. muridarum* mouse model or prospective human studies. Salvaging the Th1/Th2 narrative with adjectives such as 'multifunctional' simply masks a lack of understanding. We would be better served by trying to determine the differences between T cell subsets with complex cytokine phenotypes (for example combinatorial differentiation transcription factor patterns and/or relevant priming dendritic cell type or location, and the adjuvant systems and vaccination sites that can replicate this).
2 The Th1/Th2 paradigm appears to be misleading when used to describe mucosal immune events. In humans, it is likely that a heretical PBMC IFN-γ (Th1) + IL-13 (Th2) response to EB predicted protection from incident infections, and in mice, individual CD4 T cells produce IFN-γ and IL-13 (Th1/Th2 heresy). In the *C. muridarum* genital tract infection model, every chlamydial-specific CD4 T cell clone tested to date produces IFN-γ (Th1) and expresses the differentiation transcription factor Gata3 (Th2).
3 The *C. muridarum* model and human investigations, by definition, study a mucosal immune response. It would not be surprising to find that human chlamydial-specific CD4 T cells producing IFN-γ also express Gata3, and that the cytokine complexity currently referred to as 'multifunctional' reflects the complexity of T cell responses originating in the mucosal immune compartment.

The *C. muridarum* model will be critical for advancing understanding of mucosal T and B cell immunobiology and guide human investigations, ultimately informing a working paradigm for protection and immunopathology based on refined T cell subsets, including but not solely based on cytokines.

The role of B cells

Rodent models

Early successful vaccines like Heptavax-B against Hepatitis B virus generated potent neutralizing antibodies and sterilizing immunity. Furthermore, antibody titres could easily be used as surrogate biomarkers for protective immunity. Significant research argues against neutralizing antibody titres being the surrogate biomarker for an efficacious chlamydial vaccine. Though not likely the cornerstone, antibody responses will be an important consideration during *C. trachomatis* vaccine development. Mechanisms for antibody-mediated pathogen control include (a) direct neutralization of extracellular bacteria, (b) opsonophagocytosis, (c) antibody-dependent cell cytotoxicity and (d) transcytosis of antibody and intracellular inhibition of replication (Fig. 14.3). These mechanisms will be discussed in more detail below.

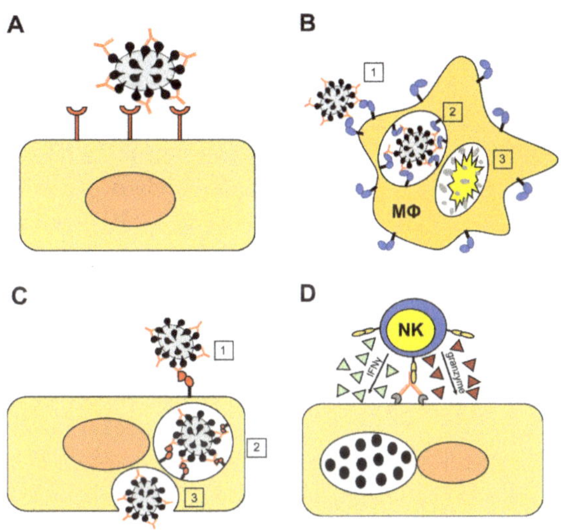

Figure 14.3 Mechanisms for antibody-mediated chlamydial control or infection. (A) Direct neutralization. Chlamydial infection is initiated by outer membrane adhesins that bind epithelial cell surface receptors. Antibodies specific for chlamydial adhesins directly prevent binding and infection. (B) Opsonophagocytosis. Antibodies coating the chlamydial surface (opsonization) enable effector cells to bind the attached antibody via Fc receptors (1) to phagocytose (2) and degrade the bacteria after lysosomal fusion (3). (C) Transcytosis. Antibodies attached to chlamydial surface proteins bind the neonatal Fc receptor (FcRn) (1) to initiate transport across the epithelium (2) into the lamina propria (3). Alternatively, antibodies specific for secreted chlamydial proteins bind the FcRn and are transported to the intracellular inclusion where they inhibit replication. (D) Antibody-dependent cellular cytotoxicity (ADCC). Antibodies attached to chlamydial antigens on the surface of infected cells cross-link Fc receptors on NK cells to trigger the release of proinflammatory cytokines and cytotoxic molecules.

In rodent models, there is evidence for chlamydial neutralization and opsonophagocytosis. Passive transfer of chlamydial immune serum or specific antibody reduces the intensity of bacterial shedding without affecting time to clearance. This result is more consistent with antibody contributing to the demise of extracellular EBs rather than elimination of bacteria within intracellular inclusions.

The guinea pig model demonstrates antibody-mediated direct EB neutralization (Fig. 14.3A). Guinea pigs are solidly protected from reinfection during the first 30 days after clearing primary guinea pig inclusion conjunctivitis (GPIC) genital tract infection, then become susceptible to low-intensity short-lived reinfection at day 77. The authors demonstrated waning genital tract antibody titres to MOMP at day 77 that may have had a role in susceptibility to reinfection (Rank *et al.*, 1988). Passive transfer of immune serum led to GPIC-specific antibodies in the genital tract detected at levels that were 20–80% of those in serum, and a reduction in genital tract infectious burden by 75% during early primary infection. Though antibody reduced the amount of GPIC in the genital tract, it did not accelerate time to clearance (Rank and Batteiger, 1989).

In the *C. muridarum* mouse model, antibody-mediated protection has been demonstrated, but it appears less potent than seen in the guinea pig model. In one study, MOMP–specific hybridomas (one IgA, the other IgG3) were implanted as subcutaneous antibody-producing tumours in BALB/c mice and generated detectable anti-MOMP monoclonal antibody levels in serum and vaginal secretions. Mice bearing chlamydial–specific hybridoma tumours had modest protection from infection with low inoculum challenges that was lost by increasing the infectious dose (Cotter *et al.*, 1995). Parallel research with a MOMP oligopeptide (LNPTIAG) variable domain 4 (VD4) vaccine generated a high titre of neutralizing antibody with similar results. Neutralizing antibody was documented in serum and vaginal secretions, but vaccinated mice showed only modest protection from infection with low inoculum challenges. There was no effect on time to clearance and protection was lost after increasing the challenge dose (Su *et al.*, 1995).

One of the most attractive vaccine candidates is a multivalent heterologous *C. trachomatis* MOMP construct that generates neutralizing antibody and a multifunctional Th1 response. The Hirep1 vaccine is a MOMP VD4 concatemer containing the LNPTIAG B cell epitope embedded within surrounding sequences from serovars D-F (Olsen *et al.*, 2015). The Hirep1 vaccine contains the mouse T cell epitope IFDTTTLNPTIAGAGDVK based on *C. trachomatis* serovar D mouse immunoproteome (Karunakaran *et al.*, 2015). Hirep1 vaccinated (C57BL/6 x C3H/HeN) F1 mice had a 2-log reduction in *C. trachomatis* serovar D shedding on day three post-infection. Passive transfer of Hirep1 immune serum into T cell-depleted mice lowered bacterial burden on day three by 1 log (Olsen *et al.*, 2015). In a follow up study with a reduced challenge inoculum, passive transfer of Hirep1 immune serum prevented infection in 48% of naïve *Rag1*$^{-/-}$ mice (Olsen *et al.*, 2017). Because these experiments were performed with human *C. trachomatis* serovar D, which demonstrates a limited duration of shedding in mice, it is unclear if passive immunization reduced the time to clearance. This multivalent vaccine is currently the most clinically progressed *C. trachomatis* vaccine. The CTH522-CAF01 Phase I trial was completed in July 2017, and results are pending. In mice and guinea pigs, protection directly attributable to neutralizing chlamydial-specific antibodies is limited. In settings where the infecting inoculum is low, pre-existing chlamydial-specific IgG/IgA antibody in the genital tract can reduce incident infections, but does not accelerate clearance.

A second mechanism for antibody-mediated control of pathogens is opsonophagocytosis that involves collaboration with innate effector cells (Fig. 14.3B). Antibody opsonizes bacteria by binding surface antigens via Fab domains. The free Fc domain binds to Fc receptors (FcR) expressed on neutrophils, monocytes, macrophages, mast cells, and basophils, leading to phagocytosis and enhanced bacterial killing. In the *C. muridarum* model, neutrophils are the dominant innate effector for opsonophagocytosis *in vivo*. Antibody-facilitated opsonophagocytosis of chlamydiae is dependent on IFN-γ (Naglak *et al.*, 2016) and the antibody Fc domain, implying FcR dependence (Naglak *et al.*, 2017). In naïve mice, neutrophil depletion has no effect on bacterial clearance from the genital tract. In primed immune mice, depletion of neutrophils and CD4 T cells resulted in prolonged bacterial shedding (two weeks), compared to CD4 depletion alone (< 1 week). Antibody collaboration with neutrophils provides a plausible explanation for the antibody-mediated T cell independent protection mechanism documented in immune mice. Immune mice that clear a primary genital tract infection are resistant to reinfection even if CD4 and CD8 T cells are depleted (Morrison and Morrison, 2001; Morrison *et al.*, 2000). B cell deficient mice that have cleared primary infection are not protected, but protection can be reconstituted by passive transfer of immune serum or antibodies specific to EB surface antigens like MOMP (Morrison and Morrison, 2005a). Why opsonophagocytosis is not functional during primary infections after passive transfer of chlamydial-specific antibody is unclear.

A third mechanism by which antibody mediates pathogen killing is via antibody-dependent cellular cytotoxicity (ADCC) (Fig. 14.3C), which typically involves natural killer (NK) cells but also myeloid effectors. Effector cells bind the Fc portion of antibodies and trigger the release of cytotoxic molecules and proinflammatory cytokines like IFN-γ. NK cell depletion with anti-asialo-GM1 antibodies causes a prolonged primary infection with *C. muridarum* and a shift towards a Th2 response (Tseng and Rank, 1998). Depletion of NK cells during secondary infection had no effect on resolution in the B-cell deficient, CD4-depleted challenge model. However, combined depletion of NK cells and CD4 T cells led to enhanced shedding over CD4 depletion alone (Naglak *et al.*, 2017). It is unclear whether negative effects of NK cell depletion on bacterial clearance reflect an ADCC mechanism or loss of early IFN-γ production that could limit cell intrinsic innate defences and upregulation of epithelial MHC class II.

The fourth antibody host defence mechanism is transcytosis. Experiments *in vitro* and *in vivo* indicate that antibody transcytosis can inhibit or enhance infection depending on the antigen (Fig. 14.3D). This mechanism raises the possibility of antibody enhancement of cell intrinsic defence by interfering with intracellular chlamydial functions mediated by effector proteins like IncA, or enhancing infection by allowing antibody-bound *C. trachomatis* to cross the epithelium and reach the basolateral surface to disseminate. The neonatal Fc receptor (FcRn) is ubiquitously expressed in the mucosal tissues of mammals on antigen-presenting cells, syncytiotrophoblasts, endothelial and epithelial cells. FcRn reversibly binds IgG under acidic conditions (pH 5–6.5) and releases it at neutral pH (pH 7.4). Therefore, FcRn facilitates bidirectional IgG delivery in and out of the lumen of the male and female reproductive tracts in a pH-dependent manner (Knee *et al.*, 2005; Li *et al.*, 2011). MOMP-specific IgG was shown to enhance uptake and translocation of chlamydiae in polarized epididymal epithelia grown on transwells, which was dependent on FcRn expression. In contrast, FcRn internalized IgG specific for IncA resulted in a 25% reduction in infection, aberrant inclusion morphology, and lysosomal-independent recruitment of autophagic proteins (Armitage *et al.*, 2014). In male mice, MOMP-IgG elicited by immunization afforded no protection against testicular chlamydial infection, whereas IncA vaccination significantly reduced testicular chlamydial burden (Roopenian and Akilesh, 2007). Other unconventional candidate vaccine antigens may include inclusion membrane proteins IncG, IncF, and CT299, which are cytoplasmic facing and could be bound by intracellular IgG (Hackstadt *et al.*, 1999). Furthermore, microinjection of CT229 antibodies into infected cells inhibited *Chlamydia* infection (Rzomp *et al.*, 2006).

The mouse model does not reveal an important role for IgA. IgA-deficient mice resolve primary and secondary chlamydial genital infections similarly to wild-type mice (Morrison and Morrison, 2005b). This is surprising since IgA is known to protect against a variety of other mucosal pathogens (Blutt *et al.*, 2012), and is secreted into genital tract secretions by local plasma cells, which are universally found in genital tract tissues of infected mice. It is possible that any contribution from IgA is masked by the redundancy of IgG, which is transported at high levels into genital tract secretions via the neonatal Fc receptor (FcRn).

Primate antibody data

Recent studies examined the development of reproductive tract immunity in rhesus monkeys inoculated cervically with the wild-type strain of *C. trachomatis*, D/UW-3/Cx or a plasmid-deficient derivative of this strain, CTD153, every other week for five weeks; followed by five challenge doses with wild-type D (Qu *et al.*, 2015). Both *C. trachomatis* strains resulted in infections with similar kinetics and induced comparable levels of protective immunity. Elevated early T cell responses were associated with protection from ascension, whereas higher antibody responses were associated with detection of oviduct infection. Levels of chlamydial-specific IgG in sera and secretions were inversely associated with the degree of infection control, implying a non-protective role for systemic and mucosal chlamydial-specific IgG.

Human antibody data

Human data support a role for antibody in limiting chlamydial burden and extent of infection, but not reinfection. In women with uncomplicated cervical infection, the highest frequency (63%) of women with detectable endocervical *C. trachomatis*-specific secretory IgA were those with the lowest bacterial burden ($< 10^1$ IFU/ml), suggesting local IgA was a contributor to reduced shedding or an indirect marker. However, 21% of women with IFU $> 10^4$ IFU/ml were also positive for sIgA. No association was found with serum levels of anti-*C. trachomatis* IgG (Brunham *et al.*, 1983).

The relationship of serum anti *C. trachomatis* IgG to burden and endometrial ascension was examined in 225 highly exposed young women in Pittsburgh (Russell *et al.*, 2016). The majority (55%) reported having previously received a diagnosis of *C. trachomatis*, and 24% reported having ≥ 2 prior infections; 83% were seropositive at enrolment. Seropositive women with midrange (1:16–1:256) and high (1:512–1:8192) titres of anti-chlamydial serum IgG, as measured by microimmunofluorescence (MIF), had lower bacterial burdens in cervical swab specimens, compared to seronegative women (seronegative vs. titres of 1:16–1:256, $P < 005$; seronegative vs. titres of 1:512–1:8192, $P = .01$). However, there were no differences in the rates of endometrial infection (adjusted RR, 1.24 [95% CI,.71–2.19]) or endometrial chlamydial burden between seronegative and seropositive women. Measurement of serum antibodies by the MIF assay included antibodies to MOMP, LPS, and other surface exposed antigens that could elicit neutralizing antibodies. Although these antibodies were associated with limitation of cervical burden, chlamydiae were still able to ascend to the upper genital tract, where chronic disease manifests.

Several studies correlated high titres of serum anti-chlamydial IgG with increased morbidity (Eckert *et al.*, 1997; El Hakim *et al.*, 2010; Ness *et al.*, 2008; Tiitinen *et al.*, 2006). Furthermore, data from human studies indicate minimal to no role for antibody in resistance to reinfection. A majority of *C. trachomatis*-infected women and at least 50% of men develop detectable levels of anti-*C. trachomatis* serum IgG and IgA (Woodhall *et al.*, 2018), and since IgG is readily transported into genital secretions by FcRn and can be produced by plasma cells infiltrating the genital mucosa, equivalent or even greater levels of specific IgG can exist in genital secretions. Albritton *et al.* recently showed that female genital tract secretions contained *C. trachomatis* EB-specific IgG and IgA antibodies more frequently than serum (Albritton *et al.*, 2017). Despite the development of serum and mucosal *C. trachomatis*-specific antibody after infection, rates of repeat infections with the same serovar vary between 30 and 60% (Batteiger *et al.*, 1989; Batteiger *et al.*, 2010; Blythe *et al.*, 1992; Brunham *et al.*, 1996). These data do not support a role for serovar-specific neutralizing antibodies in prevention from reinfection. A true test of *in vivo* neutralization in humans will require development of a human challenge model and examination of the ability of passive immunization to protect from infection. This might be possible in men screened for HLA-B27 positivity, as such studies are being carried out currently with *Campylobacter jejunii* (Maue *et al.*, 2014) which like *C. trachomatis*, carries a risk of reactive arthritis.

Analysis of a cohort of sex workers in Nairobi, Kenya by Cohen *et al.* revealed that levels of serum IgG or IgA specific for chlamydial elementary bodies (EBs) or heat shock protein 60 (HSP60) were not associated with resistance to incident infection (Cohen *et al.*, 2005). Data from the Pittsburgh cohort, described above, also revealed that seropositivity was not associated with a reduced rate of incident infection (Russell *et al.*, 2016).

It should be noted that the above studies examined only antibodies to surface exposed antigens, and as demonstrated by Armitage *et al.*, it is possible that antibodies to antigens expressed intracellularly may play an important role in host defence (Armitage *et al.*, 2014). Chlamydial protein and peptide libraries (Budrys *et al.*, 2012; Hufnagel *et al.*, 2018; Rahman *et al.*, 2018) provide a mechanism to correlate antibody responses to an array of chlamydial antigens with clinical evidence of protection from reinfection in highly exposed individuals. The discovery of chlamydial antigens associated with protection from reinfection through comprehensive proteomic analyses would demonstrate their viability as candidates for subunit chlamydial vaccines.

In summary, data from rodent models and human studies indicate anti-chlamydial antibodies are effective at reducing chlamydial burden but are inefficient at preventing reinfection. In mouse and guinea pig models, antibody-mediated protection from primary infection or reinfection can be overcome with higher inoculum challenge, and pre-existing antibody does not accelerate time to clearance. Evidence for directly neutralizing antibodies *in vivo* is limited, but data indicate that antibody can lead to enhanced antigen presentation, opsonophagocytosis and neutrophil-mediated killing, processes that are facilitated by IFN-γ.

B cells collaborating with T cells

B cells and antibody are not critical during primary infections as clearance is equivalent in C57BL/6 wild type and B cell-deficient mice. B cells are critical to immunity that prevents reinfection. Immune mice lacking B cells or FcR are markedly compromised in clearing *C. muridarum* secondary infections. Immune wild type mice have a ~3 log reduction in bacterial shedding during secondary challenge that resolves within 2 weeks. The same secondary infection in immune B cell-deficient mice demonstrated a ~2 log reduction that cleared over three weeks (Su *et al.*, 1997). Interestingly, immune mice deficient in FcRI (CD64), FcRcIII (CD16), and FccRIIB1 (CD32) lack protective immunity after secondary challenge. Immune FcR-deficient mice rechallenged with *C. muridarum* had bacterial shedding of the same intensity and duration as primary infections, demonstrating the necessity for Fc receptors in antibody mediated protection (Moore *et al.*, 2002). Working models of protective immunity should be able to explain these fundamental data on the role of B cells and FcR in secondary protective immunity.

B cells are the dominant antigen presenting cells in MLC as demonstrated by histology and immunohistochemical staining of endometrial and conjunctival tissue from *C. trachomatis* urogenital and ocular infections. In the *C. muridarum* mouse model, the dynamics of MLCs have been well studied, but complicated by a caveat related to B cell differentiation. Naïve B cells are B220 (CD45RA) positive, but as they transition to immune memory or plasma B cells they lose B220 expression (Cascalho *et al.*, 2000; Dustin *et al.*, 1995). The *in situ* investigation of c-MLC in the murine uterine horns on day 14 showed clusters of B220+ cells at the epithelial interface. By day 21, the B220+ B cell population was gone and replaced by CD4 T cell clusters in the same spatial orientation to the epithelial cell interface. CD4 T cell clusters contained a few macrophages and occasionally CD8 T cells. The majority of CD8 T cells were in stroma and intraepithelial spaces in the vagina (Morrison and Morrison, 2000). Interestingly, in B cell-deficient mice, genital tract CD4 T cells are dispersed in the genital tract stroma and CD4 MLCs are not evident (Morrison and Morrison, 2005a). The disappearance of B cell clusters between day 14 and 21 likely represents their transition from naïve B cells to memory B cells or plasma cells that lack B220 expression. This naïve B220+ to immune B220(–) B cell transition has been demonstrated by flow cytometry of genital tract B cells using CD79a, a pan-B cell marker expressed by naïve and memory B cells. In naïve mice, genital tract CD79a+ cells are 97% B220+ and 3% B220(–), but after secondary infection 22% of CD79a+ cells are B220(–) (Johnson *et al.*, 2017). In women with urogenital *C. trachomatis* infection, follicles are populated by lymphocytes and plasma cells (Kiviat *et al.*, 1985; Kiviat *et al.*, 1990a; Kiviat *et al.*, 1990b). Drawing from mouse and human genital tract data, it is likely that murine B220(–) immune B cells are localized to MLCs, based on human endometrial histology. In human trachoma follicles, elegant immunohistochemistry studies documented a prominent B cell population, with some macrophages and a mixture of CD4 and CD8 T cells (Abu el-Asrar *et al.*, 1998). The significant presence of CD8 T cells in MLCs in trachoma differs from the occasional CD8 T cell observed in human and mouse genital tract MLCs. This is a particularly interesting detail coupled with the finding that the attenuated, plasmid-deficient *C. trachomatis* vaccine generated protective CD8 T cells in the eyes of inoculated macaques. There may be a geographical difference in what constitutes protective immunity in the genital tract *versus* the conjunctiva based on the nature of Trm resident in c-MLC. In the genital tract, protective T cells residing in MLCs are CD4, while in the conjunctiva, protective T cells residing in MLC appear to be CD8. Understanding these phenomena will be important for understanding pathogenesis and facilitating vaccine development.

Why are immune B cells localized to MLCs? The most likely explanation is that they are an antigen-presenting cell that is needed to retain and maintain memory CD4 Trm within c-MLC. In infected wild type mice CD4 T cells cluster near the epithelial interface while in infected B cell deficient mice they are found scattered throughout the stroma (Morrison and Morrison, 2005a; Morrison *et al.*, 2000). This semi-quantitative post-hoc comparison implies that c-MLC formation is dependent on B cells, and that optimal protective immunity against chlamydial challenge in the genital tract requires B cell functions beyond antibody production. The finding that antigen-pulsed B cells selectively expanded a small spleen-resident chlamydial-specific Trm population suggests a unique partnership between immune B cells and CD4 Trm (Johnson *et al.*, 2017). Despite not being directly compared, FcR-deficient mice appear more compromised than B cell-deficient mice during secondary *C. muridarum* genital tract challenges (Moore *et al.*, 2002; Su *et al.*, 1997). B cells and macrophages resident in c-MLC use FcR to sample the local antibody IgG pool to 'arm' their cell surface, and may be necessary for immune protection. Immune B cells have an endogenous B cell receptor (membrane bound antibody) with a single specificity, and B cells from immunized mice activate antigen-specific T cells at 1000-fold lower antigen concentration than naïve B cells, through an antibody-antigen capture mechanism (Rock *et al.*, 1984). This antibody-antigen capture phenomenon coupled with rapid local T cell expansion is potentially a more important consideration in vaccine development than neutralizing antibody. Macrophages and dendritic cells similarly use local antibody bound to FcR to enhance antigen-presentation to T cells through antigen capture. Antibodies that bind the chlamydial surface potentially enhance antigen presentation of all EB-resident protein antigens. FcR are also needed for antibody collaboration with neutrophils to diminish bacterial shedding (Naglak *et al.*, 2016, 2017), and is dependent on the antibody Fc domain, implying FcR dependence.

How does it all work and implications for vaccine development: summary

Rational vaccine design requires consideration of memory T cell subsets. Central memory T cells (T_{CM}) are retained in the spleen and secondary lymphoid organs to facilitate rapid T cell expansion, while effector memory T cells circulate through the blood and tissues and act as early first responders. However, not all non-haematopoietic tissues are equally accessible to circulating effector memory T cells (T_{EM}). Tissues that could be subjected to pathology from a hyperactive immune response, like the skin, brain, genital tract, and conjunctiva, are considered to be immune-privileged organs to varying degrees (Shin and Iwasaki, 2013). These sites have limited vascular egress and less immune architecture for establishing resident T cells, but can create temporary vascular exit sites and establish immune inductive sites after infection.

Naïve hosts have limited pre-existing immune infrastructure in the conjunctiva of the eye and the genital tract. During primary infection, dendritic cells with captive chlamydial antigens expand antigen-specific T cells in regional draining lymph nodes. Circulating systemic T cells migrate to infected sites along epithelial chemokine gradients, utilizing transient cytokine-induced (interferons, TNFα) endothelial homing receptors (VCAM-1, MadCam-1) in stroma adjacent to the epithelial interface (Kelly and Rank, 1997). During *C. trachomatis* infections, a local T cell response ultimately eliminates the focus of infected epithelium and assembles an outpost.

Chlamydia trachomatis does not reach the upper genital tract or next ocular epithelial focus by continuous spread. Instead, infectious foci are scattered throughout the conjunctiva and the genital tract, so the focal immune response described above is simultaneously occurring at many sites. As the adaptive immune response develops, the T cells being recruited to foci of infection include an increasingly higher percentage of chlamydial-specific T cells, and the number and specificity of immune outposts steadily increases. Eventually, a critical mass of c-MLC exists, making it impossible for *C. trachomatis* to locate unprotected epithelium and the infection is eliminated (Fig. 14.4).

A working paradigm for protective immunity and immunopathology should incorporate mechanism-informative findings 1–8 below that fit comfortably within a *Trm-based narrative*:

1. Humans, mice, and guinea pigs, eventually clear chlamydial genital tract infection without antibiotics, *after assembling a critical mass of c-MLC outposts sufficient to inhibit infection.*
2. MHC class II is critical for T cell mediated clearance, based on extensive mice data and supported by strong associations with HLA class II DQ alleles in humans.
3. IFN-γ is necessary, but not sufficient, for clearance. PBMC production of IFN-γ in response to EB as a single parameter is not associated

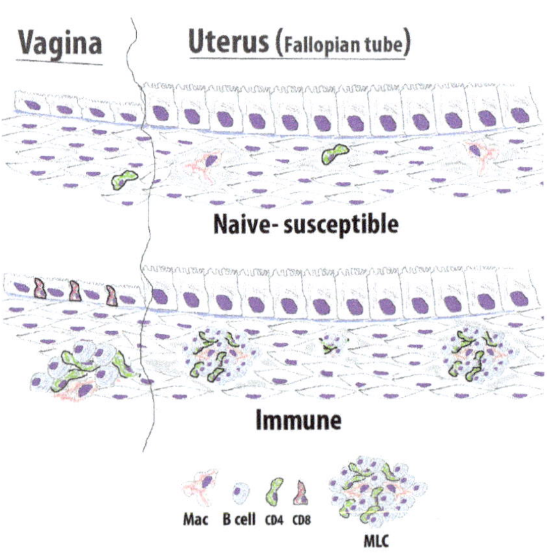

Figure 14.4 TRM-based paradigm for protective immunity. Naïve susceptible hosts have limited immune infrastructure (no outposts) underlying their reproductive tract epithelium. Immune individuals based on previous infection, and in the future successful vaccination, have a critical mass of memory lymphocyte clusters (MLC) underlying reproductive tract epithelium, thereby denying Ct safe havens (patches of epithelium to complete a replication cycle before T cell-mediated termination).

with protection in human trachoma (Bailey *et al.*, 1995) or genital tract infections (Bakshi *et al.*, 2018; Cohen *et al.*, 2005).

4 CD4 T cell responses in mouse, and likely human chlamydial genital tract infections do not adhere to the Th1/Th2 cytokine and transcription factor differentiation paradigm. *CD4 T cells in human endometrial follicles are Gata3+ (Th2 differentiation transcription factor); in C. muridarum, all* chlamydial-*specific IFN-γ producing CD4 T cell clones from mice that cleared genital tract infections appear to be Gata3+. In mucosal immune responses, Gata3 is not a reliable biomarker for Th2 polarization. While Gata3 is a Th2 marker in the systemic immune compartment, it appears to be ubiquitous in mucosal CD4 T cells regardless of their cytokine profile.* This is an example where the mouse tells us what to look for in humans.

5 CD8 T cells exist as two clearly defined subsets based on requirements for MHC class I presentation, or the lack thereof. The major chlamydial-specific CD8 response in mice and humans is not dependent on MHC class I. Trachoma vaccine data in macaques implies that there is an MHC class II restricted CD8 response.

6 Antibodies and FcR are critical to immunity for preventing reinfection but are not necessary for resolution of primary infection. (see 7).

7 B cells, the predominant antigen presenting cell in trachoma follicles and genital tract c-MLC, selectively expanded chlamydial-specific CD4 T cells with a Trm transcriptional phenotype, *implying that their role is in maintenance of c-MLC and early antigen capture, via FcR, to* accelerate T cell expansion *during recall responses.*

8 Protection in mice is potent, but not sterilizing. Secondary challenge results in a shorter duration of bacterial shedding and fewer infectious progeny. *Antibody can only neutralize a small inoculum challenge. Higher inoculums overcome this neutralization, but antibody can enhance chlamydial clearance through other mechanisms like opsonophagocytosis and enhanced antigen-presentation/T cell expansion.*

Upon resolution of infection, a subset of effector T cells present at previously infected foci become circulating T_{EM}, while others downregulate CCR7 (lymph node homing) and S1PR1 (tissue egress) to remain resident in the tissue, as intraepithelial lymphocytes (IEL) tethered to epithelial cells by CD103-mediated adherence to E-cadherin or within the MLC through interactions with B cells and/or macrophages. In mice, HSV-specific MLC have low-level CD4 T cell IFN-γ production, presumably from antigenic stimulation, that is required to drive macrophage CCL5 production necessary to maintain genital MLC (Iijima and Iwasaki, 2014). In the *C. muridarum* model, the phenomenon of persistent $PmpG_{303-311}$presentation (Johnson *et al.*, 2012b) likely reflects utilization of viable chlamydiae in antigen presenting cells for antigenic stimulation necessary to maintain the MLC. MLC are likely not created equal, with variation based on tissue location and different antigen presenting cells harbouring chlamydiae. This may be key to durable immunity. If so can a subunit vaccine generate stable c-MLC and durable local immunity? The distinction between CD4-dominated MLC in the genital tract (CD4 protective response) *versus* trachoma MLC with balanced populations of CD4 and CD8 T cells (CD8 protective response in macaques) may be another important consideration for vaccine development.

At least three nasal chlamydial vaccination strategies in mice have reduced shedding after genital challenge, with one demonstrating immune T cells localizing to the uterus in limited numbers (Jiang *et al.*, 2017a; Murthy *et al.*, 2007; Stary *et al.*, 2015). Vaccinologists would like to take advantage of T cell homing as a mucosal vaccine strategy, since the nasal passages are more accessible than the cervix or urethra. Effector memory T cells (T_{EM}) expressing the α4β7 homing receptor are programmed to routinely exit the circulation, via high endothelial venules (HEV) expressing MadCam-1, to patrol the adjacent mucosal tissue. Without local inflammation, T_{EM} follow a sphingosine phosphate concentration gradient using the S1PR1 receptor to home to the lymphatics and then the bloodstream. Tertiary lymphoid structures in the nasal passages, lungs, and gut have HEV that constitutively express the mucosal addressin MadCam-1. Infection in the respiratory tract or gut generates α4β7+ T_{EM} that circulate between the nose, gut and lungs using MadCam-1 expressing HEV. A critical question for chlamydial vaccine development is whether there is lymphoid tissue associated with the female genital tract. In mouse genital tracts and Fallopian tube organ cultures, the mucosal addressin MadCam-1 is induced by chlamydial infection (Kelly *et al.*, 2001; Kelly and Rank, 1997). However, MadCam-1 staining disappears from the mouse genital tract by day 35, suggesting that one major mechanism for maintaining reproductive tract immune privilege is by limiting lymphocyte access through transient MadCam-1 expression. Thus, the mechanism for nasal vaccines generating local immunity in the genital tract is unclear. Following *C. trachomatis* infection, circulating mucosal α4β7+ and systemic CCR7+ T cells have access to the genital tract by following chemokine gradients and using general adhesion molecules (LFA1-I, ICAM1, α4β1-VCAM1). The absence of established T cell entry sites and tertiary immune infrastructure in the mouse genital tract may explain why adoptively transferring immune T cells or T cell clones into wild type mice has a modest effect on *C. muridarum* bacterial shedding, and why 'protective' vaccines reduce shedding and infection duration but are not sterilizing. Conversely, clearance of a primary genital tract infection can generate potent immunity (c-MLC) that can prevent reinfection. After the development of natural immunity, MadCam-1 mucosal entry sites may be disassembled, but a critical mass of MLC harbouring chlamydial-specific Trm are now prepositioned to protect against subsequent infection.

It is difficult to provide the narrative for how a chlamydial vaccine would prevent primary infection without a nanoparticle vaccine (Jiang *et al.*, 2017a) or attenuated bacterial strain delivered into the genital tract, or a prime pull strategy (Shin and Iwasaki, 2012) capable of establishing local Trm/MLC-mediated immunity. A more attainable goal might be a vaccine that prevents ascension to the Fallopian tubes, thereby averting pelvic inflammatory disease and infertility, while facilitating protective c-MLC assembly to prevent subsequent infections. A better understanding of chlamydial-specific CD8 T cells and the development of surrogate biomarker(s) for protective immunity will be critical to the vaccine effort. Recent advances in understanding the complexity of the CD4 T cell response may identify biomarkers for protective immunity that can be used to assess candidate vaccine efficacy (vaccines are discussed in more detail in Chapter 15, *Chlamydia* Vaccines).

References

Abu el-Asrar, A.M., Geboes, K., and Missotten, L. (2001). Immunology of trachomatous conjunctivitis. Bull Soc Belge Ophtalmol, 73–96.

Abu el-Asrar, A.M., Geboes, K., Tabbara, K.F., al-Kharashi, S.A., Missotten, L., and Desmet, V. (1998). Immunopathogenesis of conjunctival scarring in trachoma. Eye *12 (Pt 3a)*, 453–460.

Albritton, H.L., Kozlowski, P.A., Lillis, R.A., McGowin, C.L., Siren, J.D., Taylor, S.N., Ibana, J.A., Buckner, L.R., Shen, L., and Quayle, A.J. (2017). A novel whole-bacterial enzyme linked-immunosorbant assay to quantify Chlamydia trachomatis specific antibodies reveals distinct differences between systemic and genital compartments. PLoS One *12*, e0183101.

Armitage, C.W., O'Meara, C.P., Harvie, M.C., Timms, P., Blumberg, R.S., and Beagley, K.W. (2014). Divergent outcomes following transcytosis of IgG targeting intracellular and extracellular chlamydial antigens. Immunology and cell biology *92*, 417–426.

Bailey, R.L., Holland, M.J., Whittle, H.C., and Mabey, D.C. (1995). Subjects recovering from human ocular chlamydial infection have enhanced lymphoproliferative responses to chlamydial antigens compared with those of persistently diseased controls. Infect Immun *63*, 389–392.

Bailey, R.L., Kajbaf, M., Whittle, H.C., Ward, M.E., and Mabey, D.C. (1993). The influence of local antichlamydial antibody on the acquisition and persistence of human ocular chlamydial infection: IgG antibodies are not protective. Epidemiol Infect *111*, 315–324.

Bakshi, R.K., Gupta, K., Jordan, S.J., Chi, X., Lensing, S.Y., Press, C.G., and Geisler, W.M. (2018). An Adaptive Chlamydia trachomatis-Specific IFN-gamma-Producing CD4(+) T Cell Response Is Associated With Protection Against Chlamydia Reinfection in Women. Front Immunol *9*, 1981.

Barral, R., Desai, R., Zheng, X., Frazer, L.C., Sucato, G.S., Haggerty, C.L., O'Connell, C.M., Zurenski, M.A., and Darville, T. (2014). Frequency of Chlamydia trachomatis-specific T cell interferon-gamma and interleukin-17 responses in CD4-enriched peripheral blood mononuclear cells of sexually active adolescent females. J Reprod Immunol *103*, 29–37.

Batteiger, B.E., Fraiz, J., Newhall, W.J., Katz, B.P., and Jones, R.B. (1989). Association of recurrent chlamydial infection with gonorrhea. J Infect Dis *159*, 661–669.

Batteiger, B.E., Tu, W., Ofner, S., Van Der Pol, B., Stothard, D.R., Orr, D.P., Katz, B.P., and Fortenberry, J.D. (2010). Repeated Chlamydia trachomatis genital infections in adolescent women. J Infect Dis *201*, 42–51.

Belland, R.J., Nelson, D.E., Virok, D., Crane, D.D., Hogan, D., Sturdevant, D., Beatty, W.L., and Caldwell, H.D. (2003a). Transcriptome analysis of chlamydial growth during IFN-gamma-mediated persistence and reactivation. Proc Natl Acad Sci U S A *100*, 15971–15976.

Belland, R.J., Zhong, G., Crane, D.D., Hogan, D., Sturdevant, D., Sharma, J., Beatty, W.L., and Caldwell, H.D. (2003b). Genomic transcriptional profiling of the developmental cycle of Chlamydia trachomatis. Proc Natl Acad Sci U S A *100*, 8478–8483.

Beura, L.K., Wijeyesinghe, S., Thompson, E.A., Macchietto, M.G., Rosato, P.C., Pierson, M.J., Schenkel, J.M., Mitchell, J.S., Vezys, V., Fife, B.T., *et al.* (2018). T Cells in Nonlymphoid Tissues Give Rise to Lymph-Node-Resident Memory T Cells. Immunity *48*, 327–338 e325.

Blutt, S.E., Miller, A.D., Salmon, S.L., Metzger, D.W., and Conner, M.E. (2012). IgA is important for clearance and critical for protection from rotavirus infection. Mucosal Immunol *5*, 712–719.

Blythe, M.J., Katz, B.P., Batteiger, B.E., Ganser, J.A., and Jones, R.B. (1992). Recurrent genitourinary chlamydial infections in sexually active female adolescents. J Pediatr *121*, 487–493.

Brunham, R.C., Kimani, J., Bwayo, J., Maitha, G., Maclean, I., Yang, C., Shen, C., Roman, S., Nagelkerke, N.J., Cheang, M., *et al.* (1996). The epidemiology of Chlamydia trachomatis within a sexually transmitted diseases core group. J Infect Dis *173*, 950–956.

Brunham, R.C., Kuo, C.C., Cles, L., and Holmes, K.K. (1983). Correlation of host immune response with quantitative recovery of Chlamydia trachomatis from the human endocervix. Infect Immun *39*, 1491–1494.

Budrys, N.M., Gong, S., Rodgers, A.K., Wang, J., Louden, C., Shain, R., Schenken, R.S., and Zhong, G. (2012). Chlamydia trachomatis antigens recognized in women with tubal factor infertility, normal fertility, and acute infection. Obstetrics and gynaecology *119*, 1009–1016.

Cascalho, M., Wong, J., Brown, J., Jack, H.M., Steinberg, C., and Wabl, M. (2000). A B220(–), CD19(–) population of B cells in the peripheral blood of quasimonoclonal mice. International immunology *12*, 29–35.

Cohen, C.R., Koochesfahani, K.M., Meier, A.S., Shen, C., Karunakaran, K., Ondondo, B., Kinyari, T., Mugo, N.R., Nguti, R., and Brunham, R.C. (2005). Immunoepidemiologic profile of Chlamydia trachomatis infection: importance of heat-shock protein 60 and interferon- gamma. J Infect Dis *192*, 591–599.

Cotter, T.W., Meng, Q., Shen, Z.L., Zhang, Y.X., Su, H., and Caldwell, H.D. (1995). Protective efficacy of major outer membrane protein-specific immunoglobulin A (IgA) and IgG monoclonal antibodies in a murine model of Chlamydia trachomatis genital tract infection. Infect Immun *63*, 4704–4714.

Cotter, T.W., Ramsey, K.H., Miranpuri, G.S., Poulsen, C.E., and Byrne, G.I. (1997). Dissemination of Chlamydia trachomatis chronic genital tract infection in gamma interferon gene knockout mice. Infect Immun *65*, 2145–2152.

Darville, T., Andrews, C.W., Jr., Sikes, J.D., Fraley, P.L., Braswell, L., and Rank, R.G. (2001a). Mouse strain-dependent chemokine regulation of the genital tract T helper cell type 1 immune response. Infect Immun *69*, 7419–7424.

Darville, T., Andrews, C.W., Jr., Sikes, J.D., Fraley, P.L., and Rank, R.G. (2001b). Early local cytokine profiles in strains of mice with different outcomes from chlamydial genital tract infection. Infect Immun *69*, 3556–3561.

Datta, B., Njau, F., Thalmann, J., Haller, H., and Wagner, A.D. (2014). Differential infection outcome of Chlamydia trachomatis in human blood monocytes and monocyte-derived dendritic cells. BMC Microbiol *14*, 209.

Dustin, L.B., Bullock, E.D., Hamada, Y., Azuma, T., and Loh, D.Y. (1995). Antigen-driven differentiation of naive Ig-transgenic B cells *in vitro*. J Immunol *154*, 4936–4949.

Eckert, L.O., Hawes, S.E., Wolner-Hanssen, P., Money, D.M., Peeling, R.W., Brunham, R.C., Stevens, C.E., Eschenbach, D.A., and Stamm, W.E. (1997). Prevalence and correlates of antibody to chlamydial heat shock protein in women attending sexually transmitted disease clinics and women with confirmed pelvic inflammatory disease. J Infect Dis *175*, 1453–1458.

El Hakim, E.A., Gordon, U.D., and Akande, V.A. (2010). The relationship between serum Chlamydia antibody levels and severity of disease in infertile women with tubal damage. Arch Gynecol Obstet *281*, 727–733.

Fields, K.A., McCormack, R., de Armas, L.R., and Podack, E.R. (2013). Perforin-2 restricts growth of Chlamydia trachomatis in macrophages. Infect Immun *81*, 3045–3054.

Geisler, W.M., Lensing, S.Y., Press, C.G., and Hook, E.W., 3rd (2013). Spontaneous resolution of genital Chlamydia trachomatis infection in women and protection from reinfection. J Infect Dis *207*, 1850–1856.

Gervassi, A.L., Probst, P., Stamm, W.E., Marrazzo, J., Grabstein, K.H., and Alderson, M.R. (2003). Functional characterization of class Ia- and non-class Ia-restricted Chlamydia-reactive CD8+ T cell responses in humans. J Immunol *171*, 4278–4286.

Gondek, D.C., Roan, N.R., and Starnbach, M.N. (2009). T cell responses in the absence of IFN-gamma exacerbate uterine infection with Chlamydia trachomatis. J Immunol *183*, 1313–1319.

Granstein, R.D., Staszewski, R., Knisely, T.L., Zeira, E., Nazareno, R., Latina, M., and Albert, D.M. (1990). Aqueous humour contains transforming growth factor-beta and a small (less than 3500 daltons) inhibitor of thymocyte proliferation. J Immunol *144*, 3021–3027.

Hackstadt, T., Scidmore-Carlson, M.A., Shaw, E.I., and Fischer, E.R. (1999). The Chlamydia trachomatis IncA protein is required for homotypic vesicle fusion. Cell Microbiol *1*, 119–130.

Hawkins, R.A., Rank, R.G., and Kelly, K.A. (2002). A Chlamydia trachomatis-specific Th2 clone does not provide protection against a genital infection and displays reduced trafficking to the infected genital mucosa. Infect Immun *70*, 5132–5139.

Helander, H.F., and Fandriks, L. (2014). Surface area of the digestive tract – revisited. Scand J Gastroenterol *49*, 681–689.

Hufnagel, K., Lueong, S., Willhauck-Fleckenstein, M., Hotz-Wagenblatt, A., Miao, B., Bauer, A., Michel, A., Butt, J., Pawlita, M., Hoheisel, J.D., *et al.* (2018). Immunoprofiling of Chlamydia trachomatis using whole-proteome microarrays generated by on-chip *in situ* expression. Sci Rep *8*, 7503.

Igietseme, J.U. (1996). The molecular mechanism of T-cell control of Chlamydia in mice: role of nitric oxide. Immunology *87*, 1–8.

Igietseme, J.U., He, Q., Joseph, K., Eko, F.O., Lyn, D., Ananaba, G., Campbell, A., Bandea, C., and Black, C.M.

(2009). Role of T lymphocytes in the pathogenesis of Chlamydia disease. J Infect Dis *200*, 926–934.

Igietseme, J.U., Magee, D.M., Williams, D.M., and Rank, R.G. (1994). Role for CD8+ T cells in antichlamydial immunity defined by Chlamydia-specific T-lymphocyte clones. Infect Immun *62*, 5195–5197.

Igietseme, J.U., Perry, L.L., Ananaba, G.A., Uriri, I.M., Ojior, O.O., Kumar, S.N., and Caldwell, H.D. (1998). Chlamydial infection in inducible nitric oxide synthase knockout mice. Infect Immun *66*, 1282–1286.

Igietseme, J.U., Ramsey, K.H., Magee, D.M., Williams, D.M., Kincy, T.J., and Rank, R.G. (1993). Resolution of murine chlamydial genital infection by the adoptive transfer of a biovar-specific, Th1 lymphocyte clone. Reg Immunol *5*, 317–324.

Igietseme, J.U., Uriri, I.M., Chow, M., Abe, E., and Rank, R.G. (1997). Inhibition of intracellular multiplication of human strains of Chlamydia trachomatis by nitric oxide. Biochem Biophys Res Commun *232*, 595–601.

Igietseme, J.U., Uriri, I.M., Hawkins, R., and Rank, R.G. (1996). Integrin-mediated epithelial-T cell interaction enhances nitric oxide production and increased intracellular inhibition of Chlamydia. J Leukoc Biol *59*, 656–662.

Iijima, N., and Iwasaki, A. (2014). T cell memory. A local macrophage chemokine network sustains protective tissue-resident memory CD4 T cells. Science *346*, 93–98.

Iijima, N., and Iwasaki, A. (2015). Tissue instruction for migration and retention of TRM cells. Trends in immunology *36*, 556–564.

Jayarapu, K., Kerr, M., Ofner, S., and Johnson, R.M. (2010). Chlamydia-specific CD4 T cell clones control Chlamydia muridarum replication in epithelial cells by nitric oxide-dependent and -independent mechanisms. J Immunol *185*, 6911–6920.

Jayarapu, K., Kerr, M.S., Katschke, A., and Johnson, R.M. (2009). Chlamydia muridarum-specific CD4 T-cell clones recognize infected reproductive tract epithelial cells in an interferon-dependent fashion. Infect Immun *77*, 4469–4479.

Jennings, S.R., Bonneau, R.H., Smith, P.M., Wolcott, R.M., and Chervenak, R. (1991). CD4-positive T lymphocytes are required for the generation of the primary but not the secondary CD8-positive cytolytic T lymphocyte response to herpes simplex virus in C57BL/6 mice. Cellular immunology *133*, 234–252.

Jiang, J., Liu, G., Kickhoefer, V.A., Rome, L.H., Li, L.X., McSorley, S.J., and Kelly, K.A. (2017a). A Protective Vaccine against Chlamydia Genital Infection Using Vault Nanoparticles without an Added Adjuvant. Vaccines (Basel) *5*.

Jiang, J., Maxion, H., Champion, C.I., Liu, G., and Kelly, K.A. (2017b). Expression of CXCR3 on Adaptive and Innate Immune Cells Contributes Oviduct Pathology throughout Chlamydia muridarum Infection. J Mucosal Immunol Res *1*.

Johnson, R.M. (2013). Chlamydial Diseases. In The Prokaryotes, Fourth Edition, E. Rosenberg, ed. (Berlin Heidelberg: Springer-Verlag).

Johnson, R.M., and Brunham, R.C. (2016). Tissue-Resident T Cells as the Central Paradigm of Chlamydia Immunity. Infect Immun *84*, 868–873.

Johnson, R.M., and Brunham, R.C. (2017). Reply to Vicetti Miguel *et al.*, 'Setting Sights on Chlamydia Immunity's Central Paradigm: Can We Hit a Moving Target?'. Infect Immun *85*.

Johnson, R.M., and Kerr, M.S. (2015). Modelling the transcriptome of genital tract epithelial cells and macrophages in healthy mucosa versus mucosa inflamed by Chlamydia muridarum infection. Pathogens and disease *73*, ftv100.

Johnson, R.M., Kerr, M.S., and Slaven, J.E. (2012a). Plac8-dependent and inducible NO synthase-dependent mechanisms clear Chlamydia muridarum infections from the genital tract. J Immunol *188*, 1896–1904.

Johnson, R.M., Kerr, M.S., and Slaven, J.E. (2013). Perforin is detrimental to controlling [corrected] *C. muridarum* replication *in vitro*, but not *in vivo*. PLoS One *8*, e63340.

Johnson, R.M., Kerr, M.S., and Slaven, J.E. (2014). An atypical CD8 T-cell response to Chlamydia muridarum genital tract infections includes T cells that produce interleukin-13. Immunology *142*, 248–257.

Johnson, R.M., Yu, H., Kerr, M.S., Slaven, J.E., Karunakaran, K.P., and Brunham, R.C. (2012b). PmpG303–311, a protective vaccine epitope that elicits persistent cellular immune responses in Chlamydia muridarum-immune mice. Infect Immun *80*, 2204–2211.

Johnson, R.M., Yu, H., Olivares Strank, N., Karunakaran, K., Zhu, Y., and Brunham, R.C. (2017). B cell presentation of Chlamydia antigen selects out protective CD4gamma13 T cells; implications for genital tract tissue resident memory lymphocyte clusters. Infect Immun.

Jones, C.S., Maple, P.A., Andrews, N.J., Paul, I.D., and Caul, E.O. (2003). Measurement of IgG antibodies to Chlamydia trachomatis by commercial enzyme immunoassays and immunofluorescence in sera from pregnant women and patients with infertility, pelvic inflammatory disease, ectopic pregnancy, and laboratory diagnosed Chlamydia psittaci/Chlamydia pneumoniae infection. J Clin Pathol *56*, 225–229.

Jordan, S.J., Gupta, K., Ogendi, B.M., Bakshi, R.K., Kapil, R., Press, C.G., Sabbaj, S., Lee, J.Y., and Geisler, W.M. (2017). The Predominant CD4(+) Th1 Cytokine Elicited to Chlamydia trachomatis Infection in Women Is Tumour Necrosis Factor Alpha and Not Interferon Gamma. Clin Vaccine Immunol *24*.

Kari, L., Whitmire, W.M., Olivares-Zavaleta, N., Goheen, M.M., Taylor, L.D., Carlson, J.H., Sturdevant, G.L., Lu, C., Bakios, L.E., Randall, L.B., *et al.* (2011). A live-attenuated chlamydial vaccine protects against trachoma in nonhuman primates. J Exp Med *208*, 2217–2223.

Karunakaran, K.P., Yu, H., Jiang, X., Chan, Q., Moon, K.M., Foster, L.J., and Brunham, R.C. (2015). Outer membrane proteins preferentially load MHC class II peptides: implications for a Chlamydia trachomatis T cell vaccine. Vaccine *33*, 2159–2166.

Kelly, K.A., Natarajan, S., Ruther, P., Wisse, A., Chang, M.H., and Ault, K.A. (2001). Chlamydia trachomatis infection induces mucosal addressin cell adhesion molecule-1 and vascular cell adhesion molecule-1, providing an immunologic link between the fallopian tube and other mucosal tissues. J Infect Dis *184*, 885–891.

Kelly, K.A., and Rank, R.G. (1997). Identification ofhoming receptors that mediate the recruitment of CD4 T cells to the genital tract following intravaginal infection with Chlamydia trachomatis. Infect Immun *65*, 5198–5208.

Kinnunen, A.H., Surcel, H.M., Lehtinen, M., Karhukorpi, J., Tiitinen, A., Halttunen, M., Bloigu, A., Morrison, R.P., Karttunen, R., and Paavonen, J. (2002). HLA DQ alleles and interleukin-10 polymorphism associated with Chlamydia trachomatis-related tubal factor infertility: a case–control study. Hum Reprod *17*, 2073–2078.

Kiviat, N.B., Paavonen, J.A., Brockway, J., Critchlow, C.W., Brunham, R.C., Stevens, C.E., Stamm, W.E., Kuo, C.C., DeRouen, T., and Holmes, K.K. (1985). Cytologic manifestations of cervical and vaginal infections. I. Epithelial and inflammatory cellular changes. Jama *253*, 989–996.

Kiviat, N.B., Paavonen, J.A., Wolner-Hanssen, P., Critchlow, C.W., Stamm, W.E., Douglas, J., Eschenbach, D.A., Corey, L.A., and Holmes, K.K. (1990a). Histopathology of endocervical infection caused by Chlamydia trachomatis, herpes simplex virus, Trichomonas vaginalis, and Neisseria gonorrhoeae. Human pathology *21*, 831–837.

Kiviat, N.B., Wolner-Hanssen, P., Eschenbach, D.A., Wasserheit, J.N., Paavonen, J.A., Bell, T.A., Critchlow, C.W., Stamm, W.E., Moore, D.E., and Holmes, K.K. (1990b). Endometrial histopathology in patients with culture-proved upper genital tract infection and laparoscopically diagnosed acute salpingitis. The American journal of surgical pathology *14*, 167–175.

Knee, R.A., Hickey, D.K., Beagley, K.W., and Jones, R.C. (2005). Transport of IgG across the blood–luminal barrier of the male reproductive tract of the rat and the effect of estradiol administration on reabsorption of fluid and IgG by the epididymal ducts. Biol Reprod *73*, 688–694.

Landers, D.V., Erlich, K., Sung, M., and Schachter, J. (1991). Role of L3T4-bearing T-cell populations in experimental murine chlamydial salpingitis. Infect Immun *59*, 3774–3777.

Ledford, J.G., Kovarova, M., and Koller, B.H. (2007). Impaired host defence in mice lacking ONZIN. J Immunol *178*, 5132–5143.

Li, Z., Palaniyandi, S., Zeng, R., Tuo, W., Roopenian, D.C., and Zhu, X. (2011). Transfer of IgG in the female genital tract by MHC class I-related neonatal Fc receptor (FcRn) confers protective immunity to vaginal infection. Proc Natl Acad Sci U S A *108*, 4388–4393.

Mabey, D.C., Bailey, R.L., Dunn, D., Jones, D., Williams, J.H., Whittle, H.C., and Ward, M.E. (1991). Expression of MHC class II antigens by conjunctival epithelial cells in trachoma: implications concerning the pathogenesis of blinding disease. J Clin Pathol *44*, 285–289.

Manam, S., Nicholson, B.J., and Murthy, A.K. (2013). OT-1 mice display minimal upper genital tract pathology following primary intravaginal Chlamydia muridarum infection. Pathogens and disease *67*, 221–224.

Matyszak, M.K., and Gaston, J.S. (2004). Chlamydia trachomatis-specific human CD8+ T cells show two patterns of antigen recognition. Infect Immun *72*, 4357–4367.

Maue, A.C., Poly, F., and Guerry, P. (2014). A capsule conjugate vaccine approach to prevent diarrhoeal disease caused by Campylobacter jejuni. Hum Vaccin Immunother *10*, 1499–1504.

Moore, T., Ananaba, G.A., Bolier, J., Bowers, S., Belay, T., Eko, F.O., and Igietseme, J.U. (2002). Fc receptor regulation of protective immunity against Chlamydia trachomatis. Immunology *105*, 213–221.

Morre, S.A., van den Brule, A.J., Rozendaal, L., Boeke, A.J., Voorhorst, F.J., de Blok, S., and Meijer, C.J. (2002). The natural course of asymptomatic Chlamydia trachomatis infections: 45% clearance and no development of clinical PID after one-year follow-up. Int J STD AIDS *13 Suppl 2*, 12–18.

Morrison, R.P., and Caldwell, H.D. (2002). Immunity to murine chlamydial genital infection. Infect Immun *70*, 2741–2751.

Morrison, R.P., Feilzer, K., and Tumas, D.B. (1995). Gene knockout mice establish a primary protective role for major histocompatibility complex class II-restricted responses in Chlamydia trachomatis genital tract infection. Infect Immun *63*, 4661–4668.

Morrison, S.G., Farris, C.M., Sturdevant, G.L., Whitmire, W.M., and Morrison, R.P. (2011). Murine Chlamydia trachomatis genital infection is unaltered by depletion of CD4+ T cells and diminished adaptive immunity. J Infect Dis *203*, 1120–1128.

Morrison, S.G., and Morrison, R.P. (2000). In situ analysis of the evolution of the primary immune response in murine Chlamydia trachomatis genital tract infection. Infect Immun *68*, 2870–2879.

Morrison, S.G., and Morrison, R.P. (2001). Resolution of secondary Chlamydia trachomatis genital tract infection in immune mice with depletion of both CD4+ and CD8+ T cells. Infect Immun *69*, 2643–2649.

Morrison, S.G., and Morrison, R.P. (2005a). A predominant role for antibody in acquired immunity to chlamydial genital tract reinfection. J Immunol *175*, 7536–7542.

Morrison, S.G., and Morrison, R.P. (2005b). The protective effect of antibody in immunity to murine chlamydial genital tract reinfection is independent of immunoglobulin A. Infect Immun *73*, 6183–6186.

Morrison, S.G., Su, H., Caldwell, H.D., and Morrison, R.P. (2000). Immunity to murine Chlamydia trachomatis genital tract reinfection involves B cells and CD4(+) T cells but not CD8(+) T cells. Infect Immun *68*, 6979–6987.

Mosmann, T.R., Cherwinski, H., Bond, M.W., Giedlin, M.A., and Coffman, R.L. (1986). Two types of murine helper T cell clone. I. Definition according to profiles of lymphokine activities and secreted proteins. J Immunol *136*, 2348–2357.

Murthy, A.K., Chambers, J.P., Meier, P.A., Zhong, G., and Arulanandam, B.P. (2007). Intranasal vaccination with a secreted chlamydial protein enhances resolution of genital Chlamydia muridarum infection, protects against oviduct pathology, and is highly dependent upon endogenous gamma interferon production. Infect Immun *75*, 666–676.

Murthy, A.K., Li, W., Chaganty, B.K., Kamalakaran, S., Guentzel, M.N., Seshu, J., Forsthuber, T.G., Zhong, G., and Arulanandam, B.P. (2011). Tumour necrosis factor alpha production from CD8+ T cells mediates oviduct pathological sequelae following primary genital Chlamydia muridarum infection. Infect Immun *79*, 2928–2935.

Naglak, E.K., Morrison, S.G., and Morrison, R.P. (2016). IFNgamma is Required for Optimal Antibody-Mediated Immunity against Genital Chlamydia Infection. Infect Immun.

Naglak, E.K., Morrison, S.G., and Morrison, R.P. (2017). Neutrophils Are Central to Antibody-Mediated Protection against Genital Chlamydia. Infect Immun *85*.

Nelson, D.E., Virok, D.P., Wood, H., Roshick, C., Johnson, R.M., Whitmire, W.M., Crane, D.D., Steele-Mortimer, O., Kari, L., McClarty, G., *et al.* (2005). Chlamydial IFN-{gamma} immune evasion is linked to host infection tropism. Proc Natl Acad Sci U S A *102*, 10658–10663.

Ness, R.B., Brunham, R.C., Shen, C., Bass, D.C., and Investigators, P.I.D.E.C.H.S. (2004). Associations among human leucocyte antigen (HLA) class II DQ variants, bacterial sexually transmitted diseases, endometritis, and fertility among women with clinical pelvic inflammatory disease. Sexually transmitted diseases *31*, 301–304.

Ness, R.B., Soper, D.E., Richter, H.E., Randall, H., Peipert, J.F., Nelson, D.B., Schubeck, D., McNeeley, S.G., Trout, W., Bass, D.C., *et al.* (2008). Chlamydia antibodies, chlamydia heat shock protein, and adverse sequelae after pelvic inflammatory disease: the PID Evaluation and Clinical Health (PEACH) Study. Sexually transmitted diseases *35*, 129–135.

Olivares-Zavaleta, N., Whitmire, W.M., Kari, L., Sturdevant, G.L., and Caldwell, H.D. (2014). CD8+ T cells define an unexpected role in live-attenuated vaccine protective immunity against Chlamydia trachomatis infection in macaques. J Immunol *192*, 4648–4654.

Olsen, A.W., Follmann, F., Erneholm, K., Rosenkrands, I., and Andersen, P. (2015). Protection Against Chlamydia trachomatis Infection and Upper Genital Tract Pathological Changes by Vaccine-Promoted Neutralizing Antibodies Directed to the VD4 of the Major Outer Membrane Protein. J Infect Dis *212*, 978–989.

Olsen, A.W., Lorenzen, E.K., Rosenkrands, I., Follmann, F., and Andersen, P. (2017). Protective Effect of Vaccine Promoted Neutralizing Antibodies against the Intracellular Pathogen Chlamydia trachomatis. Front Immunol *8*, 1652.

Olson, K.M., Tang, J., Brown, L., Press, C.G., and Geisler, W.M. (2018). HLA-DQB1*06 is a risk marker for chlamydia reinfection in African American women. Genes Immun *20*, 69–73.

Perry, L.L., Feilzer, K., Hughes, S., and Caldwell, H.D. (1999). Clearance of Chlamydia trachomatis from the murine genital mucosa does not require perforin-mediated cytolysis or Fas-mediated apoptosis. Infect Immun *67*, 1379–1385.

Qu, Y., Frazer, L.C., O'Connell, C.M., Tarantal, A.F., Andrews, C.W., Jr., O'Connor, S.L., Russell, A.N., Sullivan, J.E., Poston, T.B., Vallejo, A.N., *et al.* (2015). Comparable Genital Tract Infection, Pathology, and Immunity in Rhesus Macaques Inoculated with Wild-Type or Plasmid-Deficient Chlamydia trachomatis Serovar D. Infect Immun *83*, 4056–4067.

Rahman, K.S., Darville, T., Russell, A.N., O'Connell, C.M., Wiesenfeld, H.C., Hillier, S.L., Lee, D.E., and Kaltenboeck, B. (2018). Comprehensive Molecular Serology of Human Chlamydia trachomatis Infections by Peptide Enzyme-Linked Immunosorbent Assays. mSphere *3*.

Ramsey, K.H., Miranpuri, G.S., Poulsen, C.E., Marthakis, N.B., Braune, L.M., and Byrne, G.I. (1998). Inducible nitric oxide synthase does not affect resolution of murine chlamydial genital tract infections or eradication of chlamydiae in primary murine cell culture. Infect Immun *66*, 835–838.

Rank, R.G., and Batteiger, B.E. (1989). Protective role of serum antibody in immunity to chlamydial genital infection. Infect Immun *57*, 299–301.

Rank, R.G., Batteiger, B.E., and Soderberg, L.S. (1988). Susceptibility to reinfection after a primary chlamydial genital infection. Infect Immun *56*, 2243–2249.

Rank, R.G., Soderberg, L.S., and Barron, A.L. (1985). Chronic chlamydial genital infection in congenitally athymic nude mice. Infect Immun *48*, 847–849.

Rey-Ladino, J., Jiang, X., Gabel, B.R., Shen, C., and Brunham, R.C. (2007). Survival of Chlamydia muridarum within dendritic cells. Infect Immun *75*, 3707–3714.

Rock, K.L., Benacerraf, B., and Abbas, A.K. (1984). Antigen presentation by hapten-specific B lymphocytes. I. Role of surface immunoglobulin receptors. J Exp Med *160*, 1102–1113.

Roopenian, D.C., and Akilesh, S. (2007). FcRn: the neonatal Fc receptor comes of age. Nat Rev Immunol *7*, 715–725.

Russell, A.N., Zheng, X., O'Connell, C.M., Taylor, B.D., Wiesenfeld, H.C., Hillier, S.L., Zhong, W., and Darville, T. (2016). Analysis of Factors Driving Incident and Ascending Infection and the Role of Serum Antibody in Chlamydia trachomatis Genital Tract Infection. J Infect Dis *213*, 523–531.

Rzomp, K.A., Moorhead, A.R., and Scidmore, M.A. (2006). The GTPase Rab4 interacts with Chlamydia trachomatis inclusion membrane protein CT229. Infect Immun *74*, 5362–5373.

Schenkel, J.M., Fraser, K.A., and Masopust, D. (2014). Cutting edge: resident memory CD8 T cells occupy frontline niches in secondary lymphoid organs. J Immunol *192*, 2961–2964.

Shin, H., and Iwasaki, A. (2012). A vaccine strategy that protects against genital herpes by establishing local memory T cells. Nature *491*, 463–467.

Shin, H., and Iwasaki, A. (2013). Tissue-resident memory T cells. Immunol Rev *255*, 165–181.

Stary, G., Olive, A., Radovic-Moreno, A.F., Gondek, D., Alvarez, D., Basto, P.A., Perro, M., Vrbanac, V.D., Tager, A.M., Shi, J., *et al.* (2015). VACCINES. A mucosal vaccine against Chlamydia trachomatis generates two waves of protective memory T cells. Science *348*, aaa8205.

Sturdevant, G.L., and Caldwell, H.D. (2014). Innate immunity is sufficient for the clearance of Chlamydia trachomatis from the female mouse genital tract. Pathogens and disease *72*, 70–73.

Su, H., Feilzer, K., Caldwell, H.D., and Morrison, R.P. (1997). Chlamydia trachomatis genital tract infection of antibody-deficient gene knockout mice. Infect Immun *65*, 1993–1999.

Su, H., Parnell, M., and Caldwell, H.D. (1995). Protective efficacy of a parenterally administered MOMP-derived synthetic oligopeptide vaccine in a murine model of Chlamydia trachomatis genital tract infection: serum neutralizing IgG antibodies do not protect against chlamydial genital tract infection. Vaccine *13*, 1023–1032.

Tiitinen, A., Surcel, H.M., Halttunen, M., Birkelund, S., Bloigu, A., Christiansen, G., Koskela, P., Morrison, S.G., Morrison, R.P., and Paavonen, J. (2006). Chlamydia trachomatis and chlamydial heat shock protein 60-specific antibody and cell-mediated responses predict tubal factor infertility. Hum Reprod *21*, 1533–1538.

Tseng, C.T., and Rank, R.G. (1998). Role of NK cells in early host response to chlamydial genital infection. Infect Immun *66*, 5867–5875.

van den Brule, A.J., Munk, C., Winther, J.F., Kjaer, S.K., Jorgensen, H.O., Meijer, C.J., and Morre, S.A. (2002). Prevalence and persistence of asymptomatic Chlamydia trachomatis infections in urine specimens from Danish male military recruits. Int J STD AIDS *13 Suppl 2*, 19–22.

Van Voorhis, W.C., Barrett, L.K., Sweeney, Y.T., Kuo, C.C., and Patton, D.L. (1996). Analysis of lymphocyte phenotype and cytokine activity in the inflammatory infiltrates of the upper genital tract of female macaques infected with Chlamydia trachomatis. J Infect Dis *174*, 647–650.

Vicetti Miguel, R.D., and Cherpes, T.L. (2012). Hypothesis: Chlamydia trachomatis infection of the female genital tract is controlled by Type 2 immunity. Med Hypotheses *79*, 713–716.

Vicetti Miguel, R.D., Harvey, S.A., LaFramboise, W.A., Reighard, S.D., Matthews, D.B., and Cherpes, T.L. (2013). Human female genital tract infection by the obligate intracellular bacterium Chlamydia trachomatis elicits robust Type 2 immunity. PLoS One *8*, e58565.

Vlcek, K.R., Li, W., Manam, S., Zanotti, B., Nicholson, B.J., Ramsey, K.H., and Murthy, A.K. (2016). The contribution of Chlamydia-specific CD8(+) T cells to upper genital tract pathology. Immunology and cell biology *94*, 208–212.

Wang, C., Tang, J., Geisler, W.M., Crowley-Nowick, P.A., Wilson, C.M., and Kaslow, R.A. (2005). Human leucocyte antigen and cytokine gene variants as predictors of recurrent Chlamydia trachomatis infection in high-risk adolescents. J Infect Dis *191*, 1084–1092.

Wang, S., Fan, Y., Brunham, R.C., and Yang, X. (1999). IFN-gamma knockout mice show Th2-associated delayed-type hypersensitivity and the inflammatory cells fail to localize and control chlamydial infection. Eur J Immunol *29*, 3782–3792.

Woodhall, S.C., Gorwitz, R.J., Migchelsen, S.J., Gottlieb, S.L., Horner, P.J., Geisler, W.M., Winstanley, C., Hufnagel, K., Waterboer, T., Martin, D.L., *et al.* (2018). Advancing the public health applications of Chlamydia trachomatis serology. Lancet Infect Dis.

Xie, Q.W., Whisnant, R., and Nathan, C. (1993). Promoter of the mouse gene encoding calcium-independent nitric oxide synthase confers inducibility by interferon gamma and bacterial lipopolysaccharide. J Exp Med *177*, 1779–1784.

Yu, H., Jiang, X., Shen, C., Karunakaran, K.P., and Brunham, R.C. (2009). Novel Chlamydia muridarum T cell antigens induce protective immunity against lung and genital tract infection in murine models. J Immunol *182*, 1602–1608.

Yu, H., Jiang, X., Shen, C., Karunakaran, K.P., Jiang, J., Rosin, N.L., and Brunham, R.C. (2010). Chlamydia muridarum T-cell antigens formulated with the adjuvant DDA/TDB induce immunity against infection that correlates with a high frequency of gamma interferon (IFN-gamma)/tumour necrosis factor alpha and IFN-gamma/interleukin-17 double-positive CD4+ T cells. Infect Immun *78*, 2272–2282.

Yu, H., Karunakaran, K.P., Jiang, X., Shen, C., Andersen, P., and Brunham, R.C. (2012). Chlamydia muridarum T cell antigens and adjuvants that induce protective immunity in mice. Infect Immun *80*, 1510–1518.

Yu, H., Karunakaran, K.P., Kelly, I., Shen, C., Jiang, X., Foster, L.J., and Brunham, R.C. (2011). Immunization with live and dead Chlamydia muridarum induces different levels of protective immunity in a murine genital tract model: correlation with MHC class II peptide presentation and multifunctional Th1 cells. J Immunol *186*, 3615–3621.

Chlamydia Vaccines

15

Luis M. de la Maza[1], Sukumar Pal[1], Anja W. Olsen[2] and Frank Follmann[2*]

[1]Department of Pathology and Laboratory Medicine, University of California, Irvine, CA, USA.
[2]Department of Infectious Disease Immunology, Centre for Vaccine Research, Statens Serum Institut, Copenhagen, Denmark.

*Correspondence: FRF@ssi.dk

https://doi.org/10.21775/9781912530281.15

Abstract

Vaccination is the most cost effective and efficient way to control infectious diseases. Vaccines against several members of the *Chlamydia* genus are urgently needed, especially those against sexually transmitted *Chlamydia trachomatis*, where current measures have failed to control the global epidemic. Preclinical research has made significant progress over the last four decades, and vaccines are now moving into development and clinical testing. The first Phase I trial of a candidate vaccine for *C. trachomatis* sexually transmitted infections (STIs) has been completed and others are on the way. Here, we provide an overview of the status of *Chlamydia* vaccines against the principal pathogens in both humans and animals, focusing primarily on human *C. trachomatis* STIs. We discuss the approaches pursued following the trachoma clinical trials in the 1960s, including inactivated whole-cell, attenuated strain, and subunit vaccines. Novel adjuvants and delivery systems and the challenges of inducing adaptive immunity in the female genital tract are covered. Lastly, we envisage the implementation of *C. trachomatis* STI vaccines by proposing how to move them through the clinical development phases I-IV.

Chlamydia pathogens and vaccine status

Currently, 11 species of *Chlamydia* have been described, which cause a wide spectrum of diseases in humans, other mammals and birds (Bachmann *et al.*, 2014). The economic impact on humans, animal health and agricultural industries is huge, and ongoing efforts are devoted to the development of effective vaccines against many of the *Chlamydia* species (reviewed by Longbottom and Livingstone, 2006). The principal pathogens of animals include *C. psittaci*, *C. abortus*, *C. pecorum*, *C. suis*, and *C. felis*. *C. psittaci* and *C. abortus* furthermore pose a potential zoonotic risk infecting humans (Longbottom and Coulter, 2003) (see Table 15.1, for more details see Chapter 17, Veterinary *Chlamydia* Species). Commercial vaccines are available for *C. abortus* and *C. felis*. These live attenuated or inactivated whole-cell vaccines prevent disease, but they only reduce bacterial shedding in variable degree (Garcia de la Fuente *et al.*, 2004; Rocchi *et al.*, 2009; Wills *et al.*, 1987). Although similar vaccines have not been as successful against human chlamydial infections, the development of these animal vaccines is encouraging and provides hope for vaccines against *Chlamydia* species in general.

The major clinically relevant species causing disease in humans are *C. trachomatis*, responsible for ocular and genital infections, and *C. pneumoniae*, which causes community-acquired pneumonia (Hahn *et al.*, 2002) and is also associated with lung cancer, asthma and chronic inflammatory diseases such as atherosclerosis (Belland *et al.*, 2004; Chaturvedi *et al.*, 2010; Gaydos *et al.*, 1996; Kuo and Campbell, 2003; Littman *et al.*, 2005; Sessa *et al.*, 2009; Shemer-Avni and Lieberman, 1995).

Table 15.1 Principal chlamydial pathogens of animals: *C. psittaci*, *C. abortus*, *C. pecorum*, *C. suis*, *C. felis* and *C. muridarum*[a]

Pathogen	Host	Disease	Vaccine status
C. psittaci	Birds and domesticated poultry. Infects humans and other mammals (Longbottom and Coulter, 2003)	In birds, the infection results in avian chlamydiosis characterized by diarrhoea, respiratory disease and conjunctivitis (Vanrompay *et al.*, 1995). Infection in humans causes psittacosis or parrot disease (Stewardson and Grayson, 2010; Yung and Grayson, 1988)	No commercial vaccine is available. DNA major outer membrane protein (MOMP)-based vaccines have been tested in a turkey model (Loots *et al.*, 2006; Verminnen *et al.*, 2010; Verminnen *et al.*, 2005); a recombinant adenovirus vaccine expressing MOMP has been used in chickens (Zhou *et al.*, 2007); and an edible MOMP on transgenic rice seeds was tried in mice (Zhang *et al.*, 2013)
C. abortus	Sheep and goats; however, cattle, pig, horses and humans can also become infected (Entrican *et al.*, 2012; Longbottom and Coulter, 2003)	Major cause of abortion in sheep and goats (Entrican *et al.*, 2012; Longbottom and Coulter, 2003). In humans, severe life-threatening disease can occur in pregnant women and can result in spontaneous abortion or stillbirth (Buxton, 1986; Hyde and Benirschke, 1997)	Two vaccines are commercially available: a live attenuated vaccine (Rocchi *et al.*, 2009) and an inactivated whole-cell vaccine (Garcia de la Fuente *et al.*, 2004). Both vaccines are protective against disease (ovine enzootic abortion) and reduce bacterial shedding. Novel subunit vaccine approaches are ongoing (Ou *et al.*, 2013)
C. pecorum	Sheep, goats, pigs, cattle and wildlife species (Longbottom and Coulter, 2003), i.e. the koala in Australia (Kollipara *et al.*, 2013; Polkinghorne *et al.*, 2013)	In livestock, infections are typically associated with arthritis and conjunctivitis in ruminants, as well as encephalomyelitis in cattle (Anderson *et al.*, 1996; Cutlip *et al.*, 1972; Fukushi and Hirai, 1992; Meagher *et al.*, 1992). In koalas, infections causes conjunctivitis that can lead to blindness as well as pneumonia and urinary tract infections. Infections of the reproductive tract is a major cause of infertility in female koalas (Network, 2012)	No commercial vaccine is available. Vaccine research is ongoing and promising results demonstrated with MOMP- and polymorphic membrane protein G (PmpG)-based vaccines show effect on the clearance of infection in Koalas (Desclozeaux *et al.*, 2017; Kollipara *et al.*, 2012; Waugh *et al.*, 2016)
C. suis	Pigs (Longbottom and Coulter, 2003), wild boars (Di Francesco *et al.*, 2011). Transmission to humans has been reported (De Puysseleyr *et al.*, 2017)	In pigs causes conjunctivitis, enteritis and pneumonia (Englund *et al.*, 2012)	No commercial vaccine is available
C. felis	Cats. Transmission to humans has been reported in rare cases (Longbottom and Coulter, 2003)	In cats causes conjunctivitis	Two vaccines are commercially available: an inactivated adjuvanted vaccine (Fel-O-Vax Lv-K IV; Fort Dodge Laboratories, USA) and modified-live (ML) injectable vaccine (Eclipse 4+FeLV; Intervet ScheringePlough Animal Health, USA). Vaccination does not control bacterial shedding, but reduces acute disease (Wills *et al.*, 1987)

Table 15.1 Continued

Pathogen	Host	Disease	Vaccine status
C. muridarum	Mice	Causes pneumonitis in mice and was originally isolated from the lungs of mice inoculated with lung tissues from humans with respiratory infections (Nigg, 1942). Intravaginal infections in mice results in an ascending infection to the upper genital tract that can lead to hydrosalpinx and infertility	*C. muridarum* has intensively been used to study immunity, pathogenesis and vaccine protection in the mouse model (De Clercq *et al.*, 2013)

[a]*C. muridarum* is included as it is of principal importance for studies of immunity and vaccine development in the mouse.

Currently, no vaccine is available for *C. trachomatis*, and research is ongoing to justify the need for a *C. pneumoniae* vaccine since infections while common, are usually not severe and resolve without treatment. However, improving the understanding of both the biology and pathology might pave the way for a future vaccine, especially if a truly causative link between atherosclerosis or cancer and *C. pneumoniae* infection is established (Belland *et al.*, 2004; Littman *et al.*, 2005).

C. trachomatis produces several human diseases including trachoma, the world's most common cause of preventable blindness (serovars A, B, Ba, or C) and a range of STIs (serovars D-K and L1-L3). Trachoma is responsible for the blindness or visual impairment of about 1.9 million individuals worldwide (WHO Key Facts, 2019). Thanks to the improvement in sanitation and hygiene, trachoma has disappeared from most parts of the world. The implementation of the WHO's SAFE Strategy (Surgery, Antibiotic treatment, Facial cleanliness, Environmental improvement) (Wright *et al.*, 2008) attempts to eradicate trachoma from the underdeveloped countries, in particular Africa, by 2020. Data from trachoma control programs indicate that the current treatment-based approach to *C. trachomatis* transmission is unlikely to completely control the spread in endemic areas, and that an effective vaccine may be the best strategy to eliminate trachoma as a public health problem (Atik *et al.*, 2006). Active vaccine research is ongoing by several groups using either live-attenuated whole-cell vaccines (Kari *et al.*, 2011; Olivares-Zavaleta *et al.*, 2014) or subunit vaccines (Badamchi-Zadeh *et al.*, 2015; Inic-Kanada *et al.*, 2015; Kari *et al.*, 2009; Taylor *et al.*, 1988).

In contrast to the positive development in the control of trachoma, the number of *C. trachomatis* STIs continue to rise. It is estimated that 131 million people were infected in 2012 (Gottlieb *et al.*, 2016; Newman *et al.*, 2015) despite continued improvement in screening, diagnostic and treatment programs (Datta *et al.*, 2007; Johnson *et al.*, 2002). Most prevalent in adolescents and young adults, *C. trachomatis* infections often go undiagnosed as they remain asymptomatic in 70–90% of women and 30–50% of men (Haggerty *et al.*, 2010; Newman *et al.*, 2015). Most infections are believed to be transient, however, in some individuals, if left untreated, acute and chronic *C. trachomatis* infections constitute a major health concern. *C. trachomatis* in women can cause pelvic inflammatory disease (PID) and other long-term sequelae including chronic abdominal pain, infertility and ectopic pregnancy (Brunham *et al.*, 2015; Darville and Hiltke, 2010; Haggerty *et al.*, 2010; Korenromp *et al.*, 2017; Price *et al.*, 2016; Westrom, 1995; Westrom *et al.*, 1992). Globally, it is estimated that *C. trachomatis* accounts for up to 1 million new cases of infertility annually (Haggerty *et al.*, 2010). *C. trachomatis* infections also have important maternal child health consequences. Infection during pregnancy poses a risk for adverse outcomes such as miscarriage, stillbirth and premature labour (Adachi *et al.*, 2016) and may also cause neonatal conjunctivitis and/or pneumonia (Rosenman *et al.*, 2003). In young men, *C. trachomatis* is the leading cause of non-gonococcal urethritis (Brill, 2010; Wetmore *et al.*, 2011), and ascending infections may lead to acute epididymitis and orchitis (Tracy *et al.*, 2008). Reactive arthritis has also been associated with *C. trachomatis* infections (Schachter, 1980). In addition, *C. trachomatis* increases susceptibility to coinfection with other STIs, in particular, *Neisseria gonorrhoeae* and HIV-1 (Johnson and Lewis, 2008). Therefore, action is needed (Starnbach, 2018; Zhong *et al.*, 2017). This chapter will review the status of work towards a vaccine against *C. trachomatis* STIs.

Current interventions and targets for a human *C. trachomatis* vaccine

To limit the infectious burden and interrupt transmission, many countries focus on case management through screening and targeted treatment in risk populations, including adolescents, young adults, partners of *C. trachomatis* positive individuals and pregnant females. Screening for infection and treatment is recommended but has, until now, not reduced incidence rates (Gottlieb *et al.*, 2014). Infection prevalence is not declining (Gottlieb *et al.*, 2016), and reinfection rates are high (Hosenfeld *et al.*, 2009). The counter-intuitive increase in *C. trachomatis* prevalence (Brunham *et al.*, 2005; Gotz *et al.*, 2002), in light of new intervention activities, led Brunham and Rekart (Brunham and Rekart, 2008) to hypothesize that the increased prevalence is caused by the interference of early antibiotic treatment on the development of natural protective immunity (Brunham *et al.*, 2005). Experimental data in mice supports this premise (Su *et al.*, 1999). Screening remains an important tool to identify chronic asymptomatic infections and reduce PID incidence through targeted treatment (Scholes *et al.*, 1996), although modelling studies indicate that a very high screening coverage is required for incidence reduction through this intervention (Regan *et al.*, 2008; Turner *et al.*, 2006, 2011).

An effective vaccine against *C. trachomatis* should ideally cover all genital serovars and aim to curb the global epidemic by preventing infection and eliciting 'sterilizing immunity', thereby blocking horizontal and vertical transmission. However, even a partially effective vaccine could provide a cost-effective intervention by shortening the duration of infection, lowering the bacterial load, limiting transmission of co-infections and/or preventing ascending infection and PID (Grey *et al.*, 2009; Poston *et al.*, 2017). Modelling studies suggest that a *C. trachomatis* vaccine would be cost-effective (de la Maza and de la Maza, 1995; Grey *et al.*, 2009; Owusu-Edusei *et al.*, 2015) even in the context of ongoing screening programs (Owusu-Edusei *et al.*, 2015). Furthermore, the acceptance of an HPV vaccine by the public suggests that a *C. trachomatis* vaccine may be similarly well received.

Experience from human and non-human primate trachoma vaccine trials

Human *C. trachomatis* vaccine development was pioneered in the 1910s by Nicolle and collaborators using whole-organisms to protect against trachoma in humans and non-human primates (NHP) (Nicolle *et al.*, 1913). Experimental ocular infection of human volunteers demonstrated that infection could induce a partial protective immune response because re-challenge with the same serovar had a reduced clinical response and decreased bacterial shedding (Dawson *et al.*, 1966; Grayston *et al.*, 1962; Jawetz *et al.*, 1965). However, this partial protective immune response did not protect against re-challenge with a different serovar.

In the 1960s, several research groups initiated clinical trachoma vaccine trials in humans and NHP with live and inactivated whole-cell *C. trachomatis* vaccines to protect against trachoma (Clements *et al.*, 1979; Dhir *et al.*, 1967; Grayston *et al.*, 1963; Nichols *et al.*, 1969; Nichols *et al.*, 1966; Sowa *et al.*, 1969; Woolridge *et al.*, 1967). Studies used the Pasteurian principles of isolating the causative agent and then inactivating the bacteria or injecting live bacteria intramuscularly, either alone or adjuvanted with alum/mineral oil. Considerable efforts were put into characterizing the vaccines both in pre-clinical models of NHP and in mice toxicity studies (Alexander *et al.*, 1967; Wang and Grayston, 1963, 1967; Wang *et al.*, 1967a; Wang *et al.*, 1967b). There are many limitations for interpreting results across these trials i.e. variation in types of vaccines and target population. Nevertheless, these placebo-controlled studies represent the only efficacy data for human *Chlamydia* vaccines in a natural setting (Table 15.2). The majority of vaccines tested led to at least a transient reduction in incidence of infection/disease and also reduced bacterial load (Dhir *et al.*, 1967; Grayston *et al.*, 1963; Nichols *et al.*, 1969; Nichols *et al.*, 1966; Sowa *et al.*, 1969). These effects were independent of vaccine composition (live, inactivated +/– Alum or mineral oil). Sowa *et al.* even reported significantly lower conjunctival scarring in vaccinated children (Sowa *et al.*, 1969) and Collier *et al.* observed significant clinical improvement in vaccinating children who had active trachoma (Collier L.H., 1963).

Two trials, by Nichols *et al.* (1966) and Woolridge *et al.* (1967) reported higher incidence of trachoma in vaccinated individuals following exposure to *C. trachomatis* (Table 15.2). These trial results were interpreted in the light of the discouraging outcomes from challenge studies in the Taiwan NHP model showing that vaccination could lead to more severe disease upon challenge with a heterologous serovar (Mordhorst, 1967; Wang *et al.*, 1967a), possibly via hypersensitive reactions. From the clinical trials there was no convincing evidence that vaccination led to more severe disease in humans (Mabey *et al.*, 2014). The overall conclusion from the trachoma vaccine trials was that vaccination could work and that it was generally safe, even in these highly challenging

Table 15.2 Trachoma vaccine trials

Vaccine	Strain(s)	Groups	Trial design	Results	Reference
UV-inactivated EB	TRIC Tw-1	1. Adjuvanted; 2. placebo. Adjuvant = Alum	3*IM, 0–1–8 months. Pre-school siblings of children with active trachoma	Effectiveness of the vaccine was highest using 0–1–8 protocol. Combined results between the two vaccine groups showed less active trachoma in vaccinated children (8%) compared to placebo (18%)	Grayston *et al.* (1963) Taiwan
UV-inactivated EB	TRIC Tw-1 and TRIC Tw-3	1. Adjuvanted; 2. placebo. Adjuvant = Alum	4*IM, 0–3–10–21 months. Pre-school siblings of children with active trachoma		
Inactivated EB	TRIC/1/ET/Har-13/OT and TRIC/2/SAU/Har-2/OT	1. Adjuvanted 0.25 arbitrary vaccine unit (AVU); 2. un-adjuvanted 4.0 AVU; 3. un-adjuvanted 0.4 AVU. Adjuvant = water in oil Drakeol-6/Arlecel-A	2*IM, 0–6 months. Children less than 3 years of age with clinical diagnoses of active clinical trachoma, non-trachomatous conjunctivitis or no apparent conjunctival infection	Adjuvanted 0.25 AVU: reduced bacterial load (conjunctival scrapings) compared to placebo. Un-adjuvanted 4.0 AVU: reduced bacterial load (conjunctival scrapings) compared to placebo. Higher attack rate of trachoma. Un-adjuvanted 0.4 AVU: lower attack rate of trachoma (6 and 12 months) compared to placebo	Nichols *et al.* (1969); Nichols *et al.* (1966) Saudi Arabia
Inactivated EB	TRIC Tw- 1 and TRIC Tw-3	1. Adjuvanted; 2. placebo. Adjuvant = water in oil Drakeol-6/Arlecel-A	2*IM, 0–6 months. School children from second and third grade	Non-significant reduction of incidence of active trachoma from 8.8% to 5.1%	Woolridge *et al.* (1967) Taiwan
Inactivated EB	TRIC Tw- 1	1. Adjuvanted; 2. placebo. Adjuvant = Water in oil Drakeol-6/Arlecel-A	2*IM, 0–6 months. School children from second and third grade	Higher incidence of active trachoma in the vaccinated group compared to placebo. No difference in disease severity	Woolridge *et al.* (1967) Taiwan
Inactivated EB	Tw-3 and Tw-5	1. Vaccine (gradient-purified); 2. vaccine (genetron-purified); 3. placebo. No adjuvant used	2*IM, 0–3 months. Children aged less than five years. Children showing signs of trachoma were excluded	14% of the vaccinated children developed clinical trachoma compared to 37% in the placebo group ($p < 0.01$). No difference in severity between vaccinated and control children	Dhir *et al.* (1967) India

Table 15.2 Continued

Vaccine	Strain(s)	Groups	Trial design	Results	Reference
Viable EB	TRIC/WAG/ MRC-221/ OT	1. Vaccine; 2. placebo. No adjuvant used	Children with active trachoma	Significant clinical improvement in the vaccinated compared to placebo group. Prevalence of infection reduced in the vaccinated group	Collier (1963) Gambia
Viable EB	TRIC/WAG/ MRC-187/ OT	1. Adjuvanted; 2. placebo. Adjuvant = water in oil emulsion Drakeol-6/ Arlecel-A	2*IM, 0–6 months. Children aged 0–9 years. Children showing signs of trachoma and positive for culture were excluded	Six months after the second dose 17/118 children in the placebo group had acquired trachoma, compared to 7/117 in the vaccinated groups ($p = 0.053$). At 12 months there was no difference between the groups	Sowa *et al.* (1969) Gambia
Viable EB	TRIC/2/ SAU/Har-2/ OT and TRIC/USA-Cal/Cal-2/ OT	1. Vaccine; 2. placebo. No adjuvant used	2*SC, three weeks apart. Children aged 0–4 years. Children showing signs of trachoma and positive for culture were excluded	The proportion of children with conjunctival scarring was lower in the vaccine group compared to placebo (9/55 vs. 18/47, $p = 0.034$)	Sowa *et al.* (1969) Gambia

populations with varying levels of pre-existing immunity/disease and using whole-cell vaccines of varying quality/potency (Table 15.2 and Mabey *et al.*, 2014). Protection was however only short lived and possibly serovar/serogroup-specific.

Correlates of protective immunity

Determining the characteristics of a protective immune response is a major obstacle for vaccine development. With no true correlates of protective immunity against *C. trachomatis*, vaccine research is reliant on detailed studies of host–pathogen interactions and elucidating possible protective mechanisms associated with natural infection in humans and animal models. Natural immunity to *C. trachomatis* infections is often incomplete and does not provide long-term protection. However, there is evidence from both human (Bakshi *et al.*, 2018; Batteiger *et al.*, 2010; Brunham *et al.*, 1996; Cohen *et al.*, 2005; Kimani *et al.*, 1996; Molano *et al.*, 2005; Russell *et al.*, 2016) and rodent (Gondek *et al.*, 2012; Gondek *et al.*, 2009; Igietseme, 1996; Loomis and Starnbach, 2002; Morrison, 2000; Morrison and Caldwell, 2002; Morrison *et al.*, 1995; Morrison and Morrison, 2001; Su and Caldwell, 1995; Yu *et al.*, 2011) studies that cellular immunity, with Th1 responses and IFN-γ production, is needed to recover from infection and correlates with protection against reinfections. In the mouse model, CD4 T cells are required for protection against *Chlamydia muridarum* (previously called *C. trachomatis* mouse pneumonitis) infection, whereas CD8 T cells are not (Morrison *et al.*, 1995; Morrison and Morrison, 2001; Morrison *et al.*, 2000; Rank *et al.*, 1985).

Functional *C. muridarum*-specific antibodies develop during natural infection, but do not seem essential for resolution of the primary infection. However, during re-infection they control initial bacterial load (Morrison and Morrison, 2001; Rank and Batteiger, 1989) and function as facilitators for both the adaptive and the innate immune responses, thereby aiding resolution of the infection (Moore *et al.*, 2002; Moore *et al.*, 2003; Morrison and Morrison, 2005; Morrison *et al.*, 2000; Naglak *et al.*, 2016). Antibody mediated protective immunity in humans still needs to be fully elucidated. In the female genital tract (FGT) systemic cellular and humoral immunity, especially IgG (Mestecky and Fultz, 1999), is recruited upon inflammation or infection (Mestecky *et al.*, 2005) (Fig. 15.1). Brunham *et al.* found

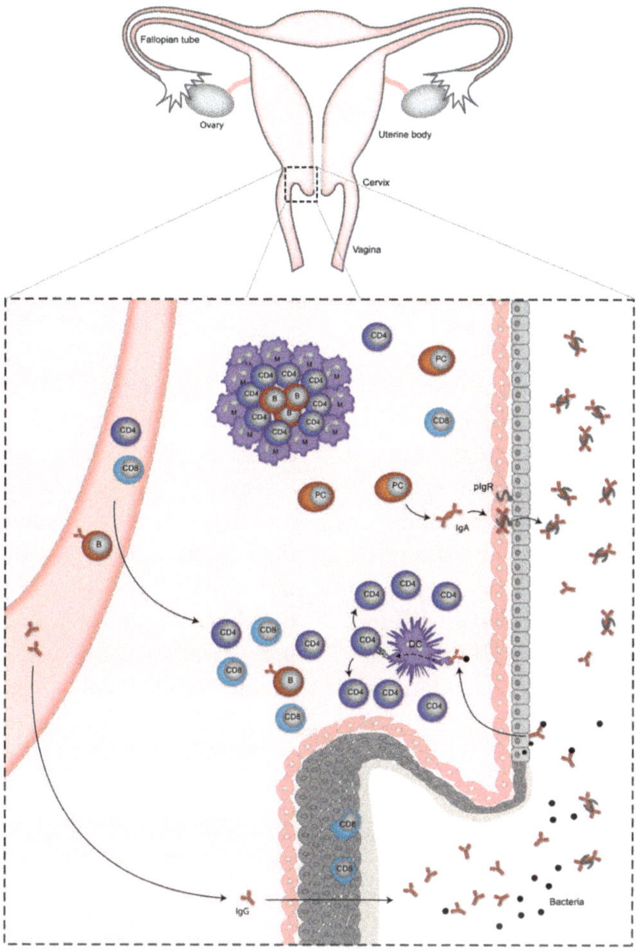

Figure 15.1 Adaptive immunity in the human female genital tract (FGT). The FGT is composed of the vagina, uterine cervix, uterus, fallopian tubes, and the ovaries. The enlarged picture shows the transition from ecto- to endocervix and endometrium. The mucosal epithelia are composed of squamous epithelial cells in the vagina and ectocervix, while columnar epithelial cells are found in the endocervix, uterus and fallopian tubes. *C. trachomatis* infects the columnar epithelium of the cervix. The vagina and ectocervix are covered by mucus. Mucus is part of the natural barrier system of the FGT and facilitates antibody mediated entrapment of microbes (Lamont, 1992; Saltzman *et al*., 1994). Both cellular and humoral systemic and local immunity can play a role in protection against a *C. trachomatis* infection. Graphical illustration by Karen Smith Korsholm.

an inverse correlation between the shedding of *C. trachomatis* and immunoglobulin A (IgA) levels in cervical secretions (Brunham *et al*., 1983). In support of this, Darville *et al.* found that in naturally infected women, serum and cervical IgG and IgA levels correlated inversely with cervical bacterial burden. Interestingly, this effect of antibodies was insufficient to prevent ascension of the infection (Darville *et al*., 2019).

Importantly, as pointed out by Plotkin (Plotkin, 2008), the mechanism of protection is not necessarily the mechanism of recovery from infection. A *C. trachomatis* vaccine needs to do better than the natural infection. *C. trachomatis* vaccine studies in animal models have demonstrated that CD4 and CD8 T cells can provide significant protection against bacterial shedding (Hansen *et al*., 2008; Igietseme *et al*., 1994; Meoni *et al*., 2009; Olivares-Zavaleta *et al*., 2014; Olsen *et al*., 2014; Olsen *et al*., 2010; Starnbach *et al*., 2003). Adoptive transfer of vaccine-induced neutralizing serum markedly reduced early bacterial shedding (O'Meara *et al*., 2016; Olsen *et al*., 2015; Olsen *et al*., 2017), pointing to a crucial role for antibodies in neutralizing/

controlling the initial inoculum. Passive transfer of serum from mice immunized with native MOMP is as protective as serum from mice infected with *C. muridarum* (Farris *et al.*, 2010). Also, inoculation of mice with neutralizing monoclonal antibodies can limit chlamydial infections (Cotter *et al.*, 1995; Pal *et al.*, 1997b). Thus, vaccines can protect via a range of immunological effector mechanisms. Recently, a consensus in the field was reached that *C. trachomatis* vaccines should aim at generating a CD4 Th1 cell response producing IFN-γ as well as a strong neutralizing antibody response (de la Maza *et al.*, 2017; Zhong *et al.*, 2017). The necessity of *Chlamydia*-specific local immunity at the time of infection is still not known. Local memory immunity will have the benefit of being able to induce a faster recall immune response upon infection than a systemic response (Fig. 15.1). Therefore many studies in animals models have focused on different immunization strategies that can lead to a strong systemic immune response in combination with the induction of local memory immunity in the genital tract (see below).

Routes of immunization

The most effective way to induce immune responses at a specific effector site is to locally administer the vaccine, or perhaps, stimulate sites with related lymph drainages (Carmichael *et al.*, 2011; Kelly *et al.*, 1996; Pal *et al.*, 1994; Parr *et al.*, 1988; Ralli-Jain *et al.*, 2010; Thapar *et al.*, 1991). Therefore, the intravaginal route may be the most appropriate for eliciting protection against *C. trachomatis*. Intravaginal immunization using gels has been used against cholera and HIV (Barnett *et al.*, 2008; Borsutzky *et al.*, 2006; Holmgren and Czerkinsky, 2005; Levine, 2010; Plotkin *et al.*, 2013; van Ginkel *et al.*, 2000). Limitations of this route are patient acceptance and possible side effects.

Several reports support the link between the mucosal immune system of the FGT, the upper respiratory airways, the gastrointestinal tract and other mucosal sites. This has prompted testing of a range of different immunization routes. Oral immunization is readily acceptable (Holmgren and Czerkinsky, 2005). A good example is the live polio vaccine. Yet, for subunit vaccines, the need for a successful delivery system and the large amounts of antigen required with oral delivery limit this approach. The intranasal, sublingual and colonic routes have also been explored (Becker *et al.*, 2006; Carmichael *et al.*, 2011; Hasegawa *et al.*, 2005; McConnell *et al.*, 2008; Pal *et al.*, 1994; Ralli-Jain *et al.*, 2010; Song *et al.*, 2008). The intranasal route seems to be the most effective in induction of mucosal antibody responses (Bergquist *et al.*, 1997; Gallichan and Rosenthal, 1995; Johansson *et al.*, 1998, 2001; Pal *et al.*, 1996). Work by Stary *et al.* (2015) supports enhanced protection using a mucosal route for immunization. These authors vaccinated mice with UV-inactivated *C. trachomatis* serovar D EBs by the intranasal or intrauterine routes using nanoparticles coupled to the adjuvant R848 (Resiquimod). Mucosal immunization elicited tissue resident memory cells that migrated to the uterine mucosa eliciting protection against a transcervical challenge. Safety and acceptability may, however, be a concern. For example, intranasal immunization using toxin-based adjuvants have been associated with a negative effects on the central nervous system (Lewis *et al.*, 2009; Mutsch *et al.*, 2004; van Ginkel *et al.*, 2000).

Combinations of parenteral and mucosal routes have also been thoroughly investigated. Using the major outer membrane protein (MOMP), Ralli-Jain *et al.* (2010) tested mucosal (sublingual and colonic) and systemic (intramuscular and subcutaneous) routes, alone and in combination, for their ability to protect mice against a *C. muridarum* respiratory challenge. The best protection was achieved when mucosal, followed by systemic immunizations, were utilized. Carmichael *et al.* (Carmichael *et al.*, 2011) also utilized combinations of mucosal (intravaginal, colonic, and intranasal) and systemic (intramuscular and subcutaneous) routes to immunize mice with MOMP against a vaginal challenge with *C. muridarum*. The strongest humoral and cell-mediated immune responses were observed in mice immunized by a combination of mucosal and systemic routes. Furthermore, based on vaginal shedding and fertility rates, mice vaccinated via the combined mucosal and systemic routes were the best protected. Other studies also showed the advantages of combining parenteral and nasal immunization to modulate immunity and cellular localization to the lung and the FGT (Christensen *et al.*, 2017; Ciabattini *et al.*, 2015; Fiorino *et al.*, 2013; Wern *et al.*, 2017). Lorenzen *et al.* used mini-pigs to study intramuscular priming with the CAF01 adjuvant followed by intranasal boosting with a subunit vaccine (Lorenzen *et al.*, 2015). This protocol induced strong genital immunity by combining systemic cellular immunity and IgG with local secretory IgA. Interestingly, in this study, nasal boosting with unadjuvanted antigen was sufficient, thus minimizing possible safety issues with adjuvanted vaccines via the nasal route.

Nevertheless, the success of the HPV vaccine challenges the concept that protection against mucosal pathogens requires induction of a local immune response at the site of entry. Compared to other mucosal secretions, the cervicovaginal area contains mostly IgG. The HPV vaccine is delivered intramuscularly, and systemic neutralizing IgG is assumed to be the main immune component that protects against this pathogen (Stanley, 2010). A similar role for antibodies was demonstrated for *C. trachomatis* by Olsen *et al.* in the murine model (Olsen *et al.*, 2015; Olsen *et al.*, 2017). A range of studies in both mice, guinea pigs and NHP also support the efficacy of parenteral vaccination against *C. trachomatis* (Batteiger *et al.*, 1993; Coler *et al.*, 2009; de Jonge *et al.*, 2011; Eko *et al.*, 2004; Kari *et al.*, 2009; Knudsen *et al.*, 2016; Olsen *et al.*, 2014; Paes *et al.*, 2016; Pal *et al.*, 2005b; Yu *et al.*, 2014) and novel adjuvants/vectors under development show the promise of eliciting long lasting protective immunity (Igietseme *et al.*, 2011). In conclusion, for a complex mucosal pathogen like *C. trachomatis*, a range of immunological responses have the potential to be protective, and only future clinical trials will enable determination of a correlate of protection in humans.

Experimental *C. trachomatis* vaccines

For an overview of vaccine antigens tested in different animal models, see Table 15.3.

Whole-cell vaccines for genital chlamydial infection

In the last five decades, following the human and NHP trachoma clinical trials, live and inactivated whole-cell vaccines have extensively been studied in the mouse model to protect against genital and respiratory challenges (Brunham and Rappuoli, 2013; de la Maza *et al.*, 2017; Farris and Morrison, 2011; Phillips *et al.*, 2019; Zhong *et al.*, 2017). Pioneering studies were performed by M. Tuffrey and D. Taylor-Robinson using various *C. trachomatis* serovars in the mouse model (Knight *et al.*, 1995; Stagg *et al.*, 1998; Taylor-Robinson and Tuffrey, 1981; Tuffrey *et al.*, 1985, 1986a b, 1990a,b, 1992). Among their conclusions, they reported that *C. trachomatis* isolates could not induce upper genital pathology in mice when delivered vaginally, but inoculation in the ovarian bursa or in the uterine horns resulted in infertility if mice were pretreated with medroxyprogesterone to stop the oestrus cycle in dioestrus. In addition, they concluded that protection against a genital challenge was associated with recruitment of MHC-Class II antigen presenting cells to the uterine tissue.

Subsequently, other investigators tested whole cell vaccines in mice. For example, Cui *et al.* used live *C. trachomatis* LGV2 EB to orally immunize mice against both respiratory and vaginal challenges and showed a decrease the bacterial load (Cui *et al.*, 1989, 1991). Intranasal vaccination with live *C. trachomatis* serovar E was also found to protect against vaginal challenge (Peterson *et al.*, 1999). Protection against infertility was first demonstrated by vaccinating mice intranasally with live *C. muridarum* (Pal *et al.*, 1996, 2005b). Mice immunized via the intranasal, oral, vaginal or subcutaneous routes using live *C. muridarum* showed significant protection against shedding following a vaginal challenge (Kelly *et al.*, 1996; Ralli-Jain *et al.*, 2010).

In the *C. muridarum* male mouse model, intranasal and urethral vaccination with live EBs have also shown robust protection against a urethral challenge (Pal *et al.*, 2004; Pal *et al.*, 2010). In vaccinated animals, cultures of the urethra, urinary bladder, epididymis and testes showed that the number of mice with positive cultures and the number of positive organs were both lower. Furthermore, the length of time that the mice had positive cultures and the total number of *C. muridarum* IFU recovered were significantly lower in vaccinated mice. Overall, the protection was more robust in mice immunized in the urethra than intranasally.

C. muridarum UV-inactivated EBs were not as protective as live EBs, and immunization with reticulate bodies was not protective (Lu *et al.*, 2012; Pal *et al.*, 1999b; Yu *et al.*, 2011). UV-inactivated *C. trachomatis* serovar E EBs, adjuvanted with AS01B,

Table 15.3 Vaccine research studies performed to develop *C. trachomatis* vaccines

Type of vaccine	Description	Animal model	Challenge strain	Reference
Live				
Viable EB	Viable EB	Mouse	*C. muridarum*	Eko *et al.* (2014); O'Connell *et al.* (2007); Pal *et al.* (1994, 1996, 1999b, 2003b, 2005a, 2010, 2015); Su *et al.* (1998, 2000); Wang *et al.* (2009); Yu *et al.* (2011)
		Mouse	*C. trachomatis* SvE	Peterson *et al.* (1999)
		Mouse	*C. trachomatis* SvL2	Cui *et al.* (1989); Cui *et al.* (1991)
		Guinea pig	*C. trachomatis* SvD	de Jonge *et al.* (2011)
Inactivated/attenuated				
UV-inactivated EB	UV-inactivated EB	Mouse	*C. trachomatis* SvK	Coler *et al.* (2009)
		Mouse	*C. trachomatis* SvL2	Stary *et al.* (2015)
		Mouse	*C. muridarum*	Lu *et al.* (2002)
Heat-killed EB	Heat-killed EB	Mouse	*C. muridarum*	Yu *et al.* (2012); Su *et al.* (1998)
Plasmid cured EB		Mouse	*C. muridarum*	Lei *et al.* (2014); O'Connell *et al.* (2007)
		Mouse	*C. trachomatis* SvD	Olivares-Zavaleta *et al.* (2010)
		Non-human primates	*C. trachomatis* SvA A2497	Kari *et al.* (2011); Olivares-Zavaleta *et al.* (2014)
Subunit vaccines				
Surface exposed				
COMC	*Chlamydia* outer membrane complex	Mouse	*C. muridarum*	Pal *et al.* (1997a)
Native MOMP:TC0052	Major outer membrane protein	Mouse	*C. muridarum*	Cheng *et al.* (2011); Farris *et al.* (2010); Pal *et al.* (1999a, 2001, 2003a, 2005b, 2006, 2015, 2017b); Sun *et al.* (2009); Tifrea *et al.* (2011, 2014)
Native MOMP:CTA0742	Major outer membrane protein	Non-human primates	*C. trachomatis* SvA	Kari *et al.* (2009)
Native MOMP:CTB_6861	Major outer membrane protein	Non-human primates	*C. trachomatis* SvB	Taylor *et al.* (1988)
Native MOMP:CCA00047	Major outer membrane protein	Guinea pig	*C. caviae*	Batteiger *et al.* (1993)

Recombinant MOMP:TC0052	Major outer membrane protein	Mouse	*C. muridarum*	Carmichael *et al*. (2011); Champion *et al*. (2009); Cheng *et al*. (2014b); Cunningham *et al*. (2009); Dong-Ji *et al*. (2000); Hansen *et al*. (2008); Hickey *et al*. (2009, 2010); O'Meara *et al*. (2013, 2016); Pal *et al*. (2017b); Ralli-Jain *et al*. (2010); Shaw *et al*. (2002); Singh *et al*. (2006); Sobinoff *et al*. (2015); Sun *et al*. (2009); Tifrea *et al*. (2011, 2103); Zhang *et al*. (1997)
Recombinant MOMP:CT681	Major outer membrane protein	Mouse	*C. trachomatis* SvF	Tuffrey *et al*. (1992)
Recombinant MOMP:CT681	Major outer membrane protein	Mouse	*C. trachomatis* SvD	Badamchi-Zadeh *et al*. (2016); Knudsen *et al*. (2016); Olsen *et al*. (2014)
Recombinant MOMP:CT681	Major outer membrane protein	Mouse	*C. trachomatis* SvK	Coler *et al*. (2009)
Recombinant MOMP:CCA00047	Major outer membrane protein	Guinea pig	*C. caviae*	Andrew *et al*. (2011)
Recombinant MOMP:CT681	Major outer membrane protein	Pig	*C. trachomatis* SvE	Schautteet *et al*. (2011)
Fragments of nine Pmps	Polymorphic membrane proteins	Mouse	*C. muridarum*	Pal *et al*. (2017a)
PmpA: CT412	Polymorphic membrane protein A	Mouse	*C. muridarum*	Muller *et al*. (2017)
PmpC:CT414	Polymorphic membrane protein C	Guinea pig	*C. caviae*	Inic-Kanada *et al*. (2015)
PmpD:CT812	Polymorphic membrane protein D	Mouse	*C. trachomatis* SvD	Paes *et al*. (2016)
PmpD:CT812	Polymorphic membrane protein D	Mouse	*C. trachomatis* SvK	Coler *et al*. (2009)
PmpE/F-1:TC0261	Polymorphic membrane protein E	Mouse	*C. muridarum*	Yu *et al*. (2012)
PmpE/F-2:TC0262	Polymorphic membrane protein F	Mouse	*C. muridarum*	Yu *et al*. (2012)
PmpG:TC0263	Polymorphic membrane protein G	Mouse	*C. muridarum*	Jiang *et al*. (2017); Yu *et al*. (2009, 2012)
PmpH:CT872	Polymorphic membrane protein H	Mouse	*C. trachomatis* SvK	Coler *et al*. (2009)
PmpH:TC0264	Polymorphic membrane protein H	Mouse	*C. muridarum*	Yu *et al*. (2012)
TC0313	Hypothetical protein	Mouse	*C. muridarum*	Meoni *et al*. (2009)
CT043	Hypothetical protein	Mouse	*C. trachomatis* SvD	Olsen *et al*. (2014)
Hirep1 (extVD4^{D,E,F}): CT681	Major outer membrane protein	Mouse	*C. trachomatis* SvD	Olsen *et al*. (2015, 2017)
CTH522 (CT681^{D}extVD4^{D,E,F,G}): CT681	Major outer membrane protein	Mouse	*C. trachomatis* SvD	Olsen *et al*. (2015)
MOMP^{D}+MOMP^{J}+MOMP^{F}:CT681	Major outer membrane proteins D, J, F	Mouse	*C. trachomatis* SvD	Karunakaran *et al*. (2015)
VD2/VD4: CT681	Major outer membrane protein	Mouse	*C. trachomatis* SvD	Hadad *et al*. (2016)
A8-VD4: CT681	Major outer membrane protein	Mouse	*C. trachomatis* SvD	Su *et al*. (1995)

Table 15.3 Continued

Type of vaccine	Description	Animal model	Challenge strain	Reference
MOMPmultiepitope:CT681	Major outer membrane protein	Mouse	*C. trachomatis* SvE	Tu *et al*. (2014)
MOMPepitope:CT681	Major outer membrane protein	Mouse	*C. trachomatis* SvE	Jiang *et al*. (2015)
Intracellular antigens				
CPAF	Chlamydial protease-like activity factor	Mouse	*C. muridarum*	Brown *et al*. (2012); Chaganty *et al*. (2010); Cong *et al*. (2007); Li *et al*. (2007, 2010; Lu *et al*. (2012); Murphey *et al*. (2006); Murthy *et al*. (2007, 2009)
		Mouse	*C. trachomatis* SvD	Olsen *et al*. (2014)
		Mouse	*C. trachomatis* SvK	Coler *et al*. (2009)
		Guinea pig	*C. trachomatis* SvD	Wali *et al*. (2017)
Other antigens				
gatB:CT004	Glu-tRNA Gln amidotransferase subunit B	Mouse	*C. trachomatis* SvD	Olsen *et al*. (2014)
lcrE:CT089	Putative regulatory molecule for type III pathway (low calcium response E)	Mouse	*C. trachomatis* SvK	Coler *et al*. (2009)
IncA:CT119	Inclusion membrane localized protein	Mouse	*C. muridarum*	Li *et al*. (2007)
CT144	Hypothetical protein	Mouse	*C. trachomatis* SvD	Picard *et al*. (2012)
rpoB:CT315	DNA-dependent RNA polymerase beta subunit	Mouse	*C. trachomatis* SvK	Coler *et al*. (2009)
tuf:CT322	Elongation factor Tu	Mouse	*C. trachomatis* SvK	Coler *et al*. (2009)
SrpA:CT442	*Chlamydia*-specific hypothetical cysteine-rich 15 kD outer membrane protein	Mouse	*C. trachomatis* SvL2	Starnbach *et al*. (2003)
OmcB:CT443	Cysteine rich outer membrane protein	Mouse	*C. trachomatis* SvD	Olsen *et al*. (2014)
Swib:CT460	SWIB (YM74) complex protein	Mouse	*C. trachomatis* SvK	Coler *et al*. (2009)
lpdA:CT557	Pyruvatedehydrogenase E3 component	Mouse	*C. trachomatis* SvK	Coler *et al*. (2009)
CT610	*Chlamydia* protein associating with death domains (CADD)	Mouse	*C. trachomatis* SvD	Olsen *et al*. (2014)
CT622	Hypothetical protein	Mouse	*C. trachomatis* SvK	Coler *et al*. (2009)
ppA:CT772	Inorganic pyrophosphatase (PPASE)	Mouse	*C. trachomatis* SvK	Coler *et al*. (2009)
htrA: CT823	Periplasmic serine protease	Mouse	*C. trachomatis* SvD	Picard *et al*. (2012)

NrdB:CT828	Ribonucleoside-diphosphate reductase, beta subunit	Mouse	*C. trachomatis SvD*	Barker *et al*. (2008)
CT875	Hypothetical protein	Mouse	*C. trachomatis* SvK	Coler *et al*. (2009)
Aasf:TC0707	Anti-anti-sigma factor	Mouse	*C. muridarum*	Yu *et al*. (2012)
TC0106	Hypothetical protein	Mouse	*C. muridarum*	Finco *et al*. (2011)
htrA:TC0210	Serine protease	Mouse	*C. muridarum*	Finco *et al*. (2011)
TC0285	Hypothetical protein	Mouse	*C. muridarum*	Yu *et al*. (2012)
TC0313	Hypothetical protein	Mouse	*C. muridarum*	Finco *et al*. (2011)
TC0420	Hypothetical protein	Mouse	*C. muridarum*	Yu *et al*. (2012)
GlgP:TC0519	Glycogen phosphorylase	Mouse	*C. muridarum*	Li *et al*. (2012)
AtpB:TC0581	ATP synthase, subunit B	Mouse	*C. muridarium*	Cheng *et al*. (2014a)
AtpA:TC0582	ATP synthase, subunit A	Mouse	*C. muridarium*	Cheng *et al*. (2014a)
AtpE:TC0584	ATP synthase, subunit E	Mouse	*C. muridarium*	Cheng *et al*. (2014a)
Tarp:TC0741	Hypothetical protein	Mouse	*C. muridarum*	Finco *et al*. (2011); Inic-Kanada *et al*. (2015); Wang *et al*. (2009); Yu *et al*. (2012)
RplF:TC0801	Ribosomal protein L6	Mouse	*C. muridarum*	Yu *et al*. (2009); Yu *et al*. (2010, 2012)
Mip:TC0828	Peptidyl-prolyl cis–trans isomerase	Mouse	*C. muridarum*	Hu *et al*. (2013)
TC0825	Hypothetical protein	Mouse	*C. muridarum*	Yu *et al*. (2012)
Plasmid antigens				
PgP3:TCA04	Plasmid virulence factor PGP3	Mouse	*C. muridarum*	Mosolygo *et al*. (2014)
PgP3:pCTD0005	Plasmid virulence factor PGP3	Mouse	*C. trachomatis* SvD	Donati *et al*. (2003)
Pgp4:TCA05	Plasmid virulence factor PGP4	Mouse	*C. muridarum*	Mosolygo *et al*. (2014)
Multisubunit vaccines				
CTH1 (rplP-omcB):CT521-CT443	5S ribosomal protein L16—cysteine rich outer membrane protein	Mouse	*C. muridarum*	Olsen *et al*. (2010)
PmpE/F-2:TC0262 PmpG-1:TC0263 MOMP:TC0052	Polymorphic membrane protein F Polymorphic membrane protein G Major outer membrane protein	Mouse	*C. muridarum*	Yu *et al*. (2010)
PmpE-H:TC0261-4 Tarp:TC0741	Polymorphic membrane protein E–H Translocated actin-recruiting phosphoprotein	Mouse	*C. muridarum*	Yu *et al*. (2012)

Table 15.3 Continued

Type of vaccine	Description	Animal model	Challenge strain	Reference
PmpE-H:TC0261-4	Polymorphic membrane proteins E–H	Mouse	*C. muridarum*	Yu *et al.* (2014)
PmpE-H:TC0261-4	Polymorphic membrane protein E–H	Mouse	*C. muridarum*	Yu *et al.* (2014)
MOMP:TC0052	Major outer membrane protein			
htrA:TC0210	Serine protease	Mouse	*C. muridarum*	Finco *et al.* (2011)
TC0313	Hypothetical protein			
TC0106	Hypothetical protein	Mouse	*C. muridarum*	Finco *et al.* (2011)
TC0431	MAC/perforin family protein			
TC0551	NADH:ubiquinone oxidoreductase	Mouse	*C. muridarum*	Finco *et al.* (2011)
TC0890	Hypothetical protein			
TC0660	Periplasmic amino acid-binding protein	Mouse	*C. muridarum*	Finco *et al.* (2011)
TC0741	Hypothetical protein			
aaxA:TC0651	Porin AaxA	Mouse	*C. muridarum*	Finco *et al.* (2011)
OmcB:TC0727	Cysteine rich outer membrane protein			
TC0106	Hypothetical protein	Mouse	*C. muridarum*	Finco *et al.* (2011)
htrA:TC0210	Serine protease			
TC0313	Hypothetical protein			
TC0741	Hypothetical protein			
TC0106	Hypothetical protein	Mouse	*C. muridarum*	Finco *et al.* (2011)
TC0431	MAC/perforin family protein			
TC0551	NADH:ubiquinone oxidoreductase			
TC0890	Hypothetical protein			
NrdB:TC0215	Ribonucleoside-diphosphate reductase, beta subunit	Mouse	*C. muridarum*	Carey *et al.* (2011)
MOMP:TC0052	Major outer membrane protein			

Aasf:TC0707	Anti-anti-sigma factor	Mouse	*C. muridarum*	Yu *et al.* (2012)
RplF:TC0801	Ribosomal protein L6			
Rec0:TC0755	DNA repair protein			
TC0420	Hypothetical protein	Mouse	*C. muridarum*	Yu *et al.* (2012)
TC0825	Hypothetical protein			
TC0285	Hypothetical protein			
atpA: TC0582	ATP synthase, subunit A	Mouse	*C. muridarum*	Cheng *et al.* (2014a)
atpD: TC0580	ATP synthase, subunit D			
atpA:TC0582	ATP synthase, subunit A	Mouse	*C. muridarum*	Cheng *et al.* (2014a)
atpB:TC0581	ATP synthase, subunit B			
atpA:TC0582	ATP synthase, subunit A	Mouse	*C. muridarum*	Cheng *et al.* (2014a)
MOMP:TC0052	Major outer membrane protein			
CopB:CT578	Type III secretion system translocator CopB	Mouse	*C. muridarum*	Bulir *et al.* (2016)
CopD:CT579	Type III secretion system translocator CopC			
CT584	Hypothetical protein			
MOMP:TC0052	Major outer membrane protein	Mouse	*C. muridarum*	O'Meara *et al.* (2017)
PmpG:TC0263	Polymorphic membrane protein			
TC0873	Hypothetical protein			
TC0500	Hypothetical protein			
MOMP-OMP2:CT681-CT443	Major outer membrane protein – cysteine rich outer membrane protein	Mouse	*C. trachomatis* SvD	Eko *et al.* (2004)
MOMP:CT681	Major outer membrane protein	Mouse	*C. trachomatis* SvK	Coler *et al.* (2009)
CT875	Hypothetical protein			
PmpD:CT812	Polymorphic membrane protein	Mouse	*C. trachomatis* SvD	Eko *et al.* (2011); Pais *et al.* (2017)
PorB:CT713	Outer membrane protein			

Table 15.3 Continued

Type of vaccine	Description	Animal model	Challenge strain	Reference
PmpE to H:CT869 to 72	Polymorphic membrane proteins E–H	Mouse	*C. trachomatis* SvD	Karunakaran *et al*. (2015)
$MOMP^{D}+MOMP^{J}+MOMP^{F}$:CT681	Major outer membrane proteins D, J, F			
Hirep1 ($extVD4^{D,E,F}$):CT681	Major outer membrane protein derivatives	Mouse	*C. trachomatis* SvD	Wern *et al*. (2017)
CTH93:CT043-CT414-CT681	Hypothetical protein – outer membrane protein C – major outer membrane protein			
Hirep1 ($extVD4^{D,E,F}$):CT681	Major outer membrane protein	Minipig	*C. trachomatis* SvD	Lorenzen *et al*. (2015)
CTH93:CT043-CT414-CT681	Hypothetical protein – outer membrane protein C – major outer membrane protein			
Hirep1 ($extVD4^{D,E,F}$):CT681	Major outer membrane protein	Minipig	*C. trachomatis* SvD	Boje *et al*. (2016)
CTH93:CT043-CT414-CT681	Hypothetical protein – outer membrane protein C – major outer membrane protein			

EB, elementary body; L2, lymphogranuloma venerum 2; Sv, serovar; UV, ultraviolet light; VD, variable domain.

were shown to induce cross protection against vaginal and intrauterine challenges with serovar K (Coler *et al.*, 2009). Adoptive transfer of dendritic cells pulsed *ex vivo* with inactivated *C. muridarum* EBs elicited strong protection against a vaginal challenge (Lu and Zhong, 1999; Su *et al.*, 1998). The ability of some whole inactivated EB preparations to induce protection has encouraged further exploration of this vaccination approach.

Live attenuated EBs have also been tested in mice. For example, mice infected vaginally with a plasmid-cured isolate of *C. muridarum*, when challenged with wild type *C. muridarum*, still shed chlamydiae but had less upper genital pathology than control animals who had been infected with wild type EBs (Lei *et al.*, 2014; O'Connell *et al.*, 2007). Plasmid-cured *C. muridarum* inoculated into the ovarian bursa gave similar results (Lei *et al.*, 2014). Vaccination with a plasmid-cured *C. trachomatis* serovar L2 decreased shedding against a vaginal challenge with serovar D (Olivares-Zavaleta *et al.*, 2010).

Whether or not these attenuated isolates can be used in humans is still under investigation. For example, Qu *et al.* (Qu *et al.*, 2015) infected, in the uterine cervix, Rhesus macaques with wild type or plasmid-deficient *C. trachomatis* serovar D (UW-3Cx). No differences in levels of immunity, infection or genital tract pathology where observed indicating that the plasmid is not required for *C. trachomatis* to colonize the genital tract of NHPs, which would argue against the use of a plasmid-deficient strain as an 'attenuated' vaccine for genital tract infections. Also, the finding of plasmid-less and plasmid-mutated strains in human *C. trachomatis* infections calls for caution in the use of these 'attenuated' isolates as vaccines (Peterson *et al.*, 1990; Sigar *et al.*, 2014).

In conclusion, live *C. trachomatis* and *C. muridarum* vaccines have been found to be highly protective against genital and respiratory challenges in mice. Some of these vaccines protect not only against vaginal shedding but also against long-term sequelae, such as infertility. The experience with live attenuated vaccines is more limited and have, so far, given mixed results depending on the animal model tested. The efficacy of immunization with whole inactivated EBs likely depends in great part on the types of adjuvants used. In our opinion, it is unlikely that any of these whole-cell vaccines will be implemented in humans to protect against genital tract infections. Whether or not they could be explored to prevent trachoma is still under debate. A live attenuated vaccine showed promising results in a recent NHP experiment, providing protection from clinical ocular disease in three out of six animals (Kari *et al.*, 2011). However, the negative effects observed in previous NHP trachoma vaccine trials, possibly linked to a detrimental delayed hypersensitivity response to the HSP60 protein (Lichtenwalner *et al.*, 2004; Morrison *et al.*, 1989a,b; Patton *et al.*, 1994), needs to be addressed before human testing. Thus, both safety concerns and challenges manufacturing a whole cell *Chlamydia* vaccine to industry standards [Good Manufacturing Practices (GMP)] are major obstacles and has prompted the search for a subunit vaccine.

Subunit vaccines

Surface exposed antigens

Major outer membrane protein

The chlamydial major outer membrane protein, (MOMP), is a primary target of vaccine development because it is highly antigenic, contains B- and T-cell epitopes and constitutes ~60% of the outer membrane protein mass (Caldwell *et al.*, 1981; de la Maza and Peterson, 2002). MOMP is considered a member of the general porin class of proteins, a group important for the passive transport of ions, sugars and nucleotides across the outer membrane of Gram-negative bacteria (Nikaido, 1992, 1994). Porins have a structural topology comprised of antiparallel β-strands spanning the outer membrane, a water-filled inner channel, tight β-turns extending into the periplasmic region and flexible loops reaching beyond the extracellular surface. Several studies have confirmed that *C. trachomatis* MOMP shares the characteristics of other Gram-negative bacterial porins; it has a monomeric molecular weight of 39.5 kDa, a hydrophobic residue content of ~40% and migration patterns on SDS-PAGE gels consistent with trimeric oligomerization (Sun *et al.*, 2007).

Results from the trachoma vaccine trials using whole-organisms indicated that protection against disease and shedding was serovar/serogroup specific. This observation supported the protection results in mice performed by Wang and Grayston

(Schachter and Dawson, 1978; Taylor, 2008; Wang and Grayston, 1984; Wang and Grayston, 1963). DNA sequencing of the *C. trachomatis* genome and phylogenetic analysis determined that MOMP probably accounted for the serovar/serogroup-specific protection observed during the trachoma vaccine trials (Fitch *et al.*, 1993; Stephens *et al.*, 1998; Stephens *et al.*, 1987; Wang and Grayston, 1984).

Tuffrey *et al.* immunized mice with rMOMP from *C. trachomatis* serovar L1, challenged them in the oviducts with serovar F (strain NI.1) and observed a decrease in vaginal shedding, but no protection against infertility (Tuffrey *et al.*, 1992). The first evidence that a subunit *Chlamydia* vaccine could protect against a genital challenge was reported in 1997 (Pal *et al.*, 1997a). Mice immunized using the *C. muridarum* chlamydial outer membrane complex (COMC) were protected against a long-term sequelae, as determined by the reduced vaginal shedding of chlamydiae and improved fertility rates (Pal *et al.*, 1997a). Subsequently, mice vaccinated with native trimeric MOMP (nMOMP) isolated from EBs, with incomplete Freund's adjuvant were shown to be protected against shedding and infertility (Pal *et al.*, 2001). Pal *et al.* vaccinated mice with nMOMP using CpG plus Montanide ISA 720 as adjuvants and showed very robust protection against an intrabursal challenge with *C. muridarum*. Animals immunized with nMOMP were as protected against shedding and infertility as those immunized intranasally with live *C. muridarum* EBs (Pal *et al.*, 2005b; Pal *et al.*, 2015). Furthermore, based on measurements of body weight, lung weight and number of IFU recovered from the lungs, nMOMP was significantly more protective than recombinant MOMP (rMOMP) against a *C. muridarum* respiratory challenge, suggesting that the tertiary structure of the antigen was important for protection. In the vaginal model, protection induced by nMOMP requires both CD4+ T cells and antibodies (Farris *et al.*, 2010; Sun *et al.*, 2009; Tifrea *et al.*, 2011). Additionally, vaccination of non-human primates with *C. trachomatis* serovar A (strain A2497) nMOMP formulated with CpG-2395 and Montanide ISA 720 protected against an ocular challenge with the same serovar, as shown by a significant decrease in ocular shedding, although no protection was observed against inflammatory responses (Kari *et al.*, 2009).

Several investigators have used *C. muridarum* rMOMP to protect against a genital challenge (Carmichael *et al.*, 2011; Hansen *et al.*, 2008; Hickey *et al.*, 2010; Sun *et al.*, 2009). For example, Hansen *et al.* (Hansen *et al.*, 2008) immunized mice with rMOMP and the cationic liposome 1 (CAF01) and, as determined by a reduced chlamydial load, obtained protection against vaginal shedding but not against pathology. Hickey *et al.* (Hickey *et al.*, 2010) administered rMOMP orally, and measured significant protection against vaginal shedding, although *C. muridarum* was still recovered at 18 days post infection. No protection against pathology was observed. The authors proposed that in the rMOMP-immunized animals the lack of protection against tissue pathology was due to the influence of the treatment with medroxyprogesterone (Depo-provera) which biases the immune responses to Th2 and therefore makes animals more susceptible to a *Chlamydia* infection (Hickey *et al.*, 2010). For the first time, Carmichael *et al.* (Carmichael *et al.*, 2011) using *C. muridarum* rMOMP, CpG-1826 plus Montanide ISA 720 as adjuvants and a combination of mucosal and systemic routes for immunization showed protection against vaginal shedding and infertility using a recombinant chlamydial antigen.

A multivalent vaccine containing the MOMP variable domain VD4 and surrounding constant regions from *C. trachomatis* serovars D, E and F (Hirep1) was tested in mice using CAF01 as adjuvant (Olsen *et al.*, 2015) (Table 15.3). Some of the immune responses were serovar-specific and others to the conserved VD4 epitope (LNPTIAG). Sera from immunized mice neutralized serovars D, E, and F *in vitro*. As determined by the number of *C. trachomatis* serovar D IFU detected in vaginal cultures and the inflammatory responses in the upper genital tract, the mice were protected. $CD4^+$ T cells and antibodies were found to be critical for protection (Olsen *et al.*, 2015). Vaccine-induced neutralizing antibodies transferred to naïve and Rag1 mice (T and B cell deficient) were able to significantly control the initial infectious load compared to control mice (Olsen *et al.*, 2015; Olsen *et al.*, 2017). A multivalent MOMP–based construct (CTH522) (Olsen *et al.*, 2015) has completed a clinical phase 1 trial (Abraham *et al.*, 2019). A vaccine combining Hirep1 and the CT043 and CT414 antigens elicited neutralizing antibodies and weak protection against vaginal shedding in mini-pigs challenged with serovar D (Boje *et al.*, 2016). Hadad *et al.* (Hadad *et al.*, 2016) used a VD2/VD4 construct of *C. trachomatis* serovar E MOMP adjuvanted with cholera toxin to protect mice against a vaginal challenge with a clinical serovar D isolate (Hadad *et al.*, 2016). The vaccine was reported to elicit protection against vaginal shedding and infertility. However, these results need to be confirmed because there is no evidence that a vaginal challenge with serovar D induces infertility.

O'Meara *et al.* vaccinated male and female mice four times intranasally at one week intervals with recombinant *C. muridarum* MOMP and the adjuvant Iscomatrix (O'Meara *et al.*, 2016). Male mice were challenged with 10^6 *C. muridarum* IFU. Depo-provera-treated females were challenged using prostate ejaculates (containing ~50 *C. muridarum* IFU) collected from *C. muridarum*-infected male mice. Immunization of female and male mice reduced the chlamydial burden and disease development in both sexes following a vaginal or a meatus urethra challenge, respectively. Importantly, only immunized female mice challenged vaginally with prostate ejaculates from immunized, infected males were 100% protected against infection and inflammatory disease of the upper genital tract. O'Meara *et al.* (O'Meara *et al.*, 2016) concluded that vaccination can induce partial protection against infection in females, but a vaccine targeting males may be required to protect females against upper genital tract pathology. This synergism between partial immunity elicited by vaccination in both females and males supports the need to implement a vaccine in both sexes.

Polymorphic membrane proteins

C. trachomatis and *C. muridarum* have nine polymorphic membrane proteins (Pmps) (Henderson and Lam, 2001; Swanson *et al.*, 2009). Chlamydial Pmps are a group of proteins with MWs ranging from ~100–150 kDa that have conserved GGA(I,L,V) and FXXN tetrapeptide motifs in their N-terminal regions (Carrasco *et al.*, 2011; Nunes *et al.*, 2007; Stephens *et al.*, 1987; Tan *et al.*, 2010; Van Lent *et al.*, 2016; Vasilevsky *et al.*, 2016). Pmps have three functional domains including a cleavable *sec*-dependent N-terminal signal for translocation to the periplasm, a passenger domain for secretion or surface localization and a C-terminal β-barrel sequence for outer membrane translocation. All nine members of the Pmp family of *C. trachomatis* serovar E (DK-20) mediate adhesion to human epithelial and endothelial cells (Becker and Hegemann, 2014). The tetrapeptide motifs are essential for Pmp adhesion to human cells (Molleken *et al.*, 2010). Henderson and Lam (Henderson and Lam, 2001) identified the similarity of Pmps with autotransporters in other species. However, the autotransporter mechanism has not been proven for *Chlamydia* Pmps. Furthermore, PmpG contains a poly G tract (nine residues for all non-LGV strains) that, like in other bacteria, could control the phase variable expression of the gene, providing a means by which these organisms can evade host immune responses (Stephens *et al.*, 1998).

Crane *et al.* (Crane *et al.*, 2006) determined that antibodies to PmpD neutralize all *C. trachomatis* serovars *in vitro*. Several laboratories have since reported protection in various animal models using Pmps as antigens. For example, mice vaccinated with PmpD from *C. trachomatis* serovar E using a TLR4 agonist as adjuvant protected against a challenge with serovar D (Paes *et al.*, 2016). Vaccination of C57BL/6, BALB/c and C3H/HeN mice with N-terminal fragments from *C. muridarum* Pmps (E, F, G and H) conferred protection against a vaginal challenge as determined by accelerated clearance of *C. muridarum* (Yu *et al.*, 2012). A polyvalent recombinant vaccine consisting of the four Pmps from serovar D and MOMPs from serovars D, F and J formulated with CAF01 protected C57BL/6 mice against vaginal shedding following a transcervical challenge with *C. trachomatis* serovar D (Karunakaran *et al.*, 2015). Individual fragments from all nine *C. trachomatis* serovar E Pmps were used to immunize BALB/c mice utilizing CpG and Montanide as adjuvants. As determined by changes in body weight and number of IFU recovered from the lungs, mice challenged intranasally with *C. muridarum* immunized with PmpC were the best protected, followed by those immunized with PmpG, E or H (Pal *et al.*, 2017a). Interestingly, Inic-Kanada *et al.* (Inic-Kanada *et al.*, 2015) reported that *C. trachomatis* serovar E PmpC elicited limited protection in guinea pigs against an ocular challenge with *Chlamydia caviae*. These two experiments suggest that Pmps may be able to elicit cross-species protection.

Intracellular antigens

Chlamydial protease-like activity factor

The chlamydial protease-like activity factor (CPAF) is highly conserved with 99% amino acid identity among the *C. trachomatis* serovars (discussed in more detail in Chapter 8, The Chlamydial Protease CPAF) (Dong *et al.*, 2005; Huang *et al.*, 2008; Tang *et al.*, 2015; Yang *et al.*, 2015, 2016a,b; Zhong *et al.*, 2001). In humans and mice, CPAF is a dominant antigen (Budrys *et al.*, 2012; Wang *et al.*, 2010). Vaccination of mice with *C. muridarum* CPAF, using IL-12 or CpG as adjuvants, shortened the duration of the infection, decreased oviduct pathology and elicited cross-serovar protection (Li *et al.*, 2007; Murthy *et al.*, 2007). The protection elicited by CPAF was not dependent on its conformation. Interestingly, no protection against infertility was observed following a primary vaginal challenge, but there was protection against a secondary *C. muridarum* challenge (Murthy *et al.*, 2011). A T-cell epitope from *C. muridarum* CPAF has also been found to be protective in HLA-DR4 transgenic mice supporting the premise that it may also be protective in humans (Li *et al.*, 2013). Mechanisms of CPAF induced protection are dependent on CD4+ T cells, producing IFN-γ and antibodies with some contribution from CD8+ T-cells (Chaganty *et al.*, 2010; Lu *et al.*, 2012; Murphey *et al.*, 2006).

Plasmid antigens

Most *C. trachomatis* isolates and *C. muridarum* have a plasmid that codes for eight proteins including plasmid glycoprotein 3 (Pgp3) (Palmer and Falkow, 1986; Peterson *et al.*, 1990; Stothard *et al.*, 1998). Crystallization of Pgp3 (MW of 28 kDa) from *C. trachomatis* serovar D showed it to be a ~84 kDa homotrimer (Galaleldeen *et al.*, 2013). Pgp3 was found in the chlamydial envelope and also in the cytoplasm of the host cell (Chen *et al.*, 2010; Li *et al.*, 2008a). Pgp3-deficient *C. trachomatis* and *C. muridarum* isolates were highly attenuated, and Pgp3 may promote chlamydial colonization by neutralizing host immune effectors (Hou *et al.*, 2015; Lei *et al.*, 2014). Sera from humans and mice identified the homotrimer but not the monomer as an immunodominant antigen (Li *et al.*, 2008c; Wang *et al.*, 2010). Donati *et al.* (Donati *et al.*, 2003) vaccinated C3H/HeN mice with a DNA plasmid expressing *C. trachomatis* serovar D Pgp3 and a control group with the same plasmid containing an irrelevant insert. As determined by the number of positive salpinx cultures, mice vaccinated with the Pgp3 plasmid were partially protected against a genital challenge with *C. trachomatis* serovar D. Vaccination of mice with *C. muridarum* Pgp3, using CpG as adjuvant, elicited protection against a genital challenge (Li *et al.*, 2008b).

Although the role of the plasmid in chlamydial pathogenesis is still under investigation, the possibility that some of the plasmid-encoded proteins may be pathogenic casts doubts about their use as a vaccine. In addition, vaccines based only on plasmid antigens such as Pgp3 may not protect against rare *C. trachomatis* strains that lack all or part of the plasmid (Peterson *et al.*, 1990; Ripa and Nilsson, 2007; Stothard *et al.*, 1998).

Multisubunit vaccines

Antigens formulated in a multisubunit vaccine may have synergistic, additive, neutral or antagonistic effects (Finco *et al.*, 2011; Qiu *et al.*, 2008; Yu *et al.*, 2010). Combining antigens increases the number of relevant epitopes, either T or B cell epitopes or a combination of both for a multifaceted vaccine. Multisubunit vaccines can also be specifically designed to target dominant antigens from each developmental form of chlamydiae (extracellular and intracellular) (O'Meara *et al.*, 2017). These so-called multistage vaccines have been exploited within tuberculosis and malaria vaccines (Theisen *et al.*, 2014; Aagaard *et al.*, 2011)

Several antigen combinations have been tested in the mouse model. For example, Finco *et al.* (Finco *et al.*, 2011) identified chlamydial antigens that elicited humoral and cell-mediated immune responses, and combinations of these proteins were tested for their ability to protect mice against an intranasal *C. muridarum* challenge. Mice immunized with the *C. muridarum* proteins TC0106, TC0210, TC0313 or TC0741 adjuvanted with LTK63 + CpG had a 0.5–0.9 $\log_{10}$ reduction in the number of IFU recovered from the lungs. Two four–antigen combinations (TC0106, TC0210, TC0313 and TC0741 and TC0106, TC0431, TC0551 and TC0890) were also tested. The first antigen combination resulted in a 4.1 $\log_{10,}$ reduction and the second combination in a 2.2 $\log_{10}$ reduction in the yield of *C. muridarum* from the lungs, indicative of additive effects. Yu *et al.* (Yu *et al.*, 2010) also found additive effects in a genital model. Three *C. muridarum* proteins, PmpE/F, PmpG, and MOMP, or a combination of them, were formulated with three different adjuvants CpG, Abisco or CAF01 and used to immunize mice. The combination of the three proteins adjuvanted with CAF01, as determined by vaginal shedding, exhibited the best degree of protection against a genital challenge. C57BL/6, BALB/c and C3H/HeN immunized with a multisubunit recombinant vaccine that included PmpE, PmpF, PmpG, PmpH and MOMP, with DDA/MPL as the adjuvant, also conferred better protection against a genital challenge than immunization with single antigens (Yu *et al.*, 2014). Coler *et al.* (Coler *et al.*, 2009) also showed that MOMP, combined with CT875 and adjuvanted with AS01B elicited better protection against a vaginal challenge with *C. trachomatis* serovar K than the individual antigens.

The combination of *C. muridarum* antigens may provide no additional protection. Yu *et al.* (2012) vaccinated mice with PmpE, PmpF, PmpG, Aasf, RplF, TC0420 or TC0825 adjuvanted with DDA-MPL and challenged the animals in the genital tract. PmpG elicited the most robust immune response and the best protection as shown by a decrease in the number of *C. muridarum* IFU recovered. However, three combinations of antigens (PmpE + Pmp F + PmpG + PmpH + Tarp; Aasf + RpIF + Rec0 and TC0420 + TC0825 + TC0285) only protected as well as the best individual protein in the formulation. Similar neutral findings were reported when testing combinations of antigens in the respiratory model (Cheng *et al.*, 2014a). Mice were immunized with components of the *C. muridarum* putative ATP synthase complex, TC0580, TC0581, TC0582, TC0584, or with MOMP. In addition, TC0582 was formulated in combination with TC0580, TC0581 or MOMP. Animals immunized with combinations of two of these three antigens were only protected as well as mice vaccinated with MOMP, the most protective protein in the formulation. Also, Li *et al.* (Li *et al.*, 2007) reported that the addition of CPAF to MOMP, or IncA, from *C. trachomatis* serovar D (UW3-Cx) did not enhance the CPAF-induced effect on *C. muridarum* clearance or oviduct pathology. These results demonstrate the need to evaluate the interactions between antigens of a multisubunit vaccine.

Additional *C. trachomatis* vaccine antigen candidates

Several additional chlamydial antigens have been tested as vaccine candidates (Table 15.3) (Farris and Morrison, 2011; Hafner *et al.*, 2008; Igietseme *et al.*, 2011; Liang *et al.*, 2017; Phillips *et al.*, 2019; Poston and Darville, 2018; Rockey *et al.*, 2009; Vasilevsky *et al.*, 2014; Yu *et al.*, 2016). In the mouse model, some of the antigens found to elicit protection include the outer membrane protein B (OmcB), putative type III secretion effector protein Tarp, macrophage infectivity potentiator (MIP), CopB, CopD, Cap1 and CT584, the inclusion membrane protein IncA, the porin protein PorB, the ribonucleoside reductase NrdB and glycogen phosphorylase (Barker *et al.*, 2008; Bulir *et al.*, 2016; Eko *et al.*, 2011; Eko *et al.*, 2004; Fling *et al.*, 2001; Kawa *et al.*, 2004; Li *et al.*, 2012; Lu *et al.*, 2013; Tsai *et al.*, 2007; Wang *et al.*, 2009).

In conclusion, based on the results of the human trachoma vaccine trials and more recent experiments in various animal models, MOMP is the leading subunit vaccine antigen. Purified nMOMP provides better protection than rMOMP, but cannot be produced in any significant quantities. The most likely difference is that rMOMP has to be correctly refolded into the native trimeric conformation to optimize protection. If this challenge can be overcome, MOMP could be the basis for a highly effective chlamydial subunit vaccine against genital infections. Since MOMP elicits serovar/serogroup protection, MOMP from two or three *C. trachomatis* serovars may be necessary to elicit broad coverage. Alternatively, making a chimeric MOMP from several serovars may also overcome the limitation. Additional antigens, such as Pmp or CPAF, could also be added to a MOMP vaccine to help provide broad cross-serovar protection against genital and ocular infections.

Adjuvants

Attenuated pathogens are very effective vaccines (Marmion, 1980). They cause a mild infection and thereby often long-lived immunity. Whole organism vaccines contain not only antigenic components but also pathogen-associated molecular patterns (PAMPs). These PAMPs activate eukaryotic cell receptors, such as toll-like receptors (TLR), stimulating innate and adaptive host immune responses. Examples of PAMPs include membrane molecules i.e. lipopolysaccharide, flagellin, lipoteichoic acid, porins and intracellular components, such as nucleic acids. These trigger different pattern recognition receptors (PRRs) on i.e. antigen presenting cells and thereby establish the setting for induction of adaptive T- and B-cell responses. Sub-unit vaccines (non-living antigens) are poorly immunogenic on their own and need help to activate the immune system. For these, adjuvants (adjuvare = help) are used. Adjuvants often incorporate synthetic analogues of PAMPs and can be tailored for the desired immune response i.e. provide specific immune responses like Th1, Th2, CD4/CD8 and specific antibody isotypes (Levine, 2010; Lowy and Schiller, 2006; Plotkin *et al.*, 2013; Zuckerman, 2006). Vaccine adjuvants are therefore key to subunit vaccines, and research and development is constantly expanding the repertoire of experimental adjuvants (O'Hagan and Fox, 2015); (Levine, 2010; Lowy and Schiller, 2006; Plotkin *et al.*, 2013; Zuckerman, 2006).

During a *C. trachomatis* infection, TLRs and nucleotide-binding oligomerization domain (NOD)1-mediated activation occur (Bas *et al.*, 2008; Nagarajan *et al.*, 2005; Prebeck *et al.*, 2001). Several chlamydial PAMPs, including LPS, MOMP, macrophage infectivity potentiator (Mip) and heat shock protein-60 (Hsp60), are involved in this process, resulting in signalling mainly through TLR2 and TLR4 (Bas *et al.*, 2008; Ingalls *et al.*, 1995; Massari *et al.*, 2013; Ouburg *et al.*, 2009). However, all these chlamydial PAMPs, including LPS, induce weak immune responses, and this may explain the lack of robust protection elicited by a primary infection (Ingalls *et al.*, 1995).

In mice, CD4 Th1 cells producing IFN-γ and neutralizing antibodies are needed to protect against a chlamydial challenge (Farris and Morrison, 2011). In consequence of this a broad range of mainly Th1 adjuvant systems have been used in vaccine testing in mice (Table 15.4). The use of adjuvants eliciting cell mediated and humoral immune responses is a general requirement for vaccines against a range of pathogens. Unfortunately, adjuvants currently licensed for human-use are mostly designed to induce antibody responses and lack the ability to induce cell-mediated immunity. Knudsen *et al.* recently made a detailed comparison of five different clinically tested adjuvants [Alhydrogel (alum), MF59®, IC31®, GLA-SE and CAF01] in challenge models of *M. tuberculosis*, influenza and *C. trachomatis*. In this study, recombinant MOMP formulated in the three strong Th1 adjuvants (IC31®, GLA-SE and CAF01) protected against intra-vaginal *C. trachomatis* serovar D challenge, whereas rMOMP in combination with mostly Th2 driven adjuvants Alum and MF59® did not protect (Knudsen *et al.*, 2016). Together with the results of Coler *et al.* using the adjuvants system 01b (AS01b) from GSK in a *C. trachomatis* serovar K challenge model (Table 15.4), this supports the use of Th1 adjuvants (Coler *et al.*, 2009). Importantly, this is also supported by results from *C. trachomatis* infection in pigs and guinea pigs (Boje *et al.*, 2016; Lorenzen *et al.*, 2015; Wali *et al.*, 2017).

Several laboratories have tested adjuvant combinations to protect against a *C. muridarum* challenge. Native MOMP formulated in MF59® was found to elicit some protection in mice against a genital *C. muridarum* challenge (Pal *et al.*, 2006). However, the most robust protection was obtained using a combination of strong Th1 (CpG-1826) plus Th1/Th2 (Montanide ISA 720) adjuvants (Pal *et al.*, 2005b). Yu *et al.* (Yu *et al.*, 2010) used recombinant PmpG-1, PmpE/F-2 and MOMP individually or in combination with three different adjuvants: CpG-1826, AbISco-100 or CAF01 (Table 15.4). The combination of PmpG-1, PmpE/F-2 and MOMP plus CAF01 was the most efficacious in protecting C57BL/6 and BALB/c mice against a vaginal challenge. These results were confirmed using four combination adjuvants: DDA-MPL, CAF01, CAF04 and Montanide ISA 720 VG plus CpG-1826, in addition to Alum, using PmpG as the antigen. The most robust protective immune responses were observed with DDA-MPL or CAF01 (Yu *et al.*, 2012).

nMOMP with the adjuvant combination CpG-1826 plus Montanide ISA 720 VG, has also been found to be very effective at eliciting protective immune responses in mice against respiratory challenges, and in non-human primates against ocular challenge (Carmichael *et al.*, 2011; Kari *et al.*, 2009; Pal *et al.*, 2005b; Tifrea *et al.*, 2018; Tifrea *et al.*, 2014). Table 15.4 summarizes adjuvants used in research/pre-clinical testing and highlights those already in clinical trials.

Table 15.4 Adjuvants tested in *Chlamydia* vaccine studies

Adjuvant	Delivery system	Immunostimulatory component(s)	Innate receptors or pathway activated	Immune responses stimulated	References
Adjuvants tested safe in human clinical trials					
Aluminium salts	Aluminium salts	–	NLRP3 inflammasome	Ab, Th2	Cheng *et al*. (2011); Knudsen *et al*. (2016); Yu *et al*. (2012); Abraham *et al.* (2019)
AS01b	Neutral liposomes	MPL and QS-21	TLR4	Ab, Th1	Coler *et al*. (2009)
CAF01 (DDA/TDB)	Cationic liposomes	TDB	Mincle	Ab, Th1/Th17	Knudsen *et al*. (2016); Olsen *et al*. (2015); Yu *et al*. (2010, 2012); Abraham *et al.* (2019)
GLA-SE	Squalene-in-water emulsion	GLA	TLR4	Ab, Th1/Th17	Knudsen *et al*. (2016); Paes *et al*. (2016)
IC31®	Cationic peptide KLK	ODN1a	TLR9	Ab, Th1	Cheng *et al*. (2011); Knudsen *et al*. (2016)
MF59®	Squalene-in-water emulsion	–	Tissue inflammation (no receptors defined), Transient ATP release	Ab, Th1/Th2	Badamchi-Zadeh *et al*. (2015); Badamchi-Zadeh *et al*. (2016); Knudsen *et al*. (2016); Pal *et al*. (2006)
Adjuvants used in preclinical testing					
CAF04 (DDA/MMG)	Cationic liposomes	MMG	Mincle	Ab, Th1/Th17	Yu *et al*. (2012)
CAF09 (DDA/MMG/Poly I:C)	Cationic liposomes	MMG, poly I:C	TLR3	Ab,Th1/Th17	Pal *et al*. (2017b)
CFA	Water-in-oil emulsion	–	NLR, inflammasome, Mincle, TLR	Ab, Th1	Pal *et al*. (1997a, 1999b, 2001)
CT	–	CT	No receptors defined	Ab, Th1/Th2	Singh *et al*. (2006)
CT-CpG	–	CpG	TLR9	Ab, Th1	Berry *et al*. (2004); Hickey *et al*. (2009, 2010)
CTA1-DD	–	CTA-subunit	No receptors defined	Ab, Th1/Th2	Cunningham *et al*. (2009); O'Meara *et al*. (2013, 2017)
CTA2B	–	CTA-subunit	No receptors defined	Ab, Th1/Th2	Ekong *et al*. (2009)
CpG-1826	–	Synthetic phosphorothioate-linked DNA oligonucleotides with optimized CpG motifs	TLR9	Th1	Cong *et al*. (2007); Li *et al*. (2010); Muller *et al*. (2017); Murthy *et al*. (2009); Skelding *et al*. (2006); Wali *et al*. (2017); Yu *et al*. (2010)

Table 15.4 Continued

Adjuvant	Delivery system	Immunostimulatory component(s)	Innate receptors or pathway activated	Immune responses stimulated	References
CpG-1826+Alum	Aluminium salts	CpG-1826	TLR9, NLRP3 inflammasome	Ab, Th1/Th2	Pal *et al*. (2002)
CpG-1826+Pam$_2$CSK$_4$	–	CpG-1826, Pam$_2$CSK$_4$	TLR9, TLR2/6	Ab, Th1/Th2	Cheng *et al*. (2014b)
CpG-1826+Montanide ISA720 VG	Water-in-oil emulsion	CpG 1826	TLR9	Ab, Th1/Th2	Carmichael *et al*. (2011); Cheng *et al*. (2014b); Pal *et al*. (2015, 2017a); Tifrea *et al*. (2014); Yu *et al*. (2012)
CpG-1826+LTK63	–	CpG-1826	TLR9	Ab, Th1	Finco *et al*. (2011)
cSAP	Polymer particles	R848	TLR7, TLR8	Ab, Th1	Stary *et al*. (2015)
DDA/MPL	Cationic liposomes	MPL	TLR4	Ab, Th1	Yu *et al*. (2012, 2014)
IFA	Water-in-oil emulsion	–	–	Ab, Th2/Th1	Blander and Amortegui (1994)
IL-12	–	IL-12	–	Ab, Th1	Chaganty *et al*. (2010); Murthy *et al*. (2006, 2007)
ISCOMATRIX and AbISCO	ISCOMS	Saponins	Mechanism undefined	Ab, Th1/Th2	Dong-Ji *et al*. (2000); Kollipara *et al*. (2012); O'Meara *et al*. (2016, 2017); Yu *et al*. (2010)
Lipid C	–	–	Mechanism undefined	Ab, Th1	Hickey *et al*. (2009, 2010)
LTK63	–	*E. coli* heat-labile enterotoxin	No receptors defined	Th1/Th2	Pal *et al*. (2006)
LTR72	–	*E. coli* mutated heat-labile enterotoxin	No receptors defined	Th1/Th2	Pal *et al*. (2006)

AS, adjuvant system; CAF, cationic adjuvant formulation; CFA, complete Freund's adjuvant; CpG, synthetic oligodeoxynucleotides containing unmethylated CpG motifs; cSAP, charge-switching synthetic adjuvant particles; CT, cholera toxin; CTA1-DD, cholera toxin A1 subunit linked to a dimer of the B cell targeting moiety D from protein A of *Staphylococcus aureus*; CTA2B, modified cholera toxin consisting of A2, the nontoxic cleavage product of the A subunit, and the nontoxic B subunit; DDA, dimethyldioctadecylammonium; GLA, glucopyranosyl; IFA, incomplete Freund´s adjuvant; IL, interleukin; ISCOMS, immune stimulating complexes; LTK63/LTR72, non-toxic mutants of heat labile *Escherichia coli* enterotoxin; MMG, mycobacterial monomycoloyl glycerol; MPL, monophosphoryl lipid; ODN1a, oligodeoxynucleotide; TLR, Toll-like receptor; poly I:C, polyinosinic:polycytidylic acid; QS-21, *Quillaja saponaria* fraction 21.

Delivery systems

Several delivery systems, including DNA plasmids, poliovirus, adenovirus, hepatitis B virus, influenza virus, *Vibrio cholerae* ghosts, *Lactobacillus rhamnosus*, and nanoparticles, have been used to vaccinate mice with various chlamydial antigens (Brown *et al.*, 2012; Champion *et al.*, 2009; Dixit *et al.*, 2014; Dong-Ji *et al.*, 2000; He *et al.*, 2007; Igietseme and Murdin, 2000; Inic-Kanada *et al.*, 2016; Inic-Kanada *et al.*, 2015; Jiang *et al.*, 2017; Jiang *et al.*, 2015; Montanaro *et al.*, 2015; Murdin *et al.*, 1993; Pal *et al.*, 1999a; Zhang *et al.*, 1997). An advantage of some of these delivery systems is the possibility of carrying antigens from several pathogens i.e. (Macmillan *et al.*, 2007). Especially DNA vaccines and recombinant viral vectors are potent inducers of cellular responses, both CD4 and CD8 T cells. Heterologous prime-boost vaccination strategies of DNA, vector and protein can also refine and enhance T-cell induction (Badamchi-Zadeh *et al.*, 2016; Hanke *et al.*, 1998; McShane *et al.*, 2001). Both DNA vaccine vectors, Human Adenovirus 5, and Modified Vaccinia Ankara (MVA) viral vectors have been tested safe in clinical trials (Hayton *et al.*, 2014; Nilsson *et al.*, 2015). Until now, the results obtained, however, do not appear to improve protection over conventional approaches using adjuvanted proteins. Also, for some of these systems the presence of not-medically relevant antigens in the delivery system raises safety concerns and the possibility of eliciting competitive irrelevant immune responses. The use of edible plants is certainly an interesting approach that requires further evaluation (Kalbina *et al.*, 2011).

Clinical development of a vaccine for *C. trachomatis* genital infections

Valid animal models mimicking human genital infection and disease are crucial for vaccine development. No single model (mouse, guinea pig, pig or NHP) is perfect for studying genital *C. trachomatis* infections and diseases (Poston *et al.*, 2017). Currently, there is consensus that animal models are relevant for selection of potential formulations but not a substitute for testing vaccine candidates in human trials early in development (Gottlieb and Johnston, 2017; Poston and Darville, 2018). To ensure the most rational and cost-effective vaccine development process, the preclinical and clinical development plans, including choice of animal models, endpoints, target populations etc. must align with the desired characteristics of the vaccine. These desired characteristics are described in a target product profile (TPP) (Singh, 2018) and include considerations about the therapeutic indication, target population, duration of protection, route of administration, safety, and efficacy requirements as outlined below.

Towards a target product profile for *Chlamydia* vaccines

In the absence of data from efficacy trials, it is difficult to qualify and rank parameters in a hypothetical target product profile for a vaccine candidate against genital *C. trachomatis*. However, it is possible to outline target ranges and desired characteristics based on presumed mechanisms of action, minimal efficacy and safety targets. The target population of a future preventative *C. trachomatis* vaccine would be older children and adolescents before sexual debut. An effective vaccine against genital *C. trachomatis* should ideally cover all genital serovars and aim to eliminate the global genital *Chlamydia* epidemic by preventing infection. However, a partially effective vaccine with ability to improve the immune control of the infection could provide a cost-effective intervention, as measured with specific parameters. These parameters are outlined in Table 15.5 and include shortened infection time, lowered bacterial load, limited transmission of co-infections and/or the prevention of ascending infections and complications such as PID, tubal infertility and extra uterine pregnancy (Grey *et al.*, 2009; Poston and Darville, 2018).

The duration of protection should ideally span the reproductive years (15–45 years of age) (Pal *et al.*, 1996). However, shorter protective duration e.g. 5–8 years and protective efficacy in only women are also cost-effective in modelling programs (Grey *et al.*, 2009). Ideally, the vaccine route and schedule would be aligned with the HPV program to reduce costs and increase coverage if rolled out in tandem in national immunization programmes.

Safety targets are cardinal for this kind of vaccine and, at a minimum, it should be safe in the target

Table 15.5 Standard phases for clinical development of a *C. trachomatis* vaccine

Phase	Description of end points	Design and powering
Phase 1	First in man study, small trial assessing safety in adults with no known prior exposure to *C. trachomatis*	Smaller groups of healthy volunteers, $n = 10–20$ per group. One recently completed trial compared two vaccines against placebo, $n = 5–15$/group. ClinicalTrials.gov Identifier: NCT02787109
Phase 1b/2a	Smaller trial(s) exploring dose and schedule, could include participants from the target population. Allows for detection of frequently occurring adverse events	Smaller groups of healthy volunteers, $n = 10–50$ per group. Selection of regimen based on safety and immunogenicity parameters
Phase 2b – prevention of infection	Prevention of infection, proof-of-concept trial powered for reduction of infection rates or chlamydial load, as a surrogate for prevention of later complications such as pelvic inflammatory disease (PID), ectopic pregnancy and infertility. The clinical end point could comprise of quantitative nucleic acid testing of urine samples or vaginal/cervical swabs, potentially guided by presence of symptoms. These trials allows for detection of common and rarer adverse events	Would include high-risk individuals, e.g. younger women aged 17–21, at-home urine sampling e.g. 2–3 times a year (Andersen *et al.*, 2002; Dirks *et al.*, 2015), one year follow-up. For example, in a setting with 5% reported infections, total annual incidence will be ~20% (incl. asymptomatic infections (Farley *et al.*, 2003; Korenromp *et al.*, 2002). Without correcting for short-lived asymptomatic and spontaneously cleared infections, sample size would be ~300, to demonstrate 60% vaccine efficacy/effectiveness (VE), at 80% power and 5% 1-sided error rate, with 10% loss to follow-up
Phase 2b/ Phase 3 – prevention of disease	Prevention of PID as an end point, indicating reduced *C. trachomatis* morbidity. PID is a clinical diagnosis and not always caused by *C. trachomatis* despite presence of the infection in the lower genital tract, calling for stringent case verification. Allows for detection of rare adverse events	One such trial, has previously been done to investigate the effect of screening and treating women for chlamydial infection on the incidence of PID. This trial assumed a sample size of 2500 and PID incidence of 3%, to demonstrate 44% risk reduction with one year follow-up, at 80% power, 5% 1-sided error and 10% loss to follow-up). ClinicalTrials.gov Identifier: NCT00115388 (Oakeshott *et al.*, 2010)
Phase 4 – post marketing	These post marketing surveillance studies allow for detection and assessment of very rare adverse events. Could allow for evaluation of rarer efficacy measures including tubal factor infertility, ectopic pregnancies, miscarriage/ stillbirth/preterm births, neonatal conjunctivitis (ophthalmia neonatorum), neonatal pneumonia, *Chlamydia*-associated reactive arthritis, epididymitis, and co-infection rates with *N. gonorrhoeae* or HIV	Post marketing surveillance, marketing studies

population of older children and adolescents and potentially also safe in risk groups including pregnant and immunocompromised individuals. An aim could be the safety profile of already marketed HPV vaccines although more adverse events might be acceptable e.g. in regions with high endemicity and high rates of vertical transmission, such as the African infertility belt (Larsen, 2003).

Clinical development strategy for a *C. trachomatis* vaccine for genital infections

Clinical development plans of a *C. trachomatis* vaccine for genital infections should follow the standard phases for pharmaceuticals and vaccines (Singh, 2018) (Table 15.5). Phase 1 and 2a trials should seek to establish safety and immunogenicity with an optimized dose/schedule and the desired route. When a candidate has proven safe and immunogenic in the target population, the next step is a direct assessment of clinical efficacy against hard clinical endpoints in phase 2b or Phase 3. These clinical endpoints remain elusive and should be driven by preclinical and modelling evidence. Ideally, phase 2b/3 trials should seek to determine the ability of the vaccines to prevent PID. Such a trial was recently attempted outside the field of vaccines by assessing the efficacy of *C. trachomatis* screening and target treatment (Oakeshott *et al.*, 2010). However, PID is a challenging endpoint because it is a clinical diagnosis, not always caused by *C. trachomatis* even if present in the lower female genital tract, and thus there is a need for stringent case verification (Brunham *et al.*, 2015). As all infected individuals can transmit the infection, but only one-in-six infected women develop PID, it seems rational to build evidence and establish early clinical proof-of-concept by assessing endpoints directly linked to the bacteria and infection, including prevention of infection, reduction of *C. trachomatis* load, length of time of vaginal shedding and reduction of rates of persistent infection. These endpoints could be addressed in smaller non-pivotal phase 2b trials in a high transmission population of young adults (Poston and Darville, 2018). The availability of kits for non-invasive urine sampling at-home, combined with quantitative nucleic acid amplification techniques, mobile phone-based prompting and real-time communication with study staff would allow for detailed and cost effective monitoring and would probably be well-suited for the trial participants (Andersen *et al.*, 2002; Dirks *et al.*, 2015; Poston and Darville, 2018). Demonstrating prevention of infection is likely a high bar for a vaccine. Reduction in ascending infection would probably suffice to limit the risk of PID, for which such proof-of-concept trials could be designed as adaptive and guided by interim analysis (Gottlieb and Johnston, 2017; Poston and Darville, 2018).

Given the global scale of the *C. trachomatis* epidemic, the development strategy could focus on both the developed and the developing world (Gottlieb and Johnston, 2017; Unemo *et al.*, 2017). To maximize the impact on global health and to piggyback on already established infrastructure from the HPV and HIV-1 field (Ramjee *et al.*, 2010), it would be rational to focus the clinical development and regulatory strategy on the developing world (Gottlieb and Johnston, 2017), i.e. in Sub Saharan Africa. Here infection rates remain very high (Torrone *et al.*, 2018), and as a result there is substantial co-morbidity, mortality and increased transmission of co-infections, including HIV and neonatal conjunctivitis as well as lower respiratory tract infection (Adachi *et al.*, 2016; Collet *et al.*, 1988; Johnson and Lewis, 2008).

Importantly, all developers involved in clinical trials should seek to build as much information as possible in terms of transmission patterns, prevalence of transient self-healing infections and risk factors for persistence in the trial design (Geisler *et al.*, 2013). Randomized controlled trials of experimental *Chlamydia* vaccines offer a unique opportunity to study such patterns and generate experience to guide future trial designs. To facilitate the development of the much-needed correlates of protection and surrogate outcomes, researchers should prioritize sampling of plasma, cells and transcriptomes, as well as data from non-invasive radiological techniques at all stages of the development process.

C. trachomatis vaccines in pre-clinical and clinical development

Chlamydia vaccine research is mainly driven by academics and governmental institutions. None of the big pharmaceutical companies have active development programs. A few small to medium enterprises (SMEs) are active (see Table 15.6). Recently, the first Phase I trial was completed (Abraham *et al.*, 2019), and the NIH/NIAD vaccine is close to clinical testing.

Statens Serum Institut (SSI) vaccine

The vaccine antigen, CTH522, is a subunit vaccine and a recombinant version of the *C. trachomatis* major outer membrane protein (MOMP). CTH522 consists of two parts; at its N-terminus, CTH522 is a recombinant linearized cysteine-free protein, covering most of serovar D MOMP, known to include conserved T helper cell epitopes (Nunes *et al.*, 2010). At its C-terminus, CTH522 has a novel antigenic stretch of extended VD4 segments from serovars D, E, F and G (designated heterologous immunorepeat = Hirep). The VD4 region of MOMP is rich in neutralizing epitopes and contains the highly conserved species-specific linear epitope LNPTIAG embedded in the central part. CTH522-specific antibodies neutralize serovars D-G, which represent up to 90% of *C. trachomatis* prevalence for human genital tract infection. CTH522 is produced recombinantly in *E. coli*, and in solution purified CTH522 antigen shows an unusual self-assembly into nanoparticles (Rose *et al.*, 2018), possibly aiding its immunogenicity.

Pre-clinical testing of CTH522 was primarily done using the cationic liposomal adjuvant CAF01

Table 15.6 *C. trachomatis* vaccines in pre-clinical and clinical development

Organization	Development stage	Indication	Vaccine
Statens Serum Institut	Phase I	*C. trachomatis* infections	Subunit vaccine (CTH522/CAF01)
NIH/NIAID	Preclinical	*C. trachomatis* infections	Live attenuated vaccine
Abera	Preclinical	*C. trachomatis* infections	Subunit vaccine (Outer Membrane Vesicles, OMVs)
BlueWillow Biologics Inc.	Preclinical	*C. trachomatis* infections	Subunit vaccine
Vault Pharma Inc.	Preclinical	*C. trachomatis* infections	Subunit vaccine (VPI-201 *Chlamydia* – Vault)
Prokarium	Preclinical	*C. trachomatis* infections	Recombinant vector vaccine (Vaxonella® platform)
Genetic Immunity Inc.	Preclinical	*C. pneumoniae* infection induced atherosclerosis	Therapeutic DNA vaccine (ChlamyDerm Immune Therapy)

(Knudsen *et al.*, 2016; Pedersen *et al.*, 2018). CAF01 has been designed for the induction of a strong cell-mediated immune (CMI) response combined with antibody production (Agger *et al.*, 2008). CTH522/CAF01 has been evaluated for immunogenicity in a range of animal models (mouse, guinea pig, pig and NHPs). Protective efficacy was found following genital *C. trachomatis* challenge in mice (Olsen *et al.*, 2015; Olsen *et al.*, 2017) and guinea pigs (Olsen *et al.* unpublished). *In vivo* and *in vitro* studies have documented the functional capacity of neutralizing antibodies (Olsen *et al.*, 2015; Olsen *et al.*, 2017) and of opsonizing antibodies (Grasse *et al.*, 2018). The depletion studies described by Olsen *et al.* suggest that there is also an important synergistic role of $CD4^+$ T cells, in addition to the neutralizing antibody response (Olsen *et al.*, 2015).

The human trial of CTH522 (Abraham *et al.*, 2019) is a parallel testing of two different adjuvant systems (CAF01 and aluminium hydroxide), with intranasal boost and assessment of both systemic and mucosal immunogenicity. The primary outcome of the trial was safety and the secondary outcome was humoral immunogenicity, given the key role of antibodies in CTH522 mediated protection. The applied immunization schedule of systemic priming followed by mucosal boost has in several studies proven effective in inducing mucosal IgA responses (Christensen *et al.*, 2017; Fiorino *et al.*, 2013; Lorenzen *et al.*, 2015). CTH522/CAF01 is planned to proceed into a Phase IIa study (2019). CTH522 has the potential to target the ocular serovar B, and the Statens Serum Institut is actively pursuing the expanded use of CTH522/CAF01 against ocular infections. In addition it is being supplemented with antigens based on immunorepeat technology to cover serovars A and C (European Union's Horizon 2020 research project TracVac, https://www.trachoma-vaccine.org/).

NIH/NIAID vaccine

The NIH/NIAID vaccine program aims to control both ocular and genital human diseases. Based on the work by Harlan Caldwell's lab NIH/NIAD is developing a plasmid-deficient live-attenuated trachoma vaccine (LATV). Pre-clinical studies found that the vaccine was highly attenuated in NHPs. The LATV had unchanged capability to replicate within epithelial cells, and animals could be repeatedly infected with the LATV without inducing ocular pathology. Vaccinated animals were protected against challenge with a virulent strain (Kari *et al.*, 2011; Olivares-Zavaleta *et al.*, 2014). Protected animals mounted a rapid inflammatory cytokine response in tears, and T cell depletion revealed that the vaccine-mediated immunity was dependent on CD8 T cells.

Summary and prospects for *Chlamydia* vaccines

The only commercial vaccines available against *Chlamydia* are against the veterinary pathogens *C. abortus* and *C. felis*. Promising results indicate that more animal vaccines will follow for species such as *C. pecorum*. There is still a critical global need for a *C. trachomatis* vaccine to protect against human genital and ocular infections. *C. trachomatis* prevalence is increasing and current control measures are unable to limit the epidemic (Gottlieb and Johnston, 2017).

Studies to develop a vaccine to protect against *C. trachomatis* infections started more than 100 years ago. At that time trachoma was still a worldwide problem. Implementation of improved sanitary conditions resulted in the eradication of trachoma from most parts of the world except from Africa and other isolated regions. Major efforts to develop a vaccine against trachoma were undertaken in the 1960s. Multiple whole cell vaccines were tested in humans and NHPs. Although those efforts did not result in the implementation of a vaccine major lessons were learned that are still guiding the research efforts today. Overall, the human trials showed that vaccination could work. Protection was, however, short-lived and serovar/serogroup specific. In a few cases, vaccinated individuals had a higher attack rate of trachoma compared with placebo. Results from NHPs suggested that vaccination could lead to a hypersensitivity reaction on re-exposure to C. trachomatis. There was no convincing evidence of this in the human clinical trials. Starting in the 1970s, the recognition that *C. trachomatis* was also a causative agent of numerous genitourinary infections reignited attempts to produce a vaccine. Extensive experimentation in animal models, in particular mice, and information from clinical studies in humans have concluded that a vaccine will have to elicit strong Th1 cellular immune responses, with production of IFN-γ and humoral responses that favour the generation of neutralizing antibodies. Determination of the DNA sequence of *C. trachomatis* and phylogenetic analysis concluded that, most likely, MOMP was the antigen that accounted for the serovar/serogroup protection observed during the trachoma vaccine. The ability of MOMP-based vaccines to protect several animal models against a genital challenge is now well-established. The search for additional protective antigens has being extensive and other proteins such as the Pmps, CPAF and Pgp3 have emerged as potential vaccine candidates. The lack of self-adjuvanticity of subunit vaccines has emphasized the need to identify adjuvants that can elicit broad systemic and mucosal cellular and humoral immune responses. The number of adjuvants currently approved for human use is very limited and therefore, there is an urgent need to advance research in this field. Like many human pathogens, the site of entry of *C. trachomatis* is mucosal. Thus it is not surprising that vaccination using combinations of mucosal and systemic routes is more effective that using a single route. Safety, efficacy, easy delivery and low costs are requirement for the successful implementation of *C. trachomatis* vaccines. Work has been restricted to preclinical testing for decades, but several recent reviews and meeting reports have addressed how the scientific community can move forward together with industry to develop a vaccine (de la Maza *et al.*, 2017; Dodet, 2014; Gottlieb and Johnston, 2017; Poston *et al.*, 2017; Yu *et al.*, 2016; Zhong *et al.*, 2017). In support of vaccine research and development, WHO, NIAID and industry have identified critical steps in 'The global roadmap for advancing development of vaccines against sexually transmitted infections' (Dodet, 2014; Gottlieb *et al.*, 2016; Gottlieb *et al.*, 2014; Poston *et al.*, 2017). In the coming years, as more vaccines enter clinical testing, the field will gain important knowledge about safety and how protective immune responses translate from pre-clinical models into humans. Future vaccine development, as seen for both the NIH/NIIAD and SSI program, could include trachoma and could fit into the global programs of WHO to eliminate trachoma in the world.

Acknowledgement

We thank Dr Morten Ruhwald for the design of the clinical development plan and for valuable discussions and comments.

References

Abraham, S., Juel, H.B., Bang, P., Cheeseman, H.M., Dohn, R.B., Cole, T., Kristiansen, M.P., Korsholm, K.S., Lewis, D., Olsen, A.W., *et al.* (2019). Safety and immunogenicity of the chlamydia vaccine candidate CTH522 adjuvanted with CAF01 liposomes or aluminium hydroxide: a first-in-human, randomised, double-blind, placebo-controlled, phase 1 trial. Lancet Infect Dis *19*, 1091–1100.10.1016/S1473-3099(19)30279-8

Adachi, K., Nielsen-Saines, K., and Klausner, J.D. (2016). *Chlamydia trachomatis* Infection in Pregnancy: The Global Challenge of Preventing Adverse Pregnancy and Infant Outcomes in Sub-Saharan Africa and Asia. Biomed Res Int 2016, 9315757.10.1155/2016/9315757.

Agger, E.M., Rosenkrands, I., Hansen, J., Brahimi, K., Vandahl, B.S., Aagaard, C., Werninghaus, K., Kirschning, C., Lang, R., Christensen, D., *et al.* (2008). Cationic liposomes formulated with synthetic mycobacterial cordfactor (CAF01): a versatile adjuvant for vaccines with different immunological requirements. PLoS One 3, e3116.10.1371/journal.pone.0003116.

Alexander, E.R., Wang, S.P., and Grayston, J.T. (1967). Further classification of TRIC agents from ocular trachoma and other sources by the mouse toxicity prevention test. Am J Ophthalmol 63, Suppl:1469–1478.

Andersen, B., Olesen, F., Moller, J.K., and Ostergaard, L. (2002). Population-based strategies for outreach screening of urogenital *Chlamydia trachomatis* infections: a randomized, controlled trial. J Infect Dis 185, 252–258.10.1086/338268.

Anderson, I.E., Baxter, S.I., Dunbar, S., Rae, A.G., Philips, H.L., Clarkson, M.J., and Herring, A.J. (1996). Analyses of the genomes of chlamydial isolates from ruminants and pigs support the adoption of the new species Chlamydia pecorum. Int J Syst Bacteriol 46, 245–251.10.1099/00207713-46-1-245.

Andrew, D.W., Hafner, L.M., Beagley, K.W., and Timms, P. (2011). Partial protection against chlamydial reproductive tract infection by a recombinant major outer membrane protein/CpG/cholera toxin intranasal vaccine in the guinea pig Chlamydia caviae model. J Reprod Immunol 91, 9–16.10.1016/j.jri.2011.06.100.

Atik, B., Thanh, T.T., Luong, V.Q., Lagree, S., and Dean, D. (2006). Impact of annual targeted treatment on infectious trachoma and susceptibility to reinfection. JAMA 296, 1488–1497.10.1001/jama.296.12.1488.

Aus, G., Adolfsson, J., Selli, C., and Widmark, A. (2002). Treatment of patients with clinical T3 prostate cancer. Scandinavian journal of urology and nephrology 36, 28–33.

Bachmann, N.L., Polkinghorne, A., and Timms, P. (2014). Chlamydia genomics: providing novel insights into chlamydial biology. Trends Microbiol 22, 464–472.10.1016/j.tim.2014.04.013.

Badamchi-Zadeh, A., McKay, P.F., Holland, M.J., Paes, W., Brzozowski, A., Lacey, C., Follmann, F., Tregoning, J.S., and Shattock, R.J. (2015). Intramuscular Immunization with Chlamydial Proteins Induces *Chlamydia trachomatis* Specific Ocular Antibodies. PLoS One 10, e0141209.10.1371/journal.pone.0141209.

Badamchi-Zadeh, A., McKay, P.F., Korber, B.T., Barinaga, G., Walters, A.A., Nunes, A., Gomes, J.P., Follmann, F., Tregoning, J.S., and Shattock, R.J. (2016). A Multi-Component Prime-Boost Vaccination Regimen with a Consensus MOMP Antigen Enhances *Chlamydia trachomatis* Clearance. Front Immunol 7, 162.10.3389/fimmu.2016.00162.

Bakshi, R.K., Gupta, K., Jordan, S.J., Chi, X., Lensing, S.Y., Press, C.G., and Geisler, W.M. (2018). An Adaptive *Chlamydia trachomatis*-Specific IFN-gamma-Producing CD4(+) T Cell Response Is Associated With Protection Against Chlamydia Reinfection in Women. Front Immunol 9, 1981.10.3389/fimmu.2018.01981.

Barker, C.J., Beagley, K.W., Hafner, L.M., and Timms, P. (2008). In silico identification and in vivo analysis of a novel T-cell antigen from Chlamydia, NrdB. Vaccine 26, 1285–1296.10.1016/j.vaccine.2007.12.048.

Barnett, S.W., Srivastava, I.K., Kan, E., Zhou, F., Goodsell, A., Cristillo, A.D., Ferrai, M.G., Weiss, D.E., Letvin, N.L., Montefiori, D., *et al.* (2008). Protection of macaques against vaginal SHIV challenge by systemic or mucosal and systemic vaccinations with HIV-envelope. AIDS 22, 339–348.10.1097/QAD.0b013e3282f3ca57.

Bas, S., Neff, L., Vuillet, M., Spenato, U., Seya, T., Matsumoto, M., and Gabay, C. (2008). The proinflammatory cytokine response to *Chlamydia trachomatis* elementary bodies in human macrophages is partly mediated by a lipoprotein, the macrophage infectivity potentiator, through TLR2/TLR1/TLR6 and CD14. J Immunol 180, 1158–1168.10.4049/jimmunol.180.2.1158.

Batteiger, B.E., Rank, R.G., Bavoil, P.M., and Soderberg, L.S. (1993). Partial protection against genital reinfection by immunization of guinea pigs with isolated outer-membrane proteins of the chlamydial agent of guinea pig inclusion conjunctivitis. J Gen Microbiol 139, 2965–2972.10.1099/00221287-139-12-2965.

Batteiger, B.E., Xu, F., Johnson, R.E., and Rekart, M.L. (2010). Protective immunity to *Chlamydia trachomatis* genital infection: evidence from human studies. J Infect Dis 201 Suppl 2, S178–189.10.1086/652400.

Becker, E., and Hegemann, J.H. (2014). All subtypes of the Pmp adhesin family are implicated in chlamydial virulence and show species-specific function. Microbiologyopen 3, 544–556.10.1002/mbo3.186.

Becker, P.D., Fiorentini, S., Link, C., Tosti, G., Ebensen, T., Caruso, A., and Guzman, C.A. (2006). The HIV-1 matrix protein p17 can be efficiently delivered by intranasal route in mice using the TLR 2/6 agonist MALP-2 as mucosal adjuvant. Vaccine 24, 5269–5276.10.1016/j.vaccine.2005.11.008.

Belland, R.J., Ouellette, S.P., Gieffers, J., and Byrne, G.I. (2004). *Chlamydia pneumoniae* and atherosclerosis. Cell Microbiol 6, 117–127.

Bergquist, C., Johansson, E.L., Lagergard, T., Holmgren, J., and Rudin, A. (1997). Intranasal vaccination of humans with recombinant cholera toxin B subunit induces systemic and local antibody responses in the upper respiratory tract and the vagina. Infect Immun 65, 2676–2684.

Berry, L.J., Hickey, D.K., Skelding, K.A., Bao, S., Rendina, A.M., Hansbro, P.M., Gockel, C.M., and Beagley, K.W. (2004). Transcutaneous immunization with combined cholera toxin and CpG adjuvant protects against Chlamydia muridarum genital tract infection. Infect Immun 72, 1019–1028.10.1128/iai.72.2.1019-1028.2004.

Blander, S.J., and Amortegui, A.J. (1994). Mice immunized with a chlamydial extract have no increase in early protective immunity despite increased inflammation following genital infection by the mouse pneumonitis agent of *Chlamydia trachomatis*. Infect Immun 62, 3617–3624.

Boje, S., Olsen, A.W., Erneholm, K., Agerholm, J.S., Jungersen, G., Andersen, P., and Follmann, F. (2016). A multi-subunit Chlamydia vaccine inducing neutralizing antibodies and strong IFN-gamma(+) CMI responses protects against a genital infection in minipigs. Immunol Cell Biol 94, 185–195.10.1038/icb.2015.79.

Borsutzky, S., Ebensen, T., Link, C., Becker, P.D., Fiorelli, V., Cafaro, A., Ensoli, B., and Guzman, C.A. (2006). Efficient systemic and mucosal responses against the HIV-1 Tat protein by prime/boost vaccination using the lipopeptide MALP-2 as adjuvant. Vaccine 24, 2049–2056.10.1016/j.vaccine.2005.11.025.

Brill, J.R. (2010). Diagnosis and treatment of urethritis in men. American family physician 81, 873–878.

Brown, T.H., David, J., Acosta-Ramirez, E., Moore, J.M., Lee, S., Zhong, G., Hancock, R.E., Xing, Z., Halperin, S.A., and Wang, J. (2012). Comparison of immune responses and protective efficacy of intranasal prime-boost immunization regimens using adenovirus-based and CpG/HH2 adjuvanted-subunit vaccines against genital Chlamydia muridarum infection. Vaccine 30, 350–360.10.1016/j.vaccine.2011.10.086.

Brunham, R.C., Gottlieb, S.L., and Paavonen, J. (2015). Pelvic inflammatory disease. N Engl J Med 372, 2039–2048.10.1056/NEJMra1411426.

Brunham, R.C., Kimani, J., Bwayo, J., Maitha, G., Maclean, I., Yang, C., Shen, C., Roman, S., Nagelkerke, N.J., Cheang, M., *et al.* (1996). The epidemiology of *Chlamydia trachomatis* within a sexually transmitted diseases core group. J Infect Dis 173, 950–956.10.1093/infdis/173.4.950.

Brunham, R.C., Kuo, C.C., Cles, L., and Holmes, K.K. (1983). Correlation of host immune response with quantitative recovery of *Chlamydia trachomatis* from the human endocervix. Infect Immun 39, 1491–1494.

Brunham, R.C., Pourbohloul, B., Mak, S., White, R., and Rekart, M.L. (2005). The unexpected impact of a *Chlamydia trachomatis* infection control program on susceptibility to reinfection. J Infect Dis 192, 1836–1844.10.1086/497341.

Brunham, R.C., and Rappuoli, R. (2013). *Chlamydia trachomatis* control requires a vaccine. Vaccine 31, 1892–1897.10.1016/j.vaccine.2013.01.024.

Brunham, R.C., and Rekart, M.L. (2008). The arrested immunity hypothesis and the epidemiology of chlamydia control. Sex Transm Dis 35, 53–54.10.1097/OLQ.0b013e31815e41a3.

Budrys, N.M., Gong, S., Rodgers, A.K., Wang, J., Louden, C., Shain, R., Schenken, R.S., and Zhong, G. (2012). *Chlamydia trachomatis* antigens recognized in women with tubal factor infertility, normal fertility, and acute infection. Obstet Gynecol 119, 1009–1016.10.1097/AOG.0b013e3182519326.

Bulir, D.C., Liang, S., Lee, A., Chong, S., Simms, E., Stone, C., Kaushic, C., Ashkar, A., and Mahony, J.B. (2016). Immunization with chlamydial type III secretion antigens reduces vaginal shedding and prevents fallopian tube pathology following live C. muridarum challenge. Vaccine 34, 3979–3985.10.1016/j.vaccine.2016.06.046.

Buxton, D. (1986). Potential danger to pregnant women of Chlamydia psittaci from sheep. The Veterinary record 118, 510–511.

Caldwell, H.D., Kromhout, J., and Schachter, J. (1981). Purification and partial characterization of the major outer membrane protein of *Chlamydia trachomatis*. Infect Immun 31, 1161–1176.

Carey, A., Cunningham, K., Andrew, D., Hafner, L., Timms, P., and Beagley, K. (2011). A comparison of the effects of a chlamydial vaccine administered during or after a C. muridarum urogenital infection of female mice. Vaccine 29, 6505–6513.10.1016/j.vaccine.2011.07.012.

Carmichael, J.R., Pal, S., Tifrea, D., and de la Maza, L.M. (2011). Induction of protection against vaginal shedding and infertility by a recombinant Chlamydia vaccine. Vaccine 29, 5276–5283.10.1016/j.vaccine.2011.05.013.

Carrasco, J.A., Tan, C., Rank, R.G., Hsia, R.C., and Bavoil, P.M. (2011). Altered developmental expression of polymorphic membrane proteins in penicillin-stressed *Chlamydia trachomatis*. Cell Microbiol 13, 1014–1025.10.1111/j.1462-5822.2011.01598.x.

Chaganty, B.K., Murthy, A.K., Evani, S.J., Li, W., Guentzel, M.N., Chambers, J.P., Zhong, G., and Arulanandam, B.P. (2010). Heat denatured enzymatically inactive recombinant chlamydial protease-like activity factor induces robust protective immunity against genital chlamydial challenge. Vaccine 28, 2323–2329.10.1016/j.vaccine.2009.12.064.

Champion, C.I., Kickhoefer, V.A., Liu, G., Moniz, R.J., Freed, A.S., Bergmann, L.L., Vaccari, D., Raval-Fernandes, S., Chan, A.M., Rome, L.H., *et al.* (2009). A vault nanoparticle vaccine induces protective mucosal immunity. PLoS One 4, e5409.10.1371/journal.pone.0005409.

Chaturvedi, A.K., Gaydos, C.A., Agreda, P., Holden, J.P., Chatterjee, N., Goedert, J.J., Caporaso, N.E., and Engels, E.A. (2010). *Chlamydia pneumoniae* infection and risk for lung cancer. Cancer Epidemiol Biomarkers Prev 19, 1498–1505.10.1158/1055-9965.EPI-09-1261.

Chen, D., Lei, L., Lu, C., Galaleldeen, A., Hart, P.J., and Zhong, G. (2010). Characterization of Pgp3, a *Chlamydia trachomatis* plasmid-encoded immunodominant antigen. J Bacteriol 192, 6017–6024.10.1128/JB.00847-10.

Cheng, C., Cruz-Fisher, M.I., Tifrea, D., Pal, S., Wizel, B., and de la Maza, L.M. (2011). Induction of protection in mice against a respiratory challenge by a vaccine formulated with the Chlamydia major outer membrane protein adjuvanted with IC31(R). Vaccine 29, 2437–2443.10.1016/j.vaccine.2011.01.031.

Cheng, C., Jain, P., Pal, S., Tifrea, D., Sun, G., Teng, A.A., Liang, X., Felgner, P.L., and de la Maza, L.M. (2014a). Assessment of the role in protection and pathogenesis of the Chlamydia muridarum V-type ATP synthase subunit A (AtpA) (TC0582). Microbes Infect 16, 123–133.10.1016/j.micinf.2013.10.012.

Cheng, C., Pal, S., Tifrea, D., Jia, Z., and de la Maza, L.M. (2014b). A vaccine formulated with a combination of TLR-2 and TLR-9 adjuvants and the recombinant major outer membrane protein elicits a robust immune response and significant protection against a Chlamydia muridarum challenge. Microbes Infect 16, 244–252.10.1016/j.micinf.2013.11.009.

Christensen, D., Mortensen, R., Rosenkrands, I., Dietrich, J., and Andersen, P. (2017). Vaccine-induced Th17 cells are established as resident memory cells in the lung and promote local IgA responses. Mucosal immunology 10, 260–270.10.1038/mi.2016.28.

Ciabattini, A., Prota, G., Christensen, D., Andersen, P., Pozzi, G., and Medaglini, D. (2015). Characterization of the Antigen-Specific CD4(+) T Cell Response Induced by Prime-Boost Strategies with CAF01 and CpG Adjuvants Administered by the Intranasal and Subcutaneous Routes. Front Immunol 6, 430.10.3389/fimmu.2015.00430.

Clements, C., Dhir, S.P., Grayston, J.T., and Wang, S.P. (1979). Long term follow-up study of a trachoma vaccine trial in villages of Northern India. Am J Ophthalmol 87, 350–353.

Cohen, C.R., Koochesfahani, K.M., Meier, A.S., Shen, C., Karunakaran, K., Ondondo, B., Kinyari, T., Mugo, N.R., Nguti, R., and Brunham, R.C. (2005). Immunoepidemiologic profile of *Chlamydia trachomatis* infection: importance of heat-shock protein 60 and interferon- gamma. J Infect Dis 192, 591–599.10.1086/432070.

Coler, R.N., Bhatia, A., Maisonneuve, J.F., Probst, P., Barth, B., Ovendale, P., Fang, H., Alderson, M., Lobet, Y., Cohen, J., *et al.* (2009). Identification and characterization of novel recombinant vaccine antigens for immunization against genital *Chlamydia trachomatis*. FEMS Immunol Med Microbiol 55, 258–270.10.1111/j.1574-695X.2008.00527.x.

Collet, M., Reniers, J., Frost, E., Gass, R., Yvert, F., Leclerc, A., Roth-Meyer, C., Ivanoff, B., and Meheus, A. (1988). Infertility in Central Africa: infection is the cause. Int J Gynaecol Obstet 26, 423–428.

Collier, L.H., S.S., Sowa, J., Blyth, W.A. (1963). Experiments with trachoma vaccines: therapeutic effect on established trachoma. Orient Arch Ophthalmol, 67–75.

Cong, Y., Jupelli, M., Guentzel, M.N., Zhong, G., Murthy, A.K., and Arulanandam, B.P. (2007). Intranasal immunization with chlamydial protease-like activity factor and CpG deoxynucleotides enhances protective immunity against genital Chlamydia muridarum infection. Vaccine 25, 3773–3780.10.1016/j.vaccine.2007.02.010.

Cotter, T.W., Meng, Q., Shen, Z.L., Zhang, Y.X., Su, H., and Caldwell, H.D. (1995). Protective efficacy of major outer membrane protein-specific immunoglobulin A (IgA) and IgG monoclonal antibodies in a murine model of *Chlamydia trachomatis* genital tract infection. Infect Immun 63, 4704–4714.

Crane, D.D., Carlson, J.H., Fischer, E.R., Bavoil, P., Hsia, R.C., Tan, C., Kuo, C.C., and Caldwell, H.D. (2006). *Chlamydia trachomatis* polymorphic membrane protein D is a species-common pan-neutralizing antigen. Proc Natl Acad Sci U S A 103, 1894–1899.10.1073/pnas.0508983103.

Cui, Z.D., LaScolea, L.J., Jr., Fisher, J., and Ogra, P.L. (1989). Immunoprophylaxis of *Chlamydia trachomatis* lymphogranuloma venereum pneumonitis in mice by oral immunization. Infect Immun 57, 739–744.

Cui, Z.D., Tristram, D., LaScolea, L.J., Kwiatkowski, T., Jr., Kopti, S., and Ogra, P.L. (1991). Induction of antibody response to *Chlamydia trachomatis* in the genital tract by oral immunization. Infect Immun 59, 1465–1469.

Cunningham, K.A., Carey, A.J., Lycke, N., Timms, P., and Beagley, K.W. (2009). CTA1-DD is an effective adjuvant for targeting anti-chlamydial immunity to the murine genital mucosa. J Reprod Immunol 81, 34–38.10.1016/j.jri.2009.04.002.

Cutlip, R.C., Smith, P.C., and Page, L.A. (1972). Chlamydial polyarthritis of lambs: a review. J Am Vet Med Assoc 161, 1213–1216.

Darville, T., Albritton, H.L., Zhong, W., Dong, L., O'Connell, C.M., Poston, T.B., Quayle, A.J., Goonetilleke, N., Wiesenfeld, H.C., Hillier, S.L., *et al.* (2019). Anti-chlamydia IgG and IgA are insufficient to prevent endometrial chlamydia infection in women, and increased anti-chlamydia IgG is associated with enhanced risk for incident infection. Am J Reprod Immunol 81, e13103.10.1111/aji.13103.

Darville, T., and Hiltke, T.J. (2010). Pathogenesis of genital tract disease due to *Chlamydia trachomatis*. J Infect Dis 201 Suppl 2, S114–125.10.1086/652397.

Datta, S.D., Sternberg, M., Johnson, R.E., Berman, S., Papp, J.R., McQuillan, G., and Weinstock, H. (2007). Gonorrhea and chlamydia in the United States among persons 14 to 39 years of age, 1999 to 2002. Ann Intern Med 147, 89–96.

Dawson, C., Jawetz, E., Hanna, L., Rose, L., Wood, T.R., and Thygeson, P. (1966). Experimental inclusion conjunctivitis in man. II. Partial resistance to reinfection. Am J Epidemiol 84, 411–425.10.1093/oxfordjournals.aje.a120654.

De Clercq, E., Kalmar, I., and Vanrompay, D. (2013). Animal models for studying female genital tract infection with *Chlamydia trachomatis*. Infect Immun 81, 3060–3067.10.1128/IAI.00357-13.

de Jonge, M.I., Keizer, S.A., El Moussaoui, H.M., van Dorsten, L., Azzawi, R., van Zuilekom, H.I., Peters, P.P., van Opzeeland, F.J., van Dijk, L., Nieuwland, R., *et al.* (2011). A novel guinea pig model of *Chlamydia trachomatis* genital tract infection. Vaccine 29, 5994–6001.10.1016/j.vaccine.2011.06.037.

de la Maza, L.M., and Peterson, E.M. (2002). Vaccines for *Chlamydia trachomatis* infections. Curr Opin Investig Drugs 3, 980–986.

de la Maza, L.M., Zhong, G., and Brunham, R.C. (2017). Update on *Chlamydia trachomatis* Vaccinology. Clin Vaccine Immunol 24.10.1128/CVI.00543-16.

de la Maza, M.A., and de la Maza, L.M. (1995). A new computer model for estimating the impact of vaccination protocols and its application to the study of *Chlamydia trachomatis* genital infections. Vaccine 13, 119–127.

De Puysseleyr, L., De Puysseleyr, K., Braeckman, L., Morre, S.A., Cox, E., and Vanrompay, D. (2017). Assessment of Chlamydia suis Infection in Pig Farmers. Transbound Emerg Dis 64, 826–833.10.1111/tbed.12446.

Desclozeaux, M., Robbins, A., Jelocnik, M., Khan, S.A., Hanger, J., Gerdts, V., Potter, A., Polkinghorne, A., and Timms, P. (2017). Immunization of a wild koala population with a recombinant Chlamydia pecorum Major Outer Membrane Protein (MOMP) or Polymorphic Membrane Protein (PMP) based vaccine: New insights into immune response, protection and clearance. PLoS One 12, e0178786.10.1371/journal.pone.0178786.

Dhir, S.P., Agarwal, L.P., Detels, R., Wang, S.P., and Grayston, J.T. (1967). Field trial of two bivalent trachoma vaccines in children of Punjab Indian villages. Am J Ophthalmol 63, Suppl:1639–1644.

Di Francesco, A., Donati, M., Morandi, F., Renzi, M., Masia, M.A., Ostanello, F., Salvatore, D., Cevenini, R., and Baldelli, R. (2011). Seroepidemiologic survey for Chlamydia suis in wild boar (Sus scrofa) populations in Italy. J Wildl Dis 47, 709–712.10.7589/0090-3558-47.3.709.

Dirks, J.A., Wolffs, P.F., Dukers-Muijrers, N.H., Brink, A.A., Speksnijder, A.G., and Hoebe, C.J. (2015). *Chlamydia trachomatis* load in population-based screening and STI-clinics: implications for screening policy. PLoS One 10, e0121433.10.1371/journal.pone.0121433.

Dixit, S., Singh, S.R., Yilma, A.N., Agee, R.D., 2nd, Taha, M., and Dennis, V.A. (2014). Poly(lactic acid)-poly(ethylene glycol) nanoparticles provide sustained delivery of a *Chlamydia trachomatis* recombinant MOMP peptide and potentiate systemic adaptive immune responses in mice. Nanomedicine: nanotechnology, biology, and medicine 10, 1311–1321.10.1016/j.nano.2014.02.009.

Dodet, B. (2014). Current barriers, challenges and opportunities for the development of effective STI vaccines: point of view of vaccine producers, biotech companies and funding agencies. Vaccine 32, 1624–1629.10.1016/j.vaccine.2013.08.032.

Donati, M., Sambri, V., Comanducci, M., Di Leo, K., Storni, E., Giacani, L., Ratti, G., and Cevenini, R. (2003). DNA immunization with pgp3 gene of *Chlamydia trachomatis* inhibits the spread of chlamydial infection from the lower to the upper genital tract in C3H/HeN mice. Vaccine 21, 1089–1093.

Dong-Ji, Z., Yang, X., Shen, C., Lu, H., Murdin, A., and Brunham, R.C. (2000). Priming with *Chlamydia trachomatis* major outer membrane protein (MOMP) DNA followed by MOMP ISCOM boosting enhances protection and is associated with increased immunoglobulin A and Th1 cellular immune responses. Infect Immun 68, 3074–3078.10.1128/iai.68.6.3074-3078.2000.

Dong, F., Zhong, Y., Arulanandam, B., and Zhong, G. (2005). Production of a proteolytically active protein, chlamydial protease/proteasome-like activity factor, by five different Chlamydia species. Infect Immun 73, 1868–1872.10.1128/IAI.73.3.1868-1872.2005.

Eko, F.O., Ekong, E., He, Q., Black, C.M., and Igietseme, J.U. (2011). Induction of immune memory by a multisubunit chlamydial vaccine. Vaccine 29, 1472–1480.10.1016/j.vaccine.2010.12.024.

Eko, F.O., He, Q., Brown, T., McMillan, L., Ifere, G.O., Ananaba, G.A., Lyn, D., Lubitz, W., Kellar, K.L., Black, C.M., *et al.* (2004). A novel recombinant multisubunit vaccine against Chlamydia. J Immunol 173, 3375–3382.10.4049/jimmunol.173.5.3375.

Eko, F.O., Mania-Pramanik, J., Pais, R., Pan, Q., Okenu, D.M., Johnson, A., Ibegbu, C., He, C., He, Q., Russell, R., *et al.* (2014). Vibrio choleraee ghosts (VCG) exert immunomodulatory effect on dendritic cells for enhanced antigen presentation and induction of protective immunity. BMC Immunol 15, 584.10.1186/s12865-014-0056-x.

Ekong, E.E., Okenu, D.N., Mania-Pramanik, J., He, Q., Igietseme, J.U., Ananaba, G.A., Lyn, D., Black, C., and Eko, F.O. (2009). A Vibrio choleraee ghost-based subunit vaccine induces cross-protective chlamydial immunity that is enhanced by CTA2B, the nontoxic derivative of cholera toxin. FEMS Immunol Med Microbiol 55, 280–291.10.1111/j.1574-695X.2008.00493.x.

Englund, S., af Segerstad, C.H., Arnlund, F., Westergren, E., and Jacobson, M. (2012). The occurrence of Chlamydia spp. in pigs with and without clinical disease. BMC Vet Res 8, 9.10.1186/1746-6148-8-9.

Entrican, G., Wheelhouse, N., Wattegedera, S.R., and Longbottom, D. (2012). New challenges for vaccination to prevent chlamydial abortion in sheep. Comp Immunol Microbiol Infect Dis 35, 271–276.10.1016/j.cimid.2011.12.001.

Farley, T.A., Cohen, D.A., and Elkins, W. (2003). Asymptomatic sexually transmitted diseases: the case for screening. Preventative medicine 36, 502–509.

Farris, C.M., and Morrison, R.P. (2011). Vaccination against Chlamydia genital infection utilizing the murine C. muridarum model. Infect Immun 79, 986–996.10.1128/IAI.00881-10.

Farris, C.M., Morrison, S.G., and Morrison, R.P. (2010). CD4+ T cells and antibody are required for optimal major outer membrane protein vaccine-induced immunity to Chlamydia muridarum genital infection. Infect Immun 78, 4374–4383.10.1128/IAI.00622-10.

Finco, O., Frigimelica, E., Buricchi, F., Petracca, R., Galli, G., Faenzi, E., Meoni, E., Bonci, A., Agnusdei, M., Nardelli, F., *et al.* (2011). Approach to discover T- and B-cell antigens of intracellular pathogens applied to the design of *Chlamydia trachomatis* vaccines. Proc Natl Acad Sci U S A 108, 9969–9974.10.1073/pnas.1101756108.

Fiorino, F., Pettini, E., Pozzi, G., Medaglini, D., and Ciabattini, A. (2013). Prime-boost strategies in mucosal immunization affect local IgA production and the type of th response. Front Immunol 4, 128.10.3389/fimmu.2013.00128.

Fitch, W.M., Peterson, E.M., and de la Maza, L.M. (1993). Phylogenetic analysis of the outer-membrane-protein genes of Chlamydiae, and its implication for vaccine development. Mol Biol Evol 10, 892–913.10.1093/oxfordjournals.molbev.a040048.

Fling, S.P., Sutherland, R.A., Steele, L.N., Hess, B., D'Orazio, S.E., Maisonneuve, J., Lampe, M.F., Probst, P., and Starnbach, M.N. (2001). CD8+ T cells recognize an inclusion membrane-associated protein from the vacuolar pathogen *Chlamydia trachomatis*. Proc Natl Acad Sci U S A 98, 1160–1165.10.1073/pnas.98.3.1160.

Fukushi, H., and Hirai, K. (1992). Proposal of Chlamydia pecorum sp. nov. for Chlamydia strains derived from ruminants. Int J Syst Bacteriol 42, 306–308.10.1099/00207713-42-2--306.

Galaleldeen, A., Taylor, A.B., Chen, D., Schuermann, J.P., Holloway, S.P., Hou, S., Gong, S., Zhong, G., and Hart, P.J. (2013). Structure of the *Chlamydia trachomatis* immunodominant antigen Pgp3. J Biol Chem 288, 22068–22079.10.1074/jbc.M113.475012.

Gallichan, W.S., and Rosenthal, K.L. (1995). Specific secretory immune responses in the female genital tract following intranasal immunization with a recombinant adenovirus expressing glycoprotein B of herpes simplex virus. Vaccine 13, 1589–1595.

Garcia de la Fuente, J.N., Gutierrez-Martin, C.B., Ortega, N., Rodriguez-Ferri, E.F., del Rio, M.L., Gonzalez, O.R., and Salinas, J. (2004). Efficacy of different commercial and new inactivated vaccines against ovine enzootic abortion. Vet Microbiol 100, 65–76.10.1016/j.vetmic.2004.01.015.

Gaydos, C.A., Summersgill, J.T., Sahney, N.N., Ramirez, J.A., and Quinn, T.C. (1996). Replication of *Chlamydia pneumoniae* in vitro in human macrophages, endothelial cells, and aortic artery smooth muscle cells. Infect Immun 64, 1614–1620.

Geisler, W.M., Lensing, S.Y., Press, C.G., and Hook, E.W., 3rd (2013). Spontaneous resolution of genital *Chlamydia trachomatis* infection in women and protection from reinfection. J Infect Dis 207, 1850–1856.10.1093/infdis/jit094.

Gondek, D.C., Olive, A.J., Stary, G., and Starnbach, M.N. (2012). CD4+ T cells are necessary and sufficient to confer protection against *Chlamydia trachomatis* infection in the murine upper genital tract. J Immunol 189, 2441–2449.10.4049/jimmunol.1103032.

Gondek, D.C., Roan, N.R., and Starnbach, M.N. (2009). T cell responses in the absence of IFN-gamma exacerbate uterine infection with *Chlamydia trachomatis*. J Immunol 183, 1313–1319.10.4049/jimmunol.0900295.

Gottlieb, S.L., Deal, C.D., Giersing, B., Rees, H., Bolan, G., Johnston, C., Timms, P., Grey-Owen, S.D., Jerse, A.E., Cameron, C.E., *et al.* (2016). The global roadmap for advancing development of vaccines against sexually transmitted infections: Update and next steps. Vaccine 34, 2939–2947.10.1016/j.vaccine.2016.03.111.

Gottlieb, S.L., and Johnston, C. (2017). Future prospects for new vaccines against sexually transmitted infections. Curr Opin Infect Dis 30, 77–86.10.1097/QCO.0000000000000343.

Gottlieb, S.L., Low, N., Newman, L.M., Bolan, G., Kamb, M., and Broutet, N. (2014). Towards global prevention of sexually transmitted infections (STIs): the need for STI vaccines. Vaccine 32, 1527–1535.10.1016/j.vaccine.2013.07.087.

Gotz, H., Lindback, J., Ripa, T., Arneborn, M., Ramsted, K., and Ekdahl, K. (2002). Is the increase in notifications of *Chlamydia trachomatis* infections in Sweden the result of changes in prevalence, sampling frequency or diagnostic methods? Scandinavian journal of infectious diseases 34, 28–34.

Grasse, M., Rosenkrands, I., Olsen, A., Follmann, F., and Dietrich, J. (2018). A flow cytometry-based assay to determine the phagocytic activity of both clinical and nonclinical antibody samples against *Chlamydia trachomatis*. Cytometry A 93, 525–532.10.1002/cyto.a.23353.

Grey, R.T., Beagley, K.W., Timms, P., and Wilson, D.P. (2009). Modelling the impact of potential vaccines on epidemics of sexually transmitted *Chlamydia trachomatis* infection. J Infect Dis 199, 1680–1688.10.1086/598983.

Grayston, J.T., Wang, S.P., Yang, Y.F., and Woolridge, R.L. (1962). The effect of trachoma virus vaccine on the course of experimental trachoma infection in blind human volunteers. J Exp Med 115, 1009–1022.10.1084/jem.115.5.1009.

Grayston, J.T., Woolridge, R.L., Wang, S.P., Yen, C.H., Yang, C.Y., Cheng, K.H., and Chang, I.H. (1963). Field studies of protection from infection by experimental trachoma virus vaccine in preschool-aged children on Taiwan. Proc Soc Exp Biol Med 112, 589–595.

Hadad, R., Marks, E., Kalbina, I., Schon, K., Unemo, M., Lycke, N., Strid, A., and Andersson, S. (2016). Protection against genital tract *Chlamydia trachomatis* infection following intranasal immunization with a novel recombinant MOMP VS2/4 antigen. APMIS 124, 1078–1086.10.1111/apm.12605.

Hafner, L., Beagley, K., and Timms, P. (2008). *Chlamydia trachomatis* infection: host immune responses and potential vaccines. Mucosal immunology 1, 116–130.10.1038/mi.2007.19.

Haggerty, C.L., Gottlieb, S.L., Taylor, B.D., Low, N., Xu, F., and Ness, R.B. (2010). Risk of sequelae after *Chlamydia trachomatis* genital infection in women. J Infect Dis 201 Suppl 2, S134–155.10.1086/652395.

Hahn, D.L., Azenabor, A.A., Beatty, W.L., and Byrne, G.I. (2002). *Chlamydia pneumoniae* as a respiratory pathogen. Front Biosci 7, e66–76.

Hanke, T., Blanchard, T.J., Schneider, J., Hannan, C.M., Becker, M., Gilbert, S.C., Hill, A.V., Smith, G.L., and McMichael, A. (1998). Enhancement of MHC class I-restricted peptide-specific T cell induction by a DNA prime/MVA boost vaccination regime. Vaccine 16, 439–445.

Hansen, J., Jensen, K.T., Follmann, F., Agger, E.M., Theisen, M., and Andersen, P. (2008). Liposome delivery of Chlamydia muridarum major outer membrane protein primes a Th1 response that protects against genital chlamydial infection in a mouse model. J Infect Dis 198, 758–767.10.1086/590670.

Hasegawa, H., Ichinohe, T., Strong, P., Watanabe, I., Ito, S., Tamura, S., Takahashi, H., Sawa, H., Chiba, J., Kurata, T., *et al.* (2005). Protection against influenza virus infection by intranasal administration of haemagglutinin vaccine with chitin microparticles as an adjuvant. J Med Virol 75, 130–136.10.1002/jmv.20247.

Hayton, E.J., Rose, A., Ibrahimsa, U., Del Sorbo, M., Capone, S., Crook, A., Black, A.P., Dorrell, L., and Hanke, T. (2014). Safety and tolerability of conserved region vaccines vectored by plasmid DNA, simian adenovirus and modified vaccinia virus ankara administered to human immunodeficiency virus type 1-uninfected adults in a randomized, single-blind phase I trial. PLoS One 9, e101591.10.1371/journal.pone.0101591.

He, Q., Martinez-Sobrido, L., Eko, F.O., Palese, P., Garcia-Sastre, A., Lyn, D., Okenu, D., Bandea, C., Ananaba, G.A., Black, C.M., *et al.* (2007). Live-attenuated influenza viruses as delivery vectors for Chlamydia vaccines. Immunology 122, 28–37.10.1111/j.1365-2567.2007.02608.x.

Henderson, I.R., and Lam, A.C. (2001). Polymorphic proteins of Chlamydia spp.--autotransporters beyond the Proteobacteria. Trends Microbiol 9, 573–578.

Hickey, D.K., Aldwell, F.E., and Beagley, K.W. (2009). Transcutaneous immunization with a novel lipid-based adjuvant protects against Chlamydia genital and respiratory infections. Vaccine 27, 6217–6225.10.1016/j.vaccine.2009.08.001.

Hickey, D.K., Aldwell, F.E., and Beagley, K.W. (2010). Oral immunization with a novel lipid-based adjuvant protects against genital Chlamydia infection. Vaccine 28, 1668–1672.10.1016/j.vaccine.2009.12.010.

Holmgren, J., and Czerkinsky, C. (2005). Mucosal immunity and vaccines. Nat Med 11, S45–53.10.1038/nm1213.

Hosenfeld, C.B., Workowski, K.A., Berman, S., Zaidi, A., Dyson, J., Mosure, D., Bolan, G., and Bauer, H.M. (2009). Repeat infection with Chlamydia and gonorrhea among females: a systematic review of the literature. Sex Transm Dis 36, 478–489.10.1097/OLQ.0b013e3181a2a933.

Hou, S., Dong, X., Yang, Z., Li, Z., Liu, Q., and Zhong, G. (2015). Chlamydial plasmid-encoded virulence factor Pgp3 neutralizes the antichlamydial activity of human cathelicidin LL-37. Infect Immun 83, 4701–4709.10.1128/IAI.00746-15.

Hu, V.H., Holland, M.J., and Burton, M.J. (2013). Trachoma: protective and pathogenic ocular immune responses to *Chlamydia trachomatis*. PLoS neglected tropical diseases 7, e2020.10.1371/journal.pntd.0002020.

Huang, Z., Feng, Y., Chen, D., Wu, X., Huang, S., Wang, X., Xiao, X., Li, W., Huang, N., Gu, L., *et al.* (2008). Structural basis for activation and inhibition of the secreted chlamydia protease CPAF. Cell Host Microbe 4, 529–542.10.1016/j.chom.2008.10.005.

Hyde, S.R., and Benirschke, K. (1997). Gestational psittacosis: case report and literature review. Mod Pathol 10, 602–607.

Igietseme, J.U. (1996). The molecular mechanism of T-cell control of Chlamydia in mice: Role of nitric oxide. Immunology 87, 1–8.

Igietseme, J.U., Eko, F.O., and Black, C.M. (2011). Chlamydia vaccines: recent developments and the role of adjuvants in future formulations. Expert Rev Vaccines 10, 1585–1596.10.1586/erv.11.139.

Igietseme, J.U., Magee, D.M., Williams, D.M., and Rank, R.G. (1994). Role for CD8+ T cells in antichlamydial immunity defined by Chlamydia-specific T-lymphocyte clones. Infect Immun 62, 5195–5197.

Igietseme, J.U., and Murdin, A. (2000). Induction of protective immunity against *Chlamydia trachomatis* genital infection by a vaccine based on major outer membrane protein-lipophilic immune response–stimulating complexes. Infect Immun 68, 6798–6806.10.1128/iai.68.12.6798-6806.2000.

Ingalls, R.R., Rice, P.A., Qureshi, N., Takayama, K., Lin, J.S., and Golenbock, D.T. (1995). The inflammatory cytokine response to *Chlamydia trachomatis* infection is endotoxin mediated. Infect Immun 63, 3125–3130.

Inic-Kanada, A., Stojanovic, M., Marinkovic, E., Becker, E., Stein, E., Lukic, I., Djokic, R., Schuerer, N., Hegemann, J.H., and Barisani-Asenbauer, T. (2016). A Probiotic Adjuvant Lactobacillus rhamnosus Enhances Specific Immune Responses after Ocular Mucosal Immunization with Chlamydial Polymorphic Membrane Protein C. PLoS One 11, e0157875.10.1371/journal.pone.0157875.

Inic-Kanada, A., Stojanovic, M., Schlacher, S., Stein, E., Belij-Rammerstorfer, S., Marinkovic, E., Lukic, I., Montanaro, J., Schuerer, N., Bintner, N., *et al.* (2015). Delivery of a Chlamydial Adhesin N-PmpC Subunit Vaccine to the Ocular Mucosa Using Particulate Carriers. PLoS One 10, e0144380.10.1371/journal.pone.0144380.

Jawetz, E., Rose, L., Hanna, L., and Thygeson, P. (1965). Experimental inclusion conjunctivitis in man: measurements of infectivity and resistance. JAMA 194, 620–632.

Jiang, J., Liu, G., Kickhoefer, V.A., Rome, L.H., Li, L.X., McSorley, S.J., and Kelly, K.A. (2017). A Protective Vaccine against Chlamydia Genital Infection Using Vault Nanoparticles without an Added Adjuvant. Vaccines (Basel) 5.10.3390/vaccines5010003.

Jiang, P., Du, W., Xiong, Y., Lv, Y., Feng, J., Zhu, S., Xue, X., Chen, S., and Zhang, L. (2015). Hepatitis B virus core antigen as a carrier for *Chlamydia trachomatis* MOMP multi-epitope peptide enhances protection against genital chlamydial infection. Oncotarget 6, 43281–43292.10.18632/oncotarget.6533.

Johansson, E.L., Rask, C., Fredriksson, M., Eriksson, K., Czerkinsky, C., and Holmgren, J. (1998). Antibodies and antibody-secreting cells in the female genital tract after vaginal or intranasal immunization with cholera toxin B subunit or conjugates. Infect Immun 66, 514–520.

Johansson, E.L., Wassen, L., Holmgren, J., Jertborn, M., and Rudin, A. (2001). Nasal and vaginal vaccinations have differential effects on antibody responses in vaginal and cervical secretions in humans. Infect Immun 69, 7481–7486.10.1128/IAI.69.12.7481-7486.2001.

Johnson, L.F., and Lewis, D.A. (2008). The effect of genital tract infections on HIV-1 shedding in the genital tract: a systematic review and meta-analysis. Sex Transm Dis 35, 946–959.10.1097/OLQ.0b013e3181812d15.

Johnson, R.E., Newhall, W.J., Papp, J.R., Knapp, J.S., Black, C.M., Gift, T.L., Steece, R., Markowitz, L.E., Devine, O.J., Walsh, C.M., *et al.* (2002). Screening tests to detect *Chlamydia trachomatis* and Neisseria gonorrhoeae infections--2002. MMWR Recomm Rep 51, 1–38; quiz CE31–34.

Kalbina, I., Wallin, A., Lindh, I., Engstrom, P., Andersson, S., and Strid, K. (2011). A novel chimeric MOMP antigen expressed in Escherichia coli, Arabidopsis thaliana, and Daucus carota as a potential *Chlamydia trachomatis* vaccine candidate. Protein Expr Purif 80, 194–202.10.1016/j.pep.2011.08.010.

Kari, L., Whitmire, W.M., Crane, D.D., Reveneau, N., Carlson, J.H., Goheen, M.M., Peterson, E.M., Pal, S., de la Maza, L.M., and Caldwell, H.D. (2009). *Chlamydia trachomatis* native major outer membrane protein induces partial protection in nonhuman primates: implication for a trachoma transmission-blocking vaccine. J Immunol 182, 8063–8070.10.4049/jimmunol.0804375.

Kari, L., Whitmire, W.M., Olivares-Zavaleta, N., Goheen, M.M., Taylor, L.D., Carlson, J.H., Sturdevant, G.L., Lu, C., Bakios, L.E., Randall, L.B., *et al.* (2011). A live-attenuated chlamydial vaccine protects against trachoma in nonhuman primates. J Exp Med 208, 2217–2223.10.1084/jem.20111266.

Karunakaran, K.P., Yu, H., Jiang, X., Chan, Q., Moon, K.M., Foster, L.J., and Brunham, R.C. (2015). Outer membrane proteins preferentially load MHC class II peptides: implications for a *Chlamydia trachomatis* T cell vaccine. Vaccine 33, 2159–2166.10.1016/j.vaccine.2015.02.055.

Kawa, D.E., Schachter, J., and Stephens, R.S. (2004). Immune response to the *Chlamydia trachomatis* outer membrane protein PorB. Vaccine 22, 4282–4286.10.1016/j.vaccine.2004.04.035.

Kelly, K.A., Robinson, E.A., and Rank, R.G. (1996). Initial route of antigen administration alters the T-cell cytokine profile produced in response to the mouse pneumonitis biovar of *Chlamydia trachomatis* following genital infection. Infect Immun 64, 4976–4983.

Kimani, J., Maclean, I.W., Bwayo, J.J., MacDonald, K., Oyugi, J., Maitha, G.M., Peeling, R.W., Cheang, M., Nagelkerke, N.J., Plummer, F.A., *et al.* (1996). Risk factors for *Chlamydia trachomatis* pelvic inflammatory disease among sex workers in Nairobi, Kenya. J Infect Dis 173, 1437–1444.10.1093/infdis/173.6.1437.

Knight, S.C., Iqball, S., Woods, C., Stagg, A., Ward, M.E., and Tuffrey, M. (1995). A peptide of *Chlamydia trachomatis* shown to be a primary T-cell epitope in vitro induces cell-mediated immunity in vivo. Immunology 85, 8–15.

Knudsen, N.P., Olsen, A., Buonsanti, C., Follmann, F., Zhang, Y., Coler, R.N., Fox, C.B., Meinke, A., D'Oro, U., Casini, D., *et al.* (2016). Different human vaccine adjuvants promote distinct antigen-independent immunological signatures tailored to different pathogens. Sci Rep 6, 19570.10.1038/srep19570.

Kollipara, A., George, C., Hanger, J., Loader, J., Polkinghorne, A., Beagley, K., and Timms, P. (2012). Vaccination of healthy and diseased koalas (Phascolarctos cinereus) with a Chlamydia pecorum multi-subunit vaccine: evaluation of immunity and pathology. Vaccine 30, 1875–1885.10.1016/j.vaccine.2011.12.125.

Kollipara, A., Polkinghorne, A., Wan, C., Kanyoka, P., Hanger, J., Loader, J., Callaghan, J., Bell, A., Ellis, W., Fitzgibbon, S., *et al.* (2013). Genetic diversity of Chlamydia pecorum strains in wild koala locations across Australia and the implications for a recombinant C. pecorum major outer membrane protein based vaccine. Vet Microbiol 167, 513–522.10.1016/j.vetmic.2013.08.009.

Korenromp, E.L., Sudaryo, M.K., de Vlas, S.J., Grey, R.H., Sewankambo, N.K., Serwadda, D., Wawer, M.J., and Habbema, J.D. (2002). What proportion of episodes of gonorrhoea and chlamydia becomes symptomatic? Int J STD AIDS 13, 91–101.10.1258/0956462021924712.

Korenromp, E.L., Wi, T., Resch, S., Stover, J., and Broutet, N. (2017). Costing of National STI Program Implementation for the Global STI Control Strategy for the Health Sector, 2016–2021. PLoS One 12, e0170773.10.1371/journal.pone.0170773.

Kuo, C.C., and Campbell, L.A. (2003). Chlamydial infections of the cardiovascular system. Front Biosci 8, e36–43.

Lamont, J.T. (1992). Mucus: the front line of intestinal mucosal defence. Ann N Y Acad Sci 664, 190–201.

Larsen, U. (2003). Infertility in central Africa. Trop Med Int Health 8, 354–367.

Lei, L., Chen, J., Hou, S., Ding, Y., Yang, Z., Zeng, H., Baseman, J., and Zhong, G. (2014). Reduced live organism recovery and lack of hydrosalpinx in mice infected with plasmid-free Chlamydia muridarum. Infect Immun 82, 983–992.10.1128/IAI.01543-13.

Levine, M. (2010). New generation vaccines. Informa Healthcare USA, Inc 4th edition.

Lewis, D.J., Huo, Z., Barnett, S., Kromann, I., Giemza, R., Galiza, E., Woodrow, M., Thierry-Carstensen, B., Andersen, P., Novicki, D., *et al.* (2009). Transient facial nerve paralysis (Bell's palsy) following intranasal delivery of a genetically detoxified mutant of Escherichia coli heat labile toxin. PLoS One 4, e6999.10.1371/journal.pone.0006999.

Li, W., Guentzel, M.N., Seshu, J., Zhong, G., Murthy, A.K., and Arulanandam, B.P. (2007). Induction of cross-serovar protection against genital chlamydial infection by a targeted multisubunit vaccination approach. Clin Vaccine Immunol 14, 1537–1544.10.1128/CVI.00274-07.

Li, W., Murthy, A.K., Guentzel, M.N., Chambers, J.P., Forsthuber, T.G., Seshu, J., Zhong, G., and Arulanandam, B.P. (2010). Immunization with a combination of integral chlamydial antigens and a defined secreted protein induces robust immunity against genital chlamydial challenge. Infect Immun 78, 3942–3949.10.1128/IAI.00346-10.

Li, W., Murthy, A.K., Lanka, G.K., Chetty, S.L., Yu, J.J., Chambers, J.P., Zhong, G., Forsthuber, T.G., Guentzel, M.N., and Arulanandam, B.P. (2013). A T cell epitope-based vaccine protects against chlamydial infection in HLA-DR4 transgenic mice. Vaccine 31, 5722–5728.10.1016/j.vaccine.2013.09.036.

Li, Z., Chen, D., Zhong, Y., Wang, S., and Zhong, G. (2008a). The chlamydial plasmid-encoded protein pgp3 is secreted into the cytosol of Chlamydia-infected cells. Infect Immun 76, 3415–3428.10.1128/IAI.01377-07.

Li, Z., Lu, C., Peng, B., Zeng, H., Zhou, Z., Wu, Y., and Zhong, G. (2012). Induction of protective immunity against Chlamydia muridarum intravaginal infection with a chlamydial glycogen phosphorylase. PLoS One 7, e32997.10.1371/journal.pone.0032997.

Li, Z., Wang, S., Wu, Y., Zhong, G., and Chen, D. (2008b). Immunization with chlamydial plasmid protein pORF5 DNA vaccine induces protective immunity against genital chlamydial infection in mice. Sci China C Life Sci 51, 973–980.10.1007/s11427-008-0130-9.

Li, Z., Zhong, Y., Lei, L., Wu, Y., Wang, S., and Zhong, G. (2008c). Antibodies from women urogenitally infected with *C. trachomatis* predominantly recognized the plasmid protein pgp3 in a conformation-dependent manner. BMC Microbiol 8, 90.10.1186/1471-2180-8-90.

Liang, S., Bulir, D., Kaushic, C., and Mahony, J. (2017). Considerations for the rational design of a Chlamydia vaccine. Human vaccines & immunotherapeutics 13, 831–835.10.1080/21645515.2016.1252886.

Lichtenwalner, A.B., Patton, D.L., Van Voorhis, W.C., Sweeney, Y.T., and Kuo, C.C. (2004). Heat shock protein 60 is the major antigen which stimulates delayed-type hypersensitivity reaction in the macaque model of *Chlamydia trachomatis* salpingitis. Infect Immun 72, 1159–1161.10.1128/iai.72.2.1159-1161.2004.

Littman, A.J., Jackson, L.A., and Vaughan, T.L. (2005). *Chlamydia pneumoniae* and lung cancer: epidemiologic evidence. Cancer Epidemiol Biomarkers Prev 14, 773–778.10.1158/1055-9965.EPI-04-0599.

Longbottom, D., and Coulter, L.J. (2003). Animal chlamydioses and zoonotic implications. J Comp Pathol 128, 217–244.

Longbottom, D., and Livingstone, M. (2006). Vaccination against chlamydial infections of man and animals. Vet J 171, 263–275.10.1016/j.tvjl.2004.09.006.

Loomis, W.P., and Starnbach, M.N. (2002). T cell responses to *Chlamydia trachomatis*. Curr Opin Microbiol 5, 87–91.

Loots, K., Loock, M.V., Vanrompay, D., and Goddeeris, B.M. (2006). CpG motifs as adjuvant in DNA vaccination against Chlamydophila psittaci in turkeys. Vaccine 24, 4598–4601.10.1016/j.vaccine.2005.08.042.

Lorenzen, E., Follmann, F., Boje, S., Erneholm, K., Olsen, A.W., Agerholm, J.S., Jungersen, G., and Andersen, P. (2015). Intramuscular Priming and Intranasal Boosting Induce Strong Genital Immunity Through Secretory IgA in Minipigs Infected with *Chlamydia trachomatis*. Front Immunol 6, 628.10.3389/fimmu.2015.00628.

Lowy, D.R., and Schiller, J.T. (2006). Prophylactic human papillomavirus vaccines. J Clin Invest 116, 1167–1173.10.1172/JCI28607.

Lu, C., Peng, B., Li, Z., Lei, L., Li, Z., Chen, L., He, Q., Zhong, G., and Wu, Y. (2013). Induction of protective immunity against Chlamydia muridarum intravaginal infection with the chlamydial immunodominant antigen macrophage infectivity potentiator. Microbes Infect 15, 329–338.10.1016/j.micinf.2013.02.001.

Lu, C., Zeng, H., Li, Z., Lei, L., Yeh, I.T., Wu, Y., and Zhong, G. (2012). Protective immunity against mouse upper genital tract pathology correlates with high IFNgamma but low IL-17 T cell and anti-secretion protein antibody responses induced by replicating chlamydial organisms in the airway. Vaccine 30, 475–485.10.1016/j.vaccine.2011.10.059.

Lu, H., Xing, Z., and Brunham, R.C. (2002). GM-CSF transgene-based adjuvant allows the establishment of protective mucosal immunity following vaccination with inactivated *Chlamydia trachomatis*. J Immunol 169, 6324–6331.10.4049/jimmunol.169.11.6324.

Lu, H., and Zhong, G. (1999). Interleukin-12 production is required for chlamydial antigen-pulsed dendritic cells to induce protection against live *Chlamydia trachomatis* infection. Infect Immun 67, 1763–1769.

Mabey, D.C., Hu, V., Bailey, R.L., Burton, M.J., and Holland, M.J. (2014). Towards a safe and effective chlamydial vaccine: lessons from the eye. Vaccine 32, 1572–1578.10.1016/j.vaccine.2013.10.016.

Macmillan, L., Ifere, G.O., He, Q., Igietseme, J.U., Kellar, K.L., Okenu, D.M., and Eko, F.O. (2007). A recombinant multivalent combination vaccine protects against Chlamydia and genital herpes. FEMS Immunol Med Microbiol 49, 46–55.10.1111/j.1574-695X.2006.00165.x.

Marmion, B.P. (1980). Prospects for new viral vaccines. Philos Trans R Soc Lond B Biol Sci 290, 395–407.10.1098/rstb.1980.0103.

Massari, P., Toussi, D.N., Tifrea, D.F., and de la Maza, L.M. (2013). Toll-like receptor 2-dependent activity of native major outer membrane protein proteosomes of *Chlamydia trachomatis*. Infect Immun 81, 303–310.10.1128/IAI.01062-12.

McConnell, E.L., Basit, A.W., and Murdan, S. (2008). Colonic antigen administration induces significantly higher humoral levels of colonic and vaginal IgA, and serum IgG compared to oral administration. Vaccine 26, 639–646.10.1016/j.vaccine.2007.11.071.

McShane, H., Brookes, R., Gilbert, S.C., and Hill, A.V. (2001). Enhanced immunogenicity of CD4(+) t-cell responses and protective efficacy of a DNA-modified vaccinia virus Ankara prime-boost vaccination regimen for murine tuberculosis. Infect Immun 69, 681–686.10.1128/IAI.69.2.681-686.2001.

Meagher, M., Quinn, W.J., and Stackhouse, L. (1992). Chlamydial-caused infectious keratoconjunctivitis in bighorn sheep of Yellowstone National Park. J Wildl Dis 28, 171–176.10.7589/0090-3558-28.2.171.

Meoni, E., Faenzi, E., Frigimelica, E., Zedda, L., Skibinski, D., Giovinazzi, S., Bonci, A., Petracca, R., Bartolini, E., Galli, G., *et al.* (2009). CT043, a protective antigen that induces a CD4+ Th1 response during *Chlamydia trachomatis* infection in mice and humans. Infect Immun 77, 4168–4176.10.1128/IAI.00344-09.

Mestecky, J., Moldoveanu, Z., and Russell, M.W. (2005). Immunologic uniqueness of the genital tract: challenge for vaccine development. Am J Reprod Immunol 53, 208–214.10.1111/j.1600-0897.2005.00267.x.

Molano, M., Meijer, C.J., Weiderpass, E., Arslan, A., Posso, H., Franceschi, S., Ronderos, M., Munoz, N., and van den Brule, A.J. (2005). The natural course of *Chlamydia trachomatis* infection in asymptomatic Colombian women: a 5-year follow-up study. J Infect Dis 191, 907–916.10.1086/428287.

Molleken, K., Schmidt, E., and Hegemann, J.H. (2010). Members of the Pmp protein family of *Chlamydia pneumoniae* mediate adhesion to human cells via short repetitive peptide motifs. Mol Microbiol 78, 1004–1017.10.1111/j.1365-2958.2010.07386.x.

Montanaro, J., Inic-Kanada, A., Ladurner, A., Stein, E., Belij, S., Bintner, N., Schlacher, S., Schuerer, N., Mayr, U.B., Lubitz, W., *et al.* (2015). Escherichia coli Nissle 1917 bacterial ghosts retain crucial surface properties and express chlamydial antigen: an imaging study of a delivery system for the ocular surface. Drug Des Devel Ther 9, 3741–3754.10.2147/DDDT.S84370.

Moore, T., Ananaba, G.A., Bolier, J., Bowers, S., Belay, T., Eko, F.O., and Igietseme, J.U. (2002). Fc receptor regulation of protective immunity against *Chlamydia trachomatis*. Immunology 105, 213–221.10.1046/j.0019-2805.2001.01354.x.

Moore, T., Ekworomadu, C.O., Eko, F.O., MacMillan, L., Ramey, K., Ananaba, G.A., Patrickson, J.W., Nagappan, P.R., Lyn, D., Black, C.M., *et al.* (2003). Fc receptor-mediated antibody regulation of T cell immunity against intracellular pathogens. J Infect Dis 188, 617–624.10.1086/377134.

Mordhorst, C.H. (1967). Experimental infections and immunogenicity of TRIC agents in monkeys. Am J Ophthalmol 63, Suppl:1603–1615.

Morrison, R.P. (2000). Differential sensitivities of *Chlamydia trachomatis* strains to inhibitory effects of gamma interferon. Infect Immun 68, 6038–6040.10.1128/iai.68.10.6038-6040.2000.

Morrison, R.P., Belland, R.J., Lyng, K., and Caldwell, H.D. (1989a). Chlamydial disease pathogenesis. The 57-kD chlamydial hypersensitivity antigen is a stress response protein. J Exp Med 170, 1271–1283.10.1084/jem.170.4.1271.

Morrison, R.P., and Caldwell, H.D. (2002). Immunity to murine chlamydial genital infection. Infect Immun 70, 2741–2751.10.1128/iai.70.6.2741-2751.2002.

Morrison, R.P., Feilzer, K., and Tumas, D.B. (1995). Gene knockout mice establish a primary protective role for major histocompatibility complex class II-restricted responses in *Chlamydia trachomatis* genital tract infection. Infect Immun 63, 4661–4668.

Morrison, R.P., Lyng, K., and Caldwell, H.D. (1989b). Chlamydial disease pathogenesis. Ocular hypersensitivity elicited by a genus-specific 57-kD protein. J Exp Med 169, 663–675.10.1084/jem.169.3.663.

Morrison, S.G., and Morrison, R.P. (2001). Resolution of secondary *Chlamydia trachomatis* genital tract infection in immune mice with depletion of both CD4+ and

CD8+ T cells. Infect Immun 69, 2643–2649.10.1128/IAI.69.4.2643-2649.2001.

Morrison, S.G., and Morrison, R.P. (2005). The protective effect of antibody in immunity to murine chlamydial genital tract reinfection is independent of immunoglobulin A. In Infect Immun, pp. 6183–6186.

Morrison, S.G., Su, H., Caldwell, H.D., and Morrison, R.P. (2000). Immunity to murine *Chlamydia trachomatis* genital tract reinfection involves B cells and CD4(+) T cells but not CD8(+) T cells. Infect Immun 68, 6979–6987.10.1128/iai.68.12.6979-6987.2000.

Mosolygo, T., Szabo, A.M., Balogh, E.P., Faludi, I., Virok, D.P., Endresz, V., Samu, A., Krenacs, T., and Burian, K. (2014). Protection promoted by pGP3 or pGP4 against Chlamydia muridarum is mediated by CD4(+) cells in C57BL/6N mice. Vaccine 32, 5228–5233.10.1016/j.vaccine.2014.07.039.

Muller, T., Becker, E., Stallmann, S., Waldhuber, A., Rommler-Dreher, F., Albrecht, S., Mohr, F., Hegemann, J.H., and Miethke, T. (2017). Vaccination with the polymorphic membrane protein A reduces Chlamydia muridarum induced genital tract pathology. Vaccine 35, 2801–2810.10.1016/j.vaccine.2017.04.017.

Murdin, A.D., Su, H., Manning, D.S., Klein, M.H., Parnell, M.J., and Caldwell, H.D. (1993). A poliovirus hybrid expressing a neutralization epitope from the major outer membrane protein of *Chlamydia trachomatis* is highly immunogenic. Infect Immun 61, 4406–4414.

Murphey, C., Murthy, A.K., Meier, P.A., Neal Guentzel, M., Zhong, G., and Arulanandam, B.P. (2006). The protective efficacy of chlamydial protease-like activity factor vaccination is dependent upon CD4+ T cells. Cellular immunology 242, 110–117.10.1016/j.cellimm.2006.10.002.

Murthy, A.K., Chaganty, B.K., Li, W., Guentzel, M.N., Chambers, J.P., Seshu, J., Zhong, G., and Arulanandam, B.P. (2009). A limited role for antibody in protective immunity induced by rCPAF and CpG vaccination against primary genital Chlamydia muridarum challenge. FEMS Immunol Med Microbiol 55, 271–279.10.1111/j.1574-695X.2008.00517.x.

Murthy, A.K., Chambers, J.P., Meier, P.A., Zhong, G., and Arulanandam, B.P. (2007). Intranasal vaccination with a secreted chlamydial protein enhances resolution of genital Chlamydia muridarum infection, protects against oviduct pathology, and is highly dependent upon endogenous gamma interferon production. Infect Immun 75, 666–676.10.1128/IAI.01280-06.

Murthy, A.K., Cong, Y., Murphey, C., Guentzel, M.N., Forsthuber, T.G., Zhong, G., and Arulanandam, B.P. (2006). Chlamydial protease-like activity factor induces protective immunity against genital chlamydial infection in transgenic mice that express the human HLA-DR4 allele. Infect Immun 74, 6722–6729.10.1128/IAI.01119-06.

Murthy, A.K., Li, W., Guentzel, M.N., Zhong, G., and Arulanandam, B.P. (2011). Vaccination with the defined chlamydial secreted protein CPAF induces robust protection against female infertility following repeated genital chlamydial challenge. Vaccine 29, 2519–2522.10.1016/j.vaccine.2011.01.074.

Mutsch, M., Zhou, W., Rhodes, P., Bopp, M., Chen, R.T., Linder, T., Spyr, C., and Steffen, R. (2004). Use of the inactivated intranasal influenza vaccine and the risk of Bell's palsy in Switzerland. N Engl J Med 350, 896–903.10.1056/NEJMoa030595.

Nagarajan, U.M., Ojcius, D.M., Stahl, L., Rank, R.G., and Darville, T. (2005). *Chlamydia trachomatis* induces expression of IFN-gamma-inducible protein 10 and IFN-beta independent of TLR2 and TLR4, but largely dependent on MyD88. J Immunol 175, 450–460.10.4049/jimmunol.175.1.450.

Naglak, E.K., Morrison, S.G., and Morrison, R.P. (2016). IFNgamma is Required for Optimal Antibody-Mediated Immunity against Genital Chlamydia Infection. Infect Immun 84, 3232–3242.10.1128/IAI.00749-16.

Network, K.R. (2012). Conserving koalas in the 21st century: synthesizing the dynamics of Australia's koala populations.

Newman, L., Rowley, J., Vander Hoorn, S., Wijesooriya, N.S., Unemo, M., Low, N., Stevens, G., Gottlieb, S., Kiarie, J., and Temmerman, M. (2015). Global Estimates of the Prevalence and Incidence of Four Curable Sexually Transmitted Infections in 2012 Based on Systematic Review and Global Reporting. PLoS One 10, e0143304.10.1371/journal.pone.0143304.

Nichols, R.L., Bell, S.D., Jr., Haddad, N.A., and Bobb, A.A. (1969). Studies on trachoma. VI. Microbiological observations in a field trial in Saudi Arabia of bivalent rachoma vaccine at three dosage levels. The American journal of tropical medicine and hygiene 18, 723–730.

Nichols, R.L., Bell, S.D., Jr., Murray, E.S., Haddad, N.A., and Bobb, A.A. (1966). Studies on trachoma. V. Clinical observations in a field trial of bivalent trachoma vaccine at three dosage levels in Saudi Arabia. The American journal of tropical medicine and hygiene 15, 639–647.

Nicolle, C., Cuenod, A., and Baizot, L. (1913). Etude experimentale du trachome. Arch Inst Pasteur de Tunis, 157–182.

Nigg, C. (1942). An Unidentified Virus Which Produces Pneumonia and Systemic Infection in Mice. Science 95, 49–50.10.1126/science.95.2454.49-a.

Nikaido, H. (1992). Porins and specific channels of bacterial outer membranes. Mol Microbiol 6, 435–442.

Nikaido, H. (1994). Porins and specific diffusion channels in bacterial outer membranes. J Biol Chem 269, 3905–3908.

Nilsson, C., Hejdeman, B., Godoy-Ramirez, K., Tecleab, T., Scarlatti, G., Brave, A., Earl, P.L., Stout, R.R., Robb, M.L., Shattock, R.J., *et al.* (2015). HIV-DNA Given with or without Intradermal Electroporation Is Safe and Highly Immunogenic in Healthy Swedish HIV-1 DNA/MVA Vaccinees: A Phase I Randomized Trial. PLoS One 10, e0131748.10.1371/journal.pone.0131748.

Nunes, A., Gomes, J.P., Mead, S., Florindo, C., Correia, H., Borrego, M.J., and Dean, D. (2007). Comparative expression profiling of the *Chlamydia trachomatis* pmp gene family for clinical and reference strains. PLoS One 2, e878.10.1371/journal.pone.0000878.

Nunes, A., Nogueira, P.J., Borrego, M.J., and Gomes, J.P. (2010). Adaptive evolution of the *Chlamydia trachomatis* dominant antigen reveals distinct evolutionary scenarios for B- and T-cell epitopes: worldwide survey. PLoS One 5.10.1371/journal.pone.0013171.

O'Connell, C.M., Ingalls, R.R., Andrews, C.W., Jr., Scurlock, A.M., and Darville, T. (2007). Plasmid-deficient

Chlamydia muridarum fail to induce immune pathology and protect against oviduct disease. J Immunol 179, 4027–4034.10.4049/jimmunol.179.6.4027.

O'Hagan, D.T., and Fox, C.B. (2015). New generation adjuvants--from empiricism to rational design. Vaccine 33 Suppl 2, B14–20.10.1016/j.vaccine.2015.01.088.

O'Meara, C.P., Armitage, C.W., Andrew, D.W., Kollipara, A., Lycke, N.Y., Potter, A.A., Gerdts, V., Petrovsky, N., and Beagley, K.W. (2017). Multistage vaccines containing outer membrane, type III secretion system and inclusion membrane proteins protects against a Chlamydia genital tract infection and pathology. Vaccine 35, 3883–3888.10.1016/j.vaccine.2017.05.063.

O'Meara, C.P., Armitage, C.W., Harvie, M.C., Timms, P., Lycke, N.Y., and Beagley, K.W. (2013). Immunization with a MOMP-based vaccine protects mice against a pulmonary Chlamydia challenge and identifies a disconnection between infection and pathology. PLoS One 8, e61962.10.1371/journal.pone.0061962.

O'Meara, C.P., Armitage, C.W., Kollipara, A., Andrew, D.W., Trim, L., Plenderleith, M.B., and Beagley, K.W. (2016). Induction of partial immunity in both males and females is sufficient to protect females against sexual transmission of Chlamydia. Mucosal immunology 9, 1076–1088.10.1038/mi.2015.125.

Oakeshott, P., Kerry, S., Aghaizu, A., Atherton, H., Hay, S., Taylor-Robinson, D., Simms, I., and Hay, P. (2010). Randomized controlled trial of screening for *Chlamydia trachomatis* to prevent pelvic inflammatory disease: the POPI (prevention of pelvic infection) trial. BMJ (Clinical research ed 340, c1642.10.1136/bmj.c1642.

Olivares-Zavaleta, N., Whitmire, W., Gardner, D., and Caldwell, H.D. (2010). Immunization with the attenuated plasmidless *Chlamydia trachomatis* L2(25667R) strain provides partial protection in a murine model of female genitourinary tract infection. Vaccine 28, 1454–1462.10.1016/j.vaccine.2009.11.073.

Olivares-Zavaleta, N., Whitmire, W.M., Kari, L., Sturdevant, G.L., and Caldwell, H.D. (2014). CD8+ T cells define an unexpected role in live-attenuated vaccine protective immunity against *Chlamydia trachomatis* infection in macaques. J Immunol 192, 4648–4654.10.4049/jimmunol.1400120.

Olsen, A.W., Andersen, P., and Follmann, F. (2014). Characterization of protective immune responses promoted by human antigen targets in a urogenital *Chlamydia trachomatis* mouse model. Vaccine 32, 685–692.10.1016/j.vaccine.2013.11.100.

Olsen, A.W., Follmann, F., Erneholm, K., Rosenkrands, I., and Andersen, P. (2015). Protection Against *Chlamydia trachomatis* Infection and Upper Genital Tract Pathological Changes by Vaccine-Promoted Neutralizing Antibodies Directed to the VD4 of the Major Outer Membrane Protein. J Infect Dis 212, 978–989.10.1093/infdis/jiv137.

Olsen, A.W., Lorenzen, E.K., Rosenkrands, I., Follmann, F., and Andersen, P. (2017). Protective Effect of Vaccine Promoted Neutralizing Antibodies against the Intracellular Pathogen *Chlamydia trachomatis*. Front Immunol 8, 1652.10.3389/fimmu.2017.01652.

Olsen, A.W., Theisen, M., Christensen, D., Follmann, F., and Andersen, P. (2010). Protection against Chlamydia promoted by a subunit vaccine (CTH1) compared with a primary intranasal infection in a mouse genital challenge model. PLoS One 5, e10768.10.1371/journal.pone.0010768.

Ou, C., Tian, D., Ling, Y., Pan, Q., He, Q., Eko, F.O., and He, C. (2013). Evaluation of an ompA-based phage-mediated DNA vaccine against Chlamydia abortus in piglets. International immunopharmacology 16, 505–510.10.1016/j.intimp.2013.04.027.

Ouburg, S., Lyons, J.M., Land, J.A., den Hartog, J.E., Fennema, J.S., de Vries, H.J., Bruggeman, C.A., Ito, J.I., Pena, A.S., Lundberg, P.S., *et al.* (2009). TLR9 KO mice, haplotypes and CPG indices in *Chlamydia trachomatis* infection. Drugs Today (Barc) 45 Suppl B, 83–93.

Owusu-Edusei, K., Jr., Chesson, H.W., Gift, T.L., Brunham, R.C., and Bolan, G. (2015). Cost-effectiveness of Chlamydia vaccination programs for young women. Emerg Infect Dis 21, 960–968.10.3201/eid2106.141270.

Paes, W., Brown, N., Brzozowski, A.M., Coler, R., Reed, S., Carter, D., Bland, M., Kaye, P.M., and Lacey, C.J.N. (2016). Recombinant polymorphic membrane protein D in combination with a novel, second-generation lipid adjuvant protects against intra-vaginal *Chlamydia trachomatis* infection in mice. Vaccine 34, 4123–4131.10.1016/j.vaccine.2016.06.081.

Pais, R., Omosun, Y., He, Q., Blas-Machado, U., Black, C., Igietseme, J.U., Fujihashi, K., and Eko, F.O. (2017). Rectal administration of a chlamydial subunit vaccine protects against genital infection and upper reproductive tract pathology in mice. PLoS One 12, e0178537.10.1371/journal.pone.0178537.

Pal, S., Barnhart, K.M., Wei, Q., Abai, A.M., Peterson, E.M., and de la Maza, L.M. (1999a). Vaccination of mice with DNA plasmids coding for the *Chlamydia trachomatis* major outer membrane protein elicits an immune response but fails to protect against a genital challenge. Vaccine 17, 459–465.

Pal, S., Davis, H.L., Peterson, E.M., and de la Maza, L.M. (2002). Immunization with the *Chlamydia trachomatis* mouse pneumonitis major outer membrane protein by use of CpG oligodeoxynucleotides as an adjuvant induces a protective immune response against an intranasal chlamydial challenge. Infect Immun 70, 4812–4817.10.1128/iai.70.9.4812-4817.2002.

Pal, S., Favaroni, A., Tifrea, D.F., Hanisch, P.T., Luczak, S.E.T., Hegemann, J.H., and de la Maza, L.M. (2017a). Comparison of the nine polymorphic membrane proteins of *Chlamydia trachomatis* for their ability to induce protective immune responses in mice against a C. muridarum challenge. Vaccine 35, 2543–2549.10.1016/j.vaccine.2017.03.070.

Pal, S., Fielder, T.J., Peterson, E.M., and de la Maza, L.M. (1994). Protection against infertility in a BALB/c mouse salpingitis model by intranasal immunization with the mouse pneumonitis biovar of *Chlamydia trachomatis*. Infect Immun 62, 3354–3362.

Pal, S., Luke, C.J., Barbour, A.G., Peterson, E.M., and de la Maza, L.M. (2003a). Immunization with the *Chlamydia trachomatis* major outer membrane protein, using the outer surface protein A of Borrelia burgdorferi as an adjuvant, can induce protection against a chlamydial genital challenge. Vaccine 21, 1455–1465. https://doi.org/10.1016/S0264-410x(02)00680-1.

Pal, S., Peterson, E.M., and de la Maza, L.M. (1996). Intranasal immunization induces long-term protection in mice against a *Chlamydia trachomatis* genital challenge. Infect Immun 64, 5341–5348.

Pal, S., Peterson, E.M., and de la Maza, L.M. (2003b). Induction of protective immunity against a *Chlamydia trachomatis* genital infection in three genetically distinct strains of mice. Immunology 110, 368–375.10.1046/j.1365-2567.2003.01748.x.

Pal, S., Peterson, E.M., and de la Maza, L.M. (2004). New murine model for the study of *Chlamydia trachomatis* genitourinary tract infections in males. Infect Immun 72, 4210–4216.10.1128/IAI.72.7.4210-4216.2004.

Pal, S., Peterson, E.M., and de la Maza, L.M. (2005a). Vaccination of newborn mice induces a strong protective immune response against respiratory and genital challenges with *Chlamydia trachomatis*. Vaccine 23, 5351–5358.10.1016/j.vaccine.2005.06.026.

Pal, S., Peterson, E.M., and de la Maza, L.M. (2005b). Vaccination with the *Chlamydia trachomatis* major outer membrane protein can elicit an immune response as protective as that resulting from inoculation with live bacteria. Infect Immun 73, 8153–8160.10.1128/IAI.73.12.8153-8160.2005.

Pal, S., Peterson, E.M., Rappuoli, R., Ratti, G., and de la Maza, L.M. (2006). Immunization with the *Chlamydia trachomatis* major outer membrane protein, using adjuvants developed for human vaccines, can induce partial protection in a mouse model against a genital challenge. Vaccine 24, 766–775.10.1016/j.vaccine.2005.08.074.

Pal, S., Rangel, J., Peterson, E.M., and de la Maza, L.M. (1999b). Immunogenic and protective ability of the two developmental forms of Chlamydiae in a mouse model of infertility. Vaccine 18, 752–761.

Pal, S., Sarcon, A.K., and de la Maza, L.M. (2010). A new murine model for testing vaccines against genital *Chlamydia trachomatis* infections in males. Vaccine 28, 7606–7612.10.1016/j.vaccine.2010.09.060.

Pal, S., Tatarenkova, O.V., and de la Maza, L.M. (2015). A vaccine formulated with the major outer membrane protein can protect C3H/HeN, a highly susceptible strain of mice, from a Chlamydia muridarum genital challenge. Immunology 146, 432–443.10.1111/imm.12520.

Pal, S., Theodor, I., Peterson, E.M., and de la Maza, L.M. (1997a). Immunization with an acellular vaccine consisting of the outer membrane complex of *Chlamydia trachomatis* induces protection against a genital challenge. Infect Immun 65, 3361–3369.

Pal, S., Theodor, I., Peterson, E.M., and de la Maza, L.M. (1997b). Monoclonal immunoglobulin A antibody to the major outer membrane protein of the *Chlamydia trachomatis* mouse pneumonitis biovar protects mice against a chlamydial genital challenge. Vaccine 15, 575–582.

Pal, S., Theodor, I., Peterson, E.M., and de la Maza, L.M. (2001). Immunization with the *Chlamydia trachomatis* mouse pneumonitis major outer membrane protein can elicit a protective immune response against a genital challenge. Infect Immun 69, 6240–6247.10.1128/IAI.69.10.6240-6247.2001.

Pal, S., Tifrea, D.F., Follmann, F., Andersen, P., and de la Maza, L.M. (2017b). The cationic liposomal adjuvants CAF01 and CAF09 formulated with the major outer membrane protein elicit robust protection in mice against a Chlamydia muridarum respiratory challenge. Vaccine 35, 1705–1711.10.1016/j.vaccine.2017.02.020.

Palmer, L., and Falkow, S. (1986). A common plasmid of *Chlamydia trachomatis*. Plasmid 16, 52–62.

Parr, E.L., Parr, M.B., and Thapar, M. (1988). A comparison of specific antibody responses in mouse vaginal fluid after immunization by several routes. J Reprod Immunol 14, 165–176.

Patton, D.L., Sweeney, Y.T., and Kuo, C.C. (1994). Demonstration of delayed hypersensitivity in *Chlamydia trachomatis* salpingitis in monkeys: a pathogenic mechanism of tubal damage. J Infect Dis 169, 680–683.10.1093/infdis/169.3.680.

Pedersen, G.K., Andersen, P., and Christensen, D. (2018). Immunocorrelates of CAF family adjuvants. Semin Immunol 39, 4–13.10.1016/j.smim.2018.10.003.

Peterson, E.M., Markoff, B.A., Schachter, J., and de la Maza, L.M. (1990). The 7.5-kb plasmid present in *Chlamydia trachomatis* is not essential for the growth of this microorganism. Plasmid 23, 144–148.

Peterson, E.M., You, J.Z., Motin, V., and de la Maza, L.M. (1999). Intranasal immunization with *Chlamydia trachomatis*, serovar E, protects from a subsequent vaginal challenge with the homologous serovar. Vaccine 17, 2901–2907.

Phillips, S., Quigley, B.L., and Timms, P. (2019). Seventy Years of Chlamydia Vaccine Research – Limitations of the Past and Directions for the Future. Front Microbiol 10, 70.10.3389/fmicb.2019.00070.

Picard, M.D., Cohane, K.P., Gierahn, T.M., Higgins, D.E., and Flechtner, J.B. (2012). High-throughput proteomic screening identifies *Chlamydia trachomatis* antigens that are capable of eliciting T cell and antibody responses that provide protection against vaginal challenge. Vaccine 30, 4387–4393.10.1016/j.vaccine.2012.01.017.

Plotkin, S., Orenstein, W.A., and Offit, P.A. (2013). Vaccines. Elsevier 6th edition.

Plotkin, S.A. (2008). Vaccines: correlates of vaccine-induced immunity. Clin Infect Dis 47, 401–409.10.1086/589862.

Polkinghorne, A., Hanger, J., and Timms, P. (2013). Recent advances in understanding the biology, epidemiology and control of chlamydial infections in koalas. Vet Microbiol 165, 214–223.10.1016/j.vetmic.2013.02.026.

Poston, T.B., and Darville, T. (2018). *Chlamydia trachomatis*: Protective Adaptive Responses and Prospects for a Vaccine. Current topics in microbiology and immunology 412, 217–237.10.1007/82_2016_6.

Poston, T.B., Gottlieb, S.L., and Darville, T. (2017). Status of vaccine research and development of vaccines for *Chlamydia trachomatis* infection. Vaccine.10.1016/j.vaccine.2017.01.023.

Prebeck, S., Kirschning, C., Durr, S., da Costa, C., Donath, B., Brand, K., Redecke, V., Wagner, H., and Miethke, T. (2001). Predominant role of toll-like receptor 2 versus 4 in *Chlamydia pneumoniae*-induced activation of dendritic cells. J Immunol 167, 3316–3323.10.4049/jimmunol.167.6.3316.

Price, M.J., Ades, A.E., Welton, N.J., Simms, I., Macleod, J., and Horner, P.J. (2016). Proportion of Pelvic

Inflammatory Disease Cases Caused by *Chlamydia trachomatis*: Consistent Picture From Different Methods. J Infect Dis 214, 617–624.10.1093/infdis/jiw178.

Qiu, S., Zhang, J., Tian, Y., Yang, Y., Huang, H., Yang, D., Lu, M., and Xu, Y. (2008). Reduced antigenicity of naturally occurring hepatitis B surface antigen variants with substitutions at the amino acid residue 126. Intervirology 51, 400–406.10.1159/000205265.

Qu, Y., Frazer, L.C., O'Connell, C.M., Tarantal, A.F., Andrews, C.W., Jr., O'Connor, S.L., Russell, A.N., Sullivan, J.E., Poston, T.B., Vallejo, A.N., *et al.* (2015). Comparable Genital Tract Infection, Pathology, and Immunity in Rhesus Macaques Inoculated with Wild-Type or Plasmid-Deficient *Chlamydia trachomatis* Serovar D. Infect Immun 83, 4056–4067.10.1128/IAI.00841-15.

Ralli-Jain, P., Tifrea, D., Cheng, C., Pal, S., and de la Maza, L.M. (2010). Enhancement of the protective efficacy of a *Chlamydia trachomatis* recombinant vaccine by combining systemic and mucosal routes for immunization. Vaccine 28, 7659–7666.10.1016/j.vaccine.2010.09.040.

Ramjee, G., Coumi, N., Dladla-Qwabe, N., Ganesh, S., Gappoo, S., Govinden, R., Guddera, V., Maharaj, R., Moodley, J., Morar, N., *et al.* (2010). Experiences in conducting multiple community-based HIV prevention trials among women in KwaZulu-Natal, South Africa. AIDS Res Ther 7, 10.10.1186/1742-6405-7-10.

Rank, R.G., and Batteiger, B.E. (1989). Protective role of serum antibody in immunity to chlamydial genital infection. Infect Immun 57, 299–301.

Rank, R.G., Soderberg, L.S., and Barron, A.L. (1985). Chronic chlamydial genital infection in congenitally athymic nude mice. Infect Immun 48, 847–849.

Regan, D.G., Wilson, D.P., and Hocking, J.S. (2008). Coverage is the key for effective screening of *Chlamydia trachomatis* in Australia. J Infect Dis 198, 349–358.10.1086/589883.

Ripa, T., and Nilsson, P.A. (2007). A *Chlamydia trachomatis* strain with a 377-bp deletion in the cryptic plasmid causing false-negative nucleic acid amplification tests. Sex Transm Dis 34, 255–256.10.1097/OLQ.0b013e31805ce2b9.

Rocchi, M.S., Wattegedera, S., Meridiani, I., and Entrican, G. (2009). Protective adaptive immunity to Chlamydophila abortus infection and control of ovine enzootic abortion (OEA). Vet Microbiol 135, 112–121.10.1016/j.vetmic.2008.09.030.

Rockey, D.D., Wang, J., Lei, L., and Zhong, G. (2009). Chlamydia vaccine candidates and tools for chlamydial antigen discovery. Expert Rev Vaccines 8, 1365–1377.10.1586/erv.09.98.

Rose, F., Karlsen, K., Jensen, P.R., Jakobsen, R.U., Wood, G.K., Rand, K.D., Godiksen, H., Andersen, P., Follmann, F., and Foged, C. (2018). Unusual Self-Assembly of the Recombinant *Chlamydia trachomatis* Major Outer Membrane Protein-Based Fusion Antigen CTH522 Into Protein Nanoparticles. J Pharm Sci 107, 1690–1700.10.1016/j.xphs.2018.02.005.

Rosenman, M.B., Mahon, B.E., Downs, S.M., and Kleiman, M.B. (2003). Oral erythromycin prophylaxis vs watchful waiting in caring for newborns exposed to *Chlamydia trachomatis*. Arch Pediatr Adolesc Med 157, 565–571.10.1001/archpedi.157.6.565.

Russell, A.N., Zheng, X., O'Connell, C.M., Wiesenfeld, H.C., Hillier, S.L., Taylor, B.D., Picard, M.D., Flechtner, J.B., Zhong, W., Frazer, L.C., *et al.* (2016). Identification of *Chlamydia trachomatis* Antigens Recognized by T Cells From Highly Exposed Women Who Limit or Resist Genital Tract Infection. J Infect Dis 214, 1884–1892.10.1093/infdis/jiw485.

Saltzman, W.M., Radomsky, M.L., Whaley, K.J., and Cone, R.A. (1994). Antibody diffusion in human cervical mucus. Biophysical journal 66, 508–515.

Schachter, J. (1980). Chlamydiae in Reiter's syndrome. Lancet 1, 40.

Schachter, J., and Dawson, C. (1978). Human chlamydial infections. PSG Publishing.

Schautteet, K., Stuyven, E., Beeckman, D.S., Van Acker, S., Carlon, M., Chiers, K., Cox, E., and Vanrompay, D. (2011). Protection of pigs against *Chlamydia trachomatis* challenge by administration of a MOMP-based DNA vaccine in the vaginal mucosa. Vaccine 29, 1399–1407.10.1016/j.vaccine.2010.12.042.

Scholes, D., Stergachis, A., Heidrich, F.E., Andrilla, H., Holmes, K.K., and Stamm, W.E. (1996). Prevention of pelvic inflammatory disease by screening for cervical chlamydial infection. N Engl J Med 334, 1362–1366.10.1056/NEJM199605233342103.

Sessa, R., Nicoletti, M., Di Pietro, M., Schiavoni, G., Santino, I., Zagaglia, C., Del Piano, M., and Cipriani, P. (2009). *Chlamydia pneumoniae* and atherosclerosis: current state and future prospectives. Int J Immunopathol Pharmacol 22, 9–14.10.1177/039463200902200102.

Shaw, J., Grund, V., Durling, L., Crane, D., and Caldwell, H.D. (2002). Dendritic cells pulsed with a recombinant chlamydial major outer membrane protein antigen elicit a CD4(+) type 2 rather than type 1 immune response that is not protective. Infect Immun 70, 1097–1105.

Shemer-Avni, Y., and Lieberman, D. (1995). *Chlamydia pneumoniae*-induced ciliostasis in ciliated bronchial epithelial cells. J Infect Dis 171, 1274–1278.10.1093/infdis/171.5.1274.

Sigar, I.M., Schripsema, J.H., Wang, Y., Clarke, I.N., Cutcliffe, L.T., Seth-Smith, H.M., Thomson, N.R., Bjartling, C., Unemo, M., Persson, K., *et al.* (2014). Plasmid deficiency in urogenital isolates of *Chlamydia trachomatis* reduces infectivity and virulence in a mouse model. Pathogens and disease 70, 61–69.10.1111/2049-632X.12086.

Singh, G. (2018). Target Product Profile and Clinical Development Plan. In Pharmaceutical Medicine and Translational Clinical Research, D. Vohora, and G. Singh, eds. (Boston: Academic Press), pp. 65–80.10.1016/b978-0-12-802103-3.00005-5.

Singh, S.R., Hulett, K., Pillai, S.R., Dennis, V.A., Oh, M.K., and Scissum-Gunn, K. (2006). Mucosal immunization with recombinant MOMP genetically linked with modified cholera toxin confers protection against *Chlamydia trachomatis* infection. Vaccine 24, 1213–1224.10.1016/j.vaccine.2005.08.097.

Skelding, K.A., Hickey, D.K., Horvat, J.C., Bao, S., Roberts, K.G., Finnie, J.M., Hansbro, P.M., and Beagley, K.W. (2006). Comparison of intranasal and transcutaneous immunization for induction of protective immunity against Chlamydia muridarum respiratory tract infection. Vaccine 24, 355–366.10.1016/j.vaccine.2005.07.104.

Sobinoff, A.P., Dando, S.J., Redgrove, K.A., Sutherland, J.M., Stanger, S.J., Armitage, C.W., Timms, P., McLaughlin, E.A., and Beagley, K.W. (2015). Chlamydia muridarum infection-induced destruction of male germ cells and sertoli cells is partially prevented by Chlamydia major outer membrane protein-specific immune CD4 cells. Biol Reprod 92, 27.10.1095/biolreprod.114.124180.

Song, J.H., Nguyen, H.H., Cuburu, N., Horimoto, T., Ko, S.Y., Park, S.H., Czerkinsky, C., and Kweon, M.N. (2008). Sublingual vaccination with influenza virus protects mice against lethal viral infection. Proc Natl Acad Sci U S A 105, 1644–1649.10.1073/pnas.0708684105.

Sowa, S., Sowa, J., Collier, L.H., and Blyth, W.A. (1969). Trachoma vaccine field trials in The Gambia. The Journal of hygiene 67, 699–717.10.1017/s0022172400042157.

Stagg, A.J., Tuffrey, M., Woods, C., Wunderink, E., and Knight, S.C. (1998). Protection against ascending infection of the genital tract by *Chlamydia trachomatis* is associated with recruitment of major histocompatibility complex class II antigen-presenting cells into uterine tissue. Infection and Immunity 66, 3535–3544.

Stanley, M. (2010). HPV – immune response to infection and vaccination. Infect Agent Cancer 5, 19.10.1186/1750-9378-5-19.

Starnbach, M.N. (2018). Action Needed on Chlamydia Vaccines. Trends Microbiol 26, 639–640.10.1016/j.tim.2018.05.006.

Starnbach, M.N., Loomis, W.P., Ovendale, P., Regan, D., Hess, B., Alderson, M.R., and Fling, S.P. (2003). An inclusion membrane protein from *Chlamydia trachomatis* enters the MHC class I pathway and stimulates a CD8+ T cell response. J Immunol 171, 4742–4749.10.4049/jimmunol.171.9.4742.

Stary, G., Olive, A., Radovic-Moreno, A.F., Gondek, D., Alvarez, D., Basto, P.A., Perro, M., Vrbanac, V.D., Tager, A.M., Shi, J., *et al.* (2015). VACCINES. A mucosal vaccine against *Chlamydia trachomatis* generates two waves of protective memory T cells. Science 348, aaa8205.10.1126/science.aaa8205.

Stephens, R.S., Kalman, S., Lammel, C., Fan, J., Marathe, R., Aravind, L., Mitchell, W., Olinger, L., Tatusov, R.L., Zhao, Q., *et al.* (1998). Genome sequence of an obligate intracellular pathogen of humans: *Chlamydia trachomatis*. Science 282, 754–759.

Stephens, R.S., Sanchez-Pescador, R., Wagar, E.A., Inouye, C., and Urdea, M.S. (1987). Diversity of *Chlamydia trachomatis* major outer membrane protein genes. J Bacteriol 169, 3879–3885.10.1128/jb.169.9.3879-3885.1987.

Stewardson, A.J., and Grayson, M.L. (2010). Psittacosis. Infect Dis Clin North Am 24, 7–25.10.1016/j.idc.2009.10.003.

Stothard, D.R., Williams, J.A., Van Der Pol, B., and Jones, R.B. (1998). Identification of a *Chlamydia trachomatis* serovar E urogenital isolate which lacks the cryptic plasmid. Infect Immun 66, 6010–6013.

Su, H., and Caldwell, H.D. (1995). CD4+ T cells play a significant role in adoptive immunity to *Chlamydia trachomatis* infection of the mouse genital tract. Infect Immun 63, 3302–3308.

Su, H., Messer, R., Whitmire, W., Fischer, E., Portis, J.C., and Caldwell, H.D. (1998). Vaccination against chlamydial genital tract infection after immunization with dendritic cells pulsed ex vivo with nonviable Chlamydiae. J Exp Med 188, 809–818.10.1084/jem.188.5.809.

Su, H., Messer, R., Whitmire, W., Hughes, S., and Caldwell, H.D. (2000). Subclinical chlamydial infection of the female mouse genital tract generates a potent protective immune response: implications for development of live attenuated chlamydial vaccine strains. Infect Immun 68, 192–196.10.1128/iai.68.1.192-196.2000.

Su, H., Morrison, R., Messer, R., Whitmire, W., Hughes, S., and Caldwell, H.D. (1999). The effect of doxycycline treatment on the development of protective immunity in a murine model of chlamydial genital infection. J Infect Dis 180, 1252–1258.10.1086/315046.

Su, H., Parnell, M., and Caldwell, H.D. (1995). Protective efficacy of a parenterally administered MOMP-derived synthetic oligopeptide vaccine in a murine model of *Chlamydia trachomatis* genital tract infection: serum neutralizing IgG antibodies do not protect against chlamydial genital tract infection. Vaccine 13, 1023–1032.

Sun, G., Pal, S., Sarcon, A.K., Kim, S., Sugawara, E., Nikaido, H., Cocco, M.J., Peterson, E.M., and de la Maza, L.M. (2007). Structural and functional analyses of the major outer membrane protein of *Chlamydia trachomatis*. J Bacteriol 189, 6222–6235.10.1128/JB.00552-07.

Sun, G., Pal, S., Weiland, J., Peterson, E.M., and de la Maza, L.M. (2009). Protection against an intranasal challenge by vaccines formulated with native and recombinant preparations of the *Chlamydia trachomatis* major outer membrane protein. Vaccine 27, 5020–5025.10.1016/j.vaccine.2009.05.008.

Swanson, K.A., Taylor, L.D., Frank, S.D., Sturdevant, G.L., Fischer, E.R., Carlson, J.H., Whitmire, W.M., and Caldwell, H.D. (2009). *Chlamydia trachomatis* polymorphic membrane protein D is an oligomeric autotransporter with a higher-order structure. Infect Immun 77, 508–516.10.1128/IAI.01173-08.

Tan, C., Hsia, R.C., Shou, H., Carrasco, J.A., Rank, R.G., and Bavoil, P.M. (2010). Variable expression of surface-exposed polymorphic membrane proteins in in vitro-grown *Chlamydia trachomatis*. Cell Microbiol 12, 174–187.10.1111/j.1462-5822.2009.01389.x.

Tang, L., Chen, J., Zhou, Z., Yu, P., Yang, Z., and Zhong, G. (2015). Chlamydia-secreted protease CPAF degrades host antimicrobial peptides. Microbes Infect 17, 402–408.10.1016/j.micinf.2015.02.005.

Taylor-Robinson, D., and Tuffrey, M. (1981). Progesterone as a key factor in the development of a mouse model for genital-tract infection with *Chlamydia trachomatis*. FEMS Microbiology Letters 12, 111–115.10.1111/j.1574-6968.1981.tb07622.x%J FEMS Microbiology Letters.

Taylor, H. (2008). Trachoma: a blinding scourge from the Bronze Age to the twenty-first century. Haddington Press Pry Ltd.

Taylor, H.R., Whittum-Hudson, J., Schachter, J., Caldwell, H.D., and Prendergast, R.A. (1988). Oral immunization with chlamydial major outer membrane protein (MOMP). Invest Ophthalmol Vis Sci 29, 1847–1853.

Thapar, M.A., Parr, E.L., Bozzola, J.J., and Parr, M.B. (1991). Secretory immune responses in the mouse vagina after parenteral or intravaginal immunization with an immunostimulating complex (ISCOM). Vaccine 9, 129–133.

Theisen, M., Roeffen, W., Singh, S.K., Andersen, G., Amoah, L., van de Vegte-Bolmer, M., Arens, T., Tiendrebeogo, R.W., Jones, S., Bousema, T., *et al.* (2014). A multi-stage malaria vaccine candidate targeting both transmission and asexual parasite life-cycle stages. Vaccine 32, 2623–2630.10.1016/j.vaccine.2014.03.020.

Tifrea, D.F., Pal, S., Le Bon, C., Giusti, F., Popot, J.L., Cocco, M.J., Zoonens, M., and de la Maza, L.M. (2018). Co-delivery of amphipol-conjugated adjuvant with antigen, and adjuvant combinations, enhance immune protection elicited by a membrane protein-based vaccine against a mucosal challenge with Chlamydia. Vaccine 36, 6640–6649.10.1016/j.vaccine.2018.09.055.

Tifrea, D.F., Pal, S., Popot, J.L., Cocco, M.J., and de la Maza, L.M. (2014). Increased immunoaccessibility of MOMP epitopes in a vaccine formulated with amphipols may account for the very robust protection elicited against a vaginal challenge with Chlamydia muridarum. J Immunol 192, 5201–5213.10.4049/jimmunol.1303392.

Tifrea, D.F., Ralli-Jain, P., Pal, S., and de la Maza, L.M. (2013). Vaccination with the recombinant major outer membrane protein elicits antibodies to the constant domains and induces cross-serovar protection against intranasal challenge with *Chlamydia trachomatis*. Infect Immun 81, 1741–1750.10.1128/IAI.00734-12.

Tifrea, D.F., Sun, G., Pal, S., Zardeneta, G., Cocco, M.J., Popot, J.L., and de la Maza, L.M. (2011). Amphipols stabilize the Chlamydia major outer membrane protein and enhance its protective ability as a vaccine. Vaccine 29, 4623–4631.10.1016/j.vaccine.2011.04.065.

Torrone, E.A., Morrison, C.S., Chen, P.L., Kwok, C., Francis, S.C., Hayes, R.J., Looker, K.J., McCormack, S., McGrath, N., van de Wijgert, J., *et al.* (2018). Prevalence of sexually transmitted infections and bacterial vaginosis among women in sub-Saharan Africa: An individual participant data meta-analysis of 18 HIV prevention studies. PLoS medicine 15, e1002511.10.1371/journal.pmed.1002511.

Tracy, C.R., Steers, W.D., and Costabile, R. (2008). Diagnosis and management of epididymitis. Urol Clin North Am 35, 101–108; vii.10.1016/j.ucl.2007.09.013.

Tsai, P.Y., Hsu, M.C., Huang, C.T., and Li, S.Y. (2007). Human antibody and antigen response to IncA antibody of *Chlamydia trachomatis*. Int J Immunopathol Pharmacol 20, 156–161.

Tu, J., Hou, B., Wang, B., Lin, X., Gong, W., Dong, H., Zhu, S., Chen, S., Xue, X., Zhao, K.N., *et al.* (2014). A multi-epitope vaccine based on *Chlamydia trachomatis* major outer membrane protein induces specific immunity in mice. Acta biochimica et biophysica Sinica 46, 401–408.10.1093/abbs/gmu016.

Tuffrey, M., Alexander, F., Conlan, W., Woods, C., and Ward, M. (1992). Heterotypic protection of mice against chlamydial salpingitis and colonization of the lower genital tract with a human serovar F isolate of *Chlamydia trachomatis* by prior immunization with recombinant serovar L1 major outer-membrane protein. J Gen Microbiol 138 Pt 8, 1707–1715.10.1099/00221287-138-8-1707.

Tuffrey, M., Alexander, F., Inman, C., and Ward, M.E. (1990a). Correlation of infertility with altered tubal morphology and function in mice with salpingitis induced by a human genital-tract isolate of *Chlamydia trachomatis*. J Reprod Fertil 88, 295–305.

Tuffrey, M., Alexander, F., and Taylor-Robinson, D. (1990b). Severity of salpingitis in mice after primary and repeated inoculation with a human strain of *Chlamydia trachomatis*. J Exp Pathol (Oxford) 71, 403–410.

Tuffrey, M., Falder, P., Gale, J., Quinn, R., and Taylor-Robinson, D. (1986a). Infertility in mice infected genitally with a human strain of *Chlamydia trachomatis*. J Reprod Fertil 78, 251–260.

Tuffrey, M., Falder, P., Gale, J., and Taylor-Robinson, D. (1986b). Salpingitis in mice induced by human strains of *Chlamydia trachomatis*. Br J Exp Pathol 67, 605–616.

Tuffrey, M., Falder, P., and Taylor-Robinson, D. (1985). Effect on *Chlamydia trachomatis* infection of the murine genital tract of adoptive transfer of congenic immune cells or specific antibody. Br J Exp Pathol 66, 427–433.

Turner, K., Adams, E., Grant, A., Macleod, J., Bell, G., Clarke, J., and Horner, P. (2011). Costs and cost effectiveness of different strategies for chlamydia screening and partner notification: an economic and mathematical modelling study. BMJ (Clinical research ed 342, c7250.10.1136/bmj.c7250.

Turner, K.M., Adams, E.J., Lamontagne, D.S., Emmett, L., Baster, K., and Edmunds, W.J. (2006). Modelling the effectiveness of chlamydia screening in England. Sex Transm Infect 82, 496–502.10.1136/sti.2005.019067.

Unemo, M., Bradshaw, C.S., Hocking, J.S., de Vries, H.J.C., Francis, S.C., Mabey, D., Marrazzo, J.M., Sonder, G.J.B., Schwebke, J.R., Hoornenborg, E., *et al.* (2017). Sexually transmitted infections: challenges ahead. The Lancet infectious diseases 17, e235-e279.10.1016/S1473-3099(17)30310-9.

van Ginkel, F.W., Nguyen, H.H., and McGhee, J.R. (2000). Vaccines for mucosal immunity to combat emerging infectious diseases. Emerg Infect Dis 6, 123–132.10.3201/eid0602.000204.

Van Lent, S., Creasy, H.H., Myers, G.S., and Vanrompay, D. (2016). The Number, Organization, and Size of Polymorphic Membrane Protein Coding Sequences as well as the Most Conserved Pmp Protein Differ within and across Chlamydia Species. J Mol Microbiol Biotechnol 26, 333–344.10.1159/000447092.

Vanrompay, D., Ducatelle, R., and Haesebrouck, F. (1995). Chlamydia psittaci infections: a review with emphasis on avian chlamydiosis. Vet Microbiol 45, 93–119.

Vasilevsky, S., Greub, G., Nardelli-Haefliger, D., and Baud, D. (2014). Genital *Chlamydia trachomatis*: understanding the roles of innate and adaptive immunity in vaccine research. Clin Microbiol Rev 27, 346–370.10.1128/CMR.00105-13.

Vasilevsky, S., Stojanov, M., Greub, G., and Baud, D. (2016). Chlamydial polymorphic membrane proteins: regulation, function and potential vaccine candidates. Virulence 7, 11–22.10.1080/21505594.2015.1111509.

Verminnen, K., Beeckman, D.S., Sanders, N.N., De Smedt, S., and Vanrompay, D.C. (2010). Vaccination of turkeys against Chlamydophila psittaci through optimized

DNA formulation and administration. Vaccine 28, 3095–3105.10.1016/j.vaccine.2010.02.064.

Verminnen, K., Loock, M.V., Cox, E., Goddeeris, B.M., and Vanrompay, D. (2005). Protection of turkeys against Chlamydophila psittaci challenge by DNA and rMOMP vaccination and evaluation of the immunomodulating effect of 1 alpha,25-dihydroxyvitamin D(3). Vaccine 23, 4509–4516.10.1016/j.vaccine.2005.04.014.

Wali, S., Gupta, R., Yu, J.J., Lanka, G.K.K., Chambers, J.P., Guentzel, M.N., Zhong, G., Murthy, A.K., and Arulanandam, B.P. (2017). Chlamydial protease-like activity factor mediated protection against *C. trachomatis* in guinea pigs. Immunol Cell Biol 95, 454–460.10.1038/icb.2016.122.

Wang, J., Chen, L., Chen, F., Zhang, X., Zhang, Y., Baseman, J., Perdue, S., Yeh, I.T., Shain, R., Holland, M., *et al.* (2009). A chlamydial type III-secreted effector protein (Tarp) is predominantly recognized by antibodies from humans infected with *Chlamydia trachomatis* and induces protective immunity against upper genital tract pathologies in mice. Vaccine 27, 2967–2980.10.1016/j.vaccine.2009.02.095.

Wang, J., Zhang, Y., Lu, C., Lei, L., Yu, P., and Zhong, G. (2010). A genome-wide profiling of the humoral immune response to *Chlamydia trachomatis* infection reveals vaccine candidate antigens expressed in humans. J Immunol 185, 1670–1680.10.4049/jimmunol.1001240.

Wang, S.-P., and Grayston, J. (1984). Microimmunofluorescence serology of *Chlamydia trachomatis*. In Peterson LMdIMaE. Medical Virology III.

Wang, S.P., and Grayston, J.T. (1963). Classification of Trachoma Virus Strains by Protection of Mice from Toxic Death. J Immunol 90, 849–856.

Wang, S.P., and Grayston, J.T. (1967). A potency test for trachoma vaccine utilizing the mouse toxicity prevention test. Am J Ophthalmol 63, Suppl:1443–1454.

Wang, S.P., Grayston, J.T., and Alexander, E.R. (1967a). Trachoma vaccine studies in monkeys. Am J Ophthalmol 63, Suppl:1615–1630.

Wang, S.P., Kenny, G.E., and Grayston, J.T. (1967b). Characterization of trachoma antigens protective against mouse toxicity. Am J Ophthalmol 63, Suppl:1454–1461.

Waugh, C., Khan, S.A., Carver, S., Hanger, J., Loader, J., Polkinghorne, A., Beagley, K., and Timms, P. (2016). A Prototype Recombinant-Protein Based Chlamydia pecorum Vaccine Results in Reduced Chlamydial Burden and Less Clinical Disease in Free-Ranging Koalas (Phascolarctos cinereus). PLoS One 11, e0146934.10.1371/journal.pone.0146934.

Wern, J.E., Sorensen, M.R., Olsen, A.W., Andersen, P., and Follmann, F. (2017). Simultaneous Subcutaneous and Intranasal Administration of a CAF01-Adjuvanted Chlamydia Vaccine Elicits Elevated IgA and Protective Th1/Th17 Responses in the Genital Tract. Front Immunol 8, 569.10.3389/fimmu.2017.00569.

Westrom, L. (1995). Effect of pelvic inflammatory disease on fertility. Venereology 8, 219–222.

Westrom, L., Joesoef, R., Reynolds, G., Hagdu, A., and Thompson, S.E. (1992). Pelvic inflammatory disease and fertility. A cohort study of 1,844 women with laparoscopically verified disease and 657 control women with normal laparoscopic results. Sex Transm Dis 19, 185–192.

Wetmore, C.M., Manhart, L.E., Lowens, M.S., Golden, M.R., Whittington, W.L., Xet-Mull, A.M., Astete, S.G., McFarland, N.L., McDougal, S.J., and Totten, P.A. (2011). Demographic, behavioural, and clinical characteristics of men with nongonococcal urethritis differ by aetiology: a case-comparison study. Sex Transm Dis 38, 180–186.10.1097/OLQ.0b013e3182040de9.

Wills, J.M., Gruffydd-Jones, T.J., Richmond, S.J., Gaskell, R.M., and Bourne, F.J. (1987). Effect of vaccination on feline Chlamydia psittaci infection. Infect Immun 55, 2653–2657.

Woolridge, R.L., Grayston, J.T., Chang, I.H., Cheng, K.H., Yang, C.Y., and Neave, C. (1967). Field trial of a monovalent and of a bivalent mineral oil adjuvant trachoma vaccine in Taiwan school children. Am J Ophthalmol 63, Suppl:1645–1650.

Wright, H.R., Turner, A., and Taylor, H.R. (2008). Trachoma. Lancet 371, 1945–1954.10.1016/S0140-6736(08)60836-3.

Yang, Z., Tang, L., Shao, L., Zhang, Y., Zhang, T., Schenken, R., Valdivia, R., and Zhong, G. (2016a). The Chlamydia-Secreted Protease CPAF Promotes Chlamydial Survival in the Mouse Lower Genital Tract. Infect Immun 84, 2697–2702.10.1128/IAI.00280-16.

Yang, Z., Tang, L., Sun, X., Chai, J., and Zhong, G. (2015). Characterization of CPAF critical residues and secretion during *Chlamydia trachomatis* infection. Infect Immun 83, 2234–2241.10.1128/IAI.00275-15.

Yang, Z., Tang, L., Zhou, Z., and Zhong, G. (2016b). Neutralizing antichlamydial activity of complement by chlamydia-secreted protease CPAF. Microbes Infect 18, 669–674.10.1016/j.micinf.2016.07.002.

Yu, H., Jiang, X., Shen, C., Karunakaran, K.P., and Brunham, R.C. (2009). Novel Chlamydia muridarum T cell antigens induce protective immunity against lung and genital tract infection in murine models. J Immunol 182, 1602–1608.10.4049/jimmunol.182.3.1602.

Yu, H., Jiang, X., Shen, C., Karunakaran, K.P., Jiang, J., Rosin, N.L., and Brunham, R.C. (2010). Chlamydia muridarum T-cell antigens formulated with the adjuvant DDA/TDB induce immunity against infection that correlates with a high frequency of gamma interferon (IFN-gamma)/tumour necrosis factor alpha and IFN-gamma/interleukin-17 double-positive CD4+ T cells. Infect Immun 78, 2272–2282.10.1128/IAI.01374-09.

Yu, H., Karunakaran, K.P., Jiang, X., and Brunham, R.C. (2014). Evaluation of a multisubunit recombinant polymorphic membrane protein and major outer membrane protein T cell vaccine against Chlamydia muridarum genital infection in three strains of mice. Vaccine 32, 4672–4680.10.1016/j.vaccine.2014.06.002.

Yu, H., Karunakaran, K.P., Jiang, X., and Brunham, R.C. (2016). Subunit vaccines for the prevention of mucosal infection with *Chlamydia trachomatis*. Expert Rev Vaccines 15, 977–988.10.1586/14760584.2016.1161510.

Yu, H., Karunakaran, K.P., Jiang, X., Shen, C., Andersen, P., and Brunham, R.C. (2012). Chlamydia muridarum T cell antigens and adjuvants that induce protective immunity in mice. Infect Immun 80, 1510–1518.10.1128/IAI.06338-11.

Yu, H., Karunakaran, K.P., Kelly, I., Shen, C., Jiang, X., Foster, L.J., and Brunham, R.C. (2011). Immunization with live and dead Chlamydia muridarum induces different levels of protective immunity in a murine genital tract model: correlation with MHC class II peptide presentation and multifunctional Th1 cells. J Immunol 186, 3615–3621.10.4049/jimmunol.1002952.

Yung, A.P., and Grayson, M.L. (1988). Psittacosis--a review of 135 cases. The Medical journal of Australia 148, 228–233.

Zhang, D., Yang, X., Berry, J., Shen, C., McClarty, G., and Brunham, R.C. (1997). DNA vaccination with the major outer-membrane protein gene induces acquired immunity to *Chlamydia trachomatis* (mouse pneumonitis) infection. J Infect Dis 176, 1035–1040.10.1086/516545.

Zhang, X.X., Yu, H., Wang, X.H., Li, X.Z., Zhu, Y.P., Li, H.X., Luo, S.J., and Yuan, Z.G. (2013). Protective efficacy against Chlamydophila psittaci by oral immunization based on transgenic rice expressing MOMP in mice. Vaccine 31, 698–703.10.1016/j.vaccine.2012.11.039.

Zhong, G., Brunham, R.C., de la Maza, L.M., Darville, T., and Deal, C. (2017). National Institute of Allergy and Infectious Diseases workshop report: 'Chlamydia vaccines: The way forward'. Vaccine.10.1016/j.vaccine.2017.10.075.

Zhong, G., Fan, P., Ji, H., Dong, F., and Huang, Y. (2001). Identification of a chlamydial protease-like activity factor responsible for the degradation of host transcription factors. J Exp Med 193, 935–942.10.1084/jem.193.8.935.

Zhou, J., Qiu, C., Cao, X.A., and Lin, G. (2007). Construction and immunogenicity of recombinant adenovirus expressing the major outer membrane protein (MOMP) of Chlamydophila psittaci in chicks. Vaccine 25, 6367–6372.10.1016/j.vaccine.2007.06.031.

Zuckerman, J.N. (2006). Protective efficacy, immunotherapeutic potential, and safety of hepatitis B vaccines. J Med Virol 78, 169–177.10.1002/jmv.20524.

Aagaard, C., Hoang, T., Dietrich, J., Cardona, P.J., Izzo, A., Dolganov, G., Schoolnik, G.K., Cassidy, J.P., Billeskov, R., and Andersen, P. (2011). A multistage tuberculosis vaccine that confers efficient protection before and after exposure. Nat Med 17, 189–194.10.1038/nm.2285

Chlamydia Infections and the Microbiota

16

Simon Graspeuntner[1], Jan Rupp[1], Peter Timms[2*] and Bonnie Quigley[2]

[1]Department of Infectious Diseases and Microbiology, University of Lübeck, Lübeck, Germany.
[2]University of the Sunshine Coast, Sippy Downs, QLD, Australia.

*Correspondence: ptimms@usc.edu.au

https://doi.org/10.21775/9781912530281.16

Abstract

The microbiota has become intensively studied in many human diseases. In this chapter, we highlight recent research about the human microbiota and its association with chlamydial infections. We introduce the vaginal microbiota in general, focusing on the microbiota composition with regard to genital *Chlamydia trachomatis* infection and provide current concepts about the exclusion of *C. trachomatis* by the most important members of the vaginal microbiota, the lactobacilli. We further provide evidence about how particular anaerobic members of the microbiota create a microenvironment supporting ongoing infection with *C. trachomatis*. The potential role of the microbiota in upper genital tract infections with *C. trachomatis* is further discussed. The limited data that are available for non-genital chlamydial infections, in particular *C. trachomatis* infection in the eye and *C. pneumoniae* infection in the lung, are also reviewed. This chapter finally discusses the recent advances in microbiota research linked to animal chlamydial diseases and provides future directions of microbiota research in chlamydial diseases.

Vaginal microbiota in genital tract infections with *C. trachomatis*

Microbiota of the female genital tract

Methods to analyse microbiota

Over many decades of clinical research, bacteria have generally been studied primarily from a pathogen-centred perspective. In more recent times, awareness has increased that microbial communities (the microbiota) which colonize human surfaces are a critical factor in human health. The term microbiota describes the complete entity of microbes at a given location (e.g. the skin, the gastrointestinal tract or the vagina). However, recent research focuses mainly on the bacteria within the microbiota and thus the term microbiota in most circumstances comprises only the bacterial component. Assessing the microbiota has been a challenging task, with only limited precision available prior to modern next-generation sequencing (NGS) techniques. NGS nowadays enables researchers to obtain a detailed picture of the microbiota in humans using high throughput sequencing-by-synthesis technologies. Thus, most modern studies on human microbiota are based on the analysis of partial 16SrRNA gene amplicons. These genes carry hypervariable regions, which enable researchers to distinguish between bacterial taxa. A complex bioinformatic data processing, followed by a range of multivariate statistical analyses, completes the technical and theoretical needs required for high quality microbiota analysis. With current sequencing-by-synthesis technology, high throughput analysis of hundreds of samples in parallel is now possible. Thus, studies on vaginal microbiota are nowadays limited by the availability of samples rather than of any subsequent analysis step. Logically, our knowledge about the microbiota from all body sites, including the vaginal microbiota, and its relation to human health and disease, has been accelerated to a completely new level of understanding in recent years.

The microbial composition of the vagina

In 2011, Ravel *et al.* published a landmark study on the urogenital microbiota of 396 asymptomatic reproductive-age women (Ravel *et al.*, 2011). They showed that different communities ofbacteria could be found in the vagina. Five different communities (Table 16.1) were initially described and numbered according to this analysis: four of them are dominated by either *Lactobacillus crispatus* (community group I), *L. gasseri* (community group II), *L. iners* (community group III) or *L. jensenii* (community group V). Lactobacilli are long known as colonizers of the vagina and these communities have a remarkably low level of diversity. Another type of bacterial community (community group IV) was described as comprising all samples, which could not be assigned to one of the other groups. This fifth group was composed of overall high bacterial diversity, with low *Lactobacillus* content consisting predominantly of various combinations of the taxa *Prevotella, Dialister, Atopobium, Gardnerella, Megasphaera, Peptoniphilus, Sneathia, Eggerthella, Aerococcus, Finegoldia* and *Mobiluncus*. This broad range of bacteria includes several strict anaerobic bacteria, bacteria that are known to be associated with vaginal diseases, and species which have not been described in the vagina before. These bacterial communities were later described as different 'community state types' (CST) (Gajer *et al.*, 2012) and this term is currently in use for describing the composition of the vaginal microbiota. In the meantime it has become evident that other CSTs exist, including a CST dominated by *Gardnerella vaginalis* (Graspeuntner *et al.*, 2018b; Nunn *et al.*, 2015). In addition, some singletons appear which represent rare bacterial communities dominated by other bacteria (Ravel *et al.*, 2011). Such bacteria include *Bifidobacterium* spp., *Enterococcus*, other lactobacilli or pathogens infecting the vagina (Graspeuntner *et al.*, 2018a; Ravel *et al.*, 2011).

While the discussion is not finally settled yet (Beamer *et al.*, 2017) there are strong hints showing that the portion of females with a diverse community is higher in Hispanics and Black compared to Asian and White females (Gajer *et al.*, 2012; Ravel *et al.*, 2011). The underlying reasons for those differences are not yet identified. However, a wide range of external and internal factors have been revealed impacting on the composition of the vaginal microbiota. Studies have shown that a varying degree of dynamic changes can occur over time, resulting in the presence of certain CSTs in the same individual. Consistent with this idea, it has been reported that the CST of women can persist over a long period of time, but might also change to different CSTs (Chaban *et al.*, 2014; Gajer *et al.*, 2012). In

Table 16.1 Most prominent bacterial community groups (community state types, CST) in the vagina and their function in *C. trachomatis* infection

Number of community group/CST following Ravel *et al.* (2011)	Characteristic bacterial species of CST	Diversity of CST	Association with *C. trachomatis* infection
I	*L. crispatus*	Low	Negative
II	*L. gasseri*	Low	Not available
III	*L. iners*	Low to high	Positive
IV	Wide range of mostly anaerobic bacteria	High	Positive
V	*L. jensenii*	Low	Not available
*	*G. vaginalis*	Low to high	Positive

*Often referred to as part of CST IV.

particular, if dominated by *L. crispatus* the CST is highly stable, but if for example, *L. iners* is the dominant species of the CST, transition to other CSTs is more likely (Gajer *et al.*, 2012). While they could not explain the major reasons for those dynamics, Gajer *et al.* demonstrated an enhanced degree of instability of the vaginal microbiota during menses, and postulated that female sex hormones influence the stability of the microbiota (Gajer *et al.*, 2012). Indeed, usage of hormonal contraceptives is suggested to change the microbial composition of the vagina (Brooks *et al.*, 2017). Smoking is associated with lower Lactobacillus abundance (Brotman *et al.*, 2014; Nelson *et al.*, 2018). Recent publications have suggested sexual behaviour as a strong impactor on the vaginal microbiota. Thus the stability of the microbial composition decreases (Gajer *et al.*, 2012) and the diversity of the microbiota increases with enhanced sexual activity (Graspeuntner *et al.*, 2018a). Thereby, sexual activity increases the likelihood of having a microbiota dominated by *L. iners* or *G. vaginalis* suggesting transmission between sexual partners playing a role (Vodstrcil *et al.*, 2017).

We are currently only beginning to understand how the vaginal microbiota is controlled, but the dynamic changes within the microbiota are clearly an important factor to be considered when addressing the role of the microbiota in infectious diseases, such as *C. trachomatis*.

The good and the bad: vaginal microbiota associated with *C. trachomatis* genital infection

For decades it has been accepted that an abnormal vaginal microbiota is a risk factor for the acquisition of sexually transmitted infections. While is it well established that a healthy vaginal microbiota consists mainly of lactobacilli, the term bacterial vaginosis (BV) is often used to describe a poorly defined polymicrobial community related to a variety of diseases. While this indicates that a broad range of bacteria play a role in the acquisition and prevention of STIs, recent advances shed a detailed light on how the composition of the entire microbiota relates to infection with *C. trachomatis*.

The microbiota in bacterial vaginosis

Bacterial vaginosis (BV) is a common vaginal syndrome connected with symptoms such as vaginal discharge, itching and burning, as well as local inflammation. Parameters such as an increased pH and analysis of vaginal fluid with wet smears (scored with Amsel criteria; Mohammadzadeh *et al.*, 2014) and Gram stain (Nugent Score; Nugent *et al.*, 1991) are used to diagnose BV. Most in use is the Nugent score, which is a health score based on the presence/absence and abundance of (i) *Lactobacillus* morphotypes (ii) *G. vaginalis* or *Bacteroides* morphotypes (small Gram-variable or Gram-negative rods) and (iii) curved Gram variable rods (Nugent *et al.*, 1991). A 'healthy vaginal microbiota' is composed of high numbers of *Lactobacillus* morphotypes with low numbers of the other above-named morphotypes. High numbers of the latter types in combination with low *Lactobacillus* content are indicative of BV (Nugent *et al.*, 1991). It is well established that BV leads to increased susceptibility to sexually transmitted infections, such as *C. trachomatis* and other STIs (Brotman *et al.*, 2010; Wiesenfeld *et al.*, 2003). Thus, it appears logical that the microbial composition behind the diagnosis of BV is of relevance to STIs.

On a community level, BV is characterized by a shift from *Lactobacillus*-dominated to a rather diverse or *Gardnerella*-dominated microbiota, with high microbiota variability (Kaambo *et al.*, 2018; Onderdonk *et al.*, 2016; Srinivasan *et al.*, 2010). When researchers have evaluated women with a diagnosed BV, the taxa *Gardnerella, Atopobium, Prevotella, Peptostreptococcus, Mobiluncus, Sneathia, Leptotrichia, Mycoplasma,* bacterial vaginosis-associated bacterium (BVAB) 1 to BVAB3, and *Lachnospiraceae* generally dominate the vaginal bacterial community (Muzny *et al.*, 2013; Onderdonk *et al.*, 2016). In addition, women with symptomatic and asymptomatic BV diagnosis display similar bacterial communities (Ravel *et al.*, 2013).

The high portion of diverse communities in some groups of females without symptoms of BV, or with other signs of urogenital disease, has led some researchers to question the dogma of a disturbed microbiota associated with increasing Nugent scores. They claim that not all diverse communities might necessarily reflect a microbiota dysbiosis (Gajer *et al.*, 2012; Ravel *et al.*, 2011). Their argument is based on the fact that for some ethnicities, it is more common to have a diverse CST with no occurrence of symptoms for BV (Gajer *et al.*, 2012; Ravel *et al.*, 2011). According to this interpretation, some samples which score high in Nugent Score test, and are diverse, might nevertheless reflect a healthy microbiota.

Taken together, this means that there is not a specific microbial composition termed BV, but rather a set of phenotypes which belong to this category and these may have different impacts on acquiring STIs, such as *C. trachomatis*. Thus, it has become clear that a detailed knowledge about the microbiota is the key to understanding how genital *C. trachomatis* infection is influenced by the bacteria colonizing the vagina.

Bacterial communities positively and negatively associated with *C. trachomatis* genital infection in women

In a pilot study, Filardo *et al.* analysed the cervical microbiota of asymptomatic reproductive age women with and without *C. trachomatis* infection. They found a notable difference of the cervical microbiota between *C. trachomatis* positive women and healthy controls, as there was a significant separation of the bacterial communities into two clusters. This analysis shows that an infection with *C. trachomatis* correlated with (i) a decrease in lactobacilli and (ii) an increase in anaerobic bacteria, such as *G. vaginalis*, *A. vaginae*, *P. amnii*, *P. timonensis* and *Leptotrichia amnionii* (Filardo *et al.*, 2017). A study of 115 Dutch women which were vaginally swabbed before and after an infection with *C. trachomatis*, revealed a significant association between the risk of infection with *C. trachomatis* and CST III (clustering after; Ravel *et al.*, 2011), which is dominated by *L. iners* (van Houdt *et al.*, 2018).

Specific statistical approaches are used to analyse the association of specific taxa within the microbiota and *C. trachomatis* infection status. Thus, they showed *L. iners*, *P. amnii*, *Faecalibacterium prausnitzii*, *Gemella*, and *Ureaplasma* to be associated with infection. In contrast, *L. crispatus*, *L. fermentum* and *Enterobacteriaceae* were associated with healthy women when analysed independent of the CSTs present (van Houdt *et al.*, 2018). In this sense, the data of van Houdt *et al.* point in the direction of specific signatures, which could be used as indicators for an increased host predisposition to acquire STIs. Of note, it cannot yet be excluded that these signatures are part of the effect of the infection.

The same group examined another cohort of Dutch females initially negative for *C. trachomatis* infection, who were exposed to a sexual partner with *C. trachomatis* infection during the study (van der Veer *et al.*, 2017). They sampled cervical swabs before and after exposure to *C. trachomatis* infected sexual partners and performed molecular detection of *C. trachomatis*. They could link the microbiota prior to exposure with disease acquisition. The prior microbiota from women who tested positive for *C. trachomatis* was generally dominated by *L. iners* or other diverse anaerobic bacteria, rather than *L. crispatus*. This study revealed *L. iners*, *Actinomyces urogenitalis*, *Gemella* and BVAB3 to be significantly associated with *C. trachomatis* infection, while *L. crispatus*, *L. vaginalis* and *Lachnospiraceae* were indicators in uninfected women (van der Veer *et al.*, 2017).

In a further study, an indicator species analysis showed a wider range of bacteria to be associated with the presence of *C. trachomatis* infection: *Streptococcus*, *Veillonella montpellierensis*, *Sutterella sanguinis*, *Atopobium vaginae*, *Ureaplasma*, *Dialister propionicifaciens*, *Granulicatella elegans*, BVAB1, *Haemophilus parainfluenzae*, *Prevotella amnii*, *Mycoplasma hominis*, *Delftia*, and BVAB2 (Graspeuntner *et al.*, 2018b). Very recent studies reproduced the consistent findings that *L. iners* and a rather diverse microbiota profile, often in concordance with high *G. vaginalis* abundance, is associated with *C. trachomatis* infection, while *L. crispatus* colonization seems to prevent *C. trachomatis* infection (Balle *et al.*, 2018; Parolin *et al.*, 2018; Tamarelle *et al.*, 2018a) (Balle *et al.*, 2018; Tamarelle *et al.*, 2018a; Ziklo *et al.*, 2018).

The microbiota in females with a history of *C. trachomatis* infection also differs in comparison to females without such a history. Thus, diverse CST and communities with high *G. vaginalis* abundance are prominent in women with high prevalence of antibodies targeting epitopes of *C. trachomatis* and self-reported previous infections (Graspeuntner *et al.*, 2018a).

Within all these analyses, caution should be given to the fact that the studies used different experimental, bioinformatic, and statistical methods (Robinson *et al.*, 2016). In particular, the choice of primers for partial amplification of the 16S gene can introduce bias, lead to dramatically different results, and complicate data interpretation (Graspeuntner *et al.*, 2018b). Further, CSTs have been defined with different methods, such as abundance thresholds of selected bacteria of interest (Nunn *et al.*, 2015; Ravel *et al.*, 2011), and unbiased hierarchical clustering options (Robinson *et al.*, 2016). Despite these methodological differences, the overall results with regard to *C. trachomatis* infection are supported by all of these studies, including a recent meta-analysis (Tamarelle *et al.*, 2018b).

In conclusion, community state type IV of the vaginal microbiota (Table 16.1), which in other studies is referred to as a diverse type (Filardo *et al.*, 2017; Graspeuntner *et al.*, 2018a; van der Veer *et al.*, 2017; van Houdt *et al.*, 2018, etc.) is associated with a higher susceptibility for acquiring *C. trachomatis*. Furthermore, the dominance of *L. iners* in the vaginal microbiota has been shown to enhance the risk of a chlamydial infection (van der Veer *et al.*, 2017; van Houdt *et al.*, 2018). Finally, a wide range of facultative and strict anaerobic bacteria is associated with *C. trachomatis* infection, mostly appearing with low overall microbiota abundance. These results indicate that there is not a single species that is associated with increased risk to *C. trachomatis* infection but the focus has to be given to the host–microbiota interactions and the metabolic/immunological properties that are being driven by the entire microbiota. While most studies focus on the microbiota composition in women, one needs additionally keep in mind, that these microbiota can be influenced by the penile microbiota of men and it was suggested the penile microbiota can serve as a reservoir for colonizing the vagina and thus influencing the microbiota composition (Liu *et al.*, 2015; Zozaya *et al.*, 2016).

While we now know a lot about associations between the bacterial composition of the vaginal microbiota and *C. trachomatis* genital infection, we are limited in the knowledge of non-bacterial members of the microbiota. However, from co-infection experiments we see that viruses such as herpes simplex virus (Vanover *et al.*, 2008) and the parasite *Toxoplasma gondii* (Romano *et al.*, 2013) are associated with differential outcome of *C. trachomatis* infection. Thus, we will need to consider the microbiota profile in its whole complexity in the future.

The anti-chlamydial effect of lactobacilli

The common thread throughout vaginal microbiota community analyses has been that the presence of lactobacilli is associated with a lower risk of *C. trachomatis* infection. This observation has led to the obvious question of how lactobacilli may inhibit *C. trachomatis* infection and whether all lactobacilli possess the mechanisms to do so. While still an active research area, we have begun to understand some of the major mechanisms by which lactobacilli protect the vaginal environment. These include acidifying the environment with lactic acid, competitive exclusion, and introducing anti-inflammatory immune properties (Fig. 16.1). Through these investigations, we are beginning to understand why *L. crispatus* is associated with health while *L. iners* appears to be associated with dysbiosis.

Lactic acid

A defining feature of lactobacilli is their ability to produce lactic acid and reduce the pH of their environment. *Lactobacillus*-dominated vaginal environments have traditionally been characterized as having physiological concentrations of lactic acid from 55–111 mM (0.2% or 22 mM total lactate) and an average pH of 4.2–4.5; conditions that can inactivate BV-associated bacteria while having no effect on vaginal lactobacilli (O'Hanlon *et al.*, 2011, 2013). The pH of the vaginal fluid is known to differ when different lactobacilli dominate the vaginal microbiota; *L. crispatus* – pH 4.0, *L. iners* – pH 4.4, *L. jensenii* – pH 4.7 and *L. gasseri* – pH 5.0 (Petrova *et al.*, 2015). More recently, an appreciation of the generally hypoxic environment of the vagina has led researchers to re-evaluate cervicovaginal fluid under more anaerobic conditions. Based on these new measurements, native fluids in the vagina range closer to pH 3.5 ± 0.2, with total lactate present at 1.0% ± 0.2% (O'Hanlon *et al.*, 2013). These updated measurements indicate the microbicidal potential of vaginal lactic acid when lactobacilli dominate the vaginal microbiota.

Once the potential benefits of both low pH and lactic acid were realized, researchers began studying the two characteristics to determine the influence of each on chlamydial EBs in tissue culture. At first, it appeared that low pH alone was all that was needed to inactivate *C. trachomatis* elementary bodies (EBs). Culture supernatants from *L. crispatus*, *L. gasseri* and *L. jensenii* (pH ~4) were all able to inactivate *C. trachomatis* EBs while the same supernatants buffered to pH 7 had no

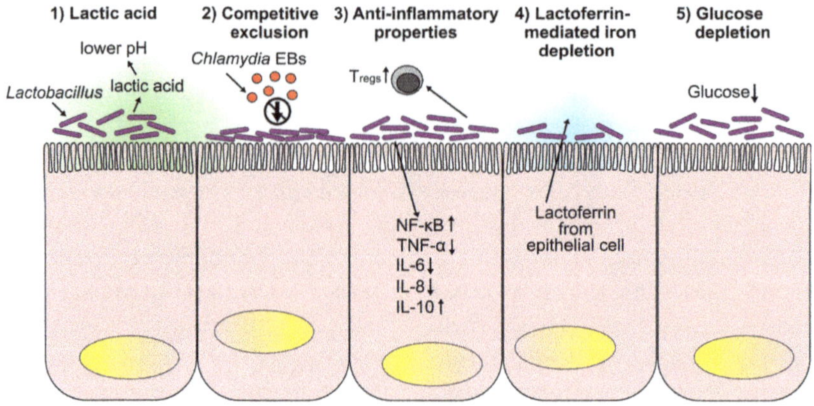

Figure 16.1 Anti-chlamydial effects of lactobacilli. Illustrations diagramming the five major activities lactobacilli have against *Chlamydia*: 1) Lactic acid, 2) Competitive exclusion, 3) Anti-inflammatory properties, 4) Lactoferrin-mediated iron depletion, and 5) Glucose depletion. EBs = elementary bodies, T_{regs} = T regulatory cells, NF-κB = nuclear factor kappa-light-chain-enhancer of activated B cells, TNF-α = tumour necrosis factor alpha, IL = interleukin.

effect on EB viability (Gong *et al.*, 2014). Interestingly, solutions of lactic acid, formic acid and acetic acid alone (prepared in water to 167 mM, pH 4.0) all had the same effect on EB viability as the cultured supernatants, suggesting that the hydrogen ions present at low pH were fully responsible for the anti-chlamydial effect (Gong *et al.*, 2014). However, further studies with *L. crispatus*, *L. gasseri* and *L. vaginalis* determined that lactic acid concentrations from culture supernatants had a mean titre of 10 mM and maximum titre of 50 mM (Nardini *et al.*, 2016). When the effects of lactic acid and hydrochloric acid solutions were tested at 10 and 50 mM (much lower than the 167 mM concentrations Gong *et al.* tested before) and pH 4 and 7, only the lactic acid solutions at pH 4 had an anti-chlamydial effect on *C. trachomatis* EBs (Nardini *et al.*, 2016). This result indicated that high concentrations of hydrogen ions were essential, but not sufficient, for inhibition of *Chlamydia* EBs. When the entire lactobacilli culture supernatant was tested, this solution was able to fully inhibit *Chlamydia* EB viability, indicating that additional metabolites present in lactobacilli supernatants could have a synergistic effect with lactic acid to inhibit *C. trachomatis* (Nardini *et al.*, 2016). Presently, it appears that low pH lactic acid, together with additional unknown lactobacilli culture components, are necessary for the full anti-chlamydial benefit.

Finally, three nonexclusive mechanisms for the effect of lactic acid on *Chlamydia* have been proposed: 1) the acid may inactivate EB surface molecules necessary for host cell attachment or entry, 2) the acid may reduce disulfides in the EB outer membrane and disrupt membrane integrity, and 3) the hydrogen ions could enter the EB and disable cellular metabolism (Gong *et al.*, 2014). Further research is needed to test these hypotheses and determine the specific active mechanisms.

Competitive exclusion

One of the simplest ways lactobacilli can interfere directly with *Chlamydia* infection is to bind to EBs and/or host cells and prevent chlamydial attachment or entry. Strains of *L. brevis* and *L. salivarius* have the ability to co-aggregate with *C. trachomatis* EBs, as well as compete directly with *Chlamydia*, while they adsorb to epithelial cells, to significantly reduce the number of chlamydial infectious units (IFUs) (Mastromarino *et al.*, 2014). This study showed that lactobacilli could have a strong anti-chlamydial effect if present during the early steps of infection when attachment and adsorption are critical. Interestingly, the addition of these lactobacilli after *C. trachomatis* had entered the epithelial cells also resulted in a significant reduction of chlamydial IFUs (Mastromarino *et al.*, 2014). The effect at this stage was less pronounced than at earlier infection stages, but was still significant. The mechanism of chlamydial inhibition after cell entry is not yet understood, but it has been proposed that a product/component of the lactobacilli may exert an intracellular effect or that lactobacilli adherence to the cell membrane may activate anti-chlamydial cell signalling.

Anti-inflammatory immune properties

When *Chlamydia* infects an epithelial cell, a pro-inflammatory response is initiated that includes the production of cytokines and chemokines such as interferon gamma (IFN-γ), interleukin 1 (IL-1), IL-6, IL-8, IL-12, and tumour necrosis factor alpha (TNF-α) (Johnson, 2004; Rasmussen *et al.*, 1997; Redgrove and McLaughlin, 2014). While this response is important for the recruitment of the innate immune system and management of the infection, prolonged activation results in chronic and intense inflammation that leads to cellular proliferation, tissue remodelling and scarring characteristic of *Chlamydia* pathology (Singer and Ouburg, 2016; Stephens, 2003).

Lactobacilli colonization, particularly by *L. crispatus*, appears to have an anti-inflammatory effect on host cells. *L. crispatus* colonization of epithelial cells has been shown to induce nuclear factor kappa B activation (NF-κB) and IL-6 production (Abramov *et al.*, 2014). Both NF-κB and IL-6 have traditionally been considered pro-inflammatory cytokines, but continued study has revealed that both also have anti-inflammatory properties that appear to be at work in this situation to down regulate inflammation (Abramov *et al.*, 2014; Lawrence *et al.*, 2001; Scheller *et al.*, 2011). *L. crispatus* has also been shown to confer an anti-inflammatory phenotype to dendritic cells through up-regulation of the anti-inflammatory cytokine IL-10 and induction of regulatory T cells (Eslami *et al.*, 2016). It is believed that these changes in dendritic cells help reduce inflammation of vaginal epithelial cells, which in turn could help prevent the scarring associated with chlamydial disease. Finally, addition of *L. crispatus* to a *C. trachomatis* cell culture infection downregulated *C. trachomatis*-induced IL-6, IL-8 and TNF-α pro-inflammatory cytokines and upregulated anti-inflammatory IL-10 cytokine expression (Rizzo *et al.*, 2015). Overall, the presence of *L. crispatus* in the vaginal environment appears to limit the inflammation caused by *Chlamydia* infection, which could have effects on the amount of cell damage and scarring that can lead to permanent complications.

Additional mechanisms

Aside from the three major mechanisms by which lactobacilli reduce infection risk in the vaginal environment during *Chlamydia* infection, there are a number of other anti–chlamydial interactions that are beginning to be investigated. Lactoferrin is a host glycoprotein, which is secreted from mucosal epithelial cells and that inhibits bacterial growth because of its ability to chelate iron (Valenti *et al.*, 2018). While the presence of either *L. brevis* organisms or lactoferrin at the epithelial cell surface had the ability to inhibit *C. trachomatis* EB adhesion and invasion and cellular IL-6 and IL-8 synthesis, the presence of both *L. brevis* and lactoferrin together had the greatest effect (Sessa *et al.*, 2017). This suggests that lactoferrin together with the iron sequestering effects of lactobacilli have additive anti-chlamydial effects that warrant further investigation. Another anti-chlamydial effect of lactobacilli growth in the vaginal environment is glucose depletion. When culture supernatants from high-glucose utilizing *L. crispatus* and *L. gasseri* strains were supplemented with glucose, there was a significant reduction in anti-*Chlamydia* activity (Nardini *et al.*, 2016). This finding revealed the importance of glucose depletion as an additional mechanism for the inhibition of *C. trachomatis* EBs by vaginal lactobacilli.

Finally, a mechanism that turned out not to be important as an anti-chlamydial factor was hydrogen peroxide (H_2O_2). For some time, it was believed that H_2O_2 produced from lactobacilli was microbicidal to pathogens in the vagina and an important property of protective lactobacilli. However, investigation revealed that physiological concentrations of H_2O_2 had no effect on BV-associated bacteria or *Chlamydia* (Gong *et al.*, 2014; O'Hanlon *et al.*, 2011). Additionally, when lactobacilli were grown under aeration conditions (with vigorous shaking) to increase the levels of H_2O_2, there was less lactic acid produced and overall, *C. trachomatis* killing was reduced (Gong *et al.*, 2014). Therefore, H_2O_2 production is no longer considered meaningful when evaluating the anti-chlamydial properties of a lactobacilli strain.

Not all lactobacilli are created equal – *L. crispatus* versus *L. iners*

In general, lactobacilli lower the risk of chlamydial infection in the vaginal environment, but there is epidemiological evidence that *L. crispatus* is more protective than *L. iners*. When the genomes of these two species are compared, a number of differences emerge that offer insight into why this is so (France *et al.*, 2016). Overall, the *L. crispatus* genome is nearly twice the size of the *L. iners* genome (2.25 Mbp versus 1.28 Mbp, respectively). This results in *L. iners* being more dependent on the vaginal environment for essential nutrients while *L. crispatus* can synthesize more of its own nutritional and energy requirements. In terms of lactic acid, *L. crispatus* produces both d- and l-lactic acid from its two copies of l-lactate dehydrogenase and one copy of d-lactate dehydrogenase, while *L. iners* can only produce l-lactic acid because it is completely missing a d-lactate dehydrogenase gene (France *et al.*, 2016; Witkin *et al.*, 2013). Interestingly, in the context of the vaginal environment, this results in a slightly lower overall pH for *L. crispatus*-dominated vaginal environments (pH 4.0) than *L. iners*-dominated vaginal environments (pH 4.4) (Petrova *et al.*, 2015). *L. crispatus* has a broader genetic capability for fermenting different sugars, which could aid in nutrient depletion against pathogens, as well as possessing the capability to breakdown putrescine, a malodorous amino acid commonly found in vaginal secretions during BV. An iron transport system present in *L. crispatus* but absent in *L. iners* may also allow *L. crispatus* to sequester iron released by the host during menses (France *et al.*, 2016). One advantage to *L. iners* is its ability to produce inerolysin, a pore-forming cytolysin, which may allow it to lyse cells in its environment and acquire the nutrients it needs. However, genetic differences combined with experimental evidence indicate that *L. crispatus* appears to be a superior anti-chlamydial lactobacillus than *L. iners* (France *et al.*, 2016). Nevertheless, there are still more experimental parameters, such as competitive exclusion properties, anti-inflammatory properties or lactoferrin production, which could be compared between these two distinct lactobacilli.

The pro-chlamydial effect of some members of the anaerobic microbiome: *Chlamydia* rescue by indole

While lactobacilli appear to reduce the risk of *C. trachomatis* infection, anaerobic vaginal communities have been implicated in increasing the risk of *Chlamydia* infection. The lack of lactobacilli in the vagina results in higher pH, the absence of lactic acid and the loss of lactobacilli-mediated effects, all of which can make the environment more conducive for *Chlamydia* infection. However, beyond simply the removal of anti-chlamydial effects, the presence of some anaerobic bacteria appears to assist genital strains of *C. trachomatis* in immune evasion and growth through the production of indole. This complicated interaction involves a host immune response to *Chlamydia*, in which there is IFN-γ-dependent depletion of the amino acid tryptophan, and a chlamydial counter-response that utilizes microbially derived indole to restore tryptophan (Loomis and Starnbach, 2002; Ziklo *et al.*, 2016a) (Fig. 16.2).

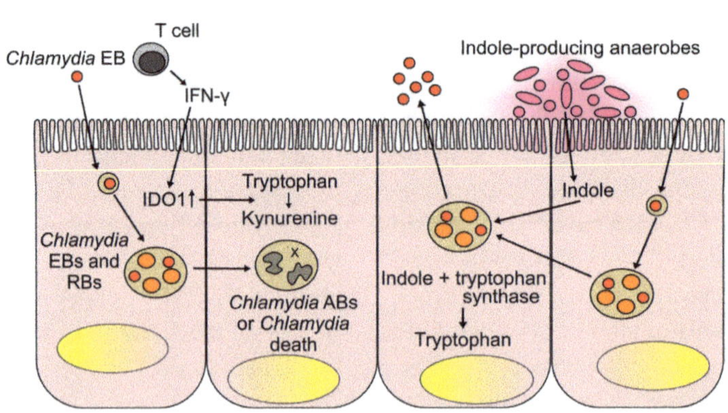

Figure 16.2 Chlamydial rescue by indole from anaerobic microbiome members. Diagram summarizing the tryptophan degradation pathway by interferon-gamma (IFN-γ) (on left side) and the rescuing of *Chlamydia* from tryptophan starvation by indole provided by the anaerobic vagina microbiome (right side). EBs = elementary bodies, RBs = reticulate bodies, ABs = aberrant bodies, IDO1 = Indoleamine 2, 3-dioxygenase 1 gene.

Setting the stage – tryptophan depletion by the immune system

It has been clearly shown that the most effective host response to a *C. trachomatis* infection is IFN-γ production by CD4+ and CD8+ cells (Loomis and Starnbach, 2002). IFN-γ can directly inhibit the replication of *C. trachomatis* within infected cells by at least three possible mechanisms; 1) up-regulation of inducible nitric oxide synthase (iNOS), which catalyses the production of various reactive nitrogen intermediates like nitric oxide (NO) (Liew, 1994), 2) down-regulation of transferrin receptors on the surface of infected cells, resulting in intracellular iron deficiency (Byrd and Horwitz, 1993), and 3) induced expression of the host enzyme idoleamine-2,3-dioxygenase (IDO), which degrades intracellular tryptophan to kynurenine (Beatty *et al.*, 1994b; Byrne *et al.*, 1986). Investigation of *C. trachomatis* growth in cell lines showed that the major mechanism by which IFN-γ exerted its effect was through nutrient deprivation by IDO-mediated tryptophan catabolism (Beatty *et al.*, 1994a). Tryptophan depletion is particularly difficult for *Chlamydia* to cope with, as none of the characterized species contains a full complement of tryptophan synthesis (*trp*) genes, making them tryptophan auxotrophs (Fehlner-Gardiner *et al.*, 2002). However, *C. trachomatis* genital strains do contain three genes from the tryptophan synthesis pathway in the plasticity zone of their genome, *trpR*, *trpA* and *trpB*, which can allow them to perform some steps in the pathway (Fehlner-Gardiner *et al.*, 2002). Interestingly, ocular strains of *C. trachomatis* lack a functional *trpA* synthase gene (Caldwell *et al.*, 2003).

Creating the opportunity – indole in the vaginal environment

In the biosynthesis of tryptophan, the last step in tryptophan synthesis is the conversion of the molecule indole to tryptophan using the *trpB* gene. Indole is widespread in the natural environment and is produced by a wide range of Gram-negative and Gram-positive bacteria as a signalling molecule (Lee and Lee, 2010). Indole has been shown to be involved in a diverse range of bacterial physiologies, including biofilm formation, drug resistance, plasmid stability and quorum signalling (Chant and Summers, 2007; Chu *et al.*, 2012; Lee *et al.*, 2009; Lee *et al.*, 2007). Interestingly, when the list of known indole-producing bacteria is compared to the list of known vaginal microbiota community members, an overlap in anaerobic bacteria is found (Ziklo *et al.*, 2016a). Prominent indole-producing vaginal bacteria include species from the genera *Prevotella*, *Peptoniphilus*, and *Porphyromonas* (Ziklo *et al.*, 2016a). Given this overlap, vaginal secretions have been screened for the presence of indole, and concentrations of 2.18–7.35 mM have been detected (Ziklo *et al.*, 2016b). Therefore, when indole-producing bacteria are present in the vaginal microbiota, they generate a source of indole that *Chlamydia* might utilize to overcome the immune depletion of tryptophan.

Rescuing *Chlamydia* – converting indole to tryptophan

There is mounting evidence that genital *C. trachomatis* use their tryptophan synthase (*trpBA*) genes to convert indole scavenged from the vaginal microbiota to tryptophan and escape immune-induced tryptophan starvation. Indole added to tryptophan-starved genital *C. trachomatis* reticulate bodies (RBs) in tissue culture showed significant recovery while the same treatment had no effect on ocular *C. trachomatis* RBs (Ziklo *et al.*, 2016b). Additionally, culture supernatants from indole-positive *Prevotella* species supported the recovery of tryptophan-starved genital *C. trachomatis* RBs while the supernatants of indole-negative *Prevotella* species had no observable benefit (Ziklo *et al.*, 2016b). Extending from *C. trachomatis*, *Chlamydia psittaci* is a pathogen of birds and can cause zoonotic infections in humans. *C. psittaci* possesses additional tryptophan biosynthesis genes to *C. trachomatis*, including *trpDCFBA*, as well as *kynU* (kynureninase) and *prsA* (phosphoribosylpyrophosphate (PRPP) synthase) (Wood *et al.*, 2004). The presence of these additional genes allowed *C. psittaci* cultured in tryptophan-free medium to be rescued by not only indole, but also by anthranilate and kynurenine (additional intermediates in the tryptophan-synthesis pathway) (Wood *et al.*, 2004). Clearly, the presence of some tryptophan biosynthesis genes allows *Chlamydia* to circumvent the immune strategy of tryptophan starvation.

The evidence that indole-mediated *Chlamydia* rescue can be linked to the microbial community members at the site of infection can be found in the differences between ocular and genital *C.*

trachomatis strains. Ocular strains of *C. trachomatis* have lost their tryptophan synthesis capability while genital strains have retained their indole-to-tryptophan conversion (Caldwell *et al.*, 2003). A proposed reason for this difference is the fact that very few ocular microbiota bacteria have indole-producing capacity while many more genital microbiota bacteria exhibit this trait (Caldwell *et al.*, 2003). When there is no selective advantage for bacteria to retain a gene, its function is often lost (Wolf and Koonin, 2013). The availability of bacterially produced indole in the vaginal environment appears to have created an advantage for genital *C. trachomatis* strains to maintain the indole-to-tryptophan conversion to overcome the IFN-γ tryptophan-depletion selective pressure (Caldwell *et al.*, 2003). Together, indole-producing vaginal microbiota members work with *Chlamydia* to circumvent the host immune system.

The role of the microbiota in developing long term sequelae of *C. trachomatis* infection

The connection between the vaginal microbiota and acute *C. trachomatis* infection raises the question of whether the microbiota also plays a role in the development of upper genital tract pathology. In this respect, both the microbiota of the lower and the upper genital tract are of particular interest and might have functions that accelerate or dampen the pathological potential of *C. trachomatis* infections.

The contribution of vaginal microbiota in upper genital tract sequelae by *C. trachomatis*

Based on the current knowledge, the vaginal microbiota may play an important role in the establishment of upper genital tract sequelae (Loeper *et al.*, 2018). As described, a microbial composition with high abundance of *L. crispatus* is associated with lower *C. trachomatis* transmission. In combination with the anti-inflammatory capacity of *L. crispatus* and its metabolic activity, this likely reflects a major component in the prevention of pathological processes induced by *C. trachomatis* and reduces subsequent ascension of the pathogen. In contrast, a diverse microbiota and a microbiota dominated by either *G. vaginalis* or *L. iners* can potentially play a role in the development of pathological sequelae by inducing a pro-inflammatory host response, rescuing *C. trachomatis* from elimination through the host, and facilitating ascension of the pathogen to the upper genital tract. This hypothesis is supported by the fact that microbiota containing BV-associated bacteria predisposes the host to pelvic inflammatory disease (Ness *et al.*, 2005; Sharma *et al.*, 2014). Along this line, a recent study on cervical microbiota in females with tubal factor infertility showing respective microbial pattern in diseased patients resembling those found to be associated with acute *C. trachomatis* infection (Graspeuntner *et al.*, 2018a).

Microbiota of the upper genital tract

While recent reports highlight the divergent role of different microbiota of the vagina in the context of *C. trachomatis* infection (Ghartey *et al.*, 2014; Nardini *et al.*, 2016; Nunn *et al.*, 2015; van der Veer *et al.*, 2017; van Houdt *et al.*, 2018), the effect of the microbiota on the upper urogenital tract is not well explored. While the upper genital tract has been thought to be sterile, there are recent reports that body habitats that are not directly exposed to the external environment are colonized by bacteria (Chen *et al.*, 2017; Mitchell *et al.*, 2015; Pelzer *et al.*, 2018).

Mitchel *et al.* found that the endometrium of most women undergoing hysterectomy is colonized by bacteria that are typical for the vagina, such as lactobacilli or *G. vaginalis* (Mitchell *et al.*, 2015). However, on a community-based level, they reported that the microbiota composition of the upper genital tract differed from the vaginal microbiota in many women (Mitchell *et al.*, 2015). A more recent report found polymicrobial communities, with *Staphylococcus* being the most abundant genus in the fallopian tube, and a noticeably low amount of *Lactobacillus* species (Pelzer *et al.*, 2018). These findings agree with a study by Chen *et al.*, which showed that lactobacilli decreased in abundance from the vagina to the fallopian tubes and was not the dominating species in the endometrium (Chen *et al.*, 2017). This study found that *Acinetobacter*, *Comamonas*, and *Pseudomonas* dominated in the upper genital tract (Chen *et al.*, 2017).

There is very limited data attributing specific upper genital tract microbiota with disease. However, BV-associated bacteria such as *Sneathia* spp., BVAB1, and *Atopobium vaginae* have been found in endometrial biopsies and shown to be associated with pelvic inflammatory disease (PID) (Haggerty *et al.*, 2016). *G. vaginalis*, *Megasphaera*, *Prevotella* spp., *A. vaginae* and *Dialister* spp. have been shown to be present in the endometrium of *C. trachomatis*-infected women (Balle *et al.*, 2018). Further, it is yet unclear if these bacteria colonize the upper genital tract and play a role in ascending infections and upper genital tract inflammation (Sharma *et al.*, 2014).

Other human chlamydial infections: role of the microbiota?

While most of the work on the relationship between the microbiota and *Chlamydia* has focused on genital infections, there is interest in the role of the microbiota in *C. trachomatis* infections of the eye, and *Chlamydia pneumoniae* lung infections.

Ocular infections

Trachoma is a blinding disease predominantly caused by *C. trachomatis* strains A-C and is the world's leading cause of preventable blindness (discussed in more detail in Chapter 1, *Chlamydia trachomatis* Infections). While it is uncommon in developed countries, it is a threat to populations in developing countries. Considering the knowledge about the microbiota in genital *C. trachomatis* infection there likely is a role of the microbiota in ocular *C. trachomatis* infection as well.

Normal conjunctival microbiota

Some recent reports have been published on the microbiota of the healthy human conjunctiva. The most prominent phyla have consistently been described to be Actinobacteria, Proteobacteria, Firmicutes, and Bacteroidetes, accounting for more than 90% of all bacterial sequences observed (Huang *et al.*, 2016; Zhou *et al.*, 2014). Little consensus between studies can be observed, however, when defining a core microbiota at the genus level. *Propionibacterium*, *Corynebacterium*, *Pseudomonas*, *Bradyrhizobium*, *Acinetobacter*, *Brevundimonas*, *Stahylococcus*, *Aquabacterium*, *Sphingomonas*, *Streptococcus*, *Streptophyta*, and *Methylobacterium* have been initially described as the potential core microbiota (Dong *et al.*, 2011) based on a limited number of samples. In contrast, Huang *et al.* described a partially different set of taxa as the core microbiota (*Corynebacterium*, *Pseudomonas*, *Staphylococcus*, *Acinetobacter*, *Streptococcus*, *Millisia*, *Anaerococcus*, *Finegoldia*, *Simonsiella*, and *Veillonella*) (Huang *et al.*, 2016). With increasing precision, the proposed core microbiota has been reduced to *Corynebacterium*, *Propionibacterium*, *Staphylococcus*, and *Streptococcus*, while other taxa occur more sporadically (Butcher *et al.*, 2017). The definition of a healthy or core conjunctival microbiota is complicated as there is a high inter-individual variation that is influenced by age and sample season, as shown in a recent study (Butcher *et al.*, 2017), and the number of studies addressing microbial composition in the eye is currently limited.

Conjunctival microbiota in trachoma disease

Trachoma is endemic in many developing countries, with one of the rare studies on microbiota in trachoma being conducted in The Gambia (Zhou *et al.*, 2014). Zhou *et al.* observed that significant differences between the bacterial communities of healthy subjects and patients with scarring conjunctiva. Patients with conjunctival scarring had increased bacterial diversity, and *Corynebacterium* and *Streptococcus* abundance, although these differences were only during the dry season. In addition, the amount of *Corynebacterium* was higher in patients with scarring and trachomatous trichiasis (i.e. eyelashes rubbing against the eyeball) compared to only scarring conjunctiva (Zhou *et al.*, 2014). Another study was conducted on patients with active trachoma cases in the Solomon Islands. In this study, Butcher *et al.* observed a broad variety of polymicrobial communities in active trachoma patients (Butcher *et al.*, 2017). They characterized differences in the abundance of *Corynebacterium*, *Propionibacterium*, *Helicobacter*, and *Paracoccus*. However, they were not able to identify specific signatures that could be associated with the disease (Butcher *et al.*, 2017).

Although these studies indicate differences between the microbiota of healthy people and patients with active trachoma or conjunctival scarring, we have limited information about the characteristics of these differences, and if and how those differences influence the disease. Large scale studies deciphering the microbiota over the different stages of trachoma will be necessary to unravel the role of the microbiota in chlamydial ocular disease.

Lung infections

Chlamydia pneumoniae infects the lung

Chlamydia pneumoniae is another human pathogen within the genus *Chlamydia,* which is different from *C. trachomatis,* not only genetically, but also by mainly infecting the airways of humans (Kuo *et al.*, 1995). *C. pneumoniae* has been linked to a wide range of respiratory diseases such as pneumonia, chronic obstructive pulmonary disease (COPD), and asthma (discussed in more detail in Chapter 2, *Chlamydia pneumoniae* Infections).

Lung microbiota

The description and analysis of the microbiota of the human airways is complicated as various sample materials can be used for molecular studies, including sputum, bronchoalveolar lavages (BAL) or lung tissue (Dickson *et al.*, 2013). The major observed phyla are Bacteroidetes and Firmicutes, with prominent genera including *Prevotella, Veillonella* and *Streptococcus.* Thus, the microbiota of the airways resembles that of the mouth, which is suggested as its primary source (Dickson *et al.*, 2016). However, the microbiota within the lung can also vary dependent on the location of the sampling (Goddard *et al.*, 2012; Willner *et al.*, 2012). In particular, taxa that are found in lower abundance, are more likely to show differences between sampling sites within the lung (Willner *et al.*, 2012).

Lung microbiota in disease

The microbiota of the airways is a particular focus of studies addressing the wide range of pulmonary diseases (Dickson *et al.*, 2016). In the field of *Chlamydia* related diseases, there has been particular attention on asthmatic diseases, as studies have demonstrated the prevalence of *C. pneumoniae* in this disease (Webley and Hahn, 2017; Webley *et al.*, 2005) and a relationship between asthma and the microbiota of the lung (see Chapter 2, *Chlamydia pneumoniae* Infections, for a more detailed discussion of this point.)

To date there is no study available that directly addresses the role of the airway microbiota in the virulence and pathogenicity of *C. pneumoniae.* Taking under consideration the defined role of the microbiota in urogenital diseases, it appears plausible to hypothesize a role of the microbiota in *C. pneumoniae* associated lung diseases. In this respect, a protective role of certain microbial communities in the lung has already been proven for respiratory infections with *Escherichia coli, Klebsiella pneumoniae, Listeria monocytogenes, Staphylococcus aureus,* and *Streptococcus pneumoniae* (Brown *et al.*, 2017). Thus, future research into *C. pneumoniae* infection in the lung should address the microbiota composition to unravel its relation to each other.

Chlamydia and microbiota in animals

Beyond human health, different species of *Chlamydia* can cause serious infection and disease in a range of animals. *Chlamydia psittaci* infection in birds, *C. pneumoniae* infection in reptiles and *Chlamydia abortus* and *Chlamydia pecorum* infection in livestock and wildlife are just a few examples of non-human chlamydial disease (Burnard and Polkinghorne, 2016) (discussed in more detail in Chapter 17, Recent Advances in Epidemiology, Pathology and Immunology of Veterinary Chlamydiae). Where it has been investigated, the local microbiomes of these different animals at the sites of chlamydial infection appear to influence the establishment and progression of chlamydial disease. While direct comparisons to human disease cannot be drawn from most of these animal-focused *Chlamydia*/microbiome studies, broadening our understanding of how *Chlamydia* interacts with different microbiomes improves our overall understanding of this complex interaction.

Koalas (*Phascolarctos cinereus*)

The koala is an iconic Australian marsupial, which is currently suffering population declines from a range of natural and man-made threats. One of these major threats is disease caused by *Chlamydia,* with *C. pecorum* causing urogenital and ocular disease while *C. pneumoniae* causes rhinitis/pneumonia complex (Polkinghorne *et al.*, 2013). Of these two pathogens, *C. pecorum* is by far the most common. While a few isolated island koala populations appear to be *Chlamydia*-free, mainland koala populations have been found to have up to a 90% *C. pecorum* infection rate (Polkinghorne *et al.*, 2013). Ocular disease manifests as keratoconjunctivitis (which can lead to scarring and blindness), while urogenital disease can involve urinary tract disease (cystitis leading to incontinence and skin ulceration) and/or reproductive tract disease (which results in infertility and reproductive loss). Chlamydial disease is diagnosed via clinical signs observed during veterinary clinical examination and ultrasonography, while chlamydial infection is established by laboratory testing of swabs taken from mucosal sites and tested for *C. pecorum* nucleic acid via PCR (Polkinghorne *et al.*, 2013). Currently, treatment of chlamydial infection and disease is achieved with antibiotics. However, antibiotic options are limited for koala use, as antibiotics commonly used to treat chlamydial infections in humans and other animals have been shown to adversely affect the gastrointestinal microbiota of koalas, preventing proper digestion of their eucalyptus leaf diet (Dahlhausen *et al.*, 2018). Recent advances in chlamydial vaccine development for koalas have shown very promising results and may offer a safer and long-term strategy for chlamydial management in koala populations in the future (Hernandez-Sanchez *et al.*, 2015; Waugh *et al.*, 2016).

As with humans, there is interest in the microbiota communities found in the eye and urogenital tracts of koalas and how these communities are altered during chlamydial infection. The ocular conjunctiva of koalas has been found to have a low abundance but a high diversity of microbes (Alfano *et al.*, 2015; Vidgen *et al.*, 2017). The dominant genera present include *Boyliae/Corynebacteria*, *Bacteroides*, *Aerococcus* and Ruminococcacae, which is similar in composition to the human conjunctiva (Alfano *et al.*, 2015; Vidgen *et al.*, 2017; Zhou *et al.*, 2014). Interestingly, during chlamydial infection, the ocular microbiota becomes dominated by *C. pecorum*, suggesting that infection overwhelms the other bacteria present in the eye (Vidgen *et al.*, 2017). Currently, sample numbers are limited for koala ocular microbiota, both with and without chlamydial infection, so conclusions about the effects of chlamydial infection on the bacteria in the eye are difficult to reach. However, a clear shift to a *C. pecorum*-dominated environment suggests that *Chlamydia* can profoundly alter the microbiota at this body site.

In humans, the female vaginal microbiota is often dominated by the Lactobacillales genera *Lactobacillus*. In the urogenital tract of female koalas, the order Lactobacillales also predominates, but the dominant genera are *Aerococcus* and *Facklamia* (Legione *et al.*, 2018; Vidgen *et al.*, 2017). Organisms from the order Clostridiales, such as *Lachnospiraceae*, *Ruminococcaceae* and *Tissierellaceae* are also prevalent (Legione *et al.*, 2018; Vidgen *et al.*, 2017). Distinguishing the female koala urogenital microbiota from the vaginal microbiota of other animals is the ubiquitous presence of *Propionibacteriaceae* in the koala microbiome (Vidgen *et al.*, 2017). The anatomical structure of the marsupial urogenital tract consists of a canal that is common to the urethra and three vaginal openings, resulting in a common site for both urinary and genital associated organisms. It is currently believed that *Propionibacteriaceae* represent part of the urinary microbiota and the unique anatomy of marsupials results in its detection in urogenital samples (Vidgen *et al.*, 2017). Finally, the composition of the male koala penile microbiota mirrored that of the female urogenital microbiota, suggesting a similar association of the microbiota of these sites between sexual partners that has been documented in human penile/vaginal comparisons (Vidgen *et al.*, 2017; Zozaya *et al.*, 2016).

During chlamydial infection, no direct correlations between *C. pecorum* infection and the presence or absence of any single microbiota member were identified. Rather, there was a correlation in overall microbiota composition, where koalas infected with a high load of *C. pecorum* (> 100 16S rRNA gene copies/μl of swab sample tested) had an overall lower microbiota diversity and richness compared to koalas with a low infection level (1–100 16S rRNA gene copies/μl of swab sample tested) or no *C. pecorum* detected (Vidgen *et al.*, 2017). Interestingly, it was noted that *Chlamydia*-negative koalas could have microbiota with similar compositions and structures to koalas with high *C. pecorum* infections. This suggested that it was unlikely that *C. pecorum* infection altered the microbiota composition but rather the microbiota composition may have influenced the potential for *C. pecorum* to establish an infection (Vidgen *et al.*, 2017). Research in this area is still in its infancy and further studies will clarify this interesting interaction between *Chlamydia* and the koala urogenital microbiota.

Animal models and their microbiota

Animal models are critical to facilitate research into complex pathogen–host interactions, especially for human diseases where experimentation cannot be done in the native host. However, an animal model is only as good as its ability to replicate the native disease situation being modelled. For studies involving human vaginal infections, such as *C. trachomatis* infection, the ability to model all aspects of the vaginal environment in laboratory animals has proven challenging. One aspect that is particularly challenging to replicate has been to find an animal system that can mimic the vaginal microbiota. Aside from humans, the common chimpanzee and the olive baboon, no other investigated animal has an average vaginal pH below 6 (Miller *et al.*, 2016). This includes other non-human primates (such as monkeys or macaques), pigs, mice, rats, rabbits, cows, sheep, goats, horses or dogs (Jacques *et al.*, 1986; Miller *et al.*, 2016; Neuendorf *et al.*, 2015; Stumpf *et al.*, 2013; Swartz *et al.*, 2014). The predominant reason for the neutral vaginal pH in these other animals is that their microbiota lack significant levels of lactobacilli. Non-human primates have on average less than 5% lactobacilli in their vaginal microbiota (Stumpf *et al.*, 2013), while guinea pigs possess less than 0.5% lactobacilli abundance (Neuendorf *et al.*, 2015). BALB/c mice, a common model organism, has a vaginal microbiota containing *Streptococcus, Acinetobacter, Sphingomonas, Enterococcus* and *Polaromonas*, making it clearly distinct from humans (Barfod *et al.*, 2013). Given the important role that the vaginal microbiota plays in protection against infection, consideration needs to be given to microbiota differences in model systems.

Where the microbiota of an animal model has been considered during *Chlamydia* research, little effect of chlamydial infection on the microbiota has been observed. In female olive baboons (*Papio anubis*), infection with *C. trachomatis* had no effect on the diversity, richness or longitudinal trends observed in the microbiota of infected versus uninfected animals (Liechty *et al.*, 2015). Similarly, *Chlamydia caviae* infection of guinea pigs affected the abundance of some bacterial phylotypes in the microbiota, but did not alter the microbiota composition (Neuendorf *et al.*, 2015). Whether this lack of effect on the animal model microbiota can be extrapolated to *C. trachomatis* infection in humans is still in need of further testing.

Outlook

Recent advances have shed considerable light on the importance of the microbiota in chlamydial infections. However, a wide range of scientific questions remains to be addressed. The research on microbial communities has been very much focused on bacteria excluding fungi and viruses. Further, we know a lot about the vaginal microbiota but our knowledge about the penile microbiota and how these both interact, are scarce. First insights beyond the sole composition of the microbiota in *C. trachomatis* infection have been provided, however a better understanding of cause and consequence of the microbiota in *C. trachomatis* infection is required. In this spite research needs to be extended to understand the interaction between microbial composition and the infection with *C. trachomatis* in large longitudinal cohort studies. This opens novel research fields such as the resulting metabolic profiles corresponding to microbiota changes that might be the major regulator for both chlamydial growth but also host immunity. Such information is desperately needed considering preventative treatment strategies on the basis of directed microbiota modulation, e.g. by probiotics.

As only limited data on host–microbiota-chlamydia interactions are available, more studies on the upper genital tract and non-genital tract infections, as well as better experimental models, are needed for better characterization of the *in vivo* situation. In addition, analyses on respective interactions in animals will foster our understanding on the impact of microbiota on chlamydial infections.

References

Abramov, V., Khlebnikov, V., Kosarev, I., Bairamova, G., Vasilenko, R., Suzina, N., Machulin, A., Sakulin, V., Kulikova, N., Vasilenko, N., *et al.* (2014). Probiotic Properties of Lactobacillus crispatus 2,029: Homeostatic Interaction with Cervicovaginal Epithelial Cells and Antagonistic Activity to Genitourinary Pathogens. Probiotics Antimicrob Proteins *6*, 165–176.

Alfano, N., Courtiol, A., Vielgrader, H., Timms, P., Roca, A.L., and Greenwood, A.D. (2015). Variation in koala microbiomes within and between individuals: effect of body region and captivity status. Sci Rep *5*, 10189.

Balle, C., Lennard, K., Dabee, S., Barnabas, S.L., Jaumdally, S.Z., Gasper, M.A., Maseko, V., Mbulawa,

Z.Z.A., Williamson, A.L., Bekker, L.G., *et al.* (2018). Endocervical and vaginal microbiota in South African adolescents with asymptomatic Chlamydia trachomatis infection. Sci Rep *8*, 11109.

Barfod, K.K., Roggenbuck, M., Hansen, L.H., Schjorring, S., Larsen, S.T., Sorensen, S.J., and Krogfelt, K.A. (2013). The murine lung microbiome in relation to the intestinal and vaginal bacterial communities. BMC Microbiol *13*, 303.

Beamer, M.A., Austin, M.N., Avolia, H.A., Meyn, L.A., Bunge, K.E., and Hillier, S.L. (2017). Bacterial species colonizing the vagina of healthy women are not associated with race. Anaerobe *45*, 40–43.

Beatty, W.L., Belanger, T.A., Desai, A.A., Morrison, R.P., and Byrne, G.I. (1994a). Tryptophan depletion as a mechanism of gamma interferon-mediated chlamydial persistence. Infect Immun *62*, 3705–3711.

Beatty, W.L., Morrison, R.P., and Byrne, G.I. (1994b). Persistent chlamydiae: from cell culture to a paradigm for chlamydial pathogenesis. Microbiol Rev *58*, 686–699.

Brooks, J.P., Edwards, D.J., Blithe, D.L., Fettweis, J.M., Serrano, M.G., Sheth, N.U., Strauss, J.F., 3rd, Buck, G.A., and Jefferson, K.K. (2017). Effects of combined oral contraceptives, depot medroxyprogesterone acetate and the levonorgestrel-releasing intrauterine system on the vaginal microbiome. Contraception *95*, 405–413.

Brotman, R.M., He, X., Gajer, P., Fadrosh, D., Sharma, E., Mongodin, E.F., Ravel, J., Glover, E.D., and Rath, J.M. (2014). Association between cigarette smoking and the vaginal microbiota: a pilot study. BMC Infect Dis *14*, 471.

Brotman, R.M., Klebanoff, M.A., Nansel, T.R., Yu, K.F., Andrews, W.W., Zhang, J., and Schwebke, J.R. (2010). Bacterial vaginosis assessed by gram stain and diminished colonization resistance to incident gonococcal, chlamydial, and trichomonal genital infection. J Infect Dis *202*, 1907–1915.

Brown, R.L., Sequeira, R.P., and Clarke, T.B. (2017). The microbiota protects against respiratory infection via GM-CSF signalling. Nature Communications *8*, 1512.

Burnard, D., and Polkinghorne, A. (2016). Chlamydial infections in wildlife-conservation threats and/or reservoirs of 'spill-over' infections? Vet Microbiol *196*, 78–84.

Butcher, R.M.R., Sokana, O., Jack, K., Kalae, E., Sui, L., Russell, C., Houghton, J., Palmer, C., Holland, M.J., Le Mesurier, R.T., *et al.* (2017). Active Trachoma Cases in the Solomon Islands Have Varied Polymicrobial Community Structures but Do Not Associate with Individual Non-Chlamydial Pathogens of the Eye. Frontiers in medicine *4*, 251.

Byrd, T.F., and Horwitz, M.A. (1993). Regulation of transferrin receptor expression and ferritin content in human mononuclear phagocytes. Coordinate upregulation by iron transferrin and downregulation by interferon gamma. J Clin Invest *91*, 969–976.

Byrne, G.I., Lehmann, L.K., and Landry, G.J. (1986). Induction of tryptophan catabolism is the mechanism for gamma-interferon-mediated inhibition of intracellular Chlamydia psittaci replication in T24 cells. Infect Immun *53*, 347–351.

Caldwell, H.D., Wood, H., Crane, D., Bailey, R., Jones, R.B., Mabey, D., Maclean, I., Mohammed, Z., Peeling, R., Roshick, C., *et al.* (2003). Polymorphisms in Chlamydia trachomatis tryptophan synthase genes differentiate between genital and ocular isolates. J Clin Invest *111*, 1757–1769.

Chaban, B., Links, M.G., Jayaprakash, P.T., Wagner, E.C., Bourque, D.K., Lohn, Z., Albert, A.Y.K., van Schalkwyk, J., Reid, G., Hemmingsen, S.M., *et al.* (2014). Characterization of the vaginal microbiota of healthy Canadian women through the menstrual cycle. Microbiome *2*, 23.

Chant, E.L., and Summers, D.K. (2007). Indole signalling contributes to the stable maintenance of Escherichia coli multicopy plasmids. Mol Microbiol *63*, 35–43.

Chen, C., Song, X., Wei, W., Zhong, H., Dai, J., Lan, Z., Li, F., Yu, X., Feng, Q., Wang, Z., *et al.* (2017). The microbiota continuum along the female reproductive tract and its relation to uterine-related diseases. Nat Commun *8*, 875.

Chu, W., Zere, T.R., Weber, M.M., Wood, T.K., Whiteley, M., Hidalgo-Romano, B., Valenzuela, E., Jr., and McLean, R.J. (2012). Indole production promotes Escherichia coli mixed-culture growth with Pseudomonas aeruginosa by inhibiting quorum signalling. Appl Environ Microbiol *78*, 411–419.

Dahlhausen, K.E., Doroud, L., Firl, A.J., Polkinghorne, A., and Eisen, J.A. (2018). Characterization of shifts of koala (Phascolarctos cinereus) intestinal microbial communities associated with antibiotic treatment. PeerJ *6*, e4452.

Dickson, R.P., Erb-Downwards, J.R., and Huffnagle, G.B. (2013). The role of the bacterial microbiome in lung disease. Expert review of respiratory medicine *7*, 245–257.

Dickson, R.P., Erb-Downwards, J.R., Martinez, F.J., and Huffnagle, G.B. (2016). The Microbiome and the Respiratory Tract. Annual review of physiology *78*, 481–504.

Dong, Q., Brulc, J.M., Iovieno, A., Bates, B., Garoutte, A., Miller, D., Revanna, K.V., Gao, X., Antonopoulos, D.A., Slepak, V.Z., *et al.* (2011). Diversity of bacteria at healthy human conjunctiva. Invest Ophthalmol Vis Sci *52*, 5408–5413.

Eslami, S., Hadjati, J., Motevaseli, E., Mirzaei, R., Farashi Bonab, S., Ansaripour, B., and Khoramizadeh, M.R. (2016). Lactobacillus crispatus strain SJ-3C-US induces human dendritic cells (DCs) maturation and confers an anti-inflammatory phenotype to DCs. APMIS *124*, 697–710.

Fehlner-Gardiner, C., Roshick, C., Carlson, J.H., Hughes, S., Belland, R.J., Caldwell, H.D., and McClarty, G. (2002). Molecular basis defining human Chlamydia trachomatis tissue tropism. A possible role for tryptophan synthase. J Biol Chem 277, 26893–26903.

Filardo, S., Di Pietro, M., Porpora, M.G., Recine, N., Farcomeni, A., Latino, M.A., and Sessa, R. (2017). Diversity of Cervical Microbiota in Asymptomatic Chlamydia trachomatis Genital Infection: A Pilot Study. Front Cell Infect Microbiol 7, 321.

France, M.T., Mendes-Soares, H., and Forney, L.J. (2016). Genomic Comparisons of Lactobacillus crispatus and Lactobacillus iners Reveal Potential Ecological Drivers of Community Composition in the Vagina. Appl Environ Microbiol *82*, 7063–7073.

Gajer, P., Brotman, R.M., Bai, G.Y., Sakamoto, J., Schuette, U.M.E., Zhong, X., Koenig, S.S.K., Fu, L., Ma, Z.S.S., Zhou, X., *et al.* (2012). Temporal Dynamics of the Human Vaginal Microbiota. Sci Transl Med *4*, 132–152.

Ghartey, J.P., Smith, B.C., Chen, Z., Buckley, N., Lo, Y., Ratner, A.J., Herold, B.C., and Burk, R.D. (2014). Lactobacillus crispatus dominant vaginal microbiome is associated with inhibitory activity of female genital tract secretions against Escherichia coli. PLoS One *9*, e96659.

Goddard, A.F., Staudinger, B.J., Dowd, S.E., Joshi-Datar, A., Wolcott, R.D., Aitken, M.L., Fligner, C.L., and Singh, P.K. (2012). Direct sampling of cystic fibrosis lungs indicates that DNA-based analyses of upper-airway specimens can misrepresent lung microbiota. Proc Natl Acad Sci U S A *109*, 13769–13774.

Gong, Z., Luna, Y., Yu, P., and Fan, H. (2014). Lactobacilli inactivate Chlamydia trachomatis through lactic acid but not H2O2. PLoS One *9*, e107758.

Graspeuntner, S., Bohlmann, M.K., Gillmann, K., Speer, R., Kuenzel, S., Mark, H., Hoellen, F., Lettau, R., Griesinger, G., Konig, I.R., *et al.* (2018a). Microbiota-based analysis reveals specific bacterial traits and a novel strategy for the diagnosis of infectious infertility. PLoS One *13*, e0191047.

Graspeuntner, S., Loeper, N., Kuenzel, S., Baines, J.F., and Rupp, J. (2018b). Selection of validated hypervariable regions is crucial in 16S-based microbiota studies of the female genital tract. Sci Rep *8*, 9678.

Haggerty, C.L., Totten, P.A., Tang, G., Astete, S.G., Ferris, M.J., Norori, J., Bass, D.C., Martin, D.H., Taylor, B.D., and Ness, R.B. (2016). Identification of novel microbes associated with pelvic inflammatory disease and infertility. Sex Transm Infect *92*, 441–446.

Hernandez-Sanchez, J., Brumm, J., Timms, P., and Beagley, K.W. (2015). Vaccination of koalas with a prototype chlamydial vaccine is safe, does not increase the incidence of lymphoma-related disease and maybe associated with increased lifespan in captive koalas. Vaccine *33*, 4459–4463.

Hilty, M., Burke, C., Pedro, H., Cardenas, P., Bush, A., Bossley, C., Davies, J., Ervine, A., Poulter, L., Pachter, L., *et al.* (2010). Disordered microbial communities in asthmatic airways. PLoS One *5*, e8578.

Huang, Y., Yang, B., and Li, W. (2016). Defining the normal core microbiome of conjunctival microbial communities. Clin Microbiol Infect *22*, 643 e647–643 e612.

Huang, Y.J., Nelson, C.E., Brodie, E.L., Desantis, T.Z., Baek, M.S., Liu, J., Woyke, T., Allgaier, M., Bristow, J., Wiener-Kronish, J.P., *et al.* (2011). Airway microbiota and bronchial hyperresponsiveness in patients with suboptimally controlled asthma. The Journal of allergy and clinical immunology *127*, 372–381 e371–373.

Jacques, M., Olson, M.E., Crichlow, A.M., Osborne, A.D., and Costerton, J.W. (1986). The normal microflora of the female rabbit's genital tract. Can J Vet Res *50*, 272–274.

Johnson, R.M. (2004). Murine oviduct epithelial cell cytokine responses to Chlamydia muridarum infection include interleukin-12-p70 secretion. Infect Immun *72*, 3951–3960.

Kaambo, E., Africa, C., Chambuso, R., and Passmore, J.S. (2018). Vaginal Microbiomes Associated With Aerobic Vaginitis and Bacterial Vaginosis. Frontiers in public health *6*, 78.

Kuo, C.C., Jackson, L.A., Campbell, L.A., and Grayston, J.T. (1995). Chlamydia pneumoniae (TWAR). Clin Microbiol Rev *8*, 451–461.

Lawrence, T., Gilroy, D.W., Colville-Nash, P.R., and Willoughby, D.A. (2001). Possible new role for NF-kappaB in the resolution of inflammation. Nat Med *7*, 1291–1297.

Lee, J., Attila, C., Cirillo, S.L., Cirillo, J.D., and Wood, T.K. (2009). Indole and 7-hydroxyindole diminish Pseudomonas aeruginosa virulence. Microb Biotechnol *2*, 75–90.

Lee, J., Jayaraman, A., and Wood, T.K. (2007). Indole is an inter-species biofilm signal mediated by SdiA. BMC Microbiol *7*, 42.

Lee, J.H., and Lee, J. (2010). Indole as an intercellular signal in microbial communities. FEMS Microbiol Rev *34*, 426–444.

Legione, A.R., Amery-Gale, J., Lynch, M., Haynes, L., Gilkerson, J.R., Sansom, F.M., and Devlin, J.M. (2018). Variation in the microbiome of the urogenital tract of Chlamydia-free female koalas (Phascolarctos cinereus) with and without 'wet bottom'. PLoS One *13*, e0194881.

Liechty, E.R., Bergin, I.L., Bassis, C.M., Chai, D., LeBar, W., Young, V.B., and Bell, J.D. (2015). The levonorgestrel-releasing intrauterine system is associated with delayed endocervical clearance of Chlamydia trachomatis without alterations in vaginal microbiota. Pathog Dis *73*, ftv070.

Liew, F.Y. (1994). Regulation of nitric oxide synthesis in infectious and autoimmune diseases. Immunol Lett *43*, 95–98.

Liu, C.M., Hungate, B.A., Tobian, A.A., Ravel, J., Prodger, J.L., Serwadda, D., Kigozi, G., Galiwango, R.M., Nalugoda, F., Keim, P., *et al.* (2015). Penile Microbiota and Female Partner Bacterial Vaginosis in Rakai, Uganda. MBio *6*, e00589.

Loeper, N., Graspeuntner, S., and Rupp, J. (2018). Microbiota changes impact on sexually transmitted infections and the development of pelvic inflammatory disease. Microbes Infect.

Loomis, W.P., and Starnbach, M.N. (2002). T cell responses to Chlamydia trachomatis. Curr Opin Microbiol *5*, 87–91.

Marri, P.R., Stern, D.A., Wright, A.L., Billheimer, D., and Martinez, F.D. (2013). Asthma-associated differences in microbial composition of induced sputum. The Journal of allergy and clinical immunology *131*, 346–352 e341–343.

Mastromarino, P., Di Pietro, M., Schiavoni, G., Nardis, C., Gentile, M., and Sessa, R. (2014). Effects of vaginal lactobacilli in Chlamydia trachomatis infection. Int J Med Microbiol *304*, 654–661.

Miller, E.A., Beasley, D.E., Dunn, R.R., and Archie, E.A. (2016). Lactobacilli Dominance and Vaginal pH: Why Is the Human Vaginal Microbiome Unique? Front Microbiol *7*, 1936.

Mitchell, C.M., Haick, A., Nkwopara, E., Garcia, R., Rendi, M., Agnew, K., Fredricks, D.N., and Eschenbach, D. (2015). Colonization of the upper genital tract by vaginal bacterial species in nonpregnant women. Am J Obstet Gynecol *212*, 611 e611–619.

Mohammadzadeh, F., Dolatian, M., Jorjani, M., and Alavi Majd, H. (2014). Diagnostic value of Amsel's clinical criteria for diagnosis of bacterial vaginosis. Global journal of health science 7, 8–14.

Muzny, C.A., Sunesara, I.R., Kumar, R., Mena, L.A., Griswold, M.E., Martin, D.H., Lefkowitz, E.J., Schwebke, J.R., and Swiatlo, E. (2013). Characterization of the vaginal microbiota among sexual risk behaviour groups of women with bacterial vaginosis. PLoS One *8*, e80254.

Nardini, P., Nahui Palomino, R.A., Parolin, C., Laghi, L., Foschi, C., Cevenini, R., Vitali, B., and Marangoni, A. (2016). Lactobacillus crispatus inhibits the infectivity of Chlamydia trachomatis elementary bodies, *in vitro* study. Sci Rep *6*, 29024.

Nelson, T.M., Borgogna, J.C., Michalek, R.D., Roberts, D.W., Rath, J.M., Glover, E.D., Ravel, J., Shardell, M.D., Yeoman, C.J., and Brotman, R.M. (2018). Cigarette smoking is associated with an altered vaginal tract metabolomic profile. Sci Rep *8*, 852.

Ness, R.B., Kip, K.E., Hillier, S.L., Soper, D.E., Stamm, C.A., Sweet, R.L., Rice, P., and Richter, H.E. (2005). A cluster analysis of bacterial vaginosis-associated microflora and pelvic inflammatory disease. American journal of epidemiology *162*, 585–590.

Neuendorf, E., Gajer, P., Bowlin, A.K., Marques, P.X., Ma, B., Yang, H., Fu, L., Humphrys, M.S., Forney, L.J., Myers, G.S., *et al.* (2015). Chlamydia caviae infection alters abundance but not composition of the guinea pig vaginal microbiota. Pathog Dis *73*.

Nugent, R.P., Krohn, M.A., and Hillier, S.L. (1991). Reliability of diagnosing bacterial vaginosis is improved by a standardized method of gram stain interpretation. J Clin Microbiol *29*, 297–301.

Nunn, K.L., Wang, Y.Y., Harit, D., Humphrys, M.S., Ma, B., Cone, R., Ravel, J., and Lai, S.K. (2015). Enhanced Trapping of HIV-1 by Human Cervicovaginal Mucus Is Associated with Lactobacillus crispatus-Dominant Microbiota. MBio *6*, e01084–01015.

O'Hanlon, D.E., Moench, T.R., and Cone, R.A. (2011). In vaginal fluid, bacteria associated with bacterial vaginosis can be suppressed with lactic acid but not hydrogen peroxide. BMC Infect Dis *11*, 200.

O'Hanlon, D.E., Moench, T.R., and Cone, R.A. (2013). Vaginal pH and microbicidal lactic acid when lactobacilli dominate the microbiota. PLoS One *8*, e80074.

Onderdonk, A.B., Delaney, M.L., and Fichorova, R.N. (2016). The Human Microbiome during Bacterial Vaginosis. Clin Microbiol Rev *29*, 223–238.

Parolin, C., Foschi, C., Laghi, L., Zhu, C., Banzola, N., Gaspari, V., D'Antuono, A., Giordani, B., Severgnini, M., Consolandi, C., *et al.* (2018). Insights Into Vaginal Bacterial Communities and Metabolic Profiles of Chlamydia trachomatis Infection: Positioning Between Eubiosis and Dysbiosis. Front Microbiol *9*, 600.

Pelzer, E.S., Willner, D., Buttini, M., Hafner, L.M., Theodoropoulos, C., and Huygens, F. (2018). The fallopian tube microbiome: implications for reproductive health. Oncotarget *9*, 21541–21551.

Petrova, M.I., Lievens, E., Malik, S., Imholz, N., and Lebeer, S. (2015). Lactobacillus species as biomarkers and agents that can promote various aspects of vaginal health. Front Physiol *6*, 81.

Polkinghorne, A., Hanger, J., and Timms, P. (2013). Recent advances in understanding the biology, epidemiology and control of chlamydial infections in koalas. Vet Microbiol *165*, 214–223.

Rasmussen, S.J., Eckmann, L., Quayle, A.J., Shen, L., Zhang, Y.X., Anderson, D.J., Fierer, J., Stephens, R.S., and Kagnoff, M.F. (1997). Secretion of proinflammatory cytokines by epithelial cells in response to Chlamydia infection suggests a central role for epithelial cells in chlamydial pathogenesis. J Clin Invest *99*, 77–87.

Ravel, J., Brotman, R.M., Gajer, P., Ma, B., Nandy, M., Fadrosh, D.W., Sakamoto, J., Koenig, S.S., Fu, L., Zhou, X., *et al.* (2013). Daily temporal dynamics of vaginal microbiota before, during and after episodes of bacterial vaginosis. Microbiome *1*, 29.

Ravel, J., Gajer, P., Abdo, Z., Schneider, G.M., Koenig, S.S.K., McCulle, S.L., Karlebach, S., Gorle, R., Russell, J., Tacket, C.O., *et al.* (2011). Vaginal microbiome of reproductive-age women. Proc Natl Acad Sci U S A *108*, 4680–4687.

Redgrove, K.A., and McLaughlin, E.A. (2014). The Role of the Immune Response in Chlamydia trachomatis Infection of the Male Genital Tract: A Double-Edged Sword. Front Immunol *5*, 534.

Rizzo, A., Fiorentino, M., Buommino, E., Donnarumma, G., Losacco, A., and Bevilacqua, N. (2015). Lactobacillus crispatus mediates anti-inflammatory cytokine interleukin-10 induction in response to Chlamydia trachomatis infection *in vitro*. Int J Med Microbiol *305*, 815–827.

Robinson, C.K., Brotman, R.M., and Ravel, J. (2016). Intricacies of assessing the human microbiome in epidemiologic studies. Ann Epidemiol *26*, 311–321.

Romano, J.D., de Beaumont, C., Carrasco, J.A., Ehrenman, K., Bavoil, P.M., and Coppens, I. (2013). Fierce competition between Toxoplasma and Chlamydia for host cell structures in dually infected cells. Eukaryotic cell *12*, 265–277.

Scheller, J., Chalaris, A., Schmidt-Arras, D., and Rose-John, S. (2011). The pro- and anti-inflammatory properties of the cytokine interleukin-6. Biochim Biophys Acta *1813*, 878–888.

Sessa, R., Di Pietro, M., Filardo, S., Bressan, A., Mastromarino, P., Biasucci, A.V., Rosa, L., Cutone, A., Berlutti, F., Paesano, R., *et al.* (2017). Lactobacilli-lactoferrin interplay in Chlamydia trachomatis infection. Pathog Dis *75*.

Sharma, H., Tal, R., Clark, N.A., and Segars, J.H. (2014). Microbiota and pelvic inflammatory disease. Semin Reprod Med *32*, 43–49.

Singer, M., and Ouburg, S. (2016). Effect of cytokine level variations in individuals on the progression and outcome of bacterial urogenital infections—a meta-analysis. Pathog Dis *74*.

Srinivasan, S., Liu, C.Z., Mitchell, C.M., Fiedler, T.L., Thomas, K.K., Agnew, K.J., Marrazzo, J.M., and Fredricks, D.N. (2010). Temporal Variability of Human Vaginal Bacteria and Relationship with Bacterial Vaginosis. PLoS One *5*.

Stephens, R.S. (2003). The cellular paradigm of chlamydial pathogenesis. Trends Microbiol *11*, 44–51.

Stumpf, R.M., Wilson, B.A., Rivera, A., Yildirim, S., Yeoman, C.J., Polk, J.D., White, B.A., and Leigh, S.R. (2013). The primate vaginal microbiome: comparative context and implications for human health and disease. Am J Phys Anthropol *152 Suppl 57*, 119–134.

Swartz, J.D., Lachman, M., Westveer, K., O'Neill, T., Geary, T., Kott, R.W., Berardinelli, J.G., Hatfield, P.G., Thomson, J.M., Roberts, A., *et al.* (2014). Characterization of the Vaginal Microbiota of Ewes and Cows Reveals a Unique Microbiota with Low Levels of Lactobacilli and Near-Neutral pH. Front Vet Sci *1*, 19.

Tamarelle, J., de Barbeyrac, B., Le Hen, I., Thiebaut, A., Bebear, C., Ravel, J., and Delarocque-Astagneau, E. (2018a). Vaginal microbiota composition and association with prevalent Chlamydia trachomatis infection: a cross-sectional study of young women attending a STI clinic in France. Sex Transm Infect.

Tamarelle, J., Thiebaut, A.C.M., de Barbeyrac, B., Bebear, C., Ravel, J., and Delarocque-Astagneau, E. (2018b). The vaginal microbiota and its association with human papillomavirus, Chlamydia trachomatis, Neisseria gonorrhoeae and Mycoplasma genitalium infections: a systematic review and meta-analysis. Clin Microbiol Infect.

Valenti, P., Rosa, L., Capobianco, D., Lepanto, M.S., Schiavi, E., Cutone, A., Paesano, R., and Mastromarino, P. (2018). Role of Lactobacilli and Lactoferrin in the Mucosal Cervicovaginal Defence. Front Immunol *9*, 376.

van der Veer, C., Bruisten, S.M., van der Helm, J.J., de Vries, H.J., and van Houdt, R. (2017). The Cervicovaginal Microbiota in Women Notified for Chlamydia trachomatis Infection: A Case-Control Study at the Sexually Transmitted Infection Outpatient Clinic in Amsterdam, The Netherlands. Clin Infect Dis *64*, 24–31.

van Houdt, R., Ma, B., Bruisten, S.M., Speksnijder, A., Ravel, J., and de Vries, H.J.C. (2018). Lactobacillus iners-dominated vaginal microbiota is associated with increased susceptibility to Chlamydia trachomatis infection in Dutch women: a case–control study. Sex Transm Infect *94*, 117–123.

Vanover, J., Sun, J., Deka, S., Kintner, J., Duffourc, M.M., and Schoborg, R.V. (2008). Herpes simplex virus co-infection-induced Chlamydia trachomatis persistence is not mediated by any known persistence inducer or anti-chlamydial pathway. Microbiology *154*, 971–978.

Vidgen, M.E., Hanger, J., and Timms, P. (2017). Microbiota composition of the koala (Phascolarctos cinereus) ocular and urogenital sites, and their association with Chlamydia infection and disease. Sci Rep *7*, 5239.

Vodstrcil, L.A., Twin, J., Garland, S.M., Fairley, C.K., Hocking, J.S., Law, M.G., Plummer, E.L., Fethers, K.A., Chow, E.P., Tabrizi, S.N., *et al.* (2017). The influence of sexual activity on the vaginal microbiota and Gardnerella vaginalis clade diversity in young women. PLoS One *12*, e0171856.

Waugh, C., Khan, S.A., Carver, S., Hanger, J., Loader, J., Polkinghorne, A., Beagley, K., and Timms, P. (2016). A Prototype Recombinant-Protein Based Chlamydia pecorum Vaccine Results in Reduced Chlamydial Burden and Less Clinical Disease in Free-Ranging Koalas (Phascolarctos cinereus). PLoS One *11*, e0146934.

Webley, W.C., and Hahn, D.L. (2017). Infection-mediated asthma: aetiology, mechanisms and treatment options, with focus on Chlamydia pneumoniae and macrolides. Respiratory research *18*, 98.

Webley, W.C., Salva, P.S., Andrzejewski, C., Cirino, F., West, C.A., Tilahun, Y., and Stuart, E.S. (2005). The bronchial lavage of paediatric patients with asthma contains infectious Chlamydia. American journal of respiratory and critical care medicine *171*, 1083–1088.

Wiesenfeld, H.C., Hillier, S.L., Krohn, M.A., Landers, D.V., and Sweet, R.L. (2003). Bacterial vaginosis is a strong predictor of Neisseria gonorrhoeae and Chlamydia trachomatis infection. Clin Infect Dis *36*, 663–668.

Willner, D., Haynes, M.R., Furlan, M., Schmieder, R., Lim, Y.W., Rainey, P.B., Rohwer, F., and Conrad, D. (2012). Spatial distribution of microbial communities in the cystic fibrosis lung. The ISME journal *6*, 471–474.

Witkin, S.S., Mendes-Soares, H., Linhares, I.M., Jayaram, A., Ledger, W.J., and Forney, L.J. (2013). Influence of vaginal bacteria and D- and L-lactic acid isomers on vaginal extracellular matrix metalloproteinase inducer: implications for protection against upper genital tract infections. MBio *4*.

Wolf, Y.I., and Koonin, E.V. (2013). Genome reduction as the dominant mode of evolution. Bioessays *35*, 829–837.

Wood, H., Roshick, C., and McClarty, G. (2004). Tryptophan recycling is responsible for the interferon-gamma resistance of Chlamydia psittaci GPIC in indoleamine dioxygenase-expressing host cells. Mol Microbiol *52*, 903–916.

Zhou, Y., Holland, M.J., Makalo, P., Joof, H., Roberts, C.H., Mabey, D.C., Bailey, R.L., Burton, M.J., Weinstock, G.M., and Burr, S.E. (2014). The conjunctival microbiome in health and trachomatous disease: a case control study. Genome Med *6*, 99.

Ziklo, N., Huston, W.M., Hocking, J.S., and Timms, P. (2016a). Chlamydia trachomatis Genital Tract Infections: When Host Immune Response and the Microbiome Collide. Trends Microbiol *24*, 750–765.

Ziklo, N., Huston, W.M., Taing, K., Katouli, M., and Timms, P. (2016b). In vitro rescue of genital strains of Chlamydia trachomatis from interferon-gamma and tryptophan depletion with indole-positive, but not indole-negative Prevotella spp. BMC Microbiol *16*, 286.

Ziklo, N., Vidgen, M.E., Taing, K., Huston, W.M., and Timms, P. (2018). Dysbiosis of the Vaginal Microbiota and Higher Vaginal Kynurenine/Tryptophan Ratio Reveals an Association with Chlamydia trachomatis Genital Infections. Front Cell Infect Microbiol *8*, 1.

Zozaya, M., Ferris, M.J., Siren, J.D., Lillis, R., Myers, L., Nsuami, M.J., Eren, A.M., Brown, J., Taylor, C.M., and Martin, D.H. (2016). Bacterial communities in penile skin, male urethra, and vaginas of heterosexual couples with and without bacterial vaginosis. Microbiome *4*, 16.

Recent Advances in Epidemiology, Pathology and Immunology of Veterinary Chlamydiae

17

Konrad Sachse[1]* and Nicole Borel[2]

[1]RNA Bioinformatics and High-Throughput Analysis, Faculty of Mathematics and Computer Science, Friedrich-Schiller University, Jena, Germany.
[2]Institute of Veterinary Pathology, Vetsuisse Faculty, University of Zurich, Zurich, Switzerland.

*Correspondence: konrad.sachse@uni-jena.de

https://doi.org/10.21775/9781912530281.17

Abstract

The veterinary chlamydiae comprise all currently known organisms of the family *Chlamydiaceae*, except *Chlamydia trachomatis*. *C. abortus* is an economically important infection in sheep and goats causing abortion, stillbirths and weak neonates. Moreover, it is a zoonotic pathogen, i.e. well known for being transmissible from animals to humans and posing a risk for pregnant women. *C. psittaci*, which causes avian chlamydiosis in birds, is the most important veterinary chlamydial agent in terms of economic impact and human health. Avian chlamydiosis has been recently extended by addition of the novel species *C. avium*, *C. gallinacea* and *Candidatus* C. ibidis. In addition, *C. felis*, causing conjunctivitis in cats, and *C. caviae*, a guinea pig pathogen, can both be zoonotic. The latter has been extensively used in experimental animal models. Similarly, *C. muridarum* infection models in mice have been in use for decades to investigate human genital chlamydial disease. *C. pecorum* has a broad host range, including livestock and wild ruminants, but also marsupials such as the koala in Australia. *C. pneumoniae*, originally thought to occur only in humans, has a broad host range, including horses, koalas, reptiles and amphibians. *C. suis*, a pathogen of pigs, is the only chlamydial species to date that can maintain stable tetracycline resistance.

General overview of veterinary chlamydiae and their importance as pathogens

The world of veterinary chlamydiae comprises all species except *C. trachomatis*, with nearly all of them being able to cause disease in their main host. An overview about host preference, infection site and genomic parameters of currently accepted *Chlamydia* species is given in Table 17.1 (for genetic relationship see Chapter 18, Thinking Outside the *Chlamydia* Box). Research in the past decade has produced a wealth of new knowledge on host range, dissemination, transmission pathways, pathogenicity mechanisms, host–pathogen interactions and genetic relationships (Borel *et al.*, 2018; Hogerwerf *et al.*, 2017; Knittler *et al.*, 2014; Knittler and Sachse, 2015; Rodolakis and Laroucau, 2015; Walker *et al.*, 2015).

In this chapter, we will describe the major etiological and genetic features of chlamydiae of veterinary interest, in alphabetical order, and comment on their economic importance. For additional discussion of geographical distribution and epidemiology, the reader is referred to published reviews (Burnard and Polkinghorne, 2016; Kaleta and Taday, 2003; Polkinghorne *et al.*, 2013; Reinhold *et al.*, 2011; Sachse *et al.*, 2015; Schautteet and Vanrompay, 2011).

Table 17.1 General characteristics of currently recognized *Chlamydia* spp.[a]

Characteristic	*C. abortus*	*C. avium*	*C. caviae*	*C. felis*	*C. gallinacea*	*C. muridarum*	*C. pecorum*	*C. pneumoniae*	*C. psittaci*	*C. suis*	*C. trachomatis*
Natural host	Sheep, goat	Pigeon, parrots	Guinea pig	Cat	Chicken	Rodents	Cattle, koala	Human, horse, koala	Birds	Swine, wild boar	Human
Other hosts	Cattle, swine, birds, horse	Mallard	Horse, human, rabbit	Dog	Other poultry	–	Sheep, goat, swine	Amphibians, reptiles	Mammals	Ruminants, human	–
Site of infection	Genital, respiratory	Respiratory	Ocular, genital	Ocular, respiratory	Respiratory	Gastro-intestinal	Gastro-intestinal, respiratory, urogenital	Respiratory, cardiovascular	Respiratory	Gastro-intestinal, genital	Ocular, urogenital tract
Genome size (Mbp)	1.14	1.05	1.18	1.17	1.04	1.08	1.11	1.23	1.17	1.08	1.04
GC content (mol-%)	39.9	36.9	39.2	39.0	37.9	40.3	41.1	40.6	39.0	42.0	41.3
Predicted proteins	932	940	1005	1013	907	911	988	1113	975	931	919
Gene content identity (%)[b]	89.5	88.8	90.0	90.9	89.0	95.5	89.1	89.7	90.2	96.0	100
16S rRNA difference (%)[c]	4.87	5.19	4.61	5.00	5.65	1.62	5.12	6.22	4.80	2.86	0
No. of rRNA operons	1	1	1	1	1	2	1	1	1	1	2
Plasmid	–	+	+	+	+	+	+	–[d]	+	+	–[d]
Bacteriophage	+	?	+	–	?	–	+	+	+	–	–

[a]Genomic parameters refer to respective type strains S26/3, 10DC88, GPIC, Fe/C-56, 08.1274/3, MoPn/Nigg, E58, TW-183, 6BC, MD56 (reference strain) and A-Har-13.
[b]Percentage of genes shared with *C. trachomatis* A-Har-13.
[c]Compared with *C. trachomatis* A-Har-13.
[d]Other strains of the species possess a plasmid.

Chlamydia abortus

C. abortus is the causative agent of ovine enzootic abortion (OEA) or enzootic abortion of ewes (EAE), an economically important disease in sheep that occurs particularly in closely congregated flocks during the parturient period (Aitken and Longbottom, 2007; Longbottom and Coulter, 2003; OIE, 2018; Rodolakis and Laroucau, 2015). Typically, infection is linked to spontaneous abortion during the last 2–3 weeks of pregnancy with the appearance of stillborn lambs and inflamed placentas. Sometimes, weak lambs are delivered but fail to survive beyond 48 hours. In flocks free of *C. abortus*, the pathogen is usually introduced through replacement animals. While the number of abortions caused in the first year after introduction of the pathogen into the flock is low, an 'abortion storm' in the second year can affect around 30% of ewes. Processes underlying pathogenesis intensify around day 90 of gestation (average gestation period is 147 days), when the foetus is in a phase of rapid growth and downregulated cellular response factors as well as pathological changes in the placenta facilitate chlamydial invasion (Livingstone *et al.*, 2017). Abortion itself probably results from a combination of impairment of materno-foetal nutrient and gaseous exchange, disruption of hormonal regulation of pregnancy and induced cytokine aggression (Buxton *et al.*, 2002; Entrican, 2002). *C. abortus* is also widespread in flocks having low or 'normal' abortion rates, with 50% of flocks testing PCR positive and 94% harbouring seropositive animals (Lenzko *et al.*, 2011). Whether such a latent and subclinical course of infection is relevant in terms of economic impact and animal health, as known from cattle (Reinhold *et al.*, 2011), needs to be investigated. Goats are affected by *Chlamydia* induced abortion in a similar fashion, and although rare, cases have also been reported for cattle (Rojas *et al.*, 2018) and swine (Salinas *et al.*, 2012).

C. abortus is genetically quite monomorphic. Its genome sequence is distinct from other chlamydiae for the absence of both toxin genes and the tryptophan operon (Thomson *et al.*, 2005). Comparative analysis of 64 strains confirmed the limited intra-species diversity while revealing distinct geographical signatures within the phylogeny and substantial variation among pmp loci (Seth-Smith *et al.*, 2017b). For the latter, see also Chapter 5, Interactions of the Chlamydial Inclusion with the Host Cell.

Chlamydia avium

C. avium is one of the recently discovered species of the genus *Chlamydia* (Sachse *et al.*, 2014) and knowledge about it is still limited. This agent has been found in pigeons and psittacine birds (Sachse and Laroucau, 2015) and more recently in mallards (Szymanska-Czerwinska *et al.*, 2017). It is likely that future surveys will extend the host range to more bird species. The pathogenicity of *C. avium* has yet to be investigated. A PCR-based survey in breeder pigeons in Germany revealed that the agent was present in 4 of 27 (14.8%) flocks (Krautwald-Junghanns *et al.*, 2013), while a study of urban pigeons in France showed that 10/125 (8%) of *Chlamydiaceae*-positive samples contained *C. avium* (Gasparini *et al.*, 2011). A recent study from the Netherlands reported prevalence rates of 36.6% and 20% in pigeons of two different cities (Heijne *et al.*, 2018).

The general prevalence of *C. avium* in psittacines is not known because only individual cases have been examined. However, pathological findings suggest that *C. avium* is capable of causing respiratory disease in both parrots and pigeons (Sachse *et al.*, 2014). In those cases, internal organs of the diseased birds were infected and showed lesions. In addition, a facultative pathogenic role in co-infection with *C. psittaci* seems possible.

Preliminary analysis of the *C. avium* genome shows close genetic relatedness to *C. gallinacea* (field strains sharing 861/891 annotated genes with it), as well as to *C. abortus* (829/891) and *C. psittaci* (804/891) (Sachse, unpublished data).

Chlamydia caviae

C. caviae got its name from its natural host, the guinea pig (classified into genus *Cavia*), where it was first recovered from the conjunctiva in a case of inclusion conjunctivitis (Murray, 1964). This agent seems to be endemic in pet guinea pigs, but can be transmitted through contact with other animals, such as rabbits (Pantchev *et al.*, 2010). More recently, it was identified in the conjunctiva and nasal mucus of young horses with signs of conjunctivitis and rhinitis (Gaede *et al.*, 2010). Several groups have used *C. caviae* infection of guinea pigs as an experimental model to investigate humoral and cellular immune responses (Belij-Rammerstorfer *et al.*, 2016; Lacy *et al.*, 2011) and to develop new chlamydial vaccines (Andrew *et al.*, 2011).

Chlamydia felis

C. felis is endemic among house cats worldwide and can cause conjunctivitis and rhinitis (Studdert *et al.*, 1981). Diagnostic surveys also identified it in dogs, where it appeared to have caused conjunctivitis as well (Pantchev *et al.*, 2010). Laroucau and colleagues conducted an epidemiological study based on variable number tandem repeat (VNTR) analysis and multilocus sequence typing (MLST) of *C. felis* field strains (Laroucau *et al.*, 2012). Using indirect haemagglutination, Wu and colleagues found seroprevalence rates of 5.9% and 12.1% in Chinese domestic cats and dogs, respectively (Wu *et al.*, 2013). While Schulz and co-workers identified 35.6% *C. felis* positives in a group of cats showing signs of upper respiratory tract disease (Schulz *et al.*, 2015), a study in the US confirmed that cats affected by upper respiratory infection were more likely to carry *C. felis* than healthy control cats (McManus *et al.*, 2014).

Chlamydia gallinacea

The main hosts of *C. gallinacea* are chicken, turkey and guinea fowl (Sachse *et al.*, 2014), but other domestic poultry and wild birds could possibly also harbour this organism.

From epidemiological studies, *C. gallinacea*, rather than *C. psittaci*, appears to be the predominant chlamydial species in poultry and turkeys. The first report indicating worldwide dissemination of this recently detected and described agent dates back to 2012 (Zocevic *et al.*, 2012). In a later survey covering poultry flocks in 24 Chinese provinces, 63.8% of all chlamydia-positive samples had *C. gallinacea* (Guo *et al.*, 2016). Furthermore, the agent was detected in 12.4% of oropharyngeal samples from backyard poultry in the state of Alabama, USA (Li *et al.*, 2017a). Likewise, 15% of the chickens sampled from Italian backyard farms were *C. gallinacea* positive (Donati *et al.*, 2018).

At present, no evidence of a pathogenic potential has been observed, but it cannot be ruled out that *C. gallinacea* contributes to avian chlamydiosis, for instance in mixed infection with *C. psittaci*. While experimentally inoculated chickens showed no signs of overt clinical disease, the average body weight gain was significantly reduced by 6.5–11.4% from three weeks post infection onwards (Guo *et al.*, 2016).

Interestingly, the analysis of the ompA locus revealed extensive intra-species diversity that could be translated into at least 14 variants or serovars (Guo *et al.*, 2016; Li *et al.*, 2017a). This highly polymorphic spectrum stands in contrast to the more homogeneous *C. avium*, but is reminiscent of the diversity within *C. psittaci*, where more than 15 ompA genotypes are known.

Genomic analysis of *C. gallinacea* has so far been based on two strains (Guo *et al.*, 2017; Holzer *et al.*, 2016). Interesting features include up to 15 predicted pmp genes and a reduced plasticity zone. Field strains of *C. gallinacea* were found to share 861/891 annotated genes with *C. avium*, its closest relative, whereas there are 830/891 common genes with *C. abortus* and 805/891 with *C. psittaci* (Sachse, unpublished data).

Chlamydia muridarum

C. muridarum was originally isolated from the lungs of albino Swiss mice with pneumonia (Nigg, 1942; Stills *et al.*, 1991) and from hamsters with proliferative ileitis. It is not certain whether this agent is widely disseminated. The only indication of a possible natural reservoir in rodents was given in the report by Ramsey and colleagues (Ramsey *et al.*, 2016).

The two strains available (MoPn, Nigg) have been used in a large number of murine infection models to investigate the course of infection, systemic dissemination, cellular immune response and transmission pathways with the aim to better understand *C. trachomatis* infections in humans (Koroleva *et al.*, 2017; Pal *et al.*, 2015; Yu *et al.*, 2014). One of the major findings of the studies involving *C. muridarum* infection models was the crucial role of chlamydial polymorphic membrane proteins in host protection (Muller *et al.*, 2017; Pal *et al.*, 2017; Yu *et al.*, 2014), which makes them promising candidate components of vaccines (for details see Chapter 15, *Chlamydia* Vaccines).

Chlamydia pecorum

C. pecorum infections have been associated with conjunctivitis, encephalomyelitis, enteritis, pneumonia, polyarthritis, abortion, and reproductive or urinary tract diseases in ruminants and swine (Mohamad and Rodolakis, 2010; Reinhold *et al.*, 2011; Storz and Kaltenboeck, 1993). Arthritis is ranked as a major cause of economic loss in Australia's sheep industry, but proper documentation of cases caused by *C. pecorum* was conducted only recently (Walker *et al.*, 2016). *C. pecorum* can also be responsible for abortions in small ruminants (Rodolakis *et al.*, 1998), but well-documented cases, such as the recently reported caprine abortion in Argentina (Giannitti *et al.*, 2016) are scarce. *C. pecorum* strains were shown to be the cause of bovine enteric disease in an endemic region of Japan (Ohtani *et al.*, 2015). In addition, the agent was involved in urogenital infections and conjunctivitis in Australian marsupials (Jackson *et al.*, 1999; Warren *et al.*, 2005).

As with other chlamydiae, the majority of infections remain asymptomatic (Li *et al.*, 2016; Reinhold *et al.*, 2011), but latently infected calves displayed detectable lung dysfunction (Reinhold *et al.*, 2008) and substantial reductions in weight gains (Poudel *et al.*, 2012). In contrast, the presence of *C. pecorum* in sheep flocks did not affect mass or growth rates (Walker *et al.*, 2018). In swine, this agent was found in mixed infections with *C. suis*, with a substantial increase in *C. pecorum* positivity at the end of the fattening period, which was associated with contact to ruminants (Hoffmann *et al.*, 2015). In summary, there is a generally held notion that *C. pecorum* infections in livestock are widespread and occasionally severe, but are underrecognized and their economic importance is underestimated (Walker *et al.*, 2015).

C. pecorum displays extensive intra-species genetic diversity, which is underpinned by the existence of at least 14 MLST types (Jelocnik *et al.*, 2014b) and pronounced branching of field strains in phylogenetic trees based on 16–23S rRNA loci. Consequently, its antigenic diversity is considerable.

To define genetic virulence markers, two groups of researchers investigated polymorphisms in the ompA, incA and ORF663 genes of *C. pecorum* strains (Mohamad *et al.*, 2014; Yousef Mohamad *et al.*, 2008). Analysis revealed that the number of certain coding tandem repeats (CTRs) was a suitable virulence marker. In the corresponding proteins, high numbers of APA or AGA amino acid repeats in IncA and 15-mer CTRs in ORF663 were associated with low virulence. Moreover, low-virulence and high-virulence clades of strains were identified in the ompA-based dendrogram. Using MLST, Jelocnik and colleagues were able to associate 'sequence type 23' with disease in sheep and cattle, and 'type 62' to host preference of sheep and koalas in Australian strains (Jelocnik *et al.*, 2014b).

Comparative genomics identified several features that are specific for *C. pecorum*. (i) the number of loci comprising the plasticity zone, (ii) the absence of essential genes of folate metabolism, and (iii) the presence of 15 genes encoding polymorphic proteins belonging to all six Pmp subtypes (Sait *et al.*, 2014). Another study found the plasticity zone with its two toxin genes toxA and toxB, as well as a 77-kbp region encoding putative type III effector proteins to be the most variable regions (Bachmann *et al.*, 2014). Further analysis based on SNPs of whole genome sequences confirmed the high degree of genetic heterogeneity within the species of *C. pecorum* (Bachmann *et al.*, 2015).

Chlamydia pneumoniae

C. pneumoniae was initially thought to occur only in humans, but epidemiological surveys since the 1990s have suggested a far broader host range. So far, the agent has been found in horses (Storey *et al.*, 1993), amphibians (Hotzel *et al.*, 2001), reptiles (Bodetti *et al.*, 2002), koalas (Wardrop *et al.*, 1999) and others. The question whether *C. pneumoniae* alone is capable of causing disease in animals is still open. It was concluded from comparative genomics that *C. pneumoniae* was originally an animal pathogen that adapted to the human host after zoonotic transmission at a later stage of evolution (Myers *et al.*, 2009).

C. pneumoniae infections of humans are discussed in detail in Chapter 2, *Chlamydia pneumoniae* Infections.

Chlamydia psittaci

C. psittaci is the causative agent of avian chlamydiosis and human psittacosis and is arguably the most important veterinary chlamydial agent in economic and health terms. The course of infection depends on many factors of both the pathogen and the host and can range from acute to chronic to subclinical. Clinical signs in affected birds include respiratory symptoms, conjunctivitis, coryza, mucopurulent discharge from nose and eyes, cough, dyspnoea, or greenish to greyish faeces, but are generally non-specific. Sub-clinical or latent infection occurs more frequently than acute cases and outbreaks (Sachse *et al.*, 2015; Vanrompay, 2013). However, the consequences on bird health and economic parameters of such long-term host–pathogen interaction are still poorly understood. There is speculation that prolonged subclinical carriage of *C. psittaci* results in chronicity, and consequently, retarded development of infected animals, as seen in cattle (Reinhold *et al.*, 2011).

C. psittaci is disseminated worldwide and was reported to occur in nine domestic fowl species and at least 460 free-living or pet bird species of 30 different orders (Kaleta and Taday, 2003). Fulminant and devastating outbreaks on poultry farms were seen in the first half of the 20th century, but are now rare. Instead, current outbreaks are characterized by reduced feed intake and respiratory signs with or without low mortality. *C. psittaci* is nearly endemic in turkey flocks, where co-infection with *Ornithobacterium rhinotracheale* can exacerbate the course (Van Loock *et al.*, 2005). Infections of domestic ducks with implications for public health and revenue of the poultry industry have been reported from Europe (Haas *et al.*, 2007; Laroucau *et al.*, 2009a) and Asia (Yin *et al.*, 2015). Reports of outbreaks in chicken production caused by *C. psittaci* or zoonotic cases ascribed to contact with *C. psittaci*-infected chickens are generally rare (Lagae *et al.*, 2014; Laroucau *et al.*, 2015), whereas economic losses and human *C. psittaci* infections linked to turkeys, ducks or mixed flocks have been reported more frequently (Dickx *et al.*, 2010; Dickx and Vanrompay, 2011; Gaede *et al.*, 2008; Hinton *et al.*, 1993). A large number of epidemiological studies was conducted in urban or feral pigeons, with *C. psittaci* prevalence ranging from 12% to nearly 100% (Magnino *et al.*, 2009).

From the genetic point of view, *C. psittaci* is a rather heterogeneous species. Genotypes based on ompA, which encodes the major outer membrane protein A, are used to represent the diversity of strains, their host preference and virulence. Fifteen genotypes are currently in use, i.e. A to F, E/B, M56, WC, 1V, 6N, Mat116, R54, YP84, and CPX0308 (Sachse *et al.*, 2015; Sachse and Ruettger, 2015).

The state of *C. psittaci* whole-genome analysis was discussed in a recent review (Knittler *et al.*, 2014), and only the major points will be mentioned here:

1. Genome sequences of 68 different strains have been deposited at NCBI and ENA databases, of which 18 are completely assembled.
2. Analysis of 20 genome sequences revealed that 911 genes (or 90%) are common among all *C. psittaci* strains (Read *et al.*, 2013).
3. The majority of strains examined by Read *et al.* were found to possess an intact tox gene and guaAB-add loci in the hypervariable plasticity zone (Read *et al.*, 2013).
4. The genome of type strain 6BC was shown to harbour 21 full-length pmp family members covering all six subtypes (Voigt *et al.*, 2012). Compared to *C. trachomatis* and *C. pneumoniae*, the number of studies on *C. psittaci* functional genome analysis is still low.

Chlamydia suis

C. suis appears to be highly adapted to swine and has been associated with clinical diseases such as conjunctivitis, rhinitis, pneumonia, enteritis and reproductive disorders. Its pathogenic potential has been demonstrated in experimental infections targeting the respiratory system (Sachse *et al.*, 2004) and the gastrointestinal tract of pigs (Guscetti *et al.*, 2009; Rogers and Andersen, 2000). Outside the natural host, there have been occasional findings by PCR in cattle (Reinhold *et al.*, 2011; Teankum *et al.*, 2007), sheep (Polkinghorne *et al.*, 2009), horses (Pantchev *et al.*, 2010) and frogs (Blumer *et al.*, 2007).

Naturally acquired infection by *C. suis* can cause reproductive problems (Kauffold *et al.*, 2006) or persistent enteric infection in swine (Hoffmann *et al.*, 2015). Nevertheless, the pathogenicity of *C. suis* in natural infections remains largely unclear as it is mainly found in their gastrointestinal tract without clinical signs. Thus, its impact on porcine health and production has yet to be substantiated (Li *et al.*, 2017b). Recent studies suggested that *C. suis* may contribute to conjunctivitis of domestic swine (Chahota *et al.*, 2018) and wild boar (Risco *et al.*, 2013). In this context, poor hygiene and inadequate management of swine herds can be predisposing factors to ocular *C. suis* infection (Becker *et al.*, 2007). While not a primary pathogen in swine, epidemiological studies indicate a correlation between the detection of *C. suis* and the occurrence of diarrhoea in affected fattening pig herds (Hoffmann *et al.*, 2015). Thus, *C. suis* may have a negative effect on animal welfare and could be considered a potential zoonotic pathogen as outlined below.

Recent genome analysis of 10 US and 29 Swiss strains revealed very high genetic diversity. Individual markers were associated with events of recombination and plasmid exchange (Seth-Smith *et al.*, 2017a). As in other *Chlamydia* spp., the plasticity zone is the most variable genomic region and includes multiple phospholipase D pld genes, toxin gene(s) and an intact trp operon. Other variable segments comprise the loci encoding Inc proteins, a locus carrying a highly repetitive helicase and putative restriction enzyme, as well as the regions around the rRNA operons.

C. suis is the only chlamydial organisms where structural elements encoding antimicrobial (i.e. tetracycline) resistance have been found. Structure and function of the Tet-island comprising the tetA(C) gene will be discussed below.

Zoonotic potential and disease manifestations in humans – an update

Some chlamydial species can be transmitted from animals to humans (zoonotic potential) and zoonotic diseases caused by *Chlamydia* spp. have been well known for many decades. The variety of manifestations in disease caused by veterinary *Chlamydia* spp. and zoonotic features are compiled in Table 17.2. The best known examples include chlamydial abortion in ruminants due to *C. abortus* (Essig and Longbottom, 2015) and psittacosis/ornithosis in birds due to *C. psittaci* (Sachse *et al.*, 2015). Recent research has uncovered new examples of zoonotic potential either from known reservoirs or from new infection sources. Thus, the recently reported zoonotic cases of *C. caviae* infection causing severe community-acquired pneumonia (Ramakers *et al.*, 2017), the detection of *C. suis* in humans (De Puysseleyr *et al.*, 2014; De Puysseleyr *et al.*, 2017) and the emerging role of *C. psittaci* in equine abortion (Polkinghorne and Greub, 2017) have extended our horizon. Such cases exemplify the emergence of new zoonotic risks for humans and point towards novel virulence properties of these microorganisms.

C. abortus

Zoonotic infections have been reported only after contact with lambing small ruminants (Essig and Longbottom, 2015; Longbottom and Coulter, 2003); (Hagemann *et al.*, 2016; Pospischil *et al.*, 2002; Walder *et al.*, 2005). During abortion, massive shedding of elementary bodies takes place leading to contamination of the environment and inhalation or ingestion by exposed individuals (Longbottom *et al.*, 2013b). Pregnant women are particularly at risk with the potential to develop life-threatening *C. abortus* infections, resulting in septic abortion and stillbirth (Essig and Longbottom, 2015). Recommendations for pregnant women include avoiding contact with sheep and goats during the lambing and kidding season. Of note, *C. abortus*-induced abortions in women may remain undiagnosed as they are not routinely submitted to the laboratory for further investigation, which leads to potential underestimation of actual case numbers. Beyond the risk to pregnant women, a recent case of atypical pneumonia due to *C. abortus* was reported in a laboratory worker involved in intranasal experimental *C. abortus* inoculation of sheep (Ortega *et al.*, 2015), thus highlighting the additional occupational risk posed by this bacterium.

C. caviae

Zoonotic transmission is known to occur occasionally and was reported to range from mild conjunctivitis (Lutz-Wohlgroth *et al.*, 2006) to severe pneumonia (Ramakers *et al.*, 2017). Recently, three unrelated cases of zoonotic *C. caviae* infections in otherwise healthy adults in their early 30s were described. All three patients suffered severe respiratory failure after exposure to ill guinea pigs, but recovered after being treated with doxycycline. In one case, transmission between the guinea pig and its owner was confirmed by identification of sequence-identical isolates in broncho-alveolar fluids from the patient and her pet guinea pig (Ramakers *et al.*, 2017). Moreover, community-acquired pneumonia (CAP) due to *C. caviae* in a patient with no direct guinea pig exposure was reported, raising questions about the zoonotic reservoirs and other

Table 17.2 Manifestations of disease caused by veterinary chlamydiae and zoonotic potential

Chlamydial species	Main hosts	Manifestations	Zoonotic potential
Chlamydia abortus	Ruminants, swine	Abortion,[a] vaginitis, endometritis, seminal vesiculitis, mastitis (latent)	Yes
Chlamydia avium	Birds	Enteritis and respiratory disease (link to apparent pathology still uncertain)	Unclear
Chlamydia caviae	Guinea pig	Conjunctivitis,[a] keratitis, pneumonia[b]	Yes
Chlamydia felis	Cat	Conjunctivitis,[a] rhinitis	Yes
Chlamydia gallinacea	Birds	No apparent pathology yet described	Unclear
Chlamydia muridarum	Mouse, hamster	Pneumonitis, ileitis	No
Chlamydia pecorum	Ruminants swine, koala	Sheep and cattle: encephalitis, polyarthritis, pneumonia, enteritis, vaginitis, endometritis; swine: polyarthritis, serositis, enteritis, pneumonia; koala: keratoconjunctivitis, vaginitis, ovarian cyst, infertility	No
Chlamydia pneumoniae	Koala, other marsupials, horse	Rhinitis, pneumonia, conjunctivitis	No
	Reptiles, amphibians	Conjunctivitis, enteritis, granulomatous inflammation of internal organs	No
Chlamydia psittaci	Birds	Conjunctivitis, pneumonia, atypical pneumonia,[b] enteritis, hepatitis	Yes
Chlamydia suis	Swine	Conjunctivitis, pneumonia, enteritis, polyarthritis	Under debate
Chlamydia poikilothermis	Snakes	No apparent pathology yet described	No
Chlamydia serpentis	Snakes	No apparent pathology yet described	No
Ca. Chlamydia corallus	Snakes	No apparent pathology yet described	No
Ca. Chlamydia ibidis	Birds	No apparent pathology yet described	No
Ca. Chlamydia sanzinia	Snakes	No apparent pathology yet described	No

[a]In both animal and human infection.
[b]In cases of zoonotic transmission.

potential transmission routes (van Grootveld *et al.*, 2018). Physicians and veterinarians should be aware of the zoonotic potential of *C. caviae*, particularly as guinea pigs are very popular pets that have frequent contact with children. This awareness is also important for researchers using *C. caviae* ocular and genital infection models in guinea pigs to study human diseases.

C. felis

The first case of a zoonotic infection in cats was reported in 1969 (Schachter *et al.*, 1969). The cat owner presented with a follicular keratoconjunctivitis, and his cat had a recent history of rhinitis and conjunctivitis. Cats may remain infected over a longer time period, maintaining an asymptomatic carrier status. Prolonged vaginal and rectal shedding in cats with conjunctivitis after experimental *C. felis* infection indicated that the genital and intestinal tracts can be reservoirs for long-term infection (Sykes, 2005) and represent a potential source of zoonotic transmission. There are no recent reports of zoonotic transmission from cats, but awareness of such infections is highly important in view of the above-mentioned case reports in guinea pigs, the popularity of cats as pets and the presence of uncontrolled, non-vaccinated stray cat populations.

C. gallinacea

An initial case report of atypical pneumonia in three French slaughterhouse workers suggested that *C.*

gallinacea could be a zoonotic pathogen (Laroucau *et al.*, 2009b). However, it could not be verified that the agent was involved in zoonotic transmission in a more recent outbreak of psittacosis among abattoir workers, where *C. gallinacea* was found alongside *C. psittaci* in slaughterhouse samples (Hulin *et al.*, 2015).

C. pecorum

C. pecorum is not thought to be a zoonotic agent. Nevertheless, it was detected in 3% of a group of rural individuals in Nepal presenting with trachoma, in both single (2 cases) and mixed infection with *C. trachomatis* (Dean *et al.*, 2013).

C. pneumoniae

As pointed out above, this organism has a broad host range with infections occurring in humans, horses, koalas and other Australian marsupials, frogs and reptiles. Phylogenetic comparison of human and animal strains showed that the latter are more diverse than isolates of human origin, e.g. genes inactivated in human isolates were intact in their corresponding animal homologues (Mitchell *et al.*, 2010; Mitchell *et al.*, 2009). The aforementioned observation led to the suggestion that *C. pneumoniae* animal strains are ancestral to human strains and that cross-species transmission may have occurred several times in the evolutionary history of this pathogen (Myers *et al.*, 2009; Roulis *et al.*, 2013). While reports confirming zoonotic transmission are still absent, careful observation regarding possible zoonotic properties of *C. pneumoniae* is necessary. Human infections caused by *C. pneumoniae* are discussed in more detail in Chapter 2, *Chlamydia pneumoniae* Infections.

C. psittaci

The zoonotic properties of this agent are well documented (see reviews in Beeckman and Vanrompay, 2009; Knittler and Sachse, 2014; Sachse *et al.*, 2015). Transmission from pet birds, poultry or wild birds occurs through inhalation of infectious aerosol or direct contact with contaminated faeces or feathers. Until the 1930s, large-scale outbreaks of human psittacosis occurred repeatedly in Europe and North America, and all of them could be traced back to shipments of infected psittacine birds from South America. Nowadays usually small groups of individuals with previous contact to psittacines, ducks, turkeys, chickens, pigeons or wild birds are affected, and fulminant manifestations are rare due to the availability of antimicrobial treatment (Ferreira *et al.*, 2017; Gaede *et al.*, 2008; Laroucau *et al.*, 2009b; Rehn *et al.*, 2013; Van Droogenbroeck *et al.*, 2009). Human-to-human transmission was believed to be irrelevant, but recently such a case was reported from Sweden (Wallensten *et al.*, 2014).

Occasionally, *C. psittaci* infections from non-avian sources can cause psittacosis as well. In this context, *C. psittaci*-induced abortion cases in horses and zoonotic infections in humans exposed to equine abortion material in New South Wales, Australia, raised increased public awareness (Jelocnik *et al.*, 2017; Polkinghorne and Greub, 2017). An outbreak of five cases of human psittacosis in veterinary students and staff exposed to equine foetal membranes was reported with involvement of a highly virulent *C. psittaci* strain of the 6BC lineage (genotype A), identical to lineages detected in Australian parrots (Branley *et al.*, 2016; Chan *et al.*, 2017). Molecular typing from another Australian equine abortion case showed close relationship to a *C. psittaci* strain detected in a dove (genotype B) and belonging to an evolutionary lineage of *C. psittaci* strains typically associated with infections in pigeons (Jelocnik *et al.*, 2018). Transmission via indirect contact, presumably via fecal environmental contamination from *C. psittaci*-infected birds was suspected as the infection source for pregnant mares (Jelocnik *et al.*, 2017). These recent findings highlight the existence of potentially novel zoonotic sources, such as *C. psittaci*-infected equine placental membranes, which pose a risk to students, veterinarians and staff during their work.

The course of human psittacosis can range from an asymptomatic illness to a flu-like syndrome to severe systemic disease with pneumonia, myocarditis and encephalitis. Affected individuals usually show mild symptoms, while immunocompromised people are more likely to develop clinical signs. Veterinarians, poultry workers, birdkeepers and pet shop employees are at an elevated risk (Deschuyffeleer *et al.*, 2012). A recent meta-analysis estimated that 1% of cases of community-acquired pneumonia in humans are caused by *C. psittaci* (Hogerwerf *et al.*, 2017). The agent was also detected by PCR in 2.2% of pharyngeal swab samples taken from 780 community-acquired pneumonia patients in Germany (Dumke *et al.*, 2015).

The number of notified human cases seems to be in decline in most countries of the Northern hemisphere. However, these numbers are likely an underestimation of the true prevalence because *C. psittaci* is usually not part of the routine laboratory diagnosis, and many cases go unreported. In these circumstances, the correct diagnosis of a *C. psittaci* infection is impeded, and it can be difficult to trace the source of a human psittacosis case.

C. suis

A zoonotic potential of *C. suis* is being discussed based on (i) detection in Nepalese trachoma patients, both as mono-infection (2 cases out of 101) and co-infection with *C. trachomatis* (5/101) (Dean *et al.*, 2013), (ii) *C. suis* positivity of ocular swabs from two Belgian abattoir employees (De Puysseleyr *et al.*, 2014), and (iii) recent isolation of viable *C. suis* from nasal, pharyngeal and stool samples taken from Belgian farmers (De Puysseleyr *et al.*, 2014, 2017). The latter finding implies the presence of live, replicating organisms and could support the hypothesis of zoonotic transmission from swine to humans. Further studies are needed to elucidate the role and pathogenic potential of such infections in humans, in particular in view of tetracycline-resistance gene transfer.

Others

The zoonotic potential of novel chlamydial species from snakes (see next section of this chapter), such as *C. serpentis* and *C. poikilothermis*, as well as *Candidatus* species *C. sanzinia* and *C. corallus*, is still unknown at this point. The same refers to the recently introduced avian chlamydiae *C. avium* and *Candidatus* C. ibidis.

In summary, there have been recent improvements in understanding the extent and impact of chlamydial zoonotic infections, but more efforts are required to document new cases and draw practical conclusions for both veterinary and human health care. Moreover, the zoonotic potential of novel members in several families of the order *Chlamydiales* are largely unknown at present. Future studies should also take into account that many chlamydial species are not restricted to one particular host, contrary to traditional belief, and that alternative reservoirs need to be considered.

Recently described new veterinary chlamydial organisms

A rapidly growing number of culture-independent studies shows a high diversity of chlamydiae in a wide range of habitats and hosts, and it is expected that various new chlamydial species will be described in the near future. Recent studies leading to descriptions of new species or *Candidatus* species of the family *Chlamydiaceae*, have focused on birds and reptiles, but this may represent only the 'tip of the iceberg' of emerging novel chlamydial species found in animal hosts (Taylor-Brown and Polkinghorne, 2017).

To report a new chlamydial species, a minimal number of five genes belonging to the core genome of the order *Chlamydiales* is recommended (Pillonel *et al.*, 2015). These genes should not be subject to horizontal transfer and should ideally be recognized taxonomic markers (Pillonel *et al.*, 2015). The panels include genes encoding for (i) DnaA, SucA, Hyp325 and FabI, which help to classify new isolates at the genus level, with cut-off similarity values of 70%, 64%, 57% and 78%, respectively, or (ii) RpoN, FtsK, PepF and Adk, which are suitable to classify isolates at the species level with 96%, 98%, 96%, and 95% cut-offs, respectively, or (iii) 16S rRNA and 23S rRNA, which are also well-recognized markers to classify isolates at the family level, with cut-offs of 92.5% and 91%, respectively (Pillonel *et al.*, 2015). These genetic cut-offs should be complemented by phenotypic characterization whenever possible. Ultrastructure description is considered mandatory to confirm the biphasic developmental stages of chlamydiae. A new species description should also include the deposition of at least one isolate into two culture collections (Greub, 2010).

Atypical *Chlamydiaceae* species in birds have been described in recent years in urban pigeons, as well as European and Asian chicken flocks and resulted in the description of two new avian chlamydial species in the reunited genus *Chlamydia: C. gallinacea* and *C. avium* (Sachse and Laroucau, 2015; Sachse *et al.*, 2014). Another chlamydial species at *Candidatus* level, *Candidatus* Chlamydia ibidis, was isolated from cloacal swabs of a healthy feral African sacred ibis (*Threskiornis aethiopicus*), which had been shot during an official culling operation carried out in Western France between 2009 and 2010 (Vorimore *et al.*, 2013).

Even more complexity was added recently, when avian *C. abortus* strains were detected in Polish wildfowl by PCR (Szymanska-Czerwinska *et al.*, 2017). The full genome of a new species intermediary between *C. psittaci* and *C. abortus* was retrieved from a red-tailed hawk (*Buteo jamaicensis*) (Joseph *et al.*, 2015). An average nucleotide identity of less than 94% compared to other *Chlamydiaceae* species suggested that the red-tailed hawk strain belongs to a new species intermediary between *C. psittaci* and *C. abortus*, possibly as a result of recombination events in the intestinal tract. Birds-of-prey or other scavengers and predators may acquire multiple species by oral uptake thus facilitating homologous recombination and subsequent emergence of new species in these hosts (Joseph *et al.*, 2015).

Reptiles also appear to harbour an unexpected diversity of novel chlamydiae (Taylor-Brown and Polkinghorne, 2017) as illustrated by recent descriptions of two novel chlamydial species (Staub *et al.*, 2018) and two *Candidatus* species in snakes (Taylor-Brown *et al.*, 2016).

Initial reports considered *C. pneumoniae* as the most widespread chlamydial species in free-ranging and captive reptilian hosts including puff adders, boas, chameleons, crocodiles, turtles and tortoises (reviewed in (Borel *et al.*, 2018). These infections manifested as granulomatous inflammation in inner organs, such as heart, liver, spleen and lungs, but also remained asymptomatic with bacterial detection at the choana and/or cloaca. Novel *C. pneumoniae*-like strains based on phylogenetic analysis of the 16S rRNA gene were detected in captive asymptomatic snake species (Taylor-Brown *et al.*, 2015). Of those, culture-independent whole-genome sequencing of genomic DNA led to the identification of the novel *Candidatus* species C. sanzinia in a clinically inapparent Madagascar tree boa (*Sanzinia madagascariensis volontany*) (Taylor-Brown *et al.*, 2016). *Candidatus* C. sanzinia falls into a phylogenetic clade with *C. pneumoniae* and *C. pecorum* and harbours a 7.5 kb cryptic plasmid. The genome of *Candidatus* C. sanzinia contains several known chlamydial virulence factors such as Tarp and Mip and further includes an expanded repertoire of outer membrane proteins, but did not show any evidence of a conventional chlamydial plasticity zone. The metagenomics approach retrieved another novel chlamydial species *Candidatus* C. corallus, obtained from the choanal sample of an asymptomatic captive Amazon Basin emerald tree boa (*Corallus batesii*) (Taylor-Brown *et al.*, 2017). The plasticity zone of *Candidatus* C. corallus is structurally most similar to the human *C. pneumoniae* isolate AR39. *Candidatus* C. corallus also contains a plasmid but lacks a tryptophan operon. Although *Candidatus* C. sanzinia and *Candidatus* C. corallus were classified as new species according to the recommended scheme (Pillonel *et al.*, 2015), they remain *Candidatus* species due to the absence of ultrastructural description for lack of successful isolation. Two novel chlamydial species, *C. serpentis* and *C. poikilothermis*, could be isolated from choanal and cloacal swabs of captive snakes and maintained in culture (Staub *et al.*, 2018). *C. serpentis* is most closely related to *C. pneumoniae*, harbours a plasmid and appears to have a less disrupted genome than *C. pneumoniae* with a more extensive plasticity zone and fewer pseudogenes. It occurred in snakes belonging to the families *Colubridae* and *Viperidae*, and type strains were isolated from captive asymptomatic snake species *Pantherophis guttatus* and *Atheris squamigera*. Phenotypic features include the typical biphasic developmental cycle, enhanced replication by adding cycloheximide, antimicrobial susceptibility to tetracycline and moxifloxacin but intermediate sensitivity to azithromycin. *C. poikilothermis* is most closely related to *C. caviae* and has unique phenotypic features: *C. poikilothermis* isolation and growth in cell culture was more successful at lower temperatures (28°C) compared to normally used incubation temperatures (37°C), which could represent an adaptation to the poikilothermic lifestyle of the reptilian host. The replication of *C. poikilothermis* is not enhanced by the host protein synthesis inhibitor cycloheximide regardless of the incubation temperature (28 or 37°C). The ultrastructural features of *C. poikilothermis* are similar to other *Chlamydiaceae* including EBs, RBs and IBs with smaller aberrant bodies containing inclusions if grown at 37°C. *C. poikilothermis* is susceptible to tetracycline and moxifloxacin but has intermediate resistance to azithromycin in the absence of typical mutations described in other species (Staub *et al.*, 2018). While all these recent findings highlight that snakes may still harbour a significant and largely uncharacterized level of diverse and novel chlamydial species, the host range, pathogenicity and zoonotic potential of such novel chlamydial species remain obscure.

Occurrence, structure and possible implications of antimicrobial resistance genes in *C. suis*

Swine can be infected by different species of *Chlamydia* including *C. psittaci, C. abortus, C. pecorum* and *C. suis*. The most common chlamydial species causing endemic infections in swine is *C. suis* (see first section of this chapter). *C. suis* is of particular importance to the genus *Chlamydia*, because it is the first and only species found to harbour stable homotypic antibiotic resistance, particularly tetracycline resistance (reviewed in Borel *et al.*, 2016; Sandoz and Rockey, 2010). Thus, *C. suis* is the only obligate intracellular bacterium to date that has acquired tetracycline resistance genes of non-chlamydial origin outside laboratory conditions. While the *C. suis* type strain S45, isolated from the faeces of an asymptomatic pig in Austria in the late 1960s, is tetracycline sensitive (Kaltenboeck *et al.*, 1992; Perez-Martinez and Storz, 1985), tetracycline-resistant *C. suis* has been detected in clinical samples of asymptomatically infected swine in several countries including the US (Dugan *et al.*, 2004; Lenart *et al.*, 2001), Switzerland (Borel *et al.*, 2012; Wanninger *et al.*, 2016), Italy (Di Francesco *et al.*, 2008), Belgium, Cyprus, Germany, the Netherlands and Israel (Schautteet *et al.*, 2013; Vanrompay *et al.*, 2018). To date, the *tetC* gene has only been retrieved from porcine, but not from human *C. suis* isolates (De Puysseleyr *et al.*, 2014; De Puysseleyr *et al.*, 2017; Suchland *et al.*, 2009). Successful *in vitro* transfer of *tetC* from tetracycline-resistant (tet^R) *C. suis* strains to a tetracycline-sensitive (tet^S) *C. trachomatis* L2 strain and *C. muridarum* by co-infection, as well as transfer from *in vitro*-generated tet^R to tet^S *C. trachomatis* strains has been demonstrated (Jeffrey *et al.*, 2013; Suchland *et al.*, 2009).

In *C. suis*, tetracycline resistance is conferred by the tetracycline resistance gene *tetA(C)*, which encodes an efflux pump and is located within a Tet-island. This Tet-island is integrated into the *C. suis* chromosome and flanked by the invasin-like protein gene [*inv*-like gene (Liu *et al.*, 2007)] and was most likely acquired by horizontal gene transfer from an external source as further outlined below (Dugan *et al.*, 2004; Joseph *et al.*, 2016; Seth-Smith *et al.*, 2017a). All Tet-islands described for *C. suis* to date contain the tetracycline resistance gene, *tetA(C)*, and the tetracycline repressor class C gene, *tetR*(C). Interestingly, *tet* operon expression in *C. suis* occurs even in the absence of tetracycline, which is unusual for Gram-negative bacteria and explains discrepancies between the detection of *tetC* gene by PCR and *in vitro* susceptibility characteristics (Donati *et al.*, 2016). Furthermore, the Tet-island contains genes coding for the plasmid stable inheritance protein (*pemK*), mobilization proteins *mobABCDE*, which are important for conjugation, and the replication factors *repA* and *repC*, which are essential for plasmid replication. Partial or complete deletion of these genes was detected in tetracycline resistant strains.

The Tet-island may also contain the insertion element *IScs605*, which is composed of two divergently oriented insertion sequences, annotated as IS200 and IS605 (Dugan *et al.*, 2004; Joseph *et al.*, 2016; Marti *et al.*, 2017; Seth-Smith *et al.*, 2017a). Dugan and colleagues were able to functionally characterize *IScs605* using an *Escherichia coli*-based mating assay, demonstrating that a transposase encoded within this element is responsible for the integration of the Tet-island into the *C. suis* genome (Dugan *et al.*, 2007).

Suchland and co-workers further advanced these initial studies by co-culturing cells with the tet^R *C. suis* strain R19 and with either the *C. trachomatis* L2 strain or *C. muridarum*, producing tet^R recombinants for both *C. trachomatis* and *C. muridarum* (Suchland *et al.*, 2009). However, the mode of transmission was different for the two species: Tetracycline resistance in *C. muridarum* was transmitted in the form of a classical double-crossover encompassing >98,000 bp from the parental *C. suis* strain, ranging from a gene encoding a helicase (*C. muridarum* gene TC_RS00430) to a hypothetical protein-encoding gene (*C. muridarum* gene TC_RS00785), and which included both ribosomal (*rrn*) operons. In contrast, recombination in *C. trachomatis* occurred at the location of the *rrn* operons of the parental strains producing a tet^R recombinant with three instead of two *rrn* operons, of which the upstream operon was from the L2 parental strain and the two downstream operons were composed of a mosaic of *C. trachomatis* L2 and *C. suis* R19. In both instances, the Tet-island was transmitted as a whole unit, including its neighbouring genes. The Suchland group exclusively investigated transmission of the Tet-island of the *C. suis* strain R19, a strain containing all identified genes of the genomic island to date, including *IScs605*.

In order to test whether the *IScs605* transposases are responsible for successful integration of the Tet-island into *C. suis*, Marti *et al.* (Marti *et al.*, 2017) co-cultured the tet^S *C. suis* type strain S45 with the tet^R *C. suis* strain Rogers132, a strain lacking the *IScs605* transposases producing two unique tet^R recombinants with the genomic backbone of S45. Thus, *IScs605* may have been essential for the initial integration of the Tet-island into the *C. suis* chromosome, but it does not appear to be necessary for gene transfer among *C. suis* strains.

Molecular-clock dating indicates that ancestral clades of tetracycline-resistant *C. suis* strains predated the discovery of tetracycline in 1947. The use of tetracycline in livestock feed for pig farming since the 1950s could be a reason for the recent bottleneck in the *C. suis* population (Joseph *et al.*, 2016). The Tet-island itself likely spread throughout *C. suis* strains by homologous recombination after acquisition by horizontal gene transfer from an external source, possibly originating from *Aeromonas salmonicida* plasmid pRAS3.2 (Dugan *et al.*, 2004) or *Snodgrassella* species (Tian *et al.*, 2012) or other *Betaproteobacteria* (Joseph *et al.*, 2016). Joseph and co-workers suggested that the Tet-island was most likely acquired by an index strain and later exchanged between *C. suis* strains via intraspecies recombination (Joseph *et al.*, 2016), which was further supported by a recent study focusing on whole-genome sequencing of 29 tet^R and tet^S *C. suis* isolates from a recent porcine cohort in Switzerland (Hoffmann *et al.*, 2015; Wanninger *et al.*, 2016). This study further showed that the *C. suis* genome is very plastic, with unprecedented intra-species diversity, highly susceptible to recombination and plasmid exchange (Seth-Smith *et al.*, 2017a). Interestingly, New World isolates had a more restricted diversity and appeared to have derived from European isolates, thus indicating historical strain transfer to the United States. In contrast, the Tet-island appeared to be a recent import into the genome of *C. suis*, with a possible American origin. The architecture of the Tet-island from Swiss *C. suis* strains was variable, but the *tetA(C)* gene was always intact and presumably functional, and the *inv* gene was inactivated in many of the strains analysed, even when not disrupted by the Tet-island. Selective pressure from tetracycline use in pig production likely led to a high number of Tet-island carrying strains. However, the Tet-island appeared to be lost in the absence of such pressure indicating strain replacement rather than loss or gain of the Tet-island. Also, in the absence of selective pressure, isolates lacking the Tet-island appear to dominate, which indicates a fitness cost of carrying this island (Borel *et al.*, 2012; Seth-Smith *et al.*, 2017a; Wanninger *et al.*, 2016).

Host–pathogen interactions in *C. psittaci* infection revealing unique features

As outlined in the first section of this chapter, the severity of disease caused by *C. psittaci* in birds and humans can vary from acute to chronic or even remain subclinical. While it is known that an asymptomatic course has measurable consequences for the host, but is difficult to investigate, the agent's capability of causing a fulminant course of infection has been of particular interest to researchers in the last decade.

The essential hallmarks of a *C. psittaci* infection in its various manifestations were verified in several infection trials in calves (Ostermann *et al.*, 2014; Ostermann *et al.*, 2013a; Ostermann *et al.*, 2013b; Reinhold *et al.*, 2012). Using this infection model, the whole variety of characteristic features observed in animals and humans in natural chlamydial infections could be reproduced. (i) The severity of disease during the acute phase of infection was shown to be dose dependent. As the infectious dose of *C. psittaci* varied from 10^6 to 10^9 inclusion-forming units (ifu) per calf, the course changed from moderate to severe along this scale. In each setting, the highest scores of respiratory and clinical signs were observed during the acute phase at two to three days post inoculation (dpi). Clinical illness subsided for the most part with resolution of symptoms by 10 dpi. (ii) Subsequently, the illness entered a protracted clinically silent stage with intermittent mild symptoms, faecal pathogen excretion, transient chlamydaemia and slightly elevated levels of monocytes and lipopolysaccharide-binding protein (LBP) in the blood. Notably, these phenomena were also seen in sentinel calves that were put in contact with the infected animals, which resembled the process of naturally acquired chlamydial infection. (iii) Regarding pathological manifestations, acute respiratory disease caused by *C. psittaci* was characterized by recruitment of inflammatory cells to the infection site, damage of the alveolar–capillary barrier, and elevated concentrations of eicosanoids and total protein in broncho-alveolar lavage fluid. Clinically, the effects manifested themselves as pulmonary dysfunction. (iv) In contrast to other bacterial infections, the humoral immune response was generally weak. Formation of specific antibodies was observed in the second week post infection, but only in two thirds of experimentally challenged calves. In sentinel animals, however, no humoral immune response could be detected. This means that the effect of disease control approaches based solely on classical immunoprophylaxis may be limited and the cellular immune response has to be addressed as well (Knittler *et al.*, 2014).

Further investigations of the cellular immune response using cell culture and a mouse model revealed a new mechanism of MHC I-cross presentation of *C. psittaci*-infected dendritic cells (Fiegl *et al.*, 2013). It was demonstrated that *C. psittaci*-infected murine dendritic cells use autophagosomal and endo-vacuolar processing to degrade bacterial compartments and proteolytically produce chlamydial peptide antigens. MHC I molecules loaded with chlamydial antigens are presented on the cell surface to $CD8^+$ T cells and subsequently recycled to the amphisomal compartment.

As part of the innate immune system, the complement system was known to become activated by surface antigens of chlamydiae. Consisting of approximately 40 serum factors, this system modulates the inflammatory response and protects against extracellular pathogens (Klos *et al.*, 2013). Early, high and long-lasting activation of the complement system was documented in a murine model of intranasal *C. psittaci* lung infection (Bode *et al.*, 2012). The crucial role of complement factor C3 and its anaphylatoxic peptide C3a for protection against chlamydial infection was shown in a challenge experiment involving C3-deficient mice. The latter developed a 100-fold higher susceptibility to *C. psittaci* infection compared to wild-type mice. In a follow-up study, it was shown that the C3a receptor was indispensable for effective protection and enhanced survival in pulmonary *C. psittaci* infection (Dutow *et al.*, 2013). In contrast, the anti-chlamydial IgG and IgM response was practically the same in both groups of mice, thus confirming that protection of the wild-type mice was not due to specific antibody production. A significant protective effect in C3-deficient mice was only attained when high-titred anti-*C. psittaci* serum was administered at an early time point, i.e. 1 dpi. These findings are of general importance for protection against infections by intracellular bacteria and provide interesting clues for vaccine development. Please see Chapter 13, *Chlamydia* and Innate Immunity, for more detail on the role of the complement during infection.

To identify pathogenicity factors and mechanisms that are specific to *C. psittaci*, host–pathogen interactions were studied in embryonated chicken eggs (Braukmann *et al.*, 2012), and young chickens (Kalmar *et al.*, 2015). In a comparative approach, infection experiments using *C. abortus*, which is closely related but prefers non-avian hosts, were conducted alongside *C. psittaci* infection. The *in ovo* chicken embryo model involves inoculation onto the chorioallantoic membrane (CAM) and, therefore, resembles natural infection across epithelial layers. It also proved a valuable tool to investigate the innate immune response. The main difference to the *in vivo* chicken model is the less developed pulmonary immune defence in the embryo, which results in high mortality rates of *C. psittaci*-infected embryos in contrast to the absence of mortalities in chickens. The main findings of both studies are summarized in Table 17.3. All differences found between the *C. psittaci* and *C. abortus* infections highlight the superior capacities of the former to establish the infection in the host organism.

Summarizing this section and recent data from the literature, the following distinctive and unique characteristics of *C. psittaci* can be stated (Knittler and Sachse, 2015):

1 *efficient dissemination* within the animal host causing systemic disease (Braukmann *et al.*, 2012; Kalmar *et al.*, 2015; Reinhold *et al.*, 2012);
2 the capability of *causing a fulminant* course of *infection* (Knittler *et al.*, 2014; Ostermann *et al.*, 2013a; Yin *et al.*, 2013);
3 timely and *strong up-regulation* of essential chlamydial genes in the face of host immune

Table 17.3 Comparison of infection course and outcome in two infection models involving *C. psittaci* and *C. abortus*

Feature	*In ovo* infection (chicken embryo)	*In vivo* infection (young chicken)
Cellular immune response in organs	Largely similar in *C. psittaci* and *C. abortus* infection	*C. psittaci* * IFN-γ, IL-1β, IL-6, IL-17, IL-22, LITAF, iNOS
Pathology	More severe in *C. psittaci* infection	More severe in *C. psittaci* infection
Systemic dissemination	*C. psittaci* more vigorous	*C. psittaci* more vigorous
mRNA expression of chlamydial genes	* incA, groEL, cpaf, ftsW in *C. psittaci*, not *C. abortus*	* groEL, cpaf, ftsW in *C. psittaci*, not *C. abortus*
Influx of immune cells (macrophage, γδ T cells)	Higher and faster in *C. psittaci*-infected sites	Higher and faster in *C. psittaci*-infected sites

* Increase of expression rate significant.

response (Braukmann *et al.*, 2012; Kalmar *et al.*, 2015);

4 *rigorous shut-down* of *C. psittaci* gene expression in intracellular persistence in cell culture (Goellner *et al.*, 2006).
5 *inhibition of apoptosis* as integral part of the *in vitro* persistence phenotype (Knittler and Sachse, 2015; Schwarzenbacher *et al.*, 2004);
6 IncA and IncB protein-mediated *utilization of host proteins* to control cell survival and intracellular transport of inclusions in order to support bacterial growth in infected cells (Bocker *et al.*, 2014; Borth *et al.*, 2011; Mital *et al.*, 2013);
7 pronounced physical *attachment* of bacterial inclusions *to mitochondria* to acquire eukaryotic substrates (Knittler *et al.*, 2014; Matsumoto *et al.*, 1991).

Pathology of *Chlamydia abortus* and *Chlamydia pecorum* in ruminants

The clinical history of *C. abortus* infection comprises abortion cases in sheep and goats, often in late pregnancy (the last 2–3 weeks) followed by expulsion of a dead foetus and foetal membranes (Longbottom and Coulter, 2003). The placenta is usually macroscopically altered by oedema, haemorrhage, necrosis and purulent exudate. The typical histopathological picture consists of placental inflammation and necrosis. In well-preserved tissue, intra-trophoblastic basophilic chlamydial inclusions might be observed (reviewed by Borel *et al.*, 2014; Sachse *et al.*, 2009). A large number of chlamydial organisms are shed in infected placental membranes leading to contamination of the environment and being the source of new infections (Longbottom and Coulter, 2003).

In sheep, *C. abortus*-induced placentitis develops late in gestation and, from then on, gradually progresses to a diffuse inflammatory response, thrombotic vasculitis and tissue necrosis (reviewed by Sammin *et al.*, 2009). The abortion event may be induced by a combination of impairment of materno-fetal nutrient and gaseous exchange, disruption of hormonal regulation of pregnancy and cytokine aggression (Essig and Longbottom, 2015). Foetal pathology can occur after thrombo-embolic infection of the foetus (Buxton *et al.*, 1990). Buxton and co-workers also showed that ewes were susceptible to infection from early gestation onwards, with infections of placenta and foetuses occurring from about 60 days of gestation, while pathological changes were detected after 90 days of gestation. The trophoblast cells are the main site for multiplication and dissemination of *C. abortus*. These cells, as well as the foetal macrophages, release pro-inflammatory mediators, such as TNF-α and IL-8, leading to tissue damage (Wheelhouse *et al.*, 2009).

The route of primary infection is presumed to be oronasal following ingestion of organisms and initial infection established in the tonsils. From there it can disseminate to other organs through blood or lymph [reviewed by (Essig and Longbottom, 2015)] until reaching the placenta at a later stage. Recent intranasal infection studies in non-pregnant ewes demonstrated that relatively low doses of infectious organisms (EBs) can induce chlamydial latency at unknown anatomical sites (Longbottom *et al.*, 2013a; Longbottom *et al.*, 2013b).

During latency the infection remains inapparent in non-pregnant animals and only becomes evident during a subsequent pregnancy. When the animals become pregnant, the infection progresses, initiates placental infection and induces abortion (Essig and Longbottom, 2015). This could explain frequent observations that initially few abortion cases are seen, whereas in the subsequent pregnancy period an abortion storm can affect 30% of animals or more.

Recent intranasal experimental infection studies in sheep using three different infection doses of *C. abortus* investigated the chlamydial latency stage (Longbottom *et al.*, 2013b). Interestingly, the highest *C. abortus* dose stimulated protective immunity resulting in a much lower abortion rate, whereas medium and low doses resulted in latent infection, which is characterized by infection of the placenta and abortion in the subsequent pregnancy. Following high-dose intranasal infection with *C. abortus*, sheep mounted a strong IFN-γ-mediated cellular immune response (Longbottom *et al.*, 2013b), which is in line with protective immunity developed after natural infection and abortion. While this experimental study involved *C. abortus* strain S26/3, which was isolated in Scotland from the placenta of a sheep that had aborted, the same group (Livingstone *et al.*, 2017) later used the genetically distinct strains LLG and POS, which were isolated in Greece from aborted goat and sheep foetuses. They showed that placental pathology, infectious load and serological response were clearly lower with LLG and POS as compared to strain S26/3.

Ovine enzootic abortion can be controlled by application of inactivated or attenuated vaccines (Longbottom and Livingstone, 2006). The latter consists of a temperature-attenuated strain 1B of *C. abortus* that was generated following N-methyl-N'-nitro-N-nitrosoguanidine (NTG) mutagenesis. Although it has been reported that this vaccine can induce abortion (Sargison *et al.*, 2015; Wheelhouse *et al.*, 2010), vaccination is still recommended to prevent ovine enzootic abortion. Whole-genome sequencing of vaccine strain 1B revealed that its genome is identical to the non-attenuated 'reverted mutant' strain 1H (Longbottom *et al.*, 2018). Thus, protection conferred by the 1B vaccine strain is unlikely to be due to NTG-induced SNPs, but more likely caused by administration of high doses of *C. abortus* EBs, which stimulate protective immunity as seen in the previously reported intranasal infection experiment (Longbottom *et al.*, 2013b).

The first reports on the pathology of livestock *C. pecorum* infections date back to the 1940s and 1960s. They described cases of encephalomyelitis (Buss disease) in cattle (McNutt and Waller, 1940) and polyarthritis in feeder lambs (Mendlowski and Segre, 1960) and calves (Storz *et al.*, 1966a,b) in the USA. Until now, chlamydial infections due to *C. pecorum* in sheep and cattle have been associated with clinical manifestations, such as infectious arthritis, conjunctivitis, encephalomyelitis, infertility, enteritis, reduced growth rates, mastitis and pneumonia (reviewed in Walker *et al.*, 2015). However, the majority of infections may remain clinically inapparent, mostly localized in the gastrointestinal tract (Reinhold *et al.*, 2011), which is comparable to *C. suis* infections in swine (Hoffmann *et al.*, 2015). In contrast to *C. abortus* in small ruminants, *C. pecorum* is considered to be endemic in cattle and sheep.

In cattle, *C. pecorum* is predominantly associated with pulmonary, articular, intestinal, ocular and central nervous disease (Reinhold *et al.*, 2011). Infection, usually at a low level of pathogen exposure, occurs early in life and repeatedly by multiple strains circulating ubiquitously in herds with mixed infections or simultaneous infection with two or more strains at different mucosal sites (Poudel *et al.*, 2012). Such recurrent *C. pecorum* infections can have a significant impact on herd performance leading to reduced growth rates in calves. Faecal shedding by carrier animals is thought to be the most important transmission route. In addition, *C. pecorum* may be shed in vaginal, ocular, and nasal discharges. The pathogenic significance of *C. pecorum* and development of overt disease seem to depend on a multitude of factors, such as co-infection with other pathogens, herd management, stress and hygiene practices on farms and host genetics (Reinhold *et al.*, 2011). In contrast to these subclinical and chronic courses of infection, *C. pecorum* is able to cause severe and fatal sporadic bovine encephalomyelitis (SBE) in cattle, previously called Buss disease (McNutt and Waller, 1940). The first SBE cases were reported in the USA and Canada between 1940 and 1960. More recently, cases of SBE in calves were reported in Western Australia (Jelocnik *et al.*, 2014a) and New Zealand (Hunt *et al.*, 2016). SBE-affected calves showed fever and depression, stiff gait, central nervous signs such as staggering, circling and opisthotonus, and eventually respiratory signs and diarrhoea if lungs and intestinal tract were also affected. Post-mortem examinations revealed encephalomyelitis, systemic infection and fibrinous polyserositis including pericarditis, pleuritis, peritonitis and arthritis. Using PCR, *C. pecorum* DNA was detected in multiple organs, and chlamydial antigen was observed by immunohistochemistry, in the cytoplasm of endothelium and macrophages of brain, spleen and lung, as well as in the mesothelium. Recent typing of *C. pecorum* strains revealed differences between strains detected in brain, heart, lung, liver and the intestine, suggesting that different strains may have a different pathogenic potential, an observation previously suggested from studies in sheep (Jelocnik *et al.*, 2014a). The predominant sequence type found in brain, heart, lung and liver (sequence type 23) of the two SBE-affected calves was identical to a sequence type previously described in SBE cases. A second sequence type was detected in an ileal tissue sample from one of these calves, which clustered with bovine faecal isolates (Jelocnik *et al.*, 2014a), thus indicating sequence types could be grouped according to anatomical, but not geographical site. Comparative genome analysis (Sait *et al.*, 2014) of four *C. pecorum* strains E58 (from calf with SBE), W73 (asymptomatic sheep), P787 (sheep with arthritis), and PV3056/3 (cow with metritis) identified differences in ORF663 and pseudogene content between strains isolated from diseased or asymptomatic ruminants. However, no other genetic differences were observed that may account for the different disease manifestations. It was also found that all analysed *C. pecorum* strains contained an almost intact tryptophan biosynthesis pathway, which might enable persistent infections in the respective host (Sait *et al.*, 2014). Two *C. pecorum* strains, the ovine polyarthritis isolate IPA and the calf SBE isolate E58, could escape IFN-γ-mediated host cell tryptophan depletion *in vitro* (Islam *et al.*, 2018) and possibly use the same mechanism to evade the immune response *in vivo* in its natural host.

In sheep, *C. pecorum*-induced polyarthritis, typically affecting growing lambs, is a debilitating disease causing economic losses (Mohamad and Rodolakis, 2010) (Rodolakis and Laroucau, 2015; Walker *et al.*, 2018; Walker *et al.*, 2015). New knowledge was recently added from Australian studies, where polyarthritis and concurrent conjunctivitis due to *C. pecorum* infections were seen in young lambs peaking at six months of age (weaning) (Walker *et al.*, 2018; Walker *et al.*, 2015). In contrast, other studies (Polkinghorne *et al.*, 2009) could not find an association between the presence of chlamydial DNA in ovine conjunctival samples and the onset of ocular disease.

C. pecorum-induced arthritis is characterized by an inflammatory and proliferative response in synovial membranes possibly progressing to chronic changes with articular erosions and fibrotic thickening. These changes result in animal welfare issues and economic losses for the farmer (Walker *et al.*, 2018). Ovine arthritis cases in Australian weaned lambs showing joint stiffness with occasional joint effusions and depression were described recently (Walker *et al.*, 2016). *C. pecorum* shedding determined by PCR was observed at multiple sites, such as the conjunctiva, rectum and vagina in this and previous studies (Lenzko *et al.*, 2011), but the sequence type 23 was commonly associated with conjunctivitis and polyarthritis in Australian sheep (Walker *et al.*, 2018).

In summary, clinical disease due to *C. pecorum* infection in ruminant hosts ranges from subclinical intestinal infection to fatal disease. The course of infection can be controlled by various factors, such as stress and co-infections on the host side, as well as strain virulence on the pathogen side, resulting in multifactorial pathogenesis and variable disease outcome.

Outlook

1. Recent research led to the discovery of more members of the family *Chlamydiaceae*. These newly defined veterinary chlamydial species need to be characterized in terms of epidemiology and pathogenicity.
2. Despite recent advances, both surveillance and documentation of zoonotic cases have to be improved to understand the extent and impact of chlamydial zoonotic infections.
3. New findings on pathogenesis of animal chlamydial infections have enabled better understanding of the specific strategies employed by the pathogens. Future projects

should focus on chlamydial latency, differences in strain virulence and the variability of host immune responses.

4 Thorough investigation of tetracycline-resistant and sensitive *C. suis* strains isolated from pigs is crucial to better understand chlamydial antibiotic resistance and its transmission through recombination events.
5 Now that genome sequence analysis has been conducted for the major chlamydial pathogens, the focus should shift to transcriptome analysis to enable comprehensive investigations of pathogen–host interactions.
6 Co-infections involving different *Chlamydia* spp. and/or other pathogens have emerged as an important research topic that should be further pursued.

References

Aitken, I.D., and Longbottom, D. (2007). Chlamydial abortion. In Diseases of sheep, I.D. Aitken, ed. (Oxford, UK: Blackwell Scientific Ltd.), pp. 105–112.

Andrew, D.W., Hafner, L.M., Beagley, K.W., and Timms, P. (2011). Partial protection against chlamydial reproductive tract infection by a recombinant major outer membrane protein/CpG/cholera toxin intranasal vaccine in the guinea pig *Chlamydia caviae* model. J Reprod Immunol *91*, 9–16 DOI: 10.1016/j.jri.2011.06.100.

Bachmann, N.L., Fraser, T.A., Bertelli, C., Jelocnik, M., Gillett, A., Funnell, O., Flanagan, C., Myers, G.S., Timms, P., and Polkinghorne, A. (2014). Comparative genomics of koala, cattle and sheep strains of *Chlamydia pecorum*. BMC Genomics *15*, 667 DOI: 10.1186/1471-2164-15-667.

Bachmann, N.L., Sullivan, M.J., Jelocnik, M., Myers, G.S., Timms, P., and Polkinghorne, A. (2015). Culture-independent genome sequencing of clinical samples reveals an unexpected heterogeneity of infections by *Chlamydia pecorum*. J Clin Microbiol *53*, 1573–1581 DOI: 10.1128/JCM.03534-14.

Becker, A., Lutz-Wohlgroth, L., Brugnera, E., Lu, Z.H., Zimmermann, D.R., Grimm, F., Grosse Beilage, E., Kaps, S., Spiess, B., Pospischil, A., *et al.* (2007). Intensively kept pigs pre-disposed to chlamydial associated conjunctivitis. J Vet Med A Physiol Pathol Clin Med *54*, 307–313 DOI: 10.1111/j.1439-0442.2007.00963.x.

Beeckman, D.S., and Vanrompay, D.C. (2009). Zoonotic Chlamydophila psittaci infections from a clinical perspective. Clin Microbiol Infect *15*, 11–17 DOI: 10.1111/j.1469-0691.2008.02669.x.

Belij-Rammerstorfer, S., Inic-Kanada, A., Stojanovic, M., Marinkovic, E., Lukic, I., Stein, E., Montanaro, J., Bintner, N., Schurer, N., Ghasemian, E., *et al.* (2016). Infectious dose and repeated infections are key factors influencing immune response characteristics in guinea pig ocular chlamydial infection. Microbes Infect *18*, 254–262 DOI: 10.1016/j.micinf.2015.12.001.

Blumer, C., Zimmermann, D.R., Weilenmann, R., Vaughan, L., and Pospischil, A. (2007). Chlamydiae in free-ranging and captive frogs in Switzerland. Vet Pathol *44*, 144–150 DOI: 10.1354/vp.44-2-144.

Bocker, S., Heurich, A., Franke, C., Monajembashi, S., Sachse, K., Saluz, H.P., and Hanel, F. (2014). *Chlamydia psittaci* inclusion membrane protein IncB associates with host protein Snapin. Int J Med Microbiol *304*, 542–553 DOI: 10.1016/j.ijmm.2014.03.005.

Bode, J., Dutow, P., Sommer, K., Janik, K., Glage, S., Tummler, B., Munder, A., Laudeley, R., Sachse, K.W., and Klos, A. (2012). A new role of the complement system: C3 provides protection in a mouse model of lung infection with intracellular Chlamydia psittaci. PLoS One *7*, e50327 DOI: 10.1371/journal.pone.0050327.

Bodetti, T.J., Jacobson, E., Wan, C., Hafner, L., Pospischil, A., Rose, K., and Timms, P. (2002). Molecular evidence to support the expansion of the host range of *Chlamydophila pneumoniae* to include reptiles as well as humans, horses, koalas and amphibians. In System Appl Microbiol, pp. 146–152.

Borel, N., Frey, C.F., Gottstein, B., Hilbe, M., Pospischil, A., Franzoso, F.D., and Waldvogel, A. (2014). Laboratory diagnosis of ruminant abortion in Europe. Vet J *200*, 218–229 DOI: 10.1016/j.tvjl.2014.03.015.

Borel, N., Leonard, C., Slade, J., and Schoborg, R.V. (2016). Chlamydial antibiotic resistance and treatment failure in veterinary and human medicine. Curr Clin Microbiol Rep 3, 10–18 DOI: 10.1007/s40588-016-0028-4.

Borel, N., Polkinghorne, A., and Pospischil, A. (2018). A review on chlamydial diseases in animals: Still a challenge for pathologists? Veterinary Pathology *55*, 374–390.

Borel, N., Regenscheit, N., Di Francesco, A., Donati, M., Markov, J., Masserey, Y., and Pospischil, A. (2012). Selection for tetracycline-resistant *Chlamydia suis* in treated pigs. Vet Microbiol *156*, 143–146 DOI: 10.1016/j.vetmic.2011.10.011.

Borth, N., Litsche, K., Franke, C., Sachse, K., Saluz, H.P., and Hanel, F. (2011). Functional interaction between type III-secreted protein IncA of *Chlamydophila psittaci* and human G3BP1. PLoS One *6*, e16692 DOI: 10.1371/journal.pone.0016692.

Branley, J., Bachmann, N.L., Jelocnik, M., Myers, G.S., and Polkinghorne, A. (2016). Australian human and parrot *Chlamydia psittaci* strains cluster within the highly virulent 6BC clade of this important zoonotic pathogen. Sci Rep *6*, 30019 DOI: 10.1038/srep30019.

Braukmann, M., Sachse, K., Jacobsen, I.D., Westermann, M., Menge, C., Saluz, H.P., and Berndt, A. (2012). Distinct intensity of host–pathogen interactions in *Chlamydia psittaci*- and *Chlamydia abortus*-infected chicken embryos. Infect Immun *80*, 2976–2988 DOI: 10.1128/IAI.00437-12.

Burnard, D., and Polkinghorne, A. (2016). Chlamydial infections in wildlife-conservation threats and/or reservoirs of 'spill-over' infections? Vet Microbiol *196*, 78–84 DOI: 10.1016/j.vetmic.2016.10.018.

Buxton, D., Anderson, I.E., Longbottom, D., Livingstone, M., Wattegedera, S., and Entrican, G. (2002). Ovine chlamydial abortion: characterization of the inflammatory immune response in placental tissues. J Comp Pathol *127*, 133–141, ISI:000178368000007.

Buxton, D., Barlow, R.M., Finlayson, J., Anderson, I.E., and Mackellar, A. (1990). Observations on the pathogenesis of *Chlamydia psittaci* infection of pregnant sheep. J Comp Pathol *102*, 221–237, 2324344.

Chahota, R., Ogawa, H., Ohya, K., Yamaguchi, T., Everett, K.D.E., and Fukushi, H. (2018). Involvement of multiple *Chlamydia suis* genotypes in porcine conjunctivitis. Transbound Emerg Dis *65*, 272–277 DOI: 10.1111/tbed.12645.

Chan, J., Doyle, B., Branley, J., Sheppeard, V., Gabor, M., Viney, K., Quinn, H., Janover, O., McCready, M., and Heller, J. (2017). An outbreak of psittacosis at a veterinary school demonstrating a novel source of infection. One Health 3, 29–33 DOI: 10.1016/j.onehlt.2017.02.003.

De Puysseleyr, K., De Puysseleyr, L., Dhondt, H., Geens, T., Braeckman, L., Morre, S.A., Cox, E., and Vanrompay, D. (2014). Evaluation of the presence and zoonotic transmission of *Chlamydia suis* in a pig slaughterhouse. BMC Infect Dis *14*, 560 DOI: 10.1186/s12879-014-0560-x.

De Puysseleyr, L., De Puysseleyr, K., Braeckman, L., Morre, S.A., Cox, E., and Vanrompay, D. (2017). Assessment of Chlamydia suis Infection in Pig Farmers. Transbound Emerg Dis *64*, 826–833 DOI: 10.1111/tbed.12446.

Dean, D., Rothschild, J., Ruettger, A., Kandel, R.P., and Sachse, K. (2013). Zoonotic Chlamydiaceae species associated with trachoma, Nepal. Emerg Infect Dis *19*, 1948–1955 DOI: 10.3201/eid1912.130656.

Deschuyffeleer, T.P., Tyberghien, L.F., Dickx, V.L., Geens, T., Saelen, J.M., Vanrompay, D.C., and Braeckman, L.A. (2012). Risk assessment and management of *Chlamydia psittaci* in poultry processing plants. The Annals of occupational hygiene *56*, 340–349 DOI: 10.1093/annhyg/mer102.

Di Francesco, A., Donati, M., Rossi, M., Pignanelli, S., Shurdhi, A., Baldelli, R., and Cevenini, R. (2008). Tetracycline-resistant *Chlamydia suis* isolates in Italy. Vet Rec *163*, 251–252 DOI: 163/8/251 [pii].

Dickx, V., Geens, T., Deschuyffeleer, T., Tyberghien, L., Harkinezhad, T., Beeckman, D.S., Braeckman, L., and Vanrompay, D. (2010). *Chlamydophila psittaci* zoonotic risk assessment in a chicken and turkey slaughterhouse. J Clin Microbiol *48*, 3244–3250 DOI: JCM.00698-10 [pii]
10.1128/JCM.00698–10.

Dickx, V., and Vanrompay, D. (2011). Zoonotic transmission of *Chlamydia psittaci* in a chicken and turkey hatchery. J Med Microbiol *60*, 775–779 DOI: 10.1099/jmm.0.030528-0.

Donati, M., Balboni, A., Laroucau, K., Aaziz, R., Vorimore, F., Borel, N., Morandi, F., Vecchio Nepita, E., and Di Francesco, A. (2016). Tetracycline susceptibility in *Chlamydia suis* pig isolates. PLoS One *11*, e0149914 DOI: 10.1371/journal.pone.0149914.

Donati, M., Laroucau, K., Guerrini, A., Balboni, A., Salvatore, D., Catelli, E., Lupini, C., Levi, A., and Di Francesco, A. (2018). Chlamydiosis in backyard chickens (*Gallus gallus*) in Italy. Vector Borne Zoonotic Dis *18*, 222–225 DOI: 10.1089/vbz.2017.2211.

Dugan, J., Rockey, D.D., Jones, L., and Andersen, A.A. (2004). Tetracycline resistance in *Chlamydia suis* mediated by genomic islands inserted into the chlamydial inv-like gene. Antimicrob Agents Chemother *48*, 3989–3995 DOI: 10.1128/AAC.48.10.3989-3995.2004.

Dumke, R., Schnee, C., Pletz, M.W., Rupp, J., Jacobs, E., Sachse, K., Rohde, G., and Capnetz Study, G. (2015). *Mycoplasma pneumoniae* and *Chlamydia* spp. infection in community-acquired pneumonia, Germany, 2011–2012. Emerg Infect Dis *21*, 426–434 DOI: 10.3201/eid2103.140927.

Dutow, P., Fehlhaber, B., Bode, J., Laudeley, R., Rheinheimer, C., Glage, S., Wetsel, R.A., Pabst, O., and Klos, A. (2013). The Complement C3a Receptor is Critical in Defense against Chlamydia psittaci in Mouse Lung Infection and Required for Antibody and Optimal T cell Response. J Infect Dis DOI: 10.1093/infdis/jit640.

Entrican, G. (2002). Immune regulation during pregnancy and host–pathogen interactions in infectious abortion. J Comp Pathol *126*, 79–94, 11944996.

Essig, A., and Longbottom, D. (2015). *Chlamydia abortus:* new aspects of infectious abortion in sheep and potential risk for pregnant women. Current Clinical Microbiology Reports 2, 22–34.

Ferreira, V.L., Silva, M.V., Bassetti, B.R., Pellini, A.C.G., and Raso, T.F. (2017). Intersectoral action for health: preventing psittacosis spread after one reported case. Epidemiol Infect *145*, 2263–2268 DOI: 10.1017/S0950268817001042.

Fiegl, D., Kagebein, D., Liebler-Tenorio, E.M., Weisser, T., Sens, M., Gutjahr, M., and Knittler, M.R. (2013). Amphisomal route of MHC class I cross-presentation in bacteria-infected dendritic cells. Journal of immunology *190*, 2791–2806 DOI: 10.4049/jimmunol.1202741.

Gaede, W., Reckling, K.F., Dresenkamp, B., Kenklies, S., Schubert, E., Noack, U., Irmscher, H.M., Ludwig, C., Hotzel, H., and Sachse, K. (2008). *Chlamydophila psittaci* infections in humans during an outbreak of psittacosis from poultry in Germany. Zoonoses Public Health *55*, 184–188 DOI: 10.1111/j.1863-2378.2008.01108.x.

Gaede, W., Reckling, K.F., Schliephake, A., Missal, D., Hotzel, H., and Sachse, K. (2010). Detection of *Chlamydophila caviae* and *Streptococcus equi* subsp. *zooepidemicus* in horses with signs of rhinitis and conjunctivitis. Vet Microbiol *142*, 440–444 DOI: 10.1016/j.vetmic.2009.10.011.

Gasparini, J., Erin, N., Bertin, C., Jacquin, L., Vorimore, F., Frantz, A., Lenouvel, P., and Laroucau, K. (2011). Impact of urban environment and host phenotype on the epidemiology of *Chlamydiaceae* in feral pigeons (*Columba livia*). Environ Microbiol *13*, 3186–3193 DOI: 10.1111/j.1462-2920.2011.02575.x.

Giannitti, F., Anderson, M., Miller, M., Rowe, J., Sverlow, K., Vasquez, M., and Canton, G. (2016). *Chlamydia pecorum*: fetal and placental lesions in sporadic caprine abortion. J Vet Diagn Invest *28*, 184–189 DOI: 10.1177/1040638715625729.

Goellner, S., Schubert, E., Liebler-Tenorio, E., Hotzel, H., Saluz, H.P., and Sachse, K. (2006). Transcriptional response patterns of *Chlamydophila psittaci* in different *in vitro* models of persistent infection. Infect Immun *74*, 4801–4808, 16861668.

Greub, G. (2010). Subcommittee on the taxonomy of the *Chlamydiae*: Minutes of the inaugural closed meeting, 21 March 2009, Little Rock, AR, USA. International

Journal of Systematic and Evolutionary Microbiology *60*, 2691–2693.

Guo, W., Jelocnik, M., Li, J., Sachse, K., Polkinghorne, A., Pannekoek, Y., Kaltenboeck, B., Gong, J., You, J., and Wang, C. (2017). From genomes to genotypes: molecular epidemiological analysis of Chlamydia gallinacea reveals a high level of genetic diversity for this newly emerging chlamydial pathogen. BMC Genomics *18*, 949 DOI: 10.1186/s12864-017-4343-9.

Guo, W., Li, J., Kaltenboeck, B., Gong, J., Fan, W., and Wang, C. (2016). *Chlamydia gallinacea*, not *C. psittaci*, is the endemic chlamydial species in chicken (*Gallus gallus*). Sci Rep *6*, 19638 DOI: 10.1038/srep19638.

Guscetti, F., Schiller, I., Sydler, T., Heinen, E., and Pospischil, A. (2009). Experimental enteric infection of gnotobiotic piglets with *Chlamydia suis* strain S45. Vet Microbiol *135*, 157–168 DOI: 10.1016/j.vetmic.2008.09.038.

Haas, W.H., Swaan, C.M., Meijer, A., Neve, G., Buchholz, U., Beer, M., van Steenbergen, J.E., and Krause, G. (2007). A Dutch case of atypical pneumonia after culling of H5N1 positive ducks in Bavaria was found infected with *Chlamydophila psittaci*. Euro Surveill *12*, E071129 071123, http://www.ncbi.nlm.nih.gov/pubmed/18053568.

Hagemann, J.B., Simnacher, U., Longbottom, D., Livingstone, M., Maile, J., Soutschek, E., Walder, G., Boden, K., Sachse, K., and Essig, A. (2016). Analysis of humoral immune responses to surface and virulence-associated *Chlamydia abortus* proteins in ovine and human abortions by use of a newly developed line immunoassay. J Clin Microbiol *54*, 1883–1890 DOI: 10.1128/JCM.00351-16.

Heijne, M., van der Goot, J.A., Fijten, H., van der Giessen, J.W., Kuijt, E., Maassen, C.B.M., van Roon, A., Wit, B., Koets, A.P., and Roest, H.I.J. (2018). A cross sectional study on Dutch layer farms to investigate the prevalence and potential risk factors for different Chlamydia species. PLoS One *13*, e0190774 DOI: 10.1371/journal.pone.0190774.

Hinton, D.G., Shipley, A., Galvin, J.W., Harkin, J.T., and Brunton, R.A. (1993). Chlamydiosis in workers at a duck farm and processing plant. Aust Vet J *70*, 174–176, 8343087.

Hoffmann, K., Schott, F., Donati, M., Di Francesco, A., Hassig, M., Wanninger, S., Sidler, X., and Borel, N. (2015). Prevalence of chlamydial infections in fattening pigs and their influencing factors. PLoS One *10*, e0143576 DOI: 10.1371/journal.pone.0143576.

Hogerwerf, L., B., D.E.G., Baan, B., and W., V.D.H. (2017). *Chlamydia psittaci* (psittacosis) as a cause of community-acquired pneumonia: a systematic review and meta-analysis. Epidemiol Infect *145*, 3096–3105 DOI: 10.1017/S0950268817002060.

Holzer, M., Laroucau, K., Creasy, H.H., Ott, S., Vorimore, F., Bavoil, P.M., Marz, M., and Sachse, K. (2016). Whole-genome sequence of *Chlamydia gallinacea* type strain 08-1274/3. Genome Announc *4* DOI: 10.1128/genomeA.00708-16.

Hotzel, H., Grossmann, E., Mutschmann, F., and Sachse, K. (2001). Genetic characterization of a *Chlamydophila pneumoniae* isolate from an African frog and comparison to currently accepted biovars. Syst Appl Microbiol *24*, 63–66, 11403400.

Hulin, V., Oger, S., Vorimore, F., Aaziz, R., de Barbeyrac, B., Berruchon, J., Sachse, K., and Laroucau, K. (2015). Host preference and zoonotic potential of *Chlamydia psittaci* and *C. gallinacea* in poultry. Pathog Dis *73*, 1–11 DOI: 10.1093/femspd/ftv005.

Hunt, H., Orbell, G., Buckle, K.N., Ha, H.J., Lawrence, K.E., Fairley, R.A., and Munday, J.S. (2016). First report and histological features of *Chlamydia pecorum* encephalitis in calves in New Zealand. N Z Vet J *64*, 364–368 DOI: 10.1080/00480169.2016.1208781.

Islam, M.M., Jelocnik, M., Huston, W.M., Timms, P., and Polkinghorne, A. (2018). Characterization of the *in vitro Chlamydia pecorum* response to gamma interferon. Infect Immun *86* DOI: 10.1128/IAI.00714-17.

Jackson, M., White, N., Giffard, P., and Timms, P. (1999). Epizootiology of Chlamydia infections in two free-range koala populations. Veterinary microbiology *65*, 225–234.

Jeffrey, B.M., Suchland, R.J., Eriksen, S.G., Sandoz, K.M., and Rockey, D.D. (2013). Genomic and phenotypic characterization of *in vitro*-generated *Chlamydia trachomatis* recombinants. BMC Microbiol *13*, 142 DOI: 10.1186/1471-2180-13-142.

Jelocnik, M., Branley, J., Heller, J., Raidal, S., Alderson, S., Galea, F., Gabor, M., and Polkinghorne, A. (2017). Multilocus sequence typing identifies an avian-like *Chlamydia psittaci* strain involved in equine placentitis and associated with subsequent human psittacosis. Emerg Microbes Infect *6*, e7 DOI: 10.1038/emi.2016.135.

Jelocnik, M., Forshaw, D., Cotter, J., Roberts, D., Timms, P., and Polkinghorne, A. (2014a). Molecular and pathological insights into *Chlamydia pecorum*-associated sporadic bovine encephalomyelitis (SBE) in Western Australia. BMC Vet Res *10*, 121 DOI: 10.1186/1746-6148-10-121.

Jelocnik, M., Jenkins, C., O'Rourke, B., Barnwell, J., and Polkinghorne, A. (2018). Molecular evidence to suggest pigeon-type *Chlamydia psittaci* in association with an equine foal loss. Transbound Emerg Dis *65*, 911–915 DOI: 10.1111/tbed.12817.

Jelocnik, M., Walker, E., Pannekoek, Y., Ellem, J., Timms, P., and Polkinghorne, A. (2014b). Evaluation of the relationship between *Chlamydia pecorum* sequence types and disease using a species-specific multi-locus sequence typing scheme (MLST). Vet Microbiol *174*, 214–222 DOI: 10.1016/j.vetmic.2014.08.018.

Joseph, S.J., Marti, H., Didelot, X., Castillo-Ramirez, S., Read, T.D., and Dean, D. (2015). *Chlamydiaceae* genomics reveals interspecies admixture and the recent evolution of *Chlamydia abortus* infecting lower mammalian species and humans. Genome Biol Evol *7*, 3070–3084 DOI: 10.1093/gbe/evv201.

Joseph, S.J., Marti, H., Didelot, X., Read, T.D., and Dean, D. (2016). Tetracycline selective pressure and homologous recombination shape the evolution of *Chlamydia suis:* A recently identified zoonotic pathogen. Genome Biol Evol *8*, 2613–2623 DOI: 10.1093/gbe/evw182.

Kaleta, E.F., and Taday, E.M. (2003). Avian host range of *Chlamydophila* spp. based on isolation, antigen detection and serology. Avian Pathol 32, 435–461 DOI: 10.1080/0307945031000159361З.

Kalmar, I., Berndt, A., Yin, L., Chiers, K., Sachse, K., and Vanrompay, D. (2015). Host–pathogen interactions in specific pathogen-free chickens following aerogenous infection with *Chlamydia psittaci* and *Chlamydia abortus*. Veterinary immunology and immunopathology *164*, 30–39 DOI: 10.1016/j.vetimm.2014.12.014.

Kaltenboeck, B., Kousoulas, K.G., and Storz, J. (1992). Two-step polymerase chain reactions and restriction endonuclease analyses detect and differentiate ompA DNA of *Chlamydia* spp. J Clin Microbiol *30*, 1098–1104, https://www.ncbi.nlm.nih.gov/pubmed/1349899.

Kauffold, J., Melzer, F., Berndt, A., Hoffmann, G., Hotzel, H., and Sachse, K. (2006). Chlamydiae in oviducts and uteri of repeat breeder pigs. Theriogenology *66*, 1816–1823., 16837032.

Klos, A., Wende, E., Wareham, K.J., and Monk, P.N. (2013). International Union of Basic and Clinical Pharmacology. [corrected]. LXXXVII. Complement peptide C5a, C4a, and C3a receptors. Pharmacological reviews *65*, 500–543, http://www.ncbi.nlm.nih.gov/pubmed/23383423.

Knittler, M.R., Berndt, A., Bocker, S., Dutow, P., Hanel, F., Heuer, D., Kagebein, D., Klos, A., Koch, S., Liebler-Tenorio, E., *et al.* (2014). *Chlamydia psittaci:* New insights into genomic diversity, clinical pathology, host–pathogen interaction and anti-bacterial immunity. Int J Med Microbiol *304*, 877–893 DOI: 10.1016/j.ijmm.2014.06.010.

Knittler, M.R., and Sachse, K. (2014). *Chlamydia psittaci*: update on an underestimated zoonotic agent (Minireview). FEMS Pathogens and Disease 73, doi: 10.1093/femspd/ftu1007.

Knittler, M.R., and Sachse, K. (2015). *Chlamydia psittaci:* update on an underestimated zoonotic agent. Pathog Dis 73, 1–15 DOI: 10.1093/femspd/ftu007.

Koroleva, E.A., Kobets, N.V., Shcherbinin, D.N., Zigangirova, N.A., Shmarov, M.M., Tukhvatulin, A.I., Logunov, D.Y., Naroditsky, B.S., and Gintsburg, A.L. (2017). Chlamydial type III secretion system needle protein induces protective immunity against *Chlamydia muridarum* intravaginal infection. Biomed Res Int *2017*, 3865802 DOI: 10.1155/2017/3865802.

Krautwald-Junghanns, M.E., Stolze, J., Schmidt, V., Böhme, J., Sachse, K., and Cramer, K. (2013). [Efficacy of doxycyclin for chlamydioses in pigeon flocks] in German. Tierärztl Praxis *41*, 392–398.

Lacy, H.M., Bowlin, A.K., Hennings, L., Scurlock, A.M., Nagarajan, U.M., and Rank, R.G. (2011). Essential role for neutrophils in pathogenesis and adaptive immunity in *Chlamydia caviae* ocular infections. Infect Immun *79*, 1889–1897 DOI: 10.1128/IAI.01257-10.

Lagae, S., Kalmar, I., Laroucau, K., Vorimore, F., and Vanrompay, D. (2014). Emerging *Chlamydia psittaci* infections in chickens and examination of transmission to humans. J Med Microbiol *63*, 399–407 DOI: 10.1099/jmm.0.064675-0.

Laroucau, K., Aaziz, R., Meurice, L., Servas, V., Chossat, I., Royer, H., de Barbeyrac, B., Vaillant, V., Moyen, J.L., Meziani, F., *et al.* (2015). Outbreak of psittacosis in a group of women exposed to *Chlamydia psittaci*-infected chickens. Euro Surveill *20*, pii: 21155, http://www.ncbi.nlm.nih.gov/pubmed/26111240.

Laroucau, K., de Barbeyrac, B., Vorimore, F., Clerc, M., Bertin, C., Harkinezhad, T., Verminnen, K., Obeniche, F., Capek, I., Bebear, C., *et al.* (2009a). Chlamydial infections in duck farms associated with human cases of psittacosis in France. Vet Microbiol *135*, 82–89 DOI: 10.1016/j.vetmic.2008.09.048.

Laroucau, K., Di Francesco, A., Vorimore, F., Thierry, S., Pingret, J.L., Bertin, C., Willems, H., Bolske, G., and Harley, R. (2012). Multilocus variable-number tandem-repeat analysis scheme for *Chlamydia felis* genotyping: comparison with multilocus sequence typing. J Clin Microbiol *50*, 1860–1866 DOI: 10.1128/JCM.00417-12.

Laroucau, K., Vorimore, F., Aaziz, R., Berndt, A., Schubert, E., and Sachse, K. (2009b). Isolation of a new chlamydial agent from infected domestic poultry coincided with cases of atypical pneumonia among slaughterhouse workers in France. Infect Genet Evol *9*, 1240–1247.

Lenart, J., Andersen, A.A., and Rockey, D.D. (2001). Growth and development of tetracycline-resistant *Chlamydia suis*. Antimicrob Agents Chemother *45*, 2198–2203 DOI: 10.1128/AAC.45.8.2198-2203.2001.

Lenzko, H., Moog, U., Henning, K., Lederbach, R., Diller, R., Menge, C., Sachse, K., and Sprague, L.D. (2011). High frequency of chlamydial co-infections in clinically healthy sheep flocks. BMC Vet Res *7*, 29 DOI: 10.1186/1746-6148-7-29.

Li, J., Guo, W., Kaltenboeck, B., Sachse, K., Yang, Y., Lu, G., Zhang, J., Luan, L., You, J., Huang, K., *et al.* (2016). *Chlamydia pecorum* is the endemic intestinal species in cattle while *C. gallinacea*, *C. psittaci* and *C. pneumoniae* associate with sporadic systemic infection. Vet Microbiol *193*, 93–99 DOI: 10.1016/j.vetmic.2016.08.008.

Li, L., Luther, M., Macklin, K., Pugh, D., Li, J., Zhang, J., Roberts, J., Kaltenboeck, B., and Wang, C. (2017a). *Chlamydia gallinacea:* a widespread emerging *Chlamydia* agent with zoonotic potential in backyard poultry. Epidemiol Infect *145*, 2701–2703 DOI: 10.1017/S0950268817001650.

Li, M., Jelocnik, M., Yang, F., Gong, J., Kaltenboeck, B., Polkinghorne, A., Feng, Z., Pannekoek, Y., Borel, N., Song, C., *et al.* (2017b). Asymptomatic infections with highly polymorphic *Chlamydia suis* are ubiquitous in pigs. BMC Vet Res *13*, 370 DOI: 10.1186/s12917-017-1295-x.

Liu, Z., Rank, R., Kaltenboeck, B., Magnino, S., Dean, D., Burall, L., Plaut, R.D., Read, T.D., Myers, G., and Bavoil, P.M. (2007). Genomic plasticity of the rrn-nqrF intergenic segment in the *Chlamydiaceae*. J Bacteriol *189*, 2128–2132 DOI: 10.1128/JB.00378-06.

Livingstone, M., Wheelhouse, N., Ensor, H., Rocchi, M., Maley, S., Aitchison, K., Wattegedera, S., Wilson, K., Sait, M., Siarkou, V., *et al.* (2017). Pathogenic outcome following experimental infection of sheep with *Chlamydia abortus* variant strains LLG and POS. PLoS One *12*, e0177653 DOI: 10.1371/journal.pone.0177653.

Longbottom, D., and Coulter, L.J. (2003). Animal chlamydioses and zoonotic implications. J Comp Pathol *128*, 217–244, https://www.ncbi.nlm.nih.gov/pubmed/12834606.

Longbottom, D., Entrican, G., Wheelhouse, N., Brough, H., and Milne, C. (2013a). Evaluation of the impact and control of enzootic abortion of ewes. Vet J *195*, 257–259 DOI: 10.1016/j.tvjl.2012.06.018.

Longbottom, D., and Livingstone, M. (2006). Vaccination against chlamydial infections of man and animals. Vet J *171*, 263–275, 16490708.

Longbottom, D., Livingstone, M., Maley, S., van der Zon, A., Rocchi, M., Wilson, K., Wheelhouse, N., Dagleish, M., Aitchison, K., Wattegedera, S., *et al.* (2013b). Intranasal infection with *Chlamydia abortus* induces dose-dependent latency and abortion in sheep. PLoS One *8*, e57950 DOI: 10.1371/journal.pone.0057950.

Longbottom, D., Sait, M., Livingstone, M., Laroucau, K., Sachse, K., Harris, S.R., Thomson, N.R., and Seth-Smith, H.M.B. (2018). Genomic evidence that the live *Chlamydia abortus* vaccine strain 1B is not attenuated and has the potential to cause disease. Vaccine *36*, 3593–3598 DOI: 10.1016/j.vaccine.2018.05.042.

Lutz-Wohlgroth, L., Becker, A., Brugnera, E., Huat, Z.L., Zimmermann, D., Grimm, F., Haessig, M., Greub, G., Kaps, S., Spiess, B., *et al.* (2006). Chlamydiales in guinea-pigs and their zoonotic potential. J Vet Med A Physiol Pathol Clin Med *53*, 185–193, 16629952.

Magnino, S., Haag-Wackernagel, D., Geigenfeind, I., Helmecke, S., Dovc, A., Prukner-Radovcic, E., Residbegovic, E., Ilieski, V., Laroucau, K., Donati, M., *et al.* (2009). Chlamydial infections in feral pigeons in Europe: Review of data and focus on public health implications. Vet Microbiol *135*, 54–67 DOI: 10.1016/j.vetmic.2008.09.045.

Marti, H., Kim, H., Joseph, S.J., Dojiri, S., Read, T.D., and Dean, D. (2017). Tet(C) gene transfer between *Chlamydia suis* strains occurs by homologous recombination after co-infection: Implications for spread of tetracycline-resistance among *Chlamydiaceae*. Front Microbiol *8*, 156 DOI: 10.3389/fmicb.2017.00156.

Matsumoto, A., Bessho, H., Uehira, K., and Suda, T. (1991). Morphological studies of the association of mitochondria with chlamydial inclusions and the fusion of chlamydial inclusions. J Electron Microsc (Tokyo) *40*, 356–363, http://www.ncbi.nlm.nih.gov/pubmed/1666645.

McManus, C.M., Levy, J.K., Andersen, L.A., McGorray, S.P., Leutenegger, C.M., Gray, L.K., Hilligas, J., and Tucker, S.J. (2014). Prevalence of upper respiratory pathogens in four management models for unowned cats in the Southeast United States. Vet J *201*, 196–201 DOI: 10.1016/j.tvjl.2014.05.015.

McNutt, S.H., and Waller, E.F. (1940). Sporadic bovine encephalomyelitis (Buss disease). Cornell Veterinarian *30*, 437–448.

Mendlowski, B., and Segre, D. (1960). Polyarthritis in sheep. I. Description of the disease and experimental transmission. American Journal of Veterinary Research *21*, 68–73.

Mital, J., Miller, N.J., Dorward, D.W., Dooley, C.A., and Hackstadt, T. (2013). Role for chlamydial inclusion membrane proteins in inclusion membrane structure and biogenesis. PLoS One *8*, e63426 DOI: 10.1371/journal.pone.0063426.

Mitchell, C.M., Hutton, S., Myers, G.S., Brunham, R., and Timms, P. (2010). *Chlamydia pneumoniae* is genetically diverse in animals and appears to have crossed the host barrier to humans on (at least) two occasions. PLoS Pathog *6*, e1000903 DOI: 10.1371/journal.ppat.1000903.

Mitchell, C.M., Mathews, S.A., Theodoropoulos, C., and Timms, P. (2009). In vitro characterisation of koala *Chlamydia pneumoniae:* morphology, inclusion development and doubling time. Vet Microbiol *136*, 91–99 DOI: 10.1016/j.vetmic.2008.10.008.

Mohamad, K.Y., Kaltenboeck, B., Rahman Kh, S., Magnino, S., Sachse, K., and Rodolakis, A. (2014). Host adaptation of *Chlamydia pecorum* towards low virulence evident in co-evolution of the ompA, incA, and ORF663 loci. PLoS One *9*, e103615 DOI: 10.1371/journal.pone.0103615.

Mohamad, K.Y., and Rodolakis, A. (2010). Recent advances in the understanding of *Chlamydophila pecorum* infections, sixteen years after it was named as the fourth species of the *Chlamydiaceae* family. Vet Res *41*, 27 DOI: 10.1051/vetres/2009075.

Muller, T., Becker, E., Stallmann, S., Waldhuber, A., Rommler-Dreher, F., Albrecht, S., Mohr, F., Hegemann, J.H., and Miethke, T. (2017). Vaccination with the polymorphic membrane protein A reduces *Chlamydia muridarum* induced genital tract pathology. Vaccine *35*, 2801–2810 DOI: 10.1016/j.vaccine.2017.04.017.

Murray, E.S. (1964). Guinea pig inclusion conjunctivitis virus. I. Isolation and identification as a member of the psittacosis-lymphogranuloma-trachoma group. J Infect Dis *114*, 1–12, https://www.ncbi.nlm.nih.gov/pubmed/14118043.

Myers, G.S., Mathews, S.A., Eppinger, M., Mitchell, C., O'Brien, K.K., White, O.R., Benahmed, F., Brunham, R.C., Read, T.D., Ravel, J., *et al.* (2009). Evidence that human *Chlamydia pneumoniae* was zoonotically acquired. J Bacteriol *191*, 7225–7233 DOI: 10.1128/JB.00746-09.

JB.00746-09 [pii].

Nigg, C. (1942). An unidentified virus which produces pneumonia and systemic infection in mice. Science *95*, 49–50 DOI: 10.1126/science.95.2454.49-a.

Ohtani, A., Kubo, M., Shimoda, H., Ohya, K., Iribe, T., Ohishi, D., Endoh, D., Omatsu, T., Mizutani, T., Fukushi, H., *et al.* (2015). Genetic and antigenic analysis of *Chlamydia pecorum* strains isolated from calves with diarrhea. J Vet Med Sci *77*, 777–782 DOI: 10.1292/jvms.14-0585.

OIE (2018). Enzootic Abortion of Ewes (ovine chlamydiosis). World Organisation for Animal Health: Manual of Diagnostic Tests and Vaccines for Terrestrial Animals *Chapter 2.7.6.*, http://www.oie.int/fileadmin/Home/eng/Health_standards/tahm/2.07.06_ENZ_ABOR.pdf.

Ortega, N., Caro, M.R., Gallego, M.C., Murcia-Belmonte, A., Alvarez, D., Del Rio, L., Cuello, F., Buendia, A.J., and Salinas, J. (2015). Isolation of *Chlamydia abortus* from a laboratory worker diagnosed with atypical pneumonia. Ir Vet J *69*, 8 DOI: 10.1186/s13620-016-0067-4.

Ostermann, C., Linde, S., Siegling-Vlitakis, C., and Reinhold, P. (2014). Evaluation of pulmonary dysfunctions and acid-base imbalances induced by *Chlamydia psittaci* in a bovine model of respiratory infection. Multidiscip Respir Med *9*, 10 DOI: 10.1186/2049-6958-9-10.

Ostermann, C., Ruttger, A., Schubert, E., Schrodl, W., Sachse, K., and Reinhold, P. (2013a). Infection, disease, and transmission dynamics in calves after experimental and natural challenge with a Bovine Chlamydia psittaci isolate. PLoS One *8*, e64066 DOI: 10.1371/journal.pone.0064066.

Ostermann, C., Schroedl, W., Schubert, E., Sachse, K., and Reinhold, P. (2013b). Dose-dependent effects of Chlamydia psittaci infection on pulmonary gas exchange, innate immunity and acute-phase reaction in a bovine respiratory model. Vet J *196*, 351–359 DOI: 10.1016/j.tvjl.2012.10.035.

Pal, S., Favaroni, A., Tifrea, D.F., Hanisch, P.T., Luczak, S.E.T., Hegemann, J.H., and de la Maza, L.M. (2017). Comparison of the nine polymorphic membrane proteins of *Chlamydia trachomatis* for their ability to induce protective immune responses in mice against a *C. muridarum* challenge. Vaccine *35*, 2543–2549 DOI: 10.1016/j.vaccine.2017.03.070.

Pal, S., Tatarenkova, O.V., and de la Maza, L.M. (2015). A vaccine formulated with the major outer membrane protein can protect C3H/HeN, a highly susceptible strain of mice, from a *Chlamydia muridarum* genital challenge. Immunology *146*, 432–443 DOI: 10.1111/imm.12520.

Pantchev, A., Sting, R., Bauerfeind, R., Tyczka, J., and Sachse, K. (2010). Detection of all *Chlamydophila* and *Chlamydia* spp. of veterinary interest using species-specific real-time PCR assays. Comp Immunol Microbiol Infect Dis *33*, 473–484 DOI: 10.1016/j.cimid.2009.08.002.

Perez-Martinez, J.A., and Storz, J. (1985). Antigenic diversity of *Chlamydia psittaci* of mammalian origin determined by microimmunofluorescence. Infect Immun *50*, 905–910, https://www.ncbi.nlm.nih.gov/pubmed/3905618.

Pillonel, T., Bertelli, C., Salamin, N., and Greub, G. (2015). Taxogenomics of the order *Chlamydiales*. Int J Syst Evol Microbiol *65*, 1381–1393 DOI: 10.1099/ijs.0.000090.

Polkinghorne, A., Borel, N., Becker, A., Lu, Z.H., Zimmermann, D.R., Brugnera, E., Pospischil, A., and Vaughan, L. (2009). Molecular evidence for chlamydial infections in the eyes of sheep. Vet Microbiol *135*, 142–146 DOI: 10.1016/j.vetmic.2008.09.034.

Polkinghorne, A., and Greub, G. (2017). A new equine and zoonotic threat emerges from an old avian pathogen, *Chlamydia psittaci*. Clin Microbiol Infect *23*, 693–694 DOI: 10.1016/j.cmi.2017.05.025.

Polkinghorne, A., Hanger, J., and Timms, P. (2013). Recent advances in understanding the biology, epidemiology and control of chlamydial infections in koalas. Vet Microbiol *165*, 214–223 DOI: 10.1016/j.vetmic.2013.02.026.

Pospischil, A., Thoma, R., Hilbe, M., Grest, P., and Gebbers, J.O. (2002). Abortion in woman caused by caprine *Chlamydophila abortus* (*Chlamydia psittaci* serovar 1). In Swiss Med Wkly, pp. 64–66.

Poudel, A., Elsasser, T.H., Rahman Kh, S., Chowdhury, E.U., and Kaltenboeck, B. (2012). Asymptomatic endemic *Chlamydia pecorum* infections reduce growth rates in calves by up to 48%. PLoS One 7, e44961 DOI: 10.1371/journal.pone.0044961.

Ramakers, B.P., Heijne, M., Lie, N., Le, T.N., van Vliet, M., Claessen, V.P.J., Tolsma, P.J.P., De Rosa, M., Roest, H.I.J., Vanrompay, D., *et al.* (2017). Zoonotic *Chlamydia caviae* presenting as community-acquired pneumonia. N Engl J Med *377*, 992–994 DOI: 10.1056/NEJMc1702983.

Ramsey, K.H., Sigar, I.M., Schripsema, J.H., Townsend, K.E., Barry, R.J., Peters, J., and Platt, K.B. (2016). Detection of *Chlamydia* infection in *Peromyscus* species rodents from sylvatic and laboratory sources. Pathog Dis *74* DOI: 10.1093/femspd/ftv129.

Read, T.D., Joseph, S.J., Didelot, X., Liang, B., Patel, L., and Dean, D. (2013). Comparative analysis of Chlamydia psittaci genomes reveals the recent emergence of a pathogenic lineage with a broad host range. MBio *4* DOI: 10.1128/mBio.00604-12.

Rehn, M., Ringberg, H., Runehagen, A., Herrmann, B., Olsen, B., Petersson, A.C., Hjertqvist, M., Kuhlmann-Berenzon, S., and Wallensten, A. (2013). Unusual increase of psittacosis in southern Sweden linked to wild bird exposure, January to April 2013. Euro Surveill *18*, 20478, http://www.ncbi.nlm.nih.gov/pubmed/23725809.

Reinhold, P., Jaeger, J., Liebler-Tenorio, E., Berndt, A., Bachmann, R., Schubert, E., Melzer, F., Elschner, M., and Sachse, K. (2008). Impact of latent infections with *Chlamydophila* species in young cattle. Vet J *175*, 202–211 DOI: 10.1016/j.tvjl.2007.01.004.

Reinhold, P., Ostermann, C., Liebler-Tenorio, E., Berndt, A., Vogel, A., Lambertz, J., Rothe, M., Ruttger, A., Schubert, E., and Sachse, K. (2012). A bovine model of respiratory *Chlamydia psittaci* infection: challenge dose titration. PLoS One *7*, e30125 DOI: 10.1371/journal.pone.0030125.

Reinhold, P., Sachse, K., and Kaltenboeck, B. (2011). *Chlamydiaceae* in cattle: commensals, trigger organisms, or pathogens? Vet J *189*, 257–267 DOI: 10.1016/j.tvjl.2010.09.003 S1090-0233(10)00294-7 [pii].

Risco, D., Garcia, A., Fernandez-Llario, P., Garcia, W.L., Benitez, J.M., Goncalves, P., Cuesta, J.M., Gomez, L., Rey, J., and Hermoso de Mendoza, J. (2013). Ocular lesions associated with *Chlamydia suis* in a wild boar piglet (*Sus scrofa*) from a semi-free range population in Spain. J Zoo Wildl Med *44*, 159–162 DOI: 10.1638/1042-7260-44.1.159.

Rodolakis, A., and Laroucau, K. (2015). *Chlamydiaceae* and chlamydial infections in sheep or goats. Vet Microbiol *181*, 107–118 DOI: 10.1016/j.vetmic.2015.07.010.

Rodolakis, A., Salinas, J., and Papp, J. (1998). Recent advances on ovine chlamydial abortion. Vet Res *29*, 275–288, https://www.ncbi.nlm.nih.gov/pubmed/9689742.

Rogers, D.G., and Andersen, A.A. (2000). Intestinal lesions caused by a strain of *Chlamydia suis* in weanling pigs infected at 21 days of age. J Vet Diagn Invest *12*, 233–239 DOI: 10.1177/104063870001200306.

Rojas, M.D.C., Fort, M., Bettermann, S., Entrocassi, C., Costamagna, S.R., Sachse, K., and Rodriguez Fermepin, M. (2018). [Detection of Chlamydia abortus in bovine reproductive losses in the province of La Pampa, Argentina]. Rev Argent Microbiol DOI: 10.1016/j.ram.2017.10.002.

Roulis, E., Polkinghorne, A., and Timms, P. (2013). *Chlamydia pneumoniae:* modern insights into an ancient pathogen. Trends Microbiol *21*, 120–128 DOI: 10.1016/j.tim.2012.10.009.

Sachse, K., Grossmann, E., Berndt, A., Schutt, C., Henning, K., Theegarten, D., Anhenn, O., and Reinhold, P. (2004).

Respiratory chlamydial infection based on experimental aerosol challenge of pigs with *Chlamydia suis*. Comp Immunol Microbiol Infect Dis *27*, 7–23 DOI: 10.1016/S0147-9571(02)00079-6.

Sachse, K., and Laroucau, K. (2015). Two more species of *Chlamydia* – does it make a difference? Pathog Dis 73, 1–3 DOI: 10.1093/femspd/ftu008.

Sachse, K., Laroucau, K., Riege, K., Wehner, S., Dilcher, M., Creasy, H.H., Weidmann, M., Myers, G., Vorimore, F., Vicari, N., *et al.* (2014). Evidence for the existence of two new members of the family *Chlamydiaceae* and proposal of *Chlamydia avium* sp. nov. and *Chlamydia gallinacea* sp. nov. Syst Appl Microbiol 37, 79–88 DOI: 10.1016/j.syapm.2013.12.004.

Sachse, K., Laroucau, K., and Vanrompay, D. (2015). Avian Chlamydiosis. Curr Clin Microbiol Reports 2, 10–21 DOI: 10.1007/s40588-014-0010-y.

Sachse, K., and Ruettger, A. (2015). Rapid microarray-based genotyping of *Chlamydia* spp. strains from clinical tissue samples. Methods Mol Biol *1247*, 391–400 DOI: 10.1007/978-1-4939-2004-4_28.

Sachse, K., Vretou, E., Livingstone, M., Borel, N., Pospischil, A., and Longbottom, D. (2009). Recent developments in the laboratory diagnosis of chlamydial infections. Vet Microbiol *135*, 2–21, 18986778.

Sait, M., Livingstone, M., Clark, E.M., Wheelhouse, N., Spalding, L., Markey, B., Magnino, S., Lainson, F.A., Myers, G.S., and Longbottom, D. (2014). Genome sequencing and comparative analysis of three *Chlamydia pecorum* strains associated with different pathogenic outcomes. BMC Genomics *15*, 23 DOI: 10.1186/1471-2164-15-23.

Salinas, J., Ortega, N., Borge, C., Rangel, M.J., Carbonero, A., Perea, A., and Caro, M.R. (2012). Abortion associated with *Chlamydia abortus* in extensively reared Iberian sows. Vet J *194*, 133–134 DOI: 10.1016/j.tvjl.2012.03.002.

Sammin, D., Markey, B., Bassett, H., and Buxton, D. (2009). The ovine placenta and placentitis-A review. Vet Microbiol *135*, 90–97 DOI: 10.1016/j.vetmic.2008.09.054.

Sandoz, K.M., and Rockey, D.D. (2010). Antibiotic resistance in Chlamydiae. Future Microbiol *5*, 1427–1442 DOI: 10.2217/fmb.10.96.

Sargison, N.D., Truyers, I.G., Howie, F.E., Thomson, J.R., Cox, A.L., Livingstone, M., and Longbottom, D. (2015). Identification of the 1B vaccine strain of *Chlamydia abortus* in aborted placentas during the investigation of toxaemic and systemic disease in sheep. N Z Vet J *63*, 284–287 DOI: 10.1080/00480169.2015.1018365.

Schachter, J., Ostler, H.B., and Meyer, K.F. (1969). Human infection with the agent of feline pneumonitis. Lancet *1*, 1063–1065, https://www.ncbi.nlm.nih.gov/pubmed/4181729.

Schautteet, K., De Clercq, E., Miry, C., Van Groenweghe, F., Delava, P., Kalmar, I., and Vanrompay, D. (2013). Tetracycline-resistant *Chlamydia suis* in cases of reproductive failure on Belgian, Cypriote and Israeli pig production farms. J Med Microbiol *62*, 331–334 DOI: 10.1099/jmm.0.042861-0.

Schautteet, K., and Vanrompay, D. (2011). *Chlamydiaceae* infections in pig. Vet Res *42*, 29 DOI: 10.1186/1297-9716-42-29.

Schulz, C., Hartmann, K., Mueller, R.S., Helps, C., and Schulz, B.S. (2015). Sampling sites for detection of feline herpesvirus-1, feline calicivirus and *Chlamydia felis* in cats with feline upper respiratory tract disease. J Feline Med Surg *17*, 1012–1019 DOI: 10.1177/1098612X15569615.

Schwarzenbacher, R., Stenner-Liewen, F., Liewen, H., Robinson, H., Yuan, H., Bossy-Wetzel, E., Reed, J.C., and Liddington, R.C. (2004). Structure of the Chlamydia protein CADD reveals a redox enzyme that modulates host cell apoptosis. J Biol Chem *279*, 29320–29324, 15087448.

Seth-Smith, H.M., Wanninger, S., Bachmann, N., Marti, H., Qi, W., Donati, M., di Francesco, A., Polkinghorne, A., and Borel, N. (2017a). The *Chlamydia suis* genome exhibits high levels of diversity, plasticity, and mobile antibiotic resistance: Comparative genomics of a recent livestock cohort shows influence of treatment regimes. Genome Biol Evol *9*, 750–760 DOI: 10.1093/gbe/evx043.

Seth-Smith, H.M.B., Buso, L.S., Livingstone, M., Sait, M., Harris, S.R., Aitchison, K.D., Vretou, E., Siarkou, V.I., Laroucau, K., Sachse, K., *et al.* (2017b). European *Chlamydia abortus* livestock isolate genomes reveal unusual stability and limited diversity, reflected in geographical signatures. BMC Genomics *18*, 344 DOI: 10.1186/s12864-017-3657-y.

Staub, E., Marti, H., Biondi, R., Levi, A., Donati, M., Leonard, C.A., Ley, S.D., Pillonel, T., Greub, G., Seth-Smith, H.M.B., *et al.* (2018). Novel *Chlamydia* species isolated from snakes are temperature-sensitive and exhibit decreased susceptibility to azithromycin. Sci Rep *8*, 5660 DOI: 10.1038/s41598-018-23897-z.

Stills, H.F., Jr., Fox, J.G., Paster, B.J., and Dewhirst, F.E. (1991). A 'new' *Chlamydia* sp. strain SFPD isolated from transmissible proliferative ileitis in hamsters. Microbiol Ecol Health Dis *4*, S99.

Storey, C., Lusher, M., Yates, P., and Richmond, S. (1993). Evidence for *Chlamydia pneumoniae* of non-human origin. J Gen Microbiol *139*, 2621–2626 DOI: 10.1099/00221287-139-11-2621.

Storz, J., and Kaltenboeck, B. (1993). Diversity of chlamydia-induced diseases. In Rickettsial and Chlamydial Diseases of Domestic Animals – Ed by Zerai Woldehiwet and Miodrag Ristic – 1st ed Pergamon Press: Oxford, New York, Seoul, Tokyo.

In: Rickettsial and Chlamydial Diseases of Domestic Animals Zerai Woldehiwet and Miodrag Ristic (Eds), Pergamon Press, Oxford, UK, pp. pp. 363–393.

Storz, J., Shupe, J.L., Smart, R.A., and Thornley, R.W. (1966a). Polyarthritis of calves: Experimental induction by a psittacosis agent. American Journal of Veterinary Research 27, 987–995.

Storz, J., Smart, R.A., Marriott, M.E., and Davis, R.V. (1966b). Polyarthritis of calves: Isolation of psittacosis agents from affected joints. American Journal of Veterinary Research 27, 633–641.

Studdert, M.J., Studdert, V.P., and Wirth, H.J. (1981). Isolation of *Chlamydia psittaci* from cats with conjunctivitis. Aust Vet J *57*, 515–517, https://www.ncbi.nlm.nih.gov/pubmed/7342937.

Suchland, R.J., Sandoz, K.M., Jeffrey, B.M., Stamm, W.E., and Rockey, D.D. (2009). Horizontal transfer of tetracycline resistance among *Chlamydia* spp. *in vitro*. Antimicrob Agents Chemother *53*, 4604–4611 DOI: 10.1128/AAC.00477-09.

Sykes, J.E. (2005). Feline chlamydiosis. Clin Tech Small Anim Pract *20*, 129–134 DOI: 10.1053/j.ctsap.2004.12.018.

Szymanska-Czerwinska, M., Mitura, A., Niemczuk, K., Zareba, K., Jodelko, A., Pluta, A., Scharf, S., Vitek, B., Aaziz, R., Vorimore, F., *et al.* (2017). Dissemination and genetic diversity of chlamydial agents in Polish wildfowl: Isolation and molecular characterisation of avian *Chlamydia abortus* strains. PLoS One *12*, e0174599 DOI: 10.1371/journal.pone.0174599.

Taylor-Brown, A., Bachmann, N.L., Borel, N., and Polkinghorne, A. (2016). Culture-independent genomic characterisation of *Candidatus* Chlamydia sanzinia, a novel uncultivated bacterium infecting snakes. BMC Genomics *17*, 710 DOI: 10.1186/s12864-016-3055-x.

Taylor-Brown, A., and Polkinghorne, A. (2017). New and emerging chlamydial infections of creatures great and small. New Microbes New Infect *18*, 28–33 DOI: 10.1016/j.nmni.2017.04.004.

Taylor-Brown, A., Ruegg, S., Polkinghorne, A., and Borel, N. (2015). Characterisation of *Chlamydia pneumoniae* and other novel chlamydial infections in captive snakes. Vet Microbiol *178*, 88–93 DOI: 10.1016/j.vetmic.2015.04.021.

Taylor-Brown, A., Spang, L., Borel, N., and Polkinghorne, A. (2017). Culture-independent metagenomics supports discovery of uncultivable bacteria within the genus *Chlamydia*. Sci Rep *7*, 10661 DOI: 10.1038/s41598-017-10757-5.

Teankum, K., Pospischil, A., Janett, F., Brugnera, E., Hoelzle, L.E., Hoelzle, K., Weilenmann, R., Zimmermann, D.R., Gerber, A., Polkinghorne, A., *et al.* (2007). Prevalence of chlamydiae in semen and genital tracts of bulls, rams and bucks. Theriogenology *67*, 303–310 DOI: 10.1016/j.theriogenology.2006.07.012.

Thomson, N.R., Yeats, C., Bell, K., Holden, M.T., Bentley, S.D., Livingstone, M., Cerdeno-Tarraga, A.M., Harris, B., Doggett, J., Ormond, D., *et al.* (2005). The *Chlamydophila abortus* genome sequence reveals an array of variable proteins that contribute to interspecies variation. Genome Res *15*, 629–640 DOI: 10.1101/gr.3684805.

Tian, B., Fadhil, N.H., Powell, J.E., Kwong, W.K., and Moran, N.A. (2012). Long-term exposure to antibiotics has caused accumulation of resistance determinants in the gut microbiota of honeybees. MBio *3* DOI: 10.1128/mBio.00377-12.

Van Droogenbroeck, C., Beeckman, D.S., Verminnen, K., Marien, M., Nauwynck, H., Boesinghe Lde, T., and Vanrompay, D. (2009). Simultaneous zoonotic transmission of *Chlamydophila psittaci* genotypes D, F and E/B to a veterinary scientist. Vet Microbiol *135*, 78–81 DOI: 10.1016/j.vetmic.2008.09.047.

van Grootveld, R., Bilsen, M.P., Boelsums, T.L., Heddema, E.R., Groeneveld, G.H., Gooskens, J., and de Boer, M.G.J. (2018). *Chlamydia caviae* causing community-acquired pneumonia: An emerging zoonosis. Vector Borne Zoonotic Dis DOI: 10.1089/vbz.2018.2304.

Van Loock, M., Geens, T., De Smit, L., Nauwynck, H., Van Empel, P., Naylor, C., Hafez, H.M., Goddeeris, B.M., and Vanrompay, D. (2005). Key role of Chlamydophila psittaci on Belgian turkey farms in association with other respiratory pathogens. Vet Microbiol *107*, 91–101 DOI: 10.1016/j.vetmic.2005.01.009.

Vanrompay, D. (2013). Avian Chlamydiosis. In: Diseases of Poultry, 13th edition Vol Section III: Bacterial Diseases (Wiley- Blackwell Publishing).

Vanrompay, D., Nguyen, T.L.A., Cutler, S.J., and Butaye, P. (2018). Antimicrobial resistance in *Chlamydiales, Rickettsia, Coxiella,* and other intracellular pathogens. Microbiol Spectr *6* DOI: 10.1128/microbiolspec.ARBA-0003-2017.

Voigt, A., Schofl, G., and Saluz, H.P. (2012). The Chlamydia psittaci genome: a comparative analysis of intracellular pathogens. PLoS One 7, e35097 DOI: 10.1371/journal.pone.0035097.

Vorimore, F., Hsia, R.C., Huot-Creasy, H., Bastian, S., Deruyter, L., Passet, A., Sachse, K., Bavoil, P., Myers, G., and Laroucau, K. (2013). Isolation of a new *Chlamydia* species from the Feral Sacred Ibis (*Threskiornis aethiopicus*): *Chlamydia ibidis*. PLoS One *8*, e74823 DOI: 10.1371/journal.pone.0074823.

Walder, G., Hotzel, H., Brezinka, C., Gritsch, W., Tauber, R., Wurzner, R., and Ploner, F. (2005). An unusual cause of sepsis during pregnancy: recognizing infection with *Chlamydophila abortus*. Obstet Gynecol *106*, 1215–1217 DOI: 10.1097/01.AOG.0000161060.69470.9c.

Walker, E., Jelocnik, M., Bommana, S., Timms, P., Carver, S., and Polkinghorne, A. (2018). Understanding the health and production impacts of endemic *Chlamydia pecorum* infections in lambs. Vet Microbiol *217*, 90–96 DOI: 10.1016/j.vetmic.2018.03.009.

Walker, E., Lee, E.J., Timms, P., and Polkinghorne, A. (2015). *Chlamydia pecorum* infections in sheep and cattle: A common and under-recognised infectious disease with significant impact on animal health. Vet J *206*, 252–260 DOI: 10.1016/j.tvjl.2015.09.022.

Walker, E., Moore, C., Shearer, P., Jelocnik, M., Bommana, S., Timms, P., and Polkinghorne, A. (2016). Clinical, diagnostic and pathologic features of presumptive cases of *Chlamydia pecorum*-associated arthritis in Australian sheep flocks. BMC Vet Res *12*, 193 DOI: 10.1186/s12917-016-0832-3.

Wallensten, A., Fredlund, H., and Runehagen, A. (2014). Multiple human-to-human transmission from a severe case of psittacosis, Sweden, January-February 2013. Euro Surveill *19*, http://www.ncbi.nlm.nih.gov/pubmed/25358043.

Wanninger, S., Donati, M., Di Francesco, A., Hassig, M., Hoffmann, K., Seth-Smith, H.M., Marti, H., and Borel, N. (2016). Selective pressure promotes tetracycline resistance of *Chlamydia suis* in fattening pigs. PLoS One *11*, e0166917 DOI: 10.1371/journal.pone.0166917.

Wardrop, S., Fowler, A., O'Callaghan, P., Giffard, P., and Timms, P. (1999). Characterization of the koala biovar of *Chlamydia pneumoniae* at four gene loci – ompAVD4 ompB, 16S rRNA, groESL spacer region. Systematic and Applied Microbiology *22*, 22–27.

Warren, K., Swan, R., Bodetti, T., Friend, T., Hill, S., and Timms, P. (2005). Ocular chlamydiales infections of western barred bandicoots (Perameles bougainville) in Western Australia. J Zoo Wildl Med *36*, 100–102 DOI: 10.1638/02-067.

Wheelhouse, N., Aitchison, K., Laroucau, K., Thomson, J., and Longbottom, D. (2010). Evidence of *Chlamydophila abortus* vaccine strain 1B as a possible cause of ovine enzootic abortion. Vaccine *28*, 5657–5663 DOI: 10.1016/j.vaccine.2010.04.114.

Wheelhouse, N., Wattegedera, S., Stanton, J., Maley, S., Watson, D., Jepson, C., Deane, D., Buxton, D., Longbottom, D., Baszler, T., *et al.* (2009). Ovine trophoblast is a primary source of TNFalpha during *Chlamydophila abortus* infection. J Reprod Immunol *80*, 49–56 DOI: 10.1016/j.jri.2008.12.003.

Wu, S.M., Huang, S.Y., Xu, M.J., Zhou, D.H., Song, H.Q., and Zhu, X.Q. (2013). *Chlamydia felis* exposure in companion dogs and cats in Lanzhou, China: a public health concern. BMC Vet Res *9*, 104 DOI: 10.1186/1746-6148-9-104.

Yin, L., Kalmar, I.D., Boden, J., and Vanrompay, D. (2015). *Chlamydia psittaci* infections in Chinese poultry: a literature review. World's Poultry Science Journal *71*, 473–482.

Yin, L., Lagae, S., Kalmar, I., Borel, N., Pospischil, A., and Vanrompay, D. (2013). Pathogenicity of low and highly virulent *Chlamydia psittaci* isolates for specific-pathogen-free chickens. Avian Dis *57*, 242–247, http://www.ncbi.nlm.nih.gov/pubmed/24689181.

Yousef Mohamad, K., Rekiki, A., Myers, G., Bavoil, P.M., and Rodolakis, A. (2008). Identification and characterisation of coding tandem repeat variants in incA gene of *Chlamydophila pecorum*. Vet Res *39*, 56 DOI: 10.1051/vetres:2008032.

Yu, H., Karunakaran, K.P., Jiang, X., and Brunham, R.C. (2014). Evaluation of a multisubunit recombinant polymorphic membrane protein and major outer membrane protein T cell vaccine against *Chlamydia muridarum* genital infection in three strains of mice. Vaccine *32*, 4672–4680 DOI: 10.1016/j.vaccine.2014.06.002.

Zocevic, A., Vorimore, F., Marhold, C., Horvatek, D., Wang, D., Slavec, B., Prentza, Z., Stavianis, G., Prukner-Radovcic, E., Dovc, A., *et al.* (2012). Molecular characterization of atypical Chlamydia and evidence of their dissemination in different European and Asian chicken flocks by specific real-time PCR. Environ Microbiol *14*, 2212–2222 DOI: 10.1111/j.1462-2920.2012.02800.x.

Thinking Outside the *Chlamydia* Box

18

Alyce Taylor-Brown[1], Tamara Halter[2], Adam Polkinghorne[3] and Matthias Horn[2*]

[1]The Wellcome Trust Sanger Institute, Cambridge, UK.
[2]Centre for Microbiology and Environmental Systems Science, University of Vienna, Vienna, Austria.
[3]University of the Sunshine Coast, Sippy Downs, QLD, Australia.

*Correspondence: horn@microbial-ecology.de

https://doi.org/10.21775/9781912530281.18

Abstract

Chlamydiae have long been studied exclusively in the context of disease. Yet, accumulating evidence over nearly three decades shows that chlamydiae are ubiquitous in the environment, thriving as symbionts of unicellular eukaryotes such as amoeba and infecting a broad range of animal hosts. These chlamydiae share the characteristic chlamydial developmental cycle and other chlamydial hallmarks. Their discovery fundamentally changed our perspective on chlamydial diversity. Instead of a single genus, *Chlamydia*, including closely related pathogens, the chlamydiae comprise hundreds of families and genera. Investigating isolates and non-cultured representatives provided insights into features that are in common with or divergent from known *Chlamydia* species, and suggested that some of these chlamydiae may also be considered pathogens. Importantly, these studies have contributed to a better understanding of the biology of all chlamydiae, and they provide a framework for investigating the evolution of the chlamydial intracellular lifestyle and pathogenicity.

Introduction: a plea to study environmental chlamydiae

The phylum Chlamydiae contains a diversity of obligate intracellular bacteria that share a common ancestor that is more than 700 million years old (Greub and Raoult, 2003; Horn *et al.*, 2004) (Fig. 18.1). Chlamydial species, and hence the genera and families that encompass them, are further united by their unique biphasic developmental cycle. Since the initial description of the first species in this phylum, that is *Chlamydia trachomatis* in humans (Halberstaedter and von Prowazek, 1907), nearly all clinical and scientific attention has focused on this organism and other species in the genus *Chlamydia*, belonging to a single family, the Chlamydiaceae.

The advent of molecular methods to complement existing culture-based approaches to characterize these difficult to culture intracellular bacteria quickly led to some hidden surprises. The first of these was the isolation of unusual obligate intracellular bacteria from an aborted bovine foetus that exhibited all the developmental stages of the unique chlamydial developmental cycle (Dilbeck *et al.*, 1990). This bacterium would much later be classified as *Waddlia chondrophila* (Rurangirwa *et al.*, 1999). Still in the early nineteen nineties, morphologically similar bacteria were found in free-living amoeba (Michel *et al.*, 1994) and were later identified as novel chlamydiae known today as *Parachlamydia acanthamoebae* and *Protochlamydia amoebophila*, respectively (Amann *et al.*, 1997; Fritsche *et al.*, 2000; Collingro *et al.*, 2005). In parallel, a cell culture contaminant, *Simkania negevensis*, was discovered to have phenotypic similarities to chlamydiae but to have gene sequences that were only distantly related to members of the genus *Chlamydia* (Everett *et al.*, 1999; Kahane *et al.*, 1993; Kahane *et al.*, 1995) (Fig. 18.1). While studies in

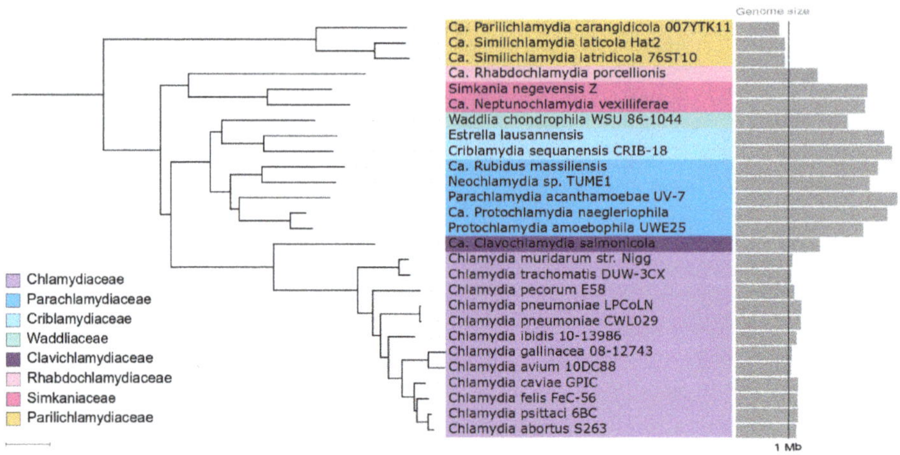

Figure 18.1 Phylogenetic diversity of the phylum Chlamydiae, based on CheckM (Price *et al*., 2009) marker protein sequences. Coloured clades indicate currently described families. The grey bars indicate the genome size. The black vertical line is a 1 Mbp size marker. The tree was constructed using FastTree (Parks *et al*., 2015), with Planctomyces and Verrucomicrobia as an outgroup. Sequences obtained from metagenome studies were included in the phylogenetic tree construction but are not shown.

humans would eventually show that *S. negevensis* is widespread, it was the environment that has proven to be the richest source for chlamydial organisms with studies of free-living amoeba quickly revealing a high diversity of new bacteria in the phylum (Horn 2008).

The detection of these bacteria led to a conundrum: how to describe this group of chlamydial families outside of what was long considered being a single-genus, single-family phylum? This challenge was first met by using ribosomal RNA gene sequence analysis to propose three new chlamydial families, the Waddliaceae, home to *W. chondrophila,* the Simkaniaceae, with its reference bacteria *S. negevensis,* and the Parachlamydiaceae, home to a growing number of chlamydial species detected in free-living amoeba, including *P. acanthamoebae* and *P. amoebophila* (Everett *et al.,* 1999). While other taxonomic changes made to this phylum would prove to be controversial, the proposal of these new families were not, and their recognition proved to be a boon to studies of chlamydial diversity. More informally, these bacterial families, genera and species have been referred to by a number of names. Some authors favour the use of '*Chlamydia*-like organisms' (CLOs) due to the historical associations with the founding genus (*Chlamydia*) in the phylum. Other authors refer to 'environmental chlamydiae' owing to the detection of many of the founding organisms in these new families from environmental sources. In reality, both terms are not ideal since a term such as 'like' is decidedly imprecise, and there is growing evidence that several of these 'environmental' chlamydiae can be detected in human and animal hosts in association with disease. While still unsatisfactory due to the *Chlamydia* genus-centric nature of the name, the term '*Chlamydia*-related bacteria' (CRB) has been suggested to describe the taxonomic diversity of bacteria outside of the family Chlamydiaceae (Tanner *et al.,* 1999). For the purpose of this review, this term, CRB, will be used to succinctly describe this diverse group of chlamydial families and the species classified within them. It should be noted that this grouping is largely for the purpose of differentiating them from *Chlamydia* species, and that there is a significant amount of biological variation between the CRB families.

So, what is the biological and clinical relevance of CRBs? The answers to this question are many and varied. From a clinical perspective, a variety of evidence provide support for a pathogenic role of many CRBs in humans and animals. Some of the strongest evidence of a pathogenic role for CRBs lies in its association with disease in fish, while for humans a pathogenic role is less convincing.

Cell biology and genomic studies of CRBs have revealed significant phenotypic and genetic diversity in the phylum Chlamydiae, surpassing anything previously described within the genus *Chlamydia*

(Taylor-Brown *et al.*, 2015). Such studies have revealed innovations in how these organisms interact with their eukaryotic hosts, an observation that is consistent with both the extended host range of these CRBs compared to their Chlamydiaceae cousins, but also with the evolutionary separation predicted to be at least 700 million years ago (Horn, 2004). While some may still question the relevance of CRBs to those interested exclusively in the genus *Chlamydia*, studies of CRBs should not be dismissed. Analyses of model CRBs are contributing to solve biological puzzles that have confounded the traditional *Chlamydia* research community for decades. Importantly, the pathogens in the genus *Chlamydia* have not just appeared at some point in animal and human history. They do originate from somewhere, namely the environment where they have evolved from a group of microbes: the CRBs. Thus, studying CRBs and the whole diversity of chlamydiae provides the context for *Chlamydia* research and can provide unprecedented insights into the evolution of the intracellular lifestyle of all chlamydiae.

In this chapter, we will provide an updated review on the taxonomic, phenotypic and genetic diversity of CRBs and their pathogenic potential. In doing so, we will also discuss the 'lessons' that can be learnt to the benefit of researchers within the traditional *Chlamydia* research field.

The expanding diversity of the phylum chlamydiae

Chlamydial taxonomy has been in a state of flux for more than 20 years, in part prompted by the discovery of new bacteria with 16S rRNA gene sequences that placed them outside of the genus *Chlamydia* (Fig. 18.1). In an attempt to improve the classification and nomenclature of this phylum, 16S rRNA gene sequence identity criteria – a convention widely used throughout bacterial taxonomy and classification – were introduced that supported the classification of CRBs on the basis of 80–90% shared nucleotide sequence identity with the family Chlamydiaceae (Everett *et al.*, 1999). Using these criteria, the first families of CRBs in the order Chlamydiales were proposed, including the Parachlamydiaceae and Simkaniaceae, and shortly after the Waddliaceae (Rurangirwa *et al.*, 1999). Several years later, the first chlamydiae were isolated from arthropods leading to the proposal of *Candidatus* Rhabdochlamydia porcellionis, the founding member of the Rhabdochlamydiaceae family (Kostanjsek *et al.*, 2004). Subsequent discoveries over the next 10 years using 16S rRNA-based sequence criteria revealed expanded family level diversity from environmental sources (Criblamydiaceae; Thomas *et al.*, 2006) and fish, the latter home to at least three family level lineages (*Ca.* Piscichlamydiaceae, *Ca.* Clavichlamydiaceae and *Ca.* Parilichlamydiaceae; Draghi *et al.*, 2004; Karlsen *et al.*, 2008; Stride *et al.*, 2013a). 16S rRNA-based phylogenies place two of these (*Ca.* Piscichlamydiaceae and *Ca.* Parilichlamydiaceae) at the root of the Chlamydiales order phylogenetic tree (Taylor-Brown *et al.*, 2015).

A rapid expansion of genomic data for cultured bacteria in the phylum but also the use of direct sequencing methods to detect and study new bacteria recently provided the opportunity for a review of the systematics of chlamydiae (Gupta *et al.*, 2015; Pillonel *et al.*, 2015). Whilst 16S rRNA gene typing has long been in use for inferring relationships between bacteria, it is increasingly understood that this method is no longer robust enough to support taxonomic classification without additional genetic data due to its short length and high level of conservation. To address this concern, phylogenetic analysis of conserved protein sequences derived from all available whole genome sequences of bacteria in the phylum was performed, which proposed a classification scheme consisting of nine phylogenetically informative proteins that can be used to identify novel chlamydial species based on their sequence identity with existing species (Pillonel *et al.*, 2015). This approach has been used to classify several new CRBs (Bou Khalil *et al.*, 2016; Benamar *et al.*, 2017; Bou Khalil *et al.*, 2017; Taylor-Brown *et al.*, 2017a; Nylund *et al.*, 2018) as well as new species within the genus *Chlamydia* (Taylor-Brown *et al.*, 2016; Taylor-Brown *et al.*, 2017b; Staub *et al.*, 2018), with and without first isolating them in culture.

Current taxonomy

This section provides a brief overview of the current taxonomy of the phylum Chlamydiae. Until recently, four accepted and five proposed families comprised the CRBs (Table 18.1), which were grouped solely on the basis of not belonging to

Table 18.1 Current CRB genus and species names, grouped by family

Family	Genus, species	Cultured?	Genome sequenced?	Reference[a]
Parachlamydiaceae	*Parachlamydia acanthamoebae*	Yes	Yes	Amann *et al.* (1997); Collingro *et al.* (2011)
	Protochlamydia amoebophila	Yes	Yes	Horn *et al.* (2004); Collingro *et al.* (2005)
	Protochlamydia naegleriophila	Yes	Yes	Casson *et al.* (2008); Bertelli *et al.* (2015)
	Protochlamydia massiliensis	Yes	Yes	Benamar *et al.* (2017)
	Protochlamydia phocaeensis	Yes	Yes	Bou Khalil *et al.* (2017)
	Neochlamydia hartmannellae	Yes	Yes	Horn *et al.* (200)0; Ishida *et al.* (2014)
	Ca. Rubidus massiliensis	Yes[b]	Yes	Bou Khalil *et al.* (2016)
	Ca. Mesochlamydia elodeae	No	Yes	Corsaro *et al.* (2013)
	Ca. Metachlamydia lacustris	No	Yes	Corsaro *et al.* (2010)
Simkaniaceae	*Simkania negevensis*	Yes	Yes	Kahane *et al.* (1999); Everett *et al.* (1999); Collingro *et al.* (2011)
	Ca. Syngnamydia venezia	No	No	Fehr *et al.* (2013)
	Ca. Syngnamydia salmonis	No	No	Nylund *et al.* (2015)
	Ca. Fritschea bemisiae	No	No	Everett *et al.* (2005)
	Ca. Fritschea eriococci	No	No	Everett *et al.* (2005)
	Neptunochlamydia vexilliferae	Yes	No	Pizzetti *et al.* (2016)
Waddliaceae	*Waddlia chondrophila*	Yes	Yes	Dilbeck *et al.* (1990); Rurangirwa *et al.* (1999); Bertelli *et al.* (2010)
	Ca. Waddlia malaysiensis	Yes[b]	No	Chua *et al.* (2005)
Criblamydiaceae	*Criblamydia sequanensis*	Yes	Yes	Thomas *et al.* (2006); Bertelli *et al.* (2014)
	Estrella lausannensis	Yes	Yes	Lienard *et al.* (2011); Bertelli *et al.* (2015)
Rhabdochlamydiaceae	*Rhabdochlamydia porcellionis*	Yes	No	Kostanjsek *et al.* (2004)
	Ca. Rhabdochlamydia crassificans	No	No	Corsaro *et al.* (2007)
	Ca. Rhabdochlamydia helvetica	No	Yes	Pillonel *et al.* (2019)
	Ca. Renichlamydia lutjani	No	No	Corsaro *et al.* (2012)
Ca. Piscichlamydiaceae	*Ca.* Piscichlamydia salmonis	No	No	Draghi *et al.* (2004)
	Ca. Piscichlamydia cyprinus	No	No	Kumar *et al.* (2013)
Ca. Parilichlamydiaceae	*Ca.* Parilichlamydia carangidicola	No	Yes	Stride *et al.* (2013a); Taylor-Brown *et al.* (2018)
	Ca. Actinochlamydia clariae	No	No	Steigen *et al.* (2013)
	Ca. Actinochlamydia pangasiae	No	No	Sood *et al.* (2018)
	Ca. Similichlamydia epinephelii	No	Yes	Taylor-Brown *et al.* (2017)
	Ca. Similichlamydia latridicola	No	Yes	Stride *et al.* (2013b); Taylor-Brown *et al.* (2018)
	Ca. Similichlamydia labri	No	Yes	Steigen *et al.* (2015)
	Ca. Similichlamydia laticola	No	Yes	Stride *et al.* (2013c); Taylor-Brown *et al.* (2018)

Table 18.1 Continued

Family	Genus, species	Cultured?	Genome sequenced?	Reference[a]
Ca. Clavochlamydiaceae	*Ca.* Clavochlamydia salmonicola	No	No	Karlsen *et al.* (2008)
Ca. Amphibiichlamydiaceae	*Ca.* Amphibiichlamydia ranarum	No	No	Martel *et al.* (2013)
	Ca. Amphibiichlamydia salamandrae	No	No	Martel *et al.* (2012)
Additional family names have been proposed to classify further chamydial species characterized solely by partial genome sequences from metagenome and single cell genome sequencing studies: *Ca.* Novochlamydiaceae, *Ca.* Pelagichlamydiaceae, *Ca.* Enkichlamydiaceae, *Ca.* Limichlamydiaceae, *Ca.* Arenachlamydiaceae, *Ca.* Kinetochlamydiaceae, *Ca.* Motilichlamydiaceae				

[a]Literature cited here includes papers in which the species were first cultured, named and genome sequenced.
[b]Cultured but not deposited in a public culture collection.

the Chlamydiaceae family. However, a number of additional names for family level lineages have been proposed based on single cell (Collingro *et al.*, 2017) and metagenome-assembled genomes (Pillonel *et al.*, 2018). A recent global proposal for a unified bacterial taxonomy supported major changes, which have not yet been discussed in the field and are thus not reflected here (Parks *et al.*, 2018). The *Candidatus* status (*Ca.*), originally introduced to tentatively classify uncultured microbes based only on their 16S rRNA gene sequence (Goebel and Stackebrandt, 1994) has been used inconsistently in chlamydial taxonomy. To enhance clarity, in this chapter, the *Candidatus* status is only used if there is no isolate available in public culture collections. Hence, any taxa that have been identified using molecular methods only, retains this status. A number of CRBs remain uncultured, however, as detailed in Table 18.1, several genomes have been characterized for these species.

Evidence for additional unclassified diversity

Metagenomic methods and various microbiome datasets hint at further uncharacterized diversity in the phylum. Lagkouvardos *et al.* (2014) used advanced, high stringency bioinformatic analyses to interrogate over 22,000 high quality, non-redundant chlamydial 16S rRNA gene sequences found in diverse databases. Whilst not offering the level of phylogenetic resolution achieved by multigene analyses, this gene is still widely used to broadly characterize the bacterial diversity of a sample or environment and to identify putative new taxa. Using near-full-length sequences and a conservative clustering approach, 17 family level lineages supported by two or more isolates were identified, and this number increased to 28 when considering families represented by only one isolate. Ecological analyses revealed that the majority of species-level taxa belong to marine and freshwater environments and only a small proportion (2%) arose from terrestrial environments (Lagkouvardos *et al.*, 2014). Single cell genome sequencing has subsequently helped to recover uncultured marine representatives (Collingro *et al.*, 2017). Metagenomics analyses of samples primarily obtained from environmental sources has recently begun to fill in sufficient sequence data to apply the current taxogenomic approaches to classify this additional diversity (Pillonel *et al.*, 2018). By collectively analyzing previously published single cell and metagenome-derived genomes, new family level lineages emerged. These supported the concurrent proposal to recognize most of the CRBs described into a new order, the Parachlamydiales, with the exception of the families *Ca.* Parilichlamydiaceae and newly proposed *Ca.* Novochlamydiaceae, which form the first families of the new orders, Parilichlamydiales and Novochlamydiales.

The underestimated host range of chlamydiae

The genetic diversity of CRBs is only matched by the diversity of hosts and environments that they have been reported in. These hosts span the full range of eukaryotic diversity from single-celled eukaryotes such as free-living amoeba to invertebrates and vertebrates in aquatic and terrestrial environments (Taylor-Brown *et al.*, 2015). Amoeba were the first hosts discovered for CRBs (Kahane *et al.*, 1993). These protists are important predators controlling microbial populations in soils, marine and freshwater habitats (Rodriguez-Zaragoza, 1994) but also inhabit artificial environments such as public water supplies, air-conditioning units, and wastewater treatment plants. The CRBs growing in amoeba differ in their host range, with some being specific for a single amoeba species whereas others, like *W. chondrophila*, can infect several different amoeba species (Michel *et al.*, 2004) or even different hosts including higher eukaryotes like humans and animals (Collingro *et al.*, 2005; Coulon *et al.*, 2012). This flexibility of CRBs led to the suggestion that CRBs may be emerging human pathogens (Corsaro and Venditti, 2004).

Arthropods appear to be important hosts of CRBs. In the last decades, emergent groups of chlamydial species were found in a variety of different arthropods including woodlice, cockroaches, spiders, ticks and fleas (Croxatto *et al.*, 2014; Hokynar *et al.*, 2017; Corsaro *et al.*, 2007; Kostanjsek *et al.*, 2004; Vanthournout and Hendrickx, 2015). In addition, CRBs have been observed in other animal hosts, such as insects and bats (Everett *et al.*, 2005; Meijer *et al.*, 2006; Hokynar *et al.*, 2016). Although terrestrial environments harbour a huge diversity of CRBs and host species, aquatic environments are putatively the largest source of CRBs (Lagkouvardos *et al.*, 2014). In addition to protists, aquatic hosts include fish, corals and clams (Milan *et al.*, 2018; Thomas *et al.*, 2008; Thurber *et al.*, 2009; Karlsen *et al.*, 2008; Fehr *et al.*, 2013; Chen *et al.*, 2011; Draghi *et al.*, 2004). There are indications that fish-associated CRBs have a narrow host spectrum as each fish species is associated with a specific CRB (Fehr *et al.*, 2013; Draghi *et al.*, 2004). This may be a result of minimal genomic content after a long period of co-evolution with a single host or simply undersampling of fish species and other hosts (Taylor-Brown *et al.*, 2018).

Amoeba-associated chlamydiae

A major focus of initial studies of CRBs was the examination of amoeba-associated bacteria within this phylum, leading to most of the early discoveries about CRBs. These bacteria appear to resist digestion by amoeba (as a food source) and to exist as symbionts that replicate within inclusions originating from modified phagosomes or within the host cytoplasm. Members of the Parachlamydiaceae appear to be ubiquitous among amoeba, with amoeba hosts including *Acanthamoeba* (Fritsche *et al.*, 2000; Greub and Raoult, 2002), *Naegleria*, and *Vermamoeba* (a.k.a. *Hartmannella*) species (Horn *et al.*, 2000). The pronounced differences of these hosts with respect to evolutionary divergence, lifestyle and cell biology are a reflection of the adaptability of CRBs (Anderson, 2017; Pánek *et al.*, 2017). Consistent with this notion, free-living amoebae have successfully been used as a bait in co-culture approaches with complex environmental samples to recover novel Parachlamydiaceae (Collingro *et al.*, 2005). Parachlamydiaceae and Parchlamdiaceae-containing amoebae have been retrieved from activated sludge (Collingro *et al.*, 2005), hot springs (Sampo *et al.*, 2014) and soil (Fritsche *et al.*, 1993). Recently, *Vermamoeba vermiformis* harbouring a novel Chlamydiales species, *Ca.* Rubidus massiliensis, was isolated from a hospital water supply in Marseille, France (Pagnier *et al.*, 2015; Bou Khalil *et al.*, 2016). In addition to environmental samples, clinical specimens, such as corneal and nasal samples, have also been found to harbour Parachlamydiaceae (Michel *et al.*, 1994; Corsaro *et al.*, 2002; Pillonel *et al.*, 2018).

The individual host range of chlamydial symbionts of amoeba appears to vary. For example, two members of the Parachlamydiaceae, *Ca.* Metachlamydia lacustris (Corsaro *et al.*, 2010) and *Ca.* Mesochlamydia elodeae (Corsaro *et al.*, 2013) were initially isolated from aquatic amoebae and demonstrate high host specificity in contrast to other Parachlamydiaceae. Interestingly, these species co-inhabit the amoebae with members of either Alphaproteobacteria or Betaproteobacteria, and such close association could offer opportunities for genetic exchange that is not provided in other eukaryotic hosts. *Neochlamydia hartmannellae* was also shown to be highly specific to its amoeba host (Horn *et al.*, 2000). Although it was not able to infect any other related amoeba species, it was able to grow in the social amoeba, *Dictyostelium discoideum* (Horn *et al.*, 2000). A similar host specificity was shown for other *Neochlamydia* spp. isolated from environmental samples (Matsuo *et al.*, 2010).

Freshwater sources have proven to be a rich source of CRB diversity, with two novel chlamydial species described in 2006 and 2011, *Criblamydia sequanensis* and *Estrella lausannensis*, respectively, both part of a novel family, the Criblamydiaceae (Thomas *et al.*, 2006; Lienard *et al.*, 2011). *E. lausannensis* and *C. sequanensis* have been studied using amoebal co-culture. *E. lausannensis* was cultured in *D. discoideum*, a genetically tractable soil amoeba (Lienard *et al.*, 2011), potentially providing a novel model system for studying CRB biology, given the vast number of mutants available. Neither species has been successfully cultured in a mammalian cell line, suggesting they are indeed not capable of infecting humans (Kebbi-Beghdadi *et al.*, 2011).

A broad diversity of chlamydiae outside of the Parachlamydiaceae and the Criblamydiaceae can also be found in amoebal hosts. Recently, *Neptunochlamydia vexillierae* was retrieved from the marine amoeba *Vexillifera* sp., which was isolated from the marine lake Lago di Paola (Pizzetti, 2016). Besides this new Simkaniaceae species, a great diversity of so far unknown chlamydial lineages, identified by 16S rRNA sequencing, could be detected in this lake (Pizzetti, 2016). Interestingly, the abundance of CRBs in Lago di Paola changed with the picoplankton availability (Pizzetti *et al.*, 2016). While repeated discoveries of novel CRBs in protist hosts suggest that these species are symbionts, it does not exempt them from also being associated with pathology in humans and animals, as will be discussed later.

Arthropod-associated chlamydiae

Arthropods are emerging as another potentially rich reservoir of CRB diversity. Early genetic analysis confirmed symbionts of the scale insect *Eriococcus spurius* and the whitefly *Bemisia tabaci* as two novel species within the Simkaniaceae, comprising a novel genus, *Ca.* Fritschea: *Ca.* F. eriococci and *Ca.* F. bemisiae (Thao *et al.*, 2003; Everett *et al.*, 2005). More recently, chlamydiae have also been found in ticks. These insects are well-known vectors for other pathogens (Karbowiak *et al.*, 2015) and as such their role as chlamydial vectors has been investigated in Switzerland, Algeria and Australia, with studies reporting prevalence rates of 6–45% in pooled ticks with individual prevalence of 0.89–28.4% (Croxatto *et al.*, 2014; Pilloux *et al.*, 2015; Burnard *et al.*, 2016; Hokynar *et al.*, 2016). The four studies report primarily Rhabdochlamydiaceae, Parachlamydiaceae, and Simkaniaceae, with Rhabdochlamydiaceae DNA being most abundant (Pilloux *et al.*, 2015). The ticks themselves appear to be unaffected by their chlamydial infection, suggesting that ticks may be 'opportunistic reservoirs' for CRBs (Karbowiak *et al.*, 2015). Whether ticks may act as vectors for CRBs is yet unclear. However, a recent study identified chlamydial sequences that were closely related to sequences previously detected in ticks in skin biopsies from tick bite patients but not controls (Hokynar *et al.*, 2016). Furthermore, the same CRB 16S rRNA sequences identified in ticks parasitizing Australian marsupials were also found in the mucosal swabs collected from the same marsupial host (Burnard *et al.*, 2017). This suggests that ticks may indeed have roles as both vectors and reservoirs, depending on the CRB species and host(s) involved.

The *Rhabdochlamydia* genus has been originally proposed to contain two species *R. porcellionis* and *R. crassificans*, although database and phylogenetic analyses show that the Rhabdochlamydiaceae is putatively the largest family in the Chlamydiae (Lagkouvardos *et al.*, 2014). *R. porcellionis* was detected in the hepatopancreas of a woodlouse, and interestingly, displays a distinctive morphology of rod-shaped elementary bodies containing translucent oblong structures (Kostanjsek *et al.*, 2004). Recently, a close relative of *R. porcellionis* (98% 16S rRNA identity) was described in the Dwarf Spider *Oedothorax gibbosus* (Vanthournout and Hendrickx, 2015). *Ca.* R. crassificans was proposed as a novel pathogen of cockroaches, where it was identified in fat body and ovary tissues (Corsaro *et al.*, 2007). So far, only *R. porcellionis* has been studied experimentally (Sixt *et al.*, 2013). This species grows well within insect cell lines (*Drosophila melanogaster* S2, *Spodoptera frugiperda* Sf9), but could not be propagated in amoebae. Characterization of the course of infection revealed the lack of apoptosis induction and a reduced sensitivity of infected cultures to experimentally induced programmed cell death. This indicates that *R. porcellionis*, like its human pathogenic relatives, has the potential to counteract this host defence mechanism (Sixt *et al.*, 2013).

The importance of *Rhabdochlamydia* species in public health remains to be explored. Of note is the fact that several human screening studies have detected *Rhabdochlamydia* species distinct from the two described in insects (Corsaro *et al.*, 2002; Corsaro *et al.*, 2003; Baud *et al.*, 2009; Hornung *et al.*, 2015), suggesting that species in this clade are specific to their hosts, and may not be able to cross host barriers, also strengthening the notion that this family may exhibit a huge diversity with respect to its phylogeny and host range.

Fish-associated chlamydiae

Fish are hosts for different chlamydiae capable of causing the gill disease epitheliocystis. Here, lesions (cysts) are formed by aggregations of infectious EBs within epithelial cells, which lead to mechanical obstruction of the lamellae in the gills of fish. To date, epitheliocystis has been reported in over 90 species of fish globally in both wild and cultured fish from marine and freshwater environments (Blandford *et al.*, 2018). Of biggest concern is the impact of mortalities caused by chlamydial epitheliocystis in cultured fish (Draghi *et al.*, 2004; Mitchell *et al.*, 2010; Stride *et al.*, 2013a; Sood *et al.*, 2018), which appear to be most affected during the larval or juvenile stages with mortalities reaching 100%, leading to economic losses.

Two species are currently proposed in the *Ca.* Piscichlamydiaceae family, which to date is the most deeply branching lineage in the phylum: *Ca.* P. salmonis, affecting wild and cultured Salmonids (Draghi *et al.*, 2004; Draghi *et al.*, 2010; Schmidt-Posthaus *et al.*, 2012) and *Ca.* P. cyprinus affecting cyprinids (Kumar *et al.*, 2013; Sellyei *et al.*, 2017). Interestingly, neither species has been described in fish from the Southern hemisphere, suggesting that geographically distinct lineages exist within these clades.

Another agent of salmonid epitheliocystis is *Ca.* C. salmonicola, which is phylogenetically closer to the Chlamydiaceae than any other chlamydial lineage (Karlsen *et al.*, 2008). *Ca.* Syngnamydia salmonis is the third chlamydial species able to infect Atlantic salmon (*Salmo salar*) and is one of two fish-pathogens belonging to the Simkaniaceae (Nylund *et al.*, 2015). The other epitheliocystis agent in this family, *Ca.* Syngnamydia venezia, was found to infect the globally distributed broad-nosed pipefish (*Syngnathus typhle*) (Fehr *et al.*, 2013).

The largest family of fish-associated chlamydiae is *Ca.* Parilichlamydiaceae. Within its three genera (*Ca.* Parilichlamydia, *Ca.* Similichlamydia and *Ca.* Actinochlamydia) lie seven *Candidatus* species (i.e. tentatively classified using molecular methods but lacking a publicly available isolate) and many more unclassified taxa. Members of this clade have been detected in geographically diverse freshwater and marine ecosystems in Africa (Steigen *et al.*, 2013), Asia (Sood *et al.*, 2018), Australia (Stride *et al.*, 2013a; Stride *et al.*, 2013b; Stride *et al.*, 2013c; Taylor-Brown *et al.*, 2017a), and Europe (Steigen *et al.*, 2015; Guevara Soto *et al.*, 2016; Seth-Smith *et al.*, 2017; Steigen *et al.*, 2018), as epitheliocystis agents of a range of ray-finned fishes (*Actinopterygii* class) (Blandford *et al.*, 2018).

An additional CRB, *Ca.* Renichlamydia lutjani (Rhabdochlamydiaceae), was identified in a blue-striped snapper from Hawaii (Corsaro and Work, 2012), and its role as an epitheliocystis agent is questionable due to its detection in internal organs as opposed to the gills or skin. On the other hand, few epitheliocystis studies have included internal organs in their investigation, an aspect needing closer attention in the future as this would have implications for disease progression and dissemination.

Apart from the known aquatic representatives of chlamydiae, marine and freshwater environments are putatively the largest source of chlamydial diversity (Lagkouvardos *et al.*, 2014). Recently, a single-cell based study reported deeply branching lineages of chlamydiae from different marine environments (Collingro *et al.*, 2015). Remarkably, all found genomes encoded flagella genes indicating that these marine CRBs are motile, although motility has yet to be demonstrated once cultured representatives of these marine become available (Collingro *et al.*, 2015). Of note, phylogenetic analyses suggest that motility is an ancient chlamydial trait that was lost in many chlamydial lineages (Collingro *et al.*, 2015). Like with any purely genome-based analysis, the natural host for these chlamydiae remains to be determined.

An expanded capacity for intracellular growth and survival

Developmental cycle differences

All CRBs described to date exhibit the biphasic developmental cycle characteristic of these bacteria, however, differences exist in cell morphology, developmental cycle time, membrane components and intracellular trafficking. These are broadly summarized in Table 18.2.

CRB developmental cycles and cell morphology

Recent advances in fixation and microscopy techniques have meant that cells can be imaged in a 'near-native' state. As such, a recent study utilizing electron cryotomography compared the developmental stages and ultrastructure of three CRBs purified from amoeba cultures (Pilhofer *et al.*, 2014). EBs and RBs of *Simkania*, *Parachlamydia* and *Protochlamydia* could be distinguished by their size, morphology and cytoplasm granularity. The coccoid EBs of these species are slightly larger than that of *Chlamydia* spp., which are in the range of 300 to 500 nm in diameter (Table 18.2). CRB EBs range in size from ~450 nm (*S. negevensis*) to ~625 nm (*P. amoebophila*), and generally exhibit regions of concentrated filamentous material, which likely is condensed DNA, while the RBs are larger, pleomorphic forms that have an irregularly shaped outer membrane and higher numbers of ribosomes (Pilhofer *et al.*, 2014).

Contrary to original reports, the crescent body described for *P. acanthamoebae* (Greub and Raoult, 2002) was deemed an artefact of the combination of chemical fixation and dehydration/embedding (Pilhofer *et al.*, 2014), as was also the case for the star shape of *E. lausannensis* and *C. sequanensis* (Lienard *et al.*, 2011; Rusconi *et al.*, 2013; Thomas *et al.*, 2006). However, several other morphological features have been described among CRBs that further differentiate them from Chlamydiaceae. For example, *Ca.* Rhabdochlamydia crassificans and *R. porcellionis* both exhibit rod-shaped mature EBs (~250–700 nm x ~100–150nm) with a five-layered cell wall, that contain one or two translucent oblong structures visible using transmission electron microscopy and chemical fixation (Corsaro *et al.*, 2007; Kostanjsek *et al.*, 2004). This cell wall feature is reminiscent of other arthropod pathogens and thus appears to be an example of specific host-adaptation strategies employed by CRBs, but

Table 18.2 Key differences between chlamydial groups

Variable	Chlamydiaceae	*Chlamydia*-related bacteria; cultured	*Chlamydia*-related bacteria; uncultured	Proposed novel single cell and metagenome-derived families; uncultured
Chromosome size (Mbp)	1.0–1.2	2.1–3.4	0.8–1.0	1.1–1.5
Plasmid size (Kbp)[a]	7.5	9.1–132	Not reported	Not reported
GC content (%)	36–42	34–43	39–43	26–49
In vitro host cells	Mammalian	Mammalian, piscine, insect, amoebal	–	–
Developmental cycle	48–72 hours	Between 48 hours and 14 days; highly host cell dependent	–	–
Morphological features	Coccoid Ø 0.3–0.5 um EBs; Ø 0.5–1.25 um RBs; conserved outer membrane protein composition	Ø 0.4–0.7 um coccoid, ovoid or rod-shaped EBs; up to Ø 1 um pleomorphic RBs; variation in outer membrane protein composition; cytosolic forms;	Variation in membrane structures: rippled, double, 5-layered; Head-and-tail forms, intermediate bodies, actinae, apical caps	–
Other distinguishing features	Highly conserved and syntenic genomes	Mitochondrial, ER and Golgi recruitment; increased metabolic capacity (e.g. Menaquinone metabolism, intact TCA)	Minimal metabolic capacity	Evidence of motility; variation in respiratory chain
Key conserved features	Biphasic developmental cycle, inclusion formation, development of persistent forms, T3SS			–

[a]Not present in all species or strains.

interestingly, it was not observed in *R. porcellionis in vitro* within insect cell lines (Sixt *et al.*, 2013).

Head-and-tail forms with central osmiophilic condensations and unipolar cytoplasmic extensions have been described for *Ca.* Piscichlamydiaceae (Draghi *et al.*, 2004) and *Ca.* Clavichlamydiaceae (Karlsen *et al.*, 2008), both of which cause epitheliocystis in Salmonids. Meanwhile, projections (actinae) (Seth-Smith *et al.*, 2017; Steigen *et al.*, 2013), rippled outer membranes (Fehr *et al.*, 2013), apical caps and electro-lucent vacuoles (Fehr *et al.*, 2013; Nylund *et al.*, 1998; Nylund *et al.*, 2015; Seth-Smith *et al.*, 2017) are just some examples of structures observed in other epitheliocystis agents from diverse families. Again, these features may be reflective of the host diversity exhibited by this phylum.

Common to all chlamydiae, both developmental forms, as well as a third form – an intermediate body – can be visualized within the inclusion concomitantly, due to the asynchronous nature of differentiation, and are generally non-uniformly distributed throughout the inclusion, depending on the growth stage (Lee *et al.*, 2018). The time taken for differentiation and replication appears to be heavily dependent on the host cell, with more efficient infection and faster replication rates in cell lines closer to the postulated or known native hosts (Horn *et al.*, 2000; Kebbi-Beghdadi *et al.*, 2011a; Lienard *et al.*, 2011; Sixt *et al.*, 2012; Sixt *et al.*, 2013; Thomas *et al.*, 2006). For example, *S. negevensis*, for which there is no known natural host, exhibits a range of growth characteristics in different cell types, all of which differ greatly from growth traits exhibited by *Chlamydia* spp. These include an eight day stationary phase after a three day exponential growth phase in Vero cells (Kahane *et al.*, 1999); a 21 hour doubling time in insect cells (Vouga *et al.*, 2017); limited cytopathic effect on human endometrial cells and pneumocytes with high adhesion rates and numbers of bacterial cells in pneumocytes (Vouga *et al.*, 2017). Further, the encystment of *S. negevensis* in *A. polyphaga* lasts up to 148 days, with infectivity remaining at 50% on day 79 (Kahane *et al.*, 2001). Whilst chlamydiae traditionally require development in a host-membrane-derived inclusion, EBs and RBs from *Neochlamydia* and *Parachlamydia* have also been observed free in the cytoplasm (in the absence of inclusion lysis) (Greub and Raoult, 2002; Horn *et al.*, 2000; Ishida *et al.*, 2014). These cells have been observed in addition to intra-vacuolar forms and could indicate a mechanism by which CRBs achieve persistent infection.

In CRBs, EB to RB conversion begins as early as 2–4 hours post infection (Bou Khalil *et al.*, 2016; Lienard *et al.*, 2011; Rusconi *et al.*, 2015). Complete infection is reached, and redifferentiation back to EBs typically begins between 30 and 42 hours post-infection (Benamar *et al.*, 2017; Bou Khalil *et al.*, 2017; Lienard *et al.*, 2011) but can take up to 14 days (Sixt *et al.*, 2013), which stands in contrast to traditional chlamydial developmental cycles that typically last 48 to 72 hours.

Effects on host organelles

Chlamydiaceae and CRBs show several variations with respect to host cell interaction. Interactions of *Chlamydia* with organelles are described in more detail in Chapter 5, Interactions of the Chlamydial Inclusion with the Host Cell. In epithelial cells, Golgi fragmentation by *C. trachomatis* generates Golgi ministacks that surround the chlamydial inclusion, promoting lipid acquisition necessary for maturation (Heuer *et al.*, 2009). The same occurs in *C. pneumoniae*, which takes up host-derived sphingomyelin as early as four hours post-infection (Wolf and Hackstadt, 2001). Golgi fragmentation is seen also in *E. lausannensis* macrophage infection, but is less pronounced in CRBs, with no accumulation detected around the inclusion (Rusconi *et al.*, 2015). Rather, actin filament disruption results in less bacterial growth suggesting that *E. lausannensis* utilizes these filaments to facilitate uptake of nutrients such as ATP and glucose. Similarly, *W. chondrophila* recruits mitochondria via microtubules and/or actin microfilaments from as early as 30 minutes following infection of macrophages derived from peripheral blood mononuclear cells, and starting at 12 hours in epithelial cell lines, resulting in continuous, complete contact with this organelle (Croxatto and Greub, 2010; Kebbi-Beghdadi *et al.*, 2011b). At around eight hours post-infection, coinciding with replication, the endoplasmic reticulum is also recruited, and the two organelles organize into two layers around the inclusion (Croxatto and Greub, 2010). Even more pronounced in amoebae and epithelial cells, *S. negevensis* sustains a very close association with the rough ER membrane, establishing ER-inclusion contact sites and forming a 'vacuolar network' tightly associated with the ER (Mehlitz *et al.*, 2014; Pilhofer *et al.*, 2016; Vouga *et al.*, 2017). Like *W. chondrophila*, the *S. negevensis* vacuole associates with mitochondria in epithelial cells, but less so in amoebae (Mehlitz *et al.*, 2014).

Apoptosis resistance is a conserved feature among Chlamydiae, however, specific families and species appear to have evolved their own mechanisms for achieving this. It is thought that the close interaction between the *Simkania*-containing vacuole and the ER helps prevent apoptosis by actively blocking ER-stress (Karunakaran *et al.*, 2011; Mehlitz *et al.*, 2014). Meanwhile the amoeba-associated *Parachlamydia acanthamoebae* can only efficiently propagate in insect cells when apoptosis is blocked experimentally (Greub *et al.*, 2003; Sixt *et al.*, 2012), suggesting that interference with programmed cell death is a key feature of chlamydiae infecting humans and animals. Chlamydial apoptosis and autophagy are discussed in more detail in Chapter 6, *Chlamydia* and Cell-autonomous Defence: Apoptosis and Autophagy.

Metabolic diversity

Chlamydial ecological and genomic diversity is paralleled by the metabolic diversity encoded by these groups. The larger genome size of the CRB families (2–3 Mb versus 1 Mb for *Chlamydia* species) partially encode a larger repertoire of biosynthetic genes, whereas all known chlamydiae remain strictly host cell dependent and the same metabolic needs are met differently by *Chlamydia* spp. (reviewed in Omsland *et al.*, 2014). What follows is not an exhaustive comparative description of metabolic diversity among Chlamydiae but rather a selection of examples that encompass the variety of metabolic differences demonstrated by CRBs. The following results are primarily derived from genomic and transcriptomic studies.

Nucleotide metabolism

Chlamydiaceae and most CRBs are auxotrophic for purine and pyrimidine nucleotides but have variable abilities to scavenge nucleotides from intermediates. The lack of a *de novo* purine synthesis pathway in the Chlamydiales is mainly owing to an inability to synthesize the central component, inosine monophosphate (IMP). However, most CRBs can utilize ADP or GDP to produce purines and deoxy-purines in various phosphorylated states. *W. chondrophila* has a unique ability to convert nucleotides to nucleobases (Bertelli *et al.*, 2010; Thomas *et al.*, 2006). Pyrimidines, on the other hand, can be synthesized from uridine monophosphate (UMP) by *E. lausannensis*. UMP is likely imported from the host cell as it cannot be synthesized from amino acid catabolism or from 5-phospho-d-ribose α–1-pyrophosphate (PRPP) produced in the pentose phosphate pathway (Bertelli *et al.*, 2015). In the case of *W. chondrophila* and *C. sequanensis* at least, pyrimidines may be generated *de novo* from l-glutamine (Bertelli *et al.*, 2010; Bertelli *et al.*, 2014).

The lack of nucleotide biosynthesis pathways in chlamydiae is compensated for by unique sets of nucleotide transporter proteins (NTTs, also referred to as Ntps). At least three distinct nucleotide transporters are encoded by CRBs (Bertelli *et al.*, 2010; Heinz *et al.*, 2014; Taylor-Brown *et al.*, 2018), which potentially enable the import of all nucleotides and ribonucleoside triphosphates required for RNA synthesis and other biosynthesis pathways (Fuchs *et al.*, 2012; Tjaden *et al.*, 1999). Some of these transport proteins exhibit significant amino acid sequence similarity but, more frequently, these are identified based on conserved motif arrangements including 9 to 12 transmembrane regions (Fisher *et al.*, 2013). Importantly, the nucleotide transporters of *P. amoebophila* and *S. negevensis* have been biochemically characterized, with different NTTs exhibiting substrate-specific antiporter and symporter functions for ATP/ADP, guanine nucleotide/ATP/H^+, nucleoside triphosphates, dCTP, UTP, GTP/ATP, and remarkably also NAD^+/ADP (Haferkamp *et al.*, 2004; Haferkamp *et al.*, 2006; Knab *et al.*, 2011).

Amino acid and cofactor metabolism

In contrast to the generally greater metabolic capacity exhibited by CRBs, many amino acid pathways are truncated or missing, resulting in patchwork pathways that generate smaller subsets of amino acids or catalyse synthesis from intermediates. These differences likely reflect the host diversity and niche adaptation exhibited throughout the phylum. *W. chondrophila*, *C. sequanensis* and *E. lausannensis* appear to possess the highest capacity for *de novo* amino acid synthesis, encoding genes to produce at least ten of the twenty essential amino acids (Bertelli *et al.*, 2015; Bertelli *et al.*, 2010; Bertelli *et al.*, 2014). Some amino acid metabolism enzymes are shared by all chlamydiae. For example, CRBs can synthesize serine from pyruvate, for later conversion to glycine by glycinehydroxymethyltransferase (*glyA*), which is encoded by all chlamydiae including the recently sequenced *Ca.* Parilichlamydiaceae (Taylor-Brown *et al.*, 2018). In some species, pyruvate or serine can also be converted to cysteine or alanine (Bertelli *et al.*, 2015).

To overcome their limited biosynthesis capabilities, a range of amino acid or oligopeptide transporters with varying substrate specificities are encoded throughout the phylum to facilitate uptake of these metabolites (Bertelli *et al.*, 2010; Collingro *et al.*, 2011; Horn *et al.*, 2004; Omsland *et al.*, 2014; Taylor-Brown *et al.*, 2018). Further, additional genetic machinery dedicated to transporting aromatic amino acids are also shared among the phylum. Tryptophan is one of the only amino acids any *Chlamydia* species can synthesize. Most strains of *C. trachomatis* encode a tryptophan synthesis operon, which is implicated in persistence and tissue tropism (Akers and Tan 2006). However these genes are – with the exception of *S. negevensis* – not conserved in CRBs. Yet, a tryptophan/tyrosine permease is present in most Chlamydiae, which may by sufficient for uptake and utilization of tryptophan and related amino acids (Taylor-Brown *et al.*, 2018).

Many CRBs, including *P. amoebophila*, but not *C. sequanensis*, (Bertelli *et al.*, 2015; Collingro *et al.*, 2011; Horn *et al.*, 2004) are able to produce electron carrier menaquinone, offering another point of difference from the Chlamydiaceae. Menaquinone is likely to be utilized in the quinone pool of the respiratory chain if host-derived ubiquinone is not available (Horn *et al.*, 2004).

P. amoebophila, like *Chlamydia* species, uses a nucleotide transporter for the uptake of host-derived nicotinamide adenine dinucleotide (NAD) (Haferkamp *et al.*, 2004; Fisher *et al.*, 2013), while other chlamydiae might use the NAD salvage pathway or, like *S. negevensis*, seem to be able to synthesize this important cofactor (Collingro *et al.*, 2011). No homologue for the NAD transporter has been identified in *W. chondrophila*, which likely synthesizes NAD from an intermediary metabolite such as quinolinate or nicotinamide (Bertelli *et al.*, 2010). These may be imported through other systems, such as membrane transporters of the major facilitator superfamily (MFS) type, which are ubiquitous in bacteria and are capable of transporting a range of molecule types.

Energy metabolism

Chlamydiae were previously referred to as 'energy parasites' (McClarty, 1994) as they obtain ATP from their host. This strategy was confirmed by initial genome sequencing efforts that showed high concentrations of ATP-binding cassette transporters and several copies of an ATP/ADP transport protein in *C. trachomatis* (Stephens *et al.*, 1998), which exchanges bacterial ADP for host ATP. Such non-mitochondrial ADP/ATP translocases were originally considered unique to intracellular pathogens and symbionts, but recent work has shown that these transport proteins are functionally diverse and occur in both free-living and obligate intracellular bacteria, including Chlamydiae and non-photosynthetic eukaryotes (Greub and Raoult, 2003; Heinz *et al.*, 2014; Major *et al.*, 2017; Schmitz-Esser *et al.*, 2004).

CRBs can also generate ATP from the glycolysis pathway, which is intact in all chlamydiae, and by oxidative phosphorylation. In this system, most CRBs, but not *Chlamydia* spp., encode the genes for a F-type ATP synthase for H^+-driven ATP synthesis, cytochromes and flavoproteins (Bertelli *et al.*, 2010; Fuchs *et al.*, 2012; Goy *et al.*, 2009; Horn *et al.*, 2004) in addition to the V-type ATP synthase, which is present in *Chlamydia* spp (Kalman *et al.*, 1999; Mojica *et al.*, 2011; Read *et al.*, 2000; Stephens *et al.*, 1998; Thomson *et al.*, 2005; Voigt *et al.*, 2012). While the TCA cycle is incomplete in Chlamydiaceae genomes (they lack three key components: citrate synthase, aconitase and isocitrate dehydrogenase, catalyzing the reactions from oxaloacetate to oxoglutarate) (Omsland *et al.*, 2014), CRBs except for the *Ca.* Parilichlamydiaceae family encode a complete TCA pathway (Collingro *et al.*, 2011; Fuchs *et al.*, 2012; Horn *et al.*, 2004; Taylor-Brown *et al.*, 2018). This suggests that highly host-specific families or species have convergently lost these pathways and thus rely more heavily on the host as a source for energy and amino acids (which are generated from TCA metabolites).

A recent study proposed a detailed model for the intracellular metabolism of the amoeba symbiont

Protochlamydia amoebophila based on global gene expression analysis (König *et al.*, 2017). During early infection these bacteria largely rely on host-derived energy (through ATP import) and amino acids, but they show a pronounced switch to a glucose-based metabolism later in the developmental cycle, during RB-EB differentiation. Interestingly, this model for a bi-phasic chlamydial metabolism is compatible with a number of similar observations for members of the Chlamydiaceae (Albrecht *et al.*, 2010; Omsland *et al.*, 2012; König *et al.*, 2017; Beder and Saluz, 2018).

Lineage-specific biological innovations

Recent comparative genome analyses have revealed that CRBs possess distinct sets of family and species-specific genes, in the range of 256 to 1340 genes, depending on the analysis, compared to 136 to 170 for Chlamydiaceae (Collingro *et al.*, 2011; Taylor-Brown *et al.*, 2018). An in-depth comparative analysis of family, genus-, or species-specific infection strategies would garner its own chapter, so we will focus again on a selection of highly relevant examples of biological innovations employed by different CRB lineages that have allowed researchers to gain further insight into chlamydial evolution and pathogenesis.

Type three secretion and effectors

The type three secretion system (T3SS) is a molecular apparatus common to many Gram-negative pathogens and symbionts. As covered in Chapter 7, Protein Secretion in *Chlamydia*, the T3SS facilitates transport of 'effectors' into the host cell cytosol (Betts-Hampikian and Fields, 2010) and is fundamental in establishing chlamydial infection via modulation and exploitation of the host cell (Coburn *et al.*, 2007; Horn *et al.*, 2004; Hueck, 1998; Peters *et al.*, 2007).

The high number of conserved structural and chaperone components of the T3SS, including inner membrane proteins, needle formation proteins and chaperones (Bertelli *et al.*, 2015; Collingro *et al.*, 2011), confirm its essential role in chlamydial pathogenesis. Its presence in all CRB genomes to date suggests it was an ancient acquisition (Collingro *et al.*, 2011; Horn *et al.*, 2004). Whilst the myriad of predicted effectors await their experimental confirmation, effectors utilized by different CRB species may further highlight host cell modulation strategies. This hypothesis is exemplified as follows. First, genomic analyses revealed an expanded family of predicted effector proteins in amoeba-associated members of the Parachlamydiaceae (Domman *et al.*, 2014), all of which contain signature domains of known effectors. Second, *W. chondrophila* Incs are more similar to *P. acanthamoebae* Incs than to *P. amoebophila* Incs, with only the two former known to be able to infect mammalian cells (Collingro *et al.*, 2011). Other notable effectors such as TARP – a 'translocated actin-recruiting protein' implicated in chlamydial invasion (Lane *et al.*, 2008), Cap-1 – 'class I accessible protein-1' that has a role in immune recognition (Balsara *et al.*, 2006) – and CADD – 'chlamydial protein associate with death domains' involved in inducing apoptosis (Stenner-Liewen *et al.*, 2002) – are absent from CRB genomes, further reflecting host diversity and range, and highlighting species-specific infection strategies.

Gene family expansions

The Chlamydiaceae possess a unique outer membrane complex (OMC), the main component of which is the major outer membrane protein (MOMP), encoded by the *ompA* gene. CRB OMCs appear to be quite distinct from their *Chlamydia* cousins. For example, *Protochlamydia* spp. appear to have replaced MOMPs with structurally similar porin proteins (Aistleitner *et al.*, 2013), whilst *Parachlamydia* spp. and *Neochlamydia* spp. are almost devoid of MOMP-like proteins (Heinz *et al.*, 2009; Ishida *et al.*, 2014). On the other hand, *S. negevensis* and *W. chondrophila* encode 35 and 11 MOMP-like proteins, respectively (Bertelli *et al.*, 2010; Collingro *et al.*, 2011).

Few examples of paralogues are seen within the genomes of obligate intracellular bacteria with their highly reduced genomes, and larger gene families are indeed rare within the Chlamydiae. Yet, 277 expansion events are described in the *S. negevensis* genome. These comprise many small gene families that are the result of many small-scale transfers or duplications (Domman *et al.*, 2014). Conversely, in the Parachlamydiaceae, fewer gene duplication events have led to larger expansions in fewer gene families, namely between *Neochlamydia* and *Protochlamydia* species. Most of these expanded gene sets (termed PEX1 and PEX2 for *Protochlamydia* species, for example) represent putative type three secretion systems effector proteins and: (a) display motifs involved in protein–protein interaction, (b) are rather common to eukaryotic than prokaryotic organisms, and (c) have domains involved in eukaryotic ubiquitin-related enzymes (Domman *et al.*, 2014). Taken together, the results of these gene family expansion events may mediate species-specific host cell modulation.

Genomic islands and plasmids

Genomic islands are regions of the chromosome that arise from horizontal gene transfer (Dobrindt *et al.*, 2004), often encoding virulence factors. Genomic islands are yet another tool chlamydiae differentially possess that confer evolutionary advantages to survival in specific host niches. Only one genomic island has been described in the Chlamydiaceae, the tetracycline resistance cassette (*tetC*) unique to *C. suis* (Dugan *et al.*, 2004), whilst different genomic islands have been acquired by Parachlamydiaceae species after the divergence of these two lineages (Horn *et al.*, 2004).

Horn *et al.* (2004) described a 100 Kbp genomic island in *P. amoebophila* encoding a putative type IV secretion system. The presence of several transposases, flanking tRNAs and the presence of other mobility genes also indicated a lateral gene transfer event (Greub *et al.*, 2004). The chromosomes of its closest relatives, *P. naegleriophila* strains Knic and Diamant, also harbour smaller genomic islands (37 Kbp and 15 Kbp, respectively). These encode genes involved in DNA damage and repair, a potassium transporter and hypothetical proteins (Bertelli *et al.*, 2016).

All *Chlamydia* species except for *C. abortus* and *Ca.* C. ibidis carry a 7.5 Kbp plasmid, which is suggested to have co-evolved with the chromosome (Jelocnik *et al.*, 2016). *Chlamydia* plasmids are non-conjugative and non-integrative, are highly conserved and highly syntenic, but mixed findings from *in vitro* and *in vivo* studies, coupled with the fact that not all strains carry it, highlight how little we understand about its role in pathogenicity and metabolism (Rockey, 2011; Zhong, 2016). In stark contrast to the *Chlamydia* plasmid, CRBs harbour plasmids ranging in size and putative function, with each carrying species-specific genes that may have roles in host adaptation. The plasmids of *S. negevensis* and *P. naegleriophila* are the largest at 132 Kbp and 145 Kbp, respectively. They are the only mobilizable F-type conjugative plasmids described in the phylum to date and carry genes associated with plasmid stability, replication and propagation and host adaptation (Bertelli *et al.*, 2016; Collingro *et al.*, 2011). Homologues of some of the plasmid-encoded elements may have already been integrated into the chromosomes of other CRBs. For example, a *Mip* homologue, which is most commonly located on the chromosome, is found on the *S. negevensis* plasmid (Collingro *et al.*, 2011). Also, parts of the *tra* subset of the type IV secretion system (on the *S. negevensis* plasmid) are encoded on the *P. amoebophila* and *P. acanthamoebae* chromosome (Greub, 2009; Greub *et al.*, 2004; Horn *et al.*, 2004, Collingro *et al.*, 2011). Interestingly, *R. massilliensis*, a member of the Parachlamydiaceae, harbours two plasmids concurrently (Bou Khalil *et al.*, 2016), offering a further point of difference from *Chlamydia* spp. This phenomenon is yet to be described in other CRBs.

Toxin, anti-toxin and detoxification genes appear to be common features among CRB plasmids. One of the two *R. massiliensis* plasmids encodes resistance or efflux genes for antibiotics, tellurite and copper (Bou Khalil *et al.*, 2016) and the other encodes a *RelF/MazF*, the endoribonuclease component of the two-part *RelEF/MazEF* toxin–antitoxin system. The latter is also encoded by the *E. lausannesis* plasmid and located on the *W. chondrophila* chromosome (Bertelli *et al.*, 2015; Bertelli *et al.*, 2010; Bou Khalil *et al.*, 2016). The *C. sequanensis* plasmid encodes cation efflux systems including one for arsenite (Bertelli *et al.*, 2014). These systems may assist the intra-amoebal lifestyle of CRBs, which is less homeostatic than other eukaryotic cells that *Chlamydia* species infect. Studying these systems may also provide insights into the evolution of mechanisms used by pathogenic chlamydiae to infect and proliferate inside other phagocytic cells such as macrophages.

Lessons from transcriptomics

A recent study employing dual-RNASeq technology explored temporal gene expression in the developmental cycle of *P. amoebophila* in its natural *Acanthamoeba* host (König *et al.*, 2017). This study provided evidence for a biphasic metabolism during the *P. amoebophila* developmental cycle as summarized above. Also, as previously shown for

Chlamydiaceae, the *P. amoebophila* developmental cycle is governed by gene sets that can be grouped roughly into genes expressed early, mid, or late during infection. As for the Chlamydiaceae, genes encoding T3SS structural proteins and effectors were found to be highly expressed very early during infection, including a *Protochlamydia*-specific orthologue of inositol phosphate phosphatase (IpgD) that facilitates host cell entry in *Shigella*. The authors suggested that this might compensate for the lack of TarP (König *et al.*, 2017). These genes are rapidly down-regulated following infection, in contrast to a second class of early genes, which remain highly expressed throughout the cycle. In *P. amoebophila*, this temporal class was dominated by putative eukaryotic-like effectors that either possess Sel1-like repeat domains or belong to the PEX1 and PEX2 expanded gene families potentially involved in modification of host ubiquitination (Domman *et al.*, 2014), showing that *P. amoebophila* may secrete species-specific effectors at this stage in addition to employing conserved mechanisms. Interestingly, the mid-cycle gene set appears highly conserved among the Chlamydiae, suggesting that whilst chlamydial species differ in their host infection strategies, RB metabolism and proliferation is highly conserved (König *et al.*, 2017).

Another study compared EB and RB transcriptomes in *W. chondrophila*, *C. psittaci* and *C. abortus*, as pathogens associated with some of the same disease manifestations in different hosts (Beder and Saluz, 2018). The number of differentially expressed genes between the two developmental stages varied from 397 for *C. abortus* (genome size: 1.14 Mbp) and 601 for *C. psittaci* (1.17 Mbp) to 998 for *W. chondrophila* (2.11 Mbp), and was also higher in RBs than EBs. The observed expression of superoxide dismutase and nitric oxide reductase in *W. chondrophila* RBs might be indicative of detoxification early on in infection. Several transporters were similarly expressed in EBs and RBs, suggesting that they are required by *W. chondrophila* throughout all stages of the developmental cycle, while virulence factors such as members of the expanded MOMP family ompA, as well as two predicted T3SS effectors, were upregulated in *W. chondrophila* EBs, suggesting they are involved in infection (Beder and Saluz, 2018).

Active transcription in EBs has been suggested by several studies (Sixt *et al.*, 2013; Beder and Saluz, 2018; König *et al.*, 2018; Grieshaber *et al.*, 2018), yet transcripts found in EBs may also represent 'carryover mRNA' that is required at the end of the developmental cycle or early during infection and loaded in EBs before the release from the host cell (Mäurer *et al.*, 2007).

Taken together, these examples show that whilst CRBs and *Chlamydia* species exhibit many lineage-specific infection and survival strategies, by further studying the biology of CRBs, we can identify conserved mechanisms and gain a better understanding of the diversity of interactions with different hosts throughout the phylum, and even among intracellular bacteria.

Chlamydia-related model organisms

Protochlamydia and *Parachlamydia*

Parachlamydia acanthamoebae was the first species described within the family Parachlamydiaceae (Amann *et al.*, 1997). This amoeba-associated species as well as *Protochlamydia amoebophila* (Collingro *et al.*, 2005), another member of the same family and the first CRB for which a genome sequence was available, have been used extensively as model organisms to study general traits of CRBs. These models also allow us to draw comparisons with infections of other phagocytic cells such as macrophages, which are used by *Chlamydia* species as a means of evading the immune system and disseminating through the body. This section will summarize the key findings about the pathogenicity and biology of these important CRBs.

Parachlamydia as a respiratory pathogen

P. acanthamoebae was originally isolated from an amoeba recovered from the nasal mucosa of a female volunteer (Amann *et al.*, 1997). After its original discovery, a number of studies provided evidence that *P. acanthamoebae* might represent a causative agent of respiratory disease in humans (Greub and Raoult, 2002; Greub *et al.*, 2003; Birtles *et al.*, 1997; Marrie *et al.*, 2001; Greub *et al.*, 2009). Although all these studies suggested the respiratory system as a target of infection, the findings are largely based on the presence of 16S rRNA sequences in samples of humans with respiratory tract infections. In order to directly correlate *P. acanthamoebae* with respiratory tract disease, animal models have been applied. In a mouse model, *P. acanthamoebae* was shown to be able to cause pneumonia in 100% of infected mice two days post-inoculation (dpi) (Casson *et al.*, 2008). Moreover, *Parachlamydia* could be recovered from infected lungs up to 42 dpi using amoebal co-culture (Casson *et al.*, 2008). It should be noted, however, that this study used immunodeficient BALB/c mice and very high intratracheal inoculations of 2.5×10^8 inclusion-forming units (IFU) of *P. acanthamoebae* per mouse. Following up on these findings, a bovine respiratory model was established (Lohr *et al.*, 2015). Male calves were challenged intra-bronchially with a dose of up to 1×10^{10} IFU per calf. Although all calves presented signs of an altered health state up to 7 dpi, *P. acanthamoebae* was not able to evade the innate immune response and could not establish a persistent infection in epithelial cells (Lohr *et al.*, 2015). A subsequent study found no evidence for severe pathophysiological consequences of *P. acanthamoebae* infection on pulmonary function in the bovine host (Lohr *et al.*, 2016). This suggests that *P. acanthamoebae* has the potential to infect mammalian hosts, but may not be able to establish an infection in the presence of a functional immune response.

So far, it has not been possible to recover *P. acanthamoebae* by co-culture from a respiratory sample originating from a human or animal field infection supporting the idea that *Parachlamydia* is not able to use higher eukaryotes as permanent hosts. However, it remains to be investigated whether *P. acanthamoebae* can infect immunodeficient humans and animals. One possible explanation for the inability of *P. acanthamoebae* to establish a sustainable infection in higher eukaryotes is its apparent inability to block apoptosis (Greub *et al.*, 2005; Sixt *et al.*, 2012). Apoptosis is a response of the innate immune system to intracellular pathogens in which damaged and infected cells are removed by programmed cell death (Chow *et al.*, 2016; Labbé and Saleh, 2008). Members of the Chlamydiaceae, and some CRBs, have evolved mechanisms to block host cell death (Herweg and Rudel, 2016; Chow *et al.*, 2016; Faherty and Maurelli, 2008; Ying *et al.*, 2007; Kahane *et al.*, 2008; Karunakaran *et al.*, 2011). This feature, discussed in more detail in Chapter 6, *Chlamydia* and Cell-autonomous Defence: Apoptosis and Autophagy, is in fact essential for the developmental cycle and a proliferative infection of *Chlamydia* species (Ying *et al.*, 2008; Dean and Powers, 2001). *P. acanthamoeba*, however, was shown to be unable to block apoptosis in insect cells and human macrophages, which restricted its growth in these and other human cell lines (Sixt *et al.*, 2012; Greub *et al.*, 2003; Hayashi *et al.*, 2010). This finding suggests that the inability of *P. acanthamoebae* to block apoptosis prevents them from escaping the immune response of higher eukaryotes and establishing a proliferative infection.

Protochlamydia and *Parachlamydia* as models to study general chlamydial biology

The type three secretion system

As reflected by its presence in all sequenced members of the Chlamydiaceae and all CRBs studied to date, the T3SS represents a key component of chlamydial biology (Read *et al.*, 2003; Thomson *et al.*, 2005; Azuma *et al.*, 2006; Horn *et al.*, 2004; Bertelli *et al.*, 2010; Collingro *et al.*, 2011; Croxatto *et al.*, 2013, Taylor-Brown *et al.*, 2017). The genomic organization of T3SS genes of *P. amoebophila*, *P. acanthamoebae* and *C. trachomatis* shares significant similarities, indicating a conservation of T3SS sequences and transcriptional expression among all chlamydiae (Horn *et al.*, 2004; Bertelli *et al.*, 2010; Croxatto *et al.*, 2013). The regulation of the chlamydial T3SS expression is clearly different from the transcriptional regulation in other bacterial species, which is under the control of *AraC*-like transcriptional activators (Francis *et al.*, 2002). As chlamydial genomes lack genes that encode *AraC*-like transcriptional activators, chlamydial T3SSs are likely regulated by a different mechanism (Croxatto *et al.*, 2013). Indeed, Case *et al.* demonstrated a mechanism in *C. trachomatis* whereby expression of some T3SS genes are regulated by promoters that are activated in response to increased DNA supercoiling, corresponding to changes in DNA supercoiling seen during the developmental cycle (Case *et al.*, 2010).

In *P. acanthamoebae*, the T3SS is mainly expressed in early stages of infection when EBs enter the host cell and establish an environment for sustainable RB replication and differentiation (Croxatto *et al.*, 2013). In the late phase of infection, when RBs start to redifferentiate to EBs, there is a second peak of T3SS expression leading to a shutdown of the apparatus and formation of silent T3SS in the EBs that enable quick activation of the complex upon infection of new host cells (Croxatto *et al.*, 2013).

In *P. amoebophila*, three steps in T3SS gene expression and assembly have been proposed (König *et al.*, 2017). The first step is the formation of the main components of novel T3S systems during RB-to-EB differentiation. In *P. amoebophila*, T3S machineries are mostly incomplete at midcycle and lack the needle protein SctF, but become fully equipped at the tail-end of RB-EB differentiation. In step two, functional T3S systems are present on extracellular EBs and facilitate host cell invasion. The third step is characterized by the peak of T3S activity occurring in early RBs, where all structural components as well as a large number of effectors are highly expressed, reflecting the major role of the T3SS during establishment of the replicative niche by early RBs (König *et al.*, 2017).

Chlamydial peptidoglycan and cell division

Two recent findings, originating from studying *P. amoebophila* illustrate the importance of investigating CRBs with respect to basic chlamydial biology. First, chlamydiae have traditionally been viewed as peptidoglycan-less microbes despite their sensitivity to beta-lactam antibiotics and the presence, in all chlamydial genomes, of genes predicted to function in the biosynthesis of this ubiquitous bacterial cell wall component (McCoy and Maurelli, 2006). Peptidoglycan was first identified in *P. amoebophila* by electron cryotomography, mass spectrometry and fluorescence labelling assays, demonstrating cell wall sacculi consisting of an unusual peptidoglycan type in these chlamydiae (Pilhofer *et al.*, 2013). Shortly thereafter, peptidoglycan was also identified in *C. trachomatis* (Liechti *et al.*, 2014; Packiam *et al.*, 2015). In parallel, new septal proteins were identified in *W. chondrophila*, together providing evidence that cell division might occur in a similar way as in other bacteria (Jacquier *et al.*, 2014; Jacquier *et al.*, 2015).

Maintenance metabolism in EBs

Due to their biological features and structure, chlamydial EBs were traditionally thought to represent a metabolically inert, spore-like stage (Hatch, 1999; Bavoil *et al.*, 1984). However, this concept has been challenged over the years (reviewed in Omsland *et al.*, 2014). The EBs of *P. amoebophila* but also members of the Chlamydiaceae possess nearly complete machineries for transcription, translation, and energy metabolism, as revealed by proteome analyses (Vandahl *et al.*, 2001; Skipp *et al.*, 2011; Sixt *et al.*, 2011). Transcripts of genes involved in glucose-based metabolism, transcription and protein synthesis are abundantly present in *P. amoebophila* EBs undergoing maturation in a time-dependent manner. Further, isotope labelling experiments revealed that EBs of *P. amoebophila* take up nutrients and stay metabolically active even when outside of a host cell (Haider *et al.*, 2010; Sixt *et al.*, 2013; Konig *et al.*, 2017). Importantly, nutrient uptake is required to maintain EB infectivity over extended time periods (Sixt *et al.*, 2013). This seems to also hold true for members of the Chlamydiaceae. For *C. trachomatis*, uptake of glucose 6-phosphate, ATP, and amino acids, in EBs has been observed (Omsland *et al.*, 2012; Grieshaber *et al.*, 2018). Together, these findings contribute to the emerging understanding that chlamydial EBs actively maintain basic metabolic processes after terminal differentiation to facilitate prolonged infectivity (Sixt *et al.*, 2013, Omsland *et al.*, 2014, Grieshaber *et al.*, 2018).

Protochlamydia and *Parachlamydia* as models to study the evolution of chlamydial pathogenicity

All chlamydiae share a common ancestor that lived around one billion years ago, and there is evidence that the emergence of their unique obligate intracellular lifestyle dates back to these ancient Precambrian times (Horn *et al.*, 2004; Greub and Raoult, 2004). Consistent with this perception, all known chlamydiae share functions and mechanisms characteristic for this intracellular lifestyle (Horn *et al.*, 2004; Greub *et al.*, 2009; Collingro *et al.*, 2011).

Despite the conservation of major biological features among all chlamydiae, Parachlamydiaceae, like other CRBs, have a larger genome that encodes mechanisms and pathways absent in the Chlamydiaceae. Due to their environmental occurrence and their primitive unicellular hosts, Parachlamydiaceae have been proposed to face similar circumstances as ancient chlamydiae and thus could serve as a model for studying the evolution of chlamydial pathogenicity (Ito *et al.*, 2012; Matsuo *et al.*, 2013). Like *P. acanthamoebae, P. amoebophila* induces apoptosis, limiting its growth in human epithelial cells (HEp-2) at 37°C (Ito *et al.*, 2012; Matsuo *et al.*, 2013; Yamane *et al.*, 2015). The discovery of the ability of a temperature-adapted *Protochlamydia* strain found in an amoeba isolate from a hot spring to grow in HEp-2 cells at body temperature suggested that temperature adaptation is yet another factor differentiating chlamydial symbionts in amoeba and classical chlamydial pathogens in the Chlamydiaceae (Sampo *et al.*, 2014; Hayashi *et al.*, 2010).

Simkania

Simkania negevensis was discovered in 1993 (Kahane *et al.*, 1993) as a contaminant in a HeLa cell culture. The original host of *S. negevensis* could never be identified (Vouga *et al.*, 2017a). Interestingly, the genome of *S. negevensis* encodes a type I intron in the 23S rRNA gene (Everett *et al.*, 1999). While absent in genomes of most prokaryotes, this structure is also present in the genome of the *A. castellanii* amoeba (Everett *et al.*, 1999), suggesting *A. castellanii* as the natural reservoir for *S. negevensis*. *S. negevensis* was the first CRB described and is still the only described member of the *Simkania* genus (Vouga *et al.*, 2017a). Since its discovery, *S. negevensis* was shown to be able to infect a broad variety of host cells including human cells, insect cells and amoeba (Sixt *et al.*, 2012; Vouga *et al.*, 2017a). Like other CRBs, *S. negevensis* has a significantly larger genome and consequently has extended metabolic capabilities that may explain its broad host range (Collingro *et al.*, 2011).

Simkania as a pathogen of humans

Although evidence of human exposure to *S. negevensis* has been reported worldwide in both healthy and diseased humans (Al-Younes and Paldanius, 2014; Friedman *et al.*, 2006; Yamaguchi *et al.*, 2005), the connection between *Simkania* exposure and disease is very weak with only exceptional cases where *Simkania* could be identified as the putative causative agent of disease (Fasoli *et al.*, 2008; Donati *et al.*, 2013). Further, recent studies using specific qPCR primers in more than 400 clinical samples also suggest a low prevalence of exposure (Vouga *et al.*, 2018; Kose *et al.*, 2015). Moreover, as older studies often lack proper negative controls and use different techniques for detection, it is now thought that S. *negevensis* is not a clinically relevant human pathogen (Vouga *et al.*, 2018).

Lessons learned from *Simkania*

S. negevensis has a wide host spectrum and was also shown to be able to grow in human endometrial cells and pneumocytes (Vouga *et al.*, 2017b). Interestingly, growth in mammalian cells was more efficient than in amoeba, even though the latter have been suggested as the natural host and main vector of transmission of *S. negevensis* (Kahane *et al.*, 2001; Vouga *et al.*, 2017b). Further, *S. negevensis* was proposed to use extrusion as an escape mechanism analogous to *C. trachomatis* (Hybiske and Stephens, 2007) as it shows release of EBs without an accompanied increase in host cell lysis (Vouga *et al.*, 2017b). These findings indicate that *S. negevensis* has the potential to infect humans and may use humans as facultative hosts. In this regard it is interesting that *S. negevensis*, like other chlamydiae, is susceptible to a variety of antimicrobials like macrolides, clindamycin, cyclines, and rifampin but resistant to β-lactams and vancomycin (Vouga *et al.*, 2017c).

Unlike other chlamydiae, *S. negevensis* lacks stabilizing cysteine-rich proteins in its outer membrane leading to EBs that are osmotically less stable (Aisteitner *et al.*, 2015). Most remarkably, *S. negevensis* forms a persistent (up to 15 days) net-like vacuolar structure upon entrance into the host cell (Kozjak-Pavlovic *et al.*, 2017; Kahane *et al.*, 2002). This so called *Simkania*-containing vacuole (SnCV) forms extensive contacts with the ER in different cell types including human cells and amoeba (Kozjak-Pavlovic *et al.*, 2017; Mehlitz *et al.*, 2014). Like other intracellular bacteria, *S. negevensis* uses the ER for the supply of growth-relevant metabolites (Herweg *et al.*, 2016). So far, *S. negevensis* is the only known intracellular bacteria that is able to inhibit ER stress, which leads to the activation of apoptosis and an innate immune response (Kozjak-Pavlovic *et al.*, 2017; Celli and Tsolis, 2015). In addition to the ER, the SnCV also makes contact with other organelles of the host cell. The Golgi is known to play a role in the provision of lipids which are used to maintain the chlamydial inclusion (Herweg *et al.*, 2016; Koch-Edelmann *et al.*, 2017). For *S. negevensis*, it was further shown that retrograde transport in the Golgi is essential for its developmental cycle by influencing its ability to establish infectious progeny (Herweg *et al.*, 2016).

Waddlia

W. chondrophila is perhaps the CRB model organism with the most striking serological and molecular evidence as a putative human and animal pathogen. This section will detail the cell biology and animal model studies that provide support for *W. chondrophila* as an abortigenic agent and respiratory pathogen in humans and animals.

Waddlia as an abortigenic agent

Since its initial isolation from bovine abortion material in the United States (Dilbeck *et al.*, 1990) and Germany (Henning *et al.*, 2002), *W. chondrophila* has emerged as a serious threat to cattle as well as humans, as evidenced by a number of serological studies in which anti-*W. chondrophila* antibodies are associated with pregnancy failure. For example, dairy cattle that had aborted showed statistically greater *W. chondrophila* antibody titres than cattle that were submitted for other reasons (Dilbeck-Robertson *et al.*, 2003). In addition, other studies have shown a strong correlation between *W. chondrophila* seropositivity and miscarriage in humans (Baud *et al.*, 2014; Baud *et al.*, 2007). Complementing these reports are molecular studies in which *W. chondrophila* DNA was detected in placental material or vaginal swabs by PCR in cattle and humans affected by pregnancy complications (Barkallah *et al.*, 2013; Baud *et al.*, 2014; Baud *et al.*, 2015; Hornung *et al.*, 2015). Importantly, *W. chondrophila* has been demonstrated to be localized to the placenta (Baud *et al.*, 2011).

Infection studies have revealed that both human endometrial cells and ovine trophoblast cells are highly permissive to *W. chondrophila*, with EB to RB conversion occurring between 6 to 12 hours post-infection (Kebbi-Beghdadi *et al.*, 2011b; Wheelhouse *et al.*, 2014). In both cell lines, genome copy numbers and the number of bacteria per infected cell peaked at 24 hours post-infection. In endometrial cells, this coincided with the absence of cytopathic effects, followed by the spontaneous development into large aberrant bodies around 72 h post infection, as visualized by immunofluorescence and electron microscopy. These features are reminiscent of the persistence stress-response stage exhibited by *Chlamydia* species, likely developed in response to starvation (Kebbi-Beghdadi *et al.*, 2011b).

The pathogenic potential of *W. chondrophila* was first established by the development of a murine model of genital *W. chondrophila* infection. It was found that only high doses (10^9 or 10^7) of live *W. chondrophila* injected into the uterine horns of mice were sufficient to stimulate strong, specific IgG responses, compared to infection with heat-inactivated bacteria (Vasilevsky *et al.*, 2015). In addition to elevated immune responses, infected mice developed organ pathology, including splenomegaly and hepatomegaly, with microscopic lesions observed in the liver, spleen and uterine horns. Macrophage infiltration was seen in the liver and histiocytic nodules developed in the spleen and endometrium of infected mice. Previously, innate immune markers, CXCL8, and TNF-a, were highly expressed in trophoblast cells in response to high infection loads of live *W. chondrophila* (Wheelhouse *et al.*, 2014), the former of which may be expressed in a MAPK-dependent manner to recruit neutrophils to the site of infection (Storrie *et al.*, 2016).

Building upon this, Wheelhouse *et al.* developed the first experimentally induced abortion model in pregnant cattle (Wheelhouse *et al.*, 2016). *W. chondrophila* EBs were isolated from a bovine placenta six months after experimental intravenous infection with *W. chondrophila*. Although superficial placental lesions were observed, the animal did not abort and no viable organisms could be recovered from other fetal tissues despite PCR detection (Wheelhouse *et al.*, 2016). This mirrored the murine model where, even though bacterial DNA was detected in other organs, no viable bacteria could be recovered from the spleen or liver seven, 14 or 21 days post-infection. Thus the ability of Waddlia to cause abortion has not yet been experimentally confirmed. Still, *Waddlia* may be considered a zoonotic pathogen as contact with animals was strongly associated with *W. chondrophila* serostatus in some cohorts (Baud *et al.*, 2007) but not others (Baud *et al.*, 2014; Hornung *et al.*, 2015).

Waddlia as a respiratory pathogen

Evidence for colonization of the respiratory tract of humans is very limited for *W. chondrophila*, which has only been detected in four respiratory samples from two studies (Buhl *et al.*, 2018; Goy *et al.*, 2009; Haider *et al.*, 2008; Lienard *et al.*, 2011; Niemi *et al.*, 2011). However, this site has been postulated as the local reservoir within humans (Baud *et al.*, 2014), from which the reproductive tract is infected via the bloodstream.

Infectivity studies and a murine model support this hypothesis. First, both heat-inactivated and live bacteria were internalized by macrophages by 24 hours post-infection (Goy *et al.*, 2008). Live bacteria replicated within vacuoles between 24 and 36 hours, which is macrophage lysis and EB release began. Next, *W. chondrophila* was successfully maintained in A549 pneumocytes, with infection plateauing around 72 hpi (Kebbi-Beghdadi *et al.*, 2011b). Pilloux *et al.* then demonstrated the pathogenicity of *W. chondrophila in vivo* (Pilloux *et al.*, 2016). Following intra-nasal inoculation of mice, *W. chondrophila* was detected in and recovered from the lungs, where it was shown to be actively replicating, from one hour to 14 days post-infection. Aligning with the studies outlined above, *W. chondrophila* DNA was detected in the spleen and kidneys. Moreover, this infection resulted in a peri-bronchial and peri-bronchiolar pattern that is characteristic of a subacute pneumonia and elicited an immune response characterized by neutrophilic and lympho-plasmacytic infiltration of alveolar spaces, macrophage recruitment and acute production of pro-inflammatory cytokines (Pilloux *et al.*, 2016).

Waddlia as a model for other chlamydial diseases

As noted earlier, many CRBs remain uncultivated, which presents a major barrier to (a) further investigating cell biology of these species, (b) investigating the growth kinetics and dynamics of these species in native and foreign host cells and (c) clarifying the infection dynamics *in vivo*. One such example is the fish respiratory disease, epitheliocystis, which is associated with an increasing diversity of deep-branching CRBs (Blandford *et al.*, 2018). The disease compromises respiratory function and has resulted in mass mortalities in cultured species, particularly affecting the juvenile stage. Despite attempts, no epitheliocystis agents have been isolated in culture, and therefore Koch's postulates have not yet been proven for those species. *W. chondrophila* has thus been used as a proxy for these agents in cell biology and animal model studies.

Initial studies showing the rapid proliferation of *W. chondrophila* in fish epithelial cell lines (Kebbi-Beghdadi *et al.*, 2011a) paved the way for a Zebrafish *W. chondrophila* infection model (Fehr *et al.*, 2016). Zebrafish models have long been in use due to their small size, ease of breeding and genetic tractability, and infection models have gained popularity due to the transparent larval stages, immune similarities to mammals, and conserved mechanisms for microbe recognition (Fehr *et al.*, 2016). Immunofluorescence confirmed infection following bath immersion, as early as four days post-fertilization, providing evidence for an oral or dermal route of infection for epitheliocystis. Fluorescence microscopy showed recruitment of neutrophils to the infection site, which is often observed in epitheliocystis and is consistent with other chlamydial infections. Microinjection of *W. chondrophila* into the swim bladder then allowed for investigation of infection at an air–epithelial interface. Inclusions were detected in the epithelium and adjacent tissues between 24 and 48 hours post infection. Infection was also characterized by thickening of the swim bladder walls, with moderate fibrosis and neutrophil migration (Fehr *et al.*, 2016).

Lastly, *W. chondrophila* was shown to be susceptible to tetracycline and macrolides, yet resistant to β-lactam and quinolone antibiotics in Vero cells (Goy and Greub, 2009), which was confirmed in the zebrafish model with inclusion formation strongly attenuated by tetracycline (Fehr *et al*, 2016). These results align with reports of successful decrease of epitheliocystis in largemouth bass after oxytetracyline bath (Goodwin *et al.*, 2005).

Conclusions

As discussed throughout this chapter, the last 30 years have seen an increase in the detection and description of novel chlamydial species from diverse hosts such as amoeba, arthropods and diverse vertebrates, and environmental niches, mainly soil and aquatic sources. The question of the limits of chlamydial diversity still remains. With an increasing risk of zoonotic disease as we encroach onto wildlife habitat, it is imperative that we characterize animal pathogens and investigate their potential for spill-over into humans. Further, as microbial diversity displays a 'greater evolutionary depth' than other kingdoms (Hugenholtz *et al.*, 1998), microorganisms, including chlamydiae, are important sources of knowledge about the strategies and limits of many forms of life, and represent potential new resources of biotechnological value. Chlamydiae, as an ecologically diverse phylum with cultivated and uncultivated representatives, offer an ideal foundation for querying the environment and novel eukaryotic hosts for novel species that may pose disease risks or answer evolutionary questions.

To aid this, an ongoing assessment of the way we describe the groups of families within the phylum is required. Terms like '*Chlamydia*-like, '*Chlamydia*-related', 'pathogenic', 'environmental' and 'traditional' do not adequately describe all species within certain groups, especially since these groupings and descriptions will presumably change as we detect and characterize more novel taxa. It is clear that there are many shared hallmarks of the Chlamydiae, in addition to a range of lineage-specific features encoded at the family, genus or species level, presumably to enable infection of, and survival in, certain hosts or niches. Steps have recently been taken towards re-classification of the phylum into three orders (Gupta *et al.*, 2015; Pillonel *et al.*, 2018), yet ideally, these efforts should be based on a unified taxonomy applicable to all lineages in the domain Bacteria (Parks *et al.*, 2018). Another hurdle to overcome will be to culture these organisms *in vitro*, which will help with properly classifying and characterizing novel bacteria and proving Koch's postulates for pathogenesis. This will also allow the confirmation of metabolic needs predicted by culture-independent genomic studies.

Chlamydia-related bacteria are a highly diverse group of intracellular bacteria that reside in the environment. Their evolutionary history as amoeba symbionts and later expansion into a range of multicellular eukaryotes provides an intriguing foundation for basic research into the evolution and mechanisms of intracellular infection strategies, virulence and symbioses.

References

Aistleitner, K., Heinz, C., Hormann, A., Heinz, E., Montanaro, J., Schulz, F., Maier, E., Pichler, P., Benz, R., and Horn, M. (2013). Identification and characterization of a novel porin family highlights a major difference in the outer membrane of chlamydial symbionts and pathogens. PloS one 8, e55010 10.1371/journal.pone.0055010.

Al-Younes H.M., Paldanius, M., M. (2014). High seroprevalence of *Simkania negevensis* in Jordan. Braz J Microbiol 45(4), 1433–1437.

Albrecht, M., Sharma, C.M., Reinhardt, R., Vogel, J., and Rudel, T. (2010). Deep sequencing-based discovery of the *Chlamydia trachomatis* transcriptome. Nucleic Acids Res, 38(3) 868–877, 10.1093/nar/gkp1032.

Amann, R., Springer, N., Schnhuber, W., Ludwig, W., Schmid, E.N., Müller, K.D., and Michel, R. (1997). Obligate intracellular bacterial parasites of acanthamoebae related to *Chlamydia* spp. Appl Environ Microbiol 63, 115–121.

Azuma, Y., Hirakawa, H., Yamashita, A., Cai, Y., Rahman, M.A., Suzuki, H., Mitaku, S., Toh, H., Goto, S., Murakami, T., *et al.*(2006). Genome sequence of the cat pathogen, *Chlamydophila felis*. DNA Res 13(1), 15–23 10.1093/dnares/dsi027.

Balsara, Z.R., Roan, N.R., Steele, L.N., and Starnbach, M.N. (2006). Developmental regulation of *Chlamydia trachomatis* class I accessible protein-1, a CD8+ T cell antigen. The Journal of infectious diseases 193, 1459–1463 10.1086/503370.

Barkallah, M., Fendri, I., Dhieb, A., Gharbi, Y., Greub, G., and Gdoura, R. (2013). First detection of *Waddlia chondrophila* in Africa using SYBR Green real-time PCR on veterinary samples. Vet microbiol 164, 101–107 10.1016/j.vetmic.2013.01.036.

Baud D., Kebbi, C., Kulling, J.P., Greub, G. (2009). Seroprevalence of different *Chlamydia*-like organisms in an asymptomatic population. Clin Microbiol Infect 15 Suppl 2:213–215 10.1111/j.1469–0691.2008.02134.x.

Baud, D., Goy, G., Osterheld, M.C., Borel, N., Vial, Y., Pospischil, A., and Greub, G. (2011). *Waddlia chondrophila*: from bovine abortion to human miscarriage. Clin Infect Dis 52, 1469–1471 10.1093/cid/cir205.

Baud, D., Goy, G., Osterheld, M.C., Croxatto, A., Borel, N., Vial, Y., Pospischil, A., and Greub, G. (2014). Role of *Waddlia chondrophila* placental infection in miscarriage. Emerg Infect Dis 20, 460–464 10.3201/eid2003.131019.

Baud, D., Goy, G., Vasilevsky, S., Osterheld, M.C., Roth-Kleiner, M., Croxatto, A., and Greub, G. (2015). Roles of bovine *Waddlia chondrophila* and *Chlamydia trachomatis* in human preterm birth. New Microbes New Infect 3, 41–45 10.1016/j.nmni.2014.11.004.

Baud, D., Thomas, V., Arafa, A., Regan, L., and Greub, G. (2007). *Waddlia chondrophila*, a potential agent of human fetal death. Emerg Infect Dis 13, 1239–1243 10.3201/eid1308.070315.

Bavoil, P., Ohlin, A., Schachter, J. (1984). Role of disulfide bonding in outer membrane structure and permeability in *Chlamydia trachomatis*. Infect Immun 44: 479–485.

Beder, T., and Saluz, H.P. (2018). Virulence-related comparative transcriptomics of infectious and non-infectious chlamydial particles. BMC genomics 19, 575 10.1186/s12864-018-4961-x.

Benamar, S., Bou Khalil, J.Y., Blanc-Tailleur, C., Bilen, M., Barrassi, L., and La Scola, B. (2017). Developmental Cycle and Genome Analysis of *Protochlamydia massiliensis* sp. nov. a New Species in the *Parachlamydiacae* Family. Front Cell Infect Microbiol 7, 385 10.3389/fcimb.2017.00385.

Bertelli, C., Collyn, F., Croxatto, A., Rückert, C., Polkinghorne, A., Kebbi-Beghdadi, C., Goesmann, A., Vaughan, L., and Greub, G. (2010). The *Waddlia* genome: a window into chlamydial biology. PloS one 5(5), e10890 10.1371/journal.pone.0010890.

Bertelli, C., Goesmann, A., and Greub, G. (2014). *Criblamydia sequanensis* Harbours a Megaplasmid Encoding Arsenite Resistance. Genome Announc 2 10.1128/genomeA.00949-14.

Bertelli, C., Aeby, S., Chassot, B., Clulow, J., Hilfiker, O., Rappo, S., Ritzmann, S., Schumacher, P., Terrettaz, C., Benaglio, P., *et al.* (2015). Sequencing and characterizing the genome of *Estrella lausannensis* as an undergraduate project: training students and biological insights. Frontiers in microbiology 6, 101 10.3389/fmicb.2015.00101.

Bertelli, C., Cisse, O.H., Rusconi, B., Kebbi-Beghdadi C., Croxatto, A., Goesmann, A., Collyn, F., Greub, G. (2016). CRISPR System Acquisition and Evolution of an Obligate Intracellular *Chlamydia*-Related Bacterium. Genome Biol Evol 8:2376–2386 10.1093/gbe/evw138.

Betts-Hampikian, H.J., and Fields, K.A. (2010). The Chlamydial Type III Secretion Mechanism: Revealing Cracks in a Tough Nut. Frontiers in microbiology 1, 114 10.3389/fmicb.2010.00114.

Birtles, R.J., Rowbotham, T.J., Storey, C., Marrie, T.J., & Raoult, D. (1997). *Chlamydia*-like obligate parasite of free-living amoebae. The Lancet 349(9056), 925–926.

Blandford, M.I., Taylor-Brown, A., Schlacher, T.A., Nowak, B., and Polkinghorne, A. (2018). Epitheliocystis in fish: An emerging aquaculture disease with a global impact. Transbound Emerg Dis 10.1111/tbed.12908.

Bou Khalil, J.Y., Benamar, S., Baudoin, J.P., Croce, O., Blanc-Tailleur C., Pagnier, I., Raoult, D., La Scola B. (2016). Developmental Cycle and Genome Analysis of '*Rubidus massiliensis*,' a New *Vermamoeba vermiformis* Pathogen. Front Cell Infect Microbiol 6:31.

Bou Khalil, J.Y., Benamar, S., Di Pinto F., Blanc-Tailleur C., Raoult, D., La Scola B. (2017). *Protochlamydia phocaeensis* sp. nov., a new Chlamydiales species with host dependent replication cycle. Microbes Infect 19:343–350 10.1016/j.micinf.2017.02.003.

Buhl, H., Eibach, D., Nagel, M., Greub, G., Borel, N., Sarpong, N., Rettig, T., Pesch, T., Aeby, S., Klockner, A., *et al.* (2018). Chlamydiae in febrile children with respiratory tract symptoms and age-matched controls, Ghana. New Microbes New Infect 22, 44–48 10.1016/j.nmni.2017.12.008.

Burnard, D, Weaver H, Gillett A, Loader J, Flanagan C, Polkinghorne, A. [Year?] A. Chlamydiales genotypes identified in ticks from Australian wildlife. Parasit Vectors. 2017 Jan 26;10(1):46. doi: 10.1186/s13071-017-1994-y.

Case, E.D., peterson, E.M., and Tan, M. (2010). Promoters for *Chlamydia* Type III Secretion genes show a differential response to DNA supercoiling that correlates with temporal expression pattern. J. Bacteriol. 192(10) 2569–2574 10.1128/JB.00068-10.

Casson, N., Michel, R., Müller, K.D., Aubert, J.D., Greub, G. (2008). *Protochlamydia naegleriophila* as etiologic agent of pneumonia. Emerg Infect Dis, 14(1):168–72 10.3201/eid1401.070980.

Casson, N., Entenza, J.M., Borel, N., Pospischil, A., Greub, G. (2008). Murine model of pneumonia caused by *Parachlamydia acanthamoebae*. Microb Pathog, 45:92–97 10.1016/j.micpath.2008.04.003.

Celli, J., & Tsolis, R.M. (2015). Bacteria, the endoplasmic reticulum and the unfolded protein response: friends or foes? Nat Rev Microbiol, 13(2), 71 10.1038/nrmicro3393.

Chen, C.P., Tseng, C.H., Chen, C.A., Tang, S.L. (2011). The dynamics of microbial partnerships in the coral *Isopora palifera*. ISME J 5:728 10.1038/ismej.2010.151.

Chow, S.H., Deo, P., & Naderer, T. (2016). Macrophage cell death in microbial infections. Cell Microbiol 18(4), 466–474 10.1111/cmi.12573.

Chua, P.K., Corkill, J.E., Hooi, P.S., Cheng, S.C., Winstanley, C., & Hart, C.A. (2005). Isolation of *Waddlia malaysiensis*, a novel intracellular bacterium, from fruit bat (*Eonycteris spelaea*). Emerging infectious diseases, 11(2), 271.

Coburn, B., Sekirov, I., & Finlay, B.B. (2007). Type III secretion systems and disease. Clin Microbiol Rev 20(4), 535–549.

Collingro, A., *et al.* (2005). Recovery of an environmental chlamydia strain from activated sludge by co-cultivation with *Acanthamoeba* sp. Microbiology 151:301–309 10.1099/mic.0.27406-0.

Collingro, A., Kostlbacher, S., Mussmann, M., Stepanauskas, R., Hallam, S.J., Horn, M. (2017). Unexpected genomic features in widespread intracellular bacteria: evidence for motility of marine.

Collingro, A., Toenshoff, E.R., Taylor, M.W., Fritsche, T.R., Wagner, M., Horn, M. (2005). '*Candidatus* Protochlamydia amoebophila', an endosymbiont of *Acanthamoeba* spp. Int J Syst Evol Microbiol 55:1863–1866 10.1099/ijs.0.63572-0.

Collingro, A., Tischler, P., Weinmaier, T., Penz, T., Heinz, E., Brunham, R.C., Read, T.D., Bavoil, P.M., Sachse, K., Kahane, S., *et al.* (2011). Unity in variety--the pan-genome of the Chlamydiae. Mol Biol Evol 28, 3253–3270 10.1093/molbev/msr161.

Collingro, A., Köstlbacher, S., Mussmann, M., Stepanauskas, R., Hallam, S.J., and Horn, M. (2017). Unexpected genomic features in widespread intracellular bacteria: evidence for motility of marine chlamydiae. ISME J. 11(10):2334–2344. 10.1038/ismej.2017.95.

Corsaro, D., Michel, R., Walochnik, J., Muller, K.D., Greub, G. (2010). *Saccamoeba lacustris*, sp. nov. (Amoebozoa: Lobosea: *Hartmannellidae*), a new lobose amoeba, parasitized by the novel chlamydia '*Candidatus* Metachlamydia lacustris' (Chlamydiae: Parachlamydiaceae). Eur J Protistol 46:86–95 10.1016/j.ejop.2009.11.002.

Corsaro, D., Muller, K.D., Wingender, J., Michel, R. (2013). '*Candidatus* Mesochlamydia elodeae' (Chlamydiae: Parachlamydiaceae), a novel chlamydia parasite of free-living amoebae. Parasitol Res 112:829–838 10.1007/s00436-012-3213-2.

Corsaro, D., Thomas, V., Goy, G., Venditti, D., Radek, R., Greub, G. (2007). '*Candidatus* Rhabdochlamydia crassificans', an intracellular bacterial pathogen of the cockroach *Blatta orientalis* (Insecta: Blattodea). Syst Appl Microbiol 30:221–228 10.1016/j.syapm.2006.06.001.

Corsaro, D., Venditti, D., Le Faou, A., Guglielmetti, P., Valassina, M. (2001). A new *Chlamydia*-like 16S rDNA sequence from a clinical sample. Microbiology Res 147, pp. 515–516 10.1099/00221287-147-3-515.

Corsaro, D., Venditti, D., Valassina, M. (2002). New parachlamydial 16S rDNA phylotypes detected in human clinical samples. Res Microbiol 153:563–567 10.1016/S0923-2508(02)01369-4.

Corsaro, D., Valassina, M., Venditti, D. (2003). Increasing diversity within Chlamydiae. Crit Rev Microbiol 29:37–78 10.1080/713610404.

Corsaro, D., Venditti, D. (2004). Emerging chlamydial infections. Crit Rev Microbiol 30:75–106 10.1080/10408410490435106.

Corsaro, D., Michel, R., Walochnik, J., Muller, K.D., Greub, G. (2010). *Saccamoeba lacustris*, sp. nov. (Amoebozoa: Lobosea: Hartmannellidae), a new lobose amoeba, parasitized by the novel chlamydia '*Candidatus* Metachlamydia lacustris' (*Chlamydiae: Parachlamydiaceae*). Eur J Protistol 46:86–95 10.1016/j.ejop.2009.11.002.

Corsaro, D., Thomas, V., Goy, G., Venditti, D., Radek, R., Greub, G. (2007). '*Candidatus* Rhabdochlamydia crassificans', an intracellular bacterial pathogen of the cockroach *Blatta orientalis* (Insecta: Blattodea). Syst Appl Microbiol 30:221–228 10.1016/j.syapm.2006.06.001.

Corsaro, D., Work, T.M. (2012). *Candidatus* Renichlamydia lutjani, a Gram-negative bacterium in internal organs of blue-striped snapper *Lutjanus kasmira* from Hawaii. Dis Aquat Organ 98:249–254 10.3354/dao02441.

Corsaro, D., Muller, K.D., Wingender, J., Michel, R. (2013). '*Candidatus* Mesochlamydia elodeae' (*Chlamydiae: Parachlamydiaceae*), a novel chlamydia parasite of free-living amoebae. Parasitol Res 112:829–838 10.1007/s00436-012-3213-2.

Coulon, C., Eterpi, M., Greub, G., Collignon, A., McDonnell G., Thomas, V. (2012). Amoebal host range, host-free survival and disinfection susceptibility of environmental Chlamydiae as compared to *Chlamydia trachomatis*. FEMS Immunol Med Microbiol 64:364–373 10.1111/j.1574-695X.2011.00919.x.

Croxatto, A., Greub, G. (2010). Early intracellular trafficking of *Waddlia chondrophila* in human macrophages. Microbiology 156:340–355 10.1099/mic.0.034546-0.

Croxatto, A., Murset, V., Chassot, B., Greub, G. (2013). Early expression of the type III secretion system of *Parachlamydia acanthamoebae* during a replicative cycle within its natural host cell *Acanthamoeba castellanii*. Pathog Dis 69:159–175 10.1111/2049-632X.12065.

Croxatto, A., Rieille, N., Kernif, T., Bitam, I., Aeby, S., Peter, O., Greub, G. (2014). Presence of Chlamydiales DNA in ticks and fleas suggests that ticks are carriers of Chlamydiae. Ticks Tick Borne Dis 5:359–365 10.1016/j.ttbdis.2013.11.009.

Dean, D., Powers, V.C. (2001). Persistent *Chlamydia trachomatis* infections resist apoptotic stimuli. Infect Immun 69: 2442–2447 10.1128/IAI.69.4.2442-2447.2001.

Dilbeck, P.M., Evermann, J.F., Crawford, T.B., Ward, A.C., Leathers, C.W., Holland, C.J., Mebus, C.A., Logan, L.L., Rurangirwa, F.R., and McGuire, T.C. (1990). Isolation of a previously undescribed rickettsia from an aborted bovine fetus. J Clin Microbiol 28, 814–816.

Dilbeck-Robertson, P., McAllister, M.M., Bradway, D., and Evermann, J.F. (2003). Results of a new serological test suggest an association of *Waddlia chondrophila* with bovine abortion. J Vet Diagn Invest 15, 568–569 10.1177/104063870301500609.

Dobrindt, U., Hochhut, B., Hentschel, U., and Hacker, J. (2004). Genomic islands in pathogenic and environmental microorganisms. Nat Rev Microbiol 2, 414–424 10.1038/nrmicro884.

Domman, D., Collingro, A., Lagkouvardos, I., Gehre, L., Weinmaier, T., Rattei, T., Subtil, A., and Horn, M. (2014). Massive expansion of ubiquitination-related gene families within the Chlamydiae. Molecular biology and evolution 10.1093/molbev/msu227.

Donati, M., Fiani, N., Di Francesco, A., Di Paolo, M., Vici, M., & Cevenini, R. (2013). IgG and IgA response to *Simkania negevensis* in sera of patients with respiratory and gastrointestinal symptoms. New Microbiol 36(3), 303–306.

Draghi, A., 2nd, Bebak, J., Daniels, S., Tulman, E.R., Geary, S.J., West, A.B., Popov, V.L., Frasca, S., Jr. (2010). Identification of '*Candidatus* Piscichlamydia salmonis' in Arctic charr *Salvelinus alpinus* during a survey of charr production facilities in North America. Dis Aquat Organ 89:39–49 10.3354/dao02171.

Draghi, A., 2nd, Popov, V.L., Kahl, M.M., Stanton, J.B., Brown, C.C., Tsongalis, G.J., West, A.B., Frasca, S., Jr. (2004). Characterization of '*Candidatus* piscichlamydia salmonis' (order Chlamydiales), a chlamydia-like bacterium associated with epitheliocystis in farmed Atlantic salmon (*Salmo salar*). J Clin Microbiol 42:5286–5297 10.1128/JCM.42.11.5286-5297.2004.

Dugan, J., Rockey, D.D., Jones, L., and Andersen, A.A. (2004). Tetracycline resistance in *Chlamydia suis* mediated by genomic islands inserted into the chlamydial *inv*-like gene. Antimicrob Agents Chemother 48, 3989–3995 10.1128/AAC.48.10.3989-3995.2004.

Everett, K.D., Bush, R.M., & Andersen, A.A. (1999). Emended description of the order Chlamydiales, proposal of Parachlamydiaceae fam. nov. and Simkaniaceae fam. nov., each containing one monotypic genus, revised taxonomy of the family Chlamydiaceae, including a new genus and five new species, and standards for the identification of organisms. Int J Syst Evol Microbiol 49(2), 415–440.

Everett, K.D., Thao, M., Horn, M., Dyszynski, G.E., Baumann, P. (2005). Novel chlamydiae in whiteflies and scale insects: endosymbionts '*Candidatus* Fritschea bemisiae' strain Falk and '*Candidatus* Fritschea eriococci' strain Elm. Int J Syst Evol Microbiol 55:1581–1587 10.1099/ijs.0.63454-0.

Faherty, C.S., Maurelli, A.T. (2008). Staying alive: bacterial inhibition of apoptosis during infection. Trends Microbiol 16: 173–180 10.1016/j.tim.2008.02.001.

Fasoli, L., Paldanius, M., Don, M., Valent, F., Vetrugno, L., Korppi, M., & Canciani, M. (2008). *Simkania negevensis* in community-acquired pneumonia in Italian children. Scand J Infect Dis 40(3), 269–272 10.1080/00365540701642146.

Fehr, A., Walther, E., Schmidt-Posthaus H., Nufer, L., Wilson, A., Svercel, M., Richter, D., Segner, H., Pospischil, A., Vaughan, L. (2013). *Candidatus* Syngnamydia venezia, a novel member of the phylum Chlamydiae from the broad nosed pipefish, *Syngnathus typhle*. PLoS One 8:e70853 10.1371/journal.pone.0070853.

Fehr, A.G., Ruetten, M., Seth-Smith H.M., Nufer, L., Voegtlin, A., Lehner, A., Greub, G., Crosier, P.S., Neuhauss, S.C., Vaughan, L. (2016). A Zebrafish Model for *Chlamydia* Infection with the Obligate Intracellular Pathogen *Waddlia chondrophila*. Front Microbiol 7:1829 10.3389/fmicb.2016.01829.

Fisher, D.J., Fernandez, R.E., and Maurelli, A.T. (2013). *Chlamydia trachomatis* transports NAD via the Npt1 ATP/ADP translocase. J Bacteriol 195, 3381–3386 10.1128/JB.00433-13.

Francis, M.S., Wolf-Watz, H., & Forsberg, Å. (2002). Regulation of type III secretion systems. Curr Opin Microbiol 5(2), 166–172 10.1016/S1369-5274(02)00301-6.

Friedman, M.G., Kahane, S., Dvoskin, B., & Hartley, J.W. (2006). Detection of *Simkania negevensis* by culture, PCR, and serology in respiratory tract infection in Cornwall, UK. J Clin Pathol 59(3), 331–333 10.1136/jcp.2004.025601.

Fritsche, T.R., Gautom, R.K., Seyedirashti, S., Bergeron, D.L., Lindquist, T.D. (1993). Occurrence of bacterial endosymbionts in *Acanthamoeba* spp. isolated from corneal and environmental specimens and contact lenses. J Clin Microbiol 31:1122–1126.

Fuchs, T.M., Eisenreich, W., Heesemann, J., and Goebel, W. (2012). Metabolic adaptation of human pathogenic and related nonpathogenic bacteria to extra- and intracellular habitats. FEMS microbiology reviews 36, 435–462 10.1111/j.1574-6976.2011.00301.x.

Goldsmith, D.B., *et al.* (2018). Comparison of microbiomes of cold-water corals *Primnoa* pacifica and *Primnoa resedaeformis*, with possible link between microbiome composition and host genotype. Sci Rep 8:12383 10.1038/s41598-018-30901-z.

Goodwin, A.E., Park, E., and Nowak, B.F. (2005). Successful treatment of largemouth bass, *Micropterus salmoides* (L.), with epitheliocystis hyperinfection. J Fish Dis 28, 623–625 10.1111/j.1365-2761.2005.00662.x.

Goy, G., Croxatto, A., and Greub, G. (2008). *Waddlia chondrophila* enters and multiplies within human macrophages. Microbes Infect 10, 556–562 10.1016/j.micinf.2008.02.003.

Goy, G., Croxatto, A., Posfay-Barbe, K.M., Gervaix, A., and Greub, G. (2009). Development of a real-time PCR for the specific detection of *Waddlia chondrophila* in clinical samples. Eur J Clin Microbiol Infect Dis 28, 1483–1486 10.1007/s10096-009-0804-7.

Greub, G., & Raoult, D. (2002). Crescent bodies of Parachlamydia acanthamoeba and its life cycle within *Acanthamoeba polyphaga*: an electron micrograph study. Appl Environ Microbiol 68:3076–3084 10.1128/AEM.68.6.3076-3084.2002.

Greub, G., & Raoult, D. (2002). Parachlamydiaceae: potential emerging pathogens. Emerg Infect Dis 8(6), 626 10.3201/eid0806.010210.

Greub, G., & Raoult, D. (2004). Microorganisms resistant to free-living amoebae. Clin Microbiol Rev 17(2), 413–433 10.1128/CMR.17.2.413-433.2004.

Greub, G., and Raoult, D. (2003). History of the ADP/ATP-translocase-encoding gene, a parasitism gene transferred from a Chlamydiales ancestor to plants 1 billion years ago. Appl Environ Microbiol 69, 5530–5535 10.1128/AEM.69.9.5530-5535.2003.

Greub, G., Boyadjiev, I., La Scola, B., Raoult, D., & Martin, C. (2003). Serological hint suggesting that Parachlamydiaceae are agents of pneumonia in polytraumatized intensive care patients. Ann N Y Acad Sci 990(1), 311–319 10.1111/j.1749-6632.2003.tb07381.x.

Greub, G., Collyn, F., Guy, L., and Roten, C.A. (2004). A genomic island present along the bacterial chromosome of the Parachlamydiaceae UWE25, an obligate amoebal endosymbiont, encodes a potentially functional F-like conjugative DNA transfer system. BMC microbiology 4, 48 10.1186/1471-2180-4-48.

Greub, G., Desnues, B., Raoult, D., & Mege, J.L. (2005). Lack of microbicidal response in human macrophages infected with *Parachlamydia acanthamoebae*. Microbes Infect, 7(4), 714–719 10.1016/j.micinf.2005.01.009.

Greub, G. (2009). *Parachlamydia acanthamoebae*, an emerging agent of pneumonia. Clin Microbiol Infect 15, 18–28 10.1111/j.1469-0691.2008.02633.x.

Greub, G., Kebbi-Beghdadi, C., Bertelli, C., Collyn, F., Riederer, B.M., Yersin, C., Croxatto, A., and Raoult, D. (2009). High throughput sequencing and proteomics to identify immunogenic proteins of a new pathogen: the dirty genome approach. PLoS One, 4(12), e8423 10.1371/journal.pone.0008423.

Grieshaber, S., Grieshaber, N., Yang, H., Baxter, B., Hackstadt, T., Omsland, A. (2018). Impact of active metabolism on *Chlamydia trachomatis* elementary body transcript profile and infectivity. J Bacteriol. 200(14), pii: e00065-18. doi:10.1128/JB.00065-18.

Guevara Soto M., Vidondo, B., Vaughan, L., Seth-Smith H.M., Nufer, L., Segner, H., Rubin, J.F., Schmidt-Posthaus H. (2016). The emergence of epitheliocystis in the upper Rhone region: evidence for Chlamydiae in wild and farmed salmonid populations. Arch Microbiol 198:315–324 10.1007/s00203-016-1192-x.

Gupta, R.S., Naushad, S., Chokshi, C., Griffiths, E., Adeolu, M. (2015). A phylogenomic and molecular markers based analysis of the phylum Chlamydiae: proposal to divide the class Chlamydiia into two orders, Chlamydiales and *Para*Chlamydiales ord. nov., and emended description of the class Chlamydiia. Antonie Van Leeuwenhoek 108:765–781 10.1007/s10482-015-0532-1.

Hackstadt, T., Fischer, E.R., Scidmore, M.A., Rockey, D.D., & Heinzen, R.A. (1997). Origins and functions of the chlamydial inclusion. Trends Microbiol 5(7), 288–293 10.1016/S0966-842X(97)01061-5.

Haferkamp, I., Schmitz-Esser, S., Linka, N., Urbany, C., Collingro, A., Wagner, M., Horn, M., and Neuhaus, H.E. (2004). A candidate NAD+ transporter in an intracellular bacterial symbiont related to Chlamydiae. Nature 432, 622–625 10.1038/nature03131.

Haferkamp, I., Schmitz-Esser, S., Wagner, M., Neigel, N., Horn, M., and Neuhaus, H.E. (2006). Tapping the nucleotide pool of the host: novel nucleotide carrier proteins of *Protochlamydia amoebophila*. Mol Microbiol 60, 1534–1545 10.1111/j.1365-2958.2006.05193.x.

Haider, S., Collingro, A., Walochnik, J., Wagner, M., and Horn, M. (2008). *Chlamydia*-like bacteria in respiratory samples of community-acquired pneumonia patients. FEMS Microbiol Lett 281, 198–202 10.1111/j.1574-6968.2008.01099.x.

Haider, S., Wagner, M., Schmid, M.C., Sixt, B.S., Christian, J.G., Häcker, G., Pichler, P., Mechtler, K., Müller, A., Baranyi, C., *et al.*(2010). Raman microspectroscopy reveals long-term extracellular activity of chlamydiae. Mol Microbiol 77(3), 687–700 10.1111/j.1365-2958.2010.07241.x.

Halberstaedter, L., Von Prowazek, S. (1907). Zur Atiologie des Trachoms. Deutsche Medizinische Wochenschrift 33, 1285–1287.

Hatch, T.P. (1999). Developmental Biology. In: Stephens RS, editor. *Chlamydia*. Washington DC: American Society for Microbiology. pp. 29–67.

Hayashi, Y., Nakamura, S., Matsuo, J., Fukumoto, T., Yoshida, M., Takahashi, K., Mizutani, Y., Yao, T., and Yamaguchi, H. (2010). Host range of obligate intracellular bacterium *Parachlamydia acanthamoebae*. Microbiol Immunol 54(11), 707–713 10.1111/j.1348-0421.2010.00265.x.

Heinz, E., Tischler, P., Rattei, T., Myers, G., Wagner, M., and Horn, M. (2009). Comprehensive *in silico* prediction and analysis of chlamydial outer membrane proteins reflects evolution and life style of the Chlamydiae. BMC genomics *10*, 634 10.1186/1471-2164-10-634.

Heinz, E., Rockey, D.D., Montanaro, J., Aistleitner, K., Wagner, M., & Horn, M. (2010). Inclusion membrane proteins of *Protochlamydia amoebophila* UWE25 reveal a conserved mechanism for host cell interaction among the Chlamydiae. J Bacteriol 192(19), 5093–5102.

Heinz, E., Hacker, C., Dean, P., Mifsud, J., Goldberg, A.V., Williams, T.A., Nakjang, S., Gregory, A., Hirt, R.P., Lucocq, J.M., *et al.* (2014). Plasma membrane-located purine nucleotide transport proteins are key components for host exploitation by microsporidian intracellular parasites. PLoS Pathog 10, e1004547 10.1371/journal.ppat.1004547.

Henning, K., Schares, G., Granzow, H., Polster, U., Hartmann, M., Hotzel, H., Sachse, K., Peters, M., and Rauser, M. (2002). *Neospora caninum* and *Waddlia chondrophila* strain 2032/99 in a septic stillborn calf. Vet Microbiol 85, 285–292 10.1016/S0378-1135(01)00510-7.

Herweg, J.A., & Rudel, T. (2016). Interaction of Chlamydiae with human macrophages. FEBS J, 283(4), 608–618 10.1111/febs.13609.

Herweg, J.A., Pons, V., Becher, D., Hecker, M., Krohne, G., Barbier, J., Berger, H., Rudel, T., and Mehlitz, A. (2016). Proteomic analysis of the *Simkania*-containing vacuole: the central role of retrograde transport. Mol Microbiol 99(1), 151–171 10.1111/mmi.13222.

Heuer, D., Rejman Lipinski, A., Machuy, N., Karlas, A., Wehrens, A., Siedler, F., Brinkmann, V., and Meyer, T.F. (2009). *Chlamydia* causes fragmentation of the Golgi compartment to ensure reproduction. Nature 457, 731–735 10.1038/nature07578.

Hokynar, K., Sormunen, J.J., Vesterinen, E.J., Partio, E.K., Lilley, T., Timonen, V., Panelius, J., Ranki, A., Puolakkainen, M. (2016). *Chlamydia*-Like Organisms (CLOs) in Finnish *Ixodes ricinus* Ticks and Human Skin. Microorganisms 4 10.3390/microorganisms4030028.

Hokynar, K., Vesterinen, E.J., Lilley, T.M., Pulliainen, A.T., Korhonen, S.J., Paavonen, J., Puolakkainen, M. (2017). Molecular evidence of *Chlamydia*-like organisms in the faeces of *Myotis daubentonii* bats. Appl Environ Microbiol 83:e02951-16 10.1128/AEM.02951-16.

Horn, M., Collingro, A., Schmitz-Esser S., Beier, C.L., Purkhold, U., *et al.* (2004). Illuminating the evolutionary history of chlamydiae. Science 304: 728–730 10.1126/science.1096330.

Horn, M., Wagner, M., Muller, K.D., Schmid, E.N., Fritsche, T.R., Schleifer, K.H., Michel, R. (2000). *Neochlamydia hartmannellae* gen. nov., sp. nov. (Parachlamydiaceae), an endoparasite of the amoeba *Hartmannella vermiformis*. Microbiology 146 (Pt 5):1231–1239 10.1099/00221287-146-5-1231.

Horn, M. (2008). Chlamydiae as symbionts in eukaryotes. Annu. Rev. Microbiol 62:113–131 10.1146/annurev.micro.62.081307.162818.

Hornung, S., Thuong, B.C., Gyger, J., Kebbi-Beghdadi C., Vasilevsky, S., Greub, G., Baud, D. (2015). Role of *Chlamydia trachomatis* and emerging *Chlamydia*-related bacteria in ectopic pregnancy in Vietnam. Epidemiol Infect 143:2635–2638 10.1017/S0950268814003616.

Hueck, C.J. (1998). Type III protein secretion systems in bacterial pathogens of animals and plants. Microbiol Mol Biol Rev 62, 379–433.

Hugenholtz, P., Goebel, B.M., and Pace, N.R. (1998). Impact of culture-independent studies on the emerging phylogenetic view of bacterial diversity. J Bacteriol *180*, 4765–4774.

Hybiske, K., & Stephens, R.S. (2007). Mechanisms of host cell exit by the intracellular bacterium *Chlamydia*. Proc Natl Acad Sci U S A 104(27), 11430–11435 10.1073/pnas.0703218104.

Ishida, K., Sekizuka, T., Hayashida, K., Matsuo, J., Takeuchi, F., Kuroda, M., Nakamura, S., Yamazaki, T., Yoshida, M., Takahashi, K., *et al.* (2014). Amoebal endosymbiont *Neochlamydia* genome sequence illuminates the bacterial role in the defence of the host amoebae against *Legionella pneumophila*. PloS one *9*, e95166 10.1371/journal.pone.0095166.

Ito, A, Matsuo J, Nakamura S, Yoshida A, Okude M, Hayashi Y, Sakai H, Yoshida M, Takahashi K, Yamaguchi H. 2012. Amoebal endosymbiont *Protochlamydia* induces apoptosis to human immortal HEp-2 cells. PLoS One 7:e30270 10.1371/journal.pone.0030270.

Jacquier, N., Frandi, A., Pillonel, T., Viollier, P.H., Greub, G. (2014). Cell wall precursors are required to organize the chlamydial division septum. Nat Commun. 5, 3578. doi: 10.1038/ncomms4578. Erratum in: Nat Commun. 2014;5:4609.

Jacquier, N., Viollier, P.H., Greub, G. (2015). The role of peptidoglycan in chlamydial cell division: towards resolving the chlamydial anomaly. FEMS Microbiol Rev. 39(2), 262–75. doi: 10.1093/femsre/fuv001.

Jelocnik, M., Bachmann, N.L., Seth-Smith, H., Thomson, N.R., Timms, P., and Polkinghorne, A.M. (2016). Molecular characterization of the *Chlamydia pecorum* plasmid from porcine, ovine, bovine, and koala strains indicates plasmid-strain co-evolution. PeerJ 4, e1661 10.7717/peerj.1661.

Kahane, S., Dvoskin, B., Friedman, M.G. (2008). The role of monocyte/macrophages as vehicles of dissemination of *Simkania negevensis*: an *in vitro* simulation model. FEMS Immunol Med Microbiol 52: 219–227 10.1111/j.1574-695X.2007.00361.x.

Kahane, S., Dvoskin, B., Mathias, M., & Friedman, M.G. (2001). Infection of *Acanthamoeba polyphaga* with *Simkania negevensis* and *S. negevensis* Survival within Amoebal Cysts. Appl Environ Microbiol, 67(10), 4789–4795 10.1128/AEM.67.10.4789-4795.2001.

Kahane, S., Gonen, R., Sayada, C., Elion, J., and Friedman, M.G. (1993). Description and partial characterization of a new *Chlamydia*-like microorganism. FEMS Microbiol Lett, 109 pp. 329–333 10.1111/j.1574-6968.1993.tb06189.x.

Kahane, S., Everett, K.D., Kimmel, N., and Friedman, M.G. (1999). *Simkania negevensis* strain ZT: growth, antigenic and genome characteristics. Int J Syst Bacteriol 49 Pt 2, 815–820.

Kahane, S., Kimmel, N., & Friedman, M.G. (2002). The growth cycle of *Simkania negevensis*. Microbiology, 148(3), 735–742 10.1099/00221287-148-3-735.

Kalman, S., Mitchell, W., Marathe, R., Lammel, C., Fan, J., Hyman, R.W., Olinger, L., Grimwood, J., Davis, R.W., and Stephens, R.S. (1999). Comparative genomes of *Chlamydia pneumoniae* and *C. trachomatis*. Nature Genet 21, 385–389 10.1038/7716.

Karbowiak, G., Biernat, B., Szewczyk, T., Sytykiewicz, H. (2015). The role of particular tick developmental stages in the circulation of tick-borne pathogens affecting humans in Central Europe, Pt 1: The general pattern. Ann Parasitol 61:221–228 10.17420/ap6104.11.

Karlsen, M., Nylund, A., Watanabe, K., Helvik, J.V., Nylund, S., Plarre, H. (2008). Characterization of *'Candidatus* Clavochlamydia salmonicola': an intracellular bacterium infecting salmonid fish. Environ Microbiol 10:208–218 10.1111/j.1462-2920.2007.01445.x.

Karunakaran, K., Mehlitz, A., Rudel, T. (2011). Evolutionary conservation of infection-induced cell death inhibition among Chlamydiales. PLoS One 6: e22528 10.1371/journal.pone.0022528.

Kebbi-Beghdadi, C., Batista, C., and Greub, G. (2011a). Permissivity of fish cell lines to three *Chlamydia*-related bacteria: *Waddlia chondrophila, Estrella lausannensis* and *Parachlamydia acanthamoebae*. FEMS Immunol Med Microbiol 63, 339–345 10.1111/j.1574-695X.2011.00856.x.

Kebbi-Beghdadi, C., Cisse, O., and Greub, G. (2011b). Permissivity of Vero cells, human pneumocytes and human endometrial cells to *Waddlia chondrophila*. Microbes and Infect 13, 566–574 10.1016/j.micinf.2011.01.020.

Knab, S, Mushak TM, Schmitz-Esser S, Horn M, Haferkamp, I. 2011. Nucleotide parasitism by *Simkania negevensis* (Chlamydiae). J Bacteriol 193:225–235 10.1128/JB.00919-10.

Koch-Edelmann, S., Banhart, S., Saied, E.M., Rose, L., Aeberhard, L., Laue, M., Doellinger, J., Arenz, C., and Heuer, D. (2017). The cellular ceramide transport protein CERT promotes *Chlamydia psittaci* infection and controls bacterial sphingolipid uptake. Cell Microbiol 19(10), e12752 10.1111/cmi.12752.

König, L., Siegl, A., Penz, T., Haider, S., Wentrup, C., Polzin, J., Mann, E., Schmitz-Esser S., Domman, D., and Horn, M. (2017). Biphasic metabolism and host interaction of a chlamydial symbiont. MSystems 2(3), e00202–16 10.1128/mSystems.00202–16.

Kose, M., Ekinci, D., Gokahmetoglu, S., Elmas, T., & Öztürk, M.K. (2015). *Simkania negevensis*: Is it a real respiratory pathogen? Acta Microbiol Immunol Hung 62(2), 161–166 10.1556/030.62.2015.2.6.

Kostanjsek, R., Strus, J., Drobne, D., Avgustin, G. (2004). *'Candidatus* Rhabdochlamydia porcellionis', an intracellular bacterium from the hepatopancreas of the terrestrial isopod *Porcellio scaber* (Crustacea: Isopoda). Int J Syst Evol Microbiol 54:543–549 10.1099/ijs.0.02802–0.

Kozjak-Pavlovic, V., Herweg, J.A., & Rudel, T. (2017). The role of host cell organelles in the development of *Simkania negevensis*. Int J Med Microbiol S1438–4221(17)30336–3 10.1016/j.ijmm.2017.10.008.

Kumar, G., Mayrhofer, R., Soliman, H., El-Matbouli M. (2013). Novel Chlamydiales associated with epitheliocystis in grass carp (*Ctenopharyngodon idella*). Vet Rec 172:47 10.1136/vr.101193.

Labbé, K., Saleh, M. (2008). Cell death in the host response to infection. Cell Death Differ 15: 1339–1349 10.1038/cdd.2008.91.

Lagkouvardos, I., Weinmaier, T., Lauro, F.M., Cavicchioli, R., Rattei, T., Horn, M. (2014). Integrating metagenomic and amplicon databases to resolve the phylogenetic and ecological diversity of the Chlamydiae. ISME J 8:115–125 10.1038/ismej.2013.

Lamoth, F., Pillonel, T., Greub, G. (2015). *Waddlia*: An emerging pathogen and a model organism to study the biology of chlamydiae. Microbes Infect 17:732–737 10.1016/j.micinf.2015.09.021.

Lane, B.J., Mutchler, C., Al Khodor, S., Grieshaber, S.S., and Carabeo, R.A. (2008). Chlamydial entry involves TARP binding of guanine nucleotide exchange factors. PLoS Pathog 4, e1000014.

Lee, J.K., Enciso, G.A., Boassa, D., Chander, C.N., Lou, T.H., Pairawan, S.S., Guo, M.C., Wan, F.Y.M., Ellisman, M.H., Sütterlin, C., and Tan., M. (2018). Replication-dependent size reduction precedes differentiation in *Chlamydia trachomatis*. Nat Comm 9(45) 10.1038/s41467–017–02432–0.

Liechti, G.W., Kuru, E., Hall, E., Kalinda, A., Brun, Y.V., VanNieuwenhze M., Maurelli, A.T. (2014). A new metabolic cell-wall labelling method reveals peptidoglycan in *Chlamydia trachomatis*. Nature. 506(7489), 507–10. doi: 10.1038/nature12892.

Lienard, J., Croxatto, A., Aeby, S., Jaton, K., Posfay-Barbe, K., Gervaix, A., and Greub, G. (2011). Development of a new Chlamydiales-specific real-time PCR and its application to respiratory clinical samples. J Clin Microbiol 49, 2637–2642 10.1128/JCM.00114–11.

Lienard, J., Croxatto, A., Prod'hom, G., and Greub, G. (2011). *Estrella lausannensis*, a new star in the

Chlamydiales order. Microbes Infect 13, 1232–1241 10.1016/j.micinf.2011.07.003.

Lohr, M., Prohl, A., Ostermann, C., Diller, R., Greub, G., Reinhold, P. (2016). Effect of *Parachlamydia acanthamoebae* on pulmonary function parameters in a bovine respiratory model. Vet J 213:9–15 10.1016/j.tvjl.2016.04.003.

Lohr, M., Prohl, A., Ostermann, C., Liebler-Tenorio E., Schroedl, W., Aeby, S., Greub, G., Reinhold, P. (2014). A bovine model of a respiratory *Parachlamydia acanthamoebae* infection. Pathog Dis 73(1):1–14 10.1111/2049-632X.12201.

Major, P., Embley, T.M., and Williams, T.A. (2017). Phylogenetic Diversity of NTT Nucleotide Transport Proteins in Free-Living and Parasitic Bacteria and Eukaryotes. Genome Biol Evol 9, 480–487 10.1093/gbe/evx015.

Marrie, T.J., Raoult, D., La Scola, B., Birtles, R.J., de Carolis, E., & Canadian Community-Acquired Pneumonia Study Group. (2001). *Legionella*-like and other amoebal pathogens as agents of community-acquired pneumonia. Emerg Infect Dis, 7(6), 1026 10.3201/eid0706.010619.

Martel, A., Adriaensen, C., Bogaerts, S., Ducatelle, R., Favoreel, H., Crameri, S., *et al.* (2012). Novel *Chlamydiaceae* disease in captive salamanders. Em. Infect. Dis. 18(6): 1020.

Martel, A., Adriaensen, C., Sharifian-Fard, M., Vandewoestyne, M., Deforce, D., Favoreel, H., *et al.* (2013). The novel 'Candidatus *Amphibiichlamydia ranarum*' is highly prevalent in invasive exotic bullfrogs (*Lithobates catesbeianus*). Environ Microbiol Rep, 5(1): 105–108.

Matsuo, J., Kawaguchi, K., Nakamura, S., Hayashi, Y., Yoshida, M., Takahashi, K., Mizutani, Y., Yao, T., and Yamaguchi, H. (2010). Survival and transfer ability of phylogenetically diverse bacterial endosymbionts in environmental *Acanthamoeba* isolates. Environ Microbiol Rep 2(4), 524–533 10.1111/j.1758-2229.2009.00094.x.

Mäurer, A.P., Mehlitz, A., Mollenkopf, H.J., Meyer, T.F. (2007). Gene expression profiles of *Chlamydophila pneumoniae* during the developmental cycle and iron depletion-mediated persistence. PLoS Pathog. 3(6): e83.

McClarty, G. (1994). Chlamydiae and the biochemistry of intracellular parasitism. Trends Microbiol 2(5), 157–164 10.1016/0966-842X(94)90665-3.

McCoy A.J., Maurelli, A.T. (2006). Building the invisible wall: updating the chlamydial peptidoglycan anomaly. Trends Microbiol. 14(2):70–7.

Mehlitz, A., Karunakaran, K., Herweg, J.A., Krohne, G., van de Linde S., Rieck, E., Sauer, M., Rudel, T. (2014). The chlamydial organism *Simkania negevensis* forms ER vacuole contact sites and inhibits ER-stress. Cell Microbiol 16:1224–1243 10.1111/cmi.12278.

Michel, R., Steinert, M., Zoeller, L., Hauroeder, B., Henning, K. (2004). Free-living amoebae may serve as hosts for the Chlamydia-like bacterium Waddlia chondrophila isolated from an aborted bovine foetus. Acta Protozool 43:37–42.

Michel, R., Hauröder-Philippczyk, B., Müller, K.-D., and Weishaar, I. (1994). *Acanthamoeba* from human nasal mucosa infected with an obligate intracellular parasite. Eur J Protistol 30, 104–110 10.1016/S0932-4739(11)80203-8.

Milan, M., Carraro, L., Fariselli, P., Martino, M.E., Cavalieri, D., Vitali, F., Boffo, L., Patarnello, T., Bargelloni, L., Cardazzo, B. (2018). Microbiota and environmental stress: how pollution affects microbial communities in Manila clams. Aquat Toxicol 194:195–207 10.1016/j.aquatox.2017.11.019.

Mitchell, S.O., Steinum, T., Rodger, H., Holland, C., Falk, K., Colquhoun, D.J. (2010). Epitheliocystis in Atlantic salmon, *Salmo salar* L., farmed in fresh water in Ireland is associated with '*Candidatus* Clavochlamydia salmonicola' infection. J Fish Dis 33:665–673 10.1111/j.1365-2761.2010.01171.x.

Mojica, S., Huot Creasy, H., Daugherty, S., Read, T.D., Kim, T., Kaltenboeck, B., Bavoil, P., and Myers, G.S. (2011). Genome sequence of the obligate intracellular animal pathogen *Chlamydia pecorum* E58. J Bacteriol 193, 3690 10.1128/JB.00454-11.

Niemi, S., Greub, G., and Puolakkainen, M. (2011). *Chlamydia*-related bacteria in respiratory samples in Finland. Microbes and Infect 824–827 10.1016/j.micinf.2011.04.012.

Nylund, A., Kvenseth, A.M., and Isdal, E. (1998). A Morphological Study of the Epitheliocystis Agent in Farmed Atlantic Salmon. J Aquat Anim Health 10, 43–55 10.1577/1548-8667.

Nylund, S., Steigen, A., Karlsbakk, E., Plarre, H., Andersen, L., Karlsen, M., Watanabe, K., Nylund, A. (2015). Characterization of '*Candidatus* Syngnamydia salmonis' (Chlamydiales, Simkaniaceae), a bacterium associated with epitheliocystis in Atlantic salmon (*Salmo salar* L.). Arch Microbiol 197:17–25 10.1007/s00203-014-1038-3.

Nylund, A., Pistone, D., Trosse, C., Blindheim, S., Andersen, L., Plarre, H. (2018). Genotyping of *Candidatus* Syngnamydia salmonis (chlamydiales; Simkaniaceae) co-cultured in *Paramoeba perurans* (amoebozoa; *Paramoebidae*). Arch Microbiol 200(6):859–867. doi: 10.1007/s00203-018-1488-0.

Omsland, A., Sager, J., Nair, V., Sturdevant, D.E., Hackstadt, T. (2012). Developmental stage-specific metabolic and transcriptional activity of *Chlamydia trachomatis* in an axenic medium. Proceedings of the National Academy of Sciences 109(48):19781–19785.

Omsland, A., Sixt, B.S., Horn, M., Hackstadt, T. (2014). Chlamydial metabolism revisited: interspecies metabolic variability and developmental stage-specific physiologic activities. FEMS Microbiol Rev 38:779–801 10.1111/1574-6976.12059.

Packiam, M., Weinrick, B., Jacobs, W.R Jr, Maurelli, A.T. (2015). Structural characterization of muropeptides from *Chlamydia trachomatis* peptidoglycan by mass spectrometry resolves 'chlamydial anomaly'. Proc Natl Acad Sci U S A 112(37), 11660–5. doi: 10.1073/pnas.1514026112.

Pagnier, I., Yutin, N., Croce, O., Makarova, K.S., Wolf, Y.I., Benamar, S., Raoult, D., Koonin, E.V., La Scola B. (2015). *Babela massiliensis*, a representative of a widespread bacterial phylum with unusual adaptations to parasitism in amoebae. Biol Direct 10:13 10.1186/s13062-015-0043-z.

Parks, D.H., Imelfort, M., Skennerton, C.T., Hugenholtz, P., & Tyson, G.W. (2015). CheckM: assessing the quality of microbial genomes recovered from isolates, single cells, and metagenomes. Genome Res 25(7):1043–55. doi: 10.1101/gr.186072.114.

Parks, D.H., Chuvochina, M., Waite, D.W., Rinke, C., Skarshewski, A., Chaumeil, P.-A., and Hugenholtz, P. (2018). A standardized bacterial taxonomy based on genome phylogeny substantially revises the tree of life. Nat Biotechnol 36, 996 10.1038/nbt.4229.

Peters, J., Wilson, D.P., Myers, G., Timms, P., and Bavoil, P.M. (2007). Type III secretion a la *Chlamydia*. Trends in microbiology 15, 241–251 10.1016/j.tim.2007.04.005.

Pilhofer, M., Aistleitner, K., Biboy, J., Grey, J., Kuru, E., Hall, E., Brun, Y.V., VanNieuwenhze, M.S., Vollmer, W., Horn, M., Jensen, G.J. (2013). Discovery of chlamydial peptidoglycan reveals bacteria with murein sacculi but without FtsZ. Nature Comm, 4, 2856.

Pilhofer, M., Aistleitner, K., Ladinsky, M.S., Konig, L., Horn, M., and Jensen, G.J. (2014). Architecture and host interface of environmental chlamydiae revealed by electron cryotomography. Environ Microbiol 16, 417–429 10.1111/1462-2920.12299.

Pillonel, T., Bertelli, C., Aeby, S., de Barsy, M., Jacquier, N., Kebbi-Beghdadi, C., *et al.* (2019). Sequencing the obligate intracellular *Rhabdochlamydia helvetica* within its tick host *Ixodes ricinus* to investigate their symbiotic relationship. Genome biology and evolution, 11(4), 1334–1344.

Pillonel, T., Bertelli, C., Greub, G. (2018). Environmental Metagenomic Assemblies Reveal Seven New Highly Divergent Chlamydial Lineages and Hallmarks of a Conserved Intracellular Lifestyle. Front Microbiol 9:79 10.3389/fmicb.2018.00079.

Pillonel, T., Bertelli, C., Salamin, N., Greub, G. (2015). Taxogenomics of the order Chlamydiales. Int J Syst Evol Microbiol 65:1381–1393 10.1099/ijs.0.000090.

Pilloux, L., Aeby, S., Gaumann, R., Burri, C., Beuret, C., Greub, G. (2015). The high prevalence and diversity of Chlamydiales DNA within *Ixodes ricinus* ticks suggest a role for ticks as reservoirs and vectors of *Chlamydia*-related bacteria. Appl Environ Microbiol 81:8177–8182 10.1128/AEM.02183-15 10.1371/journal.pone.0150909.

Pilloux, L., LeRoy, D., Brunel, C., Roger, T., and Greub, G. (2016). Mouse Model of Respiratory Tract Infection Induced by *Waddlia chondrophila*. PloS one 11(3):e0150909 10.1371/journal.pone.0150909.

Pizzetti, I., Schulz, F., Tyml, T., Fuchs, B.M., Amann, R., Horn, M., Fazi, S. (2016). Chlamydial seasonal dynamics and isolation of '*Candidatus* Neptunochlamydia vexilliferae' from a Tyrrhenian coastal lake. Environ Microbiol 18:2405–2417 10.1111/1462-2920.13111.

Price, M.N., Dehal, P.S., & Arkin, A.P. (2009). FastTree: computing large minimum evolution trees with profiles instead of a distance matrix. Mol Biol Evol 26(7), 1641–1650 10.1093/molbev/msp077.

Read, T.D., Brunham, R.C., Shen, C., Gill, S.R., Heidelberg, J.F., White, O., Hickey, E.K., Peterson, J., Utterback, T., Berry, K., *et al.* (2000). Genome sequences of *Chlamydia trachomatis* MoPn and *Chlamydia pneumoniae* AR39. Nucleic acids research 28, 1397–1406.

Read, T.D., Myers, G.S., Brunham, R.C., Nelson, W.C., Paulsen, I.T., Heidelberg, J., Holtzapple, E., Khouri, H., Federova, N.B., Carty, H.A. *et al.* (2003). Genome sequence of *Chlamydophila caviae* (*Chlamydia psittaci* GPIC): examining the role of niche-specific genes in the evolution of the Chlamydiaceae. Nucleic Acids Res 31(8), 2134–2147 10.1093/nar/gkg321.

Rockey, D.D., Heinzen, R.A., & Hackstadt, T. (1995). Cloning and characterization of a *Chlamydia psittaci* gene coding for a protein localized in the inclusion membrane of infected cells. Mol Microbiol, 15(4), 617–626 10.1111/j.1365-2958.1995.tb02371.x.

Rockey, D.D. (2011). Unravelling the basic biology and clinical significance of the chlamydial plasmid. J Exp Med 208, 2159–2162 10.1084/jem.20112088.

Rodriguez-Zaragoza S. (1994). Ecology of free living amoebae. Crit. Rev. Microbiol 20:225–41.

Rurangirwa, F.R., Dilbeck, P.M., Crawford, T.B., McGuire T.C., McElwain T.F. (1999). Analysis of the 16S rRNA gene of micro-organism WSU 86-1044 from an aborted bovine foetus reveals that it is a member of the order Chlamydiales: proposal of Waddliaceae fam. nov., *Waddlia chondrophila* gen. nov., sp. nov. Int J Syst Bacteriol 49 Pt 2:577–581 10.1099/00207713-49-2-577.

Rusconi, B., Kebbi-Beghdadi, C., and Greub, G. (2015). Trafficking of *Estrella lausannensis* in human macrophages. Pathog Dis 73 10.1093/femspd/ftv027.

Rusconi, B., Lienard, J., Aeby, S., Croxatto, A., Bertelli, C., and Greub, G. (2013). Crescent and star shapes of members of the Chlamydiales order: impact of fixative methods. Antonie van Leeuwenhoek 104, 521–532 10.1007/s10482-013-9999-9.

Sampo, A., Matsuo, J., Yamane, C., Yagita, K., Nakamura, S., Shouji, N., Hayashi, Y., Yamazaki, T., Yoshida, M., Kobayashi, M., *et al.* (2014). High-temperature adapted primitive *Protochlamydia* found in *Acanthamoeba* isolated from a hot spring can grow in immortalized human epithelial HEp-2 cells. Environ Microbiol 16:486–497 10.1111/1462-2920.12266.

Schmidt-Posthaus H., Polkinghorne, A., Nufer, L., Schifferli, A., Zimmermann, D.R., Segner, H., Steiner, P., Vaughan, L. (2012). A natural freshwater origin for two chlamydial species, *Candidatus* Piscichlamydia salmonis and *Candidatus* Clavochlamydia salmonicola, causing mixed infections in wild brown trout (*Salmo trutta*). Environ Microbiol 14:2048–2057 10.1111/j.1462-2920.2011.02670.x.

Schmitz-Esser, S., Linka, N., Collingro, A., Beier, C.L., Neuhaus, H.E., Wagner, M., and Horn, M. (2004). ATP/ADP translocases: a common feature of obligate intracellular amoebal symbionts related to Chlamydiae and *Rickettsiae*. Journal of bacteriology 186, 683–691 10.1128/JB.186.3.683-691.2004.

Sellyei, B., Molnar, K., Szekely, C. (2017). Diverse *Chlamydia*-like agents associated with epitheliocystis infection in two cyprinid fish species, the common carp (*Cyprinus carpio* L.) and the gibel carp (*Carassius auratus gibelio* L.). Acta Vet Hung 65:29–40 10.1556/004.2017.003.

Seth-Smith H.M.B., Katharios, P., Dourala, N., Mateos, J.M., Fehr, A.G.J., Nufer, L., Ruetten, M., Guevara Soto M., Vaughan, L. (2017). *Ca.* Similichlamydia in Epitheliocystis Co-infection of Gilthead Seabream Gills:

Unique Morphological Features of a Deep Branching Chlamydial Family. Front Microbiol 8:508 10.3389/fmicb.2017.00508.

Sixt, B.S., Heinz, C., Pichler, P., Heinz, E., Montanaro, J., Op den Camp H.J., Ammerer, G., Mechtler, K., Wagner, M., Horn, M. (2011). Proteomic analysis reveals a virtually complete set of proteins for translation and energy generation in elementary bodies of the amoeba symbiont Protochlamydia amoebophila. Proteomics 11:1868–1892 10.1002/pmic.201000510.

Sixt, B.S., Hiess, B., Konig, L., Horn, M. (2012). Lack of effective anti-apoptotic activities restricts growth of Parachlamydiaceae in insect cells. PLoS One 7(1): e29565 10.1371/journal.pone.0029565.

Sixt, B.S., Siegl, A., Muller, C., Watzka, M., Wultsch, A., Tziotis, D., Montanaro, J., Richter, A., Schmitt-Kopplin P., Horn, M. (2013). Metabolic features of *Protochlamydia amoebophila* elementary bodies--a link between activity and infectivity in Chlamydiae. PLoS Pathog 9(8):e1003553 10.1371/journal.ppat.1003553.

Sixt, B.S., Kostanjšek, R., Mustedanagic, A., Toenshoff, E.R., & Horn, M. (2013). Developmental cycle and host interaction of Rhabdochlamydia porcellionis, an intracellular parasite of terrestrial isopods. Environ Microbiol, *15*(11), 2980–2993 10.1111/1462–2920.

Skipp, P., Robinson, J., O'Connor C.D., Clarke, I.N. (2005). Shotgun proteomic analysis of *Chlamydia trachomatis*. Proteomics 5: 1558–1573 10.1002/pmic.200401044.

Sood, N., Pradhan, P.K., Verma, D.K., Yadav, M.K., Ravindra, Dev, A.K., Swaminathan, T.R., Sood, N.K. (2018). *Candidatus* Actinochlamydia pangasiae sp. nov. (Chlamydiales, Actinochlamydiaceae), a bacterium associated with epitheliocystis in *Pangasianodon hypophthalmus*. J Fish Dis 41:281–290 10.1111/jfd.12711.

Stackebrandt, E., and Goebel, B.M. (1994). Taxonomic Note: A Place for DNA-DNA Reassociation and 16S rRNA Sequence Analysis in the Present Species Definition in Bacteriology. Int J Syst Evol Microbiol 44, 846–849 10.1099/00207713–44–4–846.

Staub, E., Marti, H., Biondi, R., Levi, A., Donati, M., Leonard, C.A., Ley, S.D., Pillonel, T., Greub, G., Seth-Smith H.M.B., *et al.* (2018). Novel *Chlamydia* species isolated from snakes are temperature-sensitive and exhibit decreased susceptibility to azithromycin. Sci Rep 8:5660 10.1038/s41598–018–23897-z.

Steigen, A., Karlsbakk, E., Plarre, H., Watanabe, K., Overgard, A.C., Brevik, O., Nylund, A. (2015). A new intracellular bacterium, *Candidatus* Similichlamydia labri sp. nov. (Chlamydiaceae) producing epitheliocysts in ballan wrasse, *Labrus bergylta* (Pisces, Labridae). Arch Microbiol 197:311–318 10.1007/s00203–014–1061–4.

Steigen, A., Nylund, A., Karlsbakk, E., Akoll, P., Fiksdal, I.U., Nylund, S., Odong, R., Plarre, H., Semyalo, R., Skar, C., *et al.* (2013). 'Cand. Actinochlamydia clariae' gen. nov., sp. nov., a unique intracellular bacterium causing epitheliocystis in catfish (*Clarias gariepinus*) in Uganda. PLoS One 8(6):e66840 10.1371/journal.pone.0066840.

Steigen, A., Nylund, A., Plarre, H., Watanabe, K., Karlsbakk, E., Brevik, O. (2018). Presence of selected pathogens on the gills of five wrasse species in western Norway. Dis Aquat Organ 128:21–35 10.3354/dao03198.

Stenner-Liewen, F., Liewen, H., Zapata, J.M., Pawlowski, K., Godzik, A., and Reed, J.C. (2002). CADD, a *Chlamydia* Protein That Interacts with Death Receptors. Journal of Biological Chemistry 277, 9633–9636 10.1074/jbc.C100693200.

Stephens, R.S., Kalman, S., Lammel, C., Fan, J., Marathe, R., Aravind, L., Mitchell, W., Olinger, L., Tatusov, R.L., Zhao, Q., *et al.* (1998). Genome sequence of an obligate intracellular pathogen of humans: *Chlamydia trachomatis*. Science 282, 754–759 10.1126/science.282.5389.754.

Storrie, S., Longbottom, D., Barlow, P.G., and Wheelhouse, N. (2016). MAPK Activation Is Essential for *Waddlia chondrophila* Induced CXCL8 Expression in Human Epithelial Cells. PloS one 11:e0152193 10.1371/journal.pone.0152193.

Stride, M.C., Polkinghorne, A., Miller, T.L., Groff, J.M., Lapatra, S.E., Nowak, B.F. (2013a). Molecular characterization of '*Candidatus* Parilichlamydia carangidicola,' a novel *Chlamydia*-like epitheliocystis agent in yellowtail kingfish, *Seriola lalandi* (Valenciennes), and the proposal of a new family, '*Candidatus* Parilichlamydiaceae' fam. nov. (order Chlamydiales). Appl Environ Microbiol 79:1590–1597 10.1128/AEM.02899–12.

Stride, M.C., Polkinghorne, A., Miller, T.L., Nowak, B.F. (2013b). Molecular characterization of '*Candidatus* Similichlamydia latridicola' gen. nov., sp. nov. (Chlamydiales: '*Candidatus* Parilichlamydiaceae'), a novel *Chlamydia*-like epitheliocystis agent in the striped trumpeter, *Latris lineata* (Forster). Appl Environ Microbiol 79:4914–4920 10.1128/AEM.00746–13.

Stride, M.C., Polkinghorne, A., Powell, M.D., Nowak, B.F. (2013c). '*Candidatus* Similichlamydia laticola', a novel *Chlamydia*-like agent of epitheliocystis in seven consecutive cohorts of farmed Australian barramundi, *Lates calcarifer* (Bloch). PLoS One 8:e82889 10.1371/journal.pone.0082889.

Tanner, M., Harris, J., Pace, N. (1999). Molecular Phylogeny of Chlamydia and Relatives. In: Stephens R, editor. Chlamydia. Washington DC: ASM Press. 10.1128/9781555818203.ch1.

Taylor-Brown A., Bachmann, N.L., Borel, N., Polkinghorne, A. (2016.) Culture-independent genomic characterization of *Candidatus* Chlamydia sanzinia, a novel uncultivated bacterium infecting snakes. BMC Genomics 17:710 10.1186/s12864–016–3055-x.

Taylor-Brown A., Pillonel, T., Bridle, A., Qi, W., Bachmann, N.L., Miller, T.L., Greub, G., Nowak, B., Seth-Smith H.M.B., Vaughan, L., *et al.* (2017a). Culture-independent genomics of a novel chlamydial pathogen of fish provides new insight into host-specific adaptations utilized by these intracellular bacteria. Environ Microbiol 19:1899–1913 10.1111/1462–2920.13694.

Taylor-Brown A., Spang, L., Borel, N., Polkinghorne, A. (2017b). Culture-independent metagenomics supports discovery of uncultivable bacteria within the genus *Chlamydia*. Sci Rep 7(1):10661. 10.1038/s41598–017–10757–5.

Taylor-Brown A., Vaughan, L., Greub, G., Timms, P., Polkinghorne, A. (2015). Twenty years of research into *Chlamydia*-like organisms: a revolution in our understanding of the biology and pathogenicity of members of the phylum Chlamydiae. Pathog Dis 73:1–15 10.1093/femspd/ftu009.

Taylor-Brown, A., Pillonel, T., Greub, G., Vaughan, L., Nowak, B., and Polkinghorne, A. (2018). Metagenomic analysis of fish-associated *Ca.* Parilichlamydiaceae reveals striking metabolic similarities to the terrestrial Chlamydiaceae. Genome Biol Evol 10.1093/gbe/evy195.

Thao, M.L., Baumann, L., Hess, J.M., Falk, B.W., Ng, J.C., Gullan, P.J., Baumann, P. (2003). Phylogenetic evidence for two new insect-associated *Chlamydia* of the family Simkaniaceae. Curr Microbiol 47:46–50 10.1007/s00284-002-3953-9.

Thomas, V., Casson, N., Greub, G. (2006). *Criblamydia sequanensis*, a new intracellular Chlamydiales isolated from Seine river water using amoebal co-culture. Environ Microbiol 8:2125–2135 10.1111/j.1462-2920.2006.01094.x.

Thomson, N.R., Holden, M.T., Carder, C., Lennard, N., Lockey, S.J., Marsh, P., Skipp, P., O'Connor C.D., Goodhead, I., Norbertzcak, H., *et al.*(2008). *Chlamydia trachomatis*: genome sequence analysis of lymphogranuloma venereum isolates. Genome Res 18(1), 161–171 10.1101/gr.7020108.

Thomson, N.R., Yeats, C., Bell, K., Holden, M.T., Bentley, S.D., Livingstone, M., Cerdeno-Tarraga, A.M., Harris, B., Doggett, J., Ormond, D., *et al.* (2005). The *Chlamydophila abortus* genome sequence reveals an array of variable proteins that contribute to interspecies variation. Genome Res 15, 629–640 10.1101/gr.3684805.

Tjaden, J., Winkler, H.H., Schwoppe, C., Van Der Laan, M., Mohlmann, T., and Neuhaus, H.E. (1999). Two nucleotide transport proteins in *Chlamydia trachomatis*, one for net nucleoside triphosphate uptake and the other for transport of energy. J Bacteriol 181, 1196–1202.

Vandahl, B.B., Birkelund, S., Demol, H., Hoorelbeke, B., Christiansen, G., *et al.* (2001). Proteome analysis of the *Chlamydia pneumoniae* elementary body. Electrophoresis 22: 1204–1223.

Vanthournout, B., Hendrickx, F. (2015). Endosymbiont dominated bacterial communities in a dwarf spider. PLoS One 10(2):e0117297 10.1371/journal.pone.0117297.

Vasilevsky, S., Gyger, J., Piersigilli, A., Pilloux, L., Greub, G., Stojanov, M., and Baud, D. (2015). *Waddlia chondrophila* induces systemic infection, organ pathology, and elicits Th1-associated humoral immunity in a murine model of genital infection. Front Cell Infect Microbiol 5, 76 10.3389/fcimb.2015.00076.

Vega, Thurber, R., Willner-Hall D., Rodriguez-Mueller B., Desnues, C., Edwards, R.A., Angly, F., Dinsdale, E., Kelly, L., Rohwer, F. (2009). Metagenomic analysis of stressed coral holobionts. Environ Microbiol 11:2148–2163 10.1111/j.1462-2920.2009.01935.x.

Voigt, A., Schofl, G., and Saluz, H.P. (2012). The *Chlamydia psittaci* genome: a comparative analysis of intracellular pathogens. PloS one 7, e35097 10.1371/journal.pone.0035097.

Vouga, M., Baud, D., & Greub, G. (2017a). *Simkania negevensis*, an insight into the biology and clinical importance of a novel member of the Chlamydiales order. Crit Rev Microbiol 43(1), 62–80 10.3109/1040841X.2016.1165650.

Vouga, M., Baud, D., & Greub, G. (2017b). *Simkania negevensis* may produce long-lasting infections in human pneumocytes and endometrial cells. Pathog Dis 75(1) 10.1093/femspd/ftw115.

Vouga, M., Baud, D., & Greub, G. (2017c). *Simkania negevensis*, an example of the diversity of the antimicrobial susceptibility pattern among Chlamydiales. Antimicrob Agents Chemother 61(8), e00638-17 10.1128/AAC.00638-17.

Vouga, M., Baud, D., and Greub, G. (2017). *Simkania negevensis* may produce long-lasting infections in human pneumocytes and endometrial cells. Pathog Dis 75 10.1093/femspd/ftw115.

Vouga, M., Kebbi-Beghdadi, C., Liénard, J., Baskin, L., Baud, D., & Greub, G. (2018). What is the true clinical relevance of *Simkania negevensis* and other emerging Chlamydiales members? New Microbes New Infect, 23:1–5 10.1016/j.nmni.2018.01.001.

Wheelhouse, N., Coyle, C., Barlow, P.G., Mitchell, S., Greub, G., Baszler, T., Rae, M.T., and Longbottom, D. (2014). *Waddlia chondrophila* Infects and Multiplies in Ovine Trophoblast Cells Stimulating an Inflammatory Immune Response. PloS one 9, e102386 10.1371/journal.pone.0102386.

Wheelhouse, N., Flockhart, A., Aitchison, K., Livingstone, M., Finlayson, J., Flachon, V., Sellal, E., Dagleish, M.P., and Longbottom, D. (2016). Experimental challenge of pregnant cattle with the putative abortifacient *Waddlia chondrophila*. Sci Rep 6, 37150 10.1038/srep37150.

Wolf, K., and Hackstadt, T. (2001). Sphingomyelin trafficking in *Chlamydia pneumoniae*-infected cells. Cell Microbiol 3(3), 145–152 10.1046/j.1462-5822.2001.00098.x.

Yamaguchi, T., Yamazaki, T., Inoue, M., Mashida, C., Kawagoe, K., Ogawa, M., Shiga, S., Nakagawa, Y., Kishimoto, T., Kurane, I., (2005). Prevalence of antibodies against *Simkania negevensis* in a healthy Japanese population determined by the microimmunofluorescence test. FEMS Immunol Med Microbiol 43(1), 21–27 10.1016/j.femsim.2004.06.023.

Yamane, C., Yamazaki, T., Nakamura, S., Matsuo, J., Ishida, K., Yamazaki, S., Oguri, S., Shouji, N., Hayashi, Y., Yoshida, M., Yimin, Yamaguchi, H. (2015). Amoebal endosymbiont Parachlamydia acanthamoebae Bn9 can grow in immortal human epithelial HEp-2 cells at low temperature; an *in vitro* model system to study chlamydial evolution. PLoS One. 10(2), e0116486. doi: 10.1371/journal.pone.0116486.

Ying, S., Pettengill, M., Latham, E.R., Walch, A., Ojcius, D.M., *et al.* (2008). Premature apoptosis of *Chlamydia*-infected cells disrupts chlamydial development. J Infect Dis 198: 1536–1544 10.1086/592755.

Ying, S., Pettengill, M., Ojcius, D.M., Häcker, G. (2007). Host-Cell Survival and Death During *Chlamydia* Infection. Curr Immunol Rev 3: 31–40 10.2174/157339507779802179.

Zhong, G. (2016). Chlamydial Plasmid-Dependent Pathogenicity. Trends Microbiol 25(2):141–152 10.1016/j.tim.2016.09.006

Animal Models of *Chlamydia trachomatis* Infection

19

Guangming Zhong[1]*, Alison J. Quayle[2], Ashok Aiyar[2] and Tianyuan Zhang[1]

[1]Department of Microbiology, Immunology and Molecular Genetics, University of Texas Health Science Center at San Antonio, San Antonio, TX, USA.
[2]Department of Microbiology, Immunology and Parasitology, Louisiana State University Health Science Center at New Orleans, New Orleans, LA, USA.

*Correspondence: zhongg@uthscsa.edu

https://doi.org/10.21775/9781912530281.19

Abstract

Chlamydia trachomatis causes infections in the eye, urogenital, respiratory and gastrointestinal mucosa. Chronic infection can lead to significant disease; ocular infection can result in trachoma, a leading cause of preventable blindness, and genital infection can lead to the serious sequelae of ectopic pregnancy and infertility. These pathological sequelae have been proposed to result from the induction of a sustained inflammatory response, but the precise pathogenic mechanisms that lead to these diseases are not yet fully understood. Chlamydial biologists have used *in vitro* approaches that rely on biochemistry and/or cell biology, in conjunction with recently developed genetic tools, to identify bacterial factors required for interactions with host cells. Although some of the putative chlamydial virulence factors identified *in vitro* have been shown to promote chlamydial pathogenicity in animal models, it remains to be determined if this information is relevant to pathogenic mechanisms operating in humans. Furthermore, many putative virulence factors have not been evaluated in optimal animal tissue microenvironments despite the fact that various animal models with chlamydial infections have been developed. Here we describe the pathogenesis of *C. trachomatis* infections in humans and the unique features of the currently used model systems so that the most appropriate models can be chosen for addressing different aspects of *C. trachomatis* pathogenesis and evaluating vaccine candidates.

Introduction

Most human infections are caused by two species of *Chlamydia*: *C. pneumoniae* and *C. trachomatis*. The extensive and increasing global burden of *C. trachomatis*-mediated disease in particular has led to the acknowledgement that new interventional strategies are needed to supplement current public health strategies (Brunham and Rappuoli, 2013; de la Maza *et al.*, 2017; Gottlieb and Johnston, 2017; Zhong *et al.*, 2017). Excitingly, recent developments in *C. trachomatis* genetics may enable us to understand how *C. trachomatis* induces pathology and how we could prevent pathology, for example with a vaccine. Specifically, the successful transformation of *C. trachomatis* with recombinant plasmids derived from its own endogenous plasmid (Wang *et al.*, 2011) has driven a rapid dissection of plasmid-encoded factors responsible for chlamydial pathogenicity (Ding *et al.*, 2013; Gong *et al.*, 2013; Song *et al.*, 2013). The development of genetic tools has also made it possible to investigate chromosome-encoded factors in chlamydial pathogenesis (Brothwell *et al.*, 2018; Johnson and Fisher, 2013; Mueller *et al.*, 2016; Quellette, 2018). These efforts have produced large numbers of genetically defined mutants (Gong *et al.*, 2013; Johnson and Fisher, 2013; Liu *et al.*, 2014a,b; Mueller *et al.*, 2016; Putman *et al.*, 2019; Quellette, 2018; Song *et al.*, 2013). While further optimization of the genetic tools is proceeding, a discussion of the best models to evaluate these mutants is required. Although several animal models have been developed to investigate chlamydial pathogenic mechanisms and evaluate vaccine efficacy, refining and choosing the most appropriate animal models for obtaining biologically and clinically relevant information is challenging. In this context, we will describe the general pathogenic characteristics of chlamydial infections in humans, then discuss how to make the best choices of model systems, including their limitations, for obtaining biologically and clinically relevant information. Finally, we will discuss the use of human *ex vivo* models to aid in validating clinical observations. We believe that choosing robust models with appropriate infection routes, paired with appropriate mutants, should enhance our ability to identify tissue-dependent *C. trachomatis* virulence determinants. This will aid in determining the mechanisms that drive site-specific immunity and pathology and in identifying site-specific factors that may constitute, or even impede, local immunity. An ultimate outcome of the application of this knowledge would be the development of a chlamydial vaccine (refer to Chapter 15, *Chlamydia* Vaccines, for more details).

The pathology of tissue-specific *C. trachomatis* infections in humans

Trachoma

Historically, much of our knowledge of *C. trachomatis*-driven disease is derived from the *C. trachomatis* serovars that cause trachoma. Trachoma, a chronic keratoconjunctivitis, was likely the first chlamydial disease to be described in ancient Chinese writings circa 2,700 BC (Taylor, 2009). Modern chlamydial research began with the successful isolation and subsequent chick-embryo culture of trachoma-derived *C. trachomatis* organisms by Tang in 1957 (Tang *et al.*, 1957a; Tang *et al.*, 1957b). Infecting the eyes of rhesus monkeys with these organisms was found to reproduce the same disease seen in humans naturally infected with *C. trachomatis*. Subsequently, trachoma vaccine trials, with inactivated trachoma organisms, were performed in many countries (Collier and Blyth, 1966a, b; Grayston *et al.*, 1962; Grayston *et al.*, 1963; Sowa *et al.*, 1969; Woolridge *et al.*, 1967a; Woolridge *et al.*, 1967b). These historic trials revealed that the *C. trachomatis* vaccine-induced protection was generally short-lived and immunized individuals could develop more severe disease upon re-exposure.

Trachoma is the end result of inflammation and scarring sequelae initially generated by recurrent eye infection with *C. trachomatis* serovars, A-C (Mariotti *et al.*, 2009; Ramadhani *et al.*, 2016). These ocular serovars infect conjunctival epithelial cells, and organisms are shed into, and transmitted via, ocular secretions. In pandemic regions, most children are infected by the age of three, and recurrent childhood infection is common (Dawson *et al.*, 1976). The initial response is a conjunctivitis, followed by lymphoid follicle formation (Hu *et al.*, 2013). Young trachoma patients develop an epithelial hyperplasia and mixed inflammatory infiltrate (el-Asrar *et al.*, 1989). Active trachoma usually wanes by 10 years of age, suggesting that partial immunity can develop with age and exposure, and most children will have minor sequelae (Dawson *et al.*, 1976; Hu *et al.*, 2013). Individuals with moderate to severe trachoma, however, develop a scarred conjunctiva, resulting in shortening distortion and upturning of the eyelid (entropion), the lashes abrading the eye (trichiasis), and eventually blindness (Hu *et al.*, 2013) (refer to Chapter 1, *Chlamydia trachomatis* Infection, for more details).

The immunological and inflammatory correlates associated with self-limiting disease versus scarring pathology are only partially characterized. An up-regulation of proinflammatory and matrix factor expression is observed during active *C. trachomatis* infection and trachoma. Th1-type cytokines such as IFN gamma (IFNγ), and Th2-type cytokine IL-13 as well as CD8+ CTL are involved in resolution of infection. Scarring trachoma and trichiasis are typified by sustained up-regulation of proinflammatory, tissue remodelling and profibrotic genes, and a relative decrease in the anti-inflammatory IL-1 receptor antagonist (IL-1RA) (Burton *et al.*, 2011a; Burton *et al.*, 2015; Burton *et al.*, 2011b; Cohen *et al.*, 2005; Ramadhani *et al.*,

2017; Skwor *et al.*, 2008). The progression to scarring in the apparent absence of *C. trachomatis* indicates that other host determinants and exogenous factors such as co-infection could be important in driving the later stages of disease (Ramadhani *et al.*, 2016; Ramadhani *et al.*, 2017). A recent reinterpretation of the role of lymphoid follicles, specifically as potential reservoirs of tissue resident memory T (Trm) cells (Johnson and Brunham, 2016), may also hold key information that potentially could be investigated by animal modelling. The genome-wide comparison of immune response profiles have associated antibodies to a panel of *C. trachomatis* antigens with trichiasis (Lu *et al.*, 2012a). Particularly, Pgp3 antibody has been considered a risk factor for ocular pathology upon *C. trachomatis* infection in the eye (West *et al.*, 2018; Wiegand *et al.*, 2018; Winstanley *et al.*, 2017; Zambrano *et al.*, 2016).

While trachoma has decreased in many parts of the world as a result of improved sanitary conditions and implementation of the SAFE (Surgery for trichiasis, Antibiotics, Facial cleanliness and Environmental improvement) strategy, it is still the leading global cause of blindness (Jansen *et al.*, 2007; Khandekar *et al.*, 2006; Liu *et al.*, 2016; Smith *et al.*, 2013a; Smith *et al.*, 2013b; WHO, 2003; Wright *et al.*, 2008). Because of this there are renewed and promising efforts to develop a trachoma vaccine, which will rely on synergistic studies from human populations and appropriate modelling (Kari *et al.*, 2011; Mabey *et al.*, 2014).

Genitourinary infection

C. trachomatis serovars D–K are tropic for columnar and transitional epithelial cells that line the genital tract. *C. trachomatis* can be cultured from human cervico-vaginal and urethral secretions (Alani *et al.*, 1977; Barnes *et al.*, 1990; Ficarra *et al.*, 2008; Oriel *et al.*, 1976). Cervico-vaginal secretions and semen are primary horizontal vehicles of transmission for establishing genitourinary, rectal and pharyngeal infections, and rare ocular *C. trachomatis* infections. *C. trachomatis* can be visualized as free EBs in these secretions and within inclusions in epithelial cells (Dunlop *et al.*, 1989; Kiviat *et al.*, 1985; Lewis *et al.*, 2014; Swanson *et al.*, 1975). Serovars D–F are the most commonly isolated organisms from genital specimens (Geisler *et al.*, 2003; Morre *et al.*, 2000; Quint *et al.*, 2007). Approximately two-thirds of *C. trachomatis*-infected women will vertically transmit *C. trachomatis* to their neonate during a vaginal birth (Francois *et al.*, 1988; Hammerschlag *et al.*, 1979).

The urethral epithelium is the most common site of *C. trachomatis* infection in males, and 50% will experience urethritis (Stamm *et al.*, 1984). Inflammation, typified by elevated polymorphonuclear cell (PMN) numbers, is noted in symptomatic and asymptomatic men (Swartz and Kraus, 1979). It is very rare that *C. trachomatis* will ascend the male tract to cause epididymitis and possibly prostatitis (Ireton and Berger, 1984; Kiviat *et al.*, 1987). Repeat infections are extremely frequent in men (Fung *et al.*, 2007) suggesting that incomplete immunity to the organism is common.

Infections in women typically extend for months if untreated, and reinfection is common, again suggesting incomplete immunity (Burstein *et al.*, 1998). Similarly to trachoma, there is evidence that elements of immunity are serovar-specific (Brunham *et al.*, 1996). The endocervix is the primary site of *C. trachomatis* infection in women. Over 80% of women are asymptomatic, but a pelvic exam can reveal inflammation; approximately 30% will have a mucopurulent cervicitis manifested by redness, oedema and discharge, cervical friability and/or hypertrophic ectopy (Brunham *et al.*, 1984; Peipert, 2003). Proinflammatory and T cell cytokines, PMNs, CD4 and CD8 T cells, red blood cells and epithelial cell counts are significantly elevated in cervical secretions of many actively infected women (Ficarra *et al.*, 2008; Hwang *et al.*, 2015; Spear *et al.*, 2011). Further, most cervical CD4 T cells are activated effector cells and express chemokine receptors CCR5 or CXCR4, rendering them potential host cells or transmission vehicles for HIV (Ficarra *et al.*, 2008; Mittal *et al.*, 2004; Schust *et al.*, 2012). Histologically, neutrophils, macrophages, lymphocytes and plasma cells are frequently detected in *C. trachomatis*-infected endocervix and transition zone tissues (Ibana *et al.*, 2012; Kiviat *et al.*, 1990; Paavonen *et al.*, 1982). Follicular cervicitis can identify characteristic sub-epithelial lymphoid follicles similar in histological appearance to those found in trachoma (Dunlop *et al.*, 1989; Kiviat *et al.*, 1990; Paavonen *et al.*, 1982). It remains to be determined whether these lymphoid follicles are reservoirs of protective and/or pathology-inducing Trm or indicative of chronic cervical inflammation leading to permanent functional change (Johnson and Brunham, 2016).

Upper genital tract disease requires *C. trachomatis* to ascend through the cervix, leading to pelvic inflammatory disease (PID), a spectrum of disorders that includes endometritis, salpingitis, tubo-ovarian abscess and pelvic peritonitis (Brunham *et al.*, 2015). Approximately 10 to 15% women who acquire *C. trachomatis* may develop PID and 11% of women with PID eventually develop tubal infertility and 9% will experience ectopic pregnancy (Cohen and Brunham, 1999). Infertility rates can be directly associated with the number and severity of *C. trachomatis* infection (Westrom *et al.*, 1992). Similar to histological observations of conjunctival biopsies in trachoma, endometrium and fallopian tubes are infiltrated with neutrophils, lymphocytes and plasma cells, and invariably accompanied by lymphoid follicles (Kiviat *et al.*, 1990; Moller *et al.*, 1979). Chronic inflammatory damage eventually leads to tissue remodelling and fibrotic tubal blockage, by which time *C. trachomatis* organisms can no longer be detected (Hafner, 2015; Henry-Suchet *et al.*, 1996).

Synergistic human and modelling studies can be used to address key questions regarding chlamydial pathogenesis. This includes, but is not limited to identifying the inflammatory and immune events that lead to a protective versus pathology-inducing response. It is important to determine whether these responses overlap or merely differ in magnitude, are site specific (cervix versus upper tract, local versus local with a contribution from systemic) or related to antigen concentration, form or presentation. Recent human studies can help inform appropriate model and *C. trachomatis* mutant choices. First, a relatively high bacterial burden appears to be required for upper tract ascension suggesting that robust cervical immunity could be key to decreasing bacterial burden (Russell *et al.*, 2016a). Second, 'seek and treat' public health strategies indicate that PID is decreasing even as infection rates increase (Brunham and Rekart, 2009), suggesting that shortening *C. trachomatis* infection with timely treatment can reduce upper tract pathology. The identification of aberrant inclusions in human cervix also indicates that chronic infection may be aided by the ability of *C. trachomatis* to enter into a persistent growth state until environmental conditions are more favourable (Lewis *et al.*, 2014). Third, ~20% women naturally clear infection and are protected from reinfection, indicating that *C. trachomatis* infection can induce protective immunity. The correlates of protective immunity could be identified by comparative longitudinal studies in hosts that naturally clear or do not clear infection (Geisler *et al.*, 2013).

Genital epithelial cells likely initiate and sustain the local host response upon *C. trachomatis* infection by producing cytokines such as interleukin (IL)–1 and IL-8. These cytokines are also found elevated in genital secretions of actively infected women (Buckner *et al.*, 2013; Dessus-Babus *et al.*, 2000; Rasmussen *et al.*, 1997). Although these cytokines can recruit, activate and augment innate and adaptive cell responses, *C. trachomatis* may have evolved countermeasures to evade some host defence mechanisms, resulting in prolonged *C. trachomatis*–host interactions, which could lead to tissue damage. Epithelial-derived IL-1 appears key in this process as IL-1 receptor antagonist (RA) can prevent tissue damage in the *ex vivo* fallopian tube model (Hvid *et al.*, 2007). Murine modelling studies consistently identify IFNγ as the major resolving immune factor (Cain and Rank, 1995; Cotter *et al.*, 1997; Lin *et al.*, 2019; Morrison *et al.*, 1995; Morrison *et al.*, 2000; Perry *et al.*, 1997), and human *in vitro* modelling in cervical epithelial cells has also long-established that sufficient and sustained IFNγ will eradicate infection (Beatty *et al.*, 1994; Carlin *et al.*, 1989; Ouellette *et al.*, 2006; Rapoza *et al.*, 1991), though the mechanism of action in the mouse and human is different (McClarty *et al.*, 2007). Elevated cervical IFNγ concentrations are found with active infection (Arno *et al.*, 1990), and a recent analysis of high-risk young women confirmed that systemic CD4+ T cell IFNγ responses to *C. trachomatis* antigens correlated with immunity to cervical infection (Russell *et al.*, 2016a; Russell *et al.*, 2016b). Secretory IgA but not IgG, or systemic IgG, negatively correlated with decreased infectious cervical bacterial burden, suggesting a role for local IgA antibody in decreasing bacterial load (Brunham *et al.*, 1983). Using a *C. trachomatis* whole proteome array (Wang *et al.*, 2010), systemic antibody response profiles in women have been mapped to aid diagnosis of tubal infertility; antibodies to *C. trachomatis* antigens CT147 and CT875 were associated with acute *C. trachomatis* infection, while antibodies to HSP60, CT557 and CT443 were associated with tubal infertility (Budrys *et al.*, 2012; Rodgers *et al.*, 2011).

Finally, clinical studies remind us that the *in situ* effectiveness of local innate and adaptive immunity can be modulated by endogenous and exogenous environmental factors, such as seminal plasma delivery of *C. trachomatis*, hormones, antibiotics, co-infections, and host genetic differences (Aiyar *et al.*, 2014; Brotman *et al.*, 2014; Menon *et al.*, 2015; Nguyen *et al.*, 2014). *C. trachomatis* may have evolved the ability to adapt to distinct tissue environments. The primary *C. trachomatis* isolates from different human tissue sites should be used in the corresponding model system studies in order to gain knowledge that is most relevant to *C. trachomatis* infections in humans. For example, genital, but not ocular, *C. trachomatis* isolates can salvage indole from tryptophan, rendering IFNγ-mediated clearance of *C. trachomatis* problematic for the host if indole is present, which it likely is in women with bacterial vaginosis (please see Chapter 16, *Chlamydia* Infections and the Microbiota, for more details) (Aiyar *et al.*, 2014; Caldwell *et al.*, 2003; Ziklo *et al.*, 2016a; Ziklo *et al.*, 2016b). These variables could have great implications for the design of vaccines and the testing of their effectiveness.

C. trachomatis infections involving other sites

Infection by the members of the lymphogranuloma venerum (LGV) *C. trachomatis* biovar, serovars L1, L2 and L3, commonly result in proctitis, although the number of reported cases is relatively low, and are most commonly reported in sporadic outbreaks in men who have sex with men (de Vries *et al.*, 2014; Lee *et al.*, 2015; Quinn *et al.*, 1981). LGV organisms can be isolated from genital and rectal tissue and secretions and bubo pus and are generally more invasive than non-LGV *C. trachomatis* serovars, invading and disseminating in monocytes and macrophages (Schachter and Osoba, 1983). LGV infection initiated in the human genitourinary tract generally begins with self-limiting ulcers, papules or herpetic lesions and these are most in the vulva or vagina of women, and the coronal sulcus, prepuce, glans or scrotum in men. Pathology is characterized by lymphangitis; endothelial cells lining the nodes and lymph channels proliferate, nodes enlarge, necrosis occurs, neutrophils infiltrate, resulting ultimately in abscesses. Rectal infection and subsequent inflammatory events lead to bleeding and ulceration. Chronic inflammation can lead to crypt abscesses and granulomas and the subsequent fibrotic sequelae lead to rectal stricture or stenosis (Mabey and Peeling, 2002).

Genital non-LGV *C. trachomatis* serovars are also frequently detected in the human GI tract (Craig *et al.*, 2015; Gratrix *et al.*, 2014; Gratrix *et al.*, 2015; Musil *et al.*, 2016; Peters *et al.*, 2014; Quinn *et al.*, 1981). *C. trachomatis* organisms can infect human enteroendocrine cells (Dlugosz *et al.*, 2014). Oral/anal sex likely introduces *C. trachomatis* into the GI tract (Gratrix *et al.*, 2014; Musil *et al.*, 2016). However, some individuals who deny this risk behaviour also test positive for *C. trachomatis* by rectal swab (Gratrix *et al.*, 2015; Peters *et al.*, 2014), raising the possibility that genital *C. trachomatis* serovars may spread from the genital to GI tract via sexual behaviour-independent pathways perhaps possibly via the blood (Cirino *et al.*, 2006; Datta *et al.*, 2014; López-Hurtado M *et al.*, 2015). This hypothesis would be consistent with reports that sexually transmitted *C. trachomatis* are detected in synovial tissue of individuals in rare cases of *C. trachomatis*-induced reactive arthritis (Freise *et al.*, 2015) and that *C. muridarum* can spread to the GI tract via the blood stream in mice (Dai *et al.*, 2016; Zhang *et al.*, 2015). Regardless of how *C. trachomatis* gets into the GI tract, it will be important to determine whether the GI tract can serve as a vaccination route to induce transmucosal immunity (Wang *et al.*, 2018; Zhu *et al.*, 2018) and/or a reservoir to directly (Bavoil *et al.*, 2017) or indirectly (Zhong, 2018) impact chlamydial pathogenicity.

Unsurprisingly, in view of the plethora of clinical conditions associated with *C. trachomatis*, and the tropism displayed by *C. trachomatis* isolates, no single animal or cell culture model can effectively recapitulate the pathology of infection. To overcome this limitation, models have been developed that recapitulate specific clinical conditions.

Non-human primate (NHP) models of *C. trachomatis* infection

NHP models of C. trachomatis infection have historically been key to unravelling the mechanisms of pathogenesis and immunity and for evaluating vaccine efficacy in both the eye and genital tissues.

Although it is unknown whether *C. trachomatis* infects NHP naturally, *C. trachomatis* organisms can undergo productive colonization in NHP mucosal tissues following artificial inoculation. Thus, useful information has been obtained from the various NHP models of *C. trachomatis* infection, including pigtail macaques, rhesus macaques and baboons.

NHP trachoma model

Soon after Tang's successful isolation of *C. trachomatis* organisms from human ocular tissues (Tang *et al.*, 1957a; Tang *et al.*, 1957b), human isolates were used to infect monkey eyes for testing whether the isolated *C. trachomatis* organisms could meet the criteria in Koch's postulates as a cause of trachoma (Wang, 1999). Ocular inflammatory pathologies similar to those observed in humans were indeed induced in various NHP species (Collier and Mogg, 1969; Moore *et al.*, 1970; Wang and Grayston, 1967). These studies have not only confirmed the causative relationship but also demonstrated that the NHP ocular model is adequate for studying trachoma pathogenesis (Collier and Mogg, 1969; Wang and Grayston, 1967) and evaluating trachoma vaccines (Chang *et al.*, 1964; Collier and Blyth, 1966a, b; Provost and Vickers, 1972). A whole *C. trachomatis* organism-based vaccine induced robust protective immunity against conjunctivitis in monkeys when challenged within two months. However, protective immunity disappeared within a year (Chang *et al.*, 1964; Provost and Vickers, 1972).

C. trachomatis serovar A can productively colonize the ocular tissue and induce robust inflammatory pathology in rhesus monkeys while the same serovar A deficient in the chlamydial plasmid fails to do so (Kari *et al.*, 2011). Although the precise roles of the plasmid in human chlamydial ocular infection and pathogenesis require further investigations, the correlation of human trachoma pathogenesis with high titres of antibodies recognizing the plasmid-encoded protein Pgp3 (Lu *et al.*, 2012a; Wiegand *et al.*, 2018; Winstanley *et al.*, 2017; Zambrano *et al.*, 2016) suggests that the plasmid may contribute to chlamydial pathogenicity in human ocular tissue. The primate trachoma model study further revealed that live organism shedding from the ocular tissue of serovar A-infected monkeys lasted up to seven weeks and visible inflammatory pathologies were observed for up to 15 weeks (Kari *et al.*, 2011). However, the plasmid-free serovar A-infected monkeys experienced only a brief period of live organism shedding and cleared ocular infection within three weeks, without developing significant pathology. The above observation has demonstrated that the chlamydial plasmid is necessary both for *C. trachomatis* colonization and pathogenicity in ocular tissue. This conclusion is consistent with the fact that most *C. trachomatis* isolates from humans carry the plasmid. Furthermore, plasmid-free serovar A-inoculated monkeys were protected against both infection and pathology induced by wild type serovar A (Kari *et al.*, 2011) demonstrating the feasibility for developing a live attenuated *C. trachomatis* vaccine (de la Maza *et al.*, 2017; Mabey *et al.*, 2014; Wang *et al.*, 2018; Zhong *et al.*, 2017; Zhu *et al.*, 2018). Interestingly, the ocular protective immunity in the macaque was dependent on CD8+ T cells (Olivares-Zavaleta *et al.*, 2014). However, the role of CD8+ T cells against *C. trachomatis* infections at other anatomical sites remains to be determined.

NHP genital tract model

Over the years, multiple NHP species, including pigtail and rhesus macaques and baboons, have been used to model *C. trachomatis* infection in the genital tract (Bell *et al.*, 2011). There are many similarities between the genital tracts of female pigtail macaques and women, including the length of the menstrual cycle, reproductive tract anatomy, cervical tissue cellular structure, and vaginal microflora. A recent study of *C. trachomatis* pathogenicity was conducted in pigtail macaques, in which six macaques received a cervical inoculation of *C. trachomatis* and were observed and sampled at weekly intervals (Patton *et al.*, 2015). These macaques developed mild to moderate infection and disease. However, parallel animals that were inoculated with the same strain of *C. trachomatis* but deficient in the plasmid developed similar infection and disease endpoints (Patton *et al.*, 2018). These results have been reproduced in rhesus macaques (Qu *et al.*, 2015). Clearly, the plasmid is not necessary for *C. trachomatis* infection in the primate genital tracts, in contrast to its requirement in the ocular *C. trachomatis* challenge model. Although a model of baboon cervical infection with *C. trachomatis* has also been reported (Bell *et al.*, 2011), this model is not likely to be a viable option because of cost and availability issues.

Rodent models of *C. trachomatis* infection

Although *C. trachomatis* does not naturally infect rodents, mice have served as the primary models for investigating chlamydial pathogenic mechanisms (Zhong, 2017, 2018) and evaluating vaccine candidates (de la Maza *et al.*, 2017). Since *C. trachomatis* serovar L2 has been most efficiently used for creating mutations, L2 mutants have been evaluated in mice (Yang *et al.*, 2016a). Other *C. trachomatis* organisms such as serovar D with or without genetic mutations have also been evaluated in mice (Sturdevant *et al.*, 2010). However, *C. trachomatis* is not able to overcome murine innate immunity and is thus rapidly eliminated from the mouse genital tract (Sturdevant and Caldwell, 2014). In contrast, intravaginal inoculation with *C. muridarum* is sufficient for establishing productive infection and for inducing upper genital tract pathology as observed in *C. trachomatis*-infected women under laparoscopy (de la Maza *et al.*, 1994; Lei *et al.*, 2014; O'Connell *et al.*, 2007; Shah *et al.*, 2005; Sun *et al.*, 2015). This robust model in combination with genetic manipulation of *C. muridarum* (Liu *et al.*, 2014a; Liu *et al.*, 2014b) has produced useful information for significantly advancing our understanding of chlamydial pathogenesis.

Mouse model of *C. trachomatis* infection

Soon after Tang's successful isolation of *C. trachomatis* from human ocular tissues (Tang *et al.*, 1957a; Tang *et al.*, 1957b), Bell *et al.* reported that mice immunized with killed *C. trachomatis* prevented toxicity induced by intravenous injection of high dose live *C. trachomatis* organisms (Bell *et al.*, 1959). This toxicity model allowed Wang and Grayston to classify the 'trachoma virus strains' into 15 serovars (Caldwell and Perry, 1982; Stephens *et al.*, 1988; Wang and Grayston, 1963; Zhong *et al.*, 1990). It was also used to probe the role of IFNγ in chlamydial infection (Zhong and de la Maza, 1988; Zhong *et al.*, 1988; Zhong *et al.*, 1989) and to define the cellular basis of the protective immunity against chlamydial infection (Lampe *et al.*, 1998) as well as to map T cell epitopes (Grotenbreg *et al.*, 2008). However, this toxicity model failed to provide useful information on *C. trachomatis* pathogenicity in humans. In order to obtain more relevant information on *C. trachomatis* pathogenicity, it is necessary to inoculate *C. trachomatis* into mice intravaginally. Unfortunately, *C. trachomatis* fails to induce long-lasting pathology in the mouse upper genital tract although lower genital tract infections can be established (Eko *et al.*, 2014; Gondek *et al.*, 2012; Sturdevant *et al.*, 2010). This is partly due to the failure of *C. trachomatis* to overcome innate immunity in the mouse genital tract, since mice with or without adaptive immunity were similarly resistant to *C. trachomatis* infection (Sturdevant and Caldwell, 2014). Female C3H/HeJ mice, which have a natural mutation in their toll-like receptor 4 gene, are more susceptible to *C. trachomatis* infection in the genital tract (Bonner *et al.*, 2015; Sturdevant *et al.*, 2010). Improving the mouse intravaginal infection model by genetically increasing mouse susceptibility to *C. trachomatis* colonization in the genital tract is desired. The p47 GTPases Iigp2 and Irgb10 regulate innate immunity and inflammation to chlamydial infection in mice (Bernstein-Hanley *et al.*, 2006; Miyairi *et al.*, 2007). However, mice deficient in these genes still prevented *C. trachomatis* persistence in the upper genital tract (Coers *et al.*, 2011).

Since intravaginal inoculation with *C. trachomatis* seldom led to tubal pathology in mice, an intrabursal inoculation model was tested (Carmichael *et al.*, 2013). However, the intrabursal inoculation involves survival surgery (requiring anaesthetization, incision into the peritoneal cavity, injection into the ovarian bursa, stitches) and bypasses both the cervical barrier and the uterotubal junction (Zhang *et al.*, 2017). Instead, a less invasive transcervical inoculation with *C. trachomatis* was preferred (Gondek *et al.*, 2012). Transcervical inoculation with an inactivated *C. trachomatis* vaccine strain protected mouse genital tract from live *C. trachomatis* challenge infection (Stary *et al.*, 2015). In a separate study, transcervical inoculation with live *C. trachomatis* serovar D induced tubal infertility in both C57 wild type mice and HLA-DR4 transgenic mice (whose major histocompatibility complex class II antigen is replaced with the human leucocyte antigen DR4) (Pal *et al.*, 2018). The HLA-DR4 mice can be a useful model for both studying *C. trachomatis* pathogenesis and evaluating human-relevant chlamydial vaccines. It will be worth testing whether the inactivated *C. trachomatis* vaccine can prevent infertility induced by a subsequent *C. trachomatis* infection in the HLA-DR4 mice or mice swapped with more human elements (Garcia and Freitas, 2012).

Since the successful transformation of the *C. trachomatis* LGV L2 strain by Wang *et al.* (Wang *et al.*, 2011), a series of L2 mutants deficient in plasmid genes have been created (Gong *et al.*, 2013; Song *et al.*, 2013) and evaluated in mice (Ramsey *et al.*, 2014). L2 deficient in Pgp3 was attenuated in its ability to infect the mouse lower genital tracts and to induce inflammatory pathology, confirming an earlier report that plasmid-free L2 was attenuated in mice (Carlson *et al.*, 2008). A L2 strain deficient in expressing the chlamydial protease-like activity factor (CPAF) was isolated from L2 mutagenesis libraries (Kokes *et al.*, 2015; Nguyen and Valdivia, 2012; Snavely *et al.*, 2014). CPAF has been well characterized with *in vitro* assays (Chen *et al.*, 2009a; Dong *et al.*, 2004a; Dong *et al.*, 2004b; Huang *et al.*, 2008; Zhong, 2009), but some loss of function mutations in CPAF defined in a cell-free system did not significantly reduce CPAF enzymatic activity during chlamydial infection (Yang *et al.*, 2015). The discrepancy in CPAF proteolytic activity measured between cell-free and intracellular infection systems suggested that *Chlamydia* might encode compensatory co-factors for promoting CPAF functionality, which is consistent with the concept that CPAF function may be important during chlamydial infection. Indeed, a CPAF-deficient L2 strain was defective in colonizing the mouse genital tract following an intravaginal inoculation (Yang *et al.*, 2016a), demonstrating a critical role of CPAF in chlamydial survival in mice. Mechanistically, CPAF may target extracellular defence molecules (Conrad *et al.*, 2013) such as cathelicidin peptide (Tang *et al.*, 2015) and complement components (Yang *et al.*, 2016b) as well as neutrophil surface receptor (Rajeeve *et al.*, 2018) for promoting chlamydial evasion of host defence. For a more detailed description of CPAF, please see Chapter 9, The Chlamydial Protease CPAF. However, the L2 genital infection model is not robust enough for modelling *C. trachomatis* infection in humans since both the infection and pathology were transient even with wild type L2.

Mouse model of genital *C. muridarum* infection

The *C. muridarum* cervicovaginal infection mouse model has been used to study chlamydial pathogenic mechanisms and host immune responses (Cheng *et al.*, 2008; de la Maza *et al.*, 1994; Morrison and Caldwell, 2002; Nagarajan *et al.*, 2008; Shah *et al.*, 2005) because of the overall genomic similarity between *C. muridarum* and *C. trachomatis* as well as the ability of *C. muridarum* to induce upper genital tract pathology that mimics what is observed in women infected with *C. trachomatis*. Scoring gross pathologies of isolated mouse genital tracts such as uterine horn dilatation (Sun *et al.*, 2015) and hydrosalpinx (Chen *et al.*, 2014) have been described. It is worthy pointing out that only mouse hydrosalpinx, as a result of tubal fibrotic occlusion, is a medically relevant endpoint since tubal adhesion, phimosis (fimbrial adhesion) and hydrosalpinx observed in *C. trachomatis*-infected women under laparoscopy (Budrys *et al.*, 2012; Rodgers *et al.*, 2011) all have the underlying cause of fibrosis.

Upon intravaginal inoculation, *C. muridarum* ascends along the mouse genital tract by passing through both the cervical barrier and uterotubal junction to reach the oviduct and ovary tissues. The ascending infection induces severe acute inflammation in the mouse genital tracts. Pyosalpinx (dilated oviducts filled with inflammatory infiltrates, cell debris and pus) becomes detectable macroscopically between 2 and 3 weeks. By the fourth week, some pyosalpinges are converted into hydrosalpinges while others may resolve without developing hydrosalpinx. Some hydrosalpinges become long-lasting while others can resolve spontaneously. It remains unclear what determines the varied pathology outcomes. Nevertheless, it has been proposed that mouse hydrosalpinx observed eight weeks or longer after *C. muridarum* infection may be more stable (Zhang *et al.*, 2014), which may result in long-lasting infertility. It may be more medically relevant to study long-lasting infertility because human infertility is defined as the failure to conceive after 12 months or more of unprotected sex. Indeed, women with tubal factor infertility often develop fibrotic pathologies in their upper genital tracts but without any sign of acute infection (Budrys *et al.*, 2012; Rodgers *et al.*, 2011). Thus, it is logical to use the long-lasting hydrosalpinx as oviduct pathology in the *C. muridarum* mouse model (Lu *et al.*, 2012b; Peng *et al.*, 2011; Tang *et al.*, 2013, 2014).

Using this model in combination with genetic modifications for both *C. muridarum* (Liu *et al.*, 2014a) and mice (Chen *et al.*, 2010; Yang *et al.*, 2014), significant progress has been made in defining the roles of the pathogen, host, and environmental factors in chlamydial pathogenesis. Screening 11 strains of mice for their susceptibility to hydrosalpinx induction has revealed that both ascending infection and induction of tubal inflammation play critical roles in chlamydial pathogenicity (Chen *et al.*, 2014). The plasmid and plasmid-encoded Pgp3 may both promote *C. muridarum* ascension and tubal inflammatory pathology (Liu *et al.*, 2014b; O'Connell *et al.*, 2007; Zhong, 2017) by modulating both antimicrobial peptide and TNF signalling pathways (Galaleldeen *et al.*, 2013; Hou *et al.*, 2015; Hou *et al.*, 2018). On the other hand, chromosome-encoded hypothetical proteins TC0237 and TC0668 are critical for *C. muridarum* induction of hydrosalpinx but without significantly affecting ascending infection (Chen *et al.*, 2015a; Conrad *et al.*, 2015). These observations together suggest a separation of function between plasmid-encoded and chromosome-encoded factors. The distinct functions may be selected for during chlamydial adaptation to the host gastrointestinal tract (Koprivsek *et al.*, 2019; Zhang *et al.*, 2019). From the host side, mice deficient in MyD88 (Chen *et al.*, 2010; Frazer *et al.*, 2013) or STING (Zhang *et al.*, 2017) allowed more *C. muridarum* organisms to cross the cervical and uterotubal junction, which suggests that these host signalling pathways play important roles in controlling chlamydial infection in the genital tract. On the other hand, many inflammatory pathways including TNFα (Dong *et al.*, 2014; Murthy *et al.*, 2011), matrix metalloproteinase (Imtiaz *et al.*, 2006), iNOS (Ramsey *et al.*, 2001), IL-1R (Nagarajan *et al.*, 2012), caspase-1 (Cheng *et al.*, 2008; Igietseme *et al.*, 2013), IL-17 (Andrew *et al.*, 2013), CD28 (Chen *et al.*, 2009b), CXCR2 (Lee *et al.*, 2010) and C5 (Yang *et al.*, 2014) may promote chlamydial pathogenicity while CXCR5 signalling suppresses it (Jiang *et al.*, 2013; Jiang *et al.*, 2012).

Mouse genital *C. muridarum* spreading to the gastrointestinal (GI) tract

Successful transformation of *C. muridarum* (Liu *et al.*, 2014a) has provided the tool for monitoring live chlamydial infection in mice in real time (Campbell *et al.*, 2014), which has revealed both rapid vaginal *C. muridarum* ascension to the oviduct and spread to the GI tract (Zhang *et al.*, 2015). Chlamydial spreading from the genital tract to the GI tract is independent of the oral/anorectal route since mice wearing restraining Elizabethan collars and singly housed in net-bottom cages to prevent coprophagy still spread genital *C. muridarum* to their GI tract (Zhang *et al.*, 2015). Instead, genital-to-GI tract spreading seems to be dependent on the blood circulation (Dai *et al.*, 2016; Perry and Hughes, 1999; Zhang *et al.*, 2015). Once *C. muridarum* arrives in the GI tract, it does not autoinoculate extra-GI tissues (Wang *et al.*, 2016), contradicting the hypothesis that chlamydial organisms in the GI tract may serve as a reservoir for auto-inoculating the genital tract (Rank and Yeruva, 2014; Yeruva *et al.*, 2013). Despite the lack of autoinoculation, chlamydial spreading from the genital to GI tracts correlates with chlamydial pathogenicity in the upper genital tract (Shao *et al.*, 2017a; Shao *et al.*, 2017b), leading to the proposal of a two-hit model as a chlamydial pathogenic mechanism (Brothwell *et al.*, 2018). The first hit may be caused by genital *C. muridarum* ascending infection, during which, *C. muridarum* invades oviduct epithelial cells and induces tubal inflammation. The latter may both clear tubal infection and cause epithelial damage, triggering tissue repair responses including fibrosis. Normally, the fibrotic response is transient and stops upon restoration of oviduct function. However, after genital *C. muridarum* spreads to the GI tract, *C. muridarum* organisms can stay in the GI tract for long periods of time and induce pro-fibrotic lymphocytes that can be recruited into the oviduct that has experienced *C. muridarum* ascending infection. These GI tract-derived *C. muridarum*-specific profibrotic lymphocytes may act as the second hit for converting the transient tubal repairing fibrotic response into long-lasting tubal fibrotic blockade. The 2-hit model is consistent with the concept that gut bacteria-induced responses can impact extra-GI tissues (Block *et al.*, 2016; Schnupf *et al.*, 2017; Tai *et al.*, 2016; Wu *et al.*, 2010). On the contrary, if a naive mouse is exposed to *C. muridarum* in the GI tract first, the mouse essentially becomes immune to subsequent *C. muridarum* exposure in extra-GI tissues (Wang *et al.*, 2018; Zhu *et al.*, 2018). Thus, the site of first exposure to *C. muridarum* may determine the consequence. Although *C. trachomatis* is also frequently detected in the human GI tracts (Bazan *et al.*, 2015; Dukers-Muijrers *et al.*, 2013; Javanbakht *et al.*, 2012; Musil *et al.*, 2016; Steedman and McMillan, 2009), it remains unclear how human GI tract *C. trachomatis* affects *C. trachomatis* pathogenicity in the genital tract. It will be worth testing the two-hit model in human studies.

Guinea pig model

Guinea pigs have been used to study chlamydial pathogenesis mainly using genital inoculation with *Chlamydia caviae* (which was originally called the chlamydial agent of guinea pig inclusion conjunctivitis or GPIC) (Barron *et al.*, 1979; Rank *et al.*, 1979; White *et al.*, 1979). Both *C. trachomatis* (de Jonge *et al.*, 2011; Wali *et al.*, 2017) and *C. muridarum* (Kamalakaran *et al.*, 2013) have also been used to infect the genital tracts of guinea pigs. The female reproductive system of the guinea pig is closely related to humans with regard to histological features and physiology, and exhibits a 15–17-day oestrous cycle, with active secretion of corpus lutea, so that hormonal influences on infection may be assessed (Rank, 1994). Given that guinea pigs can be outbred and exhibit a greater degree of similarity than mice to the female reproductive tract of humans (Rank *et al.*, 1992), the guinea pig model can be a model for bridging the translational gap between mice, NHP and human cohorts. Nevertheless, the findings made from the guinea pig model are largely consistent with what has been observed in mice, which further validates the utility of the mouse model. In addition, the costs and availability of reagents also favour the use of the mouse over the guinea pig model.

Considerations for choosing the appropriate animal models

Overall considerations

One major challenge to all animal models is that chlamydial species have adapted to specific animal hosts. Genitourinary *C. trachomatis* serovars that can cause significant upper reproductive tract disease in humans are highly susceptible to innate host immune responses in the murine genital tract, and therefore generally fail to induce long-lasting fibrosis in the upper genital tract. Although IFNγ-mediated immunity is critical for controlling both *C. trachomatis* and *C. muridarum* infections in humans and mice, respectively, distinct mechanisms are involved. IFNγ-induced tryptophan deprivation is critical for controlling *C. trachomatis* infection in human cells while IFNγ-activated small GTPases control *C. muridarum* infection in mice. Different biovars and serovars of the same chlamydial species have developed distinct tissue tropisms in human hosts. For example, *C. trachomatis* serovars A-C are predominantly tropic for the columnar epithelium of the conjunctiva, serovars D-K for columnar and transitional epithelial cells lining the genital tract and the LGV serovars, which form their own biovar, can invade lymphoid tissue and can cause proctitis. These varied human tissue tropisms by different *C. trachomatis* serovars are likely to be shaped by tissue-specific *C. trachomatis* evolution driven by host and microbiome-encoded factors found in the local tissue environment. Patterns of chlamydial spreading within and between hosts may impact the outcome of chlamydial interactions with the hosts. Finally, tissue damage in human *C. trachomatis*-mediated disease can be caused by chronic and/or repeat or single severe infections both complicating and increasing the cost of recapitulation in models. Thus, it is important to choose the appropriate infection dose and frequency in a given model in order to obtain relevant information. A lower challenge dose may be more sensitive in detecting minor differences between treatment and control groups while a higher infection dose may not only mask the minor differences but may also activate additional pathways. A recent study (Sun *et al.*, 2017) revealed that at a high infectious dose, IL-6 was required for controlling *C. muridarum* infection but not for exacerbating *C. muridarum* induction of pathology. However, at a low infectious dose, IL-6 signalling participated in both controlling *C. muridarum* infection and exacerbating pathology. Thus, higher infectious doses of *C. muridarum* can induce pathology in an IL-6-independent manner. While the mechanisms by which low dose and high dose *C. trachomatis* infections induce pathology may differ from those of *C. muridarum*, the *C. muridarum* study has revealed subtleties that may be difficult to recapitulate in other models such as mice inoculated with *C. trachomatis*. This is because mice are highly susceptible to *C. muridarum* but resistant to *C. trachomatis* infection.

Considerations for using NHP models

Using NHP to study trachoma has been very successful. The trachoma serovars productively colonize the ocular mucosal tissue and induce inflammatory pathologies similar to those observed in human eyes upon *C. trachomatis* infection. A plasmid-deficient serovar A strain was significantly attenuated in the monkey eye. These findings demonstrate the relevance of the NHP ocular model to human ocular infection with *C. trachomatis* (since most ocular isolates from humans are plasmid-competent). They also provide information for developing the plasmid-free serovar A strain into an attenuated vaccine for preventing trachoma. In contrast, the NHP genital tract model is possibly more limited. Wild type *C. trachomatis* serovars fail to induce long-lasting pathologies in the mouse upper genital tract despite repeated inoculations, emphasizing the difficulty in employing these models to understand the pathology of *C. trachomatis* genitourinary infections, or to evaluate vaccines against the same.

Considerations for using the *C. trachomatis*–mouse model

One advantage of using the *C. trachomatis*–mouse model is that the same pathogens that cause human diseases can be directly evaluated in mice. The knowledge to be obtained on chlamydial virulence factors or vaccine candidates may be applicable to humans. However, *C. trachomatis* is rapidly cleared from the mouse genital tract, which is not suitable for investigating adaptive immunity. Nevertheless, since tissue-resident memory T cells (Trms) induced by vaccination can attack infection within a few days, the *C. trachomatis*-mouse model has been successfully used for evaluating *C. trachomatis* vaccine efficacy in terms of preventing infection.

The lack of long-lasting fibrotic pathology in the upper genital tract in the *C. trachomatis*-infected mice is another limitation of this model, which disqualifies this model for evaluating *C. trachomatis* vaccine efficacy in terms of preventing chronic pathology. Among the *C. trachomatis*-mouse models, intravenous injection was useful for serotyping *C. trachomatis* organisms, but this method is inappropriate for investigating *C. trachomatis* pathogenesis in the genital tract. Among the genital inoculation routes, transcervical inoculation seems to be most appropriate for both investigating acute pathogenicity and evaluating vaccine efficacy in the genital tract.

Considerations for using the *C. muridarum*–mouse model

The most obvious advantage of using the *C. muridarum*–mouse model is that a single inoculation with wild type *C. muridarum* to the lower genital tract can lead to long-lasting hydrosalpinx, which is an indication of tubal fibrotic occlusion. Since the underlying cause of tubal adhesion, phimosis and hydrosalpinx observed in *C. trachomatis*-infected women is fibrosis, observing the long-lasting hydrosalpinx as an endpoint in the *C. muridarum*-mouse model is medically relevant. Furthermore, many variables of the *C. muridarum*-mouse model, including inoculation doses (Sun *et al.*, 2017), routes (Chen *et al.*, 2015b; Tang *et al.*, 2014; Tang *et al.*, 2013), mouse strains (Chen *et al.*, 2014), and evaluation parameters such as infectivity, pathology and infertility (de la Maza *et al.*, 1994; Lei *et al.*, 2014; Shah *et al.*, 2005; Sun *et al.*, 2015) have been compared and optimized. Thus, the *C. muridarum*-mouse model will continue to be the model of choice for both investigating chlamydial pathogenesis and evaluating vaccine efficacy.

An added benefit of using the *C. muridarum*-mouse model is the recent finding that the tissue site of first exposure to *C. muridarum* in a mouse seems to determine the outcome of the exposure (Wang *et al.*, 2018; Zhong, 2018; Zhu *et al.*, 2018): When the first exposure occurs in the genital tract, the genital *C. muridarum* can spread to the GI tract, which may promote the induction of pathology in the upper genital tract. This observation has led to the two-hit model (discussed above) for explaining *C. muridarum* pathogenicity in the upper genital tract. However, when the first exposure is the GI tract, the mice are essentially immunized against subsequent exposures in extra-GI tissues.

The caveat is that the information obtained from the *C. muridarum* mouse model studies may not necessarily be applicable to *C. trachomatis* infection in humans. This is because both the biology of chlamydial interactions with its host and host behaviours differ significantly between mice and humans. *C. muridarum* is clearly adapted to the mouse GI tract since it can colonize the GI tract for long periods of time without causing any significant pathology. Furthermore, its genome maintains three copies of the toxin genes coding for the homologues of the large clostridium toxin that may be required for competing with gut microbiota. Although *C. muridarum* can induce long-lasting tubal fibrosis in mice when introduced to the genital tract, there is no evidence of natural transmission of *C. muridarum* among mice via the genital tract. Instead, *C. muridarum* may be more efficiently transmitted among mice via the oral-faecal route, which may have allowed *C. muridarum* to adapt to the gastrointestinal mucosal tissues. However, *C. trachomatis* has been clearly transmitted between humans via genital tract contact, suggesting that *C. trachomatis* may have started to, if not already, adapt to human genital tissues. This assumption is consistent with the observation that the *C. trachomatis* genome has accumulated numerous loss of function mutations in the genes coding for the large toxin homologue of the enteric bacterium *Clostridium*.

Considerations for using human tissue culture models

Determination of *C. trachomatis* pathogenesis mechanisms and its prevention have predominantly been the purview of animal models, complemented with *in vitro C. trachomatis* infection studies in transformed human epithelial cell lines such as HeLa or HEp-2. Transformed lines can be reproducibly infected, are economic and scalable, and circumvent the differences in *C. trachomatis* growth and host defence mechanisms inherent in non-human models. However, they have the following limitations: (a) their genetic alterations and viral oncoproteins are known to modify immune responses. (b) They generally poorly represent tissue of origin, and do not recapitulate the native tissue or establish an appropriate environmental milieu when cultured in 2D. Tissue explants (organ cultures) derived from discarded human female genital tract tissue or biopsies are the classic architecturally appropriate *ex vivo* model that can provide appropriate species, site and, to a more limited extent, environment that can be infected by *Neisseria gonorrheae, Mycoplasma genitalium* and HIV, in addition to *C. trachomatis* (Carney and Taylor-Robinson, 1973; Hvid *et al.*, 2007; Jerchel *et al.*, 2012; King *et al.*, 2009; Nicol *et al.*, 2015; Taylor-Robinson and Carney, 1974). Though limited by availability, replicate potential, and lack of cell trafficking into the tissue, they remain key tools to understand early *C. trachomatis*–epithelial interactions; explants have, for example, been used for the IL1RA fallopian tube studies described above (Hvid *et al.*, 2007), and detailed protocols are available (Jerchel *et al.*, 2012; McGee *et al.*, 1976). Patton and Kuo's elegant NHP variant on this, auto-transplanted fallopian tube, endometrial and conjunctival tissue placed in subcutaneous pockets for ease of access and allowing mononuclear trafficking and repeat *C. trachomatis* challenge, has also been key in establishing the role of immune cells in tissue damage (Patton and Kuo, 1989; Patton *et al.*, 1989; Van Voorhis *et al.*, 1997). Primary human epithelial cells grown from surgically discarded human genital tissues are a more reductionist single-cell model to address *C. trachomatis* pathogenic mechanisms. These non-immortalized, primary cells have the characteristics of their site of origin and polarize when placed on membrane inserts that can be coated with extracellular matrix, establishing unique basal and apical surfaces (Herbst-Kralovetz *et al.*, 2008; Li and Chen, 2018; Wyrick *et al.*, 1989). Such cells can be apically challenged with *C. trachomatis*, while immune cells can be added to the basal chamber and treatments (e.g. hormones, drugs, potential neutralizing antibody, microbicide) can be applied at the appropriate surface/s (Wyrick *et al.*, 1989). The utility of these cells is limited by their non-immortalized nature. They support a limited, but variable, number of divisions, before undergoing terminal senescence. Variable number of senescent cells in such cultures renders them difficult to use in long-term experiments that require reproducible responses. Approaches have been described for non-genital human epithelial cells to increase their replicative potential (Sato *et al.*, 2011), but it is unknown if these can be applied to genital-origin tissue. Such limitations in primary cells have led us and others to immortalize them using viral oncogenes, primarily E6 and E7 from human papillomavirus type 16. The first such E6/E7 immortalized line constructed using primary cervical cells was described by DiPaolo and co-workers in 1990 (Woodworth *et al.*, 1990). Numerous lines have been constructed using this approach since then to study sexually transmitted infections as well as urinary tract infections using primary epithelial cells from the vagina, cervix, transition zone, and endocervix (Deng *et al.*, 2018; Rajan *et al.*, 2000). At lower passage numbers, E6/E7 immortalized cells retain many primary-like characteristics and functions. These include growing in contact-inhibited monolayers, forming tight and *adherens* junctions, and displaying characteristics of polarized cells when grown at an air–liquid interface. They can be infected with sexually transmitted agents, and when polarized, the apical surface is complete with mucus and secreted mediators resembling some of the first barriers encountered by *C. trachomatis* (Buckner *et al.*, 2013; Buckner *et al.*, 2011; Fichorova and Anderson, 1999; Fichorova *et al.*, 2002; Fichorova *et al.*, 1997; Schust *et al.*, 2012). Limited microbiota may also be added to more accurately recapitulate the vaginal mucosa (Fichorova *et al.*, 2011; Medina-Colorado *et al.*, 2017; Rose *et al.*, 2012). The vastly increased replicative potential of these E6/E7 immortalized cell-lines, combined with their capacity to form polarized epithelia, renders them to be of considerable utility in evaluating aspects of innate and adaptive immune responses against *C. trachomatis* infections, as well as to investigate bacterial factors that drive cellular pathology. A potential problem with these cells lies in the other effects of E6 and E7, particularly at higher passage numbers. These include degradation of cellular proteins including specific cytokines, and proteins necessary to retain polarity or maintain junctional complexes (Kranjec and Banks, 2011; Nakagawa and Huibregtse, 2000; Niebler *et al.*, 2013; Storrs and Silverstein, 2007; Thomas *et al.*, 2002). To overcome these effects, newer immortalization techniques that do not rely on viral oncogenes have been developed and should further improve immortalized line variety and quality (Haga *et al.*, 2007; Kogan *et al.*, 2006; Ramirez *et al.*, 2004; Shan and Liu, 2013; Yang *et al.*, 2007).

Recent advances in three-dimensional culture techniques are predicted to expand the potential for human tissue models that can be developed or adapted for *C. trachomatis* research. Models can be imaged, studied in real time at the molecular and cellular levels and used to study specific genes in regulating cell behaviour and to determine the contribution of micro-environmental factors in the disease process (Dixon and Woodward, 2018). Examples include polarized epithelium with underlying stromal and dendritic cells and tissue organoids including those developed from stem cells (Bouschbacher *et al.*, 2008), the latter of which have successfully been used for modelling the endometrium (Boretto *et al.*, 2017; Li and Chen, 2018), a primary site for *C. trachomatis* pathogenesis.

In summary, selection of a suitable model to elucidate *C. trachomatis* pathogenesis or to evaluate the efficacy of a vaccine is complex and driven by the aspect of *C. trachomatis* pathology under investigation. NHP and rodent models to evaluate *C. trachomatis* genital tract infections have been useful but do not recapitulate human disease as closely as NHP modelling *C. trachomatis* ocular infections. Humanized mice may someday allow human *C. trachomatis* genital tract infection to be modelled in rodents. In addition, newer cell culture models that use polarized epithelia and reconstructed tissue models show great promise to evaluate the molecular mechanisms that underlie tissue damage directly induced by *C. trachomatis* infections, but currently do not permit evaluation of immune-mediated pathology or protective immune responses.

References

Aiyar, A., Quayle, A.J., Buckner, L.R., Sherchand, S.P., Chang, T.L., Zea, A.H., Martin, D.H., and Belland, R.J. (2014). Influence of the tryptophan–indole-IFNgamma axis on human genital Chlamydia trachomatis infection: role of vaginal co-infections. Front Cell Infect Microbiol 4, 72.

Alani, M.D., Darougar, S., Burns, D.C., Thin, R.N., and Dunn, H. (1977). Isolation of Chlamydia trachomatis from the male urethra. Br J Vener Dis 53, 88–92.

Andrew, D.W., Cochrane, M., Schripsema, J.H., Ramsey, K.H., Dando, S.J., O'Meara, C.P., Timms, P., and Beagley, K.W. (2013). The duration of Chlamydia muridarum genital tract infection and associated chronic pathological changes are reduced in IL-17 knockout mice but protection is not increased further by immunization. PLoS One *8*, e76664.

Arno, J.N., Ricker, V.A., Batteiger, B.E., Katz, B.P., Caine, V.A., and Jones, R.B. (1990). Interferon-gamma in endocervical secretions of women infected with Chlamydia trachomatis. J Infect Dis *162*, 1385–1389.

Barnes, R.C., Katz, B.P., Rolfs, R.T., Batteiger, B., Caine, V., and Jones, R.B. (1990). Quantitative culture of endocervical Chlamydia trachomatis. J Clin Microbiol *28*, 774–780.

Barron, A.L., White, H.J., Rank, R.G., and Soloff, B.L. (1979). Target tissues associated with genital infection of female guinea pigs by the chlamydial agent of guinea pig inclusion conjunctivitis. J Infect Dis *139*, 60–68.

Bavoil, P.M., Marques, P.X., Brotman, R., and Ravel, J. (2017). Does Active Oral Sex Contribute to Female Infertility? J Infect Dis.

Bazan, J.A., Carr Reese, P., Esber, A., Lahey, S., Ervin, M., Davis, J.A., Fields, K., and Turner, A.N. (2015). High prevalence of rectal gonorrhea and Chlamydia infection in women attending a sexually transmitted disease clinic. J Womens Health (Larchmt) *24*, 182–189.

Beatty, W.L., Belanger, T.A., Desai, A.A., Morrison, R.P., and Byrne, G.I. (1994). Role of tryptophan in gamma interferon-mediated chlamydial persistence. Ann N Y Acad Sci *730*, 304–306.

Bell, J.D., Bergin, I.L., Harris, L.H., Chai, D., Mullei, I., Mwenda, J., Dalton, V.K., Vahratian, A., Lebar, W., Zochowski, M.K., *et al.* (2011). The effects of a single cervical inoculation of Chlamydia trachomatis on the female reproductive tract of the baboon (Papio anubis). J Infect Dis *204*, 1305–1312.

Bell, S.D., Jr., Snyder, J.C., and Murray, E.S. (1959). Immunization of mice against toxic doses of homologous elementary bodies of trachoma. Science *130*, 626–627.

Bernstein-Hanley, I., Coers, J., Balsara, Z.R., Taylor, G.A., Starnbach, M.N., and Dietrich, W.F. (2006). The p47 GTPases Igtp and Irgb10 map to the Chlamydia trachomatis susceptibility locus Ctrq-3 and mediate cellular resistance in mice. Proc Natl Acad Sci U S A *103*, 14092–14097.

Block, K.E., Zheng, Z., Dent, A.L., Kee, B.L., and Huang, H. (2016). Gut Microbiota Regulates K/BxN Autoimmune Arthritis through Follicular Helper T but Not Th17 Cells. J Immunol *196*, 1550–1557.

Bonner, C., Caldwell, H.D., Carlson, J.H., Graham, M.R., Kari, L., Sturdevant, G.L., Tyler, S., Zetner, A., and McClarty, G. (2015). Chlamydia trachomatis virulence factor CT135 is stable *in vivo* but highly polymorphic *in vitro*. Pathog Dis 73, ftv043.

Boretto, M., Cox, B., Noben, M., Hendriks, N., Fassbender, A., Roose, H., Amant, F., Timmerman, D., Tomassetti, C., Vanhie, A., *et al.* (2017). Development of organoids from mouse and human endometrium showing endometrial epithelium physiology and long-term expandability. Development *144*, 1775–1786.

Bouschbacher, M., Bomsel, M., Verronese, E., Gofflo, S., Ganor, Y., Dezutter-Dambuyant, C., and Valladeau, J. (2008). Early events in HIV transmission through a human reconstructed vaginal mucosa. AIDS *22*, 1257–1266.

Brothwell, J.A., Muramatsu, M.K., Zhong, G., and Nelson, D.E. (2018). Advances and obstacles in the genetic dissection of chlamydial virulence. Curr Top Microbiol Immunol *412*, 133–158.

Brotman, R.M., Ravel, J., Bavoil, P.M., Gravitt, P.E., and Ghanem, K.G. (2014). Microbiome, sex hormones, and immune responses in the reproductive tract: challenges for vaccine development against sexually transmitted infections. Vaccine *32*, 1543–1552.

Brunham, R.C., Gottlieb, S.L., and Paavonen, J. (2015). Pelvic inflammatory disease. N Engl J Med *372*, 2039–2048.

Brunham, R.C., Kimani, J., Bwayo, J., Maitha, G., Maclean, I., Yang, C., Shen, C., Roman, S., Nagelkerke, N.J., Cheang, M., *et al.* (1996). The epidemiology of Chlamydia trachomatis within a sexually transmitted diseases core group. J Infect Dis *173*, 950–956.

Brunham, R.C., Kuo, C.C., Cles, L., and Holmes, K.K. (1983). Correlation of host immune response with quantitative recovery of Chlamydia trachomatis from the human endocervix. Infect Immun *39*, 1491–1494.

Brunham, R.C., Paavonen, J., Stevens, C.E., Kiviat, N., Kuo, C.C., Critchlow, C.W., and Holmes, K.K. (1984). Mucopurulent cervicitis—the ignored counterpart in women of urethritis in men. N Engl J Med *311*, 1–6.

Brunham, R.C., and Rappuoli, R. (2013). Chlamydia trachomatis control requires a vaccine. Vaccine *31*, 1892–1897.

Brunham, R.C., and Rekart, M.L. (2009). Considerations on Chlamydia trachomatis disease expression. FEMS Immunol Med Microbiol *55*, 162–166.

Buckner, L.R., Lewis, M.E., Greene, S.J., Foster, T.P., and Quayle, A.J. (2013). Chlamydia trachomatis infection results in a modest pro-inflammatory cytokine response and a decrease in T cell chemokine secretion in human polarized endocervical epithelial cells. Cytokine *63*, 151–165.

Buckner, L.R., Schust, D.J., Ding, J., Nagamatsu, T., Beatty, W., Chang, T.L., Greene, S.J., Lewis, M.E., Ruiz, B., Holman, S.L., *et al.* (2011). Innate immune mediator profiles and their regulation in a novel polarized immortalized epithelial cell model derived from human endocervix. J Reprod Immunol 92, 8–20.

Budrys, N.M., Gong, S., Rodgers, A.K., Wang, J., Louden, C., Shain, R., Schenken, R.S., and Zhong, G. (2012). Chlamydia trachomatis antigens recognized in women with tubal factor infertility, normal fertility, and acute infection. Obstet Gynecol *119*, 1009–1016.

Burstein, G.R., Gaydos, C.A., Diener-West, M., Howell, M.R., Zenilman, J.M., and Quinn, T.C. (1998). Incident Chlamydia trachomatis infections among inner-city adolescent females. JAMA *280*, 521–526.

Burton, M.J., Rajak, S.N., Bauer, J., Weiss, H.A., Tolbert, S.B., Shoo, A., Habtamu, E., Manjurano, A., Emerson, P.M., Mabey, D.C., *et al.* (2011a). Conjunctival transcriptome in scarring trachoma. Infect Immun *79*, 499–511.

Burton, M.J., Rajak, S.N., Hu, V.H., Ramadhani, A., Habtamu, E., Massae, P., Tadesse, Z., Callahan, K., Emerson, P.M., Khaw, P.T., *et al.* (2015). Pathogenesis of progressive scarring trachoma in Ethiopia and Tanzania and its implications for disease control: two cohort studies. PLoS Negl Trop Dis *9*, e0003763.

Burton, M.J., Ramadhani, A., Weiss, H.A., Hu, V., Massae, P., Burr, S.E., Shangali, W., Holland, M.J., Mabey, D.C., and Bailey, R.L. (2011b). Active trachoma is associated with increased conjunctival expression of IL17A and profibrotic cytokines. Infect Immun *79*, 4977–4983.

Cain, T.K., and Rank, R.G. (1995). Local Th1-like responses are induced by intravaginal infection of mice with the mouse pneumonitis biovar of Chlamydia trachomatis. Infect Immun *63*, 1784–1789.

Caldwell, H.D., and Perry, L.J. (1982). Neutralization of Chlamydia trachomatis infectivity with antibodies to the major outer membrane protein. Infect Immun *38*, 745–754.

Caldwell, H.D., Wood, H., Crane, D., Bailey, R., Jones, R.B., Mabey, D., Maclean, I., Mohammed, Z., Peeling, R., Roshick, C., *et al.* (2003). Polymorphisms in Chlamydia trachomatis tryptophan synthase genes differentiate between genital and ocular isolates. J Clin Invest *111*, 1757–1769.

Campbell, J., Huang, Y., Liu, Y., Schenken, R., Arulanandam, B., and Zhong, G. (2014). Bioluminescence Imaging of Chlamydia muridarum Ascending Infection in Mice. PLos one *9*, e101634.

Carlin, J.M., Borden, E.C., and Byrne, G.I. (1989). Interferon-induced indoleamine 2,3-dioxygenase activity inhibits Chlamydia psittaci replication in human macrophages. J Interferon Res *9*, 329–337.

Carlson, J.H., Whitmire, W.M., Crane, D.D., Wicke, L., Virtaneva, K., Sturdevant, D.E., Kupko, J.J., 3rd, Porcella, S.F., Martinez-Orengo, N., Heinzen, R.A., *et al.* (2008). The Chlamydia trachomatis plasmid is a transcriptional regulator of chromosomal genes and a virulence factor. Infect Immun *76*, 2273–2283.

Carmichael, J.R., Tifrea, D., Pal, S., and de la Maza, L.M. (2013). Differences in infectivity and induction of infertility: a comparative study of Chlamydia trachomatis strains in the murine model. Microbes Infect *15*, 219–229.

Carney, F.E., Jr., and Taylor-Robinson, D. (1973). Growth and effect of Neisseria gonorrhoeae in organ cultures. Br J Vener Dis *49*, 435–440.

Chang, H.L., Chin, H.Y., and Wang, K.C. (1964). Experimental Studies on Trachoma Vaccine in Monkeys. Chin Med J *83*, 755–762.

Chen, C., Zhou, Z., Conrad, T., Yang, Z., Dai, J., Li, Z., Wu, Y., and Zhong, G. (2015a). In vitro passage selects for Chlamydia muridarum with enhanced infectivity in cultured cells but attenuated pathogenicity in mouse upper genital tract. Infect Immun *83*, 1881–1892.

Chen, D., Chai, J., Hart, P.J., and Zhong, G. (2009a). Identifying catalytic residues in CPAF, a Chlamydia-secreted protease. Arch Biochem Biophys *485*, 16–23.

Chen, J., Yang, Z., Sun, X., Tang, L., Ding, Y., Xue, M., Zhou, Z., Baseman, J., and Zhong, G. (2015b). Intrauterine infection with plasmid-free Chlamydia muridarum reveals a critical role of the plasmid in chlamydial ascension and establishes a model for evaluating plasmid-independent pathogenicity. Infect Immun *83*, 2583–2592.

Chen, J., Zhang, H., Zhou, Z., Yang, Z., Ding, Y., Zhou, Z., Zhong, E., Arulanandam, B., Baseman, J., and Zhong, G. (2014). Chlamydial induction of hydrosalpinx in 11 strains of mice reveals multiple host mechanisms for preventing upper genital tract pathology. PLoS One *9*, e95076.

Chen, L., Cheng, W., Shivshankar, P., Lei, L., Zhang, X., Wu, Y., Yeh, I.T., and Zhong, G. (2009b). Distinct roles of CD28- and CD40 ligand-mediated costimulation in the development of protective immunity and pathology during Chlamydia muridarum urogenital infection in mice. Infect Immun 77, 3080–3089.

Chen, L., Lei, L., Chang, X., Li, Z., Lu, C., Zhang, X., Wu, Y., Yeh, I.T., and Zhong, G. (2010). Mice deficient in MyD88 Develop a Th2-dominant response and severe pathology in the upper genital tract following Chlamydia muridarum infection. J Immunol *184*, 2602–2610.

Cheng, W., Shivshankar, P., Li, Z., Chen, L., Yeh, I.T., and Zhong, G. (2008). Caspase-1 contributes to Chlamydia trachomatis-induced upper urogenital tract inflammatory pathologies without affecting the course of infection. Infect Immun *76*, 515–522.

Cirino, F., Webley, W.C., West, C., Croteau, N.L., Andrzejewski, C., Jr., and Stuart, E.S. (2006). Detection of Chlamydia in the peripheral blood cells of normal donors using *in vitro* culture, immunofluorescence microscopy and flow cytometry techniques. BMC Infect Dis *6*, 23.

Coers, J., Gondek, D.C., Olive, A.J., Rohlfing, A., Taylor, G.A., and Starnbach, M.N. (2011). Compensatory T cell responses in IRG-deficient mice prevent sustained Chlamydia trachomatis infections. PLoS Pathog *7*, e1001346.

Cohen, C.R., and Brunham, R.C. (1999). Pathogenesis of Chlamydia induced pelvic inflammatory disease. Sex Transm Infect *75*, 21–24.

Cohen, C.R., Koochesfahani, K.M., Meier, A.S., Shen, C., Karunakaran, K., Ondondo, B., Kinyari, T., Mugo, N.R., Nguti, R., and Brunham, R.C. (2005). Immunoepidemiologic profile of Chlamydia trachomatis infection: importance of heat-shock protein 60 and interferon- gamma. J Infect Dis *192*, 591–599.

Collier, L.H., and Blyth, W.A. (1966a). Immunogenicity of experimental trachoma vaccines in baboons. I. Experimental methods, and preliminary tests with vaccines prepared in chick embryos and in HeLa cells. J Hyg (Lond) *64*, 513–528.

Collier, L.H., and Blyth, W.A. (1966b). Immunogenicity of experimental trachoma vaccines in baboons. II. Experiments with adjuvants, and tests of cross-protection. J Hyg (Lond) *64*, 529–544.

Collier, L.H., and Mogg, A.E. (1969). Dissemination and immunogenicity of live TRIC agent in baboons after parenteral injection. II. Experiments with a 'slow-killing' strain. J Hyg (Lond) *67*, 449–455.

Conrad, T., Yang, Z., Ojcius, D., and Zhong, G. (2013). A path forward for the chlamydial virulence factor CPAF. Microbes Infect *15*, 1026–1032.

Conrad, T.A., Gong, S., Yang, Z., Matulich, P., Keck, J., Beltrami, N., Chen, C., Zhou, Z., Dai, J., and Zhong, G. (2015). The Chromosome-Encoded Hypothetical Protein TC0668 Is an Upper Genital Tract Pathogenicity Factor of Chlamydia muridarum. Infect Immun *84*, 467–479.

Cotter, T.W., Ramsey, K.H., Miranpuri, G.S., Poulsen, C.E., and Byrne, G.I. (1997). Dissemination of Chlamydia trachomatis chronic genital tract infection in gamma interferon gene knockout mice. Infect Immun *65*, 2145–2152.

Craig, A.P., Kong, F.Y., Yeruva, L., Hocking, J.S., Rank, R.G., Wilson, D.P., and Donovan, B. (2015). Is it time to switch to doxycycline from azithromycin for treating genital chlamydial infections in women? Modelling the impact of autoinoculation from the gastrointestinal tract to the genital tract. BMC Infect Dis *15*, 200.

Dai, J., Zhang, T., Wang, L., Shao, L., Zhu, C., Zhang, Y., Failor, C., Schenken, R., Baseman, J., He, C., *et al.* (2016). Intravenous Inoculation with Chlamydia muridarum Leads to a Long-Lasting Infection Restricted to the Gastrointestinal Tract. Infect Immun *84*, 2382–2388.

Datta, B., Njau, F., Thalmann, J., Haller, H., and Wagner, A.D. (2014). Differential infection outcome of Chlamydia trachomatis in human blood monocytes and monocyte-derived dendritic cells. BMC Microbiol *14*, 209.

Dawson, C.R., Daghfous, T., Messadi, M., Hoshiwara, I., and Schachter, J. (1976). Severe endemic trachoma in Tunisia. Br J Ophthalmol *60*, 245–252.

de Jonge, M.I., Keizer, S.A., El Moussaoui, H.M., van Dorsten, L., Azzawi, R., van Zuilekom, H.I., Peters, P.P., van Opzeeland, F.J., van Dijk, L., Nieuwland, R., *et al.* (2011). A novel guinea pig model of Chlamydia trachomatis genital tract infection. Vaccine *29*, 5994–6001.

de la Maza, L.M., Pal, S., Khamesipour, A., and Peterson, E.M. (1994). Intravaginal inoculation of mice with the Chlamydia trachomatis mouse pneumonitis biovar results in infertility. Infect Immun *62*, 2094–2097.

de la Maza, L.M., Zhong, G., and Brunham, R.C. (2017). Update on Chlamydia trachomatis Vaccinology. Clin Vaccine Immunol *24*.

de Vries, H.J., Zingoni, A., White, J.A., Ross, J.D., and Kreuter, A. (2014). 2013 European Guideline on the management of proctitis, proctocolitis and enteritis caused by sexually transmissible pathogens. Int J STD AIDS *25*, 465–474.

Deng, H., Hillpot, E., Yeboah, P., Mondal, S., and Woodworth, C.D. (2018). Susceptibility of epithelial cells cultured from different regions of human cervix to HPV16-induced immortalization. PLoS One *13*, e0199761.

Dessus-Babus, S., Knight, S.T., and Wyrick, P.B. (2000). Chlamydial infection of polarized HeLa cells induces PMN chemotaxis but the cytokine profile varies between disseminating and non-disseminating strains. Cell Microbiol 2, 317–327.

Ding, H., Gong, S., Tian, Y., Yang, Z., Brunham, R., and Zhong, G. (2013). Transformation of sexually transmitted infection-causing serovars of chlamydia trachomatis using Blasticidin for selection. PLoS One *8*, e80534.

Dixon, E.E., and Woodward, O.M. (2018). Three-dimensional *in vitro* models answer the right questions in ADPKD cystogenesis. Am J Physiol Renal Physiol *315*, F332-F335.

Dlugosz, A., Muschiol, S., Zakikhany, K., Assadi, G., D'Amato, M., and Lindberg, G. (2014). Human enteroendocrine cell responses to infection with Chlamydia trachomatis: a microarray study. Gut Pathog *6*, 24.

Dong, F., Pirbhai, M., Zhong, Y., and Zhong, G. (2004a). Cleavage-dependent activation of a chlamydia-secreted protease. Mol Microbiol *52*, 1487–1494.

Dong, F., Sharma, J., Xiao, Y., Zhong, Y., and Zhong, G. (2004b). Intramolecular dimerization is required for the chlamydia-secreted protease CPAF to degrade host transcriptional factors. Infect Immun *72*, 3869–3875.

Dong, X., Liu, Y., Chang, X., Lei, L., and Zhong, G. (2014). Signalling via tumour necrosis factor receptor 1 but not Toll-like receptor 2 contributes significantly to hydrosalpinx development following Chlamydia muridarum infection. Infect Immun *82*, 1833–1839.

Dukers-Muijrers, N.H., Speksnijder, A.G., Morre, S.A., Wolffs, P.F., van der Sande, M.A., Brink, A.A., van den Broek, I.V., Werner, M.I., and Hoebe, C.J. (2013). Detection of anorectal and cervicovaginal Chlamydia trachomatis infections following azithromycin treatment: prospective cohort study with multiple time-sequential measures of rRNA, DNA, quantitative load and symptoms. PLoS One *8*, e81236.

Dunlop, E.M., Garner, A., Darougar, S., Treharne, J.D., and Woodland, R.M. (1989). Colposcopy, biopsy, and cytology results in women with chlamydial cervicitis. Genitourin Med *65*, 22–31.

Eko, F.O., Mania-Pramanik, J., Pais, R., Pan, Q., Okenu, D.M., Johnson, A., Ibegbu, C., He, C., He, Q., Russell, R., *et al.* (2014). Vibrio choleraee ghosts (VCG) exert immunomodulatory effect on dendritic cells for enhanced antigen presentation and induction of protective immunity. BMC Immunol *15*, 584.

el-Asrar, A.M., Van den Oord, J.J., Geboes, K., Missotten, L., Emarah, M.H., and Desmet, V. (1989). Immunopathology of trachomatous conjunctivitis. Br J Ophthalmol 73, 276–282.

Ficarra, M., Ibana, J.S., Poretta, C., Ma, L., Myers, L., Taylor, S.N., Greene, S., Smith, B., Hagensee, M., Martin, D.H., *et al.* (2008). A distinct cellular profile is seen in the human endocervix during Chlamydia trachomatis infection. Am J Reprod Immunol *60*, 415–425.

Fichorova, R.N., and Anderson, D.J. (1999). Differential expression of immunobiological mediators by immortalized human cervical and vaginal epithelial cells. Biol Reprod *60*, 508–514.

Fichorova, R.N., Cronin, A.O., Lien, E., Anderson, D.J., and Ingalls, R.R. (2002). Response to Neisseria gonorrhoeae by cervicovaginal epithelial cells occurs in the absence of toll-like receptor 4-mediated signalling. J Immunol *168*, 2424–2432.

Fichorova, R.N., Rheinwald, J.G., and Anderson, D.J. (1997). Generation of papillomavirus-immortalized cell lines from normal human ectocervical, endocervical, and vaginal epithelium that maintain expression of tissue-specific differentiation proteins. Biol Reprod *57*, 847–855.

Fichorova, R.N., Yamamoto, H.S., Delaney, M.L., Onderdonk, A.B., and Doncel, G.F. (2011). Novel vaginal microflora colonization model providing new insight into microbicide mechanism of action. MBio *2*, e00168–00111.

Francois, P., Rouhan, D., Hirtz, P., and Beaudoing, A. (1988). [Chlamydia trachomatis in pediatrics]. Pediatrie *43*, 101–108.

Frazer, L.C., Sullivan, J.E., Zurenski, M.A., Mintus, M., Tomasak, T.E., Prantner, D., Nagarajan, U.M., and Darville, T. (2013). CD4+ T cell expression of MyD88 is essential for normal resolution of Chlamydia muridarum genital tract infection. J Immunol *191*, 4269–4279.

Freise, J., Bernau, I., Meier, S., Zeidler, H., and Kuipers, J.G. (2015). Optimized testing for *C. trachomatis* DNA in synovial fluid samples in clinical practice. Z Rheumatol *74*, 824–828.

Fung, M., Scott, K.C., Kent, C.K., and Klausner, J.D. (2007). Chlamydial and gonococcal reinfection among men:

a systematic review of data to evaluate the need for retesting. Sex Transm Infect *83*, 304–309.

Galaleldeen, A., Taylor, A.B., Chen, D., Schuermann, J.P., Holloway, S.P., Hou, S., Gong, S., Zhong, G., and Hart, P.J. (2013). Structure of the Chlamydia trachomatis immunodominant antigen Pgp3. J Biol Chem *288*, 22068–22079.

Garcia, S., and Freitas, A.A. (2012). Humanized mice: current states and perspectives. Immunol Lett *146*, 1–7.

Geisler, W.M., Lensing, S.Y., Press, C.G., and Hook, E.W., 3rd (2013). Spontaneous resolution of genital Chlamydia trachomatis infection in women and protection from reinfection. J Infect Dis *207*, 1850–1856.

Geisler, W.M., Suchland, R.J., Whittington, W.L., and Stamm, W.E. (2003). The relationship of serovar to clinical manifestations of urogenital Chlamydia trachomatis infection. Sex Transm Dis *30*, 160–165.

Gondek, D.C., Olive, A.J., Stary, G., and Starnbach, M.N. (2012). CD4+ T cells are necessary and sufficient to confer protection against Chlamydia trachomatis infection in the murine upper genital tract. J Immunol *189*, 2441–2449.

Gong, S., Yang, Z., Lei, L., Shen, L., and Zhong, G. (2013). Characterization of Chlamydia trachomatis Plasmid-Encoded Open Reading Frames. J Bacteriol *195*, 3819–3826.

Gottlieb, S.L., and Johnston, C. (2017). Future prospects for new vaccines against sexually transmitted infections. Curr Opin Infect Dis *30*, 77–86.

Gratrix, J., Singh, A.E., Bergman, J., Egan, C., McGinnis, J., Drews, S.J., and Read, R. (2014). Prevalence and characteristics of rectal chlamydia and gonorrhea cases among men who have sex with men after the introduction of nucleic acid amplification test screening at 2 Canadian sexually transmitted infection clinics. Sex Transm Dis *41*, 589–591.

Gratrix, J., Singh, A.E., Bergman, J., Egan, C., Plitt, S.S., McGinnis, J., Bell, C.A., Drews, S.J., and Read, R. (2015). Evidence for increased Chlamydia case finding after the introduction of rectal screening among women attending 2 Canadian sexually transmitted infection clinics. Clin Infect Dis *60*, 398–404.

Grayston, J.T., Wang, S.P., Yang, Y.F., and Woolridge, R.L. (1962). The effect of trachoma virus vaccine on the course of experimental trachoma infection in blind human volunteers. J Exp Med *115*, 1009–1022.

Grayston, J.T., Woolridge, R.L., Wang, S.P., Yen, C.H., Yang, C.Y., Cheng, K.H., and Chang, I.H. (1963). Field studies of protection from infection by experimental trachoma virus vaccine in preschool-aged children on Taiwan. Proc Soc Exp Biol Med *112*, 589–595.

Grotenbreg, G.M., Roan, N.R., Guillen, E., Meijers, R., Wang, J.H., Bell, G.W., Starnbach, M.N., and Ploegh, H.L. (2008). Discovery of CD8+ T cell epitopes in Chlamydia trachomatis infection through use of caged class I MHC tetramers. Proc Natl Acad Sci U S A *105*, 3831–3836.

Hafner, L.M. (2015). Pathogenesis of fallopian tube damage caused by Chlamydia trachomatis infections. Contraception 92, 108–115.

Haga, K., Ohno, S., Yugawa, T., Narisawa-Saito, M., Fujita, M., Sakamoto, M., Galloway, D.A., and Kiyono, T. (2007). Efficient immortalization of primary human cells by p16INK4a-specific short hairpin RNA or Bmi-1, combined with introduction of hTERT. Cancer Sci *98*, 147–154.

Hammerschlag, M.R., Anderka, M., Semine, D.Z., McComb, D., and McCormack, W.M. (1979). Prospective study of maternal and infantile infection with Chlamydia trachomatis. Pediatrics *64*, 142–148.

Henry-Suchet, J., Askienazy-Elbhar, M., and Orfila, J. (1996). Clinical consequences of immune response to CT upper genital tract infection in women. Infect Dis Obstet Gynecol *4*, 171–175.

Herbst-Kralovetz, M.M., Quayle, A.J., Ficarra, M., Greene, S., Rose, W.A., 2nd, Chesson, R., Spagnuolo, R.A., and Pyles, R.B. (2008). Quantification and comparison of toll-like receptor expression and responsiveness in primary and immortalized human female lower genital tract epithelia. Am J Reprod Immunol *59*, 212–224.

Hou, S., Dong, X., Yang, Z., Li, Z., Liu, Q., and Zhong, G. (2015). Chlamydial plasmid-encoded virulence factor Pgp3 neutralizes the antichlamydial activity of human cathelicidin LL-37. Infect Immun *83*, 4701–4709.

Hou, S., Sun, X., Dong, X., Lin, H., Tang, L., Xue, M., and Zhong, G. (2018). Chlamydial plasmid-encoded virulence factor Pgp3 interacts with human cathelicidin peptide LL-37 to modulate immune response. Microbes Infect.

Hu, V.H., Holland, M.J., and Burton, M.J. (2013). Trachoma: protective and pathogenic ocular immune responses to Chlamydia trachomatis. PLoS Negl Trop Dis *7*, e2020.

Huang, Z., Feng, Y., Chen, D., Wu, X., Huang, S., Wang, X., Xiao, X., Li, W., Huang, N., Gu, L., *et al.* (2008). Structural basis for activation and inhibition of the secreted chlamydia protease CPAF. Cell Host Microbe *4*, 529–542.

Hvid, M., Baczynska, A., Deleuran, B., Fedder, J., Knudsen, H.J., Christiansen, G., and Birkelund, S. (2007). Interleukin-1 is the initiator of Fallopian tube destruction during Chlamydia trachomatis infection. Cell Microbiol *9*, 2795–2803.

Hwang, L.Y., Scott, M.E., Ma, Y., and Moscicki, A.B. (2015). Diversity of Cervicovaginal Cytokine Response to Incident Chlamydia trachomatis Infection Among a Prospective Cohort of Young Women. Am J Reprod Immunol *74*, 228–236.

Ibana, J.A., Myers, L., Porretta, C., Lewis, M., Taylor, S.N., Martin, D.H., and Quayle, A.J. (2012). The major CD8 T cell effector memory subset in the normal and Chlamydia trachomatis-infected human endocervix is low in perforin. BMC Immunol *13*, 66.

Igietseme, J.U., Omosun, Y., Partin, J., Goldstein, J., He, Q., Joseph, K., Ellerson, D., Ansari, U., Eko, F.O., Bandea, C., *et al.* (2013). Prevention of Chlamydia-induced infertility by inhibition of local caspase activity. J Infect Dis *207*, 1095–1104.

Imtiaz, M.T., Schripsema, J.H., Sigar, I.M., Kasimos, J.N., and Ramsey, K.H. (2006). Inhibition of matrix metalloproteinases protects mice from ascending infection and chronic disease manifestations resulting from urogenital Chlamydia muridarum infection. Infect Immun *74*, 5513–5521.

Ireton, R.C., and Berger, R.E. (1984). Prostatitis and epididymitis. Urol Clin North Am *11*, 83–94.

Jansen, E., Baltussen, R.M., van Doorslaer, E., Ngirwamungu, E., Nguyen, M.P., and Kilima, P.M. (2007). An eye for inequality: how trachoma relates to poverty in Tanzania and Vietnam. Ophthalmic Epidemiol *14*, 278–287.

Javanbakht, M., Gorbach, P., Stirland, A., Chien, M., Kerndt, P., and Guerry, S. (2012). Prevalence and correlates of rectal Chlamydia and gonorrhea among female clients at sexually transmitted disease clinics. Sex Transm Dis *39*, 917–922.

Jerchel, S., Knebel, G., Konig, P., Bohlmann, M.K., and Rupp, J. (2012). A human fallopian tube model for investigation of *C. trachomatis* infections. J Vis Exp.

Jiang, J., Champion, C.I., Wei, B., Liu, G., and Kelly, K.A. (2013). CD8(+)CXCR5(+) T cells regulate pathology in the genital tract. Infect Dis Obstet Gynecol *2013*, 813238.

Jiang, J., Karimi, O., Ouburg, S., Champion, C.I., Khurana, A., Liu, G., Freed, A., Pleijster, J., Rozengurt, N., Land, J.A., *et al.* (2012). Interruption of CXCL13–CXCR5 axis increases upper genital tract pathology and activation of NKT cells following chlamydial genital infection. PLoS One 7, e47487.

Johnson, C.M., and Fisher, D.J. (2013). Site-specific, insertional inactivation of incA in Chlamydia trachomatis using a group II intron. PLoS One *8*, e83989.

Johnson, R.M., and Brunham, R.C. (2016). Tissue-Resident T Cells as the Central Paradigm of Chlamydia Immunity. Infect Immun *84*, 868–873.

Kamalakaran, S., Chaganty, B.K., Gupta, R., Guentzel, M.N., Chambers, J.P., Murthy, A.K., and Arulanandam, B.P. (2013). Vaginal chlamydial clearance following primary or secondary infection in mice occurs independently of TNF-alpha. Front Cell Infect Microbiol 3, 11.

Kari, L., Whitmire, W.M., Olivares-Zavaleta, N., Goheen, M.M., Taylor, L.D., Carlson, J.H., Sturdevant, G.L., Lu, C., Bakios, L.E., Randall, L.B., *et al.* (2011). A live-attenuated chlamydial vaccine protects against trachoma in nonhuman primates. J Exp Med *208*, 2217–2223.

Khandekar, R., Nga, N.H., and Mai, P. (2006). Blinding trachoma in the northern provinces of Vietnam--a cross sectional survey. Ophthalmic Epidemiol *13*, 183–189.

King, M., Poya, H., Rao, J., Natarajan, S., Butch, A.W., Aziz, N., Kok, S., Chang, M.H., Lyons, J.M., Ault, K., *et al.* (2009). CXCL13 expression in Chlamydia trachomatis infection of the female reproductive tract. Drugs Today (Barc) *45 Suppl B*, 125–134.

Kiviat, M.D., Kiviat, N.B., and Berger, R.E. (1987). Chlamydia trachomatis epididymitis diagnosed by fluorescent monoclonal antibody. Urology *30*, 395–397.

Kiviat, N.B., Paavonen, J.A., Brockway, J., Critchlow, C.W., Brunham, R.C., Stevens, C.E., Stamm, W.E., Kuo, C.C., DeRouen, T., and Holmes, K.K. (1985). Cytologic manifestations of cervical and vaginal infections. I. Epithelial and inflammatory cellular changes. JAMA *253*, 989–996.

Kiviat, N.B., Paavonen, J.A., Wolner-Hanssen, P., Critchlow, C.W., Stamm, W.E., Douglas, J., Eschenbach, D.A., Corey, L.A., and Holmes, K.K. (1990). Histopathology of endocervical infection caused by Chlamydia trachomatis, herpes simplex virus, Trichomonas vaginalis, and Neisseria gonorrhoeae. Hum Pathol *21*, 831–837.

Kogan, I., Goldfinger, N., Milyavsky, M., Cohen, M., Shats, I., Dobler, G., Klocker, H., Wasylyk, B., Voller, M., Aalders, T., *et al.* (2006). hTERT-immortalized prostate epithelial and stromal-derived cells: an authentic *in vitro* model for differentiation and carcinogenesis. Cancer Res *66*, 3531–3540.

Kokes, M., Dunn, J.D., Granek, J.A., Nguyen, B.D., Barker, J.R., Valdivia, R.H., and Bastidas, R.J. (2015). Integrating chemical mutagenesis and whole-genome sequencing as a platform for forward and reverse genetic analysis of Chlamydia. Cell Host Microbe *17*, 716–725.

Koprivsek, J., Zhang, T., Tian, Q., He, Y., Xu, H., Xu, Z., and Zhong, G. (2019). Distinct roles of chromosome-versus plasmid-encoded genital tract virulence factors in promoting Chlamydia muridarum colonization in the gastrointestinal tract. Infect Immun.

Kranjec, C., and Banks, L. (2011). A systematic analysis of human papillomavirus (HPV) E6 PDZ substrates identifies MAGI-1 as a major target of HPV type 16 (HPV-16) and HPV-18 whose loss accompanies disruption of tight junctions. J Virol *85*, 1757–1764.

Lampe, M.F., Wilson, C.B., Bevan, M.J., and Starnbach, M.N. (1998). Gamma interferon production by cytotoxic T lymphocytes is required for resolution of Chlamydia trachomatis infection. Infect Immun *66*, 5457–5461.

Lee, H.Y., Schripsema, J.H., Sigar, I.M., Lacy, S.R., Kasimos, J.N., Murray, C.M., and Ramsey, K.H. (2010). A role for CXC chemokine receptor-2 in the pathogenesis of urogenital Chlamydia muridarum infection in mice. FEMS Immunol Med Microbiol *60*, 49–56.

Lee, K.J., Kim, J., Shin, D.H., Jung, J.O., Koh, S., Kim, K.Y., and Lee, J.M. (2015). Chlamydial Proctitis in a Young Man Who Has Sex with Men: Misdiagnosed as Inflammatory Bowel Disease. Chonnam Med J *51*, 139–141.

Lei, L., Chen, J., Hou, S., Ding, Y., Yang, Z., Zeng, H., Baseman, J., and Zhong, G. (2014). Reduced live organism recovery and lack of hydrosalpinx in mice infected with plasmid-free Chlamydia muridarum. Infect Immun *82*, 983–992.

Lewis, M.E., Belland, R.J., AbdelRahman, Y.M., Beatty, W.L., Aiyar, A.A., Zea, A.H., Greene, S.J., Marrero, L., Buckner, L.R., Tate, D.J., *et al.* (2014). Morphologic and molecular evaluation of Chlamydia trachomatis growth in human endocervix reveals distinct growth patterns. Front Cell Infect Microbiol 4, 71.

Li, T., and Chen, Z.J. (2018). The cGAS-cGAMP-STING pathway connects DNA damage to inflammation, senescence, and cancer. J Exp Med *215*, 1287–1299.

Lin, H., He, C., Koprivsek, J.J., Chen, J., Zhou, Z., Arulanandam, B., Xu, Z., Tang, L., and Zhong, G. (2019). Antigen-Specific CD4(+) T Cell-Derived Gamma Interferon Is Both Necessary and Sufficient for Clearing Chlamydia from the Small Intestine but Not the Large Intestine. Infect Immun *87*.

Liu, T., Liang, Q., Hu, A., Feng, G., Wang, N., Peng, X., Baudouin, C., and Labbe, A. (2016). Elimination of blinding trachoma in China. J Fr Ophtalmol *39*, 836–842.

Liu, Y., Chen, C., Gong, S., Hou, S., Qi, M., Liu, Q., Baseman, J., and Zhong, G. (2014a). Transformation of Chlamydia muridarum reveals a role for Pgp5 in suppression of plasmid-dependent gene expression. J Bacteriol *196*, 989–998.

Liu, Y., Huang, Y., Yang, Z., Sun, Y., Gong, S., Hou, S., Chen, C., Li, Z., Liu, Q., Wu, Y., *et al.* (2014b). Plasmid-encoded Pgp3 is a major virulence factor for Chlamydia muridarum to induce hydrosalpinx in mice. Infect Immun *82*, 5327–5335.

López-Hurtado M., Cuevas-Recillas K.N., Flores-Salazar V.R., and F.M., G.-I. (2015). [Chlamydia trachomaatis DNA in leucocytes of peripheral blood from neonates]. Enferm Infecc Microbiol Clin *33*, 458–463.

Lu, C., Holland, M.J., Gong, S., Peng, B., Bailey, R.L., Mabey, D.W., Wu, Y., and Zhong, G. (2012a). Genome-wide identification of Chlamydia trachomatis antigens associated with trachomatous trichiasis. Invest Ophthalmol Vis Sci *53*, 2551–2559.

Lu, C., Zeng, H., Li, Z., Lei, L., Yeh, I.T., Wu, Y., and Zhong, G. (2012b). Protective immunity against mouse upper genital tract pathology correlates with high IFNgamma but low IL-17 T cell and anti-secretion protein antibody responses induced by replicating chlamydial organisms in the airway. Vaccine *30*, 475–485.

Mabey, D., and Peeling, R.W. (2002). Lymphogranuloma venereum. Sex Transm Infect *78*, 90–92.

Mabey, D.C., Hu, V., Bailey, R.L., Burton, M.J., and Holland, M.J. (2014). Towards a safe and effective chlamydial vaccine: lessons from the eye. Vaccine *32*, 1572–1578.

Mariotti, S.P., Pascolini, D., and Rose-Nussbaumer, J. (2009). Trachoma: global magnitude of a preventable cause of blindness. Br J Ophthalmol *93*, 563–568.

McClarty, G., Caldwell, H.D., and Nelson, D.E. (2007). Chlamydial interferon gamma immune evasion influences infection tropism. Curr Opin Microbiol *10*, 47–51.

McGee, Z.A., Johnson, A.P., and Taylor-Robinson, D. (1976). Human fallopian tubes in organ culture: preparation, maintenance, and quantitation of damage by pathogenic microorganisms. Infect Immun *13*, 608–618.

Medina-Colorado, A.A., Vincent, K.L., Miller, A.L., Maxwell, C.A., Dawson, L.N., Olive, T., Kozlova, E.V., Baum, M.M., and Pyles, R.B. (2017). Vaginal ecosystem modelling of growth patterns of anaerobic bacteria in microaerophilic conditions. Anaerobe *45*, 10–18.

Menon, S., Timms, P., Allan, J.A., Alexander, K., Rombauts, L., Horner, P., Keltz, M., Hocking, J., and Huston, W.M. (2015). Human and Pathogen Factors Associated with Chlamydia trachomatis-Related Infertility in Women. Clin Microbiol Rev *28*, 969–985.

Mittal, A., Rastogi, S., Reddy, B.S., Verma, S., Salhan, S., and Gupta, E. (2004). Enhanced immunocompetent cells in chlamydial cervicitis. J Reprod Med *49*, 671–677.

Miyairi, I., Tatireddigari, V.R., Mahdi, O.S., Rose, L.A., Belland, R.J., Lu, L., Williams, R.W., and Byrne, G.I. (2007). The p47 GTPases Iigp2 and Irgb10 regulate innate immunity and inflammation to murine Chlamydia psittaci infection. J Immunol *179*, 1814–1824.

Moller, B.R., Westrom, L., Ahrons, S., Ripa, K.T., Svensson, L., von Mecklenburg, C., Henrikson, H., and Mardh, P.A. (1979). Chlamydia trachomatis infection of the Fallopian tubes. Histological findings in two patients. Br J Vener Dis *55*, 422–428.

Moore, J.E., Griffiths, M.S., and Pearce, J.H. (1970). Virulence of trachoma-inclusion conjunctivitis agent after passage in baboons. Br J Exp Pathol *51*, 298–304.

Morre, S.A., Rozendaal, L., van Valkengoed, I.G., Boeke, A.J., van Voorst Vader, P.C., Schirm, J., de Blok, S., van Den Hoek, J.A., van Doornum, G.J., Meijer, C.J., *et al.* (2000). Urogenital Chlamydia trachomatis serovars in men and women with a symptomatic or asymptomatic infection: an association with clinical manifestations? J Clin Microbiol *38*, 2292–2296.

Morrison, R.P., and Caldwell, H.D. (2002). Immunity to murine chlamydial genital infection. Infect Immun *70*, 2741–2751.

Morrison, R.P., Feilzer, K., and Tumas, D.B. (1995). Gene knockout mice establish a primary protective role for major histocompatibility complex class II-restricted responses in Chlamydia trachomatis genital tract infection. Infect Immun *63*, 4661–4668.

Morrison, S.G., Su, H., Caldwell, H.D., and Morrison, R.P. (2000). Immunity to murine Chlamydia trachomatis genital tract reinfection involves B cells and CD4(+) T cells but not CD8(+) T cells. Infect Immun *68*, 6979–6987.

Mueller, K.E., Wolf, K., and Fields, K.A. (2016). Gene Deletion by Fluorescence-Reported Allelic Exchange Mutagenesis in Chlamydia trachomatis. MBio *7*, e01817–01815.

Murthy, A.K., Li, W., Chaganty, B.K., Kamalakaran, S., Guentzel, M.N., Seshu, J., Forsthuber, T.G., Zhong, G., and Arulanandam, B.P. (2011). Tumour necrosis factor alpha production from CD8+ T cells mediates oviduct pathological sequelae following primary genital Chlamydia muridarum infection. Infect Immun *79*, 2928–2935.

Musil, K., Currie, M., Sherley, M., and Martin, S. (2016). Rectal chlamydia infection in women at high risk of chlamydia attending Canberra Sexual Health Centre. Int J STD AIDS *27*, 526–530.

Nagarajan, U.M., Prantner, D., Sikes, J.D., Andrews, C.W., Jr., Goodwin, A.M., Nagarajan, S., and Darville, T. (2008). Type I interferon signalling exacerbates Chlamydia muridarum genital infection in a murine model. Infect Immun *76*, 4642–4648.

Nagarajan, U.M., Sikes, J.D., Yeruva, L., and Prantner, D. (2012). Significant role of IL-1 signalling, but limited role of inflammasome activation, in oviduct pathology during Chlamydia muridarum genital infection. J Immunol *188*, 2866–2875.

Nakagawa, S., and Huibregtse, J.M. (2000). Human scribble (Vartul) is targeted for ubiquitin-mediated degradation by the high-risk papillomavirus E6 proteins and the E6AP ubiquitin-protein ligase. Mol Cell Biol *20*, 8244–8253.

Nguyen, B.D., and Valdivia, R.H. (2012). Virulence determinants in the obligate intracellular pathogen Chlamydia trachomatis revealed by forward genetic approaches. Proc Natl Acad Sci U S A *109*, 1263–1268.

Nguyen, P.V., Kafka, J.K., Ferreira, V.H., Roth, K., and Kaushic, C. (2014). Innate and adaptive immune responses in male and female reproductive tracts in homeostasis and following HIV infection. Cell Mol Immunol *11*, 410–427.

Nicol, M.R., Emerson, C.W., Prince, H.M., Nelson, J.A., Fedoriw, Y., Sykes, C., Geller, E.J., Patterson, K.B., Cohen, M.S., and Kashuba, A.D. (2015). Models for predicting effective HIV chemoprevention in women. J Acquir Immune Defic Syndr *68*, 369–376.

Niebler, M., Qian, X., Hofler, D., Kogosov, V., Kaewprag, J., Kaufmann, A.M., Ly, R., Bohmer, G., Zawatzky, R., Rosl, F., *et al.* (2013). Post-translational control of IL-1beta via the human papillomavirus type 16 E6 oncoprotein: a novel mechanism of innate immune escape mediated by the E3-ubiquitin ligase E6-AP and p53. PLoS Pathog *9*, e1003536.

O'Connell, C.M., Ingalls, R.R., Andrews, C.W., Jr., Scurlock, A.M., and Darville, T. (2007). Plasmid-deficient Chlamydia muridarum fail to induce immune pathology and protect against oviduct disease. J Immunol *179*, 4027–4034.

Olivares-Zavaleta, N., Whitmire, W.M., Kari, L., Sturdevant, G.L., and Caldwell, H.D. (2014). CD8+ T cells define an unexpected role in live-attenuated vaccine protective immunity against Chlamydia trachomatis infection in macaques. J Immunol *192*, 4648–4654.

Oriel, J.D., Reeve, P., Wright, J.T., and Owen, J. (1976). Chlamydial infection of the male urethra. Br J Vener Dis *52*, 46–51.

Ouellette, S.P., Hatch, T.P., AbdelRahman, Y.M., Rose, L.A., Belland, R.J., and Byrne, G.I. (2006). Global transcriptional upregulation in the absence of increased translation in Chlamydia during IFNgamma-mediated host cell tryptophan starvation. Mol Microbiol *62*, 1387–1401.

Paavonen, J., Vesterinen, E., Meyer, B., and Saksela, E. (1982). Colposcopic and histological findings in cervical chlamydial infection. Obstet Gynecol *59*, 712–715.

Pal, S., Tifrea, D.F., Zhong, G., and de la Maza, L.M. (2018). Transcervical Inoculation with Chlamydia trachomatis Induces Infertility in HLA-DR4 Transgenic and Wild-Type Mice. Infect Immun *86*.

Patton, D., Sweeney, Y., Kari, L., Sturdevant, G., and Caldwell, H. (2015). Pathogenicity of plasmid positive and negative Chlamydia trachomatis in a macaque model of ocular and genital tract diseases. Sex Transm Infect *91*, A133.

Patton, D.L., and Kuo, C.C. (1989). Histopathology of Chlamydia trachomatis salpingitis after primary and repeated reinfections in the monkey subcutaneous pocket model. J Reprod Fertil *85*, 647–656.

Patton, D.L., Kuo, C.C., and Brenner, R.M. (1989). Chlamydia trachomatis oculogenital infection in the subcutaneous autotransplant model of conjunctiva, salpinx and endometrium. Br J Exp Pathol *70*, 357–367.

Patton, D.L., Sweeney, Y.C., Baldessari, A.E., Cles, L., Kari, L., Sturdevant, G.L., Yang, C., and Caldwell, H.D. (2018). The Chlamydia trachomatis Plasmid and CT135 Virulence Factors Are Not Essential for Genital Tract Infection or Pathology in Female Pig-Tailed Macaques. Infect Immun *86*.

Peipert, J.F. (2003). Clinical practice. Genital chlamydial infections. N Engl J Med *349*, 2424–2430.

Peng, B., Lu, C., Tang, L., Yeh, I.T., He, Z., Wu, Y., and Zhong, G. (2011). Enhanced upper genital tract pathologies by blocking Tim-3 and PD-L1 signalling pathways in mice intravaginally infected with Chlamydia muridarum. BMC Infect Dis *11*, 347.

Perry, L.L., Feilzer, K., and Caldwell, H.D. (1997). Immunity to Chlamydia trachomatis is mediated by T helper 1 cells through IFN-gamma-dependent and -independent pathways. J Immunol *158*, 3344–3352.

Perry, L.L., and Hughes, S. (1999). Chlamydial colonization of multiple mucosae following infection by any mucosal route. Infect Immun *67*, 3686–3689.

Peters, R.P., Dubbink, J.H., van der Eem, L., Verweij, S.P., Bos, M.L., Ouburg, S., Lewis, D.A., Struthers, H., McIntyre, J.A., and Morre, S.A. (2014). Cross-sectional study of genital, rectal, and pharyngeal Chlamydia and gonorrhea in women in rural South Africa. Sex Transm Dis *41*, 564–569.

Provost, P.J., and Vickers, J.H. (1972). Attempted immunization against trachoma infection in baboons. Am J Vet Res *33*, 599–605.

Putman, T., Hybiske, K., Jow, D., Afrasiabi, C., Lelong, S., Cano, M.A., Wu, C., and Su, A.I. (2019). ChlamBase: a curated model organism database for the Chlamydia research community. Database (Oxford) *2019*.

Qu, Y., Frazer, L.C., O'Connell, C.M., Tarantal, A.F., Andrews, C.W., Jr., O'Connor, S.L., Russell, A.N., Sullivan, J.E., Poston, T.B., Vallejo, A.N., *et al.* (2015). Comparable Genital Tract Infection, Pathology, and Immunity in Rhesus Macaques Inoculated with Wild-Type or Plasmid-Deficient Chlamydia trachomatis Serovar D. Infect Immun *83*, 4056–4067.

Quellette, P.S. (2018). Feasibility of a Conditional Knockout System for Chlamydia Based on CRISPR Interference. Front Cell Infect Microbiol *8*, doi: 10.3389/fcimb.

Quinn, T.C., Goodell, S.E., Mkrtichian, E., Schuffler, M.D., Wang, S.P., Stamm, W.E., and Holmes, K.K. (1981). Chlamydia trachomatis proctitis. N Engl J Med *305*, 195–200.

Quint, K., Porras, C., Safaeian, M., Gonzalez, P., Hildesheim, A., Quint, W., van Doorn, L.J., Silva, S., Melchers, W., Schiffman, M., *et al.* (2007). Evaluation of a novel PCR-based assay for detection and identification of Chlamydia trachomatis serovars in cervical specimens. J Clin Microbiol *45*, 3986–3991.

Rajan, N., Pruden, D.L., Kaznari, H., Cao, Q., Anderson, B.E., Duncan, J.L., and Schaeffer, A.J. (2000). Characterization of an immortalized human vaginal epithelial cell line. J Urol *163*, 616–622.

Rajeeve, K., Das, S., Prusty, B.K., and Rudel, T. (2018). Chlamydia trachomatis paralyses neutrophils to evade the host innate immune response. Nat Microbiol *3*, 824–835.

Ramadhani, A.M., Derrick, T., Holland, M.J., and Burton, M.J. (2016). Blinding Trachoma: Systematic Review of Rates and Risk Factors for Progressive Disease. PLoS Negl Trop Dis *10*, e0004859.

Ramadhani, A.M., Derrick, T., Macleod, D., Massae, P., Mtuy, T., Jeffries, D., Roberts, C.H., Bailey, R.L., Mabey, D.C.W., Holland, M.J., *et al.* (2017). Immunofibrogenic Gene Expression Patterns in Tanzanian Children with Ocular Chlamydia trachomatis Infection, Active Trachoma and Scarring: Baseline Results of a 4-Year Longitudinal Study. Front Cell Infect Microbiol *7*, 406.

Ramirez, R.D., Sheridan, S., Girard, L., Sato, M., Kim, Y., Pollack, J., Peyton, M., Zou, Y., Kurie, J.M., Dimaio,

J.M., *et al.* (2004). Immortalization of human bronchial epithelial cells in the absence of viral oncoproteins. Cancer Res *64*, 9027–9034.

Ramsey, K.H., Schripsema, J.H., Smith, B.J., Wang, Y., Jham, B.C., O'Hagan, K.P., Thomson, N.R., Murthy, A.K., Skilton, R.J., Chu, P., *et al.* (2014). Plasmid CDS5 influences infectivity and virulence in a mouse model of Chlamydia trachomatis urogenital infection. Infect Immun *82*, 3341–3349.

Ramsey, K.H., Sigar, I.M., Rana, S.V., Gupta, J., Holland, S.M., and Byrne, G.I. (2001). Role for inducible nitric oxide synthase in protection from chronic Chlamydia trachomatis urogenital disease in mice and its regulation by oxygen free radicals. Infect Immun *69*, 7374–7379.

Rank, R.G. (1994). Animal models for urogenital infections. Methods Enzymol 235, 83–93.

Rank, R.G., Ramsey, K.H., Pack, E.A., and Williams, D.M. (1992). Effect of gamma interferon on resolution of murine chlamydial genital infection. Infect Immun *60*, 4427–4429.

Rank, R.G., White, H.J., and Barron, A.L. (1979). Humoral immunity in the resolution of genital infection in female guinea pigs infected with the agent of guinea pig inclusion conjunctivitis. Infect Immun *26*, 573–579.

Rank, R.G., and Yeruva, L. (2014). Hidden in plain sight: chlamydial gastrointestinal infection and its relevance to persistence in human genital infection. Infect Immun *82*, 1362–1371.

Rapoza, P.A., Tahija, S.G., Carlin, J.P., Miller, S.L., Padilla, M.L., and Byrne, G.I. (1991). Effect of interferon on a primary conjunctival epithelial cell model of trachoma. Invest Ophthalmol Vis Sci *32*, 2919–2923.

Rasmussen, S.J., Eckmann, L., Quayle, A.J., Shen, L., Zhang, Y.X., Anderson, D.J., Fierer, J., Stephens, R.S., and Kagnoff, M.F. (1997). Secretion of proinflammatory cytokines by epithelial cells in response to Chlamydia infection suggests a central role for epithelial cells in chlamydial pathogenesis. J Clin Invest *99*, 77–87.

Rodgers, A.K., Budrys, N.M., Gong, S., Wang, J., Holden, A., Schenken, R.S., and Zhong, G. (2011). Genome-wide identification of Chlamydia trachomatis antigens associated with tubal factor infertility. Fertil Steril *96*, 715–721.

Rose, W.A., 2nd, McGowin, C.L., Spagnuolo, R.A., Eaves-Pyles, T.D., Popov, V.L., and Pyles, R.B. (2012). Commensal bacteria modulate innate immune responses of vaginal epithelial cell multilayer cultures. PLoS One *7*, e32728.

Russell, A.N., Zheng, X., O'Connell, C.M., Taylor, B.D., Wiesenfeld, H.C., Hillier, S.L., Zhong, W., and Darville, T. (2016a). Analysis of Factors Driving Incident and Ascending Infection and the Role of Serum Antibody in Chlamydia trachomatis Genital Tract Infection. J Infect Dis *213*, 523–531.

Russell, A.N., Zheng, X., O'Connell, C.M., Wiesenfeld, H.C., Hillier, S.L., Taylor, B.D., Picard, M.D., Flechtner, J.B., Zhong, W., Frazer, L.C., *et al.* (2016b). Identification of Chlamydia trachomatis Antigens Recognized by T Cells From Highly Exposed Women Who Limit or Resist Genital Tract Infection. J Infect Dis *214*, 1884–1892.

Sato, T., Stange, D.E., Ferrante, M., Vries, R.G., Van Es, J.H., Van den Brink, S., Van Houdt, W.J., Pronk, A., Van Gorp, J., Siersema, P.D., *et al.* (2011). Long-term expansion of epithelial organoids from human colon, adenoma, adenocarcinoma, and Barrett's epithelium. Gastroenterology *141*, 1762–1772.

Schachter, J., and Osoba, A.O. (1983). Lymphogranuloma venereum. Br Med Bull 39, 151–154.

Schnupf, P., Gaboriau-Routhiau, V., Sansonetti, P.J., and Cerf-Bensussan, N. (2017). Segmented filamentous bacteria, Th17 inducers and helpers in a hostile world. Curr Opin Microbiol 35, 100–109.

Schust, D.J., Ibana, J.A., Buckner, L.R., Ficarra, M., Sugimoto, J., Amedee, A.M., and Quayle, A.J. (2012). Potential mechanisms for increased HIV-1 transmission across the endocervical epithelium during *C. trachomatis* infection. Curr HIV Res *10*, 218–227.

Shah, A.A., Schripsema, J.H., Imtiaz, M.T., Sigar, I.M., Kasimos, J., Matos, P.G., Inouye, S., and Ramsey, K.H. (2005). Histopathologic changes related to fibrotic oviduct occlusion after genital tract infection of mice with Chlamydia muridarum. Sex Transm Dis *32*, 49–56.

Shan, W., and Liu, J. (2013). Transformation of the human ovarian surface epithelium with genetically defined elements. In Methods in Molecular Biology, pp. 377–392.

Shao, L., Melero, J., Zhang, N., Arulanandam, B., Baseman, J., Liu, Q., and Zhong, G. (2017a). The cryptic plasmid is more important for Chlamydia muridarum to colonize the mouse gastrointestinal tract than to infect the genital tract. PLoS One *12*, e0177691.

Shao, L., Zhang, T., Liu, Q., Wang, J., and Zhong, G. (2017b). Chlamydia muridarum with Mutations in Chromosomal Genes tc0237 and/or tc0668 Is Deficient in Colonizing the Mouse Gastrointestinal Tract. Infect Immun *85*, doi: 10.1128/IAI.00321-00317. Print 02017 Aug. PMID:28584162.

Skwor, T.A., Atik, B., Kandel, R.P., Adhikari, H.K., Sharma, B., and Dean, D. (2008). Role of secreted conjunctival mucosal cytokine and chemokine proteins in different stages of trachomatous disease. PLoS Negl Trop Dis *2*, e264.

Smith, J.L., Flueckiger, R.M., Hooper, P.J., Polack, S., Cromwell, E.A., Palmer, S.L., Emerson, P.M., Mabey, D.C., Solomon, A.W., Haddad, D., *et al.* (2013a). The geographical distribution and burden of trachoma in Africa. PLoS Negl Trop Dis *7*, e2359.

Smith, J.L., Sturrock, H.J., Olives, C., Solomon, A.W., and Brooker, S.J. (2013b). Comparing the performance of cluster random sampling and integrated threshold mapping for targeting trachoma control, using computer simulation. PLoS Negl Trop Dis *7*, e2389.

Snavely, E.A., Kokes, M., Dunn, J.D., Saka, H.A., Nguyen, B.D., Bastidas, R.J., McCafferty, D.G., and Valdivia, R.H. (2014). Reassessing the role of the secreted protease CPAF in Chlamydia trachomatis infection through genetic approaches. Pathog Dis *71*, 336–351.

Song, L., Carlson, J.H., Whitmire, W.M., Kari, L., Virtaneva, K., Sturdevant, D.E., Watkins, H., Zhou, B., Sturdevant, G.L., Porcella, S.F., *et al.* (2013). Chlamydia trachomatis plasmid-encoded Pgp4 is a transcriptional regulator of virulence-associated genes. Infect Immun *81*, 636–644.

Sowa, S., Sowa, J., Collier, L.H., and Blyth, W.A. (1969). Trachoma vaccine field trials in The Gambia. J Hyg (Lond) *67*, 699–717.

Spear, G.T., Kendrick, S.R., Chen, H.Y., Thomas, T.T., Bahk, M., Balderas, R., Ghosh, S., Weinberg, A., and Landay, A.L. (2011). Multiplex immunoassay of lower genital tract mucosal fluid from women attending an urban STD clinic shows broadly increased IL1ss and lactoferrin. PLoS One *6*, e19560.

Stamm, W.E., Koutsky, L.A., Benedetti, J.K., Jourden, J.L., Brunham, R.C., and Holmes, K.K. (1984). Chlamydia trachomatis urethral infections in men. Prevalence, risk factors, and clinical manifestations. Ann Intern Med *100*, 47–51.

Stary, G., Olive, A., Radovic-Moreno, A.F., Gondek, D., Alvarez, D., Basto, P.A., Perro, M., Vrbanac, V.D., Tager, A.M., Shi, J., *et al.* (2015). VACCINES. A mucosal vaccine against Chlamydia trachomatis generates two waves of protective memory T cells. Science *348*, aaa8205.

Steedman, N.M., and McMillan, A. (2009). Treatment of asymptomatic rectal Chlamydia trachomatis: is single-dose azithromycin effective? Int J STD AIDS *20*, 16–18.

Stephens, R.S., Wagar, E.A., and Schoolnik, G.K. (1988). High-resolution mapping of serovar-specific and common antigenic determinants of the major outer membrane protein of Chlamydia trachomatis. J Exp Med *167*, 817–831.

Storrs, C.H., and Silverstein, S.J. (2007). PATJ, a tight junction-associated PDZ protein, is a novel degradation target of high-risk human papillomavirus E6 and the alternatively spliced isoform 18 E6. J Virol *81*, 4080–4090.

Sturdevant, G.L., and Caldwell, H.D. (2014). Innate immunity is sufficient for the clearance of Chlamydia trachomatis from the female mouse genital tract. Pathog Dis *72*, 70–73.

Sturdevant, G.L., Kari, L., Gardner, D.J., Olivares-Zavaleta, N., Randall, L.B., Whitmire, W.M., Carlson, J.H., Goheen, M.M., Selleck, E.M., Martens, C., *et al.* (2010). Frameshift mutations in a single novel virulence factor alter the *in vivo* pathogenicity of Chlamydia trachomatis for the female murine genital tract. Infect Immun *78*, 3660–3668.

Sun, X., Tian, Q., Wang, L., Xue, M., and Zhong, G. (2017). IL-6-mediated signalling pathways limit Chlamydia muridarum infection and exacerbate its pathogenicity in the mouse genital tract. Microbes Infect *19*, 536–545.

Sun, X., Yang, Z., Zhang, H., Dai, J., Chen, J., Tang, L., Rippentrop, S., Xue, M., Zhong, G., and Wu, G. (2015). Chlamydia muridarum induction of glandular duct dilatation in mice. Infect Immun *83*, 2327–2337.

Swanson, J., Eschenbach, D.A., Alexander, E.R., and Holmes, K.K. (1975). Light and electron microscopic study of Chlamydia trachomatis infection of the uterine cervix. J Infect Dis *131*, 678–687.

Swartz, S.L., and Kraus, S.J. (1979). Persistent urethral leukocytosis and asymptomatic chlamydial urethritis. J Infect Dis *140*, 614–617.

Tai, N., Peng, J., Liu, F., Gulden, E., Hu, Y., Zhang, X., Chen, L., Wong, F.S., and Wen, L. (2016). Microbial antigen mimics activate diabetogenic CD8 T cells in NOD mice. J Exp Med *213*, 2129–2146.

Tang, F.F., Chang, H.L., Huang, Y.T., and Wang, K.C. (1957a). Studies on the aetiology of trachoma with special reference to isolation of the virus in chick embryo. Chin Med J *75*, 429–447.

Tang, F.F., Huang, Y.T., Chang, H.L., and Wong, K.C. (1957b). Isolation of trachoma virus in chick embryo. J Hyg Epidemiol Microbiol Immunol *1*, 109–120.

Tang, L., Chen, J., Zhou, Z., Yu, P., Yang, Z., and Zhong, G. (2015). Chlamydia-secreted protease CPAF degrades host antimicrobial peptides. Microbes Infect *17*, 402–408.

Tang, L., Yang, Z., Zhang, H., Zhou, Z., Arulanandam, B., Baseman, J., and Zhong, G. (2014). Induction of protective immunity against Chlamydia muridarum intracervical infection in DBA/1j mice. Vaccine *32*, 1407–1413.

Tang, L., Zhang, H., Lei, L., Gong, S., Zhou, Z., Baseman, J., and Zhong, G. (2013). Oviduct Infection and Hydrosalpinx in DBA1/j Mice Is Induced by Intracervical but Not Intravaginal Inoculation with Chlamydia muridarum. PLoS One *8*, e71649.

Taylor, H.R. (2009). Doyne Lecture: trachoma, is it history? Eye (Lond) *23*, 2007–2022.

Taylor-Robinson, D., and Carney, F.E., Jr. (1974). Growth and effect of mycoplasmas in Fallopian tube organ cultures. Br J Vener Dis *50*, 212–216.

Thomas, M., Laura, R., Hepner, K., Guccione, E., Sawyers, C., Lasky, L., and Banks, L. (2002). Oncogenic human papillomavirus E6 proteins target the MAGI-2 and MAGI-3 proteins for degradation. Oncogene *21*, 5088–5096.

Van Voorhis, W.C., Barrett, L.K., Sweeney, Y.T., Kuo, C.C., and Patton, D.L. (1997). Repeated Chlamydia trachomatis infection of Macaca nemestrina fallopian tubes produces a Th1-like cytokine response associated with fibrosis and scarring. Infect Immun *65*, 2175–2182.

Wali, S., Gupta, R., Yu, J.J., Lanka, G.K.K., Chambers, J.P., Guentzel, M.N., Zhong, G., Murthy, A.K., and Arulanandam, B.P. (2017). Chlamydial protease-like activity factor mediated protection against *C. trachomatis* in guinea pigs. Immunol Cell Biol *95*, 454–460.

Wang, J., Zhang, Y., Lu, C., Lei, L., Yu, P., and Zhong, G. (2010). A genome-wide profiling of the humoral immune response to Chlamydia trachomatis infection reveals vaccine candidate antigens expressed in humans. J Immunol *185*, 1670–1680.

Wang, L., Zhang, Q., Zhang, T., Zhang, Y., Zhu, C., Sun, X., Zhang, N., Xue, M., and Zhong, G. (2016). The Chlamydia muridarum Organisms Fail to Auto-Inoculate the Mouse Genital Tract after Colonization in the Gastrointestinal Tract for 70 days. PLoS One *11*, e0155880.

Wang, L., Zhu, C., Zhang, T., Tian, Q., Zhang, N., Morrison, S., Morrison, R., Xue, M., and Zhong, G. (2018). Nonpathogenic Colonization with Chlamydia in the Gastrointestinal Tract as Oral Vaccination for Inducing Transmucosal Protection. Infect Immun *86*.

Wang, S.P., and Grayston, J.T. (1963). Classification of Trachoma Virus Strains by Protection of Mice from Toxic Death. J Immunol *90*, 849–856.

Wang, S.P., and Grayston, J.T. (1967). Pannus with experimental trachoma and inclusion conjunctivitis agent infection of Taiwan monkeys. Am J Ophthalmol *63*, Suppl:1133–1145.

Wang, Y. (1999). Aetiology of trachoma: a great success in isolating and cultivating Chlamydia trachomatis. Chin Med J (Engl) *112*, 938–941.

Wang, Y., Kahane, S., Cutcliffe, L.T., Skilton, R.J., Lambden, P.R., and Clarke, I.N. (2011). Development of a transformation system for Chlamydia trachomatis: restoration of glycogen biosynthesis by acquisition of a plasmid shuttle vector. PLoS Pathog *7*, e1002258.

West, S.K., Munoz, B., Kaur, H., Dize, L., Mkocha, H., Gaydos, C.A., and Quinn, T.C. (2018). Longitudinal change in the serology of antibodies to Chlamydia trachomatis pgp3 in children residing in a trachoma area. Sci Rep *8*, 3520.

Westrom, L., Joesoef, R., Reynolds, G., Hagdu, A., and Thompson, S.E. (1992). Pelvic inflammatory disease and fertility. A cohort study of 1,844 women with laparoscopically verified disease and 657 control women with normal laparoscopic results. Sex Transm Dis *19*, 185–192.

White, H.J., Rank, R.G., Soloff, B.L., and Barron, A.L. (1979). Experimental chlamydial salpingitis in immunosuppressed guinea pigs infected in the genital tract with the agent of guinea pig inclusion conjunctivitis. Infect Immun *26*, 728–735.

WHO (2003). Report of the 2nd global scientific meeting on trachoma. World Health Organization: Geneva, Switzerland.

Wiegand, R.E., Cooley, G., Goodhew, B., Banniettis, N., Kohlhoff, S., Gwyn, S., and Martin, D.L. (2018). Latent class modelling to compare testing platforms for detection of antibodies against the Chlamydia trachomatis antigen Pgp3. Sci Rep *8*, 4232.

Winstanley, C.E., Ramsey, K.H., Marsh, P., and Clarke, I.N. (2017). Development and evaluation of an enzyme-linked immunosorbent assay for the detection of antibodies to a common urogenital derivative of Chlamydia trachomatis plasmid-encoded PGP3. J Immunol Methods *445*, 23–30.

Woodworth, C.D., Waggoner, S., Barnes, W., Stoler, M.H., and DiPaolo, J.A. (1990). Human cervical and foreskin epithelial cells immortalized by human papillomavirus DNAs exhibit dysplastic differentiation *in vivo*. Cancer Res *50*, 3709–3715.

Woolridge, R.L., Grayston, J.T., Chang, I.H., Cheng, K.H., Yang, C.Y., and Neave, C. (1967a). Field trial of a monovalent and of a bivalent mineral oil adjuvant trachoma vaccine in Taiwan school children. Am J Ophthalmol *63*, Suppl:1645–1650.

Woolridge, R.L., Grayston, J.T., Chang, I.H., Yang, C.Y., and Cheng, K.H. (1967b). Long-term follow-up of the initial (1959–1960) trachoma vaccine field trial on Taiwan. Am J Ophthalmol *63*, Suppl:1650–1655.

Wright, H.R., Turner, A., and Taylor, H.R. (2008). Trachoma. Lancet *371*, 1945–1954.

Wu, H.J., Ivanov, I.I., Darce, J., Hattori, K., Shima, T., Umesaki, Y., Littman, D.R., Benoist, C., and Mathis, D. (2010). Gut-residing segmented filamentous bacteria drive autoimmune arthritis via T helper 17 cells. Immunity *32*, 815–827.

Wyrick, P.B., Choong, J., Davis, C.H., Knight, S.T., Royal, M.O., Maslow, A.S., and Bagnell, C.R. (1989). Entry of genital Chlamydia trachomatis into polarized human epithelial cells. Infect Immun *57*, 2378–2389.

Yang, G., Rosen, D.G., Colacino, J.A., Mercado-Uribe, I., and Liu, J. (2007). Disruption of the retinoblastoma pathway by small interfering RNA and ectopic expression of the catalytic subunit of telomerase lead to immortalization of human ovarian surface epithelial cells. Oncogene *26*, 1492–1498.

Yang, Z., Conrad, T., Zhou, Z., Chen, J., Dutow, P., Klos, A., and Zhong, G. (2014). Complement factor C5 but not C3 contributes significantly to hydrosalpinx development in mice infected with Chlamydia muridarum. Infect Immun *82*, 3154–3163.

Yang, Z., Tang, L., Shao, L., Zhang, Y., Zhang, T., Schenken, R., Valdivia, R., and Zhong, G. (2016a). The Chlamydia-Secreted Protease CPAF Promotes Chlamydial Survival in the Mouse Lower Genital Tract. Infect Immun *84*, 2697–2702.

Yang, Z., Tang, L., Sun, X., Chai, J., and Zhong, G. (2015). Characterization of CPAF critical residues and secretion during Chlamydia trachomatis infection. Infect Immun *83*, 2234–2241.

Yang, Z., Tang, L., Zhou, Z., and Zhong, G. (2016b). Neutralizing antichlamydial activity of complement by chlamydia-secreted protease CPAF. Microbes Infect *18*, 669–674.

Yeruva, L., Spencer, N., Bowlin, A.K., Wang, Y., and Rank, R.G. (2013). Chlamydial infection of the gastrointestinal tract: a reservoir for persistent infection. Pathog Dis *68*, 88–95.

Zambrano, A.I., Sharma, S., Crowley, K., Dize, L., Munoz, B.E., Mishra, S.K., Rotondo, L.A., Gaydos, C.A., and West, S.K. (2016). The World Health Organization Recommendations for Trachoma Surveillance, Experience in Nepal and Added Benefit of Testing for Antibodies to Chlamydia trachomatis pgp3 Protein: NESTS Study. PLoS Negl Trop Dis *10*, e0005003.

Zhang, H., Zhou, Z., Chen, J., Wu, G., Yang, Z., Zhou, Z., Baseman, J., Zhang, J., Reddick, R.L., and Zhong, G. (2014). Lack of long lasting hydrosalpinx in A/J mice correlates with rapid but transient chlamydial ascension and neutrophil recruitment in the oviduct following intravaginal inoculation with Chlamydia muridarum. Infect Immun *82*, 2688–2696.

Zhang, Q., Huang, Y., Gong, S., Yang, Z., Sun, X., Schenken, R., and Zhong, G. (2015). In Vivo and Ex Vivo Imaging Reveals a Long-Lasting Chlamydial Infection in the Mouse Gastrointestinal Tract following Genital Tract Inoculation. Infect Immun *83*, 3568–3577.

Zhang, T., Huo, Z., Ma, J., He, C., and Zhong, G. (2019). The plasmid-encoded pGP3 promotes Chlamydia evasion of acidic barriers in both stomach and vagina. Infect Immun.

Zhang, Y., Shao, L., Li, X., and Zhong, G. (2017). Uterotubal junction prevents chlamydial ascension via innate immunity. PLoS One *12*, e0183189.

Zhong, G. (2009). Killing me softly: chlamydial use of proteolysis for evading host defences. Trends Microbiol *17*, 467–474.

Zhong, G. (2017). Chlamydial Plasmid-Dependent Pathogenicity. Trends Microbiol *25*, 141–152.

Zhong, G. (2018). Chlamydia Spreading from the Genital Tract to the Gastrointestinal Tract – A Two-Hit Hypothesis. Trends Microbiol *26*, 611–623.

Zhong, G., Brunham, R.C., de la Maza, L.M., Darville, T., and Deal, C. (2017). National Institute of Allergy and Infectious Diseases workshop report: 'Chlamydia vaccines: The way forward'. Vaccine.

Zhong, G.M., and de la Maza, L.M. (1988). Activation of mouse peritoneal macrophages *in vitro* or *in vivo* by recombinant murine gamma interferon inhibits the growth of Chlamydia trachomatis serovar L1. Infect Immun *56*, 3322–3325.

Zhong, G.M., Peterson, E.M., Czarniecki, C.W., and de la Maza, L.M. (1988). Recombinant murine gamma interferon inhibits Chlamydia trachomatis serovar L1 *in vivo*. Infect Immun *56*, 283–286.

Zhong, G.M., Peterson, E.M., Czarniecki, C.W., Schreiber, R.D., and de la Maza, L.M. (1989). Role of endogenous gamma interferon in host defence against Chlamydia trachomatis infections. Infect Immun *57*, 152–157.

Zhong, G.M., Reid, R.E., and Brunham, R.C. (1990). Mapping antigenic sites on the major outer membrane protein of Chlamydia trachomatis with synthetic peptides. Infect Immun *58*, 1450–1455.

Zhu, C., Lin, H., Tang, L., Chen, J., Wu, Y., and Zhong, G. (2018). Oral Chlamydia vaccination induces transmucosal protection in the airway. Vaccine *36*, 2061–2068.

Ziklo, N., Huston, W.M., Hocking, J.S., and Timms, P. (2016a). Chlamydia trachomatis genital tract infections: when host immune response and the microbiome collide. Trends Microbiol *24*, 750–765.

Ziklo, N., Huston, W.M., Taing, K., Katouli, M., and Timms, P. (2016b). In vitro rescue of genital strains of Chlamydia trachomatis from interferon-gamma and tryptophan depletion with indole-positive, but not indole-negative Prevotella spp. BMC Microbiol *16*, 286.

Current Books of Interest

Title	Year
Bats and Viruses: Current Research and Future Trends	2020
SUMOylation and Ubiquitination: Current and Emerging Concepts	2019
Avian Virology: Current Research and Future Trends	2019
Microbial Exopolysaccharides: Current Research and Developments	2019
Polymerase Chain Reaction: Theory and Technology	2019
Pathogenic Streptococci: From Genomics to Systems Biology and Control	2019
Insect Molecular Virology: Advances and Emerging Trends	2019
Methylotrophs and Methylotroph Communities	2019
Prions: Current Progress in Advanced Research (Second Edition)	2019
Microbiota: Current Research and Emerging Trends	2019
Microbial Ecology: Current Advances from Genomics, Metagenomics and Other Omics	2019
Porcine Viruses: From Pathogenesis to Strategies for Control	2019
Lactobacillus Genomics and Metabolic Engineering	2019
Cyanobacteria: Signaling and Regulation Systems	2018
Viruses of Microorganisms	2018
Protozoan Parasitism: From Omics to Prevention and Control	2018
Genes, Genetics and Transgenics for Virus Resistance in Plants	2018
Plant-Microbe Interactions in the Rhizosphere	2018
DNA Tumour Viruses: Virology, Pathogenesis and Vaccines	2018
Pathogenic *Escherichia coli*: Evolution, Omics, Detection and Control	2018
Postgraduate Handbook: A Comprehensive Guide for PhD and Master's Students and their Supervisors	2018
Enteroviruses: Omics, Molecular Biology, and Control	2018
Molecular Biology of Kinetoplastid Parasites	2018
Bacterial Evasion of the Host Immune System	2017
Illustrated Dictionary of Parasitology in the Post-Genomic Era	2017
Next-generation Sequencing and Bioinformatics for Plant Science	2017
Brewing Microbiology: Current Research, Omics and Microbial Ecology	2017
Metagenomics: Current Advances and Emerging Concepts	2017
The CRISPR/Cas System: Emerging Technology and Application	2017
Bacillus: Cellular and Molecular Biology (Third edition)	2017
Cyanobacteria: Omics and Manipulation	2017
Foot-and-Mouth Disease Virus: Current Research and Emerging Trends	2017
Staphylococcus: Genetics and Physiology	2016
Microbial Biodegradation: From Omics to Function and Application	2016

Full details at www.caister.com

www.ingramcontent.com/pod-product-compliance
Lightning Source LLC
LaVergne TN
LVHW070845210926
848482LV00016B/874